ARITHMETIC PROGRESSION

$a_1 + a_2 + a_3 + \cdots$ where $a_2 = a_1 + d$

nth term:

$a_n = a_1 + (n-1)d$

$a_3 = a_2 + d$

etc.

d is the common difference ...
it
neg

Sum of n terms: $S_n = \dfrac{n}{2}(a_1 + a_n)$

n is a positive integer

GEOMETRIC PROGRESSION

$a_1 + a_2 + a_3 + \cdots$ where $a_2 = r \cdot a_1$

nth term:

$a_n = a_1 \cdot r^{n-1}$

$a_3 = r \cdot a_2$

etc.

r is the common ratio ... it can be positive or negative, but $r \neq 1$, $r \neq 0$.

Sum of n terms: $S_n = \dfrac{a_1(1-r^n)}{1-r}$

Sum of infinitely many terms: $S = \dfrac{a_1}{1-r}$

FACTORIAL FUNCTION $n! = n(n-1)(n-2)\cdots 3 \cdot 2 \cdot 1$ for any positive integer n

BINOMIAL FORMULA

$$(a+b)^n = \underbrace{a^n + na^{n-1}b + \frac{n(n-1)}{2!}a^{n-2}b^2 + \frac{n(n-1)(n-2)}{3!}a^{n-3}b^3 + \cdots + b^n}_{n+1 \text{ terms}}$$

where $k > 0$

The kth term of the binomial expansion of $(a+b)^n$ is $\dfrac{n(n-1)(n-2)\cdots(n-k+1)}{k!} a^{n-k}b^k$

IMAGINARY NUMBERS

$j = \sqrt{-1}$ $j^2 = -1$ $j^3 = -j$ $j^4 = 1$ $j^5 = j$ $\sqrt{-a} = j\sqrt{a}$

n is some positive integer

★ $\sqrt{-a} \cdot \sqrt{-b} = j\sqrt{a} \cdot j\sqrt{b} = -\sqrt{ab}$ for $a > 0$, $b > 0$

if a is a positive real number

DEMOIVRE'S THEOREM

$x + jy = r(\cos\theta + j\sin\theta)$ and $\left[r(\cos\theta + j\sin\theta)\right]^n = r^n(\cos n\theta + j\sin n\theta)$

DEFINITIONS of TRIGONOMETRIC FUNCTIONS

$\sin\theta = y/r$ $\cos\theta = x/r$ $\tan\theta = y/x$

$\csc\theta = 1/\sin\theta = r/y$ $\sec\theta = 1/\cos\theta = r/x$ $\cot\theta = 1/\tan\theta = x/y$

FUNCTIONS of SPECIAL ANGLES

θ, degrees	θ, radians	sin θ	cos θ	tan θ
0	0	0	1	0
30	π/6	1/2	√3/2	√3/3
45	π/4	√2/2	√2/2	1
60	π/3	√3/2	1/2	√3
90	π/2	1	0	not defined
180	π	0	-1	0
270	3π/2	-1	0	not defined

DEGREE — RADIAN CONVERSION

$1° = \dfrac{\pi}{180}$ radians ≈ 0.017

1 radian $= \dfrac{180°}{\pi} \approx 57.3°$

Modern
Technical
Mathematics
with Calculus

Modern Technical Mathematics with Calculus

Robert A. Carman

Hal M. Saunders

Wadsworth Publishing Company

Belmont, California
A Division of Wadsworth, Inc.

Mathematics Editor: Heather Bennett
Production Editor: Greg Hubit Bookworks
Designer: John Edeen
Copy Editor: Phyllis Niklas
Technical Illustrator: Scientific Illustrators, Inc.
Cover Art: Charles Sheeler, *General Motors Research*, 1955.
Courtesy, General Motors Research Laboratories

Printed in the United States of America

1 2 3 4 5 6 7 8 9 10—89 88 87 86 85

Library of Congress Cataloging in Publication Data

Carman, Robert A.
    Modern technical mathematics with calculus.

    Includes index.
    1. Mathematics—1961–   I. Saunders, Hal M.
II. Title.
QA37.2.C29 1985   510   84-28419
ISBN 0-534-04305-4

ISBN 0-534-04305-4

To Agnes,
who taught me
that happiness is a state
of dynamic equilibrium,
and that to know means to be
learning or discovering or
teaching. The mind has no
passive state of grace.

Preface

In 1464, the German mathematician and inventor Regiomontanus wrote *De triangulis omnimodis*, a textbook of trigonometry, geometry, and algebra. This most modern of mathematics texts became one of the first printed textbooks and was used for over 200 years. In it the author writes, "Among such an abundance of things to learn, some may seem . . . hard to understand. The new student should neither be frightened or despair. . . . Where a theorem may present some problem, he may always look to the numerical examples for help." Worked examples, detailed explanations, and a table of sines were some of the "modern" features of his text.

It is our intention in this technical mathematics textbook to provide the mathematical background needed by modern students of engineering and engineering technology who will spend most of their working lives in the twenty-first century. The word *modern* in the title implies a great deal more than the existence of worked examples. Here is a list of some of the special features it includes.

- Clear, straightforward explanations emphasizing the readability of the text lead to understanding of the mathematics. The mathematics is correct and carefully stated but emphasizes an intuitive understanding rather than rigor. The emphasis here is on explaining concepts rather than simply presenting them.
- Many applications and applied problems are drawn from a great variety of modern scientific and technical areas. This is

math to be used as well as understood. These applications are summarized in a special index of occupational titles and are indicated in the text with a special heading.

- Numerous carefully worked examples with annotated step-by-step comments are keyed to the mathematical concepts and to the exercise sets.
- Color is used extensively as an instructional aid rather than simply as a decoration. The most important aspects of explanations, worked examples, and other parts of the text are highlighted in color.
- Graphic explanations complement the text and are used to highlight important concepts, common errors, helpful hints, and so on.
- The calculator is used extensively throughout the text, both for finding numerical solutions to problems and to determine the values of transcendental functions. Care is taken to first explain all concepts and problem solving without use of the calculator, and to estimate and check answers. Model problems and specific key sequences are given.
- Footnotes are used to cite references for many of the applied problems and to point out more mathematically rigorous aspects of some concepts.
- Answers to all odd-numbered problems are provided, and a Student Solutions Manual contains answers to even-numbered problems and many step-by-step solutions.
- Chapter summaries provide a list of important terms, concepts, and formulas keyed to the text. Comprehensive problem sets review all material presented in the chapter.
- Exercise sets at the end of each section of every chapter provide many problems covering the work of the text. These include both routine drill and applied technical problems.

Instructors will find the following materials available to supplement the textbook:

- A Test Manual containing three alternate forms of examinations covering the work in each chapter. These examinations are also available from Wadsworth Publishing Company on disks compatible with the Apple IIe, Apple II+ and IBM-PC.
- A Student Solutions Manual containing answers to even-numbered problems and step-by-step solutions to many problems in the textbook, prepared by Keith Wilson of Oklahoma City Community College and others.

- Appropriate microcomputer software to demonstrate and allow students to practice concepts taught in the book. Available from the publisher.

Appendices in the text provide an introduction to the electronic calculator, information on units of measurement and rounding, a review of basic geometric concepts, and a table of integrals. The organization of the text allows considerable flexibility in curriculum design. It is especially designed for students who need a review of basic concepts or who have difficulty with mathematics. An early introduction to trigonometry provides help for students who are concurrently enrolled in science and technical courses.

It is a pleasure for us to acknowledge the help of many people who have contributed to the development of this book. We are greatly indebted to the following teachers who read preliminary versions of the text and offered many helpful suggestions: Dennis Bertholf, Oklahoma State University – Stillwater; Dale E. Boye, Schoolcraft College; Gordon Brady, Erie Community College; William Brower, New Jersey Institute of Technology; Daniel L. Buchanan, Henry Ford Community College; Richard Butterworth, Massasoit Community College; Ray E. Collings, Tri-County Technical College; Robert Cournoyer, Wentworth Institute of Technology; Ronald M. Davis, Northern Virginia Community College; Henry D. Davison, St. Petersburg Junior College; Donald T. DeBonee, Ward Technical College, University of Hartford; Paul J. Eldersveld, College of Du Page; John Gill, University of Southern Colorado; Siegfried Haenisch, Trenton State College; Harold N. Hauser, Mt. Hood Community College; Michael Iannone, Trenton State College; Donald R. Ignatz, Lorain County Community College; Michael R. John, Wentworth Institute of Technology; David Johnson, Diablo Valley College; Art Kincannon, Rose State College; Ellen Kowalczyk, Madison Area Technical College; Larry R. Lance, Columbus Technical Institute; David Legg, Indiana University – Purdue University; E. P. McCravy, Midlands Technical College; Kalman Mecs, Community College of Allegheny County; Sidney Messer, City College of San Francisco; Robert Morrow, Central Piedmont Community College; Donald B. Moser, St. Louis Community College at Forest Park; Mary Beth Orrange, Erie Community College; Ronald Roblee, Michigan Technological University; Thomas Rourke, North Shore Community College; Roger Rowe, Lansing Community College; P. T. Sanjivamurthy, Cuyahoga Community College; Robert E. Seaver, Lorain County Community College; David M.

Segal, New York Institute of Technology; Gary Simundza, Wentworth Institute of Technology; J. E. Smith, New Mexico State University at Alamogordo; Richard C. Spangler, Tacoma Community College; Arlene Starwalt-Jeskey, Oscar Rose State College; Thomas Taylor, San Antonio College; Robert Watkins, Derry Institute of Technology; Keith Wilson, Oklahoma City Community College. This book has benefited greatly from their excellence as teachers.

We also wish to thank Ellen Kowalczyk, Gary Simundza, E. P. McCravy, John R. Long, Keith Dodson, and Gloria Langer for reading the galleys and checking solutions to the problems, and especially, Keith Wilson for preparing most of the Student Solutions Manual.

All graphs of specific functions were made using an HP87 computer and HP7225A plotter with software developed by Scientific Illustrators and George Morris.

Producing and publishing a textbook is a team sport, and we have been privileged to work with a highly talented team of individuals. We wish to give special thanks to Phyllis Niklas, Greg Hubit, John Edeen, Sandra Craig, and Sally Schuman for their creativity, imagination, and dedication throughout the design and shaping of the book. We are particularly fortunate to have had the constant encouragement, generous support, and wise counsel of Peter Fairchild and Heather Bennett, Wadsworth mathematics editors, and Rich Jones, Wadsworth senior editor. We are grateful for their confidence in us and for their support.

Much of the tedious and difficult work of typing this manuscript was ably performed by Chris Saunders. We appreciate her help and admire her stamina.

Finally, we are happy to acknowledge the patience and understanding of our wives Chris and Lyn. They helped us to see every insoluble problem as a shining opportunity.

Robert A. Carman

Hal M. Saunders

To the Student

Modern technology is not simply a storehouse for facts and numbers; it is an organized system of concepts, skills, and tools for understanding and description. Mathematical concepts and skills are a most important part of modern technology, and this textbook has been designed for students who will not only read it but will be working to achieve understanding and acquire skills.

Be alert for the following learning devices as you use this textbook.

Reference head Within each section of a chapter when new vocabulary is introduced or when new topics are presented, **reference headings** appear in the left margin as shown here. These signals show where the new information appears. Reference heads will be useful to you when you need to locate a word, phrase, equation, or topic, or when you need to review.

IMPORTANT ▶ Every experienced teacher knows that certain mathematical concepts and procedures will present special difficulties for students. To help you with these, special notes are included in the text. A large triangle ▶ and a warning word appear in the left margin to indicate the start of the comment, and a small triangle ◀ shows when it is completed. The word IMPORTANT, as used at the start of this paragraph, tells you that this note emphasizes an important idea or key concept. ◀

NOTE ▶ This kind of comment calls your attention to conclusions or consequences that might be overlooked. ◀

CAUTION ▶ A caution comment points out the common mistakes that you might make and shows you how to avoid them. ◀

LEARNING HINT ▶ A learning hint gives you an alternative explanation or slightly different way of thinking about and working with the concepts being presented. ◀

REMEMBER ▶ In some chapters concepts are used that were introduced earlier in the text. The REMEMBER comment serves to refresh your memory. ◀

An important feature of this textbook is the use of many worked examples. These appear in the following format.

EXAMPLE 0 In each chapter carefully worked examples are included so that you can follow through the solution of all the various types of problems that are included in the problem sets. Examples cover both routine mathematical manipulations and applications.

Solution
Step 1. The solution of each worked example is usually organized in a step-by-step format.
Step 2. Explanations for each step are provided on the left, and corresponding mathematical operations are on the right.
Step 3. Color, ◀─ Boxed comments , and other Graphical aids ↗ are used to highlight the important or tricky aspects of the solution.

The electronic calculator is an important tool for the modern technician or engineer, and we assume in this textbook that the student will make extensive use of a calculator. Problems in the exercise sets or examples in the text that involve the use of a calculator are preceded by the calculator symbol shown here.

Solutions often include a display of the proper calculator key sequences. For example, the calculation

$$\frac{85.7 + (12.9)^2}{\sqrt{71.6}}$$

would be shown as

85.7 $\boxed{+}$ 12.9 $\boxed{x^2}$ $\boxed{=}$ $\boxed{\div}$ 71.6 $\boxed{\sqrt{}}$ $\boxed{=}$ ⟶ **29.794326**

Footnotes   In the presentation of a concept, footnotes are sometimes used to provide comments or explanations on the need for making careful, mathematically correct statements.*

---

*Footnotes are also sometimes used to supply references to the source of an application so that you or your instructor can obtain more information if you wish.

Exercises 0-1

At the conclusion of each section of each chapter you will find a set of problems covering the work of that section. These will include a number of routine or drill problems as well as applications or word problems. Each applied problem begins with an indication of the technical field from which it has been taken. Many of these applications have been adapted from the textbooks you may be using in related technology or science courses, or they may have been obtained from engineers and technical workers in these areas.

0-7 | Review of Chapter 0

Each chapter concludes with a list of the important terms and concepts covered in the chapter and a list of the formulas or rules that have been presented. Page references are given for these so that you may review quickly or use this text for reference when you are doing on-the-job problem solving.

If your approach to learning mathematics is to skim the text lightly on the way to puzzling through a homework assignment, you will have difficulty with this or any textbook. If you are motivated to study mathematics so that you understand it and can use it correctly, this textbook is designed for you.

Contents

Numbers, Laws, and Operations $\quad$ 1

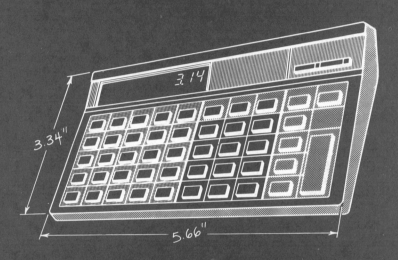

3.34″

5.66″

3.14

Over the centuries mathematics and technology have grown hand in hand. From the construction of the pyramids in ancient Egypt, through the industrial revolution of the 18th and 19th centuries, to the computer technology of today, technicians have had an ever increasing need for mathematics. As we shall demonstrate, the mathematics covered in this text provides the concepts and techniques needed to solve a great many applied problems. Furthermore, it will provide the foundation for more advanced work in mathematics and allied technical courses.

Calculators are a basic tool in modern technical work. No engineer or technician would attempt to solve today's technical problems with outdated equipment or second-rate skills. Therefore, we have designed this text to recognize the importance of the calculator in modern technical mathematics. To help you sharpen your skills we have included special examples and exercises marked with the symbol ▦ to show that a calculator is to be used. See Appendix A for a more detailed introduction to the calculator.

In this first chapter we shall review the basic operations used with real numbers and the laws governing these operations. We begin by tracing the evolution of real numbers.

1-1 | Numbers and Symbols

Numbers are the basic quantities of science, engineering, and technical work, from simple counting to complex measurements. The diagram in Fig. 1-1 shows how the numbers used in basic algebra are classified.

Integers   The familiar numbers used in counting, 1, 2, 3, and so on, are called the **counting numbers** or the **positive integers.** The **negative integers,** -1, -2, -3, and so on, are useful in describing many physical situations. Zero is also an integer, but it is neither positive nor negative. The **integers** include the positive and negative integers and zero.

Figure 1-1

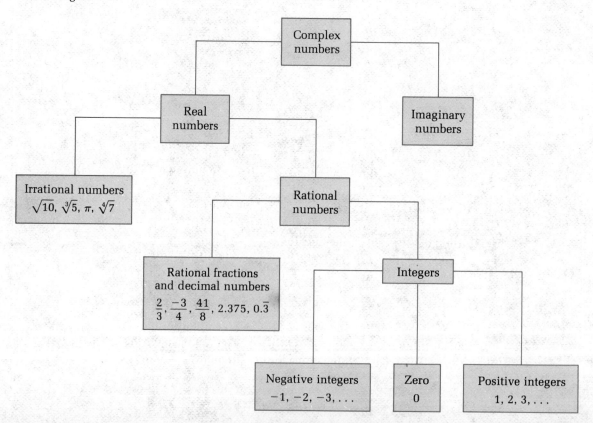

Rational numbers
In order to specify some part of a quantity, ancient mathematicians and engineers developed the idea of a rational number. We define a **rational number** as any number that can be written in the form $\frac{a}{b}$, where a is any integer and b is a nonzero integer. Thus, common fractions such as $\frac{2}{3}$, $-\frac{3}{4}$, and $\frac{41}{8}$ are rational numbers because they are written as quotients of two integers. Integers themselves are also rational numbers since they can be written as integer fractions:

$$4 = \frac{4}{1} \qquad -6 = \frac{-6}{1} \qquad 0 = \frac{0}{1}$$

Terminating decimal numbers are also rational, as you can see in the following examples:

$$0.4 = \frac{4}{10} \qquad \text{and} \qquad 2.375 = 2\frac{3}{8} \text{ or } \frac{19}{8}$$

Finally, all repeating decimals are rational numbers because they, too, can be written as integer fractions:

$$0.333\ldots = 0.\overline{3} = \frac{1}{3} \qquad \text{and} \qquad 2.272727\ldots = 2.\overline{27} = \frac{25}{11}$$

EXAMPLE 1 Show that all the following numbers are rational by writing each as the quotient of two integers:
(a) 5   (b) $-1\frac{2}{3}$ (c) 5.23   (d) $0.\overline{6}$ (e) -8

Solution (a) $5 = \dfrac{5}{1}$ (b) $-1\dfrac{2}{3} = \dfrac{-5}{3}$ (c) $5.23 = \dfrac{523}{100}$

(d) $0.\overline{6} = \dfrac{2}{3}$ (e) $-8 = \dfrac{-8}{1}$

Irrational numbers
Any number that cannot be written as the quotient of two integers is an **irrational number.** Most of us encountered our first irrational number when we used π (Greek letter "pi") to calculate the area and perimeter of a circle. In decimal form $\pi = 3.141592653579\ldots$. This decimal representation of π continues without end, and its digits follow no repeating pattern. All such nonterminating and nonrepeating decimal numbers are irrational.

Irrational numbers also arise in mathematics when we calculate the square root, or higher roots, of certain numbers. The numbers $\sqrt{10}$, $\sqrt[3]{5}$, and $\sqrt[4]{7}$ are irrational numbers because their decimal representations are nonterminating and nonrepeating:

$$\sqrt{10} = 3.162277\ldots \qquad \sqrt[3]{5} = 1.709975\ldots \qquad \sqrt[4]{7} = 1.626576\ldots$$

When the decimal form of an irrational number is used in a calculation, it must be rounded to some approximate value (see Appendix B). For example, in practical applications we can often use $\pi \approx 3.14$ (where $\approx$ means "approximately equal to").

Real numbers
In technical math we work mainly, but not exclusively, with real numbers. The **real numbers** include the rational numbers and the irrational numbers. Note that since all integers are rational numbers, they are also real numbers.

EXAMPLE 2   In the following list of real numbers, which are integers, which are rational numbers, and which are irrational numbers? (A given number may be in more than one category.)

(a) 1.7358   (b) $\dfrac{2}{3}$ (c) 0   (d) $\sqrt{6}$ (e) -9

(f) $\sqrt{16}$ (g) $\sqrt[3]{2}$ (h) $3\frac{7}{8}$ (i) $0.\overline{71}$ (j) $\dfrac{\pi}{2}$

(k) $\dfrac{\sqrt{3}}{2}$ (l) $\sqrt{729}$ (m) $\sqrt{1148}$

Solution   The integers are (c), (e), (f), and (l):

$$729 \;\boxed{\sqrt{}}\;\rightarrow\quad \boxed{27.}$$

The rational numbers are (a), (b), (c), (e), (f), (h), (i), and (l). The irrational numbers are (d), (g), (j), (k), and (m):

$$1148 \;\boxed{\sqrt{}}\;\rightarrow\quad \boxed{33.882149}$$

NOTE ▶

Example 2 (k) illustrates the fact that not all fractions are rational numbers. The number $\sqrt{3}/2$ is an irrational fraction. ◀

Imaginary numbers

To deal with some very important applications of electronics in Chapter 12, we will need numbers beyond the real numbers. There, we will study **imaginary numbers,** which require that we work with the square root of a negative number.

Complex numbers

The **complex numbers** include both the real numbers and the imaginary numbers.

Number line

It will help you to visualize and compare the magnitudes of real numbers if you represent them as points on a **number line.** Notice in Fig. 1-2 that the

Figure 1-2

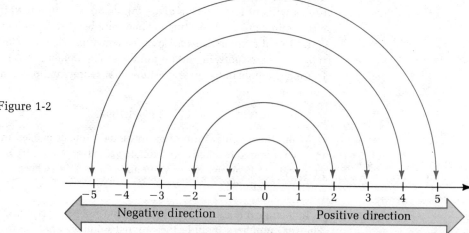

positive numbers are represented to the right of the zero point, and the negative numbers are represented to the left. Furthermore, each number is associated **Opposites** with another number called its **opposite,** which is a number the same distance from zero on the opposite side. The opposite of a number is signified by prefixing it with a "−" sign.

EXAMPLE 3 (a)   The opposite of 4 is −4.
(b)   The opposite of −2 is −(−2) = 2.

In Fig. 1-2 we have labeled only the points representing integers, but every real number can be associated with a point on the number line.

EXAMPLE 4 The real numbers 2, $-\frac{1}{2}$, −3.2, $\sqrt{7}$, π, and $-\frac{7}{3}$ are shown on the following number line:

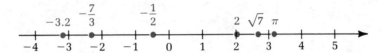

Symbols of inequality   The **symbols of inequality, < (less than)** and **> (greater than),** are used to show the relative position of real numbers on the number line. In a comparison of two numbers, the arrow-shaped symbol points toward the smaller number. Along the number line, the smaller number is to the left.

EXAMPLE 5   Since the number −3.2 is to the left of the number 2 on the number line, then

$-3.2 < 2$

The arrow shape points toward the smaller number, −3.2

This could also be written

$2 > -3.2$

The arrow shape points toward the smaller number, −3.2

To compare fractions and irrational numbers, first convert them to decimal form and then compare their decimal equivalents.

EXAMPLE 6   Insert the correct symbol, < or >, in each box:

(a) $-\dfrac{7}{62}$ ☐ -0.113 (b) $\sqrt{90}$ ☐ 3π (c) 0.09 ☐ $\dfrac{1}{11}$

Solution

(a) 7 ÷ 62 = +/- → -0.1129032 Therefore, $-\dfrac{7}{62} > -0.113$.

(b) $90 \boxed{\sqrt{}} \rightarrow$ $\boxed{9.4868330}$ and $3 \boxed{\times} \boxed{\pi} \boxed{=} \rightarrow$ $\boxed{9.4247780}$
Therefore, $\sqrt{90} > 3\pi$.

(c) $1 \boxed{\div} 11 \boxed{=} \rightarrow$ $\boxed{0.0909091}$ Therefore, $0.09 < \dfrac{1}{11}$.

The symbols $\geq$ (**greater than or equal to**) and $\leq$ (**less than or equal to**) are also frequently used in technical mathematics.

EXAMPLE 7   The statements $5 \geq 2$, $2 \leq 2$, $-7 \leq 0$, and $0 \geq 0$ are all true.

Absolute value   The **absolute value** of a number is the distance along the number line between that number and the zero point. We indicate absolute value by enclosing the number within vertical lines. The absolute value of any number is always a nonnegative number. For example,

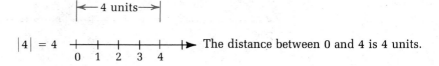

$|4| = 4$ ⟶ The distance between 0 and 4 is 4 units.

$|-5| = 5$ ⟶ The distance between 0 and -5 is 5 units.

EXAMPLE 8   Find:
(a) $|6|$ (b) $|-3|$ (c) $|\pi|$ (d) $|-0.1|$ (e) $|0|$ (f) $-|-1|$

Solution   (a) 6   (b) 3   (c) π (d) 0.1   (e) 0   (f) -1
In part (f) the negative sign is applied after taking the absolute value, because it is outside the vertical lines.

Reciprocal   Another useful concept in technical mathematics is that of the reciprocal of a number. The **reciprocal** of a nonzero* real number is defined as 1 divided by that number. When the number is written as a fraction, its reciprocal is obtained by interchanging the numerator and denominator.

EXAMPLE 9

(a)   The reciprocal of $\dfrac{3}{4}$ is $\dfrac{4}{3}$.

---

*The reciprocal of zero is undefined because, as we shall see in Section 1-3, division by zero is not defined over the real numbers.

(b) The reciprocal of 5, or $\frac{5}{1}$, is $\frac{1}{5}$.

(c) The reciprocal of $-\frac{2}{3}$ is $-\frac{3}{2}$.

(d) To find the reciprocal of a decimal number, first change it to fraction form.

$$0.6 = \frac{3}{5},$$

so the reciprocal of 0.6 is $\frac{5}{3}$.

(e) To find the reciprocal of a number using a calculator, simply enter the number and press the $\boxed{1/x}$ key. The reciprocal of 4.783 is

$$4.783 \; \boxed{1/x} \rightarrow \quad \boxed{0.2090738}$$

Denominate numbers   For many practical applications of mathematics in science and technology you will use numbers arising from physical measurement, such as 5 feet (ft), 60 miles per hour (mi/h or mph), 12.5 square inches (in.^{2} or sq in.), and 350 cubic centimeters (cm^3, cc, or cu cm). Numbers followed by specific units of measurement are called **denominate numbers.** Notice that a denominate number can be thought of as a product or a quotient:

5 ounces (oz) = 5 × 1 oz

4.2 square yards (yd^2 or sq yd) = 4.2 × 1 yd × 1 yd

1 cubic millimeter (mm^3 or cu mm) = 1 mm × 1 mm × 1 mm

$$30 \text{ meters per second (m/s)} = \frac{30 \text{ m}}{1 \text{ s}} = \frac{30 \times 1 \text{ m}}{1 \text{ s}}$$

EXAMPLE 10   Perform the indicated operations and express the answer with the proper units.

(a) 20 km × 5 km   (b) 30 mi ÷ 6 h

(c) 6 in. × 5 in. × 2 in.   (d) 3 × 22 kg

Solution

(a) 100 km^2 (or sq km)   (b) 5 mi/h

(c) 60 in.^{3} (or cu in.)   (d) 66 kg

For a more detailed discussion of denominate numbers, see Appendix B.

Literal numbers   One of the characteristics of algebra that distinguishes it from arithmetic is the use of literal numbers. A **literal number** is a letter or other symbol used to represent a real number. The use of literal numbers allows us to work with and make logical statements about entire sets or classes of numbers.

EXAMPLE 11   In an electronics course you might have learned that the electrical resistance of a wire is given by the formula

$$R = \frac{\rho L}{A}$$

where R, ρ (Greek letter "rho"), L, and A are literal numbers representing the resistance, resistivity, length, and cross-sectional area of the wire, respectively.*

Operating symbols   Addition and subtraction of literal numbers are designated with the usual $+$ and $-$ symbols. However, multiplication is normally represented by writing literal numbers without an operating symbol between them. Division of literal symbols is almost always represented by a fraction bar. Thus, the formula in Example 11 would be read: "Electrical resistance equals resistivity times length divided by area."

Variables and constants   Literal numbers that can take on many different values are called **variables.** Literal numbers that have the same value in a particular situation are called **constants.** In Example 11, R, L, and A are variables, but ρ is a constant literal number because it is always the same for any sample of a particular type of wire.

In mathematics we generally use letters near the beginning of the alphabet or Greek letters for constants and letters near the end of the alphabet for variables. This is not always true in science and technology, however, because literal numbers are often chosen for their memory value: d for distance, t for time, and so on. When two or more values of the same physical quantity appear in a formula, subscripts are used to distinguish them from each other.

EXAMPLE 12   In a physics course you might have learned that the speed of a falling body can be described by the formula

$$v_f = v_i - gt$$

where v_f, v_i, and t are variables representing final velocity, initial velocity, and time of fall, respectively. The literal number g, which represents the acceleration due to gravity, is a constant because it is approximately the same everywhere on the surface of the earth.

Exercises 1-1

Label each of the following real numbers with the correct name: integer, rational, or irrational. In some cases, more than one name may apply.

1. $-6.\overline{4}$ **2.** $0.\overline{18}$ **3.** $\sqrt{3}$ **4.** $\dfrac{\pi}{3}$ **5.** 9.6 **6.** 0.872

7. $-13\frac{1}{6}$ **8.** $\dfrac{7}{12}$ **9.** $\sqrt{25}$ **10.** 13 ▦ **11.** $-\sqrt{3721}$ ▦ **12.** $\sqrt{1371}$

*This formula is explained in more detail in George J. Angerbauer, *Principles of DC and AC Circuits* (Boston: Breton, 1978), pp. 203–204.

Insert the correct symbol of inequality, < or >, in each box.

13. 12 ☐ 18

14. −9 ☐ −13

15. $\frac{1}{2}$ ☐ $\frac{2}{3}$

16. −9.2 ☐ 3.4

17. $\sqrt{5}$ ☐ π

18. $2\frac{3}{8}$ ☐ $-\sqrt{15}$

19. $4.\overline{7}$ ☐ 4.7

20. $-0.\overline{8}$ ☐ −0.8

21. $|13|$ ☐ $|-14|$

22. $|-7|$ ☐ $|5|$

▦ 23. $\sqrt{387}$ ☐ 19

▦ 24. $-\frac{11}{27}$ ☐ −0.407

Evaluate each of the following:

25. $|13|$

26. $|-13|$

27. $|-7.2|$

28. $\left|-\frac{3}{4}\right|$

29. $-|-3|$

30. $-|0.\overline{6}|$

31. $|-33|$

32. $|58|$

Find the reciprocal of each of the following numbers:

33. −4

34. 6

35. $\frac{3}{8}$

36. $-\frac{7}{\pi}$

37. $\sqrt{7}$

38. $\frac{2}{\sqrt{3}}$

39. −0.2

▦ 40. −0.568

Perform the indicated operations and express the answers with the proper units.

41. 7 × 8.0 lb

42. 4 m × 5 m × 2 m

43. 200 km ÷ 4 h

44. 6 ft × 10 ft

45. 60 m ÷ 12 s

46. 3.14 × 14.3 in.

Solve.

47. Show the following numbers on a number line: 4, −2.2, $-\pi$, $-\frac{5}{6}$, $\sqrt{50}$, −0.8

48. Show the following numbers on a number line: $-\sqrt{15}$, $-1\frac{2}{3}$, $|-2|$, −4, −1.67, $\frac{\pi}{3}$

49. *Physics* The potential energy E of an object varies with its mass m, its height h, and the acceleration due to gravity g according to the formula

$$E = mgh$$

(a) State this formula in words.
(b) Identify the literal symbols that represent constants and those that are variables.

50. *Electronics* The current I_b in the base wire of a transistor is related to the emitter current I_e by the formula

$$I_b = I_e(1 - \alpha)$$

where α (Greek letter "alpha") is a specified value that remains the same for a given transistor.
(a) State this formula in words.
(b) Identify the literal symbols that represent constants and those that are variables.

51. Given any integer:
(a) Is its absolute value always an integer?
(b) Is its reciprocal always an integer?

52. Given any rational number:
(a) Is its absolute value always a rational number?
(b) Is its reciprocal always a rational number?

53. *Optics* An industrial optical device has two convex lenses, each with a focal length f_l of 150 mm.
(a) Using the formula

$$\frac{1}{D_i} = \frac{1}{f_l} - \frac{1}{D_o}$$

the reciprocal of the image distance for the first lens is found to be $1/D_i = 2/520$ mm^{-1}. Find D_i itself by taking the reciprocal of $1/D_i$. The answer will be in mm.*

(b) Using the same formula with the second lens, $1/D_i$ is found to be 0.0028 mm^{-1}. Find D_i for this lens.

54. *Electronics*   In a particular dc circuit, resistances of 1500 ohms (Ω) and 2000 Ω are connected in parallel so that $1/R_t$ is equal to $7/6000$ Ω^{-1}, where R_t is the total resistance. Find R_t to the nearest 10 Ω.

55. *Astronomy*   The distance to a star (measured in seconds of an arc) is defined as the reciprocal of its heliocentric parallax. What is the distance to a star whose heliocentric parallax is 0.40?

56. *Electronics*   The pulse period T in an RC circuit is the reciprocal of the pulse-repetition frequency f. Find T when f is 1480 hertz (Hz).

1-2 | Laws and Operations of Real Numbers

There are certain useful laws that we often take for granted when performing arithmetic. Since these laws are just as useful but not as obvious when dealing with literal numbers, we must explicitly define them before proceeding with algebra.

Commutative laws   When adding two real numbers such as 2 and 5, we obtain the same answer, 7, if we state the sum in the order $2 + 5$ or as $5 + 2$. Similarly, we can write the product of the numbers 4 and 6 as $4 \cdot 6$ or in reverse order as $6 \cdot 4$, and arrive at the same result, 24. Because we can reverse the *order* of two numbers in a sum or product, we say that addition and multiplication of real numbers are **commutative.**

Commutative Law of Addition $a + b = b + a$
Commutative Law of Multiplication $a \cdot b = b \cdot a$
where a and b are any two real numbers.

CAUTION ▶   Subtraction and division are **not** commutative. For example,

$5 - 2$ is *not* equal to $2 - 5$

$6 \div 2$ is *not* equal to $2 \div 6$ ◀

We normally use the commutative laws to simplify a larger calculation. For example, to calculate

$30 + 56 + 70 = ?$

rearrange the numbers using the commutative law of addition:

$30 + 70 + 56 = 100 + 56 = 156$

*Adapted from Abraham Marcus and James R. Thrower, *Introduction to Applied Physics* (Boston: Breton, 1980), p. 400.

Similarly, to find the product

$$25 \cdot 73 \cdot 4 = ?$$

rearrange the numbers using the commutative law of multiplication:

$$25 \cdot 4 \cdot 73 = 100 \cdot 73 = 7300$$

There is no need for a calculator this way.

Associative laws   Another law of real numbers may be illustrated by the calculation $37 + 80 + 20$. Without changing the order of the numbers, we can still add 80 and 20 first to simplify our computation. By doing so we are saying that

$$(37 + 80) + 20 = 37 + (80 + 20)$$

where parentheses indicate which operation to perform first.

The same sort of rule holds true for multiplication. To find the product $93 \cdot 20 \cdot 5$, it is easier to multiply $20 \cdot 5$ first than to begin with $93 \cdot 20$. Since the final result is the same, we say that

$$(93 \cdot 20) \cdot 5 = 93 \cdot (20 \cdot 5)$$

Because we can begin the sum or product of three numbers with either the first pair or the second pair, we say that addition and multiplication of real numbers are **associative.**

Associative Law of Addition $(a + b) + c = a + (b + c)$

Associative Law of Multiplication $(ab)c = a(bc)$

where a, b, and c are any three real numbers.

CAUTION ▶   Subtraction and division are **not** associative. For example,

$(8 - 4) - 9$ is *not* equal to $8 - (4 - 9)$

$(6 \div 2) \div 12$ is *not* equal to $6 \div (2 \div 12)$ ◀

EXAMPLE 13   Rewrite each of the following in a way that will simplify the calculation, and state the law used:
(a) $35 + 79 + 15$ (b) $(27 \cdot 4) \cdot 25$
(c) $86 + (14 + 39)$ (d) $200 \cdot 43 \cdot 5$

Solution
(a) $35 + 15 + 79$ Commutative law of addition
(b) $27 \cdot (4 \cdot 25)$ Associative law of multiplication
(c) $(86 + 14) + 39$ Associative law of addition
(d) $200 \cdot 5 \cdot 43$ Commutative law of multiplication

Distributive law   In algebra, one of the most useful laws of real numbers is the distributive law. The **distributive law** states that a calculation such as $5(6 + 2)$ may be rewritten $5 \cdot 6 + 5 \cdot 2$. The first statement gives us $5 \cdot 8 = 40$, while the second yields $30 + 10 = 40$.

Distributive Law

$$a(b + c) = ab + ac$$

$$a(b - c) = ab - ac$$

where a, b, and c are any three real numbers.

The distributive law will help us simplify algebraic expressions in Chapter 2. The reverse of the distributive law helps simplify arithmetic calculations. For example, since $ab + ac = a(b + c)$, then

$$3 \cdot 7 + 3 \cdot 9 = 3(7 + 9) = 3(16) = 48$$

EXAMPLE 14   Rewrite each of the following using the distributive law:
(a) $m(n - p)$ (b) $14 \cdot 37 + 14 \cdot 13$
(c) $7 \cdot 12 - 5 \cdot 12$ (d) $3(x + y + z)$

Solution
(a) $mn - mp$ (b) $14(37 + 13)$ (c) $(7 - 5)12$ (d) $3x + 3y + 3z$

Addition of signed   The laws given so far help simplify arithmetic and algebraic operations, but
numbers   we must also know the rules for performing the operations, particularly when dealing with signed numbers.

When adding two numbers of like signs: Add their absolute values and attach their common sign to the sum.

Like signs   Add their absolute values: $|5| + |2| = 5 + 2$

$$5 + 2 = 7 \quad \text{or} \quad (+5) + (+2) = +(5 + 2) = +7$$

Attach their common sign

Like signs   Add their absolute values: $|-8| + |-3| = 8 + 3$

$$(-8) + (-3) = -(8 + 3) = -11$$

Attach their common sign

When adding two numbers of unlike signs: Subtract the smaller absolute value from the larger and attach to the result the sign of the number with the larger absolute value.

Unlike signs

Subtract their absolute values:
$|+6| - |-4| = 6 - 4$

$(+6) + (-4) = +(6 - 4) = +2$

Sign of +6

Unlike signs

Subtract their absolute values:
$|-9| - |+4| = 9 - 4$

$(+4) + (-9) = -(9 - 4) = -5$

Sign of −9

NOTE ▶ Normally, the "+" sign is omitted in front of a positive number. ◀

EXAMPLE 15 Add the following numbers:

(a) $(-6) + (-8)$ (b) $(-8) + 12$ (c) $\left(-\dfrac{3}{4}\right) + \dfrac{5}{8}$

(d) $(-13.487) + 28.642 + (-19.915)$

(e) Four different stock transactions resulted in a loss of \$418, a profit of \$1265, a loss of \$1944, and a profit of \$568. What was the net result of the transactions?

Solution

(a) $(-6) + (-8) = -(6 + 8) = -14$

(b) $(-8) + 12 = +(12 - 8) = 4$

(c) For fractions, first find a common denominator:

$$\left(-\frac{3}{4}\right) + \left(\frac{5}{8}\right) = \left(-\frac{6}{8}\right) + \left(\frac{5}{8}\right)$$

Then add the numerators as signed numbers:

$$\frac{(-6) + 5}{8} = \frac{-(6 - 5)}{8} = \frac{-1}{8} = -\frac{1}{8}$$

(d) $13.487 \;[\text{+/-}]\;[\text{+}]\; 28.642 \;[\text{+}]\; 19.915 \;[\text{+/-}]\;[\text{=}] \rightarrow$ *−4.76*

(e) This problem translates as

$$(-418) + 1265 + (-1944) + 568 = ?$$

For calculations involving more than two numbers, add those with like signs first. Use the commutative law to rearrange as

$$(-418) + (-1944) + 1265 + 568 = ?$$

Find the negative total and the positive total separately:

$$(-2362) + 1833 = ?$$

Then find the difference of their absolute values and attach the proper sign:

$$= -529$$

NOTE ▶   Although you may use calculators to perform signed number arithmetic as in Example 15(d), you should not depend upon them for simple problems. To succeed in algebra, learn the rules for signed number computations and use your calculator only for problems involving decimals or very large numbers. ◀

Subtraction of signed numbers   The basic subtraction facts tell us that $11 - 8 = 3$, but technical math often involves subtracting a larger number from a smaller number, as in $8 - 11$. The addition rules imply that subtracting is the same as adding the opposite:

$$11 - 8 = 11 + (-8) = 3$$

Therefore, we may state a general rule for subtraction of signed numbers.

   When subtracting one real number from another: Change the sign of the one being subtracted and then combine them using the rules for addition of signed numbers.

To subtract		Add the opposite

$$8 - (+11) = 8 + (-11) = -3$$

To subtract		Add the opposite

$$(-5) - (-9) = (-5) + (+9) = +4 = 4$$

EXAMPLE 16   Perform the indicated operations.

(a) $5 - (-6)$ (b) $(-12) - 11$ (c) $4 - 13$

(d) $\left(-\dfrac{1}{4}\right) - \left(-\dfrac{1}{2}\right)$ ▦ (e) $6.53 - 11.16$ ▦ (f) $5378 - (-7429)$

(g)   The temperature on Venus ranges from a high of 497°C to a low of -23°C. What is the temperature range?

Solution

(a) $5 - (-6) = 5 + (+6) = 11$

(b) $(-12) - 11 = (-12) + (-11) = -23$

(c) $4 - 13 = 4 + (-13) = -9$

(d) $\left(-\dfrac{1}{4}\right) - \left(-\dfrac{1}{2}\right) = \left(-\dfrac{1}{4}\right) + \left(+\dfrac{2}{4}\right) = \dfrac{1}{4}$

(e) $6.53 \boxed{-} 11.16 \boxed{=} \rightarrow$ *-4.63*

(f) $5378 \boxed{-} 7429 \boxed{+/-} \boxed{=} \rightarrow$ *12807.*

(g)   To find the temperature range means to find the difference between the extremes. Thus, we subtract:

$$497 - (-23) = 497 + (+23) = 520°C$$

Multiplication and division of signed numbers   To multiply or divide a pair of signed numbers: First multiply or divide their absolute values; then find the sign of the result as follows:

   1.   **If the two numbers have like signs, their product or quotient is positive.**

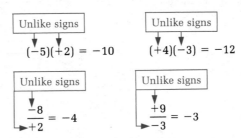

Like signs

$(+5)(+2) = +10$

Like signs

$(-4)(-3) = +12$

Like signs

$$\frac{+8}{+2} = +4$$

Like signs

$$\frac{-9}{-3} = +3$$

2. **If the two numbers have unlike signs, their product or quotient is negative.**

Unlike signs

$(-5)(+2) = -10$

Unlike signs

$(+4)(-3) = -12$

Unlike signs

$$\frac{-8}{+2} = -4$$

Unlike signs

$$\frac{+9}{-3} = -3$$

EXAMPLE 17   Perform the indicated operations.

(a) $(-9)(-11)$

(b) $\dfrac{-72}{9}$

(c) $7(-5)$

(d) $\dfrac{-48}{-8}$

(e) $\left(-\dfrac{4}{5}\right)\left(\dfrac{5}{8}\right)$

(f) $\left(\dfrac{3}{4}\right) \div \left(-\dfrac{9}{10}\right)$

(g) $(-624)(593)$

(h) $\dfrac{-108.04}{-36.5}$

Solution

(a)   99   (like signs, positive answer)

(b) -8 (unlike signs, negative result)

(c) -35 (unlike signs, negative answer)

(d)   6   (like signs, positive quotient)

(e) $\left(-\dfrac{4}{5}\right)\left(\dfrac{5}{8}\right) = \left(-\dfrac{\overset{1}{\cancel{4}}}{\underset{1}{\cancel{5}}}\right)\left(\dfrac{\overset{1}{\cancel{5}}}{\underset{2}{\cancel{8}}}\right) = -\dfrac{1}{2}$ (unlike signs, negative product)

(f) $\left(\dfrac{3}{4}\right) \div \left(-\dfrac{9}{10}\right) = \left(\dfrac{\overset{1}{\cancel{3}}}{\underset{2}{\cancel{4}}}\right)\left(-\dfrac{\overset{5}{\cancel{10}}}{\underset{3}{\cancel{9}}}\right) = -\dfrac{5}{6}$ (unlike signs, negative answer)

(g)   624 $\boxed{+/-}$ $\boxed{\times}$ 593 $\boxed{=}$ → $-370032.$

(h)   108.04 $\boxed{+/-}$ $\boxed{\div}$ 36.5 $\boxed{+/-}$ $\boxed{=}$ → 2.96

LEARNING HINT ▶ Instead of using the $\boxed{+/-}$ key for calculator products and quotients, you might find it faster to enter all numbers as positive and determine the sign of the answer yourself. ◀

When multiplying or dividing more than two real numbers, simply count the number of negative signs to determine the sign of your answer. An even number of negative signs results in a positive answer; an odd number of negative signs results in a negative answer.

EXAMPLE 18   Perform the indicated operations.

(a) $(-3)(2)(-5)(-4)$ (b) $\dfrac{(-2)(-5)(-8)}{-16}$

Solution

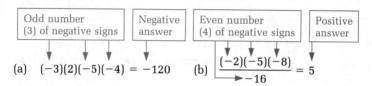

(a) $(-3)(2)(-5)(-4) = -120$ (b) $\dfrac{(-2)(-5)(-8)}{-16} = 5$

Exercises 1-2

Perform the indicated operations.

1. $(-13) + 5$
2. $7 + 9$
3. $(-6) + (-11)$
4. $10 + (-7)$
5. $5 - 14$
6. $6 - (-2)$
7. $-22 - (-13)$
8. $26 - 15$
9. $8(-2)$
10. $(-6)(-5)$
11. $(-17)(3)$
12. $8(9)$
13. $\dfrac{-12}{-3}$
14. $\dfrac{24}{-4}$
15. $\dfrac{-56}{8}$
16. $\dfrac{42}{14}$
17. $\dfrac{5}{6} + \left(-\dfrac{2}{3}\right)$
18. $-2\frac{5}{8} - 1\frac{1}{6}$
19. $(-2.7) - (-3.8)$
20. $0.8 + (-0.5)$
21. $(2.4)(0.4)$
22. $(-3\frac{1}{2})(-1\frac{1}{7})$
23. $\left(-\dfrac{7}{12}\right) \div \dfrac{2}{9}$
24. $\dfrac{-4.6}{-0.2}$
25. $12 - (-9) + 13$
26. $-8 + 14 - 22$
27. $(-4)(-5)(-7)$
28. $(-2)(5)(-6)$
29. $\dfrac{(12)(-4)}{-2}$
30. $\dfrac{(-8)(-9)}{-6}$
31. $-8.732 - 12.964 + 14.783$
32. $(-749)(-327)(-243)$
33. $\dfrac{-63.4}{17.28}$
34. $(-2846) + 3978 + (-3264) - (-4173)$

Name the mathematical law that is being used in each of the following:

35. $8(x + w) = 8x + 8w$
36. $50(2 \cdot 8) = (50 \cdot 2)8$
37. $9 \cdot 8 \cdot 7 = 9 \cdot 7 \cdot 8$
38. $5 \cdot 12 + 5 \cdot 13 = 5(12 + 13)$
39. $37 + 24 + 13 = 37 + 13 + 24$
40. $(41 + 17) + 23 = 41 + (17 + 23)$
41. $a(b + c) = (b + c)a$
42. $17(12 \cdot 8) = (12 \cdot 8)17$

Solve.

43. *Electronics*   In a certain electrostatic field, the voltage at one point is 1850 volts (V), and the voltage at another point is -860 V. Find the potential difference between the two points by subtracting the second voltage from the first.

44. *Astronomy*   The difference between apparent magnitude and absolute magnitude is used to find the distance to a star. If the apparent magnitude is $+21$ and the absolute magnitude is -4, what is this difference?

45. *Meteorology*   In Fargo, North Dakota, the temperature at 9 PM was $-7°$F. By midnight it had dropped another $6°$. What was the temperature then?

46. *Planetary science* The temperature at a certain place on Mars ranges from $-128°C$ to $20°C$. What is the difference between these extremes?
47. *Industrial management* An engineering company showed a loss of $235,000 for the first half of the year. The company's annual statement at the end of the year showed an overall profit of $523,000. What was the profit for the second half of the year?
48. If a negative number is used as a factor sixteen times in a product, what is the sign of the product?
49. If a negative number is used as a factor nine times in a product, what is the sign of the product?
50. *Business* A computer services business purchases a dozen reams of paper at $16 per ream and a dozen boxes of ribbons at $9 per box. Set up a calculation for determining the total cost of the paper and ribbons.

1-3 | Zero and the Order of Operations

Two topics concerning operations with real numbers seem to cause students considerable difficulty. They are operations with zero and the order of operations. We shall therefore devote an entire section to these topics.

Operations with zero

Addition, subtraction, and multiplication with zero cause few problems. We can summarize these operations as follows:

Operations with Zero

For any real number a:

EXAMPLES:

Addition	$a + 0 = a$	$8 + 0 = 8$
	$0 + a = a$	$0 + (-5) = -5$
Subtraction	$a - 0 = a$	$5 - 0 = 5$
	$0 - a = -a$	$0 - 9 = -9, \quad 0 - (-6) = 6$
Multiplication	$a \cdot 0 = 0$	$32 \cdot 0 = 0$
	$0 \cdot a = 0$	$0(\frac{1}{2}) = 0$

Division with zero

Division with zero causes the most problems. Dividing zero by any real number except zero itself results in zero. But dividing a real number by zero has no meaning; it is undefined. To convince yourself of this, remember that

$$\frac{8}{4} = 2 \quad \text{because} \quad 8 = 4 \cdot 2$$

Similarly,

$$\frac{0}{4} = 0 \quad \text{because} \quad 0 = 4 \cdot 0$$

But if

$$\frac{8}{0} = a \qquad \text{where } a \text{ is some real number}$$

then

$$8 = 0 \cdot a \qquad \text{which is impossible}$$

Zero multiplied by any real number is zero.

We can summarize the rules for division involving zero as follows:

Division with Zero
For any nonzero real number a: **EXAMPLES:**

$$\frac{0}{a} = 0$$ $$\frac{0}{7} = 0, \quad \frac{0}{-2.5} = 0, \quad 0 \div \frac{2}{3} = 0$$

$$\frac{a}{0} \text{ is undefined}$$ $$\frac{7}{0}, \quad \frac{-2.5}{0}, \quad \text{and} \quad \frac{2}{3} \div 0$$
are all undefined

IMPORTANT ▶   The expression $\frac{0}{0}$ is indeterminate. This means it is not a meaningful quantity in algebra. ◀

EXAMPLE 19   Perform the indicated operations.

(a) $0 + 12$ (b) $(-4) \cdot 0$ (c) $0 \div \frac{3}{5}$ (d) $0 - 20$

(e) $\dfrac{0}{1}$ (f) $17 - 0$ (g) $0(3.75)$ (h) $0 - (-48)$

(i) $\dfrac{2}{0}$ (j) $-34 + 0$ (k) $\dfrac{0}{-3.6}$ (l) $\left(-\dfrac{1}{3}\right) \div 0$

Solution

(a) 12   (b) 0   (c) 0   (d) -20
(e) 0   (f) 17   (g) 0   (h) 48
(i) Undefined   (j) -34 (k) 0   (l) Undefined

Attempting to divide by zero on a calculator will produce an error signal. For example,

$2 \; \boxed{\div} \; 0 \; \boxed{=} \; \rightarrow$ $Error$, E $0.$, $E.$

The order of operations   So far we have discussed only problems involving a single operation. Suppose two operations, such as addition and multiplication, must be performed in the same problem. Which do we do first?

Here is a summary of the rules for the **order of operations.** It is simply an agreement among all mathematicians which facilitates communication.

1. **First,** simplify within grouping symbols.
2. **Second,** perform all multiplication and division, proceeding from left to right.
3. **Third,** perform all addition and subtraction, proceeding from left to right.

EXAMPLE 20 Calculate: $6 - 2(4 + 6) \div 5 + 9$

Solution

First, simplify within parentheses.

$$6 - 2(4 + 6) \div 5 + 9 = 6 - 2(10) \div 5 + 9$$

Second, multiply and divide from left to right.

$$= 6 - 20 \div 5 + 9$$
$$= 6 - 4 + 9$$

Third, add and subtract from left to right.

$$= 2 + 9$$
$$= 11$$

Division as a fraction

Division is often indicated as a fraction in complex calculations. In this type of problem, treat the numerator and denominator as if each were enclosed in parentheses. Simplify each separately before dividing. Follow the step-by-step process in the next example.

EXAMPLE 21 Calculate: $\dfrac{3 \cdot 2 - 4 \cdot 6}{2(6 - 8)}$

Solution

Beginning with the denominator, simplify within parentheses, and then multiply.

$$\frac{3 \cdot 2 - 4 \cdot 6}{2(6 - 8)} = \frac{3 \cdot 2 - 4 \cdot 6}{2(-2)}$$

$$= \frac{3 \cdot 2 - 4 \cdot 6}{-4}$$

Then, in the numerator, multiply,

$$= \frac{6 - 24}{-4}$$

and subtract.

$$= \frac{-18}{-4}$$

Finally, divide.

$$= 4.5$$

Here are some more examples.

EXAMPLE 22 Perform all operations in the correct order.

(a) $16 + (8 - 3 \cdot 2) - 2$ ▦ (b) $\dfrac{42 - 8(17.2 - 9.6)}{5.8 + 9(-1.7)}$

(c) In determining the heat of reaction (in kilocalories) of a certain combination of chemicals, the following calculation arises:

$$H = -2(94.4) - 2(-28.0) \text{ kcal}$$

Compute this heat of reaction.*

Solution

(a)   Multiply inside parentheses. $16 + (8 - 3 \cdot 2) - 2 = 16 + (8 - 6) - 2$

Subtract inside parentheses. $= 16 + (2) - 2$

Add on the left. $= 18 - 2$

Subtract. $= 16$

(b)   Modern scientific calculators perform the order of operations automatically. However, you must be careful when entering a complex calculation such as this. **First,** compute the numerator:

$$42 \boxed{-} 8 \boxed{\times} \boxed{(} 17.2 \boxed{-} 9.6 \boxed{)} \boxed{=}$$

Then, divide by the denominator, making certain to enclose it in parentheses:

$$\boxed{\div} \boxed{(} 5.8 \boxed{+} 9 \boxed{\times} 1.7 \boxed{+/-} \boxed{)} \boxed{=} \rightarrow \quad \boxed{\textit{1.9789474}}$$

(c) $$H = -2(94.4) - 2(-28.0)$$

First, multiply. $$H = -188.8 - (-56.0)$$

Then, subtract. $$H = -132.8 \text{ kcal}$$

Exercises 1-3

Perform the indicated operations.

1. $17 + 0$ **2.** $0 + 23$ **3.** $24 - 0$ **4.** $0 - 19$ **5.** $0 - (-6)$

6. $0 - (-4.5)$ **7.** $0(8)$ **8.** $(-6)0$ **9.** $\dfrac{0}{100}$ **10.** $\dfrac{-18}{0}$

11. $\dfrac{1}{4} \div 0$ **12.** $0 \div (-2\frac{1}{8})$ **13.** $7 + 3 \cdot 2$ **14.** $8 - \dfrac{6}{2}$ **15.** $5(7) - 3(4)$

16. $17 - 12 \cdot 3 - 8 \cdot 5$ **17.** $\dfrac{8(-7)}{-4} - 12(5 - 9)$ **18.** $9(4 - 11) - \dfrac{6(-3)}{2}$

19. $26 - \dfrac{-12}{4} - 8(-3)$ **20.** $9 - \left(\dfrac{-42}{-7}\right) + 7(-4)$ **21.** $\dfrac{-28}{-4} - \dfrac{12 - 18}{0 - (-2)}$

22. $\dfrac{48}{-16} - \dfrac{6 - 20}{0 - 7}$ **23.** $\dfrac{5 \cdot 7 - 7 \cdot 11}{-5 - (-11)}$ **24.** $\dfrac{(-9)3 + 8(-6)}{5 - 6 \cdot 5}$

25. $8.65 - 7.24(3.58)$ **26.** $4273 + \dfrac{13{,}572}{-234}$ **27.** $0.2 - 0.9(5.2) - 3.7$

28. $5921 + 17(738 + 92 \cdot 56)$ **29.** $\dfrac{(243)(729) + (650)(82)}{2820 + 1883}$ **30.** $\dfrac{(1.274)18 - 12.652}{6.51 - 2(1.97)}$

*See Sidney W. Benson, *Chemical Calculations*, 2d ed. (New York: Wiley, 1963), p. 59.

Solve.

31. *Industrial management* A factory purchases twelve sheets of steel at $16 each, twenty boxes of fasteners at $8.40 per box, and fifty quarts of solvent at $2.50 per quart. Set up the calculation for determining the total cost of these items and then compute the total.

32. *Business* A computer data base charges $8.50 for a hookup and $3.50 per hour thereafter. Show the calculation necessary for a 4 h hookup and then compute this charge.

33. *Business* The Bell-Tell Co. charges $0.86 for the first minute and $0.52 for each additional minute for a long-distance call. Set up the calculation for finding the cost of an 8 min call and then compute this cost.

34. *Business* A wholesale parts supplier discounts the unit cost of an item by $0.04 for every dozen ordered. A manufacturer orders 204 items whose normal unit cost is $12.98. Set up the calculation for finding the discounted unit cost and then compute this cost.

35. *Thermodynamics* Perform the following calculation to determine the neutral temperature (in degrees Celsius) of a certain iron–copper thermocouple:

$$\frac{-(16.7 - 2.71)}{-0.0297 - 0.00790}$$

36. *Thermodynamics* Perform the following calculation to determine the inversion temperature (in degrees Celsius) of a certain nickel–platinum thermocouple:

$$\frac{-2[-19.1 - (-3.03)]}{-3.02 - (-3.25)}$$

1-4 | Exponents

So far we have concentrated on the four operations of basic arithmetic: addition, subtraction, multiplication, and division. In this section we shall discuss a convenient way to express repeated multiplication.

Exponential notation When a number is written as a product, the multipliers are called **factors** of the number. For example,

$$3 \cdot 5 = 15$$

Factors of 15 | Product

When a factor is used repeatedly in a product, we usually use **exponential notation** to indicate the product.

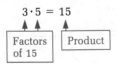

$$\underline{3 \cdot 3} = 3^2$$ Read "three to the second power," or "three squared"

Two factors of 3

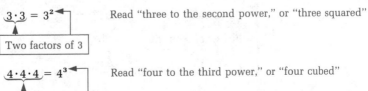

$$\underline{4 \cdot 4 \cdot 4} = 4^3$$ Read "four to the third power," or "four cubed"

Three factors of 4

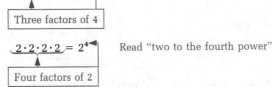

$$\underline{2 \cdot 2 \cdot 2 \cdot 2} = 2^4$$ Read "two to the fourth power"

Four factors of 2

In general, the expression a^n, for some positive integer n, indicates a real number a used as a factor n times. The real number a is called the **base,** and the number n is called the **exponent.** A number written without an exponent is understood to have an exponent of 1. Thus, $a = a^1$.

EXAMPLE 23

(a) Name the base and the exponent: 6^7 and x^8
(b) Write using exponents: $7 \cdot 7 \cdot 7 \cdot 7$ and $y \cdot y \cdot y$
(c) Express 10,000 as a power of 10.
(d) Write 8 as a power of 2.
(e) Calculate: 3^4, $(-4)^3$, $(-4)^4$
(f) Calculate: 376^2, 25^5

Solution

(a) 6^7: base is 6, exponent is 7; x^8: base is x, exponent is 8
(b) $7 \cdot 7 \cdot 7 \cdot 7 = 7^4$; $y \cdot y \cdot y = y^3$
(c) $10,000 = 10 \cdot 10 \cdot 10 \cdot 10 = 10^4$
(d) $8 = 2 \cdot 2 \cdot 2 = 2^3$
(e) $3^4 = 3 \cdot 3 \cdot 3 \cdot 3 = 81$; $(-4)^3 = (-4)(-4)(-4) = -64$;
$(-4)^4 = (-4)(-4)(-4)(-4) = 256$
(f) $376 \,\boxed{x^2} \rightarrow$ _141376._

$25 \,\boxed{y^x}\, 5 \,\boxed{=}\, \rightarrow$ _9765625._

LEARNING HINT ▶ Notice in Example 23(e) that a negative number raised to an odd power is negative, and a negative number raised to an even power is positive. ◀

Many operations can be performed on numbers written in exponential notation. The following list summarizes the rules for performing these operations. We will explain and illustrate these laws of exponents in the remainder of this section.

Laws of Exponents

For any real numbers a and b and any positive integers m and n:

Multiplication	$a^m \cdot a^n = a^{m+n}$	(1-1)
Power of a Power	$(a^m)^n = a^{mn}$	(1-2)
Power of a Product	$(ab)^m = a^m b^m$	(1-3)
Power of a Quotient	$\left(\dfrac{a}{b}\right)^m = \dfrac{a^m}{b^m}$ (where $b \neq 0$)	(1-4)
Division	$\dfrac{a^m}{a^n} = a^{m-n}$ (where $a \neq 0$)	(1-5)

Multiplication   According to the multiplication law (1-1), to multiply two numbers written in exponential notation with the same base, add the exponents and keep the original base. For example,

$$2^3 \quad \cdot \quad 2^4 = 2^{3+4} = \quad 2^7$$

Because $\boxed{2 \cdot 2 \cdot 2} \cdot \boxed{2 \cdot 2 \cdot 2 \cdot 2} = \boxed{2 \cdot 2 \cdot 2 \cdot 2 \cdot 2 \cdot 2 \cdot 2}$

This rule can be applied to calculations involving more than two numbers as long as the bases are the same.

EXAMPLE 24   If possible, express each product in exponential notation using the original base.
(a) $6^3 \cdot 6^5$ (b) $y^2 \cdot y^4$ (c) $3^7 \cdot 3 \cdot 3^5$ (d) $5^3 \cdot 9^2 \cdot 5^4$ (e) $4m^2m^3$

Solution
(a) $6^3 \cdot 6^5 = 6^{3+5} = 6^8$ (b) $y^2 \cdot y^4 = y^{2+4} = y^6$
(c) $3^7 \cdot 3 \cdot 3^5 = 3^{7+1+5} = 3^{13}$ (Remember, $3 = 3^1$.)
(d) Only the powers of 5 can be combined:

$$5^3 \cdot 9^2 \cdot 5^4 = 5^{3+4} \cdot 9^2 = 5^7 \cdot 9^2$$

(e) Only the powers of m can be combined:

$$4m^2m^3 = 4m^{2+3} = 4m^5$$

Power of a power   According to the power law (1-2), to raise a power to a power, multiply the two exponents and keep the same base. For example,

$$(2^3)^2 = 2^{3 \cdot 2} = 2^6$$

because

$$(2^3)^2 = 2^3 \cdot 2^3 = 2^{3+3} = 2^6$$

EXAMPLE 25   Express each of the following in exponential notation using a single power:
(a) $(4^3)^5$ (b) $(x^5)^4$ (c) $(y^3)(y^6)$

Solution
(a) $(4^3)^5 = 4^{3 \cdot 5} = 4^{15}$ (b) $(x^5)^4 = x^{5 \cdot 4} = x^{20}$ (c) $(y^3)(y^6) = y^{3+6} = y^9$

Did Example 25(c) fool you? You must learn to spot the difference between the "power of a power" and a straight multiplication problem such as that given in part (c). In the former, multiply exponents; in the latter, add exponents.

Power of a product   Law (1-3) shows that to determine the power of a product, first raise each number to that power, and then multiply the results. Thus,

$$(x \cdot y)^3 = x^3 \cdot y^3$$

because

$$(x \cdot y)^3 = (x \cdot y)(x \cdot y)(x \cdot y) \qquad \text{By the definition of exponents}$$

$$= (x \cdot x \cdot x)(y \cdot y \cdot y) \qquad \text{By the associative and commutative laws}$$

$$= x^3 \cdot y^3 \qquad \text{By the definition of exponents}$$

You can apply this rule in reverse to simplify arithmetic calculations. For example, the product $2^4 \cdot 5^4$ can be rewritten as $(2 \cdot 5)^4 = 10^4$, which is easier to compute than the original problem.

EXAMPLE 26   Rewrite each of the following using the power of a product rule:
(a) $(mn)^5$ (b) $3^6 \cdot 4^6$ (c) $w^2 z^2$ (d) $(2x)^3$
(e) $4^4 \cdot 5^4 \cdot 6^4$ (f) $(r^2 s^3)^4$ (g) $(-3p^5)^3$

Solution
(a) $m^5 n^5$ (b) 12^6 (c) $(wz)^2$ (d) $2^3 \cdot x^3 = 8x^3$
(e) 120^4 (f) $r^8 s^{12}$ (g) $(-3)^3 \cdot p^{15} = -27p^{15}$

Power of a quotient   Law (1-4) states that the power of the quotient of two numbers is equal to the quotient of the powers of the numbers. The following example shows how we arrive at this conclusion:

$$\left(\frac{2}{3}\right)^2 = \frac{2}{3} \cdot \frac{2}{3} = \frac{2 \cdot 2}{3 \cdot 3} = \frac{2^2}{3^2}$$

EXAMPLE 27   Rewrite each of the following using the power of a quotient rule:*

(a) $\left(\dfrac{3}{5}\right)^4$ (b) $\left(\dfrac{x}{y}\right)^3$ (c) $\left(\dfrac{w^2}{5}\right)^3$

(d)   Simplify the following expression for the centripetal acceleration a_c of the earth at the equator:

$$a_c = \left(\frac{2\pi}{T}\right)^2 R$$

Solution
(a) $\dfrac{3^4}{5^4} = \dfrac{81}{625}$ (b) $\dfrac{x^3}{y^3}$ (c) $\dfrac{w^6}{5^3} = \dfrac{w^6}{125}$ (d) $a_c = \dfrac{4\pi^2 R}{T^2}$

Division   According to the division law (1-5), to divide two numbers written in exponential notation with the same base, subtract the exponent in the denominator from the exponent in the numerator and keep the common base. For example,

---

*Since division by zero is undefined, all variables that appear as divisors from now on are assumed to be nonzero unless stated otherwise.

$$\frac{2^5}{2^2} = 2^{5-2} = 2^3 \qquad \text{because} \qquad \frac{\cancel{2} \cdot \cancel{2} \cdot 2 \cdot 2 \cdot 2}{\cancel{2} \cdot \cancel{2}} = 2 \cdot 2 \cdot 2$$

EXAMPLE 28 If possible, express each quotient in exponential notation using the original base.

(a) $\dfrac{10^7}{10^3}$ (b) $\dfrac{x^6}{x^4}$ (c) $\dfrac{z^4}{z}$ (d) $\dfrac{7^8}{6^3}$

Solution

(a) $\dfrac{10^7}{10^3} = 10^{7-3} = 10^4$ (b) $\dfrac{x^6}{x^4} = x^{6-4} = x^2$

(c) $\dfrac{z^4}{z} = z^{4-1} = z^3$ (d) $\dfrac{7^8}{6^3}$ cannot be simplified, because the bases are not the same.

Negative exponents   Suppose we apply the division law (1-5) to a problem in which the power in the denominator is greater than the power in the numerator. For example, dividing 3^2 by 3^6 we would have

$$\frac{3^2}{3^6} = 3^{2-6} = 3^{-4}$$

However, if we do this same problem by dividing out common factors in the numerator and denominator, we obtain the following result:

$$\frac{3^2}{3^6} = \frac{\cancel{3} \cdot \cancel{3}}{\cancel{3} \cdot \cancel{3} \cdot 3 \cdot 3 \cdot 3 \cdot 3} = \frac{1}{3^4}$$

To reconcile these two results, mathematicians have defined a negative exponent as follows:

$$a^{-m} = \frac{1}{a^m} \qquad \begin{array}{l} \text{for any real number } a \text{ and integer } m \\ \text{(where } a \neq 0) \end{array} \qquad \text{(1-6)}$$

EXAMPLE 29 Rewrite using positive exponents.

(a) x^{-3} (b) $\dfrac{1}{4^{-6}}$ (c) $\dfrac{3a^{-2}}{b^{-4}c^5}$

Solution

(a) $\dfrac{1}{x^3}$ (b) 4^6 (c) $\dfrac{3b^4}{a^2c^5}$

Negative exponents are allowed only in certain mathematical applications. When they are not allowed, the following amendment must be made to the definition of division:

If $m < n$ and $a \neq 0$, then

$$\frac{a^m}{a^n} = \frac{1}{a^{n-m}}$$ (1-7)

EXAMPLE 30 Express each quotient with positive exponents.

(a) $\dfrac{10^4}{10^7}$ (b) $\dfrac{x^3}{x^5}$ (c) $\dfrac{z}{z^6}$

Solution

(a) $\dfrac{10^4}{10^7} = \dfrac{1}{10^{7-4}} = \dfrac{1}{10^3}$ (b) $\dfrac{x^3}{x^5} = \dfrac{1}{x^{5-3}} = \dfrac{1}{x^2}$ (c) $\dfrac{z}{z^6} = \dfrac{1}{z^{6-1}} = \dfrac{1}{z^5}$

When negative exponents are allowed, all the laws of exponents discussed so far are valid. The next example illustrates their use.

EXAMPLE 31 Use the laws of exponents to simplify the following. Express all answers in exponential notation without fractions.

(a) $10^{-3} \cdot 10^{-5}$ (b) $\dfrac{r^2}{r^{-4}}$ (c) $4^{-3} \cdot 3^{-3}$

(d) $(x^{-2})^3$ (e) $(ab)^{-2}$ (f) $x^{-6} \cdot x^4 \cdot x^{-2}$

Solution

(a) $10^{(-3)+(-5)} = 10^{-8}$ (b) $r^{2-(-4)} = r^6$

(c) $(4 \cdot 3)^{-3} = 12^{-3}$ (d) $x^{(-2)3} = x^{-6}$

(e) $a^{-2}b^{-2}$ (f) $x^{(-6)+4+(-2)} = x^{-4}$

NOTE   Negative exponents often are used to express units. For example, meters per second might be expressed as m/s or as ms^{-1}. ◀

Zero as an exponent   So far we have defined the laws of division only for situations when the powers are different. What happens when the powers in the numerator and denominator are the same? For example, what is the value of $10^7/10^7$? Subtracting exponents, we have 10^0. On the other hand, we know that any number except zero divided by itself is 1. From this we can conclude that **any nonzero number raised to the zero power is equal to 1.**[*]

$a^0 = 1$ for any nonzero real number a (1-8)

[*]Negative exponents were defined as $a^{-n} = 1/a^n$, so a cannot equal zero. Therefore, zero to the zeroth power is not defined over the real numbers.

EXAMPLE 32 Find the value of each of the following:

(a) 10^0 (b) $x^0 y^0$ (c) $(2^5)^0$ (d) $\left(\dfrac{m}{n}\right)^0$ (e) $(5x)^0$ (f) $5x^0$

Solution

(a) 1   (b) 1   (c) 1   (d) 1   (e) 1   (f) 5

Order of operations   Did you notice the difference between parts (e) and (f) in Example 32? In part (e) the power of a product rule tells us that $(5x)^0 = 5^0 \cdot x^0 = 1 \cdot 1 = 1$. But in part (f), how did we know to evaluate the power x^0 first and then multiply by 5? Mathematicians have agreed that in a calculation, powers take precedence over multiplication and division. Adding this rule to the order of operations hierarchy, we now have the following expanded list:

Order of Operations

1. **First,** simplify within grouping symbols.
2. **Second,** simplify all powers.
3. **Third,** perform multiplication and division, left to right.
4. **Fourth,** perform addition and subtraction, left to right.

EXAMPLE 33 Calculate: $3 + (4 + 1) \cdot 2^3$

Solution

First, simplify within parentheses. $3 + (4 + 1) \cdot 2^3 = 3 + 5 \cdot 2^3$

Second, simplify powers. $= 3 + 5 \cdot 8$

Third, multiply. $= 3 + 40$

Fourth, add. $= 43$

As we mentioned earlier, most scientific calculators follow the order of operations automatically.

EXAMPLE 34 Use a calculator to simplify: $12^4 + 35(24 \cdot 13)^2$

Solution 12 $\boxed{y^x}$ 4 $\boxed{+}$ 35 $\boxed{\times}$ $\boxed{(}$ 24 $\boxed{\times}$ 13 $\boxed{)}$ $\boxed{x^2}$ $\boxed{=} \rightarrow$ 3427776.

Note how the order of operations clarifies some common misunderstandings in algebra: $2x^3$ means $2 \cdot x^3$, so only x is raised to the third power; and $-x^2$ means $-1 \cdot x^2$, so only x is raised to the second power. Similarly, $-2^4 = -16$, but $(-2)^4 = 16$.

IMPORTANT ▶   Always use parentheses when raising a negative number to a power. ◀

EXAMPLE 35   Perform the indicated operations.

(a) -3^2 (b) $(-3)^2$ (c) $2 \cdot 3^3$

(d) $(-x^3)^4$ (e) $-(x^2)^4$ (f) $x^2(x^3)^2$

Solution

(a) Square 3 first and then apply the negative sign: $-3^2 = -9$

(b) Square -3: $(-3)^2 = 9$

(c) Cube 3 first and then multiply by 2: $2 \cdot 3^3 = 2 \cdot 27 = 54$

(d) $(-x^3)^4 = (-1 \cdot x^3)^4 = (-1)^4(x^3)^4 = x^{12}$

(e) Simplify powers; then apply the negative sign: $-(x^2)^4 = -x^8$

(f) Square first; then multiply: $x^2(x^3)^2 = x^2(x^6) = x^8$

Exercises 1-4

Simplify using the laws of exponents. Express answers in exponential notation with positive exponents only.

1. $5^7 \cdot 5^4$ **2.** $x^5 \cdot x^3$ **3.** $y^4 \cdot y \cdot y^7$ **4.** $2 \cdot 2^2 \cdot 2^3$ **5.** $2^7 \cdot 5^6 \cdot 2^8$ **6.** $3x^2x^3$

7. $y^5(2y^3)$ **8.** $x^2 \cdot y^3 \cdot x^4 \cdot y^5$ **9.** $(3^6)^2$ **10.** $(y^4)^6$ **11.** $(m^2)^7$ **12.** $(x^3)^4$

13. $(ab)^4$ **14.** $(3m)^3$ **15.** $5^3 \cdot 2^3$ **16.** $3^5 \cdot 7^5 \cdot 2^5$ **17.** $(2x^2)^4$ **18.** $(-4y^3)^2$

19. $(a^2b^4)^3$ **20.** $(mn^4)^5$ **21.** $\left(\dfrac{2}{5}\right)^3$ **22.** $\left(\dfrac{3}{7}\right)^2$ **23.** $\left(\dfrac{w}{z}\right)^4$ **24.** $\left(\dfrac{k}{5}\right)^2$

25. $\left(\dfrac{2}{y^6}\right)^2$ **26.** $\left(\dfrac{x^2}{y^3}\right)^4$ **27.** $\dfrac{7^5}{7^2}$ **28.** $\dfrac{x^4}{x^5}$ **29.** 6^0 **30.** x^0

31. $\dfrac{y}{y^5}$ **32.** $\dfrac{5^9}{5^4}$ **33.** $(6p)^0$ **34.** $(mn)^0$ **35.** 5^{-3} **36.** m^{-5}

37. $4w^0$ **38.** $-6s^0$ **39.** $\dfrac{1}{y^{-3}}$ **40.** $\dfrac{1}{2^{-5}}$ **41.** $5x^{-4}$ **42.** $c^{-8}d^4$

43. $\dfrac{a^{-6}b^{-2}}{c^{-4}d^7}$ **44.** $\dfrac{2x^{-3}y^{-4}}{wz^{-2}}$

Simplify using the laws of exponents. Use negative exponents when necessary to express answers without fractions.

45. $\dfrac{1}{8^7}$ **46.** $\dfrac{1}{x^2}$ **47.** $10^{-7} \cdot 10^3$ **48.** $(x^{-5})(x^{-6})$ **49.** $(pq)^{-3}$

50. $(y^{-4})^3$ **51.** $\dfrac{10^{-8}}{10^2}$ **52.** $\dfrac{10^{-3}}{10^{-2}}$ **53.** $\dfrac{c^2}{c^5}$ **54.** $\dfrac{d^4}{d^{-6}}$

Evaluate according to the order of operations.

55. -5^2 **56.** -3^4 **57.** $(-x^2)^4$ **58.** $(-x)^8$

59. $-(x^3)^5$ **60.** $x^4(x^3)^6$ **61.** $2 + 7 \cdot 3^2$ **62.** $5(-8) - (-3)^3$

Solve.

63. *Civil engineering*   Under given conditions, calculation of the safe load (in pounds) for a beam leads to the expression below. Calculate this safe load.

$4 \cdot 18 \cdot 25^2$

64. *Physics* If you weigh 150 lb at sea level, your weight at 1600 km above sea level can be calculated by the expression below. Calculate this weight.

$$\left(\frac{6400}{6400 + 1600}\right)^2 \cdot 150$$

65. *Thermodynamics* In deriving the isothermal compressibility of a gas, the expression below may be encountered. Simplify this expression.

$$\frac{-p}{c}\left(\frac{-c}{p^2}\right)$$

66. *Astrophysics* Calculations of the angular velocity of a planet circling the sun involve the expression below. Simplify this expression.

$$\frac{GmM}{(mr)} \cdot \frac{1}{r^2}$$

67. *Mechanical engineering* In determining the moment of inertia for a section of a certain straight beam the term given below is involved. Simplify.

$$\frac{1}{12}w(2x)^3$$

68. *Mechanical engineering* The force F on a fillet weld is given by the formula below. Rewrite this formula by finding the indicated powers under the radical.

$$F = \sqrt{\left(\frac{3Rx}{2L^2}\right)^2 + \left(\frac{R}{2L}\right)^2}$$

1-5 | Scientific Notation

In technical work you will often deal with very small and very large numbers which may require a lot of space to write. For example, a certain computer function may take 5 nanoseconds (0.000000005 s), and in electricity, 1 kilowatt-hour (kWh) is equal to 3,600,000 joules (J) of work. Rather than take the time and space to write all the zeros, we use scientific notation to express such numbers.

Definition of scientific notation

In scientific notation 5 nanoseconds would be written as 5×10^{-9} s, and 1 kWh would be 3.6×10^6 J. As you can see, a number written in **scientific notation** is a product of a number between 1 and 10 and a power of 10. Stated formally, a number is expressed in scientific notation when it is in the form

$$P \times 10^k \quad \text{where } 1 \le P < 10 \text{ and } k \text{ is an integer}$$

We can use our two original examples to discover an easy method for writing numbers in scientific notation. Observe:

$$5 \times 10^{-9} = 5\left(\frac{1}{1,000,000,000}\right) = 5(0.000000001) = 0.000000005.$$

9 places left

Multiplying by 10^{-9} moved the decimal point nine places left.

$$3.6 \times 10^6 = 3.6(1,000,000) = 3,600,000.$$

6 places right

Multiplying by 10^6 moved the decimal point six places right.

Converting to scientific notation   In general, for the number $P \times 10^k$, the power of 10 moves the decimal point in P a total of k places. If k is positive, it moves the decimal point to the right; if k is negative, it moves the decimal point to the left. Therefore, to change any number from decimal notation to scientific notation, follow the procedure shown in Example 36.

EXAMPLE 36   Change to scientific notation: (a) 60,000   (b) 0.0000702

Solution

Step 1.   Form a number between 1 and 10 from the nonzero digits.

6.   7.02

Step 2.   Count the number of places you must move the decimal point to obtain the original number.

6.0000.   0.00007.02

4 places right   5 places left

Step 3.   The number from Step 2 is the power of 10. If you moved the decimal point to the right in Step 2, the power is positive; if you moved it to the left, the power is negative.

10^{+4} 10^{-5}

Step 4.   Write your final answer as a number between 1 and 10 (Step 1) multiplied by a power of 10 (Steps 2 and 3).

6×10^4 7.02×10^{-5}

EXAMPLE 37   Change each of the following to scientific notation:

(a) 0.00002   (b) 47   (c) 9   (d) $\dfrac{1}{40}$

(e) The sun's average distance from the earth is 150,000,000 km.

Solution

(a) $0.00002 = 2 \times 10^{-5}$ (b) $47 = 4.7 \times 10^1$

(c) $9 = 9 \times 10^0$ (An exponent of 0 means the number was already between 1 and 10.)

(d) **First,** change the fraction to decimal form. **Then,** convert the decimal numeral to scientific notation.

$\dfrac{1}{40} = 0.025$

$= 2.5 \times 10^{-2}$

(e) $150,000,000 \text{ km} = 1.5 \times 10^8 \text{ km}$

Converting to decimal notation To change a number from scientific notation to decimal notation, simply move the decimal point as indicated by the power of 10 and then omit the power of 10.

EXAMPLE 38 Change to decimal form:
(a) 2.72×10^4
(b) In chemistry, the relative acidity of H_2SO_3 is given as 1.2×10^{-2}.

Solution

(a) 2.72×10^4 $=$ $2.7200.$ $=$ $27,200$

Power of $+4$	Move decimal point 4 places right

(b) 1.2×10^{-2} $=$ $0.01.2$ $=$ 0.012

Power of -2	Move decimal point 2 places left

NOTE   Most scientific calculators have a means of converting back and forth between decimal and scientific notation. See your calculator instruction manual for the proper key sequences. ◀

Multiplication and division Scientific notation is especially useful when we must multiply or divide large and small numbers. To perform these operations, first change each number to scientific notation, and then multiply or divide the decimal parts and the exponential parts separately.

EXAMPLE 39 Perform the indicated operations. Express answers in scientific notation.
(a) $(26,000)(3,200,000)$
(b) Young's modulus of elasticity is equal to stress divided by strain. Find this quantity for a metal rod with a stress of 45,000 lb/in.^{2} and a strain of 0.0018 in./in.

Solution

(a) **Step 1.** Change each number to scientific notation.

$(2.6 \times 10^4)(3.2 \times 10^6)$

Step 2. Use the commutative and associative laws to regroup.

$= (2.6 \times 3.2)(10^4 \times 10^6)$

Step 3. Multiply decimal parts and exponential parts separately.

$= 8.32 \times 10^{10}$
$\approx 8.3 \times 10^{10}$, rounded

(b) **Step 1.** Change each number to scientific notation.

$\dfrac{4.5 \times 10^4}{1.8 \times 10^{-3}}$

Step 2. Rewrite the calculation as shown.

$$= \frac{4.5}{1.8} \times \frac{10^4}{10^{-3}} \Bigg\} \longrightarrow \boxed{\begin{array}{c} 10^{4-(-3)} \\ = 10^{4+3} \\ = 10^7 \end{array}}$$

Step 3. Divide decimal parts and exponential parts separately.

$$= 2.5 \times 10^7 \text{ lb/in.}^2$$

In Example 39 both answers automatically ended up in standard scientific notation. This will not always be the case. If the decimal calculation does not result in P such that $1 \leq P < 10$, it must be changed to standard scientific notation. The two resulting powers of 10 are then combined.

EXAMPLE 40 Perform the indicated operations. Express answers in standard scientific notation.

(a) To convert watts (W) to horsepower (hp), multiply by 0.00134. What is the output in horsepower of an electric utility plant generating 95,500,000 W of power?

(b) $\dfrac{0.0000072}{0.0009}$

Solution

(a) We must multiply 95,500,000 by 0.00134.

Step 1. Change each number to scientific notation.

$$(9.55 \times 10^7)(1.34 \times 10^{-3})$$

Step 2. Regroup.

$$= (9.55 \times 1.34)(10^7 \times 10^{-3})$$

Step 3. Multiply decimal parts and exponential parts separately.

$$= 12.797 \times 10^4$$

Step 4. Change the decimal part to standard scientific notation and multiply the resulting powers of 10.

$$= 1.2797 \times 10^1 \times 10^4$$
$$= 1.2797 \times 10^5$$

$$\approx 1.28 \times 10^5 \text{ hp, rounded}$$

(b) **Step 1.** Convert.

$$\frac{7.2 \times 10^{-6}}{9 \times 10^{-4}}$$

Step 2. Regroup.

$$= \frac{7.2}{9} \times \frac{10^{-6}}{10^{-4}}$$

Step 3. Divide.

$$= 0.8 \times 10^{-2}$$

Step 4. Change to standard scientific notation.

$$= 8 \times 10^{-1} \times 10^{-2}$$
$$= 8 \times 10^{-3}$$

Most calculators will display an answer in scientific notation when it is too large or too small to display in decimal form. In addition, some calculators will display an answer in scientific notation if a calculation is entered in scientific notation.

To enter a number in scientific notation, enter the decimal part, followed by the $\boxed{\text{EE}}$ or $\boxed{\text{EXP}}$ key, and then the power of 10. The calculator will display the decimal part and the power of 10 but will omit the 10 itself.

▦ **EXAMPLE 41**

(a) Calculate by entering all numbers as written: (582,000)(78,000)

(b) Calculate by entering all numbers in scientific notation:

$$\frac{(0.000423)(548,000,000)}{(7,350,000)}$$

Solution

(a) $582000 \boxed{\times} 78000 \boxed{=} \rightarrow$ ╠ 4.5396 10 ╣ $\approx 4.5 \times 10^{10}$

(b) $4.23 \boxed{\text{EE}} 4 \boxed{+/-} \boxed{\times} 5.48 \boxed{\text{EE}} 8 \boxed{\div} 7.35 \boxed{\text{EE}} 6 \boxed{=} \rightarrow$ ╠ 3.1538−02 ╣

$\approx 3.15 \times 10^{-2}$

or ╠ 0.031538 ╣

Exercises 1-5

Rewrite each number in scientific notation.

1. 7000 **2.** 8,200,000 **3.** 0.00053 **4.** 0.00000009 **5.** 40,700 **6.** 0.0000506

7. 63 **8.** 2.5 **9.** 0.7 **10.** 0.850 **11.** $\dfrac{3}{50}$ **12.** $\dfrac{1}{25}$

13. A 1 newton (N) force is equivalent to: 100,000 dynes (dyn)
14. The thermal conductivity of wood: 0.00024 cal/cm·s
15. The normal red blood cell count: 5,000,000 per mm^3
16. Young's modulus of elasticity for steel: 29,000,000 lb/in.^{2}

Rewrite each number in decimal notation.

17. 5×10^7 **18.** 4.40×10^5 **19.** 6.5×10^{-4} **20.** 9×10^{-2}
21. 7.25×10^6 **22.** 8.38×10^1 **23.** 4.9×10^0 **24.** 5.64×10^{-1}

25. 1 W is equivalent to: 2.389×10^{-4} kcal/s
26. Temperature of the solar surface (in degrees Kelvin): 6×10^3 K
27. Radius of a typical atom: 10^{-8} cm
28. The heat of fusion at 0°C: 3.33×10^5 J/kg

Rewrite each number in scientific notation and calculate the answer in scientific notation. Consider all numbers to be approximate and round accordingly.

29. (2000)(40,000) **30.** (0.00037)(0.0000024) **31.** (460,000)(0.0017)

32. (0.0000089)(3200) **33.** $\dfrac{840,000}{4200}$ **34.** $\dfrac{0.00056}{0.0020}$

35. $\dfrac{0.0000064}{80,000}$ **36.** $\dfrac{270,000}{0.000054}$ ▦ **37.** $\dfrac{(0.0502)(0.00070)}{42,000}$

38. $\dfrac{(2,780,000)(512,000)}{0.000721}$

39. $(64,000)(2800)(370,000)$

40. $(0.00075)(0.000062)(0.014)$

41. $\dfrac{0.000517}{(0.0273)(0.00469)}$

42. $\dfrac{893,000}{(5620)(387,000)}$

43. *Thermodynamics*   A 15 m × 25 m brick wall is 48 cm thick. Under particular temperature conditions, calculate the rate (in calories per second) of heat flow through the wall given by

$$(1.7 \times 10^{-4})\left(\frac{1500 \times 2500}{48}\right)$$

44. *Electrical engineering*   The lighting system for a football field consists of 60 fixtures at 1500 W each and 30 fixtures at 1650 W each. Yearly usage of the field amounts to about 250 h. At $0.000040 per watt-hour (Wh), what is the yearly cost for operating the lights?

45. *Space technology*   A detector in the Landsat satellite produces 0.5 ampere (A) of electrical current for each watt of reflected sunlight that falls on it. The detector gives a background current of 2×10^{-12} A without any sunlight. How many watts must fall on the detector to give a current 100 times greater than the background?*

46. *Nuclear physics*   If an atom has a diameter of 4×10^{-8} cm, and its nucleus has a diameter of 8×10^{-13} cm, what fraction of the diameter of the atom is the diameter of the nucleus?

47. *Electronics*   The capacitance of a certain capacitor (in microfarads, μF) is calculated using the following expression. Compute this capacitance.

$$5.75 \times 10^{-8}\left(\frac{8.00}{3.00 \times 10^{-3}}\right)$$

48. *Physics*   The energy (in joules) of a photon of visible light with a wavelength of 5.00×10^{-7} m is given by the expression below. Calculate the energy.

$$\frac{(6.63 \times 10^{-34})(3.00 \times 10^{8})}{5.00 \times 10^{-7}}$$

49. *Electronics*   The voltage V between the emitter and the base in a certain silicon transistor is given by the formula below. Calculate this voltage.

$$V = -5.0 + (5.2 \times 10^{4} \times 7.5 \times 10^{-5})$$

50. *Astronomy*   The mass of the Milky Way (in solar masses) can be calculated using the expression below. Perform this calculation.

$$\frac{(2 \times 10^{9})^{3}}{(2 \times 10^{8})^{2}}$$

51. *Chemistry*   The calculation indicated below gives the mass of one oxygen atom (in kilograms). Find this mass.

$$\frac{1.6 \times 10^{-2}}{6.0 \times 10^{23}}$$

52. *Optics*   The wavelength of blue light is approximately 4.6×10^{-7} m. Express this wavelength in angstroms (Å), given that 1 Å $= 10^{-10}$ m.

*Based on an interview with J. C. Lansing, Jr., Santa Barbara Research Center (a division of Hughes Aircraft Company), Santa Barbara, California.

1-6 | Roots and Radicals

To answer the question, "What is the length of a side of a square whose area is 16 square units?" we must find a number which, when multiplied by itself (or squared), equals 16. This number, 4, is called the **square root** of 16, and is symbolized by $\sqrt{16}$. The square root and other roots are involved in many technical applications, and this section is the first of several in which we will examine the topic in depth.

Definition of roots   In general, the **nth root of a**, $\sqrt[n]{a}$, is the number which, when used as a factor n times, gives a as a product. The $\sqrt{}$ symbol is called the **radical sign**, a is the **radicand**, and n is the **index**. When the index is not written, it is assumed to be 2, indicating square root.

EXAMPLE 42

(a) $\sqrt{49}$ (the square root of 49) $= 7$ because $7^2 = 7 \cdot 7 = 49$

(b) $\sqrt[3]{125}$ (the cube root of 125) $= 5$ because $5^3 = 5 \cdot 5 \cdot 5 = 125$

(c) $\sqrt[4]{81}$ (the fourth root of 81) $= 3$ because $3^4 = 3 \cdot 3 \cdot 3 \cdot 3 = 81$

(d) $\sqrt[5]{32}$ (the fifth root of 32) $= 2$ because $2^5 = 2 \cdot 2 \cdot 2 \cdot 2 \cdot 2 = 32$

Principal root   In Example 42(a) you may have wondered why we gave $\sqrt{49}$ as 7 and not -7 since $(-7)^2 = 49$ also. Similarly, in part (c), $(-3)^4 = 81$. To avoid this ambiguity, we define $\sqrt[n]{a}$ as the **principal nth root of a**. The principal root will always be positive if a is positive.

If we wish to indicate the negative nth root, we will place a negative sign in front of the radical. If we wish to include both the principal and the negative nth roots, we will place a $\pm$ (read "plus or minus") sign in front of the radical.

EXAMPLE 43   (a) $-\sqrt{16} = -4$ (b) $\pm\sqrt[4]{16} = \pm 2$

A calculator will always display the principal root of a number. Use the $\boxed{\sqrt{}}$ key for square roots and the $\boxed{\sqrt[x]{y}}$ or $\boxed{x^{1/y}}$ key, normally the inverse or second function of the $\boxed{y^x}$ key, for other roots.

▦ **EXAMPLE 44**   Find:

(a) $\sqrt{2209}$ (b) $\sqrt[3]{592{,}704}$

Solution

(a) $2209 \; \boxed{\sqrt{}} \rightarrow$ ▨▨▨ *47.*   (b) $592704 \; \boxed{\sqrt[x]{y}} \; 3 \; \boxed{=} \rightarrow$ ▨▨▨ *84.*

Negative radicands   So far we have discussed only positive radicands. If the radicand is negative and the index is odd, the principal root is real and negative. However, if the radicand is negative and the index is even, the result is an imaginary number.

EXAMPLE 45

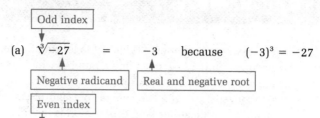

(a) $\sqrt[3]{-27}$ = -3 because $(-3)^3 = -27$

(b) $\sqrt{-4}$ is an imaginary number, because there is no real number n such that $n^2 = -4$

Negative radicand

In Chapter 12 we shall discuss imaginary numbers at length. For now, you need only be able to recognize them.

 CAUTION   Note the difference between $-\sqrt{4}$, which equals -2, and $\sqrt{-4}$, which is imaginary. ◀

EXAMPLE 46   Simplify all real numbers and identify all imaginary numbers.
(a) $\sqrt{-36}$ (b) $-\sqrt{64}$ (c) $\sqrt[3]{-8}$ (d) $-\sqrt[5]{-1}$ (e) $\sqrt[4]{-16}$

Solution
(a) Imaginary   (b) -8 (c) -2 (d) 1   (e) Imaginary

NOTE ▶   Most calculators will not accept a negative radicand, regardless of the index. Attempting to enter any such problems will result in an error signal. ◀

Simplest radical form   So far we have considered only numbers whose roots are rational, or "come out even." What if we encounter irrational numbers, such as $\sqrt{2}$ or $\sqrt{8}$? There is no integer or finite decimal number n such that $n^2 = 2$ or $n^2 = 8$. Depending on the situation, we may find approximate answers using a calculator or a table of square roots, or we may put them in **simplest radical form.** To understand simplest radical form, we must first understand one basic fact about radicals.

Square Root of a Product

For any positive real numbers a and b,

$$\sqrt{ab} = \sqrt{a \cdot b} = \sqrt{a} \cdot \sqrt{b}$$

(1-9)

That is, the square root of a product is equal to the product of the square roots. For example,

$$\sqrt{9 \cdot 4} = \sqrt{9} \cdot \sqrt{4}$$

because

$$\sqrt{9 \cdot 4} = \sqrt{36} = 6 \quad \text{and} \quad \sqrt{9} \cdot \sqrt{4} = 3 \cdot 2 = 6$$

To rewrite a radical in its simplest radical form, use this rule to "take out" the square root of any perfect square factor in the radicand. Follow the step-by-step procedure shown in Example 47.

EXAMPLE 47 Simplify $\sqrt{48}$.

Solution

16 is the largest perfect square factor of 48

$$\downarrow$$

Step 1. Rewrite the radicand as the product of two numbers using its largest perfect square factor as one of the numbers.

$$\sqrt{48} = \sqrt{16 \cdot 3}$$

Step 2. Use rule (1-9) to write the problem as the product of two square roots.

$$= \sqrt{16} \cdot \sqrt{3}$$

Step 3. Simplify the radical containing the perfect square.

$$= 4\sqrt{3}$$

If there is no perfect square factor of the radicand, the number is already in its simplest radical form.

EXAMPLE 48 Express each of the following in simplest radical form:
(a) $\sqrt{8}$ (b) $\sqrt{27}$ (c) $\sqrt{80}$ (d) $\sqrt{10}$ (e) $3\sqrt{200}$
(f) Use a calculator to compute the decimal equivalents of the radicals given in parts (a) and (e). Then find the decimal equivalents of their simplified forms and compare.

Solution
(a) $\sqrt{8} = \sqrt{4} \cdot \sqrt{2} = 2\sqrt{2}$ (b) $\sqrt{27} = \sqrt{9} \cdot \sqrt{3} = 3\sqrt{3}$
(c) $\sqrt{80} = \sqrt{16} \cdot \sqrt{5} = 4\sqrt{5}$ (d) $\sqrt{10}$ is already simplified
(e) $3\sqrt{200} = 3\sqrt{100} \cdot \sqrt{2} = 3 \cdot 10\sqrt{2} = 30\sqrt{2}$
(f) For part (a): 8 $\boxed{\sqrt{}}$ → ▨ 2.8284271

2 $\boxed{\times}$ 2 $\boxed{\sqrt{}}$ $\boxed{=}$ → ▨ 2.8284271

For part (e): 3 $\boxed{\times}$ 200 $\boxed{\sqrt{}}$ $\boxed{=}$ → ▨ 42.426407

30 $\boxed{\times}$ 2 $\boxed{\sqrt{}}$ $\boxed{=}$ → ▨ 42.426407

The quotient rule   A rule similar to (1-9) also holds for the square root of a quotient:

Square Root of a Quotient

$$\sqrt{\frac{a}{b}} = \frac{\sqrt{a}}{\sqrt{b}} \qquad \text{where } a \text{ and } b \text{ are positive real numbers} \qquad \text{(1-10)}$$

That is, the square root of a quotient is equal to the quotient of the square roots.*

EXAMPLE 49

(a) $\sqrt{\dfrac{16}{25}} = \dfrac{\sqrt{16}}{\sqrt{25}} = \dfrac{4}{5}$ **Check:** $\left(\dfrac{4}{5}\right)^2 = \dfrac{4^2}{5^2} = \dfrac{16}{25}$

(b) $\sqrt{\dfrac{8}{49}} = \dfrac{\sqrt{8}}{\sqrt{49}} = \dfrac{2\sqrt{2}}{7}$

The product and quotient rules used for simplifying square roots can be extended to other roots as well. We will discuss these problems in Chapter 11.

CAUTION ▶ Rules such as (1-9) and (1-10) do not hold for radical sums or differences. That is,

$$\sqrt{a + b} \quad \text{is } \textbf{not} \text{ equal to} \quad \sqrt{a} + \sqrt{b}$$

For example,

$$\sqrt{9 + 16} = \sqrt{25} = 5 \quad \text{but} \quad \sqrt{9} + \sqrt{16} = 3 + 4 = 7 \quad ◀$$

Order of operations   Roots occupy the same level as powers in the order of operations. In the absence of grouping symbols, roots must be simplified before multiplication, division, addition, or subtraction outside the radical. However, unless (1-9) or (1-10) is being applied, all operations inside the radical must be performed before extracting the root.

EXAMPLE 50   Simplify:

(a) $-8 + 3\sqrt{36}$ (b) $\sqrt{64 + 36}$ ▦ (c) $\dfrac{23 \pm \sqrt{23^2 - 4(8)(-23)}}{16}$

Solution

(a) **First,** find the square root. $-8 + 3\sqrt{36} = -8 + 3(6)$
   Second, multiply. $= -8 + 18$
   Third, add. $= 10$

(b) **First,** simplify inside the radical. $\sqrt{64 + 36} = \sqrt{100}$
   Then, find the root. $= 10$

(c) The $\pm$ symbol means that two answers are required. To find them both with a minimum number of keystrokes, follow this procedure:

   First, simplify inside the radical:

   23 $\boxed{x^2}$ $\boxed{-}$ 4 $\boxed{\times}$ 8 $\boxed{\times}$ 23 $\boxed{+/-}$ $\boxed{=}$

---

* Simplest radical form for a fraction requires that there be no radical in the denominator. Methods for dealing with the square root of a quotient in which the divisor is not a perfect square will be covered in Chapter 11.

Second, take the square root and store it in memory. We must use this number twice, so store it to avoid recalculating.

Third, add the square root to 23 to find the first numerator:

$\boxed{+}$ 23 $\boxed{=}$

Fourth, divide this by 16 to obtain the first answer:

$\boxed{\div}$ 16 $\boxed{=}$ → $\boxed{3.6604274}$

Fifth, to find the second answer, subtract the root in memory from 23 and divide the result by 16:

23 $\boxed{-}$ $\boxed{\text{RCL}}$ $\boxed{=}$ $\boxed{\div}$ 16 $\boxed{=}$ → $\boxed{-0.7854274}$

Exercises 1-6

Simplify all real numbers and identify all imaginary numbers.

1. $\sqrt{4}$
2. $\sqrt{25}$
3. $-\sqrt{100}$
4. $\pm\sqrt{81}$
5. $\sqrt[3]{8}$

6. $\sqrt[3]{64}$
7. $\sqrt[3]{-125}$
8. $\sqrt[3]{-343}$
9. $\sqrt[4]{256}$
10. $\sqrt[5]{1}$

11. $\sqrt{12}$
12. $\sqrt{18}$
13. $\sqrt{75}$
14. $\sqrt{72}$
15. $\sqrt{45}$

16. $\sqrt{300}$
17. $2\sqrt{98}$
18. $3\sqrt{48}$
19. $\sqrt{-20}$
20. $\sqrt{-64}$

21. $\dfrac{2\sqrt{49}}{3\sqrt{25}}$
22. $\dfrac{5\sqrt{100}}{4\sqrt{9}}$
23. $\sqrt{\dfrac{9}{25}}$
24. $\sqrt{\dfrac{36}{121}}$
25. $\pm\sqrt{\dfrac{3}{4}}$

26. $\pm\sqrt{\dfrac{5}{81}}$
27. $\sqrt{\dfrac{24}{49}}$
28. $\sqrt{\dfrac{32}{81}}$
29. $\sqrt{64+16}$
30. $\sqrt{100+25}$

31. $6+5\sqrt{64}$
32. $-9-2\sqrt{9}$

Calculate using a calculator, and round to the nearest 0.01 if necessary.

33. $\sqrt{1369}$
34. $\sqrt{277,729}$
35. $\sqrt{1856}$
36. $\sqrt{1675}$

37. $\sqrt[3]{13,824}$
38. $\sqrt[3]{29,791}$
39. $\sqrt[4]{83,521}$
40. $\sqrt[5]{371,293}$

41. $\sqrt{2(8.4)(7.2)(9.5)}$
42. $\sqrt{8(0.04)-(0.05)^2}$

43. $\dfrac{25\pm\sqrt{(-25)^2-4(14)(-17)}}{28}$
44. $\dfrac{-31\pm\sqrt{31^2-4(9)(7)}}{18}$

Solve.

45. The area of a square is 144 cm². What is the length of each side?
46. The edge of a cube is equal to the cube root of the volume. What is the edge of a cube whose volume is 216 in.³?
47. Find $8\sqrt{29}$ on a calculator and compare the result to the answer for Problem 35. What can be concluded about $1856 \div 29$?
48. Find $5\sqrt{67}$ on a calculator and compare the result to the answer for Problem 36. What can be concluded about $1675 \div 67$?

49. *Mechanical engineering*   The average normal stress in a certain $\frac{1}{2}$ in. plunger is given (in pounds per square inch, lb/in.^{2} or psi) by the expression shown below. Calculate this stress to the nearest 1000 lb/in.^{2}.

$$\frac{3000\sqrt{34}}{0.196} \text{ lb/in.}^2$$

50. *Atomic physics*   The frequency of oscillation of the atoms in the diatomic oxygen molecule is calculated using the expression below. Compute this frequency (in hertz).*

$$\sqrt{\frac{50}{1.4 \times 10^{-26}}}$$

51. *Electronics*   The optimum value of the resistance (in ohms) in a certain silicon diode is calculated using the expression below. Calculate this resistance.

$$\sqrt{15 \times 10^6 \times 24}$$

52. *Wastewater technology*   In analyzing a particular waste-handling system, the minimum retention time (in days) is given by the expression below. Calculate the retention time.

$$\frac{1}{0.21\left[1 - \sqrt{\dfrac{6400}{6400 + 55{,}000}}\right] - 0.015}$$

1-7 | Review of Chapter 1

Important Terms and Concepts

integer (p. 2)	reciprocal (p. 6)	base (p. 22)
rational number (p. 3)	denominate number (p. 7)	exponent (p. 22)
irrational number (p. 3)	literal number (p. 7)	scientific notation (p. 29)
real number (p. 3)	operating symbol (p. 8)	square root (p. 35)
imaginary number (p. 4)	variable (p. 8)	nth root (p. 35)
complex number (p. 4)	constant (p. 8)	radical sign (p. 35)
number line (p. 4)	signed number (p. 12)	radicand (p. 35)
opposite (p. 5)	operations with zero (p. 17)	index (p. 35)
symbols of inequality (p. 5)	factor (p. 21)	principal root (p. 35)
absolute value (p. 6)	exponential notation (p. 21)	simplest radical form (p. 36)

Formulas and Laws

NOTE ▶   Unless otherwise stated, a, b, and c represent any real number, and m and n represent any integer. ◀

*See Hugh D. Young, *Fundamentals of Optics and Modern Physics* (New York: McGraw-Hill, 1968), p. 310.

- commutative laws (p. 10)

    addition: $a + b = b + a$

    multiplication: $ab = ba$

- associative laws (p. 11)

    addition: $(a + b) + c = a + (b + c)$

    multiplication: $(ab)c = a(bc)$

- distributive law (p. 12)

    $a(b + c) = ab + ac$

    $a(b - c) = ab - ac$

- laws of exponents (pp. 22–26)

    multiplication: $a^m \cdot a^n = a^{m+n}$ **(1-1)**

    power of a power: $(a^m)^n = a^{mn}$ **(1-2)**

    power of a product: $(ab)^m = a^m b^m$ **(1-3)**

    power of a quotient: $\left(\dfrac{a}{b}\right)^m = \dfrac{a^m}{b^m}$ (where $b \neq 0$)   **(1-4)**

    division: $\dfrac{a^m}{a^n} = a^{m-n}$ (where $a \neq 0$)   **(1-5)**

    If $m < n$ and positive exponents are required, then $\dfrac{a^m}{a^n} = \dfrac{1}{a^{n-m}}$ (where $a \neq 0$)   **(1-7)**

- negative exponent: $a^{-m} = \dfrac{1}{a^m}$ (where $a \neq 0$) (p. 25)   **(1-6)**

- zero exponent: $a^0 = 1$ (where $a \neq 0$) (p. 26)   **(1-8)**
- square root of a product: for any positive real numbers a and b, $\sqrt{ab} = \sqrt{a} \cdot \sqrt{b}$ (p. 36)   **(1-9)**
- square root of a quotient: for any positive real numbers a and b, $\sqrt{\dfrac{a}{b}} = \dfrac{\sqrt{a}}{\sqrt{b}}$ (p. 37)   **(1-10)**

- order of operations (pp. 18, 27, and 38)
    1. Simplify within grouping symbols.
    2. Simplify all powers and roots.
    3. Perform multiplication and division from left to right.
    4. Perform addition and subtraction from left to right.

Exercises 1-7

Use the following list of numbers to answer the questions in Problems 1–8:

(a) $-\dfrac{5}{16}$ (b) 17 (c) 5π (d) -2.5 (e) $-\sqrt{5}$ (f) $-0.\overline{37}$ (g) $\sqrt{-7}$

1. Which of the numbers are integers?
2. Which of the numbers are rational?
3. Which of the numbers are irrational?
4. Which of the numbers are real?
5. Which of the numbers are imaginary?
6. Use the proper inequality symbol ($<$ or $>$) to compare (a) to (f), (b) to (c), and (e) to (f).
7. Find the absolute values of (a)–(f).
8. Find the reciprocals of (a)–(e).

Name the law that makes each of the following statements true:

9. $5(x + y) = 5x + 5y$

10. $4(5 \cdot 7) = (4 \cdot 5)7$

11. $(23 + 19) + 41 = 23 + (19 + 41)$

12. $40(11 \times 2.5) = 40(2.5 \times 11)$

13. $26 + 83 + 14 = 26 + 14 + 83$

14. $3 \cdot 7 + 3 \cdot 9 = 3 \cdot 16$

Simplify each of the following expressions:

15. $-31 + 16 \cdot$

16. $(-5) + (-18)$

17. $7 - 13$

18. $-9 - (-22)$

19. $4 - (-11) + 12$

20. $(-6) - 9 + (-13)$

21. $(-8)(11)$

22. $(-6)(-7)$

23. $\dfrac{-84}{-12}$

24. $\dfrac{90}{-15}$

25. $\dfrac{0}{-6}$

26. $\dfrac{-12}{0}$

27. $0 - 19$

28. $(6)(-8)(0)(-4)$

29. $(-6)(-5)(-3)(-2)$

30. $(-8)(4)(-2)(-3)$

31. $\dfrac{(-5)(-8)}{-4 + 2}$

32. $\dfrac{7 - (-23)}{-5 - 1}$

33. $9 - 3(-5) + \dfrac{28}{-7}$

34. $-8 - \dfrac{42}{6} - 3 \cdot 6^2(-2)$

35. $13 - 4^3$

36. $24 - (-3)^4$

37. $16 + 5(7)^2$

38. $-2^5 + (-2)^5$

39. $\sqrt{36}$

40. $\pm\sqrt{9}$

41. $\sqrt{121} - \sqrt{49}$

42. $\sqrt{54}$

43. $\sqrt{32}$

44. $3\sqrt{50}$

45. $\sqrt{\dfrac{4}{25}}$

46. $\pm\sqrt{\dfrac{27}{144}}$

47. $\sqrt[3]{-1}$

48. $-\sqrt[4]{10,000}$

49. $\sqrt{121 + 48}$

50. $15 - 2\sqrt{49}$

51. $\sqrt{3249}$

52. $\sqrt{4781}$

53. $\sqrt[3]{42,875}$

54. $\dfrac{33 \pm \sqrt{(-33)^2 - 4(12)(-19)}}{24}$

55. $\dfrac{3.74(16) - 2.98}{1.652 - 4.495}$

56. $6348 + 197[338 + 13(264 + 37^2)]$

Simplify the following using the laws of exponents. Express all answers with positive exponents.

57. $3^7 \cdot 3^6$

58. $x^4 \cdot x \cdot x^5$

59. $(3x)^3$

60. $(a^2)^8$

61. $(2x^2y^3)^3$

62. $\left(\dfrac{4}{y}\right)^2$

63. $\dfrac{m^4}{m^8}$

64. $\dfrac{y^{12}}{y^3}$

65. $(4x)^0$

66. $4x^0$

67. $7y^{-3}$

68. $x^{-2}w^4$

69. $y^2(y^3)^2$

70. $-(a^2)^3$

71. $\dfrac{1}{x^{-1}}$

72. $\dfrac{5a^{-3}}{b^6c^{-2}}$

Express all answers without fractions. Use negative exponents where necessary.

73. $\dfrac{1}{2^5}$

74. $\dfrac{a^2}{b^4}$

75. $\dfrac{10^2}{10^7}$

76. $10^5 \cdot 10^{-8}$

77. $(ax)^{-4}$

78. $(m^5)^{-2}$

Change all numbers in ordinary notation to scientific notation and all numbers in scientific notation to ordinary notation.

79. Some early projections indicated that, by the year 2000, the United States may be generating 940,000,000,000 watts of electrical energy from nuclear power.*

*Atomic Energy Commission (AEC) prediction quoted in Alvin M. Weinberg, "State of the Laboratory, 1970," *Oak Ridge National Laboratory Review,* Winter 1971, p. 1.

80. The heat of vaporization of a liquid at 100°C is 2,260,000 J/kg.
81. People have been using fire for 10^4 years.
82. One watt is equivalent to 1.34×10^{-3} hp.
83. The thermal conductivity of window glass is 0.0002 kcal/s·m·°C.
84. The specific heat of tungsten is 0.0321 cal/g·°C.
85. The earth's magnetic field at Washington, D.C., is about 5.7×10^{-5} weber per square meter (Wb/m²).
86. Jupiter has a mean diameter of 1.436×10^5 km.

Rewrite each number in scientific notation and calculate the answer in scientific notation.

87. (34,000)(250,000)
88. (74,000)(0.0000060)
89. $\dfrac{0.00088}{0.0055}$
90. $\dfrac{(85,000)(54,000)}{0.000056}$

Solve.

91. *Electronics* To obtain the voltage in a certain circuit, the calculation below must be performed. Compute this voltage to the nearest 0.1 V.

    $(0.4 + 2.8)(-1.7)$

92. *Planetary science* The temperature at a certain location on Mercury ranges from $-183°C$ to $427°C$. Find the difference between these extremes.

93. *Chemistry* The heat generated in a certain chemical reaction (in kilocalories) can be found by the calculation shown below. Find this heat of reaction.

    $4(-94.4) + 6(-68.4) - 2(-20.0)$

94. *Business* A phone call costs $1.26 for the first minute and $0.72 for each additional minute. Set up a calculation for finding the cost of a 6 min call. Then compute this cost.

95. *Refrigeration technology* The coefficient of heat transmission U of an insulating material is equal to the reciprocal of its thermal resistance R. Find U when $R = 0.92$.

96. *Physics* The fundamental dimensions of kinetic energy are ML^2T^{-2}. Express this with positive exponents only.

97. *Atomic physics* To determine the enegy equivalent E (in ergs) of the mass of an electron, the mass m (in grams) must be multiplied by the square of c, the speed of light. If $c = 3.0 \times 10^{10}$ cm/s and $m = 9.1 \times 10^{-28}$ g, find E in scientific notation.

98. *Nuclear physics* To determine the binding energy (in joules) of the hydrogen atom, the calculation shown below must be performed. Find the answer in scientific notation.

    $$-\frac{(9.11 \times 10^{-31})(6.55 \times 10^{-76})}{(6.27 \times 10^{-22})(4.40 \times 10^{-67})}$$

99. The area of a square plot of land is 4624 ft². Find the length of one side of the plot.
100. The volume of a cube is 27 ft³. Find the length of one edge of the cube.
101. *Electrical engineering* The calculation shown below is necessary for finding the velocity of an electron at an anode. Find this velocity (in meters per second).

    $$\sqrt{\frac{2(1.6 \times 10^{-19})(270)}{9.1 \times 10^{-31}}}$$

102. *Nuclear engineering* The groundwater near a certain nuclear plant is found to contain 9×10^{-8} curie (Ci) of radioactivity per liter. If the maximum allowable concentration is 3×10^{-10} Ci/liter, how many times the maximum level currently exists?

103. *Mechanical engineering* The impact factor for a certain steel step rod is given by the expression on the next page. Determine this quantity to the nearest 10 whole units.

$$1 + \sqrt{1 + 2\left(\frac{34}{40.7 \times 10^{-6}}\right)}$$

104. *Mechanical engineering*   The safety factor for a certain steel pipe used as a simply supported column is found with the calculation shown below. Compute this factor to the nearest hundredth unit.

$$\frac{5}{3} + \frac{3}{8}\left(\frac{88.67}{119.6}\right) - \frac{1}{8}\left(\frac{88.67}{119.6}\right)^3$$

Basic Concepts of Algebra

2

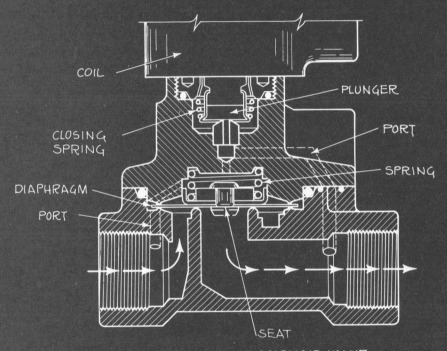

CROSS SECTION OF SOLENOID VALVE

Labels: COIL, PLUNGER, CLOSING SPRING, PORT, SPRING, DIAPHRAGM, PORT, SEAT

This chapter provides an introduction to the basic ideas and notation used in algebra. It is designed with special attention to the needs and interests of technical students. We begin by applying some of the laws and operations discussed in Chapter 1 to algebraic expressions. We will then use this knowledge to solve equations, formulas, and applied problems.

2-1 Addition and Subtraction of Algebraic Expressions

Algebraic expressions

An **algebraic expression** consists of a single number or literal symbol or a combination of numbers and literal symbols with the operations discussed in Chapter 1. Each of the following is an algebraic expression:

$$3xy \qquad x + 2 \qquad \frac{5x}{y} \qquad 3x^2 - 2y + 1 \qquad 4 \qquad x$$

All laws and rules governing operations with arithmetic expressions apply to algebraic expressions as well.

Terms and factors

Within a given algebraic expression, any quantities linked by addition are called **terms.** Any quantities linked by multiplication are called **factors.**

EXAMPLE 1 Consider the algebraic expression $5x + 6y - 2y^2$.
This expression has three terms, $5x$, $6y$, and $-2y^2$:

$$\boxed{\text{Terms}}$$
$$5x + 6y + (-2y^2)$$

The individual factors of the first term are 5 and x:

$$\boxed{\text{Factors}}$$
$$5x = 5 \cdot x$$

Note that 1 and the term itself can also be considered factors of every term.

NOTE ▶ Fractions and radical expressions are considered to be single terms. For example,

$$\frac{3a^2 + 5a + 7}{a + 6} \quad \text{is one term} \quad \text{and} \quad \frac{-2}{T}\sqrt{A^2 - X^2} \quad \text{is one term} \quad ◀$$

Coefficients

Within any given term, each factor is called a **coefficient** of the remaining factors. The nonliteral, or numerical, factor is called the **numerical coefficient.**

EXAMPLE 2 The expression $\frac{1}{2}mv^2$ gives the kinetic energy of an object. The number $\frac{1}{2}$ is the numerical coefficient, and

$\frac{1}{2}m$ is the coefficient of v^2: $\qquad \frac{1}{2}m \cdot v^2 = \frac{1}{2}mv^2$

mv is the coefficient of $\frac{1}{2}v$: $\qquad mv \cdot \frac{1}{2}v = \frac{1}{2}mv^2$

And so on.

When a numerical coefficient is not written, it is understood to be 1 or -1:

$$\text{Numerical coefficient}$$

$$a^2b = 1 \cdot a^2b \qquad -a^2b = -1 \cdot a^2b$$

Polynomials

A **polynomial** is an algebraic expression that contains only nonnegative, integral powers of the variables.* A polynomial with only one term is called a **monomial,** a polynomial with two terms is a **binomial,** and a polynomial with three terms is a **trinomial.** The word **multinomial** refers to any expression containing more than one term.

EXAMPLE 3 The expressions x, 6, $-2x^3y^2z$, and $c/3$ are all monomials.

The expressions $x + y$ and $a^2 - 4b^2$ are binomials (as well as multinomials).

The expressions $a + b - c$ and $x^2 - 2x + 2$ are trinomials (and multinomials).

Combining like terms

Under certain conditions, terms within algebraic expressions can be combined. In the expression

$$3a + 7a + b - 5b$$

$3a$ and $7a$ are called **like terms** because they contain exactly the same **literal part.**

$$\text{Same literal part}$$

$$3a + 7a$$

Similarly, b and $-5b$ are like terms because the literal part of each term, b, is exactly the same.

Like terms may be added or subtracted by combining their numerical coefficients. The distributive property shows why this is possible:

Distributive property: $ab + ac = a(b + c)$

Combining like terms: $3a + 7a = a(3 + 7) = 10a$

$$b - 5b = b(1 - 5) = -4b$$

By combining like terms, we can simplify the original algebraic expression, $3a + 7a + b - 5b$, to the equivalent, but simpler, expression, $10a - 4b$. Because these two remaining terms contain different literal parts, they are **unlike terms,** and they cannot be combined.

EXAMPLE 4 Simplify by combining like terms.

(a) $5m + 7n + 3p + 6m - 8n$ (b) $(x^2 + 5x - 3) + (3x^2 - 2x + 2)$

---

*Expressions with variables in the denominator, such as $3/x$, are not polynomials because the variables are being raised to a negative power. Expressions containing roots of variables such as $2\sqrt{x}$ are not polynomials because, as we shall see in Chapter 11, roots are considered to be fractional powers.

Solution

(a) **First,** use the commutative law to rearrange the terms as shown:

$$5m + 7n + 3p + 6m - 8n$$
$$= 5m + 6m + 7n - 8n + 3p$$

Second, use the distributive law to combine the adjacent like terms.

$$= (5 + 6)m + (7 - 8)n + 3p$$
$$= 11m - n + 3p$$

(b) With addition, the associative law allows us to drop parentheses.

$$(x^2 + 5x - 3) + (3x^2 - 2x + 2)$$
$$= x^2 + 5x - 3 + 3x^2 - 2x + 2$$

The commutative law then allows us to rearrange the terms.

$$= x^2 + 3x^2 + 5x - 2x - 3 + 2$$

Combining like terms, we obtain:

$$= 4x^2 + 3x - 1$$

Even though the first two terms of the answer to Example 4(b) contain the same variable, x, they are unlike terms because the powers of the variable differ. Like terms must have *exactly* the same variable parts. The expression in part (b) cannot be simplified further.

After a bit of practice, you will be able to combine like terms by inspection, without writing down the intermediate steps.

Subtraction with grouping symbols

Because the two original expressions in Example 4(b) were being added, the parentheses were dropped without affecting the result. However, when a grouping symbol is preceded by a subtraction sign, the signs of all terms inside must be changed in order to remove the grouping symbol.

EXAMPLE 5   Simplify by combining like terms: $(5x + 3) - (7x - 8)$

Solution

The first set of parentheses may be dropped without any changes.

$$(5x + 3) - (7x - 8)$$
$$= 5x + 3 - (7x - 8)$$

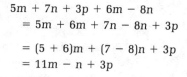
Both signs changed

But in order to remove the parentheses preceded by subtraction, all signs inside must be changed.

$$= 5x + 3 - 7x + 8$$
$$= 5x - 7x + 3 + 8$$

Then like terms may be combined.

$$= -2x + 11$$

LEARNING HINT

Some students prefer to think of subtraction with grouping symbols this way:

$$y - (7x - 8) = y + (-1)(7x - 8) = y + (-1)(7x) + (-1)(-8)$$

$$= y - 7x + 8 \blacktriangleleft$$

When several sets of grouping symbols are needed, brackets [ ] and braces { } are often used. In such problems, the innermost group should be simplified first.

EXAMPLE 6   Simplify: $-\{3 - [6a - (2a + 3)]\}$

Solution

Remove the innermost grouping symbol.

$-\{3 - [6a - (2a + 3)]\}$
$= -\{3 - [6a - 2a - 3]\}$

Both signs changed

Combine like terms.
Remove the [ ] symbols.

$= -\{3 - [4a - 3]\}$
$= -\{3 - 4a + 3\}$

Both signs changed

Combine like terms.
Remove the { } symbols.

$= -\{6 - 4a\}$
$= -6 + 4a$

Both signs changed

Exercises 2-1

Simplify each of the following algebraic expressions:

1. $8x - 6x + 5x$
2. $3y - 7y + 4y$
3. $5m + 6m - 11n + 13n$
4. $-6r + 4r - 2s + 7s$
5. $6v + 9w - 18v$
6. $8p - 7q - 12p$
7. $x + 3y - 4z - 3x + 2y$
8. $2a - 5b + c - 3b - 7c$
9. $2x^2 - 7x + 10x^2 - 11x$
10. $5y^3 - 9y^2 - 11y^2 - 6y^3$
11. $ab - 2a + 7ab + 5a$
12. $2xy - 3xy^2 - 5x^2y + 8xy^2 - 12xy$
13. $(3x + 4y - 7z) + (8x - 9y - 5z)$
14. $(4x^2 - 8xy - 9y^2) + (2x^2 - 7xy + 2y^2)$
15. $5m - (3m - 7)$
16. $8 - (2a + 5)$
17. $(3a - 4b) - (9a + 6b)$
18. $(7x + 5y) - (11x - 7y)$
19. $(2x^2 - 11x + 5) - (7x^2 + 9x - 3)$
20. $(4y^2 + 6y - 9) - (2y^2 - 8y + 7)$
21. $-(2m - 5mn) - (7mn - 8m)$
22. $-(3x + 4xy) - (-8x - 7xy)$
23. $-(a + b) - (b - 2c) - (3a + 4c)$
24. $-(c^2 - 2cd) - (cd - 3d^2) + (5c^2 - 7d^2)$
25. $-[8 - (3x - 5)]$
26. $8y - [2y - (3 - 7y)]$
27. $7m - \{3m^2 - [4m - (7m^2 + 8m)]\}$
28. $-\{8d - [5c - (6d + 5c)]\}$

29. *Civil engineering*   In determining the cost of constructing a roadway, certain cost comparisons were set up. This led to the following expression representing total cost:

$$p + \frac{1}{2}p + \frac{1}{2}p$$

Simplify by combining like terms.

30. *Construction*   A concrete mix is 1 part cement, 2 parts water, 2 parts aggregate, and 3 parts sand. This leads to the expression

$$x + 2x + 2x + 3x$$

when determining the amount of each ingredient for a particular amount of concrete. Simplify by combining like terms.

2-2 | Multiplication of Algebraic Expressions

In this section we shall discuss how to multiply a monomial by a monomial, a monomial by a multinomial, and a multinomial by a multinomial. To do this, we will use the laws and operations of real numbers (Section 1-2), the laws of exponents (Section 1-4), and the rules for adding and subtracting algebraic expressions (Section 2-1).

Monomial times a monomial   To multiply a monomial by a monomial, first use the commutative law to rearrange the factors so that like factors are adjacent. Then use the rules for signed number multiplication and the laws of exponents to multiply where possible.

EXAMPLE 7   Multiply: $(3x^2y)(-4x^3y^4)$

Solution

Step 1.   Rearrange the factors so that numerical factors are adjacent and like variables are grouped together.

$$(3x^2y)(-4x^3y^4)$$
$$= 3(-4) \cdot x^2 \cdot x^3 \cdot y \cdot y^4$$

Remember: $a^m \cdot a^n = a^{m+n}$

Step 2.   Multiply: $= -12 \cdot x^5 \cdot y^5$

The final product is $-12x^5y^5$. It has a single numerical coefficient, -12, and the powers of like bases have been combined.

EXAMPLE 8   Multiply:
(a) $(-7x^2y)(-4x^2)(-3y^6)$ (b) $(-5cd^2)(-3c^3)^2$

Solution

(a) $(-7x^2y)(-4x^2)(-3y^6) = (-7)(-4)(-3)(x^2 \cdot x^2)(y \cdot y^6)$
$$= -84x^4y^7$$

(b)   Since powers come before multiplication in the order of operations, we must square the second monomial before multiplying:

$$(-5cd^2)(-3c^3)^2 = (-5cd^2)(9c^6) \quad \text{Remember:} \quad (a^m)^n = a^{mn}$$

$$= (-5) \cdot 9 \cdot c \cdot c^6 \cdot d^2$$

$$= -45c^7d^2$$

With some experience, you will be able to arrange factors mentally and find the answer by inspection.

Monomial times a multinomial   To multiply a monomial by a multinomial, apply the distributive law. For example,

$$a\underbrace{(b + c)} = a \cdot b + a \cdot c = ab + ac$$

Monomial | Multinomial

$$x\underbrace{(x^2 + x + 2)} = x \cdot x^2 + x \cdot x + x \cdot 2 = x^3 + x^2 + 2x$$

In other words, multiply each term of the multinomial by the monomial.

EXAMPLE 9 Multiply:

(a) $-5y(2y^2 - 5xy)$ (b) $-2ab^2(5a^2 - 8ab + 4b^2)$

Solution

(a) **Step 1.** Rewrite using the distributive law.

$$-5y(2y^2 - 5xy) = (-5y)(2y^2) + (-5y)(-5xy)$$

Step 2. Find all monomial products.

$$= -10y^3 + 25xy^2$$

(b) **Step 1.**

$$(-2ab^2)(5a^2) + (-2ab^2)(-8ab) + (-2ab^2)(4b^2)$$

Step 2.

$$= -10a^3b^2 + 16a^2b^3 - 8ab^4$$

Step 1 in Example 9 can be done by inspection with practice.

Multinomial times a multinomial   To multiply a multinomial by a multinomial, again use the distributive property to multiply each term of the first multinomial by each term of the second. Then combine like terms if possible.

EXAMPLE 10 Multiply: $(2x + 3)(x - 5)$

Solution

Step 1. Rewrite using the distributive law.

$$(2x + 3)(x - 5)$$
$$= 2x(x - 5) + 3(x - 5)$$
$$= 2x \cdot x + 2x(-5) + 3 \cdot x + 3(-5)$$

Step 2. Find the indicated products.

$$= 2x^2 - 10x + 3x - 15$$

Step 3. Combine like terms.

$$= 2x^2 - 7x - 15$$

This technique can be used with multinomials containing any number of terms. The next example illustrates the product of a binomial and a trinomial.

EXAMPLE 11 Multiply: $(x - 3y)(x^2 - 5xy + 6y^2)$

Solution

Step 1. $= x(x^2 - 5xy + 6y^2) + (-3y)(x^2 - 5xy + 6y^2)$

Step 2. $= (x^3 - 5x^2y + 6xy^2) + (-3x^2y + 15xy^2 - 18y^3)$

Step 3. $= x^3 - 5x^2y - 3x^2y + 6xy^2 + 15xy^2 - 18y^3$

$= x^3 - 8x^2y + 21xy^2 - 18y^3$

LEARNING HINT ▶ Notice from Example 10 that the product of two binomials involves finding four separate products. Since this type of problem occurs so frequently in algebra, many students learn to multiply two binomials quickly by remembering the four products by name: First, Outer, Inner, Last, or **FOIL**. Here is how it works for the binomials in Example 10:

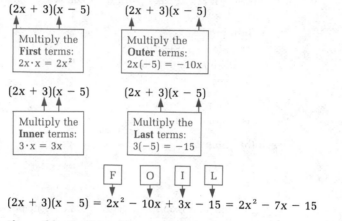

$(2x + 3)(x - 5)$

Multiply the **First** terms:
$2x \cdot x = 2x^2$

$(2x + 3)(x - 5)$

Multiply the **Outer** terms:
$2x(-5) = -10x$

$(2x + 3)(x - 5)$

Multiply the **Inner** terms:
$3 \cdot x = 3x$

$(2x + 3)(x - 5)$

Multiply the **Last** terms:
$3(-5) = -15$

| F | O | I | L |

$(2x + 3)(x - 5) = 2x^2 - 10x + 3x - 15 = 2x^2 - 7x - 15$

If possible, combine like terms as shown in the final equation. ◀

EXAMPLE 12   Use FOIL to multiply $(3x - 4y)(2x - 5y)$.

| F | O | I | L |

Solution $(3x - 4y)(2x - 5y) = 6x^2 - 15xy - 8xy + 20y^2$

$= 6x^2 - 23xy + 20y^2$

Powers of multinomials   To raise a multinomial to a power, first rewrite the expression as a product; then use the appropriate multiplication technique.

EXAMPLE 13   Find:
(a) $(x + 7)^2$ (b) $(2y - 3)^3$

Solution
(a)   **First,** rewrite the problem as a product:

$(x + 7)^2 = (x + 7)(x + 7)$

Then, proceed using FOIL:

$= x^2 + 7x + 7x + 49$

$= x^2 + 14x + 49$

(b) $(2y - 3)^3 = (2y - 3)(2y - 3)(2y - 3)$

Use FOIL to multiply the first two binomials:

$$= (4y^2 - 12y + 9)(2y - 3)$$

Use the distributive property to find the remaining product:

$$= 4y^2(2y - 3) - 12y(2y - 3) + 9(2y - 3)$$

$$= 8y^3 - 12y^2 - 24y^2 + 36y + 18y - 27$$

$$= 8y^3 - 36y^2 + 54y - 27$$

CAUTION ▶ Notice that the correct answers cannot be obtained by simply raising each term of the multinomial to the indicated power. Thus, in Example 13(a),

$(x + 7)^2$ is not equal to $x^2 + 49$

Avoid making this common mistake by always rewriting the power of a multinomial as a product. ◀

Exercises 2-2

Find each indicated product.

1. $5x(-3x^2)$

2. $(-3y^4)(-5y^2)$

3. $mn^2(m^3n^4)$

4. $(-4r^2s)3s^2$

5. $7x^3(5xy)(-2y^2)$

6. $(-6vw)(-8w)(v^4)$

7. $2a^2b(3a^3)^2$

8. $(-2xy^2)^3(3x^2y)$

9. $-4(3m - 7p)$

10. $x(x^2 + 5x)$

11. $3y(2y^2 - 7y)$

12. $-5a(3a^2 - 8ab)$

13. $-6w^2(5w^3 - 7w)$

14. $4xy(7x^2 - 5xy)$

15. $2xy(3x^2 + 5x - 8)$

16. $-7a^2(4a^2 - ab + 2b^2)$

17. $m^2n(3m - 5n + 6)$

18. $3pq^3(3p^2 + 11pq - 6q^2)$

19. $(5x)(-3)(-3x - 7y)$

20. $(2m)(3n)(3m^2 - 2n^2)$

21. $(x + 3)(x + 7)$

22. $(y - 5)(y + 8)$

23. $(x + 6)(x - 6)$

24. $(2x + 3)(2x - 3)$

25. $(2m + 5)(8m - 6)$

26. $(w - 4)(3w - 7)$

27. $(3x - 4)(2x + 7)$

28. $(6y + 1)(5y + 7)$

29. $(5a - 3b)(2a + 9b)$

30. $(4x - 7y)(3x - 2y)$

31. $(7m^2 + 5)(8m^2 + 5)$

32. $(4y^3 - 2)(y^3 + 7)$

33. $(x - 5)^2$

34. $(3y + 2)^2$

35. $(5a + 7b)^2$

36. $(6h - 7g)^2$

37. $(x - 8)(x^2 + 6x + 7)$

38. $(y + 7)(y^2 - 7y + 2)$

39. $(2p + 3)(p^3 - 2p^2 + 7p)$

40. $(3m - 4n)(5m^2 - 7mn - 8n^2)$

41. $4x(x - 7)(2x + 9)$

42. $-2y(2y - 5)(3y + 7)$

43. $(3x - 7)(x + 2)(5x - 1)$

44. $(6y + 5)(8y - 3)(y + 6)$

45. $(m - 9)^3$

46. $(5n + 3q)^3$

Solve.

47. *Finance*   The formula shown below is used to determine the amount A of an investment. Perform the multiplication on the right.

$$A = D\left(1 + \frac{pt}{100}\right)$$

48. *Physics*   In applying Newton's principles to a system of strings and pulleys, the expression $m_1(g - a)$ appears. Multiply.

49. *Physics*   The equation below describes an elastic collision. Perform the multiplication on both sides of the equation.

$$m_1(v_a - v_b)(v_a + v_b) = m_2(v_c - v_d)(v_c + v_d)$$

50. *Electronics*   The impedance Z of a Wein bridge is given by the equation below.* Simplify by multiplying in the numerator and denominator and combining like terms where possible.

$$Z = \frac{(1 + sCR)^2 + sCR}{sC(1 + sCR)}$$

51. *Computer science*   The time required to test a computer memory chip with n cells can be found using the expression $2[n(n + 2) + n]$. Find the indicated products and add like terms.

52. *Space technology*   In a 1000 channel communications system used for deep space probes, three frequencies are sent for each bit of information, and two out of the three must be correct. The error rate is given by the expression $p^2(3 + p)$. Find this product.†

53. *Finance*   After 2 years, an investment of $100 at an interest rate r compounded yearly is worth $100(1 + r)^2$. Perform the indicated operations.

54. *Aeronautical engineering*   An aircraft uses its radar to measure the direct echo range R to another aircraft. The following expression occurs in solving for x, the distance to the ground echo point: $(2R - x)^2 - x^2 - R^2$. Simplify this expression.

2-3 | Division of Algebraic Expressions

In this section we shall discuss how to divide the same types of algebraic expressions we multiplied in Section 2-2: a monomial divided by a monomial, a multinomial divided by a monomial, and a multinomial divided by a multinomial. Once again, knowledge of signed number operations and the laws of exponents will be vital to your understanding of division.

Monomial divided by a monomial   Division of a monomial by a monomial is similar to multiplication in that we can divide only like factors. Since monomial division is usually indicated by a fraction, we can separate such a problem into fractions of like factors and then simplify each one.

EXAMPLE 14   Divide: $(18x^5y^3) \div (-6xy^2) = \dfrac{18x^5y^3}{-6xy^2}$

---

*For more details, see John McWane, *Introduction to Electronics and Instrumentation* (Boston: Breton, 1981), p. 489.

†Based on an interview with William Jago, Tecolote Research Inc., Santa Barbara, California.

Solution

Step 1. Rewrite the division as the product of three quotients, each containing like variables.

$$\frac{18x^5y^3}{-6xy^2} = \frac{18}{-6} \cdot \frac{x^5}{x} \cdot \frac{y^3}{y^2}$$

Step 2. Perform each indicated division using the laws of signed numbers and exponents.

$$= -3 \cdot x^{5-1} \cdot y^{3-2}$$
$$= -3x^4y$$

EXAMPLE 15 Divide:

(a) $\dfrac{-12x^3y^2z^4}{-18x^5y^2z}$ (b) $\dfrac{(-5a^5b)(8ab^2)}{(2ab^2c)^3}$

Solution

(a) **Step 1.** $\dfrac{-12x^3y^2z^4}{-18x^5y^2z} = \dfrac{-12}{-18} \cdot \dfrac{x^3}{x^5} \cdot \dfrac{y^2}{y^2} \cdot \dfrac{z^4}{z}$

   Step 2. $= \dfrac{2}{3} \cdot x^{3-5} \cdot 1 \cdot z^{4-1} = \dfrac{2}{3}x^{-2}z^3 = \dfrac{2z^3}{3x^2}$

(b) **Step 1.** Simplify the numerator and denominator separately.

$$\frac{(-5a^5b)(8ab^2)}{(2ab^2c)^3} = \frac{-40a^6b^3}{8a^3b^6c^3}$$

   Step 2. Divide.

$$= \frac{-40}{8} \cdot \frac{a^6}{a^3} \cdot \frac{b^3}{b^6} \cdot \frac{1}{c^3} = \frac{-5a^3}{b^3c^3}$$

Multinomial divided by a monomial   To help you understand the procedure for dividing a multinomial by a monomial, we first consider the following arithmetic problem.

$$\frac{16 + 12 - 20}{4}$$

To obtain the correct answer, we would normally simplify the numerator first, and then divide.

$$= \frac{28 - 20}{4} = \frac{8}{4}$$
$$= 2$$

But we could also rewrite the problem as the sum of three quotients, divide to simplify, and then add the results.

$$= \frac{16}{4} + \frac{12}{4} - \frac{20}{4}$$
$$= 4 + 3 - 5$$
$$= 2$$

With arithmetic expressions, we normally use the first method, but with algebraic expressions, we must use the second procedure.

EXAMPLE 16 Divide: $\dfrac{12a^3 - 4a^2 + 24a}{4a^2}$

Solution

Step 1. Rewrite as three separate quotients.

$$\frac{12a^3 - 4a^2 + 24a}{4a^2}$$

$$= \frac{12a^3}{4a^2} - \frac{4a^2}{4a^2} + \frac{24a}{4a^2}$$

Step 2. Divide all monomials.

$$= 3a - 1 + \frac{6}{a}$$

In other words, to divide a multinomial by a monomial, divide each term of the multinomial separately by the monomial divisor. After some practice, you will be able to do these problems by inspection.

EXAMPLE 17   Divide:

(a) $\dfrac{15m^3 - 10m}{5m}$ (b) $\dfrac{18x^3y^2 - 24x^4y + 9x^5y^3}{-6xy^2}$

Solution

(a) $\dfrac{15m^3}{5m} - \dfrac{10m}{5m} = 3m^2 - 2$

(b) $\dfrac{18x^3y^2}{-6xy^2} - \dfrac{24x^4y}{-6xy^2} + \dfrac{9x^5y^3}{-6xy^2} = -3x^2 + \dfrac{4x^3}{y} - \dfrac{3x^4y}{2}$

Multinomial divided by a multinomial

To divide a multinomial by another multinomial, we use a procedure very similar to long division in arithmetic. Observe this similarity in the side-by-side examples given in Example 18.

EXAMPLE 18   Divide: $\dfrac{15x^2 - 46x + 16}{3x - 8}$ and $\dfrac{943}{23}$

Solution

Step 1. Rewrite in long division form:

$$3x - 8 \overline{)15x^2 - 46x + 16} \qquad 23\overline{)943}$$

Step 2. Divide the first term of the dividend by the first term of the divisor. Place this quotient above the dividend:

$$\boxed{\frac{15x^2}{3x} = 5x} \qquad\qquad \boxed{\frac{94}{23} \approx \frac{9}{2} \approx 4}$$

$$\begin{array}{r} 5x \\ 3x - 8\overline{)15x^2 - 46x + 16} \end{array} \qquad \begin{array}{r} 4 \\ 23\overline{)943} \end{array}$$

Step 3. Multiply the first term of the quotient by the entire divisor and place the result under the dividend:

$$\begin{array}{r} 5x \\ 3x - 8\overline{)15x^2 - 46x + 16} \\ 15x^2 - 40x \end{array} \qquad \begin{array}{r} 4 \\ 23\overline{)943} \\ 92 \end{array}$$

$$\boxed{5x(3x - 8)} \qquad\qquad\qquad \boxed{4(23)}$$

Step 4. Subtract this product from the dividend and bring down the next term of the dividend:

$$
\begin{array}{r}
5x \\
3x - 8\overline{)15x^2 - 46x + 16} \\
-(15x^2 - 40x) \\
\hline
-6x + 16
\end{array}
\qquad
\begin{array}{r}
4 \\
23\overline{)943} \\
-92 \\
\hline
23
\end{array}
$$

Careful: $-46x - (-40x) = -46x + 40x$

$$= -6x$$

Step 5. Repeat Steps 2, 3, and 4 until there are no terms of the dividend remaining:

$$
\dfrac{-6x}{3x} = -2
$$

$$
\begin{array}{r}
5x \quad - \quad 2 \\
3x - 8\overline{)15x^2 - 46x + 16} \\
-(15x^2 - 40x) \\
\hline
-6x + 16 \\
-(-6x + 16) \\
\hline
0
\end{array}
\qquad
\begin{array}{r}
41 \\
23\overline{)943} \\
-92 \\
\hline
23 \\
-\ 23 \\
\hline
0
\end{array}
$$

Answer: $5x - 2$ **Answer:** 41

In some cases you may need to add a place-holder to the dividend to avoid having unlike terms under each other. Also, in some cases you may have a remainder. These should be expressed as fractions. Both of these techniques are demonstrated in the next example.

EXAMPLE 19 Divide: $\dfrac{6x^3 - 28x + 9}{2x - 4}$

Solution **First,** add a place-holder for the "missing" x^2 term:

$$
2x - 4\overline{)6x^3 + 0x^2 - 28x + 9}
$$

Then, proceed as before. The completed division should look like this:

$$
\begin{array}{r}
3x^2 + 6x \quad - 2 \\
2x - 4\overline{)6x^3 + 0x^2 - 28x + 9} \\
-(6x^3 - 12x^2) \\
\hline
12x^2 - 28x \\
-(12x^2 - 24x) \\
\hline
-4x + 9 \\
-(-4x + 8) \\
\hline
1 \qquad \fbox{Remainder}
\end{array}
$$

Since there is a remainder of 1, the final answer should be expressed as

$$
3x^2 + 6x - 2 + \dfrac{1}{2x - 4} \qquad
\begin{array}{l}
\fbox{Remainder} \\
\fbox{Divisor}
\end{array}
$$

Note that this long division technique is needed only when the divisor is a multinomial.

Exercises 2-3

Perform each indicated division.

1. $\dfrac{-12x^4}{3x}$

2. $\dfrac{18y^6}{6y^2}$

3. $\dfrac{-24m^2}{-12m^5}$

4. $\dfrac{56n}{-7n^3}$

5. $\dfrac{20x^3y^4}{5xy^2}$

6. $\dfrac{-32a^5b^2}{8a^3b^2}$

7. $\dfrac{-64p^7q^4}{-8p^4q^5}$

8. $\dfrac{26r^2s^6}{-13r^4s^2}$

9. $\dfrac{(7xy^2z)(4xy^2z^3)}{10x^2yz^4}$

10. $\dfrac{-16a^5bc^3}{(-8ab^4c^2)(-3ab^3c^3)}$

11. $\dfrac{(5x^2y)^2}{-10x^3y^2z^2}$

12. $\dfrac{-12m^4n^4p^4}{-3(mn^3)^3}$

13. $\dfrac{24x^2 - 8x}{4x}$

14. $\dfrac{16y^4 + 18y^2}{2y}$

15. $\dfrac{12m^2n^3 + 18m^3n}{-6mn^2}$

16. $\dfrac{2ab^3 - 8a^2b^2}{-4ab}$

17. $\dfrac{16c^4 - 24c^3 + 8c^2}{8c^2}$

18. $\dfrac{27m^5 - 15m^3 + 9m^2}{-9m^2}$

19. $\dfrac{8x^3y^4 - 12x^2y^3 - 14xy^5}{-4x^2y^2}$

20. $\dfrac{2r^4s - 16r^3s^2 + 20r^2s^3}{4r^2s^2}$

21. $\dfrac{ax^r - bx^{r+1}}{cx^s}$ $(r > s)$

22. $\dfrac{4(x + y)^3 - 8(x + y)^2 + 6(x + y)}{2(x + y)}$

23. $\dfrac{x^2 + 9x + 20}{x + 4}$

24. $\dfrac{x^2 - x - 12}{x + 3}$

25. $\dfrac{6x^2 - 13x - 63}{3x + 7}$

26. $\dfrac{30x^2 - 49x + 20}{5x - 4}$

27. $\dfrac{6x^3 - x^2 - 29x - 21}{2x + 3}$

28. $\dfrac{8x^3 - 2x^2 - 11x + 2}{4x - 3}$

29. $\dfrac{12x^4 - 17x^3 - 8x + 9}{3x - 2}$

30. $\dfrac{6x^4 - 3x^2 - 63}{2x^2 - 7}$

31. $\dfrac{x^3 - 8}{x - 2}$

32. $\dfrac{x^3 + 27}{x + 3}$

33. $\dfrac{3x^4 - 15x^3 - 19x^2 - 25x - 40}{x^2 - 5x - 8}$

34. $\dfrac{6x^3 - 25x^2y - 39xy^2 + 35y^3}{2x - 5y}$

35. *Holography*   The formula for the radius of a reconstructed spherical wavefront is given below. Find the reciprocal of this expression, and then perform the indicated division.*

$$z_p = \dfrac{z_c z_o z_r}{z_o z_r + \mu z_c z_r - \mu z_c z_o}$$

2-4 | Linear Equations in One Variable

So far in this chapter we have discussed addition, subtraction, multiplication, and division with algebraic expressions. These operations are needed when

*For more details, see Howard M. Smith, *Principles of Holography* (New York: Wiley-Interscience, 1969), p. 200.

solving equations that lead to practical results. In this section we show how to use these algebraic operations to solve linear equations in one variable.

Linear equations in one variable

An **algebraic equation** is simply a statement that two expressions are equal. For example,

$$x + 5 = 7 \qquad x^2 - 3x = 0 \qquad 2\sqrt{x} + x = 1 \qquad 3x + 5y = 9$$

are all algebraic equations.

A **linear equation** is one in which the variables appear only to the first power. Of the above examples, only equations $x + 5 = 7$ and $3x + 5y = 9$ are linear equations, and only the first is a linear equation in one variable. The other linear equation has two variables. (Linear equations with two variables will be solved in Chapter 5.)

Solutions to linear equations

A **solution** of a linear equation in one variable is any value of the variable that makes the equation a true statement. A linear equation in one variable has either one solution, an unlimited number of solutions, or no solution.

EXAMPLE 20 The equation $x + 3 = 8$ has only one solution, 5, because 5 is the only value of x that makes the equation true:

$$5 + 3 = 8$$

Because the equation in Example 20 is true for some but not all values of the variable, it is called a **conditional equation**.

The equation $x + 3 + 6 = 2x - x + 9$ has an unlimited number of solutions. Any value of x makes the equation true. This type of equation is called an **identity.**

The equation $x - 5 = x + 6$ has no solution. There are no values of x that make the equation true. An equation with no solution is called a **contradiction.**

Solving simple linear equations

Simple equations such as

$$x = 8 \qquad x - 1 = 7 \qquad x + 5 = 13 \qquad \frac{x}{2} = 4 \qquad 3x = 24$$

can be solved by inspection. The solution to the first equation is obviously 8. A little trial and error will tell you that the solution to each of the other four equations is also 8. Equations with the same solution are called **equivalent equations.**

Trial and error is not a dependable method of solving more complex equations. Ideally, we need a method that will transform any equation into an equivalent equation of the form $x = a$, where a is the solution of the equation. To do so requires no magic, just one basic fact:

When the same arithmetic operation is performed on both sides of an equation, an equivalent equation results.

To solve an equation, perform *on both sides* the operations that will convert it into the form $x = a$. In other words, we may add the same number to both sides, subtract the same number from both sides, multiply every term on both sides by the same nonzero number, or divide every term on both sides by the same nonzero number. Let's see how this technique can be used on the four unsolved equations given earlier.

EXAMPLE 21

(a) To solve the equation,
$$x - 1 = 7$$
add 1 to both sides.
$$x - 1 + 1 = 7 + 1$$
The solution is:
$$x = 8$$

(b) To solve the equation,
$$x + 5 = 13$$
subtract 5 from both sides.
$$x + 5 - 5 = 13 - 5$$
The solution is:
$$x = 8$$

(c) To solve the equation,
$$\frac{x}{2} = 4$$
multiply both sides by 2.
$$2\left(\frac{x}{2}\right) = 4 \cdot 2$$
The solution is:
$$x = 8$$

(d) To solve the equation,
$$3x = 24$$
divide both sides by 3.
$$\frac{3x}{3} = \frac{24}{3}$$
The solution is:
$$x = 8$$

In each case in Example 21, we chose the operation that resulted in an equivalent equation of the form $x = a$.

In the final answer, the variable does not need to be on the left side of the equation. For example:

To solve the equation,
$$7 = x - 3$$
add 3 to both sides.
$$7 + 3 = x - 3 + 3$$
The solution is:
$$10 = x$$
This is exactly the same as:
$$x = 10$$

IMPORTANT ▶

Always check the solution by substituting it for the variable in the original equation. To check the solution above, substitute 10 for x in the original equation to obtain

$$7 = (10) - 3$$

$$7 = 7$$

Since this statement is true, the solution is correct. ◀

Using combined operations   In the equations encountered thus far, only one operation was needed to solve the equation. However, most often several operations must be combined to find the solution.

EXAMPLE 22 Solve:
$$3x - 8 = 19$$
Step 1. Add 8 to both sides to isolate the x term.
$$3x - 8 + 8 = 19 + 8$$
$$3x = 27$$

Step 2.   Divide both sides by 3 to obtain the solution.

$$\frac{3x}{3} = \frac{27}{3}$$

$$x = 9$$

Check: $3(9) - 8 = 19$

$27 - 8 = 19$

$19 = 19$

EXAMPLE 23   Solve and check:

(a) $\dfrac{y}{4} - 2 = -6$ (b) $9 - x = -6$

(c) $9 = 25 - 8a$ ▥ (d) $0.4m - 3.7 = 2.5$

Solution

(a)

$$\frac{y}{4} - 2 = -6$$ (b) $9 - x = -6$

Add 2. $\dfrac{y}{4} = -4$ Subtract 9. $-x = -15$

Multiply by 4. $y = -16$ Multiply by -1. $x = 15$

Check: $\dfrac{(-16)}{4} - 2 = -6$ **Check:** $9 - (15) = -6$

$-4 - 2 = -6$ $-6 = -6$

$-6 = -6$

Note that an equation in the form $-x = a$ is not yet solved. The coefficient of the variable must be positive 1, not -1. To complete the solution in such instances, multiply both sides by -1. As you will discover, this is comparable to changing the signs of both sides.

(c) $9 = 25 - 8a$

Subtract 25. $-16 = -8a$

Divide by -8. $2 = a$

Check: $9 = 25 - 8(2)$

$9 = 25 - 16$

$9 = 9$

(d) $0.4m - 3.7 = 2.5$

Add 3.7 and divide by 0.4. $2.5 \boxed{+} 3.7 \boxed{=} \boxed{\div} .4 \boxed{=} \rightarrow$ ▨ *15.5*

Solution: $m = 15.5$

Check: $.4 \boxed{\times} 15.5 \boxed{-} 3.7 \boxed{=} \rightarrow$ ▨ *2.5*

Variable on both sides   When the variable appears on both sides of an equation, first move all terms containing the variable to one side, and then proceed as usual.

EXAMPLE 24   Solve: $5x - 7 = 3x + 9$

Solution

Step 1.   Subtract $3x$ from both sides to move all variable terms to the left.

$$5x - 7 = 3x + 9$$
$$5x - 7 - 3x = 3x + 9 - 3x$$
$$2x - 7 = 9$$

Step 2.   Add 7.

$$2x = 16$$

Step 3.   Divide by 2.

$$x = 8$$

Check: $5(8) - 7 = 3(8) + 9$
$$33 = 33$$

If several like terms appear on one side of the original equation, combine them first, and then perform the necessary operations on both sides.

EXAMPLE 25   Solve: $7 + 2y - 4 = 6y - 8y - 9$

Solution

Step 1.   Combine like terms on both sides.

$$7 + 2y - 4 = 6y - 8y - 9$$
$$2y + 3 = -2y - 9$$

Step 2.   Add $2y$.

$$4y + 3 = -9$$

Step 3.   Subtract 3.

$$4y = -12$$

Step 4.   Divide by 4.

$$y = -3$$

Check: $7 + 2(-3) - 4 = 6(-3) - 8(-3) - 9$
$$-3 = -3$$

EXAMPLE 26   Solve:
(a) $4t - 12 = 7t$
(b) $2.44p - 7.68 = 1.28p + 9.46$ (Round the solution to the nearest tenth.)

Solution

(a)   Since the right side of the equation contains only a variable term, we can save a step by isolating the variable on this side.

Subtract $4t$. $-12 = 3t$

Divide by 3. $-4 = t$

Check the solution.

(b)   Combine the constant terms by adding 7.68 to both sides:

$9.46\ \boxed{+}\ 7.68\ \boxed{=}$

Divide by the coefficient of p:

$\boxed{\div}\boxed{(}\ 2.44\ \boxed{-}\ 1.28\ \boxed{)}\boxed{=} \rightarrow \boxed{\textit{14.775862}}$ $p \approx 14.8$

Check: Put the unrounded solution in memory and compute each side of the original equation separately:

Left side: $\boxed{\text{STO}}\boxed{\times}\,2.44\,\boxed{-}\,7.68\,\boxed{=}\rightarrow$ *28.373103*

Right side: $1.28\,\boxed{\times}\boxed{\text{RCL}}\boxed{+}\,9.46\,\boxed{=}\rightarrow$ *28.373103*

Grouping symbols

When parentheses or other grouping symbols appear in an equation, first remove them using the distributive law and then proceed as usual.

EXAMPLE 27   Solve: $3x - (5 - 2x) = 2 + 7(2x - 4)$

Solution
Step 1. Remove parentheses using the distributive law.

$$3x \underset{\uparrow}{-} 5 \underset{\uparrow}{+} 2x = 2 + \underset{\uparrow}{\underline{14x - 28}}$$

 Both signs change $7 \cdot 2x - 7 \cdot 4$

Step 2. Combine like terms. $5x - 5 = 14x - 26$

Step 3. Subtract $5x$. $-5 = 9x - 26$

Step 4. Add 26. $21 = 9x$

Step 5. Divide by 9. $\dfrac{7}{3} = x$ Check this.

CAUTION ▶

In Step 1 some careless students might add 2 and 7 on the right side. This is a mistake because the order of operations dictates that multiplication of 7 by $(2x - 4)$ should be performed first. ◀

Special cases

So far we have solved only conditional linear equations. However, as we mentioned earlier, it is possible for a linear equation in one variable to have an unlimited number of solutions (an identity) or no solution (a contradiction). When we attempt to solve such equations, the variable terms disappear. If a true statement remains, the equation has an unlimited number of solutions; if a false statement remains, the equation has no solution.

EXAMPLE 28   Solve:
(a) $3(4z - 7) = 8 - (6 - 12z)$ (b) $2(3m - 7) = 6m - 14$
(c) $-44 - 4(2q - 11) = q$

Solution
(a)   Remove parentheses. $12z - 21 = 8 - 6 + 12z$

Combine like terms. $12z - 21 = 2 + 12z$

Subtract $12z$. $-21 = 2$

Since the variable terms have dropped out and the resulting statement is *false*, the equation has *no* solution. No check is necessary.

(b)   Remove parentheses. $6m - 14 = 6m - 14$

Subtract $6m$. $-14 = -14$

Since the variable terms have dropped out and the resulting statement is true, the equation has an *unlimited* number of solutions. For example, for $m = 3$,

$$2(3 \cdot 3 - 7) = 6(3) - 14$$

$$4 = 4$$

(c)   Remove parentheses. $-44 - 8q + 44 = q$

Combine like terms. $-8q = q$

Keep going! The variable terms have not disappeared.

Subtract q. $-9q = 0$

Divide by -9. $q = 0$

Check: $-44 - 4(2 \cdot 0 - 11) = 0$

$$0 = 0$$

LEARNING HINT ▶ Students are often confused when only the nonliteral, or constant, terms drop out of an equation. Proceed as usual by moving all variable terms to one side. As shown by Example 28(c), the solution to such an equation will always be zero. ◀

Summary   Here is a summary of the procedure for solving any linear equation:

1.   Remove any grouping symbols using the distributive law.
2.   Combine like terms that appear on the same side of the equation.
3.   Perform the same arithmetic operations on both sides of the equation until it is in the form $x = a$.
4.   Check the solution by substituting it for the variable in the original equation.

Exercises 2-4

Solve the following equations:

1. $x + 2 = -6$

2. $y + 9 = 4$

3. $w - 4 = 11$

4. $a - 9 = -12$

5. $4x = -32$

6. $-6y = -24$

7. $\dfrac{m}{7} = -3$

8. $9 = \dfrac{m}{5}$

9. $-\dfrac{x}{3} = -12$

10. $-0.4y = 6$

11. $7 = 3m - 11$

12. $4p + 9 = -27$

13. $11 - x = 23$

14. $7 = 22 - y$

15. $\dfrac{a}{4} - 1 = 2$

16. $4.8 - 2.4c = -6$

17. $5x - 8 = x$

18. $6y = 32 - 2y$

19. $4w - 11 = 7w + 10$

20. $-5v + 13 = 29 - v$

21. $14m - 7 - 9m = 12 - 7m - 3$

22. $3n - 8n - 9 = 15 - 26 - n$

23. $3(x - 5) = 6x$

24. $7(x + 3) = -x$

25. $12x - (3 - 2x) = 25$

26. $9 - (5x + 3) = -3x$

27. $6y - 5(2y + 7) = 4y + 5$

28. $m - 3(2m - 8) = 24 - 2m$

29. $5 - 2(3w - 4) = 3w + 8$

30. $3 - 7(4v + 5) = 2 - 18v$

31. $3(2x + 4) - 7x = 5 - 9(2x - 3)$

32. $7x - 2(3x - 5) - (x - 9) = x + 11$

33. $2(3x - 9) = 7 - (5 - 6x)$

34. $5y - 3(2y - 7) = 9 - (y - 12)$

35. $6.2x - 5.9 = 21.8 + 14.6x$

36. $0.48 - 2.76y = 8.27y + 4.92$

37. *Electronics* The equation below describes the potential difference V (in volts) between two points in a circuit when it starts at 110 V and increases at a rate of 15 V per second. If t represents the time (in seconds), when is the voltage 170 V?

$$V = 110 + 15t$$

38. *Physics* The weight w (in ounces) hung on a certain spring is related to the length L (in inches) of the spring by the equation below. What is the length of the spring when a 13 oz weight is hung from it?

$$w = 10L - 87$$

39. *Civil engineering* To find one of the components of a force acting on a beam, the equation below must be solved. Find A.

$$20A - 4(15) + 20 = 40$$

40. *Electronics* Solve the equation below to find the zero level voltage of a certain clamper circuit.*

$$(8.5 \times 10^7)V = 120(8.5 - V)$$

2-5 | Evaluating and Solving Formulas

In applied technical work, the most common type of equation is the formula. A **formula** is an equation that gives a mathematical relationship between two or more measurable quantities. In this section we will discuss two of the most commonly applied skills in technical mathematics—how to evaluate and how to solve formulas.

Evaluating formulas The formula $C = \frac{5}{9}(F - 32)$ states the rule for finding the Celsius temperature C for any given Fahrenheit temperature F. If we know the Fahrenheit temperature, we can substitute it for F on the right side of the formula and calculate the corresponding Celsius temperature.

*Based on John M. Doyle, *Digital, Switching, and Timing Circuits* (Boston: Duxbury, 1976), p. 118.

EXAMPLE 29   Using the given temperature conversion formula, we can calculate C when $F = 59°$ in the following manner:

Step 1.   Rewrite the formula, replacing F with the given value. $C = \dfrac{5}{9}(59 - 32)$

Step 2.   Simplify the right side of the formula by performing the indicated operations in the proper order. **First,** simplify within parentheses. $C = \dfrac{5}{9}(27)$

Then, multiply. $C = 15°$

EXAMPLE 30   Evaluate each of the following formulas for the given values:

(a) $x = v_o t + \frac{1}{2}at^2$ when $v_o = 42$ m/s, $t = 2.0$ s, and $a = 8.0$ m/s² (The answer x will be in meters.)

(b) $L = 2C + 1.57(D + d) + \dfrac{D + d}{4C}$ when $C = 36.7$ in., $D = 24.0$ in., and $d = 4.0$ in. (L will be in inches.)

Solution

(a)   Substitute $x = \left(42 \,\dfrac{m}{s}\right)\left(2.0 \text{ s}\right) + \dfrac{1}{2}\left(8.0 \,\dfrac{m}{s^2}\right)\left(2.0 \text{ s}\right)^2$

    Simplify powers. $= \left(42 \,\dfrac{m}{\cancel{s}}\right)\left(2.0 \,\cancel{s}\right) + \dfrac{1}{2}\left(8.0 \,\dfrac{m}{\cancel{s^2}}\right)\left(4.00 \,\cancel{s^2}\right)$

    Multiply. $= 84 \text{ m} + 16 \text{ m}$

    Add. $x = 100 \text{ m}$

(b)   Substitute.

$$L = 2(36.7) + 1.57(24.0 + 4.0) + \dfrac{(24.0 + 4.0)}{4(36.7)}$$

Simplify within parentheses by inspection.

$$L = 2(36.7) + 1.57(28.0) + \dfrac{28.0}{4(36.7)}$$

Enter in your calculator as follows:

2 ⊠ 36.7 ⊞ 1.57 ⊠ 28 ⊞ 28 ⊡ 4 ⊡ 36.7 ⊟ →   *117.55074*

$L \approx 118$ in., rounded

The Pythagorean theorem   Figure 2-1 shows a **right triangle**—that is, a triangle with a 90°, or right, angle. The two **perpendicular** sides, which form the right angle, are called the **legs** (a and b in Fig. 2-1), and the side opposite the right angle is called the

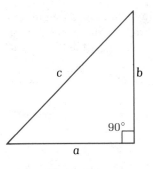

Figure 2-1

hypotenuse (*c* in the figure). The Pythagorean theorem, a formula crucial to an understanding of trigonometry, relates the three sides as follows:

Pythagorean Theorem	$c^2 = a^2 + b^2$	(2-1)

This formula may be used to find any one side of a right triangle when the other two sides are given.

EXAMPLE 31

(a) To find the missing side of the right triangle shown in Fig. 2-2, proceed as follows:

Step 1. Substitute the given values into the Pythagorean formula (2-1).

$$c^2 = 6^2 + 8^2$$

Step 2. Combine the quantities on the right.

$$c^2 = 36 + 64$$
$$c^2 = 100$$

Step 3. Take the square root of both sides to solve for *c*.

$$c = \sqrt{100} = 10 \text{ in.}$$

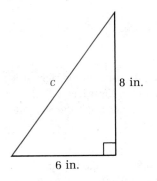

Figure 2-2

(b) To find the missing side of the right triangle shown in Fig. 2-3, proceed as follows:

Step 1. Substitute.

$$11.0^2 = a^2 + 5.0^2$$
$$121.00 = a^2 + 25.00$$

Step 2. Subtract 25 from both sides to isolate the variable term.

$$96.00 = a^2$$

Step 3. Take the square root of both sides.

$$a = \sqrt{96.00} \approx 9.8 \text{ in.}$$

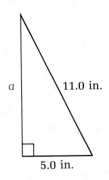

Figure 2-3

EXAMPLE 32 Find the missing side of each right triangle.

(a)

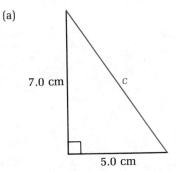

Figure 2-4

(b) A magnetic tape head with an original radius of 1.000 in. has worn down to the chord line *AB* measuring 0.250 in. across (Fig. 2-5). As the first step

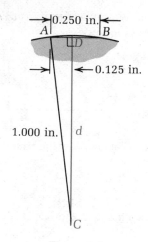

Figure 2-5

in determining the amount of useful head life remaining, an audio engineer must first find the perpendicular distance d from the center C of the head to AB. (From geometry we know that CD bisects AB and therefore $AD = 0.125$ in.) Find d.*

Solution

(a) $c^2 = 5.0^2 + 7.0^2$

$= 25.00 + 49.00$

$c = \sqrt{74.00} \approx 8.6$ cm

(b) $1.000^2 = d^2 + 0.125^2$

$1.000^2 - 0.125^2 = d^2$

$d = \sqrt{1.000^2 - 0.125^2}$

$1 \boxed{-} .125 \boxed{x^2} \boxed{=} \boxed{\sqrt{}} \rightarrow$ 0.9921567

$d \approx 0.992$ in.

Solving formulas   Technicians often need to solve for a quantity that appears on the right side of a formula. The same rules for solving equations apply in these cases. Perform the same operation on both sides of the original formula until the variable needed is isolated on one side.

EXAMPLE 33   The formula $s = \frac{1}{2}at^2$ gives the distance s traveled by an object under acceleration a for time t. If we already know the distance and time, and we wish to find the acceleration, we must solve the formula for a, as follows:

First, multiply both sides by 2: $2 \cdot s = 2\left(\frac{1}{2}at^2\right)$

$2s = at^2$

Then, divide both sides by t^2. $\frac{2s}{t^2} = \frac{at^2}{t^2}$

Write the formula with a on the left: $a = \frac{2s}{t^2}$

EXAMPLE 34   The formula $P = 2L + 2W$ gives the perimeter P of a rectangle of length L and width W. Solve for L.

Solution

Subtract $2W$ from both sides. $P - 2W = 2L$

Divide both sides by 2. $\dfrac{P - 2W}{2} = L,$ or $L = \dfrac{P - 2W}{2}$

*Provided by George Stockton, Burroughs Corporation, Piscataway, New Jersey.

Formulas containing a
fraction

To solve a formula containing a single fraction, first multiply both sides by the denominator to eliminate the fraction. Then solve for the desired variable.

EXAMPLE 35

To solve for V in the formula $\qquad I = \dfrac{E + V}{R}$

First, multiply by R to obtain: $\qquad RI = E + V$

Then, subtract E and switch sides. $\qquad V = RI - E$

Formulas containing
parentheses

To solve formulas containing parentheses, first remove parentheses using the distributive property. Then continue to solve as before.

EXAMPLE 36 Solve for t in this formula from highway engineering:

$$I = kL(T - t)$$

Solution

Remove parentheses. $\qquad\qquad\qquad\qquad I = kLT - kLt$

Subtract kLT from both sides. $\qquad\qquad I - kLT = -kLt$

Divide both sides by $-kL$. $\qquad\qquad \dfrac{I - kLT}{-kL} = t$

Switch sides and multiply the numerator and denominator by -1. $\qquad t = \dfrac{kLT - I}{kL}$

NOTE   The reason for multiplying the numerator and denominator by -1 in Example 36 is that it is preferable not to have a negative denominator in a fraction. ◀

EXAMPLE 37 Solve for R_1 in this formula from electronics: $\quad I = \dfrac{V}{R_1 + R_2}$

Solution

Multiply both sides by $(R_1 + R_2)$. $\qquad I(R_1 + R_2) = V$

Remove parentheses. $\qquad\qquad\qquad IR_1 + IR_2 = V$

Subtract IR_2 from both sides. $\qquad\qquad IR_1 = V - IR_2$

Divide both sides by I. $\qquad\qquad\qquad R_1 = \dfrac{V - IR_2}{I}$

Many students feel intimidated when trying to solve a formula, because they are working with literal symbols. Just remember that the process is the same as solving equations, and you will soon find that the lack of calculation actually makes solving formulas easier than solving equations.

Exercises 2-5

Evaluate each formula for the given values.

1. $E = IR$ (voltage, in volts) when $I = 25$ A and $R = 8$ Ω

2. $d = vt$ (distance, in feet) when $v = 6.20$ ft/s and $t = 12.0$ s

3. $E = \dfrac{F}{A}$ (illuminance, in lux) when $F = 100$ lm and $A = 5$ m²

4. $S = \dfrac{3V}{2A}$ (shearing stress of a beam, in lb/in.²) when $V = 1740$ lb and $A = 42.0$ in.²

5. $F = \frac{9}{5}C + 32$ (Fahrenheit temperature) when $C = 40°$

6. $S = \dfrac{N}{2} + 26 + 40\left(L - \dfrac{3}{4}\right)$ (horizontal range of a fire stream, in feet) when $N = 84$ lb and $L = 1\frac{1}{4}$ in.

7. $C = \frac{5}{9}(F - 32)$ (Celsius temperature) when $F = 81°$

8. $I = kL(T - t)$ (heat-caused expansion of a highway, in feet) when $k = 0.000012$, $L = 2640$ ft, $T = 102°F$, and $t = 82°F$

9. $s = \frac{1}{2}at^2$ (distance, in feet) when $a = 9.80$ ft/s² and $t = 8.00$ s

10. $V = \dfrac{4\pi}{3}r^3$ (volume of a sphere, in cubic inches) when $r = 9.00$ in.

11. $L = \dfrac{25T^4}{H^2}$ (crushing load of a square pillar, in pounds) when $T = 8.00$ in. and $H = 30.0$ ft

12. $r = 0.334v^2$ (in rapid transit, minimum comfort radius of a curve, in feet) when $v = 35.0$ mi/h

13. $x = \dfrac{R_2 L}{R_1 + R_2}$ (distance to the ground, in feet, in an underground cable) when $R_1 = 750$ Ω, $R_2 = 250$ Ω, and $L = 4000$ ft

14. $n = \dfrac{T_2}{T_1 + T_2}$ (efficiency of an engine) when $T_1 = 850$ K and $T_2 = 650$ K

15. $M = \dfrac{P(C + L)}{T}$ (taxable measure of a business, in dollars) when $P = \$55,000$, $C = \$6500$, $L = \$23,500$, and $T = \$75,000$

16. $f' = f\left(1 + \dfrac{v}{s}\right)$ (frequency of sound, in hertz) when $f = 380$ Hz, $v = 40$ mi/h, and $s = 760$ mi/h

17. $H = \sqrt{240p - p^2 - 1900} - 15$ (height of a fire stream, in feet) when $p = 55$ lb

18. $S = \sqrt{8RE - 4E^2}$ (maximum length of a line on a map, in mm) when $R = 0.25$ mm and $E = 0.15$ mm

Solve each formula for the indicated variable.

19. $F = ma$ (force) for a

20. $E = PM$ (efficiency of a pumping plant) for M

21. $H_R = H_1 - H_2$ (heat of a chemical reaction) for H_2

22. $T = \dfrac{D - d}{2}$ (thickness of tubing) for D

23. $C = \dfrac{LDU}{A}$ (level of light) for D

24. $D = \dfrac{RT}{60C}$ (in navigation, distance to a station) for C

25. $R = \dfrac{rL}{D^2}$ (resistance of a length of wire) for L

26. $V = \dfrac{D^2 H}{4}$ (volume of a cylinder) for H

27. $\dfrac{mRT}{M} = \dfrac{1}{2} mv^2$ (kinetic theory of gases) for T

28. $A = 2\pi rh + 2\pi r^2$ (surface area of a cylinder) for h

29. $s = v_o t + \dfrac{1}{2} at^2$ (distance) for v_o

30. $L = 2RH - H^2$ (length of the chord of a circle) for R

31. $L = \dfrac{1}{2}(P_1 + P_2)s$ (lateral surface area of a frustum) for P_1

32. $G = \dfrac{100(H - X)}{C}$ (center of gravity of an airplane) for X

33. $i_b = i_e(1 - \alpha)$ (in solid-state electronics, current) for i_e

34. $U = W(H + \delta)$ (potential energy of a falling object) for H

35. $L = \dfrac{R^2 N^2}{9R + 10D}$ (inductance of a coil) for D

36. $I = \dfrac{V}{R_1 + R_2}$ (current) for R_2

Find the missing side of each right triangle.

37.

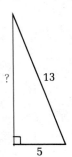

38.

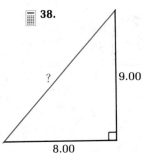

39.

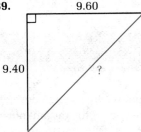

40.

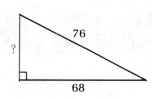

Use the Pythagorean theorem to solve for the missing quantity.

41. One end of a 12.0 ft ramp is placed on the ground, and the other end is placed on the bed of a truck 4.0 ft off the ground. What is the horizontal distance from the truck to the base of the ramp?

42. **Optics** In choosing lenses, a photographer will sometimes need to know the angle of view. The first step in determining this angle is to calculate the diagonal length of the film. Given that 35 mm film is in the shape of a rectangle measuring 36.0 mm by 24.0 mm, find the diagonal length of this type of film.

43. *Navigation*   A pilot wants to fly due east at an average ground speed of 130 mi/h when the wind is blowing from north to south at 55 mi/h. Calculate the unknown side of the right triangle in the figure to determine the air speed the pilot must maintain.

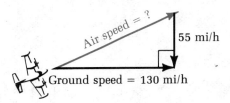

44. *Meteorology*   One hour after being launched, a weather balloon has drifted 12 mi horizontally from the launch point. The straight-line distance between the balloon and the launch point is 12.5 mi. What is the altitude of the balloon? (See the figure.)

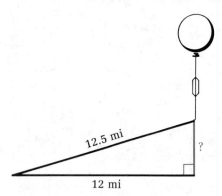

2-6 | Word Problems

In on-the-job calculations, technicians, engineers, and scientists often develop their own equations for given situations. They then solve for the unknown quantity using the techniques discussed in Section 2-4. One way to approximate this procedure is by solving word problems.

Learning to solve word problems is one of the most difficult aspects of algebra. However, certain procedures are always helpful. Read the following problem and study its five-step solution and the accompanying comments.

EXAMPLE 38   Quality control statistics have shown that 2% of the chips manufactured at an electronics hardware plant must be rejected due to defects. How many should be produced to get 900 nondefective chips?

Solution

Step 1. Read the problem carefully. It may take several careful readings to decide what is known and what must be found.

Known	**Unknown**
1. 2% of the chips are defective	The number of chips that must be produced
2. 900 chips are needed	

Step 2. Assign a variable to represent the unknown. Let

n = Number of chips to be produced

Step 3. Translate the problem statement into an equation using the variable assigned in Step 2.

The total number of chips	Minus	The 2% defective chips	Must equal	900
n	$-$	$0.02n$	$=$	900

Step 4. Solve the equation. $n - 0.02n = 900$

Combine like terms. $0.98n = 900$

Divide by 0.98. $n = 918.367\ldots$

Round. $n \approx 918$ chips

Step 5. Check your answer against the given information to be sure it is correct.

$0.02(918) \approx 18$ defective chips

This leaves 900 nondefective chips.

The remainder of this section illustrates various types of word problems and their solutions. Although each problem may seem completely different from the previous one, they can all be solved using the procedure just demonstrated. Remember the five steps of this problem-solving process:

Step 1. **Read** the problem and identify known and unknown quantities.
Step 2. **Assign** a variable.
Step 3. **Translate** the word problem into an equation.
Step 4. **Solve** the equation.
Step 5. **Check** the solution.

EXAMPLE 39 A length of wire is 48 in. long. It must be cut into two pieces, one of which is twice as long as the other. Find the length of each piece.

Solution

Step 1.   Read the problem carefully. Identify the known and unknown quantities.

Known	**Unknown**
1. Total length: 48 in.	1. Length of shorter piece
2. Longer piece is twice as long as shorter piece	2. Length of longer piece

Step 2.   Assign a variable to stand for the unknown quantity. If there are two or more unknowns, you may be able to use the known facts to express them all in terms of the same variable. Let

s = Length of the shorter piece

Then

$2s$ = Length of the longer piece

We know that one piece is twice as long as the other, so we use $2s$ to say twice s.

Step 3.   Translate the one remaining known fact into an equation.

Length = 48 in.

$s + 2s = 48$

The total length of the piece of wire consists of the shorter piece plus the longer piece.

Step 4.   Solve the equation. Make certain to give all answers requested.

$3s = 48$

$s = 16$ in.   Shorter piece

$2s = 2(16) = 32$ in.   Longer piece

Step 5.   Check your solution against the original problem statement.

The total length is 48 in.: $32 + 16 = 48$

One piece is twice as long as the other: $32 = 2 \cdot 16$

EXAMPLE 40   The sum of three voltages in a circuit is 32 V. The middle size one is 8 V more than the smallest. The largest is 4 V more than three times the smallest. What is the value of each voltage?

Solution   **Step 1.**   Read the problem and identify the known and unknown quantities.

Known	**Unknown**
1. Sum: 32	The value of each voltage
2. Middle one: 8 more than the smallest	
3. Largest: 4 more than 3 times the smallest	

Step 2.   Assign variable expressions to the unknowns. Since the two larger voltages are both compared to the smallest, let the variable itself stand for the smallest voltage:

$$x = \text{Value of the smallest voltage}$$

Then

$$x + 8 = \text{Value of the middle voltage} \qquad \textit{Hint:} \text{ "more than" means plus}$$

$$3x + 4 = \text{Value of the largest voltage}$$

Step 3.   Translate the one remaining known fact into an equation.

$$\text{Sum of the voltages} = 32$$

$$x + (x + 8) + (3x + 4) = 32$$

Step 4.   Solve the equation.

$$5x + 12 = 32$$

$$5x = 20$$

$$x = 4 \text{ V} \qquad \text{The smallest}$$

$$x + 8 = 12 \text{ V} \qquad \text{The middle one}$$

$$3x + 4 = 16 \text{ V} \qquad \text{The largest}$$

Step 5.   Check your answer.

The sum of the voltages is 32: $4 + 12 + 16 = 32$

EXAMPLE 41   A computer data base charges $6.50 for the first hour and $4.25 for each additional hour. For what length of time was it used in a certain job if the total cost was $44.75?

Solution
Step 1.   Read carefully. Identify known and unknown quantities.

Known	**Unknown**
1. First hour: $6.50	Total time
2. Each additional hour: $4.25	
3. Total cost: $44.75	

Step 2.   Assign a variable to the unknown. Let

t = Total time

Step 3.   Translate into an equation.

Cost of the first hour	Plus	Cost of $(t - 1)$ additional hours	Equals	Total cost
6.50	+	$4.25(t - 1)$	=	44.75

Step 4.   Solve.

$$6.50 + 4.25t - 4.25 = 44.75$$

$$4.25t + 2.25 = 44.75$$

$$4.25t = 42.50$$

$$t = 10 \text{ h}$$

Step 5.   Check.

$$6.50 + 4.25(9) = 44.75$$

$$44.75 = 44.75$$

An alternative approach to this problem would be to let t stand for the number of *additional* hours. This would lead to a simpler equation: $6.50 + 4.25t = 44.75$. After solving for t, you would then have to add 1 to obtain the total length of time.

Problems based on formulas   Formulas can be used to set up equations for many types of word problems. This is illustrated in the next example.

EXAMPLE 42   The width of a rectangular vent is 12 cm less than the length. The perimeter is 148 cm. Find the dimensions.

Solution

Step 1.   Read the problem. Identify known and unknown quantities.

Known	Unknown
1. Width is 12 less than length	1. Width
2. Perimeter: 148 cm	2. Length

Step 2.   Assign variable expressions to unknowns. Let

L = Length

$L - 12$ = Width   12 "less than" L means $L - 12$

Step 3. Translate. The formula for the perimeter of a rectangle is $P = 2W + 2L$. Substitute the known facts into this formula.

$$\text{Perimeter} = 2 \cdot \text{Width} + 2 \cdot \text{Length}$$

$$148 = 2(L - 12) + 2L$$

Step 4. Solve.

$$148 = 2L - 24 + 2L$$

$$148 = 4L - 24$$

$$172 = 4L$$

$$L = 43 \text{ cm} \qquad \text{Length}$$

$$L - 12 = 31 \text{ cm} \qquad \text{Width}$$

Step 5. Check.

$$2(31) + 2(43) = 148$$

$$148 = 148$$

Motion problems

Some word problems can be solved more easily if the information is organized in a chart. The next problem involves the use of the formula for uniform motion: Distance = Speed × Time, or $d = vt$.

EXAMPLE 43   A car traveling 48 mi/h leaves a certain point $1\frac{1}{2}$ h later than a bicycle rider traveling at 12 mi/h in the same direction. How long does it take the car to catch up to the bicycle?

Solution
Step 1.   Read and identify known and unknown quantities.

Known	**Unknown**
1. Car's speed: 48 mi/h	Time it takes for the car to catch the bike
2. Bike's speed: 12 mi/h	
3. Since the car left $1\frac{1}{2}$ h later than the bike, the bike's time is $1\frac{1}{2}$ h more than the car's time	

Step 2.   Assign variable expressions to the unknowns. The easiest way to do this is to organize the given information in a chart based on the motion formula:

	Speed	×	Time	=	Distance
Car	48		t		$48t$
Bicycle	12		$t + 1\frac{1}{2}$		$12(t + 1\frac{1}{2})$

The motion formula is used to fill in the Distance column by multiplying the first two columns (Speed and Time).

Step 3.   Translate. When the car overtakes the bicycle, they have traveled the same distance.

Car's distance = Bike's distance

$$48t = 12(t + 1\tfrac{1}{2})$$ From the third column of the chart

Step 4.   Solve.

$$48t = 12t + 18$$

$$36t = 18$$

$$t = \tfrac{1}{2}\,\text{h}$$

Step 5.   Check.

Car's distance $= \left(48\,\dfrac{\text{mi}}{\text{h}}\right)\cdot\tfrac{1}{2}\,\text{h} = 24$ mi

Bike's time $= \tfrac{1}{2}\,\text{h} + 1\tfrac{1}{2}\,\text{h} = 2$ h

Bike's distance $= \left(12\,\dfrac{\text{mi}}{\text{h}}\right)\cdot 2\,\text{h} = 24$ mi

EXAMPLE 44   A plane flies at a certain rate for the first 2.0 h of a trip, and then a headwind slows it down by 50 mi/h. If the total trip of 6.0 h covers 3100 mi, what is the speed of the plane during each part of the trip?

Solution
Step 1.   Read. Identify known and unknown quantities.

Known	Unknown
1. Speed during second part: 50 mi/h less than speed during first part	Speed during each part
2. Time of first part: 2.0 h	
3. Total time: 6.0 h	
4. Total distance: 3100 mi	

Step 2.   Assign variable expressions to unknowns and organize all information in a chart:

	Speed ×	Time =	Distance
First part	v	2.0	2.0v
Second part	$v - 50$	4.0	4.0($v - 50$)

Remember: The third column is simply the product of the first two.

The time for the second part (4.0 h) was obtained by subtracting the time for the first part from the total time: $6.0 - 2.0 = 4.0$.

Step 3.   Translate.

$$\text{Total distance} = 3100 \text{ mi}$$

$$2.0v + 4.0(v - 50) = 3100$$

Step 4.   Solve.

$$2.0v + 4.0v - 200 = 3100$$

$$6.0v - 200 = 3100$$

$$6.0v = 3300$$

$$v = 550 \text{ mi/h} \qquad \text{Speed for first part}$$

$$v - 50 = 500 \text{ mi/h} \qquad \text{Speed for second part}$$

Step 5.   Check.

$$\text{Distance of first part} = 2.0 \text{ h} \cdot \left(550 \, \frac{\text{mi}}{\text{h}}\right) = 1100 \text{ mi}$$

$$\text{Distance of second part} = 4.0 \text{ h} \cdot \left(500 \, \frac{\text{mi}}{\text{h}}\right) = 2000 \text{ mi}$$

$$\text{Total distance} = 1100 + 2000 = 3100 \text{ mi}$$

Percent mixture   Percent mixture problems are useful in technical work, and they, too, are more easily solved using a chart. Consider the next problem.

EXAMPLE 45   One solder alloy contains 30% lead and another contains 60% lead. How many pounds of each are needed to produce 90 lb of an alloy that is 50% lead?

Solution
Step 1.   Read. Identify known and unknown quantities.

Known	Unknown
1. First alloy: 30% lead	The number of pounds of
2. Second alloy: 60% lead	each alloy needed
3. Final alloy: 50% lead	
4. Total amount: 90 lb	

Step 2.   Assign variable expressions to unknowns and set up a chart based on the formula

(Fraction of lead) × (Amount of alloy) = Amount of lead

Again, the value of the chart lies in the fact that the third column is simply the product of the first two.

	Fraction of lead	× Amount of alloy	= Amount of lead
First alloy	0.3	x	0.3x
Second alloy	0.6	90 − x	0.6(90 − x)
Final alloy	0.5	90	0.5(90)

Many students have trouble with the expressions in the amount column in a problem like this. Think of it this way: There is a total of 90 lb of the two alloys. If x represents the amount of one alloy, then (90 − x) remains as the amount of the second alloy. The amounts of the two alloys then total 90: x + (90 − x) = 90.

Step 3. Translate into an equation.

Amount of lead in first alloy	Plus	Amount of lead in second alloy	Equals	Amount of lead in final alloy
0.3x	+	0.6(90 − x)	=	0.5(90)

Step 4. Solve.

$$0.3x + 54 - 0.6x = 45$$

$$-0.3x + 54 = 45$$

$$-0.3x = -9$$

$$x = 30 \text{ lb} \qquad \text{Amount of alloy with 30\% lead}$$

$$90 - x = 60 \text{ lb} \qquad \text{Amount of alloy with 60\% lead}$$

Step 5. Check.

Amount of lead in first alloy = 0.3(30) = 9 lb

Amount of lead in second alloy = 0.6(60) = 36 lb

Amount of lead in final alloy = 0.5(90) = 45 lb

First plus second equals final: 9 + 36 = 45

EXAMPLE 46 How many milliliters of an 85% solution of acid should be added to 32.0 mL of a 45% solution to obtain a 60% solution?

Solution
Step 1. Read and find the known and unknown quantities.

Known	**Unknown**
1. First solution: 85% acid	Necessary amount of
2. Second solution: 45% acid	the 85% solution
3. Mixture solution: 60% acid	
4. Amount of 45% solution: 32.0 mL	

Step 2.   Assign variable expressions to the unknown quantities. Put all information into a chart.

	Fraction of acid ×	Amount of solution =	Amount of acid
First solution	0.85	x	0.85x
Second solution	0.45	32.0	0.45(32.0)
Mixture solution	0.6	x + 32.0	0.6(x + 32.0)

Step 3.   Translate into an equation.

Amount of acid in the first solution	Plus	Amount of acid in the second solution	Equals	Amount of acid in the mixture solution
0.85x	+	0.45(32.0)	=	0.6(x + 32.0)

Step 4.   Solve.

$$0.85x + 14.4 = 0.6x + 19.2$$

$$0.25x = 4.8$$

$$x = 19.2 \text{ mL} \qquad \text{Amount of 85\% solution}$$

Step 5.   Check.

Amount of acid in first solution = 0.85(19.2) = 16.32 mL

Amount of acid in second solution = 0.45(32.0) = 14.4 mL

Total amount of final solution = 32.0 + 19.2 = 51.2 mL

Amount of acid in final solution = 0.6(51.2) = 30.72 mL

First plus second equals final: 16.32 + 14.4 = 30.72

Exercises 2-6

Solve.

1. *Machine technology*   Five percent of the parts tooled by a machine are rejects. How many parts must be tooled to assure 2000 acceptable ones?
2. *Industrial management*   Twenty percent of a technician's salary goes to withholding taxes. How much should his gross income be in order to take home $400 per week?
3. *Computer science*   Two computers have a combined memory of 384 kilobytes. One computer has twice as much memory as the other. What is the memory of each?
4. *Electronics*   The sum of three currents entering a circuit junction is 44 A. One current is 6 A larger than a second current, which in turn is 4 A larger than the third. What is the value of each current?

5. *Sheet metal technology*   The width of a piece of sheet metal is 8 cm less than the length. The perimeter of the sheet is 84 cm. Find the dimensions of the sheet.

6. *Industrial engineering*   The length of a rectangular machine shop is 13 yd longer than the width. The perimeter is 210 yd. What are the dimensions of the shop?

7. The third side of an isosceles triangle is 5 in. longer than twice the length of each of the two equal sides. Its perimeter is 65 in. Find its dimensions.

8. *Architecture*   The width of an office is 6 ft more than half the length. If the perimeter is 138 ft, find the dimensions of the room.

9. *Navigation*   Two planes traveling in opposite directions leave the same point at the same time. One is flying at a speed of 320 mi/h. If they are 1700 miles apart after 2.5 h, how fast is the other plane traveling?

10. Two cars headed toward each other leave towns 660 mi apart at the same time. If one car is traveling at 65 mi/h, and the other is moving at 55 mi/h, how soon do they meet?

11. A runner moving at 10 mi/h is being chased by another runner going 12 mi/h. If the first runner had a $\frac{1}{2}$ mi head start, how long does it take the second runner to overtake the first?

12. A car leaves a certain point 1 h ahead of another car. The second car is traveling 10 km/h faster than the first, and it overtakes the first car in 4 h. How fast was each car going?

13. *Navigation*   A plane flies at a certain speed for the first 3.0 h of a trip, and then a tailwind increases its speed by 70 mi/h. If the entire trip of 5.0 h covered 2440 mi, how fast was the plane going during the second half of the journey?

14. A woman rode a bike for 3 h at 20 mi/h and then walked for 2 h. If she traveled a total of 68 mi, what was her walking speed?

15. Two backpackers plan on hiking into the wilderness at 4 mi/h and hiking back to their starting point at 5 mi/h. If they have a total of $4\frac{1}{2}$ h for the round trip, how far into the wilderness can they go?

16. *Navigation*   A plane travels from A to B at 420 mi/h and returns from B to A at 360 mi/h. If the round trip took 15 h, how far is it between the two points?

17. *Chemistry*   How many liters each of a 76% acid solution and a 12% acid solution should be mixed to produce 28 L of a 60% acid solution?

18. *Chemistry*   A chemist has a 36% solution and a 90% solution of a weed killer. How many milliliters of each should be used to make 24 mL of a 72% solution?

19. *Agricultural technology*   A dairy rancher has 40 gal of a 30% butterfat solution. How much 80% butterfat should be added to produce a mixture with a 40% butterfat concentration?

20. *Auto mechanics*   An 8.0 gal radiator solution contains 20% antifreeze. How much 80% antifreeze should be added to produce a 35% solution?

21. *Chemistry*   How many liters of pure acid must be added to 15 L of a 40% acid solution to produce a 50% acid solution?

22. *Chemistry*   How much pure acid should be added to 180 mL of a 15% solution to make a 25% solution?

23. *Chemistry*   How much water must be evaporated from 80 L of a 10% salt solution to create a 16% salt solution?

24. *Chemistry*   How much water must be evaporated from 65 q of a 3% salt solution to create a 5% salt solution?

25. A phone call costs $0.38 for the first minute and $0.18 for each additional minute. How many minutes long was a call that cost $2.36?

26. *Agricultural technology*   An empty avocado crate weighs 3.6 kg. How many avocados weighing 0.4 kg each can be added before the total weight of the crate reaches 16.0 kg?

27. A salesperson receives $12.80 per hour for the first 40 h of work per week and $19.20 per hour for each hour over 40. How many overtime hours are needed in a week in order to earn $704?

28. A taxi ride costs $0.65 for the first mile and $0.45 for each additional mile. How many miles was a taxi ride that cost $8.30?

2-7 | Ratio and Proportion

The concepts of ratio and proportion are important to technical math students in two respects. First, ratios and proportions have direct practical applications to scale drawings, pulleys, gears, gas pressure, levers, and other physical situations. Second, they provide part of the foundation necessary for understanding trigonometric functions.

Ratio

A **ratio** is a number, usually written as a fraction, used to compare the magnitudes of two like quantities. Because quantities with the same units are being compared, the ratio is dimensionless—that is, it has no units.

64 teeth

16 teeth

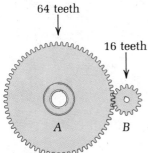

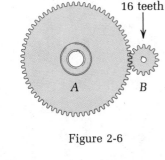

A   *B*

Figure 2-6

EXAMPLE 47 To find the gear ratio of *A* to *B* in Fig. 2-6, set up a fraction based on the number of teeth on each gear. Since *A* has 64 teeth and *B* has 16 teeth, the gear ratio of *A* to *B* is expressed as

$$\frac{A}{B} = \frac{64 \text{ teeth}}{16 \text{ teeth}} = \frac{4}{1}$$

This can also be written 4 : 1.
Both expressions are read "four to one."

The reduced fraction represents the simplest form of the ratio. Notice that the units (number of teeth) cancel out in the division, leaving the ratio a dimensionless number.

EXAMPLE 48 Express each of the following as a ratio in simplest form:
(a) The ratio of the diameter of pulley *X* to that of pulley *Y* in Fig. 2-7.
(b) 45 in. to 5 ft
(c) The specific gravity of a substance is defined as the ratio of the density of the substance to the density of water. Find the specific gravity of brass given that its density is 540 lb/ft³ and the density of water is 62.4 lb/ft³.

Y

X

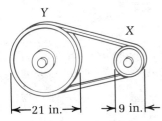

←21 in.→   →9 in.←

Figure 2-7

Solution

(a) $\frac{9 \text{ in.}}{21 \text{ in.}} = \frac{3}{7}$

(b) To get a dimensionless ratio, first change to like units:

$$\frac{45 \text{ in.}}{5 \text{ ft}} = \frac{45 \text{ in.}}{60 \text{ in.}} = \frac{3}{4}$$

(c) 540 ÷ 62.4 = → $\boxed{8.6538462}$
The specific gravity of brass is 8.7 : 1, which is usually stated simply as 8.7.

Proportion

A **proportion** is a statement that two ratios are equal. The proportion

$$\frac{a}{b} = \frac{c}{d}$$

is read "a is to b as c is to d." The quantities a, b, c, and d are called the **terms** of the proportion. Using the alternate form of each ratio, this proportion could also be written $a : b = c : d$.

Solving proportions   In this section we will be dealing with algebraic proportions containing one unknown quantity. Since a proportion is simply a linear equation, you can apply the normal equation-solving techniques to solve it.

EXAMPLE 49

To solve the proportion $$\frac{x}{4} = \frac{10}{14}$$

Multiply both sides by 4. $$4\left(\frac{x}{4}\right) = 4\left(\frac{10}{14}\right)$$

$$x = \frac{20}{7} \quad \text{or} \quad 2\frac{6}{7}$$

Cross-product rule   In the proportion

$$\frac{a}{b} = \frac{c}{d}$$

if we multiply both sides by bd,

$$bd \cdot \frac{a}{b} = \frac{c}{d} \cdot bd$$

we obtain the equivalent equation

$$ad = bc$$

Notice that the members of this equation can also be obtained from the original proportion by multiplying the numerator on one side by the denominator on the other side:

$$\frac{a}{b} \diagdown \frac{c}{d} \qquad ad = bc$$

For this reason, the following statement is known as **the cross-product rule:**

Cross-Product Rule

If $\dfrac{a}{b} = \dfrac{c}{d}$, then $ad = bc$ (where b, $d \neq 0$). $\qquad$ (2-2)

To solve any proportion, first apply the cross-product rule to eliminate fractions. Then divide both sides by the coefficient of the unknown.

EXAMPLE 50 Solve the following proportions using the cross-product rule:

(a) $\dfrac{x}{3} = \dfrac{7}{12}$ (b) $\dfrac{5}{8} = \dfrac{20}{y}$ (c) $\dfrac{2496}{3658} = \dfrac{n}{1721}$

Solution

(a) **First,** apply the cross-product rule. $12x = 21$

   Then, divide by 12. $x = \dfrac{7}{4}$ or $1\dfrac{3}{4}$

(b) $5y = 160$ (c) $2496 \boxed{\times} 1721 \boxed{\div} 3658 \boxed{=} \rightarrow$ *1174.3073* or

   $y = 32$ $n \approx 1174$, rounded

Equivalent proportions   Some quick and easy maneuvers will help you work with proportions. First, notice that an equivalent proportion results if we invert each ratio of a proportion. That is,

$$\frac{a}{b} = \frac{c}{d} \quad \text{is equivalent to} \quad \frac{b}{a} = \frac{d}{c}$$

because application of the cross-product rule to each proportion results in $ad = bc$.

An equivalent proportion also results if we equate the ratio of the numerators with the ratio of the denominators:

$$\frac{a}{b} = \frac{c}{d} \quad \text{is equivalent to} \quad \frac{a}{c} = \frac{b}{d}$$

Finally, we can also interchange the sides of any proportion. For example,

$$\frac{a}{b} = \frac{c}{d} \quad \text{is equivalent to} \quad \frac{c}{d} = \frac{a}{b}$$

EXAMPLE 51 Express the proportion $\dfrac{5}{x} = \dfrac{9}{2}$ in three other equivalent ways.

Solution

Inverting each ratio, $\dfrac{x}{5} = \dfrac{2}{9}$

Equating the ratio of the numerators with the ratio of the denominators, $\dfrac{5}{9} = \dfrac{x}{2}$

Interchanging sides, $\dfrac{9}{2} = \dfrac{5}{x}$

By continuing to invert and interchange sides, four other equivalent proportions can be created.

Solving problems   The chief reason for learning about proportions is that many practical problems can be translated into proportions and then easily solved. For example, since the dimensions of a scale drawing are proportional to the actual dimensions, we can use a proportion to solve for a missing dimension.

EXAMPLE 52   A shop drawing shows a circuit element 6.0 cm long by 2.5 cm wide. If the element is actually 0.75 cm long, find its actual width.

Solution

Step 1. Set up a proportion using ratios of like units.

$$\frac{\text{Actual length}}{\text{Drawing length}} = \frac{\text{Actual width}}{\text{Drawing width}}$$

$$\frac{0.75 \text{ cm}}{6.0 \text{ cm}} = \frac{x \text{ cm}}{2.5 \text{ cm}}$$

Step 2. Solve for the unknown using the cross-product rule.

$$6.0x = 1.875$$
$$x = 0.3125 \text{ cm} \quad \text{or}$$
$$0.31 \text{ cm, rounded}$$

Here are more problems that can be solved using proportions.

EXAMPLE 53

(a) The resistance of a wire is proportional to its length. If a wire 16 ft long has a resistance of 0.4 Ω, what is the resistance of 20 ft of the same type of wire?

(b) A metal block is allowed to slide down a rough inclined plane (Fig. 2-8). If the plane is tilted so that the block moves with constant speed, the following proportion holds:

$$\frac{F}{W} = \frac{H}{L}$$

where F is the friction force on the block and W is the weight of the block. Find the size of F if the block weighs 30 lb, $H = 1$ ft, and $L = 5$ ft.

(c) A metal cleaning compound must be diluted according to the ratio 2 parts of compound to 7 parts water. How many ounces of compound should be used to get 36 oz of diluted solution?

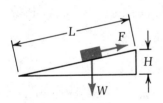

Figure 2-8

Solution

(a) We know that

$$\frac{\text{Resistance of first wire}}{\text{Resistance of second wire}} = \frac{\text{Length of first wire}}{\text{Length of second wire}}$$

If R represents the resistance of the second wire, we can substitute our given information to obtain

$$\frac{0.4}{R} = \frac{16}{20}$$

Solving, we have

$$16R = 8$$

$$R = 0.5 \ \Omega$$

(b) **First,** solve the literal proportion for F:

Use the cross-product rule. $FL = WH$

Divide both sides by L. $F = \dfrac{WH}{L}$

Then, substitute the given values:

$$F = \frac{(30 \text{ lb})(1 \text{ ft})}{5 \text{ ft}}$$

$$F = 6 \text{ lb}$$

(c) Since 36 is the total amount of diluted solution, we must set up the ratio of compound to diluted solution. From the given information, there are 2 parts of compound and 9 parts of diluted solution (2 parts compound plus 7 parts water), so we have

$$\frac{2}{9} = \frac{x}{36}$$

Solving, we obtain

$$9x = 72$$

$$x = 8 \text{ oz}$$

Similar triangles   From geometry we know that two triangles are **similar** if they have the same shape. Mathematically, this implies two important facts illustrated in Fig. 2-9:

1. Corresponding angles of similar triangles are equal.
2. Corresponding sides of similar triangles are proportional.

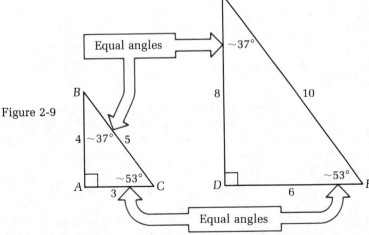

Figure 2-9

Looking at the figure, we can write fact 2 as follows:

Ratio of horizontal sides ⟹ $\dfrac{\text{Side } AC}{\text{Side } DF} = \dfrac{\text{Side } AB}{\text{Side } DE}$ ⟸ Ratio of vertical sides

$$\frac{3}{6} = \frac{4}{8}$$

We could also set up similar proportions involving one of these ratios and the ratio of the hypotenuses. The proportionality of the sides of similar triangles can be used to solve for one missing side in any given pair of similar triangles.

EXAMPLE 54   To solve for the unknown side x in the pair of similar triangles MPQ and MNO shown in Fig. 2-10, we could set up the following proportion:

$$\frac{MP}{MN} = \frac{PQ}{NO}$$

$$\frac{6 \text{ cm}}{18 \text{ cm}} = \frac{x \text{ cm}}{15 \text{ cm}}$$

Solving, we have

$$18x = 90$$

$$x = 5 \text{ cm}$$

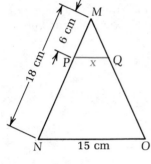

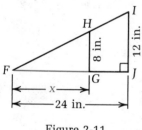

Figure 2-10

EXAMPLE 55   Solve for the unknown side x shown in Fig. 2-11.

Solution   Triangle FGH is similar to triangle FJI. Hence,

$$\frac{FG}{FJ} = \frac{GH}{JI}$$

$$\frac{x}{24} = \frac{8}{12}$$

Solving, we have

$$12x = 192$$

$$x = 16 \text{ in.}$$

Figure 2-11

In solving these problems, your proportion may not be identical to ours, but it could still be correct. As we noted earlier, the terms of a proportion can be rearranged in any one of several equivalent ways. For example, a proportion equivalent to the one used in Example 55 would be

$$\frac{FG}{GH} = \frac{FJ}{JI}$$

This method of stating the proportionality of the sides of similar right triangles will be particularly useful to us when defining the trigonometric functions.

Exercises 2-7

Express each ratio in simplest form.

1. 24 teeth to 8 teeth **2.** 16 lb to 80 lb **3.** 45 in. to 20 in. **4.** 36 V to 60 V

5. 18 ft to 12 yd **6.** 5 km to 500 m **7.** 14 qt to 5 gal **8.** 3 lb to 32 oz

9. *Optics* The magnification of a mirror or lens is the ratio of image height to object height. Find the magnification of a lens that shows an object 8 in. high to be 32 in. high.

10. *Auto mechanics* The compression ratio of an engine is the ratio of expanded volume to compressed volume. Find the compression ratio of an engine with an expanded volume of 440 cm^3 and a compressed volume of 65 cm^3.

11. *Machine technology* Find the ratio of the diameter of pulley A to that of pulley B in the figure.

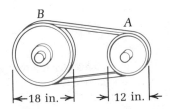

12. *Material science* Find the specific gravity [see Example 48(c)] of lead given that it has a density of 708 lb/ft^3.

13. *Electronics* The power gain of an amplifier circuit is defined as

$$\text{Power gain} = \frac{\text{Output power}}{\text{Input power}}$$

If an amplifier circuit has an input power of 0.72 W and an output power of 21.6 W, find the power gain.

14. *Environmental science* Environmental analysts have determined that the ratio of hydrocarbons to nitrogen oxide can be used as a measure of smog production. Find this ratio for an automobile that emits 6.0 g of hydrocarbons and 4.5 g of nitrogen oxide per mile.

15. *Civil engineering* The slope of a road is equal to the ratio of vertical change to horizontal change. Find the average slope of a road that drops 2500 ft in elevation over a 4.0 mi. horizontal distance. [*Careful:* The units must agree.]

16. *Land management* Many decisions are based on a cost–benefit ratio. One land-use plan has benefits worth $1.8 million and costs of $400,000. A second plan has benefits of $950,000 and costs of $210,000. Find the cost–benefit ratio of each plan and choose the better plan.

Solve each proportion for the missing variable.

17. $\dfrac{x}{4} = \dfrac{9}{16}$ **18.** $\dfrac{16}{10} = \dfrac{y}{35}$ **19.** $\dfrac{32}{12} = \dfrac{24}{a}$ **20.** $\dfrac{12}{c} = \dfrac{4.8}{32}$

21. $\dfrac{17}{33} = \dfrac{9}{e}$ **22.** $\dfrac{x}{38} = \dfrac{64}{100}$ **23.** $\dfrac{0.04}{m} = \dfrac{13.2}{16.5}$ **24.** $\dfrac{27,852}{64,962} = \dfrac{w}{5780}$

25. *Machine technology* The diameters and angular speeds of two interconnected pulleys are related by the inverse proportion given below. Find s_2 if $d_1 = 16$ in., $d_2 = 12$ in., and $s_1 = 240$ rev/min.

$$\frac{d_1}{d_2} = \frac{s_2}{s_1}$$

26. *Optics*   The lens diagram shown in the figure shows that the distances are related by the equation given below. Find q if $h_o = 56$ cm, $h_i = 16$ cm, and $p = 24$ cm.

$$\frac{h_i}{h_o} = \frac{q}{p}$$

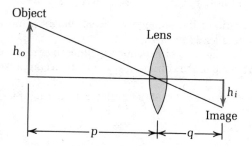

27. *Medical technology*   A lab technician uses a colorimeter to measure the light waves passing through two fluids—a standard fluid that has a known concentration of a particular substance and another fluid that has an unknown concentration. The technician then sets up the following proportion to determine the concentration in the unknown:

$$\frac{\text{Concentration in standard}}{\text{Concentration in unknown}} = \frac{\text{Absorbance of standard}}{\text{Absorbance of unknown}}$$

In a blood glucose test, the concentration of the standard was 100.0 mg/dL, the absorbance of the standard was 0.0234, and the absorbance of the unknown was 0.2060. Find the concentration of the unknown.

28. *Electrical engineering*   The electrical resistance of a given length of wire is inversely proportional to the square of its diameter, as given by the following proportion:

$$\frac{\text{Resistance of wire } A}{\text{Resistance of wire } B} = \frac{(\text{Diameter of } B)^2}{(\text{Diameter of } A)^2}$$

If a certain length of wire with diameter 34.852 mils has a resistance of 8.125 Ω, what is the resistance of the same length and type of wire with a diameter of 45.507 mils?

29. *Finance*   A computer manufacturer has just announced annual earnings of \$3.46 per share. Other computer stocks are selling at an average price–earnings ratio of 14 to 1. Using this as a guideline, what should the selling price of the stock be?

30. *Electrical engineering*   For transformers, the symbol N_p stands for the number of turns on the primary and N_s symbolizes the number of turns on the secondary. The ratio of output voltage to input voltage is equal to the ratio of N_s to N_p. If the primary consists of 200 loops and the secondary consists of 1600 loops, what will be the output voltage for an input voltage of 120 V?*

31. *Pharmacology*   If 360 mL of solution contains 330 mg of morphine, how much morphine should be in 250 mL of the same solution?

32. *Chemistry*   A certain chemical must be mixed in the ratio of 3 parts chemical to every 5 parts water. How many ounces of chemical should be used to obtain a 2 qt total mixture?

*For more information, see Frederick Bueche, *Technical Physics*, 2d ed. (New York: Harper and Row, 1981), pp. 615–617.

33. *Navigation*   If an airplane consumes 22 gal of gasoline in flying 260 mi, how many gallons will it use to fly 440 mi?

34. *Agricultural technology*   According to the United States Department of Agriculture (USDA), 3.0 oz of ground beef contains 20.0 g of protein. How many ounces of ground beef contain 48.0 g of protein?

35. An 88 ft roll of wire weighs 24 lb. If a 55 ft roll is needed, how much would it weigh?

36. If fasteners sell at 12 for $10.98, how much will 8 fasteners cost?

37. *Auto mechanics*   The transmission ratio of an engine is the ratio of the crankshaft speed to the driveshaft speed. If the transmission ratio is 4.4 to 1 and the crankshaft is turning at 1540 rev/min, what is the speed of the driveshaft?

38. The ratio of pedal turns to wheel turns in the fourth gear of a bicycle is 3 to 7. If the circumference of the wheel is 7.07 ft, how many pedal turns are needed to travel 1 mi?

39. *Construction*   The ratio of cement to sand to gravel in a dry mix of concrete is 1 : 3 : 4. How much cement and gravel are needed to mix with 420 lb of sand?

40. *Machine technology*   A standard taper pin tapers at $\frac{1}{4}$ in. per foot from one end to the other. If the large end of an 8 in. pin measures $\frac{1}{2}$ in. in diameter, what will the small end measure?

Find the unknown side for each pair of similar triangles shown below.

41.

42.

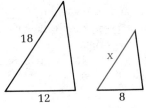

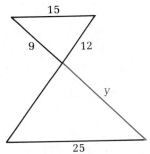

43.

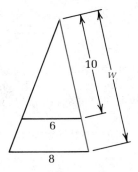

44.

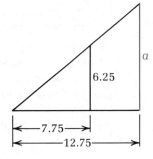

2-8 | Variation

Direct variation   Variation is a concept closely related to ratio and proportion. For example, from the study of electricity we know that the resistance of a wire is directly proportional to its length. If two lengths L_1 and L_2 of the same type and cross section of wire have resistances R_1 and R_2, this relationship can be written

$$\frac{R_1}{R_2} = \frac{L_1}{L_2}$$

As we saw in Section 2-7, this proportion can be expressed in the equivalent form

$$\frac{R_1}{L_1} = \frac{R_2}{L_2}$$

which implies that the ratio of resistance to length for a given wire is constant. If k is the value of this constant, then the proportion can be restated as the general formula

$$\frac{R}{L} = k$$

or, solving for R,

$R = kL$ where k is a constant for any given wire

This tells us that the resistance of a particular kind and cross section of wire will depend on its length. As the length of a wire increases, so will its resistance. This form of the relationship is known as **direct variation**.

In general, if a variable y is directly proportional to another variable x, then y is said to **vary directly** as x.* This direct variation can be written in the following form:

Direct Variation $y = kx$ where k is the **constant of proportionality**   (2-3)

EXAMPLE 56   Write an equation describing each of the following statements:
(a)   The volume V of a cylinder of fixed diameter varies directly as its height h.
(b)   The current i through a resistor varies directly as the voltage drop V across it.
(c)   The thermal resistance, or R value, of a wall varies directly as its thickness L.

Solution
(a) $V = kh$
(b) $i = kV$ The proportionality constant here is equal to the reciprocal of the resistance.
(c) $R = kL$ The proportionality constant depends on the composition of the wall.

If the constant of proportionality is unknown, one pair of values of the variables is needed to find k.

*The word "directly" is sometimes omitted from this statement. The phrase "y varies as x" also implies direct variation.

EXAMPLE 57 If m varies directly as w, and m = 6.0 lb when w = 4.0 ft:
(a) Find the constant of proportionality k.
(b) Find m when w = 12 ft.

Solution
(a) **First,** express the direct variation as: $m = kw$

   Second, solve for k. $k = \dfrac{m}{w}$

   Third, substitute the given values for m and w and calculate k. $k = \dfrac{6.0 \text{ lb}}{4.0 \text{ ft}} = 1.5 \text{ lb/ft}$

(b) From part (a) we have: $m = 1.5w$

   Substituting w = 12 ft, we obtain: $m = (1.5 \text{ lb/ft})(12 \text{ ft})$

   $= 18 \text{ lb}$

In some situations, one quantity will vary directly with some power of another quantity. The definition of direct variation in (2-3) can be expanded to include powers:

Direct Variation, General Form

$y = kx^n$ where n is a positive integer   (2-4)

EXAMPLE 58 The volume V of a sphere varies directly as the cube of its radius r. This relationship can be written as $V = kr^3$. (In this case, we know from geometry that $k = 4\pi/3$.)

EXAMPLE 59 Under certain conditions the flow rate Q of water through a cylindrical pipe varies directly as the square of its diameter d. If water flows through a 2.00-in.-diameter pipe at the rate of 70.0 gal/min, what is the rate of flow under the same conditions if d = 2.50 in.?

Solution
The direct variation equation is: $Q = kd^2$

Solve for k. $k = \dfrac{Q}{d^2}$

Substitute the given values of Q and d. $k = \dfrac{70.0 \text{ gal/min}}{(2.00 \text{ in.})^2}$

Calculate k. $k = 17.5 \text{ gal/min/in.}^2$

The equation now reads: $Q = 17.5d^2$

For a diameter of 2.50 in., we have: $Q = 17.5(2.50)^2$

17.5 ⨯ 2.5 x^2 = →   *109.375*

$Q \approx 109 \text{ gal/min}$

NOTE   In applied problems the proportionality constant will generally have dimensional units associated with it. ◀

Joint variation   A variable may vary directly as the product of two or more other variables. This is known as **joint variation.** If $y = kxz$, we say that y **varies jointly** as x and z. More generally, the variable y may vary with some power of x and z, and joint variation can be expressed as follows:

Joint Variation

$$y = kx^n z^m \qquad \text{where } n \text{ and } m \text{ are positive integers} \qquad (2\text{-}5)$$

EXAMPLE 60   The force F acting on a surface varies jointly as the pressure P on the surface and the square of some surface dimension d. If $F = 400$ lb when $P = 50$ lb/in.^{2} and $d = 4$ in., find the value of F when P is 80 lb/in.^{2} and d is 5 in.

Solution

The general equation is: $\qquad\qquad\qquad\qquad F = kPd^2$

Solve for k. $\qquad\qquad\qquad\qquad\qquad\qquad k = \dfrac{F}{Pd^2}$

Substitute the given values of F, P, and d. $\quad k = \dfrac{400 \text{ lb}}{(50 \text{ lb/in.}^2)(4 \text{ in.})^2}$

Calculate k. $\qquad\qquad\qquad\qquad\qquad\qquad k = \frac{1}{2}$

The equation is: $\qquad\qquad\qquad\qquad\quad F = \frac{1}{2}Pd^2$

For $P = 80$ lb/in.^{2} and $d = 5$ in.: $\qquad F = \frac{1}{2}(80)(5)^2$

Calculate F. $\qquad\qquad\qquad\qquad\qquad\quad F = 1000 \text{ lb}$

Inverse variation   **Inverse variation** occurs when the product of two variables is constant for any values of the variables.

Inverse Variation

$$\text{If} \quad y \cdot x = k, \quad \text{then} \quad y = \frac{k}{x} \quad \text{where} \quad (x \neq 0) \qquad (2\text{-}6)$$

This is read, "y **varies inversely** as x," or "y **is inversely proportional** to x."

EXAMPLE 61   At a constant temperature, the volume V of a gas varies inversely as the pressure P. If the volume of a particular gas is 15.0 L when the pressure is 250 kilopascals (kPa), what will be the volume when the pressure is 450 kPa?

Solution

The inverse variation is written: $V = \dfrac{k}{P}$

Solve for k. $k = PV$

Substitute for P and V and calculate k. $k = (250 \text{ kPa})(15.0 \text{ L})$
$= 3750 \text{ kPa} \cdot \text{L}$

The equation now reads: $V = \dfrac{3750}{P}$

Substitute P = 450 kPa: $V = \dfrac{3750 \text{ kPa} \cdot \text{L}}{450 \text{ kPa}}$

$V \approx 8.3 \text{ L}$

Inverse variation may also involve powers of the variables:

Inverse Variation, General Form

$$y = \frac{k}{x^n} \qquad x \neq 0, \text{ n a positive integer} \qquad\qquad (2\text{-}7)$$

Equation (2-7) describes a situation in which y varies inversely with some power of x.

EXAMPLE 62   The illuminance E created by a light source varies inversely as the square of its distance r from the source. If the illuminance created by a certain light source is 56.0 lx at a distance of 2.00 m, what is the illuminance from the same source at a distance of 3.50 m?

Solution

The general equation is: $E = \dfrac{k}{r^2}$

Solve for k. $k = r^2 E$

Substitute the given values of E and r. $k = (2.00 \text{ m})^2(56.0 \text{ lx})$

Calculate k. $k = 224 \text{ lx} \cdot \text{m}^2$

The equation now reads: $E = \dfrac{224}{r^2}$

Replace r with 3.50 m and calculate E: $E = \dfrac{224 \text{ lx} \cdot \text{m}^2}{(3.50 \text{ m})^2}$

$E \approx 18.3 \text{ lx}$

Combined variation   **Combined variation** occurs when a variable varies directly as another variable and inversely as a third variable. If y varies directly as x and inversely as z, the combined variation is written:

Combined Variation

$$y = \frac{kx}{z} \quad \text{or, more generally,} \quad y = \frac{kx^n}{z^m} \qquad (2\text{-}8)$$

where $z \neq 0$; n, m positive integers

EXAMPLE 63   The resistance R of a wire varies directly as its length L and inversely as its cross-sectional area A. A wire 55 ft in length with a cross-sectional area of 0.065 in.^{2} has a resistance of 0.14 Ω. What is the resistance of 180 ft of the same composition wire having a cross-sectional area of 0.040 in.^{2}?

Solution

The general equation is: $R = \dfrac{kL}{A}$

Solve for k. $k = \dfrac{RA}{L}$

Substitute the given
values of R, L, and A. $k = \dfrac{(0.14\ \Omega)(0.065\ \text{in.}^2)}{55\ \text{ft}}$

Calculate k. $.14\ \boxed{\times}\ .065\ \boxed{\div}\ 55\ \boxed{=} \rightarrow$ *0.0001655* $\Omega \cdot \text{in.}^2/\text{ft}$

Substitute the new
values of L and A
into the equation for R. $R = \dfrac{(0.0001655\ \Omega \cdot \text{in.}^2/\text{ft})(180\ \text{ft})}{0.040\ \text{in.}^2}$

With the value of k still
on the display of the
calculator, compute R: $\boxed{\times}\ 180\ \boxed{\div}\ .04\ \boxed{=} \rightarrow$ *0.7445455*

$R \approx 0.74\ \Omega$

Combined variation may include joint variation and may involve powers of the variables also. Our final example includes every type of variation discussed in this section.

EXAMPLE 64   A variable v varies jointly as a and the square of b, and inversely as the cube of c. If $v = 3.2$ when $a = 5.4$, $b = 4.6$, and $c = 1.8$, find v when $a = 7.9$, $b = 3.6$, and $c = 2.5$

Solution

The general equation is: $v = \dfrac{kab^2}{c^3}$

Solve for k. $k = \dfrac{c^3 v}{ab^2}$

Substitute the
values of v,
a, b, and c.

$$k = \frac{(1.8)^3(3.2)}{(5.4)(4.6)^2}$$

Calculate k.

$1.8 \boxed{y^x} 3 \boxed{\times} 3.2 \boxed{\div} 5.4 \boxed{\div} 4.6 \boxed{x^2} \boxed{=} \rightarrow$ ⬚ 0.1633270

Substitute the
new values for
a, b, and c.

$$v = \frac{k(7.9)(3.6)^2}{(2.5)^3}$$

Calculate v.

$\boxed{\times} 7.9 \boxed{\times} 3.6 \boxed{x^2} \boxed{\div} 2.5 \boxed{y^x} 3 \boxed{=} \rightarrow$ ⬚ 1.0702128

$v \approx 1.1$

Exercises 2-8

Write an equation describing each of the following statements:

1. x varies directly as y
2. w varies inversely as z
3. a varies jointly as b and c
4. p is directly proportional to the square of q
5. m is inversely proportional to the cube of n
6. s varies jointly as t and the fourth power of u
7. e varies jointly as f and the cube of g and inversely as h
8. x varies directly as y and inversely as the square of z

For each of the following statements, (a) find the constant of proportionality and (b) state the equation relating the variables:

9. x varies directly as y, and x = 6 ft when y = 2 s
10. a varies directly as the square of b, and a = 96 min when b = 8.0 in.
11. m varies inversely as the cube of n, and m = 6 when n = 2
12. c is inversely proportional to d, and c = 16 when d = 4
13. w varies jointly as x and the square of z, and w = 24 when x = 4 and z = 6
14. p varies jointly as q and the cube of r, and p = 128 when q = 0.2 and r = 4
15. x is directly proportional to y and inversely proportional to the square of z, and x = 14 when y = 8 and z = 2
16. a varies directly as the square of b and inversely as c, and a = 2 when b = 2 and c = 3

Solve.

17. If x is directly proportional to y, and x = 30 A when y = 12 V, find x when y = 8 V.
18. If w varies directly as the square of z, and w = 75 gal/min when z = 5.0 in., find w when z = 9.0 in.
19. If a varies inversely as the cube of b, and a = 2 when b = 3, find a when b = 2.
20. If p is inversely proportional to q, and p = $\frac{1}{4}$ when q = 8, find p when q = $\frac{1}{2}$.
21. If y varies jointly as w and the fourth power of z, and y = 45 when w = $\frac{1}{9}$ and z = 3, find y when w = 3 and z = 2.
22. If c varies jointly as d and the square of e, and c = 98 when d = 4 and e = 7, find c when d = 7 and e = 4.

23. If x varies directly as the square of y and inversely as the cube of z, and x = 18.0 when y = 12.6 and z = 4.2, find x when y = 9.2 and z = 6.4.

24. If p varies jointly as q and the fourth power of r and inversely as the square of s, and p = 10,752 when q = 24, r = 12, and s = 36, find p when q = 18, r = 8, and s = 42.

25. *Electronics*   If the current i through a resistor is 12 A when the voltage drop V across it is 30 V, find i when V = 20 V. [See Example 56(b).]

26. *Energy technology*   If the thermal resistance, or R value, of a wall 6.0 in. thick is 1.6, find R for a wall of the same composition that is 9.0 in. thick. [See Example 56(c).]

27. *Wastewater technology*   The amount of chlorine needed in a certain reservoir varies directly as the desired concentration. If 2520 lb are needed to produce a concentration of 30.0 parts per million (ppm), how much chlorine is needed to produce 25.0 ppm?

28. *Physics*   Within its elastic limit, the amount of stretch of a spring varies directly as the force applied to it. If a force of 20 lb stretches a spring 4 in., how much will the same spring be stretched by a force of 35 lb?

29. *Energy technology*   The power generated by a certain type of windmill varies directly as the cube of the wind speed. If a 40.0 mi/h wind generates 1280 W of power, how much power will a 25.0 mi/h wind produce?

30. *Wastewater technology*   The time required for an outlet pipe to empty a tank varies inversely as the square of its diameter. If a 3.0-in.-diameter pipe empties a certain tank in 6.0 h, how much time will it take a 4.0-in.-diameter pipe to empty the same tank?

31. *Machine technology*   The expansion of a metal rod varies jointly as its original length and the temperature change. If the original length of a rod is 40.0 cm and it expands 8.0×10^{-3} cm when the temperature changes 10.0°C, how much will it expand when the temperature changes 30.0°C?

32. *Energy technology*   The daily energy available from a solar collector varies jointly as the insulation and the percent of sunshine. If a given collector provides 750 British thermal units (Btu) when the insulation is rated 2500 Btu/ft^2 and the sun shines 60% of the time, how much energy will the same collector supply when the insulation is rated 2200 Btu/ft^2, and the sun shines 50% of the time?

33. *Agricultural technology*   Assuming a constant efficiency, the size of the motor required for a water-pumping system varies jointly as the discharge of the pump and the amount of lift. If a pump discharging 990 gal/min with 150 ft of lift requires a 55 horsepower (hp) motor, what size of motor is required to discharge 1100 gal/min with 160 ft of lift?*

34. *Physics*   The kinetic energy of an object varies directly as the square of its speed. If an object moving at 36 m/s has a kinetic energy of 108,000 J, what will be the kinetic energy of the same object when it is moving at 24 m/s?

35. *Acoustical engineering*   The intensity of a sound wave varies inversely as the square of the distance from its source. If the intensity is 2.0×10^{-7} W/m^2 at a distance of 8.0 m from a speaker, what is the level at a distance of 20.0 m?

36. *Physics*   At a constant temperature, the volume of a gas varies inversely as the pressure. If the volume of a certain gas is 60 ft^3 when the pressure is 18 lb/in.^{2}, what is the volume of the same mass of the gas at the same temperature when the pressure is 12 lb/in.^{2}?

37. *Electrical engineering*   The current capacity of an electrical supply using a three-phase system varies directly as the power output and inversely as the line voltage. If a current of 0.40 A is provided by the system with 120 kW of power delivered at 480 V, what is the current capacity of a similar system providing 180 kW of power at 420 V?

38. *Civil engineering*   The shearing stress of a beam varies directly as its end reaction and inversely as its cross-sectional area. A certain beam with an end reaction of 1600 lb and cross-sectional area of 40.0 in.^{2} has a shearing stress of 60.0 lb/in.^{2} What is the shearing stress of a beam of the same composition with an end reaction of 1800 lb and cross-sectional area of 64.0 in.^{2}?

*Based on Vern H. Scott, "Sprinkler Irrigation," University of California Cooperative Extension, Circular No. 456, Oct. 1956.

39. *Civil engineering* The crushing load of a square pillar varies directly as the fourth power of its thickness and inversely as the square of its height. If the crushing load of a post 0.50 ft thick and 10.0 ft high is 312.5 tons, what will be the crushing load for a post of the same composition 2.0 ft thick and 12.0 ft high?

40. *Transportation engineering*   The distance necessary for stopping a subway train varies directly as the square of its speed and inversely as its deceleration. If the train requires 180.0 ft to stop when it is traveling at 30.0 mi/h and decelerating at 0.18 g, what is the stopping distance for the train when it is traveling at 50.0 mi/h and must be decelerated at 0.16 g? (An acceleration of 1 g is the normal gravitational acceleration at the surface of the earth—about 32 ft/s².)

2-9 | Review of Chapter 2

Important Terms and Concepts

algebraic expression (p. 46)	like terms (p. 47)	equivalent equations (p. 59)
term (p. 46)	unlike terms (p. 47)	formula (p. 65)
factor (p. 46)	literal part (p. 47)	right triangle (p. 66)
coefficient (p. 46)	FOIL (p. 52)	legs (p. 66)
numerical coefficient (p. 46)	algebraic equation (p. 59)	hypotenuse (p. 67)
polynomial (p. 47)	linear equation (p. 59)	ratio (p. 83)
monomial (p. 47)	solution (p. 59)	proportion (p. 83)
binomial (p. 47)	conditional equation (p. 59)	similar triangles (p. 87)
trinomial (p. 47)	identity (p. 59)	variation (p. 91)
multinomial (p. 47)	contradiction (p. 59)	constant of proportionality (p. 92)

Formulas and Rules

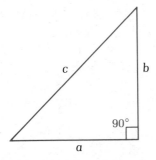

- Pythagorean theorem (p. 67): if a and b are the lengths of the legs of a right triangle and c is the length of the hypotenuse, then $c^2 = a^2 + b^2$ **(2-1)**

- cross-product rule (p. 84): if $\dfrac{a}{b} = \dfrac{c}{d}$, then $ad = bc$ (where $b, d \neq 0$)   **(2-2)**

- direct variation (p. 92):　　　$y = kx$　　　　　　　　　　　　　　**(2-3)**
  　　　　　　　　　　　　or　$y = kx^n$　　　　　　　　　　　　　**(2-4)**
- joint variation (p. 94):　$y = kx^n z^m$ **(2-5)**　　　　　　　　　**(2-5)**
- inverse variation (p. 95):　if　$y \cdot x = k$,　then　$y = \dfrac{k}{x}$　(where $x \neq 0$)　　**(2-6)**

  　　　　　　　　or　if　$y \cdot x^n = k$,　then　$y = \dfrac{k}{x^n}$　(where $x \neq 0$)　**(2-7)**

- combined variation (p. 96):　$y = \dfrac{kx^n}{z^m}$　(where $z \neq 0$)　　　　　**(2-8)**

Exercises 2-9

Simplify each expression by performing the indicated operations.

1. $2x - 11x + 4x$

2. $8w + 5v - 2z - 9w + 7v$

3. $-3ab - ab^2 + 5a^2b - 3ab^2 - 11a^2b$

4. $(3m^2 - 7m + 4) + (8m^2 - 6m - 1)$

5. $2c - (5c - 8)$

6. $(4x^2 - 11xy - 3y^2) - (12x^2 - 9xy + 5y^2)$

7. $4n - [5 - (3n + 7) - 11] - 6n$

8. $-3x(7x^3)$

9. $5ab(3a^2)(-6b^3)$

10. $6x(3x - 7)$

11. $-3mn(5m^2 - 7mn - 9n^2)$

12. $(x + 8)(x - 3)$

13. $(3y^2 - 10)(y^2 - 3)$

14. $(2x - 7y)(5x + 3y)$

15. $(5x + 7)^2$

16. $(x + 4)(2x^2 + 7x - 5)$

17. $-2y(y - 8)(3y - 7)$

18. $4n(2n - 1)(n + 6)^2$

19. $3(5x - 9y) - 7(7x + 4y)$

20. $2a[3b - 5(2a - 7b) - 8]$

21. $(5 - 3x)(2x - 5) - (x - 8)^2$

22. $\dfrac{-42x^8}{7x^2}$

23. $\dfrac{10a^2b^3c^4}{15a^5b^8c}$

24. $\dfrac{(8x^2w)(-5w^3z)}{4x^3wv}$

25. $\dfrac{28x^4 + 7x^2}{7x^2}$

26. $\dfrac{36m^7n^2 - 24mn^5 + 18m^4n}{-6m^2n}$

27. $\dfrac{15x^2 - 52x + 32}{3x - 8}$

28. $\dfrac{12x^3 - 15x^2 - 2x + 10}{4x + 3}$

29. $\dfrac{4x^4 - 11x^2 - 7x + 15}{2x - 3}$

30. $\dfrac{x^3 - 32}{x - 5}$

Solve the given equations.

31. $x + \frac{1}{2} = -\frac{1}{4}$

32. $5y - 13 = 7$

33. $12 - a = 5$

34. $3 = \dfrac{m}{3} + 7$

35. $2x - 8 = 5x$

36. $9 - 5y = y + 3$

37. $2c - 14 - 7c = 9 - 3c + 5$

38. $-2(x - 3) = 4x$

39. $6y - (5 - y) = 23$

40. $3m - 2(3m + 4) = 5m + 16$

41. $4 - 6(2 - 3x) = -(12 - 14x)$

42. $5.21x + 2.74 = 2.73x - 5.94$

Evaluate each formula for the given values.

43. $X_L = 2\pi fL$　(impedance of an inductor, in ohms) when $f = 70.0$ Hz and $L = 0.200$ henry (H)

44. $w = \dfrac{s}{f}$　(wavelength of a musical note, in meters) when $s = 344$ m/s and $f = 25.0$ Hz

45. $C = \dfrac{\varepsilon_o A}{P}$ (capacitance of parallel plates, in farads) when $\varepsilon_o = 8.85 \times 10^{-12}$ F/m, $A = 9.0$ m², and $P = 8.5 \times 10^{-4}$ m

46. $P = Ri^2$ (electrical power, in watts) when $R = 240\ \Omega$ and $i = \frac{1}{2}$ A

47. $R = \dfrac{3C^2}{2M} + \dfrac{M}{24}$ (radius of a centrifugal skid, in feet) when $C = 48.0$ ft and $M = 12.0$ ft

48. $V = C\left(1 - \dfrac{n}{N}\right)$ (linear depreciation of property, in dollars) when $C = \$15{,}000$, $n = 3$ years, and $N = 5$ years

49. $A = \dfrac{h}{2}(b_1 + b_2)$ (area of a trapezoid, in square centimeters) when $h = 34.0$ cm, $b_1 = 22.0$ cm, and $b_2 = 26.0$ cm

50. $Q = 29.7D^2\sqrt{P}$ (discharge of a fire hose, in pounds) when $D = 2.00$ in. and $P = 50.0$ lb/in.²

Solve each formula for the indicated variable.

51. $C = C_1 + C_2 + C_3$ (capacitance in parallel) for C_2

52. $P = \dfrac{W}{4A}$ (tire pressure) for W

53. $R = \dfrac{\rho L}{A}$ (resistance of a material) for L

54. $m = \dfrac{qB^2r^2}{2V}$ (mass of an ion) for V 　　　　**55.** $I = P + PRT$ (interest) for R

56. $T = t - \dfrac{h}{100}$ (temperature of the air) for h

57. $m = \dfrac{(G' - G)L}{8}$ (middle ordinate of a vertical curve) for L

58. $R = R_s + R_s \alpha_s(t - t_s)$ (resistance of materials) for t

Find the unknown side of each right triangle.

59.

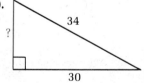

60.

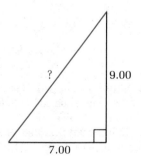

Write each ratio in simplest terms.

61. 24 A to 8 A 　　　　**62.** 8 qt to 12 qt 　　　　**63.** 6 in. to 3 ft 　　　　**64.** 2 lb to 8 oz

65. *Electrical engineering* The type of transformer needed to step up a voltage depends on the ratio of the output voltage to the input voltage. If 110 V must be raised to 16,500 V, what is this ratio?

66. *Physics* The mechanical advantage of a block and tackle is the ratio of output force to input force. Find the mechanical advantage of a block and tackle needing a force of 60 N (input) to lift 150 N (output).

67. *Electronics* The current gain of a circuit is given by the ratio $\dfrac{\alpha}{1 - \alpha}$, where α is a constant for a particular transistor. Find the current gain for $\alpha = 0.96$.

68. *Auto mechanics*   The rear axle ratio of a car is the ratio of the number of teeth on the ring gear to the number of teeth on the pinion gear. Find this ratio for a car with a ring gear that has 54 teeth and a pinion gear that has 18 teeth.

Solve.

69. $\dfrac{x}{8} = \dfrac{15}{24}$ **70.** $\dfrac{12}{21} = \dfrac{16}{y}$ ▦ **71.** $\dfrac{20.5}{10.4} = \dfrac{m}{14.5}$ ▦ **72.** $\dfrac{538}{a} = \dfrac{496}{723}$

Set up a proportion and solve for the missing quantity.

73. *Graphics technology*   A photograph 6 in. wide and 7 in. high must be reduced to fit in a space $4\frac{1}{2}$ in. wide. How much vertical space will the reduced version occupy?

74. *Pharmacology*   A certain cream mixture contains 14 g of cream A, 12 g of petrolatum, and 16 g of Unibase. A pharmacist has 21 g of cream A on hand. How many grams of each of the other ingredients are needed to prepare the cream with the given ratio?

75. *Electrical engineering*   The resistance of a wire is proportional to its length. If a wire 12 ft long has a resistance of 0.6 Ω, what is the resistance of 8 ft of the same type of wire?

76. *Chemistry*   A mixture contains 1 part chemical to 5 parts water. How many liters of chemical are needed to make a total solution of 12 L?

Solve for the indicated side in each pair of similar triangles.

77.

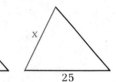

78.

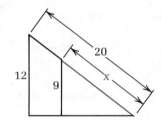

In Problems 79–86 set up equations and solve the given problems.

▦ **79.** *Industrial engineering*   On the average, 3% of the parts shipped by a manufacturer are damaged. How many parts should be shipped to ensure the arrival of 800 undamaged parts?

80. Two pieces of wire must be cut so that one is 18 in. longer than the other. If their total length is 114 in., how long should each piece be?

81. The United States Postal Service has a special 4th class library rate for shipping books. The charges are $0.35 for the first pound and $0.12 for each additional pound. How many pounds of books can be shipped for $1.19?

82. A bicycle rider traveling 16 mi/h leaves a certain point $\frac{1}{2}$ h earlier than a moped rider going 20 mi/h. How long does it take the moped to catch up to the bicycle?

83. *Navigation*   Two ships leave ports 80 nautical miles apart at the same time. As they head directly toward each other, one ship is doing 9 knots while the other is doing 7 knots. How soon do they pass each other? (1 knot = 1 nautical mile per hour).

84. The length of a rectangle is 1 in. more than three times the width. If the perimeter is 122 in., find the dimensions.

85. *Material science*   One alloy contains 70% copper, while another contains 40% copper. How many kilograms of each should be mixed to produce 60 kg of an alloy that is 50% copper?

86. *Chemistry* How much of a 35% acid solution should be added to 15.0 dL of a 10% acid solution to produce a 15% acid solution?

87. *Business* A computer shop sold m packages of diskettes on Monday for $48 each and n packages on Tuesday for the same unit price. Write two forms of an algebraic expression to indicate the total receipts from the two purchases.

88. *Electronics* The formula $V^2 = V_R^2 + (V_L - V_C)^2$ relates to the voltage across an *RCL* combination. Rewrite the formula by squaring the binomial.

89. *Thermodynamics* The internal energy of a diatomic gas having both rotational and translational motion is given by $3\mu(\frac{1}{2}RT) + 2\mu(\frac{1}{2}RT)$. Simplify this expression.

90. *Instrumentation* When a calibrated instrument containing a spring is subjected to temperature changes, a scale error results. In the calculation of scale error the following expression occurs: $[1 - (1 + ct)(1 + \alpha t)]$. Perform the indicated multiplication and simplify.*

91. *Physics* The formula below appears in the description of the elastic collision between two objects. Rewrite this equation without grouping symbols.

$$(v_{10} - v_{20}) = -(v_{1f} - v_{2f})$$

92. *Physics* The expression below occurs in the study of freely falling bodies. Perform the multiplication in the numerator.

$$\frac{(v_o + v_f)(v_f - v_o)}{2a}$$

93. *Astrophysics* The expression below is used in describing the condensation of cosmic gas and dust.[+] Perform the indicated multiplication.

$$\frac{3}{4\pi\rho s^3}\left(\frac{4\pi M\rho s^3 G}{3r^2} - \frac{L\pi s^2}{4\pi c r^2}\right)$$

94. *Computer science* In a particular computer, the storage location in memory of a number in column i of row j on page k is given by the expression below. Mutiply and combine like terms.

$$i + 5(j - 1) + 50(k - 1)$$

95. If x varies directly as y, and $x = 64$ kg when $y = 22$ N, find x when $y = 55$ N.

96. If p varies directly as the square of q, and $p = 75$ lb when $q = 5.0$ ft, find p when $q = 4.0$ ft.

97. If m varies jointly as n and d, and $m = 88$ when $n = 16$ and $d = 22$, find m when $n = 15$ and $d = 40$.

98. If r varies jointly as s and the cube of t, and $r = 54$ when $s = 6$ and $t = 3$, find r when $s = 9$ and $t = 2$.

99. If a varies inversely as b, and $a = 12$ when $b = 6$, find a when $b = 4$.

100. If x varies directly as the square of w and inversely as z, and $x = 5$ when $w = 10$ and $z = 60$, find x when $w = 16$ and $z = 40$.

101. The distance traveled by a car moving at a constant speed varies directly with the time traveled. If a car travels 420 km in 4.5 h, how far will it travel in 6.0 h at the same speed?

102. *Machine technology* The mass of a metal ball bearing varies directly as the cube of its diameter. If a ball bearing with a diameter of 0.50 cm has a mass of 0.80 g, what will be the mass of a ball bearing of the same composition with a diameter of 0.30 cm?

103. *Energy technology* The amount of energy needed hourly to heat a room varies jointly as the area of the room and the difference between the temperature outside and the thermostat setting. If a certain system requires 480 Btu/h to heat a 600 ft^2 room to a temperature 40°F warmer than the outside temperature,

*Based on Thomas G. Beckwith, N. Lewis Buck, and Roy D. Marangoni, *Mechanical Measurements* (Reading, Mass.: Addison-Wesley, 1982), p. 117.
[+]For more details, see Martin Harwit, *Astrophysical Concepts* (New York: Wiley, 1973), p. 427.

how much energy is needed for the system to heat a 500 ft^2 room to a temperature 20°F warmer than the outside temperature?

104. *Physics*   The weight of a point mass varies inversely with the square of its distance from the center of the earth. If a mass weighs 65.4 kg at the surface of the earth and the radius of the earth is approximately 6370 km, what is the weight of the mass at an altitude of 11.5 km above the earth's surface?

105. *Chemistry*   The molecular weight of a gas varies directly as its temperature and inversely as its pressure. If the molecular weight of a certain gas is 525 g/mol at 302 K under 1.40 atmosphere (atm) of pressure, what is the molecular weight of the same gas at 320 K under 1.20 atm of pressure?

106. *Lighting technology*   Assuming a constant fixture efficiency, the level of illuminance on a surface varies directly as the intensity of the light source and inversely as the area of the surface. If a light source of 5000 lm creates a light level of 50 lx on a 20 m^2 surface, what will be the level of light from a source of 8000 lm on an 8 m^2 surface?

Functions and Graphs

BOEING 707 FLAPS AND SPOILERS

Much of modern technology depends on science, and scientists describe nature by finding mathematical rules that relate one physical quantity or variable to another. For example, an electrical engineer might describe the power output of a circuit element in terms of the current flowing through it. This description may take the form of a formula or equation, a table of data, or a graph. In this chapter we will discuss these very important mathematical relations.

Definitions

A variable y is a **function** of another variable x if a relationship exists between x and y so that every value of x corresponds to *only one* value of y. A function is a mathematical description of the way one variable determines a unique value of another.

In this kind of mathematical rule, we may choose a value for x, and the rule gives the corresponding value of y. For example, if $y = 3x$ is the function, then every real number value of x gives a real number value for y:

$$x = 2 \quad \text{gives} \quad y = 3(2) = 6$$

$$x = 5 \quad \text{gives} \quad y = 3(5) = 15 \quad \text{And so on.}$$

We may choose any value for x that produces a real number for y, so x is called the **independent variable.** Because the value of y depends on the x value, y is called the **dependent variable.**

The set of numbers from which values of the independent variable may be chosen is called the **domain** of the function. The set of numbers representing all possible values of the dependent variable is called the **range** of the function. Unless otherwise noted, both the domain and range are assumed to be the set of all real numbers.

EXAMPLE 1 The current flowing in a certain circuit element during the first few seconds after a voltage is applied is given by the function

$$I = 100t^2$$

where I is the current (in milliamps) and t is time (in seconds). Here, t is the independent variable, and I is the dependent variable. Because time cannot be a negative number, the domain is the set of nonnegative real numbers. This means that the range is also the set of nonnegative real numbers. This is a function because each real value of t gives a single real value for I. The following table shows the unique values of I that result from substituting a few values for t into the formula:

t (s)	0	1	2	3	4
I (mA)	0	100	400	900	1600

Typically, a formula can be a function if it involves two variables, and if each value of the independent variable produces a unique value of the dependent variable.*

Function notation

If a variable y is some function of x, as in the equation

$$y = 4x + 1$$

* It is possible to have a function with only a dependent variable, and it is possible to have a function with more than one independent variable.

we can name it using the **function notation f(x).** Thus,

$$f(x) = 4x + 1$$

is an equivalent way of writing this function. The notation $f(x)$ is read "f of x," and it represents the value of the function f at x. It does *not* mean f times x. The letter f names the function, and the letter x in parentheses indicates the independent variable.

A function can be named this way with any letter of the alphabet, with Greek letters, or even with multiletter combinations. For example, this function might also be designated $y(x) = 4x + 1$ as a reminder that y was the dependent variable in the original equation. Similarly, the current function described in Example 1 might be written using function notation as $I(t) = 100t^2$.

EXAMPLE 2   Write the indicated functions using function notation.
(a) $y = 3x - 5$ as the function f
(b) $P = 3t^2 - t$ as the function P
(c)   Express the circumference C of a circle as a function of its diameter d.
(d)   The length of a spring that is initially 3.0 in. long increases 0.5 in. for every pound of weight on it. Express the length L of the spring as a function of the hanging weight W.

Solution
(a) $f(x) = 3x - 5$ (b) $P(t) = 3t^2 - t$
(c) $C(d) = \pi d$ (d) $L(W) = 3.0 + 0.5W$

NOTE Changing the letter that names the independent variable does not really change the function. For example,

$$f(x) = x^2 - 3x + 1 \quad \text{and} \quad f(a) = a^2 - 3a + 1$$

are the same function even though different letters are used.   ◀

Function notation is useful in naming a specific value of the dependent variable. In Example 2(a), the value of y when x is 3 might simply be written $f(3)$:

$$f(3) = 3(3) - 5 = 9 - 5 = 4$$

EXAMPLE 3   Given $f(x) = 2x^2 - 5x + 7$ find:
(a) $f(2)$ (b) $f(-3)$

Solution
(a)   **Step 1.**   Substitute 2 for x. $f(2) = 2(2)^2 - 5(2) + 7$

   Step 2.   Calculate $f(2)$ $= 2(4) - 5(2) + 7$
   according to the order of
   operations. $= 8 - 10 + 7$

                                                  $= 5$

(b) **Step 1.** Substitute. $f(-3) = 2(-3)^2 - 5(-3) + 7$

     Step 2. Calculate. $= 2(9) + 15 + 7$

                                                      $= 40$

A similar procedure is used to evaluate a function when a literal expression is substituted for the independent variable. Here, we apply the techniques for operations with algebraic expressions.

EXAMPLE 4   Given the function $g(t) = t^2 + 2t$, find $g(t + 5)$.

Solution
Step 1. Substitute $(t + 5)$ $g(t + 5) = (t + 5)^2 + 2(t + 5)$
in the equation for t.

Step 2. Simplify by performing $= t^2 + 10t + 25 + 2t + 10$
the indicated operations.
                                                $= t^2 + 12t + 35$

EXAMPLE 5   Evaluate each function as indicated.
(a)   Find $y(4)$ for

$$y(x) = \frac{4}{x} + 3\sqrt{x}$$

(b)   Given the function below, find $k(45)$ to the nearest tenth.

$$k(m) = \frac{m^2 - 50}{27 + 3m}$$

(c)   The displacement of an object with an initial velocity of 12 ft/s and an acceleration of 4 ft/s² is a function of time as given by:

$$d(t) = 12t + 2t^2$$

Find $d(3 + t) - d(t)$.

Solution

(a) $y(4) = \dfrac{4}{4} + 3\sqrt{4} = 1 + 6 = 7$

(b) $k(45) = \dfrac{(45)^2 - 50}{27 + 3(45)}$

$45\,\boxed{x^2}\,\boxed{-}\,50\,\boxed{=}\,\boxed{\div}\,\boxed{(}\,27\,\boxed{+}\,3\,\boxed{\times}\,45\,\boxed{)}\,\boxed{=}\rightarrow$ $\boxed{12.191358}$

$k(45) \approx 12.2$

(c) $d(t) = 12t + 2t^2$

$d(3 + t) = 12(3 + t) + 2(3 + t)^2$

$= 36 + 12t + 18 + 12t + 2t^2 = 2t^2 + 24t + 54$

$d(3 + t) - d(t) = 2t^2 + 24t + 54 - (12t + 2t^2)$

$= 2t^2 + 24t + 54 - 12t - 2t^2 = 12t + 54$

Restrictions on the domain   If the range of a function is restricted to real number values, we may have to limit the domain of certain functions so that undefined or imaginary values of the function do not occur.

EXAMPLE 6   In Example 5(a), the function

$$y(x) = \frac{4}{x} + 3\sqrt{x}$$

is not defined for $x = 0$, since it leads to division by zero in the term $4/x$. Furthermore, any negative values of the independent variable lead to imaginary results in the term $3\sqrt{x}$. Therefore, the domain of this function is restricted to positive real numbers, $x > 0$.

EXAMPLE 7   State the domain of each function given that the range is restricted to real number values.

(a) $f(x) = \dfrac{2}{2 - x}$ (b) $g(x) = \sqrt{x - 5}$

(c) $h(x) = \dfrac{6}{x} + \dfrac{2}{x + 1}$ (d) $y(x) = \sqrt{16 - x^2}$

Solution
(a) All real numbers except 2.
(b) All real numbers $x \geq 5$.
(c) All real numbers except 0 and -1.
(d) All real numbers between and including -4 and 4.

Exercises 3-1

Express each function using function notation.

1. If the width of a rectangle is 8 ft, write the perimeter P as a function of its length L.
2. Give the area A of a rectangle 6 m long as a function of its width W.
3. Express the side length s of an equilateral triangle as a function of its perimeter P.
4. Find the diameter d of a circle as a function of its circumference C.
5. An empty drum weighs 1.5 lb. Its weight increases 11.4 lb for every gallon g of a chemical poured into it. Write the weight W of the drum as a function of the amount of chemical it contains.
6. A full spool of wire weighs 20 kg. Its weight decreases 1 kg for every 30 m of wire removed from it. Give the weight W of the spool as a function of the length L of wire removed from it.
7. If the width of a rectangle is 4 cm, express the length L of the rectangle as a function of its perimeter P.
8. A phone call costs $0.65 for the first minute and $0.49 for each additional minute. Express the number of additional minutes M as a function of the total cost C.

Find the function values shown.

9. For $f(x) = 3x - 7$ find: (a) $f(3)$ (b) $f(-3)$ (c) $f(0)$
10. For $f(y) = 6y - 12$ find: (a) $f(2)$ (b) $f(\tfrac{1}{2})$ (c) $f(-1)$

11. For $P(w) = 3 - w$ find: (a) $P(-3)$ (b) $P(3)$ (c) $P(1.6)$
12. For $Q(m) = -6 - 2m$ find: (a) $Q(4)$ (b) $Q(-5)$ (c) $Q(3.8)$
13. For $g(y) = y^2 - 3y$ find: (a) $g(3)$ (b) $g(-2)$ (c) $g(10)$
14. For $y(x) = 3x^2 - 10x$ find: (a) $y(4)$ (b) $y(-2)$ (c) $y(0)$
15. For $\phi(a) = 2\sqrt{a} - \dfrac{a}{2}$ find: (a) $\phi(4)$ (b) $\phi(25)$ (c) $\phi\left(\frac{1}{4}\right)$
16. For $\alpha(v) = \dfrac{3}{v} + 4\sqrt{v + 1}$ find: (a) $\alpha(3)$ (b) $\alpha(15)$ (c) $\alpha(120)$
17. For $f(t) = \dfrac{8t^2 - 10}{5t}$ find: (a) $f(5)$ (b) $f(-2)$ (c) $f(10)$
18. For $f(r) = \dfrac{7r}{r^2 - 3}$ find: (a) $f(2)$ (b) $f(-1)$ (c) $f(0)$
19. For $y(x) = 3x^2 - 5x + 2$ find: (a) $y(3.7)$ (b) $y(9.72)$ (c) $y(146)$
20. For $g(y) = 7y^2 + 6y - 4$ find: (a) $g(8.2)$ (b) $g(87)$ (c) $g(350)$
21. For $L(t) = 4t - 3$ find: (a) $L(t + 2)$ (b) $L(3t)$ (c) $L(t - 1)$
22. For $M(s) = 2s^2 - 3s$ find: (a) $M(s + 3)$ (b) $M(2s)$ (c) $M(s^4)$
23. For $F(x) = x^2 - 7x + 2$ find: (a) $F(x^2)$ (b) $F(2x)$ (c) $F(2 + x) - F(x)$
24. For $G(y) = 2y^2 + 9y$ find: (a) $G(2y^3)$ (b) $G(7y)$ (c) $G(5 + y) - G(y)$

State the domain of each function, given that the range is restricted to real number values.

25. $f(x) = \dfrac{3}{x - 5}$ 26. $f(y) = \dfrac{y}{7 - y}$ 27. $g(a) = \sqrt{a - 3}$

28. $g(m) = \dfrac{1}{\sqrt{5 - m}}$ 29. $\mu(w) = \dfrac{w}{(w + 5)(w - 4)}$ 30. $\beta(n) = \dfrac{\sqrt{n^2 - 4}}{2n - 10}$

31. $Y(x) = \dfrac{1}{x} + \dfrac{1}{x - 2}$ 32. $B(y) = \dfrac{\sqrt{y + 3} + \sqrt{y - 3}}{y - 5}$

Solve.

33. The surface area of a cylinder 4.0 in. high as a function of its radius r is given by $S = 2\pi r(r + 4)$. Find the surface area of a cylinder with a radius of 5.0 in.
34. The volume of a prism with a square base and an altitude of 8.00 m is a function of the side length s of its base. The function is $V = 8.00s^2$. Find the volume of such a prism for a side length of 4.00 m.
35. *Fire science*   The pressure loss F (in pounds per square inch) of a hose is a function of the flow rate Q (in hundreds of gallons per minute). The function is $F = 2Q^2 + Q$. Find F for a flow rate of 425 gal/min.
36. *Fire management*   The fire damage potential D of a class 5 forest as a function of the average brush age A is given by $D = 2A + 5$. Find the damage potential of a forest that has an average brush age of 4.5 years.*
37. *Machine technology*   The power P (in watts) required by a metal punch machine is a function of the thickness t of the metal being punched. For 2-in.-diameter holes punched six at a time, the power is given by $P(t) = 3.17t^2$. Find the power required to drill these holes in metal 0.250 in. thick.
38. *Computer marketing*   The cost C of leasing a particular kind of computer system is a function of the number of hours of use in excess E of a certain maximum. Specifically, the cost of a 6 month lease for a certain system is given by $C(E) = \$7500 + \$35E$, where E is the number of hours exceeding 600. Find the cost of leasing this system for 745 h.
39. *Refrigeration technology*   The coefficient of performance η of a refrigeration system as a function of the cold reservoir temperature T is given by the function shown when the hot reservoir temperature is 40°C.

*Information supplied by Los Padres National Forest, Ca., USDA.

$$\eta = \frac{T}{40 - T}$$

(a) Find η for $T = 10°C$.
(b) Express η for a cold reservoir temperature of $(T + x)°C$.

40. *Hydrology* The cross-sectional area A (in square feet) of a stream may be expressed as a function of the level L of water (in feet). For a particular stream, the area is given by

$$A = 8L + \frac{5L^2}{12}$$

(a) Find the cross-sectional area for the stream when the water level is 6.0 ft.
(b) Express the function for a level of $(L - 2)$ ft.

41. Sometimes a function is defined differently for different intervals in the domain. For the function

$$f(x) = \begin{cases} x - 4 & \text{for } x < 4 \\ x^2 - 6 & \text{for } x \geq 4 \end{cases}$$

find: (a) $f(-3)$ (b) $f(4)$

42. For the function

$$g(x) = \begin{cases} 2x + 5 & \text{for } x \leq 3 \\ x^2 + 2 & \text{for } x > 3 \end{cases}$$

find: (a) $g(2)$ (b) $g(4)$

3-2 | The Rectangular Coordinate System

In describing and analyzing functions, it is often useful to have a visual representation—or **graph**—of the function. Such a graph shows how the dependent variable responds to changes in the independent variable. The graph also helps us find maximum, minimum, and zero values of the function. These values can be important in technical work, and they will become increasingly important in the development of algebra, analytic geometry, and calculus.

Definitions
For each value of x, the function $y = f(x)$ gives a corresponding value for y. The pair of numbers x and y can be used to label a point on a plane, and many such points will give a picture of the function. We can create this picture by using two number lines intersecting at right angles, as shown in Fig. 3-1 on page 112. The horizontal number line, or **x-axis**, represents values of the independent variable x, and the vertical number line, or **y-axis**, represents values of the dependent variable y. The number lines intersect at a point called the **origin,** which we label as the zero point on both number lines.

On the x-axis, the positive direction is to the right of the origin, and the negative direction is to the left. On the y-axis, the positive direction is above the origin, and the negative direction is below it.

The two axes divide the plane into four sections called **quadrants.** The usual convention is to number these I, II, III, and IV in a counterclockwise direction beginning with the upper right section.

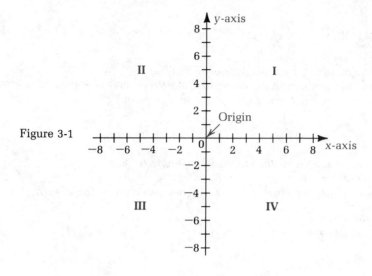

Figure 3-1

The system just described is called the **rectangular coordinate system,** because the axes are perpendicular, and because we can locate any point in the plane by specifying its horizontal and vertical **coordinates.** The coordinates are always given as an ordered pair **(x, y),** with the value of the independent variable, the **x-coordinate,** listed first, and the value of the dependent variable, the **y-coordinate,** listed second. Mathematicians often refer to the x-coordinate as the **abscissa** and to the y-coordinate as the **ordinate,** because variables other than x and y may be used.

EXAMPLE 8 Consider the ordered pairs $A(4, -2)$ and $B(-1, 3)$.
(a) What is the x-coordinate of A?
(b) What is the y-coordinate of A?
(c) Name the ordinate of B.
(d) Name the abscissa of B.

Solution
(a) 4   (b) -2 (c) 3   (d) -1

Plotting points In order to locate and plot the point corresponding to an ordered pair (x, y), begin at the origin, move horizontally x units, then vertically y units, and place a dot at the point.

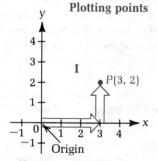

Figure 3-2

EXAMPLE 9
(a) To plot the point $P(3, 2)$, move 3 units to the right of the origin and 2 units up, as shown in Fig. 3-2. The point P is in quadrant I.
(b) To plot the point $Q(-5, -3)$, move 5 units to the left of the origin and 3 units down, as shown in Fig. 3-3. The point Q is in quadrant III.

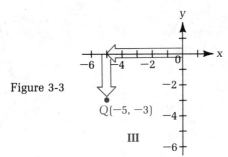

Figure 3-3

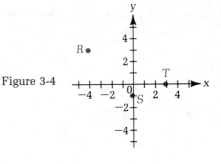

Figure 3-4

EXAMPLE 10

(a)   Plot points $M(2, -4)$ and $N(-3, 5)$ on the same graph. In which quadrants are they located?

(b)   Name the coordinates of the points R, S, and T shown in Fig. 3-4.

Solution

(a)

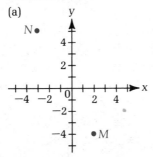

Figure 3-5

M is in quadrant IV.
N is in quadrant II.

(b) $R(-4, 3)$; $S(0, -1)$; $T(3, 0)$

From S and T we can conclude that any point on the y-axis has an x-coordinate of 0, and any point on the x-axis has a y-coordinate of 0.

Geometric figures such as lines and polygons can be described by specifying points in the plane.

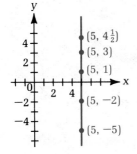

Figure 3-6

EXAMPLE 11   All points with abscissas equal to 5 are located along a vertical line 5 units to the right of the y-axis, as shown in Fig. 3-6.

EXAMPLE 12

(a)   Show the location of all points with abscissas that are equal to their ordinates.

(b)   Show the location of all points with x-coordinates that are equal to half their y-coordinates.

(c)   If $A(-2, 4)$, $B(4, 4)$, and $C(2, -1)$ are three vertices of a parallelogram, what are the coordinates of the fourth vertex, D, in quadrant III?

Solution

(a) These points are located along the diagonal line bisecting quadrants I and III.

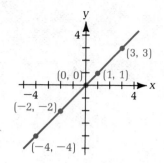

Figure 3-7

(b) The points are located along a line passing through quadrants I and III at a steeper angle than the line shown in part (a).

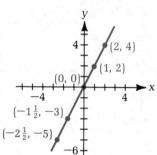

Figure 3-8

(c) D(−4, −1) is the fourth vertex of the parallelogram.*

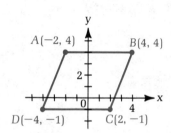

Figure 3-9

Exercises 3-2

Problems 1–6 refer to points A(3, −5), B(−2, −3), C(−7, 1), and D(4, 6).

1. What is the abscissa of B?
2. What is the y-coordinate of C?
3. Name the x-coordinate of D.
4. Name the ordinate of A.
5. In which quadrant is A? B? C? D?
6. Plot all four points on the same graph.

*Notice that the points (8, −1) in quadrant IV and (0, 9) on the y-axis could also be chosen as vertices to complete different parallelograms.

Use the figure shown here to name the coordinates of each point.

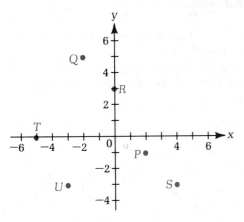

7. P   **8.** Q   **9.** R   **10.** S   **11.** T   **12.** U

Plot each set of points and connect them in the given order with a smooth curve.

13. $A(-2, 0)$, $B(-1, -3)$, $C(0, -4)$, $D(1, -3)$, $E(2, 0)$
14. $F(-2, -1)$, $G(-1, 2)$, $H(0, 3)$, $I(1, 2)$ $J(2, -1)$
15. $K(-2, -8)$, $L(-1, -1)$, $M(0, 0)$, $N(1, 1)$, $O(2, 8)$
16. $P(-3, 1)$, $Q(-2, 2)$, $R(3, 3)$, $S(12, 4)$
17. $T(-5, 0)$, $U(-4, 3)$, $V(0, 5)$, $W(4, 3)$, $Z(5, 0)$
18. $A(\frac{1}{2}, 8)$, $B(1, 4)$, $C(2, 2)$, $D(4, 1)$, $E(8, \frac{1}{2})$

Solve.

19. If $(8, 3)$, $(3, 3)$, and $(3, -2)$ are three vertices of a square, what are the coordinates of the fourth vertex?
20. If $(-1, 7)$, $(6, 7)$, and $(4, 1)$ are three vertices of a parallelogram, what are the coordinates of the fourth vertex in quadrant II?
21. In which quadrant(s) are all points with abscissas equal to -2?
22. In which quadrant(s) are all points with ordinates equal to 4?
23. In which quadrant(s) are all points with abscissas that are the opposites of their ordinates?
24. In which quadrant(s) are all points with x-coordinates that are positive?
25. In which quadrant(s) are all points with y-coordinates that are negative?
26. In which quadrant is $x > 0$ and $y < 0$?
27. In which quadrant is $x < 0$ and $y < 0$?
28. Given the point $X(a, b)$, under what conditions would $Y(b, a)$ be in a different quadrant?

3-3 | Graphing Functions

In this section we will combine the concept of a function with the idea of the rectangular coordinate system in order to plot the graphs of functions. This will enable us to visualize functions, to understand their physical meaning, and to use graphs for more advanced mathematical work.

The graph of a function   The **graph** of a function $y = f(x)$ is the collection of all points with coordinates (x, y) that satisfy the functional rule. Since it would be impossible to list and plot all such ordered pairs, we simply find a few points, connect them with a straight line or smooth curve, and then extend the curve beyond the selected points.

EXAMPLE 13 Graph the function $y = f(x) = 2x - 3$.

Solution
Step 1. Prepare a table with x and y columns, and select several values for the independent variable x.

x	y
2	
1	
0	
−1	
−2	

Step 2. Evaluate the function for each selected value of x. Fill in the y column of the table with these results.

$$f(2) = 2(2) - 3 = 1$$
$$f(1) = 2(1) - 3 = -1$$
$$f(0) = 2(0) - 3 = -3$$
$$f(-1) = 2(-1) - 3 = -5$$
$$f(-2) = 2(-2) - 3 = -7$$

x	y
2	1
1	−1
0	−3
−1	−5
−2	−7

Step 3. Plot each resulting ordered pair on a graph and connect these points with a straight line or smooth curve.

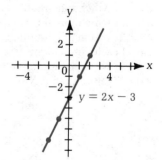

Figure 3-10

The graph of $y = 2x - 3$ is a straight line, as shown in Fig. 3-10, but not all functions produce straight-line graphs.

EXAMPLE 14 Graph the function $y = x^2 - 3$.

Solution

Step 1. Select several values for x.*

x	y
3	
2	
1	
0	
−1	
−2	
−3	

Step 2. Calculate the corresponding values of y.

$y = (3)^2 - 3 = 6$

$y = (2)^2 - 3 = 1$

$y = (1)^2 - 3 = -2$

$y = (0)^2 - 3 = -3$

$y = (-1)^2 - 3 = -2$

$y = (-2)^2 - 3 = 1$

$y = (-3)^2 - 3 = 6$

x	y
3	6
2	1
1	−2
0	−3
−1	−2
−2	1
−3	6

Step 3. Plot the points and connect with a smooth curve.

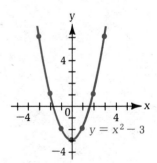

Figure 3-11

The type of curve shown in Fig. 3-11 is called a **parabola**. The point (0, −3)—the "turning point" of the curve—is the **vertex** of the parabola. Every parabola has a vertex. If your graph looks like a parabola, but you have not selected the x value generating the vertex, be sure to find the vertex as you draw the curve by estimating its location.

Restrictions   For the next two functions, be careful to select only those values of x that result in *real* values of y.

---

*For a linear function $y = f(x)$, where x appears only to the first power, a minimum of two points are needed to obtain the graph. For a nonlinear function, where x appears to a power other than 1, more points are usually needed to draw a smooth curve.

EXAMPLE 15   Graph the following functions:

(a) $y = \dfrac{2}{x} - 1$ (b) $y = \sqrt{x + 4}$

Note: In part (a) choose several values of x between −1 and 1.

Solution

(a)

x	y
3	$-\frac{1}{3}$
2	0
1	1
$\frac{1}{2}$	3
$\frac{1}{4}$	7
$-\frac{1}{4}$	−9
$-\frac{1}{2}$	−5
−1	−3
−2	−2
−3	$-1\frac{2}{3}$
−4	$-1\frac{1}{2}$

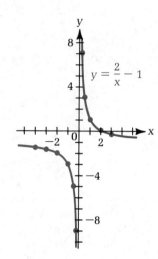

Figure 3-12

(b)

x	y
3	2.6
2	2.4
1	2.2
0	2
−1	1.7
−2	1.4
−3	1
−4	0

$7 \; \boxed{\sqrt{\,}} \rightarrow \;$ *2.6457513*

$6 \; \boxed{\sqrt{\,}} \rightarrow \;$ *2.4494897*

$5 \; \boxed{\sqrt{\,}} \rightarrow \;$ *2.2360680*

$3 \; \boxed{\sqrt{\,}} \rightarrow \;$ *1.7320508*

$2 \; \boxed{\sqrt{\,}} \rightarrow \;$ *1.4142136*

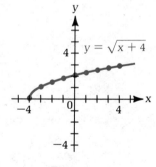

Figure 3-13

The graph shown in Fig. 3-12 is called a **hyperbola.** Notice that y is not defined for x = 0, and thus we cannot select 0 as a value for the independent variable. This produces a graph with two parts or branches, neither of which touches the y-axis. Values of x that approach 0 from the positive direction result in very large positive values of y. Values of x that approach 0 from the negative direction result in very large negative values of y.

In Example 15(b) we could not choose values of x less than −4, because they would produce imaginary values for y. This results in a graph that never goes below the x-axis (see Fig. 3-13). In graphing many practical functions, we may need to restrict the independent variables to nonnegative values because negative values are unrealistic for the given physical or technical situation. The next example further illustrates this type of restriction.

EXAMPLE 16 The lift L (in pounds) on a certain type of airplane wing is given by $L(v) = 0.083v^2$, where v is the speed of air flow over the wing (in feet per second). Graph this function for values of v up to 300 ft/s. (For this graph you must choose different scales for the horizontal and vertical axes.)

Solution First, put 0.083 in memory: .083 [STO]

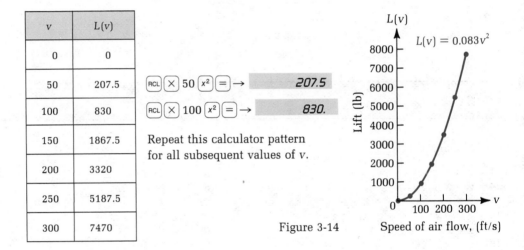

v	L(v)
0	0
50	207.5
100	830
150	1867.5
200	3320
250	5187.5
300	7470

[RCL] [×] 50 [x²] [=] → 207.5

[RCL] [×] 100 [x²] [=] → 830.

Repeat this calculator pattern for all subsequent values of v.

Figure 3-14

The function L(v) is not defined for negative values of v because a negative speed makes no physical sense in this case.

Functions defined by tables of data

As we mentioned earlier, functions may be defined by tables of data collected from observations of physical phenomena. If a clear mathematical relationship exists between two quantities, the graph of such empirical data is a smooth curve. If no mathematical function can be found to describe the relationship, straight line segments are used to join the points on the graph.

EXAMPLE 17 Measurement of the hourly temperature on a certain day in Santa Barbara gave the following data:

Time, t	6 AM	7	8	9	10	11	Noon	1 PM	2
Temperature, T(t) (°F)	58	61	66	69	73	72	75	76	75

The data are graphed in Fig. 3-15. Because no obvious mathematical formula fits the data, the points are joined by straight lines to form a broken-line graph.

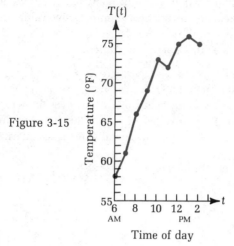

Figure 3-15

Time of day

Exercises 3-3

Graph each function.

1. $y = x$
2. $y = -3x$
3. $y = 2x - 1$
4. $y = x + 5$
5. $f(a) = 8 - a$
6. $f(b) = 4 - 2b$
7. $y = \frac{2}{3}x - 1$
8. $y = 6 - \frac{1}{2}x$
9. $y = 2x^2$
10. $y = -x^2$
11. $g(t) = t^2 + 4$
12. $g(m) = 2 - m^2$
13. $y = 3x^2 + 1$
14. $y = \frac{1}{2}x^2 - 4$
15. $y = 3x^2 - 2x$
16. $y = x - 2x^2$
17. $\alpha(n) = n^2 + 5n + 6$
18. $\beta(z) = 3 - 2z - 2z^2$
19. $y = 2x^3$
20. $y = -x^3$
21. $y = x^3 + x$
22. $y = 3x - 2x^3$
23. $E(w) = \frac{3}{w}$
24. $D(v) = \frac{1}{v - 2}$
25. $y = \frac{1}{x} - 2$
26. $y = 3 - \frac{2}{x}$
27. $y(x) = \frac{-1}{x^2}$
28. $y(x) = \frac{1}{x^2 - 1}$
29. $Q = 2\sqrt{p}$
30. $T = \sqrt{e + 2}$
31. $y = \sqrt{25 - x^2}$
32. $y = \sqrt{x^2 - 4}$

33. **Electrical engineering** The resistance (in ohms) of a copper wire 40 mils in diameter as a function of its length L (in feet) is given by $R(L) = 0.0065L$. Plot $R(L)$ for $L = 0$ to $L = 1000$ ft.
34. **Electronics** The power P (in watts) dissipated in a certain resistor depends on the current I in the resistor according to the function $P(I) = 5I^2$. Graph this function.
35. **Physics** If a space vehicle has an initial velocity of 4 ft/s and an acceleration of 2 ft/s², then its velocity at any time t is given by the equation $v = 4 + 2t$. Graph v as a function of t.
36. **Physics** The distance s traveled by an object falling freely from rest in time t is given by $s = 16t^2$. Plot s as a function of t.
37. **Physics** The position x of a particle subject to a certain force is a function of the time t of its application. Specifically, $x(t) = 3t - 4t^2 + t^3$. Graph this function.

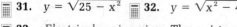

38. *Economics* A particular revenue function is defined by $R(q) = 10q - 5q^2$. Graph this function.

39. *Physics* In the study of vibration, the period T is related to the frequency f by the function $T(f) = 1/f$. Graph this function.

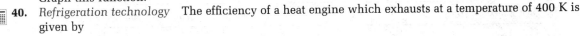 **40.** *Refrigeration technology* The efficiency of a heat engine which exhausts at a temperature of 400 K is given by

$$E = 1 - \frac{400}{T}$$

Plot E as a function of T, the temperature at which the engine uses steam. Use values of T from 450 K to 800 K.*

Graph the function represented by each table of data.

41. *Industrial management* The amount of steel used by a manufacturing company is given in the table below.

Month, M	Jan	Feb	Mar	Apr	May	Jun	Jul	Aug	Sept	Oct	Nov	Dec
Amount, $A(M)$ (10^3 lb)	25	17	20	22	26	30	31	28	25	27	24	25

42. *Meteorology* The rainfall in a certain city is represented in the table.

Year, Y	1976	1977	1978	1979	1980	1981	1982	1983
Amount, $A(Y)$ (in.)	17.5	21.2	14.6	18.8	17.2	16.5	20.4	19.3

43. *Machine technology* The surface speeds of grinding wheels of various diameters are given in the table.

Diameter, D (in.)	2	4	6	8	10	12
Surface Speed, $S(D)$ (100 ft/min)	500	1000	1500	2000	2500	3000

44. *Physics* The elongation of a spring at various force loads is given in the table.

Force, F (lb)	0	1	2	3	4	5	6	7
Length, $L(F)$ (in.)	3.0	3.5	4.0	4.5	5.0	5.5	6.5	9.0

45. *Automotive engineering* The temperature of SAE 30 oil was found to have the values given in the table at different viscosities (in centipoise, cP).

Viscosity, V (cP)	4	6	8	10	12	14	16	18
Temperature, $T(V)$ (°F)	280	240	215	199	187	178	170	163

*For further background, see Frederick Bueche, *Technical Physics*, 2d ed. (New York: Harper and Row, 1981), pp. 330–331.

46. *Energy technology*   The data in the table show the energy necessary to heat a certain object from 15°F to 210°F.

Heat Energy, E (Btu)	750	1500	2250	3000	5250	7500
Temperature, T(E) (°F)	15	75	120	165	195	210

3-4 | Solving Equations Graphically

In Section 2-4 we showed how to solve a linear equation algebraically. In this section we will discuss how to solve an equation graphically. A graphical solution is useful when the equation is difficult to solve in the usual way, or when you need to check an algebraic solution.

Graphical solution

As shown by the next example, a definite connection exists between the solution to an equation and the graph of the equation as a function.

EXAMPLE 18   Solve the equation $2x - 4 = 2$ algebraically, and then solve it graphically.

Solution

Add 4 to each side. $2x - 4 = 2$

Divide each side by 2 $2x = 6$

The solution is 3. $x = 3$

To solve the same equation graphically, follow these steps:

Step 1. Collect all terms on $2x - 6 = 0$
the same side of the equation.

Step 2. Write the expression on $y = f(x) = 2x - 6$
the left as a function of x by
replacing the zero with $f(x)$ or y.

Step 3. Graph this function.

x	y
−1	−8
0	−6
2	−2
4	2
6	6

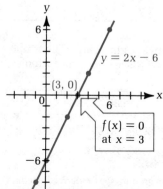

Figure 3-16   The solution is $x = 3$.

x-Intercepts and the zeros of a function

Graphically, the solution of the original equation is the x-coordinate of the point where the graph line intersects the x-axis, as demonstrated in Example 18. This point is known as the **x-intercept** of the graph. Since $f(x) = 0$ at this point, the x-intercept is called a **zero of the function.**

The zeros of a function correspond to the solutions of the equation that determines the function. Therefore, we can solve any equation graphically by finding the zeros of the corresponding function.

EXAMPLE 19 Solve the equation $5(x - 3) = x - 5$ graphically.

Solution

Step 1. Simplify and collect all terms on one side.

$5x - 15 = x - 5$

$4x - 10 = 0$

Step 2. Write the expression on the left as a function. Replace 0 with y:

$y = 4x - 10$

Step 3. Graph:

x	y
0	-10
1	-6
2	-2
3	2
4	6

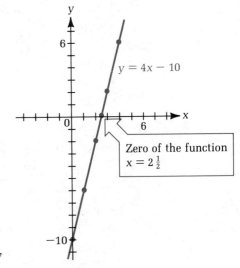

$y = 4x - 10$

Zero of the function $x = 2\frac{1}{2}$

Figure 3-17

The graph crosses the x-axis at $2\frac{1}{2}$, so $2\frac{1}{2}$ is the x-intercept of the graph, the zero of the function, and thus the solution to the original equation.

Check: Substituting $2\frac{1}{2}$ for x in the original equation, we have

$5(2\frac{1}{2} - 3) = 2\frac{1}{2} - 5$

$-2\frac{1}{2} = -2\frac{1}{2}$

This verifies that $2\frac{1}{2}$ is the solution to the equation.

Multiple solutions; estimating zeros

For some functions there may be more than one zero and therefore more than one solution to the equation. Furthermore, it may be necessary to estimate the zeros from the graph of the function.

EXAMPLE 20   Solve $4 + x = x(x - 4)$ graphically.

Solution

Step 1.   Simplify and collect all $4 + x = x^2 - 4x$
terms on one side:

$$0 = x^2 - 5x - 4$$

Step 2.   Let $y = x^2 - 5x - 4$.

Step 3.   Graph:

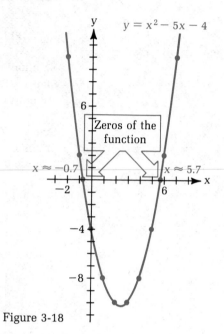

x	y
−2	10
−1	2
0	−4
1	−8
2	−10
3	−10
4	−8
5	−4
6	2
7	10

Figure 3-18

The graph of this function has two x-intercepts. We estimate these function zeros to be at $x \approx -0.7$ and $x \approx 5.7$.

Check: Substituting the estimated solutions into the original equation, we obtain

$$4 + (-0.7) \approx -0.7(-0.7 - 4)$$

$$3.3 \approx 3.29$$

and

$$4 + 5.7 \approx 5.7(5.7 - 4)$$

$$9.7 \approx 9.69$$

Our solutions are approximately correct.*

Equations with   Some equations may have no real number solution. The next example shows
no solutions   what happens when we graph the functions determined by such equations.

---

*When the solutions are approximate, the check should produce numbers that are approximately equal.

EXAMPLE 21 Solve graphically: $x^2 + 2x = -3$

Solution This is equivalent to the function

$$y = x^2 + 2x + 3$$

x	y
−4	11
−3	6
−2	3
−1	2
0	3
1	6
2	11

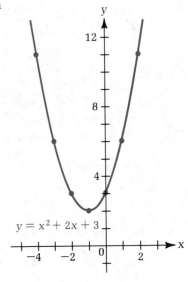

Figure 3-19

The graph of this function does not intersect the x-axis. Since the function has no real zeros, the original equation has no real solution.

Applied problems Graphical solutions may be particularly valuable in working with applied problems where the equation is difficult to solve and the solution must be estimated.

EXAMPLE 22 The range R (in kilometers) of a television station is related to the height h (in meters) of the antenna according to the formula $R = 3.56\sqrt{h}$. Find the height needed to transmit at a range of 40.0 km.

Solution **First,** substitute 40.0 for R: $40.0 = 3.56\sqrt{h}$
Then, convert to function form and solve graphically: $f(h) = 40.0 - 3.56\sqrt{h}$

h	f(h)
0	40
40	17.5
60	12.4
80	8.2
100	4.4
120	1.0
140	−2.1

Place the constant 3.56 into memory:

3.56 [STO]

40 [−] [RCL] [×] 40 [√] [=] → 17.484583

40 [−] [RCL] [×] 60 [√] [=] → 12.424359

Use the same pattern for calculating all other results.

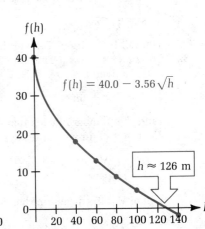

Figure 3-20

We estimate the zero of the function and therefore the solution of the equation to be $h \approx 126$ m.

Check: $40.0 = 3.56 \sqrt{126}$ $\boxed{\text{RCL}}\boxed{\times}\,126\,\boxed{\sqrt{}}\boxed{=}\rightarrow$ *39.960901*

$40.0 = 40.0$

Exercises 3-4

Solve each equation graphically. Be sure to check your solutions.

1. $4x - 8 = 0$ **2.** $2x + 7 = 0$ **3.** $-3x + 6 = 14$

4. $12 - x = 16$ **5.** $7x + 3 = x$ **6.** $4 + 3x = x$

7. $2x - 9 = 4x - 19$ **8.** $5x + 8 = 8x - 10$ **9.** $4(x - 5) = 2x$

10. $-3x = 2(2x - 21)$ **11.** $5 - (2x - 7) = 17$ **12.** $3 - 6(3x + 1) = 3$

13. $x^2 - x = 0$ **14.** $2x^2 + 6x = 0$ **15.** $x^2 - 5x + 4 = 0$

16. $2x^2 - x - 15 = 0$ **17.** $3x^2 - 7x = 3$ **18.** $x^2 = 5x + 3$

19. $(x - 3)^2 = 8$ **20.** $(x + 4)(x + 3) = 7$ **21.** $x^3 - 9x = 0$

22. $x^3 + 2x - 5 = 0$ **23.** $x^4 - 3x^2 + 1 = 0$ **24.** $x^5 = 4x$

25. $\dfrac{2}{x} - 4 = 0$ **26.** $\dfrac{1}{x^2 + 4} = 0$ **27.** $\sqrt{x + 3} - 5 = x - 8$

28. $3\sqrt{x} - x = 2$

29. *Surveying*   The elevation E (in feet) of a point on a road with a 5% grade can be found by using the formula

$$E = 50.0 + 0.05L$$

where L is the lateral distance (in feet) from the starting point. At what point would the elevation be 53.0 ft?

30. *Aeronautical engineering*   The range R (in kilometers) of a descending aircraft is given by

$$R = 13 - 0.30T$$

where T is the time of descent (in seconds). When will the aircraft arrive ($R = 0$)?

31. *Physics*   The travel distance s (in feet) of an object moving in a straight line is given by

$$s = 6 + 2t + t^2$$

where t is the time of travel (in seconds). At what time will the distance be 15 ft?

32. *Computer science*   The time T (in seconds) necessary to test a computer memory chip with $n \times 10^3$ cells is given by

$$T = 2n^2 + 6n$$

Find the number of cells that can be tested in 5 s.

33. *Electronics*   In a complex system, the coefficient of availability A of a particular component is

$$A = \frac{M}{M + 2}$$

where M is the mean time between failures (in years). When is the coefficient of availability equal to 0.2?

34. *Physics* The velocity v (in meters per second) of a falling object as it hits the ground is given by

$$v = \sqrt{19.6h}$$

where h is the starting height (in meters). If an object had an impact velocity of 10.0 ft/s, from what height was it dropped?

3-5 | Review of Chapter 3

Important Terms and Concepts

function (p. 106)	x-axis (p. 111)	abscissa (p. 112)
independent variable (p. 106)	y-axis (p. 111)	y-coordinate (p. 112)
dependent variable (p. 106)	origin (p. 111)	ordinate (p. 112)
domain (p. 106)	quadrants (p. 111)	graph of a function (p. 116)
range (p. 106)	rectangular coordinate system (p. 112)	x-intercept (p. 123)
function notation (p. 106)	x-coordinate (p. 112)	zero of a function (p. 123)

Exercises 3-5

Express each function using function notation.

1. Express the area A of a circle as a function of its diameter d.
2. Find the perimeter P of a square as a function of its side length s.
3. Write the distance d traveled by an object for 7 h as a function of its average speed v.
4. A spring 2.5 in. long increases 0.4 in. in length for every pound of weight on it. Express the length L of the spring as a function of the hanging weight W.
5. Give the radius r of a circle as a function of its circumference C.
6. Express the base b of a triangle 6 cm high as a function of its area A.

Find the following function values:

7. For $y(x) = 2x - 11$, find: (a) $y(4)$ (b) $y(-3)$ (c) $y\left(\frac{5}{2}\right)$
8. For $f(y) = 5 - 7y$, find: (a) $f(2)$ (b) $f(-5)$ (c) $f\left(-\frac{1}{7}\right)$
9. For $g(t) = t^2 - 8t$, find: (a) $g(3)$ (b) $g(-1)$ (c) $g\left(\frac{1}{2}\right)$
10. For $g(m) = 3\sqrt{m} - \dfrac{m}{8}$, find: (a) $g(4)$ (b) $g(64)$ (c) $g\left(\frac{16}{9}\right)$
11. For $\phi(a) = \dfrac{6 - 5a^2}{3a}$, find: (a) $\phi(185)$ (b) $\phi(-3.75)$ (c) $\phi(0.62)$
12. For $\mu(n) = 2n^2 - 7n + 8$, find: (a) $\mu(835)$ (b) $\mu(-2.64)$ (c) $\mu(0.65)$
13. For $h(y) = 5y + 1$, find: (a) $h(y - 2)$ (b) $h(-2y)$ (c) $h(4 + y) - h(y)$
14. For $y(x) = 3x^2 - 8x$, find: (a) $y(x + 3)$ (b) $y(3x^2)$ (c) $y(2 + x) - y(x)$

State the domain of each function, given that the range is restricted to real number values.

15. $y(x) = 2x^2 + \dfrac{3}{x}$ **16.** $f(y) = \dfrac{3}{2y - 4}$ **17.** $\alpha(a) = \sqrt{36 - a^2}$ **18.** $H(t) = \dfrac{1}{t^2 - 16}$

Graph each function.

19. $y = 2x$ **20.** $y = x - 3$ **21.** $y = \frac{1}{2}x + 3$ **22.** $y = x^2$

23. $g(t) = 3t^2 - 6$ **24.** $f(a) = a^2 + 5a$ **25.** $y = x^2$ **26.** $y = \frac{1}{2}x^3$

27. $y(x) = 2x - x^3$ **28.** $y(x) = \dfrac{4}{x}$ **29.** $y = \dfrac{1}{x^2 - 4}$ **30.** $y = \sqrt{2x}$

31. $P(t) = \sqrt{t - 1}$ **32.** $\beta(r) = \sqrt{r^2 - 1}$

Solve each equation graphically.

33. $5x - 15 = 0$ **34.** $8 - 2x = 11$ **35.** $5x - 4 = 8x$ **36.** $3x - 11 = 7x + 5$

37. $3(2x - 3) = x$ **38.** $7 - (5 - 3x) = 9$ **39.** $3x^2 = 9x$ **40.** $x^2 - 7x = -12$

41. $2x^2 - 3x = 4$ **42.** $(x + 2)^2 = 7$ **43.** $x^3 - 5x = 0$ **44.** $x^3 = 2x^2 - 3x + 5$

45. $\dfrac{6}{x} = x$ **46.** $x - \sqrt{x} = 3$

Answer questions 47–51 for these points: $A(3, 5)$, $B(7, 5)$, $C(9, -3)$, $D(-7, -2)$

47. What is: (a) The ordinate of A?   (b) The abscissa of B?

48. Name: (a) The x-coordinate of C.   (b) The y-coordinate of D.

49. In which quadrant is: (a) A?   (b) C?   (c) D?

50. Which of the given points would move to a different quadrant if its x- and y-coordinates were switched?

51. If A, B, and C are three vertices of a parallelogram, what are the coordinates of the fourth vertex in quadrant I?

52. Where are all points that have abscissas equal to their ordinates?

Solve algebraically.

53. *Finance*   The value V of an investment of \$5 at an interest rate of r after 2 years is given by

$$V = 5(1 + r)^2$$

Find V for: (a) $r = 0.06$ (b) $r = r + 0.1$

54. *Electronics*   The circuit gain G of a particular platinum wire temperature-sensing device is given by

$$G = \dfrac{R}{1200 + R}$$

Find G when: (a) $R = 250 \ \Omega$ (b) $R = 1200 \ \Omega$

Solve graphically.

55. *Economics*   A particular demand function is given by

$$p(q) = 12 - 4q$$

where p represents the maximum price a buyer will pay for a quantity q. At what quantity can a maximum price of 2 be expected?

56. *Physics* The height s (in feet) above the ground of a certain projectile is given by the formula

$$s = 40t - 16t^2$$

where t is the time (in seconds). At what times is the object 24 ft above the ground?

Graph each function.

57. *Physics* The velocity of a projectile fired upward at an initial velocity of 96 ft/s is given by

$$v = 96 - 32t$$

Plot v as a function of the time t for $t = 0$ to 6 s. At what time does the projectile begin its descent?

58. The perimeter P of a rectangle with length equal to 5 m is given by

$$P = 10 + 2w$$

Plot P as a function of the width w.

59. *Physics* The centripetal force on a 6 kg ball swung at a speed v in a curve of radius 4 m is given by

$$F = \frac{3v^2}{2}$$

Plot F as a function of the velocity v.

60. The area of a triangle whose height is 4 cm larger than its base is given by

$$A = \frac{b^2 + 4b}{2}$$

Plot A as a function of b.

61. *Transportation engineering* The ideal length L of a merging curve in a rapid transit system for a velocity of 20 mi/h is a function of the width x of the region as given by

$$L(x) = 23.2\sqrt{x}$$

Plot $L(x)$ for $x = 1$ to 12.

62. *Petroleum engineering* The pressure drop P (in pounds per square inch) in a certain oil reservoir as a function of the time t (in years) is given by

$$P(t) = 220t - 25t^2 + t^3$$

Graph $P(t)$ for $t = 0$ to 9.

Graph the function represented by each table of data.

63. *Meteorology* The hourly temperatures in Santa Barbara on a certain day are given in the table.

Time, t	6 AM	7	8	9	10	11	Noon	1 PM	2	3	4
Temperature, $T(t)$ (°F)	52	55	59	60	62	66	68	69	71	71	70

64. *Physics* The surface tension of water at various temperatures is given in the table.

Temperature, T (°C)	0	50	100	150	200	250	300	350	375
Surface Tension, $S(T)$ (dyne/cm²)	75	65	55	45	35	25	15	5	0

65. *Physics*   The elongation of a spring for various force loads is given by the set of data in the table.

Load, L (kg)	0	1	2	3	4	5	6
Length, $E(L)$ (cm)	5.8	6.3	6.8	7.3	7.8	8.6	10.1

66. *Physics*   The surface energy of water at various temperatures is given in the table.

Temperature, T (°C)	0	50	100	150	200	250	300	350	375
Surface Energy, $E(T)$ (erg/cm^2)	143	142	140	137	131	123	111	81	0

The Trigonometric Functions

4

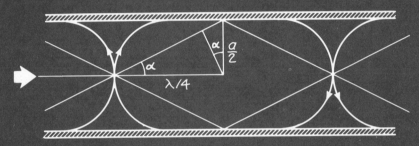

MICROWAVE ANTENNA GEOMETRY

One of the earliest engineering problems to be solved mathematically involved finding the height of an object—a mountain or pyramid, for example—that could not be measured directly. By the 6th century BC, Greek engineers used height-to-shadow-length ratios to perform these and similar calculations. By 400 AD, Hindu mathematicians had created tables of numbers that made this "shadow reckoning" easier. From this work and from even earlier calculations in astronomy, came trigonometry—the study of a special group of functions relating the sides and angles of triangles.

Over the centuries trigonometry has become an invaluable tool in almost all areas of science and technology. Today trigonometric functions are used by surveyors to calculate distances, by engineers to analyze forces, and by physicists and electronics technicians to work with electrical circuits. Perhaps the most important use of trigonometric functions is in the study of periodic phenomena such as oscillations and waves, from surf to laser beams.

In this chapter we will discuss angles and their measurement, define the basic trigonometric functions, and study some of these applications. More advanced topics of trigonometry will appear in later chapters.

4-1 | Angles and Their Measure

Definition of an angle

Figure 4-1(a) shows a **half-line,** namely that portion of a line to one side of a fixed endpoint. An **angle** is generated by rotating a half-line about its endpoint from its initial position to some terminal position. As shown in Fig. 4-1(b), the initial and terminal positions of the half-line become the **initial** and **terminal sides** of the angle. The endpoint O of the half-line becomes the **vertex** of the angle. The word *angle* refers to the amount of rotation in the direction indicated by the arrow [see Fig. 4-1(b)].

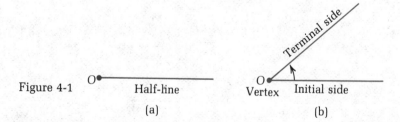

Figure 4-1

Half-line

(a)

Vertex   Initial side

(b)

Sign of an angle

The sign of an angle is determined by the direction of rotation. A counterclockwise rotation generates a **positive angle,** while a clockwise rotation produces a **negative angle.** In Fig. 4-2(a), angle α is positive. In Fig. 4-2(b), angle β is negative. Angles are usually named using lowercase Greek letters: α (alpha), β (beta), θ (theta), ϕ (phi), and so on.

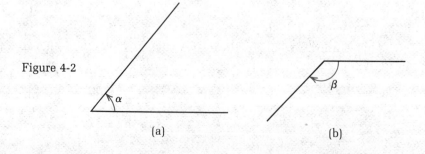

Figure 4-2

(a)

(b)

Magnitude of an angle

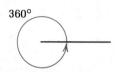

The **magnitude** of an angle may be measured in either **degrees** or **radians.** The degree unit is defined as $\frac{1}{360}$ of a complete rotation. Radian measure will be discussed in Chapter 8.

The following angles and classes of angles are often used, and their names should be memorized:

- **Straight angle** (half rotation, or 180°)

- **Right angle** (quarter rotation, or 90°)

- **Acute angle** (between 0° and 90°)

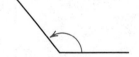

- **Obtuse angle** (between 90° and 180°)

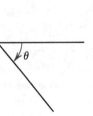

EXAMPLE 1 Figure 4-3 shows angles of the following measures:
(a) 30°   (b) 90°   (c) 120°   (d) −50°
Match each measure to one of the drawn angles.

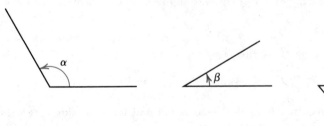

Figure 4-3

Solution
(a) β   (b) φ   (c) α   (d) θ

Divisions of a degree

For work that requires a great deal of accuracy, a degree can be divided into finer units known as **minutes** (abbreviated '). There are 60' in 1°. Minutes can be divided further into **seconds** (abbreviated "). There are 60" in 1'. When an electronic calculator is used, parts of a degree are usually expressed as decimal fractions rather than in terms of minutes and seconds. For practical work in trigonometry, you must be able to convert back and forth between decimal degrees and minutes.

EXAMPLE 2   Convert 26°36′ to decimal notation.

Solution
Step 1.   Use the fact that 1′ = 1/60° to write the minutes portion as a fraction.

$$36' = 36' \times \frac{1°}{60'}$$

So

$$26°36' = 26\frac{36°}{60}$$

Step 2.   Divide to convert the fraction to a decimal. $26\frac{36°}{60} = 26.6°$

To convert fractions of a degree from decimal form to minutes, multiply the decimal part by 60′.

EXAMPLE 3   45.4° = 45° + 0.4(60′) = 45°24′

Many calculators have a degrees/minutes/seconds key ($\boxed{\text{DMS}}$ or $\boxed{°'''}$) for performing these conversions. Even without such a key, you can use a calculator to convert in the following manner:

For Example 2:   26 $\boxed{+}$ 36 $\boxed{÷}$ 60 $\boxed{=}$ → ▓▓▓▓ *26.6*

For Example 3:   .4 $\boxed{×}$ 60 $\boxed{=}$ → ▓▓▓▓ *24.*   Note: This is only the minutes portion of the angle.

EXAMPLE 4
(a)   Convert 78°26′ to decimal form. (Round to 0.01°.)
(b)   Change 122.64° to degrees and minutes. (Round to the nearest 1′.)

Solution
(a) $78°26' = 78\frac{26°}{60} = 78.433\ldots° \approx 78.43°$
(b) $122.64° = 122° + (0.64)(60') = 122° + 38.4' \approx 122°38'$

Coterminal angles   Two angles that have the same initial and terminal sides are called **coterminal angles.** To find an angle that is coterminal to a given angle, either add a multiple of 360° to the original angle or subtract a multiple of 360° from it.

EXAMPLE 5   In Figure 4-4, 70° and 430° are coterminal angles, since 70° + 360° = 430°. Other angles coterminal to 70° are

$$70° + 2(360°) = 790° \quad \text{and} \quad 70° + 3(360°) = 1150°$$

$$70° - 360° = -290° \quad \text{and} \quad 70° - 2(360°) = -650°$$

And so on.

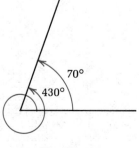

Figure 4-4

EXAMPLE 6   For each given angle, find both a positive and a negative coterminal angle.
(a)   125°   (b)   83°47′   (c)   16.25°

Solution   **Positive**   **Negative**

(a) $125° + 360° = 485°$ $125° - 360° = -235°$

(b) $83°47' + 360° = 443°47'$ $83°47' - 360° = -359°60'$
$$ + 83°47'$$
$$ \overline{-276°13'}$$

(c) $16.25° + 360° = 376.25°$ $16.25° - 360° = -343.75°$

Notice in part (b) that we used $360° = 359°60'$ in order to subtract.

Angles and rectangular coordinates

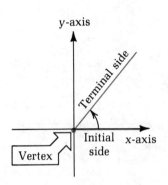

y-axis

Terminal side

Initial side   x-axis

Vertex

Figure 4-5

Figure 4-6

It often is convenient to work with angles drawn on a rectangular coordinate system. An angle that has its vertex at the origin and initial side along the positive x-axis, as shown in Fig. 4-5, is said to be in **standard position.** Instead of classifying such angles as acute or obtuse, mathematicians name them according to the quadrant in which their terminal side falls. For example, the angle in Fig. 4-5 is classified as a **first-quadrant angle.** From Fig. 4-6 we can see that first-quadrant angles have measures between 0° and 90°, second-quadrant angles have measures between 90° and 180°, third-quadrant angles have measures between 180° and 270°, and fourth-quadrant angles have measures between 270° and 360°. Right angles and straight angles, along with all other angles with terminal sides on an axis, are called **quadrantal angles.**

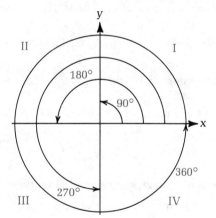

EXAMPLE 7
(a)   In what quadrants do the following angles fall?
    $10°, \quad 100°, \quad 200°, \quad 300°, \quad -50°, \quad -120°$
(b)   Name three other quadrantal angles besides 90° and 180°.

Solution
(a)   First, second, third, fourth, fourth, third
(b)   0°, 270°, and 360°; of course, there are other possibilities greater than 360°.

Naming an angle with a point

Since the initial side of an angle in standard position is fixed on the x-axis, its size is determined by the location of the terminal side. We can specify the position of the terminal side by giving the coordinates of a point anywhere

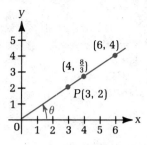

Figure 4-7

along that line, except at the origin. For example, given that the point $P(3, 2)$ is on the terminal side, we can draw the positive angle θ shown in Fig. 4-7. Notice also that many other points can be used to draw the terminal side of θ.

EXAMPLE 8 Draw an angle α in standard position with its terminal side passing through $Q(-4, -2)$.

Solution

Figure 4-8

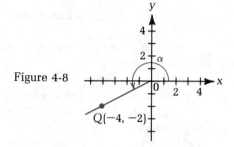

Exercises 4-1

Draw the given angles. Label the vertex V, initial side I, and terminal side T for each angle.

1. 40°	**2.** 60°	**3.** 135°	**4.** 150°	**5.** 180°	**6.** 270°
7. −90°	**8.** −360°	**9.** −30°	**10.** −165°	**11.** 225°	**12.** 340°

Find one positive and one negative coterminal angle for each of the following angles:

13. 60°	**14.** 25°	**15.** 165°	**16.** 225°	**17.** 42°15′	**18.** 310°12′
19. 137.9°	**20.** 84.2°				

Write each of the following angles in decimal form. Round to the nearest hundredth if necessary.

21. 52°18′	**22.** 125°36′	**23.** 238°45′	**24.** 17°9′	**25.** 65°43′	**26.** 325°4′
27. 98°26′	**28.** 8°17′				

Convert each of the following angles to degrees and minutes. Round to the nearest minute if necessary.

29. 58.2°	**30.** 112.5°	**31.** 227.25°	**32.** 344.65°	**33.** 8.61°	**34.** 27.48°
35. 164.83°	**36.** 271.29°				

Designate the quadrant of each angle.

37. 238°	**38.** 116°	**39.** −26°	**40.** −310°

Draw a positive angle in standard position that has a terminal side passing through the given point.

41. (3, 1)	**42.** (−2, −6)	**43.** (4, −3)	**44.** (5, 4)

4-2 | Defining the Trigonometric Functions

Right triangles in rectangular coordinates

Figure 4-9(a) shows an angle θ in standard position with a point $A(3, 4)$ on its terminal side. Dropping a perpendicular line from point A to the x-axis creates the right triangle AOC. In Fig. 4-9(b) notice that the lengths of the perpendicular sides of this triangle are $OC = 3$ and $AC = 4$, the coordinates of

Figure 4-9

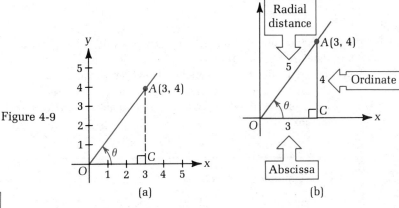

(a)   (b)

point A. Thus, for a right triangle shown in the rectangular coordinate system, the lengths of the legs are referred to as the **abscissa** and **ordinate.** The hypotenuse OA can be computed from the Pythagorean theorem:

$$OA = \sqrt{3^2 + 4^2} = \sqrt{25} = 5$$

This is also known as the **radial distance.**

We can repeat this procedure for any general point $P(x, y)$ on the terminal side of the same angle θ. For $P(x, y)$, the abscissa is x, the ordinate is y, and the radial distance is r, where $r = \sqrt{x^2 + y^2}$ (Fig. 4-10).

Triangles AOC and POQ in Figs. 4-9 and 4-10 are similar (Fig. 4-11); therefore, their side lengths are proportional. This means that

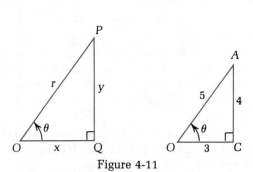

Figure 4-10

$$\frac{x}{y} = \frac{3}{4}$$

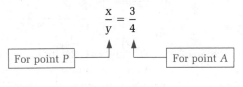

Figure 4-11

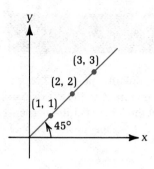

Figure 4-12

Thus, for any point on the terminal side of an angle, the ratio of x to y is a constant value that depends only on the size of the angle.

EXAMPLE 9 In Fig. 4-12 several points are shown on the terminal side of a 45° angle. What is the ratio of x to y for each of these points?

Solution For a 45° angle the ratio is always 1:

$$\frac{x}{y} = \frac{1}{1} = \frac{2}{2} = \frac{3}{3} = 1$$

The trigonometric functions

A total of six different ratios can be set up among x, y, and r for any angle in standard position (Fig. 4-13). These ratios remain constant for all points on the terminal side of a given angle θ. However, as the angle itself changes, the ratios change, so they are all functions of the angle θ. These ratios comprise a special group of functions known as the **trigonometric functions,** which are defined in the box.

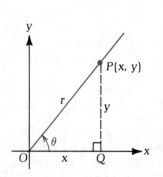

Figure 4-13

Trigonometric Functions

sine $\theta = \dfrac{y}{r}$ abbreviated sin θ

cosine $\theta = \dfrac{x}{r}$ abbreviated cos θ

tangent $\theta = \dfrac{y}{x}$ abbreviated tan θ (4-1)

cosecant $\theta = \dfrac{r}{y}$ abbreviated csc θ

secant $\theta = \dfrac{r}{x}$ abbreviated sec θ

cotangent $\theta = \dfrac{x}{y}$ abbreviated cot θ

Restrictions

The angle θ is the independent variable of the trigonometric functions. It may take on any value except those that lead to division by zero in the function values themselves (the dependent variables). For example, tan θ and sec θ are not defined for angles where x = 0. The functions cot θ and csc θ are not defined for angles where y = 0. We will discuss these cases in the next section. To eliminate this problem for sin θ and cos θ, it is assumed that r > 0 for all cases.

Reciprocal functions

The most frequently used trigonometric functions are sin θ, cos θ, and tan θ. When csc θ, sec θ, and cot θ are used, it is important to keep in mind that they are simply the reciprocals of sin θ, cos θ, and tan θ, respectively:

$$\boxed{\text{Reciprocals}}$$

$$\sin \theta = \frac{y}{r} \qquad \csc \theta = \frac{r}{y} = \frac{1}{\sin \theta}$$

$$\cos \theta = \frac{x}{r} \qquad \sec \theta = \frac{r}{x} = \frac{1}{\cos \theta}$$

$$\tan \theta = \frac{y}{x} \qquad \cot \theta = \frac{x}{y} = \frac{1}{\tan \theta}$$

This becomes especially important when finding values of csc θ, sec θ, and cot θ on a calculator.

Determining values from a given point

Except for angles that give x = 0 or y = 0, the values of the trigonometric functions for any angle are real numbers. In the remainder of this chapter we will limit our study to the trigonometric functions of acute angles. One way to calculate the values of the trigonometric functions for a given value of θ is by using a point on the terminal side of the angle.

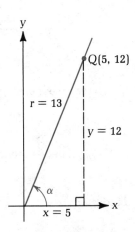

Figure 4-14

EXAMPLE 10 The point Q(5, 12) is on the terminal side of angle $\alpha \approx 67.38°$. To find the values of all six functions, first use the Pythagorean theorem to calculate r:

$$r = \sqrt{5^2 + 12^2}$$

$$r = \sqrt{169} = 13$$

Then, use the definitions and Fig. 4-14 to set up the six ratios:

$$\sin \alpha = \frac{y}{r} = \frac{12}{13} \approx 0.9231 \qquad \csc \alpha = \frac{1}{\sin \alpha} = \frac{13}{12} \approx 1.0833$$

$$\cos \alpha = \frac{x}{r} = \frac{5}{13} \approx 0.3846 \qquad \sec \alpha = \frac{1}{\cos \alpha} = \frac{13}{5} = 2.6000$$

$$\tan \alpha = \frac{y}{x} = \frac{12}{5} = 2.4000 \qquad \cot \alpha = \frac{1}{\tan \alpha} = \frac{5}{12} \approx 0.4167$$

EXAMPLE 11 Given that the terminal side of some angle β passes through the point (6, 5), find the values of all six trigonometric functions to four decimal places.

Solution Since x = 6 and y = 5, we have $r = \sqrt{6^2 + 5^2} = \sqrt{61}$.

$$\sin \beta = \frac{5}{\sqrt{61}} \approx 0.6402 \qquad 5 \div 61 \boxed{\sqrt{}} \boxed{=} \rightarrow \boxed{0.6401844}$$

$$\csc \beta = \frac{1}{\sin \beta} \approx 1.5620 \qquad \boxed{1/x} \rightarrow \boxed{1.5620499}$$

$$\cos \beta = \frac{6}{\sqrt{61}} \approx 0.7682 \qquad 6 \div 61 \boxed{\sqrt{}} \boxed{=} \rightarrow \boxed{0.7682213}$$

$$\sec \beta = \frac{1}{\cos \beta} \approx 1.3017 \qquad \boxed{1/x} \rightarrow \boxed{1.3017083}$$

$$\tan \beta = \frac{5}{6} \approx 0.8333 \qquad 5 \boxed{\div} 6 \boxed{=} \rightarrow \boxed{0.8333333}$$

$$\cot \beta = \frac{1}{\tan \beta} = 1.2000 \qquad \boxed{1/x} \rightarrow \boxed{1.2}$$

Determining values from a given value   Given the value of any single function of an angle, we can find the values of the other five functions.

EXAMPLE 12   Suppose we are given $\cos \phi = \dfrac{4}{9}$. Find the values of all other functions.

First, list any of x, y, and r that are known. Since

$$\cos \phi = \frac{4}{9} = \frac{x}{r}$$

then x = 4 and r = 9.

Second, find the length of the missing side. Using the Pythagorean theorem, we have

$$y = \sqrt{9^2 - 4^2} = \sqrt{65}$$

Third, use x, y, and r to write the sine and tangent values. Calculate the other ratios using reciprocals.

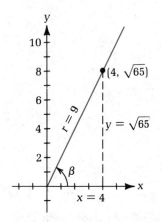

Figure 4-15

$$\sin \phi = \frac{\sqrt{65}}{9} \approx 0.8958 \qquad 65 \boxed{\sqrt{\ }} \boxed{\div} 9 \boxed{=} \rightarrow \boxed{0.8958064}$$

$$\csc \phi = \frac{1}{\sin \phi} \approx 1.1163 \qquad \boxed{1/x} \rightarrow \boxed{1.1163126}$$

$$\tan \phi = \frac{\sqrt{65}}{4} \approx 2.0156 \qquad 65 \boxed{\sqrt{\ }} \boxed{\div} 4 \boxed{=} \rightarrow \boxed{2.0155644}$$

$$\cot \phi = \frac{1}{\tan \phi} \approx 0.4961 \qquad \boxed{1/x} \rightarrow \boxed{0.4961389}$$

$$\sec \phi = \frac{1}{\cos \phi} = \frac{9}{4} = 2.25$$

A similar procedure can be used if a function value is given in decimal form.

EXAMPLE 13   Given $\sin \theta = 0.3452$, find $\cos \theta$ and $\cot \theta$ to four decimal places.

Solution **First,** write 0.3452 as a ratio to establish y and r:

$$\sin \theta = \frac{y}{r} = \frac{0.3452}{1} \qquad y = 0.3452 \text{ and } r = 1$$

Second, use the Pythagorean theorem to calculate x:

$$x = \sqrt{1^2 - 0.3452^2} \qquad 1 \boxed{-} .3452 \boxed{x^2} \boxed{=} \boxed{\sqrt{}} \rightarrow \boxed{0.9385291}$$

Third, set up the proper ratios for cos θ and cot θ:

$$\cos \theta = \frac{x}{r} \approx \frac{0.9385291}{1} \approx 0.9385$$

$$\cot \theta = \frac{x}{y} \approx \frac{0.9385291}{0.3452} \qquad \boxed{\div} .3452 \boxed{=} \rightarrow \boxed{2.7187982}$$

$$\approx 2.7188$$

Exercises 4-2

Each of the following is a point on the terminal side of an angle in standard position. Determine the exact values of all six trigonometric functions for each angle.

1. (3, 4) **2.** (24, 7) **3.** (12, 5) **4.** (5, 5)

5. (2, 5) **6.** (7, 1) **7.** $(\frac{3}{4}, 5)$ **8.** $(2, \sqrt{3})$

Each of the following is a point on the terminal side of an angle in standard position. Use your calculator to determine the approximate values for all six trigonometric functions for each angle. Round to three decimal places.

9. (3, 7) **10.** (8, 1) **11.** (0.7, 4.1) **12.** $(\sqrt{2}, \sqrt{3})$

Use the given information to find exact values for the specified functions.

13. Given $\sin \theta = \frac{3}{4}$, find: (a) $\cos \theta$ (b) $\cot \theta$
14. Given $\cos \theta = \frac{1}{2}$, find: (a) $\csc \theta$ (b) $\tan \theta$
15. Given $\tan \theta = 2$, find: (a) $\sec \theta$ (b) $\sin \theta$
16. Given $\cot \theta = 1$, find: (a) $\cos \theta$ (b) $\tan \theta$

Use the given information to find approximate values for the specified functions. Round as directed.

17. Given $\csc \theta = 2.3$, find the following to two decimal places: (a) $\sin \theta$ (b) $\cot \theta$
18. Given $\sec \theta = 1.4$, find the following to two decimal places: (a) $\csc \theta$ (b) $\cos \theta$
19. Given $\cos \theta = 0.640$, find the following to three decimal places: (a) $\sin \theta$ (b) $\tan \theta$
20. Given $\sin \theta = 0.270$, find the following to three decimal places: (a) $\tan \theta$ (b) $\sec \theta$
21. Given $\cot \theta = 1.576$, find the following to four decimal places: (a) $\sin \theta$ (b) $\sec \theta$
22. Given $\tan \theta = 3.456$, find the following to four decimal places: (a) $\cos \theta$ (b) $\csc \theta$

Solve.

23. Given A(3, 2) and B(x, 7), for what value of x are both points on the terminal side of the same angle?
24. Given P(5, 8) and Q(3, y), for what value of y are both points on the terminal side of the same angle?

4-3 | Values of the Trigonometric Functions

In Section 4-2 we showed how to find the values of the trigonometric functions from a point on the terminal side of an angle in standard position. However, in most applications, values of these functions must be found from the magnitude of the angle itself. In this section we will discuss how to find the values of the trigonometric functions of acute angles using a calculator. But first we will use some basic facts from geometry to find a few special function values. When memorized, these commonly used values will help you solve many applied problems quickly. They will also help you estimate and confirm answers obtained by using a calculator.

Functions of 45°

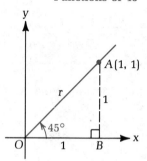

Figure 4-16

If the only tools we possessed were a protractor and a ruler, we could draw a given angle with the protractor, drop a perpendicular to the x-axis, and measure x, y, and r with a ruler. But this scale-drawing method would be tedious, and the resulting trigonometric ratios would be only approximations.

On the other hand, certain facts from geometry allow us to determine precise values for the trigonometric functions of several angles. For example, Fig. 4-16 shows right triangle AOB with perpendicular sides that each measure 1 unit long. A triangle with two sides equal in length is called an **isosceles triangle.** In an isosceles right triangle, the two acute angles are also equal and they each measure 45°. Finally, the Pythagorean theorem tells us that $r = \sqrt{1^2 + 1^2} = \sqrt{2}$. Therefore, we can calculate the values given in the box.

Special Angle: 45°

$$\sin 45° = \frac{1}{\sqrt{2}} = \frac{\sqrt{2}^*}{2} \approx 0.707$$

$$\cos 45° = \frac{1}{\sqrt{2}} = \frac{\sqrt{2}}{2} \approx 0.707$$

$$\tan 45° = \frac{1}{1} = 1$$

The figure shown in the box will help you remember these special function values. The remaining function values can be found using reciprocals.

Functions of 30° and 60°

A triangle in which all sides are equal is called an **equilateral triangle.** Figure 4-17 shows equilateral triangle POQ in which all sides are 2 units long. As

---

*As we shall see in Chapter 11, simplest radical form for a fraction requires that there be no radical in the denominator. Therefore,

$$\frac{1}{\sqrt{2}} = \frac{1}{\sqrt{2}} \cdot \frac{\sqrt{2}}{\sqrt{2}} = \frac{\sqrt{2}}{2}$$

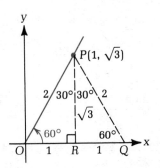

Figure 4-17

shown, all angles of an equilateral triangle measure 60°. Dropping a perpendicular from P to R on the x-axis bisects side OQ. Thus, in right triangle POR, OP = 2, OR = 1, and, using the Pythagorean theorem, $PR = \sqrt{2^2 - 1^2} = \sqrt{3}$. We can therefore calculate the function values in the box for an angle of 60°.

Special Angle: 60°

$$\sin 60° = \frac{\sqrt{3}}{2} \approx 0.866$$

$$\cos 60° = \frac{1}{2} = 0.5$$

$$\tan 60° = \frac{\sqrt{3}}{1} \approx 1.732$$

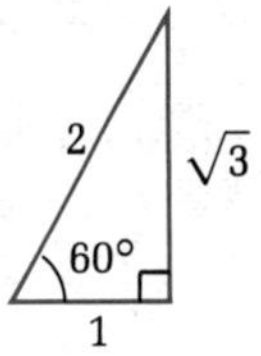

By rotating triangle POR in Fig. 4-17 to the position shown in Fig. 4-18, we can calculate the function values for an angle of 30°.

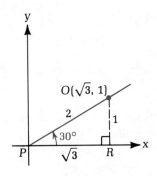

Figure 4-18

Special Angle: 30°

$$\sin 30° = \frac{1}{2} = 0.5$$

$$\cos 30° = \frac{\sqrt{3}}{2} \approx 0.866$$

$$\tan 30° = \frac{1}{\sqrt{3}} = \frac{\sqrt{3}}{3} \approx 0.577$$

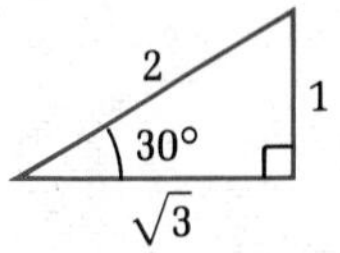

Functions of 0° and 90°

By observing the small angle θ in Fig. 4-19, we can approximate the basic trigonometric functions for 0°.

Since y is much smaller than r, the sine ratio, y/r, is also very small, and it becomes zero for $\theta = 0°$.

$$\sin \theta = \frac{y}{r}$$

$$\boxed{\sin 0° = 0}$$

Lengths x and r are approximately the same for small angles, and the ratio x/r becomes 1 for $\theta = 0°$.

$$\cos \theta = \frac{x}{r}$$

$$\boxed{\cos 0° = 1}$$

Since y is much smaller than x, the ratio y/x becomes zero for $\theta = 0°$.

$$\tan \theta = \frac{y}{x}$$

$$\boxed{\tan 0° = 0}$$

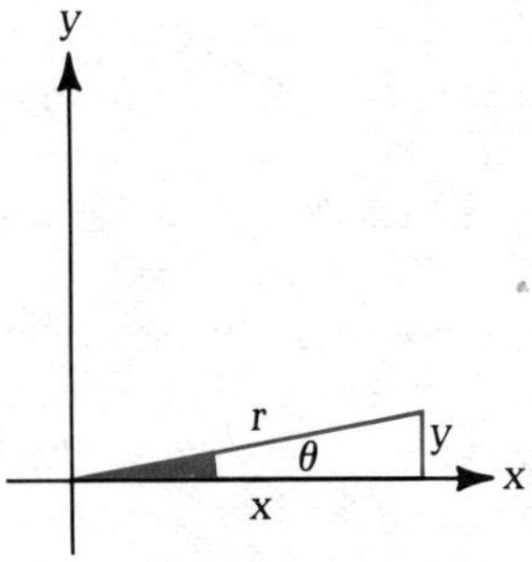

Figure 4-19

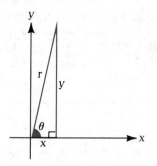

Figure 4-20

By observing the very large acute angle in Fig. 4-20, we can determine the basic trigonometric ratios for an angle of 90°.

The length of y is approximately equal to r, and the ratio y/r becomes 1 for $\theta = 90°$.

$$\sin \theta = \frac{y}{r}$$

$$\boxed{\sin 90° = 1}$$

Length x is much smaller than r, and the ratio x/r becomes zero for $\theta = 90°$.

$$\cos \theta = \frac{x}{r}$$

$$\boxed{\cos 90° = 0}$$

Since x is much smaller than y, the ratio y/x becomes equivalent to a number divided by zero for $\theta = 90°$. Thus, tan 90° is not defined.

$$\tan \theta = \frac{y}{x}$$

$$\boxed{\tan 90° \text{ is not defined}}$$

By a similar procedure we can determine the values of the reciprocal functions for 0° and 90° as follows:

θ	$\sec \theta$	$\csc \theta$	$\cot \theta$
0°	1	Undefined	Undefined
90°	Undefined	1	0

The special angle values of the basic trigonometric functions are summarized in Table 4-1. Because these special angles often arise in applied problems and in more advanced mathematics, their function values should be memorized.

TABLE 4-1
Basic Trigonometric Functions for Special Angles

	0°	30°	45°	60°	90°
$\sin \theta$	0	$\frac{1}{2} = 0.500$	$\frac{\sqrt{2}}{2} \approx 0.707$	$\frac{\sqrt{3}}{2} \approx 0.866$	1
$\cos \theta$	1	$\frac{\sqrt{3}}{2} \approx 0.866$	$\frac{\sqrt{2}}{2} \approx 0.707$	$\frac{1}{2} = 0.500$	0
$\tan \theta$	0	$\frac{\sqrt{3}}{3} \approx 0.577$	1	$\sqrt{3} \approx 1.732$	Undefined

These function values can also be helpful in checking a result obtained from a calculator. For example, suppose you unknowingly pressed the wrong button on your calculator and obtained a value of 0.9396926 for sin 20°. By referring to the special angle values, you can catch your mistake. Since sin 0° is 0 and sin 30° is 0.500, sin 20° is between 0 and 0.500—it cannot be 0.9396926.

NOTE ▶ When you use the special angle values to check answers, you can assume that the values of sin θ and tan θ continually increase between 0° and 90°, and that the values of cos θ continually decrease. ◀

EXAMPLE 14 According to the special angle values, which of the following would be reasonable values for the indicated functions?
(a)   cos 56° = 0.559   (b)   tan 75° = 0.965
(c)   sin 36° = 0.81   (d)   cos 13° = 1.37
(e)   tan 50° = 1.2   (f)   sin 68° = 0.9272

Solution The values given in parts (a), (e), and (f) would be reasonable.

Using a calculator   The geometric methods just illustrated cannot be used for all angles, and more advanced mathematics is needed to derive the values of the trigonometric functions for most other angles. However, we can determine these values from a scientific calculator.

To find the value of sin θ, cos θ, or tan θ on a calculator for an acute angle θ given in degrees, first make certain your calculator is in the degree mode. Most calculators have a [DRG] key for this purpose. (See Appendix A or your calculator instruction manual.) Then enter the value of θ, and press the correct function key. To avoid errors, use the special angle values to check your answer.

EXAMPLE 15 Find each of the following rounded to three decimal places:
(a)   sin 75°   (c)   cos 35.8°   (c)   tan 68.25°

Solution
(a)   75 [sin] →   *0.9659258*   sin 75° ≈ 0.966
The special angle values confirm that sin 75° is between 0.866 and 1.
(b)   35.8 [cos] →   *0.8110638*   cos 35.8° ≈ 0.811
The special angle values confirm that cos 35.8° is between 0.707 and 0.866.
(c)   68.25 [tan] →   *2.5065198*   tan 68.25° ≈ 2.507
The special angle values confirm that tan 68.25° is greater than 1.732.

CAUTION  On most calculators, entering 90 [tan] produces an error message. We have already determined that tan 90° is undefined, and the error display correctly conveys this fact. ◀

Angles in degrees and minutes   If an angle is given in degrees and minutes, convert it to decimal form, and then find the desired function as before.

EXAMPLE 16 Find each of the following rounded to three decimal places:
(a)   sin 48°20'   (b)   cos 81°52'   (c)   tan 26°8'

Solution
(a)   48 [+] 20 [÷] 60 [=] [sin] →   *0.7470251*   sin 48°20' ≈ 0.747
(Notice that the calculator display shows   *48.333333*   when the [=] key is pressed. Recall from Section 4-1 that this is the decimal equivalent of the angle: 48°20' = 48$\frac{20°}{60}$ = 48.333. . .°.)
(b)   81 [+] 52 [÷] 60 [=] [cos] →   *0.1414772*   cos 81°52' ≈ 0.141
(c)   26 [+] 8 [÷] 60 [=] [tan] →   *0.4906166*   tan 26°8' ≈ 0.491

If your calculator has a key for converting directly to decimal degrees, use it instead of the sequence shown in Example 16.

Finding reciprocal functions   Calculators do not have keys for the cosecant, secant, and cotangent functions. To find values of these functions on a calculator, recall that they are the reciprocals of the three basic functions, sine, cosine, and tangent. Enter the angle, followed by the corresponding basic function, and then press the $\boxed{1/x}$ key to obtain the reciprocal.

 EXAMPLE 17   Find each of the following rounded to three decimal places:
(a)   cot 28°   (b)   sec 71.4°   (c)   csc 17°15'

Solution

(a)   Since the cotangent function is the reciprocal of the tangent function, first write

$$\cot 28° = \frac{1}{\tan 28°}$$

Then enter the expression on the right into your calculator:

$$28 \boxed{\tan} \boxed{1/x} \rightarrow \boxed{1.8807265} \qquad \cot 28° \approx 1.881$$

(b)   Note that

$$\sec 71.4° = \frac{1}{\cos 71.4°}$$

Then

$$71.4 \boxed{\cos} \boxed{1/x} \rightarrow \boxed{3.1351961} \qquad \sec 71.4° \approx 3.135$$

(c)   Note that

$$\csc 17°15' = \frac{1}{\sin 17°15'}$$

Then

$$17 \boxed{+} 15 \boxed{\div} 60 \boxed{=} \sin \boxed{1/x} \rightarrow \boxed{3.3722084} \qquad \csc 17°15' \approx 3.372$$

CAUTION ▶   Be careful when using this procedure for the quadrantal angles. Since sin 0° = 0 and cos 90° = 0, your calculator will display an ▨▨▨ *Error* message when you press the $\boxed{1/x}$ key to find csc 0° and sec 90°. This correctly conveys the fact that csc 0° and sec 90° are undefined. However, when you try to find cot 90°, which is equal to 0, your calculator will *erroneously* display an error message. This is because 90 $\boxed{\tan}$ gives an error message and pressing $\boxed{1/x}$ will not change the display. In general, it is wiser to memorize the function values of the quadrantal angles than to find them with a calculator. ◀

Determining the angle from the function value   For many applications of trigonometry, you will know the value of the function, and you will need to find the angle associated with it. This reverse procedure is called *finding the inverse of a trigonometric function*. With a calculator we can find the angle when the value of the trigonometric function is known by using the (INV) or (2nd/F) key. To determine an unknown angle, enter the given function value and then press the (INV) key, followed by the function key.

EXAMPLE 18 Find θ to the nearest tenth of a degree.
(a) $\sin \theta = 0.64*$ (b) $\cos \theta = 0.2$ (c) $\tan \theta = 1.156$

Solution
(a) .64 (INV)(sin) →   *39.791820* $\theta \approx 39.8°$
(b) .2 (INV)(cos) →   *78.463041* $\theta \approx 78.5°$
(c) 1.156 (INV)(tan) →   *49.138494* $\theta \approx 49.1°$

Finding angles in degrees and minutes If an angle must be written in degrees and minutes instead of decimal degrees, subtract the whole number portion of the displayed answer and multiply the result by 60. Be sure to write down the whole number portion before subtracting or you will lose that part of your answer.

EXAMPLE 19 Find θ to the nearest minute given $\cos \theta = 0.673$.

Solution **First,** find the angle in decimal degrees:

.673 (INV)(cos) →   *47.700969*   Write down "47°."

Then, convert the decimal portion to minutes:

(−) 47 (=)(×) 60 (=) →   *42.058156*   or 42', rounded

$\theta \approx 47°42'$

Reciprocal functions To determine the value of θ given $\csc \theta$, $\sec \theta$, or $\cot \theta$ in the first quadrant, first find the corresponding value of its reciprocal function using the (1/x) key, and then use the (INV) or (2nd/F) key as before.

EXAMPLE 20
(a) Find θ to the nearest tenth of a degree given $\sec \theta = 1.34$.
(b) Find θ such that $\cot \theta = 0.468$. Round to the nearest minute.

Solution
(a) **First,** since

$$\cos \theta = \frac{1}{\sec \theta} = \frac{1}{1.34}$$

---

*Sometimes the angle θ is indicated by the expressions "arcsin 0.64" or "$\sin^{-1} 0.64$." However, we shall reserve this notation until Chapter 19, when we discuss the inverse of a trigonometric function as a function itself.

find the corresponding value of the cosine function:

$$1.34 \; \boxed{1/x} \rightarrow \boxed{0.7462687} \; \Longleftarrow \; \cos\theta$$

Then, determine the inverse of the cosine:

$$\boxed{\text{INV}} \; \boxed{\text{COS}} \rightarrow \boxed{41.731817} \qquad \theta \approx 41.7°$$

(b) Note that

$$\tan\theta = \frac{1}{\cot\theta} = \frac{1}{0.468}$$

Then,

$$.468 \; \boxed{1/x} \; \boxed{\text{INV}} \; \boxed{\text{tan}} \rightarrow \boxed{64.920406}$$

Write down "64°" and then convert the decimal part to minutes:

$$\boxed{-} \; 64 \; \boxed{=} \; \boxed{\times} \; 60 \; \boxed{=} \rightarrow \boxed{55.224355} \qquad \theta \approx 64°55'$$

Exercises 4-3

Find the value of each trigonometric function rounded to three decimal places.

1. sin 64°	**2.** cos 27°	**3.** tan 39°	**4.** sin 31°
5. cos 58°	**6.** tan 76°	**7.** sec 71°	**8.** csc 23°
9. cot 44°	**10.** sin 12.5°	**11.** cos 67.3°	**12.** tan 41.45°
13. sin 69.25°	**14.** cos 51.75°	**15.** tan 18.375°	**16.** sec 24.76°
17. csc 65.88°	**18.** cot 76.233°	**19.** cos 8.465°	**20.** tan 34.9°
21. sin 80.6°	**22.** cos 36°35'	**23.** tan 67°50'	**24.** sin 28°25'
25. csc 48°22'	**26.** sec 5°12'	**27.** cot 27°19'	**28.** tan 58°10'
29. sin 49°27'	**30.** cos 63°14'	**31.** tan 9°9'	**32.** sin 50°58'

Find θ to the nearest tenth of a degree.

33. sin θ = 0.86	**34.** cos θ = 0.76	**35.** tan θ = 0.7	**36.** sin θ = 0.3
37. cos θ = 0.6	**38.** tan θ = 1.756	**39.** sin θ = 0.568	**40.** cos θ = 0.2659
41. tan θ = 0.6712	**42.** cot θ = 2.1	**43.** sec θ = 5.3	**44.** csc θ = 1.54
45. cos θ = 0.43	**46.** tan θ = 1.992	**47.** cot θ = 0.8523	**48.** sec θ = 1.2457

Find θ to the nearest minute.

49. sin θ = 0.435	**50.** cos θ = 0.767	**51.** tan θ = 2.785	**52.** sin θ = 0.219
53. cos θ = 0.384	**54.** tan θ = 1.298	**55.** sec θ = 6.486	**56.** csc θ = 3.864
57. cot θ = 2.365	**58.** sin θ = 0.934	**59.** cos θ = 0.551	**60.** tan θ = 4.126

Solve.

61. *Surveying* To find the distance from one point to another, a surveyor uses the formula

$$D = \frac{L_2 - L_1}{\cos\theta}$$

where L_1 and L_2 are the latitudinal coordinates of the two points, and θ is the angle of departure. Find D for L_2 = 5378.0 ft, L_1 = 5144.3 ft, and θ = 21°46'.

62. *Civil engineering* In constructing roadways, calculations involving horizontal curves are often required. In the figure shown here, O is the center of the curve from B to E, R is the radius of the curve, T is the tangent length, and θ is the central angle. These quantities are interrelated by the formula below. Find R when $T = 240.0$ ft and $\theta = 80°0'$.*

$$R = T \cot \frac{\theta}{2}$$

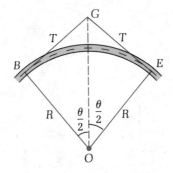

63. *Electronics* The inductive impedance of a coil (in ohms) is given by $X_L = Z \sin \theta$, where Z is the impedance and θ is the phase angle. Find X_L for $Z = 900.0 \ \Omega$ and $\theta = 65°0'$.

64. *Air traffic control* In order to track an incoming airplane on radar, the ground distance between the tracking station and the aircraft must be found. The first step in finding this distance is to find θ, the earth's central angle between the two points (see the figure). This is given by

$$\cos \theta = \frac{R}{R + H}$$

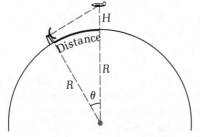

where R is the radius of the earth and H is the height of the aircraft. Find θ to the nearest tenth of a degree if R is 3440 nautical miles and H is 2.00 nautical miles.

4-4 | The Right Triangle

Many technological problems can be solved by finding the missing parts of a right triangle. In this section we will discuss how to solve such problems using the trigonometric functions.

*Adapted from Jim McCoy, *Region 5 Constructor's and Inspector's Self-Study Courses* (U.S. Forest Service, 1969).

Geometry of the right triangle   One geometric fact that will be important in solving right triangles is that *the sum of the angles in any triangle is 180°*. Since a right triangle contains one 90° angle, the other two angles must be acute angles whose sum is 90°. Any two acute angles whose sum is 90° are called **complementary angles** (Fig. 4-21).

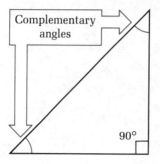

Figure 4-21

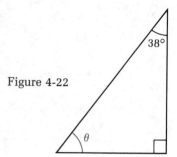

Figure 4-22

EXAMPLE 21   Find the missing angle θ in the right triangle shown in Fig. 4-22.

Solution   Since the two acute angles of a right triangle are complementary, we have

$$\theta + 38° = 90°$$

$$\theta = 90° - 38° = 52°$$

Redefining the trigonometric functions   In Section 4-2 we defined the trigonometric functions for an angle θ in standard position in the rectangular coordinate system. The functions were defined as ratios among the x- and y-coordinates of a point on the terminal side of the angle and r, the distance from the origin to that point. As shown in Fig. 4-23, x, y, and r are the three sides of a right triangle. The abscissa x and the ordinate y are the legs, r is the hypotenuse, and θ is one of the acute angles of the right triangle.

We can now redefine the trigonometric functions for a right triangle existing outside the rectangular coordinate system. As shown in Fig. 4-24, the angles of a right triangle are traditionally labeled A, B, and C, with C the right angle. The sides are labeled a, b, and c, with lowercase letters. Side c is the **hypotenuse.** With respect to angle A, side a is called the side **opposite** angle A, and side b is called the side **adjacent** to angle A. We can use this new vocabulary to define the trigonometric functions as follows:

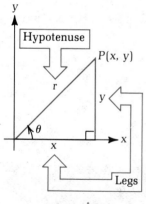

Figure 4-23

$$\sin A = \frac{\text{Opposite side}}{\text{Hypotenuse}} = \frac{a}{c} \qquad \csc A = \frac{\text{Hypotenuse}}{\text{Opposite side}} = \frac{c}{a}$$

$$\cos A = \frac{\text{Adjacent side}}{\text{Hypotenuse}} = \frac{b}{c} \qquad \sec A = \frac{\text{Hypotenuse}}{\text{Adjacent side}} = \frac{c}{b}$$

$$\tan A = \frac{\text{Opposite side}}{\text{Adjacent side}} = \frac{a}{b} \qquad \cot A = \frac{\text{Adjacent side}}{\text{Opposite side}} = \frac{b}{a}$$

We can also use this new vocabulary to define the six functions of angle B without redrawing the triangle to place B in standard position (Fig. 4-25). For

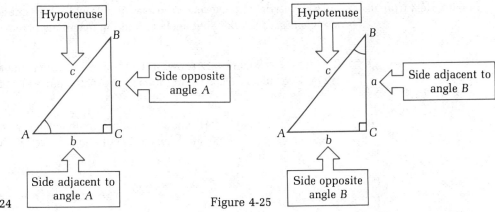

Figure 4-24

Figure 4-25

angle B, the hypotenuse remains the same, but b becomes the opposite side and a becomes the adjacent side. We can therefore list the trigonometric functions of angle B as follows:

$$\sin B = \frac{\text{Opposite side}}{\text{Hypotenuse}} = \frac{b}{c} \qquad \csc B = \frac{\text{Hypotenuse}}{\text{Opposite side}} = \frac{c}{b}$$

$$\cos B = \frac{\text{Adjacent side}}{\text{Hypotenuse}} = \frac{a}{c} \qquad \sec B = \frac{\text{Hypotenuse}}{\text{Adjacent side}} = \frac{c}{a}$$

$$\tan B = \frac{\text{Opposite side}}{\text{Adjacent side}} = \frac{b}{a} \qquad \cot B = \frac{\text{Adjacent side}}{\text{Opposite side}} = \frac{a}{b}$$

EXAMPLE 22 For the right triangle shown in Fig. 4-26, state the ratios for $\sin A$, $\cos A$, $\tan A$, $\sin B$, $\cos B$, and $\tan B$. Calculate each value to the nearest hundredth.

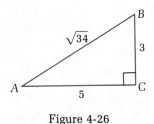

Figure 4-26

Solution $\quad \sin A = \dfrac{\text{Side opposite angle } A}{\text{Hypotenuse}} = \dfrac{3}{\sqrt{34}}$

$3 \boxed{\div} 34 \boxed{\sqrt{}} \boxed{=} \rightarrow \quad \boxed{0.5144958} \quad$ or 0.51, rounded

$\cos A = \dfrac{\text{Side adjacent to angle } A}{\text{Hypotenuse}} = \dfrac{5}{\sqrt{34}}$

$5 \boxed{\div} 34 \boxed{\sqrt{}} \boxed{=} \rightarrow \quad \boxed{0.8574929} \quad$ or 0.86, rounded

$\tan A = \dfrac{\text{Side opposite angle } A}{\text{Side adjacent to angle } A} = \dfrac{3}{5} = 0.60$

$\sin B = \dfrac{\text{Side opposite angle } B}{\text{Hypotenuse}} = \dfrac{5}{\sqrt{34}} \approx 0.86$

$\cos B = \dfrac{\text{Side adjacent to angle } B}{\text{Hypotenuse}} = \dfrac{3}{\sqrt{34}} \approx 0.51$

$\tan B = \dfrac{\text{Side opposite angle } B}{\text{Side adjacent to angle } B} = \dfrac{5}{3} \approx 1.67$

Solving right triangles   To solve a right triangle means to use one side and at least one additional side or angle other than the right angle to find the remaining parts.

▦ **EXAMPLE 23** Figure 4-27 shows a right triangle in which angle $A = 41°$ and its opposite side $a = 5.6$. To find the missing parts, proceed as follows:

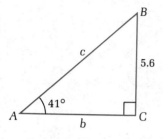

Figure 4-27

- **Angle B:** Since $A = 41°$ and A and B are complementary angles, $B = 90° - 41° = 49°$. (Check this answer by making certain all three angles sum to 180°.)

- **Side b:**

  Step 1. Form a trigonometric ratio linking the given angle and the given side with the unknown side.

  Step 2. Solve the resulting equation for b.
  Multiply by b.

  Divide by $\tan A$.

  Step 3. Substitute the known values.

  Step 4. Calculate b.

  > Given side
  >
  > $\tan A = \dfrac{a}{b}$
  >
  > Given angle | Unknown side

  $b \tan A = a$

  $b = \dfrac{a}{\tan A}$

  $b = \dfrac{5.6}{\tan 41°}$

  $5.6 \;[\div]\; 41 \;[\tan]\;[=] \rightarrow \boxed{6.4420631}$

  $b \approx 6.4$

- **Side c:** Once two sides are known, it is easiest to use the Pythagorean theorem to find the third side. Thus,

$$c = \sqrt{(5.6)^2 + (6.4)^2} \approx 8.5$$

If you do not know b yet, find c as follows:

Step 1. Use c, the given side, and the given angle to form the sine ratio.

$\sin A = \dfrac{a}{c}$

Step 2. Solve for c.

$c \sin A = a$

$c = \dfrac{a}{\sin A}$

Step 3. Substitute.

$c = \dfrac{5.6}{\sin 41°}$

Step 4. Calculate.

$5.6 \;[\div]\; 41 \;[\sin]\;[=] \rightarrow \boxed{8.5358173}$

$c \approx 8.5$

Notice that this answer agrees with the result obtained from the Pythagorean theorem. Doing it both ways serves as a way of checking the answer.

When two sides and no angles are given, use the [INV] key to determine an unknown angle and use the Pythagorean theorem to find the third side.

■ EXAMPLE 24   Solve the right triangle shown in Fig. 4-28.

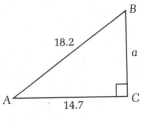

Figure 4-28

Solution

● To find angle A:

Step 1. Form a trigonometric ratio linking angle A with both of the known sides.

$$\cos A = \frac{b}{c} = \frac{14.7}{18.2}$$

Step 2. Find A.

14.7 $\div$ 18.2 $=$ INV COS → ║ *36.128928*
$$A \approx 36.1°$$

● To find angle B: Knowing $A \approx 36.1°$, we can find B by subtracting: $B \approx 90° - 36.1° \approx 53.9°$. (To solve for B without first finding A, use $\sin B = 14.7/18.2$ and proceed as before.)

● To find side a: Use the Pythagorean theorem.

$$a = \sqrt{(18.2)^2 - (14.7)^2} \approx 10.7$$

Applications   To solve applied problems, you will need to draw a right triangle from the given information. Usually, only one of the missing parts will be needed. The remainder of this section is devoted to examples of applied problems.

NOTE ▶   In applied problems, angles and side lengths are considered to be measured quantities and therefore approximate numbers. See Appendix B for information on rounding with the trigonometric functions. ◀

■ EXAMPLE 25   In a certain housing tract, solar collectors on the south-facing walls of the homes are protected from the shade by a "solar easement." That is, a local ordinance prohibits homeowners from building or planting anything that would shade any part of a neighbor's solar collectors when the sun is at an angle of 18° or greater. What is the tallest structure a homeowner could build 25 ft directly south of a neighbor's ground-level solar collector?*

Solution   First, make a drawing of the situation, as shown in Fig. 4-29.

Figure 4-29

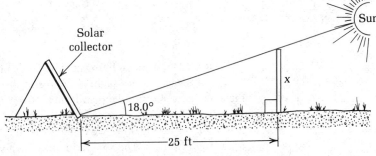

* San Diego Association of Governments, *Regional Energy Plan Implementation Program, Protecting Solar Access, Implementation Package #3* (Oct. 1981), pp. 19–20.

Then, set up a trigonometric ratio and solve for the unknown.

$$\tan 18° = \frac{x}{25}$$

$$x = 25 \tan 18°$$

$$25 \;\boxed{\times}\; 18 \;\boxed{\text{tan}}\; \boxed{=} \rightarrow \quad \boxed{8.1229924}$$

$$x \approx 8.1 \text{ ft}$$

Angles of elevation and depression

Many applied problems involve special angles known as the *angle of elevation* and the *angle of depression*.

- An **angle of elevation** is an angle formed by a line of sight above a horizontal line, as shown in Fig. 4-30(a).
- An **angle of depression** is an angle formed by a line of sight below a horizontal line, as shown in Fig. 4-30(b).

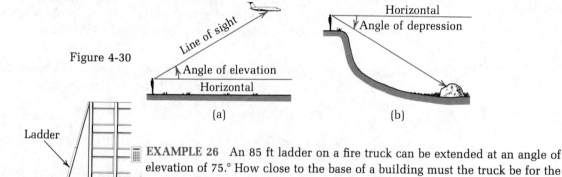

Figure 4-30

(a)   (b)

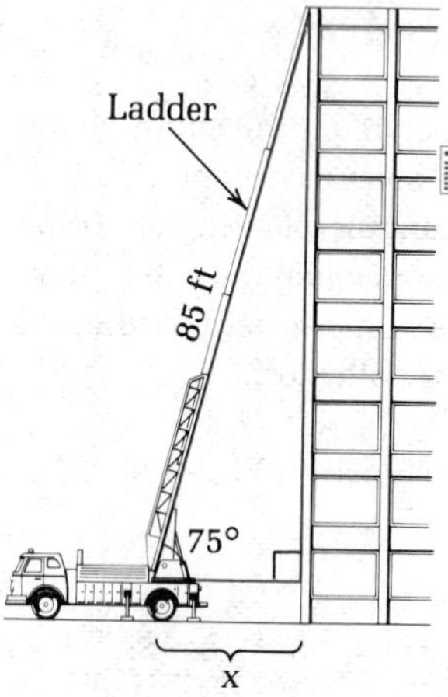

EXAMPLE 26 An 85 ft ladder on a fire truck can be extended at an angle of elevation of 75.° How close to the base of a building must the truck be for the end of the ladder to reach the building?

Solution Figure 4-31 shows the given information. From the triangle, we have

$$\cos 75° = \frac{x}{85}$$

$$x = 85 \cos 75°$$

$$85 \;\boxed{\times}\; 75 \;\boxed{\text{cos}}\; \boxed{=} \rightarrow \quad \boxed{21.999619}$$

$$x \approx 22 \text{ ft}$$

Figure 4-31

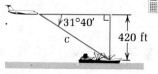

EXAMPLE 27 A plane flying at an altitude of 420 ft spots a ship in the water. The pilot measures the angle of depression to be 31°40′. Find the straight-line distance between the plane and the ship.

Solution From Fig. 4-32 we can see that the hypotenuse is the unknown part of the right triangle. The sine ratio links the given angle, the given side, and the hypotenuse. (Careful! The angle is not in standard position.)

Figure 4-32

$$\sin 31°40' = \frac{\text{Opposite side}}{\text{Hypotenuse}} = \frac{420}{c}$$

Solving for c, we have

$$c \sin 31°40' = 420$$

$$c = \frac{420}{\sin 31°40'}$$

$420 \boxed{\div} \boxed{(} 31 \boxed{+} 40 \boxed{\div} 60 \boxed{)} \boxed{\sin} \boxed{=} \rightarrow$ | 800.03569 |

$$c \approx 800 \text{ ft}$$

EXAMPLE 28   A concrete ramp must be built from a ledge 3.8 m off the ground to a point on the ground 9.1 m from the base of the ledge. At what angle will the ramp rise from the ground?

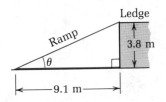

Figure 4-33

Solution   From Fig. 4-33 we see that the angle θ is the angle we want to find. We choose a trigonometric function that relates the unknown angle and the given sides:

$$\tan \theta = \frac{\text{Opposite side}}{\text{Adjacent side}} = \frac{3.8}{9.1}$$

We need to find the angle whose tangent is the given quotient:

$3.8 \boxed{\div} 9.1 \boxed{=} \boxed{\text{INV}} \boxed{\tan} \rightarrow$ | 22.664557 |

$$\theta \approx 23°$$

EXAMPLE 29   Find the head angle α of the screw shown in Fig. 4-34.

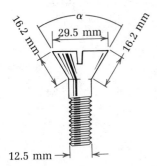

Figure 4-34

Solution   Redraw the screw head diagram as shown in Fig. 4-35.

Solve one of the triangles for $\alpha/2$.

$$\sin\left(\frac{\alpha}{2}\right) = \frac{8.5}{16.2}$$

$8.5 \boxed{\div} 16.2 \boxed{=} \boxed{\text{INV}} \boxed{\sin} \rightarrow$ | 31.647468 | $\Leftarrow \alpha/2$

Double $\alpha/2$ to get α.

$\boxed{\times} 2 \boxed{=} \rightarrow$ | 63.294936 |

$$\alpha \approx 63°$$

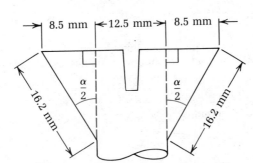

Figure 4-35

Exercises 4-4

Use the two given parts of triangle ABC with C = 90° (see the figure) to find the other three parts. Express angles to the nearest tenth of a degree in Problems 1–8 and to the nearest minute in Problems 9–16.

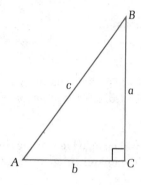

1. $B = 52.0°$, $a = 8.00$
2. $A = 24.0°$, $c = 6.00$
3. $a = 15.0$, $b = 11.0$
4. $b = 9.00$, $c = 16.0$
5. $B = 44.2°$, $b = 14.7$
6. $B = 6.4°$, $b = 22.6$
7. $c = 27.5$, $a = 18.8$
8. $a = 138$, $b = 254$
9. $A = 19°0'$, $b = 20.00$
10. $A = 63°0'$, $c = 14.00$
11. $b = 80.25$, $c = 90.75$
12. $a = 12.50$, $c = 22.50$
13. $B = 71°2'$, $c = 351.0$
14. $A = 38°7'$, $a = 5.625$
15. $a = 8.350$, $b = 2.620$
16. $b = 24.60$, $c = 33.80$

Solve.

17. *Air traffic control*   One way of measuring the aircraft visibility ceiling height h at an airport is to focus a light beam straight up at the clouds at a known distance x from an observation point (see the figure). The angle θ can then be measured. Find the ceiling height if $x = 710$ ft and $\theta = 30°$.

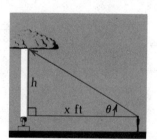

18. *Photography*   It is often useful to calculate the angle of view of a camera lens (angle α in the figure). Given the measurements shown in the figure, find α to the nearest 10'. [*Hint:*   First find $\alpha/2$.]

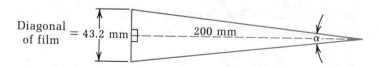

Diagonal of film = 43.2 mm   200 mm α

19. *Automotive engineering* When mounted vertically on the rear axle, the shocks on a motorcycle allow a 6.0 in. travel in the springs (see the figure). If the same shocks are mounted at an angle of 30° with the vertical, what is the vertical travel of the springs?

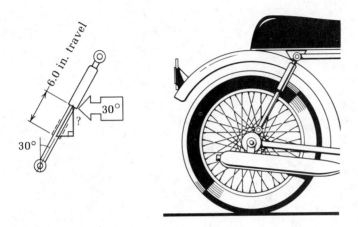

20. *Architecture* A sloped cathedral ceiling connects a wall 7 ft high with a wall 10 ft high. If the distance between the walls is 12 ft, at what angle (to the nearest degree) does the ceiling rise?

21. *Machine technology* A machinist must use a boring mill to find the holes for a bolt circle. Because the boring mill allows only "square moves," the distances x and y indicated in the figure must be found. Use the measurements shown to calculate x and y.

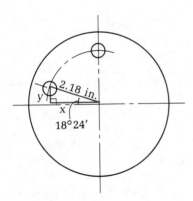

22. *Navigation* An aircraft observer flying at an altitude of 750 ft locates a small island at a 29° angle of depression. What is the horizontal distance between the plane and the island?

23. *Construction* A tunnel must be dug to the bottom of a 250 ft well from a point 40 ft away from and level with the top of the well. At what angle with respect to horizontal (to the nearest degree) must the tunnel be slanted?

24. *Meteorology*   A weather balloon rises vertically at a rate of 6.5 mi/h while moving at a 65° angle with the ground. What will be its horizontal travel after 1 h (see the figure)?

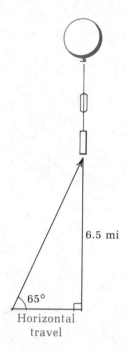

6.5 mi

65°

Horizontal
travel

25. *Forestry*   Using a clinometer and tape measure, a forest ranger locates the top of a tree at a 54° angle of elevation from a point 18 ft from the base of the tree. How tall is the tree?

26. *Machine technology*   Find the head angle α (to the nearest 0.1°) of the screw shown in the figure.

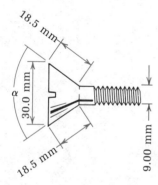

27. *Cartography*   A map maker needs to change the curve from A to C to B (see the figure) to one line from A to B. Given the measurements in the figure, find the angle θ to the nearest degree. [Hint: $\theta = 180° - \alpha$]

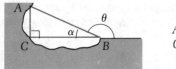

$AC = 0.4$ cm
$CB = 1.0$ cm

28. *Fire science* A hook and ladder 120 ft long is extended at a 55° angle. What vertical distance can the top of the ladder reach?

29. If a pole 40.0 ft high casts a 25.0 ft shadow, what is the angle of elevation of the sun to the nearest 10'?

30. *Drafting* A computer printer is designed with a side view as shown in the figure. Find the indicated angle β to the nearest 0.1°.

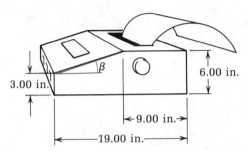

31. *Construction* A 12 ft high sign on a flat roof is held in place by a guy wire attached at an angle of 30° to the top of the sign. If the base of the sign rests on the roof, how much guy wire is needed to make the attachment?

32. *Civil engineering* A road rises at an average angle of 12.0°. What is the change in altitude for a distance along the road of 20.0 mi?

33. *Physics* A pendulum is 18.0 cm long, and the horizontal distance between its extreme positions is 4.8 cm. Through what angle (to the nearest degree) does the pendulum swing?

34. *Landscape architecture* A lot is graded at an angle of 31°0' to form an embankment, as shown in the figure. If the vertical distance between the two levels is 12.5 ft, what width x of plastic is needed to cover the slope to prevent erosion?

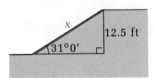

35. *Machine technology* Find the width w of the V-shaped slot shown in the figure.

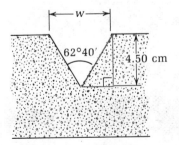

36. *Urban planning* Under the solar easement ordinance described in Example 25, what is the tallest structure a homeowner could build 37 ft directly south of a neighbor's ground level solar collector?

37. *Fire science* An 83 ft ladder on a fire truck can be extended at an angle of 65°. How close to the base of a building must the truck be for the end of the ladder to reach the building?

38. *Agricultural technology* The angle of a grain conveyor 8.5 m long can be set between 12° and 60°. What are the minimum and maximum heights the end of the conveyor can reach?*

* Based on MAA/NCTM Joint Committee, *A Sourcebook of Applications* (NCTM, 1980), p. 210.

39. As shown in the figure, the angle of elevation between a ground observer and a second story window is 47°0′. The angle of elevation between the observer and the third story window is 61°0′. If the observer is 40.0 ft from the building, what is the distance between the second and third stories?

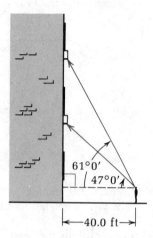

←—40.0 ft—→

40. *Navigation*   As shown in the figure, a helicopter pilot notes an angle of depression of 27° to the top of a tall building. At the same time, the angle of depression to the base of the building is 64°. If the pilot is flying over a spot at a horizontal distance 85 ft from the building, how tall is the building?

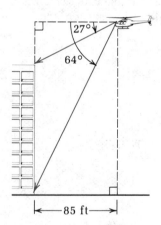

←— 85 ft —→

4-5 | Review of Chapter 4

Important Terms and Concepts

angle (p. 132)	minute (p. 133)	coterminal angles (p. 134)
initial side (p. 132)	second (p. 133)	standard position (p. 135)
terminal side (p. 132)	straight angle (p. 133)	quadrantal angle (p. 135)
vertex (p. 132)	right angle (p. 133)	complementary angles (p. 150)

positive angle (p. 132)   acute angle (p. 133)   angle of elevation (p. 154)

negative angle (p. 132)   obtuse angle (p. 133)   angle of depression (p. 154)

degree (p. 133)

Definitions of the Trigonometric Functions

For an angle θ in standard position (see p. 138):

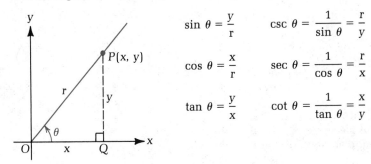

$$\sin \theta = \frac{y}{r} \qquad \csc \theta = \frac{1}{\sin \theta} = \frac{r}{y}$$

$$\cos \theta = \frac{x}{r} \qquad \sec \theta = \frac{1}{\cos \theta} = \frac{r}{x} \qquad (\textbf{4-1})$$

$$\tan \theta = \frac{y}{x} \qquad \cot \theta = \frac{1}{\tan \theta} = \frac{x}{y}$$

For a right triangle as shown in the figure (also see p. 150):

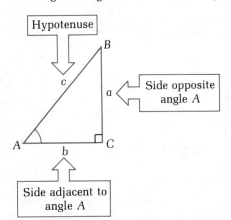

$$\sin A = \frac{\text{Opposite side}}{\text{Hypotenuse}} = \frac{a}{c} \qquad \csc A = \frac{\text{Hypotenuse}}{\text{Opposite side}} = \frac{c}{a}$$

$$\cos A = \frac{\text{Adjacent side}}{\text{Hypotenuse}} = \frac{b}{c} \qquad \sec A = \frac{\text{Hypotenuse}}{\text{Adjacent side}} = \frac{c}{b}$$

$$\tan A = \frac{\text{Opposite side}}{\text{Adjacent side}} = \frac{a}{b} \qquad \cot A = \frac{\text{Adjacent side}}{\text{Opposite side}} = \frac{b}{a}$$

Special Angle Values: see Table 4-1 (p. 144)

Exercises 4-5

Find one positive and one negative angle coterminal with each given angle.

1. 40°   **2.** 135°   **3.** 224°18′   **4.** 312.7°

In Problems 5–8 change each given angle to decimal form (rounded to the nearest hundredth), and in Problems 9–12 change each given angle to degrees and minutes (rounded to the nearest minute).

5. 24°24′ **6.** 172°15′ **7.** 256°8′ **8.** 333°53′
9. 94.5° **10.** 61.2° **11.** 285.775° **12.** 195.73°

Each of the following is a point on the terminal side of an angle in standard position. Determine the values of all six trigonometric functions for each angle. If necessary, express answers in decimal form to the nearest thousandth.

13. (8, 6) **14.** (2.5, 6) **15.** (3, 6) **16.** (10, 2)

Use the given information to find the specified functions.

17. If $\cos \theta = \frac{7}{25}$, find the exact value of: (a) $\sin \theta$ (b) $\cot \theta$
18. If $\sin \theta = \frac{1}{8}$, find the exact value of: (a) $\tan \theta$ (b) $\csc \theta$
19. If $\tan \theta = 3$, find the exact value of: (a) $\cot \theta$ (b) $\cos \theta$
20. If $\cot \theta = 1.6$, find (to two decimal places): (a) $\tan \theta$ (b) $\cos \theta$
21. If $\sec \theta = 1.04$, find (to three decimal places): (a) $\sin \theta$ (b) $\cos \theta$
22. If $\csc \theta = 2.179$, find (to four decimal places): (a) $\tan \theta$ (b) $\sin \theta$

Find the value of each trigonometric function. Round to three decimal places.

23. $\sin 38°$ **24.** $\cos 70°$ **25.** $\tan 59°$ **26.** $\sec 55°$
27. $\cos 19.8°$ **28.** $\sin 61.4°$ **29.** $\tan 49.78°$ **30.** $\csc 37.375°$
31. $\cos 53°12′$ **32.** $\sin 6°54′$ **33.** $\tan 29°25′$ **34.** $\cot 75°15′$

Find θ to the nearest tenth of a degree.

35. $\sin \theta = 0.4$ **36.** $\cos \theta = 0.82$ **37.** $\tan \theta = 3.231$
38. $\sin \theta = 0.2476$ **39.** $\sec \theta = 3.15$ **40.** $\cot \theta = 1.497$

Find θ to the nearest minute.

41. $\cos \theta = 0.667$ **42.** $\sin \theta = 0.874$ **43.** $\cot \theta = 0.919$
44. $\tan \theta = 4.213$ **45.** $\sin \theta = 0.545$ **46.** $\csc \theta = 1.112$

Use the two given parts of triangle ABC and C = 90° (see the figure) to find the other three parts. Express angles to the nearest tenth of a degree in Problems 47–54 and to the nearest minute in Problems 55–58.

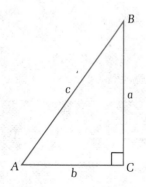

47. $A = 27.0°$, $b = 9.00$	**48.** $A = 61.0°$, $a = 25.0$	**49.** $a = 11.6$, $b = 15.9$
50. $b = 4.50$, $c = 8.10$	**51.** $B = 17.8°$, $b = 1.20$	**52.** $A = 48.5°$, $a = 4.60$
53. $a = 14.3$, $c = 23.7$	**54.** $a = 1170$, $b = 875$	**55.** $A = 48°18'$, $c = 8.870$
56. $B = 31°50'$, $a = 16.40$	**57.** $b = 1.786$, $c = 3.155$	**58.** $a = 29.62$, $c = 62.37$

 Solve.

59. *Physics* The horizontal distance x traveled by a projectile in t seconds at a velocity v is given by

$$x = vt \cos \theta$$

Find the angle θ (to the nearest degree) at which a projectile was launched if its velocity was 35 ft/s, and it had traveled 50 ft horizontally after 2 s.

60. *Physics* When light shines through a diffraction grating, the wavelength λ is given by

$$\lambda = \frac{d \sin \theta}{n}$$

where d is the width of each slit, and θ is the angle at which an nth-order image appears. Find the wavelength of the light that creates a second-order image at 68.4° if the width of a slit is 3.25×10^{-6} m.*

61. *Physics* The vibration amplitude x (in millimeters) of a mass is given by

$$x = 25 \sin (6.37t)$$

where t is the time (in seconds) and 6.37t is the angle (in degrees). Find the amplitude for $t = 0.75$ s.

62. *Physics* The maximum speed (in feet per second) of a pendulum bob is given by

$$v = \sqrt{64L(1 - \cos \theta)}$$

where L is the length of the pendulum and θ is the angle from which it is released. Find v given that a 2.0 ft pendulum is released at an angle of 34°.

63. *Electronics technology* A 14 ft antenna is held in place on a flat roof by a guy wire 21 ft long. At what angle (to the nearest degree) is the wire attached to the top of the antenna?

64. *Machine technology* Find α, the angle of taper (to the nearest 10 minutes), in the steel part shown in the figure.

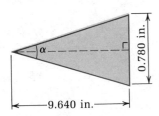

*For more information, see David Halliday and Robert Resnick, *Fundamentals of Physics* (New York: Wiley, 1981), Chap. 41.

65. *Surveying*   The latitudinal and longitudinal coordinate differences between two points are shown in the figure. To find the new angle of departure, the angle θ must be determined. Find θ to the nearest minute.

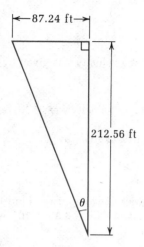

66. *Navigation*   A jet flying with an airspeed of 960 mi/h makes a change in altitude at a 17° angle of elevation. After 1 min, what horizontal distance has it covered?

67. *Navigation*   A helicopter flying at 350 ft spots a small boat in trouble. The angle of depression between the helicopter and the boat is 38°. Find the horizontal distance between the helicopter and the boat.

68. *Construction*   A rafter must be cut at an angle of 26° to join two walls 16 ft apart. What is the difference in height between the two walls?

69. *Meteorology*   A weather balloon climbs to an altitude of 6.0 mi while drifting a horizontal distance of 9.5 mi from its launch spot. At what angle θ (to the nearest degree) has the balloon risen with respect to the ground?

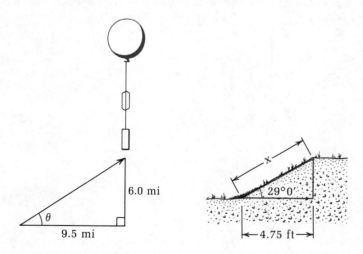

70. *Construction*   A wooden ramp must be cut to span the distance x shown in the figure at the right, above. How long should the ramp be?

Systems of Linear Equations

5

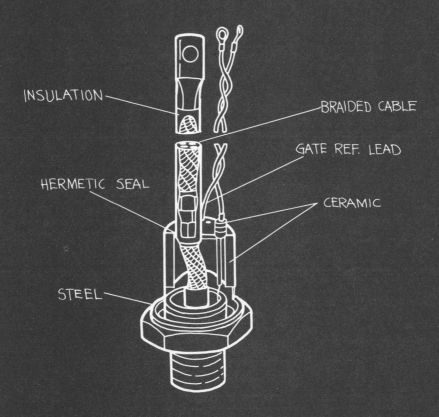

INSULATION

BRAIDED CABLE

GATE REF. LEAD

HERMETIC SEAL

CERAMIC

STEEL

SEMICONDUCTOR CONTROLLED RECTIFIER

As we saw in Chapter 2, some technical problems can be described by a single linear equation in one variable. However, many technical situations involve several conditions and unknowns simultaneously. These more complicated problems can best be represented by a system of two or more equations. In this chapter we will solve and apply systems of two and three linear equations in two and three unknowns.

Recall from Section 2-4 that a linear equation in one unknown contains a single variable raised to the first power. Thus,

$$3x + 5 = 8 \quad \text{and} \quad 3(2y - 5) = 8 - (3y + 5)$$

are both linear equations in one unknown. Linear equations in one unknown normally have a single number as a solution. We can use the techniques learned in Section 2-4 to find the solutions $x = 1$ and $y = 2$ for the equations listed above.

Any equation containing two or more variables is a linear equation if the variables are raised to the first power. The equation

$$ax + by = c \quad \text{where } a, b, \text{ and } c \text{ are real numbers } (a, b \neq 0)$$

is a linear equation in two variables, x and y.

Many situations give rise to such linear equations. For example, suppose a metallurgist combines x pounds of alloy A containing 2 oz of tin per pound with y pounds of alloy B containing 3 oz of tin per pound to form a new alloy containing a total of 24 oz of tin. Then the new alloy would have

$2x$ oz of tin from alloy A

$3y$ oz of tin from alloy B

for a total of

$2x + 3y$ oz of tin

Therefore,

$$2x + 3y = 24$$

Checking solutions   Because this linear equation has two unknowns, its solution is a pair of values of x and y that satisfy it simultaneously. The equation $2x + 3y = 24$ has an unlimited number of solutions. There are infinitely many combinations of the two alloys that will result in a new alloy with 24 oz of tin. For example, the number pair $x = 6$ lb and $y = 4$ lb is a solution, because substituting these values into the equation gives us a true statement:

$$2(6) + 3(4) = 24$$

$$12 + 12 = 24$$

EXAMPLE 1   Show by substitution that each of the following is also a solution of the linear equation $2x + 3y = 24$:
(a) $x = 9$ and $y = 2$ (b) $x = 4\frac{1}{2}$ and $y = 5$

Solution
(a) $2(9) + 3(2) = 24$ (b) $2(\frac{9}{2}) + 3(5) = 24$
        $18 + 6 = 24$ $9 + 15 = 24$

Finding solutions

So far we have simply verified given solutions to a linear equation in two variables. We can also find additional solutions using the following procedure:

Step 1.   Select any real value for one of the variables. For example, let $x = 7$.

Step 2.   Substitute this value into the equation. $2(7) + 3y = 24$

Step 3.   Solve for the other variable.
$$14 + 3y = 24$$
$$3y = 10$$
$$y = 3\tfrac{1}{3}$$

Thus, 7 lb of alloy A and $3\tfrac{1}{3}$ lb of alloy B will produce 24 oz of tin.

NOTE

Since 7 represents x and $3\tfrac{1}{3}$ represents y, we can save space by writing the solution as the ordered pair $(7, 3\tfrac{1}{3})$. As we mentioned in Chapter 3, an ordered pair is always written in the order (x, y). ◀

EXAMPLE 2   Use the equation $2x + 3y = 24$ to find:
(a) y if $x = 3$ ▦ (b) x if $y = 7.48$
State each complete solution as an ordered pair.

Solution

(a) $2(3) + 3y = 24$ (b) $2x + 3(7.48) = 24$
$\quad\quad 6 + 3y = 24$ $\quad\quad\quad\quad\quad\quad\quad 2x = 24 - 3(7.48)$
$\quad\quad\quad\quad 3y = 18$ $\quad\quad 24\ \boxed{-}\ 3\ \boxed{\times}\ 7.48\ \boxed{=}\ \boxed{\div}\ 2\ \boxed{=} \rightarrow$ $\boxed{0.78}$
$\quad\quad\quad\quad\ y = 6$ $\quad\quad$ The solution is $(0.78, 7.48)$.
$\quad$ The solution is $(3, 6)$.

Systems of equations

Now suppose we add one additional condition to our problem: "The total amount of new alloy needed is 10 lb." This can be represented by the equation

$$x + y = 10$$

Again, there are an unlimited number of solutions to this equation. The ordered pairs $(9, 1)$, $(7, 3)$, $(6, 4)$, and $(4\tfrac{1}{2}, 5\tfrac{1}{2})$ are just a few of them. However, there is only *one* solution that satisfies both of our conditions; that is, the ordered pair $(6, 4)$ satisfies $x + y = 10$ and $2x + 3y = 24$. Together the two equations are said to form the system of simultaneous linear equations,

$$2x + 3y = 24$$

$$x + y = 10$$

and $(6, 4)$ is the solution to this system.

In general, any two linear equations containing the same variables x and y are said to form a **system of simultaneous linear equations:**

$$ax + by = c$$

$$dx + ey = f \quad\quad \text{where } a, b, c, d, e, \text{ and } f \text{ are real numbers}$$

A **solution to the system** is any ordered pair (x, y) that satisfies both equations. Normally, a system of two such equations has only a single ordered pair as a solution.

EXAMPLE 3   For each system of equations, determine whether the given ordered pair represents a solution.

(a) $2x - 3y = 5$ (b) $4.5x - 3.2y = 14.03$
      $3x - 5y = 9$ $2.7x - 1.4y = 11.33$
      $(4, 1)$ $(7.1, 5.6)$

(c)   According to Kirchhoff's voltage laws,* the currents i_1 and i_2 flowing through the two resistors in the electrical circuit shown in Fig. 5-1 must satisfy the system of linear equations

$$10i_1 - 50i_2 = 0$$
$$i_1 + i_2 = 6$$

Is $i_1 = 5$ A and $i_2 = 1$ A a solution?

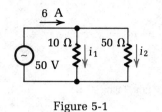

6 A

10 Ω i_1 **50 Ω** i_2

50 V

Figure 5-1

Solution

(a)   The ordered pair $(4, 1)$ is not a solution. Although it satisfies the first equation,

$$2(4) - 3(1) \stackrel{?}{=} 5$$
$$8 - 3 = 5$$
$$5 = 5$$

it does not satisfy the second equation,

$$3(4) - 5(1) \stackrel{?}{=} 9$$
$$12 - 5 \neq 9$$

(b)   The ordered pair $(7.1, 5.6)$ is a solution.

First equation: $4.5 \, \boxed{\times} \, 7.1 \, \boxed{-} \, 3.2 \, \boxed{\times} \, 5.6 \, \boxed{=} \rightarrow$ **14.03**

Second equation: $2.7 \, \boxed{\times} \, 7.1 \, \boxed{-} \, 1.4 \, \boxed{\times} \, 5.6 \, \boxed{=} \rightarrow$ **11.33**

(c)   The given pair of currents is a solution.

First equation: $10(5) - 50(1) = 0$
                         $50 - 50 = 0$

Second equation: $5 + 1 = 6$

Exercises 5-1

Determine whether each ordered pair is a solution to the given linear equation.

1. $3x - 2y = 10$:   (a) $(4, 1)$ (b) $(6, 4)$ (c) $(3, \frac{1}{2})$
2. $2x + y = -1$:   (a) $(4, 2)$ (b) $(-3, 5)$ (c) $(\frac{1}{2}, -2)$
3. $x - 4y = 18$:   (a) $(6, -3)$ (b) $(-2, -5)$ (c) $(10, -2)$

*For more information, see George J. Angerbauer, *Principles of DC and AC Circuits* (Boston: Breton, 1978), Chap. 3.

4. $-4x + 3y = 5$:   (a) $(4, \frac{1}{3})$ (b) $(0, -3)$ (c) $(-2, -1)$
5. $260x - 780y = -9360$:   (a) $(45, 27)$ (b) $(54, 36)$ (c) $(-9, 9)$
6. $6.40x + 1.60y = 1056$:   (a) $(140, 110)$ (b) $(120, 180)$ (c) $(95, 220)$

For each value of one variable, find a corresponding value of the other variable so that each pair is a solution to the given linear equation.

7. $4x - y = 6$:   (a) $x = \frac{1}{2}$ (b) $x = -3$ (c) $y = 2$
8. $3x - 6y = 18$:   (a) $x = -5$ (b) $x = 4$ (c) $y = 5$
9. $-5x + 4y = 6$:   (a) $y = 4$ (b) $y = -1$ (c) $x = -3$
10. $2x - 7y = 5$:   (a) $y = -3$ (b) $y = 5$ (c) $x = \frac{5}{4}$
11. $6.6x - 7.2y = 332.4$:   (a) $x = 22$ (b) $x = -14$ (c) $y = 84$
12. $420x + 350y = 5082$:   (a) $x = 5.60$ (b) $x = 2.50$ (c) $y = 4.68$

Determine whether each ordered pair is a solution to the given system of equations.

13. $2x - y = 9$ 14. $4x - 3y = 6$ 15. $5x - 3y = 7$
    $3x + y = 11$ $2x - 3y = 12$ $2x - 4y = 9$
    $(4, -1)$ $(-3, -6)$ $(5, 6)$
16. $6x - y = -20$ 17. $x - 9y = 7$ 18. $4x - y = -17$
    $7x - 2y = -17$ $3x - 5y = -11$ $3x - 2y = -9$
    $(-3, -2)$ $(-2, -1)$ $(-5, -3)$
19. $1200x - 1800y = 30,000$ 20. $3.7x - 1.4y = 11.61$
    $850x + 600y = 490,000$ $5.6x + 2.7y = 53.75$
    $(400, 250)$ $(5.90, 7.30)$

Solve.

21. *Electronics*   The system below results from applying Kirchhoff's laws to a certain dc circuit. Is $i_1 = 5$ A and $i_2 = 1$ A a solution of this system?

$$i_1 - 5i_2 = 0$$
$$2i_1 - 6i_2 = 4$$

22. *Material science*   The system below results from a mixture of two alloys. Is $x = 12$ oz and $y = 8$ oz a solution of this system?

$$0.45x + 0.80y = 11.8$$
$$x + y = 20$$

23. *Physics*   The system below is used to find the force on the legs of a table when it is being pulled by a known horizontal force. Are the forces $F_1 = 57$ lb and $F_2 = 23$ lb a solution to this system?

$$F_1 - F_2 = 34$$
$$F_1 + F_2 = 86$$

24. *Economics*   The equilibrium price p and quantity q for a particular item are determined by solving the system below. Is $p = \$9\frac{1}{11}$ and $q = \frac{2}{11}$ units a solution to this system?

$$p = 10 - 5q$$
$$p = 50q$$

25. *Electrical engineering*   The system of equations given on the next page results from measuring the resistance of a certain wire at two different temperatures. Is $m = 0.0779$ and $b = 24.0$ an approximate solution to this system?

$$26.8 = 36.5m + b$$
$$27.4 = 44.2m + b$$

26. *Thermodynamics* The system of equations given below results from measuring the volume of a certain gas at two different temperatures. Is $a = 4.54$ and $b = 161$ an approximate solution to this system?

$$265 = 23.0a + b$$
$$382 = 57.0a + b$$

5-2 | Graphical Solution of a System of Two Linear Equations in Two Unknowns

Locating solutions graphically

In Section 5-1 we saw that a single linear equation in two variables has an infinite number of real solutions. Since each solution is an ordered pair, each can be represented by a point on a rectangular coordinate system. The graph in Fig. 5-2 shows a few of the ordered pair solutions we found in Section 5-1 for the equation $2x + 3y = 24$. The fact that these points can be connected by a straight line tells us that the equation is linear. Furthermore, we know that every point on the line represents a solution to the equation.

In Fig. 5-3, the graph of $2x + 3y = 24$ is shown again, and the graph of the linear equation $x + y = 10$ is drawn on the same coordinate system. Notice that the two lines intersect at the point (6, 4). In Section 5-1 we showed that (6, 4) is the solution to the system formed by these two equations. The point of intersection graphically represents the only ordered pair that satisfies both equations. This tells us that we can solve any system of two linear equations by graphing both equations and locating their point of intersection.

Figure 5-2

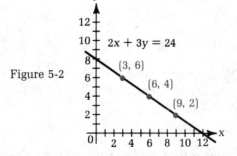

Figure 5-3

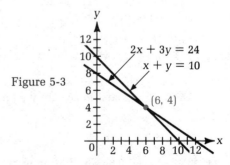

Graphing equations by the intercept method

In Chapter 3 we graphed functions by generating and plotting several ordered pairs. Since only two points are needed to draw a straight line, a linear equation can be graphed using only two ordered pair solutions. A quick and easy way to find two ordered pairs is to determine the value of y when $x = 0$ and the value of x when $y = 0$.

EXAMPLE 4   To graph the linear equation $3x + 4y = 12$, we need to find two ordered pairs that satisfy it.

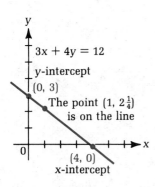

$3x + 4y = 12$

y-intercept

$(0, 3)$

The point $(1, 2\frac{1}{4})$ is on the line

$(4, 0)$

x-intercept

Figure 5-4

First, substitute $x = 0$ and solve for y. Notice that the x-term simply drops out, and y can be found easily. One ordered pair is $(0, 3)$.

$$3(0) + 4y = 12$$
$$4y = 12$$
$$y = 3$$

Second, substitute $y = 0$ and solve for x. Notice that the y-term now drops out, and x can be found easily. The second ordered pair is $(4, 0)$.

$$3x + 4(0) = 12$$
$$3x = 12$$
$$x = 4$$

To graph the line representing this equation, plot these two points and connect them with a straight line (Fig. 5-4). We know from Chapter 3 that $(4,0)$ is called the **x-intercept** of the graph. Similarly, the point $(0, 3)$ is called the **y-intercept**. To be certain the graph is correct, we should find a third ordered pair and make sure it is on the same line. For example, selecting $x = 1$ in this case yields $y = 2\frac{1}{4}$. The point $(1, 2\frac{1}{4})$ is also on the line.

EXAMPLE 5   Use the intercept method to graph the line $x + 4y = -4$.

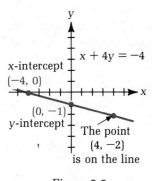

x-intercept

$(-4, 0)$

$x + 4y = -4$

$(0, -1)$

y-intercept

The point $(4, -2)$ is on the line

Figure 5-5

Solution

For $x = 0$:	For $y = 0$:	Check for $x = 4$:
$4y = -4$	$x = -4$	$4 + 4y = -4$
$y = -1$		$y = -2$
$(0, -1)$	$(-4, 0)$	$(4, -2)$

The graph is shown in Fig. 5-5.

Finding the solution to a system

Now let's combine the equations from Examples 4 and 5 to form the system

$$3x + 4y = 12$$

$$x + 4y = -4$$

To solve this system graphically, we draw the graphs of the two equations together in the same coordinate system and locate the point of intersection. From Fig. 5-6 we find the solution to be $(8, -3)$.

Check: $3(8) + 4(-3) = 12$ $(8) + 4(-3) = -4$

$24 + (-12) = 12$ $8 + (-12) = -4$

Estimating solutions

The next example shows that the graphical method does not always yield an exact solution.

EXAMPLE 6   Solve the following system graphically: $3x - 5y = 15$

$2x + y = 5$

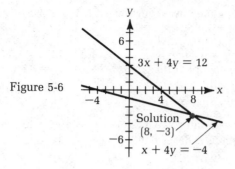

Figure 5-6

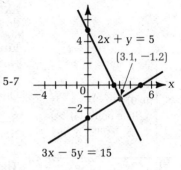

Figure 5-7

Solution   The intercepts for the graph of $3x - 5y = 15$ are $(5, 0)$ and $(0, -3)$. The intercepts for the graph of $2x + y = 5$ are $(2\frac{1}{2}, 0)$ and $(0, 5)$. Graphing both lines, we see that the point of intersection does not have integer coordinates (Fig. 5-7). Therefore, we can only estimate the solution to be approximately $(3.1, -1.2)$. [In this case, the precise solution is $(3\frac{1}{13}, -1\frac{2}{13})$.] Check your estimated answer to see if it is a reasonable solution for the system.

If the graph of an equation passes through the origin, the x- and y-intercepts are the same point: $(0, 0)$. In these cases you must choose a second nonintercept point in order to graph the line. Example 7 illustrates this.

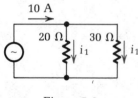

Figure 5-8

EXAMPLE 7   According to Kirchhoff's laws, the electrical circuit shown in Fig. 5-8 can be described by the following system:

$$i_1 + i_2 = 10$$
$$20i_1 - 30i_2 = 0$$

Solve this system graphically.

Solution   Using i_1 as the abscissa and i_2 as the ordinate, we determine the intercepts for the graph of $i_1 + i_2 = 10$ to be $(10, 0)$ and $(0, 10)$. The intercepts for the graph of $20i_1 - 30i_2 = 0$ are both $(0, 0)$, the origin. To find a second point for this line, we substitute $i_1 = 3$ and calculate $i_2 = 2$. Graphing both lines, we find that they intersect at the point $(6, 4)$. Therefore, $i_1 = 6$ A and $i_2 = 4$ A is the solution to the system.

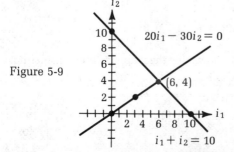

Figure 5-9

Inconsistent and dependent systems

Not every possible pair of lines intersect in exactly one point, and therefore not all systems of two linear equations in two variables have exactly one solution. Parallel lines never intersect. A system whose graph is two parallel lines has no solution. Such a system, shown in Fig. 5-10, is called an **inconsistent system of equations.**

If the two equations in a system have the same graph line, as shown in Fig. 5-11, every point on the line produces a solution. Therefore, the system has an unlimited number of solutions, and it is called a **dependent system.**

Figure 5-10

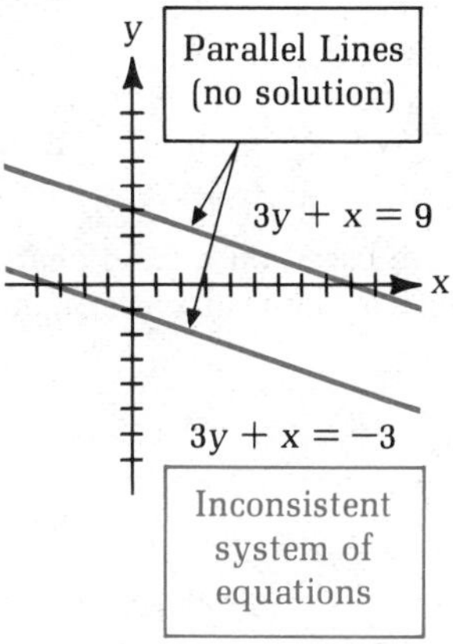

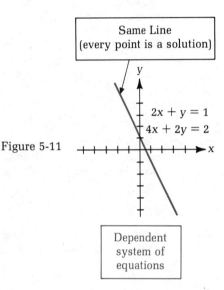

Figure 5-11

EXAMPLE 8 Solve each system graphically.

(a) $x - 2y = 4$ (b) $6x - 2y = 6$
$-2x + 4y = -8$ $-3x + y = 6$

Solution

(a) Both lines have intercepts of $(0, -2)$ and $(4, 0)$. This is a dependent system with an infinite number of solutions. Every point on the line in Fig. 5-12 gives a solution to the system.

(b) The graph of $6x - 2y = 6$ has intercepts of $(0, -3)$ and $(1, 0)$. The graph of $-3x + y = 6$ has intercepts of $(0, 6)$ and $(-2, 0)$. Figure 5-13 shows that these two lines are parallel. The system is therefore inconsistent and has no solution.

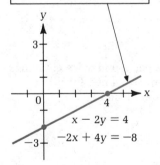

Figure 5-12

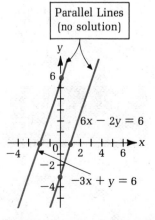

Figure 5-13

Exercises 5-2

Solve each system of equations graphically. Estimate answers to the nearest tenth if necessary.

1. $5x + 2y = 20$
   $3x - 2y = 12$

2. $x + y = 11$
   $2x - y = 7$

3. $3x - 5y = 15$
   $6x - 10y = 30$

4. $2y = 5x - 10$
   $10x - 4y = 10$

5. $y = 2x$
   $2x - 4y = -6$

6. $x = 3y$
   $2y - 4x = 30$

7. $3x + 2y = 10$
   $2x = 5y - 25$

8. $3x - 6y = 6$
   $4x - 2y = 8$

9. $x - y = 2$
   $y + x = 1$

10. $2x - 4y = -19$
    $-3x + y = -4$

11. $x - 3y = 4$
    $3y + 2x = -1$

12. $2x + 3y = 22$
    $x - y = 1$

13. $2x - y = 5$
    $2y - 4x = 3$

14. $3x - y = 5$
    $6x - 10 = 2y$

15. $6x - y = 15$
    $-x + y = -5$

16. $3x - 4y = 30$
    $4y + 3x = -6$

17. $2x - 5y = 5$
    $3x - 4y = 12$

18. $x - 4y = 4$
    $2x - 3y = 9$

19. $y = 3 - x$
    $3x + y = 11$

20. $3x = 4y + 30$
    $4y + 3x = -6$

21. $x - 2y = 3$
    $2x - 3y = 7$

22. $y + 2x = 1$
    $3y + 5x = 1$

23. $5x - 4y = 20$
    $3x + 2y = 6$

24. $x = 4 - y$
    $2x - 3y = 9$

25. Long-distance telephone rates to a certain city are based on a fixed charge x for the first minute and another fixed charge y for each additional minute. If a 7 min call costs $6.90 and a 3 min call costs $3.30, determine the two rates by solving the system

$$x + 6y = 6.90$$
$$x + 2y = 3.30$$

26. *Chemistry* A certain mixture requires 1.5 times as much carbolic acid c as water w. To find the total amount of each needed for a 7.5 mL mixture, solve the system

$$c = 1.5w$$
$$c + w = 7.5$$

27. *Physics* Two weights balance on opposite ends of a lever when the smaller weight S is 3 ft from the fulcrum and the larger weight L is 2 ft from the fulcrum. If their positions are interchanged, 5 lb must be added to the smaller weight to achieve a balance. This leads to the system below. Solve for S and L.

$$3S = 2L$$
$$2S + 10 = 3L$$

28. *Physics* The length of a spring is equal to its original length L plus P inches per pound of weight hung from it. Experimental data show the spring to be 8 in. long with an attached weight of 2 lb and 11 in. long with an attached weight of 4 lb. This leads to the system below. Solve for L and P.

$$L + 2P = 8$$
$$L + 4P = 11$$

5-3 | Algebraic Solution of a System of Two Linear Equations in Two Unknowns

By graphing a system of two linear equations, we obtain a picture of the solution and an understanding of its geometric meaning. But when a solution involves values that are not integers, this method can give us only approximate results.

In this section we will discuss two algebraic methods for solving these systems, both of which yield exact solutions.

Solving by substitution   In order to solve a system of two linear equations in two unknowns, we first need to eliminate a variable and reduce the system to one equation in one unknown. One method for achieving this is called **substitution.**

EXAMPLE 9   For the system of linear equations

$$x - 3y = 8$$
$$2x + 5y = 7$$

the substitution method works this way:

Step 1.   Solve one of the equations for one of the variables. In this case, the first equation can be solved easily for x.

$$x - 3y = 8$$
$$x = 3y + 8$$

Step 2.   Substitute the expression from Step 1 into the second equation and solve for the remaining variable.

$$2x + 5y = 7$$

Replace x with $(3y + 8)$

$$2(3y + 8) + 5y = 7$$
$$6y + 16 + 5y = 7$$
$$11y = -9$$
$$y = -\frac{9}{11}$$

Step 3.   Substitute the value from Step 2 into the equation from Step 1 and solve for the other variable.

$$x = 3\left(-\frac{9}{11}\right) + 8$$
$$x = -\frac{27}{11} + 8$$
$$x = \frac{61}{11}$$

Step 4.   Check this solution by substituting both values into each original equation:

$$\left(\frac{61}{11}\right) - 3\left(-\frac{9}{11}\right) = 8 \qquad 2\left(\frac{61}{11}\right) + 5\left(-\frac{9}{11}\right) = 7$$

$$\frac{61}{11} + \frac{27}{11} = 8 \qquad\qquad \frac{122}{11} + \left(-\frac{45}{11}\right) = 7$$

$$\frac{88}{11} = 8 \qquad\qquad\qquad \frac{77}{11} = 7$$

The solution $(\frac{61}{11}, -\frac{9}{11})$ could not have been determined exactly using the graphical method.

The substitution method works nicely for some systems but is somewhat tedious for others. Example 10 shows both situations.

EXAMPLE 10   Solve each system by substitution.

(a) $y = 6x - 2$
  $x - y = 22$

(b) $4x - 5y = 8$
  $8x + 5y = 11$

Solution

(a)   **Step 1.**   The first equation has already been solved for y.

Step 2.   Substitute $(6x - 2)$ for y in the second equation. Be sure to put this expression in parentheses. Solve for x.

$$x - (6x - 2) = 22$$
$$x - 6x + 2 = 22$$
$$-5x = 20$$
$$x = -4$$

Step 3.   Substitute -4 for x in the first equation and solve for y.

$$y = 6(-4) - 2 = -26$$

Step 4.   Check the solution by substituting both values into each original equation.

Expressed as an ordered pair, the solution is $(-4, -26)$. Although both coordinates are integers, it would take a carefully drawn graph to find this solution graphically.

(b)   **Step 1.**   Solve the first equation for x.

$$4x - 5y = 8$$
$$4x = 5y + 8$$
$$x = \frac{5y + 8}{4}$$

Step 2.   Substitute this expression for x in the second equation and solve for y.

$$8x + 5y = 11$$
$$8\left(\frac{5y + 8}{4}\right) + 5y = 11$$
$$2(5y + 8) + 5y = 11$$
$$10y + 16 + 5y = 11$$
$$15y = -5$$
$$y = -\frac{1}{3}$$

Step 3.   Replace y with $-\frac{1}{3}$ in the expression for x.

$$x = \frac{5\left(-\frac{1}{3}\right) + 8}{4}$$
$$= \frac{-\frac{5}{3} + \frac{24}{3}}{4}$$
$$= \frac{\frac{19}{3}}{4} = \frac{19}{12}$$

The solution is $\left(\frac{19}{12}, -\frac{1}{3}\right)$. Be sure to check this.

Solving by addition

Example 10(b) shows that the substitution method may be difficult when neither equation can be solved easily for a variable. In such cases we can eliminate a variable more easily by the **addition method.**

EXAMPLE 11   To solve Example 10(b) by the addition method, follow this procedure:

Step 1.   First, notice that the coefficients of y in the two equations are opposites of each other. This means we can add the two equations to eliminate y. The result will be a single equation in one unknown.

$$4x - 5y = 8$$
$$8x + 5y = 11$$
$$\overline{\hphantom{4x +} 12x + 0 = 19}$$

Step 2.   Solve for x.

$$x = \frac{19}{12}$$

Step 3.   Replace x with $\frac{19}{12}$ in either of the original two equations to solve for y.

$$4\left(\frac{19}{12}\right) - 5y = 8$$

$$\frac{19}{3} - 5y = 8$$

$$-5y = \frac{5}{3}$$

$$y = -\frac{1}{3}$$

Step 4.   Check in the usual manner.

The solution $(\frac{19}{12}, -\frac{1}{3})$ matches the solution obtained by substitution in Example 10(b); however, the addition method was easier to use in this case.

Of course, not every system is as well-suited to the addition method as the above example. Sometimes we must multiply one or both equations by a constant to produce a pair of opposite coefficients.

EXAMPLE 12   Consider the following system:

$$4x - 3y = 9$$
$$2x - 5y = 22$$

To solve this system using the addition method, follow these steps:

Step 1.   Multiply the second equation by -2 so the x-coefficients are opposites:

$$(-2)(2x - 5y) = (-2)22 \implies -4x + 10y = -44$$

(Remember, if both sides of an equation are multiplied by the same constant, the two sides remain equal.)

Step 2.   Add this multiple of the second equation to the first equation and solve for y:

$$4x - 3y = 9$$
$$-4x + 10y = -44$$
$$\overline{\hphantom{4x +} 7y = -35}$$
$$y = -5$$

Step 3.   Replace y with -5 in either of the original equations to solve for x.

$$4x - 3(-5) = 9$$
$$4x + 15 = 9$$
$$x = -\frac{3}{2}$$

Step 4.   Check the solution $(-\frac{3}{2}, -5)$ in both original equations.

In the next example both equations must be multiplied by constants to produce a pair of opposite coefficients.

EXAMPLE 13   Use the addition method to solve the system:

$$-2x - 11y = 4$$
$$5x + 9y = 27$$

Solution   Notice that the x coefficients, -2 and 5, both evenly divide 10. We will convert this pair of equations into a new pair in which the coefficients of x are $+10$ and -10.

Multiply the first equation by 5.	$-10x - 55y = 20$
Multiply the second equation by 2.	$10x + 18y = 54$
Add.	$-37y = 74$
Solve for y.	$y = -2$
Substitute -2 for y in one of the	$5x + 9(-2) = 27$
original equations and solve for x.	$5x + (-18) = 27$
	$x = 9$

The solution is $(9, -2)$. Check it.

Special cases   At the end of Section 5-2, we found that a system represented by two parallel lines has no solution (inconsistent), and a system represented by a single line has an unlimited number of solutions (dependent). It is important that you learn to recognize these two special cases when you attempt to solve systems algebraically.

EXAMPLE 14   In Example 8(b) of Section 5-2 the following system was shown to be inconsistent by the graphical method:

$$6x - 2y = 6$$
$$-3x + y = 6$$

Let's try to solve this system by the addition method:

Multiply the second equation by 2,	$6x - 2y = 6$
and add it to the first equation.	$-6x + 2y = 12$
	$0 = 18$

- *When an attempt at solving a system algebraically causes both variables to disappear, and a **false** statement results, the system is **inconsistent** and has no solution.*

EXAMPLE 15 In Example 8(a) the following system was shown to be dependent by the graphical method:

$$x - 2y = 4$$
$$-2x + 4y = -8$$

Using the substitution method, we solve the first equation for x, and substitute this expression into the second equation.

$$x = 2y + 4$$
$$-2(2y + 4) + 4y = -8$$
$$-4y - 8 + 4y = -8$$
$$-8 = -8$$

- *When an attempt at solving a system algebraically causes both variables to disappear, and a **true** statement results, the system is **dependent**.*

Applications To appreciate the value of these methods, we must examine some applied problems that lead to systems of two linear equations in two unknowns. For each of the following examples, we will explain how to set up such a system and leave their solution to you.

Percent mixture The first example is a percent mixture problem like the ones solved using one variable in Section 2-6. When the problem involves two unknown quantities, it becomes easier to solve using a system of two equations.

EXAMPLE 16 A lab technician needs 5 L of a 30% acid solution. This solution will be prepared by mixing a 60% solution with a 10% solution. How much of each solution is needed?

Solution **First,** establish the unknowns and assign a variable to each. Let

x = Amount of 60% solution

y = Amount of 10% solution

Then, look for two facts from which two equations can be formed:

1. The total amount of the two solutions is 5 L, so

$$x + y = 5$$

2.

Amount of acid in the 60% solution	+	Amount of acid in the 10% solution	=	Amount of acid in the final mixture
0.6x	+	0.1y	=	0.3(5)

Finally, combine the two equations into the system

$$x + y = 5$$
$$0.6x + 0.1y = 1.5$$

and solve for both unknowns. In this case, the addition method works nicely after multiplying the second equation by -10. The solution is $x = 2$ (liters of 60% solution) and $y = 3$ (liters of 10% solution). Be sure to check your solution against the original problem statement and the original pair of equations.

Motion problems

The next example is a simple motion problem based on the formula Speed = Distance/Time. Unlike the motion problems in Section 2-6, this type involves a speed composed of two components. Since both components are unknown, we use a system of two equations to solve.

EXAMPLE 17   An airplane travels 720 mi in 1 h 20 min flying with the wind. It returns at the same airspeed against the same wind in exactly 2 h. Find the airspeed of the plane and the speed of the wind.

Solution   First, assign variables to the unknown quantities. Let

v_p = Speed of the plane

v_w = Speed of the wind

Second, state the facts given in the problem. We use the following reasoning:

1.   With the wind, the actual ground speed is $v_p + v_w$.
2.   Against the wind, the actual ground speed is $v_p - v_w$.

Third, substitute into the speed formula to form a system of two equations:

$$\text{Speed} = \text{Distance/Time}$$

With the wind: $v_p + v_w = \dfrac{720}{\frac{4}{3}}$ $\boxed{\begin{array}{l}\text{The times must both be in hours to} \\ \text{obtain a speed in miles per hour.} \\ \text{Thus, 1 h 20 min} = 1\frac{20}{60} = 1\frac{1}{3} = \frac{4}{3}\text{h.}\end{array}}$

Against the wind: $v_p - v_w = \dfrac{720}{2}$

Or, simplifying: $v_p + v_w = 540$
$$v_p - v_w = 360$$

Finally, we solve using the addition method, and we obtain $v_p = 450$ mi/h and $v_w = 90$ mi/h. This solution checks with the original information.

Finding constants in a formula

Many formulas contain constants that vary from situation to situation. When the values of the constants are unknown for a particular situation, they can be found experimentally. First, perform as many different experimental trials as there are unknown constants. Then, set up a system of equations based on the results of the trials.

EXAMPLE 18   The French scientist Jacques Charles noted that gases expand when heated and contract when cooled. He assumed there was a linear relationship between volume V and temperature T given by

$$V = aT + b$$

where a and b are constants that depend on the particular gas involved. In one experiment Charles found that a certain gas had a volume of 500 cm³ at 27.0°C and a volume of 605 cm³ at 90.0°C. Use his data to solve for the constants a and b for this gas.*

Solution   Using the given data, we can form two equations:

$$500 = 27.0a + b$$
$$605 = 90.0a + b$$

Solving this system, we obtain $a \approx 1.67$ cm³/°C and $b = 455$ cm³. Therefore, for this gas, Charles' equation is $V \approx 1.67T + 455$.

General word problems   The next example is a general type of word problem in which two facts are stated about two unknown quantities. These statements are then converted into two equations that can be solved for the unknowns.

EXAMPLE 19   A steel beam is 26.0 ft long. If it is cut so that one piece is 9.0 ft longer than the other, how long will each piece be?

Solution   Let

$S = $ Length of the shorter piece

$L = $ Length of the longer piece

The two known facts are:

1.   The beam is 26.0 ft long: $S + L = 26.0$
2.   The longer piece is 9.0 ft longer than the shorter piece: $L = S + 9.0$

This system can be solved easily using the substitution method. The solution is $S = 8.5$ ft and $L = 17.5$ ft. Check it against the stated facts of the problem.

Exercises 5-3

Solve each system of equations using the substitution method.

1. $y = 2x$	**2.** $x = 9y$	**3.** $2x - y = 1$	**4.** $x = 3 + y$
$y = x + 6$	$x - 6y = 1$	$y = 5 - 4x$	$2y - x = -7$
5. $x = -2y - 2$	**6.** $x = 8 - 4y$	**7.** $x + y = 1$	**8.** $x - 2y = 1$
$-2x + y = 6$	$2y = 3x + 11$	$2x + y = 4$	$2x + 5y = 20$

* See: MAA/NCTM Joint Committee, *A Sourcebook of Applications of School Mathematics* (NCTM, 1980), Problem 2.71.

9. $2y - 4x = 26$
    $-y + 2x = -13$

10. $2y = 3x + 4$
    $6x - 4y = 3$

11. $2x - 5y = 8$
    $3x + 4y = -2$

12. $7x - 2y = 8$
    $4x - 3y = 6$

Solve each system of equations using elimination by addition.

13. $x + y = 11$
    $x - y = 5$

14. $2x + y = 12$
    $x - y = 6$

15. $5x - 3y = 12$
    $-3x - 3y = -4$

16. $6x - 2y = -11$
    $6x - 5y = 13$

17. $4x - 3y = 9$
    $3x + 6y = 4$

18. $4x - 5y = 11$
    $2x - 3y = -3$

19. $3x - 2y = 8$
    $6x - 4y = 4$

20. $5x + 3y = -9$
    $-10x - 6y = 18$

21. $3x + 4y = -12$
    $4x - 3y = 9$

22. $2x + 5y = 20$
    $3x + 2y = 19$

23. $2x + 7y = -21$
    $7x - 2y = 6$

24. $2x + 3y = 7$
    $3x - 2y = 4$

Solve the following systems:

25. $3x - 8y = 4$
    $x - 2y = 13$

26. $7x - 6y = 8$
    $x - y = -12$

27. $x = 1 + 2y$
    $2x + 5y = 20$

28. $x = 5 - 2y$
    $3x + 2y = 17$

29. $5x - y = 31$
    $y = 3x - 7$

30. $3y - x = -5$
    $x = 2y + 3$

31. $3y = 4x - 2$
    $5x - y - 8 = 0$

32. $5y = 2x$
    $2x = 32 - 3y$

33. $6x - y = -3$
    $-12x + 2y = 6$

34. $y = \frac{4}{3}x - 3$
    $8x - 6y = 14$

35. $3x - 5y = -11$
    $4x + 3y = 24$

36. $4x - 7y = 34$
    $6x + 5y = -11$

For each problem, set up a system of equations based on the given information and solve for both unknowns.

37. *Chemistry*   How many milliliters each of a 20% acid solution and a 50% acid solution should be combined to form a 24 mL mixture that is 25% acid?

38. *Material science*   How many kilograms each of an alloy containing 25% lead and an alloy containing 45% lead should be combined to produce 60.0 kg of an alloy that is 40% lead?

39. *Navigation*   An airplane flies 315 mi with the wind in 1 h 15 min and returns the same distance against the same wind in 1 h 30 min. Find the speeds of both the plane and the wind.

40. *Navigation*   An airplane flies 448 mi into a headwind in 2.0 h and returns the same distance with a tailwind of the same speed in 1 h 45 min. Find the speeds of both the plane and the wind.

41. *Electrical engineering*   The resistance R (in ohms) is related to the temperature T (in degrees Celsius) by the formula $R = mT + b$. If the resistance is 0.30 Ω at 30°C and 0.50 Ω at 70°C, find the constants m and b.

42. *Thermodynamics*   Given that a certain gas has a volume of 300 cm³ at 25.0°C and a volume of 425 cm³ at 75.0°C, find its constants a and b in the formula $V = aT + b$ (see Example 18).

43. A 140 ft cable is cut into two pieces so that one piece is three times as long as the other. How long is each piece?

44. *Electronics*   The sum of two voltages in a circuit is 64 V, and one is 12 V larger than the other. What are the two voltages?

5-4 | Using Determinants to Solve a System of Two Linear Equations in Two Unknowns

In mathematics if some procedure is used to solve all problems of a particular type, then a formula can usually be developed for use as a shortcut. In this section we will develop a formula for solving systems of two linear equations in two unknowns. The great value of this method is that it can be used with computers to solve systems containing three, four, or even hundreds of equations.

Determinants   A fundamental feature of this method is the use of a mathematical quantity known as a **determinant.** The quantity $\begin{vmatrix} a & b \\ c & d \end{vmatrix}$, where a, b, c, and d are any real numbers, is a **second-order determinant** because it has two rows and two columns. The numbers a, b, c, and d are the **elements** of the determinant. Elements a and d form the **principal diagonal,** while c and b are the elements of the **secondary diagonal.**

The **value** of a second-order determinant is equal to the product of the elements of the principal diagonal minus the product of the elements of the secondary diagonal.

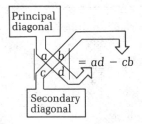

EXAMPLE 20   Calculate the value of each determinant.

(a) $\begin{vmatrix} 7 & 4 \\ 2 & 6 \end{vmatrix}$ (b) $\begin{vmatrix} -5 & 8 \\ -9 & -3 \end{vmatrix}$ (c) $\begin{vmatrix} 2.68 & 5.47 \\ 8.38 & 9.26 \end{vmatrix}$

Solution

(a) $\begin{vmatrix} 7 & 4 \\ 2 & 6 \end{vmatrix} = 7 \cdot 6 - 2 \cdot 4 = 42 - 8 = 34$

(b) $\begin{vmatrix} -5 & 8 \\ -9 & -3 \end{vmatrix} = (-5)(-3) - (-9)(8) = 15 - (-72) = 87$

(c) $2.68 \;\boxed{\times}\; 9.26 \;\boxed{-}\; 8.38 \;\boxed{\times}\; 5.47 \;\boxed{=} \rightarrow \boxed{\;-21.0218\;}$

Determinants and systems of equations   To understand the importance of determinants in solving a system of linear equations, consider the following system:

$$a_1x + b_1y = c_1$$
$$a_2x + b_2y = c_2$$

We can solve this system using the addition method from Section 5-3:

Step 1. Multiply the first equation by b_2 and the second by $-b_1$.

$$a_1b_2x + b_1b_2y = c_1b_2$$
$$-a_2b_1x - b_1b_2y = -c_2b_1$$

Step 2. Add to eliminate y.
Step 3. Apply the distributive property to the left side to combine like terms.

$$a_1b_2x - a_2b_1x = c_1b_2 - c_2b_1$$

$$(a_1b_2 - a_2b_1)x = c_1b_2 - c_2b_1$$

Step 4. Divide both sides by $(a_1b_2 - a_2b_1)$ to solve for x.

$$x = \frac{c_1b_2 - c_2b_1}{a_1b_2 - a_2b_1}$$

Step 5. Repeat this process to solve for y.

$$y = \frac{a_1c_2 - a_2c_1}{a_1b_2 - a_2b_1}$$

Notice that both the numerator and denominator of each fraction could be written as the value of a second-order determinant. For example, the common denominator

$$a_1 b_2 - a_2 b_1 \quad \text{is the value of the determinant} \quad \begin{vmatrix} a_1 & b_1 \\ a_2 & b_2 \end{vmatrix}$$

Similarly, we can express each numerator as a determinant and restate the formulas for x and y as shown in the box. This is known as **Cramer's rule**, named after its inventor, Gabriel Cramer.*

Cramer's Rule

$$x = \frac{\begin{vmatrix} c_1 & b_1 \\ c_2 & b_2 \end{vmatrix}}{\begin{vmatrix} a_1 & b_1 \\ a_2 & b_2 \end{vmatrix}} = \frac{D_x}{D} \quad \text{and} \quad y = \frac{\begin{vmatrix} a_1 & c_1 \\ a_2 & c_2 \end{vmatrix}}{\begin{vmatrix} a_1 & b_1 \\ a_2 & b_2 \end{vmatrix}} = \frac{D_y}{D} \tag{5-1}$$

where $D \neq 0$

Students often remember this rule by noticing the following:

- The determinant for the common denominator D contains the coefficients of x and y in order from the original system:

$$\begin{array}{l} a_1 x + b_1 y = c_1 \\ a_2 x + b_2 y = c_2 \end{array} \qquad \begin{vmatrix} a_1 & b_1 \\ a_2 & b_2 \end{vmatrix}$$

- The determinant for the x numerator D_x can then be found by replacing the column of x coefficients with the corresponding constants:

$$D = \begin{vmatrix} a_1 & b_1 \\ a_2 & b_2 \end{vmatrix} \qquad D_x = \begin{vmatrix} c_1 & b_1 \\ c_2 & b_2 \end{vmatrix}$$

Replace the x coefficients with the constants

- The determinant for the y numerator D_y can be found by replacing the column of y coefficients in D with the constants:

$$D = \begin{vmatrix} a_1 & b_1 \\ a_2 & b_2 \end{vmatrix} \qquad D_y = \begin{vmatrix} a_1 & c_1 \\ a_2 & c_2 \end{vmatrix}$$

Replace the y coefficients with the constants

---

* Gabriel Cramer (1704–1752), professor of mathematics at the University of Geneva in Switzerland, published this rule in 1750. However, it was actually first discovered by the Japanese mathematician Seki Kowa more than 100 years earlier.

EXAMPLE 21   Use Cramer's rule to solve:

(a) $3x - y = -7$ (b) $5x + 2y = -4$
      $x + 2y = 7$ $3x - 5y = 6$

Solution

(a) **Step 1.** Set up and calculate D using the coefficients of x and y:

$$D = \begin{vmatrix} 3 & -1 \\ 1 & 2 \end{vmatrix} = 3 \cdot 2 - (1)(-1) = 6 + 1 = 7$$

Step 2. Set up and calculate D_x by replacing the column of x coefficients in Step 1 with the constants:

$$D_x = \begin{vmatrix} -7 & -1 \\ 7 & 2 \end{vmatrix} = (-7)(2) - (7)(-1) = -14 - (-7) = -7$$

Step 3. Set up and calculate D_y by replacing the column of y coefficients in Step 1 with the constants:

$$D_y = \begin{vmatrix} 3 & -7 \\ 1 & 7 \end{vmatrix} = 3 \cdot 7 - (1)(-7) = 21 - (-7) = 28$$

Step 4. Use the determinants from the previous steps to find x and y:

$$x = \frac{D_x}{D} = \frac{-7}{7} = -1 \qquad y = \frac{D_y}{D} = \frac{28}{7} = 4$$

The solution is $(-1, 4)$. Check it by substituting these values back into the original equations.

(b) After some practice, you can proceed directly to Step 4:

$$x = \frac{D_x}{D} = \frac{\begin{vmatrix} -4 & 2 \\ 6 & -5 \end{vmatrix}}{\begin{vmatrix} 5 & 2 \\ 3 & -5 \end{vmatrix}} = \frac{20 - 12}{-25 - 6} = \frac{8}{-31}$$

$$y = \frac{D_y}{D} = \frac{\begin{vmatrix} 5 & -4 \\ 3 & 6 \end{vmatrix}}{-31} = \frac{30 + 12}{-31} = \frac{42}{-31}$$

The solution is $(-\frac{8}{31}, -\frac{42}{31})$.

Special cases   Now let's see what happens when we apply Cramer's rule to inconsistent and dependent systems.

EXAMPLE 22   Earlier in this chapter, system (a) below was found to be inconsistent and system (b) was found to be dependent. Apply Cramer's rule to each and see what happens.

(a) $6x - 2y = 6$ (b) $x - 2y = 4$
      $-3x + y = 6$ $-2x + 4y = -8$

Solution

(a) $x = \dfrac{D_x}{D} = \dfrac{\begin{vmatrix} 6 & -2 \\ 6 & 1 \end{vmatrix}}{\begin{vmatrix} 6 & -2 \\ -3 & 1 \end{vmatrix}} = \dfrac{6 - (-12)}{6 - 6} = \dfrac{18}{0}$ Not defined

$y = \dfrac{D_y}{D} = \dfrac{\begin{vmatrix} 6 & 6 \\ -3 & 6 \end{vmatrix}}{0} = \dfrac{36 - (-18)}{0} = \dfrac{54}{0}$ Not defined

(b) $x = \dfrac{D_x}{D} = \dfrac{\begin{vmatrix} 4 & -2 \\ -8 & 4 \end{vmatrix}}{\begin{vmatrix} 1 & -2 \\ -2 & 4 \end{vmatrix}} = \dfrac{16 - 16}{4 - 4} = \dfrac{0}{0}$ Not defined

$y = \dfrac{D_y}{D} = \dfrac{\begin{vmatrix} 1 & 4 \\ -2 & -8 \end{vmatrix}}{0} = \dfrac{-8 - (-8)}{0} = \dfrac{0}{0}$ Not defined

Conclusion: When solving a system of two linear equations in two unknowns using determinants:

- The system is **inconsistent** and has no solution if **only** the denominator is zero [as in Example 22 (a)].
- The system is **dependent** and has an unlimited number of solutions if **both** the numerator and denominator are zero [as in Example 22(b)].

Applications   In Section 5-3 we gave several applied problems that resulted in systems of two linear equations. Now we will examine several more applications. Any method may be used to solve the resulting systems, but for further practice with determinants, use Cramer's rule to verify our stated results.

Levers   The first example involves the concept of the **lever** as illustrated in Fig. 5-14. The pivot point of a lever is called the **fulcrum**. The distance from the fulcrum to the point where a force (F_1 or F_2) is applied is called a **lever arm**. The lever is in balance when the product of one force times its lever arm is equal to the product of the other force times its lever arm—that is,

$$F_1 L_1 = F_2 L_2$$

Assume that the weight of the lever is negligible.

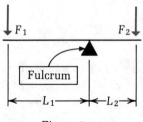

Figure 5-14

EXAMPLE 23   Two weights of 40 lb and 60 lb, respectively, balance each other on either side of the fulcrum of a lever. When a weight of 5 lb is added to the smaller weight, the larger weight must be moved 0.5 ft further from the fulcrum to maintain the balance. Find the original lengths of the lever arms.

Solution   Let L_1 = Lever arm of the smaller weight

L_2 = Lever arm of the larger weight

Figures 5-15(a) and (b) illustrate the problem. The resulting equations are shown next to the figures.

General equation: $F_1L_1 = F_2L_2$

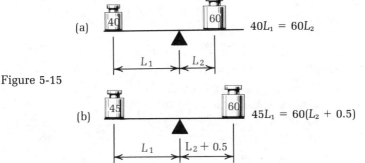

(a) $40L_1 = 60L_2$

Figure 5-15

(b) $45L_1 = 60(L_2 + 0.5)$

These two equations can be $40L_1 - 60L_2 = 0$
combined into the system $45L_1 - 60L_2 = 30$

The result is $L_1 = 6$ ft and $L_2 = 4$ ft.

Combined rates   In determining the total cost of a service, there are often two or more different rates that apply. For example, a technician's bill may be based on x hours of the shop owner's time at one cost and y hours of the apprentice's time at a lower cost. Or, the cost of mailing a package may be x cents for the first pound and y cents for each additional pound. In order to determine each individual rate or the number of units charged at each rate, a system of equations can be formed based on the given information.

EXAMPLE 24   A computer data base charges $20 per hour during prime time and $7.50 per hour during nonprime time. If an engineering firm paid $290 for 27 h, find the number of hours charged at each rate.

Solution   **First,** assign variables to the unknown quantities. Let

p = Number of hours of prime time

n = Number of hours of nonprime time

Then, set up an equation based on each fact:

1.   The total number of hours was 27: $p + n = 27$

2.   The total cost was $290: $20p + 7.5n = 290$

Solving this system, we find that there were 7 h charged at the prime-time rate and 20 h charged at the nonprime-time rate.

Investment problems   Companies often face the problem of having to decide how to split a given amount of money among two or more different investments. A system of equations may help to determine how to divide the principal.

EXAMPLE 25   A firm wants to invest $8000, some at an 8% return and the rest at a riskier 15% return. What is the least amount that can be placed at 15% and still result in a total earnings of $850 annually from the two investments?

Solution   Let x = Amount invested at 8%

y = Amount invested at 15%

The Two Known Facts	The Resulting Equations
1. The total amount of money is $8000.	$x + y = 8000$
2. The amount earned at 8% (0.08x) plus the amount earned at 15% (0.15y) equals $850. Or:	$0.08x + 0.15y = 850$ $8x + 15y = 85{,}000$

Solving the system

$$x + y = 8000$$
$$8x + 15y = 85{,}000$$

we obtain x = $5000 and y = $3000.

Exercises 5-4

Evaluate each determinant.

1. $\begin{vmatrix} 2 & 1 \\ 5 & 3 \end{vmatrix}$ **2.** $\begin{vmatrix} 5 & -1 \\ 4 & 2 \end{vmatrix}$ **3.** $\begin{vmatrix} -3 & -1 \\ 6 & 4 \end{vmatrix}$ **4.** $\begin{vmatrix} 7 & 6 \\ -5 & -3 \end{vmatrix}$

5. $\begin{vmatrix} -8 & -3 \\ 4 & -2 \end{vmatrix}$ **6.** $\begin{vmatrix} -5 & -9 \\ -1 & 7 \end{vmatrix}$ **7.** $\begin{vmatrix} 0 & -8 \\ -4 & 6 \end{vmatrix}$ **8.** $\begin{vmatrix} 12 & 0 \\ 14 & -3 \end{vmatrix}$

9. $\begin{vmatrix} -13 & -5 \\ -7 & -4 \end{vmatrix}$ **10.** $\begin{vmatrix} 5 & 11 \\ 6 & -14 \end{vmatrix}$ **11.** $\begin{vmatrix} 2.5 & 11.2 \\ 7.8 & 12.6 \end{vmatrix}$ **12.** $\begin{vmatrix} 239 & 486 \\ 517 & 812 \end{vmatrix}$

Solve each system of equations using determinants.

13. $3x + 8y = -5$
$2x + 5y = -4$

14. $3x - 4y = 20$
$3x + 5y = -7$

15. $7x + 9y = -54$
$-3x + y = -6$

16. $4x + y = -3$
$2x - 3y = -19$

17. $2x - 3y = -6$
$4x - 3y = -3$

18. $3x - 2y = -3$
$-3x + y = 1$

19. $5x + y = 15$
$x - 3y = 25$

20. $x + 5y = 0$
$3x + 10y = 10$

21. $3x - 4y = 9$
$-\frac{3}{2}x + 2y = 6$

22. $2x - 7y = 11$
$5x - y = 11$

23. $4x - 3y = 2$
$6x + y = 25$

24. $4x - 5y = 6$
$-8x + 10y = -12$

25. $9x - 5y = -8$
$4x + 6y = 17$

26. $6x - 8y = -10$
$9x + 4y = 13$

27. $14x + 11y = -64$
$-10x + 13y = 4$

28. $15x - 7y = -29$
$12x + 17y = 22$

29. $75x - 47y = 239$
$38x - 19y = 304$

30. $2.42x - 5.60y = 2.07$
$4.96x + 7.30y = 90.62$

In addition to the three types of problems illustrated in this section, the word problems in this exercise set cover the types of problems discussed in Section 5-3. Refer to these if you encounter any difficulty. Use the given information to set up a system of equations; then solve for the unknowns.

31. Physics   Two objects of mass 30 kg and 50 kg, respectively, balance each other on either side of the fulcrum of a weightless lever. When a mass of 10 kg is added to the smaller object, the larger object must be moved 3 cm further from the fulcrum to maintain the balance. Find the original lengths of the lever arms.

32. Physics   Two weights are located 16 in. and 24 in., respectively, from the fulcrum of a weightless lever. When the positions of the weights are exchanged, 15 lb must be added to the lighter one to maintain balance. What were the original weights?

33.   A phone call is charged at one rate for the first minute and at a different rate for each additional minute. If a 7 min call cost $1.20, and a 4 min call cost $0.72, find each rate.

34. Industrial management   A data base charges different hourly rates for hookups during prime time and nonprime time. During one month, a research firm was charged $577 for 22 h of prime time and 7 h of nonprime time. The next month they were charged $471 for 17 h of prime time and 9 h of nonprime time. Determine the two different hourly rates.

35. Finance   An investment counselor wants to invest $10,000, some at a 16% annual return and the rest at a 6% annual return. How much should be invested at each rate in order to earn $1400 interest per year?

36. Finance   A trust fund has $25,000 to invest in two different funds. If one pays 14% annually and the other pays 8% annually, how much should be invested in each fund in order to earn $2600 annually?

37. Chemistry   How many liters each of a 65% salt solution and a 15% salt solution should be combined to create 20.0 L of a 35% salt solution?

38. Thermodynamics   A particular gas has a volume of 200.0 cm^3 at 18.0°C and a volume of 380.0 cm^3 at 78.0°C. Use the formula $V = aT + b$ to find the constants a and b for this gas. (See Example 18.)

39.   An order of 46 printing ribbons contains 8 more black ink ribbons than two-color ribbons. How many of each kind are in the order?

40. Navigation   An airplane makes a trip of 960 mi with a tailwind in 2.0 h. Going back against the same headwind, the trip takes 2 h 40 min. Find the speeds of the wind and the plane.

5-5 | Algebraic Solution of a System of Three Linear Equations in Three Unknowns

Many technical problems involve more than two unknown quantities. In fact, aircraft designers and economists often work with systems of a thousand or more linear equations. If the number of known facts or conditions matches the number of unknowns, we can set up a system of equations to solve the problem. The techniques outlined in Sections 5-3 and 5-4 can be extended to solve these larger systems. In this section we will examine the algebraic method for solving a linear system in three unknowns.

A set of three unknown forces acting on a beam might produce the following system of three equations:

$$x + y + z = 6 \tag{1}$$

$$x - y + 2z = 2 \tag{2}$$

$$2x + y - z = 3 \tag{3}$$

where x, y, and z represent the unknown forces. A solution to this system is a set of values x, y, and z that simultaneously satisfy all three equations. If the system is consistent and independent, it will have only one solution.

To solve a system of three linear equations in three unknowns, use the addition or substitution method to reduce the system to a system of two equations in two unknowns. Then use the techniques already studied to solve this resulting system.

EXAMPLE 26   Solve the system given above using the addition method.

Solution

Step 1.   Combine two of the equations so that one of the variables is eliminated:

Adding Equations (1) and (2) results in an equation that does not contain y.

$$\begin{array}{rcl} x + y + z &=& 6 \\ x - y + 2z &=& 2 \\ \hline 2x \quad + 3z &=& 8 \end{array}$$

Step 2.   Repeat Step 1, using a different pair of the original equations to eliminate the *same* variable:

Adding Equations (2) and (3) results in another equation that does not contain y.

$$\begin{array}{rcl} x - y + 2z &=& 2 \\ 2x + y - z &=& 3 \\ \hline 3x \quad + z &=& 5 \end{array}$$

Step 3.   Use the new equations to form a system of two equations in two unknowns, and solve:

$$2x + 3z = 8$$
$$3x + z = 5$$

Using one of the techniques for solving systems of two equations, we obtain $x = 1$ and $z = 2$.

Step 4.   Substitute the values of x and z from Step 3 back into one of the original equations to solve for y:

Using Equation (1): $(1) + y + (2) = 6$
$$y = 3$$

The solution is $x = 1$, $y = 3$, and $z = 2$. This can be written as the ordered triple $(1, 3, 2)$. Check the solution by substituting it back into the original system of three equations.

EXAMPLE 27   Solve the following system using the addition method:

$$x + 2y + 3z = 4 \qquad\qquad (1)$$

$$3x + 2y + z = 16 \qquad\qquad (2)$$

$$2x - y - 5z = 15 \qquad\qquad (3)$$

Solution   By examining the coefficients of the variables, we can see that it will be easiest to eliminate y from the system.

Step 1. Multiply Equation (1) by -1 and add the result to Equation (2) to obtain:

$$-1(x + 2y + 3z = 4)$$
$$\begin{array}{r} -x - 2y - 3z = -4 \\ 3x + 2y + z = 16 \\ \hline 2x - 2z = 12 \end{array}$$

Step 2. Multiply Equation (3) by 2 and add the result to Equation (2) to obtain:

$$2(2x - y - 5z = 15)$$
$$\begin{array}{r} 4x - 2y - 10z = 30 \\ 3x + 2y + z = 16 \\ \hline 7x - 9z = 46 \end{array}$$

Step 3. Now use the results of Steps 1 and 2 to form a system of two equations in two unknowns and solve:

$$2x - 2z = 12$$
$$7x - 9z = 46$$

The solution to this sytem is $x = 4$ and $z = -2$.

Step 4. Substitute the values for x and z from Step 3 back into Equation (1) and solve for y:

$$(4) + 2y + 3(-2) = 4$$
$$2y = 6$$
$$y = 3$$

The solution is $x = 4$, $y = 3$, and $z = -2$, or $(4, 3, -2)$. Be sure to check this solution.

EXAMPLE 28   Solve the following system using substitution:

$$x - 4y - 6z = -8 \tag{1}$$
$$5x - 4y + 3z = -17 \tag{2}$$
$$-2x + 12y + 9z = 18 \tag{3}$$

Solution

Step 1. Solve Equation (1) for x: $x = 4y + 6z - 8$

Step 2. Substitute this expression for x in Equation (2) and simplify:

$$5(4y + 6z - 8) - 4y + 3z = -17$$
$$20y + 30z - 40 - 4y + 3z = -17$$
$$16y + 33z = 23 \tag{4}$$

Step 3. Substitute the same expression for x in Equation (3) and simplify:

$$-2(4y + 6z - 8) + 12y + 9z = 18$$
$$-8y - 12z + 16 + 12y + 9z = 18$$
$$4y - 3z = 2 \tag{5}$$

Step 4. Equations (4) and (5) form the system

$$16y + 33z = 23$$

$$4y - 3z = 2$$

Solve using one of the techniques discussed earlier in this chapter. The solution is $y = \frac{3}{4}$ and $z = \frac{1}{3}$.

Step 5. Substitute these values for y and z in the equation from Step 1 to solve for x:

$$x = 4\left(\frac{3}{4}\right) + 6\left(\frac{1}{3}\right) - 8$$

$$x = 3 + 2 - 8 = -3$$

The solution is $x = -3$, $y = \frac{3}{4}$, and $z = \frac{1}{3}$, or $(-3, \frac{3}{4}, \frac{1}{3})$. Check this solution by substituting it into the original system of equations.

Applied problems   We now introduce three types of applied problems that lead to linear systems in three unknowns. We will again set up the systems and leave verification of the solutions to you.

EXAMPLE 29   In order to obtain a low contract price, a manufacturer must order 3000 special crates at a time from a supplier and spend $65,000 for the order. There are three different sizes of crates with costs and capacities as follows:

	A	B	C
Cost (each crate)	$15	$25	$30
Capacity (ft³)	10	20	30

If the manufacturer projects a need of 55,000 ft³ capacity, how many of each type of crate should be ordered to meet the terms of the supplier?

Solution   **Step 1.**   Use the given variables A, B, and C to designate the unknown quantities of each.

Step 2.   Set up three equations based on the facts of the problem:

Total number: $A + B + C = 3000$

Total cost: $15A + 25B + 30C = 65{,}000$

Total capacity: $10A + 20B + 30C = 55{,}000$

Step 3.   Solve using either the addition or the substitution method and check your solution. The correct solution is $A = 1500$, $B = 500$, and $C = 1000$.

EXAMPLE 30 Solutions containing 20%, 30% and 40% sulfuric acid are mixed together to produce a 100 mL solution of 25% sulfuric acid. Twice as much of the 40% solution is used as the 30% solution. How many milliliters of each are used?

Solution **First,** assign variables to the unknowns. Let

x = Amount of 20% solution

y = Amount of 30% solution

z = Amount of 40% solution

Second, look for three conditions or facts interrelating the three unknowns:

1. There are 100 mL total. $\qquad\qquad x + y + z = 100$ **(1)**

2. The actual amounts of acid in x, y, and z add up to the amount of acid in the final solution.
$\qquad 0.2x + 0.3y + 0.4z = 0.25(100)$

  Or, multiplying by 10: $\qquad\qquad 2x + 3y + 4z = 250$ **(2)**

3. The amount of 40% solution is twice the amount of 30% solution. $\qquad\qquad z = 2y$ **(3)**

Third, combine the three equations into the system

$$x + y + z = 100 \qquad\qquad\qquad (1)$$

$$2x + 3y + 4z = 250 \qquad\qquad\qquad (2)$$

$$z = 2y \qquad\qquad\qquad (3)$$

and solve. In this case, use Equation (3) to substitute for z in (1) and (2), and you will obtain $x = 70$ mL, $y = 10$ mL, and $z = 20$ mL.

EXAMPLE 31 The total budget for a research and development project is $90,000. The money will be divided among salaries, overhead, and equipment as follows: The cost of the overhead will equal twice the combined cost of the other two items; the cost of salaries will be five times the cost of equipment. How much will be provided for each category of the budget?

Solution **First,** assign variables to the unknowns. Let

S = Cost of salaries

V = Cost of overhead

E = Cost of equipment

Second, set up three equations based on the given facts:

1. Total cost: $\qquad\qquad S + V + E = 90,000$

2. **First comparison:** $V = 2(S + E)$

3. **Second comparison:** $S = 5E$

Third, solve this system. The third equation makes the substitution method ideal. The correct solution is $S = \$25,000$, $V = \$60,000$, and $E = \$5000$.

Exercises 5-5

Solve each system of equations.

1. $x + y - z = 0$
   $x + 2y = 5$
   $2x - z = -1$

2. $x + y + z = 1$
   $x - y = -3$
   $2x + z = -2$

3. $2x - 3y + z = 4$
   $3x + 2y - 2z = -11$
   $x + 3y - 2z = -10$

4. $3x - y + 2z = 8$
   $-x + 2y - z = -6$
   $2x + 3y - 3z = -2$

5. $5x - 3y - 4z = 18$
   $3x - y - 2z = 10$
   $-x + 2y - 3z = -23$

6. $3x + 2y - z = -5$
   $2x - y + 2z = 22$
   $4x + 3y - z = -6$

7. $6x - 3y - 2z = -3$
   $2x + 9y + z = -1$
   $4x + 3y - 2z = -8$

8. $-4x + y - 2z = -5$
   $8x - 2y - 4z = -2$
   $2x - 3y - 3z = 4$

9. $3x - y - z = 5$
   $x - 5y + z = 3$
   $x + 2y - z = 1$

10. $5x + 4y - 3z = -19$
    $4x - 3y + 2z = 37$
    $3x - 2y + 4z = 31$

11. $6x + 7y - 2z = 9$
    $4x - 3y + 5z = -26$
    $3x - 2y - 3z = -25$

12. $3x - 2y - z = 3$
    $2x + 2y - 2z = 4$
    $7x + 2y - 5z = 5$

13. $5x - 3y - 2z = -2$
    $10x + 2y - z = -10$
    $-5x - 4y + 3z = 29$

14. $-x - 3y + z = -3$
    $3x + 9y + 2z = -6$
    $2x - 6y + 3z = 11$

15. $5x + 6y - 3z = -39$
    $3x - 4y - 2z = 36$
    $2x + 3y + 4z = 16$

16. $7x - 3y - 2z = -4$
    $5x + 2y - 3z = 9$
    $4x - 4y + 4z = 4$

17. $2x + 3y + 4z = -3$
    $5x - 6y - 2z = 6$
    $-4x - 9y + 8z = -5$

18. *Electronics*   Applying Kirchhoff's laws to the electric circuit shown in the figure gives the following system of equations:

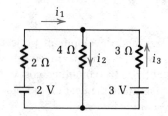

$i_1 - i_2 + i_3 = 0$

$2i_1 + 4i_2 = 2$

$4i_2 + 3i_3 = 3$

Solve for i_1, i_2, and i_3. (The results will be in amperes.)

For each problem, set up a system of linear equations and solve for the unknowns.

19. *Chemistry*   Three solutions containing 25% salt, 50% salt, and 60% salt are mixed to produce 100 mL of a solution containing 45% salt. If four times as much 25% solution is used as 50% solution, how much of each is used?

20. *Material science*   Alloy x is 50% zinc, 30% tungsten, and 20% iron. Alloy y is 40% zinc, 40% tungsten, and 20% iron. Alloy z is 20% zinc and 80% iron. How many ounces of each must be mixed to obtain 100 oz of an alloy that is 37% zinc, 25% tungsten, and 38% iron?

21. *Graphics technology* Three computer printers working together produce a total of 17 pages per minute. Printer A is twice as fast as printer B, and printer C produces 2 pages more per minute than printer A. How many pages per minute does each produce?

22. *Industrial management* The budget for the design of a new piece of computer hardware totals $75,000. The money comes from three budget centers: salaries, hardware, and on-line time. If salaries supply as much of the budget as the other two combined, and hardware provides twice as much as on-line time, how much of the budget is provided by each center?

23. *Industrial management* A trucking company owns three sizes of trucks, X, Y, and Z. These trucks are equipped to deliver three different types of machines, A, B, and C, according to the following chart:

	X	Y	Z
Machine A	1	1	1
Machine B	0	1	2
Machine C	2	1	1

Assuming each truck is fully loaded, how many trucks of each type should be sent to deliver exactly 12 type A machines, 10 type B machines, and 16 type C machines?

24. *Industrial management* A small manufacturing company must order three different size containers, X, Y, and Z, whose costs and capacities are listed in the following chart:

	X	Y	Z
Cost (each container)	$100	$150	$175
Capacity (ft³)	100	200	300

The company must purchase 2500 containers with a total capacity of 550,000 ft³. If the total cost is to be $375,000, how many of each type of container should they buy?

25. *Industrial management* In planning the output for the next quarter, a manufacturer of three models of food processors decides to produce four times as many of model A as model B and 2500 more of model C than B. If the total output will be 11,500 units, how many of each model will be produced?

26. *Industrial management* In planning a quarterly budget, a small medical lab allots twice as much money to supplies as new equipment and three times as much to salaries as supplies and new equipment combined. If $60,000 is budgeted for these three categories, how much is allotted for each?

5-6 | Using Determinants to Solve a System of Three Linear Equations in Three Unknowns

In Section 5-4 we showed how to solve a system of two linear equations in two unknowns using a second-order determinant. In this section we will discuss how to solve a system of three linear equations in three unknowns using a third-order determinant.

Third-order determinants   A **third-order determinant** contains nine elements arranged in three rows and three columns:

$$\begin{vmatrix} a_1 & b_1 & c_1 \\ a_2 & b_2 & c_2 \\ a_3 & b_3 & c_3 \end{vmatrix}$$

The **value** of a third-order determinant can be found as follows:

Step 1. Rewrite the first and second columns to the right of the determinant.

$$\begin{array}{ccc|cc} a_1 & b_1 & c_1 & a_1 & b_1 \\ a_2 & b_2 & c_2 & a_2 & b_2 \\ a_3 & b_3 & c_3 & a_3 & b_3 \end{array}$$

Step 2. Multiply the elements of the principal diagonal, and of each diagonal parallel and to the right of it, to form three separate products. **Add** these products.

Step 3. Multiply the elements of the secondary diagonal, and of each diagonal parallel and to the right of it, to form three separate products. **Subtract** each of these.

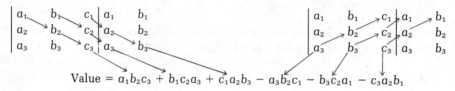

Value = $a_1 b_2 c_3 + b_1 c_2 a_3 + c_1 a_2 b_3 - a_3 b_2 c_1 - b_3 c_2 a_1 - c_3 a_2 b_1$

This method works only for third-order determinants.*

EXAMPLE 32   Find the value of each determinant.

(a) $\begin{vmatrix} 2 & 1 & -3 \\ 4 & 3 & 5 \\ -1 & -2 & -6 \end{vmatrix}$ (b) $\begin{vmatrix} 1 & 7 & 0 \\ 9 & -3 & 2 \\ 4 & 5 & -2 \end{vmatrix}$ ▦ (c) $\begin{vmatrix} 2.6 & 0.8 & 5.1 \\ -1.4 & 3.7 & 8.5 \\ 4.9 & -6.2 & 0.3 \end{vmatrix}$

Solution

(a)

$$\begin{array}{ccc|cc} 2 & 1 & -3 & 2 & 1 \\ 4 & 3 & 5 & 4 & 3 \\ -1 & -2 & -6 & -1 & -2 \end{array}$$

$-36 \quad +(-5) \quad +24$ $-9 \quad -(-20) \quad -(-24)$

Value = $-36 + (-5) + 24 - 9 - (-20) - (-24) = 18$

(b)

$$\begin{array}{ccc|cc} 1 & 7 & 0 & 1 & 7 \\ 9 & -3 & 2 & 9 & -3 \\ 4 & 5 & -2 & 4 & 5 \end{array}$$

Value = $6 + 56 + 0 - 0 - 10 - (-126) = 178$

(c) ① ② ③ ④ ⑤ ⑥

$$\begin{array}{ccc|cc} 2.6 & 0.8 & 5.1 & 2.6 & 0.8 \\ -1.4 & 3.7 & 8.5 & -1.4 & 3.7 \\ 4.9 & -6.2 & 0.3 & 4.9 & -6.2 \end{array}$$

*An alternative method that works for determinants of any order is presented in Chapter 16.

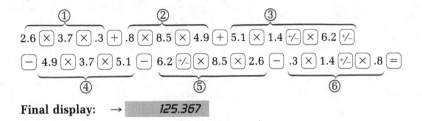

Final display: → 125.367

Using determinants to solve systems of three equations

To discover how determinants of the third order can help solve linear systems in three unknowns, we solve the general system

$$a_1 x + b_1 y + c_1 z = d_1$$

$$a_2 x + b_2 y + c_2 z = d_2$$

$$a_3 x + b_3 y + c_3 z = d_3$$

to obtain

$$x = \frac{d_1 b_2 c_3 + d_3 b_1 c_2 + d_2 b_3 c_1 - d_3 b_2 c_1 - d_1 b_3 c_2 - d_2 b_1 c_3}{a_1 b_2 c_3 + a_3 b_1 c_2 + a_2 b_3 c_1 - a_3 b_2 c_1 - a_1 b_3 c_2 - a_2 b_1 c_3}$$

$$y = \frac{a_1 d_2 c_3 + a_3 d_1 c_2 + a_2 d_3 c_1 - a_3 d_2 c_1 - a_1 d_3 c_2 - a_2 d_1 c_3}{a_1 b_2 c_3 + a_3 b_1 c_2 + a_2 b_3 c_1 - a_3 b_2 c_1 - a_1 b_3 c_2 - a_2 b_1 c_3}$$

$$z = \frac{a_1 b_2 d_3 + a_3 b_1 d_2 + a_2 b_3 d_1 - a_3 b_2 d_1 - a_1 b_3 d_2 - a_2 b_1 d_3}{a_1 b_2 c_3 + a_3 b_1 c_2 + a_2 b_3 c_1 - a_3 b_2 c_1 - a_1 b_3 c_2 - a_2 b_1 c_3}$$

Naturally, these formulas would be difficult to memorize. However, the denominator and numerators of each fraction can be written as determinants which are relatively easy to remember. Assuming the system is written in the standard form, the common denominator D is the value of the determinant whose elements are the coefficients of the variables x, y, and z:

$$a_1 x + b_1 y + c_1 z = d_1$$
$$a_2 x + b_2 y + c_2 z = d_2 \qquad D = \begin{vmatrix} a_1 & b_1 & c_1 \\ a_2 & b_2 & c_2 \\ a_3 & b_3 & c_3 \end{vmatrix}$$
$$a_3 x + b_3 y + c_3 z = d_3$$

The numerators, D_x, D_y, and D_z, can also be represented by determinants. For each variable, replace the column in D containing its coefficients with a column representing the three constants:

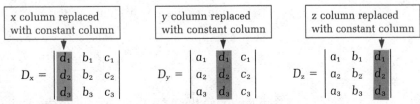

x column replaced with constant column	y column replaced with constant column	z column replaced with constant column

$$D_x = \begin{vmatrix} d_1 & b_1 & c_1 \\ d_2 & b_2 & c_2 \\ d_3 & b_3 & c_3 \end{vmatrix} \qquad D_y = \begin{vmatrix} a_1 & d_1 & c_1 \\ a_2 & d_2 & c_2 \\ a_3 & d_3 & c_3 \end{vmatrix} \qquad D_z = \begin{vmatrix} a_1 & b_1 & d_1 \\ a_2 & b_2 & d_2 \\ a_3 & b_3 & d_3 \end{vmatrix}$$

We can now restate the solution to a system of three linear equations in three unknowns as given in the box. Notice that this is simply an extension of Cramer's rule for a system of two linear equations in two unknowns.

Cramer's Rule for Three Equations in Three Unknowns

$$x = \frac{D_x}{D} \qquad y = \frac{D_y}{D} \qquad z = \frac{D_z}{D} \qquad \text{where } D \neq 0 \qquad \text{(5-2)}$$

Two special cases must be considered. If D and D_x, D_y, and D_z are all zero, the system is dependent. If D is zero, and any of D_x, D_y, and D_z are not zero, the system is inconsistent. Otherwise, the system has exactly one solution, (x, y, z).

EXAMPLE 33   Solve the following system using determinants:

$$2x + 2y + 3z = 0$$
$$-x - 3y + 7z = 15$$
$$3x + y + 4z = 21$$

Solution

Step 1.   Find D, D_x, D_y, and D_z:

$$D \longrightarrow \begin{vmatrix} 2 & 2 & 3 \\ -1 & -3 & 7 \\ 3 & 1 & 4 \end{vmatrix} \begin{matrix} 2 & 2 \\ -1 & -3 \\ 3 & 1 \end{matrix}$$

$$D = -24 + 42 + (-3) - (-27) - 14 - (-8) = 36$$

$$D_x \longrightarrow \begin{vmatrix} 0 & 2 & 3 \\ 15 & -3 & 7 \\ 21 & 1 & 4 \end{vmatrix} \begin{matrix} 0 & 2 \\ 15 & -3 \\ 21 & 1 \end{matrix}$$

$$D_x = 0 + 294 + 45 - (-189) - 0 - 120 = 408$$

$$D_y \longrightarrow \begin{vmatrix} 2 & 0 & 3 \\ -1 & 15 & 7 \\ 3 & 21 & 4 \end{vmatrix} \begin{matrix} 2 & 0 \\ -1 & 15 \\ 3 & 21 \end{matrix}$$

$$D_y = 120 + 0 + (-63) - 135 - 294 - 0 = -372$$

$$D_z \longrightarrow \begin{vmatrix} 2 & 2 & 0 \\ -1 & -3 & 15 \\ 3 & 1 & 21 \end{vmatrix} \begin{matrix} 2 & 2 \\ -1 & -3 \\ 3 & 1 \end{matrix}$$

$$D_z = -126 + 90 + 0 - 0 - 30 - (-42) = -24$$

Step 2.   Applying Cramer's rule (5-2), we have

$$x = \frac{D_x}{D} = \frac{408}{36} \qquad y = \frac{D_y}{D} = \frac{-372}{36} \qquad z = \frac{D_z}{D} = \frac{-24}{36}$$

Simplifying these fractions, we obtain the solution: $x = 11\frac{1}{3}$, $y = -10\frac{1}{3}$, and $z = -\frac{2}{3}$. Check this by substituting the values into all three of the original equations.

If a variable is missing from an equation, make certain to place a zero in the corresponding position of the determinant.

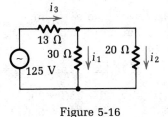

Figure 5-16

EXAMPLE 34   Using Kirchhoff's laws, the electrical circuit shown in Fig. 5-16 can be analyzed and described by the following system of equations:

$$i_1 + i_2 - i_3 = 0$$
$$30i_1 - 20i_2 = 0$$
$$30i_1 + 13i_3 = 125$$

Solve for the unknown currents (in amperes).

Solution
Step 1.   Find D, D_1, D_2, and D_3. (The subscripts refer to i_1, i_2, and i_3, respectively.)

$$D \longrightarrow \begin{vmatrix} 1 & 1 & -1 \\ 30 & -20 & 0 \\ 30 & 0 & 13 \end{vmatrix} \begin{matrix} 1 & 1 \\ 30 & -20 \\ 30 & 0 \end{matrix}$$

$$D = -260 + 0 + 0 - 600 - 0 - 390 = -1250$$

$$D_1 \longrightarrow \begin{vmatrix} 0 & 1 & -1 \\ 0 & -20 & 0 \\ 125 & 0 & 13 \end{vmatrix} \begin{matrix} 0 & 1 \\ 0 & -20 \\ 125 & 0 \end{matrix}$$

$$D_1 = 0 + 0 + 0 - 2500 - 0 - 0 = -2500$$

$$D_2 \longrightarrow \begin{vmatrix} 1 & 0 & -1 \\ 30 & 0 & 0 \\ 30 & 125 & 13 \end{vmatrix} \begin{matrix} 1 & 0 \\ 30 & 0 \\ 30 & 125 \end{matrix}$$

$$D_2 = 0 + 0 + (-3750) - 0 - 0 - 0 = -3750$$

$$D_3 \longrightarrow \begin{vmatrix} 1 & 1 & 0 \\ 30 & -20 & 0 \\ 30 & 0 & 125 \end{vmatrix} \begin{matrix} 1 & 1 \\ 30 & -20 \\ 30 & 0 \end{matrix}$$

$$D_3 = -2500 + 0 + 0 - 0 - 0 - 3750 = -6250$$

Applying formulas comparable to Cramer's rule (5-2), we have

$$i_1 = \frac{D_1}{D} = \frac{-2500}{-1250} \qquad i_2 = \frac{D_2}{D} = \frac{-3750}{-1250} \qquad i_3 = \frac{D_3}{D} = \frac{-6250}{-1250}$$

Simplifying these fractions, we obtain the solution: $i_1 = 2$ A, $i_2 = 3$ A, and $i_3 = 5$ A.

Applied problems   In Example 18, Section 5-3, we showed how to determine the constants in a formula. This same technique can be applied to a formula with three unknown constants.

Time, t (years)	ΔP (lb/in.^{2})
1	200
2	350
3	460

EXAMPLE 35   Petroleum engineers have found that the pressure drop ΔP (in pounds per square inch) in an oil reservoir is related to elapsed time t (in years) by the formula

$$\Delta P = At + Bt^2 + Ct^3$$

To determine the constants A, B, and C for a particular reservoir, three readings were taken. The results are summarized in the table. Set up a system of equations and find A, B, and C. Round to the nearest tenth.*

Solution

First, create three equations by rewriting the formula for each of the three readings.

$$\begin{array}{r} A + B + C = 200 \\ 2A + 4B + 8C = 350 \\ 3A + 9B + 27C = 460 \end{array}$$

Then, use determinants to solve for A, B, and C. The value of D is easy enough to calculate by hand:

$$D \longrightarrow \begin{vmatrix} 1 & 1 & 1 \\ 2 & 4 & 8 \\ 3 & 9 & 27 \end{vmatrix} \begin{matrix} 1 & 1 \\ 2 & 4 \\ 3 & 9 \end{matrix}$$

$$D = 108 + 24 + 18 - 12 - 72 - 54 = 12$$

Find D_A, D_B, and D_C using a calculator:

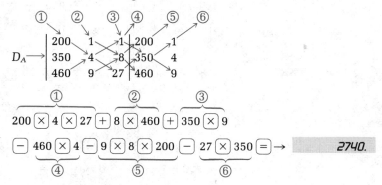

Similarly, $D_B = -360$ and $D_C = 20$.

Applying Cramer's rule (5-2), we have

$$A = \frac{D_A}{D} = \frac{2740}{12} \qquad B = \frac{D_B}{D} = \frac{-360}{12} \qquad C = \frac{D_C}{D} = \frac{20}{12}$$

Simplifying, we have $A \approx 228.3$, $B = -30$, and $C \approx 1.7$. Thus, the formula for this particular reservoir is

$$\Delta P \approx 228.3t - 30t^2 + 1.7t^3$$

An investment problem such as Example 25 in Section 5-4 might also contain three unknowns.

EXAMPLE 36   A woman wants to invest $10,000 in three different accounts, one paying 5.5%, another paying 10%, and a third paying 15%. For safety, she

* Based on an interview with John Evans, Phillips Petroleum Corp.

wants to put twice as much in the 5.5% account as in the 15% account and still earn a total of $900 interest per year. How much should she put in each account?

Solution Let

x = Amount invested at 5.5%

y = Amount invested at 10%

z = Amount invested at 15%

We have three conditions, which lead to three equations:

1. The total amount of money is
    $10,000: $x + y + z = 10,000$

2. There must be twice as much
    invested at 5.5% as at 15%: $x = 2z$

3. The total interest earned from the
    three investments must equal
    $900: $0.055x + 0.1y + 0.15z = 900$

    Or: $55x + 100y + 150z = 900,000$

The resulting system,

$$x + y + z = 10,000$$
$$x = 2z$$
$$55x + 100y + 150z = 900,000$$

has the solution $x = \$5000$, $y = \$2500$, and $z = \$2500$.

Exercises 5-6

Evaluate each determinant.

1. $\begin{vmatrix} 3 & 2 & -1 \\ -2 & 4 & 3 \\ 5 & 0 & -2 \end{vmatrix}$

2. $\begin{vmatrix} 3 & 4 & 1 \\ -5 & -2 & 0 \\ -1 & 2 & -3 \end{vmatrix}$

3. $\begin{vmatrix} 7 & 4 & -3 \\ 1 & 5 & -6 \\ -8 & 2 & 3 \end{vmatrix}$

4. $\begin{vmatrix} 1 & 6 & -1 \\ -4 & 3 & 7 \\ -2 & 5 & 3 \end{vmatrix}$

5. $\begin{vmatrix} 2 & 2 & 0 \\ 7 & -7 & 3 \\ 1 & -4 & 9 \end{vmatrix}$

6. $\begin{vmatrix} 11 & 4 & -3 \\ -2 & -6 & 7 \\ 0 & 8 & -1 \end{vmatrix}$

7. $\begin{vmatrix} 6 & 10 & 3 \\ -4 & -1 & -2 \\ -5 & 1 & -2 \end{vmatrix}$

8. $\begin{vmatrix} 9 & -3 & 5 \\ 2 & 12 & -7 \\ -3 & 2 & -4 \end{vmatrix}$

9. $\begin{vmatrix} 5 & 10 & 0 \\ -15 & -2 & 10 \\ -10 & -5 & 3 \end{vmatrix}$

10. $\begin{vmatrix} 3 & 20 & -1 \\ -3 & 0 & 3 \\ 10 & 5 & -2 \end{vmatrix}$

11. $\begin{vmatrix} 2.8 & -3.4 & 5.9 \\ 0.4 & -2.1 & 3.7 \\ -4.6 & 7.2 & -8.1 \end{vmatrix}$

12. $\begin{vmatrix} 26 & 48 & -39 \\ 75 & 37 & 86 \\ -54 & 91 & -17 \end{vmatrix}$

Solve each system of equations using determinants.

13. $x + y + z = 2$
$2x - y + z = -1$
$3x + 2y - z = 8$

14. $x - y + z = 1$
$x - 2y - z = -5$
$2x - 3y - 2z = -10$

15. $2x - y - z = 5$
$3x + y - 2z = 10$
$4x + 2y + z = 14$

16. $2x - 3y - z = 1$
$x - y + 2z = 4$
$4x - 3y + z = 11$

17. $5x - 3y - 2z = 0$
$-2x + y - z = -3$
$4x - y - 3z = -7$

18. $3x - 7y + z = 3$
$x - 5y - 4z = -11$
$-2x - 3y - z = 12$

19. $4x - y + z = -1$
$8x - 3y - 2z = 11$
$-6x + 2y - 3z = 9$

20. $5x + 2y - 9z = -7$
$-3x - 3y + 3z = 2$
$2x - 4y - 6z = -2$

21. $2x + 4y - 2z = 2$
$-2x - z = -3$
$-x + 2y - 2z = 0$

22. $x - 2y + z = 3$
$x + y - z = 2$
$3x - z = 7$

23. $2x - 9y + 4z = -4$
$-4x + 6y - 8z = 0$
$3x - 3y + 2z = 4$

24. $3x - 2y - 6z = -7$
$9x - 4y + 12z = 5$
$-2x - 6y + 4z = 1$

25. $25x - 72y + 30z = 421$
$60x + 13y - 24z = 939$
$44x - 81y + 23z = 396$

26. $2.60x - 7.20y + 1.40z = -18.66$
$3.70x + 5.50y - 4.30z = 10.65$
$8.10x - 2.50y - 6.90z = -34.45$

27. *Electronics* Applying Kirchhoff's laws to the circuit shown in the figure results in the system of equations given below. Solve for i_1, i_2, and i_3. The results will be in amperes.

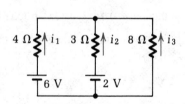

$$i_1 + i_2 + i_3 = 0$$
$$4i_1 - 8i_3 = 6$$
$$4i_1 - 3i_2 = 4$$

28. *Mechanical engineering* In determining three unknown forces acting on a structure, the system given below occurs. Solve for F_1, F_2, and F_3 (in pounds).

$$F_1 - 0.5F_2 = 0$$
$$F_2 - 0.7F_3 = 44$$
$$4F_1 - 3F_3 = 0$$

For each problem, set up a system of equations and solve for the unknowns.

29. *Physics* The displacement s of an object moving in a straight line under constant acceleration from an initial position s_0 is given by the formula

$$s = s_0 + v_0 t + \frac{1}{2}at^2$$

where t is elapsed time, v_0 is initial velocity and a is acceleration. The results of three trials in which s_0, v_0, and a were constant are shown in the chart. Find the constants s_0, v_0, and a.

Displacement, s (ft)	Time, t (s)
26.0	2.0
39.5	3.0
56.0	4.0

30. *Petroleum engineering* Use the formula for pressure drop given in Example 35 to find A, B, and C for the data listed in the table. Round to the nearest tenth if necessary.

Time, t (years)	Pressure Drop, ΔP (lb/in.^{2})
2	310
3	440
4	490

31. *Finance* A corporation wants to divide $20,000 into three different investments paying returns of 6%, 10%, and 18% annually. The company is committed to put twice as much at 6% as at 18%. How can this be achieved while still earning $2000 annually?

32. *Finance* A contractor can borrow money from three different sources, one charging 12%, one 15%, and one 20% annual interest. If the 12% source will loan only one-third as much money as the 20% source, how can the contractor borrow $27,000 and limit the total yearly payments to exactly $4500?

33. *Industrial management* Three punching machines are run so that in 1 h they can produce the following numbers of components:

	Machine X	Machine Y	Machine Z
Component A	4		3
Component B	2	6	
Component C		4	5

How many hours should each machine be run to produce 32 of component A, 28 of component B, and 32 of component C?

34. *Graphics technology* The three printing presses at Lopez Label Company produce a total of 1080 printed sheets per minute. Press A produces twice as many sheets as B, while press C produces 30 sheets more than press A. How many sheets does each press run in 1 min?

5-7 | Review of Chapter 5

Important Terms and Concepts

system of simultaneous linear equations (p. 167)

solution to a system of linear equations (p. 167)

x-intercept (p. 171)

y-intercept (p. 171)

inconsistent system (p. 173)

dependent system (p. 173)

substitution method of solving a system (p. 175)

addition method of solving a system (p. 177)

second-order determinant (p. 183)

third-order determinant (p. 196)

elements of a determinant (p. 183)

principal diagonal (p. 183)
secondary diagonal (p. 183)

Formulas and Rules

- value of a second-order determinant: $\begin{vmatrix} a & b \\ c & d \end{vmatrix} = ad - cb$ (p. 183)

- value of a third-order determinant: $\begin{vmatrix} a_1 & b_1 & c_1 \\ a_2 & b_2 & c_2 \\ a_3 & b_3 & c_3 \end{vmatrix}$

$$= a_1b_2c_3 + b_1c_2a_3 + c_1a_2b_3 - a_3b_2c_1 - b_3c_2a_1 - c_3a_2b_1 \quad \text{(p. 196)}$$

- Cramer's rule: the solution to the system $a_1x + b_1y = c_1$ is

$$a_2x + b_2y = c_2$$

$$x = \frac{\begin{vmatrix} c_1 & b_1 \\ c_2 & b_2 \end{vmatrix}}{\begin{vmatrix} a_1 & b_1 \\ a_2 & b_2 \end{vmatrix}} = \frac{D_x}{D} \qquad y = \frac{\begin{vmatrix} a_1 & c_1 \\ a_2 & c_2 \end{vmatrix}}{\begin{vmatrix} a_1 & b_1 \\ a_2 & b_2 \end{vmatrix}} = \frac{D_y}{D} \quad \text{(p. 184)}$$

(5-1)

- the solution to the system $a_1x + b_1y + c_1z = d_1$ is

$$a_2x + b_2y + c_2z = d_2$$

$$a_3x + b_3y + c_3z = d$$

$$x = \frac{\begin{vmatrix} d_1 & b_1 & c_1 \\ d_2 & b_2 & c_2 \\ d_3 & b_3 & c_3 \end{vmatrix}}{D} = \frac{D_x}{D} \qquad y = \frac{\begin{vmatrix} a_1 & d_1 & c_1 \\ a_2 & d_2 & c_2 \\ a_3 & d_3 & c_3 \end{vmatrix}}{D} = \frac{D_y}{D}$$

(5-2)

$$z = \frac{\begin{vmatrix} a_1 & b_1 & d_1 \\ a_2 & b_2 & d_2 \\ a_3 & b_3 & d_3 \end{vmatrix}}{D} = \frac{D_z}{D} \qquad \text{where} \quad D = \begin{vmatrix} a_1 & b_1 & c_1 \\ a_2 & b_2 & c_2 \\ a_3 & b_3 & c_3 \end{vmatrix} \quad \text{(p. 198)}$$

Exercises 5-7

Solve each system of equations graphically. Estimate answers to the nearest tenth if necessary.

1. $2x - y = -1$
$3x + y = -9$

2. $4x - 2y = -12$
$2x + 3y = 10$

3. $2x - 3y = 6$
$3x - 2y = 9$

4. $4x + 2y = 10$
$2x - y = -5$

5. $5x - 2y = 10$
$2x - 4y = -4$

6. $7x + 3y = 7$
$6x + 2y = 4$

7. $6x - 3y = 9$
$x - 2y = -1$

8. $4x + 3y = 12$
$3x - 2y = 6$

Solve each system of equations algebraically.

9. $x = y + 5$
$3x + 2y = 25$

10. $3x - 2y = 0$
$x = 8 - 2y$

11. $2x - y = 7$
$x + 4y = 2$

12. $4x - 3y = 8$
$6x - 3y = 9$

13. $y = 3x - 2$
$x - 2y = 4$

14. $y = 9 - 2x$
$3x - y = 11$

15. $x = \frac{1}{2}y - 5$
$3y - 6x = 4$

16. $8x - 3y = -19$
$5x + 9y = -30$

17. $2x - 7y = 26$ **18.** $4x - 6y = 8$ **19.** $3x - 7y = 29$ **20.** $4x - 9y = 21$
$\quad\ \ 6x + 5y = 0$ $\qquad\quad 6x - 9y = 12$ $\qquad\quad 2x + 5y = -29$ $\qquad\quad 3x + 6y = 3$

21. $x + y - z = -6$ **22.** $x - y - z = -1$ **23.** $4x - y + 2z = 5$
$\quad\ \ 3x - y + 2z = -5$ $\qquad 2x - 3y - 2z = 1$ $\qquad\quad 3x - 5y - 3z = -17$
$\quad\ \ 2x - 2y - 3z = -5$ $\qquad 3x - 4y + z = -4$ $\qquad\quad 5x - 3y + 4z = -9$

24. $5x + 4y - z = 5$ **25.** $2x - 7y - 6z = -9$ **26.** $2x + 3y + 6z = 16$
$\quad\ \ x + 3y - 4z = -1$ $\qquad 4x + 5y + 3z = 10$ $\qquad\quad -6x - 9y - 5z = 4$
$\quad\ \ 6x + 2y + 3z = 6$ $\qquad 6x - 3y - 3z = -1$ $\qquad\quad 4x + 6y + 4z = 0$

27. $3x + 5y - z = 4$ ▦ **28.** $40x - 50y + 20z = 620$
$\quad\ \ x + y + z = 2$ $\qquad\qquad 20x - 20y - 30z = -80$
$\quad\ \ -x - 2y + z = 3$ $\qquad\quad 30x - 30y - 25z = 120$

Evaluate each determinant.

29. $\begin{vmatrix} 3 & -2 \\ 5 & 4 \end{vmatrix}$ **30.** $\begin{vmatrix} -1 & 7 \\ 8 & -2 \end{vmatrix}$ **31.** $\begin{vmatrix} -3 & -5 \\ 6 & 9 \end{vmatrix}$ ▦ **32.** $\begin{vmatrix} 107 & 46 \\ -117 & -39 \end{vmatrix}$

33. $\begin{vmatrix} 2 & 3 & 0 \\ -5 & -1 & 4 \\ 5 & -2 & -6 \end{vmatrix}$ **34.** $\begin{vmatrix} 3 & -6 & 7 \\ 0 & 2 & 4 \\ -1 & -1 & -3 \end{vmatrix}$ **35.** $\begin{vmatrix} 8 & -5 & 3 \\ -2 & -6 & 4 \\ -3 & 7 & 2 \end{vmatrix}$ ▦ **36.** $\begin{vmatrix} 3.5 & 7.2 & 9.1 \\ -4.6 & -3.2 & -5.8 \\ 2.7 & 6.4 & 1.7 \end{vmatrix}$

Solve each system of equations using determinants.

37. $3x - 2y = 11$ **38.** $4x + 3y = -23$ **39.** $5x + y = -7$ **40.** $5x - 3y = 19$
$\quad\ \ x + 5y = -19$ $\qquad\ 3x - 2y = 4$ $\qquad\quad -3x - 2y = -7$ $\qquad\ 7x + 4y = 2$

41. $9x - 6y = 6$ **42.** $2x + 5y = 2$ **43.** $6x - 4y = 9$ **44.** $2x - 3y = 7$
$\quad\ \ 4x - 5y = 12$ $\qquad\ 3x + 8y = 5$ $\qquad\quad -3x + 2y = 7$ $\qquad\ 6x - 9y = 21$

45. $2x + 6y = -1$ **46.** $6x - 4y = -7$ ▦ **47.** $34x - 27y = 436$
$\quad\ \ 4x + 3y = 4$ $\qquad\quad -3x + 8y = -1$ $\qquad\qquad 51x - 72y = -543$

▦ **48.** $3.6x + 5.8y = 133.4$ **49.** $2x - 3y + 4z = -5$ **50.** $x - 4y + 3z = 14$
$\quad\ \ 2.7x - 4.1y = 362$ $\qquad\quad x - 2y - 2z = 7$ $\qquad\quad 3x - 2y - 2z = 8$
$\qquad\qquad\qquad\qquad\qquad -2x - y + 3z = 3$ $\qquad\quad 2x - 5y - z = 3$

51. $5x - 2y + 3z = 11$ **52.** $3x - 2y = 1$ **53.** $3x + 6y - z = -1$
$\quad\ \ 3x - 4y = -10$ $\qquad\quad 5y + 4z = 2$ $\qquad\quad -2x - 4y + z = 2$
$\quad\ \ 5y - 6z = 2$ $\qquad\quad 7x + 3y + 2z = -7$ $\qquad\ 6x - 12y + 4z = 2$

54. $6x + 3y + 4z = -2$ **55.** $2x - 3y - z = 4$ ▦ **56.** $4.7x - 1.3y - 2.8z = -23$
$\quad\ \ 3x - 2y + 6z = 1$ $\qquad\quad 4x + 3y + 2z = 1$ $\qquad\quad 6.6x - 5.2y + 1.7z = -20.8$
$\quad\ \ 4x - y + 8z = -2$ $\qquad\ 10x + 3y + 3z = 6$ $\qquad\quad 3.5x + 2.6y - 7.4z = -8.8$

Solve the following systems using any method you choose.

57. Physics In determining the tensions in the strings shown in the figure, the system of equations given below results. Solve for tensions T_1 and T_2 to the nearest pound.*

$\qquad -0.866T_1 + 0.500T_2 = 0$

$\qquad\ \cdot\ 0.500T_1 + 0.866T_2 = 500$

* Based on James R. Thrower, Jr. *Technical Statics and Strength of Materials* (Boston: Duxbury, 1976), p. 51.

58. *Electronics* Applying Kirchhoff's laws to the circuit shown in the figure produces the system of equations given below. Solve for i_1, i_2, and i_3. The results will be in amperes.

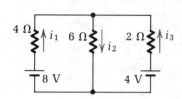

$$i_1 - i_2 + i_3 = 0$$

$$4i_1 + 6i_2 = 8$$

$$6i_2 + 2i_3 = 4$$

For each problem, set up a system of equations and solve for the unknowns.

59. *Physics* The displacement s of an object moving in a straight line under constant acceleration a is given by the formula

$$s = v_0 t + \frac{1}{2} at^2$$

where v_0 is the initial velocity (in feet per second) and t is the elapsed time (in seconds). If the displacement was 21 ft after 3 s and 32 ft after 4 s, find the initial velocity and the acceleration.

60. *Civil engineering* Five thousand feet of roadway curbing must be constructed from a combination of rock curb at $6.15 per foot and asphalt dyking at $4.30 per foot. If $25,000 is budgeted for the curbing, how many feet of each should be constructed?

61. *Chemistry* How many liters each of a 30% saline solution and a 10% saline solution should be mixed to produce 12 L of a 15% saline solution?

62. *Navigation* An airplane flies 1200 mi with a tailwind in 2 h 30 min. It returns the same distance against a headwind of the same speed in 3 h 20 min. Find the speed of the plane and of the wind.

63. A postal rate is charged at a fixed cost for the first pound and at a different rate for each additional pound. If a 3 lb package cost $1.09 to mail, and a 5 lb package cost $1.55 to mail, determine the cost of the first pound and the rate for each additional pound.

64. *Finance* A company trying to expand its plant area needs a loan of $80,000. Part of the money is secured from a bank at 14% and the rest from an investment group at 18%. If the annual loan payments are $12,000, how much is borrowed at each rate?

65. *Physics* Two weights of 60 lb and 80 lb, respectively, balance each other on either side of the fulcrum of a lever. If 20 lb are added to the larger weight, the smaller weight must be moved 4 in. further from the fulcrum to maintain the balance. Determine the original distance of each weight from the fulcrum.

66. *Pharmacology* A pharmacist has three ointments, X, Y, and Z, that contain the following percentages of creams A, B, and C:

	X	Y	Z
Cream A	60%	40%	50%
Cream B	10%	30%	50%
Cream C	30%	30%	

How many grams of each ointment should be combined to obtain a 100 g mixture containing 52% cream A, 24% cream B, and 24% cream C?

67. *Industrial management* Three different lathes, X, Y, and Z, are tooled to produce parts A, B, and C in the following daily amounts:

	X	Y	Z
Part A	8	4	
Part B	4		6
Part C	2	4	2

How many days should each lathe be run to produce 40 of part A, 24 of part B, and 26 of part C?

68. *Finance*   A company needs to divide $60,000 among three different investments, one paying a 5% annual return, another paying a 12% annual return, and the third paying a 16% annual return. The company must invest the same amount at 12% as at the other two rates combined and still earn $7300 interest per year. How should the $60,000 be divided?

69. *Industrial management*   Three machines produce 82 parts per hour. Machine A produces twice as many parts as machine B but 10 fewer parts than B and C put together. How many parts does each machine produce in an hour?

70. *Petroleum engineering*   The pressure drop ΔP (in pounds per square inch) in an oil reservoir is related to the elapsed time t (in years) by the formula

$$\Delta P = At + Bt^2 + Ct^3$$

To determine the constants A, B, and C for a particular reservoir, three experimental readings were taken. These showed that after 3 years the pressure had dropped 456 lb/in.^{2}, after 6 years the drop was 632 lb/in.^{2}, and after 9 years the drop was 692 lb/in.^{2}. Use these data to find A, B, and C to the nearest tenth.

71. *Material science*   Alloys A, B, and C contain the following percentages of metals:

	Manganese	Copper	Lead
A	20%	50%	30%
B		40%	60%
C	10%	30%	60%

In what amounts should these alloys be combined to produce 100 kg of a new alloy containing 10% manganese, 42% copper, and 48% lead?

72. *Industrial management*   A data base charges $18 per hour for prime-time use and $6.50 per hour for nonprime-time use. A research firm received a bill for $195 based on a total of 18.5 h of use. How many hours of prime time were used?

Products, Factors, and Fractions

6

CROSS-SECTION OF I-BEAM

You have already worked through many of the basic concepts and techniques of algebra. These provide the foundation necessary to understand the trigonometric functions and to solve systems of equations. In order to understand other types of equations and their technical applications, we must now develop some additional algebraic techniques.

A basic skill in algebra is the multiplication of expressions. Some of these products occur so often in mathematics that their results are worth memorizing. The list given in the box shows these special products in general form. They all may be obtained using the multiplication techniques explained in Section 2-2. The examples that follow demonstrate how to use them.

Special Products

Product of a Binomial by a Monomial	$a(x + y) = ax + ay$	(6-1)
Difference of Two Squares	$(x + y)(x - y) = x^2 - y^2$	(6-2)
Square of a Binomial	$(x + y)^2 = x^2 + 2xy + y^2$	(6-3)
Product of Two Binomials	$(x + a)(x + b) = x^2 + (a + b)x + ab$	(6-4)

The distributive law

Product (6-1) involves multiplying a binomial by a monomial. To do so, simply apply the distributive law:

$$a(x + y) = a \cdot x + a \cdot y$$

Multiply each term by a

EXAMPLE 1 Multiply: $3(x + 5)$

Solution $3 \cdot (x + 5) = 3 \cdot x + 3 \cdot 5 = 3x + 15$

Multiply each term by 3

The same technique can be used to multiply any polynomial by a monomial.

EXAMPLE 2 Multiply: $-2x^2(3x - 4y + 5)$

Solution $-2x^2 \cdot (3x - 4y + 5) = (-2x^2) \cdot (3x) + (-2x^2) \cdot (-4y) + (-2x^2) \cdot (5)$

Multiply each term by $-2x^2$

$$= -6x^3 + 8x^2y - 10x^2$$

Notice that we treat the subtraction of the first two terms as a sum: $3x - 4y = (3x) + (-4y)$.

Difference of two squares   Product (6-2) is called the **difference of two squares.** It results from the product of two special binomials. One is the sum of two terms, and the other is the difference of the same two terms. If we multiply $(x + y)$ by $(x - y)$, we have

$$(x + y)(x - y) = x^2 + xy - xy - y^2$$
$$= x^2 - y^2$$

Since the products of the inner and outer terms sum to zero, the result is the difference of two perfect squares—the square of the first term and the square of the last term in each binomial. Knowing this, we can write the results of such products by inspection.

EXAMPLE 3   Multiply: $(x + 3)(x - 3)$

Solution   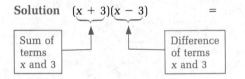   =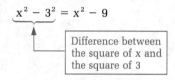

EXAMPLE 4   Multiply: $(2m - 5n)(2m + 5n)$

Solution $(2m - 5n)(2m + 5n) = (2m)^2 - (5n)^2 = 4m^2 - 25n^2$

Squaring a binomial   In product (6-3) we are squaring a binomial. Be careful to write this as a product:

$$(x + y)^2 = (x + y)(x + y) = x^2 + xy + xy + y^2$$

As we combine like terms, notice that the result contains three terms:

Square of y

$$(x + y)^2 = x^2 + 2xy + y^2$$

Square of x	Twice the product of x and y

This observation allows us to square binomials by inspection.

EXAMPLE 5
(a) $(y + 7)^2 = y^2 + 2 \cdot y \cdot 7 + 7^2 = y^2 + 14y + 49$

Twice the product of y and 7

(b) $(3a - 5b)^2 = (3a)^2 + \underbrace{2(3a)(-5b)} + (-5b)^2$

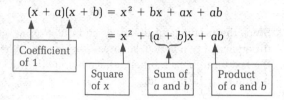

> Twice the product
> of $3a$ and $-5b$

$$= 9a^2 - 30ab + 25b^2$$

Multiplying two binomials   Product (6-4) is the result of multiplying two special binomials. Each binomial contains an x-term with a coefficient of 1. The diagram below shows how the product can be written by inspection:

$$(x + a)(x + b) = x^2 + bx + ax + ab$$

$$= x^2 + (a + b)x + ab$$

Coefficient of 1

Square of x Sum of a and b Product of a and b

EXAMPLE 6

(a) $(x + 5)(x + 3) = x^2 + (5 + 3)x + 5 \cdot 3 = x^2 + 8x + 15$

Sum of 5 and 3   Product of 5 and 3

(b) $(m - 4)(m + 2) = m^2 - 2m - 8$

Sum of -4 and 2   Product of -4 and 2

NOTE   To find a product in the form $(ax + b)(cx + d)$, where a and c are not both equal to 1, use the FOIL method discussed in Section 2-2 rather than memorize a special formula. ◄

Combined special products   All these special products should be learned well enough so that you can find them by inspection, without writing the intermediate steps. It may be necessary to write down an intermediate step in problems involving combinations of these special products, as you can see in the next examples.

EXAMPLE 7   Multiply: $5(x - 4)(x + 4)$

Solution
First, multiply the binomials. $5(x - 4)(x + 4) = 5(x^2 - 16)$

Then, multiply by the monomial. $= 5x^2 - 80$

EXAMPLE 8   Multiply: $3y(y + 2)^2$

Solution
First, square the binomial. $3y(y + 2)^2 = 3y(y^2 + 4y + 4)$

Then, multiply by the monomial. $= 3y^3 + 12y^2 + 12y$

EXAMPLE 9   Multiply: $(m + n + 3)^2$

Solution
Think of $(m + n)$ as a single term and of the whole expression as a binomial.

$$(m + n + 3)^2 = [(m + n) + 3]^2$$

First term   Second term

Square the "binomial."

$$= (m + n)^2 + 2 \cdot (m + n) \cdot 3 + 3^2$$

And simplify.

$$= m^2 + 2mn + n^2 + 6m + 6n + 9$$

Special product with cubic terms   There is an additional special product that contains cubic (third-degree) terms. Although it is not often used, it is included here for reference and for added practice in multiplying algebraic expressions:

Cube of a Binomial $(x + y)^3 = x^3 + 3x^2y + 3xy^2 + y^3$ **(6-5)**

EXAMPLE 10 $(x + 2)^3 = x^3 + 3 \cdot x^2 \cdot 2 + 3 \cdot x \cdot 2^2 + 2^3$
$$= x^3 + 6x^2 + 12x + 8$$

EXAMPLE 11 $(2x - 1)^3 = (2x)^3 + 3(2x)^2(-1) + 3(2x)(-1)^2 + (-1)^3$
$$= 8x^3 - 12x^2 + 6x - 1$$

Exercises 6-1

Find each product.

1. $15(x + 4)$
2. $x(3x - 5)$
3. $2m(m - 5)$
4. $-3a(a + 2b)$
5. $-5c^2(2c + 5)$
6. $4e^2(3e - 7)$
7. $(x + 4)(x - 4)$
8. $(y + 9)(y - 9)$
9. $(2a + 3)(2a - 3)$
10. $(3m - 4)(3m + 4)$
11. $(2x + 7y)(2x - 7y)$
12. $(x^2 + 8)(x^2 - 8)$
13. $(x + 6)^2$
14. $(y - 3)^2$
15. $(2f - 7)^2$
16. $(3a + 1)^2$
17. $(4x - 3y)^2$
18. $(5x + 2y)^2$
19. $(w + 4)(w + 6)$
20. $(v - 7)(v + 3)$
21. $(x + 8)(x - 5)$
22. $(5 + y)(3 + y)$
23. $(m + 5)(m + 3)$
24. $(n - 4)(n - 5)$
25. $(x + 5)(x - 2)$
26. $(y + 7)(y - 3)$
27. $(8 + a)(7 + a)$
28. $(m + n)(3m + n)$
29. $(x + 3y)(x - 5y)$
30. $(x - 4y)(x + 8y)$
31. $3(x + 6)(x - 6)$
32. $y(y + 2)(y - 2)$
33. $4m(3m + 1)(3m - 1)$
34. $-2n(2n + 5)(2n - 5)$
35. $5(a + 2)^2$
36. $w(w - 5)^2$
37. $-4v(2v - 3)^2$
38. $5i(3i + 7)^2$
39. $7x(x - 4)(x - 1)$
40. $-5xy(x + 9)(x + 5)$
41. $(x + y + 4)^2$
42. $(a + b - 1)^2$
43. $(x + y + 5)(x + y - 5)$
44. $(x + 2y + 7)(x + 2y - 7)$
45. $5(c + d + 2)^2$
46. $-2m(3m - n + 5)^2$

Solve.

47. *Land management*   The expression

$$\frac{V}{(1 + p)^2}$$

is used to determine the present-day value of costs that will be incurred 2 years in the future. Rewrite this expression by squaring the expression in the denominator.

48. *Physics*   The elevation E (in meters) and the boiling temperature of water T (in degrees Celsius) are related by the formula $E = 1000(100 - T) + 580(100 - T)^2$. Rewrite this formula by performing the indicated operations on the right side and combining like terms.

49. *Civil engineering*   The expression $(L + b)(2L + b)$ is found in beam deflection analysis. Multiply as indicated.

50. *Physics*   One form of the motion equation is

$$x = \frac{(v_0 + v_f)(v_f - v_0)}{2a}$$

Rewrite this equation by multiplying in the numerator.

6-2 | Factoring: Common Monomial and Difference of Two Squares

In Section 6-1 we multiplied two or more algebraic expressions to form special products. In this section and the next we will use our knowledge of these products to perform the reverse procedure, called *factoring*. We shall discuss three types of factoring, and we shall factor only polynomials with integral coefficients.

Definitions

To **factor** a polynomial means to write it as the product of two or more polynomials. We say that a polynomial is **factored completely** if each of its multinomial factors is **prime**—that is, divisible only by itself and 1.

Common monomial factoring

To factor a monomial from an expression, we must apply the distributive law [special product (6-1)] in reverse. Given a polynomial with terms that all contain a common factor, we can rewrite it as a product:

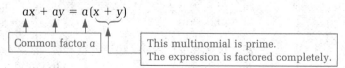

We have factored the common monomial factor a out of $ax + ay$. Since there are no other common factors, a is called the **greatest common factor,** or **GCF,** of $ax + ay$.

EXAMPLE 12   Factor completely: $3x + 6y$

Solution The greatest common factor of 3x and 6y is 3, so

$$3x + 6y = 3 \cdot x + 3 \cdot 2y = 3(x + 2y)$$

IMPORTANT ▶ Always check your answer by multiplying the factors to be certain their product is the original polynomial. This is a very important part of the factoring process. ◀

EXAMPLE 13 Factor completely: $2x^2y + x$

Solution The greatest common factor of the terms $2x^2y$ and x is x, so

$$2x^2y + x = x \cdot 2xy + x \cdot 1 = x(2xy + 1)$$

A common student error is to omit the 1 inside the parentheses. By checking your answer, you can avoid this mistake.

Failing to remove the *greatest* common factor is another frequent error. Try to avoid it in attempting to work the next example.

EXAMPLE 14 Factor completely: $12m + 18n + 30p$

Solution The terms 12m, 18n, and 30p are all divisible by 2 and 3, but their *greatest* common factor is 6. Therefore, we factor this polynomial as

$$12m + 18n + 30p = 6(2m + 3n + 5p)$$

LEARNING HINT ▶ If you had trouble determining the polynomial factor in parentheses in Example 14, try finding it this way: Once you find the greatest common factor, divide each term of the original polynomial by the GCF:

$$\frac{12m}{6} + \frac{18n}{6} + \frac{30p}{6} = 2m + 3n + 5p$$

The more complex the problem, the more useful this procedure becomes. ◀

EXAMPLE 15 Factor completely: $24ax^3 - 16a^2x^2y + 28a^3x^4$

Solution To find the greatest common factor in a complex expression like this, organize your work as follows:

Numerical coefficients: $24, -16, 28 \longrightarrow$ GCF is 4

Variable a: $a, a^2, a^3 \longrightarrow$ GCF is a

Variable x: $x^3, x^2, x^4 \longrightarrow$ GCF is x^2

Variable y: The variable y is not common to all three terms, so it will not appear in the greatest common factor.

Total GCF: $4ax^2$

By dividing each term by the greatest common factor, $4ax^2$, we find the remaining polynomial factor:

$$\frac{24ax^3}{4ax^2} - \frac{16a^2x^2y}{4ax^2} + \frac{28a^3x^4}{4ax^2} = 6x - 4ay + 7a^2x^2$$

Therefore,

$$24ax^3 - 16a^2x^2y + 28a^3x^4 = 4ax^2(6x - 4ay + 7a^2x^2)$$

Check this factorization by multiplying.

Difference of two squares   We can see how to factor the difference of two squares [special product (6-2)] by viewing the product in reverse:

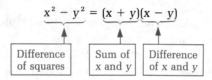

$$x^2 - y^2 = (x + y)(x - y)$$

| Difference of squares | Sum of x and y | Difference of x and y |

As you can see, the difference of two squares can be factored as the product of two binomials: one is the sum of the quantities squared, and the other is the difference of the quantities squared.

EXAMPLE 16

To factor: $x^2 - 81$

First, notice that it is the difference of two squares. $= x^2 - 9^2$

Then, write it as the product of the sum and difference shown: $= (x + 9)(x - 9)$

EXAMPLE 17   Factor completely: $4x^2 - 25$

Solution $4x^2 - 25 = (2x)^2 - (5)^2 = (2x + 5)(2x - 5)$

You will learn to factor problems of this type quickly by inspection, without writing the intermediate steps. As before, check your answer by multiplying.

In factoring the difference of two squares, remember that a variable raised to an even power is always a perfect square. For example,

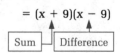

$$x^4 = (x^2)^2 \qquad x^6 = (x^3)^2 \qquad x^8 = (x^4)^2 \quad \ldots \quad \text{And so on.}$$

EXAMPLE 18 Factor completely: $16x^4 - 49y^6$

Solution $16x^4 - 49y^6 = (4x^2)^2 - (7y^3)^2 = (4x^2 + 7y^3)(4x^2 - 7y^3)$

Remember, there are two requirements for an expression to be factorable in this way: **first,** both terms must be perfect squares. **Second,** only one of them can be negative.

EXAMPLE 19 For each of the following binomials, either factor it as the difference of two squares or explain why it is unfactorable:

(a) $9x^2 + 49$ (b) $-100 - 81x^2$ (c) $36m^8 - 25n^5$ (d) $-64 + x^2$

Solution
(a) Unfactorable. This is the *sum* of squares, not the difference. There is no negative term.
(b) Unfactorable. There are two negative terms. This is not a *difference* of two terms.
(c) Unfactorable. The term $-25n^5$ is not a perfect square because the power, 5, is odd.
(d) Since there are two perfect squares and one negative term, this is factorable. However, it is easier to factor if we reverse the terms:

$$-64 + x^2 = x^2 - 64 = (x + 8)(x - 8)$$

Sometimes, the difference of two squares can be applied twice in the same problem. Always check your answer to make certain you have factored completely.

EXAMPLE 20 Factor completely: $x^4 - 16$

Solution Factoring once, we obtain

$$x^4 - 16 = (x^2 + 4)(x^2 - 4)$$

But a double check of the second binomial factor shows that it is again the difference of two squares. Factoring this binomial, we have

$$x^4 - 16 = (x^2 + 4)(x + 2)(x - 2)$$

Be sure to include the original prime factor, $x^2 + 4$, in the final answer.

Combined factoring   The factorization of many polynomials requires both of the methods learned so far.

EXAMPLE 21 Factor completely: $128x^2 - 50$

Solution $128x^2 - 50 = 2(\underbrace{64x^2 - 25}) = 2(8x + 5)(8x - 5)$

GCF   Difference of squares

In Example 21, the difference of two squares did not appear until after the common monomial factor was removed. Always remove the greatest common factor first, and then apply the difference of two squares method. This will simplify the factoring process.

EXAMPLE 22   Factor completely: $4x^4 - 16x^2$

Solution $4x^4 - 16x^2 = 4x^2(x^2 - 4) = 4x^2(x + 2)(x - 2)$

Exercises 6-2

Factor completely.

1. $2x + 8$
2. $7x - 7y$
3. $x^2 - 5x$
4. $y^3 - xy$
5. $4x^2 + 8x$
6. $5y^2 - 25y$
7. $8m^2 - 12mn$
8. $16ax - 24a^3$
9. $7x - 15y$
10. $20a^2 - 27b^2$
11. $4x - 6y + 8z$
12. $x^3 - x^2 + x$
13. $12x^2y - 24x^3 - 30x^4y^2$
14. $25m^2 + 35mn + 45m^3$
15. $18a^4b^2c - 28a^3b^3c^2 - 36a^2b$
16. $50m^2n^2p^2 + 25m^3n^3p^4 - 75mp^5$
17. $x^2 - 25$
18. $y^2 - 49$
19. $81 - x^2$
20. $1 - m^2$
21. $9a^2 - 16$
22. $36c^2 - 121$
23. $25m^2 - 36n^2$
24. $64x^2 - 9y^4$
25. $3y^2 - 48$
26. $2x^2 - 50$
27. $x^3 - 4x$
28. $y^3 - 9y$
29. $m^2 + 16$
30. $4x^2 + 36y^2$
31. $36 - 9y^2$
32. $9y^2 - 81x^2$
33. $5x^4y - 45y$
34. $-100y^8 + 25y^2$
35. $x^8 - 81$
36. $5y^4 - 5$

Solve.

37. *Cartography*   The formula below is used to find the maximum length for a line on a map. Rewrite this formula by factoring the expression under the radical.

$$s = \sqrt{8R\varepsilon - 4\varepsilon^2}$$

38. *Economics*   The expression $10q - 5q^2$ is used to determine revenue.* Factor this expression.
39. *Surveying*   The formula $d^2 = L^2 - H^2$ is used to find the difference in elevation of a slope. Rewrite this formula by factoring the right side.
40. *Thermodynamics*   When an object at temperature T_1 is surrounded by walls at a temperature of T_2, the net flow of energy by radiation is given by $e\sigma(T_1^4 - T_2^4)$. Rewrite this expression by factoring the binomial.
41. *Physics*   A factor in one of the Hermite functions is $32y^5 - 160y^3 + 120y$. Factor this expression.
42. *Civil engineering*   The equation $9x^2 - 4L^2 = 0$ occurs in determining the deflection of a beam. Factor the left side of this equation.

*Based on an interview with Anthony Marino, Professor of Economics, University of Kansas, Lawrence, Kansas.

6-3 | **Factoring Trinomials**

So far we have discussed two ways to factor polynomials based on the special products from Section 6-1. Common monomial factoring, based on special product (6-1), applies to any polynomial with terms that all contain a common factor. The difference of two squares, based on special product (6-2), applies only to certain binomials. In this section we shall examine methods of factoring trinomials like special products (6-3) and (6-4) as well as more complex trinomials.

Special product (6-4)   We begin by factoring only trinomials in the form of special product (6-4). The following diagram points out several important facts:

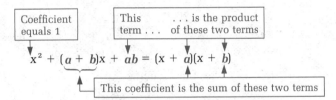

Thus, trinomials of this type can be factored as the product of two binomials. The first term of each binomial factor is x. The last terms of the binomials are the integers that meet the conditions shown above.

EXAMPLE 23   Factor: $x^2 + 5x + 6$

Solution

The x-squared term comes from the product of $x \cdot x$.

Step 1. $x^2 + 5x + 6 = (x \qquad ?)(x \qquad ?)$

Step 2.   We must replace the question marks with two integers whose product is 6, and whose sum is 5. To find them, list all pairs of integers whose product is 6 and select the pair whose sum is 5:

Pairs of integers whose product is 6	−1	−2	1	2
	−6	−3	6	3
Sum of each pair	−7	−5	7	5

The integers 2 and 3 have a product of 6 and a sum of 5.

Step 3.   Now we can factor the trinomial. The integers found in Step 2 are the last terms in the binomial factors:

$x^2 + 5x + 6 = (x + 2)(x + 3)$

Step 4.   Check your answer by multiplying the factors. Their product should be the original trinomial.

In Example 23 we were looking for a positive product and a positive sum. Our knowledge of signed numbers tells us that only a pair of positive integers will produce this result. Thus, we could have saved time by considering only 1 and 6 or 2 and 3 as possibilities in Step 2. ◀

EXAMPLE 24   Factor: $x^2 - 7x + 12$

Solution

Step 1. $x^2 - 7x + 12 = (x \quad ?)(x \quad ?)$

Step 2.   We must find two integers whose product is 12 and whose sum is -7. The possibilities are:

Integers with a product of 12	-1	-2	-3	1	2	3
	-12	-6	-4	12	6	4
Sum of each pair	-13	-8	-7	13	8	7

The integers -3 and -4 have a product of 12 and a sum of -7. Of course, you can stop testing as soon as you find the correct pair.

Step 3.   Factor the trinomial:

$$x^2 - 7x + 12 = (x - 3)(x - 4)$$

Step 4.   Check the factors by multiplying.

In Example 24 we were looking for a pair of integers with a positive product and a negative sum. You can save time by realizing that only a pair of negative integers can produce this result. ◀

EXAMPLE 25   Factor: $y^2 + 3y - 18$

Solution

Step 1. $y^2 + 3y - 18 = (y \quad ?)(y \quad ?)$

Step 2.   Find a pair of integers whose product is -18 and whose sum is 3. The possibilities are:

Pairs of integers with product of -18	1	2	3	-1	-2	-3
	-18	-9	-6	18	9	6
Sum of each pair	-17	-7	-3	17	7	3

The integers -3 and 6 have a product of -18 and a sum of 3.

Step 3. $y^2 + 3y - 18 = (y - 3)(y + 6)$

Step 4.   The answer should be checked by multiplying.

EXAMPLE 26   Factor: $x^2 - x - 30$

Solution

Step 1. $x^2 - x - 30 = (x \quad ?)(x \quad ?)$

Step 2.   We must find a pair of integers whose product is -30 and whose sum is -1:

Possible pairs of integers	1	2	3	5	−1	−2	−3	−5
	−30	−15	−10	−6	30	15	10	6
Sum of pairs	−29	−13	−7	−1	29	13	7	1

The correct pair is 5, -6.

Step 3. $x^2 - x - 30 = (x + 5)(x - 6)$

Step 4.   Check by multiplying.

With practice you will learn to find the correct pair of integers mentally and thus factor these simple trinomials by inspection.

Not all trinomials of this type can be factored.

EXAMPLE 27   Factor: $x^2 - 5x + 7$

Solution   No two integers have a product of 7 and a sum of -5. This trinomial is unfactorable.

Factoring the general trinomial   When the coefficient of the x^2-term is not 1, factoring is more difficult. Let's examine the product $(ax + b)(cx + d)$ to help discover a method:

When at least one of these terms has a coefficient other than 1 . . .	. . . the x^2-term will have a coefficient other than 1 . . .

$$(ax + b)(cx + d) = acx^2 + adx + bcx + bd$$
$$= acx^2 + (ad + bc)x + bd$$

. . . and the x-term will be the sum of the outer and inner products

The constant term bd is still the product of the last terms of the binomials.

EXAMPLE 28   Factor: $2x^2 + 7x + 3$

Solution
Step 1.   The x^2-term is the product of $2x \cdot x$. (We need only consider positive factors here.)

$$2x^2 + 7x + 3 = (2x \quad ?)(x \quad ?)$$

Step 2.   The constant term, 3, is the product of $3 \cdot 1$ or $(-3)(-1)$. Combining these possibilities with Step 1, we can now list the possible factors of the trinomial. Beside each pair of factors we list the sum of the outer and inner products.

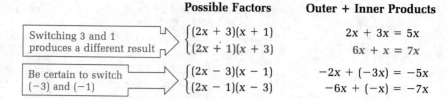

	Possible Factors	Outer + Inner Products
Switching 3 and 1 produces a different result	$\begin{cases}(2x + 3)(x + 1) \\ (2x + 1)(x + 3)\end{cases}$	$2x + 3x = 5x$ $6x + x = 7x$
Be certain to switch (-3) and (-1)	$\begin{cases}(2x - 3)(x - 1) \\ (2x - 1)(x - 3)\end{cases}$	$-2x + (-3x) = -5x$ $-6x + (-x) = -7x$

Step 3. We select the combination of factors that produces a sum of $7x$. This is the correct factorization:

$$2x^2 + 7x + 3 = (2x + 1)(x + 3)$$

Step 4. Check by multiplying.

EXAMPLE 29 Factor: $4x^2 + 3x - 10$

Solution

Step 1. The x^2-term can be the product of $4x \cdot x$ or of $2x \cdot 2x$:

$$4x^2 + 3x - 10 = (4x \quad ?)(x \quad ?)$$

or

$$(2x \quad ?)(2x \quad ?)$$

Step 2. The constant term is one of the following products:

$$10(-1) \qquad 1(-10) \qquad 5(-2) \qquad 2(-5)$$

Combining these possibilities with the possible factors for the first terms, we obtain the following list:

Possible Factors	Outer + Inner	Possible Factors	Outer + Inner
$(4x + 10)(x - 1)$	$6x$	$(4x - 10)(x + 1)$	$-6x$
$(4x + 1)(x - 10)$	$-39x$	$(4x - 1)(x + 10)$	$39x$
$(4x + 5)(x - 2)$	$-3x$	$(4x - 5)(x + 2)$	$3x$ ⟸
$(4x + 2)(x - 5)$	$-18x$	$(4x - 2)(x + 5)$	$18x$
$(2x + 10)(2x - 1)$	$18x$	$(2x - 10)(2x + 1)$	$-18x$
$(2x + 5)(2x - 2)$	$6x$	$(2x - 5)(2x + 2)$	$-6x$

Step 3. Select the combination with outer and inner products that sum to $3x$:

$$4x^2 + 3x - 10 = (4x - 5)(x + 2)$$

Step 4. Check by multiplying.

Do not be discouraged by the number of possibilities that need testing. After enough practice you will learn to find the correct factorization quickly.

LEARNING HINT ▶ There is an algebraic fact that can help reduce the number of possibilities for problems such as these. When two binomials are multiplied and one contains a common monomial factor, their product will also contain that common factor. Knowing this, you do not even need to test the following possibilities from Example 29:

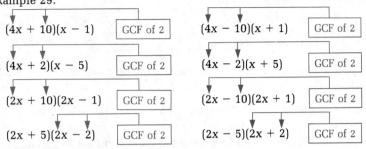

$(4x + 10)(x - 1)$ | GCF of 2 | $(4x - 10)(x + 1)$ | GCF of 2 |

$(4x + 2)(x - 5)$ | GCF of 2 | $(4x - 2)(x + 5)$ | GCF of 2 |

$(2x + 10)(2x - 1)$ | GCF of 2 | $(2x - 10)(2x + 1)$ | GCF of 2 |

$(2x + 5)(2x - 2)$ | GCF of 2 | $(2x - 5)(2x + 2)$ | GCF of 2 |

Since each of these possibilities contains a GCF in at least one of the binomials, the resulting product would contain a common factor. But the trinomial we are attempting to factor, $4x^2 + 3x - 10$, does *not* contain a common factor. Therefore, none of the above possibilities would work. ◀

Perfect squares   If the first and last terms of a trinomial are both perfect squares, then the trinomial itself might be a perfect square. We can discover how to test these by examining special product (6-3):

| Square of x | Twice the product of x and y | Square of y |

$$x^2 \;+\; 2xy \;+\; y^2 \;=\; (x + y)^2$$

EXAMPLE 30   Factor: $4x^2 + 12x + 9$

Solution   Since the first and last terms are both perfect squares, this trinomial might be a perfect square. To test this, proceed as follows:

Take the square root of the first term. $\sqrt{4x^2} = 2x$

Take the square root of the last term. $\sqrt{9} = 3$

Then find twice the product of these square roots. $2 \cdot 2x \cdot 3 = 12x$

Since the result equals the middle term, the trinomial is a perfect square:

$$4x^2 + 12x + 9 = (2x + 3)^2$$

| Square root of first term | | Square root of last term |

If you do not happen to recognize a perfect square, you will still arrive at the correct factorization using one of the other methods of factoring trinomials.

Combined factoring   Before factoring a trinomial, check to see if it contains a common monomial factor. If so, remove the GCF first.

EXAMPLE 31 Factor completely: $12x^3 - 14x^2 + 4x$

Solution

First, remove the GCF. $\qquad 12x^3 - 14x^2 + 4x = 2x(6x^2 - 7x + 2)$

Then, factor the trinomial. $\qquad\qquad\qquad = 2x(3x - 2)(2x - 1)$

Be sure to include the original GCF in the final answer.

When the first term is negative

The methods we have discussed so far work best when the first term of the trinomial is positive. If the first term is negative, it is usually best to factor out -1 from each term and continue as before.

EXAMPLE 32 Factor: $-x^2 - 2x + 24$

Solution

First, remove a common factor of -1. $\qquad -x^2 - 2x + 24 = -1(x^2 + 2x - 24)$

Then, factor the resulting trinomial. $\qquad\qquad\qquad = -1(x + 6)(x - 4)$

Finally, we can eliminate the factor of -1 by rewriting: $\qquad\qquad\qquad = (-x - 6)(x - 4)$

or $\qquad\qquad\qquad\qquad\qquad\qquad (x + 6)(-x + 4)$

Trinomials in two variables

The methods we have studied may also be used to factor certain trinomials containing two variables.

EXAMPLE 33 Factor: $x^2 + 8xy + 12y^2$

Solution

First, proceed as though the y's were not present, and factor. $\qquad x^2 + 8x + 12 = (x + 6)(x + 2)$

Then, attach a y to each constant in the factors. $\qquad x^2 + 8xy + 12y^2 = (x + 6y)(x + 2y)$

Finally, be certain to check your factorization by multiplying the factors.

Exercises 6-3

Factor completely.

1. $x^2 + 9x + 8$ $\qquad$ **2.** $x^2 - 11x + 24$ $\qquad$ **3.** $a^2 + a - 42$
4. $m^2 - 2m - 48$ $\qquad$ **5.** $x^2 - 10x + 25$ $\qquad$ **6.** $x^2 + 6x + 9$
7. $x^2 + 14xy + 33y^2$ $\qquad$ **8.** $x^2 - 6xy - 27y^2$ $\qquad$ **9.** $x^2 + 8x + 10$

10. $x^2 - 9x - 22$

11. $3x^2 + 10x + 3$

12. $2x^2 - 7x + 5$

13. $5x^2 - 2xy - 7y^2$

14. $7x^2 + 5xy - 2y^2$

15. $2n^2 + 15n + 27$

16. $5x^2 - 18x + 16$

17. $3x^2 - 2xy - 21y^2$

18. $2a^2 - 5ab - 25b^2$

19. $9x^2 + 30x + 25$

20. $4x^2 - 36xy + 81y^2$

21. $6x^2 + 11x + 3$

22. $4m^2 + 13mn + 10n^2$

23. $8x^2 - 35x + 12$

24. $6x^2 - 25x + 24$

25. $15y^2 + 4y - 4$

26. $10y^2 + 13y - 30$

27. $12m^2 - 33mn + 9n^2$

28. $8w^2 + 9w - 14$

29. $-x^2 + 8x + 20$

30. $6y^2 - 4y + 3$

31. $2x^2 - 12x + 16$

32. $-3x^2 + 21x - 18$

33. $y^4 - 5y^3 - 36y^2$

34. $6x^3 - 10x^2 - 4x$

35. $x^4 - 13x^2 + 36$

36. $8x^4 - 26x^2 + 18$

Solve.

37. *Physics* Under specified conditions, the displacement of an object in motion is given by $16t^2 + 32t - 48$. Factor this expression.

38. *Civil engineering* In solving for the dimensions of a certain open-topped channel, the following trinomial may be encountered: $H^2 - 6H + 8$. Factor this expression.

39. *Electronics* For a certain electrical circuit, the current I can be found by solving the equation $4I^2 + 120I - 396 = 0$. Factor the left side of this equation.

40. *Physics* The height h of an object thrown vertically upward from a cliff 24 ft high at an initial velocity of 8 ft/s is given by $h = 24 + 8t - 16t^2$. Factor the right side of this equation.

41. *Fire science* The flow rate Q in a certain hose can be found by solving the equation $2Q^2 + Q - 21 = 0$. Factor the left side of this equation.

42. *Civil engineering* The following equation occurs in determining the deflection of a certain beam: $9x^2 - 32Lx + 28L^2 = 0$. Factor the left side of this equation.

6-4 | Equivalent Fractions

Algebraic expressions and equations often contain fractions. Any technical student who deals with applications must be able to perform operations with algebraic fractions. These operations follow the same principles as do arithmetic fractions; however, basic operations with algebraic fractions may seem more complicated because of the presence of literal numbers.

Definitions If a and b are algebraic expressions, then the fraction a/b is called an **algebraic fraction.** An algebraic fraction is a real number for any real values of a and b except $b = 0$.

Two algebraic fractions are said to be **equivalent** if they are alternative ways of writing the same algebraic expression. Equivalent algebraic fractions name the same real number. The algebraic fractions x/y, $2x/2y$, $3x/3y$, and so on, are all equivalent algebraic fractions.

Any algebraic fraction can be rewritten in an infinite number of equivalent forms by multiplying or dividing its numerator and denominator by the same nonzero quantity.

EXAMPLE 34 Change $\dfrac{4}{5}$ to an equivalent fraction with a denominator of 10.

Solution

Since the denominator must be multiplied by 2, the numerator must also be multiplied by 2 to produce an equivalent fraction.

$$\frac{4}{5} = \frac{?}{10}$$

$$\frac{4}{5} = \frac{4 \cdot 2}{5 \cdot 2} = \frac{8}{10}$$

EXAMPLE 35   Change $\dfrac{3x}{7y}$ to an equivalent fraction with a denominator of $21y^2$.*

Solution

To change $7y$ to $21y^2$, multiply by $3y$. To create an equivalent fraction, multiply the numerator by $3y$ also.

$$\frac{3x}{7y} = \frac{?}{21y^2}$$

$$\frac{3x}{7y} = \frac{3x \cdot 3y}{7y \cdot 3y} = \frac{9xy}{21y^2}$$

Multinomial denominators   Changing to equivalent fractions with different denominators is slightly more difficult with multinomials. You may need to factor the denominators to determine the multiplier.

EXAMPLE 36   Change $\dfrac{x + 2}{x - 5}$ to an equivalent fraction with a denominator of $x^2 - 8x + 15$.

Step 1.   Factor the desired denominator:

$$\frac{x + 2}{x - 5} = \frac{?}{x^2 - 8x + 15} = \frac{?}{(x - 5)(x - 3)}$$

Step 2.   Compare the original denominator to the new denominator to determine the necessary multiplier. In this case, the original denominator already contains the factor $(x - 5)$. Thus, we need only multiply it by $(x - 3)$ to produce the desired denominator. To create an equivalent fraction, multiply the numerator by $(x - 3)$ also:

$$\frac{x + 2}{x - 5} = \frac{(x + 2)(x - 3)}{(x - 5)(x - 3)} = \frac{x^2 - x - 6}{x^2 - 8x + 15}$$

In the next example we must factor the original denominator in addition to the new one.

EXAMPLE 37   Change $\dfrac{5}{x^2 - 4x}$ to an equivalent fraction with a denominator of $4x^3 - 64x$.

---

* When variables appear in the denominator, the fraction is undefined for any values of the variable that make the denominator equal to zero. Rather than state restrictions for every problem, we will assume that these restrictions exist.

Solution **First,** factor both denominators and underline the common factors:

$$\frac{5}{x^2 - 4x} = \frac{?}{4x^3 - 64x}$$

$$\underline{x}(\underline{x - 4}) \qquad 4\underline{x}(x + 4)(\underline{x - 4})$$

We can now see that the original denominator must be multiplied by 4 and $(x + 4)$ to produce the new denominator.

Second, multiply both the numerator and denominator by $4(x + 4)$ to produce an equivalent fraction:

$$\frac{5}{x^2 - 4x} = \frac{5 \quad \cdot 4(x + 4)}{(x^2 - 4x) \cdot 4(x + 4)} = \frac{20x + 80}{4x^3 - 64x}$$

Reducing fractions

Earlier, we mentioned that equivalent fractions can be formed by multiplying or dividing the numerator and denominator by the same nonzero number. Multiplying to form an equivalent fraction is the procedure we just described. It is used to create common denominators in the first step of an addition or subtraction problem. Forming equivalent fractions through division is the procedure used to rewrite a fraction in its simplest equivalent form. This process, also known as **reducing to lowest terms,** is normally the final step in a problem involving any operation with fractions. To reduce a fraction to lowest terms, follow this procedure:

- **First,** examine both the numerator and denominator of the fraction to find their greatest common factor (GCF).
- **Second,** divide both the numerator and denominator by the GCF.

EXAMPLE 38 Reduce to lowest terms: $\dfrac{24}{42}$

Solution

The GCF of 24 and 42 is 6.

$$\frac{24}{42} = \frac{6 \cdot 4}{6 \cdot 7}$$

Dividing both the numerator and denominator by 6, we have:

$$= \frac{4}{7}$$

This process is usually written as:

$$\frac{24}{42} = \frac{\cancel{6} \cdot 4}{\cancel{6} \cdot 7} = \frac{4}{7}$$

The "cancel" marks are a reminder that we have divided both the numerator and denominator by the factor 6.

LEARNING HINT ▶

Another way to look at this procedure is to note that a separate fraction containing the GCF's can be formed:

$$\frac{24}{42} = \frac{6}{6} \cdot \frac{4}{7} = 1 \cdot \frac{4}{7} = \frac{4}{7}$$

| This fraction has a value of 1. | ◀

Now we will use this same procedure on an algebraic fraction.

EXAMPLE 39 Reduce to lowest terms: $\dfrac{20x^3y^2w}{35xy^2w^3}$

Solution

First, establish the greatest common factor of the numerator and denominator by breaking them down into parts.

$$20, 35 \longrightarrow \text{GCF is } 5$$
$$x^3, x \longrightarrow \text{GCF is } x$$
$$y^2, y^2 \longrightarrow \text{GCF is } y^2$$
$$w, w^3 \longrightarrow \text{GCF is } w$$

Total GCF: $5xy^2w$

Second, divide both the numerator and denominator by the GCF:

$$\frac{20x^3y^2w}{35xy^2w^3} = \frac{\boxed{(5xy^2w)} \cdot 4x^2}{\boxed{(5xy^2w)} \cdot 7w^2} = \frac{4x^2}{7w^2}$$

CAUTION 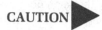 Be very careful when simplifying expressions with more than one term. Only common *factors* of the numerator and denominator can be canceled or divided out in this way. ◀

EXAMPLE 40 Reduce to lowest terms: $\dfrac{y^2 - 4}{y^2 - 5y + 6}$

Solution

First, factor both the numerator and denominator.

$$\frac{y^2 - 4}{y^2 - 5y + 6} = \frac{(y + 2)(y - 2)}{(y - 3)(y - 2)}$$

Then, divide out or cancel common factors.

$$\frac{(y + 2)\cancel{(y - 2)}}{(y - 3)\cancel{(y - 2)}} = \frac{y + 2}{y - 3}$$

Do not make the mistake of trying to reduce this further by canceling the *y*'s in the answer. Each *y* here is a *term*, not a factor, and therefore cannot be divided out of the fraction.

EXAMPLE 41 Reduce to lowest terms: $\dfrac{6m}{6m + 1}$

Solution The denominator contains a single binomial factor, $6m + 1$. The numerator contains a single monomial factor, $6m$. Since these factors are not the same, the fraction cannot be reduced.

Be careful. Complete factoring is the key to reducing an algebraic fraction correctly. Remember always to remove the greatest common monomial factor first.

EXAMPLE 42 Reduce to lowest terms: $\dfrac{6x^2 + 27x + 30}{6x^2 - 24}$

Solution Factor completely:

First, find any monomial factors.

$$\frac{6x^2 + 27x + 30}{6x^2 - 24} = \frac{3(2x^2 + 9x + 10)}{6(x^2 - 4)}$$

Second, find any binomial factors.

$$= \frac{3(2x + 5)(x + 2)}{6(x - 2)(x + 2)}$$

Divide the numerator and denominator by the common factors 3 and (x + 2):

$$\frac{\overset{1}{\cancel{3}}(2x + 5)\cancel{(x + 2)}}{\underset{2}{\cancel{6}}(x - 2)\underset{1}{\cancel{(x + 2)}}} = \frac{2x + 5}{2x - 4}$$

Notice that we cannot reduce this fraction any further because 2x is a term, not a factor.

When negative quantities appear in the numerator or denominator, it may be helpful to remember that $-x = (-1)x$. For example,

$$\frac{-3x}{3y} = \frac{(-1) \cdot \cancel{3} \cdot x}{\cancel{3} \cdot y} = \frac{(-1)x}{y} = \frac{-x}{y}$$

Also, because of the distributive law, we may write

$$x - y = (-1)(y - x)$$

and therefore,

$$\frac{x - y}{y - x} = \frac{(-1)\cancel{(y - x)}}{\cancel{y - x}} = -1$$

EXAMPLE 43 Reduce to lowest terms: $\dfrac{\overset{1}{2 - x}}{\underset{1}{x^2 - 4}}$

Solution Factoring the denominator, we obtain

$$\frac{2 - x}{(x + 2)(x - 2)}$$

Notice that the numerator (2 − x) can be written as (−1)(x − 2). Therefore,

$$\frac{2 - x}{(x + 2)(x - 2)} = \frac{(-1)\cancel{(x - 2)}}{(x + 2)\cancel{(x - 2)}} = \frac{-1}{x + 2}$$

Exercises 6-4

Change each fraction to an equivalent fraction with the indicated denominator.

1. $\dfrac{5}{8}$ to $\dfrac{?}{24}$

2. $\dfrac{3}{7}$ to $\dfrac{?}{35}$

3. $\dfrac{2x}{y}$ to $\dfrac{?}{3y^2}$

4. $\dfrac{3a}{b}$ to $\dfrac{?}{4b^3}$

5. $\dfrac{5m + n}{2n}$ to $\dfrac{?}{6n^3}$

6. $\dfrac{7x - 3y}{3x}$ to $\dfrac{?}{12x^2}$

7. $\dfrac{5}{x + 3}$ to $\dfrac{?}{3x(x + 3)}$

8. $\dfrac{x}{2x - 3}$ to $\dfrac{?}{(2x - 3)(3x + 5)}$

9. $\dfrac{x + 4}{x - 5}$ to $\dfrac{?}{2x^2 - 10x}$

10. $\dfrac{5}{y^2 + 2y}$ to $\dfrac{?}{3y^3 + 6y^2}$

11. $\dfrac{m - 3}{m + 4}$ to $\dfrac{?}{m^2 + 6m + 8}$

12. $\dfrac{5n}{n - 5}$ to $\dfrac{?}{n^2 - 25}$

13. $\dfrac{2y + 3}{3y - 2}$ to $\dfrac{?}{9y^2 - 4}$

14. $\dfrac{5x}{2x - 3}$ to $\dfrac{?}{6x^2 + x - 15}$

Reduce each fraction to lowest terms, if possible.

15. $\dfrac{18}{42}$

16. $\dfrac{30}{45}$

17. $\dfrac{4x}{6x^2}$

18. $\dfrac{10y^3}{18y}$

19. $\dfrac{24a^2b}{40ab^2}$

20. $\dfrac{54m^3n^2p}{36mp^5}$

21. $\dfrac{5a - 2b}{10a - 4b}$

22. $\dfrac{6x - 12y}{4x - 8y}$

23. $\dfrac{5x + 15y}{5x - 15y}$

24. $\dfrac{4m - 6n}{6m + 9n}$

25. $\dfrac{x^2 + 4}{x^2 - 4}$

26. $\dfrac{y^2 - 9}{3y + 9}$

27. $\dfrac{x + 5}{x^2 - 25}$

28. $\dfrac{x^2 - 9y^2}{x - 3y}$

29. $\dfrac{3x}{3x + 15}$

30. $\dfrac{x^2 + 4x}{x^2 - 4x}$

31. $\dfrac{x^2 - 16}{x^2 + 6x + 8}$

32. $\dfrac{a^2 - 2ab + b^2}{a^2 - b^2}$

33. $\dfrac{y^2 - 9y}{3y^2 + 9y}$

34. $\dfrac{4x + 6}{4x^2 - 9}$

35. $\dfrac{x^2 + 7x + 12}{x^2 - x - 12}$

36. $\dfrac{y^2 - 8y + 16}{y^2 - 9y + 20}$

37. $\dfrac{9 - x^2}{x^2 + 4x - 21}$

38. $\dfrac{x^2 - 3x + 2}{4 - x^2}$

39. $\dfrac{x^3 + 3x^2 - 28x}{x^3 + 8x^2 + 7x}$

40. $\dfrac{2x^2 + 22x + 56}{4x^2 + 28x + 48}$

41. $\dfrac{6x^2 - 17x + 5}{6x^2 - 13x - 5}$

42. $\dfrac{8x^2 - 10x + 3}{10x^2 - 19x + 6}$

43. $\dfrac{16 - 4x^2}{2x^2 - 8x + 8}$

44. $\dfrac{3x^2 + 6x - 45}{81 - 9x^2}$

6-5 Multiplication and Division of Fractions

Multiplication   Algebraic fractions can be multiplied in the same way that arithmetic fractions are multiplied. The product of two fractions is another fraction with a numerator that is the product of the original numerators and a denominator that is the product of the original denominators. In general,

$$\frac{a}{b} \cdot \frac{c}{d} = \frac{ac}{bd}$$ ⟨ Multiply the numerators
⟨ Multiply the denominators

EXAMPLE 44

(a) $\dfrac{2}{5} \cdot \dfrac{3}{7} = \dfrac{2 \cdot 3}{5 \cdot 7} = \dfrac{6}{35}$ (b) $\dfrac{4a^2}{3y} \cdot \dfrac{5x}{9y^2} = \dfrac{4 \cdot 5 \cdot a^2 \cdot x}{3 \cdot 9 \cdot y \cdot y^2} = \dfrac{20a^2x}{27y^3}$

If one of the expressions to be multiplied is not a fraction, write it as a fraction with a denominator equal to 1, and then multiply.

EXAMPLE 45 $2 \cdot \dfrac{x}{3y} = \dfrac{2}{1} \cdot \dfrac{x}{3y} = \dfrac{2x}{3y}$

When the numerator of one fraction shares a common factor with the denominator of the other, we should divide both by this common factor before multiplying. This avoids having to reduce the final answer.

EXAMPLE 46   Multiply: $\dfrac{6}{7} \cdot \dfrac{21}{8}$

Solution

$$\dfrac{6}{7} \cdot \dfrac{21}{8} = \dfrac{6 \cdot 21}{7 \cdot 8}$$

Divide the numerator and denominator by 7.

$$= \dfrac{6 \cdot \cancel{7} \cdot 3}{\cancel{7} \cdot 8} = \dfrac{6 \cdot 3}{8}$$

Divide again—this time by 2.

$$= \dfrac{\overset{3}{\cancel{6}} \cdot 3}{\underset{4}{\cancel{8}}}$$

Then multiply the remaining factors to obtain an answer in lowest terms.

$$= \dfrac{9}{4}$$

EXAMPLE 47   Multiply: $\dfrac{9m^2}{10n} \cdot \dfrac{25n^4}{18m^4}$

Solution

Combine and rearrange the factors.

$$\dfrac{9m^2}{10n} \cdot \dfrac{25n^4}{18m^4} = \dfrac{9 \cdot 25 \cdot m^2 \cdot n^4}{10 \cdot 18 \cdot m^4 \cdot n}$$

Divide out (or cancel) the common factors 9, 5, m^2, and n.

$$= \dfrac{\overset{1}{\cancel{9}} \cdot \overset{5}{\cancel{25}} \cdot \overset{1}{\cancel{m^2}} \cdot \overset{3}{\cancel{n^4}}}{\underset{2}{\cancel{10}} \cdot \underset{2}{\cancel{18}} \cdot \underset{2}{m^4} \cdot \underset{1}{\cancel{n}}}$$

Multiply the remaining factors.

$$= \dfrac{5n^3}{4m^2}$$

When multinomials appear in an algebraic fraction, they must be factored completely before you proceed.

EXAMPLE 48   Multiply: $\dfrac{2x^2 - 10x}{x^2 + 2x} \cdot \dfrac{x^2 + 9x + 14}{x^2 - 25}$

Solution

First, factor all multinomials.

$$\dfrac{2x(x - 5)}{x(x + 2)} \cdot \dfrac{(x + 7)(x + 2)}{(x + 5)(x - 5)}$$

Then, divide the numerator and denominator by the common factors x, $(x - 5)$, and $(x + 2)$.

$$= \dfrac{2x\cancel{(x - 5)}(x + 7)\cancel{(x + 2)}}{\cancel{x}\cancel{(x + 2)}(x + 5)\cancel{(x - 5)}}$$

The final product is:
$$= \frac{2(x + 7)}{x + 5} \quad \text{or} \quad \frac{2x + 14}{x + 5}$$

NOTE   In the final answer, multinomial factors do not need to be multiplied. However, when two or more monomial factors appear in the numerator or in the denominator, they should be multiplied. ◀

Division   To divide two fractions, invert the divisor and multiply. That is,

Invert the divisor
$$\frac{a}{b} \div \frac{c}{d} = \frac{a}{b} \cdot \frac{d}{c} = \frac{ad}{bc}$$
Multiply

EXAMPLE 49   Divide: (a) $\dfrac{3}{8} \div \dfrac{5}{6}$ (b) $\dfrac{12x^2}{7y} \div \dfrac{15x^3}{14y^2}$

Solution

Invert the divisor

(a) $\dfrac{3}{8} \div \dfrac{5}{6} = \dfrac{3}{8} \cdot \dfrac{6}{5} = \dfrac{3 \cdot 6}{8 \cdot 5} = \dfrac{9}{20}$

Multiply

Invert

(b) $\dfrac{12x^2}{7y} \div \dfrac{15x^3}{14y^2} = \dfrac{12x^2}{7y} \cdot \dfrac{14y^2}{15x^3}$

Multiply

$$= \frac{12 \cdot 14 \cdot x^2 \cdot y^2}{7 \cdot 15 \cdot x^3 \cdot y}$$

$$= \frac{8y}{5x}$$

In dividing, factor any existing multinomials before you invert.

EXAMPLE 50

Divide:
$$\frac{3x^2 - 12x}{9 - x^2} \div \frac{x^3 - 9x^2 + 20x}{x^2 - 10x + 21}$$

First, factor.
$$= \frac{3x(x - 4)}{(3 + x)(3 - x)} \div \frac{x(x - 5)(x - 4)}{(x - 7)(x - 3)}$$

Then, invert and multiply.
$$= \frac{3x(x - 4)}{(3 + x)(3 - x)} \cdot \frac{(x - 7)(x - 3)}{x(x - 5)(x - 4)}$$

Finally, divide by the common factors.
$$= \frac{3x(x - 4)(x - 7)(-1)(3 - x)}{(3 + x)(3 - x)x(x - 5)(x - 4)}$$

$$= \frac{-3(x - 7)}{(3 + x)(x - 5)} \quad \text{or} \quad \frac{-3x + 21}{x^2 - 2x - 15}$$

In the third step, notice that $(x - 3) = (-1)(3 - x)$.

EXAMPLE 51 Simplify: $\dfrac{2x^2 + 13x + 15}{6x} \cdot \dfrac{6x^3 - 24x^2 - 30x}{2x^2 + 5x + 3} \div (x^2 - 25)$

Solution To avoid confusion, rewrite the divisor with a denominator of 1:

$$\dfrac{2x^2 + 13x + 15}{6x} \cdot \dfrac{6x^3 - 24x^2 - 30x}{2x^2 + 5x + 3} \div \dfrac{(x^2 - 25)}{1}$$

Factor, invert, and multiply:

$$= \dfrac{(2x + 3)(x + 5)}{6x} \cdot \dfrac{6x(x - 5)(x + 1)}{(2x + 3)(x + 1)} \cdot \dfrac{1}{(x + 5)(x - 5)}$$

Divide out the common factors:

$$= \dfrac{\cancel{(2x + 3)}\cancel{(x + 5)}\cancel{6x}\cancel{(x - 5)}\cancel{(x + 1)}}{\cancel{6x}\cancel{(2x + 3)}\cancel{(x + 1)}\cancel{(x + 5)}\cancel{(x - 5)}}$$

Since all the factors cancel, the answer is 1.

CAUTION  Do not make the mistake of thinking the last answer is equal to zero. A nonzero number divided by itself is equal to 1. ◀

Exercises 6-5

Perform the indicated operations.

1. $\dfrac{5}{8} \cdot \dfrac{6}{7}$

2. $\dfrac{3}{4} \cdot \dfrac{8}{9}$

3. $\dfrac{2}{5} \cdot \dfrac{25}{16} \cdot \dfrac{4}{5}$

4. $\dfrac{8}{11} \cdot \dfrac{7}{12} \cdot 33$

5. $\dfrac{3x^4}{4y} \cdot \dfrac{10y^2}{21x^2}$

6. $\dfrac{8a^2y}{15bx^3} \cdot \dfrac{9b^3x^2}{16ay^2}$

7. $\dfrac{5m}{3m + 12} \cdot \dfrac{2m + 8}{10m^2}$

8. $\dfrac{r^2 - 9}{4c} \cdot \dfrac{c^2}{3r + 9}$

9. $\dfrac{x^2 - 4y^2}{x + 3y} \cdot (2x + 6y)$

10. $\dfrac{2a - 4b}{a^2 - b^2} \cdot \dfrac{2a + 2b}{3a - 6b}$

11. $\dfrac{x^2 - 9x + 14}{x^2 + 7x + 12} \cdot \dfrac{4x^3 + 16x^2}{3x^2 - 21x}$

12. $\dfrac{x^2 - 5x + 4}{x^2 - 1} \cdot \dfrac{(x + 1)^2}{x^2 - x - 12}$

13. $\dfrac{4 - x^2}{2 + x} \cdot \dfrac{x - 1}{x^2 - 3x + 2}$

14. $\dfrac{a^2 + 5a + 6}{9 - a^2} \cdot \dfrac{a^2 - 7a + 12}{a^2 - 4}$

15. $(x^2 - 9y^2) \cdot \dfrac{x^2 + 5xy + 6y^2}{(x + 3y)^2}$

16. $\dfrac{a^2 - a - 42}{5a^2 - 80} \cdot \dfrac{a^2 + 6a + 8}{a^2 - 10a + 21} \cdot (5a - 15)$

17. $\dfrac{2}{5} \div \dfrac{6}{11}$

18. $\dfrac{6}{25} \div \dfrac{8}{15}$

19. $\dfrac{3}{4} \div \dfrac{4}{9}$

20. $\dfrac{6}{5} \div \dfrac{15}{16}$

21. $\dfrac{7pq^2}{10m^2n} \div \dfrac{14p^3}{15mn}$

22. $\dfrac{2xy}{6b^2} \div \dfrac{4x^2}{3ab^2}$

23. $\dfrac{6w - 10}{v^2} \div \dfrac{3w - 5}{2v}$

24. $\dfrac{6t}{s^2 - 16} \div \dfrac{3t^3}{4s + 16}$

25. $\dfrac{4y + 6}{5x + 10} \div \dfrac{4y^2 - 9}{6x + 12}$

26. $(a^2 - 25) \div \dfrac{a + 5}{a - 5}$

27. $\dfrac{y - 8}{y + 4} \div (y^2 - 4y - 32)$

28. $\dfrac{3x + 9}{2x - 5} \div (2x^2 + x - 15)$

29. $\dfrac{2r^2 - 7r - 15}{r^2 - 10r + 25} \div \dfrac{2r^2 + 13r + 15}{r^2 - 25}$

30. $\dfrac{x^3 - x}{x^2 - 3x + 2} \div \dfrac{4x^2 + 8x + 4}{2x^2 - 10x + 12}$

31. $\dfrac{a^2 - 5a}{2a^3 - a^2} \div \dfrac{a^2 - 7a + 10}{2a^2 - 5a + 2}$

32. $\dfrac{ax^2 - ay^2}{x^2 - xy - 2y^2} \div \dfrac{x^2 - xy}{x^2 - 2xy}$

33. $\left(\dfrac{2m}{7x^2} \div \dfrac{4m^2}{21x}\right) \cdot \dfrac{3m^2}{x}$

34. $\dfrac{(by)^2}{2ax} \cdot \left(\dfrac{6a^2}{y} \div \dfrac{9b^2}{x^2}\right)$

35. $\left(\dfrac{2m^2 - 5m - 3}{m^2 - m - 12} \div \dfrac{2m^2 + 5m + 2}{3m + 9}\right) \cdot \dfrac{m^2 - 9}{m^2 - 2m - 8}$

36. $\left(\dfrac{r^2 + 4r - 5}{r^2 - 3r - 4} \cdot \dfrac{r^2 + 3r - 28}{(r - 3)^2} \div \dfrac{r^2 + 12r + 35}{r^2 - 2r - 3}\right)$

37. *Physics* In determining the number of lines of force in an electric field on a spherical surface, it is necessary to find the product given below. Multiply as indicated.

$$\frac{1}{4\pi\varepsilon_0} \cdot \frac{q}{r^2} \cdot 4\pi r^2$$

38. *Astrophysics* In the study of electromagnetic processes in space, it is necessary to find the quotient shown below. Divide as indicated.

$$\frac{8\pi ne^2w}{mv^3 - mvw^2} \div \frac{2mv^2 - 2\pi ne^2}{mv^2}$$

6-6 | Addition and Subtraction of Fractions

Adding or subtracting like fractions—that is, those with the same denominator—is simple. The sum of two fractions with like denominators is a fraction with a numerator that is the sum of their numerators, and with the same common denominator.

EXAMPLE 52

(a) $\dfrac{7}{8} + \dfrac{5}{8} = \dfrac{7 + 5}{8}$ ⟵ Sum of numerators

⟵ Same denominator

$= \dfrac{12}{8} = \dfrac{3}{2}$

(b) $\dfrac{4}{3x} - \dfrac{7}{3x} = \dfrac{4 - 7}{3x}$

$= \dfrac{-3}{3x} = \dfrac{-1}{x}$

(c) $\dfrac{7}{a+2} + \dfrac{3a}{a+2} - \dfrac{\overbrace{a+3}}{a+2} = \dfrac{7 + 3a - \overbrace{(a+3)}}{a+2}$

> To avoid confusion, put the expression $(a+3)$ in parentheses

> $-(a+3) = -a - 3$

$$= \dfrac{7 + 3a - a - 3}{a+2}$$

$$= \dfrac{2a+4}{a+2} = \dfrac{2(a+2)}{a+2} = 2$$

Unlike fractions   When the fractions to be added or subtracted have unlike denominators, first rewrite each as an equivalent fraction with the same denominator. This **lowest common denominator,** or **LCD,** will be the smallest number evenly divisible by both of the original denominators.

EXAMPLE 53   Subtract: $\dfrac{3}{4} - \dfrac{1}{6}$

Solution

Step 1. Note by inspection that the LCD of 4 and 6 is 12.

Step 2. Build up both fractions to equivalent fractions with denominators of 12.

$$\dfrac{3 \cdot 3}{4 \cdot 3} = \dfrac{9}{12} \qquad \dfrac{1 \cdot 2}{6 \cdot 2} = \dfrac{2}{12}$$

Step 3. Subtract the fractions with the common denominator.

$$\dfrac{3}{4} - \dfrac{1}{6} = \dfrac{9}{12} - \dfrac{2}{12} = \dfrac{7}{12}$$

EXAMPLE 54   Add: $\dfrac{x}{2} + \dfrac{2x}{3}$

Solution

Step 1. The LCD of 2 and 3 is 6.

Step 2. $\dfrac{x \cdot 3}{2 \cdot 3} = \dfrac{3x}{6} \qquad \dfrac{2x \cdot 2}{3 \cdot 2} = \dfrac{4x}{6}$

Step 3. $\dfrac{x}{2} + \dfrac{2x}{3} = \dfrac{3x}{6} + \dfrac{4x}{6} = \dfrac{7x}{6}$

Finding the lowest common denominator was easy in Examples 53 and 54, but this is not always the case. The next example illustrates a dependable method for determining the LCD and combining more complicated fractions.

EXAMPLE 55   To combine the fractions

$$\dfrac{1}{6x^2y} + \dfrac{3}{4x} - \dfrac{7}{8xy^3}$$

follow these steps:

Step 1. Factor the denominators.

$$6x^2y = 2 \cdot 3 \cdot x^2 \cdot y$$

$$4x = 2^2 \cdot \quad x$$

$$8xy^3 = 2^3 \cdot \quad x \cdot y^3$$

$$\downarrow \quad \downarrow \quad \downarrow \quad \downarrow$$

$$2^3 \cdot 3 \cdot x^2 \cdot y^3$$

Step 2. List each factor from Step 1 to its highest power. The product of these factors is the LCD.

$$\text{LCD} = 24x^2y^3$$

Step 3. Change each of the original fractions to equivalent fractions with the new LCD.

$$\frac{1}{6x^2y} = \frac{1 \cdot (4y^2)}{6x^2y \cdot (4y^2)} = \frac{4y^2}{24x^2y^3}$$

$$\frac{3}{4x} = \frac{3 \cdot (6xy^3)}{4x \cdot (6xy^3)} = \frac{18xy^3}{24x^2y^3}$$

$$\frac{7}{8xy^3} = \frac{7 \cdot (3x)}{8xy^3 \cdot (3x)} = \frac{21x}{24x^2y^3}$$

Step 4. Combine the numerators of the equivalent fractions from Step 3.

$$\frac{1}{6x^2y} + \frac{3}{4x} - \frac{7}{8xy^3} = \frac{4y^2 + 18xy^3 - 21x}{24x^2y^3}$$

In the last step, combine like terms if possible.

Use the same procedure for problems involving multinomial denominators.

EXAMPLE 56 Subtract: $\dfrac{x-3}{x^2+8x+12} - \dfrac{x}{x^2-36}$

Solution

Step 1. Factor the denominators:

$$x^2 + 8x + 12 = (x+6)(x+2) \quad \text{and} \quad x^2 - 36 = (x+6)(x-6)$$

Step 2. List each factor from Step 1 to its highest power. The product of these factors is the LCD:

$$\text{LCD} = (x+6)(x+2)(x-6)$$

Step 3. Change each of the original fractions to equivalent fractions with the LCD:

$$\frac{x-3}{x^2+8x+12} = \frac{(x-3) \cdot (x-6)}{(x+6)(x+2) \cdot (x-6)} = \frac{(x-3)(x-6)}{(x+6)(x+2)(x-6)}$$

$$\frac{x}{x^2-36} = \frac{x \cdot (x+2)}{(x+6)(x-6) \cdot (x+2)} = \frac{x(x+2)}{(x+6)(x+2)(x-6)}$$

Step 4. Subtract the equivalent fractions from Step 3:

$$\frac{x-3}{x^2+8x+12} - \frac{x}{x^2-36} = \frac{(x-3)(x-6)-x(x+2)}{(x+6)(x+2)(x-6)}$$

$$= \frac{x^2-9x+18-x^2-2x}{(x+6)(x+2)(x-6)}$$

$$= \frac{-11x+18}{(x+6)(x+2)(x-6)}$$

If the numerator can be factored, do so. This will enable you to reduce your answer if possible.

EXAMPLE 57 Combine: $\dfrac{m}{m^2+4m+4} + \dfrac{7}{m^2-2m} + \dfrac{m-5}{4-m^2}$

Solution Factoring the denominators, we have

$$m^2+4m+4 = (m+2)^2$$

$$m^2-2m = m(m-2)$$

$$4-m^2 = (2+m)(2-m)$$

Notice that the factor $(2-m)$ in the last denominator is simply the negative of $(m-2)$ from the second denominator: $(2-m) = -1(m-2)$. We will obtain a simpler LCD if we multiply the numerator and denominator of one of these fractions by (-1). In this case, we choose the second fraction, since it has the simplest numerator:

$$\frac{m}{(m+2)^2} + \frac{(-1)7}{(-1)m(m-2)} + \frac{m-5}{(2+m)(2-m)}$$

$$= \frac{m}{(m+2)^2} + \frac{(-7)}{m(2-m)} + \frac{m-5}{(m+2)(2-m)}$$

$$(2+m) = (m+2)$$

The LCD of the three denominators as now shown is $m(m+2)^2(2-m)$. Changing each fraction to an equivalent fraction with this denominator, we have

$$\frac{m}{(m+2)^2} = \frac{m}{(m+2)^2} \cdot \frac{m(2-m)}{m(2-m)} = \frac{2m^2-m^3}{m(m+2)^2(2-m)}$$

$$\frac{-7}{m(2-m)} = \frac{-7}{m(2-m)} \cdot \frac{(m+2)^2}{(m+2)^2} = \frac{-7m^2-28m-28}{m(m+2)^2(2-m)}$$

$$\frac{m-5}{(m+2)(2-m)} = \frac{(m-5)}{(m+2)(2-m)} \cdot \frac{m(m+2)}{m(m+2)} = \frac{m^3-3m^2-10m}{m(m+2)^2(2-m)}$$

Adding these equivalent fractions, we obtain

$$\frac{2m^2 - m^3 - 7m^2 - 28m - 28 + m^3 - 3m^2 - 10m}{m(m+2)^2(2-m)} = \frac{-8m^2 - 38m - 28}{m(m+2)^2(2-m)}$$

$$= \frac{-2(4m^2 + 19m + 14)}{m(m+2)^2(2-m)}$$

Factoring in the final step revealed no common factors in the numerator and denominator. The result shown is in its simplest form.

Complex fractions   A **complex algebraic fraction** is an algebraic fraction that contains one or more fractions in its numerator, its denominator, or both. To simplify a complex algebraic fraction:

- **First,** rewrite the numerator as a single fraction.
- **Second,** rewrite the denominator as a single fraction.
- **Third,** divide the numerator by the denominator.

EXAMPLE 58   Simplify: $\dfrac{\dfrac{1}{x}}{\dfrac{3}{2}}$

Solution   Since the numerator and denominator are already single fractions, we can divide:

$$\frac{1}{x} \div \frac{3}{2} = \frac{1}{x} \cdot \frac{2}{3} = \frac{2}{3x}$$

EXAMPLE 59   Simplify: $\dfrac{\dfrac{1}{3}}{\dfrac{2}{x} + \dfrac{1}{x-1}}$

Solution   The numerator is a single fraction, but the denominator must be simplified:

$$\frac{2}{x} + \frac{1}{x-1} = \frac{2(x-1) + 1 \cdot x}{x(x-1)} = \frac{3x-2}{x^2-x}$$

Dividing the original numerator by this fraction, we have

$$\frac{1}{3} \div \frac{3x-2}{x^2-x} = \frac{1}{3} \cdot \frac{x^2-x}{3x-2} = \frac{x^2-x}{9x-6}$$

EXAMPLE 60   Simplify: $\dfrac{\dfrac{1}{x} + \dfrac{1}{2}}{1 - \dfrac{2}{x}}$

Solution

First, simplify the numerator.

$$\frac{1}{x} + \frac{1}{2} = \frac{2 + x}{2x}$$

Second, simplify the denominator.

$$1 - \frac{2}{x} = \frac{x - 2}{x}$$

Third, divide the simplified numerator by the simplified denominator.

$$\frac{2 + x}{2x} \div \frac{x - 2}{x} = \frac{2 + x}{2x} \cdot \frac{\cancel{x}}{x - 2}$$

$$= \frac{2 + x}{2(x - 2)} \quad \text{or} \quad \frac{2 + x}{2x - 4}$$

Exercises 6-6

Perform the indicated operations.

1. $\dfrac{8}{9} + \dfrac{5}{9}$

2. $\dfrac{7}{12} - \dfrac{5}{12}$

3. $\dfrac{5}{x} - \dfrac{6}{x}$

4. $\dfrac{5}{2m} + \dfrac{3}{2m}$

5. $\dfrac{2a}{a + 4} + \dfrac{8}{a + 4}$

6. $\dfrac{5c}{c - 2} - \dfrac{c + 3}{c - 2}$

7. $\dfrac{x + 4}{x^2 - 9} - \dfrac{3x - 2}{x^2 - 9}$

8. $\dfrac{y + 5}{y^2 + 5y + 6} + \dfrac{2y + 1}{y^2 + 5y + 6}$

9. $\dfrac{5}{6} + \dfrac{2}{3}$

10. $\dfrac{3}{8} - \dfrac{5}{12}$

11. $\dfrac{9}{16} + \dfrac{5}{9} - \dfrac{11}{12}$

12. $\dfrac{7}{15} - \dfrac{7}{18} + \dfrac{3}{4}$

13. $\dfrac{1}{2y} - \dfrac{3}{4}$

14. $\dfrac{x}{3} + \dfrac{5}{9y}$

15. $\dfrac{5m}{a^3} + \dfrac{m}{a}$

16. $\dfrac{2}{b} - \dfrac{5}{b^2}$

17. $\dfrac{5y}{6x^2} - \dfrac{5}{12x}$

18. $\dfrac{t}{4y^3} + \dfrac{7}{6y}$

19. $\dfrac{2}{3ab} - \dfrac{5}{4a^2} + \dfrac{7}{12b}$

20. $\dfrac{1}{6mn^2} - \dfrac{5}{9m} - \dfrac{2}{3n}$

21. $\dfrac{y - 2}{y - 6} - \dfrac{5}{6 - y}$

22. $\dfrac{x + 1}{2x - 3} + \dfrac{5x + 7}{3 - 2x}$

23. $\dfrac{7}{x - 4} + \dfrac{x}{x + 4}$

24. $\dfrac{6}{x} - \dfrac{x - 3}{x + 2}$

25. $\dfrac{8}{5x + 10} - \dfrac{3x}{2x + 4}$

26. $\dfrac{x + 1}{4x - 12} + \dfrac{5}{6x - 18}$

27. $\dfrac{m - 5}{m^2 + 5m} + \dfrac{7}{2m + 10} - \dfrac{3}{2m}$

28. $\dfrac{n + 2}{3n - 6} - \dfrac{4}{3n} - \dfrac{n - 3}{n^2 - 2n}$

29. $\dfrac{5x}{x^2 - 9} - \dfrac{7}{3 - x}$

30. $\dfrac{y}{y^2 - 25} + \dfrac{4}{5 - y}$

31. $\dfrac{2y^2}{9y^2 - 1} - \dfrac{y - 7}{3y + 1}$

32. $\dfrac{2x}{x^2 + 12x + 36} + \dfrac{x + 4}{x + 6}$

33. $\dfrac{x - 5}{2x^2 - 3x - 14} - \dfrac{2x - 3}{2x^2 + 9x + 10}$

34. $\dfrac{2x}{x^2 - 16} + \dfrac{x - 7}{x^2 - 9x + 20}$

35. $\dfrac{x + 1}{x - 2} - \dfrac{x - 5}{x + 5} + \dfrac{x^2 + 3}{x^2 + 3x - 10}$

36. $\dfrac{x + 5}{4x^2 - 1} + \dfrac{3}{2x^2 + x} - \dfrac{x}{6x - 3}$

37. $\dfrac{\dfrac{5}{3}}{\dfrac{x}{6}}$

38. $\dfrac{\dfrac{2}{x}}{\dfrac{x}{4}}$

39. $\dfrac{\dfrac{1}{2}}{\dfrac{3}{x} + \dfrac{1}{x + 1}}$

40. $\dfrac{\dfrac{x - 1}{x} - \dfrac{1}{x - 2}}{\dfrac{2}{x}}$

41. $\dfrac{x - \dfrac{3}{x}}{\dfrac{1}{x} - 3}$

42. $\dfrac{\dfrac{a}{b} - 1}{\dfrac{1}{b} + a}$

43. $\dfrac{\dfrac{1}{x - 2} - x + \dfrac{1}{x}}{1 - \dfrac{1}{x^2 - 4}}$

44. $\dfrac{\dfrac{m}{m^2 + 4m + 4} - \dfrac{2}{m}}{\dfrac{1}{m^2 + 2m} - 2}$

45. *Civil engineering*   The expression given below is used in solving vertical curve problems. Simplify this fraction.

$$\dfrac{mD^2}{\dfrac{L^2}{4}}$$

46. *Fire science*   The expression below gives the amount of water lifted by a siphon. Combine these two terms.

$$\dfrac{2.304qEP}{h} - q$$

47. *Surveying*   The first two terms in a series used to find a correction factor are given below. Combine these terms.[*]

$$\dfrac{d^2}{2L} + \dfrac{d^4}{8L^3}$$

48. *Civil engineering*   The maximum stress at any point in a beam or post is given by the formula below. Combine the terms on the right side of this formula.

$$s = \dfrac{Mc}{I} + \dfrac{P}{A}$$

---

[*] For additional background, see Russell C. Brinker and Paul R. Wolf, *Elementary Surveying* (New York: Harper and Row, 1977), p. 63.

49. *Electrical engineering* In calculating the electric intensity of a field set up by a dipole, the expression below appears. Combine the two terms and simplify the numerator.

$$\frac{1}{\left(r - \dfrac{d}{2}\right)^2} - \frac{1}{\left(r + \dfrac{d}{2}\right)^2}$$

50. *Mechanical engineering* The maximum tensile stress on the cross section of a block is given by the formula below. Rewrite this formula by combining terms inside parentheses.

$$s = \frac{L}{bh}\left(\frac{6y}{b} + \frac{6z}{h} - 1\right)$$

51. *Civil engineering* In designing a reinforced concrete floor slab, an important quantity k is given by the formula below. Simplify the right side of this formula.

$$k = \frac{\dfrac{nC}{s}}{1 + \dfrac{nC}{s}}$$

52. *Electrical engineering* The flux through a certain magnetic circuit is given by the expression below. Simplify this complex fraction.

$$\frac{1}{\dfrac{x}{aM} + \dfrac{y}{bN}}$$

6-7 | Equations Involving Fractions

Many technical applications involve equations and formulas containing fractions. The concepts and techniques discussed in this chapter and in Section 2-4 allow us to solve these equations and formulas.

To solve any equation, we can perform the same basic arithmetic operation on both sides. To solve a fractional equation, first multiply both sides by the lowest common denominator. This will produce an equation without fractions.

EXAMPLE 61 Solve for x: $\dfrac{x - 3}{6} = \dfrac{x}{9} - \dfrac{5}{3}$

Solution

Step 1. Find the LCD of all fractions in the equation.

$\text{LCD} = 2 \cdot 3^2 = 18$

Step 2. Multiply every term by the LCD.

$18 \cdot \dfrac{(x - 3)}{6} = 18 \cdot \dfrac{x}{9} - 18 \cdot \dfrac{5}{3}$

Step 3. Simplify each fraction. The denominators will all cancel out, leaving an equation without fractions.

$$\frac{\overset{3}{\cancel{18}} \cdot (x - 3)}{\underset{1}{\cancel{6}}} = \frac{\overset{2}{\cancel{18}} \cdot x}{\underset{1}{\cancel{9}}} - \frac{\overset{6}{\cancel{18}} \cdot 5}{\underset{1}{\cancel{3}}}$$

$$3(x - 3) = 2x - 30$$

Step 4. Solve the resulting equation.

$$3x - 9 = 2x - 30$$

$$x = -21$$

Step 5. Check this solution by substituting it back into the original equation:

$$\frac{(-21) - 3}{6} = \frac{(-21)}{9} - \frac{5}{3}$$

$$\frac{-24}{6} = \frac{-7}{3} - \frac{5}{3}$$

$$-4 = -4$$

EXAMPLE 62 Solve for x: $\dfrac{2}{3}(x - 2) = \dfrac{1}{5} - 2x$

Solution

Step 1. The LCD is $3 \cdot 5 = 15$.

Step 2. Multiplying every term by 15, we have:

$$15 \cdot \frac{2}{3}(x - 2) = 15 \cdot \frac{1}{5} - 15 \cdot 2x$$

Step 3. Simplifying, we obtain an equation without fractions.

$$\frac{\overset{5}{\cancel{15}} \cdot 2}{\underset{1}{\cancel{3}}}(x - 2) = \frac{\overset{3}{\cancel{15}}}{\underset{1}{\cancel{5}}} - 30x$$

$$10(x - 2) = 3 - 30x$$

Step 4. Now we solve the resulting equation.

$$10x - 20 = 3 - 30x$$

$$40x = 23$$

$$x = \frac{23}{40}$$

Step 5. Check the solution.

Here is an example involving variables in the denominator. We proceed exactly the same way.

EXAMPLE 63 Solve for x: $\dfrac{2x}{x + 1} = 2 - \dfrac{5}{2x}$

Solution

Step 1. Find the LCD. $\quad$ LCD $= 2x(x + 1)$

Step 2. Multiply every term by the LCD.

$$2x(x + 1) \cdot \frac{2x}{x + 1} = 2x(x + 1) \cdot 2 - 2x(x + 1) \cdot \frac{5}{2x}$$

Step 3. Simplify.

$$\frac{2x(x + 1)2x}{x + 1} = 2x(x + 1) \cdot 2 - \frac{2x(x + 1)5}{2x}$$

Step 4. Solve.

$$4x^2 = 4x^2 + 4x - 5x - 5$$

$$0 = -x - 5$$

$$x = -5$$

The solution checks with the original equation.

Extraneous solutions

If substituting a solution back into the original equation produces a denominator of zero, the solution is **extraneous** and is not valid. Always check each denominator carefully for this possibility.

EXAMPLE 64 Solve for x: $\dfrac{2}{x^2 - 1} = \dfrac{1}{x - 1} + \dfrac{2}{x + 1}$

Solution The LCD is $(x + 1)(x - 1)$. Multiply every term by the LCD:

$$(x + 1)(x - 1) \cdot \frac{2}{x^2 - 1} = (x + 1)(x - 1) \cdot \frac{1}{x - 1} + (x + 1)(x - 1) \cdot \frac{2}{x + 1}$$

Simplify:

$$\frac{(x + 1)(x - 1)2}{(x + 1)(x - 1)} = \frac{(x + 1)(x - 1)}{x - 1} + \frac{(x + 1)(x - 1)2}{x + 1}$$

Solve:

$$2 = x + 1 + 2x - 2$$

$$2 = 3x - 1$$

$$3 = 3x$$

$$x = 1$$

Checking the solution, we find that two of the denominators, $(x^2 - 1)$ and $(x - 1)$, become zero for $x = 1$. Therefore, the solution $x = 1$ is extraneous, and this equation has no solution.

Solving formulas

Formulas containing fractions often must be rearranged to solve for a different variable. We can now apply the skills learned in Section 2-5 to formulas with fractions.

EXAMPLE 65 The formula given describes the total capacitance C of two capacitors, C_1 and C_2, connected in series. Solve this formula for C_1.

$$\frac{1}{C} = \frac{1}{C_1} + \frac{1}{C_2}$$

Solution   The LCD is $C \cdot C_1 \cdot C_2 = CC_1 C_2$.
Multiply every term of
the formula by the LCD.

$$CC_1C_2 \cdot \frac{1}{C} = CC_1C_2 \cdot \frac{1}{C_1} + CC_1C_2 \cdot \frac{1}{C_2}$$

Simplify.

$$C_1 C_2 = CC_2 + CC_1$$

To solve for C_1, add $-CC_1$ to both
sides to move all C_1-terms to one
side.

$$C_1 C_2 - CC_1 = CC_2$$

Factor the left side.

$$C_1(C_2 - C) = CC_2$$

Divide by $(C_2 - C)$.

$$C_1 = \frac{CC_2}{C_2 - C}$$

Solving rate problems   Several types of applied problems produce fractional equations. One variety of problem, called a **rate problem,** requires that you determine the amount of time needed to accomplish a task when two or more processes are working at constant rates.

EXAMPLE 66   One computer can process a company's payroll in 8 h, while a newer computer can do the same payroll in 6 h. Working together at these rates, how long would they take to complete the task?

Solution   Let x = Amount of time needed. In 1 h the first computer does 1/8 of the job. In 1 h the second computer does 1/6 of the job. In x hours they do $x/8$ and $x/6$ of the job, respectively. Working together they do one complete job. Therefore,

Part of job done by first computer		Part of job done by second computer		One job finished
$\dfrac{x}{8}$	$+$	$\dfrac{x}{6}$	$=$	1

To solve, first multiply every term by the LCD.

$$24 \cdot \frac{x}{8} + 24 \cdot \frac{x}{6} = 24 \cdot 1$$

Simplify.

$$3x + 4x = 24$$

Combine like terms.

$$7x = 24$$

Divide by 7.

$$x = 3\tfrac{3}{7} \, \text{h}$$

Check this answer by substituting it into the original problem statement.

Exercises 6-7

Solve each equation.

1. $\dfrac{x}{4} - 3 = x$

2. $10 - \dfrac{y}{3} = 3y$

3. $\dfrac{3}{4}x - \dfrac{1}{2} = \dfrac{5}{6}x$

4. $\dfrac{5}{6}x - \dfrac{1}{3} = \dfrac{1}{6}$

5. $\dfrac{2}{5}(x - 5) = 3 - \dfrac{1}{2}x$

6. $3x + \dfrac{1}{7} = \dfrac{1}{2}(x + 2)$

7. $\dfrac{y - 4}{3} - \dfrac{y}{6} = 4$

8. $\dfrac{x + 3}{8} = \dfrac{x}{2} - 3$

9. $\dfrac{m}{6} = \dfrac{m + 4}{4} + \dfrac{5}{8}$

10. $\dfrac{a}{3} - \dfrac{a - 2}{6} = \dfrac{2}{9}$

11. $\dfrac{5}{3x} - \dfrac{1}{6} = 4$

12. $\dfrac{1}{x} = 3 - \dfrac{2}{5x}$

13. $\dfrac{5}{2y} = \dfrac{3}{y + 2}$

14. $\dfrac{6}{3y + 5} = \dfrac{4}{3y}$

15. $\dfrac{10}{x - 3} = \dfrac{9}{x - 5}$

16. $\dfrac{8}{x + 4} = \dfrac{6}{x - 3}$

17. $\dfrac{6}{2m - 3} = 6$

18. $3 = \dfrac{15}{3n + 2}$

19. $\dfrac{3y}{y - 1} = 3 - \dfrac{5}{2y}$

20. $\dfrac{y}{y + 3} - 1 = \dfrac{3}{4y}$

21. $\dfrac{4}{3x + 9} = \dfrac{x + 1}{x + 3}$

22. $\dfrac{x - 3}{5x - 10} + \dfrac{1}{2} = \dfrac{4}{x - 2}$

23. $\dfrac{3}{x^2 + 3x} = \dfrac{1}{x} - \dfrac{4}{x + 3}$

24. $\dfrac{4}{2 - x} - \dfrac{x + 1}{2x - 4} + 2 = \dfrac{3}{2}$

25. $\dfrac{5}{2x - 6} + \dfrac{2}{x - 3} = \dfrac{3}{4}$

26. $3 - \dfrac{2}{3x - 12} = \dfrac{2}{3} - \dfrac{x - 3}{4 - x}$

27. $\dfrac{3}{x + 2} - \dfrac{2}{x} = \dfrac{-6}{x^2 + 2x}$

28. $\dfrac{5}{x - 3} - \dfrac{3}{2x} = \dfrac{15}{x^2 - 3x}$

29. $\dfrac{3}{3x^2 - 3x - 28} = \dfrac{5}{5x^2 - x - 20}$

30. $\dfrac{2}{3y} - \dfrac{2}{3y - 2} = \dfrac{5}{9y^2 - 4}$

31. $\dfrac{2}{a^2 + 5a + 4} - \dfrac{3}{a^2 - 1} = \dfrac{1}{a^2 + 3a - 4}$

32. $\dfrac{x - 3}{x^2 - 2x - 8} = \dfrac{1}{x - 4} + \dfrac{5}{x^2 - 5x + 4}$

Solve for the indicated variable.

33. $\dfrac{x + y}{2} - \dfrac{1}{5} = \dfrac{1}{4}$ for y

34. $\dfrac{m - 5}{n} - \dfrac{m}{3} = 4$ for m

35. $\dfrac{2a - 5}{c} - \dfrac{a}{c + 2} = \dfrac{1}{2c}$ for c

36. $\dfrac{y + 5}{x^2 + 4x} = \dfrac{y}{x} + \dfrac{3}{x + 4}$ for x

37. *Electrical engineering*   The voltage (in volts) across a resistor in a voltage divider is given by the formula below. Solve for R_1.

$$V = \frac{R_1 E}{R_1 + R_2 + R_3}$$

38. *Electrical engineering*   The total resistance (in ohms) of three resistors in parallel is given by the formula below. Solve for R_T.

$$\frac{1}{R_T} = \frac{1}{R_1} + \frac{1}{R_2} + \frac{1}{R_3}$$

39. *Accounting*   The formula at the top of the next page is used to find the gross profit percent G in terms of sales S and overhead V. Solve for S.

$$G = \frac{S - V}{S}$$

40. *Fire science*   The nozzle pressure N (in pounds per square inch) of a hose is related to engine pressure E and hose length L (in feet) by the formula below. Solve for L.

$$N = \frac{E}{1.1 + kL}$$

41. *Electronics*   If two inductors with inductances L_1 and L_2 and mutual inductance M are connected in parallel, their effective inductance L is given by the formula below. Solve for L_1.

$$L = \frac{L_1 L_2 - M^2}{L_1 + L_2 - 2M}$$

42. *Chemistry*   The concentration x (in moles per liter) of HCO_3^- in a certain solution can be found by solving the equation below. Find x.

$$\frac{x}{2(0.40 - x)} = 1.05 \times 10^{-2}$$

43. *Electronics*   When an unsymmetrical square wave signal is applied to a certain clamper circuit, the zero level V (in volts) can be found by solving the equation below. Find V.

$$\frac{1.8V}{2200(8.0 - V)} = 0.0025$$

44. *Physics*   The apparent frequency f_a of sound when the source and listener are moving toward each other can be found by solving the equation below. Solve for f_a.

$$\frac{V - V_s}{f} = \frac{V + V_L}{f_a}$$

Solve.

45. *Wastewater technology*   Working alone, one pipe can fill a tank in 10 h, while a second pipe can fill the tank in 15 h. Working together, how long does it take the two pipes to fill the tank?

46. *Graphics technology*   Working alone, one printing press can run a job in 3 h, while another press can perform the same job in 4 h. Working together, how long would it take the two machines to do the job?

47. *Business*   Working alone, one computer operator can process a set of payroll records in 60 min. Working together, she and another operator can process the same records in 24 min. How long would it take the second operator to process the records alone?

48. *Wastewater technology*   Working alone, one pipe can fill a reservoir in 12 h, a second pipe can fill the same reservoir in 15 h, and a third pipe can empty the reservoir in 20 h. If all three are operating at once, how long does it take to fill the reservoir?

6-8 | Review of Chapter 6

Important Terms and Concepts

factoring a polynomial (p. 214)

greatest common factor, GCF (p. 214)

algebraic fraction (p. 225)

equivalent fractions (p. 225)

reducing to lowest terms (p. 227)

lowest common denominator, LCD (p. 235)

complex algebraic fraction (p. 238)

extraneous solutions (p. 243)

Special Products (see Section 6-1)

- product of a binomial by a monomial: $a(x + y) = ax + ay$ **(6-1)**
- difference of two squares: $(x + y)(x - y) = x^2 - y^2$ **(6-2)**
- square of a binomial: $(x + y)^2 = x^2 + 2xy + y^2$ **(6-3)**
- product of two binomials: $(x + a)(x + b) = x^2 + (a + b)x + ab$ **(6-4)**
- cube of a binomial: $(x + y)^3 = x^3 + 3x^2y + 3xy^2 + y^3$ **(6-5)**

Exercises 6-8

Find each product.

1. $-8(x - 4y)$ **2.** $n^2(3n^2 - 4n)$ **3.** $5c(4a - 7c)$ **4.** $-3xy(x^2 - 3xy)$

5. $(y + 5)(y - 5)$ **6.** $(3x + y)(3x - y)$ **7.** $(m - 4)^2$ **8.** $(2a + 5)^2$

9. $(x + 5)(x + 7)$ **10.** $(y - 9)(y + 6)$ **11.** $(3w + 8)(5w - 2)$ **12.** $(4x - 3y)(2x - 3y)$

13. $4(x + 1)(x - 1)$ **14.** $3x(x + 3)(2x - 5)$ **15.** $-2y(2y - 7)^2$ **16.** $(x + y + 5)^2$

Factor completely.

17. $6x - 24$ **18.** $y^2 + 9y$ **19.** $8x^3 + 10x^2$

20. $7x^2 - 12y^2$ **21.** $8x - 12y + 16z$ **22.** $16a^2bc - 32a^3b^2c^2 - 48a^4b^3$

23. $x^2 - 36$ **24.** $144 - y^2$ **25.** $16a^2 - 25$

26. $49m^6 - 64n^4$ **27.** $7x^3 - 28x$ **28.** $9y^2 - 81z^2$

29. $16x^2 + 36$ **30.** $x^5 - x$ **31.** $x^2 + 4x - 12$

32. $x^2 + 5x - 24$ **33.** $m^2 + 9mn + 14n^2$ **34.** $a^2 - 11ab + 30b^2$

35. $3x^2 + 2x - 5$ **36.** $5x^2 - 16x + 3$ **37.** $9x^2 - 24x + 16$

38. $25x^2 + 20xy + 4y^2$ **39.** $2x^2 - x - 15$ **40.** $2y^2 - 11y + 12$

41. $6x^2 - 7xy - 20y^2$ **42.** $-10x^2 + 31x - 15$ **43.** $12x^3 + 23x^2 - 24x$

44. $9y^4 + 14y^2 - 8$

Perform all indicated operations and reduce answers to lowest terms.

45. $\dfrac{5x + 15}{x^2 + 3x}$ **46.** $\dfrac{x - y}{x^2 - 2xy + y^2}$ **47.** $\dfrac{a + b}{a^2 - b^2}$

48. $\dfrac{x^3 - 7x^2 + 12x}{x^3 - 16x}$

49. $\dfrac{12x^2 + 26x + 12}{18x^2 + 24x + 8}$

50. $\dfrac{24 - 6x^2}{3x^2 + 9x - 30}$

51. $\dfrac{x^2 - 1}{x^2 - 2x} \cdot \dfrac{3x - 6}{4x + 4}$

52. $\dfrac{2x}{9 - x^2} \cdot \dfrac{x^2 - 7x + 12}{x^2}$

53. $\dfrac{x^2 - 10x + 21}{x^2 + 2x - 15} \div \dfrac{x^2 - 6x - 7}{x^2 + 6x + 5}$

54. $\dfrac{x^2 - xy - 2y^2}{x^2 + xy - 2y^2} \div \dfrac{x + y}{x - y}$

55. $\dfrac{\dfrac{6}{x}}{\dfrac{3}{4}}$

56. $\dfrac{\dfrac{4}{y} - \dfrac{2}{y + 4}}{\dfrac{3}{2y}}$

57. $\dfrac{a - \dfrac{1}{a}}{\dfrac{1}{a} - 1}$

58. $\dfrac{\dfrac{1}{c^2 - 25} - \dfrac{1}{c}}{\dfrac{1}{c^2 + 5c} + 5}$

59. $\dfrac{5y - 3}{2y - 1} - \dfrac{y - 1}{2y - 1}$

60. $\dfrac{3}{8m} + \dfrac{5}{12m^2}$

61. $\dfrac{1}{6ab} - \dfrac{3}{a^2} + \dfrac{5}{2b^2}$

62. $\dfrac{x}{x + 3} + \dfrac{4}{x - 3}$

63. $\dfrac{2y}{49 - y^2} - \dfrac{5}{y - 7}$

64. $\dfrac{x + 1}{9x^2 - 16} - \dfrac{5}{3x + 4} - \dfrac{3}{x}$

65. $\dfrac{2x - 3}{2x^2 + 5x - 3} + \dfrac{x - 3}{2x^2 + 7x + 3}$

66. $\dfrac{x + 2}{x^2 - 5x} - \dfrac{x - 3}{x^2 + 5x} + \dfrac{3x}{x^2 - 25}$

Solve each equation.

67. $\dfrac{5}{8}x - 2 = \dfrac{3}{4}$

68. $\dfrac{3}{4}(x + 4) - 2x = \dfrac{5}{6}$

69. $\dfrac{x + 2}{4} - \dfrac{x}{8} = 2$

70. $\dfrac{x}{3} = \dfrac{x - 3}{6} + \dfrac{5}{9}$

71. $\dfrac{8}{5y} - \dfrac{7}{10} = \dfrac{2}{y}$

72. $\dfrac{3}{2x + 5} = \dfrac{4}{2x - 5}$

73. $\dfrac{8y}{2y - 1} = 4 - \dfrac{3}{2y}$

74. $\dfrac{-16}{x^2 - 4x} = \dfrac{3}{x} - \dfrac{4}{x - 4}$

75. $\dfrac{4}{2x^2 - x - 28} = \dfrac{5}{6x^2 + 25x + 14}$

76. $\dfrac{3}{x^2 - x - 20} = \dfrac{2}{x^2 - 9x + 20} - \dfrac{1}{x^2 - 16}$

77. $\dfrac{x - 5}{a - 5} - \dfrac{1}{2a} = \dfrac{3}{a}$ for a

78. $\dfrac{3}{xy} - \dfrac{2}{x} = \dfrac{1}{y}$ for y

Solve.

79. *Electronics* The formula below gives the sensitivity S of a circuit in terms of the resistance of a temperature sensor R_T and the resistance of another resistor R_1, where B is a constant. Rewrite this formula by squaring the denominator.

$$S = \dfrac{BR_1}{(R_1 + R_T)^2}$$

80. If the length of a rectangle is twice the width, and 3 cm are added to each dimension, the area of the enlarged rectangle is found by multiplying $(2x + 3)(x + 3)$. Find this area.

81. *Fire science* The pressure loss F in a hose is given by the formula below, where Q is the flow rate. Rewrite this formula by factoring the right side.

$$F = 2Q^2 + Q$$

82. *Astrophysics* The equilibrium pressure P within a star is given by the formula below. Factor this expression.

$$P = \dfrac{2\pi}{3}(p^2GR^2 - p^2Gr^2)$$

83. *Nuclear physics* The expression $p^2c^2 + m^2c^4$ occurs in analyzing the energy of photons. Factor this expression.

84. *Astrophysics* The total mass energy of two particles falling toward a star is given by the expression below. Combine these terms and factor the numerator.

$$mc^2 + \frac{mMG}{r}$$

85. *Civil engineering* An expression encountered in structural design is given below. Combine these terms and factor the resulting numerator.

$$\frac{9R^2a^2}{4L^4} + \frac{R^2}{4L^2}$$

86. *Planetary science* For a planet of radius R, the difference between the gravitational potential at its surface and at a height x is given by the expression below, where M is its mass and G is the gravitational constant. Combine these terms.

$$-\frac{MG}{R + x} + \frac{MG}{R}$$

87. *Electronics* The expression below is found when analyzing the phase shift of a sinusoid. Simplify this complex fraction.

$$\frac{2\pi fx}{v + \dfrac{Tt}{j}}$$

88. *Electronics* Millman's theorem, which is used for determining the common voltage V across any network, is written as shown below. Simplify the right side of this formula.

$$V = \frac{\dfrac{E_1}{R_1} + \dfrac{E_2}{R_2} + \dfrac{E_3}{R_3}}{\dfrac{1}{R_1} + \dfrac{1}{R_2} + \dfrac{1}{R_3}}$$

89. *Civil engineering* The Gordon–Rankine formula for intermediate steel columns is written as shown below. Simplify the right side of this formula.*

$$P = A\left(\frac{k}{1 + \dfrac{L^2}{kr^2}}\right)$$

90. *Electronics* The circuit gain G of a platinum wire temperature-sensing device is given by the formula below, where R_T is the resistance of the sensing device and R_1 is the resistance of a resistor. Solve for R_T.

$$G = \frac{R_T}{R_1 + R_T}$$

---

* For more information, see Charles O. Harris, *Elementary Structural Design* (Chicago: American Technical Society, 1951), p. 127.

91. *Optics*   The fundamental principle of lenses is given by the formula below, where f_1 is the distance from lens to image, f_2 is the distance from lens to object, and f is the focal length of the lens. Solve for f_2.

$$\frac{1}{f_1} + \frac{1}{f_2} = \frac{1}{f}$$

92. *Civil engineering*   The final stress s at the top of a beam is given by the formula below. Solve for I.

$$s = \frac{P}{A} - \frac{Mc}{I}$$

93. *Holography*   The formula for the radius z of a reconstructed spherical wavefront is given below. Solve for z_0.

$$z = \frac{z_c z_0 z_r}{z_0 z_r + \mu z_c z_r - \mu z_c z_0}$$

94. *Wastewater technology*   One pipe can fill a tank in 10 h, while a second pipe can fill the same tank in 8 h. If the two pipes are working together, how long does it take them to fill the tank?

95. *Graphics technology*   One printing press can run a job in 8 h working alone. When a second press is added, the same job can be done in 5 h. How long would it take the second press to run the job alone?

Quadratic Equations

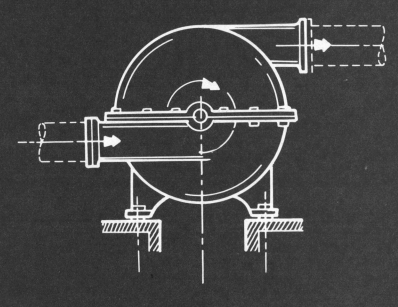

WATER PUMP

An important skill in applied mathematics is the solution of algebraic equations arising from physical situations. So far we have solved linear equations and systems of linear equations. However, many useful equations are nonlinear; that is, they contain variables with powers other than 1. Using the algebraic techniques of Chapter 6, we can solve many such nonlinear equations. We begin with quadratic equations.

Definitions A **quadratic equation** is any equation that can be put in the standard form

$$ax^2 + bx + c = 0 \quad \text{where } a, b, \text{ and } c \text{ are real numbers and } a \neq 0$$

The presence of the **quadratic term** ax^2 is what makes the equation quadratic, and therefore a cannot be zero. The presence of the **linear term** bx and the **constant term** c are not required. Therefore, b or c, or both, may equal zero.

Not all quadratic equations appear initially in this standard form. Normally, though, you can identify a quadratic equation by inspection.

EXAMPLE 1 Which of the following equations are quadratic?
(a) $3x^2 - 7x + 2 = 0$ (b) $5x^2 - 7 = 0$
(c) $2x = 9x^2$ (d) $4x - 7 = 0$
(e) $(x + 2)^2 = 0$ (f) $x^3 - 5x^2 + 9x = 6$
(g) $x + 5 = 7$ (h) $2x^2 - 5x = 2x^2 - 8$

Solution Equations (a), (b), (c), and (e) are all quadratic. In (e) notice that the quadratic term appears when the binomial is squared. Equations (d) and (g) have no quadratic term and are linear. Equation (f) contains a cubic term in addition to a quadratic term. In a quadratic equation, the quadratic term must be the highest-degree term, so (f) is not quadratic. Equation (h) is not quadratic, because the quadratic terms drop out when like terms are combined; this equation is linear.

Roots of quadratic equations A **solution,** or **root,** of any equation is a value of the variable that makes the equation a true statement. The degree of the equation indicates the number of roots it may have. A quadratic, or second-degree, equation has two roots. These roots may be a pair of real numbers (in special cases, these numbers may be equal), or they may be a pair of imaginary numbers. Any root may be checked by substituting it into the equation to be certain that a true statement results.

EXAMPLE 2 For each equation, determine whether the given values are roots.
(a) $x^2 - 7x + 12 = 0$ for $x = 3$ and $x = 4$
(b) $x^2 + 4x + 4 = 0$ for $x = 2$ and $x = -2$

Solution
(a) Both values are roots:

$$(3)^2 - 7(3) + 12 \stackrel{?}{=} 0 \qquad (4)^2 - 7(4) + 12 \stackrel{?}{=} 0$$

$$9 - 21 + 12 \stackrel{?}{=} 0 \qquad 16 - 28 + 12 \stackrel{?}{=} 0$$

$$0 = 0 \qquad\qquad 0 = 0$$

(b) Only -2 is a root:

$$(2)^2 + 4(2) + 4 \overset{?}{=} 0 \qquad\qquad (-2)^2 + 4(-2) + 4 \overset{?}{=} 0$$

$$4 + 8 + 4 \overset{?}{=} 0 \qquad\qquad\qquad 4 - 8 + 4 \overset{?}{=} 0$$

$$16 \neq 0 \qquad\qquad\qquad\qquad 0 = 0$$

The zero-product rule Several methods can be used to solve quadratic equations. Those with rational roots can be solved by factoring. This method depends on one simple arithmetic fact called the **zero-product rule:**

If the product of two or more factors is equal to zero, then at least one of those factors must equal zero.

The following examples show how to apply the zero-product rule to factorable quadratic equations.

EXAMPLE 3

To solve the quadratic equation:

$$x^2 + 9x + 8 = 0$$

First, factor the left side.

$$(x + 8)(x + 1) = 0$$

Second, since the zero-product rule tells us that one of the two factors must be equal to zero, set each factor equal to zero.

Either	Or
$x + 8 = 0$	$x + 1 = 0$

Third, solve the resulting equations.

$$x = -8 \quad \text{or} \quad x = -1$$

The roots are -8 and -1.

Always check both roots independently by substituting them back into the original equation as in Example 2:

For x = -8:   **For x = -1:**

$$(-8)^2 + 9(-8) + 8 \overset{?}{=} 0 \qquad\qquad (-1)^2 + 9(-1) + 8 \overset{?}{=} 0$$

$$64 - 72 + 8 \overset{?}{=} 0 \qquad\qquad\qquad 1 - 9 + 8 \overset{?}{=} 0$$

$$0 = 0 \qquad\qquad\qquad\qquad 0 = 0$$

Both roots satisfy the original quadratic equation.

To use the zero-product rule, the quadratic equation must be in standard form, with zero on one side of the equation.

EXAMPLE 4   Solve for x: $3x^2 + 5x = 28$

Solution

First, subtract 28 from both sides to put the equation in standard form.

$$3x^2 + 5x - 28 = 0$$

Second, factor the expression on the left.

$$(3x - 7)(x + 4) = 0$$

Third, set each factor equal to zero and solve the resulting equations.

$$3x - 7 = 0 \quad \text{or} \quad x + 4 = 0$$
$$3x = 7$$

$$x = \frac{7}{3} \quad \text{or} \quad x = -4$$

The roots are $\frac{7}{3}$ and -4. Check both values in the original equation.

The next example illustrates the special case of an equation that has only one unique root.

EXAMPLE 5 Solve for x: $x^2 = 8x - 16$

Solution

First, put the equation in standard form. $x^2 - 8x + 16 = 0$

Second, factor. $(x - 4)(x - 4) = 0$

Third, set each factor equal to zero. $x - 4 = 0 \quad \text{or} \quad x - 4 = 0$

Solve. $x = 4 \quad \text{or} \quad x = 4$

Since the trinomial is a perfect square, each factor is identical. This produces a "double root" of 4. Check it.

Incomplete quadratics   An **incomplete quadratic equation** is one that is missing either the linear or constant term. These can also be solved by factoring if the roots are rational. As the next examples show, only simple monomial factoring or difference of squares methods are needed.

EXAMPLE 6 Solve: (a) $2x^2 + 7x = 0$ (b) $x^2 - 9 = 0$

Solution

(a) **First,** factor out the greatest common factor. $x(2x + 7) = 0$

Second, set each factor equal to zero.

Finally, solve the equations to find the roots.

$$x = 0 \quad \text{or} \quad 2x + 7 = 0$$

$$x = 0 \quad \text{or} \quad x = -\frac{7}{2}$$

The roots are 0 and $-\frac{7}{2}$. Check them.

(b) You should recognize this binomial as the difference of two squares.

Factor it. $(x + 3)(x - 3) = 0$

Then, set each factor equal to zero and solve.

$$x + 3 = 0 \quad \text{or} \quad x - 3 = 0$$
$$x = -3 \quad \text{or} \quad x = 3$$

The roots are -3 and 3. Check them.

Fractional equations　　Fractional equations similar to those discussed in Chapter 6 can sometimes lead to quadratic equations.

EXAMPLE 7　Solve for x:　$\dfrac{1}{x} + \dfrac{6}{x-2} = -3$

Solution　**First,** multiply every term by the LCD:　$x(x-2)$

$$x(x-2)\cdot\frac{1}{x} + x(x-2)\cdot\frac{6}{x-2} = -3\cdot x(x-2)$$

Then, reduce each fraction, simplify, and solve as before:

$$x - 2 + 6x = -3x^2 + 6x$$

$$3x^2 + x - 2 = 0$$

$$(3x - 2)(x + 1) = 0$$

$$3x - 2 = 0 \quad \text{or} \quad x + 1 = 0$$

$$x = \frac{2}{3} \quad \text{or} \quad x = -1$$

The roots are $\frac{2}{3}$ and -1.

Finally, check:

$$\frac{1}{\left(\frac{2}{3}\right)} + \frac{6}{\left(\frac{2}{3}\right) - 2} = -3 \qquad \frac{1}{(-1)} + \frac{6}{(-1) - 2} = -3$$

$$\frac{3}{2} - \frac{9}{2} = -3 \qquad\qquad -1 - 2 = -3$$

CAUTION 　When checking solutions to fractional equations, remember to reject any that produce a denominator of zero. ◀

Of course, the zero-product rule is useful only if the quadratic expression can be factored. In Sections 7-2 and 7-3, we shall discuss how to solve quadratic equations that are not factorable.*

Word problems　　Many applied problems lead to quadratic equations, and we shall introduce one new type of word problem in each section of this chapter.

EXAMPLE 8　A rectangular open-topped water channel must be formed from a strip of sheet steel 12 ft wide. If the cross-sectional area must be 16 ft², what should be the dimensions of the channel?

---

*Given enough ingenuity we can factor *any* quadratic expression, although the coefficients may be nonintegral or even irrational. When we say that an expression is "factorable," we mean that it can be written as a product of factors with *integer* coefficients.

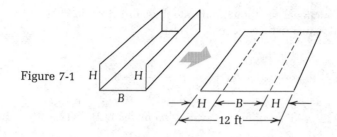

Figure 7-1

Solution   The three sides of the cross section consist of two heights and a base (Fig. 7-1).

The total width of the sheet is:

$$H + H + B = 12 \text{ ft}$$

$$2H + B = 12$$

Solving for B:

$$B = 12 - 2H$$

Now, we use the formula for area, $A = BH$, and substitute $A = 16$ and $B = 12 - 2H$.

$$16 = (12 - 2H)H$$

This is a quadratic equation.

$$16 = 12H - 2H^2$$

Solving, we obtain:

$$2H^2 - 12H + 16 = 0$$

$$2(H - 4)(H - 2) = 0$$

Notice that we can ignore the constant factor 2 when using the zero-product rule

$$H - 4 = 0 \quad \text{or} \quad H - 2 = 0$$

$$H = 4 \text{ ft} \quad \text{or} \quad H = 2 \text{ ft}$$

Substituting each value into $B = 12 - 2H$, we have:

$$B = 12 - 2(4) \quad \text{or} \quad B = 12 - 2(2)$$

$$B = 4 \text{ ft} \quad \text{or} \quad B = 8 \text{ ft}$$

The channel must either have a square cross-section, 4 ft by 4 ft, or have a height of 2 ft and a base of 8 ft.

Exercises 7-1

Determine whether each equation is quadratic. Put those that are quadratic in standard form and identify the constants a, b, and c.

1. $5x^2 - 9x = 4$ **2.** $3 - 7x^2 = 6x$ **3.** $2x^2 = 5$ **4.** $7x^2 - 9 = 0$
5. $5x - 3 = 6$ **6.** $4x^3 - 7x^2 + 6x = 8$ **7.** $3x = 8x^2$ **8.** $5x^2 - x = 0$
9. $\sqrt{x^2 + 5x + 6} = 0$ **10.** $x(x - 5) = x^2 - 7$ **11.** $2x(x - 3) = x^2 - 9$ **12.** $(x - 5)^2 = 6$

Solve each quadratic equation by factoring.

13. $x^2 - 12x + 32 = 0$ **14.** $x^2 + 9x + 20 = 0$ **15.** $x^2 - 2x - 35 = 0$
16. $x^2 + 3x - 40 = 0$ **17.** $x^2 = 4x + 45$ **18.** $x^2 + 13x = -36$

19. $2x^2 - 9x + 10 = 0$ **20.** $3x^2 - 7x - 6 = 0$ **21.** $5x^2 + 16x = -3$

22. $7x^2 = 4 - 12x$ **23.** $x^2 - 16 = 0$ **24.** $4x^2 - 25 = 0$

25. $9x^2 = 49$ **26.** $x^2 = 121$ **27.** $x^2 + 9x = 0$

28. $5x^2 - 15x = 0$ **29.** $x^2 = 3x$ **30.** $3x^2 = 8x$

31. $x^2 + 6x + 9 = 0$ **32.** $9x^2 - 12x + 4 = 0$ **33.** $4x^2 + 4x = 3$

34. $6x^2 = 13x - 6$ **35.** $2x^2 - 11x + 15 = 3$ **36.** $3x^2 - 7x + 5 = 25 - 14x$

37. $(x - 6)^2 = 3x - 20$ **38.** $(x - 8)(x + 4) = (2x + 3)(x - 7) + 1$ **39.** $\dfrac{x + 2}{x} - \dfrac{9x}{x + 4} = 2$

40. $\dfrac{2x}{x + 2} - 2 = \dfrac{x - 8}{x - 2}$ **41.** $\dfrac{x - 4}{x + 5} = 3 - \dfrac{x + 3}{x - 5}$ **42.** $\dfrac{3}{x - 2} - 2 = \dfrac{2}{x - 1}$

43. Physics   The height h (in feet) of an object dropped from a helicopter at an altitude of 1600 ft is given by the equation $h = 1600 - 16t^2$, where t is the elapsed time (in seconds). When will the object hit the ground?

44. Physics   The height h (in feet) of an object thrown vertically downward from a rooftop 56 ft high at an initial velocity of 4 ft/s is given by the equation $h = 56 - 4t - 16t^2$, where t is the elapsed time (in seconds). When will the object hit the ground?

45. Fire science   The pressure loss P due to friction (in pounds per square inch per 100 ft length) in a $2\frac{1}{2}$ in. hose is given by $P = 2Q^2 + Q$, where Q is the flow rate (in hundreds of gallons per minute). Find the flow rate that produces a pressure loss of 15 lb/in².

46. Economics   A certain revenue function is given by $R = 10q - 5q^2$, where q is the quantity of goods sold. Find q when $R = 0$.

Set up quadratic equations and solve.

47. A rectangular tank must be exactly 4 m longer than it is wide. If its area must be 117 m², find its dimensions.

48. Construction   A rectangular concrete pipe is constructed with an 8 in. by 14 in. interior channel, as shown in the figure. What uniform width w of concrete must be formed on all sides if the total cross-sectional area of the pipe must be 216 in.²?

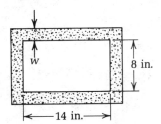

49. Civil engineering   A rectangular open-topped water channel must be constructed using a 10 ft wide piece of sheet steel. What dimensions will result in a cross-sectional area of 12 ft²? (See Example 8.)

50. Construction   A secured storage area is formed using one wall of a building and three sides of barbed-wire fencing. Using exactly 40 ft of fencing, what dimensions will produce a rectangular yard with an area of 192 ft²?

7-2   Solution by Completing the Square

Solving a quadratic equation by factoring works well in some cases, but it is very limited in its usefulness. This method works only for equations with rational roots. Unfortunately, many quadratic equations have irrational roots, and these equations must be solved using other methods.

To discover one such method, let's examine a quadratic equation that is not factorable.

EXAMPLE 9

To solve the quadratic equation: $x^2 - 4 = 8$

First, rewrite it with the quadratic term on the left.

$$x^2 = 12$$

Second, take the square root of both sides.

$$x = \pm\sqrt{12} = \pm 2\sqrt{3}$$

$$x \approx \pm 3.46$$

The roots are $+2\sqrt{3}$ and $-2\sqrt{3}$.

In Example 9 we have used the fact that performing the same operation on both sides of an equation results in an equivalent equation. Notice that in order to obtain both roots of the equation, we must include the negative as well as the positive, or principal, square root. In some situations, this results in two equations that require additional steps to solve.

EXAMPLE 10 Solve by the square root method of Example 9.

(a) $4x^2 - 25 = 0$ (b) $(x - 7)^2 = 64$ (c) $\left(x + \dfrac{1}{2}\right)^2 = \dfrac{3}{4}$

Solution

(a) Isolate the quadratic term on the left.

$$4x^2 = 25$$

Take the square root of both sides.

$$2x = 5 \quad \text{or} \quad 2x = -5$$

Solve each resulting equation.

$$x = \frac{5}{2} \quad \text{or} \quad x = -\frac{5}{2}$$

(b) The quadratic term is isolated on the left:

$$(x - 7)^2 = 64$$

Take the square root of both sides.

$$x - 7 = 8 \quad \text{or} \quad x - 7 = -8$$

Solve each resulting equation.

$$x = 15 \quad \text{or} \quad x = -1$$

(c) The quadratic term is isolated on the left.

$$\left(x + \frac{1}{2}\right)^2 = \frac{3}{4}$$

Take the square root of both sides.

$$x + \frac{1}{2} = \frac{\sqrt{3}}{2} \quad \text{or} \quad x + \frac{1}{2} = -\frac{\sqrt{3}}{2}$$

Solve each resulting equation.

$$x = -\frac{1}{2} + \frac{\sqrt{3}}{2} \quad \text{or} \quad x = -\frac{1}{2} - \frac{\sqrt{3}}{2}$$

Combined into single fractions, the roots are

$$\frac{-1 + \sqrt{3}}{2} \quad \text{and} \quad \frac{-1 - \sqrt{3}}{2}$$

We can write these as

$$\frac{-1 \pm \sqrt{3}}{2} \quad \text{where the symbol "±" is read "plus or minus"}$$

It is easy to use substitution to check the roots in Examples 10(a) and (b), but considerably more difficult in part (c). You will find it quicker to use your calculator to check irrational roots. To check (c):

Calculate the first root.   1 $\boxed{+/-}$ $\boxed{+}$ 3 $\boxed{\sqrt{}}$ $\boxed{=}$ $\boxed{\div}$ 2 $\boxed{=}$ → ▓ *0.3660254* ▓

Substitute in $\left(x + \dfrac{1}{2}\right)^2 = \dfrac{3}{4}$. $\boxed{+}$ .5 $\boxed{=}$ $\boxed{x^2}$ → ▓ *0.75* ▓   It checks.

Then calculate the
second root.   1 $\boxed{+/-}$ $\boxed{-}$ 3 $\boxed{\sqrt{}}$ $\boxed{=}$ $\boxed{\div}$ 2 $\boxed{=}$ → ▓ *−1.3660254* ▓

Substitute. $\boxed{+}$ .5 $\boxed{=}$ $\boxed{x^2}$ → ▓ *0.75* ▓   This one
checks too.

In the preceding examples, each equation could be solved quickly because the expression containing the variable was a perfect square. If the expression is not a perfect square, we can modify the equation to make it a perfect square. This process is called **completing the square.**

EXAMPLE 11   To solve: $x^2 + 4x - 6 = 0$

First, check to be certain the equation is not factorable. Then follow these steps:

Step 1. Isolate all
x-terms on the left.

$$x^2 + 4x - 6 = 0$$
$$x^2 + 4x = 6$$

Step 2. Determine
what constant will make
the left side a perfect
square trinomial. Add
this to both sides.

$$x^2 + 4x + 4 = 6 + 4$$

Step 3. Factor, and
solve as before.

$$(x + 2)^2 = 10$$
$$x + 2 = \sqrt{10} \quad \text{or} \quad x + 2 = -\sqrt{10}$$
$$x = -2 + \sqrt{10} \quad \text{or} \quad x = -2 - \sqrt{10}$$

The roots are $-2 \pm \sqrt{10}$.

Step 2 of this procedure is not usually as easy as it seems. The key to finding the correct constant for completing the square can be discovered by examining special product (6-3) from Section 6-1:

| Half of this squared . . . equals this |

$$(x + a)^2 = x^2 + 2ax + a^2$$

Notice that the constant a^2 is equal to the square of half the linear coefficient. Half of $2a$ is a, and a^2 is equal to the constant term. Use this fact to solve the quadratic equation in the next example.

EXAMPLE 12   Solve: $x^2 - 10x + 7 = 0$

Solution

Step 1.   Isolate the variable terms on the left side.

$$x^2 - 10x + 7 = 0$$
$$x^2 - 10x = -7$$

Step 2.   Complete the square on the left.

$$x^2 - 10x + 25 = -7 + 25$$

[*Hint:*   Half of -10 is -5 and $(-5)^2 = 25$, so add 25 to both sides.]

Step 3.   Factor the expression on the left.

$$(x - 5)^2 = 18$$

Step 4.   Take the square root of both sides and solve.

$$x - 5 = \sqrt{18} \quad \text{or} \quad x - 5 = -\sqrt{18}$$
$$x = 5 + \sqrt{18} \quad \text{or} \quad x = 5 - \sqrt{18}$$
$$x = 5 \pm 3\sqrt{2}$$

These roots are irrational numbers expressed in simplest radical form. If a decimal solution is desired, the irrational numbers may be approximated using a calculator before converting them to simplest radical form:

The first root is: $5\ \boxed{+}\ 18\boxed{\sqrt{}}^*\ \boxed{=} \rightarrow$ 9.2426407

The second root is: $5\ \boxed{-}\ 18\ \boxed{\sqrt{}}\ \boxed{=} \rightarrow$ 0.7573593

When the linear coefficient is an odd number, the problem becomes a bit more tricky; however, the method remains the same.

EXAMPLE 13   Solve $x^2 + 7x - 5 = 4$. Express the roots in decimal form rounded to the nearest thousandth.

Solution

Step 1.   Isolate the variable terms on the left side.

$$x^2 + 7x - 5 = 4$$
$$x^2 + 7x = 9$$

---

* By pressing $\boxed{\text{STO}}$ at this point, you can place $\sqrt{18}$ in memory and save a keystroke by using $\boxed{\text{RCL}}$ instead of $18\ \boxed{\sqrt{}}$ in calculating the second root.

Step 2. To complete the square, take half of 7 and square it: $(\frac{7}{2})^2 = \frac{49}{4}$. Add this quantity to both sides.

$$x^2 + 7x + \frac{49}{4} = 9 + \frac{49}{4}$$

Step 3. Factor the left side and simplify the right side.

$$\left(x + \frac{7}{2}\right)^2 = \frac{85}{4}$$

(Notice that the constant in the binomial, $\frac{7}{2}$, is the square root of the constant that completed the square, $\frac{49}{4}$.)

Step 4. Solve as before.

$$x + \frac{7}{2} = \frac{\sqrt{85}}{2} \quad \text{or} \quad x + \frac{7}{2} = -\frac{\sqrt{85}}{2}$$

$$x = \frac{-7 \pm \sqrt{85}}{2}$$

The first root is: 7 $\boxed{+/-}$ $\boxed{+}$ 85 $\boxed{\sqrt{\ }}$ $\boxed{=}$ $\boxed{\div}$ 2 $\boxed{=}$ → $\boxed{\mathit{1.1097722}}$

The second root is: 7 $\boxed{+/-}$ $\boxed{-}$ 85 $\boxed{\sqrt{\ }}$ $\boxed{=}$ $\boxed{\div}$ 2 $\boxed{=}$ → $\boxed{\mathit{-8.1097722}}$

The roots are approximately 1.110 and −8.110, rounded.

Notice that in this method the quadratic term must have a coefficient of 1. If it does not, divide all terms by the coefficient of the x^2-term and continue as before.

EXAMPLE 14 Solve $3x^2 - 10x + 12 = 4$ by completing the square.

Solution

Subtract 12 from both sides.

$$3x^2 - 10x = -8$$

Divide all terms by 3.

$$x^2 - \frac{10}{3}x = -\frac{8}{3}$$

Complete the square by taking half the linear coefficient $[\frac{1}{2}(-\frac{10}{3}) = -\frac{5}{3}]$ and squaring it: $(-\frac{5}{3})^2 = \frac{25}{9}$.

Add this number to both sides.

$$x^2 - \frac{10}{3}x + \frac{25}{9} = -\frac{8}{3} + \frac{25}{9}$$

Factor the left side while simplifying the right side.

$$\left(x - \frac{5}{3}\right)^2 = \frac{1}{9}$$

Solve as before.

$$x - \frac{5}{3} = \frac{1}{3} \quad \text{or} \quad x - \frac{5}{3} = -\frac{1}{3}$$

$$x = \frac{5}{3} + \frac{1}{3} \quad \text{or} \quad x = \frac{5}{3} - \frac{1}{3}$$

$$x = 2 \quad \text{or} \quad x = \frac{4}{3}$$

The roots are 2 and $\frac{4}{3}$. Since these roots are rational numbers, we know that the original equation could have been solved by factoring.

Rate problems   Earlier, we solved rate problems that resulted in linear equations. However, some rate problems may produce quadratic equations.

EXAMPLE 15   One pump can fill a tank in 2 h less time than a second pump. Together they fill the tank in 3 h. How long does it take each pump to fill the tank alone? (Round to the nearest 0.1 h.)

Solution Let

$$x = \text{Time for the slower pump to fill the tank}$$

$$x - 2 = \text{Time for the faster pump to fill the tank}$$

- In 1 h the slower pump fills $\dfrac{1}{x}$ of the tank.

- In 1 h the faster pump fills $\dfrac{1}{x - 2}$ of the tank.

- In 3 h they fill $\dfrac{3}{x}$ and $\dfrac{3}{x - 2}$ of the tank, respectively.

The standard rate equation given in Section 6-7 becomes

$$\frac{3}{x} + \frac{3}{x - 2} = 1$$

To solve, multiply all terms by the LCD: $x(x - 2)$

$$x(x - 2) \cdot \frac{3}{x} + x(x - 2) \cdot \frac{3}{x - 2} = x(x - 2) \cdot 1$$

Eliminate fractions.
Then solve the resulting
quadratic equation by
completing the square.

$$3x - 6 + 3x = x^2 - 2x$$

$$-6 = x^2 - 8x$$

$$-6 + 16 = x^2 - 8x + 16$$

$$10 = (x - 4)^2$$

$$x - 4 = \sqrt{10} \quad \text{or} \quad x - 4 = -\sqrt{10}$$

$$x = 4 + \sqrt{10} \quad \text{or} \quad x = 4 - \sqrt{10}$$

$4 \boxed{+} 10 \boxed{\sqrt{}} \boxed{=} \rightarrow$ *7.1622777*   or $4 \boxed{-} 10 \boxed{\sqrt{}} \boxed{=} \rightarrow$ *0.8377223*

or 7.2 h, rounded   or 0.8 h, rounded

These represent the possible times for the slower pump. The faster pump fills the tank in 2 h less time. For the first root this would mean 5.2 h. But the second root produces a negative answer for the faster pump, -1.2 h. Since time cannot be a negative quantity, we must reject the second solution. The only acceptable solution is 7.2 h for the slower pump and 5.2 h for the faster pump.

Check: If the slower pump fills the tank in 7.2 h, then in 1 h it fills $\frac{1}{7.2}$, or about 0.14, of a tank. The faster pump fills it in 5.2 h, so it fills $\frac{1}{5.2}$, or about 0.19, of a tank in 1 h. This is a total of $0.14 + 0.19 = 0.33$ of a tank per hour. In 3 h the two pumps fill 3(0.33), or 0.99, of a tank—approximately a full tank.

Exercises 7-2

Solve each quadratic equation by the square root method demonstrated in Example 9.

1. $x^2 = 16$
2. $x^2 - 64 = 0$
3. $9x^2 - 121 = 0$
4. $36x^2 = 49$
5. $x^2 = 8$
6. $x^2 - 3 = 0$
7. $4x^2 - 7 = 0$
8. $16x^2 = 41$
9. $(x - 5)^2 = 16$
10. $(x + 3)^2 = 81$
11. $(x + 2)^2 = 8$
12. $(x - \frac{1}{4})^2 = \frac{7}{16}$
13. $x^2 + 12x + 36 = 5$
14. $x^2 - 10x + 25 = 12$

Solve each quadratic equation by completing the square. Round calculator answers to the nearest thousandth. Otherwise, express all irrational roots in radical form.

15. $x^2 + 10x + 16 = 0$
16. $x^2 - 2x - 48 = 0$
17. $x^2 + x = 42$
18. $x^2 = 11x - 28$
19. $x^2 + 6x + 6 = 0$
20. $x^2 + 2x - 7 = 0$
21. $x^2 - 5x - 3 = 5$
22. $x^2 - 9x + 9 = 2$
23. $3x^2 = 14x - 15$
24. $2x^2 + 9x + 10 = 0$
25. $5x^2 - 8x - 5 = 0$
26. $7x^2 + 14x + 6 = 0$
27. $6x^2 + x = 3$
28. $4x^2 - 9x + 8 = 4$
29. $\dfrac{2}{x - 5} + \dfrac{x}{x + 3} = 2$
30. $\dfrac{x}{x + 4} - \dfrac{5}{x} = 3$
31. $x^2 + 4mx + n = 0$
32. $x^2 - 3bx - 2c = 0$

Set up quadratic equations and solve by completing the square. Round to the nearest tenth.

33. *Wastewater technology* One outlet takes 3 h longer to empty a water tank than another outlet. Together they empty the tank in 4 h. How long would it take each outlet to empty the tank alone?

34. *Graphics technology* One printer can do a typesetting job in 10 min less than another. Together they can do the job in 25 min. How long does it take each printer to do the job alone?

35. A rectangular lot is 12 ft longer than it is wide. Its area is 78 ft². Find the dimensions.

36. *Construction* A rectangular storage yard is bordered on three sides by fencing and on the fourth side by an existing building. Using 60 ft of fencing, what are the dimensions of a 420 ft² storage yard?

7-3 | The Quadratic Formula

Deriving the quadratic formula   A formula usually can be developed to replace any repetitive mathematical procedure. Since completing the square is just such a technique, there must be a formula that represents the solution to a quadratic equation. To develop this formula, we can apply the method for completing the square to the general quadratic equation

$$ax^2 + bx + c = 0$$

Step 1. Isolate the variable terms on one side.

$$ax^2 + bx + c = 0$$
$$ax^2 + bx = -c$$

Step 2. Divide every term by a.

$$x^2 + \frac{b}{a}x = -\frac{c}{a}$$

Step 3. Complete the square. Take half the linear coefficient, $\frac{1}{2} \cdot \frac{b}{a} = \frac{b}{2a}$, and square it: $\left(\frac{b}{2a}\right)^2 = \frac{b^2}{4a^2}$.

Add this to both sides.

$$x^2 + \frac{b}{a}x + \frac{b^2}{4a^2} = -\frac{c}{a} + \frac{b^2}{4a^2}$$

Step 4. Factor the left side while simplifying the right side.

$$\left(x + \frac{b}{2a}\right)^2 = -\frac{c \cdot 4a}{a \cdot 4a} + \frac{b^2}{4a^2}$$

$$\left(x + \frac{b}{2a}\right)^2 = \frac{b^2 - 4ac}{4a^2}$$

Step 5. Take the square root of both sides.

$$x + \frac{b}{2a} = \frac{\pm\sqrt{b^2 - 4ac}}{2a}$$

Step 6. Solve for x. The result is the **quadratic formula:**

The Quadratic Formula $x = \dfrac{-b \pm \sqrt{b^2 - 4ac}}{2a}$ (7-1)

Using the quadratic formula   If we know the coefficients a, b, and c of the terms of a quadratic equation in standard form, we can use this formula to calculate the two roots.

EXAMPLE 16   To solve $x^2 - 10x + 8 = 6$, use the quadratic formula.

First, put the equation in standard form.

$$x^2 - 10x + 2 = 0$$

$$a = 1 \qquad b = -10 \qquad c = 2$$

Second, note the coefficients a, b, and c:

Third, substitute these values into the formula and simplify:

$$x = \frac{-(-10) \pm \sqrt{(-10)^2 - 4(1)(2)}}{2(1)}$$

$$x = \frac{10 \pm \sqrt{92}}{2} = \frac{10 \pm 2\sqrt{23}}{2}$$

$$= \frac{2(5 \pm \sqrt{23})}{2} = 5 \pm \sqrt{23}$$

Finally, be sure to check your answer.

CAUTION  Always express an exact, irrational answer in simplest radical form and, if possible, reduce it to lowest terms. In the final step of Example 16 we were able to divide the numerator and denominator by a common factor of 2. ◀

If $b^2 - 4ac$ is a perfect square, you will need to carry your solution one step further.

EXAMPLE 17 Solve $2x^2 + 3x = 27$ using the quadratic formula.

Solution

Write in standard form.

$$2x^2 + 3x - 27 = 0$$

Identify a, b, and c.

$$a = 2 \qquad b = 3 \qquad c = -27$$

Substitute into the formula.

$$x = \frac{-3 \pm \sqrt{3^2 - 4(2)(-27)}}{2(2)}$$

Simplify.

$$x = \frac{-3 \pm \sqrt{225}}{4}$$

Since 225 is a perfect square, continue.

$$x = \frac{-3 \pm 15}{4}$$

This results in two rational roots:

$$x = \frac{-3 + 15}{4} \quad \text{or} \quad x = \frac{-3 - 15}{4}$$

$$x = 3 \quad \text{or} \qquad x = -\frac{9}{2}$$

Do not leave the solution in "$\pm$" form when the roots are rational.

EXAMPLE 18 Solve: $4x^2 - 20x + 25 = 0$

Solution Since the equation already is in standard form, begin by identifying the coefficients: $a = 4$, $b = -20$, $c = 25$. Then substitute:

$$x = \frac{-(-20) \pm \sqrt{(-20)^2 - 4(4)(25)}}{2(4)}$$

$$x = \frac{20 \pm \sqrt{0}}{8} = \frac{5}{2}$$

Because the radicand $b^2 - 4ac$ is zero, the two roots have the *same value*. That value is called a *double root*.

Also, notice that in Examples 16 and 17, $b^2 - 4ac$ was *positive*, resulting in *two unequal real roots*. If $b^2 - 4ac$ is *negative*, the quadratic equation has *no real roots*. The roots are imaginary.*

---

*Because the quantity $b^2 - 4ac$ can be used in this way to determine the nature of the roots of a quadratic equation, it is given a special name, the **discriminant**.

EXAMPLE 19   For a certain temperature-sensitive electronic device, the voltage output is related to the temperature according to the formula

$$V = 3.1T - 0.014T^2$$

where V is the output voltage (in millivolts) and T is the temperature (in degrees Celsius). What temperature is needed to produce a voltage of 200 mV?

Solution

Substitute the given value of V. $200 = 3.1T - 0.014T^2$

Rewrite in standard form. $0.014T^2 - 3.1T + 200 = 0$

Substitute $a = 0.014$, $b = -3.1$, and $c = 200$ into the quadratic formula.

$$T = \frac{3.1 \pm \sqrt{(-3.1)^2 - 4(0.014)(200)}}{2(0.014)}$$

Calculate T: 3.1 [+/−] [x²] [−] 4 [×] .014 [×] 200 [=] [√] →   *Error*

Because $b^2 - 4ac$ is negative (-1.59), the calculator displays an error message as soon as we press the [√] key. This means that the roots are imaginary numbers, so the equation has no real solution. There is no temperature corresponding to a voltage of 200 mV.

The quadratic formula also can be used to solve quadratic equations with literal coefficients.

EXAMPLE 20   The general formula for the surface area of a cylinder is $A = 2\pi r^2 + 2\pi rh$, where r is the radius and h is the altitude. Solve the formula for r.

Solution

Rewrite in standard form. $2\pi r^2 + 2\pi rh - A = 0$

Identify the coefficients. $\boxed{a = 2\pi}$ $\boxed{b = 2\pi h}$ $\boxed{c = -A}$

(Remember, r is the designated variable.)

Substitute into the quadratic formula:

$$r = \frac{-2\pi h \pm \sqrt{(2\pi h)^2 - 4(2\pi)(-A)}}{2(2\pi)}$$

$$= \frac{-2\pi h \pm \sqrt{4\pi^2 h^2 + 8\pi A}}{4\pi}$$

Simplify:

$$r = \frac{-2\pi h \pm 2\sqrt{\pi^2 h^2 + 2\pi A}}{4\pi} = \frac{-\pi h \pm \sqrt{\pi^2 h^2 + 2\pi A}}{2\pi}$$

But the radius r must be a positive quantity. Therefore, the only valid root is

$$r = \frac{-\pi h + \sqrt{\pi^2 h^2 + 2\pi A}}{2\pi}$$

This new formula now can be used to find the radius whenever the altitude and surface area are given.

Motion problems   Certain problems involving motion may lead to quadratic equations.

EXAMPLE 21   An airplane traveled the first 640 mi of a trip at one speed and then decreased its average speed by 50 mi/h for the last 220 mi. If the whole trip took $2\frac{1}{2}$ h, what was the average speed for the first part of the journey?

Solution Let

$$v = \text{Average speed for the first part}$$

$$v - 50 = \text{Average speed for the second part}$$

For uniform motion,

$$\text{Time} = \frac{\text{Distance}}{\text{Average speed}}$$

Time for first part	+	Time for second part	=	Total time
$\dfrac{640}{v}$	+	$\dfrac{220}{v-50}$	=	$\dfrac{5}{2}$

Multiply by the LCD: $2v(v-50)$

$$2v(v-50)\cdot\frac{640}{v} + 2v(v-50)\cdot\frac{220}{v-50} = 2v(v-50)\cdot\frac{5}{2}$$

Reduce. $2(v-50)\cdot 640 + 2v\cdot 220 = v(v-50)\cdot 5$

Simplify. $1280v - 64{,}000 + 440v = 5v^2 - 250v$

Put in standard form. $5v^2 - 1970v + 64{,}000 = 0$

Divide by the common factor 5. $v^2 - 394v + 12{,}800 = 0$

Substitute into the quadratic formula.

$$v = \frac{394 \pm \sqrt{(-394)^2 - 4(1)(12{,}800)}}{2}$$

394 $\boxed{x^2}$ $\boxed{-}$ 4 $\boxed{\times}$ 12800 $\boxed{=}$ $\boxed{\sqrt{\ }}$ $\boxed{\text{STO}}$ $\boxed{+}$ 394 $\boxed{=}$ $\boxed{\div}$ 2 $\boxed{=}$ → ▆ *358.27306*

Or 394 $\boxed{-}$ $\boxed{\text{RCL}}$ $\boxed{=}$ $\boxed{\div}$ 2 $\boxed{=}$ → ▆ *35.726940*

The second root is unacceptable, because it would produce a negative speed for the second part of the journey. The rounded answer is 360 mi/h.

Exercises 7-3

Solve each quadratic equation using the quadratic formula. Round all calculator answers to the nearest thousandth. Otherwise, express all irrational roots in simplest radical form.

1. $x^2 + 5x + 4 = 0$

2. $x^2 - 8x + 12 = 0$

3. $y^2 - 3y = 28$

4. $m^2 + 5m = 14$

5. $x^2 + 6x + 9 = 0$

6. $x^2 + 14x + 49 = 0$

7. $v^2 + 5v - 9 = 0$

8. $t^2 - 6t + 7 = 0$

9. $2x^2 - x - 28 = 0$

10. $3x^2 + 13x - 30 = 0$

11. $5n^2 - 3n - 4 = 0$

12. $4p^2 - 8p + 1 = 0$

13. $2x^2 + 3x + 8 = 0$

14. $3x^2 - x + 2 = 0$

15. $2c^2 - 7c = 5$

16. $6e^2 - 2e = 7$

17. $5x^2 - 12 = 0$

18. $3x^2 - 13 = 0$

19. $2x^2 - 7x = 0$

20. $4x^2 + 9x = 0$

21. $3r^2 + 10r + 5 = 7$

22. $2p^2 - 9p - 14 = 6$

23. $2 - 7x - 8x^2 = 0$

24. $17 - 4x - 9x^2 = 0$

25. $\dfrac{x}{x^2 - 9} + \dfrac{5}{x + 3} = 2$

26. $\dfrac{x + 2}{x + 5} + \dfrac{x}{x^2 + 7x + 10} = \dfrac{2}{x + 2}$

27. $x^2 + 5tx + 3 = 0$

28. $x^2 - 8x + 2q = 0$

Solve.

29. *Electronics*   For the formula given in Example 19, what temperature is needed to produce a voltage of 150 mV?

30. *Fire science*   Under certain conditions, the nozzle pressure p (in pounds per square inch) and the height h (in feet) of a fire hose stream are related by the formula $(h + 17)^2 = 240p - p^2 - 1900$. For a pressure of 40.0 lb/in.², what is the height of the fire hose stream?*

31. *Physics*   The distance s traveled by an object is described by the formula $s = v_0 t + \frac{1}{2}at^2$, where v_0 is its initial velocity, t is the elapsed time, and a is the acceleration. Solve the formula for t.

32. *Electronics*   For a certain electrical circuit, the power P is given by $P = RI^2 + VI$, where V is the voltage, R is the resistance, and I is the current. Solve this formula for I.

Set up quadratic equations and solve.

33. *Navigation*   A jet traveled the first 1260 km of a trip at one speed and then decreased its speed by 40 km/h for the last 875 km. If the entire trip took 2.2 h, what was its speed during the first half of the trip?

34. *Navigation*   On Tuesday the Central Alaska mail plane flew 2450 mi at a certain speed. On Wednesday the plane made the return trip in $\frac{1}{2}$ h less time by increasing its average speed by 85 mi/h. What was its average speed on Tuesday?

35. *Sheet metal technology*   A rectangular heating duct must be constructed from a piece of sheet metal 68 in. wide. What dimensions will produce a cross-sectional area of 288 in.²?

36. *Wastewater technology*   Working together two pumps, X and Y, can empty a tank in 5.0 h. Working alone pipe X takes 1.0 h longer to fill the tank than pipe Y. How long does it take each pipe to fill the tank alone?

---

* For more information, see Fred Shepperd, *Fire Service Hydraulics* (New York: Fire Engineering Book Dept., Division of Technical Publishing Co., 1970).

| 7-4 | **Review of Chapter 7** |

Important Terms and Concepts

quadratic equation (p. 252)

quadratic term (p. 252)

linear term (p. 252)

constant term (p. 252)

zero-product rule (p. 253)

incomplete quadratic equation (p. 254)

completing the square (p. 259)

The Quadratic Formula (p. 264)

For a quadratic equation in the form $ax^2 + bx + c = 0$, where $a \neq 0$: $x = \dfrac{-b \pm \sqrt{b^2 - 4ac}}{2a}$ **(7-1)**

Exercises 7-4

Solve by factoring.

1. $x^2 - 81 = 0$
2. $4x^2 - 49 = 0$
3. $3y^2 + 10y = 0$
4. $m^2 = 6m$
5. $x^2 + 10x + 25 = 0$
6. $9x^2 + 42x + 49 = 0$
7. $x^2 + 9x = -18$
8. $x^2 = 12x - 35$
9. $(a - 3)^2 = 41 - 2a$
10. $(n - 4)(n + 6) = n + 18$
11. $5x^2 - 22x - 15 = 0$
12. $6x^2 - 49x + 8 = 0$

Solve using any method. Express all irrational roots in radical form.

13. $x^2 = 25$
14. $16x^2 = 9$
15. $49x^2 - 12 = 0$
16. $x^2 - 7 = 0$
17. $(c - 3)^2 = 100$
18. $n^2 - 16n + 64 = 10$
19. $x^2 - 2x - 48 = 0$
20. $x^2 + 11x + 30 = 0$
21. $x^2 + 12x + 36 = 0$
22. $x^2 - 14x + 49 = 0$
23. $x^2 + 10x - 3 = 0$
24. $x^2 - 8x - 5 = 0$
25. $y^2 + 5y + 2 = 0$
26. $m^2 - 2m + 11 = 0$
27. $ax^2 - 2x - 3a = 0$
28. $x^2 - 6mx + 7m = 0$

Solve. Express all irrational roots in decimal form rounded to three places.

29. $5t^2 - 8t + 2 = 0$
30. $7v^2 + 6v - 3 = 0$
31. $3x^2 + 9x - 4 = 0$
32. $2x^2 - 7x - 13 = 0$
33. $12y^2 = 3y - 5$
34. $10z^2 + 7 = 5z$
35. $4x^2 - 3x - 2 = 0$
36. $6x^2 + 5x - 6 = 0$
37. $x^2 + 6x = 11$
38. $x^2 = 4x + 7$
39. $\dfrac{x - 5}{x + 3} = \dfrac{4}{x}$
40. $\dfrac{x + 4}{x - 6} = \dfrac{2x + 7}{x - 3}$
41. $\dfrac{x + 5}{2} - 1 = \dfrac{3}{x}$
42. $\dfrac{2}{x} - \dfrac{x - 5}{x + 2} = \dfrac{7}{x^2 + 2x}$

Solve.

43. **Electronics** In a particular electrical circuit, the current I, voltage V, and resistance R are related by the formula $8000 = RI^2 + VI$. Find I (in amperes) when $R = 8.0\ \Omega$ and $V = 120$ V.

44. The surface area A of a right circular cylinder is given by the formula $A = 2\pi r^2 + 2\pi rh$, where r is the radius and h is the height. Find r if $A = 625$ in.^{2} and $h = 12.0$ in.

45. *Physics*   The height h (in meters) of an object dropped from an altitude of 750 m is given by the formula $h = 750 - 4.9t^2$, where t is the elapsed time (in seconds). After how many seconds will the object hit the ground?

46. *Physics*   If air friction is ignored, the height h (in meters) of a projectile fired vertically upward is given by the formula $h = 36t - 4.9t^2$, where t is the elapsed time (in seconds). When will the object strike the ground?

47. *Electronics*   For the formula in Example 19, what temperature is needed to produce a voltage of 120 mV?

48. *Fire science*   Under certain conditions, the nozzle pressure p (in pounds per square inch) and the height h (in feet) of a fire hose stream are related by the formula $(h + 15)^2 = 240p - p^2 - 1900$. Find the minimum pressure needed to produce a stream 30.0 ft high.

49. *Electrical engineering*   When two resistances, R_1 and R_2, are connected in series, an equivalent resistance of 18 Ω results. This can be expressed by the equation $R_1 + R_2 = 18$. When the same resistances are connected in parallel, an equivalent resistance of 4 Ω results. This can be expressed by the equation

$$\frac{R_1 R_2}{R_1 + R_2} = 4$$

Solve this pair of equations for R_1 and R_2.

50. *Police science*   The stopping distance d (in feet) of an automobile traveling x miles per hour can be approximated by the formula below. At what speed will the stopping distance be approximately 110 ft?

$$d = x + \frac{x^2}{20}$$

Set up quadratic equations and solve.

51. *Sheet metal technology*   A rectangular, open-topped water channel must be constructed using a 30 in. wide piece of sheet metal. What dimensions will produce a cross-sectional area of 108 in.^{2}?

52. A box must be constructed using a rectangular piece of cardboard measuring 28 in. by 16 in. If the area of the bottom of the box must be 189 in.^{2}, what uniform width w must be cut from each corner (see the figure)?

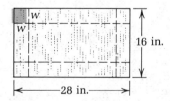

53. *Wastewater technology*   Working together two pipes, A and B, can fill a tank in 4.0 h. Working alone pipe A takes 1.0 h less time to fill the tank than pipe B. How long does it take each pipe to fill the tank alone?

54. *Navigation*   A jet flies for 1850 mi at a certain speed. The next day it makes the return trip in $\frac{1}{2}$ h less time by increasing its speed by 125 mi/h. What was its speed the second day?

55. *Navigation* In order to find the altitude of another aircraft, a jet first measures the direct echo range and the ground echo range between itself and the aircraft. Suppose these measurements produce the triangle shown in the figure. Using the Pythagorean theorem, write a formula involving R and X, and then solve for R in terms of X.

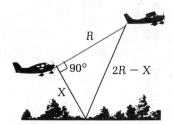

56. *Electronics* In choosing a resistor with resistance R that produces a particular sensitivity in a circuit, the following formula is used: $(R + a)^2 = 3(R + b)^2$, where a and b are constants. Solve for R.

57. *Chemistry* In calculating the concentration of a certain compound in a solution, the equation below must be solved. Solve for the concentration x (in moles per liter).

$$7.4 \times 10^{-4} = \frac{x^2}{0.50 - x}$$

58. *Machine technology* The diameter d of the circle shown here is given by the formula

$$d = \frac{\left(\dfrac{w}{2}\right)^2 + h^2}{h}$$

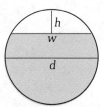

(a) Solve this formula for h. Which of the two answers represents the value of h shown in the figure?

(b) Find the depth of cut (h) needed to produce a 0.25 in. flat (w) on the cross section of a cylindrical shaft 0.75 in. in diameter (d).

Trigonometric Functions of Any Angle; Radian Measure

8

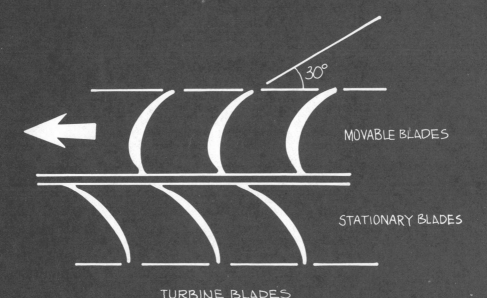

30°

MOVABLE BLADES

STATIONARY BLADES

TURBINE BLADES

In Chapter 4 we discussed the trigonometric functions for acute angles measured in degrees. In some applications, the values of these functions must be determined for angles larger than 90° and for angles measured in *radians*. These topics are the subjects of this chapter.

Nonacute angles

The values of all trigonometric functions are positive for acute, or first-quadrant, angles. However, some experimentation with your calculator will reveal that this is not the case for angles larger than 90°. For example, cos 120° and sin 220° are both negative:

$$120 \boxed{\cos} \rightarrow \boxed{-0.5} \qquad 220 \boxed{\sin} \rightarrow \boxed{-0.6427876}$$

Although a scientific calculator will display the function values for an angle of any size, many applications involving nonacute angles require you to know the sign of the function value by inspection. Since these signs vary according to the quadrant in which the terminal side of the angle is located, you must know the range of angle measures associated with each quadrant. If you need to review these, study Fig. 8-1 before proceeding.

Figure 8-1

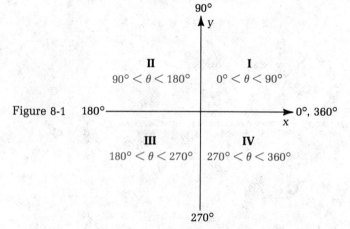

In order to determine the signs of the trigonometric functions in each quadrant, we must also review the definitions of the trigonometric functions. As stated in Chapter 4, for any point (x, y) on the terminal side of an angle θ (Fig. 8-2), these values are given in the box:

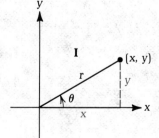

Figure 8-2

Definitions of the Trigonometric Functions

$$\sin \theta = \frac{y}{r} \qquad \cos \theta = \frac{x}{r} \qquad \tan \theta = \frac{y}{x}$$

$$\csc \theta = \frac{r}{y} \qquad \sec \theta = \frac{r}{x} \qquad \cot \theta = \frac{x}{y}$$

(8-1)

As we showed in Chapter 4, r is always positive: therefore, the signs of the trigonometric functions depend on the signs of x and y. As shown in Fig. 8-2, x and y are both positive in quadrant I, and, consequently, so are all six trigonometric functions. However, for angles in quadrant II (Fig. 8-3), x is negative and y is positive. Referring to the ratios in definition (8-1), we see that sin θ and

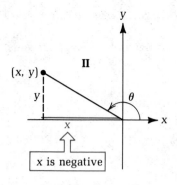

Figure 8-3

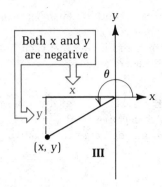

Figure 8-4

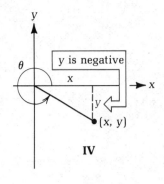

Figure 8-5

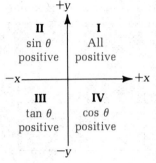

Figure 8-6

csc θ both remain positive because they depend on y. But cos θ, sec θ, tan θ, and cot θ are now negative because they depend on ratios involving x.

In quadrant III (Fig. 8-4), x and y are both negative. Thus, sin θ, cos θ, csc θ, and sec θ are all negative. However, since tan θ and cot θ both involve the ratio of these two negative quantities, they are positive in the third quadrant.

In quadrant IV (Fig. 8-5), x is positive and y is negative. Therefore, sin θ, tan θ, cot θ, and csc θ are all negative, while cos θ and sec θ are both positive.

If you want to memorize these signs, notice that only one of the three basic functions is positive in each of quadrants II, III, and IV. Once you know these three functions, you then know that the reciprocal functions are also positive, and that all other functions are negative. These results are summarized in Fig. 8-6.

LEARNING HINT ▶

If you have trouble learning these signs, try a memory device such as the phrase "All students take care":

Quadrant:	I	II	III	IV
	All	**Students**	**Take**	**Care**
Positive functions:	All	sin	tan	cos

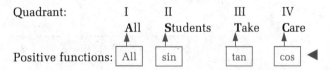

EXAMPLE 1   Determine the sign of each of the following:

(a)   sin 237°   (b)   cos 308°   (c)   tan 337°   (d)   cot 189°
(e)   cos 148°   (f)   csc 280°   (g)   tan 95.6°   (h)   sec 240°45′
(i)   sin (−190°)   (j)   cos (−24°)

Solution

	Quadrant	r	x	y	Function Sign		Quadrant	r	x	y	Function Sign
(a)	III	+	−	−	−	(b)	IV	+	+	−	+
(c)	IV	+	+	−	−	(d)	III	+	−	−	+
(e)	II	+	−	+	−	(f)	IV	+	+	−	−
(g)	II	+	−	+	−	(h)	III	+	−	−	−
(i)	II	+	−	+	+	(j)	IV	+	+	−	+

For parts (i) and (j) remember that a negative angle is measured *clockwise* from the positive x-axis.

Notice from Fig. 8-6 that each function is positive in two quadrants and negative in two quadrants.

EXAMPLE 2   Name the quadrant or quadrants in which:
(a) $\sin \theta$ is negative   (b) $\tan \theta$ is positive
(c) $\cos \theta$ and $\cot \theta$ are both negative

Solution
(a)   III and IV   (b)   I and III   (c)   II only

Quadrantal angles   As we mentioned in Chapter 4, when the vertex is at the origin and the terminal side of an angle is along one of the axes, the angle is a *quadrantal angle*: 0°, 90°, 180°, or 270°. We have already determined the function values for 0° and 90° given in the table.

θ	$\sin \theta$	$\cos \theta$	$\tan \theta$	$\csc \theta$	$\sec \theta$	$\cot \theta$
0°	0	1	0	Undef.	1	Undef.
90°	1	0	Undef.	1	Undef.	0

To determine the values of the functions for 180° and 270°, we need to find x, y, and r for each angle. For 180° (Fig. 8-7), the terminal side is along the negative x-axis. By choosing the point $(-1, 0)$ on the terminal side, we have $x = -1$, $y = 0$, and $r = 1$. Referring to definition (8-1), we obtain the values given in the table.

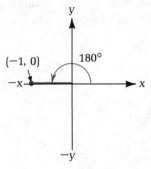

Figure 8-7

θ	$\sin \theta$	$\cos \theta$	$\tan \theta$	$\csc \theta$	$\sec \theta$	$\cot \theta$
180°	0	-1	0	Undef.	-1	Undef.

For 270° (Fig. 8-8), the terminal side is along the negative y-axis. By choosing the point $(0, -1)$ on the terminal side, we have $x = 0$, $y = -1$, and $r = 1$. Substituting these values into definition (8-1), we obtain the values given in the table.

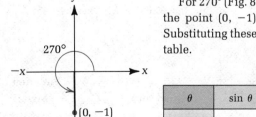

Figure 8-8

θ	$\sin \theta$	$\cos \theta$	$\tan \theta$	$\csc \theta$	$\sec \theta$	$\cot \theta$
270°	-1	0	Undef.	-1	Undef.	0

EXAMPLE 3   Find: (a) $\sin (-90°)$ (b) $\cos 450°$ (c) $\tan (-180°)$

Solution

(a)   sin (−90°) = sin 270° = −1   (b)   cos 450° = cos 90° = 0

(c)   tan (−180°) = tan 180° = 0

Finding function values from a given point

To determine the value of a trigonometric function given a point on the terminal side of the angle, calculate r using the Pythagorean theorem and set up the proper ratio. The correct sign will be part of the answer automatically.

EXAMPLE 4   Find the values of all six trigonometric functions given (−2, 5) as a point on the terminal side of the angle θ. Round to four decimal places if necessary.

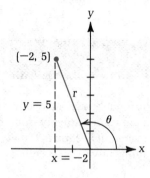

Figure 8-9

Solution   First, determine r, the hypotenuse of the right triangle in Fig. 8-9:

$$r = \sqrt{x^2 + y^2} = \sqrt{(-2)^2 + 5^2} = \sqrt{29}$$

Then, substitute in definition (8-1) to obtain

$$\sin \theta = \frac{5}{\sqrt{29}} \qquad \cos \theta = \frac{-2}{\sqrt{29}} \qquad \tan \theta = \frac{5}{-2}$$

$$\approx 0.9285 \qquad\qquad \approx -0.3714 \qquad\qquad = -2.5$$

$$\csc \theta = \frac{\sqrt{29}}{5} \qquad \sec \theta = \frac{\sqrt{29}}{-2} \qquad \cot \theta = \frac{-2}{5}$$

$$\approx 1.0770 \qquad\qquad \approx -2.6926 \qquad\qquad = -0.4$$

As you should have expected for this second-quadrant angle, sin θ and csc θ are positive, and the other functions are negative.

Exercises 8-1

Determine the sign of each of the following. Do not use a calculator.

1. sin 156°	**2.** sin 243°	**3.** cos 287°	**4.** cos 54°	**5.** tan 136°
6. tan 218°	**7.** sin 72.8°	**8.** sin 309°	**9.** cos 260°	**10.** cos 99°
11. tan 244°	**12.** tan (−71°)	**13.** csc (−38°)	**14.** csc 42°	**15.** sec 111°
16. sec 215°	**17.** cot 327°	**18.** cot 88°	**19.** sin 196°	**20.** sin 170°19′
21. cos 15°51′	**22.** cos 352°	**23.** tan 42°	**24.** tan 94.1°	

Name the quadrant or quadrants for which each of the following is true:

25. cos θ is positive   **26.** cos θ is negative   **27.** sin θ is positive

28. csc θ is negative   **29.** cot θ is positive   **30.** tan θ is negative

31. sin θ is negative, cos θ is positive   **32.** sin θ is positive, cos θ is negative

33. tan θ is positive, sec θ is negative   **34.** cot θ is negative, cos θ is positive

35. sin θ and tan θ are both negative   **36.** cos θ and tan θ are both negative.

Find all values of θ, $0° \le \theta < 360°$, for which each of the following is true:

37. $\sin \theta = 1$ **38.** $\cos \theta = 1$ **39.** $\sin \theta = 0$ **40.** $\cos \theta = 0$
41. $\tan \theta = 0$ **42.** $\tan \theta$ is undefined   **43.** $\sin \theta = -1$ **44.** $\cos \theta = -1$

Determine the values of all six trigonometric functions for the angles with terminal sides that pass through the given points. If necessary, round to four decimal places.

45. $(3, 1)$ **46.** $(2, 3)$ **47.** $(-4, -3)$ **48.** $(-5, -12)$
49. $(2, -1)$ **50.** $(3, -5)$ **51.** $(-6, 2)$ **52.** $(-5, 4)$

8-2 | Trigonometric Functions of Any Angle

In Section 8-1 we found the values of the trigonometric functions given a point on the terminal side of θ. In this section we will find the values directly from the angle itself. Because the quadrantal angles were covered in Section 8-1, the definitions and techniques developed in this section apply only to angles other than the quadrantal angles.

Using a calculator   Finding the value of any trigonometric function for any angle given in degrees is quick and simple using a scientific calculator. First, make sure your calculator is in the degree mode. Then, for $\sin \theta$, $\cos \theta$, and $\tan \theta$, enter the angle measure followed by the appropriate function key. For $\csc \theta$, $\sec \theta$, and $\cot \theta$, enter the angle, then the key of the reciprocal function, and finally the $\boxed{1/x}$ key.

EXAMPLE 5   Use a calculator to find:
(a) $\sin 145°$ (b) $\cos 97.4°$ (c) $\tan 220°$
(d) $\cot (-15°)$ (e) $\sec 250°14'$ (f) $\csc 420°$

Solution
(a) 145 $\boxed{\sin}$ $\rightarrow$ 0.5735764 (b) 97.4 $\boxed{\cos}$ $\rightarrow$ -0.1287956
(c) 220 $\boxed{\tan}$ $\rightarrow$ 0.8390996

(d) $\cot (-15°) = \dfrac{1}{\tan (-15°)}$ 15 $\boxed{+/-}$ $\boxed{\tan}$ $\boxed{1/x}$ $\rightarrow$ -3.7320508

(e) $\sec 250°14' = \dfrac{1}{\cos 250°14'}$

250 $\boxed{+}$ 14 $\boxed{\div}$ 60 $\boxed{=}$ $\boxed{\cos}$ $\boxed{1/x}$ $\rightarrow$ -2.9569135

(f) $\csc 420° = \dfrac{1}{\sin 420°}$ 420 $\boxed{\sin}$ $\boxed{1/x}$ $\rightarrow$ 1.1547005

Check the signs by noting the quadrant in which the terminal side of each angle falls.

Coterminal angles   Even though we can use a calculator to find the values of the trigonometric functions for nonacute angles, it is important that you be able to relate these

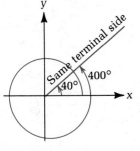

Figure 8-10

Reference angle

values to those for acute angles. This information will be useful for finding certain function values and for finding the angles corresponding to a given value of a function. Using a calculator, it is easy to verify that

$$\sin 400° = 0.6427876 \quad \text{and} \quad \sin 40° = 0.6427876$$

To see why these values are equal, look at Fig. 8-10. Both angles have the same terminal side; therefore, the values of x, y, and r are the same. Because the trigonometric functions depend only on x, y, and r, the value of a given function is the same for any pair of coterminal angles.

Again, using a calculator, it is easy to verify that

$$\sin 145° = \sin 35° \approx 0.5735764$$

Figure 8-11(a) shows why these values are equal:

$$\sin 145° = \frac{y}{r} \quad \text{and} \quad \sin 35° = \frac{y}{r}$$

A calculator will also confirm that

$$\sin 325° = -\sin 35° \approx -0.5735764$$

Figure 8-11(b) verifies that

$$\sin 35° = \frac{y}{r} \quad \text{and} \quad \sin 325° = -\frac{y}{r}$$

Figure 8-11

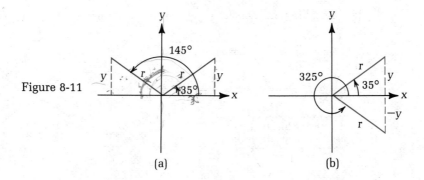

(a) (b)

In general, the *absolute value* of a trigonometric function for an angle in any quadrant is equal to the same function of a first-quadrant angle, called its *reference angle,* defined as follows:

For an angle θ, the **reference angle, Ref θ,** is the positive, acute angle formed by the terminal side of θ and the x-axis.

EXAMPLE 6 Find the reference angle for each of the following:
(a) 98° (b) 35° (c) 190.8° (d) 285°20′

Solution

(a)   82°

(b)   35°

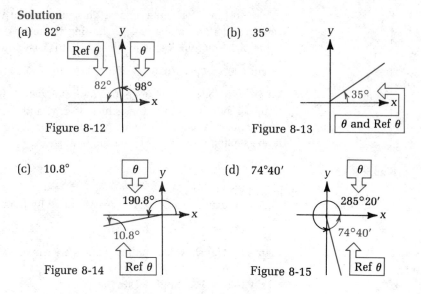

Figure 8-12

Figure 8-13

(c)   10.8°

(d)   74°40′

Figure 8-14

Figure 8-15

We can find the value of any trigonometric function if we know the corresponding value of its reference angle. The sign of this value is found according to the quadrant in which the angle falls, as we saw in Section 8-1. Although you will probably use a calculator for finding most function values, the reference angle method is useful for finding the *exact* function values of angles that have reference angles of 30°, 45°, and 60°.

EXAMPLE 7

(a)   Recalling that $\cos 30° = \dfrac{\sqrt{3}}{2}$, we can conclude that

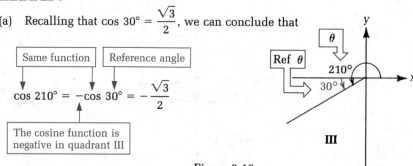

| Same function | Reference angle |

$$\cos 210° = -\cos 30° = -\frac{\sqrt{3}}{2}$$

The cosine function is
negative in quadrant III

Figure 8-16

(b)   Because $\sin 45° = \dfrac{\sqrt{2}}{2}$,

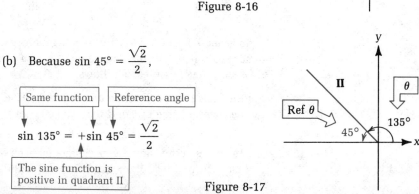

| Same function | Reference angle |

$$\sin 135° = +\sin 45° = \frac{\sqrt{2}}{2}$$

The sine function is
positive in quadrant II

Figure 8-17

(c)   Given tan 60° = $\sqrt{3}$,

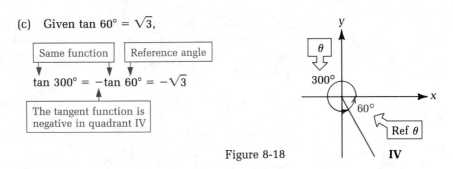

Figure 8-18

Summary   For any given problem, the reference angle can easily be determined by inspection or by making a quick sketch. If you prefer to use a formula for determining the reference angle, see Table 8-1. The values of the three basic functions for an angle θ, $0° < \theta < 360°$, are listed in terms of the reference angle, Ref θ. The signs of csc θ, sec θ, and cot θ correspond to the signs of their reciprocal functions.

TABLE 8-1

For θ in Quadrant	Ref θ	sin θ =	cos θ =	tan θ =
I	Ref $\theta = \theta$	sin (Ref θ)	cos (Ref θ)	tan (Ref θ)
II	Ref $\theta = 180° - \theta$	sin (Ref θ)	$-$cos (Ref θ)	$-$tan (Ref θ)
III	Ref $\theta = \theta - 180°$	$-$sin (Ref θ)	$-$cos (Ref θ)	tan (Ref θ)
IV	Ref $\theta = 360° - \theta$	$-$sin (Ref θ)	cos (Ref θ)	$-$tan (Ref θ)

Finding the angles from the function value   Recall from Chapter 4 that the [INV] or [2nd F] key on a calculator can be used to find the angle θ corresponding to a given value of the trigonometric function. For $0° < \theta < 360°$, excluding the quadrantal angles, there are two values of θ corresponding to each value of the trigonometric function. To find both values with a calculator, first use the absolute value of the given function value to find the reference angle and store Ref θ in memory. Then use the sign of the function to determine the actual values of θ. Figure 8-19 illustrates how to use Ref θ to find θ once its quadrant has been determined.

Figure 8-19

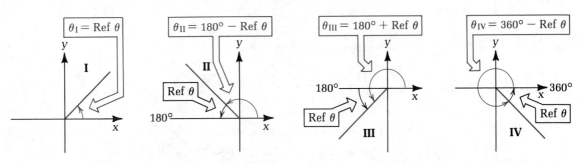

For θ in quadrant I

(a)

For θ in quadrant II

(b)

For θ in quadrant III

(c)

For θ in quadrant IV

(d)

EXAMPLE 8 Find all values of θ, $0° < \theta < 360°$, for which:
(a) $\sin \theta = 0.6$ (b) $\cos \theta = -0.425$ (c) $\cot \theta = -1.4877$
Round answers to parts (a) and (c) to the nearest 0.1° and answers to part (b) to the nearest minute.

Solution

(a) **Step 1.** $|0.6| = 0.6$ $\sin (\text{Ref } \theta) = 0.6$

.6 [INV] [sin] [STO] → 36.869898 Ref $\theta \approx 36.9°$

Step 2. Since $\sin \theta$ is positive, the two values of θ are in quadrants I and II. The first-quadrant angle is the same as Ref θ from Step 1. As shown in Fig. 8-19(b), θ in quadrant II can be calculated as follows:

180 [−] [RCL] [=] → 143.13010

Therefore, $\theta_I \approx 36.9°$ and $\theta_{II} \approx 143.1°$.

(b) **Step 1.** $|-0.425| = 0.425$ $\cos (\text{Ref } \theta) = 0.425$

.425 [INV] [cos] [STO] → 64.849337 Ref $\theta \approx 64.8°$

Step 2. Since $\cos \theta$ is negative, the two values of θ are in quadrants II and III. Referring to Fig. 8-19(b) and (c), we can calculate these values of θ to the nearest minute as follows:

Quadrant II: 180 [−] [RCL] [=] → 115.15066 Record 115°

[−] 115 [=] [×] 60 [=] → 9.0398048 $\theta_{II} \approx 115°9'$

Quadrant III: 180 [+] [RCL] [=] → 244.84934 Record 244°

[−] 244 [=] [×] 60 [=] → 50.960195 $\theta_{III} \approx 244°51'$

(c) **Step 1.** $|-1.4877| = 1.4877$ $\cot (\text{Ref } \theta) = 1.4877$

Step 2. Rewrite this in terms of the tangent function:

$$\tan (\text{Ref } \theta) = \frac{1}{1.4877}$$

Then calculate and store Ref θ:

1.4877 [1/x] [INV] [tan] [STO] → 33.908147

Step 3. Since $\cot \theta$ is negative, θ can be in quadrant II or IV. As shown in Fig. 8-19(b) and (d), we can determine these values of θ as follows:

Quadrant II: 180 [−] [RCL] [=] → 146.09185 $\theta_{II} \approx 146.1°$

Quadrant IV: 360 [−] [RCL] [=] → 326.09185 $\theta_{IV} \approx 326.1°$

If the given function value is a familiar special angle value, θ can be determined without the use of a calculator.

EXAMPLE 9 Find θ, $0° < \theta < 360°$, for which $\sin \theta = -\dfrac{\sqrt{2}}{2}$.

Solution
Step 1.

$$\sin (\text{Ref } \theta) = \left| -\frac{\sqrt{2}}{2} \right| = \frac{\sqrt{2}}{2}$$

From the special angle values, recall that Ref $\theta = 45°$.

Step 2. Since sine is negative, θ is located in quadrants III and IV. Referring to Fig. 8-19(c) and (d), we have

$$\theta_{\text{III}} = 180° + \text{Ref } \theta = 180° + 45° = 225°$$

$$\theta_{\text{IV}} = 360° - \text{Ref } \theta = 360° - 45° = 315°$$

In applied work the conditions of the problem will dictate which value of θ is needed.

EXAMPLE 10
(a) Find $\cos \theta$ when $\sin \theta = 0.5299$ and $\tan \theta < 0$.
(b) Find $\tan \theta$ when $\cos \theta = -0.78$ and $\sin \theta < 0$.

Solution

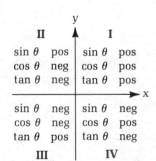

Figure 8-20

(a) Since $\sin \theta$ is positive and $\tan \theta$ is negative, θ must be in quadrant II (see Fig. 8-20). Finding Ref θ on a calculator, we have

.5299 $\boxed{\text{INV}}$ $\boxed{\text{sin}}$ $\boxed{\text{STO}}$ $\rightarrow$ $\boxed{31.998699}$

The corresponding second-quadrant angle is

180 $\boxed{-}$ $\boxed{\text{RCL}}$ $\boxed{=}$ $\rightarrow$ $\boxed{148.00130}$

Finally, we obtain the cosine of this angle:

$\boxed{\text{cos}}$ $\rightarrow$ $\boxed{-0.8480601}$ $\cos \theta \approx -0.8481$, rounded

(b) Since $\cos \theta$ and $\sin \theta$ are both negative, θ must be quadrant III (see Fig. 8-20). Finding Ref θ from $\cos (\text{Ref } \theta) = |-0.78| = 0.78$, we obtain

.78 $\boxed{\text{INV}}$ $\boxed{\text{cos}}$ $\boxed{\text{STO}}$ $\rightarrow$ $\boxed{38.739425}$

The corresponding third-quadrant angle is

180 $\boxed{+}$ $\boxed{\text{RCL}}$ $\boxed{=}$ $\rightarrow$ $\boxed{218.73942}$

The tangent of this angle is

$\boxed{\text{tan}}$ $\rightarrow$ $\boxed{0.8022814}$ $\tan \theta \approx 0.80$, rounded

In some applications you must find an angle from a point on its terminal side. Again, determine the reference angle first and then locate θ from the signs of the coordinates.

EXAMPLE 11
(a)   Find θ if $(-3.7, 2.4)$ is a point on its terminal side.
(b)   Find θ if $(7, -8)$ is a point on its terminal side.

Solution   In each case we are given x and y, so we must use

$$\tan (\text{Ref } \theta) = \left| \frac{y}{x} \right|$$

to find the reference angle.

(a)   Since

$$\tan (\text{Ref } \theta) = \left| \frac{2.4}{-3.7} \right| = \frac{2.4}{3.7}$$

find Ref θ as follows:

2.4 $\boxed{\div}$ 3.7 $\boxed{=}$ $\boxed{\text{INV}}$ $\boxed{\tan}$ $\boxed{\text{STO}}$ → $\boxed{32.969404}$ = Ref θ

Then, since $(-3.7, 2.4)$ is in quadrant II [see Fig. 8-21(a)], find the corresponding value of θ:

180 $\boxed{-}$ $\boxed{\text{RCL}}$ $\boxed{=}$ → $\boxed{147.03060}$ $\theta \approx 147°$

(b)   First note that

$$\tan (\text{Ref } \theta) = \left| \frac{-8}{7} \right| = \frac{8}{7}$$

To find Ref θ enter

8 $\boxed{\div}$ 7 $\boxed{=}$ $\boxed{\text{INV}}$ $\boxed{\tan}$ $\boxed{\text{STO}}$ → $\boxed{48.814075}$ = Ref θ

Then, since $(7, -8)$ is in quadrant IV [see Fig. 8-21(b)], find θ as follows:

360 $\boxed{-}$ $\boxed{\text{RCL}}$ $\boxed{=}$ → $\boxed{311.18593}$ $\theta \approx 311°$

Figure 8-21 shows that our answers are reasonable. Always check your answer with a similar sketch.

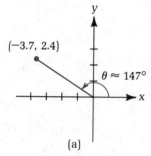

(a)

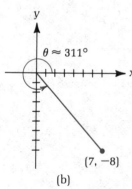

(b)

Figure 8-21

Exercises 8-2

Use a calculator to find each of the following. Round to the nearest 0.001.

1. cos 242°	**2.** cos 166°	**3.** sin 317°	**4.** sin 87.4°	**5.** tan 44°38′
6. tan 346°	**7.** cos 295°	**8.** cos $(-26.57°)$	**9.** sin 228°25′	**10.** sin 106°
11. tan 219.8°	**12.** tan 108°	**13.** sec 53.75°	**14.** sec 213°	**15.** csc 171.28°
16. csc $(-68°)$	**17.** cot 141°	**18.** cot 26.75°	**19.** cos 137°12′	**20.** cos 74°

21. sin 31°   **22.** sin 260°   **23.** tan (−51.5°)   **24.** tan 185°   **25.** sin 440°

26. cos 568°   **27.** tan 396.4°   **28.** cot 627°28′

Express each of the following in terms of the same function of the corresponding reference angle:

29. sin 260°   **30.** sin 157°   **31.** tan 298°37′   **32.** tan 214°   **33.** cos 247°

34. cos 313°16′   **35.** sec (−41°)   **36.** csc 188.5°   **37.** cot 155.6°   **38.** cot 420°

Find the exact value of each of the following without using a calculator:

39. cos 120°   **40.** sin 330°   **41.** tan 225°   **42.** cos 225°   **43.** sin 240°

44. tan 150°

Find all values of θ, $0° < \theta < 360°$, rounded to the nearest $0.1°$.

45. $\sin \theta = 0.4$ **46.** $\sin \theta = -0.84$ **47.** $\cos \theta = 0.564$ **48.** $\cos \theta = -0.2$

49. $\tan \theta = -1.4$ **50.** $\tan \theta = 0.752$ **51.** $\sec \theta = -2.53$ **52.** $\csc \theta = 1.3$

53. $\cot \theta = 3.1$ **54.** $\sin \theta = 0.736$ **55.** $\cos \theta = -0.7$ **56.** $\tan \theta = -2.14$

Find all values of θ, $0° < \theta < 360°$, to the nearest minute.

57. $\sin \theta = -0.804$ **58.** $\sin \theta = 0.539$ **59.** $\cos \theta = 0.643$ **60.** $\cos \theta = -0.268$

61. $\tan \theta = 1.842$ **62.** $\tan \theta = -4.393$ **63.** $\cot \theta = -3.747$ **64.** $\sec \theta = 4.165$

Determine θ, $0° < \theta < 360°$, without using a calculator.

65. $\sin \theta = \dfrac{1}{2}$ **66.** $\sin \theta = -\dfrac{\sqrt{3}}{2}$ **67.** $\cos \theta = -\dfrac{\sqrt{3}}{2}$

68. $\cos \theta = \dfrac{\sqrt{2}}{2}$ **69.** $\tan \theta = -\dfrac{\sqrt{3}}{3}$ **70.** $\tan \theta = 1$

Determine the function value, given the following conditions:

71. Find $\sin \theta$ when $\cos \theta = 0.4365$ and $\tan \theta < 0$.

72. Find $\cos \theta$ when $\tan \theta = -1.46$ and $\sin \theta > 0$.

73. Find $\tan \theta$ when $\sec \theta = -2.459$ and $\csc \theta < 0$.

74. Find $\cot \theta$ when $\sin \theta = -0.7243$ and $\cos \theta > 0$.

Find to the nearest degree the angle that has the given point on its terminal side.

75. $(2, -5)$ **76.** $(-3.6, -2.9)$ **77.** $(-5.4, 3.1)$ **78.** $(3, 7)$

79. *Electronics* The instantaneous value i of a sinusoidal current in an ac circuit is given by $i = i_m \sin \theta$, where i_m is the maximum current. Find i when $i_m = 22.5$ mA and $\theta = 125.0°$.

80. *Electronics* The instantaneous value V of a sinusoidal voltage in an ac circuit is given by $V = V_m \sin \theta$, where V_m is the maximum voltage. Find V when $V_m = 74$ V and $\theta = 162°$.

81. *Navigation* Since the British nautical mile is defined as the length of a minute of arc on a meridian, its length D (in feet) varies with latitude according to the formula

$$D = 6077 - 31 \cos 2\theta$$

where θ is the latitude (in degrees). What is the length of a British nautical mile at a latitude of 74°0′ N?

82. *Optics* According to the law of Malus, the intensity I of the light transmitted through a pair of polaroid filters is given by the equation $I = I_m(\cos \theta)^2$, where I_m is the maximum intensity and θ is the angle between the polarizing directions of the filters. Find the ratio I/I_m for $\theta = 155°0′$.*

* For more information see David Halliday and Robert Resnick, *Fundamentals of Physics* (New York: Wiley, 1981), p. 682.

8-3 | Radian Measure

In daily use, in geometry, and especially in most technical applications, angles are measured in units of decimal degrees. The degree, defined as $\frac{1}{360}$th of a circle, is an arbitrary unit that dates from ancient times when Babylonian engineers used a base 60 number system. In modern scientific work, a much more useful and more natural unit is the *radian*, abbreviated "rad."

Definition

To define the radian unit of measure, consider a circle of radius r, as shown in Fig. 8-22.

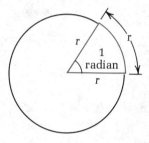

A **radian** is the measure of an angle whose vertex is at the center of a circle of radius r and which intercepts an arc equal to length r.

To determine the number of radians in a full circle, first recall that the formula for the circumference C of a circle of radius r is $C = 2\pi r$. If we divide both sides of this equation by r, we get

$$\frac{C}{r} = 2\pi$$

Figure 8-22

This tells us that there are 2π radians in a full circle. We now can obtain the following values for quadrantal angles:

Quadrantal Angles

$90° = \dfrac{\pi}{2}$ rad ≈ 1.57 rad $180° = \pi$ rad ≈ 3.14 rad

$270° = \dfrac{3\pi}{2}$ rad ≈ 4.71 rad $360° = 2\pi$ rad ≈ 6.28 rad

(8-2)

In working with radian measure, it is important to be able to determine by inspection the quadrant in which an angle is located. Try to memorize the information in Fig. 8-23 before attempting Example 12.

EXAMPLE 12   In which quadrant is each of the following angles located?

(a) $\dfrac{5\pi}{4}$ rad   (b) $\dfrac{2\pi}{3}$ rad   (c) 0.217 rad

(d) 5.86 rad   (e) 2.0 rad   (f) -0.4636476 rad

Solution

(a) III   (b) II   (c) I   (d) IV   (e) II   (f) IV

In part (f) recall that a negative angle is measured clockwise from the positive x-axis.

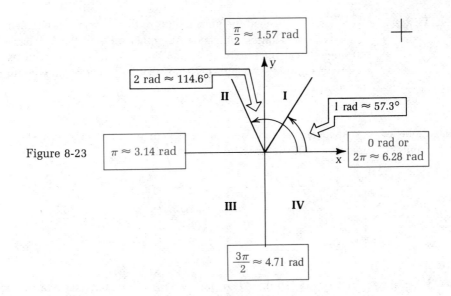

Figure 8-23

Conversion To work with radian measure you must also be able to convert back and forth between radian measure and degree measure. Using the fact that $360° = 2\pi$, we can divide both sides of this equation by 360 to obtain the following conversion formula:

$$\textbf{Degree–Radian Conversion} \qquad 1° = \frac{\pi}{180}\ \text{rad} \approx 0.017\ \text{rad} \qquad (8\text{-}3)$$

Degrees to radians   This means that to convert from degrees to radians, we must multiply degrees by: $\dfrac{\pi}{180}\dfrac{\text{rad}}{\text{degree}}$

EXAMPLE 13 Convert each of the following angles measured in degrees to radian measure:

(a) 20°   (b) 145°   ▦(c) 68.7°   ▦(d) 234°17′

Express answers to parts (a) and (b) in terms of π and answers to parts (c) and (d) in decimal form to the nearest 0.01 rad.

Solution

(a) $\left(\dfrac{\pi}{180}\dfrac{\text{rad}}{\text{degree}}\right)(20°) = \dfrac{\pi}{9}\ \text{rad}$ (b) $\left(\dfrac{\pi}{180}\dfrac{\text{rad}}{\text{degree}}\right)(145°) = \dfrac{29\pi}{36}\ \text{rad}$

(c) 68.7 ⊠ π ÷ 180 ⊟ → *1.1990412*   or approximately 1.20 rad

(d) 234 ⊞ 17 ÷ 60 ⊟ ⊠ π ÷ 180 ⊟ → *4.0890155* ≈ 4.09 rad

NOTE

Although we attached the unit *radians* to these angle measurements, they are actually dimensionless numbers. Any other units in the problem will be unaffected by calculations with radian measure. In many problems, the word *radian* is omitted, so you may assume that any angle given without units is in radian measure. ◄

Radian equivalents of 30°, 45° and 60°

Angles of 30°, 45°, and 60° are so common in technical work, their radian measure equivalents should be memorized:

Degrees	30°	45°	60°
Radians	$\dfrac{\pi}{6}$	$\dfrac{\pi}{4}$	$\dfrac{\pi}{3}$

Radians to degrees

Using the fact that $2\pi = 360°$, we can divide both sides of this equation by 2π to obtain the following conversion formula:

$$\textbf{Radian–Degree Conversion} \qquad 1 \text{ rad} = \frac{180°}{\pi} \approx 57.3° \qquad \text{(8-4)}$$

Therefore, to convert radians to degrees, we multiply radians by:

$$\frac{180}{\pi} \frac{\text{degree}}{\text{rad}}$$

EXAMPLE 14   Convert each radian measure to degrees. Round answers to parts (c) and (d) to the nearest 0.1° and answer to part (e) to the nearest minute.

(a) $\dfrac{3\pi}{4}$ rad   (b) $\dfrac{\pi}{8}$ rad ▦ (c) 2.00 rad ▦ (d) 0.860 rad

▦ (e) 3.057 rad

Solution

(a) $\left(\dfrac{\overset{45}{\cancel{180°}}}{\cancel{\pi}}\right)\left(\dfrac{3\cancel{\pi}}{\underset{1}{\cancel{4}}}\right) = 135°$ (b) $\left(\dfrac{\overset{45}{\cancel{180°}}}{\cancel{\pi}}\right)\left(\dfrac{\cancel{\pi}}{\underset{2}{\cancel{8}}}\right) = 22.5°$

(c) $2 \boxed{\times} 180 \boxed{\div} \boxed{\pi} \boxed{=} \rightarrow$ `114.59156`   or 114.6°, rounded

(d) $.86 \boxed{\times} 180 \boxed{\div} \boxed{\pi} \boxed{=} \rightarrow$ `49.274370`   or 49.3°, rounded

(e) $3.057 \boxed{\times} 180 \boxed{\div} \boxed{\pi} \boxed{=} \rightarrow$ `175.15320`   Record 175°

$\boxed{-} 175 \boxed{=} \boxed{\times} 60 \boxed{=} \rightarrow$ `9.1918782`   3.057 rad ≈ 175°9′

Finding function values

To find the trigonometric functions for an angle given in radians on your calculator, first put your calculator in radian mode. (We will often indicate this with the ⟨RAD⟩ symbol, but see your calculator instruction manual or Appendix A for the proper key sequence.) Then proceed as you do with degrees. Enter the angle followed by the key of the desired function.

EXAMPLE 15 Find each of the following rounded to three decimal places:

(a) $\sin 0.8$ rad   (b) $\cos 2.493$ rad   (c) $\tan \dfrac{\pi}{6}$ rad

(d) $\sec \dfrac{5\pi}{12}$ rad   (e) $\sin \dfrac{4\pi}{3}$ rad

Solution **First,** put your calculator in radian mode. Then:

(a) .8 [sin] →   *0.7173561* $\sin 0.8 \approx 0.717$

(b) 2.493 [cos] →   *−0.7969347* $\cos 2.493 \approx -0.797$

(c) [π][÷]6[=][tan] →   *0.5773503* $\tan \dfrac{\pi}{6} \approx 0.577$

You can find the *exact* value of $\tan\left(\frac{\pi}{6}\right)$ without a calculator by recalling that $\frac{\pi}{6} = 30°$, and

$$\tan \frac{\pi}{6} = \tan 30° = \frac{\sqrt{3}}{3}$$

(d) $\sec \dfrac{5\pi}{12} = \dfrac{1}{\cos (5\pi/12)}$ 5[×][π][÷]12[=][cos][⅟ₓ] →   *3.8637033*

$\sec \dfrac{5\pi}{12} \approx 3.864$

(e) 4[×][π][÷]3[=][sin] →   *−0.8660254* $\sin \dfrac{4\pi}{3} \approx -0.866$

You can find the exact value of this function without a calculator by noting that the angle $4\pi/3$ is in quadrant III and its reference angle is

$$\frac{4\pi}{3} - \pi = \frac{\pi}{3}$$

Then, recall that

$$\sin \frac{\pi}{3} = \sin 60° = \frac{\sqrt{3}}{2}$$

Since sine is negative in quadrant III,

$$\sin \frac{4\pi}{3} = -\frac{\sqrt{3}}{2}$$

CAUTION ▶ On the display of some calculators, the indication of radian mode is directly over the negative sign. Be certain to read the display carefully so the negative sign is not overlooked. ◀

Finding the angles from the function value   As we discussed in Section 8-2, for $0 < \theta < 2\pi$, excluding quadrantal angles, there are two possible values of θ for each value of a trigonometric function. To determine the value of an angle in radians when the value of the function is known, first make certain the calculator is in radian mode. Then find the reference angle using the absolute value of the given function value and

Figure 8-24

store Ref θ in memory. Finally, use the sign of the function to determine both values of θ. Figure 8-24 shows how to use Ref θ to determine θ once you locate its quadrant.

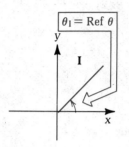

For θ in quadrant I

(a)

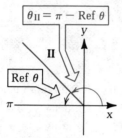

For θ in quadrant II

(b)

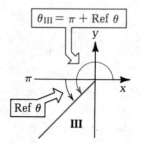

For θ in quadrant III

(c)

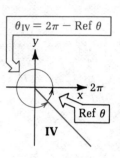

For θ in quadrant IV

(d)

EXAMPLE 16 Find all values of θ, $0 < \theta < 2\pi$, for which:
(a) $\sin \theta = 0.8$ (b) $\tan \theta = -1.78$ (c) $\sec \theta = -2.1361$
Round to the nearest 0.01 rad.

Solution **First,** make certain your calculator is in radian mode.
Then, proceed:

(a) **Step 1.** $|0.8| = 0.8$ so $\sin (\text{Ref } \theta) = 0.8$

.8 [INV] [sin] [STO] → ▨ 0.9272952 ▨ Ref $\theta \approx 0.93$ rad

Step 2. Since $\sin \theta$ is positive, θ lies in quadrant I or quadrant II. The first-quadrant value of θ is the same as Ref θ—that is, approximately 0.93 rad. To find θ in quadrant II, make a sketch similar to Fig. 8-24(b) and continue as follows:

[π] [−] [RCL] [=] → ▨ 2.2142974 ▨

Thus, $\theta_I \approx 0.93$ rad and $\theta_{II} \approx 2.21$ rad.

(b) **Step 1.** $|-1.78| = 1.78$ so $\tan (\text{Ref } \theta) = 1.78$

1.78 [INV] [tan] [STO] → ▨ 1.0589405 ▨ Ref $\theta \approx 1.06$

Step 2. Since $\tan \theta$ is negative, θ lies in quadrant II or quadrant IV. Referring to Fig. 8-24(b) and (d), we obtain

Quadrant II: [π] [−] [RCL] [=] → ▨ 2.0826522 ▨

Quadrant IV: 2 [×] [π] [−] [RCL] [=] → ▨ 5.2242448 ▨

Therefore, $\theta_{II} \approx 2.08$ rad and $\theta_{IV} \approx 5.22$ rad.

(c) **First,** rewrite sec θ in terms of cos θ: If $\sec \theta = -2.1361$, then

$$\cos \theta = \frac{1}{-2.1361}$$

Then, proceed as before:

$$\left|\frac{1}{-2.1361}\right| = \frac{1}{2.1361} \qquad \text{so} \qquad \cos(\text{Ref } \theta) = \frac{1}{2.1361}$$

2.1361 $\boxed{1/x}$ $\boxed{\text{INV}}$ $\boxed{\text{cos}}$ $\boxed{\text{STO}}$ $\rightarrow$ $\boxed{\textit{1.0836084}}$ Ref $\theta \approx 1.08$ rad

The secant function, like its reciprocal the cosine function, is negative in quadrants II and III. Referring to Fig. 8-24(b) and (c), we have

Quadrant II: $\boxed{\pi}$ $\boxed{-}$ $\boxed{\text{RCL}}$ $\boxed{=}$ $\rightarrow$ $\boxed{\textit{2.0579843}}$

Quadrant III: $\boxed{\pi}$ $\boxed{+}$ $\boxed{\text{RCL}}$ $\boxed{=}$ $\rightarrow$ $\boxed{\textit{4.2252010}}$

Therefore, $\theta_{II} \approx 2.06$ rad and $\theta_{III} \approx 4.23$ rad.

If the given function value is one of the familiar special angle values, θ can be determined without the use of a calculator.

EXAMPLE 17 Find all values of θ, $0 < \theta < 2\pi$, such that $\cos\theta = -\frac{1}{2}$. Express θ as a rational multiple of π.

Solution

Step 1. $\left|-\frac{1}{2}\right| = \frac{1}{2}$ so $\cos(\text{Ref } \theta) = \frac{1}{2}$

Recall that

$$\cos 60° = \cos\frac{\pi}{3} = \frac{1}{2}$$

Therefore, Ref $\theta = \pi/3$.

Step 2. Since cosine is negative in quadrants II and III,

$$\theta_{II} = \pi - \frac{\pi}{3} = \frac{2\pi}{3} \qquad \text{and} \qquad \theta_{III} = \pi + \frac{\pi}{3} = \frac{4\pi}{3}$$

CAUTION 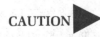 As you can see, determining the values of an angle from the value of a function requires a thorough knowledge of the signs of the trigonometric functions. Be certain to review Fig. 8-6 (in Section 8-1) before attempting the exercises. ◀

Exercises 8-3

In which quadrant is each angle located? (All angles are given in radians.)

1. $\dfrac{7\pi}{4}$ 2. $\dfrac{5\pi}{12}$ 3. $\dfrac{4\pi}{3}$ 4. $\dfrac{5\pi}{6}$ 5. $-\dfrac{5\pi}{3}$ 6. $\dfrac{\pi}{12}$

7. 2.7 8. −1.437 9. 1.83 10. 3.2 11. 0.84 12. 6.037

Convert each angle measured in degrees to radians, expressed as a rational multiple of π.

13. 60°   **14.** 150°   **15.** 275°   **16.** 240°   **17.** 72°   **18.** 350°
19. 200°   **20.** 25°   **21.** 50°   **22.** 216°   **23.** 315°   **24.** 120°

Convert each angle measured in degrees to decimal radians. Round to the nearest 0.01 rad.

25. 56°0′   **26.** 38.6°   **27.** 115.7°   **28.** 166°0′   **29.** 224°28′   **30.** 245°50′
31. 292°0′   **32.** 328°0′   **33.** −42.6°   **34.** 139.58°   **35.** 75°10′   **36.** 207.8°

Convert each angle measured in radians to decimal degrees. Use a calculator to find values for Problems 49–56. Round to the nearest 0.1° if necessary.

37. $\dfrac{5\pi}{4}$ **38.** $\dfrac{2\pi}{3}$ **39.** $\dfrac{7\pi}{6}$ **40.** $\dfrac{\pi}{6}$ **41.** $\dfrac{7\pi}{12}$ **42.** $\dfrac{5\pi}{3}$

43. $\dfrac{5\pi}{8}$ **44.** $\dfrac{3\pi}{8}$ **45.** $\dfrac{11\pi}{8}$ **46.** $\dfrac{31\pi}{36}$ **47.** $\dfrac{17\pi}{12}$ **48.** $\dfrac{7\pi}{4}$

49. 1.40   **50.** 3.80   **51.** 4.62   **52.** 0.970   **53.** 5.20   **54.** 2.40
55. 3.06   **56.** −1.06

Convert each angle measured in radians to degrees. Round to the nearest minute.

57. 4.027   **58.** 1.249   **59.** 6.038   **60.** 2.740

Find the following function values. Find the exact values for Problems 61–66, and use a calculator to find values for Problems 67–84 to the nearest 0.001.

61. $\sin\dfrac{5\pi}{3}$ **62.** $\sin\dfrac{3\pi}{4}$ **63.** $\cos\dfrac{7\pi}{6}$ **64.** $\cos\dfrac{5\pi}{3}$ **65.** $\tan\dfrac{2\pi}{3}$ **66.** $\tan\dfrac{4\pi}{3}$

67. $\cot\dfrac{11\pi}{12}$ **68.** $\sec\dfrac{5\pi}{4}$ **69.** $\csc\dfrac{11\pi}{6}$ **70.** $\tan\dfrac{\pi}{8}$ **71.** $\sin\dfrac{7\pi}{4}$ **72.** $\cos\dfrac{3\pi}{8}$

73. $\cos 0.5$ **74.** $\cos 3.25$ **75.** $\tan 2.49$ **76.** $\tan 4.68$ **77.** $\sin 5.62$ **78.** $\sin 1.8$
79. $\csc 1.2$ **80.** $\cot 0.79$ **81.** $\sec 1.73$ **82.** $\sin 3.754$ **83.** $\cos 4.986$ **84.** $\tan(-1.215)$

Find all values of θ, $0 < \theta < 2\pi$, that satisfy the following. In Problems 85–90 express θ as a rational multiple of π, and in Problems 91–102 use a calculator to find θ to the nearest 0.01 rad.

85. $\cos\theta = \dfrac{\sqrt{3}}{2}$ **86.** $\sin\theta = \dfrac{\sqrt{2}}{2}$ **87.** $\tan\theta = -1$ **88.** $\cos\theta = -\dfrac{\sqrt{2}}{2}$

89. $\sin\theta = -\dfrac{1}{2}$ **90.** $\tan\theta = \dfrac{\sqrt{3}}{3}$ **91.** $\cos\theta = 0.4$ **92.** $\cos\theta = -0.63$

93. $\sin\theta = -0.516$ **94.** $\sin\theta = 0.84$ **95.** $\tan\theta = 3.72$ **96.** $\tan\theta = -0.965$
97. $\cot\theta = -2.4$ **98.** $\tan\theta = 4.1638$ **99.** $\csc\theta = 3.6$ **100.** $\sin\theta = -0.7135$
101. $\cos\theta = -0.2146$ **102.** $\sec\theta = 1.75$

Solve.

103. *Electronics*   Under certain conditions, the instantaneous sinusoidal current in an ac circuit is given by the formula $i = i_m \sin 2\pi ft$, where i_m is the maximum current, f is the frequency of the wave, and t is the elapsed time. Find i when $i_m = 8.0$ A, $f = 60$ Hz, and $t = 0.010$ s.

104. *Electronics*   The instantaneous sinusoidal voltage in a certain ac circuit is given by the formula $V = V_m \sin 2\pi ft$, where V_m is the maximum voltage, f is the frequency, and t is elapsed time. Find V when $V_m = 150$ V, $f = 200$ Hz, and $t = 0.018$ s.

105. *Physics*   The vibration of a certain mechanical system is given by $y = 0.25 \cos 1.3t$, where t is the time (in seconds) and y is the displacement (in meters). Find y at $t = 1.5$ s.

106. *Physics*   The maximum speed of a pendulum bob is given by $v = \sqrt{2gL(1 - \cos\theta)}$, where g is the acceleration due to gravity, L is the length of the pendulum, and θ is the angle from which it is released. Find v when $g = 9.8$ m/s², $L = 0.15$ m, and $\theta = 0.42$ rad.
107. *Physics*   In Problem 105 find the smallest value of t for which $y = 0.14$ m.
108. *Physics*   In Problem 104 find the smallest value of t for which $f = 200$ Hz, $V = 105$ V, and $V_m = 140$ V.

8-4 | Applications of Radian Measure

There are a great many technical applications of trigonometry in which the angles are expressed in radian measure. In this section we will look at three important formulas that use radian measure. These involve the length of a circular arc, the area of a sector of a circle, and the relationship between linear and angular velocity.

Length of a circular arc

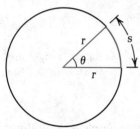

Figure 8-25

The circumference C of a full circle of radius r is given by $C = 2\pi r$. From plane geometry we know that the length of an arc on any circle is proportional to the central angle (Fig. 8-25):

$$\frac{\text{Arc length, } s}{\text{Circumference, } C} = \frac{\text{Central angle of arc, } \theta}{\text{Central angle of circle}}$$

$$\frac{s}{2\pi r} = \frac{\theta}{2\pi}$$

Solving this proportion for s, we obtain the following formula:

Length of Circular Arc	$s = r\theta$ where θ is in radians	**(8-5)**

That is, arc length = radius times central angle (in radians).

EXAMPLE 18   Find the length of an arc on a circle with radius 8.00 in. and central angle $\frac{\pi}{3}$.

Solution $s = r\theta$

$$= (8.00)\frac{\pi}{3} \qquad 8\boxed{\times}\boxed{\pi}\boxed{\div}3\boxed{=}\rightarrow \qquad \boxed{8.3775804}$$

$$s \approx 8.38 \text{ in., rounded}$$

REMEMBER ▶   Radian units are actually a dimensionless ratio. Therefore, in. × rad = in. If the angle is given in degrees, you must convert it to radians before substituting into the arc length formula (8-5). ◀

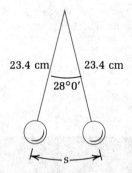

23.4 cm   23.4 cm

28°0′

s

Figure 8-26

EXAMPLE 19   A pendulum 23.4 cm long swings through an angle of 28°0′. What is the length of the arc covered by the end of the pendulum?

Solution   **First,** convert 28°0′ to radians:

$$28 \; \boxed{\times} \; \boxed{\pi} \; \boxed{\div} \; 180 \; \boxed{=} \; \rightarrow \quad \boxed{0.4886922}$$

Then, apply the formula for arc length (8-5):

$$\boxed{\times} \; 23.4 \; \boxed{=} \; \rightarrow \quad \boxed{11.435397} \qquad s \approx 11.4 \text{ cm}$$

The next example demonstrates how we may use the arc length formula (8-5) to find the central angle when the arc length and radius are known.

EXAMPLE 20   Find the central angle of an arc of length 52.0 cm on a circle with a radius of 18.5 cm.

Solution

First, solve (8-5) for θ. $\theta = \dfrac{s}{r}$

Then, substitute for s and r $\theta = \dfrac{52.0 \text{ cm}}{18.5 \text{ cm}} \approx 2.81 \text{ rad}$
and calculate θ.

Formula (8-5) can also be used to find the radius when the arc length and central angle are known. As shown in the next example, the central angle often is given in revolutions in applied problems.

EXAMPLE 21   A wheel travels 18 ft while making five complete revolutions. What is the radius of the wheel?

Solution   **First,** convert θ to radians. Since there are 2π radians in each revolution,

$$5 \text{ rev} = 5 \, \cancel{\text{rev}} \cdot \frac{2\pi \text{ rad}}{1 \, \cancel{\text{rev}}} = 10\pi \text{ rad}$$

Second, solve the formula $s = r\theta$ for r. $r = \dfrac{s}{\theta}$

Finally, substitute for s and θ and $r = \dfrac{18 \text{ ft}}{10\pi} \approx 0.57 \text{ ft}$
calculate r.

Area of a sector   Radian measure also allows us to find the area of a sector of a circle. Plane geometry tells us that the area of a sector is proportional to its central angle (Fig. 8-27):

$$\frac{\text{Area of sector, } A}{\text{Area of circle}} = \frac{\text{Central angle of sector, } \theta}{\text{Central angle of circle}}$$

$$\frac{A}{\pi r^2} = \frac{\theta}{2\pi}$$

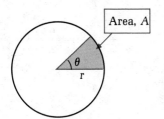

Area, A

θ

r

Figure 8-27

Solving for A, we have the following formula:

Area of Sector	$A = \dfrac{1}{2}r^2\theta$	(8-6)

EXAMPLE 22 Find the area of the sector of a circle with:
(a) A central angle of 1.60 rad and a radius of 6.00 ft
(b) A central angle of 56°0′ and a diameter of 44.8 cm

Solution

(a) $A = \dfrac{1}{2}r^2\theta = \dfrac{1}{2}(6.00 \text{ ft})^2(1.60) = 28.8 \text{ ft}^2$

(b) **First,** change 56°0′ to radians:

$$56 \boxed{\times}\boxed{\pi}\boxed{\div} 180 \boxed{=} \rightarrow \quad \boxed{0.9773844}$$

Second, use the diameter of 44.8 cm to find the radius:

$$r = \frac{44.8}{2} = 22.4 \text{ cm}$$

Third, use the area formula (8-6) to calculate A:

$$\boxed{\times}.5 \boxed{\times} 22.4 \boxed{x^2}\boxed{=} \rightarrow \quad \boxed{245.20619} \quad \text{or } 245 \text{ cm}^2, \text{ rounded}$$

We can also use the area of a sector formula (8-6) to find r or θ given the other two quantities.

EXAMPLE 23 Given a sector with an area of 400 m² and a radius of 20 m, find the central angle in radians.

Solution

First, solve the formula $A = \frac{1}{2}r^2\theta$ for θ. $\theta = \dfrac{2A}{r^2}$

Then, substitute $A = 400$ m² and $\theta = \dfrac{2(400 \text{ m}^2)}{(20 \text{ m})^2} = 2 \text{ rad}$
$r = 20$ m and calculate θ.

Linear and angular velocity   Earlier, we showed that an object moving in a circular arc travels a distance s when it goes through an angle θ, where $s = r\theta$. Since average velocity equals the distance traveled divided by the time of travel, we obtain

$$v = \frac{s}{t} \quad \text{or} \quad v = \frac{r\theta}{t}$$

The quantity s/t is defined as the **average linear velocity** v, and the quantity θ/t is defined as the **average angular velocity** ω (Greek letter "omega"). We can rewrite the formula as

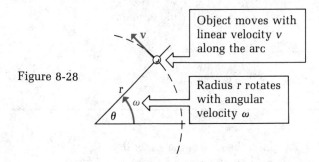

Figure 8-28

Object moves with linear velocity v along the arc

Radius r rotates with angular velocity ω

Angular Motion	$v = r\omega$	(8-7)

---

to describe the relationship between linear and angular velocity (see Fig. 8-28).

EXAMPLE 24 A flywheel rotates at an angular velocity of 2.0 rad per second (rad/s). If the radius of the flywheel is 16 in., find the linear velocity of a point on the rim in feet per second.

Solution

First, substitute in (8-7) to calculate v.

$$v = r\omega = (16 \text{ in.})\left(2.0 \frac{\text{rad}}{\text{s}}\right)$$

$$= 32 \frac{\text{in.}}{\text{s}}$$

Then, convert to feet per second.*

$$= \left(32 \frac{\text{in.}}{\text{s}}\right)\left(\frac{1 \text{ ft}}{12 \text{ in.}}\right) \approx 2.7 \text{ ft/s}$$

Notice that the radian units drop out of the calculation.

In practical applications, the rotation rate may be given in revolutions per unit of time. Multiply such a quantity by 2π rad/rev to convert it to angular velocity.

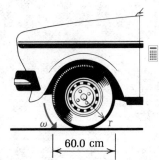

EXAMPLE 25 A car wheel 60.0 cm in diameter is rotating at an angular velocity of 168 rev/min (Fig. 8-29). Find the speed of the car in meters per second.

Solution **First,** find the radius in meters:

$$r = \left(\frac{60.0 \text{ cm}}{2}\right)\left(\frac{1 \text{ m}}{100 \text{ cm}}\right) = 0.300 \text{ m}$$

60.0 cm

Figure 8-29

* See Appendix B for a discussion of unit conversion.

Second, convert revolutions per minute to radians per second:

$$\omega = \left(168\frac{\text{rev}}{\text{min}}\right)\left(\frac{2\pi\text{ rad}}{1\text{ rev}}\right)\left(\frac{1\text{ min}}{60\text{ s}}\right)$$

168 ⊠ 2 ⊠ π ⊡ 60 = →   17.592919

Third, use (8-7) to calculate v. Multiplying r = 0.300 m by the value of ω in the calculator, we have

⊠ .3 = →   5.2778757   or   v ≈ 5.28 m/s, rounded

Formula (8-7) can be used to solve for either ω or r given the other two quantities.

EXAMPLE 26 A race car is driven around a circular track at an average speed of 80.0 mi/h. If the radius of the track is 800.0 ft, find the angular velocity of the car in revolutions per minute.

Solution

First, solve the formula
v = rω for ω.

$$\omega = \frac{v}{r}$$

Second, convert the linear velocity to feet per minute to agree with the other units.

$$\left(80.0\frac{\text{mi}}{\text{h}}\right)\left(\frac{5280\text{ ft}}{1\text{ mi}}\right)\left(\frac{1\text{ h}}{60\text{ min}}\right) = 7040\text{ ft/min}$$

Third, use ω = v/r to calculate ω.

$$\omega = \frac{7040\text{ ft/min}}{800.0\text{ ft}} = 8.80\text{ rad/min}$$

Fourth, convert ω to revolutions per minute.

$$\omega = \left(8.80\frac{\text{rad}}{\text{min}}\right)\left(\frac{1\text{ rev}}{2\pi\text{ rad}}\right) \approx 1.40\text{ rev/min}$$

Using a calculator, all four steps are entered as follows:

80 ⊠ 5280 ⊡ 60 ⊡ 800 ⊡ 2 ⊡ π = →   1.4005635

Exercises 8-4

For an arc of length s and central angle θ on a circle of radius r or diameter d, find the unknown quantity from the given information. Give θ in decimal radians rounded to 0.01.

1. r = 6.50 in., $\theta = \frac{\pi}{6}$,   s = ?
     2. r = 22.6 cm, θ = 3.04,   s = ?

3. d = 8.6 m, θ = 120°0′,   s = ?
     4. r = 3.05 ft, θ = 38°0′,   s = ?

5. s = 17.5 in.,   r = 4.75 in., θ = ?
     6. s = 43.6 ft,   d = 61.2 ft, θ = ?

7. s = 225 mm, θ = 2.45,   r = ?
     8. s = 162 mm, θ = 54°0′,   r = ?

For a sector with area A and central angle θ on a circle of radius r or diameter d, find the unknown quantity from the given information. Give θ in decimal radians, rounded to 0.01.

9. r = 12.6 in., $\theta = \dfrac{3\pi}{4}$,   A = ?   10. r = 7.00 cm,   θ = 0.820,   A = ?

11. r = 38.2 m,   θ = 18°0′,   A = ?   12. d = 438 ft,   θ = 135°0′,   A = ?

13. A = 4650 ft², r = 65.0 ft,   θ = ?   14. A = 428 in², d = 44.5 in., θ = ?

15. A = 2750 cm²,   θ = 3.02,   r = ?   16. A = 138.2 m²,   θ = 75°0′,   r = ?

Solve.

17. *Acoustical engineering*   The central angle formed by a section of tape passing over a circular magnetic tape head is 115°0′. If the radius of the head is 1.50 in., how long is the section of tape? (See the figure.)

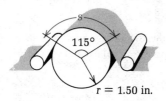

18. *Physics*   A pendulum bob 9.5 in. long swings through a central angle of 14°. What is the length of the arc through which the end of the bob moves?

19. *Construction*   The fencing for a certain Little League baseball field is constructed to create a uniform radius of 205 ft from home plate. If the boundaries of the fencing create a 100°0′ angle with home plate, how many feet of fencing are needed? (See the figure.)

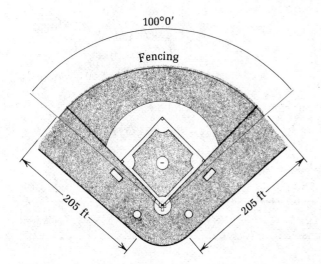

20. *Planetary science*   The moon's orbit around the earth moves it through a central angle of about 13.2° per day relative to the center of the earth. If the distance from the earth's center to the moon is approximately 3.84 × 10⁵ km, how far (in kilometers) does the moon actually move in a day?

21. *Transportation engineering*   A subway train travels 0.10 mi on a circular section of track. If the central angle of the curve is 172°, what is the radius of the curve in feet?

22. *Sheet metal technology*   In order to make a conical duct, a manufacturer of air-conditioning units orders a piece of sheet metal in the shape of a sector of a circle. If the radius of the sector (the slant height of the duct) is 28.5 in. and the arc length (the circumference of the duct opening) is 26.4 in., what is the central angle θ of the metal piece in radians? (See the figure.)

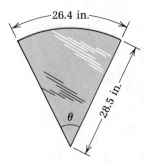

23. *Sheet metal technology*   What is the area of the piece of sheet metal in Problem 22?

24. *Construction*   Top soil must be brought in to prepare the baseball field described in Problem 19. What is the area of the sector bounded by home plate and the fence?

25. A plot of land in the shape of a sector of a circle has an area of 2.25 acres. If the radius of the plot is 314 ft, what is the measure of the central angle in degrees? (1 acre $\approx$ 43,560 ft^2)

26. *Construction*   A sector-shaped window has a central angle of 120° and an area of 5800 cm^2. Find the radius of the window.

27. *Mechanical engineering*   The blades of a food processor rotate at an angular velocity of 1400 rev/min. If the blades are 3.5 in. long, what is the linear velocity of the end of a blade in inches per second?

28. *Energy technology*   A waterwheel rotates at an angular velocity of 11.4 rev/min, and the diameter of the wheel is 12.5 ft. Calculate the linear velocity of the buckets on the edge of the wheel in feet per second. (See the figure.)

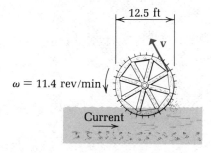

29. *Physics*   A flywheel rotates at an angular velocity of 3.50 rad/s. If the radius of the flywheel is 18.5 in., find the linear velocity of a point on the rim in feet per second.

30. *Aeronautical engineering*   The propeller of an airplane is 2.5 m in diameter and it rotates at 2200 rev/min. Find the linear velocity of the tip of the propeller in meters per second.

31. *Physics*   A car is traveling at a speed of 45 mi/h. If the wheels of the car are 24 in. in diameter, find their angular velocity in radians per second.

32. *Industrial engineering*   Two pulleys, one 32.0 cm in diameter and the other 48.0 cm in diameter, are connected by a belt 358 cm long. If the belt makes one revolution every 5.25 s, what is the angular velocity of each pulley?

8-5 | Review of Chapter 8

Important Terms and Concepts

reference angle (p. 279)   central angle (p. 293 and Appendix C)   average linear velocity v (p. 295)

radian (p. 286)   sector (p. 294 and Appendix C)   average angular velocity ω (p. 295)

arc (p. 293 and Appendix C)

Formulas and Definitions

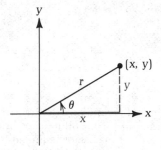

- definitions of the trigonometric functions (p. 274):

$$\sin \theta = \frac{y}{r} \qquad \cos \theta = \frac{x}{r} \qquad \tan \theta = \frac{y}{x}$$

$$\csc \theta = \frac{r}{y} \qquad \sec \theta = \frac{r}{x} \qquad \cot \theta = \frac{x}{y}$$

$(8\text{-}1)$

- signs of the trigonometric functions (p. 274)   The figure shows by quadrants the signs of the three basic trigonometric functions; the signs of the reciprocal functions, csc θ, sec θ, and cot θ, are the same as the signs of their corresponding basic functions.

```
         y
  II    ▲    I
sin θ pos │ sin θ pos
cos θ neg │ cos θ pos
tan θ neg │ tan θ pos
──────────┼──────────► x
sin θ neg │ sin θ neg
cos θ neg │ cos θ pos
tan θ pos │ tan θ neg
  III     │    IV
```

- function values for the quadrantal angles (p. 276):

θ, degrees	θ, radians	$\sin \theta$	$\cos \theta$	$\tan \theta$	$\csc \theta$	$\sec \theta$	$\cos \theta$
0°	0	0	1	0	Undefined	1	Undefined
90°	$\dfrac{\pi}{2}$	1	0	Undefined	1	Undefined	0
180°	π	0	−1	0	Undefined	−1	Undefined
270°	$\dfrac{3\pi}{2}$	−1	0	Undefined	−1	Undefined	0

- degree–radian and radian–degree conversion (pp. 287–288). $1° = \dfrac{\pi}{180} \text{ rad} \approx 0.017 \text{ rad}$ **(8-3)**

$$1 \text{ rad} = \dfrac{180°}{\pi} \approx 57.3° \qquad \textbf{(8-4)}$$

- radian equivalents of special angles (p. 288):

Degrees	30°	45°	60°
Radians	$\dfrac{\pi}{6}$	$\dfrac{\pi}{4}$	$\dfrac{\pi}{3}$

- length of a circular arc (p. 293): $s = r\theta$ **(8-5)**
- area of a sector (p. 294): $A = \dfrac{1}{2}r^2\theta$ **(8-6)**
- relationship between linear velocity v and angular velocity ω (p. 295): $v = r\omega$ **(8-7)**

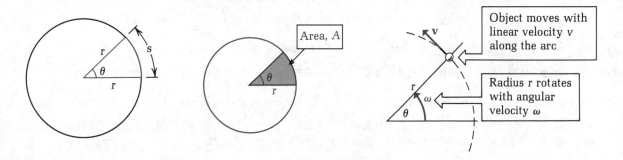

Exercises 8-5

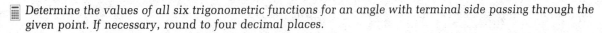 *Determine the values of all six trigonometric functions for an angle with terminal side passing through the given point. If necessary, round to four decimal places.*

1. $(-12, 5)$ **2.** $(8, 14)$ **3.** $(6, -7)$ **4.** $(-1, -2)$

Express each of the following in terms of the same function of the corresponding reference angle:

5. cos 152°   **6.** sin 246°   **7.** tan 324°   **8.** cot 142°
9. sin 101°15′   **10.** tan 196.7°   **11.** sec 222°   **12.** cos (−34°)

Convert each of the following angles measured in degrees to radian measure expressed as a rational multiple of π:

13. 275°   **14.** 36°   **15.** 220°   **16.** 135°

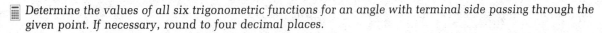 *Convert each of the following angles measured in degrees to decimal radians. Round to the nearest 0.01.*

17. 49°0′   **18.** 318°0′   **19.** 216.5°   **20.** 172°40′

Convert each of the following angles measured in radians to decimal degrees rounded to the nearest 0.1° if necessary:

21. $\dfrac{\pi}{4}$
22. $\dfrac{11\pi}{12}$
23. $\dfrac{13\pi}{8}$
24. $\dfrac{7\pi}{3}$
25. 1.10
26. 1.70
27. 2.45
28. −0.820

Use a calculator to find each of the following. Round to the nearest 0.001.

29. sin 237°
30. cos 159°
31. tan 312°
32. cot 117°
33. cos 218.4°
34. tan 227.7°
35. sec (−38°16′)
36. sin 153°50′
37. csc 329.25°
38. tan 93.75°
39. $\cos \dfrac{15\pi}{8}$
40. $\sin \dfrac{17\pi}{18}$
41. $\tan \dfrac{\pi}{6}$
42. $\cot \dfrac{5\pi}{4}$
43. $\sin \dfrac{7\pi}{8}$
44. cos 2.2
45. sec 0.63
46. csc 3.715
47. cos (−1.44)
48. sin 4.682

Find exact values for each of the following without the use of a calculator:

49. sin 210°
50. cos 315°
51. tan 120°
52. $\sin \dfrac{2\pi}{3}$
53. $\cos \dfrac{11\pi}{6}$
54. $\tan \dfrac{5\pi}{4}$

Find all values of θ in decimal degrees, $0° < \theta < 360°$, that satisfy the following. Round to the nearest 0.1°.

55. cos θ = −0.662
56. sin θ = 0.59
57. tan θ = 0.381
58. cot θ = −3.571

Find all values of θ, $0° < \theta < 360°$, that satisfy the following. Round to the nearest minute.

59. sin θ = −0.41
60. tan θ = 4.26
61. sec θ = −2.5
62. cos θ = 0.388

Find all values of θ in decimal radians, $0 < \theta < 2\pi$, that satisfy the following. Round to the nearest 0.01.

63. tan θ = −2.57
64. sin θ = 0.678
65. cos θ = 0.49
66. csc θ = −1.7
67. sin θ = −0.28
68. cos θ = −0.724
69. tan θ = 3.1
70. cot θ = 0.8635

Find θ without the use of a calculator. For Problems 71–73 give θ in degrees, $0° < \theta < 360°$, and for Problems 74–76 give θ as a rational multiple of π radians, $0 < \theta < 2\pi$.

71. $\sin \theta = \dfrac{\sqrt{3}}{2}$
72. $\cos \theta = \dfrac{1}{2}$
73. $\tan \theta = -\sqrt{3}$
74. $\sin \theta = -\dfrac{\sqrt{2}}{2}$
75. $\cos \theta = \dfrac{\sqrt{2}}{2}$
76. $\tan \theta = -\dfrac{\sqrt{3}}{3}$

Solve.

77. *Electronics* Use the formula $i = i_m \sin \theta$ to find the instantaneous sinusoidal current, given i_m = 28.2 mA and θ = 118°0′. (See Problem 79 in Section 8-2.)

78. *Electronics* Use the formula $V = V_m \sin 2\pi ft$ to find the instantaneous voltage V given V_m = 139 V, f = 1000 Hz, and t = 0.0085 s.

79. Find the length of an arc on a circle with radius 125 ft and central angle 0.625 rad.

80. Find the central angle in radians of an arc of length 0.145 cm on a circle with a radius of 0.105 cm.

81. Find the radius of a circle for which a central angle of 65°0′ subtends an arc of 12.0 in.

82. Find the area of the sector of a circle with a radius of 4.5 cm and a central angle of 120°.

83. Given a sector with an area of 2.5 m² and a radius of 1.2 m, find the central angle in radians.

84. Given a sector with an area of 1.75 cm² and a central angle of $\dfrac{3\pi}{5}$, find the radius.

85. *Physics* A flywheel rotates at an angular velocity of 1.75 rad/s. If the radius of the flywheel is 33.5 cm, find the linear velocity of a point on the rim in meters per second.

86. *Physics*   If an automobile is traveling at a linear velocity of 55 mi/h, and the tires are 27 in. in diameter, find the angular velocity of the tires in revolutions per minute.

87. *Construction*   A circular rotating restaurant at the top of a building makes one complete revolution every 20.0 min. If the diameter of the restaurant is 60.0 ft, find the linear velocity of the outer surface in miles per hour.

88. *Transportation engineering*   A subway train travels through a circular curve at 18.5 mi/h. If the radius of the curve is 68.2 ft, what is the angular velocity of the train in radians per second?

89. *Physics*   A potter's wheel rotates at an angular velocity of 110 rev/min. If the wheel is 12 in. in diameter, what is the linear velocity of a point on the rim of the wheel in feet per second?

90. *Construction*   The raised stopboard at the front of a shot-put circle is an arc 122 cm in length. If the radius of the arc is 106 cm, find the central angle.

91. *Electronics*   The needle on a voltmeter deflects through an angle of 85.0° as it passes from one extreme to the other. If the needle is 14.5 cm long, find the length of the scale.

92. *Optics*   If an object is seen with a visual angle of 2.35 rad measured from the center of an eye, and the eye is 24.18 mm in diameter, calculate the arc length of the image on the retina.

93. Fully opened, a paper fan in the shape of a sector of a circle has a central angle of 110° and a radius of 8.4 in. How many square inches of paper were used to make the fan?

94. A conical cup is formed from a piece of paper cut in the shape of the sector of a circle. If the radius of the sector is 4.2 in., and the central angle measures 124°, what is the area of the paper (in square inches) used to make the cup?

95. *Forestry*   During aerial spraying of insecticide to combat the gypsy moth, a helicopter crew is assigned a sector of forest with a central angle of 145° and a radius of 3.8 mi. How many square miles of forest will the crew spray?

96. *Space technology*   The *Landsat*, a satellite used to study the earth's resources, circles the earth at an altitude of 750 km. The smallest detail to be observed is a square 30 m on each side. What angle does this square subtend at the satellite?

[*Hint:* Use the side of the square as an approximation for arc length.]

Vectors and Oblique Triangles

9

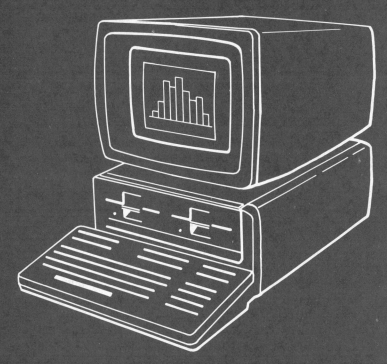

COMPUTER WITH KEYBOARD, DISC DRIVE, AND MONITOR

Certain quantities in physics have associated with them a direction as well as a size. In order to use these quantities in scientific and technical work, we must be able to deal with both of these characteristics.

Scalar   Most quantities of interest in scientific and technical work are either *vectors* or *scalars*. A **scalar** is a quantity that has only size, or *magnitude*. It can be described completely by specifying a single number and a unit of measurement. For example, mass, length, time, density, energy, and temperature are all scalar quantities.

Vector   A **vector** is a quantity that has both *magnitude* and *direction*. Mathematically, a vector must be described by two or more numbers and an appropriate unit of measurement. Force, velocity, acceleration, momentum, and electric and magnetic fields are all vector quantities.

The usual rules of arithmetic and algebra can be used to add, subtract, multiply, and divide scalars, but, as you will see in this chapter, special rules and procedures must be developed to allow the addition, subtraction, or multiplication of vectors.

EXAMPLE 1 Determine whether each of the following is a scalar or a vector:
(a)   46°C   (b)   170 mi/h at 240° SSW   (c)   250 kg
(d)   26 lb applied at an angle of 30° to horizontal

Figure 9-1

Solution The quantities in parts (a) and (c), temperature and mass, are scalars; they involve only magnitude. The quantities in parts (b) and (d), velocity and force, are vectors; they involve direction (240° SSW, 30° angle) as well as magnitude.

To understand the vector nature of some quantities, consider the following example. Figure 9-1 shows the spaceship *Columbia* moving above the earth. For simplicity, let us assume that there are only two forces acting on it: its weight **W**, the gravitational force exerted on it by the earth, and **F**, the propulsive force exerted by its rocket engines. Near the earth at takeoff the ship weighs about 100 tons, and the engines could provide about 250 tons of thrust in the direction shown. How can we add these two force vectors to find the size and direction of the total force causing the rocket ship to accelerate? We *cannot* simply add the forces to get **F** + **W** = 350 tons. Somehow the direction of the force must be taken into account. In this section and the next we will explain two methods for adding such vectors.

Notation   In printing, a vector is represented by a boldface letter: **F** for force, **v** for velocity, **a** for acceleration, and so on. In handwriting, a vector is usually indicated by placing an arrow over the letter: $\vec{F}, \vec{v}, \vec{a}$, and so on. The magnitude of the vector is represented by the letter itself. For example, the magnitude of the electric field **E** is simply E.

Geometric representation of a vector   Vector quantities are often represented in diagrams by a directed line segment or arrow, as shown in Fig. 9-2. Vector **v** is directed from point O to point

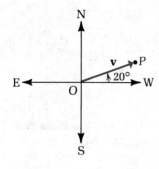

Figure 9-2

P. Point O is called the **initial point,** or tail of the vector, and point P is called the **terminal point,** or head of the vector.

The direction of the arrow shows the direction of the vector, and the length of the arrow is proportional to its magnitude. Vector **v** in Fig. 9-2 represents a velocity of 30 ft/s directed 20° north of east. Notice that we have used a scale of 1 cm on the drawing equals 20 ft/s, and the angle is drawn exactly.

EXAMPLE 2   Use a scale of 1 cm = 20 ft/s to draw the following velocity vectors:

(a)   **v₁**: 40 ft/s, north   (b)   **v₂**: 15 ft/s, 210° from the positive x-axis

Solution

(a) v_1 = 40 ft/s   (b) v_2 = 15 ft/s

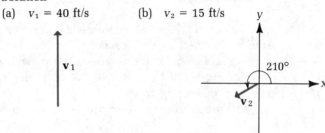

Addition of vectors   To add or subtract vectors graphically, we may need to move a vector arrow from place to place in a drawing. This is allowed only if we take care to move it parallel to itself, without changing its direction.

To add vectors graphically, follow the steps outlined in Example 3.

EXAMPLE 3   If **A** = 10 lb at an angle of 60° from horizontal, and **B** = 20 lb at an angle of 0°, find **C** = **A** + **B**. Estimate the magnitude (to the nearest 5 lb) and the direction (to the nearest 5°) of **C**.

Solution

Step 1.   Draw an arrow representing vector **A**. Choose a convenient scale; here, we have used 1 in. = 20 lb.

Step 2.   Using the same scale, draw an arrow representing vector **B**. Start the tail of **B** at the head of **A** and give **B** its correct direction.

Step 3.   Draw an arrow **C** from the tail of **A** to the head of **B**. This arrow represents the vector sum **A** + **B** = **C**.

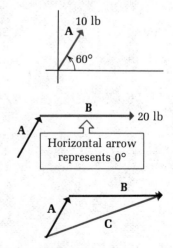

Step 4. Find the magnitude and direction of **C** by measuring its length and direction.

$C \approx 25$ lb
$\theta \approx 20°$

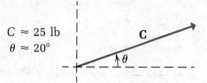

Resultant

The vector sum of two or more vectors is often called their **resultant** and denoted **R**. The resultant is a vector that has the same physical effect as the original vectors acting together.

NOTE ▶

The resultant has *both* magnitude and direction. An answer is not complete if either is omitted. ◀

Parallelogram method

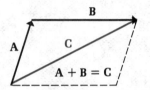

Figure 9-3

Notice in Fig. 9-3 that the resultant is the diagonal of a parallelogram having the original vectors as sides. For this reason the graphical way of adding two vectors is often called the **parallelogram method.** Also notice by comparing Fig. 9-3 and Fig. 9-4 that **A** + **B** = **B** + **A**. The order in which the vectors are added does not change the resultant.

This graphical procedure can easily be extended to add any number of vectors.

Figure 9-4

EXAMPLE 4 Find **C** + **D** + **E** and estimate the magnitude (to the nearest 0.5 unit) and direction (to the nearest 5°) of the resultant (scale: $\frac{1}{4}$ in. = 1 unit):

$C = 2.0$
$D = 4.0$
$E = 3.5$

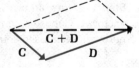

Solution

Step 1. Move vector **D** parallel to itself until its tail coincides with the head of **C**.

Step 2. Repeat Step 1 until all vectors being added have been placed head to tail.

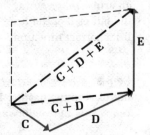

Step 3. The resultant is the vector with initial point at the tail of the first vector and terminal point at the head of the last vector.

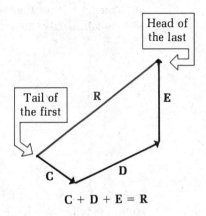

$$C + D + E = R$$

Step 4. Measure the length and direction of the resultant **R**.

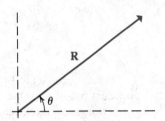

$R \approx 6.5$ at an angle of approximately 40° with the horizontal

A similar method is used to add vectors acting in the same or opposite directions.

EXAMPLE 5   Add the following pairs of vectors:
(a) **F + G** = ?   (b) **H + I** = ?

Solution
(a) **Step 1.** Place the vectors head to tail.

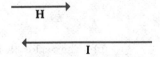

Step 2. Draw the resultant alongside, from the tail of the first to the head of the last.

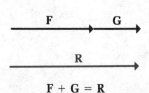

$$F + G = R$$

(b) **Step 1.** Place the vectors head
to tail, parallel to each other.

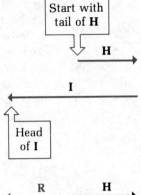

Step 2. The resultant **R** extends
from the tail of the first to the
head of the last.

$$H + I = R$$

Multiplying a vector by a scalar

Vectors can be multiplied in a variety of ways. In the simplest of these, the product $n\mathbf{A}$ of a vector **A** and a positive scalar n is a vector having the same direction as **A** and a magnitude of nA. The next example shows this product geometrically.

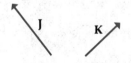

Figure 9-5

EXAMPLE 6   Given **J** and **K** in Fig. 9-5, find:
(a)   2**J**   (b)   2**J** + 3**K**

Solution
(a)   The vector 2**J** has the same
direction and twice the
magnitude of **J**.

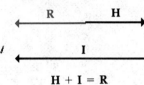

(b)   The vector 3**K** has the same
direction and three times the
magnitude of **K**. Using the
parallelogram method, we find

$$2\mathbf{J} + 3\mathbf{K} = \mathbf{R}$$

The product of a vector **A** and a negative scalar n is a vector having direction opposite to **A** and magnitude $|n|A$.

EXAMPLE 7 Given **L** in Fig. 9-6, find -2**L**.

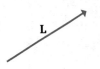

Figure 9-6

Solution
The vector -2**L** has the opposite direction and twice the magnitude of the vector **L**.

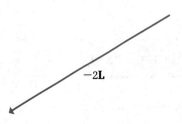

Subtraction of vectors The difference of two vectors can be written as a sum and found geometrically. For example,

A − **B** = **A** + (−**B**)

where the vector −**B** has the same magnitude but is opposite in direction to the vector **B**.

EXAMPLE 8 Given **M** and **N** in Fig. 9-7, find:
(a) **M** − **N**   (b) 2**M** − **N**

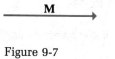

Figure 9-7

Solution
(a) **First,** draw the vector −**N**, with the same magnitude as **N** and in the opposite direction.

Then, add **M** + (−**N**) graphically as before.

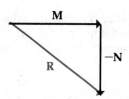

M − **N** = **M** + (−**N**) = **R**

(b) **First,** draw the vector 2**M** with double the magnitude and in the same direction as **M**.

Then, add 2**M** + (−**N**) as before.

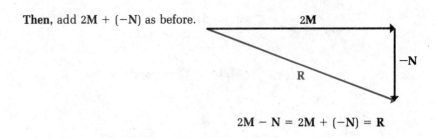

$$2M - N = 2M + (-N) = R$$

Exercises 9-1

Determine *which of the following are scalars and which are vectors:*

1. 202 kg   **2.** 65 in.   **3.** 17 mi at 170° WNW
4. 15 lb at 22° to horizontal   **5.** 1.7 lb/in³   **6.** 165 mi/h due east
7. 2.3 ft/s² at 70° NNE   **8.** 46°F   **9.** 414 N, horizontal
10. 350,000 Btu   **11.** 3 × 10⁴ J   **12.** 5.8 × 10⁻³ m/s, 67° SSE

Draw each vector using the scale 1 cm = 10 units. All angles are in standard position unless stated otherwise.

13. 35 ft/s, south   **14.** 10 m/s², east   **15.** 15 N, west   **16.** 40 mi, north
17. 25 lb, 230°   **18.** 30 N, 85°   **19.** 5 ft/s², 30° N of E   **20.** 20 mi/h, 15° W of S

Add the given vectors using the parallelogram method.

21.   **22.**   **23.**

24.   **25.**   **26.**

Find each vector sum or difference using the parallelogram method on vectors **A, B, C, D,** *and* **E.** *Estimate the magnitude (to the nearest 0.5 unit) and direction (to the nearest 5°) of each resultant. Use a scale of* $\frac{1}{4}$ *in. = 1 unit of magnitude, and measure all angles from standard position.*

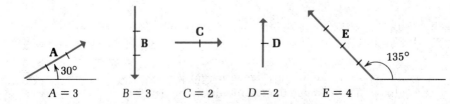

27. $A + B$	28. $B + C$	29. $C + E$	30. $A + E$
31. $A + C + D$	32. $B + D + E$	33. $A + B + C + E$	34. $B + C + D + E$
35. $B + 2D$	36. $E + 3D$	37. $B - D$	38. $A - D$
39. $E - A$	40. $D - C$	41. $A - 2B$	42. $E - C$
43. $A + 3C - 2E$	44. $D - 2B + 4C$		

9-2 | Adding Vectors Using Components

The graphical method of adding vectors is easy to visualize, but it is not very useful if the vectors are acting in three-dimensional space, and it gives only approximate answers. The magnitude and direction of the resultant vector must be measured rather than calculated exactly. In this section we will show how to calculate the sum of two or more vectors exactly using trigonometry.

Components

Draw a vector **A** on a coordinate plane with its initial point at the origin (Fig. 9-8). If perpendicular lines are drawn from the head of **A** to the axes, the two new vectors formed along the axes are called the **components** of **A**. The original vector **A** can be written as the sum of these components:

$$A = A_x + A_y$$

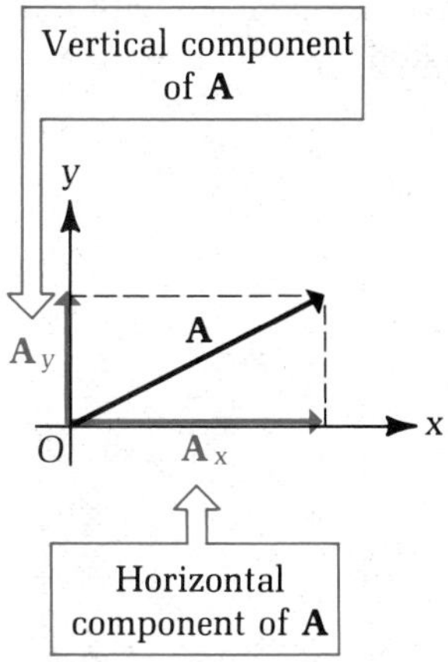

The process of finding component vectors A_x and A_y when you are given **A** is called **resolving A into its components.** If you are given the magnitude A and direction θ of vector **A**, the magnitudes of the components can be found using trigonometric ratios. In triangle OPQ shown in Fig. 9-9, the magnitudes of the components may be found as follows:

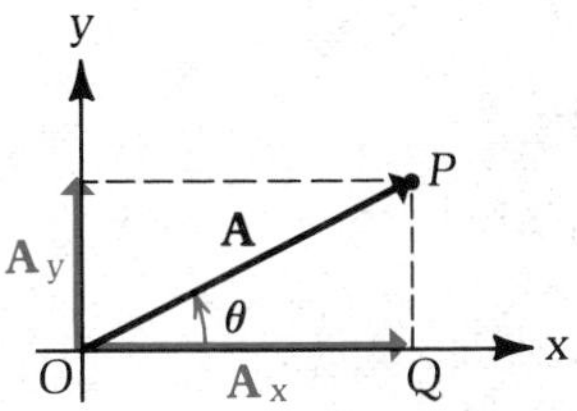

Figure 9-8

Figure 9-9

Components of a Vector

$$\cos \theta = \frac{A_x}{A} \quad \text{or} \quad A_x = A \cos \theta$$

$$\sin \theta = \frac{A_y}{A} \quad \text{or} \quad A_y = A \sin \theta$$

(9-1)

Figure 9-10

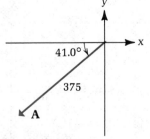

Figure 9-11

EXAMPLE 9 Resolve the following vectors into their horizontal and vertical components:

(a) $v = 3.85$ ft/s, $\theta = 56°0'$ (b) $W = 28.5$ lb, $\theta = 147.3°$

Solution

(a) $v_x = (3.85 \text{ ft/s})(\cos 56°0')$ $3.85 \times 56 \boxed{\cos} \boxed{=} \rightarrow$ 2.1528927
     $v_x \approx 2.15$ ft/s
     $v_y = (3.85 \text{ ft/s})(\sin 56°0')$ $3.85 \times 56 \boxed{\sin} \boxed{=} \rightarrow$ 3.1917947
     $v_y \approx 3.19$ ft/s

(b) $W_x = (28.5 \text{ lb})(\cos 147.3°)$ $28.5 \times 147.3 \boxed{\cos} \boxed{=} \rightarrow$ -23.983057
     $W_x \approx -24.0$ lb
     $W_y = (28.5 \text{ lb})(\sin 147.3°)$ $28.5 \times 147.3 \boxed{\sin} \boxed{=} \rightarrow$ 15.396849
     $W_y \approx 15.4$ lb

Notice in Example 9(b) that the angle could be given as an angle with the negative x-axis as shown in Fig. 9-12. We would then find the components this way:

W_x is negative in quadrant II

$$W_x = (-28.5 \text{ lb})(\cos 32.7°) \approx -24.0 \text{ lb}$$

$$W_y = (+28.5 \text{ lb})(\sin 32.7°) \approx 15.4 \text{ lb}$$

W_y is positive in quadrant II

Figure 9-12

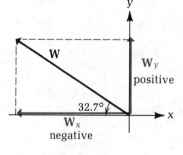

EXAMPLE 10 Resolve into components the vector **A** shown in Fig. 9-13.

Solution **First,** note that both components are negative. **Then,** apply formulas (9-1) using the triangle shown. Attach the proper signs.

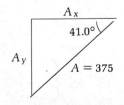

$$A_x = (-375)(\cos 41.0°) \approx -283$$

$$A_y = (-375)(\sin 41.0°) \approx -246$$

Figure 9-13

Using components to find magnitude and direction

To add vectors algebraically, we must not only be able to resolve a vector into its components, we must also be able to perform the reverse procedure. That is, given the components of a vector, determine its magnitude and direction.

Figure 9-9 shows the vector **A** with its components A_x and A_y. We can find the magnitude A in terms of A_x and A_y using the Pythagorean theorem:

$$A^2 = A_x^2 + A_y^2$$

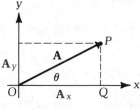

Figure 9-9

Taking the square root of both sides of this equation, we obtain the general formula:

Magnitude of a Vector	$A = \sqrt{A_x^2 + A_y^2}$	**(9-2)**

To find the direction θ in terms of A_x and A_y, we use the tangent function to find the reference angle Ref θ:

Direction of a Vector	$\tan (\text{Ref } \theta) = \left\lvert \dfrac{A_y}{A_x} \right\rvert$	**(9-3)**

The signs of A_x and A_y will tell you the quadrant in which **A** is found, and the proper value of the direction θ can then be determined.

EXAMPLE 11 Given the following pairs of components, determine the magnitude and direction of each vector.

(a) $v_x = 2.2$ m/s, $v_y = 4.3$ m/s   (b) $R_x = -820$ lb, $R_y = -460$ lb

Solution

(a) Use the magnitude formula (9-2) to find the magnitude of **v**:

$$v = \sqrt{2.2^2 + 4.3^2} \qquad 2.2\;\boxed{x^2}\;\boxed{+}\;4.3\;\boxed{x^2}\;\boxed{=}\;\boxed{\sqrt{}} \rightarrow \quad \textit{4.8301139}$$

$$v \approx 4.8 \text{ m/s}$$

Use the direction formula (9-3) to find the direction of **v**:

$$\tan (\text{Ref } \theta) = \left\lvert \frac{4.3}{2.2} \right\rvert \qquad 4.3\;\boxed{\div}\;2.2\;\boxed{=}\;\boxed{\text{INV}}\;\boxed{\tan} \rightarrow \quad \textit{62.904448}$$

Ref $\theta \approx 63°$

Because both v_x and v_y are positive, we know that **v** is in the first quadrant. Thus, $\theta \approx 63°$ and $v \approx 4.8$ m/s.

(b) Magnitude: $R = \sqrt{(-820)^2 + (-460)^2} \approx 940$ lb

Direction: $\tan (\text{Ref } \theta) = \left\lvert \dfrac{-460}{-820} \right\rvert$ Ref $\theta \approx 29°$

Because both R_x and R_y are negative, we know that **R** is in the third quadrant. Therefore, in standard position,

$$\theta \approx 180° + 29° \approx 209°$$

Adding vectors using components   To use this component method to add vectors with known magnitudes and directions, follow the steps listed below.

Step 1.   Resolve each vector into its horizontal or x-component and its vertical or y-component. Use the component formulas (9-1).
Step 2.   Find R_x, the sum of all x-components, and R_y, the sum of all y-components; R_x and R_y represent the x- and y-components, respectively, of the resultant **R**.
Step 3.   Find the magnitude of the resultant. Use the magnitude formula (9-2).
Step 4.   Find the direction of the resultant. Use the direction formula (9-3) and the signs from Step 2.

IMPORTANT ▶   When recording the intermediate results in Steps 1 and 2, always round to at least one digit more than the desired accuracy of the answer. Premature rounding may cause your final answer to be incorrect. See Appendix B for more information on rounding. ◀

EXAMPLE 12   Given vector forces **F** and **P** with $F = 15.0$ lb, $\theta_F = 24°0'$, $P = 28.0$ lb, and $\theta_P = 73°0'$, find the resultant force, $\mathbf{R} = \mathbf{F} + \mathbf{P}$ (see Fig. 9-14).

Solution

Step 1. $F_x = (15.0 \text{ lb})(\cos 24°0') \approx 13.70 \text{ lb}$

$\qquad\quad F_y = (15.0 \text{ lb})(\sin 24°0') \approx 6.101 \text{ lb}$

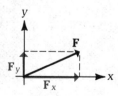

$\qquad\quad P_x = (28.0 \text{ lb})(\cos 73°0') \approx 8.186 \text{ lb}$

$\qquad\quad P_y = (28.0 \text{ lb})(\sin 73°0') \approx 26.78 \text{ lb}$

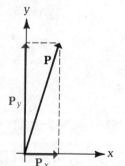

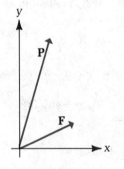

Figure 9-14

Step 2. $R_x = F_x + P_x = 13.70 \text{ lb} + 8.186 \text{ lb}$
$$\approx 21.89 \text{ lb}$$

$$R_y = F_y + P_y = 6.101 \text{ lb} + 26.78 \text{ lb}$$
$$\approx 32.88 \text{ lb}$$

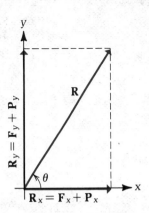

Step 3. $R = \sqrt{R_x^2 + R_y^2} = \sqrt{(21.89)^2 + (32.88)^2}$
$$\approx 39.5 \text{ lb}$$

Step 4. $\tan (\text{Ref } \theta) = \left| \dfrac{R_y}{R_x} \right| = \left| \dfrac{32.88 \text{ lb}}{21.89 \text{ lb}} \right|$ $\text{Ref } \theta \approx 56.3°$

Since R_x and R_y are both positive, **R** is in the first quadrant, and the resultant vector force is approximately 39.5 lb at an angle of 56.3° with the positive x-axis.

As shown in Example 10, if an angle is not given in standard position, we have to adjust the signs of the magnitude of each component accordingly.

EXAMPLE 13 Use the information shown in Fig. 9-15 to find the sum of vectors **C** and **D**.

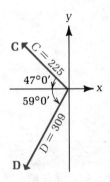

Figure 9-15

Solution **First,** note the signs:

C_x: Negative C_y: Positive

D_x: Negative D_y: Negative

Then, proceed as in Example 12:

Step 1. $C_x = (-225)(\cos 47°0') \approx -153.4$

$C_y = (225)(\sin 47°0') \approx 164.6$

$D_x = (-309)(\cos 59°0') \approx -159.1$

$D_y = (-309)(\sin 59°0') \approx -264.9$

Step 2. $R_x = C_x + D_x = (-153.4) + (-159.1) = -312.5$

$R_y = C_y + D_y = 164.6 + (-264.9) = -100.3$

Step 3. Magnitude $R = \sqrt{(-312.5)^2 + (-100.3)^2} \approx 328$

Step 4. $\tan (\text{Ref } \theta) = \left| \dfrac{-100.3}{-312.5} \right|$ $\text{Ref } \theta \approx 17.8°$

From the signs of R_x and R_y, we know that the vector **R** is in the third quadrant, as shown in the figure. Therefore, $R \approx 328$ and $\theta \approx 197.8°$ as measured from the positive x-axis.

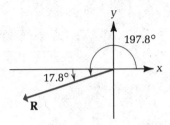

Using a calculator   Once you become proficient at adding vectors using components, you may want to find the magnitude R of the resultant with one continuous calculation using a calculator. It may be helpful to prepare a table summarizing the necessary operations.

▦ **EXAMPLE 14** Add the vectors shown in Fig. 9-16 given $A = 24.7$, $B = 38.2$, and $C = 51.3$.

Figure 9-16

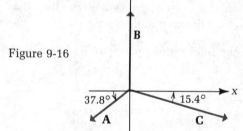

Solution

	Vector	x-Component	y-Component
Step 1.	A	$A_x = (-24.7)(\cos 37.8°)$	$A_y = (-24.7)(\sin 37.8°)$
	B	$B_x = (38.2)(\cos 90°)$	$B_y = (38.2)(\sin 90°)$
	C	$C_x = (51.3)(\cos 15.4°)$	$C_y = (-51.3)(\sin 15.4°)$
Step 2.	R	$R_x = A_x + B_x + C_x$	$R_y = A_y + B_y + C_y$

Step 3. $R = \sqrt{R_x^2 + R_y^2}$

R_x: 24.7 [+/-] [×] 37.8 [cos] [+] 38.2 [×] 90 [cos] [+] 51.3 [×] 15.4 [cos] [=] [STO] → $\boxed{29.941265}$

R_y: 24.7 [+/-] [×] 37.8 [sin] [+] 38.2 [×] 90 [sin] [+] 51.3 [+/-] [×] 15.4 [sin] [=] → $\boxed{9.4381669}$

R: [x²] [+] [RCL] [x²] [=] [√] → $\boxed{31.393604}$ ⟋ Record this

Step 4. $\tan(\text{Ref } \theta) = \left| \dfrac{R_y}{R_x} \right|$ 9.4381669 [÷] [RCL] [=] [INV] [tan] → $\boxed{17.496035}$

Therefore, $R \approx 31.4$ and, since R_x and R_y are both positive, $\theta = \text{Ref } \theta \approx 17.5°$.

LEARNING HINT ▶

If we examine the x- and y-components of vector **B** in Example 14, we find that $B_x = 0$ and $B_y = 38.2$. Notice that any vector **P** directed along the positive y-axis has x-component $P_x = P \cos 90° = 0$ and y-component $P_y = P \sin 90° = P$. Similarly, any vector **Q** directed along the positive x-axis has $Q_y = Q \sin 0° = 0$ and $Q_x = Q \cos 0° = Q$. Furthermore, a vector **R** with direction $\theta = 270°$ has $R_x = 0$ and $R_y = -R$, and a vector **S** with direction $\theta = 180°$ has $S_y = 0$ and $S_x = -S$. These facts are summarized in Fig. 9-17.

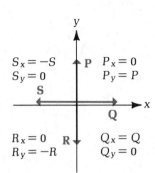

$S_x = -S$ **P** $P_x = 0$
$S_y = 0$ $P_y = P$
S

$R_x = 0$ **R** $Q_x = Q$
$R_y = -R$ $Q_y = 0$

Figure 9-17

EXAMPLE 15 For the vectors in Fig. 9-18, $A = 12.0$ and $B = 10.0$. Find:

(a) **R = A − B**   (b) **P = 2B − 3A**

Solution

(a) $A_x = (-12.0)(\cos 60.0°) = -6.000$

$A_y = (12.0)(\sin 60.0°) \approx 10.39$

$B_x = (-10.0)(\cos 27.0°) \approx -8.910$

$B_y = (-10.0)(\sin 27.0°) \approx -4.540$

$R_x = A_x - B_x = -6.000 - (-8.910) = 2.910$

$R_y = A_y - B_y = 10.39 - (-4.540) = 14.930$

$R = \sqrt{(2.91)^2 + (14.93)^2} \approx 15.2$

$\tan (\text{Ref } \theta) = \left| \dfrac{14.93}{2.91} \right|$ $\text{Ref } \theta \approx 79.0°$

The signs tell us that θ is therefore approximately 79.0°.

(b) $P_x = 2B_x - 3A_x = 2(-8.910) - 3(-6.000) \approx 0.1800$

$P_y = 2B_y - 3A_y = 2(-4.540) - 3(10.39) \approx -40.25$

$P = \sqrt{(0.1800)^2 + (-40.25)^2} \approx 40.3$

$\tan (\text{Ref } \theta) = \left| \dfrac{-40.25}{0.1800} \right|$ $\text{Ref } \theta \approx 89.7°$

The signs of P_x and P_y place θ in quadrant IV. Therefore, $P \approx 40.3$ and $\theta \approx 360° - 89.7° \approx 270.3°$.

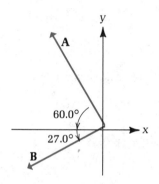

A

60.0°
27.0°

B

Figure 9-18

Exercises 9-2

Resolve each vector into its horizontal and vertical components.

1. $A = 1.74$ ft/s, $\theta = 27°0'$
2. $B = 16.5$ lb, $\theta = 72°0'$
3. $C = 285$ mi, $\theta = 127°0'$
4. $D = 146$ km, $\theta = 217°0'$
5. $E = 5.7$ lb, $\theta = 260°$
6. $F = 6.2$ m/s, $\theta = 165°$
7. $G = 1250$ N, $\theta = -37°15'$
8. $H = 0.0525$ m/s^2, $\theta = 54°40'$

9.

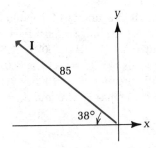

10.

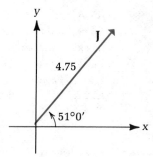

11.

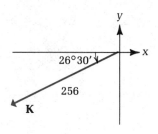

12.

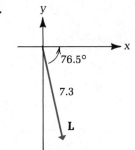

Given the following pairs of components, determine the magnitude and direction of each vector. Give the direction as an angle in standard position.

13. $R_x = 5.8$ m/s, $R_y = 3.2$ m/s
15. $T_x = 28.5$ ft/s², $T_y = -18.2$ ft/s²
17. $V_x = -854$ lb, $V_y = 275$ lb
19. $L_x = -75.4$ mi, $L_y = -51.6$ mi

14. $S_x = 625$ lb, $S_y = 856$ lb
16. $U_x = 3.65$ ft/s, $U_y = -8.12$ ft/s
18. $W_x = -0.068$ N, $W_y = 0.014$ N
20. $M_x = -1280$ m, $M_y = -5310$ m

Add each pair of vectors.

21. $A = 62.0$ lb, $\theta_A = 0.0°$
     $B = 48.0$ lb, $\theta_B = 90.0°$
23. $E = 485$ mi/h, $\theta_E = 90.0°$
     $F = 86.0$ mi/h, $\theta_F = 315.0°$
25. $T = 1.80$ m/s², $\theta_T = 195.5°$
     $U = 2.30$ m/s², $\theta_U = -48.5°$
27. $L = 81.4$ cm, $\theta_L = 15.5°$
     $M = 16.5$ cm, $\theta_M = 70.0°$

22. $C = 265$ lb, $\theta_C = 90.0°$
     $D = 375$ lb, $\theta_D = 180.0°$
24. $G = 6.65$ ft/s, $\theta_G = 75.0°$
     $H = 4.25$ ft/s, $\theta_F = 125.0°$
26. $V = 5850$ N, $\theta_V = 56.4°$
     $W = 2150$ N, $\theta_W = 345.6°$
28. $R = 0.250$ mi, $\theta_R = 145.4°$
     $S = 0.125$ mi, $\theta_S = 240.2°$

Use the figures below to find the indicated resultants.

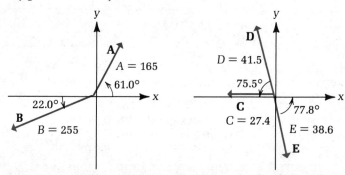

29. 2A   30. 3D   31. A + B   32. A − 2B

33. C + D + E   34. 2D − E + 3C   35. A − B   36. C − D

9-3 | Applications of Vectors

The techniques developed in this chapter can be applied to many practical situations in science and technology. In this section we shall concentrate on applications involving the vector quantities displacement, velocity, and force. Since the directions of the compass will be used in many of these applications, Fig. 9-19 is given here for reference.

Figure 9-19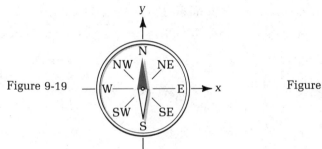

Figure 9-20

Displacement

Displacement is a vector quantity describing the change in position of an object as it moves from one point to another. In Fig. 9-20, for example, the cyclist moves from point A to point B along the curved road. The distance traveled is the length of path, 10 mi, between A and B, along the road. The displacement of the cyclist from A to B is the vector $\mathbf{D}$ connecting the points. The magnitude of $\mathbf{D}$ is $D = 2$ mi.

EXAMPLE 16 A ship heading out of a harbor travels 1.8 mi due east and then turns up the coast and travels 2.0 mi due north. At the end of this trip, what is its displacement from the harbor?

Solution The problem calls for the sum of the two given displacements. First, represent them in the coordinate plane as shown in Fig. 9-21. Since the given vectors are parallel to the coordinate axes, they represent the x- and y-components of the resultant displacement:

$$R_x = 1.8 \text{ mi} \quad \text{and} \quad R_y = 2.0 \text{ mi}$$

$$R = \sqrt{(1.8)^2 + (2.0)^2} \approx 2.7 \text{ mi}$$

$$\tan (\text{Ref } \theta) = \left| \frac{2.0}{1.8} \right| \qquad \theta = \text{Ref } \theta \approx 48°$$

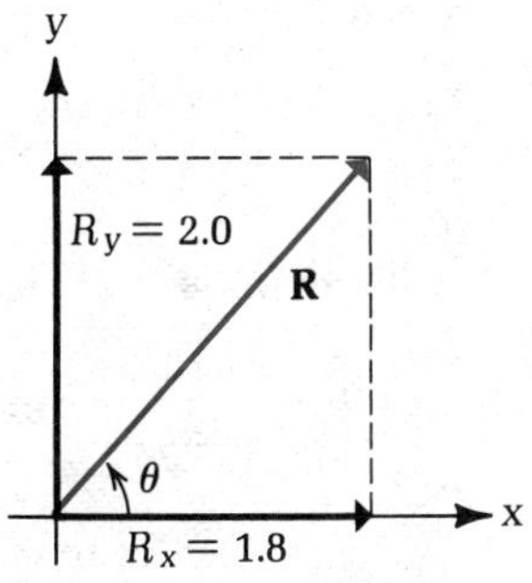

Figure 9-21

The ship is approximately 2.7 mi from the harbor in a direction 48° north of east.

Velocity

The **velocity** of a moving object is the rate at which its displacement changes with time. In Fig. 9-20, for example, if the cyclist travels from A to B in 1 h, and if the magnitude of the displacement D is 2 mi, the cyclist's average velocity is 2 mi/h along the direction of **D**. Of course, the average *speed* of the cyclist along the road is much higher, since 10 mi have actually been traveled in the hour.

EXAMPLE 17 An airplane is flying in a wind blowing at a velocity of 75 mi/h from the north. If the velocity of the plane is 550 mi/h northwest with respect to the air, what is the resultant velocity of the plane with respect to the ground?

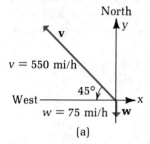

(a)

Solution

First, represent each vector in the coordinate plane with its initial point at the origin, as shown in Fig. 9-22(a).

Second, note that v_x and w_y are negative.

Third, use the component method to add the velocities:

$$v_x = (-550)(\cos 45°) \approx -389 \text{ mi/h} \qquad v_y = (550)(\sin 45°) \approx 389 \text{ mi/h}$$

$$w_x = 0 \qquad\qquad w_y = -75 \text{ mi/h}$$

If $\mathbf{u} = \mathbf{v} + \mathbf{w}$, then

$$u_x = v_x + w_x \approx -389 \text{ mi/h} \qquad u_y = v_y + w_y \approx 314 \text{ mi/h}$$

$$u = \sqrt{(-389)^2 + (314)^2} \approx 500 \text{ mi/h}$$

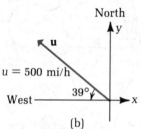

(b)

Figure 9-22

$$\tan (\text{Ref } \theta) = \left| \frac{314}{-389} \right| \qquad \text{Ref } \theta \approx 39°$$

The signs of u_x and u_y tell us that θ is in quadrant II. Therefore, the velocity is approximately 500 mi/h in a direction 39° north of west, as shown in Fig. 9-22(b).

EXAMPLE 18 As a plane flies at a velocity **v** of 125 m/s in a direction 15.0° below horizontal, a parachutist jumps out. For the first 10 s or so, the parachutist falls at a velocity of magnitude $u = 9.80t$ m/s. Ignoring the effects of air resistance, what is the velocity of the parachutist after exactly 8 s of free fall?

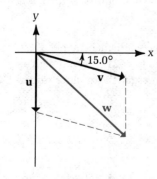

Figure 9-23

Solution The resultant velocity of the parachutist is the sum of the falling velocity **u** and the velocity **v** received from the plane: $\mathbf{w} = \mathbf{u} + \mathbf{v}$.

First, represent the vectors as shown in Fig. 9-23.

Second, note that the signs of v_y and u_y are negative.

Then, use the component method to find the resultant velocity **w**. Since $u_x = 0$,

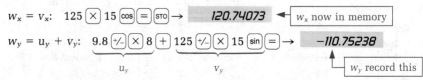

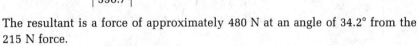

$$w = \sqrt{w_y^2 + w_x^2}: \quad \boxed{x^2}\boxed{+}\boxed{\text{RCL}}\boxed{x^2}\boxed{=}\boxed{\sqrt{}} \rightarrow \boxed{163.84265} \leftarrow \boxed{w}$$

$$\tan (\text{Ref } \theta) = \left| \frac{w_y}{w_x} \right|: \quad 110.75238 \boxed{\div}\boxed{\text{RCL}}\boxed{=}\boxed{\text{INV}}\boxed{\tan} \rightarrow \boxed{42.529363}$$

$$\uparrow \quad \boxed{\text{Ref } \theta}$$

Since w_x is positive and w_y is negative, θ is in quadrant IV. After 8 s, the parachutist's velocity is approximately 164 m/s in a direction 42.5° below horizontal.

Force   **Force** is a vector quantity associated with a push or pull of one object on another. The net effect of a force is to produce a change in the velocity of the object on which it acts. If several forces act simultaneously on an object, the change in velocity produced is proportional to the vector sum of all the forces acting.

EXAMPLE 19   Two forces, one of 215 N and the other of 325 N, pull on an object. If the angle between the forces is 56.0°, what is the resultant of the forces?

Solution   The direction of the forces is not specified. Therefore, it is most convenient to represent one on the positive x-axis and the other at a positive angle of 56.0° with the x-axis (Fig. 9-24). Then,

$$F_x = 215 \text{ N} \qquad\qquad\qquad F_y = 0$$

$$\underline{G_x = 325 \cos 56.0° \approx 181.7 \text{ N}} \qquad \underline{G_y = 325 \sin 56.0° \approx 269.4 \text{ N}}$$

$$R_x \approx 396.7 \text{ N} \qquad\qquad\qquad R_y \approx 269.4 \text{ N}$$

$$R = \sqrt{(396.7)^2 + (269.4)^2} \approx 480 \text{ N}$$

$$\tan (\text{Ref } \theta) = \left| \frac{269.4}{396.7} \right| \qquad \theta = \text{Ref } \theta \approx 34.2°$$

The resultant is a force of approximately 480 N at an angle of 34.2° from the 215 N force.

Figure 9-24

Inclined planes   One variety of physics problem involves the forces acting on an object at rest or moving on an inclined plane. As shown in Fig. 9-25, there are three forces operating on such an object. The weight **W** of the object due to gravity acts vertically downward. The normal force **N** is exerted by the plane on the object and is perpendicular to the inclined surface. The friction force **F** is exerted by the surface of the plane on the surface of the object and is parallel to the surface of the plane in a direction opposing any motion of the block.

According to Newton's first law, if the object is at rest or moving at a constant speed down the plane, the resultant of all three forces is zero. This means that the sum of the x-components is zero, and the sum of the y-components is zero. Assume that the block is small enough so that all forces act at its center.

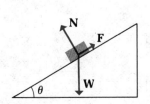

Figure 9-25

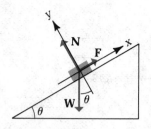

Figure 9-26

Since two of the forces, **N** and **F**, are perpendicular, it is most convenient to position the coordinate axes to coincide with these forces (Fig. 9-26). Notice that the angle between **W** and the negative y-axis is also equal to θ.

EXAMPLE 20   Suppose an object weighing 15.0 lb is at rest on a board inclined at an angle of 25°0'.
(a)   Find the static friction force exerted by the surface of the plane.
(b)   Find the normal force of the plane on the object.

Solution   **Step 1.**   Position the coordinate axes as shown in Fig. 9-27 and label the angles.

Step 2.   Make up a table of components.

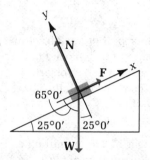

Figure 9-27

Vector	x-Component	y-Component
N	0	N
W	$-15.0 \cos 65°0'$	$-15.0 \sin 65°0'$
F	F	0
Resultant **R**	R_x	R_y

Step 3.   If the block is at rest or moving at constant speed, the resultant will equal zero.

(a) $R_x = F - 15.0 \cos 65°0' = 0$ or $F = 15.0 \cos 65°0' \approx 6.34$ lb
(b) $R_y = N - 15.0 \sin 65°0' = 0$ or $N = 15.0 \sin 65°0' \approx 13.6$ lb

Another typical physics problem requires that you find the tension force in ropes or cables supporting a static object.

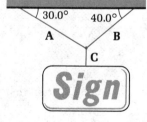

Figure 9-28

EXAMPLE 21   A sign is hung on ropes from a ceiling, as shown in Fig. 9-28. If the sign weighs 65.0 lb, what is the tension in each of the three ropes?

Solution   **Step 1.**   Position the coordinate axes and label the angles as shown in Fig. 9-29. Because the three tension forces act on the knot, we set up the origin there.

Step 2.   Make up a table of components.

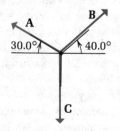

Figure 9-29

Vector	x-Component	y-Component
A	$-A \cos 30.0°$	$A \sin 30.0°$
B	$B \cos 40.0°$	$B \sin 40.0°$
C	0	$-C = -65.0$
Resultant **R**	R_x	R_y

Step 3. Since the knot is at rest under the action of the three forces, $\mathbf{R} = 0$.

$$R_x = -A \cos 30.0° + B \cos 40.0° = 0$$

$$R_y = A \sin 30.0° + B \sin 40.0° - C = 0$$

$$-0.866A + 0.766B = 0$$

$$0.500A + 0.643B - 65.0 = 0$$

Solving this system of two equations in two unknowns, we obtain $A \approx 53.0$ lb and $B \approx 59.9$ lb.

Exercises 9-3

1. *Navigation* A small airplane takes off and flies 6.0 mi due west before turning and flying 7.8 mi due north. What is its displacement from the takeoff spot?
2. *Navigation* A ship leaves a harbor traveling due south. After 12 miles it turns and heads due east for 27 miles. What is its displacement from the harbor?
3. *Physics* Two forces, one of 355 N and the other of 185 N, act on an object at right angles to each other. What is the resultant of these forces?
4. *Physics* A force of 88.0 lb acts on an object while simultaneously a second force of 53.0 lb acts on the same object at a right angle to the first force. Find the resultant of these two forces.
5. *Navigation* An airplane traveling at 675 mi/h due east with respect to the air is flying in a wind blowing at 55.0 mi/h from the south. Find the resultant velocity of the plane with respect to the ground.
6. *Navigation* A sailboat heads due west at 8.0 knots with respect to the water and is subjected to a current of 2.0 knots from the north. What is the resultant velocity of the boat?
7. *Navigation* A jet flies 1450 km west, then turns 35.0° north of west and travels an additional 1820 km. Find its displacement from the original position.
8. *Navigation* During a cruise, a ship travels 18 mi south, turns and sails 24 mi at 16° south of east, and makes a final turn and travels 11 mi in a direction 31° east of south. Calculate its total displacement from the starting point.
9. *Navigation* An airplane heads due northwest at a speed of 580 mi/h with respect to the air. If a wind blows at 85 mi/h from a direction 12° south of west, what is the velocity of the plane with respect to the ground?
10. *Physics* If an object weighing 8.0 lb is at rest on a board inclined at an angle of 18° with the horizontal, what must be the static friction force exerted by the surface of the plane?
11. *Physics* A crate weighing 2250 lb is resting on an incline sloping at an angle of 19.0° with the horizontal. What static friction force must exist in order to keep the crate from slipping?
12. *Navigation* As a plane flies at a velocity of 175 m/s in a direction 23.5° below horizontal, a parachutist jumps out. Ignoring the effects of air resistance, what is the velocity of the parachutist after 5.0 s of free fall? (See Example 18.)
13. *Physics* A projectile is fired at an angle of 65.0° with respect to horizontal at a speed of 18.2 ft/s. What horizontal distance has it traveled after 0.500 s? (Assume that it maintains its horizontal velocity component for the entire time.)
14. *Navigation* A plane climbs at an angle of 38° with respect to the horizontal at a speed of 480 mi/h. What is the altitude of the plane after 1.0 min?

15. *Physics*   A light fixture weighing 24.0 lb is hung from a ceiling, as shown in the figure. Find the tension in each of the support cables.

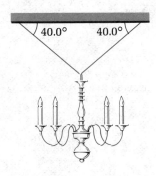

16. *Physics*   A 12.5 lb weight suspended from a rope attached to a wall is pulled outward from the wall by a horizontal force **F**, as shown in the figure. If the rope makes an angle of 25.0° with the wall, what is the tension in the rope?

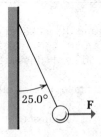

17. *Physics*   In Problem 16, what is the magnitude of force **F**?

18. *Construction*   A crate is attached to a rope and pulled at a constant speed along a rough horizontal surface. If the friction force between the crate and the surface is 275 lb, and if the rope is pulled at an angle of 32.0° upward from the horizontal, what force must be exerted by the rope?

19. *Machine technology*   Using a metal saw, a machinist makes consecutive cuts of 4.8 in. at 15°, 3.5 in. at 35°, and 1.6 in. at 57° in a metal plate. What is the total displacement of the cuts? (All angles are with respect to horizontal.)

20. *Navigation*   If a boat moves due southwest across a lake at 15 mi/h with respect to the water, where the current is moving at 2.0 mi/h in a direction 35° east of north, find the velocity of the boat.

21. *Physics*   A 65 lb sign is hung from three ropes, as shown in the figure. Find the tension in each rope.

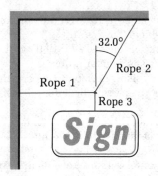

22. *Physics* A child weighing 62 lb sits in a swing and is pushed by a horizontal force of 44 lb. What is the tension in each of the two supporting ropes?

9-4 | The Law of Sines

We have already seen how the trignometric functions can be used to solve *right* triangles. However, it is often necessary to solve for the missing parts of an **oblique triangle**—that is, a triangle with no right angle. In the next two sections we shall derive and illustrate two formulas for solving oblique triangles. We begin with the *law of sines*.

Oblique triangles

There are two types of oblique triangles: acute and obtuse. In an **acute triangle** [Fig. 9-30(a)], all three angles are acute. In an **obtuse triangle** [Fig. 9-30(b)], one angle is obtuse. In each figure the angles have been labeled A, B, and C, and the sides opposite them have been labeled a, b, and c, respectively.

Figure 9-30

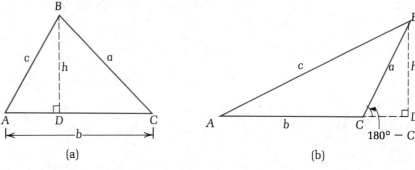

(a)   (b)

The height h has been drawn as a dashed perpendicular line from B to some point D. For each triangle, h can be expressed in two ways. From Fig. 9-30(a), we have

$h = c \sin A$ or $h = a \sin C$

From Fig. 9-30(b), we have

$h = c \sin A$ or $h = a \sin (180° - C) = a \sin C *$

The expressions for h are the same for both triangles. Equating the two expressions for h, we have

$a \sin C = c \sin A$

Dividing both sides by $\sin A \cdot \sin C$, we obtain

$$\frac{a}{\sin A} = \frac{c}{\sin C}$$

*Because C is an obtuse angle, $(180° - C)$ is its reference angle. The sine is positive for both angles, and therefore $\sin (180° - C) = \sin C$.

If we had drawn h from A to its opposite side a, a similar derivation would show that

$$\frac{b}{\sin B} = \frac{c}{\sin C}$$

Law of sines   Combining the two results gives us the **law of sines:**

Law of Sines	$\dfrac{a}{\sin A} = \dfrac{b}{\sin B} = \dfrac{c}{\sin C}$	**(9-4)**

In words, this means that *the side lengths of a triangle are proportional to the sines of the opposite angles.* The law of sines allows us to solve any triangle in which either Case 1 or Case 2 is true:

Case 1.   Two angles and a side are known:

Case 2.   Two sides and the angle opposite one of them are known:

Case 1   Example 22 illustrates Case 1, in which two angles and a side of the triangle are known.

EXAMPLE 22   Given $A = 47°$, $B = 38°$, and $a = 8.0$, find C, b, and c.

Solution
Step 1.   Find C. Since the sum of the angles of any triangle is 180°,

$$C = 180° - 47° - 38° = 95°$$

Step 2.   Apply the law of sines to find b. Given A and a, we use the following form of the law:

This ratio contains the unknown $\Longrightarrow$ $\dfrac{b}{\sin B} = \dfrac{a}{\sin A}$ $\Longleftarrow$ Both terms of this ratio are known

Substituting the given values,

$$\frac{b}{\sin 38°} = \frac{8.0}{\sin 47°}$$

$$b = \frac{8.0(\sin 38°)}{\sin 47°}$$

8 ⊠ 38 (sin) (÷) 47 (sin) (=) →   *6.7344867* $b \approx 6.7$

Step 3. Find c. Using A and a once again, we have

$$\frac{c}{\sin 95°} = \frac{8.0}{\sin 47°}$$

$$c = \frac{8.0(\sin 95°)}{\sin 47°}$$

8 ⊠ 95 (sin) (÷) 47 (sin) (=) →   *10.896995* $c \approx 11$

Case 2   When two sides and the angle opposite one of them are known in an oblique triangle, there may be one solution or two solutions to the triangle. For this reason, Case 2 is often called the *ambiguous case*. The next example illustrates the possibilities.

EXAMPLE 23   Given $A = 30.0°$ and $b = 10.0$ solve the triangle in which:
(a) $a = 12.0$ ▦   (b) $a = 8.00$

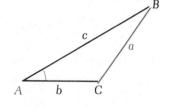

Solution   In each case we know A, a, and b. First, make a quick sketch (see the figure) and notice that the given angle (A) is opposite one of the given sides (a). We must therefore solve first for B using

$$\frac{b}{\sin B} = \frac{a}{\sin A}$$

$$\sin B = \frac{b \sin A}{a}$$

Then we can find C and c.

(a) $\sin B = \dfrac{10.0 \sin 30.0°}{12.0} \approx 0.417$

Solving for B, we obtain a reference angle of $24.6°$. But sine is positive in quadrants I and II, so this leads to two possibilities:

$$B_1 \approx 24.6° \quad \text{or} \quad B_2 \approx 180° - 24.6° \approx 155.4°$$

For the acute value B_1,

$$C_1 \approx 180° - 30.0° - 24.6° \approx 125.4°$$

$$\frac{c_1}{\sin 125.4°} = \frac{12.0}{\sin 30.0°}$$

$$c_1 \approx 19.6$$

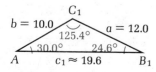

Figure 9-31

This triangle is illustrated in Fig. 9-31.

For the obtuse value B_2,

$$C_2 \approx 180° - 30.0° - 155.4° \approx -5.4°$$

This negative value means that B_2 does not result in a triangle. Figure 9-32 shows that an arc of length 12.0 from C strikes side c at only one point. The given values result in only one solution: $B \approx 24.6°$, $C \approx 125.4°$, $c \approx 19.6$.

Figure 9-32

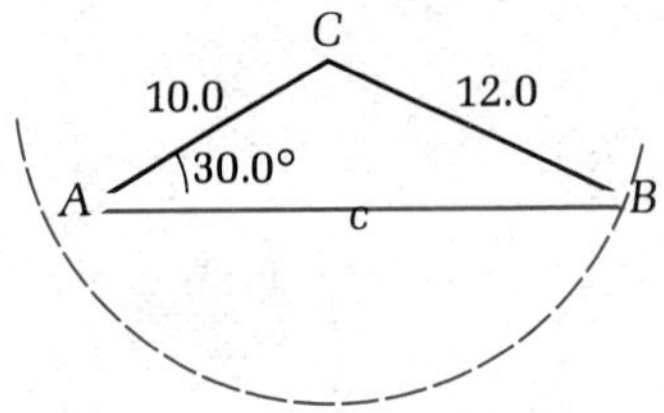

(b) $\sin B = \dfrac{10.0 \sin 30.0°}{8.00}$ $10 \boxed{\times} 30 \boxed{\sin} \boxed{\div} 8 \boxed{=} \rightarrow$ $\boxed{0.625}$

$= 0.625$

Solving for B, we again have two possibilities:

$\boxed{\text{INV}}\boxed{\sin}\boxed{\text{STO}} \rightarrow$ $\boxed{38.682187}$

$B_1 \approx 38.7°$ or $B_2 \approx 180° - 38.7° \approx 141.3°$

For the acute value of angle B,

$C_1 \approx 180° - 30.0° - 38.7°$ $180 \boxed{-} 30 \boxed{-} \boxed{\text{RCL}} \boxed{=} \rightarrow$ $\boxed{111.31781}$

$\approx 111.3°$

$\dfrac{c_1}{\sin 111.3°} = \dfrac{8.00}{\sin 30.0°}$ $\boxed{\sin} \boxed{\times} 8 \boxed{\div} 30 \boxed{\sin} \boxed{=} \rightarrow$ $\boxed{14.905252}$

$c_1 \approx 14.9$

This triangle is illustrated in Fig. 9-33.

Figure 9-33 $b = 10.0$ C_1 $111.3°$ $a = 8.00$

$30.0°$ $38.7°$

A $c_1 \approx 14.9$ B_1

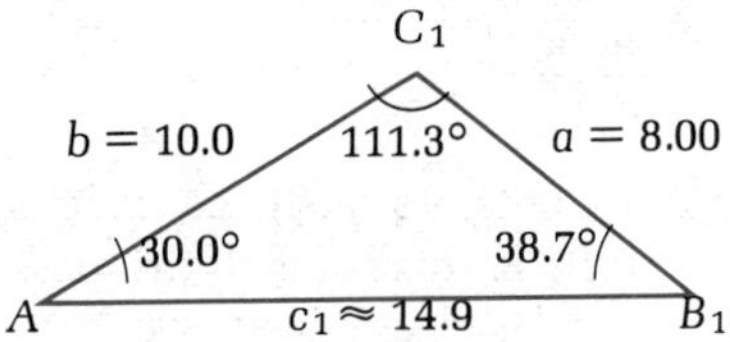

For the obtuse value of angle B,

$C_2 \approx 180° - 30.0° - 141.3°$ $180 \boxed{-} 30 \boxed{-} \boxed{(} 180 \boxed{-} \boxed{\text{RCL}} \boxed{)} \boxed{=}$

$\approx 8.7°$ $\rightarrow$ $\boxed{8.6821875}$

$\dfrac{c_2}{\sin 8.7°} = \dfrac{8.00}{\sin 30.0°}$ $\boxed{\sin} \boxed{\times} 8 \boxed{\div} 30 \boxed{\sin} \boxed{=} \rightarrow$ $\boxed{2.4152560}$

$c_2 \approx 2.42$

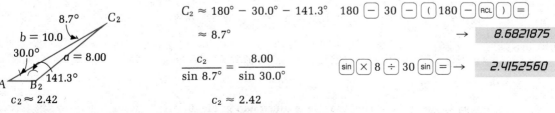

$8.7°$ C_2
$b = 10.0$
$30.0°$ $a = 8.00$
A $141.3°$
B_2
$c_2 \approx 2.42$

Figure 9-34

This triangle is illustrated in Fig. 9-34.

As shown in Fig. 9-35, the given values resulted in two solutions because an arc of length 8.00 from C strikes side c at two points, B_1 and B_2.

Figure 9-35

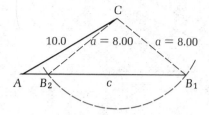

Summary   Example 23 shows that we can determine the number of solutions for Case 2 by working through the problem. However, we can also make this determination beforehand by drawing the following conclusions from the given information. Given sides a and b and angle A opposite a, an oblique triangle will have:

- One solution if $a > b$ or $A > 90°$ Example 23(a)
- Two solutions if $a > b \sin A$ and $a < b$ Example 23(b)

NOTE ▶   If $a = b \sin A$, the result is a *right* triangle, *not* an oblique triangle (Fig. 9-36), and the right triangle may be solved using the methods of Section 4-4. If $a < b \sin A$, then, according to the law of sines, $\sin B$ would have to be greater than 1, which is impossible. As shown in Figure 9-37, no triangle exists for $a < b \sin A$. ◀

Figure 9-36

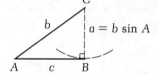

Figure 9-37

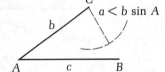

Applications   The law of sines is useful in many applications. For example, airplane pilots use it in solving "wind triangles" to determine the proper heading of an airplane.

EXAMPLE 24   An airplane pilot wants to fly from point A to point B, 30.0° north of east from A. If the wind is blowing from the west at 45.0 mi/h, and the plane flies at 375 mi/h with respect to the air, what should the pilot's heading be?

Solution   First, draw a diagram, as shown in Fig. 9-38. Since the wind will carry the plane east, the pilot must aim for point C, west of B. We are given sides a and b, the velocity vectors of the wind and the plane. Angles ABC and BAD are alternate interior angles. Since $\angle BAD$ is given as 30.0°, we know that $\angle ABC = 30.0°$ also. Now we can apply the law of sines. Notice that this is an application of Case 2, and that because $375 > 45$, it will have only one solution. The problem asks for $\angle CAD$, which means we must first find $\angle CAB$ as follows:

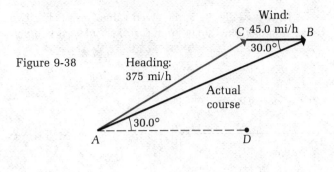

Figure 9-38

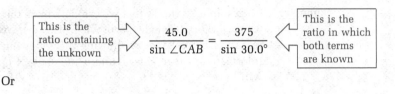

Or

$$\sin \angle CAB = \frac{45.0 \sin 30.0°}{375}$$

$$\angle CAB \approx 3.4° \qquad 45 \boxed{\times} 30 \boxed{\sin} \boxed{\div} 375 \boxed{=} \boxed{\text{INV}} \boxed{\sin} \rightarrow \boxed{3.4398128}$$

The pilot's heading should be 30.0° + 3.4° = 33.4° north of east.

Exercises 9-4

Solve each triangle.

1. $a = 6.5$, $A = 43°$, $B = 62°$ **2.** $a = 17.2$, $A = 71.0°$, $C = 44.0°$

3. $b = 165$, $B = 31.0°$, $C = 110.0°$ **4.** $b = 2300$, $B = 120°$, $A = 35°$

5. $a = 96.0$, $b = 58.0$, $B = 30.0°$ **6.** $a = 4.7$, $b = 2.8$, $A = 65°$

7. $b = 17.5$, $c = 7.50$, $B = 112.0°$ **8.** $b = 265$, $c = 172$, $C = 27.0°$

9. $a = 73.6$, $B = 27.0°$, $C = 41.0°$ **10.** $b = 440$, $A = 48°$, $C = 38°$

11. $c = 7.80$, $C = 105.6°$, $A = 46.5°$ **12.** $c = 0.450$, $C = 29.1°$, $B = 98.7°$

13. $a = 521$, $c = 512$, $C = 77.4°$ **14.** $a = 187$, $c = 275$, $A = 30.0°$

15. $a = 3.80$, $b = 9.70$, $B = 38.8°$ **16.** $a = 18$, $b = 25$, $A = 45°$

17. $c = 0.0650$, $A = 36°50'$, $B = 78°40'$ **18.** $a = 4.280$, $A = 125°15'$, $C = 22°35'$

19. $b = 5.80$, $c = 7.50$, $B = 44°40'$ **20.** $b = 592.0$, $c = 758.0$, $C = 107°25'$

Solve.

21. *Navigation* An airplane pilot wants to fly from a starting point to a second point 27° west of south from the start. If the wind is blowing from the west at 65 mi/h, and the plane flies at a speed of 450 mi/h with respect to the air, what heading should the pilot take?

22. *Navigation* On the first leg of a cruise, a boat must travel to an island 41° south of east. If there is a 4.0 knot current from the south, and the boat travels at 9.5 knots with respect to the water, what heading should it take?

23. *Navigation*   A boat sailing at 9.0 knots with respect to the water takes a heading of 16° south of east. A swell from the east alters the actual heading to 20° south of east. What is the speed of the current? What is the resultant speed of the boat?

24. *Navigation*   A jet plane flying at a speed of 580 mi/h with respect to the air heads 41°0′ north of west. Because the wind is blowing from the east, the actual heading is 37°0′ north of west. Find the speed of the wind and the resultant speed of the plane.

25. *Forestry*   In measuring the height of a tree, a forest ranger measures the angle of elevation of the treetop as 46°. The ranger then moves 15 ft toward the tree to a second point and finds the angle of elevation to be 59°. (See the figure.) How tall is the tree?

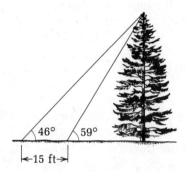

26. *Navigation*   An object on the ground is sighted from an airplane at an angle of depression of 56°. After flying another 2.0 km, the plane remains on the same side of the object, and the angle of depression is 86°. (See the figure.) At that point, what is the distance from the plane to the object?

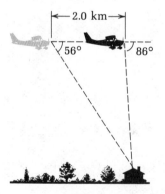

27. *Civil engineering*   Engineers cementing the surface of a creek have made the measurements shown on the cross section in the figure. What is the length of slope *AB*?

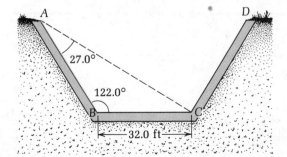

28. *Navigation*   An airplane leaves an airport at A and flies due west to point B, 35 mi from A. The plane then turns 25° north of west and flies directly to a second airport at point C. If A and C are 48 mi apart, how far apart are points B and C?

9-5 | The Law of Cosines

The law of sines can be used to solve oblique triangles for which (1) two angles and a side are known, or (2) two sides and the angle opposite one of the sides are known. However, there are two other cases for which an oblique triangle must sometimes be solved:

Case 3.   When two sides and the angle formed by them are known.
Case 4.   When three sides are known.

Law of cosines

These cases can be solved using a series of three formulas known as the *law of cosines*. To derive this law, consider the acute triangle ABC shown in Fig. 9-39. The height h divides the triangle into two right triangles and splits AC into lengths x and $b - x$, as shown.

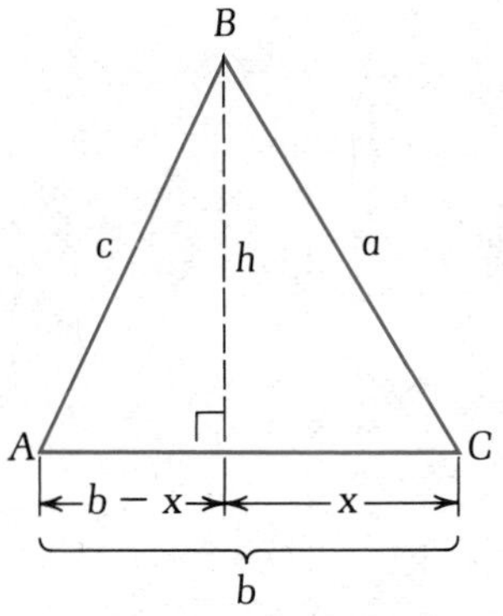

Applying the Pythagorean theorem to each right triangle, we get

$$c^2 = h^2 + (b - x)^2 \qquad \text{and} \qquad a^2 = h^2 + x^2$$

Then, solving each equation for h^2, we obtain

$$h^2 = c^2 - b^2 + 2bx - x^2 \qquad \text{and} \qquad h^2 = a^2 - x^2$$

Equating the two expressions for h^2 gives

$$c^2 - b^2 + 2bx - x^2 = a^2 - x^2$$

And solving this equation for c^2, we get

$$c^2 = a^2 + b^2 - 2bx$$

Now, since

$$\frac{x}{a} = \cos C \qquad \text{then} \qquad x = a \cos C$$

and substituting this into the above equation gives us the **law of cosines:**

Figure 9-39

| **Law of Cosines** | $c^2 = a^2 + b^2 - 2ab \cos C$ | **(9-5)** |

The law of cosines enables us to find the third side c of any triangle given the other two sides a and b and the angle C between them. It can also be used to find an angle C of any triangle if all three sides are known.

Since our original naming of the sides was arbitrary, the law of cosines can also be stated in two other ways:

$$a^2 = b^2 + c^2 - 2bc \cos A \qquad b^2 = a^2 + c^2 - 2ac \cos B$$

Although we derived the law of cosines only for an acute triangle, it can be found in a similar way for an obtuse triangle. This derivation is left for the student.

Case 3   In solving Case 3 problems, first use the law of cosines to determine the missing side and one of the missing angles. The final angle can then be found by subtracting.

EXAMPLE 25   Solve the triangle with $a = 22.8$, $b = 15.2$, and $C = 42.0°$.

Solution   First, make a sketch (Fig. 9-40) and notice that the given angle C is the angle formed by the given sides a and b. Then proceed as follows:

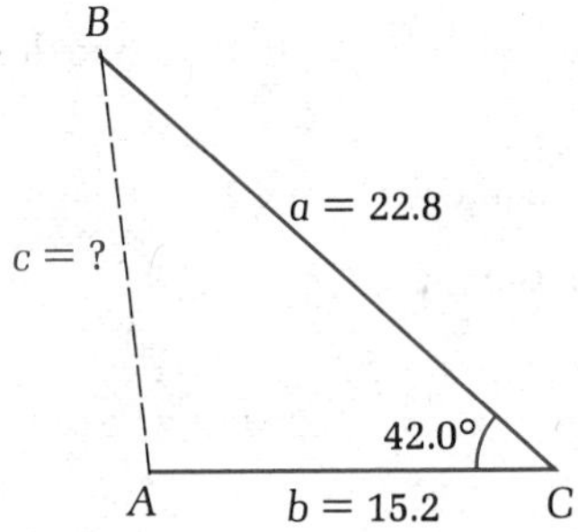

Figure 9-40

Step 1.   Use the law of cosines to find the third side, c:

$$c^2 = (22.8)^2 + (15.2)^2 - 2(22.8)(15.2) \cos 42.0°$$

$$c = \sqrt{(22.8)^2 + (15.2)^2 - 2(22.8)(15.2) \cos 42.0°}$$

22.8 $\boxed{x^2}$ $\boxed{+}$ 15.2 $\boxed{x^2}$ $\boxed{-}$ 2 $\boxed{\times}$ 22.8 $\boxed{\times}$ 15.2 $\boxed{\times}$ 42 $\boxed{\cos}$ $\boxed{=}$ $\boxed{\sqrt{\ }}$ $\boxed{\text{STO}}$ → *15.355503*

$$c \approx 15.4$$

Step 2.   We now have all three sides and one angle, so we use the law of cosines again to find one of the missing angles. Since

$$a^2 = b^2 + c^2 - 2bc \cos A$$

we can solve for $\cos A$:

$$\cos A = \frac{b^2 + c^2 - a^2}{2bc} = \frac{(15.2)^2 + c^2 - (22.8)^2}{2(15.2)c}$$

To find A, evaluate this expression and then press the $\boxed{\text{INV}}\boxed{\text{cos}}$ keys. Use the unrounded value of c that was stored in memory in Step 1:

$15.2\;\boxed{x^2}\;\boxed{+}\;\boxed{\text{RCL}}\;\boxed{x^2}\;\boxed{-}\;22.8\;\boxed{x^2}\;\boxed{=}\;\boxed{\div}\;2\;\boxed{\div}\;15.2\;\boxed{\div}\;\boxed{\text{RCL}}\;\boxed{=}\;\boxed{\text{INV}}\;\boxed{\text{cos}}\;\rightarrow$ *96.520317*

$A \approx 96.5°$

Step 3. Since all three angles must sum to 180°, find the missing angle by subtracting:

$B \approx 180° - 96.5° - 42.0° \approx 41.5°$

Case 4   In solving Case 4 problems, first use the law of cosines to find one missing angle. Then use the law of sines to determine a second angle. Again, the third angle can be calculated by subtracting.

NOTE ▶   To avoid having to deal with ambiguous cases, always find the largest angle first. The largest angle is always opposite the largest side. ◀

EXAMPLE 26   Solve the triangle with $a = 34.6$, $b = 21.7$, and $c = 39.1$

Solution
Step 1.   Use the law of cosines to find the *largest* missing angle C (see Fig. 9-41). Since

$$c^2 = a^2 + b^2 - 2ab \cos C$$

we can solve for C as follows:

$$\cos C = \frac{a^2 + b^2 - c^2}{2ab} = \frac{(34.6)^2 + (21.7)^2 - (39.1)^2}{2(34.6)(21.7)}$$

$$C \approx 84.7°$$

Step 2.   Use the law of sines to find one of the other unknown angles:

$$\frac{34.6}{\sin A} = \frac{39.1}{\sin 84.7°}$$

$$A \approx 61.8°$$

Step 3.   Subtract to find the third angle:

$$B \approx 180° - 61.8° - 84.7° \approx 33.5°$$

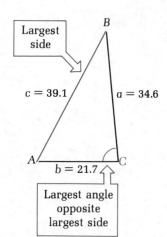

Figure 9-41

Applications   There are many applied situations in which the law of cosines is useful.

EXAMPLE 27   A cargo ship heads due west from port for 8.0 mi. It then turns 24° toward the southwest and travels another 22.0 mi. At this point, how far is the ship from port?

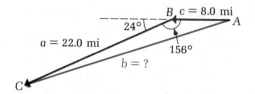

Figure 9-42

Solution   First, draw a diagram, as shown in Fig. 9-42. The displacement vectors form a triangle in which sides a and c are known. As shown in Fig. 9-42, we can also calculate angle B as $180° - 24° = 156°$.

Then, apply the law of cosines:

$$b = \sqrt{(22.0)^2 + (8.0)^2 - 2(22.0)(8.0)\cos 156°}$$

$$b \approx 29 \text{ mi}$$

Exercises 9-5

Solve each triangle.

1. $a = 8.5$, $b = 6.2$, $C = 41°$
2. $a = 19.3$, $b = 28.7$, $C = 57.0°$
3. $a = 625$, $c = 189$, $B = 102.0°$
4. $b = 1150$, $c = 3110$, $A = 125.0°$
5. $a = 27.2$, $b = 33.4$, $c = 44.6$
6. $a = 4.8$, $b = 1.6$, $c = 4.2$
7. $a = 2420$, $b = 1970$, $c = 4130$
8. $a = 571$, $b = 724$, $c = 273$
9. $a = 77$, $c = 64$, $B = 66°$
10. $b = 812$, $c = 654$, $A = 39.3°$
11. $a = 8150$, $b = 3250$, $C = 27.8°$
12. $a = 0.85$, $c = 0.71$, $B = 71°$
13. $a = 1.25$, $b = 5.08$, $c = 3.96$
14. $a = 6.95$, $b = 9.33$, $c = 7.24$
15. $a = 66.2$, $b = 58.4$, $c = 27.1$
16. $a = 46.5$, $b = 31.3$, $c = 29.4$
17. $b = 0.385$, $c = 0.612$, $A = 118°30'$
18. $a = 7.42$, $c = 5.96$, $B = 99°40'$
19. $a = 137$, $b = 419$, $c = 303$
20. $a = 517$, $b = 701$, $c = 476$

21. *Navigation*   Starting from port, a ship heads due north for 6.5 mi. It then heads 31° east of north and travels another 2.5 mi. At this point, how far from port is the ship?

22. *Machine technology*   A machinist makes a cut 13.8 cm long in a piece of metal. Then, another cut is made that is 18.6 cm long and at an angle of 62°0' with the first cut. How long a cut must be made to join the two endpoints?

23. *Civil engineering*   The figure shows a triangular roadway divider constructed where one lane of a street merges with another. Given the dimensions shown, find angle A.

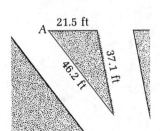

24. *Drafting* A side view of a microcomputer is shown in the figure. Find the size of angle x.

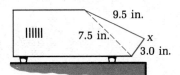

25. The lot shown in the figure is split along a diagonal, as indicated. What length of fencing is needed for the boundary line?

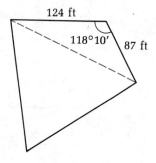

26. *Construction* Two sides of a sloped ceiling meet at an angle of 105.5°. If the distances along the sides to the opposite walls are 11.0 ft and 13.0 ft, what length of beam is needed to join the walls?

27. *Navigation* A boat moving with a speed of 7.0 knots with respect to the water heads due northeast. If a swell develops with a velocity of 1.5 knots from 22° west of north, what will be the actual heading and speed of the boat?

28. *Navigation* An airplane traveling at a speed of 860 km/h with respect to the air is headed 28° west of south. If the wind is blowing from the east at 75 km/h what is the actual velocity of the plane?

9-6 | Review of Chapter 9

Important Terms and Concepts

vector (p. 306)

scalar (p. 306)

initial point (p. 307)

terminal point (p. 307)

resultant (p. 308)

parallelogram method for adding vectors (p. 308)

components of a vector (p. 313)

adding vectors using components (p. 316)

displacement (p. 321)

velocity (p. 322)

force (p. 323)

inclined plane (p. 323)

oblique triangle (p. 327)

Formulas

- resolving a vector into its components (p. 313)

horizontal: $A_x = A \cos \theta$

vertical: $A_y = A \sin \theta$ **(9-1)**

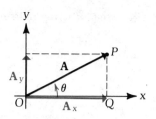

- magnitude A and direction θ of a vector (p. 315): $A = \sqrt{A_x^2 + A_y^2}$ **(9-2)**

$$\tan (\text{Ref } \theta) = \left| \frac{A_y}{A_x} \right| \quad \textbf{(9-3)}$$

- summary of procedure for solving oblique triangles (pp. 327–332)

When Given:		Use:
Case 1.	Two angles and a side.	Law of sines: $\dfrac{a}{\sin A} = \dfrac{b}{\sin B} = \dfrac{c}{\sin C}$ **(9-4)**
Case 2.	Two sides and the angle opposite one of them. (There will be either 1 or 2 solutions.)	
Case 3.	Two sides and the angle formed by them.	Law of cosines: $a^2 = b^2 + c^2 - 2bc \cos A$
Case 4.	Three sides.	$b^2 = a^2 + c^2 - 2ac \cos B$ **(9-5)** $\quad c^2 = a^2 + b^2 - 2ab \cos C$

Exercises 9-6

Resolve each vector into its horizontal and vertical components.

1. $A = 15$ m/s, $\theta = 49°$

2. $B = 1.65$ ft/s², $\theta = 113.0°$

3. $C = 245$ lb, $\theta = 245°30'$

4. $D = 1760$ mi, $\theta = 317°40'$

5.

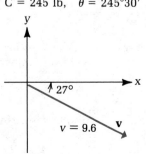

6.

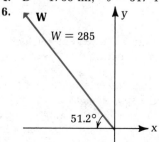

Given the following pairs of components, determine the magnitude and direction of each vector.

7. $E_x = 14.6$ lb, $E_y = 27.9$ lb

8. $F_x = -165$ N, $F_y = 332$ N

9. $G_x = -0.56$ m/s^2, $G_y = -0.32$ m/s^2

10. $H_x = 1420$ ft, $H_y = -1860$ ft

Add each pair of vectors.

11. $L = 75.0$ lb, $\theta_L = 0.0°$
    $M = 49.0$ lb, $\theta_M = 90.0°$

12. $N = 538$ m, $\theta_N = 90.0°$
    $P = 865$ m, $\theta_P = 180.0°$

13. $A = 4.63$ ft/s, $\theta_A = 270.0°$
    $B = 1.72$ ft/s, $\theta_B = 168.0°$

14. $S = 12.5$ cm/s^2, $\theta_S = 26.5°$
    $T = 75.6$ cm/s^2, $\theta_T = 81.8°$

15. $U = 0.950$ in., $\theta_U = 325.5°$
    $V = 0.610$ in., $\theta_V = 246.5°$

16. $W = 375$ mi/h, $\theta_W = 155.4°$
    $Z = 56.5$ mi/h, $\theta_Z = 72.2°$

17. $\mathbf{A} + \mathbf{B}$ in the figure below

18. $\mathbf{C} + 2\mathbf{D} - \mathbf{E}$ in the figure below

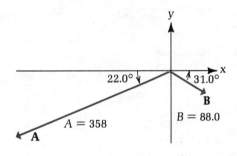

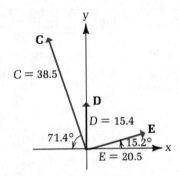

Solve each triangle.

19. $a = 17.9$, $A = 65.0°$, $B = 39.0°$

20. $a = 39.7$, $b = 51.6$, $c = 42.3$

21. $a = 7.9$, $b = 5.1$, $C = 96°$

22. $a = 7.6$, $b = 4.8$, $B = 30°$

23. $a = 166$, $c = 259$, $B = 47.0°$

24. $b = 87.5$, $c = 23.4$, $A = 118°30'$

25. $b = 65.8$, $B = 41.0°$, $C = 32.0°$

26. $b = 65.7$, $c = 89.2$, $B = 25.4°$

27. $a = 721$, $b = 444$, $c = 293$

28. $b = 4.78$, $A = 125.2°$, $B = 27.0°$

29. $a = 1160$, $c = 2470$, $C = 116.2°$

30. $a = 6.58$, $b = 9.32$, $c = 4.15$

31. $a = 260$, $c = 340$, $A = 37°$

32. $a = 0.352$, $b = 0.478$, $C = 63°20'$

33. $a = 1450$, $b = 3450$, $c = 3850$

34. $c = 0.62$, $C = 104°$, $A = 39°$

Solve.

35. *Navigation*   A motor boat crosses a river at a velocity of 8.0 mi/h with respect to the water, heading perpendicular to the stream. If a current of 2.5 mi/h is moving down the river, what is the velocity of the boat?

36. *Navigation*   An airplane takes off due north and flies for 16 km before turning due east and flying another 24 km. What is its displacement from the takeoff point?

37. *Physics*   The crate shown in the figure is being pulled by two forces. What is the resultant force on the crate? State the direction with respect to the dashed line perpendicular to the crate.

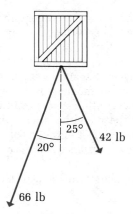

38. *Physics*   A projectile is fired with a velocity of 850 m/s at an angle of 70° with respect to the horizontal. Assuming its horizontal velocity is constant for the first 5.0 s, what horizontal distance does it travel?

39. *Physics*   If an object weighing 16 lb is at rest on a board inclined at an angle of 16°, what must be the static friction force exerted by the surface on the object?

40. *Machine technology*   A machinist makes the following cuts in a piece of metal: 6.20 cm at 12.0°, 4.80 cm at 31.0°, and 2.70 cm at 46.0°. All angles are measured with respect to horizontal. What is the total displacement of the cuts?

41. *Navigation*   An airplane flies at a velocity of 640 mi/h, 42° north of west with respect to the air. If a wind blows at 72 mi/h from 27° east of south, what is the resultant velocity of the plane?

42. *Physics*   The figure shows a 58.0 lb weight hung from several wires. Find the tension in each wire.

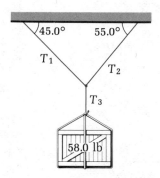

43. *Navigation*   While a plane is flying at a velocity of 880 ft/s climbing at a direction 14° above horizontal, an object is dropped from the plane. Assuming air resistance can be neglected, what is the velocity of the object after 3.5 s? (Assume that the velocity due to gravity is 32t ft/s. See Example 18.)

44. *Navigation*   An airplane pilot wants to fly from point A to a second point B, 38° north of west from A. If the wind is blowing from the west at 55 km/h, and the plane is flying at 520 km/h with respect to the air, what heading should the pilot take?

45. *Navigation*   A boat moving at 14.0 knots with respect to the water takes a heading 21.0° west of north. A current from the north causes the actual course to be 23.5° west of north. Find the velocity of the current and the resultant velocity of the boat.

46. *Construction*   To accommodate a stained glass panel, a window is built in the shape of an isoceles triangle. If the base has a length of 14 in. and the two equal sides are each 18 in. long, what is the angle at the top of the window?

47. *Carpentry*   Two walls meeting at an angle of 95° form the sides of a triangular corner cupboard. If the sides of the cupboard along each wall measure 28 in. and 32 in., what is the length of the front of the cupboard?

48. *Navigation*   While a plane is flying at an altitude of 6500 ft and rising at an angle of 12° with respect to horizontal, an object is dropped from the plane. If the plane travels 18,000 ft by the time the object hits the ground, what is the distance of the plane from the object at the moment of impact?

49. *Navigation*   A boat travels 31° south of west with a speed of 11 knots with respect to still water. If the current has a velocity of 3.5 knots from the north, find the actual speed and heading of the boat.

50. *Navigation*   Two ground observers 1.5 mi apart spot a plane at angles of elevation of 51° and 72°, respectively. If both observers are on the same side of the plane, find the distance of the second observer from the plane.

51. *Drafting*   Determine the center-to-center measurement x in the adjustment bracket shown in the figure.*

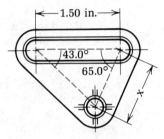

52. *Drafting*   Determine the center-to-center distance x in the layout illustrated in the figure.

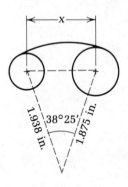

Graphs of the Trigonometric Functions

10

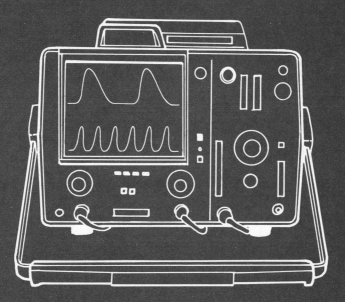

OSCILLOSCOPE

Graphs of the trigonometric functions are useful in two ways. First, they help us understand the functions by showing how function values vary in relation to the size of the angle. The graph is a picture of the function. Second, the graphs have direct applications in technical areas such as electronics and physics.

It is especially important that you understand Chapters 3 (Functions and Graphs) and 8 (Trigonometric Functions of Any Angle; Radian Measure) before you study the material of this chapter.

When plotting the graphs of the trigonometric functions in the rectangular coordinate system, it is conventional to express the value of the angle in radians.

Sine function

For example, to graph the function

$$y = \sin x$$

first select several values of the independent variable x in radians. In order to get a symmetric arrangement of graph points, choose values that are multiples of $\frac{\pi}{6}$, or 30°. Then determine the corresponding values of y. Use either your knowledge of the special angle values or a calculator to complete the table as shown.

x	$-\dfrac{\pi}{6}$	0	$\dfrac{\pi}{6}$	$\dfrac{\pi}{3}$	$\dfrac{\pi}{2}$	$\dfrac{2\pi}{3}$	$\dfrac{5\pi}{6}$	π
y = sin x	−0.5	0	0.5	0.87	1	0.87	0.5	0

x	$\dfrac{7\pi}{6}$	$\dfrac{4\pi}{3}$	$\dfrac{3\pi}{2}$	$\dfrac{5\pi}{3}$	$\dfrac{11\pi}{6}$	2π	$\dfrac{13\pi}{6}$
y = sin x	−0.5	−0.87	−1	−0.87	−0.5	0	0.5

Plotting the ordered pairs (x, y), we can sketch the graph of y = sin x as shown in Fig. 10-1(a). As indicated in Fig. 10-1(b), the graph of y = sin x continues indefinitely in both directions. Notice that the function values repeat

Figure 10-1

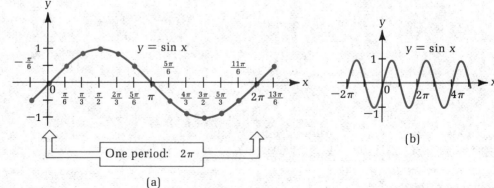

One period: 2π

(a)

themselves. The value of sin x is exactly equal to the value of sin $(x + 2\pi)$.

Periodic function   Because of this, the sine function is a **periodic function,** with a period of 2π. The

Cycle   graph of the function over the interval of one period is called a **cycle.**

Characteristics of the   The following characteristics of the graph of the sine function will help you

sine function   plot variations of the sine function quickly, without the need for detailed tables
of values:

- For one complete cycle, starting at x = 0, the graph crosses the x-axis at the beginning of the cycle and at its midpoint. These are the **zeros** of the function, as noted in Fig. 10-2(a).
- A **maximum** or **minimum** value occurs midway between each consecutive pair of zeros, as noted in Fig. 10-2(b).

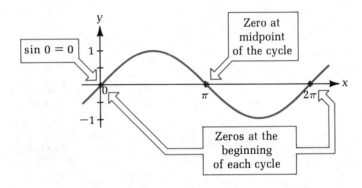

(a)

Figure 10-2

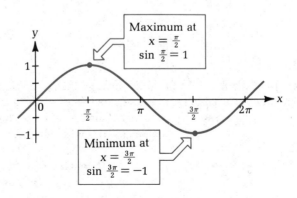

(b)

Cosine function   To see how the graph of the cosine function compares to that of the sine function, we first prepare a table of values.

x	$-\dfrac{\pi}{6}$	0	$\dfrac{\pi}{6}$	$\dfrac{\pi}{3}$	$\dfrac{\pi}{2}$	$\dfrac{2\pi}{3}$	$\dfrac{5\pi}{6}$	π
$y = \cos x$	0.87	1	0.87	0.5	0	-0.5	-0.87	-1

x	$\dfrac{7\pi}{6}$	$\dfrac{4\pi}{3}$	$\dfrac{3\pi}{2}$	$\dfrac{5\pi}{3}$	$\dfrac{11\pi}{6}$	2π	$\dfrac{13\pi}{6}$
$y = \cos x$	-0.87	-0.5	0	0.5	0.87	1	0.87

The graph of $y = \cos x$ is shown in Fig. 10-3.

Figure 10-3

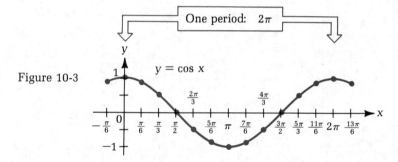

Characteristics of the cosine function   Like the sine function, the cosine function is periodic, and it has the same period (2π) as the sine function. The graph of the cosine function also has the same shape as that of the sine function, but it is displaced $\frac{\pi}{2}$ units to the left. The following characteristics will help you graph variations of the cosine function:

- Starting at $x = 0$, a **maximum** value of the cosine function occurs at the beginning of each cycle, and a **minimum** value occurs halfway through the cycle, as shown in Fig. 10-4(a).
- The graph crosses the x-axis midway between each maximum and minimum. These are the **zeros** of the function, as noted in Fig. 10-4(b).

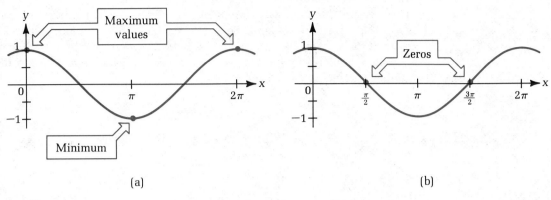

(a)   (b)

Figure 10-4

Now we shall examine what happens to the graph when we multiply the sine or cosine function by a constant.

▦ **EXAMPLE 1**   Graph $y = 4 \sin x$.

Solution   **First,** we prepare a table of values using the following calculator key sequence:

$4 \sin\left(-\dfrac{\pi}{6}\right)$: $\boxed{\text{RAD}}\,\boxed{\pi}\,\boxed{\div}\,6\,\boxed{=}\,\boxed{+/-}\,\boxed{\sin}\,\boxed{\times}\,4\,\boxed{=} \rightarrow$ *−2.*

$4 \sin\dfrac{\pi}{3}$: $\boxed{\pi}\,\boxed{\div}\,3\,\boxed{=}\,\boxed{\sin}\,\boxed{\times}\,4\,\boxed{=} \rightarrow$ *3.4641016*

Again, it is easiest if we choose multiples of $\frac{\pi}{6}$ for values of x.

x	$-\dfrac{\pi}{6}$	0	$\dfrac{\pi}{6}$	$\dfrac{\pi}{3}$	$\dfrac{\pi}{2}$	$\dfrac{2\pi}{3}$	$\dfrac{5\pi}{6}$	π
y	−2	0	2	3.5	4	3.5	2	0

x	$\dfrac{7\pi}{6}$	$\dfrac{4\pi}{3}$	$\dfrac{3\pi}{2}$	$\dfrac{5\pi}{3}$	$\dfrac{11\pi}{6}$	2π	$\dfrac{13\pi}{6}$
y	−2	−3.5	−4	−3.5	−2	0	2

Then, we plot the graph, as shown in Fig. 10-5. This curve has the same general shape as the graph of $y = \sin x$. The zeros, maximum, and minimum points all occur at the same values of x as for $y = \sin x$. However, since every value of $y = \sin x$ has been multiplied by 4, the maximum value is now 4 and the minimum value is -4.

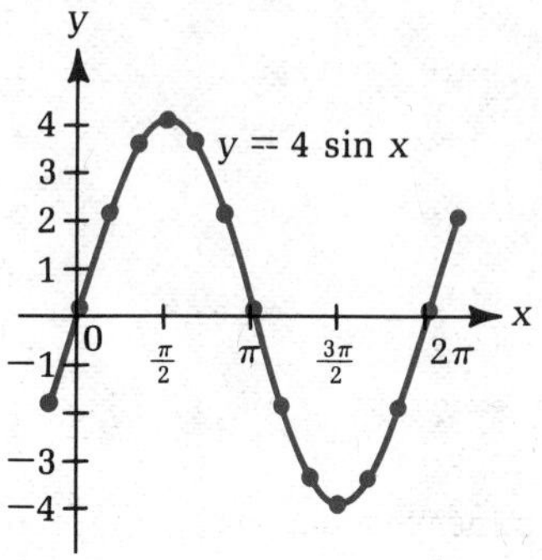

Figure 10-5

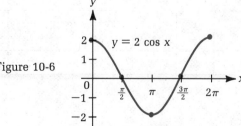

Amplitude   In general, the graphs of $y = a \sin x$ and $y = a \cos x$ will have a maximum value $|a|$ and a minimum value $-|a|$. The constant $|a|$ is called the **amplitude** of the curve. The graphs of $y = a \sin x$ and $y = a \cos x$ will lie entirely between $|a|$ and $-|a|$.

EXAMPLE 2   Graph $y = 2 \cos x$ over the interval $0 \le x \le 2\pi$.

Solution   The amplitude of this curve is $|a| = 2$. The curve is identical to that of $y = \cos x$, except that the maximum value is 2 and the minimum value is -2 (Fig. 10-6). Notice that the maximum value still occurs at the beginning of each cycle, at $x = 0$ and 2π. The minimum value occurs halfway through the cycle, at $x = \pi$. The zeros occur halfway between the maximum and minimum points, at $x = \frac{\pi}{2}$ and $\frac{3\pi}{2}$.

Figure 10-6

Example 3 shows the graph of a cosine function when a is negative.

EXAMPLE 3   Graph $y = -2 \cos x$ over the interval $0 \le x \le 2\pi$.

Solution   The amplitude $|a| = |-2| = 2$, which means the graph should have a maximum value of 2 and a minimum value of -2. The following table of five key points shows this to be true:

x	0	$\dfrac{\pi}{2}$	π	$\dfrac{3\pi}{2}$	2π
$y = -2 \cos x$	-2	0	2	0	-2

If we compare the graph of $y = 2 \cos x$ in Fig. 10-6 to the graph of $y = -2 \cos x$ in Fig. 10-7, we see that a negative value of a has the effect of inverting the curve. That is, the x values of the maximum and minimum points simply interchange while the positions of the zero points remain the same. We can now sketch the graph of $y = a \sin x$ or $y = a \cos x$ without preparing a table of values.

Figure 10-7

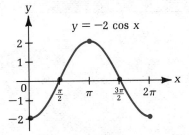

EXAMPLE 4   Graph $y = -1.6 \sin x$ for $0 \le x \le 2\pi$.

Solution   The amplitude of the function is $|-1.6| = 1.6$, so the maximum value of y is 1.6 and the minimum value is -1.6. The zeros are still at $x = 0$, π, and 2π. Since a is negative, a *minimum* occurs at $\dfrac{\pi}{2}$, while a *maximum* occurs at $\dfrac{3\pi}{2}$. Check this by using your calculator to find y when $x = \dfrac{\pi}{2}$:

$$\boxed{\text{RAD}}\;\boxed{\pi}\;\boxed{\div}\;\boxed{2}\;\boxed{=}\;\boxed{\sin}\;\boxed{\times}\;1.6\;\boxed{+/-}\;\boxed{=}\;\rightarrow \quad \text{-1.6}$$

The final graph is shown in Fig. 10-8.

Figure 10-8

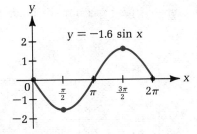

Table 10-1 summarizes the key features of the graphs of $y = a \sin x$ and $y = a \cos x$.

Table 10-1

Key Points	$y = a \sin x$ $a > 0$	$y = a \sin x$ $a < 0$	$y = a \cos x$ $a > 0$	$y = a \cos x$ $a < 0$
Zeros at:	$x = 0, \pi, 2\pi, \ldots$	$x = 0, \pi, 2\pi, \ldots$	$x = \dfrac{\pi}{2}, \dfrac{3\pi}{2}, \ldots$	$x = \dfrac{\pi}{2}, \dfrac{3\pi}{2}, \ldots$
Maximum Values at:	$x = \dfrac{\pi}{2}, \dfrac{5\pi}{2}, \ldots$	$x = \dfrac{3\pi}{2}, \dfrac{7\pi}{2}, \ldots$	$x = 0, 2\pi, \ldots$	$x = \pi, 3\pi, \ldots$
Minimum Values at:	$x = \dfrac{3\pi}{2}, \dfrac{7\pi}{2}, \ldots$	$x = \dfrac{\pi}{2}, \dfrac{5\pi}{2}, \ldots$	$x = \pi, 3\pi, \ldots$	$x = 0, 2\pi, \ldots$

Exercises 10-1

Sketch the graph of each function over the interval $0 \le x \le 2\pi$.

1. $y = 2 \sin x$
2. $y = 3 \sin x$
3. $y = 3 \cos x$
4. $y = 4 \cos x$
5. $y = -3 \sin x$
6. $y = -\sin x$
7. $y = -\cos x$
8. $y = -6 \cos x$
9. $y = \frac{1}{2} \sin x$
10. $y = -\frac{5}{2} \sin x$
11. $y = -\frac{3}{4} \cos x$
12. $y = \frac{7}{2} \cos x$
13. $y = 0.7 \sin x$
14. $y = -5.5 \sin x$
15. $y = 1.8 \cos x$
16. $y = -3.6 \cos x$

10-2 | Graphs of the Sine and Cosine Functions: Period

In Section 10-1 we showed that the sine and cosine functions are periodic functions—that is, their values repeat with a certain fixed period. Functions in the form $y = a \sin x$ and $y = a \cos x$ have a period of 2π. However, not all sine and cosine functions have periods of 2π.

EXAMPLE 5   Graph $y = \sin 2x$.

Solution   **First,** prepare a table of values listing x, 2x, and sin 2x.

x	0	$\dfrac{\pi}{4}$	$\dfrac{\pi}{2}$	$\dfrac{3\pi}{4}$	π	$\dfrac{5\pi}{4}$	$\dfrac{3\pi}{2}$	$\dfrac{7\pi}{4}$	2π
2x	0	$\dfrac{\pi}{2}$	π	$\dfrac{3\pi}{2}$	2π	$\dfrac{5\pi}{2}$	3π	$\dfrac{7\pi}{2}$	4π
$y = \sin 2x$	0	1	0	-1	0	1	0	-1	0

Then, plot the ordered pairs (x, y) and sketch the curve, as shown in Fig. 10-9.

Figure 10-9

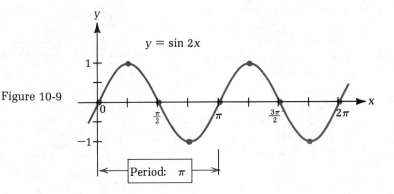

Comparing the graph of $y = \sin 2x$ with Fig. 10-1, we see that it has the same shape, but the function values repeat every π units rather than every 2π units. Thus, $y = \sin 2x$ has a period of π.

Period   In general, a function of the form $y = a \sin bx$ or $y = a \cos bx$ has a **period** equal to $\dfrac{2\pi}{b}$.

The zeros and the maximum and minimum points still occur at the same *relative* positions within each cycle. For example, the zeros of $y = \sin 2x$ occur at the beginning of each cycle $(0, \pi, 2\pi, \ldots)$ and halfway through each cycle $\left(\frac{\pi}{2}, \frac{3\pi}{2}, \ldots\right)$. The maximum and minimum values occur halfway between the zeros $\left(\frac{\pi}{4}, \frac{3\pi}{4}, \frac{5\pi}{4}, \ldots\right)$.

EXAMPLE 6   Sketch the graph of $y = \cos \dfrac{x}{2}$ over one period.

Solution

Step 1. Find the period and amplitude.

$$\text{Period} = \frac{2\pi}{b} = \frac{2\pi}{\frac{1}{2}} = 4\pi$$

$$\text{Amplitude} = 1$$

Step 2. Mark the amplitude and period on a set of coordinate axes, and construct a box that will enclose one cycle, as shown in Fig. 10-10(a).

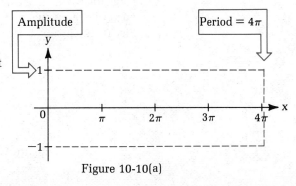

Figure 10-10(a)

Step 3. The maximum is at the beginning of each cycle. The minimum is midway between these points on the bottom edge of the box. The zero points are midway between each maximum and minimum. Plot these five points and sketch dashed lines, as shown in Fig. 10-10(b).

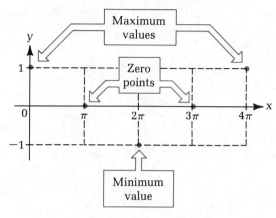

Figure 10-10(b)

Step 4. Sketch the curve, as shown in Fig. 10-11.

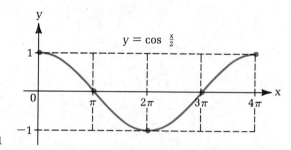

Figure 10-11

EXAMPLE 7   Sketch the graph of $y = 2 \sin 3x$ for one period.

Solution

Step 1. Find the period and amplitude.

$$\text{Period} = \frac{2\pi}{b} = \frac{2\pi}{3}$$
$$\text{Amplitude} = 2$$

Step 2. Draw the rectangle that will enclose one cycle, as shown in Fig. 10-12(a).

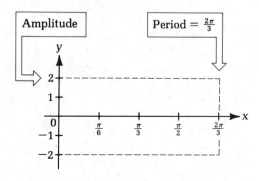

Figure 10-12(a)

Step 3. For a sine function
the zeros are at the beginning
of each cycle $\left(0 \text{ and } \frac{2\pi}{3}\right)$ and
at its midpoint $\left(\frac{\pi}{3}\right)$. Midway
between the first two zeros
$\left(\text{at } \frac{\pi}{6}\right)$ y reaches its maximum
value of 2. Midway between
the last two zeros, y drops to
its minimum value of -2.
Plot these five points, as
shown in Fig. 10-12(b).

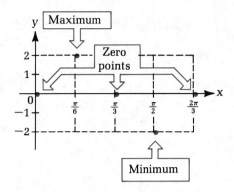

Figure 10-12(b)

Step 4. Sketch the curve (Fig. 10-13).

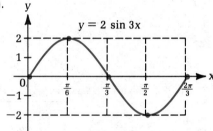

Figure 10-13

EXAMPLE 8 Sketch the graph of $y = -3 \cos \pi x$ for three periods.

Solution
Step 1. The period is

$$\frac{2\pi}{b} = \frac{2\pi}{\pi} = 2$$

The amplitude is 3.

Steps 2 and 3. Draw the rectangle that will enclose one cycle and plot the key
points of the curve, as shown in Fig. 10-14. Since a is negative, the x values of
the maximum and minimum points interchange. The *minimum* is now at the
beginning of each cycle ($x = 0$ and 2) and the *maximum* is midway between
these points (at $x = 1$). The zeros remain midway between each maximum and
minimum ($x = 0.5$ and 1.5).

Step 4. Sketch the curve and repeat it for three cycles (Fig. 10-15).

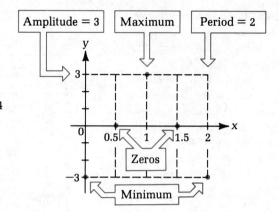

Figure 10-14

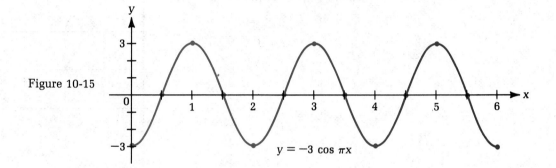

Figure 10-15

$$y = -3 \cos \pi x$$

Table 10-2 summarizes the important features of the graphs of $y = a \sin bx$ and $y = a \cos bx$.

Table 10-2

	$y = a \sin bx$	$y = a \cos bx$
Period, p	$\dfrac{2\pi}{b}$	$\dfrac{2\pi}{b}$
Zeros at:	$x = 0, \dfrac{p}{2}, p, \ldots$	$x = \dfrac{p}{4}, \dfrac{3p}{4}, \ldots$
Maximum/Minimum Values at:	$x = \dfrac{p}{4}, \dfrac{3p}{4}, \ldots$	$x = 0, \dfrac{p}{2}, p, \ldots$

Exercises 10-2

State the period and amplitude for each function, and sketch the graph.

1. $y = \sin 4x$

2. $y = \sin \dfrac{x}{2}$

3. $y = \cos \dfrac{2x}{3}$

4. $y = \cos 3x$

5. $y = 3 \sin 6x$

6. $y = 4 \sin \pi x$

7. $y = 2 \cos 2\pi x$

8. $y = 2 \cos 4x$

9. $y = -\sin 8x$

10. $y = -2 \sin 3\pi x$

11. $y = -3 \cos \dfrac{\pi x}{2}$

12. $y = -6 \cos \dfrac{x}{3}$

13. $y = \dfrac{1}{2} \sin 5x$

14. $y = -\dfrac{7}{4} \sin \dfrac{2\pi x}{3}$

15. $y = -\dfrac{3}{4} \cos 4\pi x$

16. $y = \dfrac{3}{2} \cos \dfrac{x}{2}$

17. $y = -2.8 \sin \dfrac{3x}{4}$

18. $y = 4.5 \sin \dfrac{\pi x}{4}$

19. $y = 0.1 \cos 10\pi x$

20. $y = -3.7 \cos \dfrac{x}{4}$

Solve.

21. *Electronics*   The current i (in amperes) in an ac circuit is given by $i = 4.0 \sin 120\pi t$, where t is the time in seconds. Plot i as a function of t for $0 \le t \le 0.1$ s.

22. *Physics*   The displacement y (in centimeters) of a mass oscillating on a spring is given by $y = 2.5 \cos 2\pi t$, where t is the time in seconds. Plot y as a function of t for $0 \le t \le 1$ s.

23. *Energy technology*   The horizontal position of a bucket on a waterwheel is given by $x = 5 \sin 4\pi t$, where t is the time in minutes. Plot x as a function of t for $0 \le t \le 2$ min.

24. *Physics*   The equation of motion of a floating cylinder as it bobs up and down is

$$y = y_0 \sin t \sqrt{\frac{\rho g A}{m}}$$

where y and y_0 are measured in meters, t is time in seconds, $g = 9.8$ m/s^2, A is the horizontal cross-sectional area in square meters, m is the mass in kilograms, and ρ is the density of water (a constant). Plot y as a function of t for a 75 kg floating buoy with a cross-sectional area of 1.2 m^2, bobbing with an amplitude of $y_0 = 0.5$ m. Use $\rho = 1000$ kg/m^3.

10-3 | Graphs of the Sine and Cosine Functions: Phase Shift

Phase constant   In many technical applications of trigonometric graphs, the angle has a quantity added to it called the **phase constant.** For example, in the function

$$y = a \cos (bx + c)$$

c is the phase constant. Example 9 illustrates its effect on the graph.

EXAMPLE 9   Graph the following function: $y = \sin \left(2x + \dfrac{\pi}{3} \right)$

Solution   **First,** prepare a table of values. Find these with a calculator as follows:

For $x = \dfrac{\pi}{6}$: $\boxed{\text{RAD}}\ 2\ \boxed{\times}\ \boxed{\pi}\ \boxed{\div}\ 6\ \boxed{+}\ \boxed{\pi}\ \boxed{\div}\ 3\ \boxed{=}\ \boxed{\text{sin}}\ \rightarrow$ **0.8660254**

x	$-\dfrac{\pi}{6}$	0	$\dfrac{\pi}{12}$	$\dfrac{\pi}{6}$	$\dfrac{\pi}{3}$	$\dfrac{\pi}{2}$	$\dfrac{7\pi}{12}$	$\dfrac{2\pi}{3}$	$\dfrac{5\pi}{6}$	π
y	0	0.87	1	0.87	0	−0.87	−1	−0.87	0	0.87

Then, plot the points and sketch the curve, as shown in Fig. 10-16.

Figure 10-16

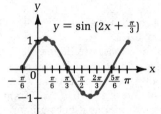

$$y = \sin\left(2x + \tfrac{\pi}{3}\right)$$

A careful look at this graph reveals that it is identical to the graph of $y = \sin 2x$ in Example 5, but it is shifted $\frac{\pi}{6}$ units to the left. For example, a zero occurs at $x = -\frac{\pi}{6}$ instead of at $x = 0$.

Phase shift   To graph any function in the form $y = a \sin (bx + c)$ or $y = a \cos (bx + c)$, shift the curve so that the beginning of the cycle (previously $x = 0$) is now the value of x where $bx + c = 0$. This value of x is equal to $-c/b$ and is called the **phase shift**, or displacement, of the curve. If the phase shift is **positive,** the curve is shifted to the **right.** If the phase shift is **negative,** the curve is shifted to the **left.** The period and amplitude are unaffected by the phase shift.

With this knowledge, we can now sketch the graph of any function containing a phase constant without preparing a table of values.

EXAMPLE 10   Sketch the graph of the function: $y = 2 \cos \left(\dfrac{x}{2} - \dfrac{\pi}{4}\right)$

Solution

Step 1.   Find the period, amplitude, and phase shift.

$$\text{Period} = \frac{2\pi}{b} = \frac{2\pi}{\frac{1}{2}} = 4\pi$$

$$\text{Amplitude} = 2$$

To find the phase shift, let:

$$\frac{x}{2} - \frac{\pi}{4} = 0$$

And solve for x:

$$\frac{x}{2} = \frac{\pi}{4}$$

$$x = \frac{\pi}{2}$$

Shift the beginning of the cycle to $x = \frac{\pi}{2}$.

Step 2. Sketch the cosine curve with a period of 4π, an amplitude of 2, and using $x = \frac{\pi}{2}$ as the beginning of the cycle. Then extend the curve back to the y-axis (Fig. 10-17).

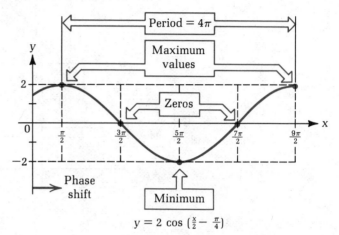

Figure 10-17

$$y = 2 \cos \left(\frac{x}{2} - \frac{\pi}{4}\right)$$

Notice from the graph that the maximum still occurs at the beginning of each cycle $\left(\frac{\pi}{2} \text{ and } \frac{9\pi}{2}\right)$. The minimum occurs midway between these values $\left(\text{at } \frac{5\pi}{2}\right)$. Finally, there are zeros midway between each maximum and minimum $\left(\text{at } \frac{3\pi}{2} \text{ and } \frac{7\pi}{2}\right)$. The graph of

$$y = 2 \cos \left(\frac{x}{2} - \frac{\pi}{4}\right)$$

is identical to that of

$$y = 2 \cos \frac{x}{2}$$

but it is shifted $\frac{\pi}{2}$ units to the right.

EXAMPLE 11 Sketch the graph of the function: $y = -\frac{1}{2} \sin \left(x + \frac{\pi}{2}\right)$

Solution
Step 1.

(Note that a is negative, so the curve will be inverted.)

$$\text{Period} = \frac{2\pi}{b} = \frac{2\pi}{1} = 2\pi$$

$$\text{Amplitude} = 0.5$$

To find the phase shift, let: $x + \frac{\pi}{2} = 0$

And solve for x:

$$x = -\frac{\pi}{2}$$

Shift the starting point to $x = -\frac{\pi}{2}$.

Step 2. Sketch one cycle of the inverted sine curve with a period of 2π, an amplitude of 0.5, and using $x = -\frac{\pi}{2}$ as the beginning of the cycle. Then extend the curve so that an entire cycle appears to the right of the y-axis (Fig. 10-18).

Figure 10-18

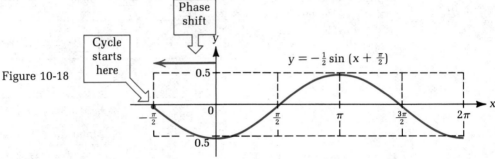

EXAMPLE 12 The motion of a certain oscillating system is described by the equation

$$y = a \cos\left(\frac{\pi x}{L} - \pi\right)$$

where x and y are measured in centimeters, $a = 5$ cm, and $L = \frac{1}{4}$ cm. Find the period, amplitude, and phase shift of the motion, and sketch the graph for $0 \le x \le \frac{3}{4}$.

Solution Substituting the given values of a and L, we obtain

$$y = 5 \cos (4\pi x - \pi)$$

Since $b = 4\pi$,

$$\text{Period} = \frac{2\pi}{b} = \frac{2\pi}{4\pi} = \frac{1}{2}$$

Since a is given, we know that the amplitude is 5 cm. To find the phase shift, we solve

$$4\pi x - \pi = 0$$

$$x = \frac{1}{4}$$

To graph, sketch the cosine curve with a period of $\frac{1}{2}$, an amplitude of 5, and using $x = \frac{1}{4}$ as the start of the cycle (Fig. 10-19).

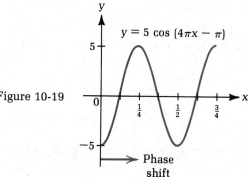

Figure 10-19

Summary

To graph a function (fn) of the form

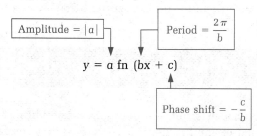

where "fn" can be either sin or cos, shift the beginning of the cycle to $x = -c/b$ and then sketch the basic curve of the function with the given amplitude and period.

Exercises 10-3

State the period, amplitude, and phase shift for each function, and sketch the graph.

1. $y = \sin\left(x + \dfrac{\pi}{3}\right)$

2. $y = \sin\left(x - \dfrac{\pi}{4}\right)$

3. $y = \cos\left(x - \dfrac{\pi}{8}\right)$

4. $y = \cos\left(x + \dfrac{\pi}{2}\right)$

5. $y = 2\sin(x - \pi)$

6. $y = -\sin\left(2x + \dfrac{\pi}{6}\right)$

7. $y = -\cos(x + \pi)$

8. $y = 3\cos\left(3x - \dfrac{\pi}{4}\right)$

9. $y = -\dfrac{3}{2}\sin\left(\dfrac{x}{2} + \dfrac{\pi}{8}\right)$

10. $y = -3\sin\left(\dfrac{x}{4} - \dfrac{\pi}{2}\right)$

11. $y = -4\cos\left(\dfrac{x}{3} - \dfrac{\pi}{6}\right)$

12. $y = \dfrac{5}{4}\cos\left(\dfrac{x}{2} + \dfrac{\pi}{3}\right)$

13. $y = \sin\left(\pi x + \dfrac{\pi}{2}\right)$

14. $y = -\sin\left(2\pi x - \dfrac{\pi}{4}\right)$

15. $y = -\dfrac{1}{3}\cos(3\pi x + \pi)$

16. $y = \cos\left(4\pi x - \dfrac{\pi}{6}\right)$

17. $y = \dfrac{2}{3}\sin\left(\dfrac{\pi x}{2} - 1\right)$

18. $y = -4\sin\left(\dfrac{\pi x}{3} + \dfrac{1}{2}\right)$

19. $y = -2\cos(2\pi x - 4)$

20. $y = \dfrac{1}{2}\cos(\pi x + 2)$

21. $y = 0.4\sin\left(3x + \dfrac{\pi}{8}\right)$

22. $y = -2.2\sin(4\pi x - \pi)$

23. $y = -2.6\cos\left(\dfrac{\pi x}{2} + 1\right)$

24. $y = 150\cos\left(\dfrac{2x}{3} - \dfrac{\pi}{4}\right)$

Solve.

25. *Physics*   Sketch the function given in Example 12 for $a = 8$ cm and $L = 0.4$ cm.

26. *Electronics*   The current i (in amperes) in a certain ac circuit is given by the function below, where t is the time in seconds. Sketch two cycles of this curve.

$$i = 45 \cos \left(120\pi t + \frac{\pi}{2} \right)$$

27. *Physics*   The equation of a certain sinusoidal wave on a string is $y = 0.04 \sin [2\pi(0.1x + 3.0t)]$. Graph two cycles of y as a function of x for $t = \frac{1}{8}$ s.

28. *Oceanography*   At a certain point in the ocean the vertical displacement of the water due to wave action is given by

$$y = 4 \sin \left[\frac{\pi}{4}(t + 2) \right]$$

where y is measured in meters and t is time in seconds. Graph y as a function of t for two cycles.

10-4 | Graphs of Other Trigonometric Functions

Although technical applications are most likely to involve the graphs of the sine and cosine functions, it is worthwhile to examine briefly the graphs of the tangent, cotangent, secant, and cosecant functions.

Tangent function   To sketch the graph of $y = \tan x$, first we use a calculator or our knowledge of the special angle values to prepare a table of sample points.*

x	$-\dfrac{\pi}{6}$	0	$\dfrac{\pi}{6}$	$\dfrac{\pi}{3}$	$\dfrac{\pi}{2}$	$\dfrac{2\pi}{3}$	$\dfrac{5\pi}{6}$	π
$y = \tan x$	-0.6	0	0.6	1.7	†	-1.7	-0.6	0

x	$\dfrac{7\pi}{6}$	$\dfrac{4\pi}{3}$	$\dfrac{3\pi}{2}$	$\dfrac{5\pi}{3}$	$\dfrac{11\pi}{6}$	2π	$\dfrac{13\pi}{6}$
$y = \tan x$	0.6	1.7	†	-1.7	-0.6	0	0.6

† Undefined.

Plotting these ordered pairs, we obtain the graph shown in Fig. 10-20. Notice that the tangent function is also periodic and has a period of π. The dashed vertical lines at $\frac{\pi}{2}$ and $\frac{3\pi}{2}$ are called **asymptotes.** They show that the function is undefined for these x values, and that as these values are approached, the function value becomes either a very large positive number or a very large

* Remember that $\tan \left(\frac{\pi}{2}\right)$ and $\tan \left(\frac{3\pi}{2}\right)$ are undefined. Most calculators will indicate this by displaying an error message when the [tan] key is pressed after entering these angles.

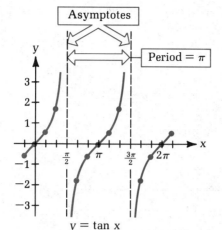

Figure 10-20

$y = \tan x$

negative number. There is no point on the graph at $\frac{\pi}{2}, \frac{3\pi}{2}$, and so on. This means that the function has no maximum or minimum values.

 **Reciprocal functions**   The cotangent, secant, and cosecant curves can be sketched easily if we recall that

$$\cot x = \frac{1}{\tan x} \qquad \sec x = \frac{1}{\cos x} \qquad \csc x = \frac{1}{\sin x}$$

To graph any of these functions, first select appropriate values of x and find the values of the corresponding basic function, tangent, cosine, or sine. Then determine the reciprocals of these values, plot them, and sketch the graph. Figure 10-21 shows the graph of $y = \cot x$.

REMINDER ▶ If you use a calculator to help determine the points on this graph, remember that $\cot x = 0$ at $x = \frac{\pi}{2}, \frac{3\pi}{2}$, etc. Most calculators will display an error message when the $\boxed{\tan}\boxed{1/x}$ sequence is used after entering these angles. ◀

Figure 10-21

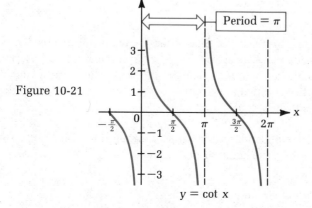

$y = \cot x$

Notice that the cotangent function has a period of π and asymptotes at $x = 0$, π, 2π, and so on.

The graphs of $y = \sec x$ and $y = \csc x$ are shown in Figs. 10-22 and 10-23.

Figure 10-22

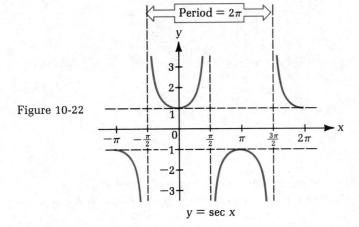

$y = \sec x$

Figure 10-23

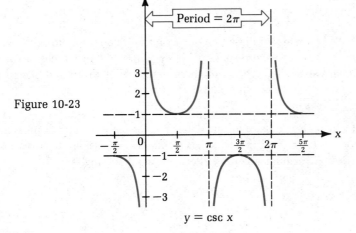

$y = \csc x$

Notice that the period of each of these functions is 2π. For the secant function, the asymptotes are at $-\frac{\pi}{2}, \frac{\pi}{2}, \frac{3\pi}{2}$, and so on; and for the cosecant function, the asymptotes are at 0, π, 2π, and so on. The functions are not defined at these values. Notice also that the y values for both functions are greater than or equal to 1 or less than or equal to -1.

Variations in tan, cot, sec, and csc graphs   Various modifications of the tangent, cotangent, secant, and cosecant functions leave the basic graph form unchanged.

EXAMPLE 13   Graph $y = 2 \tan x$.

Solution The function may be graphed using a table of values obtained with a calculator. Figure 10-24 shows the graph of $y = 2 \tan x$ compared with the graph of $y = \tan x$. Notice that multiplying the function by a constant does not change the period of the graph or the positions of the asymptotes.

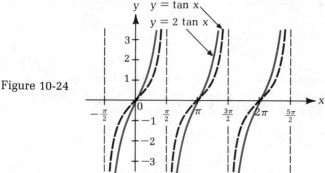

Figure 10-24

If the constant is negative, the curve is inverted.

EXAMPLE 14 Figure 10-25 shows the graph of $y = -\csc x$. The period and positions of the asymptotes are the same as for the graph of $y = \csc x$, but the curve is inverted. The sections of the graph that are above the x-axis in Fig. 10-23 are below the x-axis here, and vice versa.

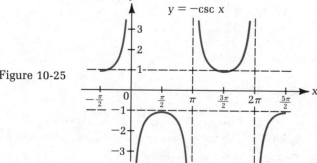

Figure 10-25

For the tangent, cotangent, secant, and cosecant functions, modifications of the angle have the same effect on period and phase shift as for the sine and cosine functions. Since the graphs of these four functions occur infrequently in applied work, it is usually best to plot their variations from a table of values rather than memorizing and modifying the basic curves.

EXAMPLE 15 Graph $y = \cot 2x$.

Solution   The following table of values is all we need to make a sketch of the curve:

x	0	$\dfrac{\pi}{8}$	$\dfrac{\pi}{4}$	$\dfrac{3\pi}{8}$	$\dfrac{\pi}{2}$	$\dfrac{5\pi}{8}$	$\dfrac{3\pi}{4}$	$\dfrac{7\pi}{8}$	π
y = cot 2x	†	1	0	−1	†	1	0	−1	†

† Undefined.

The graph is shown in Fig. 10-26. Notice that the coefficient of x can be used to determine the period of the function. From our study of the sine and cosine functions, we would expect that for y = a cot bx,

$$\text{Period} = \frac{\text{Basic period}}{b} = \frac{\pi}{b}$$

From the graph of y = cot 2x, we see that the period is $\frac{\pi}{2}$. The shape of the graph remains the same, but the asymptotes are now at x = 0, $\frac{\pi}{2}$, π, and so on.

Figure 10-26

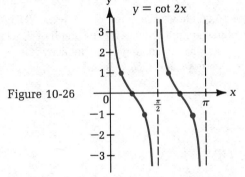

y = cot 2x

EXAMPLE 16   Graph y = sec (2x − π).

Solution   Prepare a table of values using a calculator. Since the secant function is the reciprocal of the cosine function, the calculation is done as follows:

For x = −$\dfrac{\pi}{8}$:   [RAD] 2 [×] [π] [÷] 8 [+/−] [−] [π] [=] [cos] [1/x] →   *−1.4142136*

x	$-\dfrac{\pi}{4}$	$-\dfrac{\pi}{8}$	0	$\dfrac{\pi}{8}$	$\dfrac{\pi}{4}$	$\dfrac{3\pi}{8}$	$\dfrac{\pi}{2}$	$\dfrac{5\pi}{8}$	$\dfrac{3\pi}{4}$	$\dfrac{7\pi}{8}$	π
y	†	−1.4	−1	−1.4	†	1.4	1	1.4	†	−1.4	−1

† Undefined.

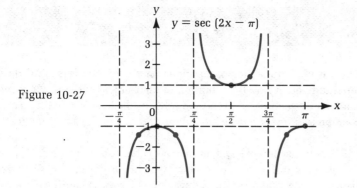

Figure 10-27

As we would expect, the graph in Fig. 10-27 has

$$\text{Period} = \frac{2\pi}{2} = \pi$$

The phase shift should be the solution to the equation $2x - \pi = 0$, or $x = \frac{\pi}{2}$. A sketch will verify that this curve is shifted $\frac{\pi}{2}$ units to the right of $y = \sec 2x$.

In general, the graphs of the modifications of the tangent, cotangent, secant, and cosecant functions will retain the general shape of the basic unmodified functions. The variations may be shifted in position, inverted, expanded, or compressed.

Exercises 10-4

Sketch the graph of each function.

1. $y = 3 \tan x$

2. $y = 2 \cot x$

3. $y = \frac{3}{2} \sec x$

4. $y = 0.8 \csc x$

5. $y = -\frac{1}{2} \cot x$

6. $y = -\tan x$

7. $y = -2 \csc x$

8. $y = -4 \sec x$

9. $y = \tan 2x$

10. $y = -\cot \frac{1}{2}x$

11. $y = -1.8 \sec 3x$

12. $y = 4 \csc 2x$

13. $y = \sec \left(x - \frac{\pi}{4} \right)$

14. $y = \tan \left(x + \frac{\pi}{2} \right)$

15. $y = \csc \left(x + \frac{\pi}{6} \right)$

16. $y = \sec (x - \pi)$

17. $y = -2 \tan \left(2x - \frac{\pi}{2} \right)$

18. $y = 3.5 \cot \left(\pi x + \frac{\pi}{4} \right)$

19. $y = \frac{1}{2} \csc (2\pi x - 1)$

20. $y = -\sec \left(3\pi x + \frac{\pi}{2} \right)$

10-5 | Applications

A very useful application of the trigonometric functions is in the description of periodic phenomena: the vibration of mechanical systems, wave motion, current flow in electrical circuits, and other cyclic changes. In this section we will examine two important technical applications of the sine function.

Simple harmonic motion

For any mechanical system or object that can be made to vibrate, there is always an equilibrium position in which the object can sit at rest. If it is displaced from that position, the system will exert a restoring force on it tending to return it to equilibrium. If the size of the restoring force is directly proportional to the amount it is displaced from equilibrium, the vibrating object is said to perform **simple harmonic motion.**

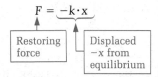

In Fig. 10-28, a block is shown oscillating up and down under the action of a spring. At rest in Fig. 10-28(a), the block sits at the equilibrium position A. When the block is pulled down to position B and then released in Fig. 10-28(b), the spring is stretched and pulls the block back up toward position A. In Fig. 10-28(c), the spring has pulled the block past the equilibrium position and

Figure 10-28

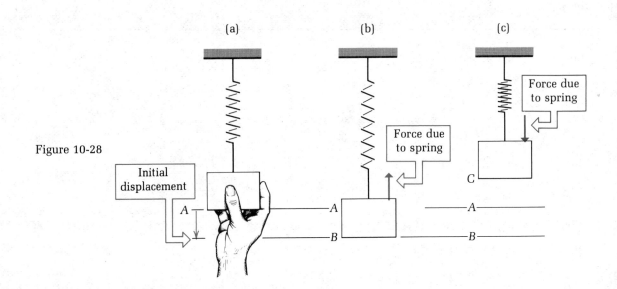

moved it to point C. This causes the spring to compress. The spring now pushes the block down toward the equilibrium position A.

If the spring is very light and is not overstretched initially, and if there are no friction forces acting on the block, it will oscillate up and down endlessly in simple harmonic motion.

If a marking pen is attached to the block and allowed to trace a record of its movement on a paper tape moving past the apparatus as shown in Fig. 10-29, we see that the motion can be described with a sine function.

Figure 10-29

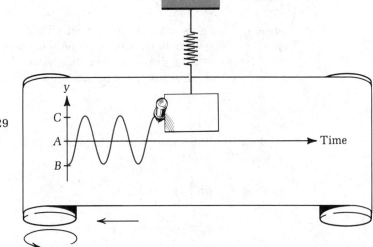

The displacement y of any object in simple harmonic motion is described over time t (in seconds) by the following equation:*

Simple Harmonic Motion	$y = a \sin (2\pi ft + \phi)$	(10-1)

Here, $|a|$ is the **amplitude** of the motion, the maximum displacement of the mass from equilibrium; f is the **frequency** of the motion (in hertz), the number of oscillations the object makes in 1 s; and ϕ is the **phase constant.**

REMINDER ▶ The phase *constant* is *not* the same as the phase *shift*. The phase shift of the graph of y is equal to $-\phi/2\pi f$, the value of t when $2\pi ft + \phi = 0$. ◀

The frequency depends on the mass of the block and the stiffness of the spring. The phase constant depends on the position and speed of the block when it is set in motion at time $t = 0$. The **period** T of the motion (in seconds) is defined as

* This may also be expressed as the cosine function, $y = a \cos (2\pi ft + \phi')$, where $\phi' = \phi - \left(\frac{\pi}{2}\right)$.

$$\text{Period} = T = \frac{1}{f}$$

This is the time needed for one complete oscillation.

EXAMPLE 17 A system vibrating in simple harmonic motion is described by the equation

$$y = 4 \sin\left(4\pi t - \frac{\pi}{2}\right)$$

where y is the displacement (in centimeters) and t is time (in seconds).
(a) Find the amplitude, frequency, period, and phase constant for this motion.
(b) What will be the displacement of the system 2 s after it begins vibrating?
(c) When will the system be at its maximum positive displacement?

Solution
(a) Rewriting this equation in the form of the general simple harmonic motion equation (10-1), we see that

$$y = 4 \sin\left[2\pi \cdot 2 \cdot t + \left(-\frac{\pi}{2}\right)\right]$$

$$\text{Amplitude} = |a| = 4 \text{ cm}$$

$$\text{Frequency} = f = 2 \text{ Hz}$$

$$\text{Period} = \frac{1}{f} = \frac{1}{2} \text{ s} = 0.5 \text{ s}$$

$$\text{Phase constant} = -\frac{\pi}{2}$$

(b) At $t = 2$ s, $y = 4 \sin\left(4\pi \cdot 2 - \frac{\pi}{2}\right)$

$$= 4 \sin\left(\frac{15\pi}{2}\right) = -4 \text{ cm}$$

(c) Since $\sin x$ is maximum at $x = \frac{\pi}{2}$, the displacement y is maximum when

$$4\pi t - \frac{\pi}{2} = \frac{\pi}{2}$$

Solving for t, we get

$$4\pi t = \pi$$

$$t = \frac{\pi}{4\pi} = \frac{1}{4} \text{ s} = 0.25 \text{ s}$$

The displacement will be a maximum at 0.25 s and at intervals of one period thereafter: $t = 0.25, 0.75, 1.25, 1.75$, and so on.

EXAMPLE 18 Write an equation describing the simple harmonic motion of a system with a period of 3 s, amplitude of 4 ft, and phase constant of $-\frac{7\pi}{6}$ rad.

Solution We are given $|a| = 4$ ft, $\phi = -\frac{7\pi}{6}$, and $T = 3$ s. Since $T = 1/f$,

$$f = \frac{1}{T} = \frac{1}{3} \text{ Hz}$$

Using $a = +4$ and substituting these values into the simple harmonic motion equation (10-1), $y = a \sin(2\pi f t + \phi)$, we have

$$y = 4 \sin\left[2\pi \cdot \frac{1}{3}t + \left(-\frac{7\pi}{6}\right)\right]$$

$$y = 4 \sin\left(\frac{2\pi t}{3} - \frac{7\pi}{6}\right)$$

Alternating current

An alternating current (ac) generator is a voltage source with output that varies in some periodic way. Modern generators can produce a variety of voltage patterns or wave forms, but the simplest is the **sinusoidal voltage** that varies with time according to the following equation:

Sinusoidal Voltage	$V = V_0 \sin(2\pi f t)$	**(10-2)**

Here, V_0 is the **peak voltage**—that is, the maximum voltage across the generator during each cycle—and f is its frequency. In the United States the voltage generated by power plants has a frequency f of 60 Hz and a peak voltage of 170 V.

EXAMPLE 19 Sketch two cycles of the function describing the voltage across a home wall socket where $V_0 = 170$ V and $f = 60$ Hz.

Solution The equation has zeros at

$$2\pi f t = 0 \qquad 2\pi f t = \pi \qquad 2\pi f t = 2\pi$$

Thus,

$$t = 0 \text{ s} \qquad t = \frac{\pi}{2\pi f} \qquad t = \frac{2\pi}{2\pi f}$$

$$= \frac{1}{2f} = \frac{1}{120} \text{ s} \qquad = \frac{1}{f} = \frac{1}{60} \text{ s}$$

The graph is shown in Fig. 10-30.

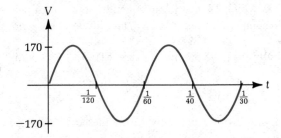

Figure 10-30

When an ac generator is connected to a resistor, the current i in the resistor varies with time t in the same pattern as the voltage:

Alternating Current	$i = i_0 \sin(2\pi ft)$	**(10-3)**

Here, i_0 is called the **peak current.** The current is alternately positive and negative, so that electric charges alternately move in one direction through the resistor and then in the opposite direction.

If more complex electrical elements are placed in an ac circuit, the voltage across these devices and the current flowing in the circuit may not necessarily peak at the same time. They are then said to be **out of phase** with each other, and the difference in phase appears in the sine function as a **phase angle** ϕ. This is usually expressed by writing the voltage across the circuit element as follows:

	$V = V_0 \sin(2\pi ft + \phi)$	**(10-4)**

If ϕ is a negative angle, the voltage is said to **lag** the current; and if ϕ is a positive angle, the voltage is said to **lead** the current.

EXAMPLE 20   Sketch two cycles of the following equations for $f = 60$ Hz:
(a) $i = i_0 \sin(2\pi ft)$ (b) $V = V_0 \sin(2\pi ft + \phi)$ when $\phi = \frac{\pi}{2}$
(c) $V = V_0 \sin(2\pi ft + \phi)$ when $\phi = -\frac{\pi}{2}$

Solution
(a)   Figure 10-31 shows the graph of $i = i_0 \sin(120\pi t)$.

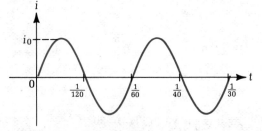

Figure 10-31

(b)   Figure 10-32 shows the graph of

$$V = V_0 \sin \left(120\pi t + \frac{\pi}{2} \right)$$

The voltage has already completed $\frac{1}{4}$ cycle by $t = 0$.

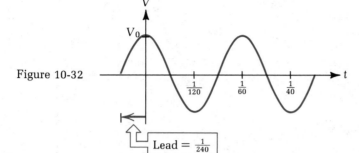

Figure 10-32

(c)   Figure 10-33 shows the graph of

$$V = V_0 \sin \left(120\pi t - \frac{\pi}{2} \right)$$

The voltage is $\frac{1}{4}$ cycle behind the current at $t = 0$.

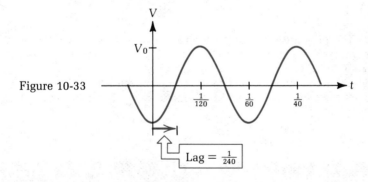

Figure 10-33

Exercises 10-5

Problems 1–4 state the equations of four different systems vibrating in simple harmonic motion. For each equation answer questions (a)–(d) below. Assume that y is measured in feet and t is time in seconds.

(a)   *Find the amplitude, frequency, period, and phase constant.*
(b)   *What is the displacement of the object 3 s after it begins vibrating?*
(c)   *When will the system be at its maximum positive displacement?*
(d)   *Sketch two cycles of the equation.*

1. $y = 6.0 \sin \left(2\pi t - \dfrac{\pi}{4} \right)$ 2. $y = 2.0 \sin \left(\pi t - \dfrac{\pi}{2} \right)$

3. $y = 0.80 \sin (3\pi t + \pi)$ 4. $y = 3.5 \sin \left(6\pi t + \dfrac{\pi}{2} \right)$

Write an equation describing the simple harmonic motion of each system.

5. Period = 2 s;   Amplitude = 3 cm;   Phase constant = $\dfrac{\pi}{3}$ rad

6. Period = 0.40 s;   Amplitude = 1.5 in.;   Phase constant = $\dfrac{5\pi}{6}$ rad

7. Period = 0.75 s;   Amplitude = 0.60 m;   Phase constant = $-\dfrac{\pi}{6}$ rad

8. Period = 5 s;   Amplitude = 6 ft;   Phase constant = $-\dfrac{2\pi}{3}$ rad

Sketch two cycles of $V = V_0 \sin (2\pi ft)$ for:

9. $V_0 = 6.0$ V   and $f = 1000$ Hz   10. $V_0 = 30.0$ mV   and $f = 400$ Hz

Sketch two cycles of $i = i_0 \sin (2\pi ft)$ for:

11. $i_0 = 5.0$ A   and $f = 1000$ Hz   12. $i_0 = 0.80$ A   and $f = 60$ Hz

Sketch two cycles of $V = V_0 \sin (2\pi ft + \phi)$ for:

13. $V_0 = 170$ V, $f = 60$ Hz,   and $\phi = -\dfrac{\pi}{2}$ 14. $V_0 = 88$ V, $f = 60$ Hz,   and $\phi = \dfrac{\pi}{2}$

Solve.

15. *Acoustical engineering*   The pressure variation in a sound wave can be described by the equation $P = A \sin 2\pi ft$. Plot the pressure P as a function of the time t (in seconds) for a sound wave with a maximum pressure A of 4.0 lb/in.² and a frequency f of 4500 Hz.

16. *Electronics*   An electronic component test table vibrates according to the equation $y = A \sin 2\pi ft$. Plot the vibration displacement y as a function of the time t for a vibration frequency of 2000 Hz. Mark the y-axis in units of A.

10-6 | Combining Graphs of Trigonometric Functions

In many technical applications, especially in electronics, complex functions appear that are a combination of two or more trigonometric functions. In this section, we will show how you can make a reasonably accurate sketch of these combined functions by a method known as **addition of ordinates**. This method is demonstrated in the next example.

EXAMPLE 21   Graph the following function for one cycle:

$$y = \cos 2x + 2 \sin x$$

Solution   First, plot $y = \cos 2x$ (light solid curve in Fig. 10-34) and $y = 2 \sin x$ (dashed curve) on the same set of coordinate axes.

Second, locate positions along the x-axis where either function has a maximum, minimum, or zero, or where the graphs intersect. Mark these. (We have indicated them with the letters A, B, C, and so on in Fig. 10-34.)

Third, add the y-coordinates of the two graphs at each of these points and place a dot at this new y-coordinate. For example,

At A, $0 + 1 = 1$

At B,   approximately $0.75 + 0.75 = 1.5$

At C, $0 + 1.4 = 1.4$

At D, $2 - 1 = 1$

And so on.

Figure 10-34

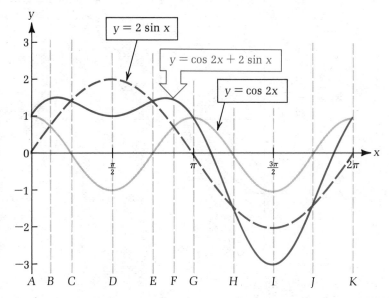

CAUTION ▶ Be sure to add negative coordinates as negative numbers. To find the point I, we add -2 to -1 to get -3. ◀

Finally, draw a smooth curve through the points you have found. The result is shown in Fig. 10-34.

The period of the new function is the least common multiple of the periods of the two component functions. The period of $y = 2 \sin x$ is 2π, and the period of $y = \cos 2x$ is π, so the period of the function $y = \cos 2x + 2 \sin x$ is 2π. When plotting a function of this kind, be certain to extend the graph for its complete period.

When a combined function is formed by subtraction, plot the negative of the subtracted function. In this way the combined ordinates may still be found by addition.

EXAMPLE 22   Graph the following function for one cycle:

$$y = \sin 2x - 2 \cos \frac{x}{2}$$

Solution   **First,** plot $y = \sin 2x$ (light solid curve in Fig. 10-35) and $y = -2 \cos \frac{x}{2}$ (dashed curve) on the same set of coordinate axes.

Second, locate maximum and minimum values, zeros, and intersections.

Third, add the y-coordinates.

Finally, sketch the curve, as shown in Fig. 10-35.

Figure 10-35

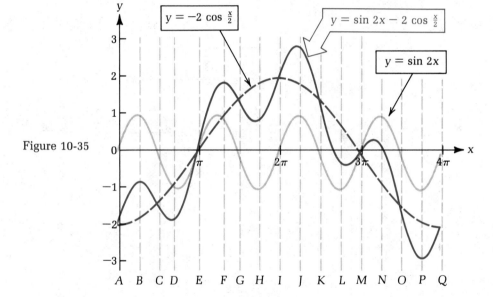

LEARNING HINT

A calculator can be used either to find or to check points on graphs involving addition of ordinates. In Example 22, we can find or check the value of y at $x = \frac{7\pi}{4}$ (point H) as follows:

Plotting graphs may involve the addition of a trigonometric function to a nontrigonometric function. The same procedure is used to determine the graph.

EXAMPLE 23   Graph $y = x + 2 \sin x$ for $0 \le x \le 2\pi$.

Solution   In Fig. 10-36, we have graphed $y = x$ (dashed line) and $y = 2 \sin x$ (light solid curve). Then, adding their y-coordinates gives the graph of $y = x + 2 \sin x$.

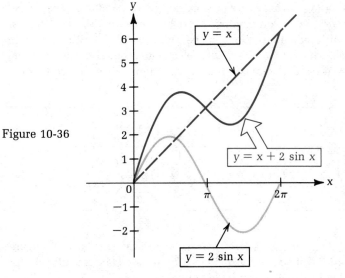

Figure 10-36

NOTE ▶ When a nontrigonometric function is involved, make certain that the scales used to calibrate the two axes agree. ◀

Lissajous figures   Many technicians and scientists use an electronic instrument called an **oscil-loscope** in their work. This device can convert an electrical signal into a graphic display on a televisionlike screen. In particular, two voltage signals, each of the pattern

$$V = V_0 \sin (2\pi ft + \phi)$$

may be combined and compared by associating one with the x-axis of the display and the other with the y-axis, and plotting the succession of x and y values for each instant of time through the period of the functions. The resulting curve is called a **Lissajous figure** (named after Jules Antoine Lissajous, who first studied such figures in 1857).

If the signals are equal in frequency and are in phase, the figure will be a straight line. If the signals are equal in frequency and are out of phase, the figure displayed will be either a circle or an ellipse, depending on the signal amplitude and the phase angle. For example, the combined functions

$$y = 2 \sin (2\pi ft) \quad \text{and} \quad x = 2 \sin \left(2\pi ft + \frac{\pi}{2} \right)$$

will be displayed as a circle.

If the frequencies are not equal, the combined function will be a much more complex curve, as shown in the next example.

EXAMPLE 24   Plot the Lissajous figure determined by the functions

$$x = \cos \left(\pi t + \frac{3\pi}{2} \right) \qquad y = 2 \sin 2\pi t$$

Solution   When two equations describe separate variables (x and y in this case) in terms of a common third variable (time t in this case), they are called **parametric equations**. To find the points needed to plot the graph, use t as an independent variable. Then select values of t and calculate x and y. The table shows a few points on the graph, rounded to the nearest tenth.

t (s)	0	$\dfrac{1}{4}$	$\dfrac{1}{2}$	$\dfrac{3}{4}$	1	$\dfrac{5}{4}$	$\dfrac{3}{2}$	$\dfrac{7}{4}$	2
$x = \cos\left(\pi t + \dfrac{3\pi}{2}\right)$	0	0.7	1	0.7	0	−0.7	−1	−0.7	0
$y = 2\sin 2\pi t$	0	2	0	−2	0	2	0	−2	0

If we plot the ordered pairs (x, y), we have the Lissajous figure shown in Fig. 10-37. Notice that this Lissajous figure cannot be considered a function because every x value corresponds to two y values.

Figure 10-37

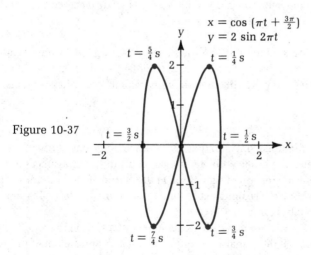

$$x = \cos\left(\pi t + \tfrac{3\pi}{2}\right)$$
$$y = 2\sin 2\pi t$$

EXAMPLE 25   Plot the figure given by

$$x = 2\sin t \qquad y = \sin\left(t + \frac{\pi}{4}\right)$$

Solution   Since the coefficient of t is not a multiple of π, it is convenient to choose values of t that are multiples of π. Using a calculator we can determine values of x and y as shown by the following samples for $t = \frac{\pi}{6}$.

Value of x: [RAD] [π] [÷] 6 [=] [STO] [sin] [×] 2 [=] → ▨▨▨ *1.*

Value of y: [π] [÷] 4 [+] [RCL] [=] [sin] → *0.9659258*

Rounding all values to the nearest tenth, we obtain the following table:

t	0	$\frac{\pi}{6}$	$\frac{\pi}{4}$	$\frac{\pi}{3}$	$\frac{\pi}{2}$	$\frac{3\pi}{4}$	π	$\frac{5\pi}{4}$	$\frac{3\pi}{2}$	$\frac{5\pi}{3}$	$\frac{7\pi}{4}$	2π
x	0	1	1.4	1.7	2	1.4	0	−1.4	−2	−1.7	−1.4	0
y	0.7	1.0	1	1.0	0.7	0	−0.7	−1	−0.7	−0.3	0	0.7

The graph is shown in Fig. 10-38.

Figure 10-38

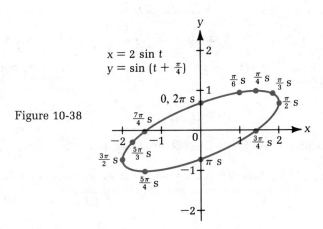

Exercises 10-6

Sketch the graph of each function.

1. $y = \sin x + \cos x$
2. $y = 2 \cos x + \sin x$
3. $y = \sin 2x - \cos x$
4. $y = \cos x - \cos 3x$
5. $y = 2 \cos \pi x + \sin 4\pi x$
6. $y = 2 \sin 2\pi x + \sin \pi x$
7. $y = \frac{1}{2} \sin 2x - \cos \frac{x}{2}$
8. $y = 2 \sin 2x - \cos 4x$
9. $y = x + \cos x$
10. $y = x + \sin 2x$
11. $y = \frac{1}{2}x - \sin \pi x$
12. $y = -x + \cos 2\pi x$
13. $y = \frac{x}{3} + 2 \cos 3x$
14. $y = \frac{1}{2}x + 2 \sin x$
15. $y = \frac{x^2}{4} + \sin x$
16. $y = -\frac{x^2}{2} + 2 \cos 2x$
17. $y = \cos \left(x + \frac{\pi}{4} \right) + 2 \sin 2x$
18. $y = 2 \sin \left(x - \frac{\pi}{3} \right) - \sin x$
19. $y = 3 \sin \left(2x - \frac{\pi}{2} \right) - \cos \left(2x - \frac{\pi}{4} \right)$
20. $y = 2 \cos (3\pi x - 1) + \sin \left(\pi x + \frac{1}{2} \right)$

Plot the graph of each Lissajous figure.

21. $x = \cos 2t, \quad y = \cos 2t$
22. $x = \sin \pi t, \quad y = 2 \sin \pi t$

23. $x = \sin 2\pi t$, $y = \cos 2\pi t$ **24.** $x = \cos t$, $y = 2 \sin t$

25. $x = 2 \sin t$, $y = \sin 2t$ **26.** $x = \cos 2\pi t$, $y = 3 \sin \pi t$

27. $x = \sin\left(\pi t + \dfrac{\pi}{4}\right)$, $y = 2 \cos \pi t$ **28.** $x = \cos 2t$, $y = \sin\left(2t - \dfrac{\pi}{4}\right)$

Solve.

29. *Surveying*   When using an external focusing instrument, the horizontal stadia multiplier H is given by

$$H = 1 + 100S \cos^2\theta$$

where S is the stadia intercept and θ is the vertical angle. Sketch H as a function of θ for $S = 1.0$ ft and $0 \le \theta \le \dfrac{\pi}{2}$.

30. *Optics*   The intensity of a light source is given by the equation $I = 4.0 + 4.0 \cos \theta$. Plot I as a function of θ for $0 \le \theta \le 2\pi$.

31. *Environmental science*   The number of tons of pollutant released weekly into the atmosphere in a large city can be approximated by the equation

$$P = 1.2 + \cos \frac{\pi x}{26}$$

where x is the number of weeks elapsed from the beginning of the year. Sketch the graph of this equation over a 2 year period $(0 \le x \le 104)$.*

32. *Physics*   Fourier series are used to study heat flow, sound, electrical fields, and other technical applications. The first two terms of a certain Fourier series are given by the equation below. Sketch the graph of this equation.

$$y = \sin x + \frac{\sin 3x}{3}$$

33. *Navigation*   In tracking the course of a ship, it is necessary to label the date and time of reported ship positions directly above the plotted ship-position triangle on navigation charts. The label must begin at a point $P(x, y)$ related to the rotation angle θ of the triangle and the offset distances D_1 and D_2 from the center of the triangle. The coordinates of P are given by the parametric equations

$$x = D_1 \cos \theta + D_2 \sin \theta \qquad y = D_1 \sin \theta - D_2 \cos \theta$$

Sketch the figure generated by these equations for $D_2 = 2D_1$.†

34. *Instrumentation*   The time variation of a certain mechanical strain is given by the equation below. Sketch the graph of this equation.

$$E = 80 + 60 \sin (15t) - 45 \cos (30t)$$

10-7 | Review of Chapter 10

Important Terms and Concepts

periodic function (p. 345)   amplitude (p. 348)   alternating current, ac (p. 369)

cycle (p. 345)   period (p. 351)   addition of ordinates (p. 372)

---

* Based on Raymond A. Barnett, *Analytic Trigonometry with Applications* (Belmont, Ca.: Wadsworth, 1980) pp. 84–85.

† Courtesy of Bud Selfridge, Ocean Routes, Inc., Palo Alto, Ca.

zeros (p. 345)

maximum (p. 345)

minimum (p. 345)

phase constant (p. 355)

phase shift (p. 356)

simple harmonic motion (p. 366)

Lissajous figure (p. 375)

parametric equations (p. 376)

Formulas

- simple harmonic motion (p. 367): $y = a \sin (2\pi ft + \phi)$ **(10-1)**
- sinusoidal voltage (p. 369): $V = V_0 \sin (2\pi ft)$ **(10-2)**
- alternating current (p. 370): $i = i_0 \sin (2\pi ft)$ **(10-3)**
- voltage out of phase with current (p. 370): $V = V_0 \sin (2\pi ft + \phi)$ **(10-4)**

Exercises 10-7

Sketch the graph of each function.

1. $y = 5 \cos x$

2. $y = -2 \sin x$

3. $y = \dfrac{3}{2} \sin x$

4. $y = -0.8 \cos x$

5. $y = \cos 2x$

6. $y = \sin 3x$

7. $y = \sin \dfrac{x}{3}$

8. $y = \cos 2\pi x$

9. $y = -\cos \dfrac{\pi x}{4}$

10. $y = 3 \sin 4\pi x$

11. $y = 2.2 \sin \dfrac{3\pi x}{4}$

12. $y = -\dfrac{5}{4} \cos \dfrac{3x}{2}$

13. $y = \cos \left(x + \dfrac{\pi}{3} \right)$

14. $y = \sin \left(3x - \dfrac{\pi}{2} \right)$

15. $y = \sin \left(\pi x + \dfrac{\pi}{6} \right)$

16. $y = \cos \left(2\pi x - \dfrac{1}{2} \right)$

17. $y = 2 \cos \left(\dfrac{x}{2} - \pi \right)$

18. $y = 1.3 \sin \left(x + \dfrac{\pi}{8} \right)$

19. $y = -4 \sin \left(\dfrac{\pi x}{4} + 1 \right)$

20. $y = -\dfrac{1}{2} \cos \left(4\pi x - \dfrac{\pi}{2} \right)$

21. $y = -\dfrac{1}{2} \tan x$

22. $y = 3 \sec x$

23. $y = \cot 3x$

24. $y = -2 \csc \dfrac{x}{2}$

25. $y = \tan \left(x + \dfrac{\pi}{4} \right)$

26. $y = -\dfrac{1}{2} \csc \left(\pi x - \dfrac{\pi}{2} \right)$

27. $y = \cos x + 2 \sin x$

28. $y = \sin x - \cos 2x$

29. $y = 2 \sin \pi x - \cos 2\pi x$

30. $y = \sin \left(x - \dfrac{\pi}{4} \right) + 2 \cos 2x$

31. $y = x + \cos 2x$

32. $y = \dfrac{x}{2} - \dfrac{1}{4} \sin 4\pi x$

Plot each figure.

33. $x = \sin t, \quad y = 3 \sin t$

34. $x = 2 \sin \pi t, \quad y = 3 \cos \pi t$

35. $x = \cos \dfrac{t}{2}, \quad y = \sin \left(t + \dfrac{\pi}{6} \right)$

36. $x = \cos \left(\pi t - \dfrac{\pi}{4} \right), \quad y = \cos 2\pi t$

Solve.

37. *Physics*   The displacement y (in centimeters) of a mass oscillating on a spring is given by

$$y = 0.75 \sin \left(2\pi t + \dfrac{\pi}{6} \right)$$

where t is time in seconds. Sketch two cycles of y as a function of t and determine the displacement after 2.0 s.

38. **Physics**   A system vibrating in simple harmonic motion is described by the equation

$$y = 3.0 \sin \left(\pi t - \frac{\pi}{4} \right)$$

where y is the displacement (in inches) and t is time in seconds. Sketch two cycles of y as a function of t and determine when the system reaches its maximum displacement.

39. **Electronics**   The equation $i = i_0 \sin (2\pi ft)$ describes the current in a resistor connected to an ac generator. Sketch two cycles of i as a function of t when $i_0 = 2.5$ A and $f = 60$ Hz. What is the current after $\frac{1}{90}$ s?

40. **Electronics**   The equation $V = V_0 \sin (2\pi ft + \phi)$ gives the voltage across the elements in an ac circuit. Sketch two cycles of V as a function of t when $V_0 = 4.0$ V, $f = 1000$ Hz, and $\phi = \frac{\pi}{2}$. What is the voltage after 0.010 s?

41. **Physics**   A mechanical system vibrating in simple harmonic motion has a period of 4 s, an amplitude of 0.8 ft, and a phase constant of $\frac{\pi}{4}$. Write an equation describing the motion.

42. **Physics**   An oscillating mass moves through its full range of motion in 1.5 s, reaches a maximum displacement of 2.5 cm, and has a phase constant of $-\frac{3\pi}{4}$. Write an equation describing this simple harmonic motion.

43. **Electronics**   The voltage in an ac circuit is given by $V = 50 \cos 30\pi t + 80 \cos 60\pi t$. Sketch the graph of the voltage as a function of time t.

44. **Physics**   The first two terms of a certain Fourier series are given by the equation below. Sketch one cycle of the graph.

$$y = \sin \pi x + \frac{\sin 2\pi x}{2}$$

45. **Surveying**   The formula $D = 100S \cos \theta$ is used to find the direct distance to an inclined sight for a given angle θ and stadia intercept S. Sketch $\frac{1}{4}$ cycle of this equation for $S = 2.50$ ft.

46. **Acoustical engineering**   A sound wave is described by the equation $P = A \sin 2\pi ft$, where P is the pressure (in pounds per square inch), A is the maximum pressure, f is the frequency (in hertz), and t is the time (in seconds). Plot P as a function of time for a sound wave with $A = 0.0034$ lb/in.^{2} and a frequency of 3000 Hz.

47. **Oceanography**   The vertical displacement y of a water wave can be described by

$$y = A \sin \left[2\pi \left(ft - \frac{r}{\lambda} \right) \right]$$

where A is the maximum height, f is the frequency, t is time, r is the distance from the source, and λ is the wavelength. Plot two cycles of y as a function of t for a wave 23.0 ft from the source, with a maximum height of 4.0 ft, a frequency of $\frac{1}{6}$ s, and a wavelength of 184 ft.

48. **Electronics**   An electron in an oscilloscope is deflected by perpendicular electric fields so that its displacement components are given by the equations below. Plot the path of this electron for $A = 4.0$ cm and $f = 120$ Hz.

$$x = A \cos 2\pi ft \quad \text{and} \quad y = A \cos \left(2\pi ft + \frac{\pi}{6} \right)$$

Exponents and Radicals

11

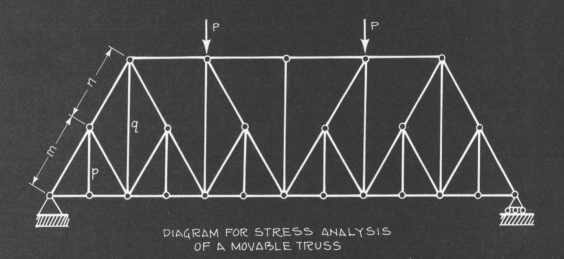

DIAGRAM FOR STRESS ANALYSIS
OF A MOVABLE TRUSS

In Sections 1-4 and 1-6 we discussed the basic laws of exponents and presented some of the definitions and properties of roots and radicals. In this chapter we will explore these topics in greater depth in order to prepare you for more advanced work in mathematics.

11-1 | Integral Exponents

First, we need to review the laws of exponents as they apply to integral exponents.

Laws of Exponents

For any real numbers a and b and any integers m and n:

Multiplication	$a^m \cdot a^n = a^{m+n}$	(11-1)
Power of a Power	$(a^m)^n = a^{mn}$	(11-2)
Power of a Product	$(ab)^n = a^n b^n$	(11-3)
Power of a Quotient	$\left(\dfrac{a}{b}\right)^n = \dfrac{a^n}{b^n}$ (where $b \neq 0$)	(11-4)
Division	$\dfrac{a^m}{a^n} = a^{m-n}$ (where $a \neq 0$)	(11-5)
Negative Exponent	$a^{-m} = \dfrac{1}{a^m}$ and $\dfrac{1}{a^{-m}} = a^m$ (where $a \neq 0$)	(11-6)
Zero Exponent	$a^0 = 1$ (where $a \neq 0$)	(11-7)

Try the following problems for a warmup.

EXAMPLE 1   Simplify each expression using one of the laws of exponents. Your answers may contain negative exponents.

(a) $x^4 \cdot x^{-1}$ (b) $y^{-5} \cdot y^2$ (c) $(m^{-2}n)^4$

(d) $(2x^3)^3$ (e) $\left(\dfrac{2}{y^2}\right)^4$ (f) $\dfrac{y^3}{y^7}$

(g) $\dfrac{z^8}{z^5}$ (h) $6x^0$ (i) $(6x)^0$

(j) $(-6)^0$ (k) -6^0

Solution

(a) $x^4 \cdot x^{-1} = x^{4+(-1)} = x^3$

(b) $y^{-5} \cdot y^2 = y^{-5+2} = y^{-3}$

(c) $(m^{-2}n)^4 = m^{(-2)(4)}n^{1 \cdot 4} = m^{-8}n^4$

(d) $(2x^3)^3 = 2^3 \cdot (x^3)^3 = 8x^9$

(e) $\left(\dfrac{2}{y^2}\right)^4 = \dfrac{2^4}{(y^2)^4} = \dfrac{16}{y^8}$

(f) $\dfrac{y^3}{y^7} = y^{3-7} = y^{-4}$

(g) $\dfrac{z^8}{z^5} = z^{8-5} = z^3$

(h) $6x^0 = 6 \cdot 1 = 6$

(i) $(6x)^0 = 1$

(j) $(-6)^0 = 1$

(k) $-6^0 = -1 \cdot 6^0 = -1 \cdot 1 = -1$

CAUTION ▶ Study Example 1(h)–(k) carefully. In parts (i) and (j), the parentheses mean that the entire quantity is raised to the zero power, resulting in an answer of 1. But in parts (h) and (k), the order of operations tells us to simplify the power first and then multiply. ◀

Notice that the answers to several parts of Example 1 contain negative exponents. Many mathematical applications require positive exponents in all answers. Such problems can be rewritten with positive exponents by applying (11-6):

For Example 1(b): $y^{-3} = \dfrac{1}{y^3}$ For Example 1(c): $m^{-8}n^4 = \dfrac{n^4}{m^8}$

And so on. In technical work, you should write the expression with positive exponents unless you are using scientific notation or you have been directed otherwise.

LEARNING HINT ▶ To change the sign of the exponent in any exponential expression of the form a^n, invert the expression. ◀

EXAMPLE 2 Simplify each expression using the laws of exponents. Express all answers with positive exponents only.

(a) $m^5 \cdot m^{-7}$ (b) $(3ab^{-3})^2$ (c) $(2x^2y^{-1})^{-3}$

(d) $\left(\dfrac{3}{x^4}\right)^{-2}$ (e) $\left(\dfrac{x^2}{y^{-2}}\right)^{-3}$ (f) $\dfrac{x^2y^3}{x^5y}$

Solution

To change the sign of the exponent, invert

(a) $m^5 \cdot m^{-7} = m^{5+(-7)} = m^{-2} = \dfrac{1}{m^2}$

(b) $(3ab^{-3})^2 = 3^2 \cdot a^2 \cdot b^{-6} = 9a^2 \cdot \dfrac{1}{b^6} = \dfrac{9a^2}{b^6}$

$(-1)(-3) = 3$

(c) $(2x^2y^{-1})^{-3} = 2^{-3} \cdot x^{-6} \cdot y^3 = \dfrac{1}{8} \cdot \dfrac{1}{x^6} \cdot y^3 = \dfrac{y^3}{8x^6}$

$$\boxed{3^{-2} = \frac{1}{3^2}}$$

(d) $\left(\dfrac{3}{x^4}\right)^{-2} = \dfrac{3^{-2}}{x^{-8}} = \dfrac{x^8}{3^2} = \dfrac{x^8}{9}$

$$\boxed{\dfrac{1}{x^{-8}} = x^8}$$

(e) $\left(\dfrac{x^2}{y^{-2}}\right)^{-3} = \dfrac{x^{-6}}{y^6} = \dfrac{1}{x^6 y^6}$

(f) $\dfrac{x^2 y^3}{x^5 y} = x^{2-5} \cdot y^{3-1} = x^{-3} y^2 = \dfrac{y^2}{x^3}$

Multinomial expressions   The laws of exponents are also useful in simplifying multinomial expressions. However, you must be careful to avoid certain common mistakes. Pay special attention to the CAUTION arrows in the next example.

EXAMPLE 3   Simplify $(2x^{-1} + y)^{-2}$. Express the answer with positive exponents.

Solution   First, rewrite the expression with only positive exponents:

$$(2x^{-1} + y)^{-2} = \frac{1}{(2x^{-1} + y)^2} = \frac{1}{\left(\dfrac{2}{x} + y\right)^2}$$

Now, simplify:

$$\frac{1}{\left(\dfrac{2 + xy}{x}\right)^2} = \frac{1}{\dfrac{(2 + xy)^2}{x^2}} = \frac{x^2}{(2 + xy)^2}$$

We do not have to expand the final denominator.

CAUTION #1 ▶ $(2x^{-1} + y)^{-2}$ is **not** equal to $(2x^{-1})^{-2} + y^{-2}$

$(a + b)^n$ is **not** equal to $a^n + b^n$ ◀

CAUTION #2 ▶   Remember that

$$\boxed{\text{The exponent is on this factor}}$$

$$2x^{-1} = 2 \cdot x^{-1} = 2 \cdot \frac{1}{x} = \frac{2}{x} ◀$$

$$\boxed{\text{Separate factors}}$$

EXAMPLE 4   Simplify: $\dfrac{mn^{-1} - m^{-1}}{m^{-2} + n^{-2}}$

Solution

First, rewrite with positive exponents.

$$\frac{mn^{-1} - m^{-1}}{m^{-2} + n^{-2}} = \frac{\dfrac{m}{n} - \dfrac{1}{m}}{\dfrac{1}{m^2} + \dfrac{1}{n^2}}$$

Second, simplify the numerator and denominator separately.

$$= \frac{\dfrac{m^2 - n}{mn}}{\dfrac{n^2 + m^2}{m^2 n^2}}$$

Third, perform the indicated division.

$$= \frac{m^2 - n}{mn} \div \frac{n^2 + m^2}{m^2 n^2}$$

$$= \frac{m^2 - n}{mn} \cdot \frac{m^2 n^2}{n^2 + m^2}$$

$$= \frac{mn(m^2 - n)}{n^2 + m^2}$$

EXAMPLE 5   Simplify: $2(x - 3)^2(2x + 5)^{-2} - (x - 3)(2x + 5)^{-1}$

Solution

First, rewrite with positive exponents.

$$2(x - 3)^2(2x + 5)^{-2} - (x - 3)(2x + 5)^{-1}$$

$$= \frac{2(x - 3)^2}{(2x + 5)^2} - \frac{x - 3}{2x + 5}$$

Then, subtract.

$$= \frac{2(x - 3)^2 - (x - 3)(2x + 5)}{(2x + 5)^2}$$

$$= \frac{2(x^2 - 6x + 9) - (2x^2 - x - 15)}{(2x + 5)^2}$$

$$= \frac{2x^2 - 12x + 18 - 2x^2 + x + 15}{(2x + 5)^2}$$

$$= \frac{-11x + 33}{(2x + 5)^2}$$

Exercises 11-1

Simplify and express all answers with positive exponents.

1. $a^6 a^{-2}$
2. $x^{-4} x^5$
3. $y^{-7} y^3$
4. $m^2 m^{-5}$
5. $(2x^{-2} y)^3$
6. $(3w^2 v^{-3})^4$
7. $(4mn^{-2})^{-2}$
8. $(-2x^{-2} y^3)^{-4}$

9. $\dfrac{a^2}{a^6}$ **10.** $\dfrac{b^3}{b^8}$ **11.** $\dfrac{c^3 d}{cd^4}$ **12.** $\dfrac{6m^2 n^5}{2m^6 n^2}$

13. $\dfrac{a^{-3} b^4 c^{-2}}{a^{-4} b^{-2} c^3}$ **14.** $\dfrac{5s^{-3} t^2 u}{s^{-2} t^5 u^{-4}}$ **15.** $\left(\dfrac{x^3}{y^2}\right)^{-2} \div \left(\dfrac{x^2}{y^4}\right)^2$ **16.** $\left(\dfrac{a^{-2}}{b}\right)^{-2} \div \left(\dfrac{a^4}{b^{-2}}\right)^{-3}$

17. $\left(\dfrac{2}{x^2}\right)^{-2}$ **18.** $\left(\dfrac{3}{y^4}\right)^{-1}$ **19.** $\left(\dfrac{w^3}{v^{-2}}\right)^{-5}$ **20.** $\left(\dfrac{2x^{-3}}{y^{-2}}\right)^{-4}$

21. $5x^0$ **22.** $(5x)^0$ **23.** $(-5)^0$ **24.** -5^0

25. $(x + 2)^{-1}$ **26.** $(3x - 4)^{-2}$ **27.** $3^{-1} + y^{-1}$ **28.** $a^{-2} + 4^{-2}$

29. $\left(\dfrac{a}{2b}\right)^{-2} \left(\dfrac{b^{-2}}{a}\right)^3$ **30.** $\left(\dfrac{x^2}{y}\right)^{-3} \left(\dfrac{x^{-1}}{y}\right)^2$ **31.** $x^{-2} y^3 + x^{-1} y$ **32.** $m^2 n^{-1} - n^2 m^{-1}$

33. $(2x^{-3} y^2)^{-1} - x^{-2}$ **34.** $a^{-3} + (3a^2 b^{-1})^{-2}$ **35.** $(m^{-1} + n^{-1})^{-1}$ **36.** $(x^{-1} + y)^{-1}$

37. $(3a + b^{-1})^{-2}$ **38.** $(w^{-1} + 2v^{-1})^2$ **39.** $\dfrac{x^{-1} + y^{-1}}{x^{-1} - y^{-1}}$ **40.** $\dfrac{a^{-2} - b^{-2}}{a^{-2} + b^{-2}}$

41. $\dfrac{2m^{-2} + n^{-2}}{m^{-1} - n^{-1}}$ **42.** $\dfrac{x^{-3} + x^{-2} y}{yx^{-1} - x^{-2}}$ **43.** $(x - 4)^{-2} + 3(x + 1)(x - 4)^{-3}$

44. $2(y + 1)(y - 1)^{-1} - (y - 1)^{-2}(y + 1)^2$

45. *Dimensional analysis*   In a certain formula for power output, the equation $P = \text{kg} \cdot \text{m}^{-3} \cdot \text{m}^2 \cdot (\text{ms}^{-1})^3$ is found. Simplify the right side of this equation and express it with positive exponents only. The result gives the dimension of the power unit, the watt.

46. *Astrophysics*   In the study of astrophysics the expression below appears. Simplify this expression and write it with positive exponents only.

$$\dfrac{-MG}{rc^2} \left(1 + \dfrac{MG}{rc^2}\right)^{-1}$$

47. *Electrical engineering*   The ratio of the conductivities of a certain material at two different temperatures is given below. Rewrite the ratio on the right side with positive exponents only.

$$\dfrac{n_{325}}{n_{295}} = \dfrac{e^{-17.9}}{e^{-19.7}}$$

48. *Thermodynamics*   The number of molecules in states with energy E is given by $n = ge^{-a - bE}$. Assuming that a, b, and E are all positive, rewrite the expression on the right with positive exponents only.

11-2 | Fractional Exponents

Meaning of a fractional exponent   In previous work with exponents or factoring, you may have noticed that in order to find the square root of a number in exponential notation, you simply divide the exponent by 2. For example,

$$\sqrt{2^6} = \sqrt{64} = 8 \quad \text{which is} \quad 2^3$$

In exponential form, we write

$$\sqrt{2^6} = 2^{6/2} = 2^3$$

This technique can be extended to other roots as well. The *cube* root of a number in exponential notation can be found be dividing the exponent by 3. For example,

$$\sqrt[3]{2^6} = \sqrt[3]{64} = 4 \qquad \text{which is} \qquad 2^2$$

In exponential form,

$$\sqrt[3]{2^6} = 2^{6/3} = 2^2$$

In general, **fractional exponents** have the following meaning:

Definition of Fractional Exponents

$$a^{m/n} = \sqrt[n]{a^m} = \left(\sqrt[n]{a}\right)^m \qquad \text{where } m \text{ and } n \text{ are integers} \qquad \textbf{(11-8)}$$
$$\text{and } n \text{ is positive}$$

Since the special case where $m = 1$ occurs so often, we state it below as a separate definition:

$$a^{1/n} = \sqrt[n]{a} \qquad \text{where } n \text{ is a positive integer} \qquad \qquad \textbf{(11-9)}$$

NOTE

From Chapter 1 we know that when a is negative and n is even, $\sqrt[n]{a}$ is not a real number, and therefore we will not apply definitions (11-8) and (11-9) in these cases. ◄

EXAMPLE 6   Find: (a) $16^{1/4}$ (b) $(-8)^{1/3}$ (c) $2 \cdot 27^{2/3}$

Solution
(a)   From definition (11-9), $16^{1/4} = \sqrt[4]{16}$, and since $16 = 2^4$, we have $\sqrt[4]{16} = 2$.
(b) $(-8)^{1/3} = \sqrt[3]{-8} = -2$
(c)   According to the order of operations, we must first simplify exponents. Notice that the definition of fractional exponents (11-8) gives us two ways to do this. Using the first method,

$$2 \cdot 27^{2/3} = 2\sqrt[3]{27^2} = 2\sqrt[3]{729} \qquad \text{Since } 729 = 9^3,$$

$$= 2 \cdot 9 = 18$$

Here, we squared 27 first and then took the cube root. But, using the second method of (11-8), we could have obtained the same answer by taking the cube root first and then squaring the result:

$$2\sqrt[3]{27^2} = 2\left(\sqrt[3]{27}\right)^2 = 2 \cdot 3^2 = 2 \cdot 9 = 18$$

In general, the second method is simpler, because it avoids large numbers.

EXAMPLE 7   Find $81^{-5/4}$.

Solution

$$81^{-5/4} = \frac{1}{81^{5/4}} = \frac{1}{\sqrt[4]{81^5}} = \frac{1}{\left(\sqrt[4]{81}\right)^5} = \frac{1}{3^5} = \frac{1}{243}$$

$$81 = 3^4 \text{ and } \sqrt[4]{3^4} = 3$$

Fractional exponents with a calculator

In Section 1-6 we showed how to find roots other than square roots on a calculator using the $\boxed{\sqrt[x]{y}}$ key.* For example, to find $\sqrt[4]{38,416}$, enter

38416 $\boxed{\sqrt[x]{y}}$ 4 $\boxed{=}$ → ▨▨▨▨▨ *14.*

However, since

$$\sqrt[4]{38,416} = 38,416^{1/4} = 38416^{0.25}$$

we can also enter

38416 $\boxed{y^x}$ .25 $\boxed{=}$ → ▨▨▨▨▨ *14.*

Therefore, any fractional power can be calculated directly using the $\boxed{y^x}$ key, but first the exponent must be written in decimal form.

EXAMPLE 8
(a)   Find $48^{3/4}$.
(b)   The formula $A = 0.5^{t/h}$ gives the fraction A of a radioactive substance remaining after t years, given that its half-life is h years. If strontium-90 has a half-life of 28 years, what fraction of a given amount will remain after 15 years?

Solution
(a)   Since $\frac{3}{4} = 0.75$, we have
   48 $\boxed{y^x}$ .75 $\boxed{=}$ → ▨ *18.236056*
(b)   **First,** substitute $h = 28$ and $t = 15$ into the formula:

   $$A = 0.5^{15/28}$$

   Then, calculate A:

   .5 $\boxed{y^x}$ $\boxed{(}$ 15 $\boxed{\div}$ 28 $\boxed{)}$ $\boxed{=}$ → ▨ *0.6898171*

   $A \approx 0.69$ or 69%

In Example 8(b), notice the use of parentheses to calculate the decimal equivalent of the exponent.

Operations with fractional exponents

All the previously defined laws of exponents are valid for fractional exponents when used with positive bases. The next example illustrates their use.

---

* Some calculators have a key marked $\boxed{x^{1/y}}$ for finding roots, while others use the $\boxed{\text{INV}}$ $\boxed{y^x}$ sequence. See your owner's manual for details.

EXAMPLE 9   Simplify each expression using the laws of exponents.

(a) $3^{1/2} \cdot 3^{3/4}$ (b) $(8^{2/3})^{1/2}$ (c) $(32m^{1/3})^{3/5}$

(d) $\left(\dfrac{x^{3/4}}{x^{3/2}}\right)^{2/3}$ (e) $\dfrac{49^{5/4}}{49^{3/4}}$ (f) $\dfrac{m^{5/6}}{m}$

(g)   Enter the original expression given in part (a) on a calculator. Then, enter its simplified answer and compare the results.

Solution

(a) $3^{1/2} \cdot 3^{3/4} = 3^{1/2+3/4} = 3^{5/4}$

(b) $(8^{2/3})^{1/2} = 8^{(2/3) \cdot (1/2)} = 8^{1/3} = 2$

$$\boxed{32^{3/5} = \left(\sqrt[5]{32}\right)^3 = \left(\sqrt[5]{2^5}\right)^3 = 2^3 = 8}$$

(c) $(32m^{1/3})^{3/5} = 32^{3/5}m^{(1/3) \cdot (3/5)} = 8m^{1/5}$

(d) $\left(\dfrac{x^{3/4}}{x^{3/2}}\right)^{2/3} = (x^{3/4-3/2})^{2/3} = (x^{-3/4})^{2/3} = x^{-1/2} = \dfrac{1}{x^{1/2}}$

(e) $\dfrac{49^{5/4}}{49^{3/4}} = 49^{5/4-3/4} = 49^{1/2} = 7$

(f) $\dfrac{m^{5/6}}{m} = m^{5/6-1} = m^{-1/6} = \dfrac{1}{m^{1/6}}$

(g) $3 \boxed{y^x} .5 \boxed{\times} 3 \boxed{y^x} .75 \boxed{=} \rightarrow$ `3.9482220`

$3 \boxed{y^x} 1.25 \boxed{=} \rightarrow$ `3.9482220`

$$\boxed{\dfrac{5}{4} = 1.25}$$

The two answers agree.

NOTE ▶

Example 9(g) shows that the laws of exponents can be valuable even when using a calculator. Most students can arrive at a numerical answer more quickly by combining the exponents first and then entering the result on a calculator. ◀

Fractional exponents are particularly useful when simplifying complex algebraic expressions. For example, it is much easier to simplify Example 9(d) when it is stated with fractional exponents than when it is stated in radical form as

$$\sqrt[3]{\left(\dfrac{\sqrt[4]{x^3}}{\sqrt{x^3}}\right)^2}$$

Here are more illustrations of the use of fractional exponents in algebraic expressions.

EXAMPLE 10   Simplify each expression and give all answers with positive exponents.

(a) $\left(\dfrac{7^2 m^3 n^{-7/2}}{2^{4/3} m^{1/3} n^{-1/2}}\right)^{3/2}$ (b) $(8x^3)^{-2/3} - 2x^{-3}$

(c) $(2x + 3)(x - 4)^{-1/3} - (x - 4)^{2/3}$

Solution

(a) **Step 1.** Working inside the parentheses, combine powers of like bases. Notice that each variable is moved to the part of the fraction where the combined powers are positive.

$$\left(\frac{7^2 m^{3-1/3}}{2^{4/3} n^{-1/2+7/2}}\right)^{3/2} = \left(\frac{7^2 m^{8/3}}{2^{4/3} n^3}\right)^{3/2}$$

Step 2. Raise the entire fraction to the indicated power.

$$= \frac{7^3 m^{(8/3)\cdot(3/2)}}{2^{(4/3)\cdot(3/2)} n^{3\cdot(3/2)}}$$

$$= \frac{343 m^4}{4 n^{9/2}}$$

(b) **Step 1.** To rewrite with positive exponents, raise the first product to the indicated power.

$$(8x^3)^{-2/3} - 2x^{-3}$$

$$= 8^{-2/3} x^{-2} - 2x^{-3}$$

$$\boxed{x^{3(-2/3)} = x^{-2}}$$

Step 2. Rewrite both terms with positive exponents.

$$= \frac{1}{4x^2} - \frac{2}{x^3}$$

Step 3. Combine the two fractions. (The LCD is $4x^3$.)

$$= \frac{x - 8}{4x^3}$$

(c) **Step 1.** Rewrite the expression with positive exponents.

$$(2x + 3)(x - 4)^{-1/3} - (x - 4)^{2/3}$$

$$= \frac{2x + 3}{(x - 4)^{1/3}} - (x - 4)^{2/3}$$

Step 2. Combine both terms into one fraction. (The LCD is $(x - 4)^{1/3}$.)

$$= \frac{2x + 3 - (x - 4)^{2/3}(x - 4)^{1/3}}{(x - 4)^{1/3}}$$

Step 3. Perform the indicated multiplication.

$$= \frac{2x + 3 - (x - 4)}{(x - 4)^{1/3}}$$

$$\boxed{(x - 4)^{2/3}(x - 4)^{1/3} = (x - 4)^1}$$

Step 4. Subtract in the numerator. Remember to change both signs of $(x - 4)$.

$$= \frac{x + 7}{(x - 4)^{1/3}}$$

Exercises 11-2

Evaluate each expression. Simplify first using the laws of exponents if necessary.

1. $16^{1/2}$ 2. $144^{1/2}$ 3. $8^{1/3}$ 4. $81^{1/4}$
5. $3(-64)^{2/3}$ 6. $2 \cdot 16^{3/4}$ 7. $100^{-1/2}$ 8. $(-27)^{-4/3}$

9. $5^{1/2} \cdot 5^{3/2}$ **10.** $3^{2/3} \cdot 3^{7/3}$ **11.** $(2^8)^{3/4}$ **12.** $(3^6)^{-2/3}$

13. $\dfrac{1000^{1/3}}{1000^{2/3}}$ **14.** $\dfrac{81^{-1/2}}{49^{1/2}}$ **15.** $64^{-1/2} - 8^{-2/3}$ **16.** $\dfrac{36^{1/2}}{25^{-1/2}} + \dfrac{4^{-1/2}}{25^{1/2}}$

Calculate and round to the nearest 0.01.

17. $56^{1/2}$ **18.** $137^{1/2}$ **19.** $175^{2/3}$ **20.** $66^{5/6}$

21. $2.85^{0.64}$ **22.** $3(24.8^{3.12})$ **23.** $2(0.2064^{-4/3})$ **24.** $580^{-2/5}$

25. $4^{3/5} \cdot 4^{1/5}$ **26.** $6^{-2/3} \cdot 6^{1/3}$ **27.** $3^{1/4} \cdot 4^{2/3}$ **28.** $5^{1/3} \cdot 2^{3/4}$

29. $(6^{2/3})^{-1/4}$ **30.** $(12^{-1/3})^{3/4}$ **31.** $(2.4^{1.6})^{0.5}$ **32.** $(0.185^{3.2})^{-1.5}$

33. $\dfrac{7^{4/5}}{7^{6/5}}$ **34.** $\dfrac{32^{5/8}}{24^{-3/8}}$ **35.** $\dfrac{500^{1/2}}{200^{1/2}} - \dfrac{80^{1/4}}{60^{3/4}}$ **36.** $120^{3/4} + 80^{2/3}$

Simplify each expression and give all answers with exponents that are positive rational numbers.

37. $x^{1/4} x^{1/2}$ **38.** $y^{2/3} y^{4/3}$ **39.** $a^{-3/4} a^{1/4}$ **40.** $c^{-1/3} c^{-4/3}$

41. $w^{1/2} w$ **42.** $x^{-1/3} x$ **43.** $\dfrac{m^{3/4}}{m^{-1/4}}$ **44.** $\dfrac{n^{-2/3}}{n^{1/3}}$

45. $\dfrac{y^{3/4}}{y^{1/2}}$ **46.** $\dfrac{b^{-1/4}}{b^{1/3}}$ **47.** $\dfrac{m}{m^{-2/3}}$ **48.** $\dfrac{a^{5/3}}{a}$

49. $\dfrac{x^{1/3} x^{1/2}}{x^{1/4}}$ **50.** $\dfrac{y^{-1/3}}{y^{3/4} y^{3/2}}$ **51.** $(b^{3/4})^{1/3}$ **52.** $(c^{5/3})^{3/4}$

53. $(27x^{-3/4})^{2/3}$ **54.** $(64a^{-2/3} b^{1/2})^{3/2}$ **55.** $\left(\dfrac{m^{5/4}}{m^{1/3}}\right)^{2/3}$ **56.** $\left(\dfrac{16v^{-3}}{v^{2/3}}\right)^{3/4}$

57. $\left(\dfrac{x^{5/3} y^{1/4}}{x^{1/6} y^{1/2}}\right)^{-2/3}$ **58.** $\left(\dfrac{4m^6 n^{-2/3}}{3^{2/3} m^{3/4} n^2}\right)^{1/2}$ **59.** $(16x^4)^{-3/4} + 4x^{-2}$ **60.** $(2y^{-1/2})^2 - (8y^3)^{-1/3}$

61. $(x + 5)^{-1/2}(x + 2) - (x + 5)^{1/2}$ **62.** $(3y + 1)(y - 6)^{-2/3} - (y - 6)^{1/3}$

63. *Marine biology* In the study of photosynthesis in a lake it is found that the light intensity is reduced 15% through a depth of 20 cm. The formula $x = (0.85)^{d/20}$ will give the approximate fraction of surface light intensity at a depth d. Find x at a depth of 50 cm.

64. *Nuclear engineering* In Example 8(b), what fraction of a given sample of strontium-90 will remain after 20 years?

65. *Cost analysis* The formula $C = aQ^b$ gives the cost of the Qth item manufactured when a is the cost of the first and b is a constant. Find the cost of the 15th airplane built by a company if the first airplane cost $35,100,000 and $b = -0.248$.*

66. *Hydrology* Under certain conditions, the equation for soil erosive loss by water is $X = 0.065S^{1.49}$, where X is the number of metric tons of soil lost per year per hectare and S is the percent slope being eroded. Calculate X for $S = 25\%$.

67. *Civil engineering* The formula for determining the flow rate Q of water passing through a certain culvert is $Q = 99.3AR^{2/3} S^{1/2}$ ft³/s, where A is the end area of water in the culvert, R is the perimeter of the culvert exposed to water, and S is the slope of the culvert. Find Q when $A = 0.20$ ft², $R = 1.1$ ft, and $S = 0.020$ ft/ft.

68. *Astrophysics* The ratio below is used to find the number of stars that are brighter than a given one:

$$\frac{r_0}{r_1} = 2.512^{1/2}$$

Calculate $(r_0/r_1)^3$.

*Courtesy of Arthur Kluge, Tecolote Research, Inc., Santa Barbara, California.

11-3 | Simplest Radical Form

The roots of numbers can be expressed both with fractional exponents and with radical symbols. One means of expression is not *always* preferable to the other, so we must be able to simplify expressions written in either form. The remainder of this chapter will deal with expressions in radical form.

Laws of exponents in radical form

Several of the laws of exponents can be rewritten with radical symbols and used to help us simplify radical expressions. They are listed in the box.

Exponential Form	Radical Form		
$(a^n)^{1/n} = a$	$\sqrt[n]{a^n} = a$	nth Root of nth Power	(11-10)
$(ab)^{1/n} = a^{1/n} \cdot b^{1/n}$	$\sqrt[n]{ab} = \sqrt[n]{a} \cdot \sqrt[n]{b}$	Root of a Product	(11-11)
$(a^{1/n})^{1/m} = a^{1/mn}$	$\sqrt[m]{\sqrt[n]{a}} = \sqrt[mn]{a}$	Root of a Root	(11-12)
$\left(\dfrac{a}{b}\right)^{1/n} = \dfrac{a^{1/n}}{b^{1/n}}$	$\sqrt[n]{\dfrac{a}{b}} = \dfrac{\sqrt[n]{a}}{\sqrt[n]{b}}$	Root of a Quotient	(11-13)

Here, a and b are assumed to represent positive real numbers, and m and n are positive integers.

Criteria for simplest radical form

Four basic requirements must be satisfied for a radical expression to be in **simplest radical form:**

- There must be no perfect nth-root factor under a radical with index n.
- Multiple radicals—that is, radicals within radicals—must be combined.
- The denominator of a fraction must not contain a radical.
- The order of the radical must, if possible, be reduced.

The following examples show how each of these requirements is met.

Removing perfect nth-root factors

In Section 1-6 we simplified $\sqrt{48}$ by removing the largest factor that was a perfect square (16) from the radicand as follows:

$$\sqrt{48} = \sqrt{16 \cdot 3} = \sqrt{16} \cdot \sqrt{3} = 4\sqrt{3}$$

The same procedure can be used to simplify radicals of any order n with radicands that have perfect nth-root factors.

EXAMPLE 11 Simplify $\sqrt[3]{54}$. Use a calculator to compare the decimal equivalents for the original expression and the simplified answer.

Solution

Step 1. The largest perfect *cube* factor of 54 is 27. $\sqrt[3]{54} = \sqrt[3]{27 \cdot 2}$

$$27 = 3^3$$

Step 2. Use the root of a product law (11-11) to write: $= \sqrt[3]{27} \cdot \sqrt[3]{2}$

Step 3. Use law (11-10) to obtain: $= \sqrt[3]{3^3} \cdot \sqrt[3]{2}$

$$= 3\sqrt[3]{2}$$

Check this answer as shown below:

Original problem: $54 \;\boxed{\sqrt[x]{y}}\; 3 \;\boxed{=}\; \rightarrow$ *3.7797631*

Simplified answer: $3 \;\boxed{\times}\; 2 \;\boxed{\sqrt[x]{y}}\; 3 \;\boxed{=}\; \rightarrow$ *3.7797631*

NOTE ▶

Notice in Example 11 that it requires more steps to evaluate the simplified version with a calculator. The simplified version is not always quicker to calculate, but it is the exact equivalent, and it is the standard form when the radicand involves literal numbers. ◀

EXAMPLE 12 Simplify: (a) $\sqrt{x^3 y^7}$ (b) $\sqrt[4]{48x^5}$

Solution

(a) $\sqrt{x^3 y^7} = \sqrt{x^2 \cdot x \cdot y^6 \cdot y} = \sqrt{x^2 y^6} \cdot \sqrt{xy} = xy^3 \sqrt{xy}$

(b) $\sqrt[4]{48x^5} = \sqrt[4]{16 \cdot 3 \cdot x^4 \cdot x} = \sqrt[4]{16x^4} \cdot \sqrt[4]{3x} = 2x\sqrt[4]{3x}$

Perfect 4th-root factors

Radicals within radicals

The root of a root law (11-12) is used to simplify expressions containing radicals within radicals. These multiple radicals must be combined into one radical with an index that is the product of the original indexes.

EXAMPLE 13 Simplify: (a) $\sqrt[4]{\sqrt[3]{7}}$ (b) $\sqrt[3]{\sqrt{2y}}$ (c) $\sqrt{2\sqrt{x}}$

Solution

(a) $\sqrt[4]{\sqrt[3]{7}} = \sqrt[12]{7}$

Index of 2 $3 \cdot 2 = 6$

(b) $\sqrt[3]{\sqrt{2y}} = \sqrt[6]{2y}$

(c) $\quad \sqrt{2\sqrt{x}} = \sqrt{\sqrt{4x}} = \sqrt[4]{4x}$

Square 2 to place
it under the innermost radical

**Rationalizing the
denominator**

The third requirement for simplest radical form is that the denominator of a fraction must not contain a radical. To achieve this, we multiply the numerator and denominator by the smallest quantity that results in a rational denominator. Thus, the procedure is called **rationalizing the denominator.**

EXAMPLE 14   Simplify: $\dfrac{\sqrt{5}}{\sqrt{3x}}$

Solution

Multiply both the numerator and denominator by $\sqrt{3x}$ to rationalize the denominator.
$$\frac{\sqrt{5}}{\sqrt{3x}} = \frac{\sqrt{5}\cdot\sqrt{3x}}{\sqrt{3x}\cdot\sqrt{3x}} = \frac{\sqrt{15x}}{3x}$$

Notice that the root of a product law (11-11) is used in reverse to combine the two radicals in the numerator.

This same procedure must be used when a fraction is under a radical.

EXAMPLE 15   Simplify: (a) $\sqrt{\dfrac{5}{2}}$ (b) $\sqrt[3]{\dfrac{y}{x^2}}$

Solution

(a)   Use the root of a quotient law (11-13) to rewrite.
$$\sqrt{\frac{5}{2}} = \frac{\sqrt{5}}{\sqrt{2}}$$

Multiply the numerator and denominator by $\sqrt{2}$.
$$\frac{\sqrt{5}}{\sqrt{2}}\cdot\frac{\sqrt{2}}{\sqrt{2}} = \frac{\sqrt{10}}{2}$$

(b)   Rewrite.
$$\sqrt[3]{\frac{y}{x^2}} = \frac{\sqrt[3]{y}}{\sqrt[3]{x^2}}$$

Multiply the numerator and denominator by $\sqrt[3]{x}$.
$$\frac{\sqrt[3]{y}}{\sqrt[3]{x^2}}\cdot\frac{\sqrt[3]{x}}{\sqrt[3]{x}} = \frac{\sqrt[3]{xy}}{\sqrt[3]{x^3}} = \frac{\sqrt[3]{xy}}{x}$$

The product of
the radicands is
a perfect cube

**Reducing the order of a
radical**

The final requirement for simplest radical form is that the order of a radical be reduced if possible. Fractional exponents are useful in this procedure. A numerical example is given below to demonstrate.

EXAMPLE 16 Simplify: $\sqrt[4]{9}$

Solution

First, rewrite the radicand in exponential notation. $\sqrt[4]{9} = \sqrt[4]{3^2}$

Then, rewrite this expression using fractional exponents. $= 3^{2/4}$

If the fractional exponent can be reduced, then the order of $= 3^{1/2}$
the radical is also reduced.

$= \sqrt{3}$

EXAMPLE 17 Simplify: (a) $\sqrt[4]{4x^6}$ (b) $\sqrt[6]{9x^2y^4}$

Solution
(a) $\sqrt[4]{4x^6} = \sqrt[4]{x^4 \cdot 4x^2} = \sqrt[4]{x^4} \cdot \sqrt[4]{2^2x^2}$

$= x \cdot 2^{2/4} \cdot x^{2/4}$

$= x \cdot 2^{1/2} \cdot x^{1/2}$

$= x\sqrt{2x}$

(b) $\sqrt[6]{9x^2y^4} = \sqrt[6]{3^2x^2y^4} = 3^{2/6} \cdot x^{2/6} \cdot y^{4/6}$

$= 3^{1/3} \cdot x^{1/3} \cdot y^{2/3}$

$= \sqrt[3]{3xy^2}$

Multinomial radicands   The requirements for simplest radical form also apply to radicands that are
multinomials.

EXAMPLE 18 Simplify: $\sqrt{2x^2 + 6x + \dfrac{9}{2}}$

Solution

Step 1. Combine all terms
into one fraction.

$\sqrt{2x^2 + 6x + \dfrac{9}{2}} = \sqrt{\dfrac{4x^2 + 12x + 9}{2}}$

Step 2. Factor the
numerator.

$= \sqrt{\dfrac{(2x + 3)^2}{2}}$

Step 3. Remove the largest
perfect square factor. If it is a
multinomial, it must be
placed in parentheses.

$= (2x + 3)\sqrt{\dfrac{1}{2}}$

Step 4. Rationalize the
denominator.

$= (2x + 3)\dfrac{\sqrt{1} \cdot \sqrt{2}}{\sqrt{2} \cdot \sqrt{2}}$

$= \dfrac{(2x + 3)\sqrt{2}}{2}$

CAUTION ▶   As shown in Steps 1 and 2 of Example 18, we must always combine terms and factor *before* simplifying. Remember, $\sqrt[n]{a + b}$ is *not* equal to $\sqrt[n]{a} + \sqrt[n]{b}$. ◀

Exercises 11-3

Express each of the following in simplest radical form:

1. $\sqrt{18}$
2. $\sqrt{60}$
3. $\sqrt{125}$
4. $\sqrt{72}$
5. $\sqrt{x^5 y^2}$
6. $\sqrt{m^4 n^3}$

7. $\sqrt{v^5 w^5}$
8. $\sqrt{a^3 b^9}$
9. $\sqrt{7c^6}$
10. $\sqrt{36q^3}$
11. $\sqrt{27m^5 n^6}$
12. $\sqrt{72a^4 b^7}$

13. $\sqrt[3]{24}$
14. $\sqrt[3]{-81}$
15. $\sqrt[5]{-64}$
16. $\sqrt[4]{32}$
17. $\sqrt[3]{125m^4}$
18. $\sqrt[3]{16n^5}$

19. $\sqrt[4]{48a^3 b^6}$
20. $\sqrt[5]{32x^2 y^6}$
21. $\sqrt{\sqrt{3}}$
22. $\sqrt[4]{\sqrt{5}}$
23. $\sqrt[4]{3\sqrt{x}}$
24. $\sqrt[3]{2\sqrt{2y}}$

25. $\dfrac{3}{\sqrt{2y}}$
26. $\dfrac{\sqrt{5x}}{\sqrt{6}}$
27. $\dfrac{\sqrt{2}}{\sqrt[3]{3m}}$
28. $\dfrac{\sqrt[4]{5a}}{\sqrt{2}}$
29. $\sqrt{\dfrac{3}{5}}$
30. $\sqrt{\dfrac{7}{2}}$

31. $\sqrt{\dfrac{x}{5}}$
32. $\sqrt{\dfrac{6}{y}}$
33. $\sqrt[3]{\dfrac{2}{3}}$
34. $\sqrt[4]{\dfrac{3}{2}}$
35. $\sqrt[5]{\dfrac{x^3}{7}}$
36. $\sqrt[3]{\dfrac{a^2}{b}}$

37. $\sqrt[4]{36}$
38. $\sqrt[4]{100}$
39. $\sqrt[9]{8}$
40. $\sqrt[8]{16}$
41. $\sqrt[4]{x^2 y^2}$
42. $\sqrt[8]{49a^2 b^6}$

43. $\sqrt{\dfrac{1}{4} + \dfrac{3}{2}}$
44. $\sqrt{\dfrac{a}{b} - \dfrac{b}{a}}$
45. $\sqrt{x^2 + 4x + 4}$
46. $\sqrt{y^2 - 8y + 16}$

47. $\sqrt{2x^2 - 2x + \dfrac{1}{2}}$
48. $\sqrt{3x^2 + 4x + \dfrac{4}{3}}$
49. $\sqrt{\dfrac{7x^{-2} y^2 z}{27x^3 y^{-1} z^3}}$
50. $\sqrt{\dfrac{8a^5 b^2 c^{-3}}{49a^2 b^6 c^{-5}}}$

51. $\sqrt{\dfrac{x + 1}{x - 1}}$
52. $\sqrt{\dfrac{y - 2}{y}}$

53. **Physics**   The frequency of oscillation f of a simple pendulum is given by the formula

$$f = \frac{1}{2\pi} \sqrt{\frac{g}{L}}$$

where $g \approx 9.8 \text{ m/s}^2$ (the acceleration due to gravity) and L is the length of the pendulum (in meters). Express f in simplest radical form when $L = 0.49$ m. Then calculate the answer to the nearest hundredth.

54. **Physics**   Under ideal conditions, the velocity v of an object falling freely from rest from a height h is given by $v = \sqrt{2gh}$, where g is the acceleration due to gravity. Express this velocity in simplest radical form for an object dropped from a height of 5 m. Use $g = 9.8 \text{ m/s}^2$.

55. **Mechanical engineering**   In calculating the impact factor for dynamic loading, the formula given below is useful. Rewrite the right side of the formula in simplest radical form.

$$k = 1 + \sqrt{1 + \frac{2h}{m}}$$

56. **Optics**   According to the Doppler effect, the perceived frequency of a signal depends on the relative motion of the source and the observer. This frequency f' when the observer is moving away from the source with velocity v is given by the formula below, where c is the velocity of sound and f is the actual frequency of the signal.* Express this in simplest radical form.

$$f' = \frac{fc}{c - v} \sqrt{1 - \frac{v^2}{c^2}}$$

*Based on Hugh D. Young, *Fundamentals of Optics and Modern Physics* (New York: McGraw-Hill, 1968), p. 56.

11-4 | Addition and Subtraction of Radicals

Like radicals   The distributive property allows us to combine like radicals in the same way we combine like terms in an algebraic expression. **Like radicals** are defined as radicals that have the same index and the same radicand. Their numerical coefficients may differ. In general:

$$b\sqrt[n]{a} + c\sqrt[n]{a} = (b + c)\sqrt[n]{a} \qquad\qquad \text{(11-14)}$$

Same index (pointing to index)
Same radicand (pointing to radicand)

EXAMPLE 19   Simplify:
(a) $3\sqrt{3} + 5\sqrt{3}$ (b) $2\sqrt[3]{x} - 7\sqrt[3]{x} + \sqrt[3]{x}$
(c) $5\sqrt{7} + 2\sqrt{5} - 4\sqrt{7}$ (d) $7\sqrt[3]{2y} - 5\sqrt{2y} + 2\sqrt[3]{2y}$

Solution
(a) $3\sqrt{3} + 5\sqrt{3} = (3 + 5)\sqrt{3} = 8\sqrt{3}$
(b) $2\sqrt[3]{x} - 7\sqrt[3]{x} + \sqrt[3]{x} = (2 - 7 + 1)\sqrt[3]{x} = -4\sqrt[3]{x}$

Coefficient is 1

(c) $5\sqrt{7} + 2\sqrt{5} - 4\sqrt{7} = \sqrt{7} + 2\sqrt{5}$

Only these two are like radicals   Answer contains two terms

(d) $7\sqrt[3]{2y} - 5\sqrt{2y} + 2\sqrt[3]{2y} = 9\sqrt[3]{2y} - 5\sqrt{2y}$

The index of 2 makes this unlike the other terms

It may not always be obvious that two terms are like terms until all radicals have been simplified.

EXAMPLE 20   Simplify: (a) $2\sqrt{10} - 3\sqrt{40} + \sqrt{90}$ (b) $\sqrt{12} - \sqrt{\dfrac{1}{3}}$

Solution

(a)   **First,** simplify all radicals. $2\sqrt{10} - 3\sqrt{40} + \sqrt{90} = 2\sqrt{10} - 6\sqrt{10} + 3\sqrt{10}$

Then, combine like radicals. $= -\sqrt{10}$

(b) $\sqrt{12} - \sqrt{\dfrac{1}{3}} = 2\sqrt{3} - \dfrac{\sqrt{3}}{3} = \dfrac{6\sqrt{3} - \sqrt{3}}{3} = \dfrac{5\sqrt{3}}{3}$

From the examples above, we can see that by first combining like radicals, we get a simpler form that is easier to calculate than the original, and the result may have fewer rounding errors.

Literal radicands   When the radicands are algebraic expressions, the final coefficient may be a multinomial.

EXAMPLE 21   Simplify: (a) $2\sqrt{x^5} - \sqrt{xy^4}$ (b) $\sqrt{\dfrac{x-y}{x}} + \sqrt{\dfrac{x}{x-y}}$

Solution

(a) **First,** simplify each radical.

$$2\sqrt{x^5} - \sqrt{xy^4} = 2x^2\sqrt{x} - y^2\sqrt{x}$$

Then, combine like radicals using the distributive property.

$$= (2x^2 - y^2)\sqrt{x}$$

(b) **First,** rationalize each denominator.

$$\sqrt{\dfrac{x-y}{x}} + \sqrt{\dfrac{x}{x-y}} = \dfrac{\sqrt{x-y}}{\sqrt{x}} + \dfrac{\sqrt{x}}{\sqrt{x-y}}$$

$$= \dfrac{\sqrt{x-y}}{\sqrt{x}} \cdot \dfrac{\sqrt{x}}{\sqrt{x}} + \dfrac{\sqrt{x}}{\sqrt{x-y}} \cdot \dfrac{\sqrt{x-y}}{\sqrt{x-y}}$$

$$= \dfrac{\sqrt{x^2 - xy}}{x} + \dfrac{\sqrt{x^2 - xy}}{x-y}$$

Then, combine fractions.

$$= \dfrac{(x-y)\sqrt{x^2 - xy} + x\sqrt{x^2 - xy}}{x(x-y)}$$

$$= \dfrac{(2x-y)\sqrt{x^2 - xy}}{x(x-y)}$$

Exercises 11-4

Perform the indicated operations and express all answers in simplest radical form.

1. $2\sqrt{2} + 4\sqrt{2}$
2. $\sqrt{3} + 3\sqrt{3}$
3. $\sqrt{5} - 2\sqrt{5}$
4. $3\sqrt{7} - 5\sqrt{7}$
5. $8\sqrt{x} - 3\sqrt{x}$
6. $2\sqrt[3]{y} + 3\sqrt[3]{y}$
7. $3\sqrt{3} - 2\sqrt{2} - 2\sqrt{3}$
8. $4\sqrt{5} - \sqrt{5} + 5\sqrt{7}$
9. $2\sqrt{5} - 3\sqrt[3]{5} - 4\sqrt[3]{5}$
10. $4\sqrt{x} + 2\sqrt[3]{x} + 5\sqrt{x}$
11. $2\sqrt{y} + \sqrt{2y} + \sqrt{y}$
12. $\sqrt[3]{x^2} - 2\sqrt[3]{x} - 3\sqrt[3]{x^2}$
13. $\sqrt{2} + \sqrt{8}$
14. $\sqrt{3} - \sqrt{75}$
15. $2\sqrt{5} - 3\sqrt{20}$
16. $3\sqrt{7} + 2\sqrt{28}$
17. $5\sqrt{6} + \sqrt{24} - 2\sqrt{54}$
18. $3\sqrt{18} - 2\sqrt{8} + 4\sqrt{72}$

19. $3\sqrt{27} + 6\sqrt{50} - \sqrt{12}$ **20.** $5\sqrt{5} - 2\sqrt{45} + 3\sqrt{60}$ **21.** $\sqrt{40} - \sqrt{\dfrac{1}{10}}$

22. $\sqrt{72} + \sqrt{\dfrac{1}{2}}$ **23.** $2\sqrt{24} + 3\sqrt{\dfrac{2}{3}}$ **24.** $5\sqrt{\dfrac{3}{5}} - 4\sqrt{60}$

25. $\sqrt{\dfrac{4}{3}} + \sqrt{12} + 2\sqrt{300}$ **26.** $\sqrt{\dfrac{5}{2}} - \sqrt{200} + \sqrt{90}$ **27.** $\sqrt[3]{16} + \sqrt[3]{54}$

28. $\sqrt[4]{32} - \sqrt[4]{162}$ **29.** $\sqrt{\sqrt{3}} - \sqrt[4]{48}$ **30.** $\sqrt[3]{25} + \sqrt{80}$

31. $\sqrt{x^2y^3} + \sqrt{4y^5}$ **32.** $2\sqrt{ab^2} - \sqrt{a^3}$ **33.** $\sqrt{12vw^4} + \sqrt{27v^3} - \sqrt{75v}$

34. $\sqrt[3]{16m} - \sqrt[3]{2mn^3} + \sqrt[3]{54m}$ **35.** $\sqrt{\dfrac{y}{3}} - 4\sqrt{\dfrac{y}{12}}$

36. $6\sqrt[3]{\dfrac{a^2}{3}} + \sqrt[3]{24a^5}$ **37.** $\sqrt{\dfrac{a+b}{a}} - \sqrt{\dfrac{a}{a+b}}$ **38.** $\sqrt{1 - \dfrac{1}{x}} + \sqrt{\dfrac{x}{x-1}}$

39. *Civil engineering* In solving for the horizontal and vertical displacements of a truss, an engineer finds it necessary to combine the terms below. Find this sum.

$$\frac{\sqrt{3}}{8}x + \frac{3\sqrt{3}}{8}x$$

40. *Mechanical engineering* In determining the stress on a mechanical plunger, it is necessary to solve the equation below. Find x in simplest radical form.

$$\frac{15\sqrt{34}}{17}x - \frac{15\sqrt{34}}{34}x - 15 = 0$$

11-5 | Multiplication of Radicals

By restating the root of a product law (11-11) in reverse as

$$\sqrt[n]{a} \cdot \sqrt[n]{b} = \sqrt[n]{ab}$$

we can see that radicals of the same order can be multiplied by expressing the product of their radicands under the same radical sign.

EXAMPLE 22 Multiply: (a) $\sqrt{7} \cdot \sqrt{3}$ (b) $\sqrt[3]{2x}\sqrt[3]{5x}$

Solution

(a) $\sqrt{7} \cdot \sqrt{3} = \sqrt{7 \cdot 3} = \sqrt{21}$ (b) $\sqrt[3]{2x}\sqrt[3]{5x} = \sqrt[3]{2x \cdot 5x} = \sqrt[3]{10x^2}$

All factors outside the radical must be multiplied separately.

EXAMPLE 23 Multiply: (a) $(3\sqrt{3})(4\sqrt{5})$ (b) $(2\sqrt{x})(3\sqrt{2})$

Solution

(a) $(3\sqrt{3})(4\sqrt{5}) = 3 \cdot 4\sqrt{3 \cdot 5} = 12\sqrt{15}$

(b) $(2\sqrt{x})(3\sqrt{2}) = 2 \cdot 3\sqrt{x \cdot 2} = 6\sqrt{2x}$

In many cases, the product will have to be simplified. Always check for this possibility.

EXAMPLE 24 Multiply: (a) $\sqrt{6}\sqrt{2}$ (b) $\sqrt[4]{6y^3}\cdot\sqrt[4]{8y^3}$ (c) $\sqrt{\dfrac{2}{3}}\sqrt{\dfrac{x}{2}}$

Solution

(a) $\sqrt{6}\sqrt{2} = \sqrt{12} = \sqrt{4}\sqrt{3} = 2\sqrt{3}$

(b) $\sqrt[4]{6y^3}\cdot\sqrt[4]{8y^3} = \sqrt[4]{48y^6} = \sqrt[4]{16\cdot3\cdot y^4\cdot y^2} = \sqrt[4]{16y^4}\sqrt[4]{3y^2} = 2y\sqrt[4]{3y^2}$

(c) $\sqrt{\dfrac{2}{3}}\sqrt{\dfrac{x}{2}} = \sqrt{\dfrac{x}{3}} = \dfrac{\sqrt{x}\cdot\sqrt{3}}{\sqrt{3}\cdot\sqrt{3}} = \dfrac{\sqrt{3x}}{3}$

When raising a monomial radical expression to a power, be certain to raise the coefficient, as well as the radical, to that power.

EXAMPLE 25 Simplify: (a) $(2\sqrt{3})^2$ (b) $(2x\sqrt[3]{y})^3$ (c) $(3\sqrt{x+2})^2$

Solution

(a) $(2\sqrt{3})^2 = 2^2(\sqrt{3})^2 = 4\cdot3 = 12$

(b) $(2x\sqrt[3]{y})^3 = (2x)^3(\sqrt[3]{y})^3 = 8x^3y$

(c) $(3\sqrt{x+2})^2 = 3^2(\sqrt{x+2})^2 = 9(x+2) = 9x + 18$

Multinomial expressions   When multinomial expressions are involved, follow the procedure you learned for multiplying algebraic expressions. It will help if you can recognize when the special products appear.

EXAMPLE 26 Multiply:

(a) $2\sqrt{3}(\sqrt{6} + 3\sqrt{3})$ (b) $(\sqrt{5} + 2)(\sqrt{5} - 2)$

(c) $(2\sqrt{x} - \sqrt{y})(3\sqrt{x} - 5\sqrt{y})$ (d) $(3 - \sqrt{6})^2$

(e) $(3 + 5\sqrt{x+2})^2$

Solution

(a) *Monomial times a binomial.* Use the distributive property:

$$2\sqrt{3}(\sqrt{6} + 3\sqrt{3}) = 2\sqrt{3}\cdot\sqrt{6} + 2\sqrt{3}\cdot3\sqrt{3}$$

$$= 2\sqrt{18} + 6\sqrt{9}$$

$$= 6\sqrt{2} + 18 \text{ or } 18 + 6\sqrt{2}$$

(b) *Sum of two terms times the difference of the same terms; difference of two squares:*

$$(\sqrt{5} + 2)(\sqrt{5} - 2) = (\sqrt{5})^2 - (2)^2 = 5 - 4 = 1$$

(c) *Binomial times a binomial.* Use FOIL:

$$(2\sqrt{x} - \sqrt{y})(3\sqrt{x} - 5\sqrt{y})$$

$$= \underbrace{2\sqrt{x}\cdot3\sqrt{x}}_{F} + \underbrace{2\sqrt{x}(-5\sqrt{y})}_{O} + \underbrace{(-\sqrt{y})(3\sqrt{x})}_{I} + \underbrace{(-\sqrt{y})(-5\sqrt{y})}_{L}$$

$$= 6x - 10\sqrt{xy} - 3\sqrt{xy} + 5y$$

$$= 6x - 13\sqrt{xy} + 5y$$

(d) *Binomial squared.* Use $(a + b)^2 = a^2 + 2ab + b^2$:

$$\left(3 - \sqrt{6}\right)^2 = 3^2 + 2(3)\left(-\sqrt{6}\right) + \left(-\sqrt{6}\right)^2$$

$$= 9 - 6\sqrt{6} + 6$$

$$= 15 - 6\sqrt{6}$$

(e) *Binomial squared:*

$$\left(3 + 5\sqrt{x + 2}\right)^2 = 3^2 + 2(3)\left(5\sqrt{x + 2}\right) + \left(5\sqrt{x + 2}\right)^2$$

$$= 9 + 30\sqrt{x + 2} + 25(x + 2)$$

$$= 59 + 30\sqrt{x + 2} + 25x$$

Multiplying radicals of different orders   Earlier, we emphasized that only radicals of like order can be multiplied using the root of a product law (11-11). However, by using fractional exponents, it is possible to multiply radicals of unlike orders.

EXAMPLE 27 Multiply: $\sqrt{3} \cdot \sqrt[3]{2x^2}$

Solution

Step 1. Express each radical in exponential form. $\sqrt{3} \cdot \sqrt[3]{2x^2} = 3^{1/2}(2x^2)^{1/3}$

Step 2. Change the fractional exponents to equivalent fractions with a common denominator. $= 3^{3/6}(2x^2)^{2/6}$

Step 3. Express the product under one radical and simplify. $= \sqrt[6]{3^3 \cdot (2x^2)^2}$

$= \sqrt[6]{108x^4}$

Exercises 11-5

Find each product and express the answer in simplest radical form.

1. $\sqrt{2}\sqrt{5}$
2. $\sqrt{11}\sqrt{6}$
3. $\sqrt{2x}\sqrt{3y}$
4. $\sqrt[3]{4y}\sqrt[3]{3y}$
5. $\left(4\sqrt{7}\right)\left(3\sqrt{5}\right)$
6. $\left(2\sqrt{x}\right)\left(5\sqrt{10y}\right)$
7. $\sqrt{5}\sqrt{10}$
8. $\sqrt{6}\sqrt{15}$
9. $\sqrt[3]{4}\sqrt[3]{6}$
10. $\sqrt[4]{2}\sqrt[4]{8}$
11. $\sqrt[3]{3m^2}\sqrt[3]{9m}$
12. $\sqrt[5]{16n^3}\sqrt[5]{4n^4}$
13. $\left(6\sqrt{5}\right)^2$
14. $\left(4\sqrt{x}\right)^2$
15. $\left(3x\sqrt[3]{2x}\right)^3$
16. $\left(2\sqrt[3]{7}\right)^3$
17. $\left(2\sqrt{x + 5}\right)^2$
18. $\left(3\sqrt{2y} - 3\right)^2$
19. $\sqrt{\dfrac{2}{5}}\sqrt{30}$
20. $\sqrt{\dfrac{x}{3}}\sqrt{\dfrac{5}{6}}$
21. $\sqrt{5}\left(\sqrt{3} - \sqrt{2}\right)$
22. $2\sqrt{6}\left(\sqrt{5} + 3\sqrt{7}\right)$
23. $2\sqrt{10}\left(4\sqrt{2} + 3\sqrt{5}\right)$

24. $\sqrt{3}\left(3\sqrt{3} - 2\sqrt{15}\right)$

25. $5\sqrt{x}\left(\sqrt{xy} - 2\sqrt{y}\right)$

26. $\sqrt{2y}\left(3\sqrt{6x} - 5\sqrt{y}\right)$

27. $\left(\sqrt{5} + 3\right)\left(\sqrt{5} - 3\right)$

28. $\left(2\sqrt{3} - 2\right)\left(2\sqrt{3} + 2\right)$

29. $\left(\sqrt{x} - 2\sqrt{y}\right)\left(\sqrt{x} + 2\sqrt{y}\right)$

30. $\left(m + \sqrt{mn}\right)\left(m - \sqrt{mn}\right)$

31. $\left(\sqrt{5} + 3\right)\left(2\sqrt{5} + 2\right)$

32. $\left(2\sqrt{3} - 7\right)\left(\sqrt{3} + 4\right)$

33. $\left(3\sqrt{2} + \sqrt{7}\right)\left(4\sqrt{2} - 3\sqrt{7}\right)$

34. $\left(5\sqrt{2} - 2\sqrt{10}\right)\left(2\sqrt{2} - 3\sqrt{10}\right)$

35. $\left(2\sqrt{x} - 3\right)\left(3\sqrt{x} - 5\right)$

36. $\left(5\sqrt{vw} + \sqrt{v}\right)\left(2\sqrt{vw} + 3\sqrt{v}\right)$

37. $\left(\sqrt{5} - 4\right)^2$

38. $\left(2\sqrt{3} + 7\right)^2$

39. $\left(3 + 2\sqrt{x + 1}\right)^2$

40. $\left(\sqrt{3x + 2} - 4\right)^2$

41. $\sqrt{5}\sqrt[3]{3}$

42. $\sqrt[4]{2}\sqrt[3]{6}$

43. $\sqrt{3x}\sqrt[4]{5x}$

44. $\sqrt[3]{mn}\sqrt[5]{2m}$

45. $\left(\sqrt{\sqrt{3} + 2}\right)\left(\sqrt{\sqrt{3} - 2}\right)$

46. $\left(2 + \sqrt{x} + \sqrt{5}\right)\left(\sqrt{2x} - \sqrt{3}\right)$

47. $\left(\sqrt{\dfrac{x}{3}} + \sqrt{\dfrac{2}{5}}\right)\left(\sqrt{\dfrac{x}{3}} + \sqrt{10}\right)$

48. $\left(2 + \sqrt{x - 1}\right)\left(3 + \sqrt{x - 1}\right)$

49. *Civil engineering*   The formula below is used to find the potential energy in a truss. Multiply as indicated and express the result in simplest radical form.

$$C = \frac{3}{8\sqrt[3]{2}} \cdot \frac{KA}{\sqrt[3]{L}}$$

11-6 | Division with Radicals

To divide one radical expression by another, we use the root of a quotient law (11-13) in reverse:

$$\frac{\sqrt[n]{a}}{\sqrt[n]{b}} = \sqrt[n]{\frac{a}{b}} \quad \text{(where } b \neq 0\text{)}$$

and then rationalize the denominator if necessary.

EXAMPLE 28   Divide: (a) $\dfrac{\sqrt{24}}{\sqrt{8}}$ (b) $\dfrac{9\sqrt{8x}}{3\sqrt{6x^2}}$

Solution

(a) $\dfrac{\sqrt{24}}{\sqrt{8}} = \sqrt{\dfrac{24}{8}} = \sqrt{3}$

(b) $\dfrac{9\sqrt{8x}}{3\sqrt{6x^2}} = \dfrac{9}{3}\sqrt{\dfrac{8x}{6x^2}} = 3\sqrt{\dfrac{4}{3x}} = \dfrac{3\sqrt{4} \cdot \sqrt{3x}}{\sqrt{3x} \cdot \sqrt{3x}} = \dfrac{6\sqrt{3x}}{3x} = \dfrac{2\sqrt{3x}}{x}$

As you can see, it is not necessary to rewrite the entire quotient under one radical. In Example 28(b), we could have first reduced the fraction,

$$\frac{\overset{3}{\cancel{9}}\sqrt{\overset{4}{\cancel{8}}\,\overset{1}{\cancel{x}}}}{\underset{1}{\cancel{3}}\sqrt{\underset{3}{\cancel{6}}\,\underset{x}{x^{\cancel{2}}}}} = \frac{3\sqrt{4}}{\sqrt{3x}}$$

and then rationalized the denominator.

REMEMBER ▶ Dividing or reducing in this way is possible only if both the numerator and denominator contain a radical of the same order. Expressions such as $\sqrt{14}/7$ or $\sqrt[3]{26}/\sqrt{13}$ cannot be simplified in this manner. ◀

If you cannot divide or reduce the radicals, simply rationalize the denominator.

EXAMPLE 29 Simplify: (a) $\dfrac{\sqrt{2x}}{\sqrt{3y}}$ (b) $\dfrac{\sqrt{6m^2}}{\sqrt[3]{2m}}$ (c) $\dfrac{\sqrt{15} + 2\sqrt{3}}{\sqrt{5}}$

Solution

(a) $\dfrac{\sqrt{2x}}{\sqrt{3y}} = \dfrac{\sqrt{2x} \cdot \sqrt{3y}}{\sqrt{3y} \cdot \sqrt{3y}} = \dfrac{\sqrt{6xy}}{3y}$ ◀

Note that this answer cannot be reduced further, because 6xy is under a radical and 3y is not.

(b) Although the radicals in the numerator and denominator contain common factors, they are of different orders, and division cannot be performed.

First, rewrite using fractional exponents with a common denominator.

$$\dfrac{\sqrt{6m^2}}{\sqrt[3]{2m}} = \dfrac{(6m^2)^{1/2}}{(2m)^{1/3}} = \dfrac{(6m^2)^{3/6}}{(2m)^{2/6}}$$

Then, change back to radical form.

$$= \sqrt[6]{\dfrac{(6m^2)^3}{(2m)^2}}$$

Finally, simplify the radicand and, if necessary, rationalize the denominator.

$$= \sqrt[6]{\dfrac{216m^6}{4m^2}}$$

$$= \sqrt[6]{54m^4}$$

(c) We must be careful not to cancel $\sqrt{15}$ with $\sqrt{5}$, because $\sqrt{15}$ is a term in the numerator—*not* a factor.

First, multiply by $\dfrac{\sqrt{5}}{\sqrt{5}}$ to rationalize the denominator. Notice the use of parentheses in the numerator.

$$\dfrac{\left(\sqrt{15} + 2\sqrt{3}\right)}{\sqrt{5}} \cdot \dfrac{\sqrt{5}}{\sqrt{5}} = \dfrac{\sqrt{75} + 2\sqrt{15}}{5}$$

Then, simplify the numerator.

$$= \dfrac{5\sqrt{3} + 2\sqrt{15}}{5}$$

Again, note that we cannot cancel the 5s, because 5 is not a common factor of the entire numerator.

Binomial denominators   To rationalize a binomial denominator containing a radical, we must use our knowledge of the difference of two squares. Two binomials whose product is the difference of two squares are called **conjugates** of each other. For example, $(a + b)$ and $(a - b)$ are conjugates of each other, $(2x + 7)$ and $(2x - 7)$ are conjugates of each other, and $(2\sqrt{3} + 5)$ and $(2\sqrt{3} - 5)$ are conjugates of each other. When a pair of conjugates are multiplied, the cross-product terms drop out and the difference of two squares results. For example,

$$(\sqrt{2} + 3)(\sqrt{2} - 3) = (\sqrt{2})^2 - (3)^2$$
$$= 2 - 9 = -7$$

We can use this fact to eliminate the radicals in a denominator that is a binomial expression. To rationalize a binomial denominator, multiply the numerator and denominator by the conjugate of the denominator.

EXAMPLE 30 Simplify: $\dfrac{2}{\sqrt{3} - 4}$

Solution

First, multiply the numerator and denominator by $\sqrt{3} + 4$, the conjugate of $\sqrt{3} - 4$.

$$\frac{2}{(\sqrt{3} - 4)} \cdot \frac{(\sqrt{3} + 4)}{(\sqrt{3} + 4)} = \frac{2\sqrt{3} + 8}{(\sqrt{3})^2 - (4)^2}$$

Then, simplify.

$$= \frac{2\sqrt{3} + 8}{3 - 16}$$

$$= \frac{2\sqrt{3} + 8}{-13} \text{ or } \frac{-2\sqrt{3} - 8}{13}$$

> The denominator now has no radical term

EXAMPLE 31 Simplify: $\dfrac{4\sqrt{x} - 7}{3 - 2\sqrt{x}}$

Solution   The fraction cannot be reduced as is.

Step 1. The conjugate of the denominator is $3 + 2\sqrt{x}$, so multiply the numerator and denominator by this quantity.

$$\frac{(4\sqrt{x} - 7) \cdot (3 + 2\sqrt{x})}{(3 - 2\sqrt{x}) \cdot (3 + 2\sqrt{x})}$$

> Note: $4\sqrt{x} \cdot 2\sqrt{x} = 8 \cdot \sqrt{x^2} = 8x$

Step 2. Use FOIL in the numerator and the difference of two squares in the denominator.

$$= \frac{12\sqrt{x} + 8x - 21 - 14\sqrt{x}}{3^2 - (2\sqrt{x})^2}$$

Step 3. Simplify.

$$= \frac{8x - 21 - 2\sqrt{x}}{9 - 4x}$$

> Denominator has no radical term

EXAMPLE 32 Simplify: $\dfrac{\sqrt{x} - 2\sqrt{y}}{\sqrt{x} + \sqrt{y}}$

Solution

Step 1. The conjugate of $\sqrt{x} + \sqrt{y}$ is $\sqrt{x} - \sqrt{y}$.

$\dfrac{\sqrt{x} - 2\sqrt{y}}{\sqrt{x} + \sqrt{y}} \cdot \dfrac{\sqrt{x} - \sqrt{y}}{\sqrt{x} - \sqrt{y}}$

Step 2. Multiply.

$= \dfrac{x - \sqrt{xy} - 2\sqrt{xy} + 2y}{\left(\sqrt{x}\right)^2 - \left(\sqrt{y}\right)^2}$ ← Difference of squares

Step 3. Simplify.

$= \dfrac{x + 2y - 3\sqrt{xy}}{x - y}$ ← Radical term has been eliminated

Exercises 11-6

Simplify and express each of the following in simplest radical form:

1. $\dfrac{\sqrt{15}}{\sqrt{5}}$

2. $\dfrac{\sqrt{30x}}{\sqrt{6x}}$

3. $\dfrac{\sqrt{y^5}}{\sqrt{y}}$

4. $\dfrac{\sqrt{32}}{\sqrt{2}}$

5. $\dfrac{\sqrt{40}}{\sqrt{5}}$

6. $\dfrac{\sqrt{8x^4}}{\sqrt{x}}$

7. $\dfrac{\sqrt[3]{x^5 y}}{\sqrt[3]{xy}}$

8. $\dfrac{\sqrt[3]{108}}{\sqrt[3]{2}}$

9. $\dfrac{\sqrt{6}}{\sqrt{48}}$

10. $\dfrac{\sqrt{3x}}{\sqrt{15x^2}}$

11. $\dfrac{\sqrt{10y}}{\sqrt{8y^4}}$

12. $\dfrac{\sqrt{8}}{\sqrt{12x}}$

13. $\dfrac{\sqrt{11}}{\sqrt{6}}$

14. $\dfrac{6\sqrt{7}}{2\sqrt{2}}$

15. $\dfrac{8\sqrt{3x}}{4\sqrt{5y}}$

16. $\dfrac{\sqrt{13m}}{\sqrt{2n}}$

17. $\dfrac{\sqrt[3]{12}}{\sqrt{2}}$

18. $\dfrac{\sqrt{3x}}{\sqrt[3]{4x^2}}$

19. $\dfrac{\sqrt{x} + \sqrt{2}}{\sqrt{5}}$

20. $\dfrac{2\sqrt{3} - \sqrt{7}}{\sqrt{10}}$

21. $\dfrac{\sqrt{6} - \sqrt{10}}{\sqrt{3}}$

22. $\dfrac{\sqrt{10} + \sqrt{2x}}{\sqrt{2}}$

23. $\dfrac{3}{\sqrt{5} - 1}$

24. $\dfrac{1}{2 + \sqrt{x}}$

25. $\dfrac{5}{\sqrt{m} + \sqrt{n}}$

26. $\dfrac{2}{\sqrt{2} - \sqrt{10}}$

27. $\dfrac{\sqrt{6}}{2\sqrt{6} - 5}$

28. $\dfrac{\sqrt{a}}{a - b\sqrt{a}}$

29. $\dfrac{3\sqrt{x}}{2\sqrt{x} + 1}$

30. $\dfrac{4\sqrt{3}}{3 + 2\sqrt{3}}$

31. $\dfrac{2\sqrt{5} + 1}{2\sqrt{5} - 1}$

32. $\dfrac{2 - 3\sqrt{7}}{1 + 2\sqrt{7}}$

33. $\dfrac{3 - \sqrt{2y}}{2 + \sqrt{2y}}$

34. $\dfrac{\sqrt{m} - 5}{2\sqrt{m} - 1}$

35. $\dfrac{3\sqrt{10} + \sqrt{5}}{2\sqrt{10} + \sqrt{5}}$

36. $\dfrac{5\sqrt{3} - 2\sqrt{2}}{3\sqrt{3} - 4\sqrt{2}}$

37. $\dfrac{\sqrt{x} - \sqrt{y}}{2\sqrt{x} + \sqrt{y}}$

38. $\dfrac{\sqrt{x} + y}{\sqrt{x} + \sqrt{y}}$

39. *Machine technology* The formula below gives the diagonal length d of a regular hexagon, where f is the distance across the flats. Express this in simplest radical form.

$$d = \dfrac{2f}{\sqrt{3}}$$

40. *Statistics* One way of calculating standard deviation is to use *Peter's formula*. The value of the coefficient in this formula is given below, where N is the number of trials. Write this coefficient in simplest radical form when $N = 25$.

$$\dfrac{5}{4N\sqrt{N - 1}}$$

11-7 | Review of Chapter 11

Important Terms and Concepts

fractional exponent (p. 387)   rationalizing the denominator (p. 394)   conjugates (p. 403)
simplest radical form (p. 392)   like radicals (p. 397)

Definitions and Laws (pp. 382–402)

- multiplication: $a^m \cdot a^n = a^{m+n}$ **(11-1)**
- power of a power: $(a^m)^n = a^{mn}$ **(11-2)**
- power of a product: $(ab)^n = a^n b^n$ **(11-3)**
- power of a quotient: $\left(\dfrac{a}{b}\right)^n = \dfrac{a^n}{b^n}$ (where $b \neq 0$)   **(11-4)**
- division: $\dfrac{a^m}{a^n} = a^{m-n}$ (where $a \neq 0$)   **(11-5)**
- negative exponent: $a^{-m} = \dfrac{1}{a^m}$ and $\dfrac{1}{a^{-m}} = a^m$ (where $a \neq 0$)   **(11-6)**
- zero exponent: $a^0 = 1$ (where $a \neq 0$)   **(11-7)**
- fractional exponents: $a^{m/n} = \sqrt[n]{a^m} = \left(\sqrt[n]{a}\right)^m$ **(11-8)**
  $a^{1/n} = \sqrt[n]{a}$ **(11-9)**

  For a and b positive real numbers:
- nth root of nth power: $\sqrt[n]{a^n} = a$ **(11-10)**
- root of a product: $\sqrt[n]{ab} = \sqrt[n]{a} \cdot \sqrt[n]{b}$ **(11-11)**
- root of a root: $\sqrt[m]{\sqrt[n]{a}} = \sqrt[mn]{a}$ **(11-12)**
- root of a quotient: $\sqrt[n]{\dfrac{a}{b}} = \dfrac{\sqrt[n]{a}}{\sqrt[n]{b}}$ **(11-13)**
- combining like radicals: $b\sqrt[n]{a} + c\sqrt[n]{a} = (b + c)\sqrt[n]{a}$ **(11-14)**
- multiplication of radicals: $\sqrt[n]{a} \cdot \sqrt[n]{b} = \sqrt[n]{ab}$
- division of radicals: $\dfrac{\sqrt[n]{a}}{\sqrt[n]{b}} = \sqrt[n]{\dfrac{a}{b}}$

Exercises 11-7

Simplify each expression and give all answers with positive exponents.

1. $m^{-5} m^3$

2. $n^{-1/2} n^{3/2}$

3. $\dfrac{x^5}{x^6}$

4. $\dfrac{a^{-2/3}}{a^{1/2}}$

5. $\dfrac{12a^3 b^5}{4a^5 b}$

6. $(y^{-1/2})^{2/5}$

7. $(x^{-3} y)^2$

8. $(8a^{-1/4} b^{5/3})^{-2/3}$

9. $-3x^{-5} x^0$

10. $\left(\dfrac{3c^{-2}}{d^3}\right)^{-4}$

11. $\left(\dfrac{m^{3/4} n^{1/2}}{m^{2/3} n^{1/6}}\right)^{-4/3}$

12. $x^{-2} + y^{-2}$

13. $(x + y)^{-2}$ **14.** $\left(\dfrac{a^{-2}}{b}\right)^{-3}\left(\dfrac{b^{-1}}{a}\right)^2$ **15.** $(8x^6)^{-1/3} - (2x^{-1/2})^2$ **16.** $v^{-3}w^2 + (2v^{-2}w)^2$

17. $(a^{-1} + 2b)^{-1}$ **18.** $\dfrac{3r^{-1}}{r^{-1} - s^{-1}}$ **19.** $(x^2 + 4x + 4)^{-1/2}$

20. $(2m - 3)(m + 4)^{-1/4} - (m + 4)^{3/4}$

Calculate. Round all irrational answers to the nearest 0.01.

21. $27^{1/3}$ **22.** $(-32)^{3/5}$ **23.** $2(81^{3/4})$ **24.** $28^{4/3}$

25. $4.72^{1.28}$ **26.** $3.5(180^{-0.85})$ **27.** $7^{3/4}7^{1/2}$ **28.** $(2^{-1/2})^6$

29. $(12.5^{1.5})^{2.0}$ **30.** $\dfrac{36^{1/4}}{36^{3/4}}$ **31.** $\dfrac{40^{2/3}}{30^{3/4}}$ **32.** $16^{-1/2} - 27^{-1/3}$

Perform the indicated operations. Express all answers in simplest radical form.

33. $\sqrt{98}$ **34.** $\sqrt{a^3b^4}$ **35.** $\sqrt{80x^5y}$

36. $\sqrt[3]{-128}$ **37.** $\sqrt[3]{16n^7}$ **38.** $\sqrt[3]{\sqrt{7}}$

39. $\sqrt{5\sqrt{y}}$ **40.** $\dfrac{5}{\sqrt{2xy}}$ **41.** $\dfrac{\sqrt{11}}{\sqrt{5x}}$

42. $\sqrt{\dfrac{5}{2}}$ **43.** $\sqrt[3]{\dfrac{7}{x}}$ **44.** $\sqrt[6]{125}$

45. $\sqrt[8]{25x^2y^4}$ **46.** $\sqrt{x^2 - 8x + 16}$ **47.** $\sqrt{7} + 5\sqrt{7}$

48. $3\sqrt[3]{10} - 8\sqrt[3]{10}$ **49.** $3\sqrt{y} - \sqrt{3y} + 4\sqrt{3y}$ **50.** $2\sqrt{5} + \sqrt{20}$

51. $2\sqrt{12} + 5\sqrt{48} - 4\sqrt{18}$ **52.** $3\sqrt{56} - \sqrt{\dfrac{2}{7}}$ **53.** $2\sqrt{a^4b^5} - a\sqrt{a^2b}$

54. $\sqrt[3]{54x^4} + \sqrt[3]{y^3x} - \sqrt[3]{x^7}$ **55.** $\sqrt{5m}\sqrt{3n}$ **56.** $\sqrt{6}\sqrt{12}$

57. $\sqrt[4]{8n^3}\sqrt[4]{4n^2}$ **58.** $(4\sqrt{7})^2$ **59.** $(2\sqrt{3x} + 1)^2$

60. $\sqrt{\dfrac{x}{2}}\sqrt{\dfrac{x}{6}}$ **61.** $3\sqrt{2}(\sqrt{3} - 2\sqrt{10})$ **62.** $\sqrt{5x}(2\sqrt{3x} + 3\sqrt{5y})$

63. $(\sqrt{7} + 2)(\sqrt{7} - 2)$ **64.** $(3\sqrt{6} - 4)(\sqrt{6} + 3)$ **65.** $(\sqrt{x} + 2\sqrt{y})(2\sqrt{x} - \sqrt{y})$

66. $(3\sqrt{2} - 5)^2$ **67.** $(2 + \sqrt{x - 5})^2$ **68.** $\sqrt[3]{2}\sqrt{2}$

69. $\dfrac{\sqrt{48}}{\sqrt{2}}$ **70.** $\dfrac{\sqrt{5y}}{\sqrt{30y^2}}$ **71.** $\dfrac{\sqrt{15} - \sqrt{10}}{\sqrt{3}}$

72. $\dfrac{\sqrt{x}}{3\sqrt{x} + 2}$ **73.** $\dfrac{4\sqrt{3} + 2}{2\sqrt{3} - 4}$ **74.** $\dfrac{\sqrt{2x} - 2\sqrt{y}}{\sqrt{2x} + \sqrt{y}}$

Solve.

75. *Dimensional analysis* The standard pressure unit, the pascal, is derived as shown below. Simplify the expression on the right, and state it with positive exponents only.

$$Pa = \dfrac{kg \cdot m^{-1} \cdot s^{-2} \cdot m^3 \cdot s^{-1}}{m^3 \cdot s^{-1}}$$

76. *Hydrology* The erosiveness of rain E (in tons of soil lost per hectare) is approximated by $E = x^r$, where x is the yearly erosion rate of the land and r is the maximum rainfall per hour. Calculate E when $x = 7.25$ tons/ha and $r = 4.75$ in./h.

77. *Energy technology*   The Weir formula for the flow rate Q (in cubic feet per second) over a rectangular-notched, watertight dam is $Q = CLH^{3/2}$, where C is the discharge coefficient, L is the width of the opening, and H is the head above the opening.* Find Q for $C = 3.4$, $L = 2.6$ ft, and $H = 0.70$ ft.

78. *Lighting technology*   The ideal height (in inches) of a light under certain conditions is given by the expression below. Determine this height to the nearest inch.

$$\left(\frac{38^2 - 25 \cdot 3^{2/3}}{3^{2/3} - 1} \right)^{1/2}$$

79. The formula below gives the length s of a side of an isosceles right triangle with hypotenuse c. Rewrite this formula in simplest radical form.

$$s = \frac{c}{\sqrt{2}}$$

80. *Civil engineering*   The horizontal displacement x of the joints of a pin-connected, plane truss is given by the formula below. Rewrite this formula with the numerical coefficient in simplest radical form.

$$x = \frac{1}{2(1 + \sqrt{2})} \cdot \frac{PL}{AE}$$

81. *Aerodynamics*   The ideal keel length L (in feet) for a hang glider weighing 75 lb is

$$L = \sqrt{\sqrt{2}(75 + w)}$$

where w is the weight of the pilot (in pounds). Express this length in simplest radical form for a pilot weighing 165 lb, and then calculate the length to the nearest foot.

82. *Electronics*   For a certain electrical circuit, the current oscillates at a frequency given by the expression below. Express this in simplest radical form.

$$\sqrt{\frac{1}{LC} - \frac{R^2}{4L^2}}$$

---

* For more information, see J. Leckie, G. Masters, H. Whitehouse, and L. Young, *Other Homes and Garbage* (San Francisco: Sierra Club, 1975), p. 63.

Complex Numbers

12

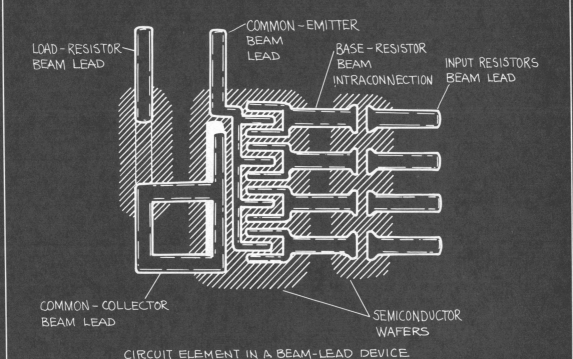

LOAD-RESISTOR
BEAM LEAD

COMMON-EMITTER
BEAM
LEAD

BASE-RESISTOR
BEAM
INTRACONNECTION

INPUT RESISTORS
BEAM LEAD

COMMON-COLLECTOR
BEAM LEAD

SEMICONDUCTOR
WAFERS

CIRCUIT ELEMENT IN A BEAM-LEAD DEVICE
BEFORE INVERTING AND MOUNTING

The complex numbers include both the real numbers and the imaginary numbers, but so far we have dealt almost exclusively with real numbers. In this chapter we will examine the properties and forms of imaginary numbers and discuss how to perform the basic operations with them. We will then explore some important applications of imaginary numbers in electronics.

Imaginary numbers   The need for imaginary numbers arises when we must take the square root of a negative number. For example, to solve the equation $x^2 = -a$, where a is a positive real number, take the square root of both sides to obtain $x = \pm\sqrt{-a}$. But there exists no real number whose square is $-a$ for $a > 0$. In order to solve problems involving the square root of a negative number, we need the following definitions:

$$\sqrt{-a} = j\sqrt{a} \quad \text{for } a > 0 \tag{12-1}$$

where

$$j^2 = -1 \tag{12-2}$$

The number j* is called the **imaginary unit,** and numbers of the form $\sqrt{-a}$ (where $a > 0$) are called **pure imaginary numbers.** The imaginary unit is symbolized by either the letter i or j, but since one of the major technical applications of imaginary numbers is in electronics where the letter i is reserved to represent electric current, we shall use the letter j for imaginary numbers.

EXAMPLE 1 Rewrite each expression in terms of j.
(a) $\sqrt{-4}$ (b) $\sqrt{-7}$ (c) $\sqrt{-8}$
(d) Express the answer to part (b) in decimal form (to the nearest hundredth).

Solution
(a) By definition (12-1),

$$\sqrt{-4} = j\sqrt{4}$$

$$= 2j$$

(b) $\sqrt{-7} = j\sqrt{7}$

> Write it with the j in front of the radical. If it is written as $\sqrt{7}j$, the j may appear to be under the radical sign.

(c) $\sqrt{-8} = j\sqrt{8} = j \cdot 2\sqrt{2} = 2j\sqrt{2}$
(d) $7\ \boxed{\sqrt{}} \rightarrow \boxed{2.6457513} \qquad \sqrt{-7} \approx 2.65j$

As we noted earlier, attempting to find the square root of a negative number directly on a calculator results in an error message. However, Example 1(d)

---

* Because $j^2 = -1$, j is often written as $\sqrt{-1}$. But we need to be careful with this, because up to now the square root has been defined only for nonnegative real numbers. Definition (12-1) allows us to extend the definition of the square root to negative numbers.

illustrates how we may use a calculator to express imaginary numbers as decimal multiples of j.

Operations with imaginary numbers

The square root of a product law (1-9) $\left(\sqrt{ab} = \sqrt{a} \cdot \sqrt{b}\right)$ used up to now with positive real numbers is *not* true when both a and b are negative real numbers. The definition of j^2 (12-2) allows us to work with radicals whose radicands are negative numbers.

Before performing operations involving negative radicands, first write each radical as the product of j and a positive real number, as in Example 1. Then, use the properties of roots of positive real numbers to perform the required operations.

For example, to multiply

$$\sqrt{-2} \cdot \sqrt{-18}$$

first, write both radicals in terms of j:

$$\sqrt{-2} \cdot \sqrt{-18} = j\sqrt{2} \cdot j\sqrt{18}$$

Next, multiply:

$$= j^2 \cdot \sqrt{2} \cdot \sqrt{18}$$
$$= (-1)\sqrt{2 \cdot 18}$$
$$= -\sqrt{36} = -6$$

CAUTION ▶ Do *not* write

$$\sqrt{-2} \cdot \sqrt{-18} = \sqrt{(-2)(-18)} = \sqrt{36} = 6$$

> This is not allowed if *both* radicands are negative.

Always write the square root of a negative number in terms of j before using it in an algebraic operation. ◀

EXAMPLE 2 Perform the indicated operations.
(a) $\sqrt{-3} \cdot \sqrt{-6}$ (b) $\left(\sqrt{-5}\right)^2$ (c) $\left(3\sqrt{-2}\right)^2 + 3\sqrt{(-2)^2}$

Solution
(a) **First,** rewrite each number in terms of j:

$$\sqrt{-3} = j\sqrt{3} \qquad \sqrt{-6} = j\sqrt{6}$$

Then,

$$\sqrt{-3} \cdot \sqrt{-6} = j\sqrt{3} \cdot j\sqrt{6} = j^2 \cdot \sqrt{3} \cdot \sqrt{6}$$
$$= (-1)\sqrt{18}$$
$$= -3\sqrt{2}$$

> The radicands are positive, so we can multiply them according to the square root of a product law (1-9).

(b) $\sqrt{-5} = j\sqrt{5}$ so $\left(\sqrt{-5}\right)^2 = \left(j\sqrt{5}\right)^2 = j^2\left(\sqrt{5}\right)^2$

$$= (-1)(5) = -5$$

(c) $3\sqrt{-2} = 3j\sqrt{2}$ so $\left(3\sqrt{-2}\right)^2 = \left(3j\sqrt{2}\right)^2$

$$= 9j^2\left(\sqrt{2}\right)^2 = 9(-1)(2) = -18$$

Then,

$$\left(3\sqrt{-2}\right)^2 + 3\sqrt{(-2)^2} = -18 + 3\sqrt{4} = -18 + 6 = -12$$

Powers of j   Many problems involve powers of j greater than the second power. These can all be simplified, and their simplification follows a pattern:

$j^1 = j$ $j^5 = j^4 \cdot j = j$

$j^2 = -1$ $j^6 = j^4 \cdot j^2 = -1$

$j^3 = j^2 \cdot j = (-1)j = -j$ $j^7 = j^6 \cdot j = -j$

$j^4 = j^2 \cdot j^2 = (-1)(-1) = 1$ $j^8 = j^4 \cdot j^4 = 1$

Notice that the powers of j repeat in the sequence j, -1, $-j$, 1. Furthermore, if the power is a multiple of 4, the number is equal to 1. These facts can be used to simplify large powers of j.

EXAMPLE 3 Simplify:
(a) j^{26} (b) j^{12} (c) $j^{77} - j^{323}$ (d) $3j^{40} - j^{122}$

Solution
(a)   **Step 1.**   Rewrite as a product of powers. $j^{26} = j^{24} \cdot j^2$
Use the largest multiple of 4 as one of the powers.

Step 2.   Simplify. $= 1 \cdot (-1) = -1$

(b) $j^{12} = 1$ (because 12 is a multiple of 4)
(c) $j^{77} - j^{323} = j^{76} \cdot j - j^{320} \cdot j^3 = 1 \cdot j - 1(-j) = j + j = 2j$
(d) $3j^{40} - j^{122} = 3 \cdot 1 - j^{120} \cdot j^2 = 3 - 1 \cdot j^2 = 3 - 1(-1) = 3 + 1 = 4$

Complex numbers in rectangular form   So far, the complex numbers we have dealt with have been either pure real numbers, such as -4, 2.5, and $\sqrt{3}$, or pure imaginary numbers, such as $-4j$, $2.5j$, and $j\sqrt{3}$. But a complex number is often a combination of the two. The **rectangular form** of a complex number is written as

$a + bj$ where a and b are real numbers

The number a is called the **real part** of $a + bj$, while bj is known as the **imaginary part**. If $a = 0$, it becomes a pure imaginary number.

EXAMPLE 4 For each of the following complex numbers, name the real part and the imaginary part:
(a) $5 + 6j$ (b) -27 (c) $j\sqrt{3}$ (d) $\sqrt{3} - j$

Solution
(a) The real part is 5 and the imaginary part is $6j$.
(b) In rectangular form, $-27 = -27 + 0j$. Thus, -27 is the real part and $0j$ is the imaginary part. This is a pure real number.
(c) $j\sqrt{3} = 0 + j\sqrt{3}$
The real part is 0 and the imaginary part is $j\sqrt{3}$. This is a pure imaginary number.
(d) $\sqrt{3} - j = \sqrt{3} + (-j)$
The real part is $\sqrt{3}$ and the imaginary part is $-j$.

Equality of complex numbers   Two complex numbers are equal if and only if both their real parts are equal and their imaginary parts are equal.

EXAMPLE 5
(a) The numbers $3 + 4j$ and $x + yj$ are equal only if $x = 3$ and $y = 4$.
(b) The numbers $a + bj$ and $\sqrt{2} - j$ are equal only if $a = \sqrt{2}$ and $b = -1$.

Equations involving complex numbers   Equations involving complex numbers can be solved using the equality principle shown in Example 5. To solve, move all terms not containing the variable to one side of the equation and all terms containing the variable to the other side. Then equate the real parts and the imaginary parts separately.

EXAMPLE 6   Solve for x and y: $2x - 3j + 4 = yj + 8 - j$

Solution

Step 1. Move the terms not containing variables to the right and the variable terms to the left.

$2x - yj = 8 - j + 3j - 4$

| Imaginary parts |

$2x - yj = 4 + 2j$

| Real parts |

Step 2. Equate the real parts and solve.

$2x = 4$

$x = 2$

Step 3. Equate the imaginary parts and solve.

$-yj = 2j$

$y = -2$

In some cases, solving a single equation may lead to a system of equations.

EXAMPLE 7   Solve for x and y: $-7j + 2x + xj = 2yj - y - 1$

Solution

First, move all terms containing the variables to the left, and move all others to the right.

$2x + xj - 2yj + y = -1 + 7j$

Then, equate the real parts and the imaginary parts separately.

Real: $2x + y = -1$

Imaginary: $xj - 2yj = 7j$

$x - 2y = 7$

This leads to a system of equations.

$$\begin{cases} 2x + y = -1 \\ x - 2y = 7 \end{cases}$$

Finally, solve the system of equations to obtain $x = 1$ and $y = -3$.

Complex conjugates   The **conjugate** of the complex number $a + bj$ is defined to be the number $a - bj$. Conjugate pairs are useful in the division of complex numbers and in the solution of many polynomial equations. To find the conjugate of a complex number, change the sign of the imaginary part:

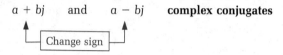

$a + bj$ and $a - bj$ **complex conjugates**

Change sign

This is true for any complex number, even those where $a = 0$ or $b = 0$.

EXAMPLE 8   Name the conjugate of each complex number.
(a) $3 + 4j$ (b) $-2 - 7j$ (c) $2j$ (d) $-j\sqrt{3}$ (e)   3

Solution   (a) $3 - 4j$ (b) $-2 + 7j$ (c) $-2j$ (d) $j\sqrt{3}$ (e)   3
In part (c), $2j$ is the complex number $a + bj$ with $a = 0$ and $b = 2$. Its conjugate has $a = 0$ and $b = -2$. In part (e), notice that any real number is the complex conjugate of itself.

Exercises 12-1

Rewrite each of the following in terms of j:

1. $\sqrt{-9}$ 2. $\sqrt{-49}$ 3. $\sqrt{-5}$ 4. $\sqrt{-23}$ 5. $\sqrt{-12}$ 6. $\sqrt{-50}$

7. $\sqrt{-\dfrac{3}{4}}$ 8. $\sqrt{-\dfrac{2}{3}}$ 9. $\sqrt{-0.04}$ 10. $\sqrt{-1.21}$ ▦ 11. $\sqrt{-853}$ ▦ 12. $\sqrt{-0.742}$

Perform the indicated operations and simplify.

13. $\sqrt{-10}\cdot\sqrt{-2}$ 14. $\sqrt{-3}\cdot\sqrt{-15}$ 15. $\left(\sqrt{-2}\right)^2$ 16. $\left(\sqrt{-3}\right)^2$
17. $\left(2\sqrt{-5}\right)^2 + 2\sqrt{(-5)^2}$ 18. $5\sqrt{(-6)^2} - \left(3\sqrt{-6}\right)^2$

Simplify.

19. j^{14} 20. j^{10} 21. j^{23} 22. j^{39}
23. j^{16} 24. j^{40} 25. j^{21} 26. j^{43}
27. $j^{18} - j^{32}$ 28. $2j^{16} - j^{42}$ 29. $j^{19} - 3j^{33}$ 30. $4j^{45} - j^{59}$

Name the real part and the imaginary part of each complex number.

31. $2 - 3j$ **32.** $5 + j\sqrt{2}$ **33.** -31 **34.** $7j$

35. $3 + \sqrt{-16}$ **36.** $\sqrt{-24} - \sqrt{27}$ **37.** $2j^6 - 3j^5$ **38.** $\left(\sqrt{-11}\right)^2 + \sqrt{-11}$

Find the conjugate of each complex number.

39. $5 - j$ **40.** $3 + 7j$ **41.** $-6 + 2j$ **42.** $-9 - 7j$

43. $-5j$ **44.** $3j\sqrt{3}$ **45.** 17 **46.** -5

Solve for x and y.

47. $2 - 11j = 2x + yj$ **48.** $-3 + 24j = x - 4yj$

49. $4j - 12 = x - 7 - 2yj$ **50.** $-2 - 3x + 3yj = 7 - 9j$

51. $3x + 4j - 5 = yj - 3 + 2j$ **52.** $2yj - 7 + j = 8 - 3j + 2x$

53. $3x - xj + 4j = 2yj - 8 - y$ **54.** $6j - 3x - yj = 9 - 2xj + y$

55. Electronics   The impedance Z in a series RCL circuit is given by

$$Z = R + jX_L - jX_C$$

Determine the impedance if $R = 24 \ \Omega$, $X_L = 14 \ \Omega$, and $X_C = 38 \ \Omega$.

56. Electronics   The impedance Z can be expressed in rectangular form as $Z = R \pm jX$, where R represents the resistance and X represents the reactance in the circuit. Using the vocabulary learned in this section, what kind of complex number represents the impedance in a:

    (a)   Circuit with no resistance?   (b)   Circuit with no reactance?

57. Electronics   A certain parallel LC circuit has an equivalent impedance Z_{eq} of

$$Z_{eq} = \frac{sL}{1 + s^2LC}$$

Given that $s = j\omega$, substitute for s in the above formula and rewrite and simplify the formula for Z_{eq}.*

12-2 | Operations with Complex Numbers

In this section we shall learn how to perform the four basic operations of addition, subtraction, multiplication, and division with complex numbers. As we emphasized in Section 12-1, the imaginary part of a complex number always must be expressed in terms of j before performing these operations.

Addition and subtraction   To add or subtract complex numbers, simply combine their real and imaginary parts separately:

$$(a + bj) + (c + dj) = (a + c) + (b + d)j$$

EXAMPLE 9   Perform the indicated operations.

(a) $(2 + 5j) + (3 - 8j)$ (b) $\left(5 - \sqrt{-49}\right) - \left(2\sqrt{-64} - 3\right)$

---

* Based on John McWane, *Introduction to Electronics and Instrumentation* (Boston: Breton, 1981), p. 482.

Solution

(a)   **Step 1.**   Use the associative and commutative
laws to rearrange terms.

$\boxed{\text{Real parts}}$ $\boxed{\text{Imaginary parts}}$

$$(2 + 5j) + (3 - 8j) = (2 + 3) + (5j - 8j)$$

Step 2.   Combine.

$$= 5 + (-3j) = 5 - 3j$$

(b)   **Step 1.**   Rewrite in terms
of j.

$$\left(5 - \sqrt{-49}\right) - \left(2\sqrt{-64} - 3\right)$$

$$= (5 - 7j) - (2 \cdot 8j - 3)$$

Step 2.   Use the distributive law to remove pa-
rentheses.

$$= 5 - 7j - 16j + 3$$

Step 3.   Rearrange.

$$= (5 + 3) + (-7j - 16j)$$

Step 4.   Combine.

$$= 8 - 23j$$

Multiplication   Addition and subtraction of complex numbers is similar to addition and subtraction of like terms in algebraic expressions. This similarity also can be seen in multiplication. To multiply complex numbers, apply the same techniques used in multiplying algebraic expressions. As the final step, simplify all powers of j.

EXAMPLE 10   Multiply as indicated.
(a) $2j(4 + 3j)$ (b) $\left(5 + 2\sqrt{-4}\right)\left(3 - \sqrt{-25}\right)$ (c) $(2 + 3j)(2 - 3j)$

Solution

(a)   Use the distrib-
utive property.

$$2j(4 + 3j) = 2j \cdot 4 + 2j \cdot 3j$$

$$= 8j + 6(-1)$$

$$= -6 + 8j$$

(b)   **First,** rewrite in
terms of j.

$$\left(5 + 2\sqrt{-4}\right)\left(3 - \sqrt{-25}\right) = (5 + 4j)(3 - 5j)$$

Then, multiply
the binomial ex-
pressions.

$$= 15 - 25j + 12j - 20j^2$$

$$= 15 - 13j + 20$$

$$= 35 - 13j$$

(c)   Multiplying a
pair of conju-
gates produces
the difference of
two squares.

$$(2 + 3j)(2 - 3j) = 2^2 - (3j)^2$$

$$= 4 - 9j^2$$

$$= 4 + 9 = 13$$

Division   Dividing complex numbers is similar to dividing radical expressions. Find the quotient by rationalizing the denominator. Example 10(c) shows that multiplying a complex number by its conjugate results in a rational number.

EXAMPLE 11 Simplify each quotient.

(a) $\dfrac{-2j}{3 + 4j}$ (b) $\dfrac{2 + \sqrt{-81}}{\sqrt{-9}}$ (c) $\dfrac{7 - 4j}{6 + 5j^7}$

Solution

(a) Multiply the numerator and denominator by $3 - 4j$, the conjugate of the denominator.

$$\dfrac{-2j}{(3 + 4j)} \cdot \dfrac{(3 - 4j)}{(3 - 4j)}$$

Simplify.

$$= \dfrac{-6j + 8j^2}{9 - 16j^2} = \dfrac{-8 - 6j}{25}$$

This answer may be left as is, or it may be expressed as

$$-\dfrac{8}{25} - \dfrac{6}{25}j$$

(b) **First,** rewrite in terms of j.

$$\dfrac{2 + \sqrt{-81}}{\sqrt{-9}} = \dfrac{2 + 9j}{3j}$$

Then, multiply the numerator and denominator by $-3j$, the conjugate of $3j$.

$$= \dfrac{(2 + 9j)}{3j} \cdot \dfrac{(-3j)}{(-3j)}$$

$$= \dfrac{-6j - 27j^2}{-9j^2}$$

$$= \dfrac{27 - 6j}{9}$$

$$= \dfrac{9 - 2j}{3} \text{ or } 3 - \dfrac{2}{3}j$$

Notice that this answer was simplified by reducing the fraction to lowest terms.

(c) **First,** simplify the power of j:
$j^7 = j^4 \cdot j^3 = 1(-j) = -j$

$$\dfrac{7 - 4j}{6 + 5j^7} = \dfrac{7 - 4j}{6 - 5j}$$

Then, multiply by the conjugate of $6 - 5j$.

$$= \dfrac{(7 - 4j)}{(6 - 5j)} \cdot \dfrac{(6 + 5j)}{(6 + 5j)}$$

$$= \dfrac{42 + 35j - 24j - 20j^2}{36 - 25j^2}$$

$$= \dfrac{62 + 11j}{61}$$

Exercises 12-2

Perform the indicated operations. Express answers in the form a + bj.

1. $(3 + 7j) + (5 - 2j)$

2. $(-5 - 6j) + (7 - 11j)$

3. $(-4 + 5j) - (8 - 3j)$

4. $(2 - 9j) - (6 - j)$

5. $\left(7 + 2\sqrt{-16}\right) + \left(12 - \sqrt{-100}\right)$ 6. $\left(-13 - \sqrt{-36}\right) + \left(-9 + \sqrt{-4}\right)$

7. $\left(11 - \sqrt{-25}\right) - \left(-8 + \sqrt{-1}\right)$ 8. $\left(-7 - 3\sqrt{-81}\right) - \left(23 - \sqrt{-9}\right)$

9. $(3j - 5) - (2j - 15) + (4j - 12)$ 10. $(8 - j) + 6j - (-13 - 7j)$

11. $\left(8 - \sqrt{-49}\right) - \sqrt{-64} + \left(3 - \sqrt{-4}\right)$ 12. $\left(-6 - \sqrt{-121}\right) - \left(-5 + \sqrt{-144}\right) - \sqrt{-9}$

13. $(3j)(-2j)(-5j)$ 14. $(-3j)(4j)(6j)$ 15. $3\sqrt{-25}\sqrt{-24}\sqrt{-16}$

16. $\sqrt{-18}\sqrt{-36}\sqrt{-64}$ 17. $\left(\sqrt{-3}\right)^3$ 18. $\left(\sqrt{-100}\right)^4$

19. $5j(3 - 8j)$ 20. $(-6 - 4j)(-2j)$ 21. $\left(10 - \sqrt{-36}\right)\sqrt{-49}$

22. $\sqrt{-9}\left(-7 + \sqrt{-144}\right)$ 23. $(3 - 4j)(2 + 3j)$ 24. $(5 - 7j)(6 - j)$

25. $(3 + 5j)(3 - 5j)$ 26. $(4 - j)(4 + j)$ 27. $(2 + 5j)^2$

28. $(7 - j)^2$ 29. $(-2 + 4j)^3$ 30. $(-1 - 2j)^4$

31. $\left(3 - 5\sqrt{-64}\right)\left(2 + 3\sqrt{-9}\right)$ 32. $\left(-1 + 3\sqrt{-4}\right)\left(5 - 2\sqrt{-16}\right)$

33. $\left(2 + \sqrt{-18}\right) - \left(3 - \sqrt{-8}\right)$ 34. $\left(5 - \sqrt{-75}\right) + \left(6 - \sqrt{-48}\right)$

35. $6j^2 - 5j^3 - 2\sqrt{-81}$ 36. $2j^5 - 8j^4 + 4j\sqrt{-36}$ 37. $\dfrac{3j}{3 - 5j}$

38. $\dfrac{-4j}{2 + 3j}$ 39. $\dfrac{7}{5 + j}$ 40. $\dfrac{-6}{2 - 7j}$

41. $\dfrac{8}{6 - \sqrt{-4}}$ 42. $\dfrac{\sqrt{-9}}{8 + \sqrt{-16}}$ 43. $\dfrac{2 - j}{5j}$

44. $\dfrac{3 + 7j}{j}$ 45. $\dfrac{5 - 4j}{2 - 9j}$ 46. $\dfrac{11 - 2j}{3 + 7j}$

47. $\dfrac{-2 + \sqrt{-25}}{3 - \sqrt{-36}}$ 48. $\dfrac{3 - \sqrt{-5}}{2 + \sqrt{-5}}$ 49. $\dfrac{8 - 6j}{5 + 3j^5}$

50. $\dfrac{2 + 4j}{6j - 3j^6}$

Use a calculator to perform the indicated operations. Express answers in rectangular form.

51. $(32.78 - 24.61j) + (27.94 + 87.21j)$ 52. $(5370 + 6420j) - (2890 - 3840j)$

53. $(2.78j)(3.65j)$ 54. $(24.3 - 16.7j)(87.8 + 21.2j)$

55. $(28 - 72j)^3$ 56. $\dfrac{3.2 + 5.6j}{2.8 - 3.7j}$

Given the complex number $a + bj$ and its conjugate $a - bj$, find:

57. Their sum 58. Their difference: $(a + bj) - (a - bj)$

59. Their product 60. The sum of their reciprocals

61. Find: (a) $(a + bj)^2 + (a - bj)^2$ (b) $(a + bj)^3 + (a - bj)^3$

62. Show that: $\left[\left(\sqrt{3} + 1\right) + \left(\sqrt{3} - 1\right)j\right]^3 = 16(1 + j)$

12-3 | Representing Complex Numbers Graphically

In this section we shall use the rectangular coordinate system to represent complex numbers graphically and to show the addition of complex numbers. This will enable us to develop an alternative method for writing complex numbers in the next section.

The complex plane

By using the vertical axis as the **imaginary axis** and the horizontal axis as the **real axis,** we can use the traditional rectangular coordinate system to represent points on the **complex plane.** In this plane, every complex number of the form $x + yj$ is represented by the point with coordinates (x, y), as shown in Fig. 12-1.

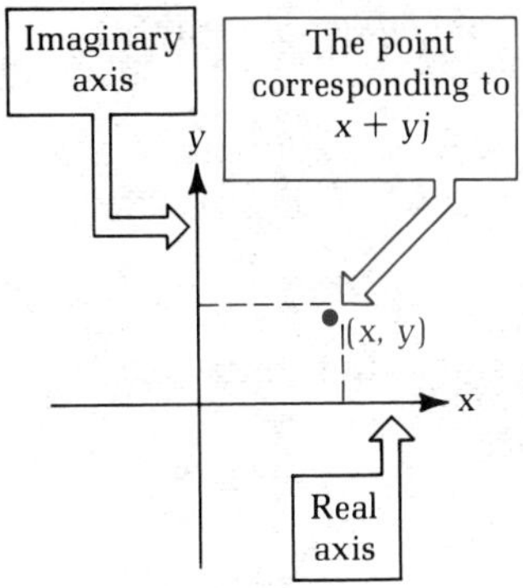

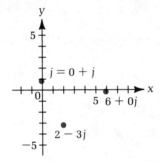

Figure 12-1

EXAMPLE 12 Plot the points that represent the following numbers:
(a) $2 - 3j$ (b) j (c) 6

Solution
(a) As shown in Fig. 12-2, the complex number $2 - 3j$ is represented by the point $(2, -3)$.
(b) Similarly, since $j = 0 + 1j$, the complex number j is represented by the point $(0, 1)$.
(c) Finally, since $6 = 6 + 0j$, the complex number 6 is represented by the point $(6, 0)$.

Figure 12-2

From Example 12(b) we can see that points representing pure imaginary numbers are on the y-, or imaginary, axis. Example 12(c) shows that points representing real numbers are on the x-, or real, axis.

Using vectors to add complex numbers

A related concept of importance in technical applications involves the use of vectors to represent complex numbers in the complex plane. If we define the real part x of a complex number $x + yj$ as the horizontal component and the coefficient y as the vertical component, then $x + yj$ represents a vector drawn from the origin to the point (x, y) on the complex plane. This vector representation allows us to add complex numbers graphically in the same way that we add vectors.

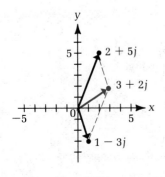

Figure 12-3

EXAMPLE 13 Add the complex numbers $2 + 5j$ and $1 - 3j$ graphically.

Solution **First,** draw vectors to the points $(2, 5)$ and $(1, -3)$. These points represent the two given numbers.

Then, using the two vectors as adjacent sides, complete the parallelogram as shown in Fig. 12-3. The fourth vertex $(3, 2)$ represents the sum $3 + 2j$ of the given numbers.

Subtraction   Since subtraction has been defined as addition of the opposite, we can also use vectors to subtract one complex number from another.

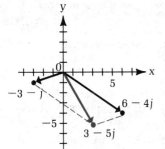

EXAMPLE 14   Perform the subtraction $(-3 - j) - (-6 + 4j)$ graphically.

Solution

Step 1. Rewrite as an addition problem.

$$(-3 - j) - (-6 + 4j) = (-3 - j) + (6 - 4j)$$

Step 2. Plot these complex numbers as vectors and add as in Example 13. From Fig. 12-4:

$$= 3 - 5j$$

Figure 12-4

EXAMPLE 15   Add $2 - 5j$ to its conjugate graphically.

Solution   The conjugate of $2 - 5j$ is $2 + 5j$. Their sum is 4, as shown graphically in Fig. 12-5. A conjugate pair will always be symmetrically placed with respect to the x-axis, and their sum will always be represented by a point on the x-, or real, axis.

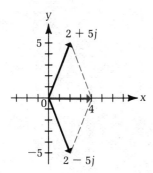

Figure 12-5

Exercises 12-3

Plot the points representing the following complex numbers:

1. $3 - 2j$ **2.** $5 + 4j$ **3.** $-6 + j$ **4.** $-1 - 3j$ **5.** $2j$ **6.** -3

Perform the indicated operations graphically.

7. $(3 - 2j) + (5 + j)$ **8.** $(-1 + 4j) + (-5 - 2j)$ **9.** $(-6 - 3j) + (1 - 3j)$
10. $(-8 + 3j) + (5 + 4j)$ **11.** $3j + (7 - 4j)$ **12.** $-7 + (2 + 6j)$
13. $(6 - 6j) - (4 + 3j)$ **14.** $(-2 - 3j) - (-5 + 3j)$ **15.** $(7 + 8j) - (3 + 2j)$
16. $(-1 + j) - (3 - 4j)$ **17.** $8 - (2 - j)$ **18.** $-2j - (-5 + 6j)$
19. $2j - (5 - 3j) + (3 + j)$ **20.** $(6 - 4j) - (5 + 2j) + 3$ **21.** $1 - (3 - 7j) - (-6 + 8j)$
22. $(-3 + 5j) - (5 - j) - 3j$ **23.** $(3 - 7j) + (3 + 7j)$ **24.** $(2 + 4j) - (2 - 4j)$

If P is a complex number $a + bj$ and Q is its conjugate $a - bj$, show each of the following graphically:

25. $P - Q$ **26.** $2P + Q$ **27.** $P + j$ **28.** $P + a$

| 12-4 | **Polar Form of a Complex Number** |

In this section we shall combine the graphical representation of complex numbers with our knowledge of trigonometry to develop an alternative form for writing complex numbers. This new form can be used to simplify certain operations performed with complex numbers.

Definitions

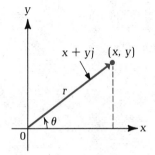

Figure 12-6

The complex number $x + yj$ can be represented by a vector drawn from the origin to the point (x, y). As shown in Fig. 12-6, this vector makes an angle θ with the real axis. If r is the distance from the origin to point (x, y), then from Chapter 4 we know that

$$x = r \cos \theta \quad \text{and} \quad y = r \sin \theta \tag{12-3}$$

Furthermore, the Pythagorean theorem gives us

$$r = \sqrt{x^2 + y^2} \tag{12-4}$$

To find the angle θ, apply the methods learned in Chapters 8 and 9. **First,** find Ref θ from

$$\tan (\text{Ref } \theta) = \left| \frac{y}{x} \right| \tag{12-5}$$

Then, use the graphical representation of $x + yj$ to place θ in the proper quadrant.

Using (12-3), we can rewrite the rectangular form of a complex number $x + yj$ as

$$x + yj = r \cos \theta + (r \sin \theta)j$$

Factoring r from each term on the right, we obtain the **polar form** of the complex number $x + yj$:

Polar Form of a Complex Number

$$x + yj = r(\cos \theta + j \sin \theta) \tag{12-6}$$

The number r is called the **modulus** or **absolute value** of the complex number, and the angle θ is the **argument** of the complex number.

Changing from rectangular to polar form

To change a complex number from rectangular form, $x + yj$, to polar form, $r(\cos\theta + j\sin\theta)$:

First, use equations (12-4) and (12-5) to find r and Ref θ.
Next, graph the number to determine in which quadrant θ lies and use Ref θ to find θ.
Finally, substitute these values in equation (12-6).

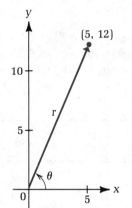

Figure 12-7

▦ **EXAMPLE 16**   Change the complex number $5 + 12j$ to polar form.

Solution   Step 1.   Use the Pythagorean theorem to find r:

$$r = \sqrt{5^2 + 12^2} = \sqrt{169} = 13$$

Step 2.   Use (12-5) to find Ref θ:

$$\tan(\text{Ref }\theta) = \left|\frac{12}{5}\right| \qquad 12\ \boxed{÷}\ 5\ \boxed{=}\ \boxed{\text{INV}}\ \boxed{\tan}\ \boxed{\text{STO}} \rightarrow \boxed{67.380135}$$

Step 3.   Graph the number to determine the quadrant in which it lies and use Ref θ to find θ. Figure 12-7 shows that the given number is in quadrant I. Therefore, $\theta = \text{Ref }\theta \approx 67.4°$.

Step 4.   Use (12-6) to write the given number in polar form:

$$5 + 12j \approx 13(\cos 67.4° + j\sin 67.4°)*$$

IMPORTANT ▶

In various applications in electronics, the notation $r\underline{/\theta}$ is used as an abbreviation for the polar form $r(\cos\theta + j\sin\theta)$. For example, the solution to Example 16 may be written as $13\underline{/67.4°}$ in abbreviated form. Conversely, the notation $2.8\underline{/1.7}$ is equivalent to $2.8(\cos 1.7 + j\sin 1.7)$ in standard polar form. We will use this notation often so that you may become familiar with it.   ◀

▦ **EXAMPLE 17**   Rewrite the complex number $-4.20 + 3.35j$ both in standard polar form and in $r\underline{/\theta}$ notation.

Solution   Step 1.   Find r:

$$r = \sqrt{(-4.20)^2 + (3.35)^2}$$

$$4.2\ \boxed{x^2}\ \boxed{+}\ 3.35\ \boxed{x^2}\ \boxed{=}\ \boxed{\sqrt{\ }} \rightarrow \boxed{5.3723831}$$

$$r \approx 5.37$$

---

*Unless a problem is stated in an applied context, assume that all given numbers are exact. If necessary, express your answer with r rounded to three significant digits and θ rounded to the nearest tenth of a degree.

(−4.20, 3.35)

Figure 12-8

Step 2. Calculate Ref θ:

$$\tan (\text{Ref } \theta) = \left| \frac{3.35}{-4.20} \right| = \frac{3.35}{4.20}$$

3.35 $\boxed{÷}$ 4.2 $\boxed{=}$ $\boxed{\text{INV}}$ $\boxed{\text{tan}}$ $\boxed{\text{STO}}$ → $\boxed{38.576530}$

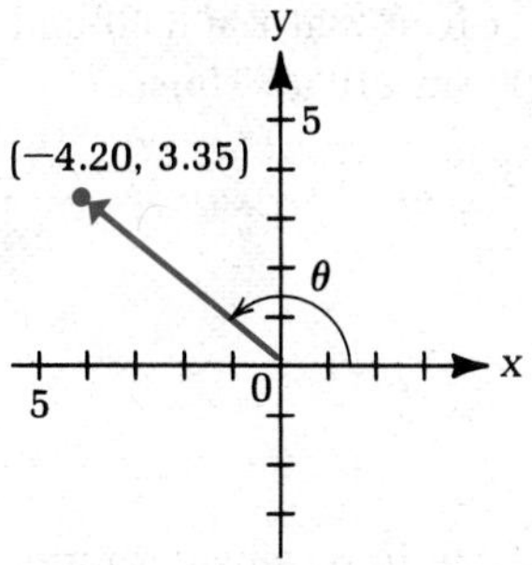

Step 3. Figure 12-8 shows that θ is in quadrant II, so

$$\theta = 180 - \text{Ref } \theta \qquad 180 \,\boxed{-}\,\boxed{\text{RCL}}\,\boxed{=}\, → \boxed{141.42347}$$

Therefore, $\theta \approx 141.4°$

Step 4. $-4.20 + 3.35j \approx 5.37(\cos 141.4° + j \sin 141.4°)$
In $r\underline{/\theta}$ notation, we have

$$\approx 5.37\underline{/141.4°}$$

Polar form of real numbers and pure imaginary numbers

Since real numbers lie on the x-axis ($\theta = 0°$ or $180°$) and pure imaginary numbers lie on the y-axis ($\theta = 90°$ or $270°$), these numbers can be put into polar form without calculation. The value of r will be the absolute value of the real or imaginary coefficient.

EXAMPLE 18 Rewrite the following numbers in polar form:
(a)   5   (b) $-2j$ (c) $4j$ (d) -3

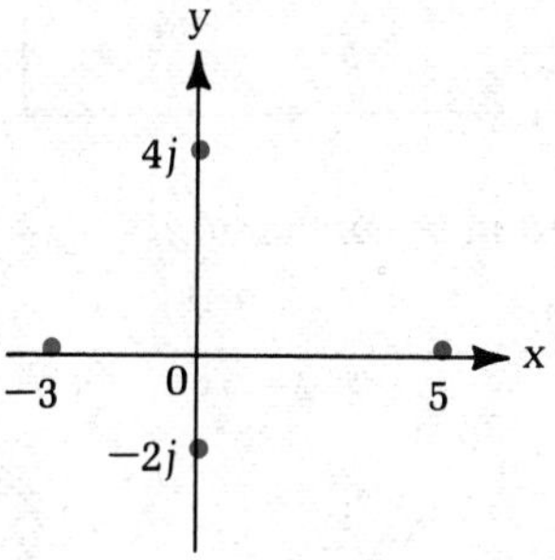

Figure 12-9

Solution
(a)   Since 5 is on the positive x-axis (Fig. 12-9), $\theta = 0°$, and $r = |5| = 5$. Therefore, the polar form is

$$5 = 5(\cos 0° + j \sin 0°) = 5\underline{/0°}$$

(b)   Because $-2j$ is on the negative y-axis, $\theta = 270°$ and $r = |-2| = 2$. Thus,

$$-2j = 2(\cos 270° + j \sin 270°) = 2\underline{/270°}$$

(c)   Given that $4j$ is on the positive y-axis, we obtain $\theta = 90°$, and $r = |4| = 4$. Therefore,

$$4j = 4(\cos 90° + j \sin 90°) = 4\underline{/90°}$$

(d)   Noting that -3 is on the negative x-axis, we have $\theta = 180°$, and $r = |-3| = 3$. Thus,

$$-3 = 3(\cos 180° + j \sin 180°) = 3\underline{/180°}$$

Changing from polar to rectangular form

To convert a complex number from polar form to rectangular form, use equation (12-3) to find x and y from r and θ.

EXAMPLE 19 Convert the complex number $3.00(\cos 210.0° + j \sin 210.0°)$ to rectangular form.

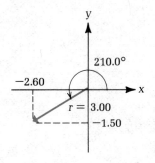

Figure 12-10

Solution Determine x and y from equation (12-3): Given that $r = 3.00$ and $\theta = 210.0°$, we have $x = 3.00 \cos 210.0°$ and $y = 3.00 \sin 210.0°$. Thus,

$$3 \; \boxed{\times} \; 210 \; \boxed{\cos} \; \boxed{=} \to \quad \boxed{-2.5980762} \qquad x \approx -2.60$$

$$3 \; \boxed{\times} \; 210 \; \boxed{\sin} \; \boxed{=} \to \quad \boxed{-1.5} \qquad y = -1.50$$

$3.00(\cos 210.0° + j \sin 210.0°) \approx -2.60 - 1.50j$

Check: Figure 12-10 will help to verify this answer.

EXAMPLE 20 Express the complex number $18.5 \underline{/320.0°}$ in rectangular form.

Solution **First**, note that $r = 18.5$ and $\theta = 320.0°$.
Then, from equation (12-3) we have

$$x = 18.5 \cos 320.0° \approx 14.2 \qquad \text{and} \qquad y = 18.5 \sin 320.0° \approx -11.9$$

Therefore,

$$18.5 \underline{/320.0°} \approx 14.2 - 11.9j$$

NOTE   Some scientific calculators have keys especially designed for converting back and forth between rectangular and polar form. See the owner's manual for your calculator for details. ◄

Exercises 12-4

Rewrite each complex number in polar form. Give the answers to Problems 11–18 in $r\underline{/\theta}$ notation.

1. $3 - 4j$ 2. $7 + 24j$ 3. $-8 + 15j$ 4. $-6 - 8j$
5. $4 + 7j$ 6. $-2 + 3j$ 7. $-8 - 5j$ 8. $5 - 3j$
9. $3.40 - 2.50j$ 10. $-6.50 + 4.20j$ 11. $1.80 + 1.80j$ 12. $-7.25 - 3.75j$
13. $-2.18 - 6.68j$ 14. $\sqrt{5} - j\sqrt{7}$ 15. 6 16. -4
17. $-3j$ 18. $7j$

Express each complex number in rectangular form.

19. $2.00(\cos 65.0° + j \sin 65.0°)$ 20. $6.00(\cos 235.0° + j \sin 235.0°)$
21. $3.50\underline{/317.0°}$ 22. $4.30\underline{/142.0°}$
23. $17.2(\cos 285.3° + j \sin 285.3°)$ 24. $11.4(\cos 291.6° + j \sin 291.6°)$
25. $135\underline{/98.1°}$ 26. $275\underline{/18.5°}$
27. $9(\cos 0° + j \sin 0°)$ 28. $14(\cos 90° + j \sin 90°)$
29. $12\underline{/270°}$ 30. $7\underline{/180°}$
31. $\cos 341.6° + j \sin 341.6°$ 32. $\underline{/156.7°}$
33. $13.7\underline{/-57°}$ 34. $8.50[\cos(-33.5°) + j \sin(-33.5°)]$

35. Find r and $\tan$ (Ref θ) for: (a) $a^b(a - bj)$ (b) $ba(b - aj) + ab(a - bj)$
36. If $y > 0$, then $\sqrt{x + yj}$ is the complex number $p + qj$ where

$$p = \sqrt{\frac{\sqrt{x^2 + y^2} + x}{2}} \qquad \text{and} \qquad q = \sqrt{\frac{\sqrt{x^2 + y^2} - x}{2}}$$

If $y < 0$, $\sqrt{x + yj}$ is the complex number $p - qj$. Find:

(a) $\sqrt{8 + 6j}$ (b) $\sqrt{1 + j}$ (c) $\sqrt{1 - j}$ (d) $\sqrt{-j}$

12-5 | Exponential Form of a Complex Number

In addition to rectangular form and polar form, there is a third way to write complex numbers known as **exponential form.** This form of the number is used in physics and electronics applications. The exponential form of a complex number is defined as follows:

Exponential Form of a Complex Number

$$re^{j\theta} = r(\cos \theta + j \sin \theta) = r\underline{/\theta} \qquad (12\text{-}7)$$

Here, r is the **modulus** and θ is the **argument in radians.** The base e is an irrational number with an approximate value of 2.71828. The angle θ is expressed in radians so that the exponent $j\theta$ is a dimensionless number and obeys all the laws of exponents. The exponential form of complex quantities is very convenient for multiplication and division, just as the rectangular form $x + yj$ is especially suited to addition and subtraction.

Polar to exponential form   Converting a complex number from polar form to exponential form simply involves changing the argument θ from degrees to radians. The exponential form of the number can then be written by inspection.

EXAMPLE 21   Change the complex number $2.45(\cos 87.0° + j \sin 87.0°)$ to exponential form.

Solution   From Chapter 8 we know that

$$87.0° = 87.0°\left(\frac{\pi}{180°}\right)\text{rad} \qquad 87 \boxed{\times}\boxed{\pi}\boxed{\div} 180 \boxed{=} \rightarrow \quad \boxed{1.5184364}$$

We can see from the given polar form that $r = 2.45$, and so we write the complex number as

$$re^{j\theta} \approx 2.45e^{1.52j}$$

Rectangular to exponential form   To change a complex number from rectangular form to exponential form, first use the Pythagorean theorem (12-4) to find r. Then, with your calculator in radian mode, determine θ from its reference angle Ref θ as before, and proceed as in Example 21.

▦ **EXAMPLE 22**   Express $-2 - 3j$ in exponential form.

Solution   First, find r:

$$r = \sqrt{x^2 + y^2} = \sqrt{(-2)^2 + (-3)^2} = \sqrt{13} \approx 3.61$$

Then, find θ:

$$\tan (\text{Ref } \theta) = \left| \frac{-3}{-2} \right| = 1.5$$

[RAD] 1.5 [INV] [tan] [STO] → ‖ *0.9827937* ◁ Ref θ

Because $-2 - 3j$ is in the third quadrant,

$$\theta = \pi + \text{Ref } \theta$$ [π] [+] [RCL] [=] → ‖ *4.1243864* ◁ θ in radians

Finally, write the exponential equivalent:

$$-2 - 3j \approx 3.61e^{4.12j}$$

EXAMPLE 23   In an ac circuit, suppose the current is represented by the complex number $i = 0.160 - 0.300j$ A. Write this number in exponential form $i = i_0 e^{j\theta}$, and find the magnitude i_0 of the current in the circuit.

Solution

$$i_0 = \sqrt{(0.160)^2 + (-0.300)^2} \approx 0.340 \text{ A}$$

$$\tan (\text{Ref } \theta) = \left| \frac{-0.300}{0.160} \right| \qquad \text{Ref } \theta \approx 1.08 \text{ rad}$$

Since θ is in quadrant IV,

$$\theta \approx 2\pi - 1.08 \approx 5.20 \text{ rad}$$

Therefore, $i \approx 0.340e^{5.20j}$. The magnitude of the current in the circuit is approximately 0.340 A.

Exponential to polar form   In polar form the argument θ is usually given in degrees in technical applications. Therefore, to convert a complex number from exponential form to polar form, first convert θ to degrees and then write the number as shown in (12-6).

▦ **EXAMPLE 24**   Convert $6.00e^{3.30j}$ to polar form.

Solution

$$6.00e^{3.30j} = 6.00(\cos 3.30 \text{ rad} + j \sin 3.30 \text{ rad})$$

Convert θ to degrees:

3.3 [×] 180 [÷] [π] [=] → ‖ *189.07607*

Thus,

$$6.00e^{3.30j} \approx 6.00(\cos 189.1° + j \sin 189.1°) \approx 6.00\underline{/189.1°}$$

Exponential to	To convert a complex number from exponential to rectangular form, first
rectangular form	rewrite it in polar form keeping θ in radians. Then convert this to the rectangular equivalent.

▦ **EXAMPLE 25** Express the complex number $1.78e^{0.620j}$ in rectangular form.

Solution

Step 1. Convert to polar form:

$$1.78e^{0.620j} = 1.78(\cos 0.620 \text{ rad} + j \sin 0.620 \text{ rad})$$

Step 2. Calculate the rectangular equivalent using (12-3):

$x = 1.78 \cos 0.620$ $\quad$ [RAD] 1.78 [×] .62 [cos] [=] → $\boxed{1.4487037}$

$y = 1.78 \sin 0.620$ $\quad$ 1.78 [×] .62 [sin] [=] → $\boxed{1.0342426}$

Thus,

$$1.78e^{0.620j} \approx 1.45 + 1.03j$$

Summary

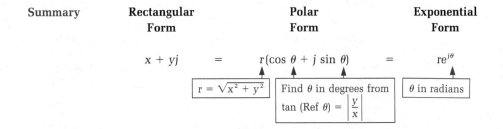

Rectangular Form		Polar Form		Exponential Form
$x + yj$	=	$r(\cos \theta + j \sin \theta)$	=	$re^{j\theta}$

$r = \sqrt{x^2 + y^2}$

Find θ in degrees from $\tan (\text{Ref } \theta) = \left| \dfrac{y}{x} \right|$

θ in radians

Exercises 12-5

Express each complex number in exponential form.

1. $3.60(\cos 55.0° + j \sin 55.0°)$ $\qquad$ **2.** $5.80(\cos 175.0° + j \sin 175.0°)$

3. $12.7\underline{/248.0°}$ $\qquad\qquad\qquad$ **4.** $23.9\underline{/317.0°}$

5. $0.640(\cos 97.0° + j \sin 97.0°)$ $\qquad$ **6.** $0.750(\cos 195.0° + j \sin 195.0°)$

7. $485[\cos(-48.5°) + j \sin (-48.5°)]$ $\qquad$ **8.** $1650(\cos 16.8° + j \sin 16.8°)$

9. $-6 + 8j$ $\qquad$ **10.** $8 + 15j$ $\qquad$ **11.** $-3 - 2j$ $\qquad$ **12.** $5.10 - 3.80j$

13. $12.6 + 19.4j$ $\qquad$ **14.** $-157 + 286j$ $\qquad$ **15.** $1860 - 1450j$ $\qquad$ **16.** $-3.58 - 5.75j$

Express each complex number in polar form.

17. $15.0e^{1.60j}$ $\qquad$ **18.** $17.0e^{4.64j}$ $\qquad$ **19.** $12.6e^{0.453j}$ $\qquad$ **20.** $17.5e^{1.97j}$

21. $2.60e^{3.21j}$ $\qquad$ **22.** $1.45e^{5.43j}$

Express each complex number in rectangular form.

23. $4.00e^{2.70j}$ $\qquad$ **24.** $6.00e^{0.250j}$ $\qquad$ **25.** $4.75e^{2.85j}$ $\qquad$ **26.** $82.4e^{5.38j}$

27. $1.97e^{6.85j}$ $\qquad$ **28.** $2.38e^{4.78j}$

Solve.

29. *Electronics*   In an ac circuit the current is represented by the complex number $i = 0.250 - 0.400j$ A. Write this number in exponential form and determine the magnitude of the current.

30. *Electronics*   Determine the magnitude of an ac voltage represented by $V = -97.3 + 61.3j$ V.

31. *Electronics*   For a current or voltage represented by a complex number in rectangular form, the real part of the number is the in-phase component. Find the in-phase component of a voltage represented by $V = 110e^{3.7j}$ V.

32. *Electronics*   Find the in-phase component of a current represented by $i = 0.68e^{1.4j}$ A. (See Problem 31.)

12-6 | Products, Quotients, Powers, and Roots of Complex Numbers

One of the main advantages of both the polar and exponential forms of complex numbers is that they enable you to perform certain operations with complex numbers more easily than when they are written in rectangular form. In this section we shall show how to multiply, divide, and take powers and roots of complex numbers using exponential and polar form. The final results of these operations are usually given in either polar or rectangular form.

Multiplication

When complex numbers are written in exponential form, we can use the laws of exponents to express their product:

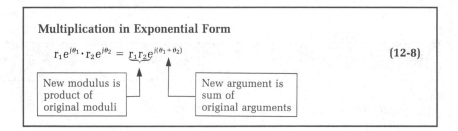

> **Multiplication in Exponential Form**
>
> $$r_1 e^{j\theta_1} \cdot r_2 e^{j\theta_2} = r_1 r_2 e^{j(\theta_1 + \theta_2)} \qquad \text{(12-8)}$$
>
> New modulus is product of original moduli
>
> New argument is sum of original arguments

We can express this in polar form as follows:

> **Multiplication in Polar Form**
>
> $$[r_1(\cos \theta_1 + j \sin \theta_1)] \cdot [r_2(\cos \theta_2 + j \sin \theta_2)]$$
> $$= r_1 r_2 [\cos (\theta_1 + \theta_2) + j \sin (\theta_1 + \theta_2)] \qquad \text{(12-9)}$$
>
> or
>
> $$r_1 \underline{/\theta_1} \cdot r_2 \underline{/\theta_2} = r_1 r_2 \underline{/\theta_1 + \theta_2}$$

NOTE ▶

In cases where the argument of the product $(\theta_1 + \theta_2)$ is greater than or equal to 360°, it is conventional to rewrite the argument as a nonnegative coterminal angle less than 360°. We can do this because, as we discussed in Chapter 8, a trigonometric function of an angle θ is equal to the same function of any angle coterminal to θ, and therefore,

$$r(\cos \theta + j \sin \theta) = r[\cos (\theta + 360k°) + j \sin (\theta + 360k°)] \quad \textbf{(12-10)}$$

where k represents any integer. ◀

EXAMPLE 26 Multiply: $2e^{3j} \cdot 7e^{4j}$

Solution

$$2e^{3j} \cdot 7e^{4j} = 2 \cdot 7e^{(3+4)j}$$
$$= 14e^{7j} \approx 14e^{0.72j}$$

EXAMPLE 27 Multiply: $[3(\cos 30° + j \sin 30°)] \cdot [2(\cos 40° + j \sin 40°)]$

Solution The product of the moduli $r_1 \cdot r_2$ is 6. The sum of the arguments θ_1 and θ_2 is 70°. The final product can be written

$$6(\cos 70° + j \sin 70°) \quad \text{or} \quad 6\underline{/70°}$$

If the numbers to be multiplied are given in rectangular form, first convert each to polar form, and then multiply.

EXAMPLE 28 Multiply $(2 - 5j)$ by $(-3 + 4j)$ using polar form. Express the answer in polar form.

Solution
Step 1. Convert each number to polar form. For $2 - 5j$:

$$r_1 = \sqrt{2^2 + (-5)^2} = \sqrt{29} \approx 5.385*$$

$$\tan (\text{Ref } \theta_1) = \left| \frac{-5}{2} \right| \qquad \text{Ref } \theta_1 \approx 68.20°*$$

Since $2 - 5j$ is in quadrant IV, $\theta_1 \approx 360° - 68.20° \approx 291.80°$

---

* To avoid errors due to premature rounding, round r and θ to one more digit than the desired accuracy of the answer.

For $-3 + 4j$:

$$r_2 = \sqrt{(-3)^2 + (4)^2} = \sqrt{25} = 5$$

$$\tan (\text{Ref } \theta_2) = \left| \frac{4}{-3} \right| \qquad \text{Ref } \theta_2 \approx 53.13°$$

Since $-3 + 4j$ is in quadrant II, $\theta_2 \approx 180° - 53.13° \approx 126.87°$
Therefore, using $r\underline{/\theta}$ notation, we have

$$2 - 5j \approx 5.385\underline{/291.80°} \qquad \text{and} \qquad -3 + 4j \approx 5\underline{/126.87°}$$

Step 2. Now we can use (12-9) to write the product:

$$(2 - 5j)(-3 + 4j) \approx (5.385)(5)\underline{/291.80° + 126.87°}$$

$$\approx 26.9\underline{/418.7°} \approx 26.9\underline{/58.7°} \qquad \boxed{418.7° - 360° = 58.7°}$$

$$\text{or } 26.9(\cos 58.7° + j \sin 58.7°)$$

Note that the argument θ was written as a positive coterminal angle less than 360°.

Division   To express the quotient of two complex numbers in polar form, first note that when we perform the operation in exponential form, we have:

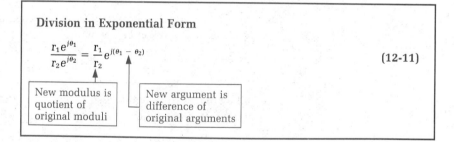

Division in Exponential Form

$$\frac{r_1 e^{j\theta_1}}{r_2 e^{j\theta_2}} = \frac{r_1}{r_2} e^{j(\theta_1 - \theta_2)} \qquad\qquad (12\text{-}11)$$

New modulus is quotient of original moduli

New argument is difference of original arguments

In polar form, this is:

Division in Polar Form

$$\frac{r_1(\cos \theta_1 + j \sin \theta_1)}{r_2(\cos \theta_2 + j \sin \theta_2)} = \frac{r_1}{r_2}[\cos (\theta_1 - \theta_2) + j \sin (\theta_1 - \theta_2)] \qquad (12\text{-}12)$$

or

$$\frac{r_1\underline{/\theta_1}}{r_2\underline{/\theta_2}} = \frac{r_1}{r_2}\underline{/\theta_1 - \theta_2}$$

As with multiplication, the final argument should be expressed as a non-negative angle less than 360°.

EXAMPLE 29   Divide: $\dfrac{6e^{5j}}{2e^{3j}}$

Solution

$$\frac{6e^{5j}}{2e^{3j}} = \frac{6}{2}[e^{(5-3)j}] = 3e^{2j}$$

EXAMPLE 30   Divide: $\dfrac{8(\cos 120° + j \sin 120°)}{2(\cos 75° + j \sin 75°)}$

Solution   Since

$$\frac{r_1}{r_2} = \frac{8}{2} = 4 \quad\text{and}\quad \theta_1 - \theta_2 = 120° - 75° = 45°$$

we have

$$\frac{8(\cos 120° + j \sin 120°)}{2(\cos 75° + j \sin 75°)} = 4(\cos 45° + j \sin 45°)$$

If the original problem is given in rectangular form, the answer can be obtained in the same manner by first changing each number to polar form.

EXAMPLE 31   Use polar form to divide: $\dfrac{-1 - 2j}{4 - 5j}$

Express the answer in both polar form and rectangular form.

Solution
Step 1.   Convert each number to polar form:

$$-1 - 2j \approx 2.236\underline{/243.43°} \qquad 4 - 5j \approx 6.403\underline{/308.66°}$$

Step 2.   Use equation (12-12) to express the quotient:

$$\frac{r_1}{r_2} \approx \frac{2.236}{6.403} \approx 0.3492 \qquad \theta_1 - \theta_2 \approx 243.43° - 308.66°$$

$$\approx -65.23° \approx 294.77°$$

Thus, in polar form, we have

$$\frac{-1 - 2j}{4 - 5j} \approx 0.349\underline{/294.8°} \quad\text{or}\quad 0.349(\cos 294.8° + j \sin 294.8°)$$

Step 3.   Use the more accurate values from Step 2 to convert the result to rectangular form:

$$x \approx 0.3492 \cos 294.77° \approx 0.146$$

$$y \approx 0.3492 \sin 294.77° \approx -0.317$$

Therefore, in rectangular form, we obtain

$$\frac{-1 - 2j}{4 - 5j} \approx 0.146 - 0.317j$$

Powers and roots   Using exponential form it is easy to raise a complex number in polar form to a power. Applying the power of a power law, we have:

$$(re^{j\theta})^n = r^n e^{jn\theta} \qquad\qquad \textbf{(12-13)}$$

New modulus   New argument

If we rewrite both sides in polar form, we have:

$$[r(\cos\theta + j\sin\theta)]^n = r^n(\cos n\theta + j\sin n\theta) \qquad\qquad \textbf{(12-14)}$$

or

$$(r\underline{/\theta})^n = r^n\underline{/n\theta}$$

As before, we express the final argument $n\theta$ as a nonnegative angle less than 360°.

Equation (12-14) is known as **DeMoivre's theorem** (named after the Frenchman Abraham DeMoivre, who first used it in 1707). DeMoivre's theorem is true for any positive integer n.

EXAMPLE 32   Find: $[4(\cos 45° + j\sin 45°)]^3$

Solution   Using DeMoivre's theorem, we have

$$[4(\cos 45° + j\sin 45°)]^3 = 4^3(\cos 3\cdot 45° + j\sin 3\cdot 45°)$$

$$= 64(\cos 135° + j\sin 135°)$$

This same procedure can be used to find powers of a complex number in rectangular form if we first find r and θ.

EXAMPLE 33   Find $(-4.60 + 2.50j)^5$. Express the result in both polar form and rectangular form.

Solution   The power n = 5; therefore, to apply DeMoivre's theorem, we must find r^5 and 5θ.

Step 1. Find r:

$$r = \sqrt{(-4.60)^2 + (2.50)^2}$$ 4.6 $\boxed{x^2}$ $\boxed{+}$ 2.5 $\boxed{x^2}$ $\boxed{=}$ $\boxed{\sqrt{}}$ → ▨ *5.2354560*

Step 2. Find r^5:

$\boxed{y^x}$ 5 $\boxed{=}$ → ▨ *3933.4405*   ◁ Record this for later use

Step 3. Find θ:

$$\tan (\text{Ref } \theta) = \left| \frac{2.50}{-4.60} \right|$$ 2.5 $\boxed{\div}$ 4.6 $\boxed{=}$ $\boxed{\text{INV}}$ $\boxed{\text{tan}}$ $\boxed{\text{STO}}$ → ▨ *28.523119*

Since θ is in quadrant II, $\theta = 180 - \text{Ref } \theta$:

180 $\boxed{-}$ $\boxed{\text{RCL}}$ $\boxed{=}$ → ▨ *151.47688*

Step 4. Find 5θ:

$\boxed{\times}$ 5 $\boxed{=}$ → ▨ *757.38441*

Convert this to a coterminal angle less than 360° and store for use in Step 5:

$\boxed{-}$ 720 $\boxed{=}$ $\boxed{\text{STO}}$ → ▨ *37.384407*

Rounding the results of Steps 2 and 4, we have the result in polar form:

$$(-4.60 + 2.50j)^5 \approx 3930(\cos 37.4° + j \sin 37.4°)$$

Step 5. Use the unrounded versions of r^5 and 5θ to find the answer in rectangular form:

3933.4405 $\boxed{\times}$ $\boxed{\text{RCL}}$ $\boxed{\cos}$ $\boxed{=}$ → ▨ *3125.4327*

3933.4405 $\boxed{\times}$ $\boxed{\text{RCL}}$ $\boxed{\sin}$ $\boxed{=}$ → ▨ *2388.2262*

Therefore, $(-4.60 + 2.50j)^5 \approx 3130 + 2390j$.

Roots of complex numbers   Although DeMoivre's theorem was originally proposed for positive *integral* powers, it can be extended to apply to any positive *rational power* n. Therefore, by using fractional exponents, we are able to find the *roots* of a complex number. As shown in the next example, every complex number has n distinct nth roots.

EXAMPLE 34 Find all possible roots for: $\sqrt[3]{3(\cos 150° + j \sin 150°)}$

Solution

Step 1. Since the problem is in polar form, we rewrite it using fractional exponents and apply (12-14) as follows:

$$\sqrt[3]{3(\cos 150° + j \sin 150°)} = [3(\cos 150° + j \sin 150°)]^{1/3}$$

$$= 3^{1/3}(\cos \frac{1}{3} \cdot 150° + j \sin \frac{1}{3} \cdot 150°)$$

$$\approx 1.44(\cos 50° + j \sin 50°)$$

Step 2. By equation (12-10), we know that two complex numbers in polar form are equivalent if their absolute values are equal and if their arguments are coterminal angles. Therefore, by adding 360° to the original argument, we can find a second root of the original number:

$$\overbrace{150° + 360°}$$

$$[3(\cos 150° + j \sin 150°)]^{1/3} = [3(\cos 510° + j \sin 510°)]^{1/3}$$

$$= 3^{1/3}(\cos \frac{1}{3} \cdot 510° + j \sin \frac{1}{3} \cdot 510°)$$

$$\approx 1.44(\cos 170° + j \sin 170°)$$

We now have two roots.

Step 3. Repeat Step 2 until the argument of the root exceeds 360°. This will not occur until all n of the nth roots have been found. Since this problem involves cube roots, there should be one more root:

$$\overbrace{510° + 360°}$$

$$[3(\cos 150° + j \sin 150°)]^{1/3} = 3(\cos 870° + j \sin 870°)^{1/3}$$

$$= 3^{1/3}(\cos \frac{1}{3} \cdot 870° + j \sin \frac{1}{3} \cdot 870°)$$

$$\approx 1.44(\cos 290° + j \sin 290°)$$

We now have all three cube roots. We can confirm this by adding 360° to 870° to obtain 1230° and noting that $\frac{1}{3}(1230°) = 410°$, which is greater than 360°. Since 410° is coterminal to 50°, the argument of the first root, attempting the procedure again will result in a repeat of the first root.

General case for roots We can generalize the procedure used in Example 34 for finding all the nth roots of a complex number as follows: If a is a nonzero complex number such that

$$a = x + yj = r(\cos \theta + j \sin \theta) = r\underline{/\theta}$$

then all the nth roots of a are also complex numbers such that:

Roots of a Complex Number

$$\sqrt[n]{a} = \sqrt[n]{r}\left[\cos\left(\frac{\theta + 360k°}{n}\right) + j \sin\left(\frac{\theta + 360k°}{n}\right)\right] \qquad (12\text{-}15)$$

$$= \sqrt[n]{r}\ \underline{/\left(\frac{\theta + 360k°}{n}\right)}$$

for n a positive integer and $k = 0, 1, 2, \ldots, n - 1$

As we saw in Example 34, the roots repeat themselves for $k > n - 1$. For example, for $k = n$, the argument becomes

$$\frac{\theta + n \cdot 360°}{n} = \frac{\theta}{n} + 360°$$

But this angle is coterminal to θ/n, which is the argument of the first root where $k = 0$.

EXAMPLE 35   Find $\sqrt[4]{5 - 8j}$. Express the four roots in rectangular form.

Solution   First, find r and θ using the methods learned earlier:

$$r = \sqrt{89} \approx 9.434 \qquad \text{and} \qquad \theta \approx 302.0°$$

Then, use equation (12-15) for $k = 0, 1, 2,$ and 3 to find all four fourth roots.

$$\sqrt[4]{(5 - 8j)} \approx \sqrt[4]{9.434}\left[\cos\left(\frac{302.0° + 360k°}{4}\right) + j \sin\left(\frac{302.0° + 360k°}{4}\right) \right]$$

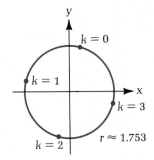

Figure 12-11

For $k = 0$: $\approx 1.753(\cos 75.5° + j \sin 75.5°) \approx 0.439 + 1.70j$

For $k = 1$: $\approx 1.753(\cos 165.5° + j \sin 165.5) \approx -1.70 + 0.439j$

For $k = 2$: $\approx 1.753(\cos 255.5° + j \sin 255.5°) \approx -0.439 - 1.70j$

For $k = 3$: $\approx 1.753(\cos 345.5° + j \sin 345.5°) \approx 1.70 - 0.439j$

If we graph these roots as shown in Fig. 12-11, we see that they are spaced at equal intervals about the origin at a radius $r \approx 1.753$.

In the next example we illustrate the same procedure using $r\underline{/\theta}$ notation.

EXAMPLE 36   Find the three exact values of $\sqrt[3]{j}$.

Solution
Step 1.   Since $j = 1(\cos 90° + j \sin 90°)$, we have

$$r = 1 \qquad \text{and} \qquad \theta = 90°$$

Step 2.   Using the $r\underline{/\theta}$ form of (12-15),

$$j = \sqrt[3]{1}\;\underline{\Big/\dfrac{90° + 360k°}{3}}$$

For $k = 0$, $\sqrt[3]{j} = 1\underline{/30°} = \dfrac{\sqrt{3}}{2} + \dfrac{j}{2}$

For $k = 1$, $\sqrt[3]{j} = 1\underline{/150°} = -\dfrac{\sqrt{3}}{2} + \dfrac{j}{2}$

For $k = 2$, $\sqrt[3]{j} = 1\underline{/270°} = -j$

These roots may be checked by cubing each to see if the result is equal to j.

Complex roots of real numbers   Since real numbers are also complex numbers, the nth root of a real number should also have n possible complex values. The next example shows how to find them.

EXAMPLE 37   Find: $\sqrt[5]{-1}$

Solution   We note first that the real root of $\sqrt[5]{-1}$ is -1. Using (12-15) we should be able to find this root and four additional imaginary roots.

Step 1.   Since $-1 = 1(\cos 180° + j \sin 180°)$,

$$\theta = 180° \quad \text{and} \quad r = 1$$

Step 2.   Using (12-15),

$$\sqrt[5]{(-1)} = 1\left[\cos\left(\frac{180° + 360k°}{5}\right) + j \sin\left(\frac{180° + 360k°}{5}\right)\right]$$

For $k = 0$: $= 1(\cos 36° + j \sin 36°) \approx 0.809 + 0.588j$

For $k = 1$: $= 1(\cos 108° + j \sin 108°) \approx -0.309 + 0.951j$

For $k = 2$: $= 1(\cos 180° + j \sin 180°) = -1$

For $k = 3$: $= 1(\cos 252° + j \sin 252°) \approx -0.309 - 0.951j$

For $k = 4$: $= 1(\cos 324° + j \sin 324°) \approx 0.809 - 0.588j$

Exercises 12-6

Perform the indicated operations. Leave the results in exponential form.

1. $3e^{4j} \cdot 5e^{2j}$ **2.** $5e^{3j} \cdot 2e^{j}$ **3.** $(2.5e^{1.4j})(3.8e^{3.2j})$ **4.** $(2.7e^{0.48j})(6.5e^{0.27j})$

5. $\dfrac{9e^{4j}}{3e^{3j}}$ **6.** $\dfrac{42e^{2j}}{7e^{5j}}$ **7.** $\dfrac{5.87e^{1.64j}}{3.24e^{2.78j}}$ **8.** $\dfrac{38.7e^{4.39j}}{72.4e^{2.05j}}$

Perform the indicated operations. Leave the results in polar form.

9. $[3(\cos 20° + j \sin 20°)][3(\cos 40° + j \sin 40°)]$

10. $[6(\cos 265° + j \sin 265°)][5(\cos 72° + j \sin 72°)]$

11. $(1.50\underline{/142.0°})(3.70\underline{/306.0°})$

12. $(16\underline{/173°})(15\underline{/204°})$

13. $\dfrac{6\underline{/150°}}{3\underline{/85°}}$

14. $\dfrac{24\underline{/170°}}{6\underline{/40°}}$

15. $\dfrac{20(\cos 315° + j \sin 315°)}{8.0(\cos 204° + j \sin 204°)}$

16. $\dfrac{2.4(\cos 238° + j \sin 238°)}{3.2(\cos 99° + j \sin 99°)}$

17. $[3(\cos 35° + j \sin 35°)]^3$

18. $[4(\cos 65° + j \sin 65°)]^3$

19. $(1.70\underline{/113°})^4$

20. $(0.780\underline{/156°})^6$

Perform the indicated operations using polar form. Express the answers in both polar form and rectangular form.

21. $(6 - 8j)(12 + 5j)$

22. $(-3 + 4j)(-12 - 9j)$

23. $(3.40 + 7.60j)(-5.20 + 2.30j)$

24. $(-1.30 - 8.70j)(4.80 - 5.30j)$

25. $\dfrac{9}{2 - 3j}$

26. $\dfrac{-6}{4 + 7j}$

27. $\dfrac{1.60 + 3.10j}{-8.70 - 9.40j}$

28. $\dfrac{4.30 - 9.20j}{-2.50 + 6.90j}$

29. $(-6 - 3j)^4$

30. $(2.30 - 7.10j)^3$

Use DeMoivre's theorem to find all possible roots. Express the results in the same form as the original radicand.

31. $\sqrt{2(\cos 30° + j \sin 30°)}$

32. $\sqrt{16(\cos 45° + j \sin 45°)}$

33. $\sqrt[3]{8(\cos 135° + j \sin 135°)}$

34. $\sqrt[4]{12.7(\cos 165.0° + j \sin 165.0°)}$

35. $\sqrt[3]{-6 + 8j}$

36. $\sqrt{5 - 4j}$

37. $\sqrt[5]{1}$

38. $\sqrt[3]{5}$

39. $\sqrt[4]{j}$

40. $\sqrt[3]{-j}$ (exact values)

41. $\sqrt[3]{-8}$ (exact values)

42. $\sqrt[5]{-32}$

43. $\sqrt{7.30 + 4.60j}$

44. $\sqrt[4]{-1.80 - 9.50j}$

12-7 | Applying Complex Numbers to ac Circuits

In this section we shall explore one of the most important applications of complex numbers: the description of voltage and current changes in an ac (alternating current) circuit. Certain elements in an ac circuit modify the in-phase relationship between current and voltage, causing them to peak at different times in the different circuit elements. Because a phase difference may exist between the current and the voltage drops at different points in a circuit, it is very helpful to use complex numbers to represent these quantities.

Circuit elements   We need to consider three basic circuit elements. Most metals are conductors of electric current, so that when a potential difference, or voltage, V is applied across a conducting wire, a current i will flow in it. As electrons move through the conductor, they undergo collisions with atoms and molecules in the wire. **Resistance**   The **resistance** R of the conductor, or **resistor** element, is a measure of this opposition to current flow. These variables are related by **Ohm's law:**

> **Ohm's Law** $R = \dfrac{V_R}{i}$ (12-16)
>
> where R is measured in ohms (Ω), and is shown in a circuit diagram by the symbol —⋀⋀—, V_R is the voltage drop across the resistor, and i is the current flowing in the resistor in amperes

Capacitance   A second circuit element, the **capacitor,** is a device that allows electric charge to build up momentarily on a conducting surface, creating an electric field in which energy may be stored. In its simplest form, the capacitor is a pair of conducting surfaces separated by an insulator. When the voltage across the surface changes in an ac circuit, the electric charge on the capacitor surfaces alternately increases and decreases. The **capacitance** C of the device is a measure of the charge stored per volt of potential difference across it. Capacitance is measured in farads (F), and is shown in a circuit diagram by the symbol —⊣⊢—.

Like a resistor, a capacitor also acts as an impedance to current flow. As the charge accumulates on one surface, the voltage across the surfaces builds up and opposes the further accumulation of charge. When the capacitor is fully charged, current stops.

Inductance   When a current flows in a conductor, it generates a magnetic field proportional in strength to the current. If the current varies, as in an ac circuit, changes in this magnetic field produce a voltage that acts to oppose the change in current. This effect is small for a straight wire, but it can be very large when the wire is arranged in a coil or **inductor.** The **inductance** L of a coil is a measure of the countervoltage induced for a given rate of change of current. The inductance depends on the number of turns of wire, the inside area of the coil, and the type of material inside the coil. Inductance is measured in henries (H), and is shown in a circuit diagram by the symbol ——ᴍᴍ——.

In an ac circuit, the voltage source delivers energy to the circuit elements. Some of this energy is dissipated as heat in the resistors, some is stored in the electric field of the capacitance elements, and some is stored in the magnetic field of the inductance elements. As the current changes in magnitude and direction, the stored energy oscillates back and forth from the electric field of the capacitor to the magnetic field of the inductor.

Reactance   Because both capacitors and inductors offer opposition to current flow in a circuit, it is useful to define a quantity analogous to resistance for them. The **capacitative reactance** X_C is the apparent resistance of a capacitor to current flow:

Capacitive Reactance $$X_C = \frac{1}{2\pi f C}$$ (12-17)

where f is the frequency of the voltage and X_C is measured in ohms

The **inductive reactance** X_L is the apparent resistance of an inductor to current flow:

Inductive Reactance $$X_L = 2\pi f L$$ (12-18)

where f is again the frequency of the voltage and X_L is measured in ohms

Reactance is related to the current through the circuit and the voltage drop across the circuit elements by a rule analogous to Ohm's law (12-16).

Voltage Drop Across a Circuit Element

For a resistor: $V_R = iR$ (12-19)

For a capacitor: $V_C = iX_C$ (12-20)

For an inductor: $V_L = iX_L$ (12-21)

EXAMPLE 38   An ac series circuit contains an inductor with $L = 0.0200$ H, a resistor with $R = 3.50\ \Omega$, and a capacitor with $C = 1.20 \times 10^{-6}$ F (see Fig. 12-12). A 15.0 V, 1000 Hz power source is connected across the circuit. Calculate X_L and X_C. Round to the nearest ohm.

Figure 12-12

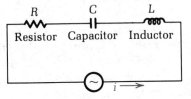

Solution

$$X_L = 2\pi f L = 2 \cdot \pi \cdot 1000 \cdot 0.0200$$

$$2\ \boxed{\times}\ \boxed{\pi}\ \boxed{\times}\ 1000\ \boxed{\times}\ .02\ \boxed{=} \rightarrow \quad \textit{125.66371}$$

$$X_L \approx 126\ \Omega$$

$$X_C = \frac{1}{2\pi f C} = \frac{1}{2 \cdot \pi \cdot 1000(1.20 \times 10^{-6})}$$

$$2\ \boxed{\times}\ \boxed{\pi}\ \boxed{\times}\ 1000\ \boxed{\times}\ 1.2\ \boxed{\text{EE}}\ 6\ \boxed{+/-}\ \boxed{=}\ \boxed{1/x} \rightarrow \quad \textit{1.3263 02}$$

$$X_C \approx 133\ \Omega$$

Phase   The voltage across a resistor is always in phase with the current when it is alone in a circuit. Both current and voltage will peak at the same time. However, the voltage across both capacitor and inductor elements will be out of phase. The voltage across a capacitor when it is alone in a circuit will *lag* the current by 90°; that is, the voltage peaks 90° after the current. For an inductor alone in a circuit, the voltage across the inductor *leads* the current by 90°; that is, the voltage peaks 90° before the current peaks. These phase relationships are shown in Fig. 12-13.

If a circuit contains all three elements, we must take into account the phase differences in order to calculate the current and voltage in the circuit. Complex

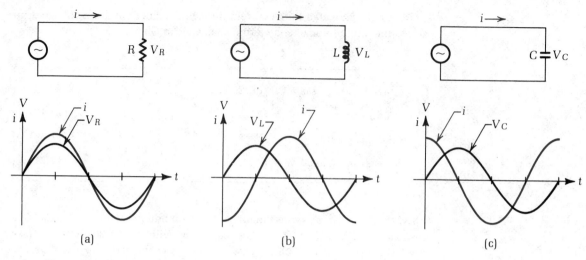

Figure 12-13

numbers are valuable in describing these quantities and their phase relationships.

Because the voltage across a resistor is in phase with the current, it is convenient to represent the resistance by a real number R. In the complex plane, R is shown along the positive real axis [Fig. 12-14(a)].

The voltage across an inductor leads the current by 90°, so the inductive reactance is represented by the pure imaginary number $-X_L j$ and is shown along the positive imaginary axis [Fig. 12-14(b)].

Because the voltage across a capacitor lags the current by 90°, it is represented by the pure imaginary number $-X_C j$ and is shown along the negative imaginary axis [Fig. 12-14(c)].

Figure 12-14

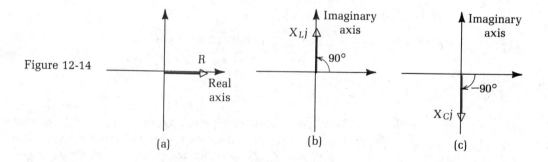

Phasors   Representing these quantities by arrows, as shown in Fig. 12-14, will probably remind you of vector notation, but resistance and reactance are *not* vector quantities. To distinguish them, these arrows, called **phasors,** are drawn with open arrowheads. Phasors rotate counterclockwise about the origin with the frequency f of the applied voltage to show phase relationships.

Impedance   For a circuit containing all three elements, the voltage across the whole circuit is given by

$$V = V_R + V_L + V_C$$

$$V = iR + iX_L j - iX_C j = i[R + j(X_L - X_C)]$$

Impedance $V = iZ$ (12-22)

where the impedance Z is the effective resistance of the entire circuit:

$$Z = R + j(X_L - X_C)$$

The magnitude of the impedance is therefore given by

Magnitude of Impedance $|Z| = \sqrt{R^2 + (X_L - X_C)^2}$ (12-23)

The phase angle ϕ between the current and the voltage across the circuit is given by

---

Phase Angle $\tan \phi = \dfrac{X_L - X_C}{R}$ (12-24)

EXAMPLE 39   For the circuit in Fig. 12-12, assume that $X_L = 126\ \Omega$, $X_C = 133\ \Omega$, $V = 15$ V, and $R = 3.8\ \Omega$. Calculate the phase angle ϕ, the magnitude of the impedance, the current in the circuit, and the voltage across each of the circuit elements.

Solution $\tan \phi = \dfrac{X_L - X_C}{R} = \dfrac{126 - 133}{3.8}$

Therefore, $\phi \approx -62°$. The voltage lags the current by approximately 62°.

$$|Z| = \sqrt{3.8^2 + (126 - 133)^2}$$

3.8 $\boxed{x^2}$ $\boxed{+}$ $\boxed{(}$ 126 $\boxed{-}$ 133 $\boxed{)}$ $\boxed{x^2}$ $\boxed{=}$ $\boxed{\sqrt{}}$ $\boxed{\text{STO}}$ →   *7.9649231* $|Z| \approx 8.0\ \Omega$

$i = \dfrac{V}{|Z|} = \dfrac{15}{|Z|}$ 15 $\boxed{\div}$ $\boxed{\text{RCL}}$ $\boxed{=}$ $\boxed{\text{STO}}$ →   *1.8832574* $i \approx 1.9$ A

$V_R = iR = i(3.8)$ $\boxed{\text{RCL}}$ $\boxed{\times}$ 3.8 $\boxed{=}$ →   *7.1563779* $V_R \approx 7.2$ V

$V_L = iX_L = i(126)$ $\boxed{\text{RCL}}$ $\boxed{\times}$ 126 $\boxed{=}$ →   *237.29043* $V_L \approx 240$ V

$V_C = iX_C = i(133)$ $\boxed{\text{RCL}}$ $\boxed{\times}$ 133 $\boxed{=}$ →   *250.47323* $V_C \approx 250$ V

Notice that the voltage across the circuit is 15 V, and this is *not* equal to the sum of the voltage drops across the three circuit elements. Because these voltages are out of phase, we must add them using phasors.

EXAMPLE 40 An electric motor with an inductance of 0.0400 H and a resistance of 20.0 Ω is connected to a 120 V, 60 Hz power source.
(a) Calculate the magnitude of the impedance of the motor.
(b) Calculate the current i through the motor.
(c) If a 480 μF (= 480 $\times$ 10^{-6} F) capacitance is added to the circuit, what current will it draw?
(d) Find the phase angle in part (c).

Solution

(a) $X_L = 2\pi fL = 2\cdot\pi\cdot(60)\cdot(0.0400) \approx 15.08$ Ω
$X_C = 0$
$|Z| = \sqrt{R^2 + (X_L - X_C)^2} = \sqrt{20.0^2 + (15.08 - 0)^2} \approx 25.0$ Ω

(b) $i = \dfrac{V}{|Z|} = \dfrac{120\text{ V}}{25.0\ \Omega} \approx 4.8$ A

(c) $X_C = \dfrac{1}{2\pi fC} = \dfrac{1}{2\cdot\pi\cdot60\cdot480\times10^{-6}} \approx 5.53$ Ω
$|Z| = \sqrt{20.0^2 + (15.08 - 5.53)^2} \approx 22.2$ Ω
$i \approx \dfrac{120\text{ V}}{22.2\ \Omega} \approx 5.4$ A

(d) $\tan\phi \approx \dfrac{15.08 - 5.53}{20.0} \approx \dfrac{9.55}{20.0}$
$\phi \approx 26°$

The voltage leads the current by 26°.

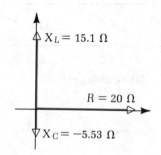

$X_L = 15.1$ Ω

$R = 20$ Ω

$X_C = -5.53$ Ω

Figure 12-15

Complex current

The phase angle ϕ tells us the constant phase difference between the current and the circuit voltage. Since current and the voltage across the resistor are in phase, by choosing R along the real axis we have represented current along this direction also. If some other angle is chosen for current, we must use complex numbers to represent current and voltage as well as impedance. For example, in Fig. 12-16, the complex current $8 - 6j$ would have magnitude given by the length of the phasor arrow and the phase direction shown.

Imaginary axis

8

Real axis

−6

Figure 12-16

EXAMPLE 41 In a certain circuit, if the complex current is $3.00 - 4.00j$ A and the complex impedance is $5.00 + 2.00j$ Ω, calculate the magnitude of the voltage.

Solution

$$V = iZ = (3.00 - 4.00j)(5.00 + 2.00j)$$

$$= 15.0 + 6.00j - 20.0j - 8.00j^2$$

$$= 23.0 - 14.0j \text{ V}$$

The magnitude of the voltage is

$$|V| = \sqrt{23.0^2 + (-14.0)^2} = \sqrt{725.00} \approx 26.9 \text{ V}$$

Resonance   Because X_L and X_C depend on the frequency of the voltage source [equations (12-17) and (12-18)], the impedance [equations (12-22) and (12-23)] is also dependent on the frequency. When $X_L = X_C$, the magnitude of the impedance will be a minimum ($|Z| = R$), and therefore the current in the circuit will be a maximum. The frequency f_0 at which $X_L = X_C$ is called the **resonant frequency.** If $X_L = X_C$, then

$$2\pi f_0 L = \frac{1}{2\pi f_0 C}$$

and, solving for f_0, we obtain the following formula:

Resonant Frequency	$f_0 = \dfrac{1}{2\pi\sqrt{LC}}$	(12-25)

EXAMPLE 42   Find the resonant frequency of the circuit described in Example 38.

Solution

$$f_0 = \frac{1}{2\pi\sqrt{LC}} = \frac{1}{2\cdot\pi\sqrt{(0.0200)(1.20 \times 10^{-6})}}$$

.02 ⊠ 1.2 ᴇᴇ 6 ⁺∕₋ ═ √ ⊠ 2 ⊠ π ═ ¹∕ₓ → $\boxed{1.0273 \quad 03}$

$$f_0 \approx 1030 \text{ Hz}$$

At its resonant frequency, an RLC series circuit is very sensitive to small input voltages. This kind of circuit is called a *tuning circuit* and is used in radio and television receivers. Usually, the frequency is changed by varying the capacitance C using a variable capacitor.

EXAMPLE 43   The tuning circuit of a radio consists of a variable capacitor and a 25 mH (= 25×10^{-3} H) coil with 12 Ω resistance.
(a)   Calculate the capacitance needed to tune the circuit to a radio station broadcasting at 950 kHz (= 950×10^3 Hz).
(b)   If the radio wave arriving at the antenna produces an ac signal of 2.4×10^{-5} V, what would be the current in the antenna circuit?
(c)   What is the voltage across the coil?*

---

* Adapted from Alan Van Heuvelan, *Physics: A General Introduction*, p. 584, Copyright © 1982 by Alan Van Heuvelan. By permission of Little, Brown and Company (Inc.)

Solution

(a) Solve the resonant frequency equation (12-25) for C and substitute the given values.

$$2\pi f_0 = \frac{1}{\sqrt{LC}}$$

$$LC = \frac{1}{4\pi^2 f_0^2}$$

$$C = \frac{1}{4\pi^2 f_0^2 L} = \frac{1}{4\pi^2 (950 \times 10^3)^2 (25 \times 10^{-3})}$$

$4\;[\times]\;[\pi]\;[x^2]\;[\times]\;950\;[\text{EE}]\;3\;[x^2]\;[\times]\;25\;[\text{EE}]\;3\;[+/-]\;[=]\;[1/x]\;\rightarrow\;\boxed{1.1227 \quad -12}$

$$C \approx 1.1 \times 10^{-12}\ \text{F}$$

(b) At resonance, $X_L = X_C$, so $|Z| = R$ and the current is

$$i = \frac{V}{R} = \frac{2.4 \times 10^{-5}\ \text{V}}{12} = 2.0 \times 10^{-6}\ \text{A}$$

(c) $V_L = iX_L = i(2\pi f L) = (2.0 \times 10^{-6})(2\pi)(950 \times 10^3)(25 \times 10^{-3})$

$V_L \approx 0.30\ \text{V}$

This is the input signal to the radio amplifier.

Exercises 12-7

1. An ac series circuit contains an inductor, a resistor, and a capacitor with the following values:

   $L = 0.0250\ \text{H}$ $\qquad$ $R = 6.70\ \Omega$ $\qquad$ $C = 7.50 \times 10^{-6}\ \text{F}$

   If a 15.0 V, 1000 Hz power source is connected across the circuit, find:
   (a) Inductive reactance
   (b) Capacitive reactance
   (c) Magnitude of the impedance
   (d) Current
   (e) Phase angle between current and voltage
   (f) Voltage across the resistor
   (g) Voltage across the inductor
   (h) Voltage across the capacitor

2. Answer parts (a)–(h) from Problem 1 for a similar circuit with

   $L = 6.25 \times 10^{-3}\ \text{H}$ $\qquad$ $R = 11.0\ \Omega$ $\qquad$ $C = 1.25 \times 10^{-5}\ \text{F}$

   and a 115 V, 60 Hz power source connected across the circuit.

3. An electric motor has an inductance of 0.0750 H and a resistance of 16.0 Ω. If it is connected to a 120 V, 60 Hz power source, calculate the magnitude of the impedance of the motor and of the current through the motor.

4. In Problem 3, find the current when a capacitance of 24×10^{-5} F is added to the circuit. What is the phase angle with the capacitance added?

5. In a given circuit, $R = 4.00\ \Omega$, $X_C = 9.00\ \Omega$, and $X_L = 8.00\ \Omega$. Find the magnitude of the impedance and the phase angle between the current and voltage.

6. In Problem 3, find the phase angle between the current and voltage.

7. In a certain circuit, if the complex current is $6.00 + 8.00j$ A and the complex impedance is $4.00 - 5.00j\ \Omega$, calculate the magnitude of the voltage.

8. If the complex current in a certain circuit is $2.00 - 5.00j$ A and the complex impedance is $7.00 - 2.00j\ \Omega$, what is the magnitude of the voltage?

9. The voltage across a certain circuit is $8.00 - 5.00j$ V and the impedance is $6.00 + 2.00j$ Ω. Calculate the magnitude of the current.

10. Given a current of $3.00 - 4.00j$ A and a voltage of $5.00 - 6.00j$ V, find the magnitude of the impedance in the circuit.

11. Find the resonant frequency of the circuit in Problem 1.

12. Find the resonant frequency of the circuit in Problem 2.

13. The tuning circuit of a radio consists of a variable capacitor and a 35 mH (35×10^{-3} H) coil with a 10.0 Ω resistance.
    (a) Determine the capitance needed to tune the circuit to a station broadcasting at 12.4×10^5 Hz.
    (b) If the radio wave arriving at the antenna produces an ac signal of 4.5×10^{-5} V, what is the current in the antenna circuit?
    (c) What is the voltage across the coil?

14. The tuning circuit of a radio contains a 7.0 Ω resistance, a 25 mH coil, and a variable capacitor.
    (a) What is the capacitance needed to tune the circuit to a station broadcasting at 6.8×10^5 Hz?
    (b) Determine the current produced in the antenna circuit by a radio wave arriving with an ac signal of 2.5×10^{-5} V.
    (c) Find the voltage across the coil.

15. How much capacitance must be inserted in series with a 35 Ω load resistor to produce a current of 1.4 A from a 60 V radio transmitter with a frequency of 2.5×10^{-3} Hz?

16. A 2.2 H inductance is connected in series with a resistor. If the combination is connected to a 220 V, 60 Hz power supply, what is the resistance necessary to produce a current of 0.25 A?

12-8 │ Review of Chapter 12

Important Terms and Concepts

complex numbers (p. 409)

imaginary unit (p. 410)

pure imaginary number (p. 411)

rectangular form (p. 412)

real part (p. 412)

imaginary part (p. 412)

complex conjugates (p. 414)

imaginary axis (p. 419)

real axis (p. 419)

complex plane (p. 419)

polar form (p. 421)

modulus (p. 422)

argument (p. 422)

exponential form (p. 425)

resistance R (in ohms, Ω)

resistor (p. 437)

capacitance C (in farads, F)

capacitor (p. 437)

inductance L (in henries, H)

inductor (p. 438)

reactance (p. 438)

phase (p. 439)

impedance Z (p. 441)

resonant frequency (p. 443)

Formulas and Definitions

- $\sqrt{-a} = j\sqrt{a}$ for $a > 0$ (p. 410)   **(12-1)**
- $j^2 = -1$ (p. 410)   **(12-2)**
- relationship between rectangular and polar form of a complex number (p. 421):

$$x + yj = r(\cos \theta + j \sin \theta)$$ **(12-6)**

$$\text{where} \quad r = \sqrt{x^2 + y^2}$$ **(12-4)**

$$\text{and} \quad \tan (\text{Ref } \theta) = \left| \frac{y}{x} \right|$$ **(12-5)**

- exponential form of a complex number (p. 425): $re^{j\theta} = r(\cos \theta + j \sin \theta) = r\underline{/\theta}$ **(12-7)**
- product of complex numbers (p. 428)

  exponential form: $r_1 e^{j\theta_1} \cdot r_2 e^{j\theta_2} = r_1 r_2 e^{j(\theta_1 + \theta_2)}$ **(12-8)**

  polar form: $[r_1(\cos \theta_1 + j \sin \theta_1)] \cdot [r_2(\cos \theta_2 + j \sin \theta_2)] = r_1 r_2[\cos (\theta_1 + \theta_2) + j \sin (\theta_1 + \theta_2)]$ **(12-9)**

  $$\text{or} \quad r_1\underline{/\theta_1} \cdot r_2\underline{/\theta_2} = r_1 r_2\underline{/\theta_1 + \theta_2}$$
- equivalent complex numbers (p. 429):

  $$r(\cos \theta + j \sin \theta) = r[\cos (\theta + 360k°) + j \sin (\theta + 360k°)] \quad \textbf{(12-10)}$$
- quotient of complex numbers (p. 430)

  exponential form: $\dfrac{r_1 e^{j\theta_1}}{r_2 e^{j\theta_2}} = \dfrac{r_1}{r_2} e^{j(\theta_1 - \theta_2)}$ **(12-11)**

  polar form: $\dfrac{r_1(\cos \theta_1 + j \sin \theta_1)}{r_2(\cos \theta_2 + j \sin \theta_2)} = \dfrac{r_1}{r_2}[\cos (\theta_1 - \theta_2) + j \sin (\theta_1 - \theta_2)]$

  $$\text{or} \quad \dfrac{r_1\underline{/\theta_1}}{r_2\underline{/\theta_2}} = \dfrac{r_1}{r_2}\underline{/\theta_1 - \theta_2} \quad \textbf{(12-12)}$$
- DeMoivre's theorem (p. 432): $[r(\cos \theta + j \sin \theta)]^n = r^n(\cos n\theta + j \sin n\theta)$ **(12-14)**

  $$(r\underline{/\theta})^n = r^n\underline{/n\theta}$$
- roots of a complex number (p. 434): for $a = x + yj = r(\cos \theta + j \sin \theta) = r\underline{/\theta}$,

  $$\sqrt[n]{a} = \sqrt[n]{r}\left[\cos \left(\dfrac{\theta + 360k°}{n}\right) + j \sin \left(\dfrac{\theta + 360k°}{n}\right)\right] \quad \textbf{(12-15)}$$

  $$= \sqrt[n]{r}\,\underline{/\dfrac{\theta + 360k°}{n}}$$

  where n is a positive integer and $k = 0, 1, 2, \ldots, n - 1$
- Ohm's law (p. 437): $R = \dfrac{V_R}{i}$ **(12-16)**
- capacitive reactance (p. 438): $X_C = \dfrac{1}{2\pi f C}$ **(12-17)**
- inductive reactance (p. 438): $X_L = 2\pi f L$ **(12-18)**
- voltage across a circuit element (p. 439)

  for a resistor, $V_R = iR$ **(12-19)**

  for a capacitor, $V_C = iX_C$ **(12-20)**

  for an inductor, $V_L = iX_L$ **(12-21)**
- impedance formulas (p. 441): for a circuit containing all three elements,

  $$V = iZ$$

  where $Z = R + j(X_L - X_C)$ **(12-22)**

  $$|Z| = \sqrt{R^2 + (X_L - X_C)^2} \quad \textbf{(12-23)}$$

  $$\tan \phi = \dfrac{X_L - X_C}{R} \quad \textbf{(12-24)}$$
- resonant frequency (p. 443): $f_0 = \dfrac{1}{2\pi\sqrt{LC}}$ **(12-25)**

Exercises 12-8

Solve for x and y.

1. $4 - 15j = 2x + 5yj$ 2. $6j + 14 = 8 - x - 3yj$
3. $5x - 2j + 3 = yj - 6 - 7j$ 4. $7j + 4x - 2yj = 6 - 3xj - y$

Perform the indicated operations. Express all answers in rectangular form.

5. $(2 - 6j) + (3 + 2j)$
6. $(-5 - 9j) + (12 - 13j)$
7. $(4 - 4j) - (8 - 7j)$
8. $(-7.24 + 2.96j) - (3.78 + 10.36j) + (11.49 - 1.65j)$
9. $(3 + \sqrt{-49}) - (-7 - \sqrt{-81})$
10. $\sqrt{-32}\sqrt{-64}\sqrt{-9}$
11. $(5j)(-7j)(-2j)$
12. $-4j(4 - 6j)$
13. $(\sqrt{-7})^4$
14. $(3 - \sqrt{-16})(2 + \sqrt{-4})$
15. $(7 + 4j)(2 + 5j)$
16. $(1.20 - 6.60j)(-2.50 + 3.90j)(5.60j)$
17. $(-3 - 4j)^3$
18. $\dfrac{2j}{5 - j}$
19. $\dfrac{-4}{3 + 7j}$
20. $\dfrac{-8 + 3j}{6j}$
21. $\dfrac{\sqrt{-36}}{7 - \sqrt{-100}}$
22. $\dfrac{4 - 7j}{9 + 2j}$
23. $4j^7 - 6j^4 + 5j\sqrt{-49}$
24. $\dfrac{(22 + 85j)(54 - 76j)}{-53 + 47j}$

Perform the indicated operations graphically. Check your answers algebraically.

25. $(2 + 4j) + (-3 + j)$
26. $(-5 - 3j) + (6 - 2j)$
27. $(5 - 4j) - (2 - j) + 2j$
28. $6 - (7 + 4j) + (-6 + 2j)$

Express each complex number in polar form and in exponential form.

29. $9 + 12j$
30. $16 - 30j$
31. $2 - 3j$
32. $-6.80 + 3.60j$
33. $15.9 - 27.7j$
34. $375 + 428j$
35. 7
36. $-5j$

Express each complex number in rectangular form.

37. $3.00(\cos 40.0° + j \sin 40.0°)$
38. $12.7(\cos 345.0° + j \sin 345.0°)$
39. $\underline{/164.8°}$
40. $125\underline{/250.4°}$
41. $6.00e^{2.30j}$
42. $4.50e^{0.710j}$
43. $5.25e^{5.04j}$
44. $37.5e^{3.75j}$

Perform the indicated operations. Leave the results in exponential form.

45. $4e^{3j} \cdot 5e^{5j}$
46. $(4.7e^{1.6j})(1.4e^{2.7j})$
47. $\dfrac{54e^{5j}}{9e^{3j}}$
48. $\dfrac{3.75e^{4.72j}}{5.25e^{2.84j}}$

Perform the indicated operations. Leave the results in polar form.

49. $[4(\cos 38° + j \sin 38°)][5(\cos 27° + j \sin 27°)]$
50. $(2.60\underline{/147.0°})(3.30\underline{/107.0°})$
51. $\dfrac{32\underline{/75°}}{8\underline{/15°}}$
52. $\dfrac{4.5(\cos 267° + j \sin 267°)}{6.0(\cos 114° + j \sin 114°)}$
53. $[5(\cos 55° + j \sin 55°)]^3$
54. $[3.80(\cos 104.0° + j \sin 104.0°)]^4$

Perform the indicated operations using polar form. Express the answers in both polar form and rectangular form.

55. $(4 - 3j)(6 + 8j)$
56. $(-5.70 + 2.60j)(-3.10 - 1.90j)$
57. $\dfrac{7}{3 - 2j}$
58. $\dfrac{-3.80 - 4.60j}{7.50 + 5.80j}$
59. $(-3 + j)^4$
60. $\dfrac{(3.50 - 7.40j)^3}{8.00}$

Use DeMoivre's theorem to find all possible roots.

61. $\sqrt{16(\cos 120° + j \sin 120°)}$
62. $\sqrt[3]{7.20(\cos 50.0° + j \sin 50.0°)}$
63. $\sqrt[3]{6.00 + 7.00j}$
64. $\sqrt{12.5 - 18.6j}$
65. $\sqrt[5]{2}$
66. $\sqrt{-j}$ (exact values)
67. $\sqrt[3]{-27}$ (exact values)
68. $\sqrt[4]{16}$

Solve.

69. *Electronics*   In a certain ac circuit the complex current is $i = 1.50 - 2.50j$ A. Write this number in exponential form and determine the magnitude of the current.

70. *Electronics*   Find the in-phase component of a voltage represented by $V = 12e^{2.4j}$ (see Problem 31 in Exercises 12-5).

71. *Electronics*   An ac circuit with a 24.0 V, 1000 Hz power source contains an inductor, a resistor, and a capacitor with the following values:

$$L = 2.50 \times 10^{-3} \text{ H} \qquad R = 8.50 \ \Omega \qquad C = 8.75 \times 10^{-6} \text{ F}$$

Find:
(a) Inductive reactance
(b) Capacitive reactance
(c) Magnitude of the impedance
(d) Current
(e) Phase angle between current and voltage
(f) Voltage across the resistor
(g) Voltage across the inductor
(h) Voltage across the capacitor

72. *Electronics*   An electric motor has an inductance of 0.0500 H and a resistance of 18.5 Ω. If it is connected to a 120 V, 60 Hz power source, find:
(a) Magnitude of the impedance of the motor
(b) Current through the motor
(c) Current when an 8.50×10^{-5} F capacitance is added to the circuit
(d) Phase angle in part (c)

73. *Electronics*   In a given circuit the current is $5.00 - 4.00j$ A and the impedance is $3.00 + 8.00j \ \Omega$. Find the magnitude of the voltage.

74. *Electronics*   If the voltage across an ac circuit is $7.50 + 2.50j$ V and the impedance is $3.50 - 4.50j \ \Omega$, find the magnitude of the current.

75. *Electronics*   Find the resonant frequency of the circuit in Problem 71.

76. *Electronics*   The tuning circuit of a radio station contains a resistance of 8.5 Ω, an inductance of 18 mH, and a variable capacitor.
(a) Find the capacitance needed to tune the circuit to a station broadcasting at 1380 kHz.
(b) Determine the current produced in the antenna circuit by a radio wave arriving with an ac signal of 3.2×10^{-5} V.

77. *Electronics*   In a particular circuit, $R = 7.00 \ \Omega$, $X_C = 11.0 \ \Omega$, and $X_L = 6.00 \ \Omega$. Find the magnitude of the impedance and the phase angle between the current and voltage.

78. *Electronics*   A 60 Volt radio transmitter operates at a frequency of 2.8×10^{-3} Hz. How much capacitance must be inserted in series with a 38 Ω load resistor to produce a current of 1.2 A?

Exponential and Logarithmic Functions

13

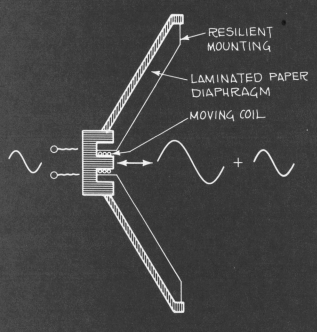

RESILIENT MOUNTING

LAMINATED PAPER DIAPHRAGM

MOVING COIL

HARMONIC OSCILLATIONS
OF AN ORDINARY SPEAKER

Many scientific and technical phenomena, especially those involving growth and decay, are best described using a nonalgebraic function called the *exponential function* and a related function called the *logarithmic function*. A short list of applications includes: population growth and decline in biological systems, changes in current in an electrical circuit, capacitor discharge, radioactive decay of unstable isotopes, loudness of a sound, acidity of a chemical solution, growth of financial investments, and the damping of vibrations in a mechanical system. We will illustrate a few of these applications in the examples of this chapter.

13-1 | The Exponential Function

Definition   An **exponential function** is a function in which the independent variable appears in the exponent. Such a function is written in the following form:

Exponential Function $y = b^x$	(13-1)

where the base b may be any positive real number other than 1. The independent variable x can be any real number, and the dependent variable y is a positive real number. If b were allowed to take on negative values, expressions such as $(-2)^{1/2}$ and $(-3)^{1/4}$ would produce imaginary values for y. If $b = 1$ were allowed, y would never vary—it would always equal 1.

NOTE ▶

There are two important differences between the exponential function as defined in equation (13-1) and the exponential expressions we have dealt with so far. First, in Chapters 1 and 11 we worked with expressions of the form x^n, where the base x is a real number variable and the exponent n is a rational number constant. In an exponential *function*, the *exponent* is any real number variable, and the *base* is a positive real number constant other than 1.

Second, by allowing x to be any *real* number in equation (13-1), we are assuming that a unique value of b^x exists for *irrational* values of x as well as rational values. The validation of this assumption is beyond the scope of this text, but Example 1 will show how b^x might be defined for an irrational value of x. ◀

▨ **EXAMPLE 1**   Consider the number $3^{\sqrt{2}}$. Although $\sqrt{2}$ is an irrational number whose decimal equivalent 1.41421 . . . is nonterminating and nonrepeating, we can use rational approximations for $\sqrt{2}$ to determine roughly the value of $3^{\sqrt{2}}$:

$$3^1 = 3$$
$$3^{1.4} = 3^{7/5} = \sqrt[5]{3^7} \approx 4.6555367$$
$$3^{1.41} \approx 4.7069650$$
$$3^{1.414} \approx 4.7276950$$

And so on. Continuing this procedure, we see that the value of 3^x seems to approach a fixed number as x approaches $\sqrt{2}$. Using a calculator, we obtain

3 $\boxed{y^x}$ 2 $\boxed{\sqrt{\ }}$ $\boxed{=}$ → $\quad$ *4.7288044*

Graphs of exponential functions   By graphing examples of exponential functions, we can discover some of their important properties.

EXAMPLE 2   Graph the exponential function $y = 3^x$.

Solution   **First,** prepare a table of values:

x	−3	−2	−1	0	1	2	3
y	$\frac{1}{27}$	$\frac{1}{9}$	$\frac{1}{3}$	1	3	9	27

Then, plot these points and, since $y = b^x$ is defined for all real values of x, connect the points with a smooth curve (Fig. 13-1).

Figure 13-1

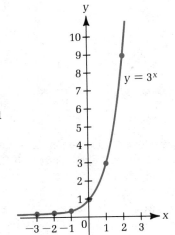

In Example 2 the base was greater than 1. As the next example will show, the exponential function behaves much differently when the base is between 0 and 1.

EXAMPLE 3   Plot the graph of the exponential function $y = 0.4^x$.

Solution   **First,** use a calculator to find y for several values of x:

x	y
−3	15.6
−2	6.3
−1	2.5
0	1
1	0.4
2	0.2
3	0.1

.4 $\boxed{y^x}$ 3 $\boxed{+/-}$ $\boxed{=}$ → 15.625

.4 $\boxed{y^x}$ 2 $\boxed{+/-}$ $\boxed{=}$ → 6.25

.4 $\boxed{y^x}$ 1 $\boxed{+/-}$ $\boxed{=}$ → 2.5

Any nonzero number raised to the zero power is 1.

Any number raised to the first power is equal to itself.

.4 $\boxed{y^x}$ 2 $\boxed{=}$ → 0.16

.4 $\boxed{y^x}$ 3 $\boxed{=}$ → 0.064

Then, graph as before (Fig. 13-2).

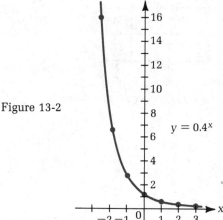

Figure 13-2

$y = 0.4^x$

Properties of exponential functions   The curves in Examples 2 and 3 demonstrate the following properties of exponential functions and their graphs. For $y = b^x$:

- If $b > 1$, as shown in Example 1, this is an increasing function—that is, y increases as x increases, and the curve slopes up and to the right.
- If $0 < b < 1$, as shown in Example 2, this is a decreasing function—that is, y decreases as x increases, and the curve slopes down and to the right.
- Exponential functions are positive for all values of x. The curve approaches the x-axis as an asymptote.
- All functions of the form $y = b^x$ intersect the y-axis at the point $(0, 1)$.

Base e   The base of an exponential function can be any positive real number other than 1. So far we have considered only functions with rational number bases. However, one of the most widely used bases in applications of the exponential function is the irrational number e. As noted in Section 12-5, e has an approximate value of 2.71828. Because functions of the form $y = e^x$ are so common in scientific and technical work, all scientific calculators automatically compute powers of e using either an $\boxed{e^x}$ key or the $\boxed{\text{INV}}\,\boxed{\text{In}x}$ sequence.

EXAMPLE 4   The current i in an RC circuit during charging is given by the equation

$$i(t) = \frac{V}{R}\, e^{-t/RC}$$

where V is the impressed voltage, R is the resistance, and C is the capacitance. Plot $i(t)$ for $V = 3200$ V, $R = 1600\ \Omega$, and $C = 1.25 \times 10^{-3}$ F.

Solution   Substituting the given values into the equation, we obtain

$$RC = 1600(1.25 \times 10^{-3}) = 2$$

$$i(t) = \left(\frac{3200}{1600}\right) e^{-t/2} = 2e^{-t/2}$$

Now we select several values of t and calculate i, as shown in the table. The graph is shown in Fig. 13-3.

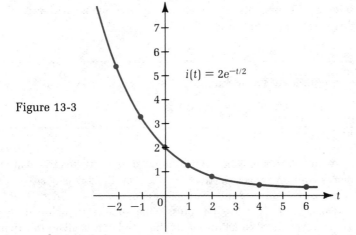

t	$-t/2$	$i(t)$
-2	1.0	5.4
-1	0.5	3.3
0	0	2.0
1	-0.5	1.2
2	-1.0	0.7
4	-2.0	0.3
6	-3.0	0.1

1 [e^x] [×] 2 [=] → 5.4365637

.5 [e^x] [×] 2 [=] → 3.2974425

0 [e^x] [×] 2 [=] → 2.

.5 [+/-] [e^x] [×] 2 [=] → 1.2130613

1 [+/-] [e^x] [×] 2 [=] → 0.7357589

2 [+/-] [e^x] [×] 2 [=] → 0.2706706

3 [+/-] [e^x] [×] 2 [=] → 0.0995741

Figure 13-3

$$i(t) = 2e^{-t/2}$$

Exercises 13-1

Evaluate each of the following:

1. $7.5^{4.9}$ **2.** $\left(\frac{2}{3}\right)^{-0.8}$ **3.** $5^{\sqrt{5}}$ **4.** $12^{\sqrt{3}}$ **5.** $\left(\frac{5}{7}\right)^{\pi}$ **6.** $0.7^{2\pi}$

7. $7^{-\sqrt{2}}$ **8.** $0.25^{-\pi}$ **9.** $e^{2.6}$ **10.** $e^{-0.3}$ **11.** $3(6^{-0.4})$ **12.** $\pi(e^{\pi})$

Plot the graph of each function.

13. $y = 2^x$ **14.** $y = 5^x$ **15.** $y = \left(\frac{1}{3}\right)^x$ **16.** $y = \left(\frac{1}{6}\right)^x$

17. $y = 2.5^x$ **18.** $y = 0.7^x$ **19.** $y = 2(e^x)$ **20.** $y = 4(1.5^x)$
21. $y = 4.5(0.25^x)$ **22.** $y = 0.2(8^x)$ **23.** $y = 3.8^{2x}$ **24.** $y = 5e^{-3x}$

25. *Lighting technology*   Each 1 mm thickness of a certain translucent material reduces the intensity of a light beam passing through it by 15%. Therefore, the intensity I is a function of the thickness of the material t as given by $I = 0.85^t$. Plot I as a function of t.

26. *Finance*   If an amount of money P is invested at $r\%$ interest compounded continuously, then after t years the amount has grown to $A = Pe^{rt/100}$. Plot A as a function of t for $0 \le t \le 20$ if $P = \$1$ and $r = 18\%$.

27. *Electronics*   In the equation given in Example 4 plot $i(t)$ for $V = 1500$ V, $R = 1200$ Ω, and $C = 1.5 \times 10^{-3}$ F.

28. *Physics*   The formula $C = a(1 - e^{bx})$, known as the **time constant curve,** has many uses in science and technology. Plot C as a function of x for $a = 75$ and $b = -0.18$. Use values of x in the range $0 \le x \le 50$.

13-2 | The Logarithmic Function

Definition   When working with the exponential function $y = b^x$, it is often necessary to express the exponent x in terms of the base b and the number y. Since no existing algebraic techniques allow us to solve for x in the exponential function (13-1), we must define a new mathematical entity called a *logarithm* to assist in solving for exponents. Exponent x is equal to the **logarithm** of y to the base b; this is written as follows:

Definition of Logarithm

If $y = b^x$, then $x = \log_b y$ (13-2)

As in (13-1), x may be any real number, b is any positive real number other than 1, and thus y is a positive real number. Since y is positive, the logarithm of a negative number and the logarithm of 0 are undefined.

The equation $x = \log_b y$ expresses the same relationship among x, y, and b as the exponential function $y = b^x$, but the latter is given in exponential form and solved explicitly for y, while (13-2) is stated in logarithmic form and solved explicitly for x.

Many mathematical and technical applications require that you be able to translate from one of these forms to the other. Keep in mind that a logarithm is directly related to an exponent.

 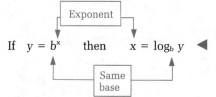

REMEMBER   If $y = b^x$ then $x = \log_b y$

Exponent

Same base

EXAMPLE 5 Rewrite each equation in logarithmic form.
(a) $y = 5^x$ (b) $0.5^m = n$ (c) $\frac{1}{9} = 3^{-2}$ (d) $16^{1/4} = 2$
(e) $e^x = 5.5$

Solution
(a) $x = \log_5 y$ (b) $m = \log_{0.5} n$ (c) $-2 = \log_3(\frac{1}{9})$
(d) $\frac{1}{4} = \log_{16} 2$ (e) $x = \log_e 5.5$

EXAMPLE 6 Rewrite each equation in exponential form.
(a) $\log_{10} y = x$ (b) $\log_7 49 = 2$ (c) $\log_4(\frac{1}{64}) = -3$
(d) $\log_{64} 2 = \frac{1}{6}$

Solution
(a) $y = 10^x$ (b) $49 = 7^2$ (c) $\frac{1}{64} = 4^{-3}$ (d) $2 = 64^{1/6}$

Solving logarithmic equations   You will often need to solve equations in logarithmic form for one missing variable. This can be done more easily by expressing the equations in exponential form.

EXAMPLE 7 Solve for the variable.
(a) $\log_5 y = 3$ (b) $\log_2(\frac{1}{2}) = x$ (c) $\log_b(\frac{1}{81}) = -4$

Solution
(a) $\log_5 y = 3$ means $y = 5^3$, so $y = 125$
(b) $\log_2(\frac{1}{2}) = x$ means $2^x = \frac{1}{2}$, therefore, $x = -1$
(c) $\log_b(\frac{1}{81}) = -4$ means $b^{-4} = \frac{1}{81}$
    But this is equivalent to $b^4 = 81$, so $b = 3$.

EXAMPLE 8 The psychological loudness L of a sound (in decibels, dB) is related to the physical intensity I of the sound wave by the logarithmic equation

$$L = 10 \log_{10}\left(\frac{I}{I_0}\right)$$

where $I_0 = 10^{-12}$ W/m^2, the lowest intensity of a 1000 Hz sound that can be heard by the average person.
(a) Solve this equation for I.
(b) Find I for a whisper of loudness, 20 dB.
(c) Find I for a rock concert at 120 dB.

Solution
(a) If

$$L = 10 \log_{10}\left(\frac{I}{I_0}\right)$$

then, dividing both sides by 10,

$$\frac{L}{10} = \log_{10}\left(\frac{I}{I_0}\right)$$

and, by the definition of a logarithm (13-2), this is equivalent to a corresponding exponential function

$$\frac{I}{I_0} = 10^{L/10}$$

$$I = I_0 10^{L/10}$$

(b) For $L = 20$ dB: $\quad I = (10^{-12} \text{ W/m}^2)(10^{20/10})$

$$= (10^{-12})(10^2) \text{ W/m}^2 = 10^{-10} \text{ W/m}^2$$

(c) For $L = 120$ dB: $\quad I = (10^{-12} \text{ W/m}^2)(10^{120/10})$

$$= (10^{-12})(10^{12}) \text{ W/m}^2 = 1 \text{ W/m}^2$$

The rock concert has 10^{10} (10 billion) times the intensity of the whisper (see Fig. 13-4).

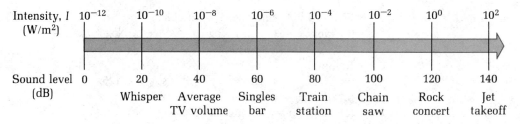

Figure 13-4

Logarithmic function   The logarithmic relation expressed in equation (13-2) is a function because each value of the independent variable corresponds to only one value of the dependent variable. However, when a function is written, the convention is to label the dependent variable as y and the independent variable as x. By switching x and y in equation (13-2), we can write the logarithmic function in the traditional way as follows:

Logarithmic Function $y = \log_b x$ where $x > 0, b > 0, b \neq 1$ (13-3)

Graphing logarithmic functions   Graphing the logarithmic function enables us to discover some of its important properties. When graphing logarithmic functions, it is often easier to change the given logarithmic equation to its equivalent exponential form in order to calculate the graph points.

EXAMPLE 9   Graph: $y = \log_2 x$

Solution   Changing this to exponential form, we have $x = 2^y$. Substituting several y values into this equation and solving for x, we obtain the following table of values:

x	$\frac{1}{8}$	$\frac{1}{4}$	$\frac{1}{2}$	1	2	4	8
y	-3	-2	-1	0	1	2	3

$$2^{-3} = \frac{1}{8}$$ $$2^0 = 1$$ $$2^2 = 4$$

Plotting these ordered pairs, we draw the graph shown in Fig. 13-5.

Figure 13-5

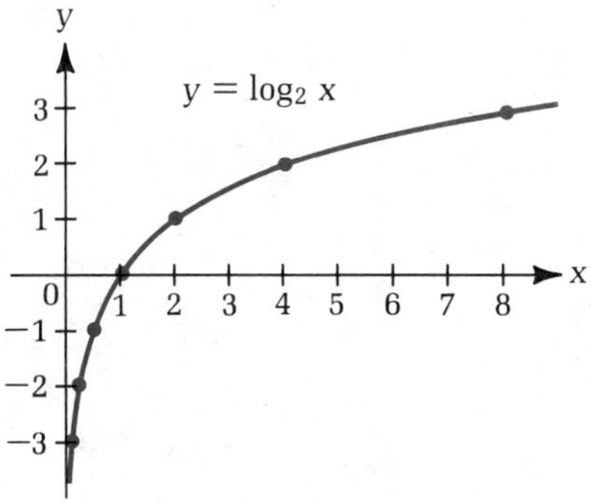

The exponential curve was reversed when the base was less than 1, so we expect that the logarithmic curve will change in a similar way for a fractional base. The next example shows this.

EXAMPLE 10   Graph: $y = 2 \log_{0.5} x$

Solution   To change this function to exponential form, first divide both sides by 2 to obtain

$$\frac{y}{2} = \log_{0.5} x$$

Then write it as

$$x = 0.5^{y/2}$$

Choosing several values for y and using a calculator to find x, we compile the following table of values:

y	y/2	x
−6	−3	8
−4	−2	4
−2	−1	2
−1	−0.5	1.4
0	0	1
1	0.5	0.7
2	1.0	0.5

.5 $\boxed{y^x}$ 3 $\boxed{+/-}$ $\boxed{=}$ →   *8.*

.5 $\boxed{y^x}$ 2 $\boxed{+/-}$ $\boxed{=}$ →   *4.*

.5 $\boxed{y^x}$ 1 $\boxed{+/-}$ $\boxed{=}$ →   *2.*

.5 $\boxed{y^x}$ .5 $\boxed{+/-}$ $\boxed{=}$ →   *1.4142136*

.5 $\boxed{y^x}$ .5 $\boxed{=}$ →   *0.7071068*

Plotting the ordered pairs (x, y), we draw the curve in Fig. 13-6.

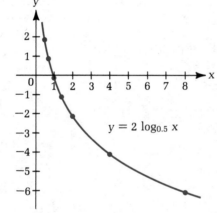

Figure 13-6

$$y = 2 \log_{0.5} x$$

Examples 9 and 10 demonstrate the following properties of the logarithmic function $y = \log_b x$:

- If $0 < b < 1$, it is a decreasing function (Example 10).
- If $b > 1$, it is an increasing function (Example 9).
- The y-axis is an asymptote of the curve.
- The x-intercept is at the point $(1, 0)$.
- The function is undefined for $x \le 0$.

Exercises 13-2

Rewrite each equation in logarithmic form.

1. $2^4 = 16$ **2.** $125 = 5^3$ **3.** $\frac{1}{27} = 3^{-3}$ **4.** $4^{-2} = \frac{1}{16}$

5. $256^{1/4} = 4$ **6.** $2 = 8^{1/3}$ **7.** $\frac{1}{4} = \left(\frac{1}{2}\right)^2$ **8.** $\left(\frac{1}{3}\right)^{-4} = 81$
9. $y = 4^x$ **10.** $1.2^w = v$ **11.** $a^5 = c$ **12.** $6 = x^y$

Rewrite each equation in exponential form.

13. $\log_2 32 = 5$ **14.** $\log_6 36 = 2$ **15.** $\log_8 8 = 1$ **16.** $\log_7 1 = 0$
17. $\log_5\left(\frac{1}{5}\right) = -1$ **18.** $\log_3\left(\frac{1}{9}\right) = -2$ **19.** $\log_{1/2} 4 = -2$ **20.** $\log_{1/5}\left(\frac{1}{25}\right) = 2$
21. $\log_{10} m = n$ **22.** $\log_x y = 4$ **23.** $\log_a 12 = y$ **24.** $\log_4 w = 22$

Solve for the variable.

25. $\log_7 49 = x$ **26.** $\log_2 64 = x$ **27.** $\log_6\left(\frac{1}{216}\right) = x$ **28.** $\log_4\left(\frac{1}{64}\right) = x$
29. $\log_{1/8}\left(\frac{1}{64}\right) = x$ **30.** $\log_{0.1} 1000 = x$ **31.** $\log_{10} N = 4$ **32.** $\log_3 N = 5$
33. $\log_{16} N = \frac{1}{4}$ **34.** $\log_9 N = \frac{1}{2}$ **35.** $\log_2 N = -7$ **36.** $\log_8 N = -\frac{1}{3}$
37. $\log_b 100 = 2$ **38.** $\log_b\left(\frac{1}{121}\right) = -2$ **39.** $\log_b 9 = \frac{2}{3}$ **40.** $\log_b 8 = \frac{3}{4}$
41. $\log_b 7 = -1$ **42.** $\log_b 12 = 1$

Plot the graph of each function.

43. $y = \log_3 x$ **44.** $y = \log_6 x$ **45.** $y = \log_{1/2} x$ **46.** $y = \log_{3/4} x$
47. $y = \log_{0.6} x$ **48.** $y = \log_{2.2} x$ **49.** $y = 3 \log_2 x$ **50.** $y = 10 \log_{1/4} x$
51. $y = 4 \log_{0.8} x$ **52.** $y = 0.5 \log_8 x$

Solve.

53. *Electronics*   The current in an RL series circuit is given by the equation below. Rewrite this in logarithmic form and solve for t.

$$I = \frac{A}{L} e^{-Rt/L}$$

54. *Medical technology*   The diastolic pressure in an artery during a heartbeat is given by the equation below. Rewrite this in logarithmic form and solve for t.

$$P = P_0 e^{-Rt}$$

55. *Physics*   In describing vertical motion in a resisting medium, the time t is given by the equation below.* Rewrite this in exponential form and solve for v.

$$t = \frac{-m}{c} \log_e\left(\frac{mg + cv}{mg + cv_0}\right)$$

56. *Astronomy*   The apparent magnitude m of a star is given by

$$m = M + 5 \log_{10}\left(\frac{r}{10}\right)$$

where M is the absolute magnitude of the star and r is its distance from the earth. Solve this formula for r.

---

* For additional background, see Grant R. Fowles, *Analytical Mechanics*, 3d ed. (New York: Holt, Rinehart and Winston, 1977), p. 39.

13-3 | Laws of Logarithms

In Section 13-2 we showed that a logarithm is an exponent. This fact allows us to establish several important laws of logarithms based on the laws of exponents. The three laws of exponents we shall use are given in the box.

Multiplication Law for Exponents	$a^m \cdot a^n = a^{m+n}$	(13-4)
Division Law for Exponents	$\dfrac{a^m}{a^n} = a^{m-n}$ $(a \neq 0)$	(13-5)
Power of a Power Law for Exponents	$(a^m)^n = a^{mn}$	(13-6)

To develop the laws of logarithms, we first define variables u and v as follows:

$$u = \log_b x \quad \text{and} \quad v = \log_b y$$

Changing these to exponential form gives us

$$x = b^u \quad \text{and} \quad y = b^v$$

Logarithm of a product   The first law of logarithms concerns the logarithm of a product. Using the expressions for x and y, we can write

$$xy = b^u \cdot b^v$$

Now, applying the multiplication law (13-4), we have

$$xy = b^{u+v}$$

Rewriting this in logarithmic form,

$$\log_b xy = u + v$$

Finally, substituting the defined values for u and v into this equation, we have the following important law concerning the logarithm of a product:

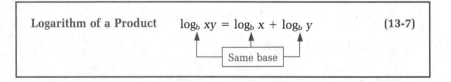

Logarithm of a Product	$\log_b xy = \log_b x + \log_b y$	(13-7)

Same base

Stated in words, this means that the logarithm of the product of two numbers is equal to the sum of the logarithms of the numbers.

We can use this law in two ways. In the next example we use it to rewrite the logarithm of a product as the sum of the logarithms of its factors.

EXAMPLE 11

(a) $\log_3 4x = \log_3 4 + \log_3 x$

(b) $\log_7 30 = \log_7(5 \cdot 6) = \log_7 5 + \log_7 6$

(c) $\log_{10} 27 = \log_{10}(9 \cdot 3) = \log_{10} 9 + \log_{10} 3$

(d) $\log_e 3ab = \log_e 3 + \log_e a + \log_e b$

Notice that all these examples could have been rewritten in other ways. For example, 11(b) is also equal to $\log_7 2 + \log_7 15$ or $\log_7 3 + \log_7 10$.

We can also use the logarithm of a product law (13-7) to rewrite the sum of logarithms as a single logarithm.

EXAMPLE 12

(a) $\log_2 5 + \log_2 7 = \log_2(5 \cdot 7) = \log_2 35$

(b) $\log_e 9 + \log_e w = \log_e 9w$

(c) $\log_{10} 4 + \log_{10} 3 + \log_3 2 = \log_{10} 12 + \log_3 2$

In Example 12(c) notice that only logarithms of *like bases* may be combined in this way.

Logarithm of a quotient   The second law concerns the logarithm of a quotient. Using u, v, x, and y as before, we can write

$$\frac{x}{y} = \frac{b^u}{b^v}$$

Applying the division law for exponents (13-5), we obtain

$$\frac{x}{y} = b^{u-v}$$

Rewriting this equation in logarithmic form, we have

$$\log_b\left(\frac{x}{y}\right) = u - v$$

Substituting for u and v gives us the second important law concerning logarithms:

Logarithm of a Quotient $\log_b\left(\dfrac{x}{y}\right) = \log_b x - \log_b y$ (13-8)

Stated in words, this means that the logarithm of the quotient of two numbers is equal to the logarithm of the numerator (dividend) minus the logarithm of the denominator (divisor).

This law also may be used in two ways. First, we show how to write the logarithm of a division as the difference of the logarithms of the dividend and the divisor.

EXAMPLE 13

(a) $\log_{10}\left(\dfrac{x}{4}\right) = \log_{10} x - \log_{10} 4$

(b) $\log_5 6 = \log_5\left(\dfrac{12}{2}\right) = \log_5 12 - \log_5 2$

(c) $\log_2\left(\dfrac{3x}{5}\right) = \log_2 3x - \log_2 5 = \log_2 3 + \log_2 x - \log_2 5$

Again, there are many other ways of writing these. Example 13(b) could have been written $\log_5 30 - \log_5 5$ or $\log_5 24 - \log_5 4$, for example.

The next example shows how we may use the logarithm of a quotient law (13-8) to rewrite the difference of two logarithms as a single logarithm.

EXAMPLE 14

(a) $\log_6 11 - \log_6 5 = \log_6\left(\dfrac{11}{5}\right)$

(b) $-\log_e x + \log_e 2 = \log_e 2 - \log_e x = \log_e\left(\dfrac{2}{x}\right)$

(c) $\log_{10} 26 - \log_{10} 2 = \log_{10}\left(\dfrac{26}{2}\right) = \log_{10} 13$

CAUTION    Do not confuse the sum or difference of two logarithms in laws (13-7) and (13-8) with expressions such as $\log_b(x + y)$ or $\log_b(x - y)$. These expressions cannot be further simplified. ◄

Logarithm of a power   The third logarithmic law concerns the logarithm of a power. Using our earlier definitions of x and u, we note that since $x = b^u$, then

$$x^n = (b^u)^n$$

Using the power of a power law for exponents (13-6),

$$x^n = b^{un}$$

Rewriting this in logarithmic form,

$$\log_b x^n = un = nu$$

Substituting for u, we obtain the third important law of logarithms:

Logarithm of a Power $\log_b x^n = n \log_b x$ (13-9)

Stated in words, this means that the logarithm of the nth power of a number is equal to n times the logarithm of the number.

We can use (13-9) to rewrite the logarithm of a power as the multiple of a logarithm.

EXAMPLE 15

(a) $\log_5 y^2 = 2\log_5 y$

(b) $\log_7 \sqrt{5} = \log_7 5^{1/2} = \frac{1}{2}\log_7 5$

(c) $\log_{10} 27 = \log_{10} 3^3 = 3\log_{10} 3$

(d) $\log_e\left(\dfrac{x^2 y^3}{z^4}\right) = \log_e x^2 + \log_e y^3 - \log_e z^4$

$$= 2\log_e x + 3\log_e y - 4\log_e z$$

Using law (13-9), we also can rewrite the multiple of a logarithm as the logarithm of a power.

EXAMPLE 16

(a) $3\log_2 x = \log_2 x^3$

(b) $2\log_3 5 = \log_3 5^2 = \log_3 25$

(c) $\frac{1}{3}\log_{10} 8 = \log_{10} 8^{1/3} = \log_{10} 2$

(d) $4\log_e x + 3\log_e 2 - 5\log_e y = \log_e x^4 + \log_e 2^3 - \log_e y^5$

$$= \log_e\left(\frac{8x^4}{y^5}\right)$$

Notice that in part (d) we used all three laws of logarithms.

Exact values of logarithms   Several additional properties of logarithms allow us to determine exact values of certain logarithms. For example, since $b = b^1$, in logarithmic form we can write

$$\log_b b = 1 \qquad\qquad \textbf{(13-10)}$$

Using the logarithm of a power law (13-9), we can also write

$$\log_b b^n = n\log_b b = n \qquad\qquad \textbf{(13-11)}$$

Finally, since $b^0 = 1$, then in logarithmic form,

$$\log_b 1 = 0 \qquad\qquad \textbf{(13-12)}$$

EXAMPLE 17 Find the exact value of each of the following:

(a) $\log_4 16$ 　　(b) $\log_2 2^{5.7}$ 　　(c) $\log_3(\frac{1}{9})$

(d) $\log_{10} \sqrt{10}$ 　　(e) $\log_4 \sqrt[5]{64}$

Solution

(a)   Since $16 = 4^2$, we have: $\log_4 16 = \log_4 4^2$

Then, applying (13-11): $= 2$

(b)   Using (13-11), we have: $\log_2 2^{5.7} = 5.7$

(c)   Because $\frac{1}{9} = 3^{-2}$, we can write: $\log_3(\frac{1}{9}) = \log_3 3^{-2}$

Then, using (13-11): $= -2$

(d)   Using fractional exponents: $\log_{10} \sqrt{10} = \log_{10} 10^{1/2}$

Then, applying (13-11): $= \frac{1}{2}$

(e)   As in part (d), we have: $\log_4 \sqrt[5]{64} = \log_4 64^{1/5}$

Rewriting 64 as 4^3 and simplifying: $= \log_4 (4^3)^{1/5}$

$= \log_4 4^{3/5} = \frac{3}{5}$

In the next example laws (13-7), (13-8), and (13-9) are used to write a single logarithm as a sum, difference, or multiple of logarithms. Then (13-10), (13-11), and (13-12) are used to obtain an exact value for part of each answer.

EXAMPLE 18
(a) $\log_3 12 = \log_3(3 \cdot 4) = \log_3 3 + \log_3 4 = 1 + \log_3 4 = 1 + 2 \log_3 2$
(b) $\log_5(\frac{3}{25}) = \log_5 3 - \log_5 25 = \log_5 3 - \log_5 5^2 = \log_5 3 - 2$
(c) $\log_2(\frac{1}{24}) = \log_2 1 - \log_2 24 = 0 - \log_2(8 \cdot 3) = -(\log_2 8 + \log_2 3)$
$= -3 - \log_2 3$

Exercises 13-3

Express each of the following as a sum, difference, or multiple of logarithms:

1. $\log_2 3x$
2. $\log_3 ab$
3. $\log_5\left(\frac{a}{b}\right)$
4. $\log_7\left(\frac{x}{5}\right)$

5. $\log_6 x^3$
6. $\log_8 3^4$
7. $\log_4 \sqrt{11}$
8. $\log_{10} \sqrt[3]{m}$

9. $\log_{16} 3xy$
10. $\log_2\left(\frac{x^2}{y}\right)$
11. $\log_5 m\sqrt[4]{n}$
12. $\log_3\left(\frac{a^4\sqrt{b}}{c^5}\right)$

Express each of the following as the logarithm of a single quantity:

13. $\log_3 w + \log_3 y$
14. $\log_5 2 + \log_5 m$
15. $\log_7 3 - \log_7 p$

16. $\log_2 r - \log_2 5$
17. $2 \log_6 x$
18. $\frac{1}{5} \log_4 u$

19. $4 \log_2 3$
20. $\frac{1}{2} \log_5 36$
21. $\log_8 5 + \log_8 3$

22. $\log_e 12 + \log_e 5$
23. $-\log_c 14 + \log_c 7$
24. $-\log_3 5 + \log_3 20$

25. $3 \log_2 x - \log_2 5 + \log_2 y$
26. $\log_{10} m + \log_{10} 3 - 4 \log_{10} n$
27. $\log_2 9 - \log_2 3 + 2 \log_2 5$

28. $\log_5 8 - 2 \log_5 4 + \log_5 V$

Find the exact value of each of the following:

29. $\log_2 8$
30. $\log_3 81$
31. $\log_5(\frac{1}{25})$
32. $\log_{10} 0.01$

33. $\log_4 4^{3.2}$
34. $\log_7 7^{6.8}$
35. $\log_3 \sqrt{3}$
36. $\log_2 \sqrt[3]{2}$

37. $\log_3 \sqrt[3]{81}$
38. $\log_2 \sqrt{32}$
39. $\log_5 \sqrt[4]{5} - \frac{1}{2}\log_{10} 1000$
40. $3 \log_7 49 - 2 \log_2(\frac{1}{32})$

Simplify each of the following so that an exact value can be found for at least part of the answer:

41. $\log_2 10$
42. $\log_3 54$
43. $\log_4(\frac{3}{4})$
44. $\log_7(\frac{49}{2})$

45. $\log_5 \sqrt{15}$
46. $\log_3 \sqrt[4]{12}$
47. $\log_6(\frac{1}{72})$
48. $\log_5(\frac{1}{75})$

49. $\log_{10} 500$
50. $\log_{10} 2000$
51. $\log_4 2^5$
52. $\log_9(3^4 \cdot 4^3)$

13-4 | Finding Logarithms of Any Base

In Section 13-3 we showed that the exact numerical values of certain logarithms can be determined using the laws of logarithms. However, most applications require that we find numerical values for other logarithms as well. As you will see in this section, these values can be found using a scientific calculator.

Base 10   The two most commonly used bases for the logarithms that appear in technical applications are 10 and e. The logarithm of some number x to base 10 is called a **common logarithm** and is normally written **log x** without the base; log x is the power to which 10 must be raised to obtain x.

To find log x using a calculator, simply enter x and then press the $\boxed{\text{log}}$ key.*

EXAMPLE 19 Find: (a) log 56   (b) log $\sqrt{0.0247}$ (c) log sin 27°40′

Solution

(a) 56 $\boxed{\text{log}}$ → $\quad$ *1.7481880* $\qquad$ log 56 ≈ 1.75
This means that $10^{1.75} \approx 56$.

(b) .0247 $\boxed{\sqrt{}}$ $\boxed{\text{log}}$ → $\quad$ *−0.8036515* $\qquad$ log $\sqrt{0.0247}$ ≈ −0.804
Therefore, $10^{-0.804} \approx \sqrt{0.0247}$.

(c) 27 $\boxed{+}$ 40 $\boxed{\div}$ 60 $\boxed{=}$ $\boxed{\sin}$ $\boxed{\text{log}}$ → $\quad$ *−0.3331762* $\qquad$ log sin 27°40′ ≈ −0.333
So $10^{-0.333} \approx \sin 27°40′$.

Base e   The logarithm of a number x to base e is known as a **natural logarithm** and is normally abbreviated **ln x**; ln x is the power to which e must be raised to obtain x.

To find ln x using a calculator, enter x followed by the $\boxed{\text{ln x}}$ key.

---

* For applied problems, where approximate numbers are involved, round log x to the same accuracy as x. Otherwise, round to three significant digits.

▦ **EXAMPLE 20** Find: (a) $\ln 3700$ (b) $\ln 0.07726$ (c) $\ln \sqrt[3]{96,800}$

Solution

(a) 3700 $\boxed{\ln x}$ →   ‖ *8.2160881* ‖ $\ln 3700 \approx 8.22$

This means that $e^{8.22} \approx 3700$

(b) $.07726$ $\boxed{\ln x}$ →   ‖ *−2.5605789* ‖ $\ln 0.07726 \approx -2.56$

Therefore, $e^{-2.56} \approx 0.07726$

(c) 96800 $\boxed{\sqrt[x]{y}}$ 3 $\boxed{=}$ $\boxed{\ln x}$ →   ‖ *3.8268008* ‖ $\ln \sqrt[3]{96,800} \approx 3.83$

So $e^{3.83} \approx \sqrt[3]{96,800}$

▦ **EXAMPLE 21** Medical lab technicians use the formula

$$A = 2 - \log T$$

to determine the light absorbence A of a substance, where T, the percent of transmission of light, is known. If 45% of the light is transmitted through a certain solution, what is its absorbence?

Solution

First, substitute the given value for T. $A = 2 - \log 45$

Then, calculate A. 2 $\boxed{-}$ 45 $\boxed{\log}$ $\boxed{=}$ →   ‖ *0.3467875* ‖

$A \approx 0.35$

▦ **EXAMPLE 22** Oceanographers are concerned with the irradiance I (the amount of available light, measured in lux) at certain depths in the ocean. Bodies of water are classified according to an **attenuation coefficient** k that indicates the rate at which the irradiance decreases as the depth increases. This coefficient is given by the formula

$$k = \frac{\ln I_1 - \ln I_2}{L}$$

where I_1 is the irradiance at one depth, I_2 is the irradiance at a deeper point, and L is the distance between the two depths (in meters).* Find k to the nearest thousandth if the irradiance at 20 m is 41.5 lx, and the irradiance at 50 m is 12.0 lx.

Solution We are given $I_1 = 41.5$, $I_2 = 12.0$, and $L = 50 - 20 = 30$ m. Substituting into the formula, we obtain

$$k = \frac{\ln 41.5 - \ln 12.0}{30}$$

41.5 $\boxed{\ln x}$ $\boxed{-}$ 12 $\boxed{\ln x}$ $\boxed{=}$ $\boxed{÷}$ 30 $\boxed{=}$ →   ‖ *0.0413596* ‖ $k \approx 0.041$ m^{-1}

---

* Based on Robert G. Wetzel and Gene E. Likens, *Limnological Analyses* (Philadelphia: W. B. Saunders, 1979), p. 18.

Logarithms of any base   In some applications the logarithm of a number to a base other than 10 or e is needed. Calculators are not designed to find these values directly. However, we can express a logarithm to any base in terms of a base 10 or base e logarithm and then determine its value.

We develop a formula for changing the base of a logarithm from b to a as follows:

First, let: $y = \log_b x$

In exponential form this is equivalent to: $b^y = x$

Taking the logarithms to base a of both sides: $\log_a b^y = \log_a x$

Using the power rule for logs (13-9): $y \log_a b = \log_a x$

Dividing both sides by $\log_a b$: $y = \dfrac{\log_a x}{\log_a b}$

Finally, substituting $\log_b x$ for y, we have the following formula:

Change-of-Base Formula $\log_b x = \dfrac{\log_a x}{\log_a b}$ ◄————┐   (13-13)

The bases on this side are both 10 or both e.

The next example demonstrates how to use this formula to find the logarithm of a number given to a base other than 10 or e.

EXAMPLE 23   Find:
(a) $\log_2 11$ by changing it to base 10.
(b) $\log_5 0.417$ by changing it to base e.

Solution

(a)   **Step 1.**   Change the base using equation (13-13): $\log_2 11 = \dfrac{\log 11}{\log 2}$

Step 2.   Perform the indicated calculation: $11 \boxed{\text{log}} \boxed{\div} 2 \boxed{\text{log}} \boxed{=} \rightarrow$ 3.45943I6

$\log_2 11 \approx 3.46$

(b)   **Step 1.** $\log_5 0.417 = \dfrac{\ln 0.417}{\ln 5}$

Step 2. $.417 \boxed{\text{ln}x} \boxed{\div} 5 \boxed{\text{ln}x} \boxed{=} \rightarrow$ −0.5434624

$\log_5 0.417 \approx -0.543$

It makes no difference whether you change to base 10 or base e to do this kind of calculation—the final results will be the same.

Antilogarithms   If the value y of $y = \log_b N$ is known, then N can be calculated. The value of N is called the **antilogarithm** of $\log_b N$. To find an antilogarithm using a calculator, rewrite the problem in exponential form and use the $\boxed{y^x}$ key to obtain a numerical solution.

EXAMPLE 24   Find the antilogarithm N for each of the following:
(a) $\log_3 N = 26.2$ (b) $\log_2 N = -0.714$

Solution

(a)   **Step 1.**   Rewrite in exponen- $N = 3^{26.2}$
tial form:

   Step 2.   Calculate: $3 \boxed{y^x} 26.2 \boxed{=} \rightarrow$ $\boxed{3.1665 \quad 12}$

   $N \approx 3.17 \times 10^{12}$

(b)   **Step 1.** $N = 2^{-0.714}$

   Step 2. $2 \boxed{y^x} .714 \boxed{+/-}\boxed{=} \rightarrow$ $\boxed{0.6096275}$

   $N \approx 0.610$

Antilogarithms involving base 10 and base e may be determined using the same procedure, but there is an even easier method. To find N for $y = \log_{10} N$, enter y followed by the $\boxed{10^x}$ key or $\boxed{INV}\boxed{\log}$ keys; for $y = \ln N$, enter y followed by the $\boxed{e^x}$ key or $\boxed{INV}\boxed{\ln x}$ keys.

EXAMPLE 25   Find N: (a) $\log N = 2.57$ (b) $\ln N = -5.73$

Solution
(a) $2.57 \boxed{10^x} \rightarrow$ $\boxed{371.53523}$ $N \approx 372$
(b) $5.73 \boxed{+/-}\boxed{e^x} \rightarrow$ $\boxed{0.0032471}$ $N \approx 0.00325$

EXAMPLE 26   Since hydrochloric acid constitutes a part of the exhaust of the space shuttle, environmental scientists in California were concerned with the effect of its launch on the acidity of rainfall. Acidity, or pH, is given by the formula

$$pH = -\log [H^+]$$

where $[H^+]$ is the hydrogen ion concentration (in moles per liter). If the pH of rain near the launch site was measured to be 1.35, what was the concentration of hydrogen ions?

Solution

First, substitute for pH. $1.35 = -\log [H^+]$

Second, multiply by -1. $\qquad\qquad -1.35 = \log[H^+]$

Third, calculate the antilogarithm $[H^+]$. $\qquad 1.35\ \boxed{+/-}\ \boxed{10^x} \rightarrow\ \boxed{0.0446684}$

$$[H^+] \approx 4.47 \times 10^{-2}\ \text{mole/L}$$

Exercises 13-4

Evaluate each of the following:

1. $\log 38$
2. $\log 7$
3. $\log 550$
4. $\log 1275$
5. $\ln 15$
6. $\ln 3$
7. $\ln 2460$
8. $\ln 718$
9. $\log 0.084$
10. $\log 0.59$
11. $\log (4.23 \times 10^{13})$
12. $\log (5.81 \times 10^{-4})$
13. $\ln 0.02335$
14. $\ln 0.6$
15. $\ln (7.24 \times 10^{-6})$
16. $\ln (8.75 \times 10^8)$
17. $\log \sin 38°20'$
18. $\log \cos 59°40'$
19. $\ln \tan 126°18'$
20. $\ln \cot 12°51'$
21. $3 \log 84{,}900$
22. $-\log (1.75 \times 10^{-3})$
23. $-12 \ln (2.84 \times 10^{-5})$
24. $50 \ln 620$

Convert the following to logarithms with base 10 or base e and evaluate:

25. $\log_2 62$
26. $\log_5 85$
27. $\log_3 113$
28. $\log_{16} 875$
29. $\log_8 0.026$
30. $\log_7 0.543$
31. $\log_{50} 3750$
32. $\log_{20}(3.7 \times 10^{-3})$

Solve for N.

33. $\log N = 2.78$
34. $\log N = -12.9$
35. $\ln N = -37.2$
36. $\ln N = 1.84$
37. $\log_2 N = 5.63$
38. $\log_3 N = -15.3$
39. $\log N = -0.249$
40. $\log N = 0.865$
41. $\ln N = 0.493$
42. $\ln N = -0.00726$
43. $\log_8 N = -0.583$
44. $\log_{16} N = 2.14$

Solve.

45. **Economics** The yearly increase of energy use in the United States between the years 1880 and 1960 can be approximated by the expression below.* Calculate this rate of increase, and express it as a percent.

$$\frac{\ln 10}{80}$$

46. **Energy technology** The depth d (in centimeters) of an oxidation pond is given by

$$d = \frac{\ln I}{C\alpha}$$

where I is the incident light intensity (in footcandles), C is the concentration of algal matter (in parts per million), and α is the absorption coefficient. Find d when $I = 820$ fc, $C = 180$ ppm, and $\alpha = 0.0015$.

47. **Nuclear engineering** The time t required for a radioactive substance to decay to a safe level is given by

$$t = \frac{T}{0.301} \log\left(\frac{A}{y}\right)$$

*Based on David Ritterhouse Inglis, *Nuclear Energy: Its Physics and Its Social Challenge* (Reading, Mass.: Addison-Wesley, 1979), p. 3.

where T is the half-life of the substance, A is the initial radiation level, and y is the level considered safe. If the half-life of strontium-90 is 28 years, find the ratio A/y for $t = 16$ years. (This will indicate what multiple of y will decay to a safe level in 16 years.)

48. *Environmental science* After a launch of the space shuttle, scientists measured the pH to be 1.4 near the launch site and 3.2 a short distance from the site. How many times higher was the hydrogen ion concentration at the point nearer the launch site? (See Example 26.)

49. *Medical technology* In Example 21 if the light absorbence of a solution is found to be 0.49, what percent of the light is transmitted through it?

50. *Biology* Enzymes will deactivate if the pH of their environment is not kept within certain limits. The pH of a certain type of enzyme solution is given by

$$pH = pK + \log\left(\frac{[Salt]}{[Acid]}\right)$$

where pK is an ionization constant, [Salt] denotes the concentration of a salt, and [Acid] denotes the concentration of an acid. Find the ratio of salt concentration to acid concentration for a solution in which pH = 4.95 and pK = 4.76.

13-5 | Exponential and Logarithmic Equations

Exponential equations An **exponential equation** is an equation in which the variable is an exponent. In some cases these equations can be solved by inspection.

EXAMPLE 27 Solve: (a) $3^x = 27$ (b) $2^y = \dfrac{1}{8}$

Solution

(a) Since $27 = 3^3$, then $x = 3$

(b) Since $8 = 2^3$, then $\dfrac{1}{8} = \dfrac{1}{2^3} = 2^{-3}$ and $y = -3$

If an exponential equation cannot be solved by inspection, take either the common or natural logarithm of both sides and then solve.

EXAMPLE 28 If oil consumption in the United States increases at the rate of 5% per year, then the number of years y before the annual consumption doubles is given by $(1.05)^y = 2$. Solve for y.

Solution

First, take the logarithm of both sides.

$\log 1.05^y = \log 2$

Second, apply the power law (13-9).

$y \log 1.05 = \log 2$

Third, divide by log 1.05.

$$y = \frac{\log 2}{\log 1.05}$$

Finally, calculate y.   2 $\boxed{\log}$ $\boxed{\div}$ 1.05 $\boxed{\log}$ $\boxed{=}$ → $\quad$ *14.206699*

At a 5% annual rate of increase, consumption will double in approximately 14 years.

To check, store y and use it to calculate $(1.05)^y$:

$\boxed{\text{STO}}$ 1.05 $\boxed{y^x}$ $\boxed{\text{RCL}}$ $\boxed{=}$ → $\quad$ *2.*

▦ **EXAMPLE 29**   Solve $4^{x+3} = 20$.

Solution

First, take the logarithm of both sides.

$$\log 4^{x+3} = \log 20$$

Then, in order to solve, apply the power law (13-9).

$$(x + 3) \log 4 = \log 20$$

Divide by log 4.

$$x + 3 = \frac{\log 20}{\log 4}$$

Subtract 3.

$$x = \frac{\log 20}{\log 4} - 3$$

Finally, calculate x.   20 $\boxed{\log}$ $\boxed{\div}$ 4 $\boxed{\log}$ $\boxed{-}$ 3 $\boxed{=}$ → $\quad$ *−0.8390360*

$$x \approx -0.839$$

Check: $\boxed{\text{STO}}$ 4 $\boxed{y^x}$ $\boxed{(}$ $\boxed{\text{RCL}}$ $\boxed{+}$ 3 $\boxed{)}$ $\boxed{=}$ → $\quad$ *20.*

If the variable appears in more than one exponent, additional algebraic manipulation may be needed.

▦ **EXAMPLE 30**   Solve $2(3^y) = 7^{y+1}$.

Solution

Take the logarithm of both sides.

$$\log 2(3^y) = \log 7^{y+1}$$

To solve, apply the product law (13-7) to the left side.

$$\log 2 + \log 3^y = \log 7^{y+1}$$

Apply the power law (13-9) to both sides.

$$\log 2 + y \log 3 = (y + 1) \log 7$$

Multiply on the right.

$$\log 2 + y \log 3 = y \log 7 + \log 7$$

Isolate all y-terms on one side.

$$y \log 3 - y \log 7 = \log 7 - \log 2$$

| Use the distributive property to combine y-terms. | $y(\log 3 - \log 7) = \log 7 - \log 2$ |

| Divide by $(\log 3 - \log 7)$. | $y = \dfrac{\log 7 - \log 2}{\log 3 - \log 7}$ |

| Use the quotient law (13-8) to simplify the calculation. | $y = \dfrac{\log \left(\frac{7}{2}\right)}{\log \left(\frac{3}{7}\right)}$ |

Calculate.
$$3.5 \boxed{\log} \boxed{\div} \boxed{(} \boxed{(} 3 \boxed{\div} 7 \boxed{)} \boxed{\log} \boxed{=} \rightarrow \boxed{-1.4785390}$$
$$y \approx -1.48$$

To check, first put the answer in memory: $\boxed{\text{STO}}$
Then, calculate each side of the original equation:

Left side: $2 \boxed{\times} 3 \boxed{y^x} \boxed{\text{RCL}} \boxed{=} \rightarrow \boxed{0.3940829}$

Right side: $7 \boxed{y^x} \boxed{(} 1 \boxed{+} \boxed{\text{RCL}} \boxed{)} \boxed{=} \rightarrow \boxed{0.3940829}$

NOTE ▶ Examples 28–30 also could have been solved by taking the *natural* logarithm of both sides. ◀

Equations with e^x   To solve an exponential equation in which e is the only base that appears, change the equation to logarithmic form directly and then solve. Equations of this kind are used in many technical applications.

▦ EXAMPLE 31   Solve $2e^{3x} = 8$.

Solution

First, divide both sides by 2. $e^{3x} = 4$

Then, rewrite the equation in logarithmic form using (13-2).

$3x = \ln 4 \leftarrow$ | If $e^{3x} = 4$ then $3x = \log_e 4 = \ln 4$ |

Finally, solve.

$$x = \frac{\ln 4}{3}$$

$$4 \boxed{\ln x} \boxed{\div} 3 \boxed{=} \rightarrow \boxed{0.4620981}$$

$$x \approx 0.462$$

Check: $\boxed{\text{STO}} 3 \boxed{\times} \boxed{\text{RCL}} \boxed{=} \boxed{e^x} \boxed{\times} 2 \boxed{=} \rightarrow \boxed{8.}$

▦ EXAMPLE 32   A manufacturer of electronic parts has found that the fraction of parts failing in t weeks of use is roughly

$$F(t) = 1 - e^{-0.04t}$$

This kind of function is known as a **reliability function** in product engineering. After how many weeks will 5% of the parts fail?

Solution

First, change 5% to a decimal fraction, substitute into the formula, and simplify.

$$0.05 = 1 - e^{-0.04t}$$

$$e^{-0.04t} = 0.95$$

Then, rewrite the equation in logarithmic form.

$$-0.04t = \ln 0.95$$

Finally, solve for t.

$$t = \frac{\ln 0.95}{-0.04}$$

.95 $\boxed{\ln x}$ $\boxed{\div}$ .04 $\boxed{+/-}$ $\boxed{=}$ → $\boxed{1.2823324}$

$$t \approx 1.3 \text{ weeks}$$

Therefore, 5% of the parts will have failed early in the second week of use.

In some formulas the power of e can be derived experimentally for each specific situation.

EXAMPLE 33 Newton's law of cooling describes how objects are warmed or cooled by their surroundings. If the temperature difference between an object and its surroundings is not too great, then its temperature T at time t is given by

$$T = T_s + T_0 e^{-kt}$$

where T_s is the constant temperature of the surroundings, T_0 is the temperature of the object at $t = 0$, and k is a constant that depends on the characteristics of the object.

A hot liquid in a container is allowed to cool in a room at 60°F. If the liquid cools from 180°F to 130°F after 7.8 min:
(a) Calculate k to the nearest thousandth.
(b) How long would it take the liquid to cool from 130°F to 95°F under the same conditions?

Solution

(a) **First,** substitute the experimentally derived values into the formula.

$$130 = 60 + 180e^{-k(7.8)}$$

Then, solve for k.

$$70 = 180e^{-7.8k}$$

$$\frac{70}{180} = e^{-7.8k}$$

$$\ln\left(\frac{70}{180}\right) = -7.8k$$

$$k = \frac{\ln\left(\frac{70}{180}\right)}{-7.8}$$

$$k \approx 0.121$$

(b)   The formula now reads:

$$T = T_s + T_0 e^{-0.121t}$$

Substituting the new values:

$$95 = 60 + 130 e^{-0.121t}$$

Now, solve for t.

$$\frac{35}{130} = e^{-0.121t}$$

$$\ln\left(\frac{35}{130}\right) = -0.121t$$

$$t = \frac{\ln\left(\frac{35}{130}\right)}{-0.121}$$

$$t \approx 11 \text{ min}$$

Logarithmic equations   In Section 13-2 we solved simple logarithmic equations by rewriting them in exponential form. We can use the same procedure to solve more complex logarithmic equations if we first simplify them using the laws of logarithms.

EXAMPLE 34   Solve: $\log (x + 1) + \log 2 = -1$

Solution

Step 1.   Use the product law (13-7) to combine the terms on the left into a single expression.

$$\log 2(x + 1) = -1$$

Step 2.   Rewrite in exponential form using (13-2), and simplify.

$$2(x + 1) = 10^{-1}$$

$$2x + 2 = \frac{1}{10}$$

Step 3.   Solve.

$$20x + 20 = 1$$

$$20x = -19$$

$$x = \frac{-19}{20} = -0.95$$

Step 4.   Check.

$$\log (-0.95 + 1) + \log 2 = -1$$

$$\log 0.05 + \log 2 = -1$$

$.05\ \boxed{\log}\ \boxed{+}\ 2\ \boxed{\log}\ \boxed{=} \rightarrow$ ▨▨▨▨▨ $-1.$

CAUTION ▶   Since the logarithm of a negative number is undefined, it is very important always to check your answer. ◀

EXAMPLE 35   Solve: $2 \ln (x - 1) - \ln (2x + 4) = 0$

Solution

Step 1.   Use the power law (13-9) and the quotient law (13-8) to combine logarithmic terms.

$$\ln (x - 1)^2 - \ln (2x + 4) = 0$$

$$\ln \left[\frac{(x - 1)^2}{2x + 4}\right] = 0$$

Step 2. Rewrite in exponential form.

$$\frac{(x-1)^2}{2x+4} = e^0$$

$$\frac{(x-1)^2}{2x+4} = 1$$

Step 3. Solve.

$$(x-1)^2 = 2x+4$$

$$x^2 - 2x + 1 = 2x + 4$$

$$x^2 - 4x - 3 = 0$$

$$x = \frac{4 \pm \sqrt{16+12}}{2} = 2 \pm \sqrt{7}$$

Step 4. Check for $x = 2 + \sqrt{7}$:

$$2 \ln \left(2 + \sqrt{7} - 1\right) - \ln \left[2\left(2 + \sqrt{7}\right) + 4\right] \overset{?}{=} 0$$

$$2 \ln \left(1 + \sqrt{7}\right) - \ln \left(8 + 2\sqrt{7}\right) \overset{?}{=} 0$$

$2 \,\boxed{\times}\, \boxed{(}\, 1 \,\boxed{+}\, 7 \,\boxed{\sqrt{\ }}\, \boxed{)}\, \boxed{\ln x}\, \boxed{-}\, \boxed{(}\, 8 \,\boxed{+}\, 2 \,\boxed{\times}\, 7 \,\boxed{\sqrt{\ }}\, \boxed{)}\, \boxed{\ln x}\, \boxed{=}\, \rightarrow$ | **2** | **−10** |

$$\approx 0$$

Check for $x = 2 - \sqrt{7}$:

$$2 \ln \left(2 - \sqrt{7} - 1\right) - \ln \left[2\left(2 - \sqrt{7}\right) + 4\right] \overset{?}{=} 0$$

$$2 \ln \underbrace{\left(1 - \sqrt{7}\right)} - \ln \left(8 - 2\sqrt{7}\right) \overset{?}{=} 0$$

| $1 - \sqrt{7} \approx -1.65$, a negative number |

The first term invoves the logarithm of a negative number $\left(1 - \sqrt{7}\right)$; therefore, this solution is not valid. The only solution is $x = 2 + \sqrt{7} \approx 4.65$.

The equation in Example 35 could have been solved more easily using the method explained next.

If each side of a logarithmic equation can be combined into a single logarithmic expression, then the equation can be solved using the following principle:

| If $\log_a x = \log_a y$, then $x = y$. | (13-14) |

Notice that the bases must be equal in (13-14).

EXAMPLE 36 Solve:

(a) $3 \log 2 = \log 5 + \log x$ (b) $\log_5(3x + 2) - \log_5 x = \log_5 2x$

Solution

(a) **Step 1.** Rewrite the equation in the form $\log_a x = \log_a y$.

$\log 2^3 = \log 5x$

Step 2. Since the bases are equal, write the equation without logarithms.

$$2^3 = 5x$$
$$8 = 5x$$

Step 3. Solve.

$$x = \frac{8}{5} = 1.6$$

Always check your answer in the original equation.

(b) **Step 1.** $\log_5(3x + 2) - \log_5 x = \log_5 2x$

$$\log_5 \left(\frac{3x + 2}{x} \right) = \log_5 2x$$

Step 2.

$$\frac{3x + 2}{x} = 2x$$

Step 3.

$$3x + 2 = 2x^2$$
$$2x^2 - 3x - 2 = 0$$
$$(2x + 1)(x - 2) = 0$$
$$x = -\frac{1}{2} \quad \text{or} \quad x = 2$$

Checking these roots, we find that $x = -\frac{1}{2}$ results in the logarithm of a negative number in two terms of the original equation. The only valid solution is $x = 2$.

Exercises 13-5

Solve.

1. $3^x = \frac{1}{9}$
2. $2^x = 128$
3. $7^x = 40$
4. $2(6^x) = 4$
5. $3e^x = 22$
6. $e^x = 0.5$
7. $0.04(5^x) = 15,625$
8. $4^x = \frac{1}{16,384}$
9. $3^{x+1} = 16$
10. $0.8^{2x} = 6$
11. $6(2^{3x}) = 15$
12. $0.15(9^{x-2}) = 600$
13. $e^{5x} = 120$
14. $e^{x+3} = 16$
15. $0.8e^{x-1} = 2$
16. $4e^{0.5x} = 0.3$
17. $4^{x+2} = 7^x$
18. $5^{x-1} = 3^{2x}$
19. $3(4.8^x) = 7.7^{x-2}$
20. $0.7(2.9^{x+3}) = 1.6^{3x}$
21. $2 \log_3 x = 4$
22. $\frac{1}{2} \log_2 x = 3$
23. $2 \log (x + 4) = 1$
24. $2 \ln (x - 3) = 0$
25. $\log (x - 2) + \log 3 = 0$
26. $\log 5 - \log (x + 2) = 1$
27. $\ln (x + 3) - \ln (x + 4) = 0$
28. $\ln 3 + \ln (x - 6) = 0$
29. $\log (x + 1) + \log (x + 2) = 1$
30. $\log_5 x + \log_5(2x - 3) = 1$
31. $2 \ln (x + 3) - \ln (x + 5) = 0$
32. $\log_2(3x + 3) - 2 \log_2(2x - 1) = 0$
33. $\log_4(x - 6) = \log_4 3x$
34. $\log (2x + 3) = \log (3x - 7)$
35. $\ln 5 + \ln (x - 2) = \ln x$
36. $\log_2(2x + 5) - 2 \log_2 3 = \log_2(x - 1)$
37. $\log (3x - 1) + \log (x + 2) = \log (4 - 2x)$
38. $2 \ln x = \ln (x + 6) - \ln 5$
39. $\ln (3x - 2) = 2 \ln (x - 3) - \ln (x - 7)$
40. $2 \log (x + 4) - \log (2x - 1) = \log (x + 5)$

41. *Space technology* A detector in the *Landsat* satellite must be able to discern 250 shades of brightness and transmit what it samples as a radio wave carrying a series of binary digits. Solve the equation $2^x = 250$ to find the number of digits required for each sample. Round to the nearest whole number.

42. *Physics* A certain vacuum pump removes 4% of the gas from a chamber with every stroke. Solve the equation

$$0.02 = (0.96)^x$$

to calculate the number of strokes necessary to remove 98% of the gas from the chamber.*

43. *Environmental science* The formula

$$\frac{A}{C} = \frac{[(1 + \frac{n}{100})^t - 1]}{\frac{n}{100}}$$

relates the initial amount A and the rate of consumption C of a resource to the time t it will last if consumption C increases at $n\%$ per year. If there are currently 8×10^{13} tons of a certain mineral resource remaining, and the rate of consumption is 4×10^8 tons per year, how long will the resource last if consumption increases at 3% per year? (Round to the nearest year.)

44. *Medical technology* According to the U.S. Public Health service, the number N of cases of a certain contagious disease in an epidemic tended to follow the **sigmoidal function**

$$\frac{N}{1000} = \frac{15}{1 + 49e^{-kt}}$$

where k is a constant and t is the elapsed time in weeks. If 13,400 new cases of the disease were reported by the end of the fourth week of study:
(a) Find the value of k to the nearest tenth.
(b) Calculate the total number of cases expected by the end of the sixth week.

45. *Physics* A hot object, initially at 160°F, is allowed to cool in a room at 70°F. If the temperature of the object is 120°F after 11 min, use Newton's law of cooling (Example 33) to:
(a) Find k to the nearest thousandth.
(b) Determine the amount of time needed to cool the same object from 120°F to 90°F in the same room.

46. *Environmental science* The population of a certain city in the United States in 1982 was 450,000, and it was doubling every 20 years. The formula

$$P = 4.5 \times 10^5(2^{t/20})$$

gives the population P after an additional t years. At this rate, when will the population be 1.5 million?

47. *Nuclear engineering* We can use the equation $0.8 = (\frac{1}{2})^{15/h}$ to determine the half-life h (in years) of a substance when 20% of it has decayed in 15 years. Solve for h.

48. *Computer science* In order to speed up a computer simulation, a process called *data packing* is used. The formula

$$n(2^m - 1) = 2^x - 1$$

gives the relationship among the number of items n to be added, the number of bits m used to represent each item, and the number of bits x required to represent the maximum sum.[†]
(a) Solve for x using natural logarithms.
(b) Find x when $m = 6$ and $n = 10$.

49. *Atomic physics* Solve the equation below for n, the number of molecules with energy state E in a certain substance.

$$\ln g - \ln n - a - bE = 0$$

50. *Marine biology* The irradiance formula discussed in Example 22 of Section 13-4 is often expressed in terms of I_2 instead of k. Solve the formula for I_2.

---

*Based on MAA/NCTM Joint Committee, *A Sourcebook of Applications of School Mathematics* (NCTM, 1980), p. 244.
†Courtesy of Jim Wahl, Flow General Corp., Santa Barbara, California.

13-6 | Graphs on Logarithmic and Semilogarithmic Graph Paper

While constructing graphs of exponential and logarithmic functions in Sections 13-1 and 13-2, you may have noticed that one variable changed in value much faster than the other. This makes it very difficult to graph the functions over a wide range of values or to read values accurately from the graph. For example, as x goes from 1 to 5, $y = 10^x$ goes from 10 to 100,000. However, two special types of graph paper are available that allow us to avoid such difficulties when graphing these and similar functions.

On **semilogarithmic,** or **semilog,** graph paper (Fig. 13-7), distances along the vertical axis are uneven because they are spaced in proportion to the logarithms of the numbers. The distances along the horizontal axis remain evenly spaced.

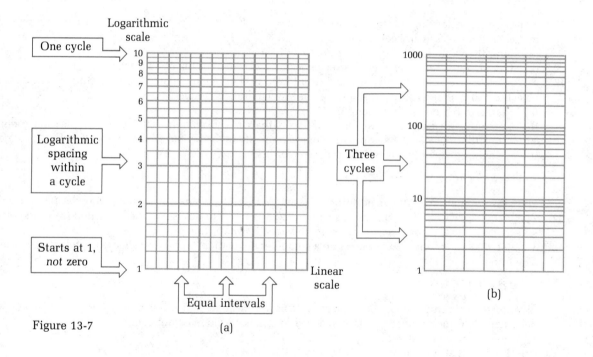

Figure 13-7   (a)

On **logarithmic,** or **log–log,** graph paper (Fig. 13-8), distances along both axes are marked in proportion to the logarithms of the numbers.

Semilog graphs   Semilogarithmic graph paper is normally used for exponential and logarithmic functions where the exponent or logarithm is a variable. Plotting such functions on semilog paper will often produce a linear graph. Use of this paper increases the readability of the graphs and makes it possible to discover exponential or logarithmic functions describing experimental measurements.

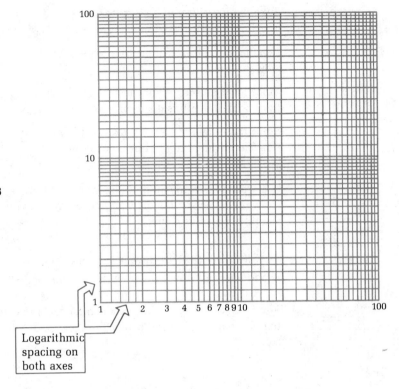

Figure 13-8

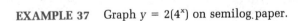

Logarithmic
spacing on
both axes

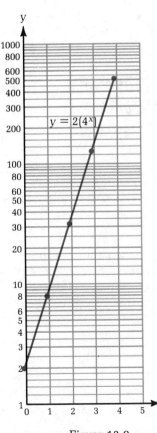

$y = 2(4^x)$

Figure 13-9

EXAMPLE 37   Graph $y = 2(4^x)$ on semilog paper.

Solution
Step 1.   Prepare a table of values:

x	0	1	2	3	4
y	2	8	32	128	512

Step 2.   Label the axes and plot the ordered pairs. As shown in Fig. 13-9, the x-axis is marked in the usual manner. The y-axis of the graph paper is pre-marked with the numbers 1–9 in repetitive cycles. These are used as multiples of increasing powers of 10. In this case, the y values from the table cover the range from 2 to 512. Therefore, the first cycle must represent the values from 1 to 10. The second cycle covers the values from 10 to 100, and the third cycle represents the values from 100 to 1000.

   In Example 37 the graph of $y = 2(4^x)$ was shown to be a straight line on semilog graph paper. The following explains why:

Take the logarithm of both sides of the equation.	$\log y = \log 2(4^x)$
Use the laws of logarithms to rewrite the right side as shown.	$\log y = \log 2 + \log 4^x$ $\log y = \log 2 + x \log 4$
Since log 2 and log 4 are constants, replace them with the letters a and b.	$\log y = a + bx$
Since the numbers on the y-axis represent log y, this is equivalent to the linear equation:	$y = a + bx$

Because an exponential function of the form $y = ab^x$ will always produce a straight line on semilog paper, we need to plot only two points. A third point should be used as a check.

EXAMPLE 38 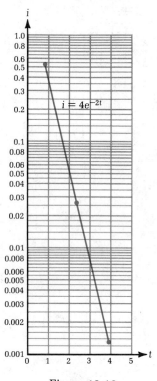 The current i in an RC circuit is given by $i = 4e^{-2t}$. Plot i as a function of time t (in seconds) on semilog paper for $1 \le t \le 4$.

Solution   Since this is an exponential function, its graph will be linear on semilog paper. We should plot three ordered pairs to be certain our line is correct.

t	$-2t$	i
1	-2	0.54
2.5	-5	0.027
4	-8	0.0013

$4 \boxed{\times} 2 \boxed{+/-} \boxed{e^x} \boxed{=} \rightarrow$ *0.5413411*

$4 \boxed{\times} 5 \boxed{+/-} \boxed{e^x} \boxed{=} \rightarrow$ *0.0269518*

$4 \boxed{\times} 8 \boxed{+/-} \boxed{e^x} \boxed{=} \rightarrow$ *0.0013419*

The range of values for i indicates that the first cycle of the vertical axis should represent multiples of 10^{-3}, and that it should continue through multiples of 10^{-1}. The graph is shown in Fig. 13-10.

Figure 13-10

$i = 4e^{-2t}$

Graphs on log–log paper

Logarithmic graph paper is used when the values of both the abscissa and the ordinate cover a wide range. It is especially useful in graphing power functions of the form $y = ax^b$. These differ from exponential functions in that the variable appears as the base, and not as the exponent.

EXAMPLE 39   Graph $x^2y^3 = 1$ on logarithmic paper.

Solution
Step 1.   Solve for y:

$$x^2y^3 = 1$$

$$y^3 = \frac{1}{x^2}$$

$$y = \sqrt[3]{\frac{1}{x^2}}$$

Or, using fractional exponents:

$$y = x^{-2/3}$$

Step 2. Prepare a table of values.

x	y
0.5	1.59
1	1
2	0.63
5	0.34
10	0.22
20	0.14

2 $\boxed{+/-}$ $\boxed{÷}$ 3 $\boxed{=}$ $\boxed{STO}$ ⟵ Store the power in memory first

.5 $\boxed{y^x}$ $\boxed{RCL}$ $\boxed{=}$ → _1.5874011_

1 raised to any power equals 1

2 $\boxed{y^x}$ $\boxed{RCL}$ $\boxed{=}$ → _0.6299605_

5 $\boxed{y^x}$ $\boxed{RCL}$ $\boxed{=}$ → _0.3419952_

10 $\boxed{y^x}$ $\boxed{RCL}$ $\boxed{=}$ → _0.2154435_

20 $\boxed{y^x}$ $\boxed{RCL}$ $\boxed{=}$ → _0.1357209_

Step 3. Label the axes and plot the ordered pairs on log–log graph paper (Fig. 13-11). Notice that the x-axis is marked in logarithmic proportions. For an x range of 0.5 to 20 we need three cycles: 0.1 to 1, 1 to 10, and 10 to 100. For a y range of 0.14 to 1.59 we need only two cycles: 0.1 to 1 and 1 to 10.

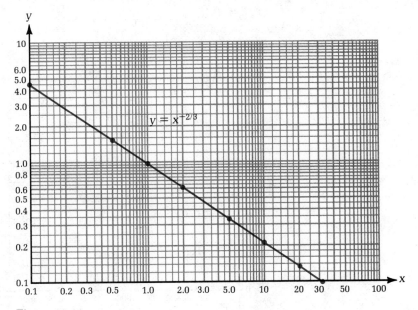

Figure 13-11

A power function in the form $y = ax^b$ will always produce a straight line graph on log–log graph paper, as shown in Example 39. Here is the reason why:

Take the logarithm of both sides. $\log\ y = \log\ ax^b$

Use the laws of logarithms to rewrite the right side as shown.

$\log y = \log a + \log x^b$

$\log y = \log a + b \log x$

Since the numbers on the y-axis represent $\log y$ and the numbers on the x-axis represent $\log x$, then on logarithmic paper this is equivalent to the equation

$$y = \log a + bx$$

Since $\log a$ is a constant, this is a linear equation.

Semilog and log–log graph papers are also useful in graphing experimental data in which one or both variables cover a wide range of values.

EXAMPLE 40   The data in the table show the energy required to heat a certain substance from 10°F to 600°F. Plot the temperature as a function of heat energy on log–log paper.

Heat Energy, H (Btu)	200	400	600	1000	2000	3300
Temperature, T (°F)	10	28	50	105	285	600

Solution

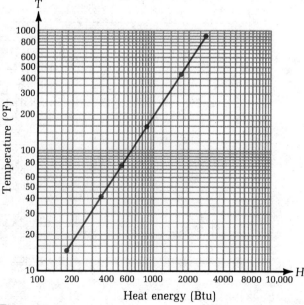

Heat energy (Btu)

Figure 13-12

Exercises 13-6

Graph each function on semilog paper.

1. $y = 5^x$ 2. $y = 6^x$ 3. $y = 8(2^x)$ 4. $y = 4(9^x)$
5. $y = 4^{-x}$ 6. $y = 2^{-x}$ 7. $y = 10^{x/2}$ 8. $y = 7^{x-1}$
9. $y = e^x$ 10. $y = 3e^{2x}$ 11. $y = e^{-x}$ 12. $y = e^{-3x}$

Graph each function on log–log paper.

13. $y = \frac{1}{2}x^3$ 14. $y = 0.06x^4$ 15. $y = 10x^{-2}$ 16. $y = 25x^{-3}$
17. $y = x^{3/4}$ 18. $y = 2\sqrt{x}$ 19. $xy = 10$ 20. $xy = 50$
21. $x^2y = 5$ 22. $xy^2 = 1$ 23. $x^3y^2 = 8$ 24. $xy^3 = 1$

25. Physics Under certain conditions, a wave pulse on a stretched string at a certain time can be described by

$$y = 0.2e^{-(0.4x-0.3)^2}$$

where x is the position (measured in meters) of a point on the string and y is the displacement (in meters) of that point. Plot y as a function of x, $0 \le x \le 7$, on semilog paper.
[*Caution:* The graph is not a straight line.]

26. Biology The growth of a colony of bacteria follows the equation $y = 2e^{1.8t}$, where y is the number of bacteria at some time t (in hours). Plot y as a function of t on semilog paper over the range $0 \le t \le 5$ h.

27. Physics The moment of inertia I of an irregular object is given by $I = kMx^2$, where M is its mass, x is a linear dimension, and k is a constant. Plot I as a function of x when $k = 0.30$ and $M = 10$ kg. Use log–log paper and values of x in the range $0.2 \le x \le 1.5$ m.

28. Thermodynamics For a gas undergoing adiabatic expansion, pressure P (in atmospheres) and volume V (in liters) are related by the equation $PV^a = k$, where a and k are constants based on the type of gas. Use log–log paper to plot V as a function of P for nitrogen gas, where $a = 1.4$ and $k = 50$. Use values of P in the range $0.1 \le P \le 4$ atm.

29. Thermodynamics The boiling temperature of water increases with pressure. The chart shows the boiling temperature T at various values of pressure P. Plot these values on semilog paper, using the horizontal axis for temperature. Connect the points with a smooth curve.

Temperature, T (°F)	328	401	467	545	596	636	668	695
Pressure, P (lb/in.²)	100	250	500	1000	1500	2000	2500	3000

30. Electronics The table shows the values of the current I in a varistor at different voltages V. Plot I as a function of V on log–log paper.*

V	25.0	46.7	62.1	82.9	100	122	187	230
I	0.0028	0.024	0.063	0.17	0.33	0.71	3.5	9.2

*Based on Ernest Zebrowski, Jr., *Fundamentals of Physical Measurement* (Boston: Duxbury, 1979), p. 179.

13-7 | Review of Chapter 13

Important Terms and Concepts

exponential function (p. 450)

logarithm (p. 454)

logarithmic function (p. 456)

common logarithm (p. 465)

natural logarithm (p. 465)

antilogarithm (p. 468)

exponential equation (p. 470)

logarithmic equation (p. 474)

semilogarithmic graph paper (p. 478)

logarithmic (log–log) graph paper (p. 478)

Formulas and Laws

- exponential function (p. 450): $y = b^x$ $(y > 0, \quad b > 0, \quad b \neq 1)$ **(13-1)**

- logarithmic function (p. 454): $y = \log_b x$ $(x > 0, \quad b > 0, \quad b \neq 1)$ **(13-3)**

- laws of logarithms (Section 13-3)

  logarithm of a product: $\log_b xy = \log_b x + \log_b y$ **(13-7)**

  logarithm of a quotient: $\log_b\left(\dfrac{x}{y}\right) = \log_b x - \log_b y$ **(13-8)**

  logarithm of a power: $\log_b x^n = n \log_b x$ **(13-9)**

  $\log_b b = 1$ **(13-10)**

  $\log_b b^n = n$ **(13-11)**

  $\log_b 1 = 0$ **(13-12)**

- change-of-base formula (p. 467): $\log_b x = \dfrac{\log_a x}{\log_a b}$ **(13-13)**

- If $\log_a x = \log_a y,$ then $x = y.$ **(13-14)**

Exercises 13-7

Solve for the variable in each equation without using a calculator.

1. $\log_2 16 = x$ **2.** $\log_3(\frac{1}{27}) = x$ **3.** $\log_5 5^{3.2} = x$
4. $\log_6 \sqrt{6} = x$ **5.** $\log_4 N = 3$ **6.** $\log_2 N = -2$
7. $\log_{25} N = \frac{1}{2}$ **8.** $\log_{1/6} N = 2$ **9.** $\log_b 9 = 1$
10. $\log_b 4 = -2$ **11.** $\log_b 125 = \frac{3}{2}$ **12.** $\log_b 343 = 3$

Solve for the variable in each equation. Apply the change-of-base formula (13-13) first if necessary.

13. $\log 57 = x$ **14.** $\log \cos 37°22' = x$
15. $\ln 480 = x$ **16.** $\ln (6.42 \times 10^5) = x$
17. $\log_2 80 = x$ **18.** $\log_5 320 = x$

19. $\log N = 5.24$ 20. $\log N = -2.13$
21. $\ln N = -9.64$ 22. $\log_2 N = 0.528$
23. $\log (5.73 \times 10^{-3}) = x$ 24. $\log 6500 = x$
25. $\ln 6 = x$ 26. $\ln 0.583 = x$
27. $\log_3 5000 = x$ 28. $\log_4(7.84 \times 10^{-4}) = x$
29. $\log N = 14.7$ 30. $\log N = -0.265$
31. $\log_3 N = -1.62$ 32. $\ln N = 3.24$

Express each of the following as a sum, difference, or multiple of logarithms:

33. $\log_3 xyz$ 34. $\log\left(\dfrac{m}{2}\right)$ 35. $\ln c^4$

36. $\log_5 \sqrt[6]{n}$ 37. $\log\left(\dfrac{a^3}{b^2}\right)$ 38. $\ln w^2\sqrt{v}$

Solve each equation.

39. $3^x = 81$ 40. $4^x = 12$ 41. $e^x = 0.478$
42. $3(2^x) = \frac{1}{6}$ 43. $5^{x+2} = 22$ 44. $0.4(e^{2x}) = 20$
45. $2^{x+3} = 5^x$ 46. $4(2.7^x) = 5.9^{x-2}$
47. $2 \log_2 x = 6$ 48. $2 \log (x - 5) = 1$
49. $\ln 6 + \ln (x + 3) = 0$ 50. $\log_3 8 - \log_3(x - 2) = 1$
51. $\log (5x - 3) - \log (2x - 1) = 0$ 52. $2 \log x + \log 5 = \log 20$
53. $\ln (x - 5) + \ln (x + 6) = \ln (3x - 6)$
54. $\log_7(x + 2) + \log_7(2x + 1) = \log_7(x + 3)$

Graph each function. Use normal graph paper for Problems 55–58, semilog paper for Problems 59 and 60, and log–log paper for Problems 61 and 62.

55. $y = 4^x$ 56. $y = 3(0.2^x)$ 57. $y = \log_4 x$
58. $y = 3 \log_{0.7} x$ 59. $y = 7^x$ 60. $y = 5e^{3x}$
61. $y = 0.03x^2$ 62. $xy^3 = 8$

Solve.

63. **Electronics**   Charge decays from the plates of a capacitor in an *RC* circuit according to the formula below. Solve this formula for *t*.

$$q = Qe^{-t/RC}$$

64. **Electronics**   The resistance of a cylindrical insulator is given by the formula below. Solve for *b*.

$$R = \frac{\rho \ln (b/a)}{2\pi L}$$

65. **Acoustical engineering**   The loudness *L* of a sound (in decibels) is given by the equation

$$L = 10 \log\left(\frac{I}{I_0}\right)$$

where *I* is the intensity of the sound and $I_0 = 10^{-12}$ W/m². If the intensity of the sound of a rocket takeoff is 10^3 W/m², calculate the loudness of this sound.

66. **Electronics**   In determining the potential difference *V* (in volts) in a cell with a certain type of electrode, the calculation below must be made. Find *V*.

$$V = 0.010 + 0.025 \ln\left(\frac{\sqrt{0.025}}{\sqrt{0.065}}\right)$$

67. *Chemistry*   An experiment is designed to find the specific reaction rate constant at a certain temperature. Substitution of the known values into the Arrhenius equation gives

$$\log\left(\frac{k_1}{0.45}\right) = \frac{4.2 \times 10^6}{(4.6)(775)(675)}$$

Solve for k_1.

68. *Seismology*   The Richter scale magnitude R of an earthquake is related to its intensity I by the formula

$$R = \log\left(\frac{I}{I_0}\right)$$

where I_0 is a standard minimum level. Very often, seismologists know the magnitudes of two earthquakes and must compare their intensities. How many times more intense is an earthquake measuring 7.8 on the Richter scale than one measuring 6.1?
[*Hint:*   Substitute for R and solve for I for each earthquake. Then set up a ratio of the intensities.]

69. *Meteorology*   The atmospheric pressure P (in pounds per square inch) at an altitude h (in miles) above sea level can be calculated approximately from the formula

$$P \approx 14.7e^{-0.21h} \qquad \text{for} \quad 0 \leq h \leq 12 \text{ mi}$$

At what altitude will the pressure be 2.0 lb/in.^{2}?

70. *Economics*   In cost analysis theory, it is often assumed that as the total quantity of units produced doubles, the cost per unit declines by a constant percentage. This leads to the equation $S = 2^b$, where S is the fraction of decrease and b is a constant used to determine the actual cost. Find b when $S = 0.85$.

71. *Nuclear engineering*   A certain radioactive isotope decays according to the formula $A = Ce^{-0.0017t}$, where A is the amount present at any time t and C is the amount present at $t = 0$. After how many years will a 20.0 g sample of this isotope decay to 10.0 g? (This is the half-life of the isotope.)

72. *Biology*   The growth of a colony of bacteria under optimum conditions follows the exponential equation $y = Ae^{kt}$, where y is the number of bacteria at some time t and A is the number at time $t = 0$. A single cholera bacterium divides every $\frac{1}{2}$ h. That is, $A = 1$, and $y = 2$ at time $t = \frac{1}{2}$. Use this information to solve for k. Then determine the number of bacteria present after 24 h if $A = 100$.

73. *Electrical engineering*   The potential V at any point between two infinitely long parallel line charges is given by the equation below. Solve for a.*

$$V = -\left(\frac{\lambda}{2\pi\epsilon_0}\right)(\ln a - \ln b)$$

74. *Economics*   In a study of the market penetration of a new kind of energy generator for electric utilities, it was expected that the cost of the changeover would be approximated by $C = a(1 - e^{bx})$, where C is the cost (in millions of dollars), a and b are constants, and x is the percent of penetration. Suppose $C = \$100$ million when $x = 5\%$, and a is found to be 166.7. Use this information to solve for b, and then find the cost of a 15% penetration.

75. *Meteorology*   Refer to Problem 69. Plot P as a function of h for $0 \leq h \leq 12$ mi. Use normal graph paper.

76. *Electronics*   The current rise i (in amperes) in a certain RC circuit is given by the equation below. Plot i as a function of t on semilog paper for $0.01 \leq t \leq 0.05$ s.

$$i = 2.5e^{-150t}$$

77. *Energy technology*   The power P (in watts) generated by a windmill is given by $P = 0.01v^3$, where v is the velocity of the wind in miles per hour. Plot P as a function of v on semilog paper for $10 \leq v \leq 40$. (The graph of this function is not a straight line.)

---

*For more information, see John R. Reitz and Frederick J. Milford, *Foundations of Electromagnetic Theory*, 3d ed. (Reading, Mass.: Addison-Wesley, 1979), p. 61.

78. *Planetary science* According to Kepler's third law of planetary motion, the period T and mean distance d from the sun for a planet are related. Use log–log paper to plot T as a function of d from the following table of data:

	Mercury	Venus	Earth	Mars
Mean Distance from Sun, d	0.39	0.72	1.00	1.52
Period, T (years)	0.24	0.62	1.00	1.88

	Jupiter	Saturn	Uranus	Neptune
Mean Distance from Sun, d	5.20	9.54	19.18	30.07
Period, T (years)	11.86	29.46	84.01	164.79

Nonlinear Equations and Systems

14

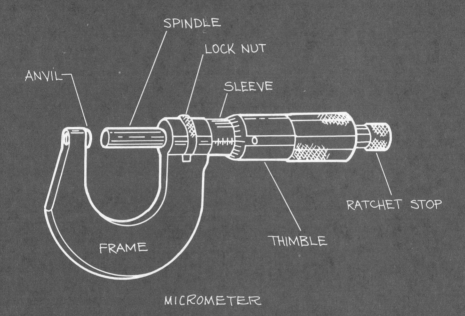

SPINDLE

LOCK NUT

ANVIL

SLEEVE

RATCHET STOP

FRAME

THIMBLE

MICROMETER

So far, we have solved many types of equations in one variable, including linear equations, quadratic equations, and exponential and logarithmic equations. We have also solved systems of linear equations in two and three variables. In this chapter we will apply both graphic and algebraic techniques to solve systems involving nonlinear equations in two variables and some new types of nonlinear equations in one variable.

Definition   Earlier, a linear equation was defined as an equation in which only first powers of the variable(s) appear. A **nonlinear equation** contains at least one term involving either a variable raised to a power other than 1 or the product of two or more variables. The following equations are all nonlinear:

Power of 2

$$x^2 + y = 5$$

Power of -1:
$$\frac{1}{x} = x^{-1}$$

$$y = \frac{1}{x}$$

Product of two variables

$$xy = 4$$

Quadratic equations in two variables   Certain kinds of technical applications lead to systems involving nonlinear equations. In this section we shall use graphic techniques to solve nonlinear systems of two equations in two variables. First, we will examine the characteristic forms of some special quadratic equations in two variables.

EXAMPLE 1 Graph the equation: $x^2 + y^2 = 36$

Solution

Step 1. Solve for y. $x^2 + y^2 = 36$

$$y^2 = 36 - x^2$$

$$y = \pm\sqrt{36 - x^2}$$

Step 2. Prepare a table of ordered pairs. Use positive and negative pairs of values of x:

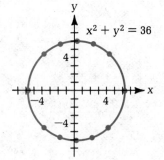

Figure 14-1

x	−6	−4	−2	0	2	4	6
y	0	±4.5	±5.7	±6	±5.7	±4.5	0

Notice that imaginary values for y result when x > 6 or x < −6.

Step 3. Plot the points and connect them with a smooth curve. The result shown in Fig. 14-1 is a *circle*.

Circle   Any equation in the form $x^2 + y^2 = r^2$ produces a **circle** with center at the origin. The constant r is the radius of the circle.

In the next example notice what happens when the x^2- and y^2-terms have coefficients that are different but have the same sign.

EXAMPLE 2   Graph the equation: $x^2 + 2y^2 = 4$

Solution

Step 1.   Solve for y. $x^2 + 2y^2 = 4$

$$2y^2 = 4 - x^2$$

$$y^2 = \frac{4 - x^2}{2}$$

$$y = \pm\sqrt{\frac{4 - x^2}{2}}$$

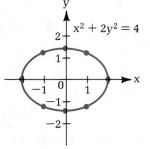

Step 2.

x	−2	−1	0	1	2
y	0	±1.2	±1.4	±1.2	0

Imaginary values for y result when $x < -2$ or $x > 2$.

Figure 14-2

Step 3.   The result shown in Fig. 14-2 is an *ellipse*.

Ellipse   Any equation in the form $ax^2 + cy^2 = f$, where a, c, and f are all positive numbers or all negative numbers, and $a \ne c$, produces an **ellipse** centered at the origin.

The next example examines the effect of replacing the plus sign in the ellipse equation with a minus sign.

EXAMPLE 3   Graph the equation: $x^2 - 4y^2 = 10$

Solution

Step 1.   Solve for y. $x^2 - 4y^2 = 10$

$$-4y^2 = 10 - x^2$$

$$y^2 = \frac{10 - x^2}{-4}$$

$$= \frac{x^2 - 10}{4}$$

$$y = \pm\frac{1}{2}\sqrt{x^2 - 10}$$

Step 2.

x	y
$\pm\sqrt{10} \approx \pm 3.2$	0
±5	$\pm\frac{1}{2}\sqrt{15} \approx \pm 1.9$
±7	$\pm\frac{1}{2}\sqrt{39} \approx \pm 3.1$
±9	$\pm\frac{1}{2}\sqrt{71} \approx \pm 4.2$

Step 3.   The result shown in Fig. 14-3 is a *hyperbola*.

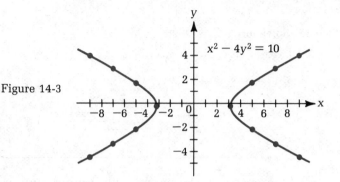

Figure 14-3

Hyperbola   Any equation in the form $ax^2 - cy^2 = f$, where a, c, and f are all positive numbers or all negative numbers, produces a **hyperbola** centered at the origin. A hyperbola also results from any equation in the form $xy = k$, where k is a constant.

In each example given so far, the equation contained second-degree terms in both variables. Another distinctive curve results if one of the variables appears only to the first power.

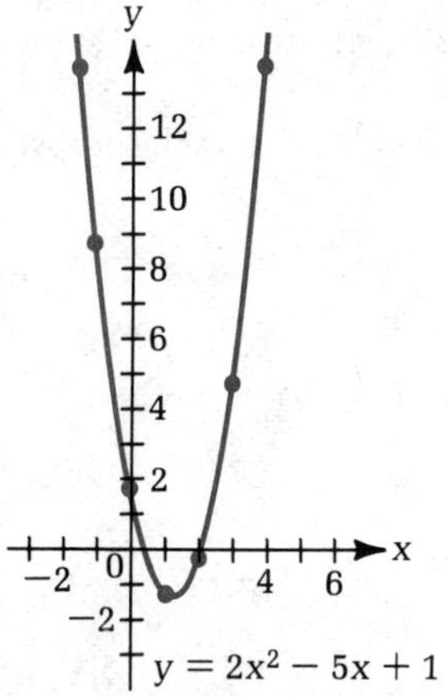

EXAMPLE 4   Graph the equation: $y = 2x^2 - 5x + 1$

Solution   Since the equation is already solved for y, we can immediately prepare a table of values:

x	−1.5	−1	0	1	2	3	4
y	13	8	1	−2	−1	4	13

$y = 2x^2 - 5x + 1$

Figure 14-4

The curve shown in Fig. 14-4 is a *parabola*.

Parabola   Any equation in the form $y = ax^2 + dx + f$ (where $a \neq 0$)   or $x = by^2 + ey + f$ (where $b \neq 0$) produces a **parabola.**

The four special curves shown in Figs. 14-1–14-4 are known as **conic sections.** We will examine these in much greater detail in Chapter 20.

Nonlinear systems   A **nonlinear system of equations** is a system in which at least one equation is nonlinear. As we saw in Chapter 5, real solutions to a system of equations are the coordinates of the points of intersection of their graphs.

EXAMPLE 5   Graphically solve the system: $y = 2^x$

$$x^2 + y^2 = 4$$

Solution

Step 1. Note that the first equation gives an exponential curve, and the second is a circle with radius 2. We have the following tables of values:

$$y = 2^x$$

x	-2	-1	0	1	2	3
y	$\frac{1}{4}$	$\frac{1}{2}$	1	2	4	8

$$x^2 + y^2 = 4$$

$$y = \pm\sqrt{4 - x^2}$$

x	±2	±1	0
y	0	±1.7	±2

Step 2. Plot the graphs on the same set of coordinate axes (Fig. 14-5).

Figure 14-5

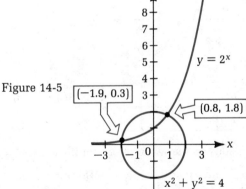

Step 3. Estimate the points of intersection. These are the solutions to the system. For this system, the solutions are $x \approx -1.9$, $y \approx 0.3$ or $x \approx 0.8$, $y \approx 1.8$.

Step 4. Check both solutions by substituting them back into the original system of equations.

Check for $x \approx -1.9$, $y \approx 0.3$:

First equation gives: $0.3 \approx 2^{-1.9}$

$0.3 \approx 0.27$

Second equation gives: $(-1.9)^2 + (0.3)^2 \approx 4$

$3.7 \approx 4$

Check for $x \approx 0.8$, $y \approx 1.8$:

First equation gives: $1.8 \approx 2^{0.8}$

$1.8 \approx 1.74$

Second equation gives: $(0.8)^2 + (1.8)^2 \approx 4$

$3.88 \approx 4$

NOTE   When the solutions are approximate, the check should produce numbers that are equal when rounded. ◀

EXAMPLE 6   Graphically solve the system: $y = x^2 - 1$
$y = \log_2 x$

Solution

Step 1.   The graph of the first equation is a parabola, and the second is a logarithmic curve.

$$y = x^2 - 1$$

x	0	±1	±2	±3
y	−1	0	3	8

$$y = \log_2 x$$
$$2^y = x$$

x	$\frac{1}{2}$	1	2	4	8
y	−1	0	1	2	3

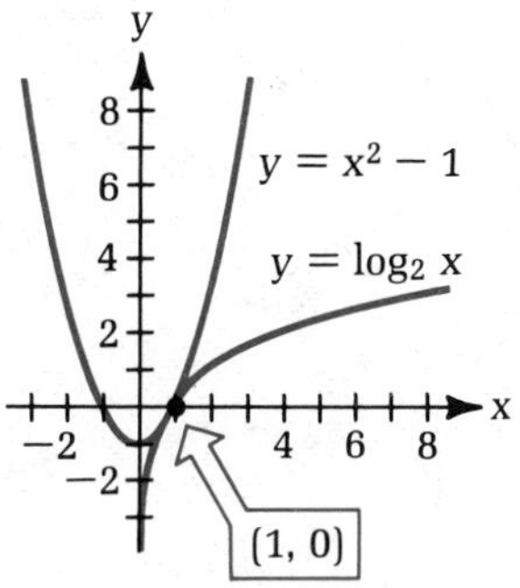

Step 2.   Plot the graphs on the same set of coordinate axes (Fig. 14-6).

Step 3.   The graph shows that the only point of intersection is $(1, 0)$. Thus, $x = 1$, $y = 0$ is the only solution to this system.

Step 4.   Check this solution by substituting in both equations in the original system.

Figure 14-6

EXAMPLE 7   A farmer wants to fence off a 225 yd^2 rectangular plot for growing beans. If 56 yd of fencing are available, what are the possible dimensions of the plot? Solve graphically.

Solution   First, set up a system of equations based on the given information concerning area and perimeter. Let

L = Length of field W = Width of field

Area: $LW = 225$

Perimeter: $2L + 2W = 56$

Then, since it does not matter which variable is independent, we arbitrarily select L as the independent variable and W as the dependent variable. Solving each equation for W and preparing the tables of values, we obtain the following:

$$LW = 225$$

$$W = \frac{225}{L}$$

L	10	15	20	25
W	22.5	15	11.25	9

$$2L + 2W = 56$$

$$W = 28 - L$$

L	5	10	15	20
W	23	18	13	8

Notice that since neither length nor width can be a negative quantity, we selected only values of L between 0 and 28 yd. As we would expect, the graph of the first equation is part of a hyperbola, and the graph of the second equation is a straight line (Fig. 14-7). Since the two graphs never intersect, there is no real solution to the system. The farmer does not have enough fence for a plot of 225 yd².

Figure 14-7

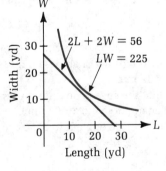

EXAMPLE 8 The kinetic energy (K.E.) of an object of mass m (in kilograms) moving with speed v (in meters per second) is $\frac{1}{2}mv^2$ J (joules). Two objects with masses of 1.5 kg and 2.4 kg are moving at different speeds so that their total kinetic energy is 2260 J. If the smaller object is moving 2.0 m/s faster than the larger object, how fast is each moving?

Solution **First,** assign variables to the unknown quantities. The speeds of the two objects are unknown, so let

v_1 = Speed of smaller object v_2 = Speed of larger object

Next, set up a system of two equations based on the given information. The masses corresponding to v_1 and v_2 are given as $m_1 = 1.5$ kg and $m_2 = 2.4$ kg. Using the formula K.E. $= \frac{1}{2}mv^2$, we can express the individual kinetic energies of m_1 and m_2 as follows:

For m_1:   K.E. $= \frac{1}{2}(1.5)v_1^2$ For m_2:   K.E. $= \frac{1}{2}(2.4)v_2^2$

$= 0.75v_1^2$ $= 1.2v_2^2$

We can now write a system of two equations:

The total kinetic energy is 2260 J: $0.75v_1^2 + 1.2v_2^2 = 2260$

The smaller object moves 2.0 m/s
faster than the larger object: $v_1 - v_2 = 2.0$

Then, continue as before. We arbitrarily select v_1 as the independent variable and v_2 as the dependent variable. Solving for v_2 and preparing tables of positive values (speed cannot be negative), we have the following:

$0.75v_1^2 + 1.2v_2^2 = 2260$ $v_1 - v_2 = 2.0$

$v_2 = \sqrt{\dfrac{2260 - 0.75v_1^2}{1.2}}$ $v_2 = v_1 - 2.0$

v_1	0	10	20	30	40	50
v_2	43.4	42.7	40.4	36.3	29.7	17.9

v_1	2	10	20	30	40
v_2	0	8	18	28	38

For the formula on the left, use the following calculator sequence:

For $v_1 = 50$:   2260 $\boxed{-}$ .75 $\boxed{\times}$ 50 $\boxed{x^2}$ $\boxed{=}$ $\boxed{\div}$ 1.2 $\boxed{=}$ $\boxed{\sqrt{}}$ $\rightarrow$ *17.911821*

As we would expect, the graph of the first equation is a portion of an ellipse, and the graph of the second is a straight line (Fig. 14-8). The only solution is at $v_1 \approx 35$ m/s and $v_2 \approx 33$ m/s.

Checking, we see that the speeds do differ by 2 m/s. Furthermore, we calculate the total kinetic energy to be 2226 J, which, considering the approximate nature of the solution, is close enough to 2260 J to verify our answer.

Figure 14-8

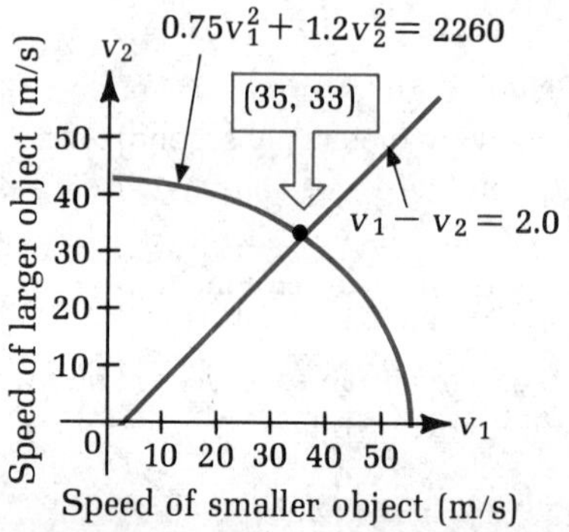

Exercises 14-1

Without plotting a graph, identify the type of curve represented by each equation.

1. $3x^2 - 7y^2 = 6$
2. $x + 2y = 5$
3. $x^2 + y^2 = 81$
4. $6x^2 + y^2 = 12$
5. $y = 3x^2 - 7x + 2$
6. $x^2 + y^2 = 100$
7. $2y^2 + 6x^2 = 9$
8. $x = 3y^2 - 5y$

Solve each system of equations graphically. Estimate your answers to the nearest tenth.

9. $y = 2x - 1$
   $y = x^2$
10. $x = y$
    $x^2 + y^2 = 9$
11. $2x + y = 5$
    $xy = 4$
12. $x - y = 4$
    $x^2 + 2y^2 = 10$
13. $2y - x = 6$
    $x^2 - 4y^2 = 4$
14. $y^2 - 2x = 6$
    $x^2 - y^2 = 4$
15. $y = x^2 - 2$
    $x^2 + 4y^2 = 9$
16. $y = x^2 - 3x$
    $xy = 9$
17. $x^2 + y^2 = 4$
    $2y = x^2$
18. $y = -x^2$
    $y = 2x^2 - 1$
19. $x^2 + y^2 = 16$
    $x^2 - y^2 = 4$
20. $x^2 + 4y^2 = 16$
    $x^2 - 2y^2 = 4$
21. $x^2 + y^2 = 10$
    $x^2 - 4y^2 = 1$
22. $xy = 1$
    $x^2 + y^2 = 2$
23. $y = \sin x$
    $y = 2x - 1$
24. $y = 3^x$
    $y = x^2 + 5$
25. $y = \log_3 x$
    $y = x - 1$
26. $y = \cos x$
    $x = y^2 - 1$
27. $2x^2 - y^2 = 2$
    $xy = 2$
28. $3x^2 + y^2 = 6$
    $x^2 + 2y^2 = 12$

Solve.

29. *Sheet metal technology* A rectangular piece of sheet metal has an area of 20.0 in.^{2} and a perimeter of 22.0 in. Set up two equations in two unknowns and solve this system graphically to find the dimensions of the piece.
30. *Agricultural technology* A farmer has a rectangular pasture with an area of 120.0 m^2 that requires 48.0 m of fencing to enclose it. Find the dimensions of the lot graphically.
31. The sum of the squares of two positive numbers is 14. The difference between the numbers is 6. Solve for the numbers graphically.
32. *Industrial design* In designing the top of a cylindrical container, an engineer plans for a hole x inches in radius to be drilled through a larger circular surface y inches in radius. If 160 in.^{2} of the larger surface remain after the hole is drilled, and if $y - x = 7.0$ in., find x and y graphically.
    [Hint: The difference between the areas of the two circles is 160 in.^{2}.]
33. *Electronics* When a current i passes through a resistor R, the power dissipated in the resistor is i^2R. In a circuit containing resistors of 120 Ω and 85 Ω connected in parallel, current i_1 passes through the 120 Ω resistor and current i_2 passes through the 85 Ω resistor. Find i_1 and i_2 if the total power dissipated is 310 W, and the total current drawn by the circuit is 2.2 A.
34. *Physics* Solve Example 8 if the two masses are 2.2 kg and 3.7 kg, the total kinetic energy is 1870 J, and the sum of the speeds of the masses is 51 m/s.

14-2 | Solving Nonlinear Systems Algebraically

As we showed in Chapter 5 and again in Section 14-1, the graphical solution to a system of equations yields only an approximate answer. When it is possible to find one, an algebraic solution is much more useful.

The algebraic methods developed for solving linear systems in Chapter 5 can be applied to nonlinear systems in two variables. The two methods we used were substitution and addition. With both methods, a variable is eliminated, and a system of two equations in two variables can be reduced to a single equation in one variable.

Solving by substitution   The substitution method should always be used when one equation is linear and the other is nonlinear.

EXAMPLE 9   Use the substitution method to solve the nonlinear system given below. Round your answer to the nearest hundredth.

$$xy = 6$$

$$2x - 3y = 12$$

Solution

Step 1. Solve either equation for one of the variables. $xy = 6$

Dividing both sides of the first equation by y, we obtain: $x = \dfrac{6}{y}$

Step 2. Substitute for x in the second equation. $2\left(\dfrac{6}{y}\right) - 3y = 12$

Step 3. Solve for y. $\dfrac{12}{y} - 3y = 12$

$$12 - 3y^2 = 12y$$

$$-3y^2 - 12y + 12 = 0$$

$$y^2 + 4y - 4 = 0$$

First, attempt to factor the left side of this equation to find a solution. Since it cannot be factored, we use the quadratic equation.

$$y = \frac{-4 \pm \sqrt{16 - 4(1)(-4)}}{2} = \frac{-4 \pm \sqrt{32}}{2} = -2 \pm 2\sqrt{2}$$

$2 \boxed{+/-} \boxed{+} 2 \boxed{\times} 2 \boxed{\sqrt{}} \boxed{=} \rightarrow$ *0.8284271*

$2 \boxed{+/-} \boxed{-} 2 \boxed{\times} 2 \boxed{\sqrt{}} \boxed{=} \rightarrow$ *-4.8284271*

$y \approx -4.83$ or $y \approx 0.83$

Step 4. Substitute each unrounded y value into the equation in Step 1 and solve for the corresponding values of x:

For the first value of y,

find $x = \dfrac{6}{y}$: $6 \boxed{\div} 4.8284271 \boxed{+/-} \boxed{=} \rightarrow$ *-1.2426407*

$$x \approx -1.24$$

For the second value of y,

find $x = \frac{6}{y}$: 6 [÷] 0.8284271 [=] → 7.2426409

$x \approx 7.24$

The approximate solutions are $(-1.24, -4.83)$ or $(7.24, 0.83)$.

Step 5. Check for $(-1.24, -4.83)$:

First equation: 1.24 [+/-] [×] 4.83 [+/-] [=] → 5.9892 ≈ 6

Second equation: 2 [×] 1.24 [+/-] [−] 3 [×] 4.83 [+/-] [=] → 12.01 ≈ 12

Check for $(7.24, 0.83)$:

First equation: 7.24 [×] .83 [=] → 6.00092 ≈ 6

Second equation: 2 [×] 7.24 [−] 3 [×] .83 [=] → 11.99 ≈ 12

Solving by addition

When both equations contain squared terms, the addition method may usually be used.

EXAMPLE 10 Solve by addition:

$$x^2 + y^2 = 25$$
$$-x^2 + 2y^2 = 2$$

Solution

Step 1. Notice that the coefficients of the x^2-terms are opposites, so we can add the two equations to eliminate these terms.

$$\begin{array}{r} x^2 + y^2 = 25 \\ -x^2 + 2y^2 = 2 \\ \hline 3y^2 = 27 \end{array}$$

Step 2. Solve for y.

$$y^2 = 9$$
$$y = 3 \quad \text{or} \quad y = -3$$

Step 3. Substitute each value for y back into the first equation and find the corresponding values for x:

For $y = 3$:

$$x^2 + (3)^2 = 25$$
$$x^2 + 9 = 25$$
$$x^2 = 16$$
$$x = 4 \quad \text{or} \quad x = -4$$

For $y = -3$:

$$x^2 + (-3)^2 = 25$$
$$x^2 + 9 = 25$$
$$x^2 = 16$$
$$x = 4 \quad \text{or} \quad x = -4$$

Since there are two x values for each y value, there are a total of four solutions: $(4, 3)$, $(-4, 3)$, $(4, -3)$, $(-4, -3)$.

Step 4. Check each solution by substituting it back into both of the original equations.

As with linear systems, it may be necessary to multiply one or both equations by a constant before using addition to solve nonlinear systems.

EXAMPLE 11 Use addition to solve the following system:

$$2x^2 - y^2 = 5$$

$$3x^2 + 2y^2 = 11$$

Solution

Step 1. Multiply the top equation by 2: And add it to the second equation to eliminate y.

$$4x^2 - 2y^2 = 10$$
$$3x^2 + 2y^2 = 11$$
$$\overline{\,7x^2\; = 21}$$

Step 2. Solve for x.

$$x^2 = 3$$

$$x = \sqrt{3} \quad \text{or} \quad x = -\sqrt{3}$$

Step 3. Substitute these values back into the first of the original equations and solve for y:

For $x = \sqrt{3}$:	**For $x = -\sqrt{3}$:**
$2(\sqrt{3})^2 - y^2 = 5$	$2(-\sqrt{3})^2 - y^2 = 5$
$6 - y^2 = 5$	$6 - y^2 = 5$
$y^2 = 1$	$y^2 = 1$
$y = 1 \quad \text{or} \quad y = -1$	$y = 1 \quad \text{or} \quad y = -1$

Again, we have four solutions:

$$(\sqrt{3}, 1), \quad (\sqrt{3}, -1), \quad (-\sqrt{3}, 1), \quad (-\sqrt{3}, -1).$$

Step 4. Be sure to check all solutions.

Applications **EXAMPLE 12** An aircraft initially at 13,000 m is losing altitude so that its height h_a at any time t (in seconds) is

$$h_a = 13{,}000 - 300t$$

When the plane is at 13,000 m, a missile is fired from the ground so that its height is given by

$$h_m = 3000t - 5t^2$$

When will the plane and the missile be at the same height? Round to the nearest second.*

---

*Courtesy of Gaylen Sandwisch, Santa Barbara, CA.

Solution
$$h_a = h_m$$

$$13{,}000 - 300t = 3000t - 5t^2$$

$$5t^2 - 3300t + 13{,}000 = 0$$

$$t^2 - 660t + 2600 = 0$$

$$t = \frac{660 \pm \sqrt{(-660)^2 - 4(2600)}}{2}$$

$$t \approx 656 \text{ s} \quad \text{or} \quad t \approx 4 \text{ s}$$

Checking these answers, we find that the height of both objects at $t = 4$ s is roughly 12,000 m. However, at $t = 656$ s, the height of both objects is approximately $-184{,}000$ m. Since neither craft will ever go lower than ground level, $h = 0$, the only acceptable solution is $t = 4$ s (see Fig. 14-9).

Figure 14-9

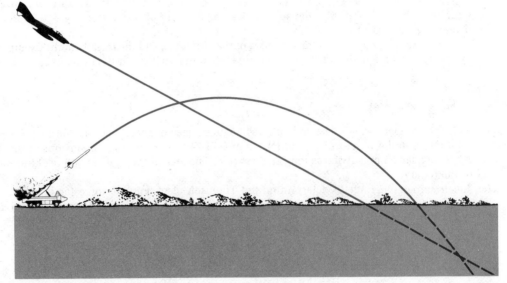

Exercises 14-2

Solve each system of equations algebraically. Give exact answers.

1. $y = x - 1$
$y = x^2 - 3$

2. $y = x + 2$
$y = x^2 - 5x + 10$

3. $x^2 + 4y^2 = 4$
$y = x + 1$

4. $x^2 + y^2 = 6$
$y = 2x$

5. $x^2 + 4y^2 = 6$
$2x^2 - y^2 = 3$

6. $x^2 + y^2 = 16$
$x^2 - 3y^2 = 4$

7. $xy = 4$
$4x - 3y = 2$

8. $x^2 - y^2 = 4$
$x + 2y = 2$

9. $y = x^2$
$x^2 + 4y = 10$

10. $y = x^2 - 2$
$3x^2 - y = 4$

Solve each system of equations algebraically. Round irrational answers to the nearest hundredth.

11. $x^2 + y^2 = 9$
$x^2 - y^2 = 4$

12. $x^2 + y^2 = 4$
$x^2 - 2y^2 = -4$

13. $x = y^2$
$x = 2y^2 - 3y$

14. $y = x^2 + 1$
$y = x^2 - 5x$

15. $x^2 + y^2 = 9$
$x^2 - y = 4$

16. $x^2 - 2y^2 = 5$
$y^2 = x - 6$

17. $x^2 + 2y^2 = 3$
$x^2 - 2y^2 = 9$

18. $3x^2 - y^2 = 4$
$x^2 + 2y^2 = 4$

19. $x^2 + y^2 = 8$
$3x - y = 2$

20. $xy = -2$
$2x + y = 1$

Solve.

21. A rectangular lot 1440 ft^2 in area requires 154 ft of fence to enclose it. Find the dimensions of the lot.

22. The sum of the squares of two positive numbers is 1217, and the difference of their squares is 705. Find the numbers.

23. *Industrial design*   One of the surfaces of a surgical tool is a circle with a radius of x in. A square hole y in. on each side is cut from the surface, leaving 0.47 in.^{2} of metal. If $x - y = 0.22$ in., find x and y. [Note: Use $\pi \approx 3.14$.]

24. *Aeronautical engineering*   If the altitudes of the aircraft h_a and the missile h_m in Example 12 are given by the system of equations below, determine when they will be at the same height.

$$h_a = 10{,}000 - 250t$$

$$h_m = 3200t - 4t^2$$

25. *Sheet metal technology*   The area of a rectangular piece of sheet metal is 285 in.^{2}. A second piece is 3 in. longer and 4 in. narrower and has an area of 242 in.^{2}. Find the dimensions of the first piece.

26. A rectangular room is twice as long as it is wide. The area of the room is 648 ft^2. Find the dimensions of the room.

27. *Electrical engineering*   In determining the Thomson electromotive force (emf) of a certain thermocouple, the system of equations given below is found. Solve for temperatures T_1 and T_2 (in degrees Kelvin).

$$0.02(T_2^2 - T_1^2) = 4280$$

$$T_2 - T_1 = 250$$

28. *Transportation engineering*   In the analysis of a rapid transit track, the total length L of the merge and demerge regions is related to the radius r of the curve and the offset distance x by the equation

$$L^2 = 4rx$$

In addition, the radius is related to the velocity v of the vehicle by

$$r = 0.334v^2$$

Combine these two equations to express L in terms of v and x only.

29. *Physics*   The height y (in meters) of a projectile as a function of time t (in seconds) from lift-off is given by the equation

$$y = 20 + 80t - 10t^2$$

Its horizontal distance of travel is given by

$$x = 10 + 50t$$

Find x and y when the horizontal distance is exactly four times the height.

30. *Space technology* A space probe is designed to separate into two parts when a spring-loaded connecting bolt is fired. The motion of the two parts, with mass $m_1 = 48$ kg and $m_2 = 72$ kg, satisfies the following energy and momentum equations:

Energy: $$\left(\frac{1}{2}\right)(48)(v_1^2) + \left(\frac{1}{2}\right)(72)(v_2^2) = 18,000$$

Momentum: $$48v_1 + 72v_2 = 0$$

where v_1 and v_2 are the velocities of the parts (in meters per second). Find v_1 and v_2.

14-3 | Equations Involving Radicals

Many technical formulas contain radical expressions, and you will often have to solve these formulas for different variables.

Equations containing radicals may be solved by first raising both sides of the equation to a power that causes the radicals to disappear. This technique may lead to either a linear or a quadratic equation to be solved.

CAUTION  Raising both sides of an equation to a power sometimes leads to extraneous solutions. Always check your answers. ◄

EXAMPLE 13 Solve: $\sqrt{2x - 3} = 5$

Solution

First, square both sides.
$$\left(\sqrt{2x - 3}\right)^2 = (5)^2$$
$$2x - 3 = 25$$

Then, solve the resulting equation.
$$2x = 28$$
$$x = 14$$

Check your answer.
$$\sqrt{2(14) - 3} \overset{?}{=} 5$$
$$\sqrt{25} \overset{?}{=} 5$$
$$5 = 5$$

EXAMPLE 14 Solve: $\sqrt[4]{x + 3} = 2$

Solution

First, raise both sides to the fourth power.
$$\left(\sqrt[4]{x + 3}\right)^4 = 2^4$$
$$x + 3 = 16$$

Then, solve.
$$x = 13$$

Check.
$$\sqrt[4]{(13) + 3} \overset{?}{=} 2$$
$$\sqrt[4]{16} \overset{?}{=} 2$$
$$2 = 2$$

Always try to isolate the radical term on one side of the equation before raising each side to a power.

EXAMPLE 15   Solve: $\sqrt{x - 4} + 4 = x$

Solution

First, isolate the radical term on one side.

$$\sqrt{x - 4} = x - 4$$

Then, solve as before.
Square both sides.
The equation is quadratic.

$$\left(\sqrt{x - 4}\right)^2 = (x - 4)^2 \qquad \text{Square the entire binomial}$$

$$x - 4 = x^2 - 8x + 16$$

$$0 = x^2 - 9x + 20$$

Factor.

$$0 = (x - 5)(x - 4)$$

Solve.

$$x = 5 \quad \text{or} \quad x = 4$$

Check both solutions.

$$\sqrt{(5) - 4} + 4 \stackrel{?}{=} (5) \quad \text{or} \quad \sqrt{(4) - 4} + 4 \stackrel{?}{=} (4)$$

$$5 = 5 \qquad\qquad\qquad 4 = 4$$

If there is more than one radical term, it may be necessary to raise the equation to a power more than once to remove all radicals.

EXAMPLE 16   Solve: $2\sqrt{x + 3} - \sqrt{2x + 3} = 3$. Round all roots to the nearest hundredth.

Solution

First, move one radical to the right side.

$$2\sqrt{x + 3} = 3 + \sqrt{2x + 3}$$

Second, square both sides.

$$\left(2\sqrt{x + 3}\right)^2 = \left(3 + \sqrt{2x + 3}\right)^2$$

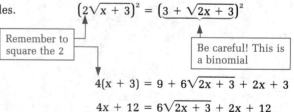

Remember to square the 2

Be careful! This is a binomial

$$4(x + 3) = 9 + 6\sqrt{2x + 3} + 2x + 3$$

$$4x + 12 = 6\sqrt{2x + 3} + 2x + 12$$

Third, isolate the remaining radical on one side and again square both sides.

$$6\sqrt{2x + 3} = 2x$$

$$\left(6\sqrt{2x + 3}\right)^2 = (2x)^2$$

$$36(2x + 3) = 4x^2$$

Finally, since there are no remaining radicals, we can solve this quadratic equation for x.

$$4x^2 = 72x + 108$$

$$4x^2 - 72x - 108 = 0$$

$$x^2 - 18x - 27 = 0$$

$$x = \frac{18 \pm \sqrt{324 - 4(1)(-27)}}{2}$$

$$= \frac{18 \pm \sqrt{432}}{2} = 9 \pm 6\sqrt{3}$$

$$x \approx 19.39 \quad \text{or} \quad x \approx -1.39$$

Checking, we find that the positive root satisfies the original equation; however, substituting -1.39 for x in the original equation, we have

$$2 \boxed{\times} \boxed{(} 1.39 \boxed{+/-} \boxed{+} 3 \boxed{)} \boxed{\sqrt{}} \boxed{-} \boxed{(} 2 \boxed{\times} 1.39 \boxed{+/-} \boxed{+} 3 \boxed{)} \boxed{\sqrt{}} \boxed{=} \rightarrow$$

$$\underbrace{\hspace{3.5cm}}_{2\sqrt{x+3}} \qquad \underbrace{\hspace{3.5cm}}_{\sqrt{2x+3}} \qquad \boxed{2.0686739}$$

Since the right side of the equation is 3, this root is not correct and is therefore extraneous. The only solution is $x \approx 19.39$.

Applications   We can use these same techniques for solving formulas and literal equations involving radicals.

EXAMPLE 17   In the theory of special relativity, the formula

$$m = \frac{m_0}{\sqrt{1 - \dfrac{v^2}{c^2}}}$$

relates the mass m of an object of rest mass m_0 as measured from a frame of reference moving with velocity v. The number c is the speed of light. Solve for v.

Solution

Step 1.   Multiply both sides by $\sqrt{1 - (v^2/c^2)}$.

$$m\sqrt{1 - \frac{v^2}{c^2}} = m_0$$

Step 2.   Square both sides.

$$m^2\left(1 - \frac{v^2}{c^2}\right) = m_0^2$$

Step 3.   Solve.

$$m^2 - \frac{m^2 v^2}{c^2} = m_0^2$$

$$c^2 m^2 - m^2 v^2 = c^2 m_0^2$$

$$m^2 v^2 = c^2 m^2 - c^2 m_0^2$$

$$v^2 = \frac{c^2(m^2 - m_0^2)}{m^2}$$

Physicists normally write this formula as $v = c \cdot k$, where the constant k involves the ratio of m_0/m, so now we rearrange the formula to put it in this form.

$$v^2 = c^2\left[1 - \left(\frac{m_0}{m}\right)^2\right]$$

$$v = c\sqrt{1 - \left(\frac{m_0}{m}\right)^2}$$

Certain geometric problems also lead to equations involving radicals.

EXAMPLE 18   The perimeter of a rectangular plate is 34 cm, and its diagonal length is 13 cm. Find the width and length of the plate.

Solution
Step 1.   Let

W = Width of plate L = Length of plate

Step 2.   Use the given information to write two equations (see Fig. 14-10).

Figure 14-10

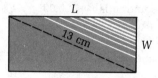

Using the Pythagorean theorem:

$$L^2 + W^2 = 169$$

The formula for perimeter gives:

$$2L + 2W = 34$$
$$L + W = 17$$

Step 3.   Solve the first equation for L and substitute the result into the second equation.

$$L^2 = 169 - W^2$$
$$L = \sqrt{169 - W^2}$$
$$\sqrt{169 - W^2} + W = 17$$

Step 4.   Solve for W by isolating the radical on one side.

$$\sqrt{169 - W^2} = 17 - W$$

Squaring both sides:

$$169 - W^2 = 289 - 34W + W^2$$

Collecting all terms on one side:

$$0 = 2W^2 - 34W + 120$$
$$0 = W^2 - 17W + 60$$

Factoring:

$$0 = (W - 5)(W - 12)$$

Solving:

$$W = 5 \text{ cm} \quad \text{or} \quad W = 12 \text{ cm}$$

Step 5.   Substitute each solution back into one of the original equations and solve for L.

For W = 5 cm: **For W = 12 cm:**

$L + 5 = 17$ $L + 12 = 17$

$L = 12$ cm $L = 5$ cm

Since L and W are interchangeable, the two solutions are the same, and the dimensions of the plate are 5 cm by 12 cm. Checking, we find that the perimeter is $2(5) + 2(12) = 34$ cm, and the diagonal length is $\sqrt{5^2 + 12^2} = 13$ cm.

Exercises 14-3

Solve each equation. Round all irrational roots to the nearest hundredth.

1. $\sqrt{x + 2} = 3$ **2.** $\sqrt{x - 5} = 7$ **3.** $\sqrt{2x - 3} = 4$

4. $\sqrt{3x + 1} = 2$ **5.** $\sqrt[3]{x + 4} = 2$ **6.** $\sqrt[4]{5x - 11} = 3$

7. $\sqrt{x + 6} = x$ **8.** $\sqrt{7x - 12} = x$ **9.** $\sqrt{x^2 - 4} = x + 2$

10. $\sqrt{4x^2 - 15} = 2x + 3$ **11.** $\sqrt{x - 1} = x - 1$ **12.** $\sqrt{2x + 6} = x + 3$

13. $\sqrt{2x + 3} = x$ **14.** $\sqrt{x + 1} = 2x$ **15.** $3\sqrt{x + 4} = x$

16. $2\sqrt{2x + 6} = x$ **17.** $\sqrt{x + 2} - \sqrt{2x - 1} = 2$ **18.** $\sqrt{x - 3} + \sqrt{x + 4} = 5$

19. $\sqrt{2x + 1} + 3\sqrt{x} = 9$ **20.** $2\sqrt{x} - \sqrt{x - 5} = 4$ **21.** $\sqrt{x + 3} + 1 = \sqrt{2x + 1}$

22. $\sqrt{3x - 1} - 1 = \sqrt{x + 2}$

23. *Electrical engineering* A bar magnet oscillates about its position of equilibrium with a period T given by the equation below. Solve for the magnetic field intensity B.

$$T = 2\pi\sqrt{\frac{I}{MB}}$$

24. *Cartography* The formula below is used to find the maximum line length S on a map from the radius R of a curve and the tolerance limit E. Solve this formula for E.

$$S = \sqrt{8RE - 4E^2}$$

25. *Electronics* In the study of multielectrode devices, the equation below is found. Solve for J.

$$V^{3/4} = \frac{3d}{4}\sqrt{\frac{4J}{E}\sqrt{\frac{m}{2e}}}$$

26. *Navigation* The formula below is used in calculating the range d of ship-to-shore radio communications. Solve for h.

$$d = \sqrt{2rh + h^2}$$

27. *Sheet metal technology* The perimeter of a rectangular plate of sheet metal is 124 cm, and the length of its diagonal is 50.0 cm. Find the dimensions of the plate.

28. A rectangular field has a diagonal length of 153 m and a perimeter of 414 m. Find the dimensions of the field.

29. *Navigation* Solve the formula in Problem 26 for r.

30. *Physics* The "length contraction" result of the special theory of relativity states that the length L of an object measured by an observer moving with speed v relative to the object and along its length is

$$L = L_0\sqrt{1 - \frac{v^2}{c^2}}$$

where c is the speed of light in a vacuum and L_0 is the length of the object as measured by an observer at rest with respect to it. For what speed will $L = \frac{1}{4}L_0$?

14-4 | Solving Equations in Quadratic Form

Transforming equations into quadratic form

A quadratic equation has the form

$$ax^2 + bx + c = 0 \qquad \text{where} \quad a \neq 0$$

Notice that the quadratic term, ax^2, has degree 2, and this is twice the degree of the linear term bx.

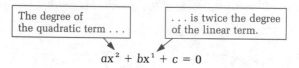

$$ax^2 + bx^1 + c = 0$$

Using this fact, we can transform many equations into quadratic form and then use quadratic methods to solve them. For example, to transform the equation

$$x^4 - 5x^2 + 9 = 0$$

into quadratic form, notice that the degree of the first term is 4, which is twice the degree of the second term. If we rewrite the equation as

$$(x^2)^2 - 5(x^2) + 9 = 0$$

and let $y = x^2$, we have

$$y^2 - 5y + 9 = 0$$

which is quadratic.

EXAMPLE 19 Transform each equation into quadratic form.

(a) $3x^{-6} + 2x^{-3} + 7 = 0$ (b) $x^{2/3} - 7x^{1/3} = 0$

(c) $2x - 3\sqrt{x} = 9$ (d) $5(x^2 + 2x)^2 - 2(x^2 + 2x) + 8 = 0$

Solution

(a) Since -6 is twice -3, rewrite the equation as

$$3(x^{-3})^2 + 2(x^{-3}) + 7 = 0$$

Let $y = x^{-3}$ and substitute:

$$3y^2 + 2y + 7 = 0$$

(b) The power $\frac{2}{3}$ is twice $\frac{1}{3}$, so we rewrite the equation as

$$(x^{1/3})^2 - 7(x^{1/3}) = 0$$

Let $y = x^{1/3}$; then the equation is

$$y^2 - 7y = 0$$

(c)   Rewrite $\sqrt{x}$ as $x^{1/2}$.

The power 1 is twice $\frac{1}{2}$.

Let $y = x^{1/2}$; then the equation becomes:

$$2x^1 - 3x^{1/2} = 9$$
$$2(x^{1/2})^2 - 3(x^{1/2}) = 9$$
$$2y^2 - 3y = 9$$
$$\text{or} \quad 2y^2 - 3y - 9 = 0$$

(d)   The powers are 2 and 1:

$$5(x^2 + 2x)^2 - 2(x^2 + 2x)^1 + 8 = 0$$

If we let $y = x^2 + 2x$, then the equation becomes

$$5y^2 - 2y + 8 = 0$$

Equations in quadratic form   After transforming a nonquadratic equation into quadratic form, solving the new equation may lead to a solution of the original equation.

EXAMPLE 20   Solve: $x^6 - 9x^3 + 8 = 0$

Solution

Step 1. Rewrite the equation in quadratic form:

The power 6 is twice 3.

$$(x^3)^2 - 9(x^3) + 8 = 0$$

Let $y = x^3$.

$$y^2 - 9y + 8 = 0$$

Step 2. Solve for y by factoring.

$$(y - 8)(y - 1) = 0$$
$$y = 8 \quad \text{or} \quad y = 1$$

Step 3. Replace y with x^3 and solve for x.

$$x^3 = 8 \quad \text{or} \quad x^3 = 1$$
$$x = \sqrt[3]{8} \qquad x = \sqrt[3]{1}$$
$$x = 2 \quad \text{or} \quad x = 1$$

Step 4. Check your answers in the original equation.

For $x = 2$:	For $x = 1$:
$(2)^6 - 9(2)^3 + 8 \stackrel{?}{=} 0$	$(1)^6 - 9(1)^3 + 8 \stackrel{?}{=} 0$
$64 - 72 + 8 \stackrel{?}{=} 0$	$1 - 9 + 8 \stackrel{?}{=} 0$
$0 = 0$	$0 = 0$

It is very important to check every solution in this type of problem, since **extraneous roots** may be introduced during the solution process. If *all* potential roots prove to be extraneous, there is *no* solution to the system.

EXAMPLE 21 Solve: $\sqrt{x} - 2\sqrt[4]{x} = 3$

Solution

Step 1. Convert the equation into quadratic form.

Rewrite with fractional exponents.

$$x^{1/2} - 2x^{1/4} = 3$$

$$(x^{1/4})^2 - 2(x^{1/4}) = 3$$

Let $y = x^{1/4}$.

$$y^2 - 2y = 3$$

Step 2. Solve for y.

$$y^2 - 2y - 3 = 0$$

$$(y - 3)(y + 1) = 0$$

$$y = 3 \quad \text{or} \quad y = -1$$

Step 3. Replace y with $x^{1/4}$ and solve for x.

$$x^{1/4} = 3 \quad \text{or} \quad x^{1/4} = -1$$

Raise each side to the fourth power.

$$(x^{1/4})^4 = 3^4 \quad \text{or} \quad (x^{1/4})^4 = (-1)^4$$

$$x = 81 \quad \text{or} \quad x = 1$$

Step 4. Check.

For x = 81:   **For x = 1:**

$\sqrt{(81)} - 2\sqrt[4]{(81)} \stackrel{?}{=} 3$ $\sqrt{(1)} - 2\sqrt[4]{(1)} \stackrel{?}{=} 3$

$9 - 6 \stackrel{?}{=} 3$ $1 - 2 \stackrel{?}{=} 3$

$3 = 3$ $-1 \neq 3$

Since $x = 1$ is not a valid solution, $x = 81$ is the only solution.

EXAMPLE 22 Solve: $2x^{-2} + 7x^{-1} + 4 = 0$. Round all roots to the nearest hundredth.

Solution

Step 1. Rewrite in quadratic form.

$$2(x^{-1})^2 + 7(x^{-1}) + 4 = 0$$

Let $y = x^{-1}$.

Step 2. Solve for y.

$$y = \frac{-7 \pm \sqrt{49 - 4(2)(4)}}{4}$$

$$= \frac{-7 \pm \sqrt{17}}{4}$$

$7 \boxed{+/-} \boxed{+} 17 \boxed{\sqrt{}} \boxed{=} \boxed{\div} 4 \boxed{=} \rightarrow \boxed{-0.7192236}$ $y \approx -0.72$

$7 \boxed{+/-} \boxed{-} 17 \boxed{\sqrt{}} \boxed{=} \boxed{\div} 4 \boxed{=} \rightarrow \boxed{-2.7807764}$ $y \approx -2.78$

Step 3. Replace y with x^{-1} and solve for x. For greater accuracy, retain the unrounded values and round after the last calculation.

$$x^{-1} = \frac{1}{x} \approx -0.7192236 \quad \text{or} \quad x^{-1} = \frac{1}{x} \approx -2.7807764$$

Enter $\boxed{1/x}$. $x \approx -1.3903882$ $x \approx -0.3596118$

Round. $x \approx -1.39$ or $x \approx -0.36$

Step 4.   Check both solutions. In this case, both solutions are valid.

EXAMPLE 23   Solve: $(x^2 + 3x)^2 - 2(x^2 + 3x) - 8 = 0$

Solution

Step 1.   Let $y = x^2 + 3x$ and $y^2 - 2y - 8 = 0$
rewrite in quadratic form.

Step 2.   Solve for y. $(y - 4)(y + 2) = 0$

$$y = 4 \quad \text{or} \quad y = -2$$

Step 3.   Replace y with $x^2 + 3x$ and solve for x.

$x^2 + 3x = 4$ or $x^2 + 3x = -2$

$x^2 + 3x - 4 = 0$ $x^2 + 3x + 2 = 0$

$(x + 4)(x - 1) = 0$ $(x + 2)(x + 1) = 0$

$x = -4 \quad \text{or} \quad x = 1 \quad$ or $x = -2 \quad \text{or} \quad x = -1$

Step 4.   Check all solutions by substituting them into the original equation. In this case, all four are valid solutions.

The next example illustrates a geometric problem that leads to an equation in quadratic form.

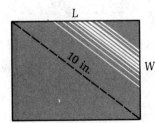

Figure 14-11

EXAMPLE 24   A rectangular piece of sheet metal has a diagonal length of 10 in. and an area of 48 in.^{2} (Fig. 14-11). Find the width and length of the piece.

Solution

Step 1.   Let W = Width and L = Length.

Step 2.   Write equations representing each fact in the problem.

The area is: $LW = 48$

The Pythagorean theorem gives: $L^2 + W^2 = 100$

Step 3.   Solve the first equation for L and substitute into the second equation. $L = \dfrac{48}{W}$

Therefore: $\left(\dfrac{48}{W}\right)^2 + W^2 = 100$

Step 4.   Solve for W.

$$\frac{2304}{W^2} + W^2 = 100$$

$$2304 + W^4 = 100W^2$$

$$W^4 - 100W^2 + 2304 = 0$$

Let $x = W^2$.

$$x^2 - 100x + 2304 = 0$$

Solve for x.

$$(x - 36)(x - 64) = 0$$

$$x = 36 \quad \text{or} \quad x = 64$$

Replace x with W^2.

$$W^2 = 36 \quad \text{or} \quad W^2 = 64$$

$$W = \pm 6 \quad \text{or} \quad W = \pm 8$$

Since a negative width makes no physical sense, the only acceptable widths are 6 in. or 8 in.

Step 5.   Substitute these values back into the equation in Step 3. Since

$$L = \frac{48}{W}$$

we have

$$L = \frac{48}{6} = 8 \quad \text{or} \quad L = \frac{48}{8} = 6$$

The two solutions, $W = 6$ in. and $L = 8$ in. or $W = 8$ in. and $L = 6$ in., are actually the same, since we can call either dimension W or L. The dimensions of the piece of sheet metal are 6 in. by 8 in.

Step 6.   Check this solution.

Exercises 14-4

Solve each equation. Round all irrational answers to the nearest hundredth.

1. $x^6 - 26x^3 - 27 = 0$
2. $x^4 - 10x^2 + 9 = 0$
3. $x^{-2} - x^{-1} - 42 = 0$
4. $x^{-4} - 13x^{-2} + 36 = 0$
5. $x^4 - 6x^2 + 7 = 0$
6. $x^6 + 2x^3 - 24 = 0$
7. $x^{-6} - 4x^{-3} - 21 = 0$
8. $x^{-2} - 2x^{-1} - 6 = 0$
9. $2x^4 + 3x^2 - 4 = 0$
10. $3x^6 - 2x^3 - 3 = 0$
11. $x - 3\sqrt{x} + 2 = 0$
12. $2x - 5\sqrt{x} - 3 = 0$
13. $3\sqrt{x} + 5\sqrt[4]{x} - 2 = 0$
14. $\sqrt[3]{x} - 4\sqrt[6]{x} + 4 = 0$
15. $x^{2/3} - 4x^{1/3} - 12 = 0$
16. $x^{-2/3} - x^{-1/3} - 6 = 0$
17. $(x^2 + 2x)^2 - 2(x^2 + 2x) - 3 = 0$
18. $(x^2 - 3x)^2 - 8(x^2 - 3x) - 20 = 0$
19. $(x^2 - 3)^{-2} - 3(x^2 - 3)^{-1} + 2 = 0$
20. $(x^2 - 2x)^{2/3} - (x^2 - 2x)^{1/3} - 2 = 0$

21. The lid for a rectangular box has a diagonal length of 13 in. and an area of 60 in.^{2}. Find its width and length.
22. If a commercially zoned lot is in the shape of a rectangle with area 3000 ft^2 and diagonal 85 ft, what are its dimensions?

14-5 | Review of Chapter 14

Important Terms and Concepts

nonlinear equation (p. 490)   ellipse (p. 491)   parabola (p. 492)

circle (p. 490)   hyperbola (p. 492)   nonlinear system of equations (p. 492)

Summary of Procedures

- To solve a nonlinear system of two equations graphically: Plot the graph of each equation and estimate the point(s) of intersection.
- To solve a nonlinear system of two equations algebraically: **First,** use the substitution or addition method to eliminate a variable. **Second,** solve the resulting equation for the remaining variable. **Third,** substitute this value back into either of the original equations and solve for the other variable.
- To solve an equation containing radicals: Isolate the radical term on one side and raise both sides to a power that causes the radical to disappear. Repeat the procedure if there is more than one radical term.
- To solve an equation in quadratic form: **First,** substitute a new variable to transform the equation into a quadratic equation and use quadratic methods to solve. **Then,** replace the substituted variable with the original expression and solve for the original variable.

Exercises 14-5

Solve each system of equations graphically. Estimate answers to the nearest tenth.

1. $y = 2x$
$x^2 + y^2 = 10$

2. $x - y = -3$
$y = 3x^2$

3. $x + 3y = 5$
$x^2 - y^2 = 6$

4. $2x + y = 1$
$x^2 + 3y^2 = 5$

5. $y = x^2 - 2$
$y = 2x^2 - 5x$

6. $y = 2x^2$
$4x^2 + y^2 = 1$

7. $2x^2 + 3y^2 = 6$
$3x^2 - y^2 = 4$

8. $x^2 + y^2 = 16$
$xy = 4$

9. $y = \cos x$
$y = \dfrac{x}{2}$

10. $y = 2^x$
$y = 3x$

Solve each system of equations algebraically. Give exact answers.

11. $y = 4x$
$y = x^2$

12. $2x - y = 4$
$xy = -2$

13. $x^2 + y^2 = 12$
$x - y = 8$

14. $2x^2 + y^2 = 4$
$x = 2y - 3$

15. $x^2 + y^2 = 6$
$x^2 - y^2 = 2$

16. $x^2 + y^2 = 36$
$x^2 + 8y^2 = 64$

Solve each system of equations algebraically. Round all irrational answers to the nearest hundredth.

17. $y^2 = x + 4$
$3y^2 - x^2 = 2$

18. $4x^2 + 2y^2 = 7$
$x^2 - 3y = 4$

19. $y = 3x - 2$
$y = x^2 - 2$

20. $x^2 + 3y^2 = 7$
$x^2 - 3y^2 = 3$

Solve. Round all irrational answers to the nearest hundredth.

21. $x^4 - 17x^2 + 16 = 0$

22. $3x^{-2} + 5x^{-1} - 6 = 0$

23. $4x^{-4} - 5x^{-2} + 1 = 0$

24. $x^{2/3} + 6x^{1/3} + 8 = 0$

25. $\sqrt{x} - \sqrt[4]{x} - 6 = 0$

26. $2x - \sqrt{x} - 10 = 0$

27. $9x^{-4/3} - 37x^{-2/3} + 4 = 0$

28. $(x^2 + 5x)^2 + 10(x^2 + 5x) + 24 = 0$

29. $\sqrt{x - 3} = 4$

30. $\sqrt[3]{x + 2} = 3$ $\qquad$ **31.** $\sqrt{7x - 10} = x$ $\qquad$ **32.** $\sqrt{x^2 + 3} = x - 1$

33. $\sqrt{x + 5} = x - 2$ $\qquad$ **34.** $\sqrt{x - 3} = 2x$ $\qquad$ **35.** $\sqrt{x} + \sqrt{2x + 3} = 2$

36. $\sqrt{3x + 2} - \sqrt{x + 2} = 1$

Solve.

37. *Machine technology* The formula below is used to calculate the depth of cut needed to make the flat of an arbor. Solve for D.

$$H = \frac{1}{2}\left(D - \sqrt{D^2 - W^2}\right)$$

38. *Civil engineering* The formula below is found in the stress analysis of a concrete beam.[*] Solve for np. [*Hint:* Let $x = np$; then solve for x.]

$$K = -np + \sqrt{(np)^2 + 2np}$$

39. A rectangular piece of steel has an area of 252 cm² and a perimeter of 66 cm. Find its dimensions.

40. *Agricultural technology* A cylindrical storage tank y ft in radius is located on a square pasture x ft on each side. If 870 ft² of usable land remains around the tank, and if $x - y = 25$ ft, find x and y to the nearest foot.

41. *Aeronautical engineering* If the altitudes h_a and h_m of the aircraft and missile in Example 12 are given by the system below, find the time t (to the nearest second) when their altitudes are the same.

$$h_a = 15,000 - 325t \qquad h_m = 2800t - 5t^2$$

42. A rectangular table has a diagonal length of 6 ft 3 in. and an area of 10.5 ft². Find the dimensions of the table.

43. The perimeter of a rectangular lot is 868 ft, and the diagonal length is 310 ft. Find the dimensions of the lot.

44. *Civil engineering* In the analysis of the reactions at two points in a certain beam, the following equations appear:

$$2RL^2 - ML + C = 0 \qquad 8RL^2 - 3ML + 12C = 0$$

Eliminate R from these two equations and solve for C in terms of M and L.

45. *Physics* The total kinetic energy of two masses, $m_1 = 1.8$ kg and $m_2 = 3.5$ kg, after a collision is given by the equation

$$\left(\frac{1}{2}\right)(1.8)(v_1^2) + \left(\frac{1}{2}\right)(3.5)(v_2^2) = 80.0$$

where v_1 and v_2 are the speeds of the colliding objects (in meters per second). If v_1 is four times as large as v_2, find speeds v_1 and v_2.

46. *Physics* The special theory of relativity predicts that time intervals will be measured differently by moving and stationary observers according to the equation

$$t = \frac{t_0}{\sqrt{1 - \dfrac{v^2}{c^2}}}$$

where c is the speed of light, v is the relative speed of the observers, t is a time interval as recorded by an observer moving with respect to the clock, and t_0 is the same time interval as recorded by an observer at rest with respect to the clock. How fast must a spaceship travel for each month on the ship ($t_0 = 1$) to correspond to 1 year on earth ($t = 12$)? (*Hint:* Solve for v/c.)

---

[*] For additional information, see Nelson R. Bauld, Jr., *Mechanics of Materials* (Monterey, Ca.: Brooks/Cole Engineering Division, Wadsworth, 1982), p. 246.

Solving Equations of a Higher Degree

15

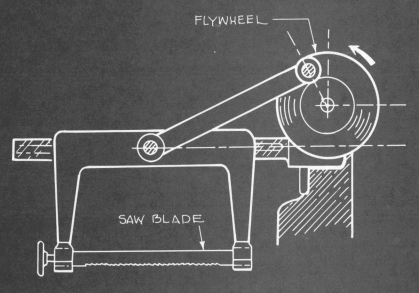

FLYWHEEL

SAW BLADE

FLYWHEEL-POWERED SAW

Most of the equation-solving techniques we have discussed so far are useful primarily for linear and quadratic equations. With the exception of some special cases we encountered in Chapters 7 and 14, we have no methods for solving equations of the third degree or higher. In this chapter we shall develop some techniques for solving these higher-degree equations.

The Remainder Theorem and the Factor Theorem

When you first learned how to perform long division in arithmetic, you might have stated the answer to $23 \div 6$ as "3, remainder 5." You would then check your answer by multiplying the divisor, 6, by the quotient, 3, and adding the remainder, 5. The result is equal to the dividend, 23. This same method of checking can be used to verify the division of a polynomial by another polynomial.

EXAMPLE 1   Divide $3x^2 - 7x + 1$ by $x - 3$ and check your answer.

Solution   Using the long division technique learned in Section 2-3, we have

$$
\begin{array}{r}
3x + 2 \\
x - 3 \overline{) 3x^2 - 7x + 1} \\
\underline{3x^2 - 9x} \\
2x + 1 \\
\underline{2x - 6} \\
7
\end{array}
$$

Check: Divisor·Quotient + Remainder

$3x^2 - 7x + 1 = (x - 3)(3x + 2) + 7$

$3x^2 - 7x + 1 = 3x^2 - 7x - 6 + 7$

$3x^2 - 7x + 1 = 3x^2 - 7x + 1$

This method of checking a division is the basis of an important theorem that we need to state in algebraic terms. In general, a **polynomial function** $f(x)$ can be written in the form

$$ f(x) = a_n x^n + a_{n-1} x^{n-1} + \cdots + a_1 x + a_0 $$

where n is a nonnegative integer.

As we saw in Example 1, if a polynomial function is divided by a binomial of the form $x - r$, resulting in a quotient $q(x)$ and a remainder R, then

$$ f(x) = (x - r)q(x) + R \qquad\qquad \text{(15-1)} $$

In other words, the original function is equal to the divisor times the quotient plus the remainder.

If we replace x with r in (15-1), we have

$$f(r) = (r - r)q(x) + R$$
$$= 0 + R = R \qquad\qquad (15\text{-}2)$$

This result leads us to the **remainder theorem:**

The remainder theorem   *If a polynomial f(x) is divided by (x − r) until a constant remainder R is obtained, then f (r) = R.*

Now, let us return to Example 1 to see how the remainder theorem is used.

EXAMPLE 2   To apply the remainder theorem to Example 1, first define

$$f(x) = 3x^2 - 7x + 1$$

Since the divisor x − r was x − 3, then r = 3. According to the remainder theorem, f(3) must be equal to the remainder. To find the remainder when f(x) is divided by x − 3, proceed as follows:

$$f(3) = 3(3)^2 - 7(3) + 1$$
$$= 3(9) - 7(3) + 1$$
$$= 27 - 21 + 1$$
$$= 7$$

This is the remainder we obtained in Example 1.

This means that we no longer need to divide a polynomial by a factor of the form x − r in order to determine the remainder. If we simply find f(r), it will be equal to the remainder.

EXAMPLE 3   What is the remainder when $4x^3 - 7x^2 + 5x + 2$ is divided by x + 2?

Solution
Step 1.   Since the remainder theorem is stated for a divisor of the form x − r, we must write x + 2 as x − (−2) to find that r = −2

Step 2. $R = f(-2) = 4(-2)^3 - 7(-2)^2 + 5(-2) + 2$
$$= 4(-8) - 7(4) + 5(-2) + 2$$
$$= -32 - 28 - 10 + 2$$
$$R = -68$$

The factor theorem   At this point, you may be wondering why we would be interested in finding the remainder when some polynomial $f(x)$ is divided by a binomial $x - r$. The reason is that if $x - r$ divides $f(x)$ exactly, then $R = 0$, and therefore $x - r$ is a factor of $f(x)$. This is called the **factor theorem**:

If $f(r) = R = 0$, then $x - r$ is a factor of $f(x)$.

EXAMPLE 4   Determine whether the second expression is a factor of the first.
(a) $x^3 + 3x^2 - 9x - 2$; $x - 2$
(b) $2x^4 + 3x^3 - 2x^2 + 4x + 5$; $x + 3$

Solution
(a)   From $x - 2$ we note that $r = 2$, so we have

$$R = f(2) = (2)^3 + 3(2)^2 - 9(2) - 2$$

$$= 8 + 12 - 18 - 2 = 0$$

Since $R = 0$, the binomial $x - 2$ is a factor of the given polynomial.
(b)   Rewriting $x + 3$ as $x - (-3)$, we have $r = -3$, so

$$R = f(-3) = 2(-3)^4 + 3(-3)^3 - 2(-3)^2 + 4(-3) + 5$$

$$= 162 + (-81) - 18 - 12 + 5 = 56$$

Since $R \neq 0$, the binomial $x + 3$ is not a factor of the given polynomial.

Factors, zeros, and roots   We have just established that if the quantity $f(r)$ is 0, then the binomial $x - r$ is a **factor of the polynomial** $f(x)$. Also recall from Chapter 3 that if $f(r) = 0$, then r is a **zero of the function** $f(x)$ and therefore r is a **root of the equation** $f(x) = 0$. An understanding of the interrelationship among factors, zeros, and roots will be extremely important in our discussion of roots of polynomial equations in Section 15-3.

EXAMPLE 5
(a)   Is 4 a zero of the function given below?

$$f(x) = 2x^3 - 11x^2 + 19x - 28$$

(b)   Is -14 a root of the equation given below?

$$x^4 + 13x^3 - 15x^2 + 196 = 0$$

Solution
(a) $f(4) = 2(4)^3 - 11(4)^2 + 19(4) - 28$

$$= 128 - 176 + 76 - 28 = 0$$

Thus, 4 is a *zero* of the given function. This also means that 4 is a *root* of the equation

$$2x^3 - 11x^2 + 19x - 28 = 0$$

and that $x - 4$ is a *factor* of the polynomial

$$2x^3 - 11x^2 + 19x - 28$$

(b)   With many calculators the $\boxed{y^x}$ key gives an error message when a negative base is used. To evaluate the left side for $x = -14$, calculate each term using $x = +14$. Then, since $(-14)^n$ is negative for any odd integer n, change the sign of any term containing an odd power of x.

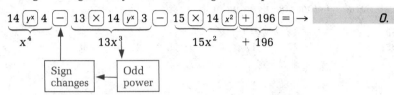

Since $x = -14$ gives us 0 on the left side, we know that -14 is a *root* of the equation $x^4 + 13x^3 - 15x^2 + 196 = 0$. It is also a *zero* of the function $f(x) = x^4 + 13x^3 - 15x^2 + 196$, and the binomial $x - (-14)$, or $x + 14$, is a *factor* of the polynomial $x^4 + 13x^3 - 15x^2 + 196$.

REMINDER ▶   Although most calculators automatically perform the order of operations as shown in Example 5(b), you must remember the proper order when calculating by hand as in Example 5(a). In the absence of grouping symbols, simplify powers first, multiplication and division next, and addition and subtraction last. Return to Section 1-4 if you need to review this procedure.   ◀

Exercises 15-1

Find each remainder using both long division and the remainder theorem.

1. $(x^3 - 2x^2 + 3x - 4) \div (x - 2)$
2. $(x^3 + 3x^2 - 6x + 1) \div (x - 1)$
3. $(x^4 + 4x^3 + 2x^2 + x + 2) \div (x + 1)$
4. $(2x^5 + 7x^4 - 2x^3 - 6x^2 + x - 3) \div (x + 4)$
5. $(2x^4 + 6x^3 - 3x^2 - 7) \div (x - 5)$
6. $(3x^3 - x - 6) \div (x - 3)$

Find each remainder using the remainder theorem.

7. $(x^3 + 5x^2 - 3x - 2) \div (x + 1)$
8. $(2x^4 - x^3 - 5x^2 + 2x - 3) \div (x + 2)$
9. $(x^5 + 4x^4 - 6x^2 + 8x - 3) \div (x - 5)$
10. $(x^4 - 2x^3 - 5x^2 + 6x - 1) \div (x + 2)$
11. $(x^5 - x^3 + 3x^2 - 4x + 1) \div (x - 1)$
12. $(x^3 - 4x^2 + 2x + 3) \div (x - 3)$
13. $(3x^3 + 5x^2 - x + 6) \div (x + 3)$
14. $(2x^5 - 5x^4 - x^2 + 4x - 6) \div (x - 4)$
15. $(2x^4 - x^3 - 7x^2 + 8) \div (x - 2)$
16. $(2x^3 + 6x^2 + 17) \div (x + 5)$
17. $(x^5 - 11x^4 - 11x^3 - 22x - 1500) \div (x - 12)$
18. $(2x^4 + 35x^3 - 41x^2 - 84x + 5900) \div (x + 18)$

Determine whether the second expression is a factor of the first.

19. $x^3 - x^2 - 3x + 2;\ \ x - 2$
20. $x^3 - 2x^2 - 5x + 3;\ \ x - 3$
21. $2x^3 + 7x^2 - 3x + 1;\ \ x + 4$
22. $3x^3 + 5x^2 - 4x - 6;\ \ x + 1$
23. $x^4 - 6x^2 - 8x - 3;\ \ x - 3$
24. $x^5 - 4x^3 - 7x^2 + 16;\ \ x + 2$

25. $x^6 - 7x^4 - 8x^2 - 80$; $x + 3$ **26.** $x^7 - x^5 - x^3 - x + 2$; $x - 1$

27. $x^6 - 67x^4 - 23x^3 + 500$; $x + 8$ **28.** $x^7 - 6x^6 - 30x^5 + 28x^4 - 8x^3 - 729$; $x - 9$

Determine whether the given number is a zero of the function.

29. $f(x) = x^3 + 3x^2 - 6x - 15$; -3 **30.** $f(x) = 2x^3 + 7x^2 - 17x + 78$; -6

31. $f(x) = 4x^4 - 2x^3 - 4x^2 + 1$; $\frac{1}{2}$ **32.** $f(x) = 3x^5 - 22x^4 - 29x^3 - 62x^2 - 16x - 6500$; 9

Determine whether the given value is a root of the equation.

33. $3x^3 - 5x^2 - 4x + 4 = 0$; 2 **34.** $x^4 - 10x^3 - 18x^2 - 864 = 0$; 12

35. $x^5 + 17x^4 + 27x^3 - 17x^2 - 5850 = 0$; -15 **36.** $x^3 + 6x^2 + 15x + 60 = 0$; -5

15-2 | Synthetic Division

In Section 15-1 we showed that the remainder theorem provides a quick way to determine the remainder when a polynomial function $f(x)$ is divided by the binomial $x - r$. However, in many cases we must also find the quotient of the division. Fortunately, in these cases we do not have to rely on long division because there is a shortcut method called **synthetic division** that simplifies the division process.

To understand how synthetic division works, we must first examine the result of a sample long division problem in algebra. To divide $3x^4 - 5x^3 + 2x^2 - 3x + 1$ by $x - 2$, we normally set it up this way:

$$
\begin{array}{r}
3x^3 + x^2 + 4x + 5 \\
x - 2 \overline{)3x^4 - 5x^3 + 2x^2 - 3x + 1} \\
\underline{3x^4 - 6x^3} \\
x^3 + 2x^2 \\
\underline{x^3 - 2x^2} \\
4x^2 - 3x \\
\underline{4x^2 - 8x} \\
5x + 1 \\
\underline{5x - 10} \\
11
\end{array}
$$

Notice that the powers of x in the quotient begin with one less than the power of x in the dividend and continue in descending order. The literal parts of the quotient are x^3, x^2, x and $x^0 = 1$. Because this is so, we can omit writing the literal portion of each term while dividing and work only with the coefficients. Notice, too, that the terms shown in color are duplicates of the terms above them. It will save time if we do not write these. Finally, we do not have to write the x-term in the divisor, since we will use the procedure only when the coefficient of x in the divisor is 1. The following is what remains after we make these deletions:

$$
\begin{array}{r}
3 + 1 + 4 + 5 \\
-2\overline{)3 - 5 + 2 - 3 + 1} \\
-\,6 \\
\hline
1 \\
-\,2 \\
\hline
4 \\
-\,8 \\
\hline
5 \\
-\,10 \\
\hline
11
\end{array}
$$

We can now compress this by moving all terms up and eliminating the addition signs:

$$
\begin{array}{rrrrr}
3 & 1 & 4 & 5 & \\
-2\overline{)3} & -5 & 2 & -3 & 1 \\
& -6 & -2 & -8 & -10 \\
\hline
& 1 & 4 & 5 & 11
\end{array}
$$

Notice now that coefficients of the quotient in the top row are repeated in the bottom row with the exception of the first term. By bringing the first term down to the bottom, we can omit the top row:

$$
\begin{array}{rrrrrl}
-2\overline{)3} & -5 & 2 & -3 & 1 & \longleftarrow \text{Row 1} \\
& -6 & -2 & -8 & -10 & \longleftarrow \text{Row 2} \\
\hline
3 & 1 & 4 & 5 & 11 & \longleftarrow \text{Row 3}
\end{array}
$$

Finally, note that each number in the bottom row is the difference of the numbers in the first two rows. Because addition is easier than subtraction, we can change the signs in Row 2 and use the actual value of r as the divisor.

Here is the final form for synthetic division:

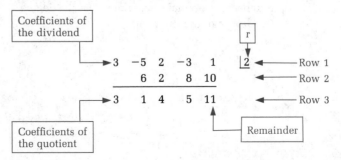

Notice that each number in the middle row is the product of r and the preceding number from Row 3. Example 6 shows the step-by-step procedure for synthetic division.

EXAMPLE 6   Use synthetic division to divide $4x^3 - 7x^2 + 3x - 5$ by $x - 3$.

Solution
Step 1.   With all terms arranged in descending order of powers, write the coefficients of the dividend in a row. Place r to the right and bring down the first coefficient to the answer row.

$$
\begin{array}{rrrr|r}
4 & -7 & 3 & -5 & \underline{3} \\
\big\downarrow & & & & \\
\hline
4
\end{array}
$$

Step 2.   Multiply the number in the bottom row by r and place the product in the middle row one space to the right.

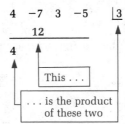

Step 3.   Add and place the result in the answer row.

$$
\begin{array}{rrrr|r}
4 & -7 & 3 & -5 & \underline{3} \\
 & 12 & & & \\
\hline
4 & 5 & & &
\end{array}
$$

Add

Step 4.   Repeat Steps 2 and 3 until all numbers in the first row have been used.

$$
\begin{array}{rrrr|r}
4 & -7 & 3 & -5 & \underline{3} \\
 & 12 & 15 & 54 & \\
\hline
4 & 5 & 18 & 49 &
\end{array}
$$

Step 5.   Write the quotient using the coefficients in the answer row. The last number is the remainder. Remember, the degree of the quotient is one less than the degree of the dividend.

$$
\begin{array}{cccc}
\boxed{x^3} & \boxed{x^2} & \boxed{x} & \boxed{\text{Constant}} \\
\downarrow & \downarrow & \downarrow & \downarrow \\
4 & -7 & 3 & -5 \\
 & 12 & 15 & 54 \\
\hline
4 & 5 & 18 & 49 \\
\uparrow & \uparrow & \uparrow & \\
\boxed{x^2} & \boxed{x} & \boxed{\text{Constant}} &
\end{array}
$$

Quotient: $4x^2 + 5x + 18$
Remainder: 49

EXAMPLE 7   Use synthetic division to divide $x^5 - 2x^3 + 5x - 4$ by $x + 3$.

Solution

First, insert 0's for each missing power of the variable:

Term: x^5 x^4 x^3 x^2 x^1 x^0

Coefficient: 1   0 -2 0   5 -4

Then, proceed as before. Note that for $x + 3$, $r = -3$

$$
\begin{array}{rrrrrr|r}
1 & 0 & -2 & 0 & 5 & -4 & \underline{-3} \\
  & -3 & 9 & -21 & 63 & -204 & \\
\hline
1 & -3 & 7 & -21 & 68 & -208 &
\end{array}
$$

Since the dividend has degree 5, the quotient has degree 4. The quotient is $x^4 - 3x^3 + 7x^2 - 21x + 68$, and the remainder is -208.

Testing factors   To determine whether $x - r$ is a factor of the polynomial $f(x)$, we can (a) use the factor theorem and test to see if $f(x) = 0$, as shown in Section 15-1; or (b) use synthetic division to divide $f(x)$ by $x - r$ to see if the remainder is 0.

EXAMPLE 8   Use synthetic division to determine whether $x + 4$ is a factor of $2x^4 + 3x^3 - 17x^2 + 6x - 24$.

Solution Using synthetic division, we have

$$
\begin{array}{rrrrr|r}
2 & 3 & -17 & 6 & -24 & \underline{-4} \\
  & -8 & 20 & -12 & 24 & \\
\hline
2 & -5 & 3 & -6 & 0 &
\end{array}
$$

Since the remainder is **0**, $x + 4$ is a factor of the polynomial.

Factors in the form   The synthetic division procedure is valid only for divisors in the form $x - r$.
$ax - b$ However, not all binomial factors of a polynomial are in this form. To test a possible factor of the form $ax - b$, first rewrite it as $a\left(x - \dfrac{b}{a}\right)$. Then use synthetic division to test $x - \dfrac{b}{a}$. If this remainder is 0 and the resulting quotient is divisible by a, then $ax - b$ is a factor of the original polynomial.

EXAMPLE 9   Use synthetic division to test whether $2x - 1$ is a factor of $6x^3 - 13x^2 + 13x - 4$.

Solution
Step 1. Factor out the coefficient of x from the binomial:

$$2x - 1 = 2\left(x - \frac{1}{2}\right)$$

Step 2.   Test $x - \frac{1}{2}$ using synthetic division:

$$
\begin{array}{rrrr|l}
6 & -13 & 13 & -4 & \dfrac{1}{2} \\[4pt]
  & 3 & -5 & 4 & \\
\hline
6 & -10 & 8 & 0 &
\end{array}
$$

Since the remainder is 0, $x - \frac{1}{2}$ is a factor.

Step 3.   Test the quotient coefficients in Step 2 to see if they are divisible by 2. Since 6, -10, and 8 are all exactly divisible by 2, then $2(x - \frac{1}{2})$, or $2x - 1$, is a factor of the original polynomial.

Testing zeros and roots

In Section 15-1 we showed that if r is a zero of $f(x)$, and therefore a root of the equation $f(x) = 0$, then $x - r$ is a factor of $f(x)$. It is much quicker to test zeros and roots by seeing if $f(r) = 0$ than by dividing $f(x)$ by $x - r$ using long division (see Example 5). But, using a calculator and synthetic division, we may find it easier to test a zero or root r by dividing the polynomial than by testing $f(r)$, especially if the degree of $f(x)$ is large, or if the coefficients are very large numbers.

EXAMPLE 10   Use synthetic division to test whether -5.75 is a root of $24x^3 + 182x^2 + 221x - 184 = 0$.

Solution   If -5.75 is a root of the given equation, then $x - (-5.75)$ is a factor of $f(x)$. We test this using synthetic division as follows:

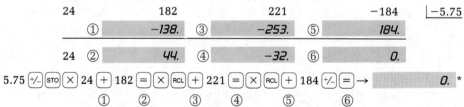

The circled numbers indicate when the corresponding calculator displays appear.

The remainder is 0, so -5.75 is a root of the original equation.

Exercises 15-2

Use synthetic division to divide.

1. $(x^3 - x^2 - 3x + 5) \div (x - 1)$

2. $(x^3 + 3x^2 - 5x - 2) \div (x - 2)$

3. $(x^4 + 3x^3 + 3x^2 + x - 2) \div (x + 2)$

4. $(x^5 - 2x^4 + 5x^3 + 7x^2 - 2x - 1) \div (x + 1)$

---

* In some problems of this type you may get a small remainder due to rounding errors with a specific calculator.

5. $(x^3 + 5x^2 - 4) \div (x + 3)$

6. $(x^4 - 8x^2 - 7x - 6) \div (x - 4)$

7. $(x^5 - 5x^3 - 9x + 2) \div (x - 3)$

8. $(x^3 - 12x + 17) \div (x + 4)$

9. $(2x^3 - 6x^2 - 5x + 9) \div (x + 1)$

10. $(3x^4 + 7x^3 + 8x^2 - 6x + 1) \div (x + 2)$

11. $(2x^5 - 5x^4 - 2x^3 + 3x^2 + 4) \div (x - 1)$

12. $(4x^3 - 5x^2 - 7x - 6) \div (x - 2)$

13. $(2x^4 + 8x^3 + 12x + 5) \div (x + 4)$

14. $(3x^5 - 5x^4 - 2x^3 + 6x - 3) \div (x - 2)$

15. $(14x^3 - 67x^2 - 112x - 8000) \div (x - 11)$

16. $(22x^3 + 187x^2 - 106x + 4000) \div (x + 13)$

Use synthetic division to determine whether the second expression is a factor of the first.

17. $x^3 - 5x^2 + 9x - 13$; $x - 3$

18. $x^3 + 6x^2 + 6x - 2$; $x + 4$

19. $x^4 - 6x^3 - 12x^2 + 4x - 8$; $x + 2$

20. $3x^4 - 11x^3 - 16x^2 - 18x - 10$; $x - 5$

21. $2x^3 + 3x^2 - 3x + 2$; $2x + 1$

22. $6x^4 + 7x^3 - 4x + 1$; $3x - 1$

23. $32x^4 + 56x^3 - 8x^2 + 37x - 57$; $4x - 3$ **24.** $24x^3 + 156x^2 - 1200x + 4600$; $x + 12$

Use synthetic division to determine whether the given number is a zero of the function.

25. $f(x) = x^3 - 4x^2 - 15x + 18$; 6

26. $f(x) = x^4 + 6x^3 + 7x^2 - 2x + 8$; -4

27. $f(x) = 16x^4 + 220x^3 - 622x^2 + 750x + 225$; -16

28. $f(x) = 3x^3 - 11x^2 + 12x - 2$; $\frac{2}{3}$

Use synthetic division to determine whether the given value is a root of the equation.

29. $x^5 - 2x^3 - x + 14 = 0$; -2

30. $x^3 + 7x^2 + 6x + 8 = 0$; -3

31. $4x^4 - 6x^3 + 8x^2 - 5x + 4 = 0$; $\frac{1}{2}$ **32.** $2x^3 - 35x^2 + 130x + 209 = 0$; 9.5

15-3 | The Roots of a Polynomial Equation

So far, we have shown that if r is a root of a polynomial equation $f(x) = 0$, then $x - r$ is a factor of the polynomial $f(x)$. We also found that synthetic division enables us to test a factor very quickly. In this section we shall provide several theorems that allow us to use our skills and knowledge to solve polynomial equations.

Number of roots   The **fundamental theorem of algebra** states that every polynomial equation has at least one root. The root may be real or imaginary. Since the proof of this theorem requires more advanced mathematics, we accept it here without proof. Two other interrelated theorems are based on this one:

- Every polynomial $f(x)$ of degree n has n linear factors.
- Every polynomial equation $f(x) = 0$ of degree n has exactly n roots.

We can prove these statements as follows: Let $f(x) = 0$ represent a polynomial equation of degree n. According to the fundamental theorem, this equation has at least one root. If this root is r_1, then $x - r_1$ is a factor of $f(x)$. We can now write $f(x)$ as

$$f(x) = (x - r_1)f_1(x)$$

where $f_1(x)$ is the quotient of $f(x) \div (x - r_1)$. The degree of $f_1(x)$ is $n - 1$, one less than the degree of $f(x)$ because we divided $f(x)$ by a linear factor. If $f_1(x) = 0$

is a polynomial equation, then it must have at least one root. If this root is r_2, then $x - r_2$ is a factor of $f_1(x)$. We can now write $f(x)$ as the product

$$f(x) = (x - r_1)(x - r_2)f_2(x)$$

where the degree of $f_2(x)$ is $n - 2$. We can continue this process until a constant quotient a (degree 0) is found, and therefore,

$$f(x) = a(x - r_1)(x - r_2) \cdot \cdots \cdot (x - r_n)$$

Thus, the polynomial $f(x)$ of degree n has n linear factors, and the equation $f(x) = 0$ has n roots.

EXAMPLE 11   Consider the polynomial equation $f(x) = x^4 - 13x^2 + 36 = 0$. According to the theorems we have stated, $f(x)$ should have four linear factors, and the equation $f(x) = 0$ should have four roots. Since $f(x)$ is quadratic in form, we can factor it as follows:

$$x^4 - 13x^2 + 36 = (x^2 - 9)(x^2 - 4)$$

$$= (x + 3)(x - 3)(x + 2)(x - 2)$$

As predicted, there are four linear factors. There are also four roots to the equation: $-3, 3, -2, 2$.

Multiple roots   The four roots found in Example 11 were all distinct. However, a polynomial may have two or more identical factors, and thus the corresponding equation may have two or more identical roots. These are known as **multiple roots.**

EXAMPLE 12   Solve the equation $f(x) = x^2 - 4x + 4 = 0$.

Solution   Because it is a second-degree equation, there should be two factors and therefore two roots. Factoring $f(x)$, we obtain

$$f(x) = x^2 - 4x + 4 = (x - 2)^2$$

As we expect of a second-degree function, $x^2 - 4x + 4$ has two factors, $x - 2$ and $x - 2$. The corresponding roots to the equation are both $x = 2$, a double root.

Reducing equations to quadratic form   The equations in Examples 11 and 12 were quadratic in form and could be solved using quadratic techniques. We also can solve a higher-degree equation, even if it is not in quadratic form, if we know all but two of its roots. Using synthetic division and the appropriate quadratic technique, we can find the two remaining roots.

EXAMPLE 13   Find the remaining roots of $x^3 - 3x^2 - 10x + 24 = 0$, given that one of the roots is 4.

Solution   Since one of the roots of this equation is 4, one of the factors is $x - 4$. Using synthetic division, we can find the remaining quadratic factor:

$$
\begin{array}{rrrr|r}
1 & -3 & -10 & 24 & \underline{4} \\
 & 4 & 4 & -24 & \\
\hline
1 & 1 & -6 & 0 &
\end{array}
$$

The bottom row tells us that the remaining quadratic factor is
$1 \cdot x^2 + 1 \cdot x - 6 = x^2 + x - 6$. The original equation can now be rewritten as

$$(x - 4)(x^2 + x - 6) = 0$$

and the two remaining roots of the original equation are the roots of
$x^2 + x - 6 = 0$. Factoring this expression, we have

$$(x + 3)(x - 2) = 0$$

Therefore, -3 and 2 are the two remaining roots.

EXAMPLE 14   Find the remaining roots of

$$2x^4 + 17x^3 + 44x^2 + 21x - 36 = 0$$

given that -3 is a double root. Round to the nearest hundredth.

Solution   Since -3 is a double root, $x + 3$ must be a double factor. Therefore,
we use synthetic division twice:

First
division
$$
\begin{array}{rrrrr|r}
2 & 17 & 44 & 21 & -36 & \underline{-3} \\
 & -6 & -33 & -33 & 36 & \\
\hline
2 & 11 & 11 & -12 & 0 & \underline{-3} \\
\end{array}
$$
Second
division
$$
\begin{array}{rrrr}
 & -6 & -15 & 12 \\
\hline
2 & 5 & -4 & 0 \\
\end{array}
$$

The remaining quadratic factor is $2x^2 + 5x - 4$, so the original equation can
now be rewritten as

$$(x + 3)^2(2x^2 + 5x - 4) = 0$$

The two remaining roots of the original equation are the roots of

$$2x^2 + 5x - 4 = 0$$

Since this is not factorable with integer coefficients, we use the quadratic
formula to obtain

$$x = \frac{-5 \pm \sqrt{25 - 4(2)(-4)}}{4} = \frac{-5 \pm \sqrt{57}}{4}$$

First root: 5 $\boxed{+/-}$ $\boxed{+}$ 57 $\boxed{\sqrt{}}$ $\boxed{=}$ $\boxed{\div}$ 4 $\boxed{=}$ → *0.6374586*

Second root: 5 $\boxed{+/-}$ $\boxed{-}$ 57 $\boxed{\sqrt{}}$ $\boxed{=}$ $\boxed{\div}$ 4 $\boxed{=}$ → *-3.1374586*

Thus, the two remaining roots are $x \approx 0.64$ and $x \approx -3.14$.

Imaginary roots   The fundamental theorem of algebra indicates that some of the roots of a
polynomial equation may be imaginary. Imaginary roots will always occur in

pairs, so that if $a + bj$, with $b \neq 0$, is a root of some polynomial equation $f(x) = 0$, then its conjugate $a - bj$ will also be a root of that equation.

EXAMPLE 15   Find the remaining roots of the equation

$$3x^4 + 7x^3 + x^2 - 23x - 20 = 0$$

given that $\frac{5}{3}$ and -1 are both roots.

Solution   We divide synthetically by $x - \frac{5}{3}$ and $x + 1$ as follows:

$$
\begin{array}{rrrrr|l}
3 & 7 & 1 & -23 & -20 & \dfrac{5}{3} \\[1mm]
& 5 & 20 & 35 & 20 & \\[1mm]
\hline
3 & 12 & 21 & 12 & 0 & \underline{\;-1\;} \\[1mm]
& -3 & -9 & -12 & & \\[1mm]
\hline
3 & 9 & 12 & 0 & &
\end{array}
$$

The remaining quadratic factor is $3x^2 + 9x + 12$

To solve the equation: $3x^2 + 9x + 12 = 0$

Divide both sides by 3. $x^2 + 3x + 4 = 0$

Then use the quadratic formula. $x = \dfrac{-3 \pm \sqrt{9 - 4(1)(4)}}{2}$

$$= \dfrac{-3 \pm \sqrt{-7}}{2}$$

In rectangular form, the two remaining roots are

$$\dfrac{-3}{2} + j\dfrac{\sqrt{7}}{2} \quad \text{and} \quad \dfrac{-3}{2} - j\dfrac{\sqrt{7}}{2}$$

As predicted, the imaginary roots found in Example 15 are conjugates. This fact can help us solve problems such as the next example.

EXAMPLE 16   Find the remaining roots of $2x^4 - 9x^3 - 81x - 162 = 0$, given that one of its roots is $3j$.

Solution   Since imaginary roots always occur in conjugate pairs, and $3j$ is one imaginary root, then $-3j$ must be the second imaginary root. Using synthetic division twice, we obtain

$$
\begin{array}{rrrrr|l}
2 & -9 & 0 & -81 & -162 & \underline{\;3j\;} \\
& 6j & -27j - 18 & 81 - 54j & 162 & \\
\hline
2 & -9 + 6j & -27j - 18 & -54j & 0 & \underline{\;-3j\;} \\
& -6j & 27j & 54j & & \\
\hline
2 & -9 & -18 & 0 & &
\end{array}
$$

The remaining quadratic factor is $2x^2 - 9x - 18$. Solving the equation $2x^2 - 9x - 18 = 0$, we have

$$(2x + 3)(x - 6) = 0$$

So the two remaining roots are $x = -\frac{3}{2}$ and $x = 6$.

Exercises 15-3

The numbers in parentheses are roots of the given equations. Find the remaining roots. Round all irrational roots to the nearest hundredth.

1. $x^3 + 4x^2 + x - 6 = 0$ (1)
2. $x^3 - 9x^2 + 26x - 24 = 0$ (2)
3. $x^3 + 3x^2 - 10x - 24 = 0$ (−2)
4. $x^3 - x^2 - 37x - 35 = 0$ (−1)
5. $2x^3 - 5x^2 - 11x - 4 = 0$ $(-\frac{1}{2})$
6. $3x^3 + 5x^2 - 26x + 8 = 0$ $(\frac{1}{3})$
7. $2x^3 + 9x^2 - 2x - 24 = 0$ (−4)
8. $4x^3 + x^2 - 36x - 9 = 0$ (3)
9. $6x^3 + 25x^2 + 12x - 18 = 0$ $(-\frac{3}{2})$
10. $2x^3 - 3x^2 - 33x - 10 = 0$ (5)
11. $2x^3 + 8x^2 - 19x + 30 = 0$ (−6)
12. $3x^3 + 4x^2 + 14x - 12 = 0$ $(\frac{2}{3})$
13. $x^3 - 4x^2 + x - 4 = 0$ (j)
14. $x^3 + 5x^2 + 8x + 6 = 0$ (−1 + j)
15. $x^4 - 9x^3 + 19x^2 + 9x - 20 = 0$ (4, −1)
16. $2x^4 - 5x^3 - 39x^2 + 153x - 135 = 0$ (3 is a double root)
17. $x^4 + 10x^3 + 26x^2 + 16x - 8 = 0$ (−2 is a double root)
18. $6x^4 - 5x^3 - 21x^2 - x + 6 = 0$ $(\frac{1}{2}, -\frac{2}{3})$　19. $24x^4 - 22x^3 + 8x^2 + 3x - 1 = 0$ $(\frac{1}{4}, -\frac{1}{3})$
20. $x^4 + 29x^2 + 100 = 0$ (2j)
21. $3x^5 + 5x^4 - 13x^3 - 23x^2 - 16x - 28 = 0$ (−j, −2)
22. $x^6 + 9x^5 + 17x^4 - 26x^3 - 49x^2 + 57x - 9 = 0$ (1 and −3 are both double roots)
23. $2x^3 - 47x^2 - 1558x + 17{,}328 = 0$ (9.5)
24. $2x^3 + 27x^2 - 2759x + 18{,}330 = 0$ (7.5)

Solve.

25. Name all roots of the equation: $(x - 5)(x^2 - 64)(x^2 - 49) = 0$
26. Name all roots of the equation: $(x + 4)(x^2 + 6x + 9)(x^2 - 10x + 25) = 0$
27. Find a polynomial equation with roots −1, 3, and 4.
28. Find a polynomial equation having a double root of −2, and with a third root of 5.

15-4 | Rational Roots

In Section 15-3 we showed that if we knew all but two roots of a polynomial equation, we could find the remaining roots by factoring or by using the quadratic formula. In this section we shall explain how to find the rational roots that may reduce the equation to a quadratic equation.

Integer roots　　Finding all but two roots of a polynomial equation of degree greater than 2 generally requires an intelligent use of trial and error. By utilizing certain clues, you can limit the number of trials necessary to discover the correct solution.

EXAMPLE 17   Consider the polynomial equation

$$(x - 2)(x + 3)(x + 5) = 0$$

Because it is in factored form, its roots can be found easily by inspection. They are 2, -3, and -5. If we multiply the factors, we can rewrite the equation as

$$x^3 + 6x^2 - x - 30 = 0$$

Notice that all three roots are integral factors of the constant term -30.

Example 17 leads to the following theorem:

For any polynomial equation $f(x) = 0$ whose highest-power term has a coefficient of 1, all integral roots of the equation are factors of the constant term of $f(x)$.

Thus, if

$$f(x) = x^n + a_{n-1}x^{n-1} + \cdots + a_1x + a_0 = 0$$

can be factored as

$$f(x) = (x - r_1)(x - r_2) \cdots \cdots (x - r_n)f_1(x) = 0$$

where $f_1(x)$ is a polynomial factor representing the irrational roots, then the integral roots r_1, r_2, r_3, and so on, must be factors of a_0.

We can use this theorem to solve the equation in the next example.

EXAMPLE 18   Solve the equation $x^3 + 4x^2 - 20x - 48 = 0$.

Solution   To reduce this to a quadratic equation, we must somehow find one root.

Step 1.   All integral roots are factors of the constant term -48. The possible factors of the constant term are: $\pm 1, \pm 2, \pm 3, \pm 4, \pm 6, \pm 8, \pm 12, \pm 16, \pm 24, \pm 48$.

Step 2.   We can use synthetic division to test each of the factors as possible roots. Proceeding from left to right, we soon discover that -2 is a root:

$$
\begin{array}{rrrr|r}
1 & 4 & -20 & -48 & \underline{-2} \\
  & -2 & -4 & 48 & \\
\hline
1 & 2 & -24 & 0 & \\
\end{array}
$$

The resulting quotient, $x^2 + 2x - 24$, is quadratic.

Step 3.   Set the quadratic quotient equal to 0 and solve:

$$x^2 + 2x - 24 = 0$$

$$(x + 6)(x - 4) = 0$$

The roots are $x = -6$ and $x = 4$, so we now have all three roots: $-2, -6, 4$.

Step 4.   Check your answer by substituting each root back into the original equation.

Rational roots   Unfortunately, the highest-power term of a polynomial may not always have a coefficient of 1. Therefore, a similar technique must be developed for solving equations involving polynomials such as

$$a_n x^n + a_{n-1} x^{n-1} + \cdots + a_0 = 0 \qquad \text{where} \quad a_n \neq 1, a_n \neq 0$$

If we divide each term by a_n, we have

$$x^n + \frac{a_{n-1}}{a_n} x^{n-1} + \cdots + \frac{a_0}{a_n} = 0$$

and because x^n now has a coefficient of 1, we can apply the previous theorem to this equation and state a more general theorem. The roots of the original equation are now factors of a_0/a_n, the constant term a_0 divided by the highest-power coefficient a_n. Since this is a rational number but not necessarily an integer, the following theorem applies to rational roots:

For the polynomial equation

$$f(x) = a_n x^n + a_{n-1} x^{n-1} + \cdots + a_1 x + a_0 = 0$$

every rational root has the form

$$\frac{\text{Integer factor of } a_0}{\text{Integer factor of } a_n}$$

EXAMPLE 19   Solve the equation $2x^3 - 5x^2 - 15x - 6 = 0$.

Solution
Step 1.   List all possible rational roots. Since the integer factors of -6 are ± 1, ± 2, ± 3, and ± 6, and the integer factors of 2 are ± 1 and ± 2, then the possible rational roots must be: $\pm 1, \pm 2, \pm 3, \pm 6, \pm \dfrac{1}{2}, \pm \dfrac{3}{2}$.

Step 2.   Use synthetic division to test the factors as possible roots. Repeat this until a remainder of 0 is obtained. Proceeding from left to right, we first find that $-\frac{1}{2}$ is a root.

$$
\begin{array}{rrrr|r}
2 & -5 & -15 & -6 & \;-\dfrac{1}{2} \\[2mm]
  & -1 & 3 & 6 & \\
\hline
2 & -6 & -12 & 0 &
\end{array}
$$

The remaining quotient, $2x^2 - 6x - 12$, is quadratic.

Step 3.   Solve the corresponding quadratic equation:

$$2x^2 - 6x - 12 = 0$$

$$x^2 - 3x - 6 = 0$$

Use the quadratic formula:

$$x = \frac{3 \pm \sqrt{9 - 4(1)(-6)}}{2} = \frac{3 \pm \sqrt{33}}{2}$$

In decimal form, $x \approx 4.37$ or $x \approx -1.37$. We now have all three roots: $-\frac{1}{2}$, 4.37, -1.37. Notice that only one of the roots is rational.

Step 4. Check.

LEARNING HINT ▶ Through more advanced mathematics, it can be shown that the sum of the roots of the polynomial equation

$$a_n x^n + a_{n-1} x^{n-1} + \cdots + a_0 = 0$$

is equal to $-a_{n-1}/a_n$. This fact provides a quick, alternate way of checking your answer. For Example 19,

$$-\frac{a_{n-1}}{a_n} = -\left(\frac{-5}{2}\right) = \frac{5}{2}$$

and the sum of the roots is

$$-\frac{1}{2} + 4.37 + (-1.37) = \frac{5}{2} \quad ◀$$

Descartes' rule of signs   While the theorem concerning rational roots certainly reduces the amount of trial and error needed to find these roots, there still may be a large number of possibilities to test. Two additional rules can be used to reduce these possibilities. The first of these is **Descartes' rule of signs,** which states: For any polynomial equation $f(x) = 0$, the **maximum number of real positive roots** is equal to the number of sign changes in the terms of $f(x)$. The **maximum number of negative real roots** is equal to the number of sign changes in the terms of $f(-x)$.

EXAMPLE 20   Find the maximum number of positive roots and the maximum number of negative roots of the equation:

$$x^5 - 3x^4 - 2x^3 + 7x^2 + 6x - 4 = 0$$

Solution   There are *three* sign changes in $f(x)$:

$$f(x) = x^5 \underset{①}{-} 3x^4 \underset{②}{-} 2x^3 + 7x^2 \underset{③}{+} 6x - 4$$

Therefore, at most, there are three positive roots.
   There are *two* sign changes in $f(-x)$:

$$f(-x) = (-x)^5 - 3(-x)^4 - 2(-x)^3 + 7(-x)^2 + 6(-x) - 4$$

$$= -x^5 - 3x^4 \underset{①}{+} 2x^3 + 7x^2 \underset{②}{-} 6x - 4$$

There are no more than two negative roots.

CAUTION  Because there may be imaginary roots, you cannot find the number of negative roots by subtracting the number of positive roots from the number of total roots. ◀

The next example shows how to use this information to help solve an equation.

EXAMPLE 21   Solve the equation $2x^4 + 9x^3 - 11x^2 - 78x - 72 = 0$.

Solution
Step 1. Establish the maximum number of positive roots and the maximum number of negative roots:

$$f(x) = 2x^4 + 9x^3 - 11x^2 - 78x - 72$$

①

One sign change means that there is at most one positive root. Now,

$$f(-x) = 2x^4 - 9x^3 - 11x^2 + 78x - 72$$

①　　②　③

Three sign changes indicate that there are a maximum of three negative roots.

Step 2. List all possible rational roots. Dividing integral factors of -72 by integral factors of 2, we have: $\pm1, \pm2, \pm3, \pm4, \pm6, \pm8, \pm9, \pm12, \pm18, \pm24, \pm36, \pm72, \pm\frac{1}{2}, \pm\frac{3}{2}, \pm\frac{9}{2}$.

Step 3. To reduce the equation to a quadratic, we must find two roots by trial and error. The number of possibilities may seem overwhelming, but if we can find the one positive root (assuming it does exist), then we will have to try only negative possibilities for a second root. Testing the positive possibilities, we find that $x = 3$ is a root:

$$
\begin{array}{rrrrr|r}
2 & 9 & -11 & -78 & -72 & \underline{3} \\
  & 6 & 45 & 102 & 72 & \\
\hline
2 & 15 & 34 & 24 & 0 &
\end{array}
$$

Now we test only the negative possibilities on the resulting quotient, $2x^3 + 15x^2 + 34x + 24$, and we quickly discover that -2 is a root.

$$
\begin{array}{rrrr|r}
2 & 15 & 34 & 24 & \underline{-2} \\
  & -4 & -22 & -24 & \\
\hline
2 & 11 & 12 & 0 &
\end{array}
$$

Step 4. The resulting quotient is quadratic, and the remaining roots can be found by solving in the usual way:

$$2x^2 + 11x + 12 = 0$$

$$(2x + 3)(x + 4) = 0$$

$$x = -\frac{3}{2} \quad \text{or} \quad x = -4$$

The four roots are 3, -2, $-\frac{3}{2}$, and -4.

Step 5.   These roots all satisfy the original equation.

Maximum and minimum values

If, in Example 21, we had used a polynomial equation with two roots of each sign, and if the first successful possibility was far down the list, solving the problem would still require a great deal of testing. A two-part rule that can further simplify the trial and error process is stated below:

- If the test of a possible **positive** root is unsuccessful but yields a quotient with only **positive** coefficients, there cannot be a root any **larger** than the one tested.
- If the test of a possible **negative** root is unsuccessful but yields a quotient with **alternating signs,** there cannot be a root any **smaller** than the one tested.

EXAMPLE 22   Solve the equation $3x^4 + x^3 - 17x^2 - 29x - 30 = 0$.

Solution

Step 1.   Determine the maximum number of positive roots and the maximum number of negative roots.

$$f(x) = 3x^4 + x^3 - 17x^2 - 29x - 30$$

①   No more than one positive root

$$f(-x) = 3x^4 - x^3 - 17x^2 + 29x - 30$$

①   ②   ③   No more than three negative roots

Step 2.   List the possible roots. Dividing the factors of -30 by the factors of 3, we have the following distinct possibilities: ± 1, ± 2, ± 3, ± 5, ± 6, ± 10, ± 15, ± 30, $\pm \frac{1}{3}$, $\pm \frac{2}{3}$, $\pm \frac{5}{3}$, $\pm \frac{10}{3}$.

Step 3.   Test the possibilities until all but two roots are found. Since there is at most one positive root, begin testing these first. With our new rules, it is best to begin in the middle of the list. Testing 6, we have

3	1	-17	-29	-30	$\underline{6}$
	18	114	582	3318	
3	19	97	553	3288	

Although 6 does not work, the coefficients of the quotient above are all positive, so we know there can be no root larger than 6. In trying those possibilities less than 6, we soon discover that 3 is the one positive root:

3	1	-17	-29	-30	$\underline{3}$
	9	30	39	30	
3	10	13	10	0	

Now we use the quotient, $3x^3 + 10x^2 + 13x + 10$, to test a negative value somewhere near the middle of the list. Testing -6, we have

$$
\begin{array}{rrrrl}
3 & 10 & 13 & 10 & \underline{\smash{|}\,-6} \\
 & -18 & 48 & -366 & \\
\hline
3 & -8 & 61 & -356 &
\end{array}
$$

Although -6 is unsuccessful, the alternating signs in the bottom row tell us that there is no root smaller than -6. Testing negative roots larger than -6, we soon find that -2 is a root:

$$
\begin{array}{rrrrl}
3 & 10 & 13 & 10 & \underline{\smash{|}\,-2} \\
 & -6 & -8 & -10 & \\
\hline
3 & 4 & 5 & 0 &
\end{array}
$$

Step 4. Set the quadratic quotient equal to 0 and solve:

$$3x^2 + 4x + 5 = 0$$

Therefore,

$$x = \frac{-4 \pm \sqrt{16 - 4(3)(5)}}{6} = \frac{-4 \pm \sqrt{-44}}{6} = \frac{-2 \pm j\sqrt{11}}{3}$$

The remaining two roots are imaginary. We now have all four solutions:

$$3, \quad -2, \quad -\frac{2}{3} + j\frac{\sqrt{11}}{3}, \quad -\frac{2}{3} - j\frac{\sqrt{11}}{3}$$

Step 5. Check these roots.

Summary   Here is a summary of the information that should be used to solve a polynomial equation of degree n.

- There are n roots, some of which may be identical.
- Any rational root must be equal to a factor of the constant term divided by a factor of the highest-power coefficient.
- The maximum number of positive roots is equal to the number of sign changes in $f(x)$. The maximum number of negative roots is equal to the number of sign changes in $f(-x)$.
- If an unsuccessful test with a positive number yields a quotient with only positive coefficients, there can be no root larger than the number tested. If an unsuccessful test with a negative number yields a quotient with alternating signs, there can be no root smaller than the number tested.
- Once all but two roots have been found by trial and error, the remaining roots can be found using the quadratic formula.
- Imaginary roots occur as conjugate pairs.

Exercises 15-4

For each equation, (a) list all possible rational roots, (b) state the maximum number of positive roots, and (c) state the maximum number of negative roots.

1. $x^3 - 3x^2 - 8x + 4 = 0$

2. $2x^4 - 7x^3 + 5x^2 - 2x + 8 = 0$

3. $3x^4 + 6x^3 - 2x^2 + x + 20 = 0$

4. $4x^5 + 3x^4 - 2x^3 + 7x^2 + 5x - 30 = 0$

Solve each equation. Round irrational roots to the nearest hundredth.

5. $x^3 - x^2 - 4x + 4 = 0$

6. $x^3 + 5x - 6 = 0$

7. $x^3 - 5x - 4 = 0$

8. $x^3 - x^2 - 7x + 3 = 0$

9. $2x^3 - 3x^2 - 18x - 8 = 0$

10. $3x^3 - 5x^2 - 16x + 12 = 0$

11. $4x^3 + 7x - 4 = 0$

12. $2x^3 - 3x^2 - 13x - 6 = 0$

13. $x^4 + x^3 - 7x^2 - x + 6 = 0$

14. $x^4 - 9x^2 - 4x + 12 = 0$

15. $x^4 - 2x^3 - 14x^2 - 7x + 10 = 0$

16. $x^4 + 3x^3 - 5x^2 - 6x - 8 = 0$

17. $2x^4 + 5x^3 - 5x^2 - 20x - 12 = 0$

18. $3x^4 - 14x^3 + 16x^2 - 2x - 3 = 0$

19. $4x^4 + 12x^3 + 9x^2 + 12x + 5 = 0$

20. $6x^4 - 10x^3 - 13x^2 + 15x + 6 = 0$

21. $x^5 - 4x^4 - 20x^3 - 10x^2 + 19x + 14 = 0$

22. $x^6 + x^5 - 12x^4 - 10x^3 + 29x^2 + 9x - 18 = 0$

23. $3x^6 + 4x^5 - 5x^4 - 5x^3 - 4x^2 + 5x + 2 = 0$

24. $2x^5 + 2x^4 - 31x^3 - 9x^2 + 82x + 24 = 0$

Solve.

25. *Civil engineering* The deflection $V(x)$ of the centroidal line of a certain beam is given by $V(x) = k(x^4 - 108x + 243)$, where x is the position along the beam. Find the position x where the deflection V is 0.

26. *Mechanical engineering* In determining the principal normal stresses in a certain material the equation below appears. Find the magnitudes S of these stresses in kips per square inch (ksi).*

$$S^3 - 6S^2 + 3S + 10 = 0$$

27. *Physics* The horizontal displacement x (in meters) of a certain object varies with time t (in seconds) according to the equation below. Find t when $x = 4$ m.

$$x = 3t^2 - t^3$$

28. *Physics* The angle θ (in radians) through which the flywheel of a certain generator turns is related to the time t (in seconds) by

$$\theta = t + 6t^3 - 2t^4$$

Find t when $\theta = 3$ rad. Round to the nearest tenth if necessary.

29. *Sheet metal technology* A rectangular piece of sheet metal was made into a box by cutting identical squares from the four corners and bending up the sides, as shown in the figure. If the piece of sheet metal originally measured 10.0 in. by 15.0 in. and the volume of the box is 63 in.³, what was the length of each side of the squares that were cut out?

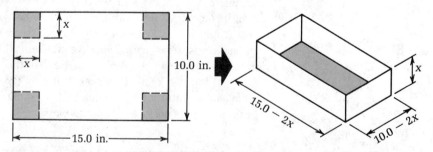

30. A rectangular box is constructed so that the width is 2 cm longer than the depth, and the length is 4 cm longer than the width. If the volume of the box is 576 cm³, what are its dimensions?

---

* Based on Nelson R. Bauld, Jr., Mechanics of Materials (Monterey, Ca.: Brooks/Cole Engineering Division, Wadsworth, 1982), p. 606.

15-5 | **Irrational Roots**

So far the only method we have for finding the irrational or imaginary roots of a polynomial equation is the quadratic formula. This means that we can completely solve such an equation only if no more than two of its roots are irrational or imaginary. In this section we shall discuss another method for finding irrational roots. This will help us find the real roots of any polynomial equation.

Locating roots graphically
We know that all real roots of the polynomial equation $f(x) = 0$ are zeros of the function $f(x)$. This suggests that we can locate the real roots of such an equation by graphing $f(x)$ and finding the x-intercept(s).

EXAMPLE 23 Given that $f(x) = x^3 - 2x^2 + 5x - 1 = 0$ has one real root, locate it by graphing.

Solution
Step 1. Prepare a table of values:

x	−1	0	1	2	3
$f(x)$	−9	−1	3	9	23

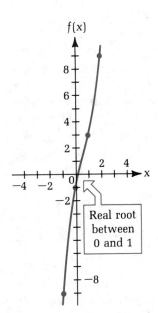

Figure 15-1

Step 2. Plot the ordered pairs and sketch the curve of the function (Fig. 15-1).

Step 3. Locate the real root of the equation by finding the x-intercept of the graph. Since this function crosses the x-axis between 0 and 1, its real root must lie in that interval.

By reexamining the table of values from Step 1, we can conclude that drawing the actual graph was unnecessary in locating the real root of the equation. The table itself shows that $f(x)$ changes sign between $x = 0$ and $x = 1$, and thus a zero of the function must lie in this interval.

CAUTION ▶ When the sign of $f(x)$ changes between $x = a$ and $x = b$, there is an *odd* number of real roots between a and b. If we had not been told that there was only one real root in Example 23, from the table we could have guessed only that there was either one or three real roots between 0 and 1 (see Fig. 15-2). Furthermore, if there is *no* sign change in $f(x)$ between $x = a$ and $x = b$, there is either *no* real root or an *even* number of real roots in the interval. For example, in Example 23 since $f(1) = 3$ and $f(2) = 9$, the graph could have crossed the x-axis twice within this interval (see Fig. 15-3). ◀

Linear interpolation
To determine a real root to a greater degree of accuracy, we use a method called **linear interpolation.** This method assumes that if two points on a graph are close together, the curve joining the points may be approximated by a straight line.

Figure 15-2

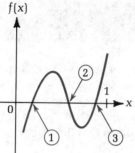

Figure 15-3

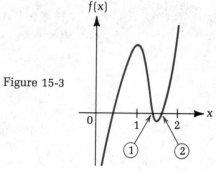

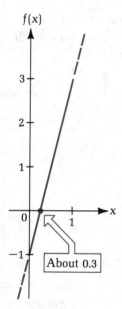

Figure 15-4

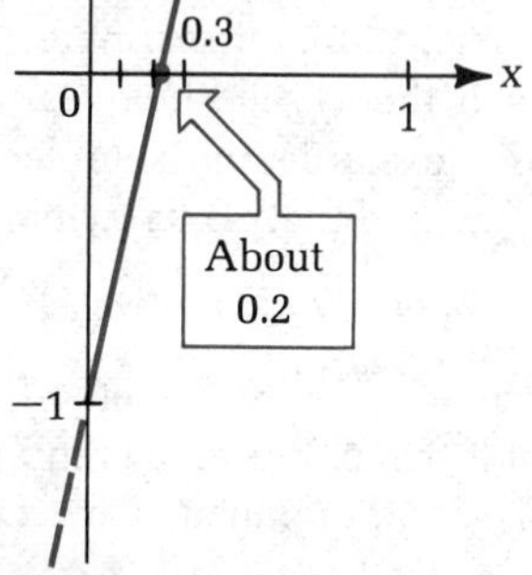

EXAMPLE 24   Find the root in Example 23 to the nearest hundredth.

Solution

Step 1.   Make an approximation within the given interval and test it. The table from Step 1 of Example 23 shows that $f(0) = -1$ and $f(1) = 3$. Using linear interpolation, we assume that the curve between these two points approximates a straight line (Fig. 15-4). Therefore, the root is probably closer to 0 than to 1. We make an estimate of 0.3 and find $f(0.3)$ using synthetic division. Using a calculator, we have the following result:

```
1   -2      5      -1     | 0.3
        0.3   -0.51   1.347
   _____
1   -1.7   4.49    0.347
```

Step 2.   Establish a new interval and make another approximation. Repeat this step until the desired degree of precision is obtained.

We now know that $f(0) = -1$, $f(0.3) = 0.347$, and $f(1) = 3$. Since a sign change occurs between $f(0.3)$ and $f(0)$, the root must lie between 0.3 and 0 (Fig. 15-5). Furthermore, since 0.347 is closer to 0 than is -1, the root should be closer to 0.3. Testing **0.2**, we have the following:

```
1   -2      5      -1     | 0.2
        0.2   -0.36   0.928
   _____
1   -1.8   4.64   -0.072
```

We now know that $f(0.2) = -0.072$ and $f(0.3) = 0.347$. The sign change tells us that the root lies between 0.2 and 0.3 (Fig. 15-6). The size of the remainders indicates that the root is much closer to 0.2. Testing **0.21** and **0.22**, we have the following:

```
1   -2        5        -1        | 0.21
        0.21   -0.3759   0.971061
   _____
1   -1.79    4.6241   -0.028939
```

```
1   -2        5        -1        | 0.22
        0.22   -0.3916   1.013848
   _____
1   -1.78    4.6084    0.013848
```

Figure 15-5

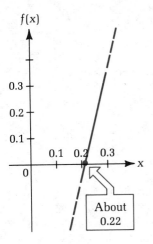

$f(x)$

0.3

0.2

0.1

0.1 0.2 0.3

0

→ x

About
0.22

Figure 15-6

The sign change tells us that the root lies between these values, and the remainders indicate that the root is probably closer to 0.22.

Step 3. To verify this, we make a final test halfway between 0.21 and 0.22 and find that $f(0.215) \approx -0.0075$. Because this is negative, the sign change occurs between 0.215 and 0.22, confirming that the root is closer to 0.22 than to 0.21. Therefore, $x \approx 0.22$, rounded to the nearest hundredth.

Step 4. Check this root by substituting it back into the original equation.

LEARNING HINT ▶

Through advanced mathematics, it can be shown that if a sign change occurs between $f(a) = b$ and $f(c) = d$, then the best linear approximation of the root is given by

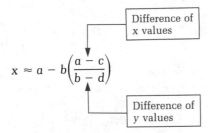

Difference of x values

$$x \approx a - b\left(\frac{a - c}{b - d}\right)$$

Difference of y values

This is often expressed in determinant form (see Chapter 5) as

$$x \approx \frac{\begin{vmatrix} a & b \\ c & d \end{vmatrix}}{d - b}$$

For Example 24, since

$$f(0) = -1 \qquad \text{and} \qquad f(0.3) = 0.347$$

a b c d

then

$$x \approx 0 - (-1)\left(\frac{0 - 0.3}{-1 - 0.347}\right) \approx \frac{-0.3}{-1.347} \approx 0.2227171 \ldots$$

Rounded to the nearest hundredth, this agrees with the result found through three additional synthetic divisions. ◀

Here is another example illustrating linear interpolation using a calculator.

EXAMPLE 25 To the nearest hundredth, find the irrational root of $2x^4 + 3x^3 + 5x - 2 = 0$ that lies between -3 and -2.

Solution

Step 1.   Use $f(-3)$ and $f(-2)$ to make a first approximation.

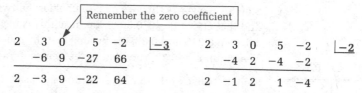

Remember the zero coefficient

$$
\begin{array}{rrrrrr|r}
2 & 3 & 0 & 5 & -2 & & -3 \\
  & -6 & 9 & -27 & 66 & \\
\hline
2 & -3 & 9 & -22 & 64 &
\end{array}
\qquad
\begin{array}{rrrrr|r}
2 & 3 & 0 & 5 & -2 & -2 \\
  & -4 & 2 & -4 & -2 & \\
\hline
2 & -1 & 2 & 1 & -4 &
\end{array}
$$

Step 2.   Make a new approximation to bracket the root more closely. Repeat this step until the desired degree of precision is obtained.

Since $f(-3) = 64$ and $f(-2) = -4$, the root is probably much closer to -2 than to -3 (Fig. 15-7). Testing -2.1 with a calculator, we have

2.1 $\boxed{+/-}$ $\boxed{\text{STO}}$ $\boxed{\times}$ 2 $\boxed{+}$ 3 $\boxed{=}$ $\boxed{\times}$ $\boxed{\text{RCL}}$ $\boxed{\times}$ $\boxed{\text{RCL}}$ $\boxed{+}$ 5 $\boxed{=}$ $\boxed{\times}$ $\boxed{\text{RCL}}$ $\boxed{-}$ 2 $\boxed{=}$ $\rightarrow$

-1.3868

We now know that $f(-3) = 64$, $f(-2.1) \approx -1.39$, and $f(-2) = -4$. The sign change between $f(-3)$ and $f(-2.1)$ means that the root lies between -3 and -2.1. The remainders indicate that the root is much closer to -2.1. Testing -2.2,

2.2 $\boxed{+/-}$ $\boxed{\text{STO}}$ $\boxed{\times}$ 2 $\boxed{+}$ 3 $\boxed{=}$ $\boxed{\times}$ $\boxed{\text{RCL}}$ $\boxed{\times}$ $\boxed{\text{RCL}}$ $\boxed{+}$ 5 $\boxed{=}$ $\boxed{\times}$ $\boxed{\text{RCL}}$ $\boxed{-}$ 2 $\boxed{=}$ $\rightarrow$

1.9072

The root lies between -2.2 and -2.1. Noting the relative magnitudes of the remainders, we then test -2.14 and -2.15 and find that $f(-2.14) \approx -0.1555597$ and $f(-2.15) = 0.1698875$. The root lies between -2.14 and -2.15, and it appears to be closer to -2.14.

Step 3.   Having reached our desired degree of precision, we make one final test midway between -2.14 and -2.15 and discover that $f(-2.145) \approx 0.0062662$. The sign change indicates that the root lies between -2.14 and -2.145, and thus $x \approx -2.14$ to the nearest hundredth.

We can now combine the techniques for finding irrational roots with those for finding rational roots in order to determine all the real roots of a polynomial equation.

EXAMPLE 26   Find all three real roots of the equation

$$x^5 - x^4 - 5x^3 - x^2 + x + 5 = 0$$

Round irrational roots to the nearest hundredth.

Solution   **First,** try to find any rational roots. Using the techniques discussed in Section 15-4, we test the possibilities ± 5 and ± 1 and find that 1 is the only rational root:

$$
\begin{array}{rrrrrr|r}
1 & -1 & -5 & -1 & 1 & 5 & 1 \\
  & 1 & 0 & -5 & -6 & -5 & \\
\hline
1 & 0 & -5 & -6 & -5 & 0 &
\end{array}
$$

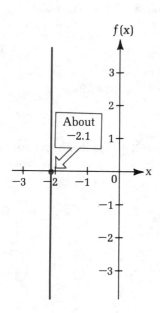

$f(x)$

About -2.1

Figure 15-7

Then, try to find any irrational roots. Using the quotient from the successful test above, we now can solve

$$f(x) = x^4 - 5x^2 - 6x - 5 = 0$$

Prepare a table of values as follows:

x	−3	−2	−1	0	1	2	3	4
f(x)	49	3	−3	−5	−15	−21	13	147

Sign changes

The sign changes indicate irrational roots between −2 and −1 and between 2 and 3 (Fig. 15-8). Using linear interpolation, we find these roots to be x ≈ −1.79 and x ≈ 2.79. Thus, the real roots are 1, −1.79, and 2.79. Since the equation has a total of five roots, it must have two imaginary roots. This is reasonable, since we know such roots must occur in pairs.

Check these roots.

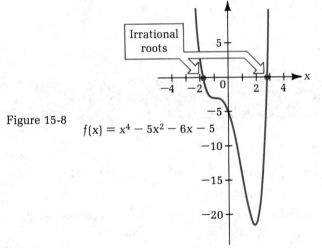

Figure 15-8

$$f(x) = x^4 - 5x^2 - 6x - 5$$

Exercises 15-5

Use linear interpolation to find the irrational root located between the given values of x to the nearest hundredth.

1. $x^3 + 2x^2 + 3x - 4 = 0$; 0 and 1
3. $x^3 + 3x^2 + 3x - 4 = 0$; 0 and 1
5. $2x^3 + 3x^2 - 7x + 3 = 0$; −3 and −2

2. $x^3 - 4x^2 + 7x - 2 = 0$, 0 and 1
4. $x^3 - 5x^2 + 2x - 3 = 0$; 4 and 5
6. $3x^3 - x^2 - 12x + 8 = 0$; −3 and −2

7. $x^4 + 5x^3 - 6x^2 - 2x - 7 = 0$;   1 and 2
8. $x^4 - x^3 + 2x^2 - 6x - 12 = 0$;   2 and 3
9. $3x^4 - 2x^3 - 2x^2 - 4x - 3 = 0$;   1 and 2
10. $2x^4 + 3x^3 - x^2 + x + 1 = 0$;   −1 and −2
11. $x^5 - 2x^3 - 3x^2 + 2x + 3 = 0$;   −1 and 0
12. $x^5 - 3x^4 - 8x^2 + x - 2 = 0$;   3 and 4
13. $x^3 - 2.41x^2 - 4.87x - 3.95 = 0$;   3 and 4
14. $x^3 + 7.21x^2 - 6.12x - 4.97 = 0$;   1 and 2

The number in parentheses tells the number of real roots to each equation. Find the real roots to the nearest hundredth.

15. $x^3 + 2x^2 + 2x - 8 = 0$ (1)
16. $3x^3 + 5x^2 - 2x + 6 = 0$ (1)
17. $x^4 + x^3 - 4x^2 - 29x - 35 = 0$ (2)
18. $x^5 + x^4 + 2x^3 - x^2 - 21x + 18 = 0$ (3)

Solve. Round to the nearest hundredth if necessary.

19. *Petroleum engineering*   The pressure drop P (in pounds per square inch) in a certain oil reservoir as a function of time t (in years) is approximated by the equation

$$P = 140t - 15t^2 + t^3$$

Find the time for which the pressure drop is 500 lb/in.².

20. *Civil engineering*   To determine the location of maximum deflection of a certain beam, the equation $5x^4 - 6L^2x^2 + L^4 = 0$ is used.* Find x (in feet) when L is 2.0 ft. The solution must be a positive number less than 2.

21. *Industrial design*   The length of a rectangular container is 5.00 m more than its width. The depth is 3.00 m less than the width. If the volume of the container is 200.0 m³, find its dimensions.

22. *Industrial design*   The interior of a rectangular packing crate measures 9.0 in. by 12.0 in. by 15.0 in. How thick must the walls be if the exterior volume of the crate is 2600 in.³? (Round this answer to the nearest tenth.)

15-6 | Review of Chapter 15

Important Terms and Concepts

polynomial function (p. 516)

the remainder theorem (p. 517)

the factor theorem (p. 518)

factor of a polynomial (p. 518)

zero of a function (p. 518)

root of an equation (p. 518)

synthetic division (p. 520)

fundamental theorem of algebra (p. 525)

multiple roots (p. 526)

imaginary roots (p. 527)

Descartes' rule of signs (p. 532)

maximum and minimum values (p. 534)

linear interpolation (p. 537)

Summary

• For a summary of the information that should be used to solve a polynomial equation of degree n, see p. 535.

---

*Based on Nelson R. Bauld, Jr., *Mechanics of Materials* (Monterey, Ca.: Brooks/Cole Engineering Division, Wadsworth, 1982), p. 386.

Exercises 15-6

Find the remainder of each division.

1. $(x^3 - 3x^2 - 7x + 5) \div (x - 3)$ **2.** $(4x^3 + 7x^2 - 13x - 15) \div (x + 2)$
3. $(2x^4 - 3x^2 - 5x + 2) \div (x - 1)$ **4.** $(x^4 + 3x^3 - 3x^2 + 3x - 4) \div (x + 4)$

Determine whether the second expression is a factor of the first.

5. $3x^3 + 4x^2 - x + 6$; $x + 2$ **6.** $x^3 + 2x^2 - 5x + 6$; $x + 3$
7. $x^4 - 8x^3 + 3x^2 + 24x + 120$; $x - 6$ **8.** $4x^4 - 10x^3 + 7x + 3$; $2x - 3$

Determine whether the given number is a zero of the function.

9. $f(x) = 2x^3 - 5x^2 + 2$; $\frac{1}{2}$ **10.** $f(x) = x^4 + 7x^3 + 10x^2 - 4x + 6$; -3
11. $f(x) = x^3 - 9x^2 + 18x + 10$; 5 **12.** $f(x) = 4x^4 - 18x^3 - 64x^2 - 72x - 37$; 6.5

Determine whether the given value is a root of the equation.

13. $x^3 + 5x^2 + 6x + 8 = 0$; -4 **14.** $2x^4 + 9x^3 + 8x^2 - x + 10 = 0$; $-\frac{5}{2}$
15. $x^4 - 6x^2 - 8x + 3 = 0$; 3 **16.** $8x^3 + 78x^2 + 83x - 125 = 0$; -8.25

Use synthetic division to divide.

17. $(x^3 - 6x^2 + 5x - 9) \div (x - 3)$ **18.** $(x^3 - 4x - 2) \div (x + 2)$
19. $(4x^3 + 9x^2 - 9x - 3) \div (x + 4)$ **20.** $(13x^3 - 230x^2 + 315x + 525) \div (x - 16)$
21. $(x^4 - 3x^3 - 8x^2 + 7x + 3) \div (x - 1)$ **22.** $(2x^4 - 6x^3 - 5x^2 + 7x + 3) \div (x - 2)$
23. $(x^5 - 9x^3 - 13x^2 - 20x + 2) \div (x - 4)$ **24.** $(x^6 + 4x^5 - 3x^4 + 7x^3 - 12x^2 + 28x + 6) \div (x + 5)$

The numbers given in parentheses are roots of the given equations. Find the remaining roots. Round irrational roots to the nearest hundredth.

25. $x^3 - 4x^2 - 20x + 48 = 0$ (2) **26.** $x^3 - 4x^2 - 17x + 12 = 0$ (-3)
27. $2x^3 - 5x^2 - 6x - 24 = 0$ (4) **28.** $4x^3 - 8x^2 - 37x + 20 = 0$ $\left(-\frac{5}{2}\right)$
29. $x^4 + 2x^3 - 3x^2 - 4x + 4 = 0$ (1 is a double root)   **30.** $3x^4 + 2x^3 - 2x^2 + 4x - 16 = 0$ $\left(\frac{4}{3}, -2\right)$
31. $x^5 + x^4 + x^3 + x^2 - 12x - 12 = 0$ **32.** $6x^5 - 5x^4 - 93x^3 - 4x^2 + 252x + 144 = 0$
     $(2j, -1)$ $\left(-\frac{2}{3}, 2, -\frac{3}{2}\right)$

Solve each equation.

33. $x^3 + 3x^2 - x - 3 = 0$ **34.** $x^3 - x^2 - 11x - 4 = 0$
35. $2x^3 + 9x^2 - 4x + 5 = 0$ **36.** $6x^3 + 13x^2 - 3x - 10 = 0$
37. $x^4 - 3x^3 - 26x^2 + 44x + 24 = 0$ **38.** $3x^4 + 2x^3 - 17x^2 - 8x + 20 = 0$
39. $4x^5 - 23x^3 - 66x^2 - 47x - 8 = 0$ **40.** $x^5 - 6x^4 + 12x^3 - 12x^2 + 11x - 6 = 0$

Find the irrational root located between the given values of x to the nearest hundredth.

41. $x^3 - x^2 - 3x + 1 = 0$;   0 and 1   **42.** $x^3 + 4x^2 - x - 2 = 0$; -1 and 0
43. $2x^3 - 3x^2 - 2x + 2 = 0$;   1 and 2   **44.** $3x^4 + 3x^3 - 2x^2 + 5x + 4 = 0$; -1 and -2

Solve. Round irrational roots to the nearest hundredth.

45.   *Physics*   The angular acceleration α of a wheel is related to the time t (in seconds) by

$$\alpha = 4t^3 - 9t^2$$

Find t for $\alpha = 3.00$ rad/s². [*Note:* There is only one positive real root.]

46.   *Mechanical engineering*   The principal normal strains for a certain state of strain can be found by solving the equation below. Find these strain values.

$$x^3 - 16x^2 + 80x - 128 = 0$$

47.   *Mechanical engineering*   The characteristic stress polynomial for a certain material is

$$S^3 - 6S^2 - 78S + 108 = 0$$

Find the irrational value of the stress that lies between 1 and 2 ksi.

48.   *Petroleum engineering*   The pressure drop P (in pounds per square inch) in a certain oil reservoir is given by

$$P = 220t - 18t^2 + t^3$$

where t is the time in years. After how much time does the pressure drop by exactly 800 lb/in.²? [*Note:* There is only one positive real root.]

49.   A rectangular piece of cardboard 12.0 in. by 20.0 in. is formed into a box by cutting a square from each corner and folding up the sides. If the volume of the box is 153 in.³, what is the length of each side of the squares that were cut out?

50.   A steel storage tank measures 10.0 in. by 12.0 in. by 18.0 in. on the outside. How thick can the walls be and still provide an interior volume of 1280 in.³? (Assume a uniform thickness.)

Determinants and Matrices

16

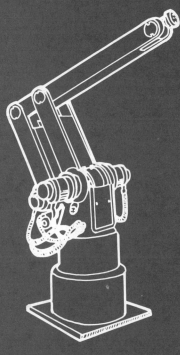

ROBOT ARM

In Chapter 5 we introduced the concept of a determinant and showed how to use determinants to solve systems of linear equations. There, we limited our discussion to second- and third-order determinants and, consequently, to systems of two and three linear equations. However, many applications involve systems of four or more linear equations. As we shall see in this chapter, these larger systems of equations can be solved using determinants. In this chapter we will also introduce matrices, a related concept that is particularly useful in solving systems of equations with a computer.

In order to solve systems of four or more linear equations using determinants, we must first develop techniques for evaluating higher-order determinants. In this section we will discuss a technique called **expansion by cofactors.** This method can be used to evaluate a determinant of any order.

Minors

The **minor** $M(a)$ of an element a of a determinant is the determinant that remains after deleting the elements in the row and column containing a. For example, in the determinant

$$\begin{vmatrix} a_1 & b_1 & c_1 \\ a_2 & b_2 & c_2 \\ a_3 & b_3 & c_3 \end{vmatrix}$$

the minor of a_1 is found by deleting the elements of the first row and first column:

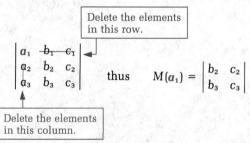

Delete the elements in this row.

$$\begin{vmatrix} a_1 & b_1 & c_1 \\ a_2 & b_2 & c_2 \\ a_3 & b_3 & c_3 \end{vmatrix} \qquad \text{thus} \qquad M(a_1) = \begin{vmatrix} b_2 & c_2 \\ b_3 & c_3 \end{vmatrix}$$

Delete the elements in this column.

Similarly, the minor of c_2 is found by deleting the second row and third column:

$$\begin{vmatrix} a_1 & b_1 & c_1 \\ a_2 & b_2 & c_2 \\ a_3 & b_3 & c_3 \end{vmatrix} \qquad \text{thus} \qquad M(c_2) = \begin{vmatrix} a_1 & b_1 \\ a_3 & b_3 \end{vmatrix}$$

EXAMPLE 1

(a) Find the minor of 4 in

$$\begin{vmatrix} 1 & 2 & 3 \\ 4 & 5 & 6 \\ 7 & 8 & 9 \end{vmatrix}$$

(b) Find the minor of -11 in

$$\begin{vmatrix} -1 & -2 & -3 & -4 \\ -5 & -6 & -7 & -8 \\ -9 & -10 & -11 & -12 \\ -13 & -14 & -15 & -16 \end{vmatrix}$$

Solution

(a) Since 4 is the element in the second row and first column, delete these:

$$\begin{vmatrix} 1 & 2 & 3 \\ 4 & 5 & 6 \\ 7 & 8 & 9 \end{vmatrix} \qquad \text{thus} \qquad M(4) = \begin{vmatrix} 2 & 3 \\ 8 & 9 \end{vmatrix}$$

(b) Since -11 is the element in the third row and third column, delete these:

$$\begin{vmatrix} -1 & -2 & -3 & -4 \\ -5 & -6 & -7 & -8 \\ -9 & -10 & -11 & -12 \\ -13 & -14 & -15 & -16 \end{vmatrix} \qquad \text{thus} \qquad M(-11) = \begin{vmatrix} -1 & -2 & -4 \\ -5 & -6 & -8 \\ -13 & -14 & -16 \end{vmatrix}$$

Cofactors The **cofactor** of an element is simply its minor with a sign attached. This sign is found as follows:

1. Determine the numbers of the row and the column in which the element lies and add these numbers.
2. If this sum is *even*, the sign of the cofactor is *positive*; if the sum is *odd*, the sign is *negative*.

LEARNING HINT ▶ Another way to find the sign of the cofactor of an element is to memorize the following array of signs:

$$\begin{vmatrix} + & - & + \\ - & + & - \\ + & - & + \end{vmatrix}$$

The "+" sign in the upper left-hand corner means that the element in Row 1, Column 1 has a positive cofactor. Then, reading either across or down, the signs of the cofactors alternate with the corresponding elements. This alternating pattern can be extended to a determinant of any order. ◀

EXAMPLE 2 In the determinant below find the cofactor of: (a) 7   (b) 4

$$\begin{vmatrix} 9 & 8 & 7 \\ 6 & 5 & 4 \\ 3 & 2 & 1 \end{vmatrix}$$

Solution

(a) **Step 1.** Determine the sign of the cofactor:

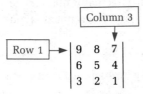

The element 7 is located in Row 1, Column 3. The sum of the row and column numbers is $1 + 3 = 4$, an even number; therefore, the sign of the cofactor is *positive*.

Step 2. Attach this sign to the minor of the given element.

Eliminating the row and column containing 7, we have:

$$\begin{vmatrix} 9 & 8 & 7 \\ 6 & 5 & 4 \\ 3 & 2 & 1 \end{vmatrix}$$

Attaching the proper sign to the minor gives us the cofactor of 7.

$$+ \begin{vmatrix} 6 & 5 \\ 3 & 2 \end{vmatrix}$$

(b) **Step 1.**   The element 4 is located in Row 2, Column 3:

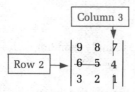

The sum of the row and column numbers is 5, an *odd* number. Thus, the sign of the cofactor is *negative*.

Step 2.   We now find the minor of 4 and attach the sign from Step 1. The cofactor of 4 is

$$- \begin{vmatrix} 9 & 8 \\ 3 & 2 \end{vmatrix}$$

Expansion by cofactors

To see how cofactors can be used to evaluate determinants, consider the expansion of the general third-order determinant as stated in Section 5-6:

$$\begin{vmatrix} a_1 & b_1 & c_1 \\ a_2 & b_2 & c_2 \\ a_3 & b_3 & c_3 \end{vmatrix} = a_1 b_2 c_3 + a_3 b_1 c_2 + a_2 b_3 c_1 - a_3 b_2 c_1 - a_1 b_3 c_2 - a_2 b_1 c_3 \quad \textbf{(16-1)}$$

If we rearrange the terms on the right and factor a_1, $-b_1$, and c_1 from the terms containing them, we obtain

$$a_1 (b_2 c_3 - b_3 c_2) - b_1 (a_2 c_3 - a_3 c_2) + c_1 (a_2 b_3 - a_3 b_2)$$

Now each binomial represents the expansion of a second-order determinant:

$$a_1 \begin{vmatrix} b_2 & c_2 \\ b_3 & c_3 \end{vmatrix} - b_1 \begin{vmatrix} a_2 & c_2 \\ a_3 & c_3 \end{vmatrix} + c_1 \begin{vmatrix} a_2 & b_2 \\ a_3 & b_3 \end{vmatrix}$$

Notice that this expression is actually the sum of products in which each element in Row 1 of the original determinant is multiplied by its cofactor.

Cofactor Expansion of a Third-Order Determinant

$$\begin{vmatrix} a_1 & b_1 & c_1 \\ a_2 & b_2 & c_2 \\ a_3 & b_3 & c_3 \end{vmatrix} = a_1 \underbrace{\begin{vmatrix} b_2 & c_2 \\ b_3 & c_3 \end{vmatrix}}_{\substack{\text{Cofactor} \\ \text{of } a_1}} + b_1 \underbrace{\left(- \begin{vmatrix} a_2 & c_2 \\ a_3 & c_3 \end{vmatrix} \right)}_{\substack{\text{Cofactor} \\ \text{of } b_1}} + c_1 \underbrace{\begin{vmatrix} a_2 & b_2 \\ a_3 & b_3 \end{vmatrix}}_{\substack{\text{Cofactor} \\ \text{of } c_1}} \quad \textbf{(16-2)}$$

Notice that the second cofactor is negative while the first and third are positive.

Actually, the terms in (16-1) could have been rearranged and factored in any one of six different ways. For example, if we factor $-b_1$, b_2, and $-b_3$ from the terms containing them, the value of this same determinant would be

$$\begin{vmatrix} a_1 & b_1 & c_1 \\ a_2 & b_2 & c_2 \\ a_3 & b_3 & c_3 \end{vmatrix} = b_1\left(-\begin{vmatrix} a_2 & c_2 \\ a_3 & c_3 \end{vmatrix}\right) + b_2\begin{vmatrix} a_1 & c_1 \\ a_3 & c_3 \end{vmatrix} + b_3\left(-\begin{vmatrix} a_1 & c_1 \\ a_2 & c_2 \end{vmatrix}\right)$$

This is similar to (16-2), but now each term is the product of an element from Column 2 and its cofactor. By determining all six possibilities, we find that a third-order determinant can be expanded in this manner using *any* row or column. This leads to the following general procedure for expanding determinants of any order:

*The value of a determinant of any order is the sum of the products obtained when each element in any row or column is multiplied by its cofactor.**

EXAMPLE 3   Evaluate the determinant below using cofactors and expanding in terms of:
(a)   Row 1   (b)   Column 2

$$\begin{vmatrix} 2 & 3 & -1 \\ 4 & 0 & 5 \\ -5 & 0 & 1 \end{vmatrix}$$

Solution

(a)   **Step 1.**   Write the sum of the products of each element of Row 1 by its cofactor:

$$\begin{vmatrix} 2 & 3 & -1 \\ 4 & 0 & 5 \\ -5 & 0 & 1 \end{vmatrix} = 2\begin{vmatrix} 0 & 5 \\ 0 & 1 \end{vmatrix} - 3\begin{vmatrix} 4 & 5 \\ -5 & 1 \end{vmatrix} + (-1)\begin{vmatrix} 4 & 0 \\ -5 & 0 \end{vmatrix}$$

Row 1, Column 2:
$1 + 2 = 3$ (odd),
so the sign is negative

Step 2.   Simplify (remember that $\begin{vmatrix} a & b \\ c & d \end{vmatrix} = ad - cb$):

$$= 2(0\cdot 1 - 0\cdot 5) - 3[4\cdot 1 - (-5)5] + (-1)[4\cdot 0 - (-5)0]$$

$$= 2(0) - 3(29) + (-1)(0)$$

$$= -87$$

(b)   **Step 1.**   Using the second column, we have

$$\begin{vmatrix} 2 & 3 & -1 \\ 4 & 0 & 5 \\ -5 & 0 & 1 \end{vmatrix} = -3\begin{vmatrix} 4 & 5 \\ -5 & 1 \end{vmatrix} + 0\begin{vmatrix} 2 & -1 \\ -5 & 1 \end{vmatrix} - 0\begin{vmatrix} 2 & -1 \\ 4 & 5 \end{vmatrix}$$

Notice that we have only one nonzero term when we expand by Column 2.

*The cofactor method of evaluating determinants is the procedure used in computer programs.

Step 2. Simplifying:

$$= -3[4\cdot1 - (-5)5] = -3(29) = -87$$

As expected, the two answers agree. However, expanding by Column 2 proved to be much easier since it contained two zero elements. This observation can be especially helpful in evaluating determinants of a higher order and in checking your work.

EXAMPLE 4 Evaluate:
$$\begin{vmatrix} 2 & -1 & 5 & 4 \\ 3 & 1 & 7 & -3 \\ 0 & 6 & 0 & -2 \\ 4 & 1 & -5 & 3 \end{vmatrix}$$

Solution

Step 1. Since Row 3 contains the most 0's, it should be used to expand the determinant.

$$\begin{vmatrix} 2 & -1 & 5 & 4 \\ 3 & 1 & 7 & -3 \\ 0 & 6 & 0 & -2 \\ 4 & 1 & -5 & 3 \end{vmatrix} = -6\begin{vmatrix} 2 & 5 & 4 \\ 3 & 7 & -3 \\ 4 & -5 & 3 \end{vmatrix} - (-2)\begin{vmatrix} 2 & -1 & 5 \\ 3 & 1 & 7 \\ 4 & 1 & -5 \end{vmatrix}$$

Step 2. Expand each minor by cofactors:

Using Row 1

$$= -6\left[2\begin{vmatrix} 7 & -3 \\ -5 & 3 \end{vmatrix} - 5\begin{vmatrix} 3 & -3 \\ 4 & 3 \end{vmatrix} + 4\begin{vmatrix} 3 & 7 \\ 4 & -5 \end{vmatrix}\right]$$

$$+ 2\left[-(-1)\begin{vmatrix} 3 & 7 \\ 4 & -5 \end{vmatrix} + 1\begin{vmatrix} 2 & 5 \\ 4 & -5 \end{vmatrix} - 1\begin{vmatrix} 2 & 5 \\ 3 & 7 \end{vmatrix}\right]$$

Using Column 2

Step 3. Simplify:

$$= -6[2(21 - 15) - 5(9 + 12) + 4(-15 - 28)]$$

$$+ 2[(-15 - 28) + (-10 - 20) - (14 - 15)]$$

$$= -6[12 - 105 + (-172)] + 2[-43 + (-30) - (-1)]$$

$$= 1590 + (-144) = 1446$$

Solving linear systems   As we mentioned earlier, Cramer's rule allows us to solve systems of linear equations using determinants. With our new method of expansion, we can now apply Cramer's rule to systems involving four or more linear equations. You may wish to review Sections 5-4 and 5-6 before attempting the next example.

EXAMPLE 5   Solve the following system of equations:

$$a + b + c + d = 6$$

$$2a - b - c = 1$$

$$2b + 3c = 0$$

$$3a - 4b + 2d = -1$$

Solution

Step 1.   Find D, D_a, D_b, D_c, and D_d.

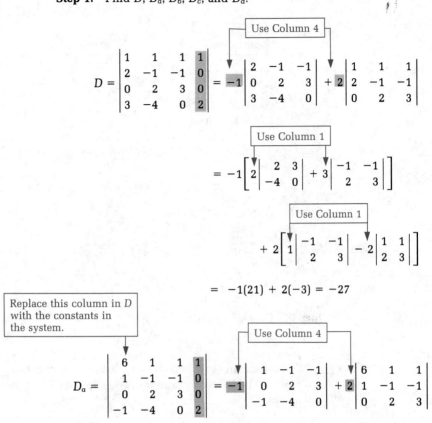

$$D = \begin{vmatrix} 1 & 1 & 1 & 1 \\ 2 & -1 & -1 & 0 \\ 0 & 2 & 3 & 0 \\ 3 & -4 & 0 & 2 \end{vmatrix} = -1\begin{vmatrix} 2 & -1 & -1 \\ 0 & 2 & 3 \\ 3 & -4 & 0 \end{vmatrix} + 2\begin{vmatrix} 1 & 1 & 1 \\ 2 & -1 & -1 \\ 0 & 2 & 3 \end{vmatrix}$$

Use Column 4

$$= -1\left[2\begin{vmatrix} 2 & 3 \\ -4 & 0 \end{vmatrix} + 3\begin{vmatrix} -1 & -1 \\ 2 & 3 \end{vmatrix} \right]$$

Use Column 1

$$+ 2\left[1\begin{vmatrix} -1 & -1 \\ 2 & 3 \end{vmatrix} - 2\begin{vmatrix} 1 & 1 \\ 2 & 3 \end{vmatrix} \right]$$

Use Column 1

$$= -1(21) + 2(-3) = -27$$

Replace this column in D with the constants in the system.

$$D_a = \begin{vmatrix} 6 & 1 & 1 & 1 \\ 1 & -1 & -1 & 0 \\ 0 & 2 & 3 & 0 \\ -1 & -4 & 0 & 2 \end{vmatrix} = -1\begin{vmatrix} 1 & -1 & -1 \\ 0 & 2 & 3 \\ -1 & -4 & 0 \end{vmatrix} + 2\begin{vmatrix} 6 & 1 & 1 \\ 1 & -1 & -1 \\ 0 & 2 & 3 \end{vmatrix}$$

Use Column 4

Expanding about Column 1 of each third-order determinant:

$$= -1[12 + 1] + 2[-6 - 1] = -27$$

Replace with constants

$$D_b = \begin{vmatrix} 1 & 6 & 1 & 1 \\ 2 & 1 & -1 & 0 \\ 0 & 0 & 3 & 0 \\ 3 & -1 & 0 & 2 \end{vmatrix} = 3\begin{vmatrix} 1 & 6 & 1 \\ 2 & 1 & 0 \\ 3 & -1 & 2 \end{vmatrix}$$

Use Row 3

Using Column 3 of the third-order determinant:

$$= 3[-5 + (-22)] = -81$$

Replace with constants

$$D_c = \begin{vmatrix} 1 & 1 & 6 & 1 \\ 2 & -1 & 1 & 0 \\ 0 & 2 & 0 & 0 \\ 3 & -4 & -1 & 2 \end{vmatrix} = -2\begin{vmatrix} 1 & 6 & 1 \\ 2 & 1 & 0 \\ 3 & -1 & 2 \end{vmatrix}$$

Use Row 3

Using Column 3:

$$= -2[-5 + (-22)] = 54$$

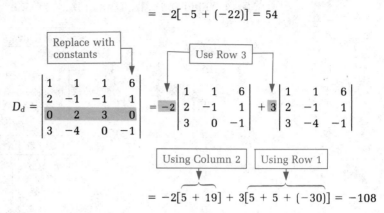

$$= -2[5 + 19] + 3[5 + 5 + (-30)] = -108$$

Step 2. Apply Cramer's rule.

$$a = \frac{D_a}{D} = \frac{-27}{-27} = 1 \qquad b = \frac{D_b}{D} = \frac{-81}{-27} = 3$$

$$c = \frac{D_c}{D} = \frac{54}{-27} = -2 \qquad d = \frac{D_d}{D} = \frac{-108}{-27} = 4$$

Step 3. Check by substituting these values into all original equations.

Applications   We have already solved applied problems involving three equations in three unknowns. The following example results in a system of four equations in four unknowns.

EXAMPLE 6   Four models of a calculator are carried by a retail store. The store manager wants the inventory to contain twenty calculators costing a total of $310. The four models, $W, X, Y,$ and $Z,$ cost $10, $15, $20, and $25, respectively. The manager wants to stock three times as many of model W as $Z,$ and twice as many of model X as $Y.$ How many of each must the store stock?

Solution   Let W, X, Y, Z represent the number of calculators of each type to be stocked. There are four unknowns, so we must have four conditions in order to solve. The store wants

1. 20 calculators total: $W + X + Y + Z = 20$
2. A total cost of $310: $10W + 15X + 20Y + 25Z = 310$
3. Three times as many of W as Z: $W = 3Z$
4. Twice as many of X as Y: $X = 2Y$

To solve this system of four equations in four unknowns, first divide every term of the second equation by 5 to obtain

$$2W + 3X + 4Y + 5Z = 62$$

Then use determinants and Cramer's rule as in Example 5 to obtain

$$D = -7 \qquad D_W = -42 \qquad D_X = -56 \qquad D_Y = -28 \qquad D_Z = -14$$

Therefore, $W = 6$, $X = 8$, $Y = 4$, and $Z = 2$.

Another way to solve this system is to substitute $3Z$ for W (equation 3) and $2Y$ for X (equation 4) into the first two equations to obtain the reduced system

$$3Y + 4Z = 20$$

$$10Y + 11Z = 62$$

Then solve using determinants or by elimination. The answers will be the same using either method.

Exercises 16-1

Evaluate each determinant using expansion by cofactors.

1. $\begin{vmatrix} 1 & 2 & 0 \\ 3 & -1 & 1 \\ 0 & 4 & -2 \end{vmatrix}$
2. $\begin{vmatrix} 5 & 0 & 1 \\ 2 & 2 & -1 \\ 0 & 3 & -2 \end{vmatrix}$
3. $\begin{vmatrix} -4 & 1 & 1 \\ 2 & 0 & 0 \\ -3 & -5 & 1 \end{vmatrix}$

4. $\begin{vmatrix} 1 & 6 & 0 \\ -2 & 0 & 4 \\ 7 & 2 & 1 \end{vmatrix}$
5. $\begin{vmatrix} 8 & 0 & -1 \\ 2 & 2 & 1 \\ 5 & -3 & 0 \end{vmatrix}$
6. $\begin{vmatrix} 1 & 2 & -3 \\ 6 & 3 & 1 \\ -1 & -2 & -4 \end{vmatrix}$

7. $\begin{vmatrix} 125 & 85 & 12 \\ 64 & 160 & 72 \\ 18 & 80 & 100 \end{vmatrix}$
8. $\begin{vmatrix} 1.6 & 2.4 & 3.8 \\ 5.1 & 3.3 & 2.7 \\ 4.4 & 7.2 & 1.6 \end{vmatrix}$
9. $\begin{vmatrix} 1 & -1 & 0 & 2 \\ 3 & 1 & 4 & 0 \\ 0 & 0 & 2 & -3 \\ 1 & 3 & 0 & -2 \end{vmatrix}$

10. $\begin{vmatrix} 2 & 0 & -3 & -1 \\ -4 & 1 & 0 & 5 \\ 1 & 3 & 4 & -1 \\ 2 & 0 & -1 & 2 \end{vmatrix}$
11. $\begin{vmatrix} -3 & 5 & 0 & 2 \\ 1 & 4 & 1 & 0 \\ -2 & 2 & 1 & 1 \\ 3 & 0 & -1 & 2 \end{vmatrix}$
12. $\begin{vmatrix} 6 & 0 & 1 & -1 \\ 3 & 1 & 5 & 2 \\ -1 & 4 & -2 & -3 \\ 0 & -5 & -6 & 1 \end{vmatrix}$

13. $\begin{vmatrix} 1 & 0 & 1 & -3 & 0 \\ 2 & 0 & -1 & -4 & -1 \\ 3 & -2 & 0 & 1 & -2 \\ 1 & 3 & 0 & -3 & 0 \\ 0 & 1 & 0 & 0 & -3 \end{vmatrix}$
14. $\begin{vmatrix} 4 & 0 & 1 & -2 & 0 \\ -3 & -2 & 2 & 3 & 0 \\ -3 & 6 & 0 & 0 & -1 \\ -1 & 1 & 1 & -4 & 0 \\ 0 & 0 & 2 & -1 & 2 \end{vmatrix}$

15. $\begin{vmatrix} 44 & 16 & 81 & 12 \\ 75 & 23 & 94 & 18 \\ 66 & 87 & 33 & 25 \\ 19 & 64 & 51 & 36 \end{vmatrix}$
16. $\begin{vmatrix} 1.3 & 5.8 & 0.6 & -2.9 \\ -2.7 & 3.1 & -1.4 & 0.9 \\ -0.5 & 4.2 & 3.6 & -1.8 \\ 3.7 & 1.4 & 4.6 & 2.6 \end{vmatrix}$

Solve each system of equations using Cramer's rule. Expand all determinants by cofactors.

17. $x + y + z = 2$
$x - 2y + 3z = -3$
$3x + 2z = 4$

18. $3x - 4y = -1$
$2y + 3z = -1$
$x - 2y - z = 0$

19. $a - 2b + c = -7$
$-2a + b - 2c = 2$
$-3a + 4b + 6c = -5$

20. $3r + 4s + 7t = 5$
$6r - 2s - 5t = -5$
$-9r + 2s - 4t = 6$

21. $2w - 5y = 7$
$4x - 3z = 12$
$x + y + z = 2$
$w + 2y - 3z = -1$

22. $3w + 2x = 2$
$w - x - 4y = 0$
$3y + z = 0$
$w + x - y + z = -4$

23. $3a + 4b - d = 0$
$2b + 3c - 2d = -2$
$6a - 2b - 5c = 4$
$-3a + 2b - 3c + 4d = 0$

24. $3r + 2s - t + u = 10$
$-r - 3s + 2t - 4u = -3$
$2r + 3s + t - u = -5$
$r + s + t + u = 2$

25. *Electronics*   Applying Kirchoff's laws to the electrical circuit shown in the figure results in the following system of equations. Solve for the three currents (in amperes).

$$5I_1 - 3I_3 = 2$$

$$2I_1 - 4I_2 + 8I_3 = 0$$

$$3I_1 - 6I_2 - 17I_3 = 0$$

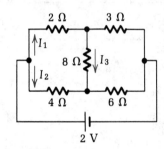

26. *Industrial management*   A research and development firm submits a $64,000 budget for a government project. The money is to be divided among four categories as follows: Supplies will cost as much as travel expenses and on-line computer time combined; salaries will cost twice as much as supplies and on-line time combined; on-line time will cost $4000 less than supplies. How much money is allotted to each category?

27. *Chemistry*   Four different compounds, A, B, C, and D, contain different percentages of ingredients X, Y, and Z, as shown in the table. How many grams of each compound should be combined to produce a 100 g mixture containing 35% X, 24% Y, and 41% Z?

INGREDIENT	A	B	C	D
X	50%	40%	20%	
Y	50%		20%	50%
Z		60%	60%	50%

28. *Business*   Three types of disk drives are carried by a computer shop. A single drive costs $300, a dual drive costs $500, and a hard-disk drive costs $1200. The manager of the store has allotted $8000 for purchasing the drives and wants to stock 15 drives, including twice as many single drives as dual drives. How many of each can be ordered?

16-2 | Important Properties of Determinants

Evaluating higher-order determinants can be tricky and time-consuming, especially if the determinant contains only nonzero elements. However, certain properties of determinants can be used to simplify this process. In this section we shall state these properties and illustrate each with an example.* Then, we shall show how a combination of the properties can be used to simplify a determinant before evaluating it.

Property 1

If all elements in a particular row or column are 0, the value of the determinant is 0.

$$\begin{vmatrix} 0 & 0 & 0 \\ d & e & f \\ g & h & i \end{vmatrix} = \begin{vmatrix} 0 & b & c \\ 0 & e & f \\ 0 & h & i \end{vmatrix} = 0$$

EXAMPLE 7 Evaluate: $\begin{vmatrix} 0 & 1 & 2 \\ 0 & 3 & 4 \\ 0 & 5 & 6 \end{vmatrix}$

Solution Property 1 tells us that since all the elements of Column 1 are 0, the value of the determinant is 0. An expansion using Column 1 shows why this is true:

$$\begin{vmatrix} 0 & 1 & 2 \\ 0 & 3 & 4 \\ 0 & 5 & 6 \end{vmatrix} = 0\begin{vmatrix} 3 & 4 \\ 5 & 6 \end{vmatrix} - 0\begin{vmatrix} 1 & 2 \\ 5 & 6 \end{vmatrix} + 0\begin{vmatrix} 1 & 2 \\ 3 & 4 \end{vmatrix} = 0$$

Property 2

If all elements above or below the principal diagonal are 0, the value of the determinant is the product of the elements of the principal diagonal.

$$\begin{vmatrix} a & 0 & 0 \\ d & e & 0 \\ g & h & i \end{vmatrix} = \begin{vmatrix} a & b & c \\ 0 & e & f \\ 0 & 0 & i \end{vmatrix} = aei$$

---

*We shall not prove these properties but rather demonstrate their validity with examples.

EXAMPLE 8   Evaluate:
$$
\begin{vmatrix}
2 & 5 & 8 \\
0 & 3 & -9 \\
0 & 0 & -5
\end{vmatrix}
$$

Solution   According to Property 2,

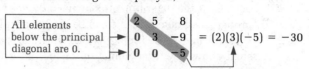

$$
\boxed{\begin{array}{l}\text{All elements} \\ \text{below the principal} \\ \text{diagonal are 0.}\end{array}} \rightarrow
\begin{vmatrix}
2 & 5 & 8 \\
0 & 3 & -9 \\
0 & 0 & -5
\end{vmatrix} = (2)(3)(-5) = -30
$$

We verify that this is true by expanding using Column 1:

$$
2\begin{vmatrix} 3 & -9 \\ 0 & -5 \end{vmatrix} = 2(-15 - 0) = -30
$$

Property 3

If two rows or two columns of a determinant are identical, the value of the determinant is 0.

$$
\begin{vmatrix}
a & b & c \\
a & b & c \\
g & h & i
\end{vmatrix} = 0
\qquad
\begin{vmatrix}
a & a & c \\
d & d & f \\
g & g & i
\end{vmatrix} = 0
$$

EXAMPLE 9   Applying Property 3,

$$
\begin{vmatrix}
2 & 3 & 1 \\
-1 & 0 & 2 \\
2 & 3 & 1
\end{vmatrix} = 0
$$

because the first and third rows are the same. Verify that the value of this determinant is 0 by expanding about Row 2:

$$
\begin{vmatrix}
2 & 3 & 1 \\
-1 & 0 & 2 \\
2 & 3 & 1
\end{vmatrix} = -(-1)\begin{vmatrix} 3 & 1 \\ 3 & 1 \end{vmatrix} - 2\begin{vmatrix} 2 & 3 \\ 2 & 3 \end{vmatrix}
$$

$$
= 1(3 - 3) - 2(6 - 6) = 0 - 0 = 0
$$

Property 4

If any two columns or any two rows of a determinant are interchanged, the value of the resulting determinant is the negative of the value of the original determinant.

$$
\begin{vmatrix}
a & b & c \\
d & e & f \\
g & h & i
\end{vmatrix} = (-1)\begin{vmatrix}
d & e & f \\
a & b & c \\
g & h & i
\end{vmatrix}
$$

EXAMPLE 10 Property 4 tells us that

$$\begin{vmatrix} 5 & -2 & 3 \\ 2 & 0 & 4 \\ 0 & 0 & -6 \end{vmatrix} = (-1) \begin{vmatrix} -2 & 5 & 3 \\ 0 & 2 & 4 \\ 0 & 0 & -6 \end{vmatrix}$$

These columns have been interchanged.

Expand each determinant to verify this property:

$$\begin{vmatrix} 5 & -2 & 3 \\ 2 & 0 & 4 \\ 0 & 0 & -6 \end{vmatrix} = 2 \begin{vmatrix} 2 & 4 \\ 0 & -6 \end{vmatrix} = 2(-12) = -24 \qquad \text{Expanding on Column 2}$$

$$\begin{vmatrix} -2 & 5 & 3 \\ 0 & 2 & 4 \\ 0 & 0 & -6 \end{vmatrix} = (-2)(2)(-6) = +24 \qquad \text{Using Property 2}$$

Property 5

If all corresponding rows and columns of a determinant are interchanged, the value of the determinant is unchanged.

$$\begin{vmatrix} a & b & c \\ d & e & f \\ g & h & i \end{vmatrix} = \begin{vmatrix} a & d & g \\ b & e & h \\ c & f & i \end{vmatrix}$$

EXAMPLE 11 Property 5 tells us that

$$\begin{vmatrix} 2 & 5 & 3 \\ 1 & -1 & 0 \\ 0 & 2 & 4 \end{vmatrix} = \begin{vmatrix} 2 & 1 & 0 \\ 5 & -1 & 2 \\ 3 & 0 & 4 \end{vmatrix}$$

Column 3 becomes Row 3
Column 2 becomes Row 2
Column 1 becomes Row 1

The second determinant was formed from the first one by interchanging Row 1 with Column 1, Row 2 with Column 2, and Row 3 with Column 3. Verify this by evaluating both determinants:

$$\begin{vmatrix} 2 & 5 & 3 \\ 1 & -1 & 0 \\ 0 & 2 & 4 \end{vmatrix} = 2 \begin{vmatrix} -1 & 0 \\ 2 & 4 \end{vmatrix} - 1 \begin{vmatrix} 5 & 3 \\ 2 & 4 \end{vmatrix} = -22$$

$$\begin{vmatrix} 2 & 1 & 0 \\ 5 & -1 & 2 \\ 3 & 0 & 4 \end{vmatrix} = 2 \begin{vmatrix} -1 & 2 \\ 0 & 4 \end{vmatrix} - 1 \begin{vmatrix} 5 & 2 \\ 3 & 4 \end{vmatrix} = -22$$

Property 6

If all the elements of a row or column are multiplied by the same nonzero number k, the value of the resulting determinant will be k times the value of the original determinant.

$$\begin{vmatrix} a & b & kc \\ d & e & kf \\ g & h & ki \end{vmatrix} = k \begin{vmatrix} a & b & c \\ d & e & f \\ g & h & i \end{vmatrix} \qquad \begin{vmatrix} ka & kb & kc \\ d & e & f \\ g & h & i \end{vmatrix} = k \begin{vmatrix} a & b & c \\ d & e & f \\ g & h & i \end{vmatrix}$$

EXAMPLE 12   According to Property 6,

$$\begin{vmatrix} 1 & -3 & -2 \\ 4 & 6 & -8 \\ 2 & 1 & 3 \end{vmatrix} = 2 \begin{vmatrix} 1 & -3 & -2 \\ 2 & 3 & -4 \\ 2 & 1 & 3 \end{vmatrix}$$

because the elements of Row 2 in the first determinant contain the common factor 2. By expanding, we show that

$$\begin{vmatrix} 1 & -3 & -2 \\ 4 & 6 & -8 \\ 2 & 1 & 3 \end{vmatrix} = 126 \qquad \text{and} \qquad 2 \begin{vmatrix} 1 & -3 & -2 \\ 2 & 3 & -4 \\ 2 & 1 & 3 \end{vmatrix} = 2(63) = 126$$

This property is useful because the factored determinant may be easier to evaluate than the original determinant.

Property 7

If the same multiple of each element of a row (or column) is added to the corresponding element of another row (or column), the value of the determinant is unchanged.

$$\begin{vmatrix} a & b & c \\ d & e & f \\ g & h & i \end{vmatrix} = \begin{vmatrix} a + kb & b & c \\ d + ke & e & f \\ g + kh & h & i \end{vmatrix}$$

EXAMPLE 13   Form a determinant equivalent to

$$\begin{vmatrix} 3 & 2 & -2 \\ -6 & -4 & 2 \\ 5 & 6 & 7 \end{vmatrix}$$

by adding 2 times each element in the first row to the corresponding element in the second row.

$$\boxed{2\cdot 3 = 6} \qquad \boxed{2\cdot 2 = 4} \qquad \boxed{2(-2) = -4}$$

Solution $\begin{vmatrix} 3 & 2 & -2 \\ -6 & -4 & 2 \\ 5 & 6 & 7 \end{vmatrix} = \begin{vmatrix} 3 & 2 & -2 \\ -6+6 & -4+4 & 2+(-4) \\ 5 & 6 & 7 \end{vmatrix}$

$$= \begin{vmatrix} 3 & 2 & -2 \\ 0 & 0 & -2 \\ 5 & 6 & 7 \end{vmatrix}$$

As you can see, the new determinant is much easier to evaluate than the original, because there are two 0's in Row 2. If you are clever, you can use this property to simplify higher-order determinants by writing an equivalent determinant containing 0's in all but one position of a row or column. In some cases, you may be able to obtain 0's in all positions below the principal diagonal so that the determinant can be evaluated using Property 2.

Properties 4–7 can be used to simplify a determinant before attempting to evaluate it. Use these to transform the determinant so that only one nonzero element remains in one row or column. Sometimes, the result will enable you to apply Property 1, 2, or 3.

EXAMPLE 14 Use the properties developed in this section to simplify and evaluate the following determinant:

$$A = \begin{vmatrix} 1 & 3 & -2 & 4 \\ 0 & 2 & -1 & -3 \\ -5 & 4 & 2 & 1 \\ 4 & 2 & 1 & -1 \end{vmatrix}$$

Solution Our goal is to reduce the order of the determinant by obtaining 0's in all but one position of a row or column. There are many ways to do this, but it is probably best to work with either Column 1 or Row 2 since they already contain a zero element. Here is a sample of how determinant A might be simplified.

Step 1. Multiply each element of Column 3 by 2, and add it to the corresponding element of Column 2. This use of Property 7 creates a 0 in Column 2, Row 2. Column 3 is unchanged.

$$A = \begin{vmatrix} 1 & -1 & -2 & 4 \\ 0 & 0 & -1 & -3 \\ -5 & 8 & 2 & 1 \\ 4 & 4 & 1 & -1 \end{vmatrix}$$

Step 2. Multiply each element of Column 3 by -3, and add this to the corresponding element of Column 4. This application of Property 7 creates a 0 in Column 4, Row 2. Column 3 is unchanged.

$$= \begin{vmatrix} 1 & -1 & -2 & 10 \\ 0 & 0 & -1 & 0 \\ -5 & 8 & 2 & -5 \\ 4 & 4 & 1 & -4 \end{vmatrix}$$

Step 3. Expand this determinant about Row 2, and reduce it to a third-order determinant.

$$= -(-1)\begin{vmatrix} 1 & -1 & 10 \\ -5 & 8 & -5 \\ 4 & 4 & -4 \end{vmatrix}$$

Step 4. Use Property 6 to factor 4 from each element of Row 3.

$$= 4\begin{vmatrix} 1 & -1 & 10 \\ -5 & 8 & -5 \\ 1 & 1 & -1 \end{vmatrix}$$

Step 5. Use Property 7 to add the elements of Column 2 to Column 3. This produces a 0 in Row 3, Column 3.

$$= 4\begin{vmatrix} 1 & -1 & 9 \\ -5 & 8 & 3 \\ 1 & 1 & 0 \end{vmatrix}$$

Step 6. Multiply the elements of Column 1 by -1, and add them to the corresponding elements of Column 2. This produces a 0 in Row 3, Column 2.

$$= 4\begin{vmatrix} 1 & -2 & 9 \\ -5 & 13 & 3 \\ 1 & 0 & 0 \end{vmatrix}$$

Step 7. Expand the determinant about Row 3.

$$= 4(1)\begin{vmatrix} -2 & 9 \\ 13 & 3 \end{vmatrix}$$

$$A = 4(-6 - 117) = -492$$

LEARNING HINT  Notice that in Steps 1, 2, 5, and 6 we created 0's in a certain *row* of each determinant by combining *columns*. Similarly, to create 0's in a *column*, it is usually best to combine *rows*. Keep this in mind so that you do not lose one 0 while creating another. ◄

Exercises 16-2

Evaluate each determinant by inspection.

1. $\begin{vmatrix} 2 & 7 & -3 \\ 0 & 0 & 0 \\ 4 & -5 & 6 \end{vmatrix}$
  2. $\begin{vmatrix} 3 & -5 & 2 \\ 0 & 2 & -1 \\ 0 & 0 & -1 \end{vmatrix}$
  3. $\begin{vmatrix} 2 & 0 & 0 \\ 5 & 0 & 3 \\ -1 & -4 & -2 \end{vmatrix}$

4. $\begin{vmatrix} 3 & 2 & 0 \\ -1 & 6 & 0 \\ -5 & 1 & 0 \end{vmatrix}$
  5. $\begin{vmatrix} 1 & 3 & -2 \\ 0 & 4 & 6 \\ 1 & 3 & -2 \end{vmatrix}$
  6. $\begin{vmatrix} 1 & 3 & 4 \\ 0 & 0 & 5 \\ 0 & 2 & -2 \end{vmatrix}$

7. $\begin{vmatrix} 2 & 0 & 0 & 0 \\ -3 & -1 & 0 & 0 \\ 4 & 5 & 3 & 0 \\ -1 & -3 & 2 & 4 \end{vmatrix}$
  8. $\begin{vmatrix} -3 & 2 & 1 & 2 \\ -1 & 7 & 0 & 7 \\ 0 & 5 & 3 & 5 \\ 3 & 6 & 2 & 6 \end{vmatrix}$

9. $\begin{vmatrix} 3 & -1 & 4 \\ -1 & 0 & 2 \\ 2 & 1 & 6 \end{vmatrix}$ given that $\begin{vmatrix} 3 & 4 & -1 \\ -1 & 2 & 0 \\ 2 & 6 & 1 \end{vmatrix} = 20$

10. $\begin{vmatrix} -3 & 7 & 1 \\ 4 & 8 & 2 \\ 1 & 3 & 0 \end{vmatrix}$ given that $\begin{vmatrix} -3 & 7 & 1 \\ 2 & 4 & 1 \\ 1 & 3 & 0 \end{vmatrix} = 18$

Use Properties 1–7 to evaluate each determinant.

11. $\begin{vmatrix} 2 & -1 & 1 \\ 3 & 0 & 2 \\ 1 & 1 & -3 \end{vmatrix}$

12. $\begin{vmatrix} 5 & 1 & 0 \\ -2 & 1 & 1 \\ -4 & 2 & -1 \end{vmatrix}$

13. $\begin{vmatrix} 3 & 0 & 2 \\ 5 & 2 & 3 \\ -4 & -1 & -2 \end{vmatrix}$

14. $\begin{vmatrix} 3 & 2 & -3 \\ 2 & -3 & 4 \\ -6 & -5 & -2 \end{vmatrix}$

15. $\begin{vmatrix} 4 & -2 & 2 \\ 3 & 1 & -1 \\ 2 & -3 & 5 \end{vmatrix}$

16. $\begin{vmatrix} -2 & 6 & 1 \\ -4 & 3 & 2 \\ 1 & -9 & -1 \end{vmatrix}$

17. $\begin{vmatrix} 3 & -1 & 0 & 1 \\ 0 & 2 & -2 & 0 \\ -1 & 4 & 2 & -3 \\ 2 & 3 & 0 & -6 \end{vmatrix}$

18. $\begin{vmatrix} 1 & 0 & 1 & 2 \\ -2 & -3 & 0 & 4 \\ -1 & 2 & 0 & -3 \\ 3 & -1 & -3 & -2 \end{vmatrix}$

19. $\begin{vmatrix} -2 & 1 & 2 & 4 \\ 3 & 2 & -3 & -1 \\ -4 & 1 & 5 & 2 \\ 1 & 2 & 2 & -1 \end{vmatrix}$

20. $\begin{vmatrix} 4 & 2 & -2 & -4 \\ -1 & 1 & 3 & 6 \\ -2 & -1 & -3 & 8 \\ 5 & -2 & -3 & 2 \end{vmatrix}$

21. $\begin{vmatrix} 2 & 1 & 0 & -1 & 1 \\ 0 & -1 & 2 & -3 & 0 \\ 1 & 4 & -1 & 0 & 2 \\ -1 & 0 & 0 & 2 & -2 \\ -2 & 1 & 1 & 0 & 0 \end{vmatrix}$

22. $\begin{vmatrix} 2 & 3 & 1 & 0 & -1 \\ 5 & 1 & 0 & 2 & -2 \\ -3 & 0 & 1 & 1 & 1 \\ 0 & 6 & 1 & 3 & -1 \\ 1 & 2 & 3 & 0 & 0 \end{vmatrix}$

Solve each system of equations using Cramer's rule. Use Properties 1–7 to evaluate the resulting determinants.

23. $2x + 3y - z = -6$
 $x - y + 2z = 2$
 $2x + 2y - 3z = -13$

24. $x + y + z = 1$
 $x + 2y + 2z = -2$
 $2x + 3y + 4z = -2$

25. $2a - 3b + c = 2$
 $4a + 3b - 2c = -1$
 $6a - 6b - c = -1$

26. $5d - e + f = 0$
 $10d - e - 2f = 5$
 $-5d + 2e + 3f = 5$

27. $2w - x + y - z = -6$
 $w + y + 2z = 4$
 $w - 2x + 2y = -9$
 $2w - 3x - y + z = 2$

28. $3x + t = -3$
 $2y - z = -4$
 $x + y - z = -5$
 $3x - 4y + 2z + t = 3$

29. $2a - 6b - c + d = 4$
 $4a + 3b + 5c + 2d = 5$
 $6a - 3b - 2c + 2d = 1$
 $-4a + 9b + c + 3d = 3$

30. $r + s + t + u = 2$
 $2r - s - t + 3u = -1$
 $-r + 2s + 2t - 2u = 3$
 $4r + 3s - t - 3u = 1$

31. *Physics*   Solve for forces F_1, F_2, F_3, and F_4:

$$F_1 + F_2 + F_3 + F_4 = 6$$

$$F_1 + F_3 - F_4 = 9$$

$$F_2 + F_4 = -1$$

$$F_3 - F_4 = 6$$

32. *Electronics*   Applying Kirchhoff's laws to the circuit shown in the figure results in the following system of equations. Solve for the three currents (in amperes).

$$I_1 + I_2 + I_3 = 0$$

$$3I_1 - 6I_2 = 6$$

$$6I_2 - 2I_3 = 4$$

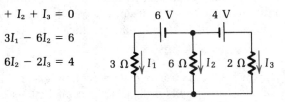

33. *Material science* Alloys containing 30% lead, 40% lead, and 60% lead are combined to produce 100 kg of an alloy containing 51% lead. If twice as much of the 60% lead alloy is used as the 40% alloy, how much of each is used?

34. *Industrial management* A company has three vans, A, B, and C, equipped to hold the following numbers of machines X, Y, and Z:

Machine	Van A	Van B	Van C
X	2	1	3
Y	1	0	1
Z	0	1	2

How many complete loads of each van are required to ship 15 type X machines, 5 type Y machines, and 8 type Z machines?

16-3 | Introduction to Matrices

As we have seen, determinants are useful for solving systems of linear equations by hand. A similar mathematical device, known as a *matrix*, can also be used to solve systems of linear equations. Matrices are especially valuable for today's technology because they are ideal for solving large systems of linear equations with a computer.

Definition

A **matrix** is a rectangular array of numbers that can be manipulated according to certain rules. For example, the following are all matrices:

$$\begin{bmatrix} 2 & 3 \\ 7 & 8 \\ 5 & 6 \end{bmatrix} \quad [2 \ 0 \ 1 \ -3] \quad \begin{bmatrix} 3 & 5 \\ 7 & 1 \end{bmatrix}$$

Notice that the matrix is indicated by using brackets.

Matrices and determinants

A matrix differs from a determinant in two important ways:

1. A matrix does not necessarily contain the same number of rows as columns, but a determinant must. In the above examples, the first matrix contains three rows but only two columns. The second matrix has one row and four columns. The third matrix contains two rows and two columns.

2. The word *matrix* refers to the array itself, whereas the word *determinant* refers to the numerical value calculated according to the rules we have listed.

To further distinguish a matrix from a determinant, we enclose a matrix in square brackets. Matrices are usually designated by capital letters.

Matrix notation The entries, or **elements,** of a matrix may be any real or imaginary numbers or algebraic expressions. The **order** of a matrix is specified by a pair of numbers telling the number of rows and the number of columns. A matrix with m rows and n columns is said to be of order $m \times n$, read "m by n."

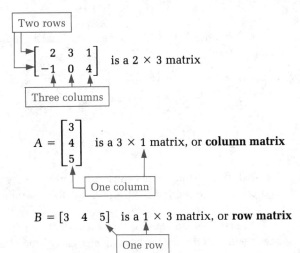

Note that A and B are considered different matrices even though they contain the same elements in the same sequence.

Square matrices A **square matrix** has the same number of rows as columns. The **principal diagonal** of a square matrix is the diagonal running from upper left to lower right:

$$\begin{bmatrix} 2 & 3 & 0 & 4 \\ 4 & 0 & 1 & -1 \\ 0 & 1 & 1 & 2 \\ 3 & 2 & 2 & 4 \end{bmatrix}$$

Principal diagonal

Several square matrices are used so often in matrix algebra that they are given special names. The **zero matrix** is a square matrix in which every element is equal to 0:

$$0_2 = \begin{bmatrix} 0 & 0 \\ 0 & 0 \end{bmatrix} \qquad 0_3 = \begin{bmatrix} 0 & 0 & 0 \\ 0 & 0 & 0 \\ 0 & 0 & 0 \end{bmatrix} \qquad \text{And so on}$$

The **unit,** or **identity, matrix** is a square matrix with only 1's along the principal diagonal and 0's elsewhere:

$$I_2 = \begin{bmatrix} 1 & 0 \\ 0 & 1 \end{bmatrix} \qquad I_3 = \begin{bmatrix} 1 & 0 & 0 \\ 0 & 1 & 0 \\ 0 & 0 & 1 \end{bmatrix} \qquad \text{And so on}$$

The **J matrix** is a square matrix in which all elements are 1's:

$$J_2 = \begin{bmatrix} 1 & 1 \\ 1 & 1 \end{bmatrix} \qquad J_3 = \begin{bmatrix} 1 & 1 & 1 \\ 1 & 1 & 1 \\ 1 & 1 & 1 \end{bmatrix} \qquad \text{And so on}$$

When naming a specific element in a matrix, **double-subscript notation** is usually used:

$$A = \begin{bmatrix} a_{11} & a_{12} & a_{13} \\ a_{21} & a_{22} & a_{23} \\ a_{31} & a_{32} & a_{33} \end{bmatrix}$$

a_{ij}

Row number   Column number

The first digit of the double subscript represents the row where the element is located, and the second digit represents the column.

Definition of equality   Two matrices are **equal** if they have the same order, and if each element of one is equal to the corresponding element of the other.

EXAMPLE 15   Find the value of each variable:

(a) $\begin{bmatrix} u & v & -w \\ -x & 2y & \dfrac{z}{2} \end{bmatrix} = \begin{bmatrix} 5 & 0 & 1 \\ -4 & -6 & 3 \end{bmatrix}$

(b) $[a \quad 2a + b \quad a - b + c] = [4 \quad 6 \quad 2]$

Solution

(a) Since the matrices are equal, corresponding elements must be equal. Thus, $u = 5$ and $v = 0$. Solving for the remaining elements, we have

$$-w = 1 \qquad -x = -4 \qquad 2y = -6 \qquad \frac{z}{2} = 3$$

$$w = -1 \qquad x = 4 \qquad y = -3 \qquad z = 6$$

(b) Equating corresponding elements, we have

$$a = 4 \qquad 2a + b = 6 \qquad a - b + c = 2$$

Substituting 4 for a in the second equation, we obtain

$$2(4) + b = 6$$

$$b = -2$$

Substituting for a and b in the third equation gives

$$(4) - (-2) + c = 2$$

$$6 + c = 2$$

$$c = -4$$

Addition of matrices   Two matrices may be added if they are of the same order—that is, if they contain the same number of rows and the same number of columns. Their sum is the matrix with elements that are the sums of corresponding elements.

EXAMPLE 16 Add:

(a) $\begin{bmatrix} 2 & 3 & 5 \\ 1 & 6 & 4 \end{bmatrix} + \begin{bmatrix} 3 & -7 & 8 \\ 2 & 5 & 9 \end{bmatrix} = ?$ (b) $\begin{bmatrix} 4 & 2 \\ 2 & 0 \\ 7 & 1 \end{bmatrix} + \begin{bmatrix} 3 \\ 9 \\ 5 \end{bmatrix} = ?$

Solution

(a) $\begin{bmatrix} 2 & 3 & 5 \\ 1 & 6 & 4 \end{bmatrix} + \begin{bmatrix} 3 & -7 & 8 \\ 2 & 5 & 9 \end{bmatrix} = \begin{bmatrix} 2+3 & 3+(-7) & 5+8 \\ 1+2 & 6+5 & 4+9 \end{bmatrix}$

$$= \begin{bmatrix} 5 & -4 & 13 \\ 3 & 11 & 13 \end{bmatrix}$$

(b) These two matrices are not of the same order; therefore, they cannot be added.

Multiplication by a constant   **Scalar multiplication** is defined as the multiplication of a matrix by a constant. The product of such a multiplication is a new matrix of the same order, each element of which is the product of the original element and the constant.

EXAMPLE 17 If $Q = \begin{bmatrix} 4 & 1 & -2 \\ 7 & -6 & 5 \end{bmatrix}$, find: (a) $3Q$ (b) $-Q$

Solution

(a) $3Q = \begin{bmatrix} 3(4) & 3(1) & 3(-2) \\ 3(7) & 3(-6) & 3(5) \end{bmatrix} = \begin{bmatrix} 12 & 3 & -6 \\ 21 & -18 & 15 \end{bmatrix}$

(b) $-Q = (-1)(Q) = \begin{bmatrix} -4 & -1 & 2 \\ -7 & 6 & -5 \end{bmatrix}$

Note that the commutative and associative laws of *addition* and the distributive law are all valid for matrices. Examples involving these laws are found in Exercises 16-3, Problems 19–22.

Subtraction of matrices   One matrix can be subtracted from another by restating subtraction as addition. If A and B represent matrices of the same order, then

$$A - B = A + (-B)$$

EXAMPLE 18 Given $A = [2 \quad 3 \quad 7 \quad -1]$ and $B = [5 \quad -2 \quad 0 \quad 4]$, find:
(a) $A - B$ (b) $2A - 3B$

Solution

(a) $A - B = A + (-B) = [2 \quad 3 \quad 7 \quad -1] + [-5 \quad 2 \quad 0 \quad -4]$

$$= [-3 \quad 5 \quad 7 \quad -5]$$

(b) $2A - 3B = 2A + (-3B) = [4 \quad 6 \quad 14 \quad -2] + [-15 \quad 6 \quad 0 \quad -12]$

$= [-11 \quad 12 \quad 14 \quad -14]$

Exercises 16-3

Find the value of each variable.

1. $\begin{bmatrix} x & -y & z \\ 2u & \dfrac{v}{3} & -w \end{bmatrix} = \begin{bmatrix} 1 & 3 & -5 \\ -4 & 9 & -2 \end{bmatrix}$

2. $\begin{bmatrix} a & 2b \\ -3c & -d \\ -\dfrac{e}{2} & 3f \end{bmatrix} = \begin{bmatrix} 5 & 3 \\ 12 & -4 \\ 8 & -6 \end{bmatrix}$

3. $\begin{bmatrix} p \\ 2p + q \\ p - q + r \end{bmatrix} = \begin{bmatrix} -3 \\ 4 \\ 5 \end{bmatrix}$

4. $[2x \quad 3x - 2y \quad x - y + 2z] = [-6 \quad -13 \quad 7]$

Perform the indicated operations.

5. $\begin{bmatrix} 2 & -1 & 3 \\ 5 & -3 & 2 \end{bmatrix} + \begin{bmatrix} -3 & 4 & 2 \\ -6 & -2 & 1 \end{bmatrix}$

6. $\begin{bmatrix} 2 \\ 3 \\ 5 \end{bmatrix} + \begin{bmatrix} -3 \\ 6 \\ -4 \end{bmatrix}$

7. $\begin{bmatrix} 5 & 7 \\ -1 & -4 \end{bmatrix} - \begin{bmatrix} -3 & 6 \\ 2 & 7 \end{bmatrix}$

8. $\begin{bmatrix} -11 & 6 \\ 4 & 2 \\ -3 & 9 \end{bmatrix} - \begin{bmatrix} 5 & 4 \\ -1 & 0 \\ 2 & 6 \end{bmatrix}$

Use the following matrices to perform the indicated operations:

$$A = \begin{bmatrix} 3 & 2 & 4 \\ -1 & -3 & 5 \end{bmatrix} \qquad B = \begin{bmatrix} -1 & -3 & 6 \\ 0 & 4 & 8 \end{bmatrix} \qquad C = \begin{bmatrix} 2 & 5 \\ -3 & -6 \end{bmatrix}$$

9. $2A$ **10.** $3C$ **11.** $-B$ **12.** $-4A$ **13.** $A - B$

14. $B - A$ **15.** $A + 2B$ **16.** $3B - 2A$ **17.** $A + 2C$ **18.** $C - B$

Verify the following laws using matrices M, N, and P:

$$M = \begin{bmatrix} -1 & -3 & 2 & 4 \\ 2 & 5 & -2 & 3 \end{bmatrix} \qquad N = \begin{bmatrix} 3 & 1 & 4 & -1 \\ -2 & 2 & 0 & 4 \end{bmatrix} \qquad P = \begin{bmatrix} -5 & 0 & 1 & 1 \\ 2 & 6 & 0 & 2 \end{bmatrix}$$

19. $M + N = N + M$ **20.** $M + (N + P) = (M + N) + P$

21. $M + 0 = M$ **22.** $2(M + P) = 2M + 2P$

23. *Business*   Two branches of an auto supply franchise in the same city used the following matrices to inventory the amount (in quarts) of three different brands and four different weights of motor oil:

		Brand				Brand		
		X	Y	Z		X	Y	Z
	10W–30	12	8	19		16	22	5
Weight	10W–40	6	13	7		11	12	9
	20W–40	4	3	8		3	8	6
	30W	7	2	4		5	1	0
		Branch *A*				Branch *B*		

If one branch went out of business and the inventories were merged, what matrix would represent the total amount of each type and weight of oil?

24. *Business* The following matrix represents the normal monthly order of a retail shoe store for three models of running shoes in eight different sizes:

$$
\begin{array}{c}
\\
\textbf{Model} \quad
\begin{array}{c}
\text{Jogger} \\
\text{Trainer} \\
\text{Racer}
\end{array}
\end{array}
\begin{array}{c}
\textbf{Size} \\
\begin{array}{cccccccc}
8 & 9 & 9\frac{1}{2} & 10 & 10\frac{1}{2} & 11 & 11\frac{1}{2} & 12
\end{array} \\
\begin{bmatrix}
3 & 4 & 4 & 7 & 8 & 6 & 2 & 1 \\
2 & 3 & 3 & 6 & 5 & 5 & 1 & 1 \\
4 & 6 & 6 & 8 & 10 & 8 & 4 & 2
\end{bmatrix}
\end{array}
$$

During their annual sale the store expects to do three times the usual volume of business. What matrix represents their order prior to the sale?

25. *Optics* The **coherency matrix*** is given as

$$
C = \begin{bmatrix}
\sin^2 \alpha & -\sin \alpha \cos \alpha \\
-\sin \alpha \cos \alpha & \cos^2 \alpha
\end{bmatrix}
$$

Write this matrix for $\alpha = \pi/4$.

26. *Optics* In Problem 25, are there any values of α that will produce the identity matrix?

16-4 | Multiplication of Matrices

Matrices are very common mathematical arrays. A cash register receipt is a simple column matrix, the inning-by-inning score of a baseball game is a 2×9 matrix, the daily stock market report is a very large matrix with hundreds of rows, and so on. Because we want to apply matrices to the solution of systems of linear equations, we must first define the product of two matrices.

Definition of multiplication
Two matrices may be multiplied if and only if the number of *columns* in the first matrix equals the number of *rows* in the second matrix.

If A is an $m \times n$ matrix and B is an $n \times p$ matrix, then their product is $AB = C$, where C is an $m \times p$ matrix.

$$
\begin{array}{ccccc}
A & \cdot & B & = & C \\
m \times n & & n \times p & & m \times p
\end{array}
$$

Same

This means, for example, that if

$$
A = \begin{bmatrix} 1 & 2 & 0 \\ 0 & -1 & 2 \end{bmatrix} \quad \text{and} \quad B = \begin{bmatrix} 1 & -1 & 2 \\ 0 & 2 & 3 \\ 1 & 1 & -1 \end{bmatrix}
$$

*Based on Miles V. Klein, *Optics* (New York: Wiley, 1970), p. 513.

then the product AB is possible, but the product BA is not possible:

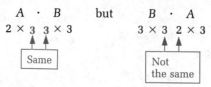

Obviously, $A \cdot B \neq B \cdot A$. Multiplication of matrices is *not* commutative.

When two matrices are multiplied, the elements of the product matrix are obtained by multiplying the elements of the rows of the first matrix by the corresponding elements of the columns of the second matrix and finding the sum of these products.

EXAMPLE 19 Multiply: $\begin{bmatrix} 1 & 2 & 0 \end{bmatrix} \begin{bmatrix} 2 & 0 \\ 3 & 1 \\ 1 & 1 \end{bmatrix} = ?$

Solution

Step 1. Check to be certain the multiplication is allowed.

Number of columns of first = Number of rows of second

$$3 = 3$$

Step 2. Multiply elements of the first row of the left-hand matrix and the first column of the right-hand matrix, and add these products.

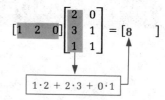

Step 3. Multiply elements of the first row of the left-hand matrix and the second column of the right-hand matrix, and add these products.

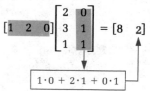

LEARNING HINT ▶ Remember each row-by-column multiplication as a "lazy T" that produces *one element* of the product matrix. ◀

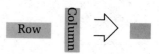

EXAMPLE 20 Multiply: $\begin{bmatrix} 1 & 2 \\ 3 & 0 \\ 1 & -1 \end{bmatrix} \begin{bmatrix} 2 & 3 \\ 0 & 1 \end{bmatrix} = ?$

Solution

Step 1. Check the number of rows and columns.

Number of columns of first = Number of rows of second

$$2 = 2$$

Step 2. Multiply elements of the first row on the left by the columns on the right, and add these products.

$$\begin{bmatrix} 1 & 2 \\ 3 & 0 \\ 1 & -1 \end{bmatrix} \begin{bmatrix} 2 & 3 \\ 0 & 1 \end{bmatrix} = \begin{bmatrix} 2 & \\ & \end{bmatrix}$$

$$1 \cdot 2 + 2 \cdot 0$$

$$\begin{bmatrix} 1 & 2 \\ 3 & 0 \\ 1 & -1 \end{bmatrix} \begin{bmatrix} 2 & 3 \\ 0 & 1 \end{bmatrix} = \begin{bmatrix} 2 & 5 \\ & \end{bmatrix}$$

$$1 \cdot 3 + 2 \cdot 1$$

Step 3. When the first row of the left-hand matrix has been used with all columns of the right-hand matrix, repeat the sequence with the second row.

$$\begin{bmatrix} 1 & 2 \\ 3 & 0 \\ 1 & -1 \end{bmatrix} \begin{bmatrix} 2 & 3 \\ 0 & 1 \end{bmatrix} = \begin{bmatrix} 2 & 5 \\ 6 & \end{bmatrix}$$

$$3 \cdot 2 + 0 \cdot 0$$

$$\begin{bmatrix} 1 & 2 \\ 3 & 0 \\ 1 & -1 \end{bmatrix} \begin{bmatrix} 2 & 3 \\ 0 & 1 \end{bmatrix} = \begin{bmatrix} 2 & 5 \\ 6 & 9 \end{bmatrix}$$

$$3 \cdot 3 + 0 \cdot 1$$

Step 4. Continue until all rows on the left have multiplied all columns on the right.

$$\begin{bmatrix} 1 & 2 \\ 3 & 0 \\ 1 & -1 \end{bmatrix} \begin{bmatrix} 2 & 3 \\ 0 & 1 \end{bmatrix} = \begin{bmatrix} 2 & 5 \\ 6 & 9 \\ 2 & \end{bmatrix}$$

$$1 \cdot 2 + (-1)0$$

$$\begin{bmatrix} 1 & 2 \\ 3 & 0 \\ 1 & -1 \end{bmatrix} \begin{bmatrix} 2 & 3 \\ 0 & 1 \end{bmatrix} = \begin{bmatrix} 2 & 5 \\ 6 & 9 \\ 2 & 2 \end{bmatrix}$$

$$1 \cdot 3 + (-1)1$$

REMEMBER ▶ The "lazy T" multiplication of the ith row and the jth column produces the element in the corresponding position in the product matrix:

If the matrices A and B being multiplied are square matrices, with the same number of rows as columns, then both AB and BA can be found, but they are generally not equal. For any square matrix A, we can find the integer powers of A, A^2, A^3, A^4, and so on, by multiplying.

EXAMPLE 21 If $A = \begin{bmatrix} 2 & 3 & 1 \\ 0 & 1 & 1 \\ -1 & 2 & 2 \end{bmatrix}$ and $B = \begin{bmatrix} 0 & 0 & 2 \\ 1 & 1 & 4 \\ 0 & 2 & 3 \end{bmatrix}$, find:

(a) AB (b) BA (c) A^2

Solution

(a) $AB = \begin{bmatrix} 2 & 3 & 1 \\ 0 & 1 & 1 \\ -1 & 2 & 2 \end{bmatrix} \begin{bmatrix} 0 & 0 & 2 \\ 1 & 1 & 4 \\ 0 & 2 & 3 \end{bmatrix} = \begin{bmatrix} 3 & 5 & 19 \\ 1 & 3 & 7 \\ 2 & 6 & 12 \end{bmatrix}$

(b) $BA = \begin{bmatrix} 0 & 0 & 2 \\ 1 & 1 & 4 \\ 0 & 2 & 3 \end{bmatrix} \begin{bmatrix} 2 & 3 & 1 \\ 0 & 1 & 1 \\ -1 & 2 & 2 \end{bmatrix} = \begin{bmatrix} -2 & 4 & 4 \\ -2 & 12 & 10 \\ -3 & 8 & 8 \end{bmatrix}$

(c) $A^2 = A \cdot A = \begin{bmatrix} 2 & 3 & 1 \\ 0 & 1 & 1 \\ -1 & 2 & 2 \end{bmatrix} \begin{bmatrix} 2 & 3 & 1 \\ 0 & 1 & 1 \\ -1 & 2 & 2 \end{bmatrix} = \begin{bmatrix} 3 & 11 & 7 \\ -1 & 3 & 3 \\ -4 & 3 & 5 \end{bmatrix}$

Identity matrix

A matrix of particular importance in solving systems of equations is the **identity matrix** defined earlier. The product of the identity matrix and any other matrix A of the same order is A. That is,

$$AI = IA = A \qquad \text{where} \quad I = \begin{bmatrix} 1 & 0 & 0 & 0 & \cdots \\ 0 & 1 & 0 & 0 & \cdots \\ 0 & 0 & 1 & 0 & \cdots \\ 0 & 0 & 0 & 1 & \cdots \\ \cdot & \cdot & \cdot & \cdot \\ \cdot & \cdot & \cdot & \cdot \end{bmatrix}$$

EXAMPLE 22 If $A = \begin{bmatrix} 3 & 4 \\ -2 & 1 \end{bmatrix}$, show that $AI = IA = A$.

Solution Since A is a 2×2 matrix, I must be the 2×2 identity matrix,

$I = \begin{bmatrix} 1 & 0 \\ 0 & 1 \end{bmatrix}$. Then

$$AI = \begin{bmatrix} 3 & 4 \\ -2 & 1 \end{bmatrix} \begin{bmatrix} 1 & 0 \\ 0 & 1 \end{bmatrix} = \begin{bmatrix} 3 \cdot 1 + 4 \cdot 0 & 3 \cdot 0 + 4 \cdot 1 \\ (-2)1 + 1 \cdot 0 & (-2)0 + 1 \cdot 1 \end{bmatrix} = \begin{bmatrix} 3 & 4 \\ -2 & 1 \end{bmatrix} = A$$

$$IA = \begin{bmatrix} 1 & 0 \\ 0 & 1 \end{bmatrix} \begin{bmatrix} 3 & 4 \\ -2 & 1 \end{bmatrix} = \begin{bmatrix} 1 \cdot 3 + 0(-2) & 1 \cdot 4 + 0 \cdot 1 \\ 0 \cdot 3 + 1(-2) & 0 \cdot 4 + 1 \cdot 1 \end{bmatrix} = \begin{bmatrix} 3 & 4 \\ -2 & 1 \end{bmatrix} = A$$

EXAMPLE 23 The Brain Co. produces two models, A and B, of a personal microcomputer. In a given month, the company produces 1250 units of model A and 875 units of model B. The Super Chip Co. provides three types of chips, x, y, and z, used in these computers. Model A requires 8 type x chips, 4 type y chips, and 3 type z chips, while model B requires 12 type x chips, 4 type y chips, and 5 type z chips. Use matrices to determine the total number of each kind of chip needed for the monthly output of the two models.

Solution The monthly production of the two models can be represented by the production matrix P:

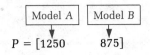

$$P = [1250 \qquad 875]$$

The number of chips required by the two models can be represented by the components matrix C:

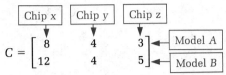

$$C = \begin{bmatrix} 8 & 4 & 3 \\ 12 & 4 & 5 \end{bmatrix}$$

The product PC of the two matrices gives a matrix N listing the total number of chips of each type required for the proposed output:

$$N = PC = [1250 \quad 875]\begin{bmatrix} 8 & 4 & 3 \\ 12 & 4 & 5 \end{bmatrix}$$

Using a calculator, the elements of the product matrix are found as follows:

Row	Column	Calculation	Result
1	1	1250 ⊗ 8 ⊕ 875 ⊗ 12 ⊜ →	*20500.*
1	2	1250 ⊗ 4 ⊕ 875 ⊗ 4 ⊜ →	*8500.*
1	3	1250 ⊗ 3 ⊕ 875 ⊗ 5 ⊜ →	*8125.*

Therefore,

$$N = PC = [20{,}500 \quad 8500 \quad 8125]$$

The company will need 20,500 type x chips, 8500 type y chips, and 8125 type z chips.

EXAMPLE 24   If the cost of the chips in Example 23 is 40¢ for chip x, 50¢ for chip y, and 60¢ for chip z, then we can form a cost matrix Q and use it to calculate the total cost of the chips:

$$Q = \begin{bmatrix} 0.4 \\ 0.5 \\ 0.6 \end{bmatrix}$$

$$\text{Total cost} = NQ = [20{,}500 \quad 8500 \quad 8125]\begin{bmatrix} 0.4 \\ 0.5 \\ 0.6 \end{bmatrix} = [17{,}325] \quad \text{or } \$17{,}325$$

The usefulness of working with matrices becomes more evident in actual commercial problems where hundreds of components and their costs may appear in various combinations in several dozen products. Computers can be programmed to manipulate such large matrices very quickly.

Exercises 16-4

Consider the order of each of the following matrices, and determine whether the products indicated in Problems 1–12 can be found.

A: 3×3 B: 2×3 C: 3×2 D: 1×3 E: 3×1

1. AB   **2.** BC   **3.** CD   **4.** DE   **5.** BA   **6.** CB
7. DC   **8.** ED   **9.** AE   **10.** EA   **11.** BE   **12.** EB

Compute each product if possible.

13. $\begin{bmatrix} 2 & -3 & 1 \end{bmatrix} \begin{bmatrix} 3 & 5 \\ 1 & 2 \\ -4 & 3 \end{bmatrix}$

14. $\begin{bmatrix} -1 & -3 \end{bmatrix} \begin{bmatrix} 1 & 4 & -1 \\ 3 & -1 & 0 \end{bmatrix}$

15. $\begin{bmatrix} -3 & 5 \\ 4 & 2 \end{bmatrix} \begin{bmatrix} 7 & -1 \\ 0 & -2 \end{bmatrix}$

16. $\begin{bmatrix} 5 & -1 \\ 3 & 3 \end{bmatrix} \begin{bmatrix} 3 & 4 & 2 \\ -1 & -1 & -5 \end{bmatrix}$

17. $\begin{bmatrix} -6 & 2 & 1 \\ 4 & 0 & 7 \end{bmatrix} \begin{bmatrix} -11 & 0 \\ 5 & 4 \\ 1 & 2 \end{bmatrix}$

18. $\begin{bmatrix} 6 & 2 \\ -5 & 10 \\ 4 & -3 \end{bmatrix} \begin{bmatrix} 3 & 4 & 6 \\ 1 & -1 & -3 \\ 4 & 2 & 6 \end{bmatrix}$

19. $\begin{bmatrix} 1 & 0 & 9 \\ -7 & 3 & 2 \end{bmatrix} \begin{bmatrix} 5 & 1 \\ 3 & -2 \end{bmatrix}$

20. $\begin{bmatrix} 5 & 1 & 0 \\ -6 & -8 & 4 \\ 2 & 3 & 3 \end{bmatrix} \begin{bmatrix} 1 & 0 & 3 \\ 2 & -6 & -1 \\ 3 & 7 & 0 \end{bmatrix}$

21. $\begin{bmatrix} 1 & 2 & 4 & -3 \end{bmatrix} \begin{bmatrix} 2 \\ -3 \\ -2 \\ 5 \end{bmatrix}$

22. $\begin{bmatrix} 3 & 7 & 5 & -4 \\ -2 & -6 & 1 & -3 \end{bmatrix} \begin{bmatrix} 4 \\ 0 \\ 6 \\ -3 \end{bmatrix}$

23. $\begin{bmatrix} 22 & 18 & 14 \\ 30 & 26 & 12 \end{bmatrix} \begin{bmatrix} 125 & 180 \\ 140 & 96 \\ 72 & 192 \end{bmatrix}$

24. $\begin{bmatrix} 3.7 & 4.6 & 5.9 & 2.8 & 6.1 \end{bmatrix} \begin{bmatrix} 54 & 17 \\ 26 & 19 \\ 41 & 33 \\ 17 & 25 \\ 15 & 18 \end{bmatrix}$

Use matrices A and B to perform the indicated operations.

$$A = \begin{bmatrix} 3 & 4 & -2 \\ 0 & 5 & -1 \\ 2 & -3 & -4 \end{bmatrix} \qquad B = \begin{bmatrix} 1 & 2 & 0 \\ -2 & 3 & -5 \\ -1 & 4 & 2 \end{bmatrix}$$

25. AB   **26.** BA   **27.** A^2 **28.** B^2

Use each of the following to show that $AI = IA = A$.

29. $\begin{bmatrix} 1 & 3 \\ 4 & -2 \end{bmatrix}$ **30.** $\begin{bmatrix} 5 & 1 \\ -6 & 4 \end{bmatrix}$ **31.** $\begin{bmatrix} 1 & 0 & -5 \\ 3 & 6 & 0 \\ 2 & 4 & -3 \end{bmatrix}$ **32.** $\begin{bmatrix} -7 & -1 & -1 \\ 2 & 3 & -2 \\ 1 & 1 & 5 \end{bmatrix}$

33. *Automotive engineering* A matrix can be used to describe the rotation of tires on a car.

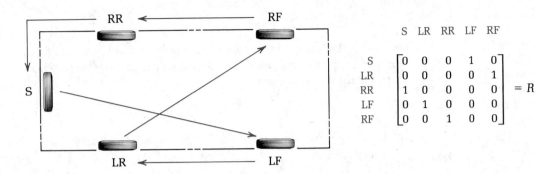

In matrix R, a 1 means that the tire in that column is replaced by the tire in that row. Find R^2, the position of the tires after two successive rotations.

34. *Industrial management* In Example 23 determine the number of chips and the total cost if the company reduces production to 850 units of model A and 625 units of model B.

35. *Cartography* A figure defined by the points $P_1(1, 3)$, $P_2(2, 1)$, $P_3(6, 3)$, $P_4(3, 2)$ is rotated through an angle θ using the product*

$$\begin{bmatrix} 1 & 3 \\ 2 & 1 \\ 6 & 3 \\ 3 & 2 \end{bmatrix} \begin{bmatrix} \cos\theta & \sin\theta \\ -\sin\theta & \cos\theta \end{bmatrix}$$

Find the resulting matrix for a rotation of $\theta = \pi$.

36. Show that

$$\begin{bmatrix} 2 & 3 \\ -1 & -2 \end{bmatrix}^2 = I$$

and, in general, that

$$\begin{bmatrix} n & 1+n \\ 1-n & -n \end{bmatrix}^2 = I$$

for any real number n. (Does it surprise you that the square root of the identity matrix I is not unique?)

16-5 | The Inverse of a Matrix

The final concept needed to solve systems of linear equations with matrices is that of the inverse of a matrix. In this section we shall define the inverse of a matrix and demonstrate two methods for finding it. We will then be prepared to solve systems of linear equations using matrices in Section 16-6.

* Based on an interview with Jill Skoog, Cartographer, United States Army.

Definition   The **inverse** of the square matrix A is another square matrix A^{-1} of the same order, such that

$$AA^{-1} = A^{-1}A = I$$

where I is the identity matrix of the same order as A. Only a square matrix can have an inverse. It can never have more than one inverse, but, as we shall see, it may have no inverse at all.

Finding an inverse   If we are given both a matrix A and its inverse A^{-1}, then it is easy to show
matrix   that they are inverses.

EXAMPLE 25   If $A = \begin{bmatrix} 3 & 1 \\ 5 & 2 \end{bmatrix}$,   then $A^{-1} = \begin{bmatrix} 2 & -1 \\ -5 & 3 \end{bmatrix}$.

Because

$$\begin{bmatrix} 3 & 1 \\ 5 & 2 \end{bmatrix}\begin{bmatrix} 2 & -1 \\ -5 & 3 \end{bmatrix} = \begin{bmatrix} 1 & 0 \\ 0 & 1 \end{bmatrix}$$

It is much more difficult to *find* the inverse of a matrix than to verify that $AA^{-1} = I$. The first method we shall give for finding the inverse of a matrix is valid for 2×2 matrices only, and the second is a general method that will enable you to find the inverse of any square matrix.

Determinant method   If we are given a 2×2 matrix A, then we know that its inverse is another 2×2 matrix A^{-1}, and they will satisfy the equation $AA^{-1} = I$. If

$$A = \begin{bmatrix} a & b \\ c & d \end{bmatrix}   \text{we can assume}   A^{-1} = \begin{bmatrix} x & y \\ z & w \end{bmatrix}$$

Multiplying leads to the following system of equations:

$$AA^{-1} = \begin{bmatrix} a & b \\ c & d \end{bmatrix}\begin{bmatrix} x & y \\ z & w \end{bmatrix} = \begin{bmatrix} 1 & 0 \\ 0 & 1 \end{bmatrix} \Longrightarrow \begin{array}{l} ax + bz = 1 \\ ay + bw = 0 \\ cx + dz = 0 \\ cy + dw = 1 \end{array}$$

Solving, we obtain

$$x = \frac{d}{ad - bc} = \frac{d}{D}   \qquad y = \frac{-b}{D}   \qquad z = \frac{-c}{D}   \qquad w = \frac{a}{D}$$

where $D = \begin{vmatrix} a & b \\ c & d \end{vmatrix}$

This gives us the following formula for finding the inverse of a 2 × 2 matrix:

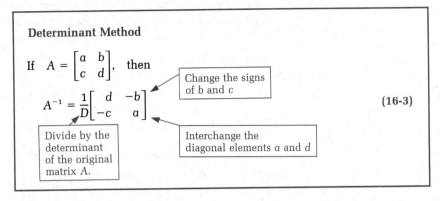

Determinant Method

If $A = \begin{bmatrix} a & b \\ c & d \end{bmatrix}$, then

$A^{-1} = \dfrac{1}{D}\begin{bmatrix} d & -b \\ -c & a \end{bmatrix}$

Change the signs of b and c

Divide by the determinant of the original matrix A.

Interchange the diagonal elements a and d

(16-3)

This expression for A^{-1}, the inverse of A, can be used to find the inverse of any 2 × 2 matrix whose determinant D is not equal to 0. If $D = 0$, the inverse does not exist.

EXAMPLE 26 Find the inverse of: $P = \begin{bmatrix} 3 & -4 \\ 2 & -2 \end{bmatrix}$

Solution
Step 1. The determinant is

$$D = \begin{vmatrix} 3 & -4 \\ 2 & -2 \end{vmatrix} = 3(-2) - 2(-4) = -6 + 8 = 2$$

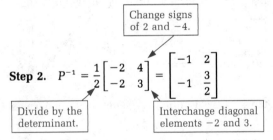

Change signs of 2 and −4.

Step 2. $P^{-1} = \dfrac{1}{2}\begin{bmatrix} -2 & 4 \\ -2 & 3 \end{bmatrix} = \begin{bmatrix} -1 & 2 \\ -1 & \dfrac{3}{2} \end{bmatrix}$

Divide by the determinant.

Interchange diagonal elements −2 and 3.

Step 3. Check:

$$\begin{bmatrix} 3 & -4 \\ 2 & -2 \end{bmatrix}\begin{bmatrix} -1 & 2 \\ -1 & \dfrac{3}{2} \end{bmatrix} = \begin{bmatrix} 1 & 0 \\ 0 & 1 \end{bmatrix}$$

Gauss–Jordan method   A second, and much more general, procedure for finding an inverse is known as the *Gauss–Jordan method* (named after the 19th century mathematicians who first developed it). This procedure is particularly useful because it can be programmed for computers.

The **Gauss–Jordan method** consists of a sequence of steps that simultaneously transform the given matrix into an identity matrix and an identity matrix into the inverse. The steps used are based on the following operations, or **row transformations:**

- Interchange any two rows.
- Multiply or divide the elements in any row by a nonzero number.
- Change a row by adding to it the corresponding elements of another row.

EXAMPLE 27   Find the inverse of: $Q = \begin{bmatrix} 3 & 4 \\ -2 & -2 \end{bmatrix}$

Solution   **First,** write an identity matrix of the same order alongside Q:

$$\left[\begin{array}{cc|cc} 3 & 4 & 1 & 0 \\ -2 & -2 & 0 & 1 \end{array} \right]$$

Then, use row transformations to transform Q to an identity matrix. Perform the *same* operations on the adjacent identity matrix. Use the following order of operations:

Step 1.   Start with Column 1. Be sure the first entry in Row 1 is not 0. If it is a 0, interchange the first row with another row.

Step 2.   Use row transformations to get a 1 in Row 1, Column 1:

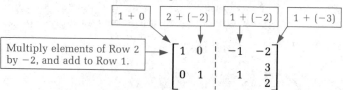

Step 3.   Transform all other elements in Column 1 to 0's:

Multiply each element of Row 1 by 2, and add to Row 2.

$$\left[\begin{array}{cc|cc} 1 & 2 & 1 & 1 \\ 0 & 2 & 2 & 3 \end{array} \right]$$

$-2 + 2$ $-2 + 4$ $0 + 2$ $1 + 2$

Step 4.   Go to Column 2. Again transform the element in the principal diagonal to 1:

Divide each element by 2.

$$\left[\begin{array}{cc|cc} 1 & 2 & 1 & 1 \\ 0 & 1 & 1 & \dfrac{3}{2} \end{array} \right]$$

$0 \div 2$ $2 \div 2$ $3 \div 2$

Step 5.   Transform the remaining elements of Column 2 to 0's:

$1 + 0$ $2 + (-2)$ $1 + (-2)$ $1 + (-3)$

Multiply elements of Row 2 by -2, and add to Row 1.

$$\left[\begin{array}{cc|cc} 1 & 0 & -1 & -2 \\ 0 & 1 & 1 & \dfrac{3}{2} \end{array} \right]$$

Continue until all columns have been transformed and the matrix on the left is the identity matrix, as shown above. Then the inverse is the matrix on the right.

$$Q^{-1} = \begin{bmatrix} -1 & -2 \\ 1 & \dfrac{3}{2} \end{bmatrix}$$

Step 6. Check:

$$QQ^{-1} = \begin{bmatrix} 3 & 4 \\ -2 & -2 \end{bmatrix} \begin{bmatrix} -1 & -2 \\ 1 & \dfrac{3}{2} \end{bmatrix} = \begin{bmatrix} 1 & 0 \\ 0 & 1 \end{bmatrix}$$

LEARNING HINT  Notice that we work column-by-column, transforming the principal diagonal element to 1 first, and then transforming the other elements in the column to 0. ◀

EXAMPLE 28 Find the inverse of: $C = \begin{bmatrix} 0 & 2 & -1 \\ 2 & 3 & -2 \\ 1 & 0 & 1 \end{bmatrix}$

Solution

First, attach the 3×3 identity matrix to C.

$$\left[\begin{array}{ccc|ccc} 0 & 2 & -1 & 1 & 0 & 0 \\ 2 & 3 & -2 & 0 & 1 & 0 \\ 1 & 0 & 1 & 0 & 0 & 1 \end{array} \right]$$

Step 1. Interchange Rows 1 and 3 to get a nonzero digit in Row 1, Column 1.

$$\left[\begin{array}{ccc|ccc} 1 & 0 & 1 & 0 & 0 & 1 \\ 2 & 3 & -2 & 0 & 1 & 0 \\ 0 & 2 & -1 & 1 & 0 & 0 \end{array} \right]$$

Step 2. Use the transformation rules to get 0's elsewhere in Column 1.

$$\left[\begin{array}{ccc|ccc} 1 & 0 & 1 & 0 & 0 & 1 \\ 0 & 3 & -4 & 0 & 1 & -2 \\ 0 & 2 & -1 & 1 & 0 & 0 \end{array} \right]$$

Multiply Row 1 by -2, and add to Row 2.

Step 3. Go to Column 2. Transform the principal diagonal element to 1.

$$\left[\begin{array}{ccc|ccc} 1 & 0 & 1 & 0 & 0 & 1 \\ 0 & 1 & -3 & -1 & 1 & -2 \\ 0 & 2 & -1 & 1 & 0 & 0 \end{array} \right]$$

Multiply Row 3 by -1, and add to Row 2.

Step 4. Transform all other elements in Column 2 to 0's.

$$\left[\begin{array}{ccc|ccc} 1 & 0 & 1 & 0 & 0 & 1 \\ 0 & 1 & -3 & -1 & 1 & -2 \\ 0 & 0 & 5 & 3 & -2 & 4 \end{array} \right]$$

Multiply Row 2 by -2, and add to Row 3.

Step 5. Go to Column 3. Transform the principal diagonal element to 1.

$$\left[\begin{array}{ccc|ccc} 1 & 0 & 1 & 0 & 0 & 1 \\ 0 & 1 & -3 & -1 & 1 & -2 \\ 0 & 0 & 1 & \dfrac{3}{5} & -\dfrac{2}{5} & \dfrac{4}{5} \end{array} \right]$$

Divide Row 3 by 5.

Step 6. Transform all other elements in Column 3 to 0's.

$$\left[\begin{array}{ccc|ccc} 1 & 0 & 0 & -\dfrac{3}{5} & \dfrac{2}{5} & \dfrac{1}{5} \\ 0 & 1 & -3 & -1 & 1 & -2 \\ 0 & 0 & 1 & \dfrac{3}{5} & -\dfrac{2}{5} & \dfrac{4}{5} \end{array} \right]$$

Multiply Row 3 by -1, and add to Row 1.

$$\begin{bmatrix} 1 & 0 & 0 & \vdots & -\dfrac{3}{5} & \dfrac{2}{5} & \dfrac{1}{5} \\ 0 & 1 & 0 & \vdots & \dfrac{4}{5} & -\dfrac{1}{5} & \dfrac{2}{5} \\ 0 & 0 & 1 & \vdots & \dfrac{3}{5} & -\dfrac{2}{5} & \dfrac{4}{5} \end{bmatrix}$$

Multiply Row 3 by 3, and add to Row 2.

Step 7.

$$C^{-1} = \begin{bmatrix} -\dfrac{3}{5} & \dfrac{2}{5} & \dfrac{1}{5} \\ \dfrac{4}{5} & -\dfrac{1}{5} & \dfrac{2}{5} \\ \dfrac{3}{5} & -\dfrac{2}{5} & \dfrac{4}{5} \end{bmatrix}$$

Check by multiplying: $CC^{-1} = I$

Not all square matrices have inverses. If a matrix does not have an inverse, the transformation process will produce a row or column in the original matrix consisting entirely of 0's. Also, if a matrix has no inverse, its determinant will be equal to 0.

EXAMPLE 29 Find the inverse of: $T = \begin{bmatrix} 1 & 0 & 1 \\ 0 & 2 & 1 \\ 2 & 4 & 4 \end{bmatrix}$

Solution
$$\begin{bmatrix} 1 & 0 & 1 & \vdots & 1 & 0 & 0 \\ 0 & 2 & 1 & \vdots & 0 & 1 & 0 \\ 2 & 4 & 4 & \vdots & 0 & 0 & 1 \end{bmatrix} \Longrightarrow \begin{bmatrix} 1 & 0 & 1 & \vdots & 1 & 0 & 0 \\ 0 & 2 & 1 & \vdots & 0 & 1 & 0 \\ 0 & 4 & 2 & \vdots & -2 & 0 & 1 \end{bmatrix}$$

$$\Longrightarrow \begin{bmatrix} 1 & 0 & 1 & \vdots & 1 & 0 & 0 \\ 0 & 1 & \dfrac{1}{2} & \vdots & 0 & \dfrac{1}{2} & 0 \\ 0 & 4 & 2 & \vdots & -2 & 0 & 1 \end{bmatrix}$$

$$\Longrightarrow \begin{bmatrix} 1 & 0 & 1 & \vdots & 1 & 0 & 0 \\ 0 & 1 & \dfrac{1}{2} & \vdots & 0 & \dfrac{1}{2} & 0 \\ 0 & 0 & 0 & \vdots & -2 & -2 & 1 \end{bmatrix}$$

But on this step we obtain a row of all 0's.

The matrix has no inverse. Check by evaluating the determinant of the matrix: $D_T = 0$

Exercises 16-5

Use the method of Example 26 to find the inverse of each matrix.

1. $\begin{bmatrix} 3 & 1 \\ -1 & 2 \end{bmatrix}$ **2.** $\begin{bmatrix} 4 & 0 \\ 1 & -2 \end{bmatrix}$ **3.** $\begin{bmatrix} -3 & 5 \\ 2 & 0 \end{bmatrix}$ **4.** $\begin{bmatrix} 4 & 5 \\ 2 & 3 \end{bmatrix}$

5. $\begin{bmatrix} -2 & 5 \\ 3 & -8 \end{bmatrix}$ **6.** $\begin{bmatrix} 10 & -3 \\ 6 & -2 \end{bmatrix}$ ▦ **7.** $\begin{bmatrix} 4.02 & 9.14 \\ -4.05 & -6.50 \end{bmatrix}$ ▦ **8.** $\begin{bmatrix} -1.06 & 3.05 \\ -4.42 & 8.30 \end{bmatrix}$

Use the Gauss–Jordan method to find the inverse of each matrix.

9. $\begin{bmatrix} 3 & 4 \\ 2 & 1 \end{bmatrix}$ **10.** $\begin{bmatrix} -2 & -2 \\ 5 & 4 \end{bmatrix}$ **11.** $\begin{bmatrix} 6 & -3 \\ -3 & 2 \end{bmatrix}$ **12.** $\begin{bmatrix} 4 & 2 \\ -3 & -1 \end{bmatrix}$

13. $\begin{bmatrix} 5 & -1 \\ -8 & 3 \end{bmatrix}$ **14.** $\begin{bmatrix} 4 & -6 \\ 3 & -4 \end{bmatrix}$ **15.** $\begin{bmatrix} 1 & 8 \\ 1 & 7 \end{bmatrix}$ **16.** $\begin{bmatrix} 5 & 2 \\ 1 & -2 \end{bmatrix}$

17. $\begin{bmatrix} 1 & 4 & 2 \\ -1 & 1 & 0 \\ 0 & 3 & -3 \end{bmatrix}$ **18.** $\begin{bmatrix} 2 & -3 & 0 \\ 1 & -1 & 2 \\ 0 & 3 & 2 \end{bmatrix}$ **19.** $\begin{bmatrix} 1 & 6 & 5 \\ -3 & 2 & -1 \\ -1 & 1 & 0 \end{bmatrix}$ **20.** $\begin{bmatrix} 1 & 3 & 2 \\ 0 & -2 & -3 \\ -2 & 1 & 6 \end{bmatrix}$

21. $\begin{bmatrix} 2 & -2 & -3 \\ -1 & 5 & 1 \\ 1 & -4 & 0 \end{bmatrix}$ **22.** $\begin{bmatrix} 3 & -3 & 0 \\ -1 & -4 & 2 \\ 2 & -6 & 1 \end{bmatrix}$ **23.** $\begin{bmatrix} -2 & 1 & 1 \\ 3 & 0 & 4 \\ -6 & -1 & -12 \end{bmatrix}$ **24.** $\begin{bmatrix} -4 & 6 & -7 \\ 2 & 3 & 4 \\ 1 & -2 & 2 \end{bmatrix}$

25. Show that $J = \begin{bmatrix} 1 & 1 \\ 1 & 1 \end{bmatrix}$ has no inverse.

26. *Electronics* In the analysis of three-phase circuits, it is sometimes necessary to find the inverse of the matrix shown below. Find S^{-1}.

$$S = \begin{bmatrix} 1 & 1 & 1 \\ 1 & a^2 & a \\ 1 & a & a^2 \end{bmatrix}$$

16-6 | Solving Systems of Linear Equations Using Matrices

We shall now use our knowledge of matrices to solve systems of linear equations. As you learn this method, keep in mind that, although it may not be the easiest method for a paper and pencil solution of a small set of equations, this matrix solution is ideal for solving large linear systems, particularly with a computer. Because of the ever-increasing importance of computers in technical occupations and the large-order systems of equations that appear, all technical students should be familiar with this method.

Background Consider the following general system of two linear equations in two unknowns:

$$a_1x + b_1y = c_1$$

$$a_2x + b_2y = c_2$$

Using matrix notation, we can write this system of equations as a single matrix equation:

$$\begin{bmatrix} a_1 & b_1 \\ a_2 & b_2 \end{bmatrix} \begin{bmatrix} x \\ y \end{bmatrix} = \begin{bmatrix} c_1 \\ c_2 \end{bmatrix} \quad \text{or} \quad AX = C \qquad (16\text{-}4)$$

where $A = \begin{bmatrix} a_1 & b_1 \\ a_2 & b_2 \end{bmatrix}$ is the **coefficient matrix**

$X = \begin{bmatrix} x \\ y \end{bmatrix}$ is the **solution matrix**

$C = \begin{bmatrix} c_1 \\ c_2 \end{bmatrix}$ is the **constant matrix**

The system of equations is solved when we find the matrix X. To find X from the matrix equation, we proceed as follows:

First, multiply both sides of (16-4) by A^{-1}. $A^{-1}AX = A^{-1}C$

Next, since $A^{-1}A = I$, we obtain: $IX = A^{-1}C$

Finally, since $IX = X$: $X = A^{-1}C$ (16-5)

This last equation shows us the matrix method for solving a system of linear equations.

Step 1.   Form the coefficient matrix A and find its inverse A^{-1}.

Step 2.   Multiply A^{-1} by the constant matrix C. The result is the solution matrix X, and the elements of X are the values of the variables.

EXAMPLE 30   Use the matrix method to solve the following system:

$3x - 2y = -8$

$4x - 5y = -6$

Solution

Step 1.   The coefficient matrix is: $A = \begin{bmatrix} 3 & -2 \\ 4 & -5 \end{bmatrix}$.   Using the methods of Section 16-5,

$$A^{-1} = \begin{bmatrix} \dfrac{5}{7} & -\dfrac{2}{7} \\[2mm] \dfrac{4}{7} & -\dfrac{3}{7} \end{bmatrix}$$

Step 2.   Multiply A^{-1} by $C = \begin{bmatrix} -8 \\ -6 \end{bmatrix}$ to find X:

$$X = A^{-1}C = \begin{bmatrix} \dfrac{5}{7} & -\dfrac{2}{7} \\[2mm] \dfrac{4}{7} & -\dfrac{3}{7} \end{bmatrix} \begin{bmatrix} -8 \\ -6 \end{bmatrix} = \begin{bmatrix} -4 \\ -2 \end{bmatrix}$$

Since $X = \begin{bmatrix} x \\ y \end{bmatrix} = \begin{bmatrix} -4 \\ -2 \end{bmatrix}$, then $x = -4$ and $y = -2$.

Step 3.   Check your solution by substituting into the original system:

$$3(-4) - 2(-2) = -8 \qquad 4(-4) - 5(-2) = -6$$

$$-8 = -8 \qquad\qquad -6 = -6$$

Application: two equations in two unknowns

To solve an applied problem, first form a system of equations from the given information and then solve as before. (See Chapter 5 for a review of applications with linear systems.)

EXAMPLE 31   How many liters each of a 40% saline solution and a 10% saline solution should be combined to produce 12 L of a 15% saline solution?

Solution   **First,** form two equations from the given information. If we let

x = Amount of 40% solution

y = Amount of 10% solution

then there is a total of 12 L:

$x + y = 12$

The amount of salt in the 40% and 10% solutions equals the amount of salt in the 15% solution:

$0.4x + 0.1y = 0.15(12)$

$4x + y = 18$

We now have a system of two equations:

$x + y = 12$

$4x + y = 18$

Thus,

$$A = \begin{bmatrix} 1 & 1 \\ 4 & 1 \end{bmatrix} \quad \text{and} \quad A^{-1} = \begin{bmatrix} -\dfrac{1}{3} & \dfrac{1}{3} \\ \dfrac{4}{3} & -\dfrac{1}{3} \end{bmatrix}$$

Since $C = \begin{bmatrix} 12 \\ 18 \end{bmatrix}$,

$$X = A^{-1}C = \begin{bmatrix} -\dfrac{1}{3} & \dfrac{1}{3} \\ \dfrac{4}{3} & -\dfrac{1}{3} \end{bmatrix} \begin{bmatrix} 12 \\ 18 \end{bmatrix} = \begin{bmatrix} 2 \\ 10 \end{bmatrix}$$

Therefore, $x = 2$ and $y = 10$. We must combine 2 L of 40% solution with 10 L of 10% solution. Check this.

Systems of three equations   The matrix method can be used to solve a system of linear equations of any size. Here is an example involving three equations in three unknowns.

EXAMPLE 32   If we apply Kirchoff's laws to the circuit shown in Figure 16-1, we have the system of equations given below. Find the three currents, I_1, I_2, and I_3.

$$I_1 - I_2 + I_3 = 0$$

$$2I_1 + 8I_2 = 4$$

$$8I_2 + 3I_3 = 17$$

Figure 16-1

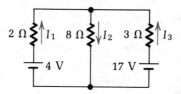

Solution   Be careful to represent the missing variables with 0's.

$$A = \begin{bmatrix} 1 & -1 & 1 \\ 2 & 8 & 0 \\ 0 & 8 & 3 \end{bmatrix}$$

and, using the Gauss–Jordan method,

$$A^{-1} = \begin{bmatrix} \dfrac{12}{23} & \dfrac{11}{46} & -\dfrac{4}{23} \\ -\dfrac{3}{23} & \dfrac{3}{46} & \dfrac{1}{23} \\ \dfrac{8}{23} & -\dfrac{4}{23} & \dfrac{5}{23} \end{bmatrix}$$

Solving for X, we have

$$X = A^{-1}C = \begin{bmatrix} \dfrac{12}{23} & \dfrac{11}{46} & -\dfrac{4}{23} \\ -\dfrac{3}{23} & \dfrac{3}{46} & \dfrac{1}{23} \\ \dfrac{8}{23} & -\dfrac{4}{23} & \dfrac{5}{23} \end{bmatrix} \begin{bmatrix} 0 \\ 4 \\ 17 \end{bmatrix} = \begin{bmatrix} -2 \\ 1 \\ 3 \end{bmatrix}$$

Thus, $I_1 = -2$ A, $I_2 = 1$ A, and $I_3 = 3$ A. Check this solution.

Application: three equations in three unknowns   The next example involves finding the constants in a formula based on experimental data. See Example 35 in Section 5-6 for review.

EXAMPLE 33   Petroleum engineers have found that the pressure drop P in an oil reservoir is related to elapsed time t by the formula

$$P = at + bt^2 + ct^3$$

To determine the constants a, b, and c for a particular reservoir, three readings were taken. The results are summarized in the table. Find a, b, and c.

Time, t (years)	P (lb/in.²)
1	275
2	425
3	525

Solution **First,** set up three equations based on the given data:

t t^2 t^3

$1\,a + 1\,b + 1\,c = 275$ First year

$2\,a + 4\,b + 8\,c = 425$ Second year

$3\,a + 9\,b + 27\,c = 525$ Third year

Then, proceed as before. The coefficient matrix and its inverse are

$$A = \begin{bmatrix} 1 & 1 & 1 \\ 2 & 4 & 8 \\ 3 & 9 & 27 \end{bmatrix} \qquad A^{-1} = \begin{bmatrix} 3 & -\dfrac{3}{2} & \dfrac{1}{3} \\ -\dfrac{5}{2} & 2 & -\dfrac{1}{2} \\ \dfrac{1}{2} & -\dfrac{1}{2} & \dfrac{1}{6} \end{bmatrix}$$

Solving for X, we have

$$X = A^{-1}C = \begin{bmatrix} 3 & -\dfrac{3}{2} & \dfrac{1}{3} \\ -\dfrac{5}{2} & 2 & -\dfrac{1}{2} \\ \dfrac{1}{2} & -\dfrac{1}{2} & \dfrac{1}{6} \end{bmatrix}\begin{bmatrix} 275 \\ 425 \\ 525 \end{bmatrix} = \begin{bmatrix} a \\ b \\ c \end{bmatrix}$$

$a\Rightarrow 3\;\boxed{\times}\;275\;\boxed{+}\;3\;\boxed{+/-}\boxed{\times}\;425\;\boxed{\div}\;2\;\boxed{+}\;525\;\boxed{\div}\;3\;\boxed{=}\;\rightarrow$ *362.5*

$b\Rightarrow 5\;\boxed{+/-}\boxed{\times}\;275\;\boxed{\div}\;2\;\boxed{+}\;2\;\boxed{\times}\;425\;\boxed{-}\;525\;\boxed{\div}\;2\;\boxed{=}\;\rightarrow$ *−100.*

$c\Rightarrow 275\;\boxed{\div}\;2\;\boxed{-}\;425\;\boxed{\div}\;2\;\boxed{+}\;525\;\boxed{\div}\;6\;\boxed{=}\;\rightarrow$ *12.5*

Therefore, $a = 362.5$ lb/in.^{2}/year, $b = -100$ lb/in.^{2}/year^2, and $c = 12.5$ lb/in.^{2}/year^3.

NOTE ▶ If a system is dependent or inconsistent (as defined in Chapter 5), its coefficient matrix will have *no* inverse. ◀

Exercises 16-6

Solve each system of equations using the matrix method.

1. $x - y = 3$
   $2x + y = 3$

2. $x + y = -1$
   $2x + 3y = 0$

3. $2x + y = 9$
   $5x - 3y = 17$

4. $4x + 3y = 1$
   $3x + 2y = 0$

5. $2x - y = 3$
   $4x - 3y = 8$

6. $5x - 6y = 3$
   $-2x + 3y = -1$

7. $3x - 2y = -12$
   $6x + y = 1$

 8. $4.20x + 6.80y = 67.38$
   $9.10x - 2.70y = 51.85$

9. $x + y + z = 1$
   $2x - 2z = -18$
   $3x - y - 4z = -33$

10. $x + 2y + z = -6$
$2x + z = 5$
$-5y - z = 21$

11. $x + y + 2z = 2$
$4x - 2y + z = 17$
$-2x + 3y + 2z = -12$

12. $-3x + 2y + z = 7$
$2x + y - 2z = 11$
$-4x - y + 3z = -14$

13. $3x + y + 2z = 3$
$6x - 2y + z = 15$
$-3x + y - z = -10$

14. $2x + y - z = -4$
$4x - y + z = 13$
$6x - 3y + 2z = 27$

15. $5x + 2y + z = 711$
$x + 5y + z = 725$
$2x + 3y + 2z = 610$

16. $-2x + 5y + 2z = 22$
$3x + 4y + z = 25$
$x - 2y - z = -9$

17. *Civil engineering*   In determining the reactions at the supports for a statically indeterminate beam, it is necessary to solve the system of equations given below. Solve for V and M in terms of P and L.*

$$4VL + 4M = PL$$

$$16VL + 12M = 5PL$$

18. *Civil engineering*   To test the elastic deformation of a beam, weights w_1, w_2, and w_3 are applied to it, and the total deflection is measured. The following equations are obtained. Solve for w_1, w_2, and w_3.

$$0.01w_1 + 0.02w_2 + 0.04w_3 = 2.0$$

$$0.02w_1 + 0.01w_2 + 0.02w_3 = 2.5$$

$$0.04w_1 + 0.02w_2 + 0.01w_3 = 3.0$$

[*Hint:*   Factor out 0.01 from the coefficient matrix to simplify finding the inverse. Then multiply the resulting solutions by 100.]

19. *Physics*   The displacement s of an object under constant acceleration a for an elapsed time t is given by

$$s = s_0 + v_0 t + \frac{1}{2}at^2$$

where s_0 is the initial displacement and v_0 is the initial velocity. Experimental results on a certain object showed a displacement of 7 ft after 1 s, 15 ft after 2 s, and 25 ft after 3 s. Solve for s_0, v_0, and a.

20. *Chemistry*   Solutions of 15% saline and 25% saline are combined to obtain 100 mL of 18% saline solution. How many milliliters of each solution are used?

16-7 | Review of Chapter 16

Important Terms and Concepts

minor (p. 546)

cofactor (p. 547)

expansion by cofactors (p. 548)

Cramer's rule (p. 550)

matrix (p. 562)

square matrix (p. 563)

principal diagonal (p. 563)

zero matrix (p. 563)

unit, or identity, matrix (p. 563)

J matrix (p. 564)

multiplication of matrices (p. 567)

inverse of a matrix (p. 574)

determinant method (p. 575)

Gauss–Jordan method (p. 575)

---

* Based on Nelson R. Bauld, Jr., *Mechanics of Materials* (Monterey, Ca.: Brooks/Cole Engineering Division, Wadsworth, 1982), p. 511.

elements of a matrix (p. 563)   equality of matrices (p. 564)   row transformations (p. 576)

order of a matrix (p. 563)   addition of matrices (p. 565)   coefficient matrix (p. 580)

column matrix (p. 563)   scalar multiplication (p. 565)   solution matrix (p. 580)

row matrix (p. 563)   subtraction of matrices (p. 565)   constant matrix (p. 580)

Properties of Determinants (Section 16-2)

1. If all elements in a particular row or column are 0, the value of the determinant is 0.
2. If all elements above or below the principal diagonal are 0, the value of the determinant is the product of the elements of the principal diagonal.
3. If two rows or two columns of a determinant are identical, the value of the determinant is 0.
4. If any two columns or any two rows of a determinant are interchanged, the value of the resulting determinant is the negative of the value of the original determinant.
5. If all corresponding rows and columns of a determinant are interchanged, the value of the determinant is unchanged.
6. If all the elements of a row or column are multiplied by the same nonzero number k, the value of the resulting determinant will be k times the value of the original determinant.
7. If the same multiple of each element of a row (or column) is added to the corresponding element of another row (or column), the value of the determinant is unchanged.

Exercises 16-7

Evaluate each determinant using expansion by cofactors.

1. $\begin{vmatrix} 4 & 1 & 2 \\ -1 & -2 & 0 \\ 3 & -4 & 1 \end{vmatrix}$
2. $\begin{vmatrix} -1 & 5 & 3 \\ 0 & 2 & 4 \\ 1 & -2 & -3 \end{vmatrix}$
3. $\begin{vmatrix} 3 & 4 & -1 \\ 2 & 3 & -6 \\ -1 & 4 & 0 \end{vmatrix}$

4. $\begin{vmatrix} 3 & 7 & 1 \\ -2 & -5 & 3 \\ 4 & 1 & -1 \end{vmatrix}$
5. $\begin{vmatrix} 2 & 1 & 1 & 0 \\ -3 & 2 & 1 & 4 \\ 3 & 0 & 4 & 2 \\ 1 & 5 & -3 & -2 \end{vmatrix}$
6. $\begin{vmatrix} 3 & -2 & 5 & 1 \\ 0 & 4 & -1 & 2 \\ 1 & -3 & 0 & 4 \\ -4 & 1 & -5 & -2 \end{vmatrix}$

7. $\begin{vmatrix} 2.7 & 3.1 & -4.2 & 5.9 \\ -0.6 & 4.7 & 7.1 & 2.6 \\ -5.8 & 4.8 & -3.2 & 1.9 \\ 1.8 & -2.4 & 3.6 & 2.5 \end{vmatrix}$
8. $\begin{vmatrix} 44 & 68 & 83 \\ 71 & 19 & 46 \\ 51 & 73 & 26 \end{vmatrix}$

Use Properties 1–7 of determinants to help evaluate each determinant.

9. $\begin{vmatrix} 2 & 3 & 7 \\ 1 & 5 & 1 \\ 2 & 3 & 7 \end{vmatrix}$
10. $\begin{vmatrix} 7 & -3 & 4 \\ 0 & 0 & 0 \\ 3 & -2 & 5 \end{vmatrix}$
11. $\begin{vmatrix} 6 & 4 & 2 \\ 3 & -3 & 1 \\ 3 & 2 & 1 \end{vmatrix}$
12. $\begin{vmatrix} 5 & 1 & 8 \\ 0 & 2 & 3 \\ 0 & 0 & -3 \end{vmatrix}$

13. $\begin{vmatrix} 3 & 0 & 0 & 0 \\ -2 & 4 & 0 & 0 \\ 8 & 6 & -2 & 0 \\ 7 & 3 & -4 & 1 \end{vmatrix}$
14. $\begin{vmatrix} 5 & 0 & 0 & 0 \\ -2 & 0 & 4 & 0 \\ 7 & 3 & 6 & 0 \\ 4 & 1 & -2 & 5 \end{vmatrix}$
15. $\begin{vmatrix} 3 & -3 & 1 \\ 4 & -2 & 0 \\ 2 & 6 & 2 \end{vmatrix}$
16. $\begin{vmatrix} 1 & 5 & -3 \\ 0 & 2 & 6 \\ -1 & -4 & 9 \end{vmatrix}$

17. $\begin{vmatrix} 5 & 1 & 2 \\ 7 & -4 & 8 \\ 3 & 1 & 6 \end{vmatrix}$
18. $\begin{vmatrix} 1 & 2 & 0 & 5 \\ -1 & -4 & 3 & 0 \\ 2 & -5 & 1 & 2 \\ -3 & 1 & 4 & -1 \end{vmatrix}$
19. $\begin{vmatrix} 1 & -2 & 3 & -4 \\ -1 & -4 & 0 & 6 \\ -2 & 5 & 1 & 3 \\ -3 & 2 & -1 & 1 \end{vmatrix}$
20. $\begin{vmatrix} 8 & 1 & 2 & 3 \\ 4 & -1 & -1 & -3 \\ 12 & 0 & -2 & -4 \\ 16 & -3 & 5 & 2 \end{vmatrix}$

Use Cramer's rule to solve each system of equations. Apply Properties 1–7 of determinants wherever possible.

21. $x + y - z = 4$
$2x - y + z = -1$
$3x + 2z = 1$

22. $x - 3y - 2z = -2$
$4x + y + 3z = 11$
$-2x - 3y + z = -3$

23. $a + b - c - 2d = 5$
$3a - 2b + c + d = -8$
$-a + 3b - 2c - d = 1$
$2a - 4b + 5c + d = 6$

24. $-2w + x - 3z = -5$
$2x - 5y + 3z = 1$
$4w + x = 7$
$2w + x - y + z = 3$

Use the following matrices to perform the indicated operations:

$A = \begin{bmatrix} 2 & -3 \\ 0 & 1 \\ 5 & -2 \end{bmatrix}$
$B = \begin{bmatrix} -1 & 4 \\ 6 & -2 \\ -3 & 1 \end{bmatrix}$
$C = \begin{bmatrix} 1 & -4 & -1 \\ -2 & 6 & 3 \\ 1 & 4 & 0 \end{bmatrix}$

25. $A + B$
26. $B + C$
27. $A - B$
28. $B - A$
29. $2A$
30. $-4C$
31. $3A + 2B$
32. $3B - 4A$
33. C^2
34. AB
35. BC
36. CA

Find the following products:

37. $[1 \quad -2 \quad 3] \begin{bmatrix} 2 & 1 & -2 \\ 5 & -1 & 0 \\ 0 & 3 & 4 \end{bmatrix}$
38. $\begin{bmatrix} 3 & 6 \\ -1 & 1 \end{bmatrix}\begin{bmatrix} -1 & 1 & 2 \\ 5 & 0 & 4 \end{bmatrix}$
39. $\begin{bmatrix} 2 & 6 \\ -3 & 0 \\ 5 & 4 \end{bmatrix}\begin{bmatrix} 3 & -7 & -5 \\ -4 & 4 & 1 \end{bmatrix}$

40. $\begin{bmatrix} 22 & 13 & 75 & 29 \\ 43 & 66 & 28 & 51 \end{bmatrix}\begin{bmatrix} 140 \\ 88 \\ 220 \\ 155 \end{bmatrix}$

Find the inverse of each matrix.

41. $\begin{bmatrix} 3 & -2 \\ 1 & -1 \end{bmatrix}$
42. $\begin{bmatrix} 1 & 4 \\ 2 & 6 \end{bmatrix}$
43. $\begin{bmatrix} -5 & -4 \\ 6 & 8 \end{bmatrix}$

44. $\begin{bmatrix} 1 & 1 & 0 \\ 3 & -2 & 1 \\ 2 & -4 & 1 \end{bmatrix}$
45. $\begin{bmatrix} 2 & -3 & 1 \\ 0 & 4 & -2 \\ -5 & -1 & 2 \end{bmatrix}$
46. $\begin{bmatrix} -2 & -4 & 3 \\ 4 & 1 & -1 \\ 0 & -2 & 2 \end{bmatrix}$

Solve each system of equations using the matrix method.

47. $4x - 3y = 3$
$-x + y = -2$

48. $5x + 4y = -6$
$3x + 2y = -5$

49. $-3.70x + 7.40y = 9.62$
$7.50x - 3.50y = 51.8$

50. $-6x + y = 4$
$9x - 2y = -4$

51. $4x + y - z = -3$
$2x - y = 5$
$-3x - 2y + z = 7$

52. $x - 2y - 4z = -5$
$-2x + 2y + 5z = 4$
$-4x - y + 2z = -8$

53. $x + y - 2z = 11$
$3x + 2y + 4z = 9$
$2x + y + 3z = 4$

54. $2x + 7y - 9z = 181$
$x + 5y - 3z = 197$
$x + 2y - 4z = 38$

Solve.

55. *Mechanical engineering* The direction cosines for a certain state of strain can be found by solving the following system. Find n_1 and n_2.

$$-3n_1 - n_2 = 1$$

$$-n_1 - 3n_2 = -1$$

56. *Mechanical engineering* The forces exerted by three rivets on a metal cover plate are given by the system of equations below. Solve for the three forces.

$$F_1 - F_3 = 0$$

$$F_2 - 1.58F_3 = 0$$

$$4.50F_1 + 2.80F_2 + 4.50F_3 = 17{,}900$$

57. *Civil engineering* The equilibrium configuration for a truss is given by the system of equations below. Solve for the displacements x, y, and z in terms of k.

$$5.613x - 0.284y - 0.423z = 0$$

$$0.059x + 3.064y - 0.714z = 0$$

$$-0.766x - 0.714y + 0.714z = k$$

58. *Industrial management* The Write-Rite Typewriter Co. produced four models of typewriters during four different years as represented by the equations

$$2.5a + 0.4b + 4c = 22.2$$

$$2.3a + 0.7b + 4c = 22.7$$

$$1.8a + 1.9b + 1.5c + 2.5d = 27.8$$

$$1.7a + 2.6b + 3d = 26.2$$

where a, b, c, and d are the cost of each model sold (in hundreds of dollars), and the numbers on the right give total sales (in units of $100,000). Find a, b, c, and d.

59. *Cartography* A map projection is achieved by multiplying matrix A below, which represents the coordinates of the original figure, by matrix B, which represents the type of projection. Find the projection matrix AB.

$$A = \begin{bmatrix} 0 & 0 & 1 \\ 3 & 0 & 1 \\ 3 & 3 & 1 \end{bmatrix} \qquad B = \begin{bmatrix} 1 & 0 & 0.1 \\ 0 & 1 & 0.3 \\ 3 & 3 & 2 \end{bmatrix}$$

60. *Optics* In optics theory, it is necessary to multiply the matrices below.* Find this product.

$$\begin{bmatrix} 1 & 0 \\ \dfrac{-p}{x} & \dfrac{n}{x} \end{bmatrix} \begin{bmatrix} \dfrac{n - pd}{n} & \dfrac{md}{n} \\ \dfrac{-p}{n} & \dfrac{m}{n} \end{bmatrix}$$

* Based on Miles V. Klein, *Optics* (New York: Wiley, 1970), p. 88.

Set up a system of equations and solve.

61. *Electronics*   For a simple circuit element, resistance varies with temperature according to the formula

$$R = aT + b$$

If the resistance is 0.7 Ω at 60°C and 0.9 Ω at 80°C, find a and b.

62. *Material science*   Alloys containing 25%, 40%, and 60% copper are combined to produce 50 kg of an alloy containing 38% (19 kg) copper. If 10 kg more of the 60% alloy are used than of the 40% alloy, how many kilograms of each are used?

63. *Industrial management*   Three different factories, *A*, *B*, and *C*, produce the following numbers of industrial robots *X*, *Y*, and *Z* per day:

Robot	Factory A	Factory B	Factory C
X	2		5
Y	3	2	
Z		4	1

How many days should each factory be operated to produce 30 *X* robots, 23 *Y* robots, and 20 *Z* robots?

64. *Business*   Four different types of paper, *A*, *B*, *C*, and *D*, are ordered for a company's high-speed printers. The costs per box are *A*, $20; *B*, $24; *C*, $30; *D*, $45. The company places an order for 28 boxes totaling $750. The order contains three times as many boxes of *A* as *D*, and two more boxes of *C* than *B*. How many boxes of each are in the order?

Inequalities | 17

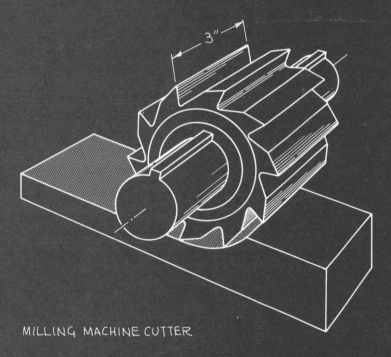

3″

MILLING MACHINE CUTTER

An equation is a mathematical statement of the equality of two quantities. But many mathematical relationships or models of real situations cannot be described easily using equations. For example, it may be necessary to write a statement showing when an object is traveling *faster* than a given velocity, when the combined resistance in a circuit is *lower* than a certain value, or how production should be adjusted to *maximize* profit. Inequalities are used to describe all these situations.

Definitions

An **inequality** is a statement that one quantity is **greater than (>)** or **less than (<)** another quantity. The following statements are all inequalities:

1. $2 < 8$ **2.** $x > -3$ **3.** $y < x + 2$

Two is   x is   y is
less than   greater than   less than
eight   negative three x plus two

The inequality symbols > and < define the **sense** of the inequality. The expressions on either side of the inequality symbol are called its **members.** Statements 1 and 3 above have the same sense, while statements 2 and 3 have the opposite sense.

A number line provides an effective means for visualizing an inequality. On a number line, the smaller number in an inequality is always to the *left* of the larger numbers, as shown in Fig. 17-1.

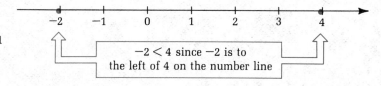

Figure 17-1

$-2 < 4$ since -2 is to the left of 4 on the number line

LEARNING HINT

Think of the inequality symbol as an arrowhead pointing toward the smaller number. ◄

The **solution** of an inequality in one variable consists of all real values of the variable that make the inequality a true statement. A **conditional inequality** is an inequality that is true for some, but not all, allowed values of the variable. An **unconditional inequality** is true for all real values of the variable.

EXAMPLE 1
(a) The statement $x + 2 > 7$ is a conditional inequality. It is true only for values of x greater than 5. Its solution can be expressed by the inequality $x > 5$.
(b) The statement $x^2 + 3 > 0$ is an unconditional inequality. It is true for all real values of x.

An inequality can be combined with an equality using the symbols ≤ (less than or equal to) or ≥ (greater than or equal to).

EXAMPLE 2
(a) $x \leq 4$ is read "x is less than or equal to 4." This inequality is true for values of x less than 4 or for $x = 4$.

(b) x ≥ −2 is read "x is greater than or equal to −2." This inequality is true
for values of x greater than −2 or for x = −2.

Number line graphs   It is often helpful to graph the solution to an inequality in one variable on a
number line. This procedure can be summarized as shown in Fig. 17-2.

Figure 17-2

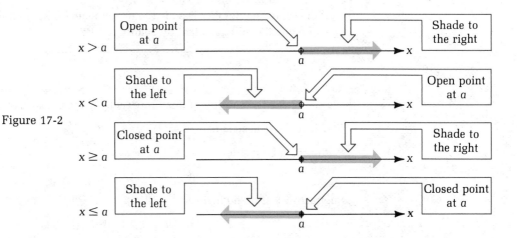

Notice that an open point at *a* indicates that *a* itself is *not* part of the solution,
as in > and <. A closed point at *a* indicates that *a* *is* part of the solution, as in
≥ and ≤. Furthermore, the shading indicates the direction in which the solutions lie: to the right of *a* for >, and to the left of *a* for <.

EXAMPLE 3   Graph: (a) x < 6   (b) x ≥ −3

Solution

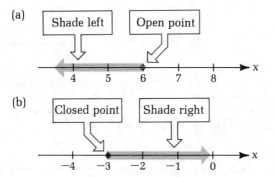

Solving linear   A **linear inequality** is an inequality in which the variable terms appear only
inequalities   to the first power. The process of solving a linear inequality in one variable is
very much like the process of solving a linear equation in one variable. On both
members of the inequality, perform whatever legal operations are needed to
convert the inequality into the form x ☐ *a*, where ☐ represents an inequality

symbol. *Legal operations* are any operations that produce **equivalent inequalities**—that is, inequalities with the same solution. Three properties govern these operations. The first is stated below.

---

1. **The Addition/Subtraction Property**
    If the same real number is added to or subtracted from both members of an inequality, an equivalent inequality results.

EXAMPLE 4

We can check the validity of this property by using actual numbers.

Consider the true statement:	$8 > 2$
Adding 5 to both sides:	$8 + 5 > 2 + 5$

$$13 > 7 \qquad \text{This is also a true statement.}$$

Subtracting 5 from both sides: $8 - 5 > 2 - 5$

$$3 > -3 \qquad \text{This is also true.}$$

We can use the addition/subtraction property to solve certain one-step inequalities.

EXAMPLE 5   Solve and graph: (a) $x - 7 > 3$ (b) $y + 4 \leq 1$

Solution

(a) To isolate x on one side, add 7 to both sides. The solution is:

$$x - 7 > 3$$
$$x - 7 + 7 > 3 + 7$$
$$x > 10$$

Check by substituting any number greater than 10 for x in the original inequality:

$$(11) - 7 > 3$$
$$4 > 3$$

(b) To isolate y on one side, subtract 4 from both sides. The solution is:

$$y + 4 \leq 1$$
$$y + 4 - 4 \leq 1 - 4$$
$$y \leq -3$$

Check by substituting -3 and a number less than -3 in the original inequality:

$$(-3) + 4 \leq 1 \qquad (-4) + 4 \leq 1$$

$$1 \leq 1 \qquad\qquad 0 \leq 1$$

---

2. The Multiplication/Division Property: Positive Numbers

If both members of an inequality are multiplied or divided by the same **positive** real number, an equivalent inequality results.

EXAMPLE 6

We can demonstrate this property using actual numbers.

Given the true statement: $\qquad\qquad -2 < 6$

If we multiply both sides by 3: $\qquad -2 \cdot 3 < 6 \cdot 3$

A true statement results. $\qquad\qquad -6 < 18$

If we divide both sides by 2: $\qquad \dfrac{-2}{2} < \dfrac{6}{2}$

Again, a true statement results. $\qquad -1 < 3$

This property can be used to solve some one-step inequalities.

EXAMPLE 7 Solve and graph: (a) $\dfrac{w}{4} \geq -3$ (b) $3z < 12$

Solution

(a) To isolate w with a coefficient of 1, multiply both sides by 4.

$$\dfrac{w}{4} \geq -3$$

$$4 \cdot \dfrac{w}{4} \geq 4 \cdot (-3)$$

The solution is:

$$w \geq -12$$

(b) To isolate z with a coefficient of 1, divide both sides by 3.

$$3z < 12$$

$$\dfrac{3z}{3} < \dfrac{12}{3}$$

The solution is:

$$z < 4$$

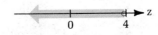

Check each solution by substituting a number from the shaded area of the graph into the original equation.

So far, solving linear inequalities seems much the same as solving linear equations. However, the third property is very different.

---

3.   The Multiplication/Division Property: Negative Numbers
If both members of an inequality are multiplied or divided by the same **negative** real number, the sense of the inequality is **reversed.**

EXAMPLE 8

Consider the true statement: $12 > -6$

$-6 \quad 0 \qquad 12$

If we multiply both sides by -2, the sense of the inequality sign must be reversed.

$(-2) \cdot (12) < (-2) \cdot (-6)$

$-24 < 12$

$-24 \qquad 0 \quad 12$

Similarly, if we divide both sides by -3, we must reverse the inequality sign.

$\dfrac{12}{-3} < \dfrac{-6}{-3}$

$-4 < 2$

$-4 \qquad 0 \quad 2$

EXAMPLE 9   Solve and graph: (a) $-\dfrac{2}{3}x \le 6$ (b) $-4y > -24$

Solution

(a)   To isolate x with a coefficient of 1, multiply both sides by $-\frac{3}{2}$. Because we are multiplying by a negative number, we must reverse the inequality symbol.

$$-\frac{2}{3}x \le 6$$

| $\le$ becomes $\ge$ |

$$\left(-\frac{3}{2}\right)\left(-\frac{2}{3}x\right) \ge \left(-\frac{3}{2}\right)(6)$$

The solution is

$$x \geq -9$$

Check it:

$$-\frac{2}{3}(-9) \leq 6 \qquad\qquad -\frac{2}{3}(0) \leq 6$$

$$6 \leq 6 \qquad\qquad\qquad 0 \leq 6$$

(b) To isolate y with a coefficient of 1, divide both sides by -4 and reverse the inequality symbol.

$$-4y > -24$$

$$\frac{-4y}{-4} < \frac{-24}{-4}$$

The solution is

Check it.

$$y < 6$$

Inequalities with solutions that require more than one step may be solved using a combination of these properties.

REMEMBER ▶ Solving a linear inequality is very similar to solving a linear equation, *except* that the sense of an inequality must be reversed when both sides are multiplied or divided by a **negative** number. ◀

EXAMPLE 10 Solve the following inequalities:

(a) $2x - 3 > 7$ (b) $-5x + 4 \geq 14$

(c) $2.78y - 9.26 < 3.49y + 6.08$ (d) $-3(y - 7) \leq 6 - (7y + 2)$

Solution

(a) Add 3: $2x > 10$ (b) Subtract 4: $-5x \geq 10$

    Divide by 2: $x > 5$ Divide by -5: $x \leq -2$

> Dividing by a negative number reverses the sense.

(c) **Step 1.** Combine the coefficients of the y-terms on the left side, note the sign of the result, and store in memory:

2.78 $\boxed{-}$ 3.49 $\boxed{=}$ $\boxed{\text{STO}}$ →   *−0.71*   Coefficient of y is negative

Step 2. Combine the constant terms on the right side, and divide by the coefficient of y in memory:

6.08 $\boxed{+}$ 9.26 $\boxed{=}$ $\boxed{÷}$ $\boxed{\text{RCL}}$ $\boxed{=}$ →   *−21.605634*

Step 3. State the result, making certain to reverse the sense of the inequality since the sign in Step 1 was negative:

$$y > -21.6 \quad \text{rounded}$$

The sense of the inequality reverses in this case.

CAUTION ▶   To avoid any confusion that might be caused by reading the solution back-wards, it is usually best to move the variable to the left side when solving an inequality. ◀

(d)   Use the distributive property:

$$-3(y - 7) \le 6 - (7y + 2)$$

$$-3y + 21 \le 6 - 7y - 2$$

Combine like terms on the right: $-3y + 21 \le -7y + 4$

Add 7y: $4y + 21 \le 4$

Subtract 21: $4y \le -17$

Divide by 4: $y \le \dfrac{-17}{4}$

EXAMPLE 11   The engineers at the Ramrom Computer Co. know that it will cost them $6000 to design a new component and $3.50 per unit to manufacture it. Therefore, the total cost is $C = \$3.50N + \$6000$, where N is the number of units sold. The amount of money they will take in is $R = \$4.00N - \$2000$. For what values of N will Ramrom make a profit? That is, when is $R > C$?

Solution   Since we want $R > C$, then

$$4.00N - 2000 > 3.50N + 6000$$

$$0.50N > 8000$$

$$N > 16,000$$

They will break even at 16,000 units sold and make a profit for more than that.

Compound inequalities   When a statement consists of two or more inequalities connected by either the word *or* or the word *and*, it is known as a **compound inequality.**

When two inequalities are connected by the word *and*, the solution consists of only those values of the variable that make *both* inequalities true.

EXAMPLE 12   For the compound inequality

$$x > 3 \quad and \quad x < 7$$

the solution consists of all real numbers between 3 and 7. We can write this compound inequality using the shorthand notation

$3 < x < 7$

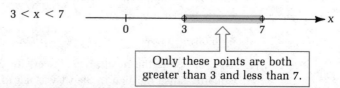

Only these points are both greater than 3 and less than 7.

When two inequalities are connected by the word *or*, the solution consists of any values of the variable that make *either* of the inequalities true.

EXAMPLE 13   For the compound inequality

x < 2   or   x > 8

the solution consists of all real numbers less than 2 or greater than 8:

x < 2   or   x > 8

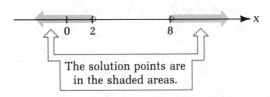

The solution points are in the shaded areas.

There is no shorthand notation for writing a compound inequality containing the word *or*.

EXAMPLE 14   Express in simplest form and graph:
(a)   (x > 4 and x < 9)   or   (x ≤ 0 or x ≥ 12)
(b)   (5 < x < 13)   and   (x > 10 or x < 7)

Solution
(a)   Simplifying the first compound statement, we have

(4 < x < 9)   or   (x ≤ 0 or x ≥ 12)

The *or* between the expressions in parentheses means that values satisfying *either* compound statement will make up the solution.

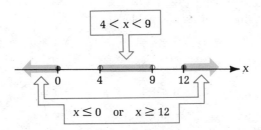

4 < x < 9

x ≤ 0   or   x ≥ 12

There is no simpler way to express the solution.
(b)   The word *and* between the expressions in parentheses means that only values satisfying *both* compound statements will make up the solution. We can best determine these values by graphing each compound statement and finding where they overlap:

$5 < x < 13$

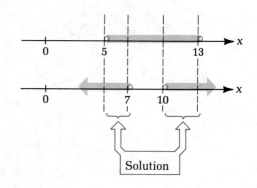

$x > 10$ or $x < 7$

The dashed lines indicate that the solution is

$(5 < x < 7)$ or $(10 < x < 13)$

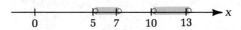

If the original problem is stated as a compound inequality, solve each half of the statement and join the solutions with the original connecting word.

EXAMPLE 15   Solve the following compound inequalities:
(a) $3x - 2 \leq -5$ or $2x + 3 > 7$ (b) $3x - 3 \leq x - 1 \leq 2x + 5$
(c) $-7 < 5 - 2x < 3$

Solution
(a)   Solve each half separately:

$$3x - 2 \leq -5 \qquad \text{or} \qquad 2x + 3 > 7$$

Add 2: $3x \leq -3$ Subtract 3: $2x > 4$
Divide by 3: $x \leq -1$ Divide by 2: $x > 2$

The solution is $x \leq -1$ or $x > 2$

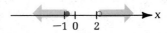

(b)   Rewrite the problem as

$$3x - 3 \leq x - 1 \quad \text{and} \quad x - 1 \leq 2x + 5$$

Then solve each part:

$2x - 3 \leq -1$ $-x - 1 \leq 5$

$2x \leq 2$ $-x \leq 6$

$x \leq 1$ $x \geq -6$

The solution is $x \leq 1$ and $x \geq -6$, which may be written as

$$-6 \leq x \leq 1$$

(c) With the variable in the middle member only, the inequality may be solved as given.

$$-7 < 5 - 2x < 3$$

Subtract 5 from all three members. $$-12 < -2x < -2$$

Divide by -2 and reverse both inequality signs: $$6 > x > 1$$

In more conventional form, the solution is $$1 < x < 6$$

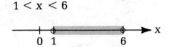

Exercises 17-1

Solve and graph the following inequalities:

1. $x + 5 > 0$ **2.** $x - 2 < 0$ **3.** $y + 4 \le 2$

4. $y - 6 \ge -9$ **5.** $4x < -12$ **6.** $-\frac{3}{2}x > 18$

7. $-\dfrac{m}{5} \le 1$ **8.** $6n \ge 21$ **9.** $-2x + 5 > 17$

10. $8 - x < 12$ **11.** $\dfrac{3a}{4} - 2 \le 7$ **12.** $4b + 8 \ge 3$

13. $3x \le x + 9$ **14.** $2x + 6 < 5x$ **15.** $w + 8 \ge 2w + 6$

16. $\dfrac{z}{3} - 5 > \dfrac{z}{6} + 1$ **17.** $\frac{1}{2}(2 - x) \ge 3$ **18.** $2 - (5 - 3x) < 6$

19. $4x - (x - 6) > x + 2$ **20.** $4 - 2(x - 3) \le 5x$ **21.** $x < 3$ and $x > -1$

22. $x \le 9$ or $x \ge 12$ **23.** $2x - 3 \ge 1$ or $-x + 2 > 5$

24. $3x - 8 < -5$ and $5 - x < 6$ **25.** $-2 < 5x - 2 < 3$

26. $3 \le -2x - 1 \le 17$ **27.** $4x + 6 \le x - 3 \le 3x + 5$

28. $2x - 5 < x - 6 < 5x + 6$ **29.** $(6 < x < 11)$ or $(x > 15$ or $x < 1)$

30. $(x < -2$ or $x > 6)$ or $(x > 0$ and $x < 4)$ **31.** $(x \le 7$ or $x \ge 10)$ and $(5 < x < 12)$

32. $(x < 8$ and $x > 3)$ and $(x > 6$ or $x < 4)$ **33.** $2.73x - 4.69 < 8.31$

34. $1.36x + 2.48 \le -8.12x + 56.8$ **35.** $7492x - 2758 > 7563x + 3428$

36. $46{,}852 - (27{,}934 + 568x) \ge 2496$

37. *Aeronautical engineering* The speed of a supersonic aircraft is often given in terms of the **Mach** number, which is the ratio of the speed of the aircraft to the speed of sound. For example, since the speed of sound is approximately 740 mi/h, Mach 1.2 would be 1.2(740 mi/h) $\approx$ 890 mi/h. Use a compound inequality statement to express the range of speed v of an aircraft designed to fly between 1.5 and 2.2 Mach.*

38. *Meteorology* The formula $T = t - \dfrac{11h}{2000}$ gives the approximate air temperature T (in degrees Fahrenheit) at a given height h (in feet) if the ground temperature is $t°F$. If the ground temperature is 80°F, for what values of h above that location will the air temperature remain above 32°F? (Round to the nearest hundred feet.)

* Based on MAA/NCTM Joint Committee, *A Sourcebook of Applications of School Mathematics* (NCTM, 1980), p. 89.

39. *Business*   Salespeople for the Solar Engineering Corp. are paid a commission of 15% on all sales over $4000. What amount must be sold in order to earn a commission over $1200?

40. *Industrial management*   The engineers at the Ramrom Computer Co. have designed a component costing $8000 for research and development plus $2.80 per unit for production. The revenue from the component is $R = \$5.20N - \$2500$. For what values of N does Ramrom make a profit? (See Example 11.)

17-2 | Solving Nonlinear Inequalities

In Section 17-1, linear, or first-degree, inequalities in one variable were solved using the same techniques previously applied to linear equations. However, the techniques used for solving higher-degree inequalities are somewhat different from those applied to higher-degree equations. In this section we shall consider two techniques for solving higher-degree inequalities: one for factorable inequalities and another for nonfactorable inequalities.

Critical value   The first technique is based on the following statement, which we will accept without proof: If $x = a$ is a zero of the function $y = f(x)$, or if the function is not defined at $x = a$, then a is called a **critical value.** The sign of the function can change only at a critical value.

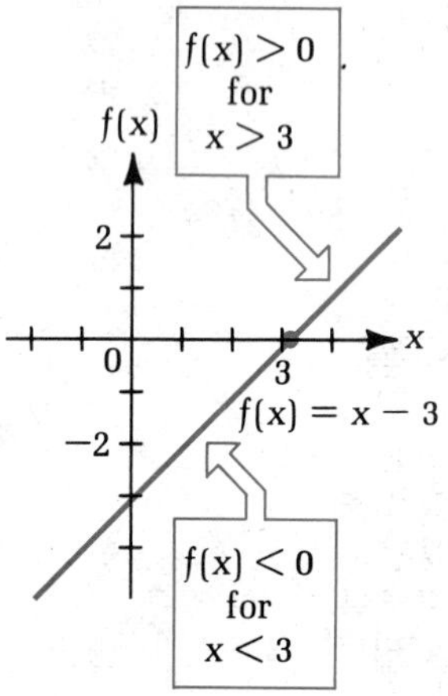

Figure 17-3

EXAMPLE 16   Consider the linear function $f(x) = x - 3$. By inspection we see that there is one critical value, the zero $x = 3$. If we substitute for x any value less than 3, $f(x)$ is negative. For example,

$$f(2) = (2) - 3 = -1 \qquad f(0) = (0) - 3 = -3 \qquad \text{and so on}$$

If we substitute for x any value greater than 3, $f(x)$ is positive. For example,

$$f(3.5) = (3.5) - 3 = 0.5 \qquad f(7) = (7) - 3 = 4 \qquad \text{and so on}$$

Therefore, the sign of $f(x)$ changes at the critical value $x = 3$.

Solving factorable inequalities   The concept of a critical value is useful particularly in solving factorable, nonlinear inequalities. The step-by-step procedure is illustrated in Example 17.

EXAMPLE 17   Solve: $x^2 - 3x > 10$

Solution

Step 1.   Move all terms to the left and factor the left member:

$$x^2 - 3x - 10 > 0$$

$$(x - 5)(x + 2) > 0$$

Thus, for the function $x^2 - 3x - 10$, the critical values are the zeros $x = 5$ and $x = -2$.

Step 2.   Draw a number line and mark an open point at each critical value. This circle would be darkened if the inequality sign was $\le$ or $\ge$. Notice that this creates one more interval on the number line than the number of critical values:

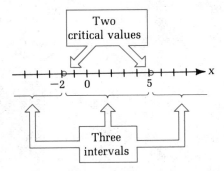

Step 3.   Choose a sample point in each interval and see if it satisfies the factored inequality. This can be done by simply testing signs:

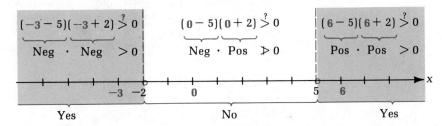

Step 4.   Because the sign of a function can change only at a critical value, the sign will be the same everywhere within the intervals defined by the critical values. Therefore, if a sample point in Step 3 satisfies the inequality, all points within that interval will also satisfy the inequality. Shade in all "yes" intervals on the graph and state the solution in inequality form:

$x < -2$ or $x > 5$

This technique for solving nonlinear inequalities can also be used when the left member of the inequality is written as a quotient. Care must be taken not to include in the solution any values that cause the denominator to be 0.

EXAMPLE 18   Solve: $\dfrac{x + 2}{x - 3} \le 0$

Solution
Step 1.   All terms are on the left and no factoring is necessary.

Step 2.   Graph the critical values. Since $x = -2$ is a zero of the function, it is a critical value. Because $x = 3$ causes the function to be undefined, it is also a critical value. Normally, the ≤ symbol would result in closed points at the critical values. However, because x = 3 causes the denominator to be 0, we must use an open point at x = 3 to indicate that it is not part of the solution:

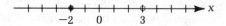

Step 3. Test a sample point in each interval:

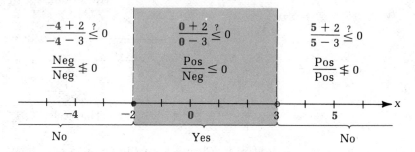

Step 4. Make a final graph of the "yes" interval and state the solution in inequality form. The boundary points stay exactly as they were in Steps 2 and 3, closed at x = 2 and open at x = 3:

$-2 \le x < 3$ 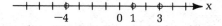

In a more complex problem, there may be several critical values.

EXAMPLE 19   Solve the nonlinear inequality: $\dfrac{(x + 4)(x - 1)^2}{3 - x} < 0.$

Solution

Step 1. The equation is already written in factored form.

Step 2. There are three unique critical values, −4, 1, and 3. Although x = 3 creates a denominator of 0, this point is open anyway because of the < symbol:

Step 3. Test the four intervals as before. Notice that $(x - 1)^2$ is always positive:

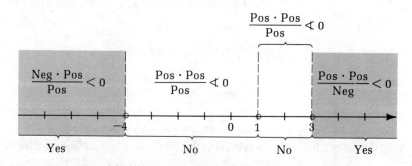

Step 4.   Graph and state the final solution:

$$x < -4 \quad \text{or} \quad x > 3$$

Solving unfactorable inequalities

So far, we have considered only inequalities in which the left member was factorable. If an unfactorable expression results after all terms are moved to the left, the inequality should be solved by graphing in the coordinate plane.

EXAMPLE 20   Solve the inequality: $x^2 > 2x + 5$

Solution

Step 1.   Move all terms to the left side:

$$x^2 - 2x - 5 > 0$$

Step 2.   Since the expression on the left is unfactorable, set it equal to $f(x)$ and graph $f(x) = x^2 - 2x - 5$, as shown in Fig. 17-4(a).

Step 3.   The solution of $f(x) > 0$ consists of all values of x for which the graph is *above* the x-axis. For $f(x) < 0$, the solution consists of all values of x for which the graph is *below* the x-axis. In this case, we want $f(x) > 0$. If we estimate the zeros of the function to be $x \approx 3.4$ and $x \approx -1.4$, then the curve

x	f(x)
-3	10
-2	3
-1	-2
0	-5
1	-6
2	-5
3	-2
4	3
5	10

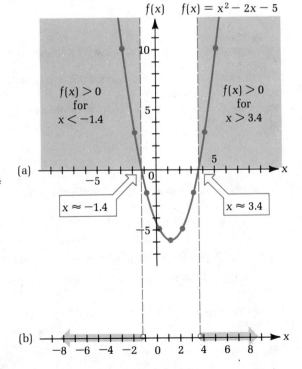

Figure 17-4

is above the x-axis to the left of -1.4 and to the right of 3.4.* Thus, written in inequality form, the approximate solution is

$$x < -1.4 \quad \text{or} \quad x > 3.4$$

The number line graph of this solution is shown in Fig. 17-4(b).

Exercises 17-2

Solve the following inequalities and graph each solution on a number line:

1. $x^2 - 4 > 0$
2. $x^2 - 9 \leq 0$
3. $4x^2 - 25 < 0$
4. $16 - 9x^2 \geq 0$
5. $x^2 + 2 \geq 0$
6. $x^2 + 9 < 0$
7. $x^2 + 7x < 0$
8. $x^2 - 5x \geq 0$
9. $x^2 + 6x \geq -8$
10. $x^2 - x \leq 30$
11. $2x^2 + 15 < 11x$
12. $3x^2 + 2 > -7x$
13. $x^3 + 3x^2 - 28x \geq 0$
14. $x^3 - 9x^2 + 18x < 0$
15. $x^2 + 4x + 2 < 0$
16. $5x^2 - 3x - 3 \geq 0$
17. $x^4 - 8x^2 - 9 < 0$
18. $x^3 - 2x^2 - 13x - 10 \geq 0$
19. $x^3 - 5x^2 + x - 1 \leq 0$
20. $x^4 + 3x^3 - 2x^2 + x - 5 > 0$
21. $\dfrac{x - 3}{x + 4} > 0$
22. $\dfrac{x + 5}{x + 1} < 0$
23. $\dfrac{x - 1}{2 - x} \leq 0$
24. $\dfrac{2x + 1}{x - 1} \geq 0$
25. $\dfrac{3}{x^2 + 5x + 6} < 0$
26. $\dfrac{-1}{x^2 - 3x - 40} \leq 0$
27. $\dfrac{x + 4}{x^2 - 2x - 35} \geq 0$
28. $\dfrac{x^2 - 10x + 24}{x - 1} > 0$
29. $\dfrac{(x - 3)^2(x + 2)}{5 - x} \geq 0$
30. $\dfrac{(x + 1)(x - 4)}{(x + 3)^2} \leq 0$
31. $\dfrac{x(x - 5)(x + 3)}{(x + 2)^3} < 0$
32. $\dfrac{x^2(x + 6)(x - 3)}{(x - 5)(x + 1)^5} > 0$

33. *Electronics*   The transient current i (in amperes) in a circuit varies with the time t (in milliseconds) according to $i = 6 + 3t^2$. Express as an inequality the time required for the current to exceed 9 A.

34. *Energy technology*   The inequality below gives the desired depth b (in feet) of the buckets of a certain waterwheel when the maximum stream flow rate is Q.† Find b (to the nearest 0.1 ft) when $Q = 8$ ft³/s. [Note: If two positive roots are obtained, the smaller root represents the solution for an undershot wheel.]

$$\frac{2b(12 - b)}{3} \leq Q$$

35. *Physics*   If we neglect air resistance, the range R of a projectile is given by

$$R = \frac{v_x^2 + v_y^2}{g}$$

where v_x and v_y are the horizontal and vertical components of its initial velocity. Given $v_x = 85$ ft/s and $g = 32$ ft/s², for what values of v_y will the projectile have a range of at least 320 ft?

---

*A graphical solution generally yields only approximate critical values. If more exact solutions are required, use the quadratic formula or linear interpolation.

†For more information, see Leckie, Jim; Masters, Gil; Whitehouse, Harry; Young, Lily. *Other Homes and Garbage* (San Francisco: Sierra Club Books, 1975), p. 69.

36. *Petroleum engineering* The pressure–time equation for a certain oil reservoir is

$$P = t^3 - 12t^2 + 120t$$

Express as an inequality the minimum number of years t necessary for the pressure drop P to exceed 310 lb/in.².

17-3 | Inequalities Involving Absolute Value

Absolute value   The **absolute value** $|a|$ is the distance along the number line from the origin to point a. The solution of the inequality $|x| > 3$ therefore consists of all numbers on the number line located *more* than 3 units from the origin. The number line in Fig. 17-5 shows that all numbers greater than 3 or less than -3 satisfy this requirement. Thus, the single statement $|x| > 3$ is equivalent to the pair of inequalities $x < -3$ or $x > 3$.

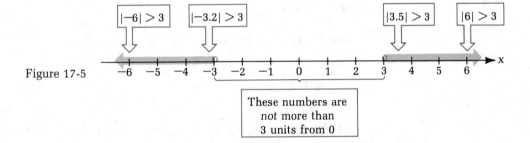

Figure 17-5

Similarly, the solution of the inequality $|x| < 3$ consists of all numbers located *less* than 3 units from the origin. All numbers between -3 and 3 satisfy this requirement, as shown in Fig. 17-6. Therefore, the single inequality statement $|x| < 3$ is equivalent to the compound statement $-3 < x < 3$.

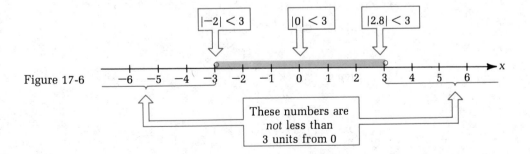

Figure 17-6

Solving absolute value   The preceding examples have shown that an inequality involving absolute
inequalities   value can be rewritten as an equivalent compound inequality without absolute value. The compound inequality is the solution to the absolute value inequality.

Solution to Absolute Value Inequalities

If $|f(x)| > a$, then $f(x) < -a$ or $f(x) > a$.   **(17-1)**

If $|f(x)| < a$, then $-a < f(x) < a$.   **(17-2)**

Solutions (17-1) and (17-2) also apply when the symbols $\geq$ and $\leq$ are used.

EXAMPLE 21   Solve: $|x + 2| > 5$

Solution

Step 1.   Use (17-1) to rewrite the problem as a compound inequality:

If $|x + 2| > 5$, then $x + 2 < -5$ or $x + 2 > 5$.

Step 2.   Solve this compound inequality to obtain

$x < -7$ or $x > 3$

Step 3.   Check it:

For $x = -8$:   **For $x = 4$:**

$|-8 + 2| > 5$ $|4 + 2| > 5$

$\quad\quad 6 > 5$ $\quad\quad 6 > 5$

EXAMPLE 22   Solve: $3|2x - 3| < 15$

Solution

Step 1.   Divide both sides by 3:

$|2x - 3| < 5$

Step 2.   Use (17-2) to state the problem as a compound inequality:

If $|2x - 3| < 5$, then $-5 < 2x - 3 < 5$.

Step 3.   Solve: $-2 < 2x < 8$

$\quad\quad\quad\quad\quad -1 < x < 4$

Step 4.   Check it.

EXAMPLE 23   Solve: $\left|\dfrac{-3x + 4}{2}\right| \leq 4$

Solution

Step 1.   By (17-2):

$$-4 \leq \frac{-3x + 4}{2} \leq 4$$

Step 2.   Solve: $-8 \le -3x + 4 \le 8$

$$-12 \le -3x \le 4$$

$$4 \ge x \ge -\frac{4}{3}$$

Note the reversal of sense

$$-\frac{4}{3} \le x \le 4$$

Step 3.   Check it.

When the absolute value of a higher-degree expression is involved, the same method is used, but the solution may be more complicated.

EXAMPLE 24   Solve: $\left| x^2 - 5x + 5 \right| > 1$

Solution

Step 1.   Use (17-1) to rewrite this as two inequalities:

$$x^2 - 5x + 5 < -1 \quad \text{or} \quad x^2 - 5x + 5 > 1$$

Step 2.   Solve the first inequality:

Move all terms to the left. $x^2 - 5x + 6 < 0$

Factor. $(x - 3)(x - 2) < 0$

Graph the critical values and test the intervals:

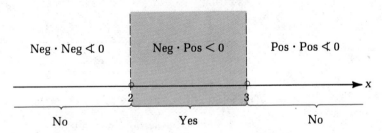

The solution to the first inequality is $2 < x < 3$.

Step 3.   Solve the second inequality:

Move all terms to the left. $x^2 - 5x + 4 > 0$

Factor. $(x - 4)(x - 1) > 0$

Use the critical values, as before:

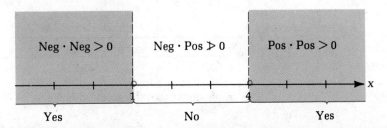

The solution to the second inequality is $x < 1$ or $x > 4$.

Step 4. Combine the two solutions using the original connecting word:

$(2 < x < 3)$ or $(x < 1$ or $x > 4)$

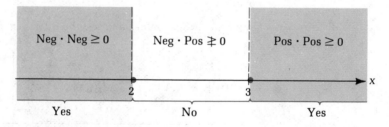

There is no simpler way to state this.

EXAMPLE 25 Solve: $|x^2 - 5x + 5| \le 1$

Solution
Step 1. Use (17-2) to rewrite this as

$$-1 \le x^2 - 5x + 5 \le 1$$

This means:

$$x^2 - 5x + 5 \ge -1 \quad \text{and} \quad x^2 - 5x + 5 \le 1$$

Step 2. Solve the first inequality:

Move all terms to the left. $\qquad x^2 - 5x + 6 \ge 0$

Factor. $\qquad\qquad\qquad (x - 3)(x - 2) \ge 0$

Use the critical values:

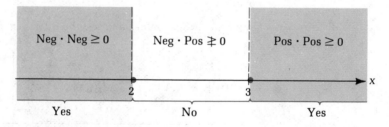

The solution is $x \le 2$ or $x \ge 3$.

Step 3. Solve the second inequality:

Move all terms to the left. $\qquad x^2 - 5x + 4 \le 0$

Factor. $\qquad\qquad\qquad (x - 4)(x - 1) \le 0$

Use the critical values:

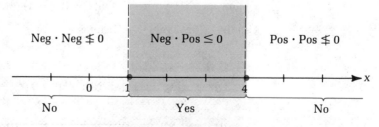

The graph shows that the solution is $1 \le x \le 4$.

Step 4. Combine the two solutions using the original connecting word:

$$(x \le 2 \text{ or } x \ge 3) \quad \text{and} \quad (1 \le x \le 4)$$

The word *and* means that *both* conditions must be satisfied simultaneously. Using the method of Example 14(b) in Section 17-1, we see that only values of x between and including 1 and 2 or between and including 3 and 4 will meet these conditions. The solution is

$$(1 \le x \le 2) \quad \text{or} \quad (3 \le x \le 4)$$

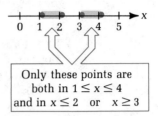

Only these points are
both in $1 \le x \le 4$
and in $x \le 2$ or $x \ge 3$

Exercises 17-3

Solve.

1. $|x + 3| > 7$ 2. $|x - 2| \ge 5$ 3. $|x - 5| \le 1$ 4. $|x + 6| < 2$

5. $|2x - 3| \ge 1$ 6. $|2x + 5| > 11$ 7. $|3x - 5| < 8$ 8. $|4x + 1| \le 7$

9. $|2 - x| > 2$ 10. $|7 - 2x| \le 1$ 11. $|-2x + 7| < 3$ 12. $|-5x - 4| \ge 1$

13. $3|x + 5| \ge 9$ 14. $2|x - 6| < 8$ 15. $\left|\dfrac{x}{3} - 2\right| > 3$ 16. $\left|\dfrac{3x - 1}{4}\right| \le 6$

17. $|x^2 + x - 9| < 3$ 18. $|x^2 - 4x - 1| \le 4$ 19. $|x^2 - 7x + 11| \ge 1$ 20. $|x^2 - 2x - 1| \ge 2$

21. $|3.4x - 8.5| < 20.4$ 22. $|74x - 925| \ge 3885$

23. *Computer science* A certain diskette must be kept within a temperature range given by $|x - 85°F| \le 35°F$. What are the minimum and maximum temperatures of the range?

24. *Machine technology* The specifications for the length of a certain machine part can be expressed as $|x - 3.52 \text{ in.}| \le 0.02$ in. What are the minimum and maximum acceptable lengths?

25. *Electronics* The inequality $|R - 10.2| < 0.4$ expresses the range of resistance (in ohms) of a certain resistor under given temperature conditions. Solve this inequality for R to find the range of resistance.

26. *Material science* By solving the inequality

$$|-2.8T^2 + 820T - 44,000| < 10,000$$

for the temperature T (in degrees Kelvin), we obtain the range of temperatures of 100.0 g of silver when it receives 2.0×10^4 cal of heat. Solve for T to the nearest 10 K.

17-4 Inequalities in Two Variables and Systems of Inequalities

So far, we have solved only inequalities involving one variable. In this section we will solve graphically single inequalities in two variables and systems of two inequalities in two variables. Finally, we will explore an important application involving systems of inequalities.

Inequalities in two variables   In Chapter 5 we showed that a single equation in two variables has an infinite number of ordered-pair solutions. The solutions cannot be listed, but they can be illustrated by a graph in the rectangular coordinate system. An inequality in two variables also has an infinite number of solutions. These can be represented by a region in the coordinate plane.

EXAMPLE 26   Consider the equation $y = 2x$. A graph of its ordered-pair solutions results in the straight line shown in Fig. 17-7(a). As the sample solutions show, every point on this line has a y-coordinate equal to twice its x-coordinate.

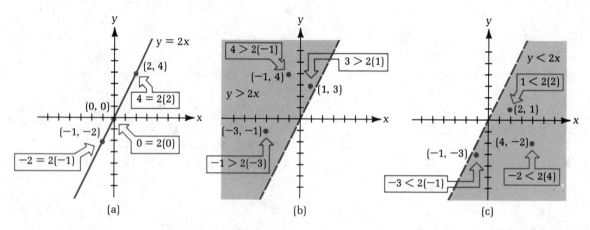

Figure 17-7

Next, consider the inequality $y > 2x$. Its solution consists of every point with a y-coordinate *greater than* twice its x-coordinate. The three sample points shown in Fig. 17-7(b) all satisfy this requirement. Notice that they are all in the region *above* the line $y = 2x$. Further experimentation will show that any point in the shaded region above this line satisfies the given inequality.

Now consider the inequality $y < 2x$. The sample points shown in Fig. 17-7(c) all satisfy this statement because their y-coordinates are all *less than* twice their x-coordinates. Every point in the shaded area *below* this line will satisfy the given inequality.

Finally, notice that the points on the line $y = 2x$ do not satisfy either $y > 2x$ or $y < 2x$. This fact is shown by the use of the dashed lines in Figs. 17-7(b) and (c).

Example 26 leads us to the following method for graphing a single inequality in one variable.

To graph an inequality in the form $y > f(x)$ or $y < f(x)$:

- **First,** sketch the graph of $y = f(x)$. Use a dashed line to show that it is not part of the solution.

- **Second,** indicate $y > f(x)$ by shading the region of the plane *above* the curve, or indicate $y < f(x)$ by shading the region of the plane *below* the curve.
- **Finally,** always check the solution by choosing two points on the plane—one in the shaded region and one in the unshaded region—and substituting their coordinates into the original inequality to be certain the graph is correct.

EXAMPLE 27 Sketch the graph of $y < x - 2$.

Solution First, prepare a table of ordered pairs and sketch the graph of $y = x - 2$ with a dashed line (Fig. 17-8).

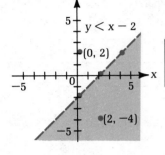

x	0	2	4
y	-2	0	2

For a linear function find the y-intercept, the x-intercept, and a third point to check.

Figure 17-8

Second, shade the region *below* the line.

Finally, check the solution by testing two points, one in the shaded region and one in the unshaded region. For example:

Selecting (2, −4) **in the shaded region:**	**Selecting (0, 2)** **in the unshaded region:**
$y < x - 2$	$y < x - 2$
$(-4) < (2) - 2$	$(2) < (0) - 2$
$-4 < 0$ True	$2 < -2$ False

As we would expect from the graph, $(2, -4)$ is part of the solution and $(0, 2)$ is not.

If the inequality symbol is $\leq$ or $\geq$, use a *solid* graph line to show that the points on $y = f(x)$ *are* part of the solution.

EXAMPLE 28 Sketch the graph of $y + 2x \geq 3$.

Solution First, isolate y on the left side:

$$y \geq -2x + 3$$

Then prepare a table of values and graph $y = -2x + 3$ with a solid line (Fig. 17-9).

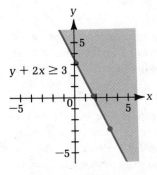

Figure 17-9

x	0	$\frac{3}{2}$	3
y	3	0	-3

Second, shade the region of the plane *above* the line.

Finally, check the solution by substituting the coordinates of two points in the original inequality.

This method can also be used for nonlinear inequalities.

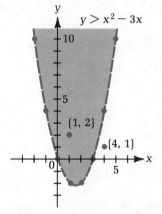

$y > x^2 - 3x$

Figure 17-10

EXAMPLE 29   Sketch the graph of $y > x^2 - 3x$.

Solution   **First,** prepare a table of values and sketch the graph of $y = x^2 - 3x$ using a dashed curve. In this case, the graph is a parabola (Fig. 17-10).

x	−2	−1	0	1	2	3	4	5
y	10	4	0	−2	−2	0	4	10

Second, shade the region above the curve (inside the parabola).

Finally, check two points.

   Inside: (1, 2)   **Outside: (4, 1)**

$(2) > (1)^2 - 3$ True $(1) > (4)^2 - 12$ False

Systems of inequalities   Problems involving more than one condition on the variable may result in a system of inequalities. To solve such a system, we must locate all points that simultaneously satisfy all inequalities in the system. These points may be found by graphing the inequalities in the same plane and visually determining where their solutions intersect.

EXAMPLE 30   Solve the system: $y > x^2$

$$y \le x + 2$$

Solution   The graph of $y > x^2$ is shown in Fig. 17-11(a), and the graph of $y \le x + 2$ is shown in Fig. 17-11(b). The solution to the system consists of all points in the region of the plane where the two shaded regions overlap, as shown in Fig. 17-11(c). The points on the portion of the line $y = x + 2$ inside the parabola are also part of the solution.

Linear programming   One of the most important applications of systems of inequalities is in a mathematical procedure known as **linear programming.** Linear programming uses graphs of systems involving two or more variables to determine how to maximize or minimize some quantity such as profits, costs, materials used, or

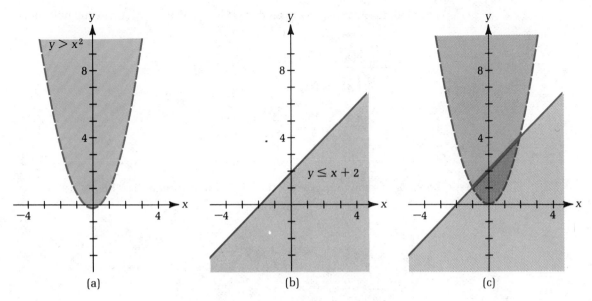

(a)   (b)   (c)

Figure 17-11

time spent when these variables are subject to certain constraints. The best way to show linear programming in action is through an example.

EXAMPLE 31   A computer equipment company has two factories, X and Y, each able to produce two different models, A and B, of a printer. Factory X produces 20 units of model A and 15 units of model B per day. Factory Y produces 10 units of model A and 25 units of model B per day. The operating cost is \$1500 per day for factory X and \$1000 per day for factory Y. The company must produce at least 200 units of model A and 300 units of model B. How many days should each factory operate to minimize production cost?

Solution
Step 1.   Assign variables to the unknown quantities. Let

x = Number of days factory X operates

y = Number of days factory Y operates

Step 2.   Formulate an equation for the quantity you want to maximize or minimize. In this case, the quantity to be minimized is production cost, and it is given by the equation, or **objective function,**

$C = 1500x + 1000y$

Step 3.   Form inequalities to represent the constraints, or limitations, on the variables.

First, no factory can operate $x \geq 0$
for a negative number of days: $y \geq 0$

Second, together they must produce at least 200 units of A and 300 units of B:

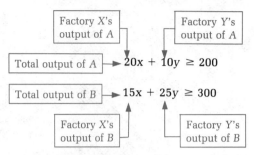

Step 4.   Graph the system formed by all four inequalities (Fig. 17-12).

Figure 17-12

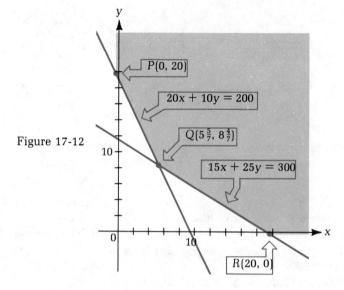

Step 5.   The points in the shaded region are known as **feasible points.** Any point in the feasible region is a solution to the system of inequalities in Step 3. However, we want to determine which of the feasible points will yield a minimum value of the cost function in Step 2. In this type of problem, the particular point that will yield the desired maximum or minimum for the objective function is always one of the vertices of the feasible region. To determine which point this is, substitute the coordinates of each vertex into the objective function. In this case, the vertices are points P, Q, and R, as shown in Fig. 17-12.

For P: $C = 1500(0) + 1000(20) = \$20,000$

For R: $C = 1500(20) + 1000(0) = \$30,000$

The coordinates of point $Q(5\frac{5}{7},\ 8\frac{4}{7})$ are found by solving the system of equations

$$20x + 10y = 200$$

$$15x + 25y = 300$$

For Q: $C = 1500(\frac{40}{7}) + 1000(\frac{60}{7})$

$1500 \boxed{\times} 40 \boxed{\div} 7 \boxed{+} 1000 \boxed{\times} 60 \boxed{\div} 7 \boxed{=} \rightarrow$ 17142.857

$C \approx \$17{,}143$

Since the coordinates of point Q result in a smaller value of C than the coordinates of P and R, the minimum cost is achieved by operating factory X for $5\frac{5}{7}$ days and factory Y for $8\frac{4}{7}$ days.

Using a computer, it is possible to solve very complex linear programming problems involving many variables and limiting conditions.

Exercises 17-4

Sketch the graph of each inequality.

1. $y > x + 1$ **2.** $y \geq 2x$ **3.** $y \leq 3x - 1$ **4.** $y < 2 - x$

5. $y + 2x < 1$ **6.** $x + 2y \leq 6$ **7.** $x + 3y \geq 9$ **8.** $y + 4x > 4$

9. $3x - 2y - 8 \leq 0$ **10.** $5x + 3y - 6 > 0$ **11.** $y \leq x^2$ **12.** $y > -\dfrac{x^2}{2}$

13. $y < -x^2 - 2$ **14.** $y \leq 2x^2 + 1$ **15.** $y < x^2 + 2x$ **16.** $y \geq -\dfrac{x^2}{2} + x$

17. $4x - x^2 + 4y \leq 0$ **18.** $-5x + x^2 - y > 0$ **19.** $y > x^3$ **20.** $y \leq x^3 - 3x$

Solve each system of inequalities graphically.

21. $y < x + 1$
$y > 2 - x$

22. $y < \dfrac{x}{2}$
$y < 3x + 1$

23. $y \geq 2x^2$
$y \geq x + 3$

24. $y \geq x - 3$
$y \leq -x^2 + 2$

25. $y - x^2 + 2 > 0$
$y + x^2 \leq 2$

26. $y - 2x^2 + 3 \geq 0$
$y + \dfrac{x^2}{2} > 1$

Solve the following linear programming problems:

27. *Industrial management*   A small warehouse for a furniture store contains 800 ft² of usable floor space. The store manager wants to use the warehouse to stock a special purchase of tables and chairs. Each table costs \$150 and requires 10 ft² of floor space, while each chair costs \$200 and requires 8 ft² of space. If the manager can spend no more than \$15,000 on the two items, how many of each should be ordered to maximize the total number of items?

28. *Nutrition*   A brand X multivitamin tablet contains 15 mg of iron and 10 mg of zinc. Brand Y contains 8 mg of iron and 10 mg of zinc. A nutritionist advises a patient to take at least 70 mg of each mineral per day. If brand X pills cost 4¢ apiece and brand Y pills cost 3¢ apiece, how many of each should the patient take per day to fulfill the prescription at minimum cost?

29. *Industrial management*   In Example 31, what combination would minimize production cost if the operating cost for factory Y dropped to \$700 per day?

30. *Nutrition*   In Problem 28, how many of each pill would minimize cost if brand X went on sale for 2.5¢ apiece?

17-5 | Review of Chapter 17

Important Terms and Concepts

inequality (p. 590)

sense of an inequality (p. 590)

members of an inequality (p. 590)

solution of an inequality (p. 590)

conditional inequality (p. 590)

unconditional inequality (p. 590)

linear inequality (p. 591)

equivalent inequality (p. 592)

compound inequality (p. 596)

nonlinear inequality (p. 600)

critical value (p. 600)

absolute value (p. 605)

linear programming (p. 612)

objective function (p. 613)

feasible points (p. 614)

Properties for Solving Linear Inequalities

- addition/subtraction property (p. 592)
  If the same real number is added to or subtracted from both members of an inequality, an equivalent inequality results.
- multiplication/division property: positive numbers (p. 593)
  If both members of an inequality are multiplied or divided by the same **positive** real number, an equivalent inequality results.
- multiplication/division property: negative numbers (p. 594)
  If both members of an inequality are multiplied or divided by the same **negative** real number, the sense of the inequality is reversed.

Exercises 17-5

Solve the following inequalities:

1. $x + 4 < 0$

2. $3x > -9$

3. $-\dfrac{x}{6} \geq -1$

4. $3x - 6 \leq -12$

5. $2 - 7x > 16$

6. $-x + 5 \leq 12$

7. $5x < 7x - 4$

8. $\dfrac{x}{2} - 3 \geq \dfrac{x}{8} + 2$

9. $x - 2(x + 5) \leq 6$

10. $3 - (5 - 4x) > 2x + 7$

11. $3x + 2 \geq 5$ or $-\dfrac{2x}{3} > 6$

12. $-5 < 2x - 7 < 3$

13. $x^2 - 16 \geq 0$

14. $x^2 + 4 > 0$

15. $x^2 - x < 0$

16. $x^2 - 12x + 32 \leq 0$

17. $2x^2 + 3x > 14$

18. $3x^2 - 5x + 1 < 0$

19. $x^3 + 3x^2 - 4x - 12 \leq 0$

20. $x^4 + 4x^2 - 12 > 0$

21. $\dfrac{x - 5}{3 - x} \geq 0$

22. $\dfrac{3x - 5}{x + 1} < 0$

23. $\dfrac{1}{x^2 - 3x - 40} > 0$

24. $\dfrac{(x - 6)(x + 3)^2}{x - 4} \leq 0$

25. $|x - 7| < 4$

26. $|4x - 3| > 9$

27. $|5 - 2x| \geq 1$

28. $|x^2 - 6x + 7| \leq 2$

Sketch the graph of each of the following:

29. $y < x$ **30.** $y \geq 2x - 3$ **31.** $x + 2y \leq 4$ **32.** $3x - y + 2 > 0$
33. $y < 2x^2$ **34.** $y > x^2 - x$ **35.** $3x - x^2 - y \geq 0$ **36.** $y < \frac{1}{2}x^3$

Solve each system graphically.

37. $y > -\dfrac{x}{2} - 2$ **38.** $y > \dfrac{x^2}{4}$

$\quad\ \ y > x + 1$ $\qquad\qquad y < \dfrac{x}{2} + 5$

39. *Chemistry*   The temperature in degrees Fahrenheit is related to the temperature in degrees Celsius by the formula $F = \frac{9}{5}C + 32$. Express as a compound inequality the temperature range in degrees Fahrenheit for a chemical solution that must be kept between $-5°C$ and $10°C$ inclusive.

40. *Electronics*   The equation $V = 110 + 15t$ describes the changing potential difference V (in volts) between two points in a certain circuit with an initial voltage of 110 V. Express as an inequality the minimum amount of time t (in seconds) needed for the voltage to exceed 200 V.

41. *Physics*   The length L (in inches) of a certain spring is given by

$$L = \frac{w}{10} + 8$$

where w is the weight (in ounces) hanging on it. Express as an inequality the maximum amount of weight that will keep the length under 10.5 in.

42. *Fire science*   Firefighters use the formula $S = 40L + 33$ to find the horizontal range S of a fire stream (in feet) when the hose has 80 lb/in.^{2} of pressure. Express as an inequality the minimum nozzle diameter L necessary to achieve a range of a least 125 ft (L will be in inches).

43. *Hydrology*   The cross-sectional area A (in square feet) of a certain trapezoid-shaped stream is related to the depth d by $A = 8d + d^2$. For what values of d will A exceed 84 ft^2?

44. *Physics*   The distance s (in feet) traveled by an object under certain conditions is given by $s = 16 + 8t + 3t^2$, where t is the time in seconds. For what values of t is $s > 54$ ft?

45. *Industrial engineering*   The projected number of acceptable parts P tooled by a certain machine is given by $|P - 2000| \leq 500$. Solve for P and give the projected minimum and maximum number of acceptable parts.

46. *Industrial management*   Two types of industrial lamps are produced by a manufacturer. Lamp X requires 3 h of labor for the body and 1 h for wiring. Lamp Y requires 2 h for the body and 2 h for wiring. The profit on lamp X is \$10, and the profit on lamp Y is \$12. The machine shop (body) provides 120 h of time per week, and the electrical shop (wiring) provides 80 h. How many of each type of lamp should be manufactured per week in order to maximize profit?

Sequences, Series, and the Binomial Formula

18

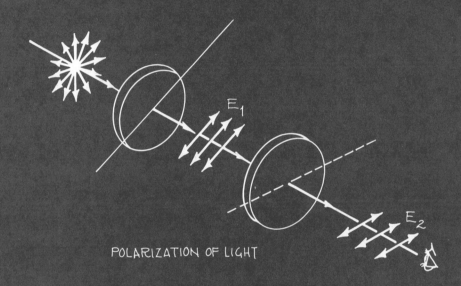

POLARIZATION OF LIGHT

In many applications of mathematics, numerical results follow certain patterns. For example, the velocity of a falling object will increase by a constant amount every second, and the level of radioactivity of a fission by-product from a nuclear test will decrease by one-half in each successive half-life. It is sometimes useful to find the sums of these patterns or to use the patterns to project some future result. In this chapter we will discuss mathematical techniques for doing both.

Definitions   A **sequence** is a collection of real or complex numbers arranged in a particular order.* For example,

2, 5, 9, 14, . . .   and   10, 11, 12

are both sequences. The three dots in the first sequence are used to indicate that the pattern continues.

The numbers in the sequence are called the **terms** of the sequence. If the sequence has a last term, it is a **finite** sequence. Any sequence that is not finite but continues without end is called an **infinite** sequence.

Usually, we use a letter with a subscript to name the terms of a sequence:

$a_1 = $ First term, $a_2 = $ Second term,   . . .

and, in general,

$a_n = $ nth term

Arithmetic progression   An **arithmetic** sequence or **arithmetic progression** (abbreviated AP) is a sequence in which each successive term differs from the previous term by a constant amount. This constant amount is called the **common difference d** of the terms.

EXAMPLE 1   The sequence 3, 7, 11, 15, . . . is an arithmetic progression with a common difference of 4. Each term after the first can be obtained by adding 4 to the previous term:

3,   7,   11,   15, . . .

+4   +4   +4

In general, if a_1 is the first term of an arithmetic progression with a common difference d, then the terms of the progression are

a_1, $a_1 + d$, $a_1 + 2d$, $a_1 + 3d$,   . . . , $a_1 + (n - 1)d$

First   Second   Third   Fourth   nth
term $= a_1$ term $= a_2$ term $= a_3$ term $= a_4$ term $= a_n$

The nth term a_n can be found from the following formula:

The nth Term of an Arithmetic Progression

$$a_n = a_1 + (n - 1)d$$

(18-1)

---

* In this chapter we will be concerned only with sequences of real numbers.

EXAMPLE 2 Find the 12th term of the AP: 3, 7, 11, 15, . . .

Solution
Step 1. Find a_1, n, and d:

$a_1 = 3$ $\qquad$ n = 12 $\qquad$ d = 4

Step 2. Use the nth-term formula to calculate a_{12}:

$a_{12} = 3 + (12 - 1)4 = 3 + 44 = 47$

The 12th term is 47.

EXAMPLE 3 An object moving in a straight line starts with velocity 7.2 m/s and constant acceleration 1.4 m/s². Find its velocity at the end of each of the first 3 s and after 45 s.

Solution The velocities form an AP:

Velocity at start = v_0 = 7.2 m/s

Velocity after 1 s = v_1 = 7.2 m/s + 1.4 m/s = 8.6 m/s

Velocity after 2 s = v_2 = 8.6 m/s + 1.4 m/s = 10.0 m/s

Velocity after 3 s = v_3 = 8.6 m/s + 2(1.4) m/s = 11.4 m/s

Velocity after n s = v_n = v_1 + (n - 1)d

Velocity after 45 s = v_{45} = 8.6 + (45 - 1)(1.4)

8.6 $\boxed{+}$ 44 $\boxed{\times}$ 1.4 $\boxed{=}$ → $\qquad$ *70.2* = 70.2 m/s

The nth-term formula (18-1) can also be used to find the first term a_1, the number of terms n, or the difference between terms d if the other quantities are known.

EXAMPLE 4
(a) Given $a_1 = 6$, d = 4, and $a_n = 62$, find n.
(b) If the first term of an AP is 11 and the 13th term is −25, find the common difference.

Solution

(a) **Step 1.** Substitute the given values into (18-1).

$\qquad\qquad\qquad\qquad\qquad\qquad\qquad\qquad$ 62 = 6 + (n - 1)4

$\qquad$ **Step 2.** Solve. $\qquad\qquad\qquad\qquad\qquad\qquad$ 62 = 6 + 4n - 4

$\qquad\qquad\qquad\qquad\qquad\qquad\qquad\qquad\qquad\qquad\qquad$ 4n = 60

$\qquad\qquad\qquad\qquad\qquad\qquad\qquad\qquad\qquad\qquad\qquad$ n = 15

Thus, 62 is the 15th term of the progression.

(b) **Step 1.**   We are given $a_1 = 11$, $n = 13$, and $a_{13} = a_n = -25$.
Substitute these values into (18-1).

$$-25 = 11 + (13 - 1)d$$

Step 2.   Solve.

$$-25 = 11 + 12d$$

$$12d = -36$$

$$d = -3$$

In this case, d is negative. The progression is a decreasing one: 11, 8, 5, 2, −1, −4, . . .

Series

For every sequence we can write a corresponding **series** which is the indicated sum of the terms of the sequence. For example, the finite sequence

1, 3, 5, 7, 9   has the associated series $1 + 3 + 5 + 7 + 9$

and the infinite sequence

1, 2, 4, 8, 16, . . .

has the associated series $1 + 2 + 4 + 8 + 16 + \cdots$

A series is an expression showing a sum, but an infinite series associated with an infinite sequence may or may not sum up to a real number. For example, the series

$1 + 2 + 4 + 8 + 16 + \cdots$

does not have a real number sum.

Sum of an arithmetic progression

The sum S_n of the first n terms of an arithmetic progression can always be calculated. To develop a formula for the sum, we list the series forward and backward and then add the two equations:

$$S_n = a_1 + (a_1 + d) + (a_1 + 2d) + \cdots + (a_n - 2d) + (a_n - d) + a_n$$

$$\underline{S_n = a_n + (a_n - d) + (a_n - 2d) + \cdots + (a_1 + 2d) + (a_1 + d) + a_1}$$

$$2S_n = (a_1 + a_n) + (a_1 + a_n) + (a_1 + a_n) + \cdots + (a_1 + a_n) + (a_1 + a_n) + (a_1 + a_n)$$

Since a_n is the nth term, $(a_1 + a_n)$ appears n times in this sum. Therefore, we can rewrite this as $2S_n = n(a_1 + a_n)$. Solving for S_n, we have the following formula for the sum of n terms of an arithmetic progression:

Sum of *n* Terms of an AP $S_n = \dfrac{n(a_1 + a_n)}{2}$ (18-2)

LEARNING HINT

Many students think of the sum of an arithmetic progression as the number of terms times the average of the first and last terms:

$$S_n = n\left(\frac{a_1 + a_n}{2}\right) = \frac{n(a_1 + a_n)}{2} \quad \blacktriangleleft$$

Number of terms

Average of the first and last terms

EXAMPLE 5 Find the sum of the first 14 terms of an arithmetic progression if the first term is 6 and the 14th term is 162.

Solution We are given $n = 14$, $a_1 = 6$, and $a_{14} = a_n = 162$. Substituting into (18-2), we have:

$$S_{14} = \frac{14(6 + 162)}{2} = \frac{\overset{7}{\cancel{14}}(168)}{\underset{1}{\cancel{2}}} = 1176$$

In some cases, we may need to use formula (18-1) before we can find the sum.

EXAMPLE 6 Find the sum of the first 300 positive odd integers:
$1 + 3 + 5 + \cdots$

Solution
Step 1. We know that $a_1 = 1$ (the first positive odd integer), $n = 300$, and $d = 2$ (the difference between consecutive odd integers). We need to find the nth term a_{300} using (18-1):

$$a_{300} = 1 + (300 - 1)2 = 599$$

Step 2. Now use formula (18-2) to find the sum S_{300}:

$$S_{300} = \frac{300(1 + 599)}{2} = 90,000$$

EXAMPLE 7 In a series of measurements, the current in a temperature-sensitive circuit element decreases steadily. The current was 1280 mA at the start and decreased 15% by the first measurement, 13% at the second, 11% at the third, and so on. Assuming each decrease applies to the original value, what is the size of the current at the seventh measurement?

Solution
Step 1. The percentage drop is an arithmetic progression, where $a_1 = 15\%$, $d = -2\%$, and $n = 7$. We must find S_7, but first we need to determine a_7. Applying (18-1),

$$a_7 = 15\% + (7 - 1)(-2\%) = 3\%$$

Step 2. Substituting in (18-2), we find S_7:

$$S_7 = \frac{7(15\% + 3\%)}{2} = 63\%$$

Step 3. Since the circuit current has decreased 63%, it has retained 100% − 63% = 37% of its original value.

$$\text{Current} = (0.37)(1280) \approx 474 \text{ mA}$$

If the sum and two of the quantities a_1, a_n, and n are known, we can use the sum formula (18-2) to solve for the third quantity.

EXAMPLE 8
(a) If the first term of an AP is 16 and the sum of the first 12 terms is 246, find the 12th term.
(b) If $a_1 = 4.6$, $a_n = 45.2$, and $S_n = 1444.2$, find n.

Solution

(a) **Step 1.** First, note that $a_1 = 16$, $n = 12$, and $S_{12} = 246$. Then, substitute these values into (18-2) to find a_{12}.

$$246 = \frac{12(16 + a_{12})}{2}$$

Step 2. Solve.

$$246 = 6(16 + a_{12})$$
$$246 = 96 + 6a_{12}$$
$$6a_{12} = 150$$
$$a_{12} = 25$$

(b) **Step 1.** Substitute in (18-2).

$$1444.2 = \frac{n(4.6 + 45.2)}{2}$$

Step 2. Solve.

4.6 ⊞ 45.2 ⊜ ÷ 2 ⊜ →   *24.9* $1444.2 = 24.9n$

STO 1444.2 ÷ RCL ⊜ →   *58.* $n = 58$

In some problems, formulas (18-1) and (18-2) must be solved simultaneously in order to find the missing quantity.

EXAMPLE 9   A student takes a job as a part-time lab assistant to pay off a $7750 student loan. If the salary is $250 for the first month, and it is raised $5 per month for each successive month, how long will it take the student to pay off the loan?

Solution   We are given $a_1 = \$250$, $d = \$5$, and $S_n = \$7750$.

Step 1. Substitute a_1 and d into (18-1): $a_n = 250 + (n - 1)5$
$$a_n = 5n + 245$$

Step 2. Substitute a_1 and S_n into (18-2): $7750 = \frac{n(250 + a_n)}{2}$

Step 3.   Solve the system of two equations resulting from Steps 1 and 2. Substitute the expression for a_n from Step 1 into the equation from Step 2:

$$7750 = \frac{n[250 + (5n + 245)]}{2}$$

Solve for n:

$$7750 = \frac{n(5n + 495)}{2}$$

$$15,500 = 5n^2 + 495n$$

$$n^2 + 99n - 3100 = 0$$

$$(n - 25)(n + 124) = 0$$

$$n = 25 \quad \text{or} \quad n = -124$$

Since n must be positive, n = 25. The student must work 25 months to earn enough money to pay off the loan.

Exercises 18-1

Find the indicated term of each arithmetic progression.

1. The 7th term of: 2, 5, 8, . . .
2. The 8th term of: 3, 8, 13, . . .
3. The 10th term of: 5, 3, 1, . . .
4. The 15th term of: 1, −3, −7, . . .
5. The 20th term of: $1, \frac{5}{2}, 4, \ldots$
6. The 25th term of: $2, \frac{7}{4}, \frac{3}{2}, \ldots$
7. The 6th term, where $a_1 = -5$ and $d = 6$
8. The 16th term, where $a_1 = 9$ and $d = 8$
9. The 13th term, where $a_1 = 4$ and $d = -5$
10. The 40th term, where $a_1 = -2$ and $d = -\frac{2}{3}$

Find the sum of the first n terms of each arithmetic progression.

11. $n = 12$, $a_1 = 2$, $a_{12} = 5$
12. $n = 14$, $a_1 = 6$, $a_{14} = -80$
13. $n = 32$, $a_1 = -2$, $a_{32} = -124$
14. $n = 28$, $a_1 = -\frac{1}{2}$, $a_{28} = 27$
15. $n = 20$, $a_1 = 3$, $d = -7$
16. $n = 16$, $a_1 = 5$, $d = 8$

Use the information given in Problems 17–34 to find the indicated missing quantities for each arithmetic progression.

17. If $a_1 = 2$, $d = 4$, and $a_n = 38$, find n and S_n.
18. If $a_1 = -5$, $d = -3$, and $a_n = -65$, find n and S_n.
19. If $a_1 = 8$ and $a_{24} = -38$, find d and S_{24}.
20. If $a_1 = 4$ and $a_{50} = 123$, find d and S_{50}.
21. If $a_1 = -8$, $n = 40$, and $d = \frac{5}{2}$, find a_{40} and S_{40}.
22. If $a_1 = 6$, $n = 30$, and $d = 0.3$, find a_{30} and S_{30}.
23. If $a_1 = 22$, $a_n = -17$, and $S_n = 35$, find d and n.
24. If $a_1 = 3$, $a_n = 32$, and $S_n = 1032.5$, find d and n.
25. If the first term is 32, the common difference is −3, and the nth term is −10, find n and the sum of the first n terms.
26. If the first term is 44.6 and the common difference is 2.8, find the 30th term and the sum of the first 30 terms.

27. If the first term is $\frac{3}{4}$ and the sum of the first 45 terms is $\frac{3735}{8}$, find the common difference and the 45th term.

28. If the sum of the first 60 terms is -4680 and the first term is 40, find the common difference and the 60th term.

29. If the sum of the first 26 terms is 2938 and the common difference is -2, find the first term and the 26th term.

30. If the common difference is -4, the nth term is -25, and the sum of n terms is 1085, find n and the first term.

31. If $a_1 = 4$, $d = 2$, and $S_n = 378$, find n and a_n. (See Example 9.)

32. If $a_1 = -12$, $d = 3$, and $S_n = 105$, find n and a_n. (See Example 9.)

33. If the 4th term is 3 and the 7th term is 9, find a_1, d, and S_{12}.

34. If the 9th term is 34 and the 15th term is 10, find a_1, d, and S_{30}.

Solve.

35. Find the sum of the first 400 positive even integers.

36. Find the sum of the first 250 positive multiples of 3.

37. *Physics*   An object moving in a straight line has an initial velocity of 9.4 m/s and a constant acceleration of 1.8 m/s². Find its velocity at the end of 16 s. (See Example 3.)

38. *Environmental science*   An oceanside cliff is eroding at the rate of 4 in./year. If it is currently 16 ft 0 in. from the nearest property line, how close will it be after 40 years?
[*Hint:* $a_1 = 16$ ft $- 4$ in. Find a_{40}.]

39. *Physics*   In Problem 37, what is the total distance traveled by the object?
[*Hint:*   Find the sum of the AP formed by the *average* velocity during each second.]

40. *Business*   A technical researcher is hired at a starting salary of $1400 per month. If the salary is raised $15 per month, how much will the researcher have earned after 3 years?

41. *Business*   A software development company reported sales of a certain program to be 22 units at the end of the first month. If sales are projected to increase at the rate of 4 units per month, after how many months will total sales reach 20,000 units?

42. *Business*   A commercial product had peak sales of 1680 units per month, after which sales began decreasing by approximately 8 units per month. If this rate of decrease continues, how many years after the peak month will sales reach a level of 650 units per month?

18-2 | Geometric Progressions

Definitions   A **geometric progression** (abbreviated GP) is a sequence in which each term is a constant multiple of the previous term. This multiple is called the **common ratio r** of the progression.

EXAMPLE 10

(a)   The sequence 2, 6, 18, 54, . . . is a geometric progression with a common ratio of 3. Each term after the first can be obtained by multiplying the preceding term by 3:

$$r = 3$$

(b) The sequence 16, 8, 4, 2, . . . is a GP with a common ratio of $\frac{1}{2}$:

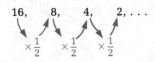

16,   8,   4,   2, . . .

$\times\frac{1}{2}$ $\times\frac{1}{2}$ $\times\frac{1}{2}$

$$r = \frac{1}{2}$$

(c) The sequence −3, 6, −12, 24, . . . is a GP with a common ratio of −2:

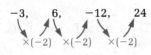

−3,   6,   −12,   24

$\times(-2)$ $\times(-2)$ $\times(-2)$

$$r = -2$$

NOTE

As you can see, the terms of a GP increase if $r > 1$ [Example 10(a)], decrease if $0 < r < 1$ [Example 10(b)], and alternately increase and decrease when $r < 0$ [Example 10(c)]. For $r = 0$, all terms after the first are 0, and for $r = 1$, the terms are identical. ◄

The *n*th term

In general, if a_1 is the first term of a geometric progression and the common ratio is r, the first n terms of the progression are

$$a_1, \quad a_1r, \quad a_1r^2, \quad a_1r^3, \quad \ldots, \quad a_1r^{n-1}$$

First   Second   Third   Fourth   nth term
term   term   term   term

The nth term a_n is given by the following formula:

The *n*th Term of a Geometric Progression	$a_n = a_1r^{n-1}$ **(18-3)**

EXAMPLE 11   Find the 10th term of the GP: 3, 12, 48, . . .

Solution
Step 1.   Find a_1, r, and n:

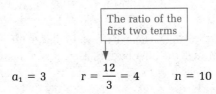

The ratio of the
first two terms

$$a_1 = 3 \qquad r = \frac{12}{3} = 4 \qquad n = 10$$

Step 2. Substitute these values into (18-3) and solve for a_{10}.

$$a_{10} = 3(4)^{10-1} = 3(4)^9 \longleftarrow \boxed{\begin{array}{l}\text{Caution:}\\ \text{This is}\\ \text{not } 12^9\end{array}}$$

$$3\,\boxed{\times}\,4\,\boxed{y^x}\,9\,\boxed{=}\rightarrow \boxed{786432.}$$

The 10th term of the GP is 786,432.

Formula (18-3) can also be used to solve for a_1, r, or n.

EXAMPLE 12 The level of cadmium metal pollution in a certain lake was found to be increasing geometrically. If it was 4.00 parts per billion (ppb) in January and 20.25 ppb in May, then:
(a)   By what ratio is the level increasing each month?
(b)   What will be the level in December?

Solution
(a)   **Step 1.**   Determine the given quantities.

$$a_1 = 4.00 \qquad n = 5 \qquad a_5 = 20.25$$

$$\boxed{\text{May is the 5th month}}$$

Step 2.   Substitute these values into (18-3) and solve for the ratio r:

$$20.25 = 4.00r^{5-1}$$

$$r^4 = \frac{20.25}{4.00}$$

$$r = \sqrt[4]{\frac{20.25}{4.00}} \qquad 20.25\,\boxed{\div}\,4\,\boxed{=}\,\boxed{\sqrt[x]{y}}\,4\,\boxed{=}\rightarrow \boxed{1.5}$$

The pollutant level is increasing by a ratio of 1.5 per month.

(b)   **Step 1.**   December is the 12th month, so $n = 12$, and we know that $a_1 = 4.00$ and $r = 1.5$.

Step 2.   The level in December is

$$a_{12} = 4.00(1.5)^{12-1} = 4.00(1.5)^{11}$$

$$4\,\boxed{\times}\,1.5\,\boxed{y^x}\,11\,\boxed{=}\rightarrow \boxed{345.99023}$$

The pollutant level will be approximately 346 ppb in December.

EXAMPLE 13 Each sheet of a certain plastic filter reduces the intensity of light passing through the filter by 15%. How many sheets will reduce a light intensity of 100 candelas (cd) to 45 cd?

Solution

Step 1. Determine the given quantities:

$$a_1 = 100 \qquad a_n = 45 \qquad r = 0.85 \leftarrow \boxed{\text{If the intensity is reduced by 15\%, then 85\% of it remains.}}$$

Step 2. Substitute these values into (18-3) and solve for $n - 1$, where n, the number of terms, is 1 more than the number of filters (Fig. 18-1):

$$45 = 100(0.85)^{n-1}$$

Divide by 100.

$$0.45 = (0.85)^{n-1}$$

Take the logarithm of both sides.

$$\log 0.45 = \log 0.85^{n-1}$$

Use the power rule for logs.

$$\log 0.45 = (n - 1) \log 0.85$$

Divide by log 0.85.

$$n - 1 = \frac{\log 0.45}{\log 0.85}$$

$$.45 \boxed{\text{log}} \boxed{\div} .85 \boxed{\text{log}} \boxed{=} \rightarrow \boxed{4.9133212} \approx 5$$

Five sheets of the filter are needed.

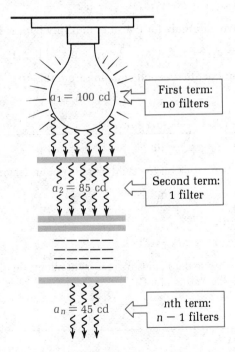

First term: no filters

$a_1 = 100$ cd

Second term: 1 filter

$a_2 = 85$ cd

nth term: $n - 1$ filters

$a_n = 45$ cd

Figure 18-1

EXAMPLE 14 A radioactive by-product of a nuclear test has a half-life of 4.8 days. This means that the radioactivity level decreases by one-half every 4.8 days. Thirty days after the test, the radioactivity level of the site is measured to be 12 microcuries (μCi). What was the original radioactivity level?

Solution To find the original level a_1 using (18-3), we must know a_n, r, and n. We are given $a_n = 12$ and $r = \dfrac{1}{2}$. To calculate n, first note that the number of half-lives elapsed *since* the start is:

$$\frac{30 \text{ days}}{4.8 \text{ days/half-life}} \approx 6.25$$

Therefore, counting a_1, the number of *terms* $n \approx 7.25$. Substituting into (18-3),

$$12 = a_1 \left[\left(\frac{1}{2} \right)^{6.25} \right] \qquad \boxed{n-1}$$

$$a_1 = \frac{12}{\left(\frac{1}{2} \right)^{6.25}}$$

$$12 \; \boxed{\div} \; .5 \; \boxed{y^x} \; 6.25 \; \boxed{=} \rightarrow \qquad \boxed{913.31106}$$

$$a_1 \approx 910 \; \mu\text{Ci}$$

The sum of a GP The sum S_n of the first n terms of a geometric progression is given by

$$S_n = a_1 + a_1 r + a_1 r^2 + \cdots + a_1 r^{n-2} + a_1 r^{n-1}$$

To derive a formula for the sum, multiply both sides of this equation by r:

$$r S_n = a_1 r + a_1 r^2 + a_1 r^3 + \cdots + a_1 r^{n-1} + a_1 r^n$$

Then subtract the second equation from the first:

$$
\begin{array}{r}
S_n = a_1 + \;\; a_1 r + a_1 r^2 + \cdots + a_1 r^{n-2} + a_1 r^{n-1} \\
-r S_n = \quad\; - (a_1 r + a_1 r^2 + \cdots + a_1 r^{n-2} + a_1 r^{n-1} + a_1 r^n) \\
\hline
S_n - r S_n = a_1 + \;\; 0 \;\; + \;\; 0 \;\; + \cdots + \;\; 0 \;\; + \;\; 0 \;\; - a_1 r^n
\end{array}
$$

All but two terms on the right side drop out, leaving

$$S_n - r S_n = a_1 - a_1 r^n$$

Factor both sides:

$$S_n (1 - r) = a_1 (1 - r^n)$$

Finally, divide both sides by $(1 - r)$ to obtain the desired formula for the sum of the first n terms of a GP:

Sum of n Terms of a GP $\qquad S_n = \dfrac{a_1 (1 - r^n)}{1 - r} \qquad$ for $r \neq 1 \qquad$ (18-4)

REMEMBER ▶ When evaluating the expression $(1 - r^n)$ in (18-4), *first* raise r to the nth power; *then* subtract the result from 1. ◀

EXAMPLE 15 Find the sum of the first 8 terms of the GP: $3, \dfrac{3}{2}, \dfrac{3}{4}, \ldots$

Solution

Step 1. Establish a_1, r, and n:

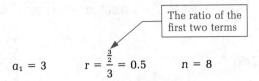

The ratio of the first two terms

$$a_1 = 3 \qquad r = \frac{\frac{3}{2}}{3} = 0.5 \qquad n = 8$$

Step 2. Substitute these values into (18-4):

$$S_8 = \frac{3(1 - 0.5^8)}{1 - 0.5}$$

and calculate S_8:

$$1 \; \boxed{-} \; .5 \; \boxed{y^x} \; 8 \; \boxed{=} \; \boxed{\times} \; 3 \; \boxed{\div} \; .5 \; \boxed{=} \to \quad \boxed{5.9765625}$$

EXAMPLE 16 At the beginning of each year, the owner of a machine shop puts $2000 into an IRA account paying 12% interest compounded annually. How much money will be in the account when he retires after 15 years? (Assume that he does *not* make the 16th deposit.)

Solution

|Principal | Interest|

After 1 year, the account contains: $\qquad 2000 + 2000(0.12)$

Or, by factoring: $\qquad = 2000(1.12)$

At the end of the second year, 12% interest will be added to this amount once more.

Interest calculated twice

It will then be worth: $\qquad [2000(1.12)](1.12) = 2000(1.12)^2$

Meanwhile, a second deposit has earned a full year's interest, so it is worth:

Interest calculated once

$$2000(1.12)^1$$

There is now a total of $2000(1.12) + 2000(1.12)^2$. Continuing this pattern, the total after 15 years is

$$2000(1.12) + 2000(1.12)^2 + 2000(1.12)^3 + \cdots + 2000(1.12)^{15}$$

$$= 2000[1.12 + (1.12)^2 + (1.12)^3 + \cdots + (1.12)^{15}]$$

The expression in brackets is a geometric progression with

$$a_1 = 1.12 \qquad r = \frac{(1.12)^2}{1.12} = 1.12 \qquad n = 15$$

The sum is

$$S_{15} = \frac{1.12(1 - 1.12^{15})}{1 - 1.12} \approx 41.753280$$

Multiplying by 2000 gives the total amount in the account:

$$2000(41.753280) = \$83,506.56 \approx \$83,507$$

Exercises 18-2

Find the indicated term of each geometric progression.

1. The 9th term of: 2, 6, 18, . . .
2. The 8th term of: 3, −6, 12, . . .
3. The 7th term of: 12, −6, 3, . . .
4. The 10th term of: 5, 3, 1.8, . . .
5. The 6th term, where $a_1 = -4$ and $r = -3$
6. The 7th term, where $a_1 = 2$ and $r = 5$
7. The 8th term, where $a_1 = 60$ and $r = \frac{1}{2}$
8. The 6th term, where $a_1 = 120$ and $r = -\frac{1}{3}$

Find the sum of the first n terms of each geometric progression.

9. −8, −4, −2, . . . $(n = 7)$
10. 3, 9, 27, . . . $(n = 6)$
11. 4, −6, 9, . . . $(n = 8)$
12. −50, 40, −32, . . . $(n = 7)$

13. $n = 6$, $a_1 = 6$, $r = 2$
14. $n = 5$, $a_1 = -12$, $r = -\dfrac{1}{3}$

15. $n = 5$, $a_1 = -4$, $a_5 = -324$
16. $n = 7$, $a_1 = 256$, $a_7 = \dfrac{1}{16}$

Use the information given in Problems 17–26 to find the indicated missing quantities for each geometric progression.

17. If $a_1 = 3$, $r = 4$, and $a_n = 192$, find n and S_n.
18. If $a_8 = 512$ and $r = -2$, find a_1 and S_8.
19. If $r = -3$ and $S_7 = -1094$, find a_1 and a_7.
20. If $a_1 = 2$, $r = -2$, and $S_n = -170$, find n and a_n.

21. If the 6th term is −160 and the common ratio is 2, find the first term and the sum of the first 6 terms.
22. If the first term is 1000, the common ratio is 0.1, and the nth term is 0.001, find n and the sum of the first n terms.
23. If the first term is 1, the common ratio is 3, and the sum of the first n terms is 3280, find n and the nth term.
24. If the sum of the first 8 terms is 318.75 and the common ratio is $\dfrac{1}{2}$, find the first term and the 8th term.
25. If the 3rd term is 4 and the 7th term is 16, find the 9th term.
26. If the 5th term is 36 and the 9th term is 324, find the 3rd term.

Solve.

27. *Finance*   A certain tax-sheltered annuity pays 9% interest compounded annually. What is the value (to the nearest dollar) after 20 years of an initial investment of $6000?
[*Careful:* This is not the same question posed in Example 16.]
28. *Finance*   In Problem 27, suppose that $6000 is invested *each* year. What is the total value of the account after 20 years?

29. *Environmental science*   In Example 12, another lake in the vicinity contained 2.00 ppb in January and 3.92 ppb in March. What will be the level of cadmium metal pollution in that lake in December? (Assume that the level is increasing geometrically.)

30. *Finance*   The median price of homes in a certain area is increasing at the rate of 4% per year. At this rate, what will be the value in 20 years of a home currently worth $160,000?

31. *Environmental science*   The level of formaldehyde gas present in a room insulated by formaldehyde foam is found to decrease by 3% per week. If the current level is 0.25 ppm, how long will it take the level to reach 0.10 ppm?
    [*Caution:*   See Example 13.]

32. *Nuclear physics*   A radioactive by-product of a nuclear test has a half-life of 6.4 days. If the radioactivity level was measured to be 1.4 μCi 60 days after the test, what was the original radioactivity level of this by-product? (See Example 14.)

33. *Optics*   The intensity of light passing through a 3 in. filter was measured to be 75 cd. If each 1 in. thickness of filter reduces the intensity by 12%, what is the intensity of the original source?

34. *Biology*   At the beginning of an experiment, three bacteria are present in a culture. If each bacteria divides once in every hour, how many will there be at the *end* of 12 h?

35. Solve equation (18-3) for n.

36. Solve equation (18-4) for n.

18-3 | Infinite Geometric Series

Sum of an infinite geometric series   In Section 18-2 we developed a formula for summing the first n terms of any geometric progression. In this section we will show that, for certain infinite geometric progressions, the sum of *all* terms can be found, even though there are infinitely many terms.

Consider the following geometric progressions:

(a)   2, 8, 32, 128, . . . $r = 4$

(b)   2, 1, $\dfrac{1}{2}$, $\dfrac{1}{4}$, . . . $r = \dfrac{1}{2}$

By inspection, we can see that the sum of the terms of progression (a) will increase without bound. But, if we examine the partial sums for progression (b) listed in Table 18-1, we see that they seem to approach a limiting value. As n increases, the sum S_n approaches closer and closer to 4.

TABLE 18-1

For n =	Series	S_n
2	$2 + 1$	3
3	$2 + 1 + \frac{1}{2}$	$3\frac{1}{2}$
4	$2 + 1 + \frac{1}{2} + \frac{1}{4}$	$3\frac{3}{4}$
5	$2 + 1 + \frac{1}{2} + \frac{1}{4} + \frac{1}{8}$	$3\frac{7}{8}$
6	$2 + 1 + \frac{1}{2} + \frac{1}{4} + \frac{1}{8} + \frac{1}{16}$	$3\frac{15}{16}$
20	$2 + 1 + \frac{1}{2} + \frac{1}{4} + \frac{1}{8} + \cdots + \frac{1}{2^{18}}$	≈ 3.9999962

Notice that for progression (a) above, $|r| > 1$, and for progression (b), $|r| < 1$. If we reexamine formula (18-4),

$$S_n = \frac{a_1(1 - r^n)}{1 - r}$$

we see that if $|r| < 1$, then r^n becomes smaller and smaller as n increases. If n is extremely large, r^n becomes effectively 0. In mathematical notation, we say that

$$\lim_{n \to \infty} r^n = 0 \qquad \text{for } |r| < 1$$

This is read "the limit of r^n, as n increases without bound, is zero."

If we now rewrite formula (18-4) for a GP with an infinite number of terms and $|r| < 1$, the sum S is

$$S = \frac{a_1(1 - 0)}{1 - r}$$

or:

Sum of an Infinite GP $S = \dfrac{a_1}{1 - r}$ for $|r| < 1$ (18-5)

EXAMPLE 17   Applying formula (18-5) to the GP $2, 1, \dfrac{1}{2}, \ldots$ that we examined in Table 18-1, we have $a_1 = 2$ and $r = \dfrac{1}{2}$, so

$$S = \frac{2}{1 - \frac{1}{2}} = 4$$

EXAMPLE 18   Find the sum of the infinite geometric progression:

$-6, 2, -\dfrac{2}{3}, \ldots$

Solution
Step 1. Find a_1 and r:

$$a_1 = -6 \qquad r = \frac{2}{-6} = -\frac{1}{3}$$

Step 2. Substitute a_1 and r into (18-5). $S = \dfrac{-6}{1 - (-\frac{1}{3})}$

Solve. $S = \dfrac{-6}{\frac{4}{3}} = -\dfrac{9}{2}$

Repeating decimals   A geometric progression can be used to find the fractional equivalent of a repeating decimal.

EXAMPLE 19 Find the fractional equivalent of the repeating decimal: $0.\overline{5} = 0.5555\ldots$

Solution
Step 1. Rewrite as a series:

$$0.5555\ldots = 0.5 + 0.05 + 0.005 + 0.0005 + \cdots$$

Step 2. The terms of this series form an infinite GP with

$$a_1 = 0.5 \qquad r = \frac{0.05}{0.5} = 0.1$$

Step 3. Use the infinite sum formula (18-5) to find the sum and express it as a fraction in lowest terms:

$$S = \frac{0.5}{1 - 0.1} = \frac{0.5}{0.9} = \frac{5}{9}$$

EXAMPLE 20 Find the fractional equivalent of: $0.\overline{5137} = 0.5137137137\ldots$ Check your answer with a calculator.

Solution
Step 1. Rewrite as

$$0.5 + 0.0137 + 0.0000137 + 0.0000000137 + \cdots$$

Step 2. Find a_1 and r for the GP comprised of all terms but the first one:

$$a_1 = 0.0137 \qquad r = \frac{0.0000137}{0.0137} = 0.001$$

Step 3. Find the sum of the GP:

$$S = \frac{0.0137}{1 - 0.001} = \frac{0.0137}{0.999} = \frac{137}{9990}$$

The repeating decimal is

$$0.5 + S = \frac{1}{2} + \frac{137}{9990} = \frac{5132}{9990} = \frac{2566}{4995}$$

Check: $2566 \boxed{\div} 4995 \boxed{=} \rightarrow$ $\boxed{0.5137137}$

All answers should be reduced to lowest terms if possible.

EXAMPLE 21 An object suspended on the end of a spring oscillates up and down. If the total distance it travels during its initial oscillation is 70 mm, and if the distance it travels during each successive oscillation is 80% of the previous distance, what total vertical distance does the object travel before coming to rest?

Solution

Step 1. $a_1 = 70$ $r = 0.8$

Step 2. $S = \dfrac{70}{1 - 0.8} = \dfrac{70}{0.2} = 350$ mm

Exercises 18-3

Find the sum of each infinite geometric progression.

1. 12, 6, 3, . . .
2. 9, 3, 1, . . .
3. 4, −2, 1, . . .
4. −8, −2, −$\frac{1}{2}$, . . .
5. 20, 4, $\frac{4}{5}$, . . .
6. 18, 3, $\frac{1}{2}$, . . .
7. 18, 12, 8, . . .
8. 8, 6, $\frac{9}{2}$, . . .
9. −48, 36, −27, . . .
10. 5, −4, $\frac{16}{5}$, . . .
11. −9, 3$\sqrt{3}$, −3, . . .
12. 4, 2$\sqrt{2}$, 2, . . .
13. 8a^2, 4a, 2, . . .
14. 6x^3, −2x, $\frac{2}{3x}$, . . .

Find the fractional equivalent of each repeating decimal.

15. $0.\overline{7}$
16. $0.\overline{6}$
17. $0.\overline{39}$
18. $0.\overline{07}$
19. $0.\overline{127}$
20. $0.8\overline{31}$
21. $0.42\overline{16}$
22. $0.7\overline{89}$
23. $2.3\overline{49}$
24. $9.8\overline{54}$

25. *Physics*   An object oscillating up and down on the end of a spring has an initial oscillation of 50.0 mm, and each successive oscillation is 70% of the previous one. What total vertical distance does the object travel before coming to rest?

26. *Physics*   A pendulum bob travels 20.0 in. during its first swing, 14.0 in. during its second swing, 9.8 in. during its third swing, and so on. How far will it travel before coming to rest?

27. Physics   A moving object decelerates in such a way that it travels 80.0 m during the first second, 64.0 m during the second, 51.2 m during the third, and so on. How far does it travel before coming to rest?

28. Physics   A bouncing ball rebounds $\frac{3}{4}$ as far as it falls. If it is initially dropped from a height of 25 ft, what total vertical distance does it travel before coming to rest?

18-4 | The Binomial Formula

In our study of algebra so far, we have had occasion to raise binomial expressions to the second and third powers, but rarely to any higher powers. However, as you proceed with more advanced mathematics and its applications, such calculations will arise. In this section we develop a formula for raising a binomial to any power.

Expanding $(a + b)^n$ Consider the general binomial expression $(a + b)^n$. Using direct multiplication, we can expand this expression for several integral values of n:

$$(a + b)^0 = \qquad\qquad\qquad 1$$

$$(a + b)^1 = \qquad\qquad\qquad a + b$$

$$(a + b)^2 = \qquad\qquad a^2 + 2ab + b^2$$

$$(a + b)^3 = \qquad\quad a^3 + 3a^2b + 3ab^2 + b^3$$

$$(a + b)^4 = \qquad a^4 + 4a^3b + 6a^2b^2 + 4ab^3 + b^4$$

$$(a + b)^5 = a^5 + 5a^4b + 10a^3b^2 + 10a^2b^3 + 5ab^4 + b^5$$

Patterns of the expansions   This lengthy multiplication is a very time-consuming way to expand $(a + b)^n$. However, the following patterns will guide us to an easier method for expansion:

- Each expansion of $(a + b)^n$ contains **$n + 1$** terms.
  Example: The expansion of $(a + b)^5$ contains **6** terms.
- The first term is always a^n, and the last term is always b^n.
  Example: The expansion of $(a + b)^3$ begins with a^3 and ends with b^3.
- Proceeding from left to right, the power of a *decreases* by 1 in each successive term, and the power of b *increases* by 1 in each successive term. The sum of the exponents in each term is therefore always equal to n.

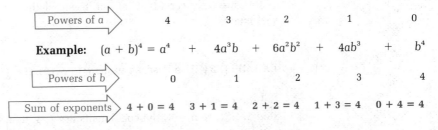

| Powers of a | 4 | 3 | 2 | 1 | 0 |

Example: $(a + b)^4 = a^4 \quad + \quad 4a^3b \quad + \quad 6a^2b^2 \quad + \quad 4ab^3 \quad + \quad b^4$

| Powers of b | 0 | 1 | 2 | 3 | 4 |

| Sum of exponents | $4 + 0 = 4$ | $3 + 1 = 4$ | $2 + 2 = 4$ | $1 + 3 = 4$ | $0 + 4 = 4$ |

- The coefficients of the terms form a symmetrical triangular pattern:

n = 0   1

n = 1   1   1

n = 2   1   2   1

n = 3   1   3   3   1

n = 4   1   4   6   4   1

n = 5   1   5   10   10   5   1

Pascal's triangle   The pattern shown above is known as **Pascal's triangle** (named after the 17th century French mathematician, Blaise Pascal). Notice that each row begins and ends with 1, and that each number is the sum of the two numbers above it on either side:

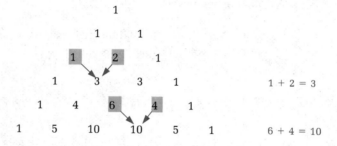

EXAMPLE 22 Use Pascal's triangle and the other patterns listed to expand: $(x + y)^6$

Solution

Step 1. To determine the coefficients, find the row of the triangle corresponding to n = 6:

n = 5:   [1]   [5]   [10]   [10]   [5]   [1]

n = 6:   1   6   15   20   15   6   1

Step 2. Combine these coefficients with the patterns for the exponents listed earlier, and write the expression:

$(x + y)^6 = x^6 + 6x^5y + 15x^4y^2 + 20x^3y^3 + 15x^2y^4 + 6xy^5 + y^6$

Be especially careful when the terms of the binomial themselves contain coefficients.

EXAMPLE 23 Use Pascal's triangle to expand: $(3m - 2n)^4$

Solution

Step 1. Since n = 4, the coefficients are 1, 4, 6, 4, 1.

Step 2. Referring to our original expansion of $(a + b)^4$, we note that $a = 3m$ and $b = -2n$. Therefore,

$$[3m + (-2n)]^4 = (3m)^4 + 4(3m)^3(-2n) + 6(3m)^2(-2n)^2$$
$$+ 4(3m)(-2n)^3 + (-2n)^4$$

Step 3. Simplify:

$$= 81m^4 - 216m^3n + 216m^2n^2 - 96mn^3 + 16n^4$$

The factorial function

Pascal's triangle is a useful tool for smaller powers of a binomial, but it is cumbersome for larger powers. For these situations, a quicker method has been developed which involves a special function called the **factorial function.** The number n! (read "n factorial") is defined as the product of the first n integers.

The Factorial Function

$$n! = n(n - 1)(n - 2)(n - 3) \cdot \cdots \cdot 3 \cdot 2 \cdot 1 \qquad (18\text{-}6)$$

where n is any positive integer

For example,

$$2! = 2 \cdot 1 = 2$$
$$3! = 3 \cdot 2 \cdot 1 = 6$$
$$4! = 4 \cdot 3 \cdot 2 \cdot 1 = 24$$

And, by definition:

$$1! = 1 \qquad \text{and} \qquad 0! = 1$$

NOTE ▶ Most scientific calculators have a factorial key marked $\boxed{x!}$ or $\boxed{n!}$ for evaluating large factorials. ◀

EXAMPLE 24 Calculate: (a) 5! (b) 25!

Solution
(a) $5! = 5 \cdot 4 \cdot 3 \cdot 2 \cdot 1 = 120$ (b) $25 \boxed{x!} \rightarrow$ | *1.5511 25* |

$$25! \approx 1.6 \times 10^{25}$$

The binomial formula

The expansion of $(a + b)^n$, where n is any positive integer, can now be written using factorial notation. This equation is known as the **binomial formula.**

The Binomial Formula

$$(a + b)^n = a^n + na^{n-1}b + \frac{n(n-1)}{2!}a^{n-2}b^2 + \frac{n(n-1)(n-2)}{3!}a^{n-3}b^3 + \cdots + b^n \qquad (18\text{-}7)$$

where n is any positive integer

EXAMPLE 25   Use the binomial formula to expand: $(x + y)^7$

Solution

$$(x + y)^7 = x^7 + 7x^6y + \left(\frac{7\cdot6}{2\cdot1}\right)x^5y^2 + \left(\frac{7\cdot6\cdot5}{3\cdot2\cdot1}\right)x^4y^3 + \left(\frac{7\cdot6\cdot5\cdot4}{4\cdot3\cdot2\cdot1}\right)x^3y^4$$

$$+ \left(\frac{7\cdot6\cdot5\cdot4\cdot3}{5\cdot4\cdot3\cdot2\cdot1}\right)x^2y^5 + \left(\frac{7\cdot6\cdot5\cdot4\cdot3\cdot2}{6\cdot5\cdot4\cdot3\cdot2\cdot1}\right)xy^6 + y^7$$

$$= x^7 + 7x^6y + 21x^5y^2 + 35x^4y^3 + 35x^3y^4 + 21x^2y^5 + 7xy^6 + y^7$$

Be especially careful when either term of the binomial contains a coefficient.

EXAMPLE 26   Use the binomial formula to expand: $(2x - 5)^4$

Solution

$$(2x - 5)^4 = (2x)^4 + 4(2x)^3(-5) + \frac{4\cdot3}{2\cdot1}(2x)^2(-5)^2 + \frac{4\cdot3\cdot2}{3\cdot2\cdot1}(2x)(-5)^3 + (-5)^4$$

$$= 16x^4 - 160x^3 + 600x^2 - 1000x + 625$$

Finding a given term   It is possible to use the binomial formula to find one specific term of an expansion without writing the others. The $(k + 1)$st term of $(a + b)^n$ is given by

k factors

$$\underbrace{\frac{n(n-1)(n-2)\cdot\,\cdots\,\cdot(n-k+1)}{k!}a^{n-k}b^k}$$

$\longleftarrow$ k is 1 less than the number of the term

For example, for the 3rd term of $(x + y)^5$,

$$n = 5 \qquad \text{and} \qquad k = 2 \longleftarrow \text{k is 1 less than the number of the term}$$

$$n - k = 5 - 2 = 3$$

$k = 2$ factors k

$$\left(\frac{5\cdot4}{2\cdot1}\right)x^3y^2 = 10x^3y^2$$

$2!$

EXAMPLE 27 Find the 5th term in the expansion of: $(x + y)^9$

Solution For the 5th term, $k = 4$. Thus,

$$\left(\frac{9 \cdot 8 \cdot 7 \cdot 6}{4 \cdot 3 \cdot 2 \cdot 1}\right) x^{9-4} y^4 = 126 x^5 y^4$$

Expanding $(1 + b)^n$ Using advanced mathematics, we can show that, if $a = 1$ and $|b| < 1$, the binomial formula is valid for any real values of n. Substituting into (18-7), we obtain the following binomial series:

$$(1 + b)^n = 1 + nb + \frac{n(n-1)}{2!} b^2 + \frac{n(n-1)(n-2)}{3!} b^3 + \cdots \quad (18\text{-}8)$$

where n is any real number and $|b| < 1$

For negative and fractional values of n, (18-8) becomes an infinite series. For these values of n, the formula is useful because the first few terms provide a reasonable approximation for $(1 + b)^n$, and the expansion gives a representation of the binomial power that cannot be found using normal multiplication.

EXAMPLE 28 Write the first four terms in the binomial expansion of $(1 + x)^{-2}$, where $|x| < 1$.

Solution Applying (18-8), we obtain

$$(1 + x)^{-2} = 1 + (-2)x + \frac{(-2)(-3)}{2 \cdot 1} x^2 + \frac{(-2)(-3)(-4)}{3 \cdot 2 \cdot 1} x^3 + \cdots$$

$$= 1 - 2x + 3x^2 - 4x^3 + \cdots$$

Exercises 18-4

Write the expansion of each expression using Pascal's triangle.

1. $(m + 4)^4$ **2.** $(x - 3)^3$ **3.** $(y - 2)^5$ **4.** $(a + 5)^7$
5. $(2x - 1)^6$ **6.** $(3y + 1)^4$ **7.** $(4m + 3n)^5$ **8.** $(7x - 6y)^3$

Write the expansion of each expression using the binomial formula.

9. $(x + 5)^5$ **10.** $(x - 6)^4$ **11.** $(x - y)^6$ **12.** $(a + 3)^7$
13. $(m - 2n)^6$ **14.** $(5x + 4y)^4$ **15.** $(8a + 7b)^3$ **16.** $(3m - 2n)^5$
17. $(x^2 + 3y)^6$ **18.** $(2x^3 - 5y^2)^4$

Write the first four terms in the binomial expansion of each expression.

19. $(y - 3)^9$ **20.** $(x + 4)^8$ **21.** $(1 + x)^{-3}$ $(|x| < 1)$ **22.** $(1 - x)^{-1/2}$ $(|x| < 1)$

Solve.

23. Find the 4th term in the expansion of $(x + 3)^{10}$.
24. Find the 6th term in the expansion of $(y - 2)^{13}$.
25. Find the y^4 term in the expansion of $(x - y)^{11}$.
26. Find the x^7 term in the expansion of $(x + y)^{12}$.

Solve.

27. *Electronics*   In complex electronic systems it is important to determine the availability A of a component as defined by

$$A = \frac{F}{F + R}$$

where F is the mean time between failures and R is the mean time to repair. When $R << F$ (read "R is much less than F"), it is desirable to approximate A by using the first few terms of its binomial expansion. Divide the numerator and denominator by F and write the first three terms of this expansion using (18-8).*

28. *Hydraulics*   The velocity profile for turbulent flow in a smooth pipe is given by the expression

$$\left(1 - \frac{2r}{D}\right)^{1/n} \qquad \text{where} \quad \left|\frac{2r}{D}\right| < 1$$

Use (18-8) to write the first four terms in the expansion of this expression for $n = 6$.

18-5 | Review of Chapter 18

Important Terms and Concepts

sequence (p. 620)	arithmetic progression, AP (p. 620)	common ratio (p. 626)
terms of a sequence (p. 620)	common difference (p. 620)	Pascal's triangle (p. 638)
finite sequence (p. 620)	series (p. 622)	factorial (p. 639)
infinite sequence (p. 620)	geometric progression, GP (p. 626)	

Formulas

- nth term of an AP (p. 620): $a_n = a_1 + (n - 1)d$ **(18-1)**

- sum of the first n terms of an AP (p. 622): $S_n = \dfrac{n(a_1 + a_n)}{2}$ **(18-2)**

- nth term of a GP (p. 627): $a_n = a_1 r^{n-1}$ **(18-3)**

---

* Courtesy of William Jago, Tecolote Research, Inc., Santa Barbara, California.

- sum of the first n terms of a GP (p. 630): $S_n = \dfrac{a_1(1 - r^n)}{1 - r}$ where $r \neq 1$ **(18-4)**

- sum of an infinite GP (p. 634): $S = \dfrac{a_1}{1 - r}$ for $|r| < 1$ **(18-5)**

- factorial function (p. 639): $n! = n(n - 1)(n - 2)(n - 3) \cdots 3 \cdot 2 \cdot 1$ **(18-6)**
  where n is any positive integer

- binomial formula (p. 640):

  $$(a + b)^n = a^n + na^{n-1}b + \frac{n(n - 1)}{2!}a^{n-2}b^2 + \frac{n(n - 1)(n - 2)}{3!}a^{n-3}b^3 + \cdots + b^n \qquad \textbf{(18-7)}$$

- finding a given term in a binomial expansion (p. 640): The $(k + 1)$st term of $(a + b)^n$ is given by
  $$\frac{n(n - 1)(n - 2) \cdots (n - k + 1)}{k!}a^{n-k}b^k$$

- for any real number n where $|b| < 1$ (p. 641),

  $$(1 + b)^n = 1 + nb + \frac{n(n - 1)}{2!}b^2 + \frac{n(n - 1)(n - 2)}{3!}b^3 + \cdots \qquad \textbf{(18-8)}$$

Exercises 18-5

Find the indicated term of each progression.

1. The 10th term of: $-1, 4, 9, \ldots$
2. The 11th term of: $17, 13, 9, \ldots$
3. The 9th term of: $3, -9, 27, \ldots$
4. The 8th term of: $16, 4, 1, \ldots$
5. The 22nd term, where $a_1 = 8$ and $d = -3$
6. The 14th term, where $a_1 = -28$ and $d = 7$
7. The 10th term, where $a_1 = 2$ and $r = -\frac{5}{2}$
8. The 9th term, where $a_1 = 5$ and $r = 2$

Find the sum of the first n terms of each progression.

9. $4, 12, 36, \ldots$ $(n = 9)$
10. $-2, 3, -4.5, \ldots$ $(n = 13)$
11. $15, 5, \frac{5}{3}, \ldots$ (all terms)
12. $80, -60, 45, \ldots$ (all terms)
13. $n = 14, \quad a_1 = -3, \quad a_{14} = 36$ (AP)
14. $n = 6, \quad a_1 = 76, \quad a_6 = \frac{19}{8}$ (AP)
15. $n = 12, \quad a_1 = -1, \quad d = -3$
16. $n = 11, \quad a_1 = 7, \quad d = 5$

Use the given information in Problems 17–24 to find the indicated value.

17. $a_1 = 5, \quad d = -6, \quad a_{14} = ?$
18. $a_1 = 4, \quad r = 2, \quad a_n = 2048, \quad n = ?$
19. $a_1 = 7, \quad d = -3, \quad a_n = -20, \quad S_n = ?$
20. $a_1 = 13, \quad a_{17} = 77, \quad d = ?$

21. If the 9th term of a GP is 512 and the common ratio is 2, find the first term.

22. If the sum of the first seven terms of a GP is $79\dfrac{3}{8}$ and the common ratio is $\dfrac{1}{2}$, find the 7th term.

23. If the first term of an AP is 21, the nth term is 57, and the sum of the first n terms is 390, find n.

24. If the first term of an AP is 18, the common difference is -3, and the sum of the first n terms is -21, find n.

Find the fractional equivalent of each repeating decimal.

25. $0.\overline{2}$
26. $0.\overline{43}$
27. $0.7\overline{627}$
28. $0.6\overline{18}$

Write the binomial expansion of each expression.

29. $(x - 4)^5$
30. $(y + 2)^6$
31. $(3x + 5y)^4$
32. $(2x - 3)^7$

Find the first four terms in the binomial expansion of each expression.

33. $(x + 2)^{11}$ **34.** $(y - 5)^{10}$ **35.** $(1 + x)^{-4}$ $(|x| < 1)$ **36.** $(1 - x)^{-1/3}$ $(|x| < 1)$

Solve.

37. Find the sum of the first 150 multiples of 4.

38. If the 3rd term of an AP is 15 and the 7th term is 375, find the first term.

39. Find the 5th term in the expansion of $(x - 6)^9$.

40. Find the a^8 term in the expansion of $(3 - a)^{13}$.

41. *Wastewater technology*   The amount of corrosion accumulating on the inside of a water pipe decreases the inside diameter by 0.3 cm per year. If the diameter of a new pipe is 15.0 cm, what will be the diameter after 12 years?

42. *Physics*   If the initial velocity of an object moving in a straight line is 12.6 ft/s and the object is under a constant acceleration of 2.5 ft/s², how far will it travel in 60 s? (See Problem 37 in Exercises 18-1.)

43. *Finance*   If $10,000 is invested in a fixed-rate 20 year term savings account paying 14% compounded annually, find the value of the investment at the end of the term.

44. *Finance*   A self-employed person deposits $1500 each year into an IRA account paying 11% compounded annually. What is the total value of the account after 14 years?

45. *Environmental science*   A river contaminated with a certain pollutant is being cleaned up. If the level of pollution was found to be 32.0 ppb initially and is 13.5 ppb *after* 3 years of cleanup ($n = 4$), what will the level be after 8 years of cleanup? (Assume that the pollution level is decreasing geometrically.)

46. *Electronics*   In determining the intensity in the interference pattern among N antennas, it becomes necessary to sum

$$E = E_0 e^{i\omega t}[1 + e^{i\theta} + e^{2i\theta} + e^{i(n-1)\theta}]$$

Write an expression for E by finding the sum of the expression in brackets.*

47. *Physics*   A ball is dropped from a height of 10 ft. If each successive bounce reaches two-thirds the height of the previous one, through what total vertical distance does the ball travel before coming to rest?

48. *Instrumentation*   In the study of obstruction meters, the formula below is found. Use (18-8) to find the first three terms of the expansion of E.

$$E = \frac{1}{\sqrt{1 - \beta^4}} \qquad \text{where } |\beta| < 1$$

---

* For more information, see Hugh D. Young, *Fundamentals of Optics and Modern Physics* (New York: McGraw-Hill, 1968), p. 79.

Special Topics in Trigonometry

19

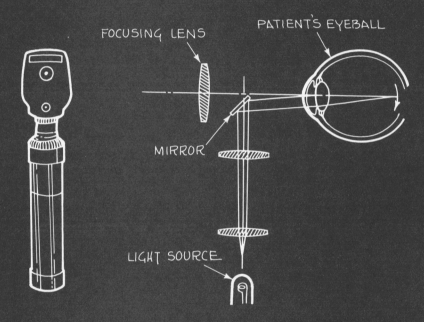

FOCUSING LENS

PATIENT'S EYEBALL

MIRROR

LIGHT SOURCE

OPTHALMOSCOPE

RAY DIAGRAM FOR OPTHALMOSCOPE

In Chapters 4, 8, and 10 we defined and applied the basic trigonometric functions, discussed radian measure, and explored the graphs of the trigonometric functions. In Chapters 9 and 12 we also applied trigonometric functions to the study of vectors and complex numbers. In this chapter we shall show how to rewrite trigonometric expressions in equivalent forms, solve trigonometric equations, and work with the inverse trigonometric functions.

Because of the interrelationships among the trigonometric functions, it is possible to rewrite any trigonometric expression in several equivalent ways. This process of rewriting can be useful when we want to simplify trigonometric expressions, solve equations, develop useful formulas, or solve a great many problems in analytic geometry, calculus, and other advanced mathematical topics.

Identities

An **identity** is a relationship between two expressions that is true for all permissible values of the variables involved. For example, the equation

$$x^2 - 7x + 12 = (x - 3)(x - 4)$$

is an identity, because substituting any value of x produces a true statement. On the other hand, the equation

$$x^2 - 7x + 12 = 0$$

is *not* an identity, because it is true only for $x = 3$ or $x = 4$.

Basic trigonometric identities

In Chapter 4 we defined the six trigonometric functions in terms of x, y, and r and then developed the following trigonometric identities in defining the reciprocal functions:

$$\csc \theta = \frac{1}{\sin \theta} \tag{19-1}$$

$$\text{or} \quad \sin \theta = \frac{1}{\csc \theta} \quad \text{or} \quad \sin \theta \csc \theta = 1$$

$$\sec \theta = \frac{1}{\cos \theta} \tag{19-2}$$

$$\text{or} \quad \cos \theta = \frac{1}{\sec \theta} \quad \text{or} \quad \cos \theta \sec \theta = 1$$

$$\cot \theta = \frac{1}{\tan \theta} \tag{19-3}$$

$$\text{or} \quad \tan \theta = \frac{1}{\cot \theta} \quad \text{or} \quad \tan \theta \cot \theta = 1$$

Furthermore, tan θ and cot θ can both be defined in terms of sin θ and cos θ (see Fig. 19-1):

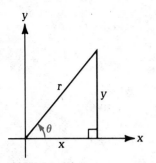

y

Figure 19-1

$$\tan \theta = \frac{y}{x} = \frac{y/r}{x/r} = \frac{\sin \theta}{\cos \theta} \qquad \tan \theta = \frac{\sin \theta}{\cos \theta} \qquad \text{(19-4)}$$

$$\cot \theta = \frac{1}{\tan \theta} = \frac{\cos \theta}{\sin \theta} \qquad \cot \theta = \frac{\cos \theta}{\sin \theta} \qquad \text{(19-5)}$$

Several other basic identities can be developed using the Pythagorean theorem, $x^2 + y^2 = r^2$ (refer to Fig. 19-1). First, if we divide both sides by r^2,

$$\left(\frac{x}{r}\right)^2 + \left(\frac{y}{r}\right)^2 = 1$$

and then substitute for x/r and y/r, we obtain:

$$\cos^2 \theta + \sin^2 \theta = 1 \qquad \text{(19-6)}$$

On the other hand, if we divide both sides of the Pythagorean theorem by x^2,

$$1 + \left(\frac{y}{x}\right)^2 = \left(\frac{r}{x}\right)^2$$

and then substitute for y/x and r/x, we get:

$$1 + \tan^2 \theta = \sec^2 \theta \qquad \text{(19-7)}$$

Or, if we divide both sides of the Pythagorean theorem by y^2,

$$\left(\frac{x}{y}\right)^2 + 1 = \left(\frac{r}{y}\right)^2$$

and then substitute for x/y and r/y, we have:

$$\cot^2 \theta + 1 = \csc^2 \theta \qquad \text{(19-8)}$$

NOTE   The expressions $\cos^2 \theta$, $\sin^2 \theta$, and so on, are convenient ways of writing $(\cos \theta)^2$, $(\sin \theta)^2$, etc. The exponent 2 in $\cos^2 \theta$ means that the entire *function* is squared—*not* the angle. ◀

For ease of reference we now summarize the eight basic trigonometric identities. These formulas should be memorized.

$$\csc \theta = \frac{1}{\sin \theta} \qquad \textbf{(19-1)} \qquad \sec \theta = \frac{1}{\cos \theta} \qquad \textbf{(19-2)}$$

$$\cot \theta = \frac{1}{\tan \theta} \qquad \textbf{(19-3)} \qquad \tan \theta = \frac{\sin \theta}{\cos \theta} \qquad \textbf{(19-4)}$$

$$\cot \theta = \frac{\cos \theta}{\sin \theta} \qquad \textbf{(19-5)} \qquad \sin^2 \theta + \cos^2 \theta = 1 \qquad \textbf{(19-6)}$$

$$1 + \tan^2 \theta = \sec^2 \theta \qquad \textbf{(19-7)} \qquad 1 + \cot^2 \theta = \csc^2 \theta \qquad \textbf{(19-8)}$$

The argument θ may stand for any angle, real number, or algebraic expression.

EXAMPLE 1

(a)   Applying (19-1): $\csc (x + 3) = \dfrac{1}{\sin (x + 3)}$

(b)   Applying (19-4): $\tan 29° = \dfrac{\sin 29°}{\cos 29°}$

We can check this using a calculator:

$$\boxed{\text{DEG}}\ 29\ \boxed{\tan} \rightarrow \quad 0.5543091$$
$$29\ \boxed{\sin}\ \boxed{\div}\ 29\ \boxed{\cos}\ \boxed{=} \rightarrow \quad 0.5543091$$

(c)   Applying (19-6): $\sin^2\left(\dfrac{\pi}{4}\right) + \cos^2\left(\dfrac{\pi}{4}\right) = 1$

To check this, recall that

$$\frac{\pi}{4} = 45° \qquad \text{and} \qquad \sin 45° = \cos 45° = \frac{\sqrt{2}}{2}$$

Substituting into the identity gives

$$\left(\frac{\sqrt{2}}{2}\right)^2 + \left(\frac{\sqrt{2}}{2}\right)^2 = 1$$

$$\frac{1}{2} + \frac{1}{2} = 1$$

One of the main values of knowing the eight basic identities is that they allow you to rewrite any trigonometric expression in several equivalent forms. To practice this skill, you will use them to prove additional identities. These proofs often require you to recognize variations of the eight basic identities.

EXAMPLE 2
(a)   Use identity (19-6) to write expressions for $\sin^2 \theta$ and for $\cos^2 \theta$.
(b)   Use identity (19-7) to write an expression for $\tan^2 \theta$.
(c)   Use identity (19-8) to write an expression for $\cot^2 \theta$.

Solution

(a) Subtracting $\cos^2\theta$ from both sides of (19-6), we have: $\sin^2\theta = 1 - \cos^2\theta$.
Subtracting $\sin^2\theta$ from both sides of (19-6) gives: $\cos^2\theta = 1 - \sin^2\theta$.

(b) Subtracting 1 from both sides, we obtain: $\tan^2\theta = \sec^2\theta - 1$.

(c) Subtracting 1 from both sides results in: $\cot^2\theta = \csc^2\theta - 1$.

The simple variations found in Example 2 will prove to be just as important as the original identities.

Proving identities

We can now use the eight basic identities and their variations to prove more complex trigonometric identities. There is no set method for performing these proofs, and many of them can be done in more than one way. The general approach is to use some of the basic identities to transform one member until it is identical to the other member.

EXAMPLE 3   Prove that: $\tan x = \sin x \sec x$

Solution   Generally, the best procedure is to transform the more complicated member until it is identical to the simpler one. In this case, we transform the right member.

Use (19-2). $\sin x \sec x = \sin x \cdot \dfrac{1}{\cos x}$

Combine. $= \dfrac{\sin x}{\cos x}$

Use (19-4). $\sin x \sec x = \boxed{\tan x}$

In the first step we chose to substitute for $\sec x$ rather than $\sin x$, because one of the eight basic identities defines $\tan\theta$ in terms of $\sin\theta$, but none of them defines $\tan\theta$ in terms of $\sec\theta$.

EXAMPLE 4   Show that: $\sec y = \dfrac{\csc y}{\cot y}$

Solution   We will transform the right side because it is the more complicated member. Since $\sec\theta$ is defined in terms of $\cos\theta$ in (19-2), we should first substitute for both $\csc y$ and $\cot y$.

Use (19-1). $\dfrac{\csc y}{\cot y} = \dfrac{\dfrac{1}{\sin y}}{\cot y}$

Use (19-5). $= \dfrac{\dfrac{1}{\sin y}}{\dfrac{\cos y}{\sin y}}$

Simplify. $= \dfrac{1}{\sin y} \cdot \dfrac{\sin y}{\cos y} = \dfrac{1}{\cos y}$

Use (19-2). $\dfrac{\csc y}{\cot y} = \boxed{\sec y}$

In some cases, it is best to simplify an expression algebraically before substituting.

EXAMPLE 5   Prove the identity: $\sin u + \dfrac{\cos^2 u}{\sin u} = \csc u$

Solution

Begin by combining the left side into a single fraction.

$$\sin u + \dfrac{\cos^2 u}{\sin u} = \dfrac{\sin^2 u}{\sin u} + \dfrac{\cos^2 u}{\sin u} = \dfrac{\sin^2 u + \cos^2 u}{\sin u}$$

Use (19-6). $= \dfrac{1}{\sin u}$

Use (19-1). $\sin u + \dfrac{\cos^2 u}{\sin u} = \boxed{\csc u}$

If both sides are equally complex, other considerations must dictate which side to simplify.

EXAMPLE 6   Prove that: $\tan y + \cot y = \dfrac{\tan y}{\sin^2 y}$

Solution   Here, the two sides appear to be equally complex. However, it is easier to combine two terms into one than to split a single fraction into a sum. We therefore choose to work on the left side.

Use (19-4) and (19-5) to replace $\tan y$ and $\cot y$.

$$\tan y + \cot y = \dfrac{\sin y}{\cos y} + \dfrac{\cos y}{\sin y}$$

Combine the two fractions. $= \dfrac{\sin^2 y + \cos^2 y}{\cos y \sin y}$

Use (19-6). $= \dfrac{1}{\cos y \sin y}$

The expression we are trying to obtain has $\tan y$ in the numerator. Therefore, multiply the numerator and denominator by $\tan y$.

$$= \dfrac{\tan y}{\tan y \cos y \sin y}$$

Use (19-4). $= \dfrac{\tan y}{\dfrac{\sin y}{\cos y} \cdot \cos y \sin y}$

Simplify the denominator.

$$\tan y + \cot y = \frac{\tan y}{\sin^2 y}$$

EXAMPLE 7 Prove the identity: $\sec x(\sec x + \tan x) = \dfrac{1}{1 - \sin x}$

Solution The left side is more complicated. First, apply the distributive property:

$$\sec x(\sec x + \tan x) = \sec^2 x + \sec x \tan x$$

Now, rewrite the entire expression in terms of sin x and cos x.

Use (19-2) to replace $\sec^2 x$ and sec x and (19-4) to replace tan x.

$$= \frac{1}{\cos^2 x} + \frac{1}{\cos x} \cdot \frac{\sin x}{\cos x}$$

Multiply and combine fractions.

$$= \frac{1}{\cos^2 x} + \frac{\sin x}{\cos^2 x}$$

$$= \frac{1 + \sin x}{\cos^2 x}$$

Use a variation of (19-6).

$$= \frac{1 + \sin x}{1 - \sin^2 x}$$

Factor the denominator.

$$= \frac{1 + \sin x}{(1 + \sin x)(1 - \sin x)}$$

Reduce.

$$\sec x(\sec x + \tan x) = \frac{1}{1 - \sin x}$$

EXAMPLE 8 Prove that: $\cot^4 m + \cot^2 m = \csc^4 m - \csc^2 m$

Solution Both sides seem equally difficult to transform, so we arbitrarily choose to work on the left member.

Factor $\cot^2 m$ from each term.　$\cot^4 m + \cot^2 m = \cot^2 m(1 + \cot^2 m)$

Use (19-8).　$= \cot^2 m$

Use a variation of (19-8).　$= (\csc^2 m - 1)\csc^2 m$

Multiply.　$\cot^4 m + \cot^2 m = \csc^4 m - \csc^2 m$

Exercises 19-1

Use a calculator to verify each of the following identities for the angle given:

1. (19-5) for $\theta = 71°$

2. (19-6) for $\theta = 38°$

3. (19-7) for $\theta = \dfrac{5\pi}{6}$

4. (19-8) for $\theta = 2.58$

Prove the following identities:

5. $\sin \theta = \dfrac{\cos \theta}{\cot \theta}$ **6.** $\cos \theta = \dfrac{\sin \theta}{\tan \theta}$ **7.** $\cot x = \cos x \csc x$ **8.** $\csc x = \sec x \cot x$

9. $\cos \theta \tan \theta \cot \theta \sec \theta = 1$ **10.** $\dfrac{\sin \theta - \cos \theta}{\cos \theta} = \tan \theta - 1$

11. $1 - 2 \sin x \cos x = (\cos x - \sin x)^2$ **12.** $\dfrac{1 - \sin x}{\cos x} = \dfrac{\cos x}{1 + \sin x}$

13. $\csc x = \sin x + \cos x \cot x$ **14.** $\tan^2 x - \sin^2 x = \tan^2 x \sin^2 x$

15. $\dfrac{\tan \theta + 1}{\tan \theta} = \dfrac{\cos \theta + \sin \theta}{\sin \theta}$ **16.** $\tan x \sec x = \dfrac{1 + \tan^2 x}{\csc x}$

17. $\dfrac{1 - \csc x}{2 \csc x} = \dfrac{\sin x - 1}{2}$ **18.** $\dfrac{\cot^2 \theta - \csc^2 \theta}{1 - \sec^2 \theta} = \cot^2 \theta$

19. $\csc \theta = \dfrac{\cot \theta + \tan \theta}{\sec \theta}$ **20.** $(\csc x + \cot x)^2 = \dfrac{1 + \cos x}{1 - \cos x}$

21. $\dfrac{1 - \sin y}{1 + \sin y} = (\tan y - \sec y)^2$ **22.** $\sin^4 x - \cos^4 x = \sin^2 x - \cos^2 x$

23. $\sin \theta (\csc \theta - \sin \theta) = \cos^2 \theta$ **24.** $\dfrac{1}{\csc \theta - \cot \theta} - \dfrac{1}{\csc \theta + \cot \theta} = 2 \cot \theta$

25. $\cos x + \cot x = \dfrac{\csc x + 1}{\sec x}$ **26.** $\csc x = \dfrac{\sin x}{1 - \cos x} - \cot x$

27. $\cot^2 \theta = \dfrac{\cos \theta}{\sec \theta - 1} - \dfrac{\cos \theta}{\tan^2 \theta}$ **28.** $2 \cos \theta = \dfrac{\sin \theta + \tan \theta \cos \theta}{\tan \theta}$

29. $\csc \theta - \cot \theta = \dfrac{1}{\cot \theta + \csc \theta}$ **30.** $\dfrac{\sin \theta}{1 + \sin \theta} = \dfrac{1}{1 + \csc \theta}$

31. $\dfrac{\sin^2 x}{1 + \cos x} = \dfrac{\tan x - \sin x}{\tan x}$ **32.** $\csc^4 x = 1 + \cot^2 x (1 + \csc^2 x)$

33. $\dfrac{1 + \sin n}{\cos n} + \dfrac{\cos n}{1 + \sin n} = 2 \sec n$ **34.** $\sec^4 x - \sec^2 x = \tan^4 x + \tan^2 x$

35. $\cot m - \dfrac{\sin m}{1 + \cos m} = 2 \cot m - \csc m$ **36.** $2 + \tan^2 a = \dfrac{\sec^4 a - 1}{\tan^2 a}$

37. $\dfrac{1 + \cos y}{\sin y} + \dfrac{\sin y}{1 + \cos y} = 2 \csc y$ **38.** $\dfrac{1 - \sec x}{1 + \sec x} = \dfrac{\sin x - \tan x}{\sin x + \tan x}$

39. $\dfrac{\tan \theta}{2 \tan \theta - \sin \theta} = \dfrac{1}{2 - \cos \theta}$ **40.** $\dfrac{1}{\cos \theta - \sin \theta} = \dfrac{\sec \theta + 2 \tan \theta \sec \theta}{1 + \tan \theta - 2 \tan^2 \theta}$

41. *Astrophysics*   The formula below appears in calculating the orbit of a meteorite as it approaches the earth:*

$$\sin \theta = \left[1 + \left(\frac{2sE}{MG} \right)^2 \right]^{-1/2}$$

Show that this can be written as: $\cot \theta = \dfrac{2sE}{MG}$

[*Hint:*   Let $x = \dfrac{2sE}{MG}$ to simplify your work.]

42. *Atomic physics*   One of the angular functions for the hydrogen atom is

$$P_{20} = \frac{\sqrt{10}}{4} (3 \cos^2 \theta - 1)$$

Express this in terms of $\sin \theta$.

* Additional background can be found in Martin Harwit, *Astrophysical Concepts* (New York: Wiley, 1973), p. 41.

43. *Physics* In describing the motion of a conical pendulum, physicists use the following formula:

$$\frac{g \sin \theta}{L} - \frac{h^2 \cos \theta}{\sin^3 \theta} = 0$$

Solve for h^2 in terms of $\cos \theta$ only.

44. *Physics* When two particles, each of mass m carrying charge q, are suspended from strings of length L, the angle θ each string makes with the vertical is given by

$$\frac{\tan^3 \theta}{1 + \tan^2 \theta} = \frac{q^2}{kmgL^2}$$

Show that the left side of this equation is equal to $\tan \theta - \cos \theta \sin \theta$.

19-2 | Sum and Difference Identities

In this section we shall develop equivalent expressions for the sine, cosine, and tangent of the sum and difference of two angles. Aside from having direct applications themselves, these identities also lead to the even more important double-angle and half-angle formulas discussed in the next section.

Sine and cosine of a sum or difference Our first objective is to find equivalent expressions for $\sin (\alpha \pm \beta)$ and $\cos (\alpha \pm \beta)$ in terms of functions of α and β separately. We will first dispel a common misconception by showing that $\sin (\alpha + \beta)$ is *not* equal to $\sin \alpha + \sin \beta$.

▦ **EXAMPLE 9** Show that $\sin (40° + 30°)$ is not equal to $\sin 40° + \sin 30°$.

Solution $\sin (40° + 30°) = \sin 70°$:   70 ⟨sin⟩ → ░0.9396926░

$\sin 40° + \sin 30°$:   40 ⟨sin⟩ + 30 ⟨sin⟩ = → ░1.1427876░

We can derive expressions for $\sin (\alpha \pm \beta)$ and $\cos (\alpha \pm \beta)$ by using the exponential form of a complex number, as defined in Chapter 12:

$$e^{j\theta} = \cos \theta + j \sin \theta$$

For angles α and β,

$$e^{j\alpha} = \cos \alpha + j \sin \alpha \quad \text{and} \quad e^{j\beta} = \cos \beta + j \sin \beta$$

Multiplying, we get

$$e^{j\alpha} \cdot e^{j\beta} = (\cos \alpha + j \sin \alpha)(\cos \beta + j \sin \beta)$$

$$e^{j(\alpha+\beta)} = (\cos \alpha \cos \beta - \sin \alpha \sin \beta) + j(\sin \alpha \cos \beta + \cos \alpha \sin \beta)$$

But, substituting $\alpha + \beta$ for θ in the definition of the exponential form,

$$e^{j(\alpha+\beta)} = \cos (\alpha + \beta) + j \sin (\alpha + \beta)$$

Therefore,

$$\cos(\alpha + \beta) + j\sin(\alpha + \beta) = (\cos\alpha\cos\beta - \sin\alpha\sin\beta)$$

$$+ j(\sin\alpha\cos\beta + \cos\alpha\sin\beta)$$

By equating real and imaginary parts, we obtain the **sum identities:**

$$\cos(\alpha + \beta) = \cos\alpha\cos\beta - \sin\alpha\sin\beta \qquad \text{(19-9)}$$

$$\sin(\alpha + \beta) = \sin\alpha\cos\beta + \cos\alpha\sin\beta \qquad \text{(19-10)}$$

A similar derivation using division of $e^{j\alpha}$ by $e^{j\beta}$ gives us the following difference identities:*

$$\cos(\alpha - \beta) = \cos\alpha\cos\beta + \sin\alpha\sin\beta \qquad \text{(19-11)}$$

$$\sin(\alpha - \beta) = \sin\alpha\cos\beta - \cos\alpha\sin\beta \qquad \text{(19-12)}$$

These formulas hold for any angles α and β. We can write these sum and difference identities in a more compact form as

$$\sin(\alpha \pm \beta) = \sin\alpha\cos\beta \pm \cos\alpha\sin\beta$$

$$\cos(\alpha \pm \beta) = \cos\alpha\cos\beta \mp \sin\alpha\sin\beta$$

EXAMPLE 10   Find $\sin 15°$ in simplest radical form by using $15° = 60° - 45°$.

Solution $\sin 15° = \sin(60° - 45°) = \sin 60° \cos 45° - \cos 60° \sin 45°$

$$= \frac{\sqrt{3}}{2} \cdot \frac{\sqrt{2}}{2} - \frac{1}{2} \cdot \frac{\sqrt{2}}{2} = \frac{\sqrt{6} - \sqrt{2}}{4}$$

EXAMPLE 11   Given that $\sin\alpha = \dfrac{5}{13}$ (α in quadrant I) and $\cos\beta = -\dfrac{4}{5}$ (β in quadrant II), find $\cos(\alpha + \beta)$ using the sum identity (19-9).

Solution   To use (19-9), we must first find $\cos\alpha$ and $\sin\beta$ by applying the identity $\sin^2\theta + \cos^2\theta = 1$.

---

* This derivation is left for the student in Problem 41 of Exercises 19-2.

Step 1. To find $\cos \alpha$: $\left(\dfrac{5}{13}\right)^2 + \cos^2 \alpha = 1$

$$\cos^2 \alpha = 1 - \left(\dfrac{5}{13}\right)^2$$

$$= 1 - \dfrac{25}{169} = \dfrac{144}{169}$$

$$\cos \alpha = \pm \dfrac{12}{13}$$

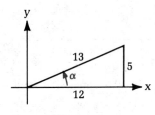

Figure 19-2

But α is in quadrant I where cosine is positive (Fig. 19-2); therefore, $\cos \alpha = +\dfrac{12}{13}$.

Step 2. To find $\sin \beta$: $\sin^2 \beta + \left(-\dfrac{4}{5}\right)^2 = 1$

$$\sin^2 \beta = 1 - \left(-\dfrac{4}{5}\right)^2$$

$$= 1 - \dfrac{16}{25} = \dfrac{9}{25}$$

$$\sin \beta = \pm \dfrac{3}{5}$$

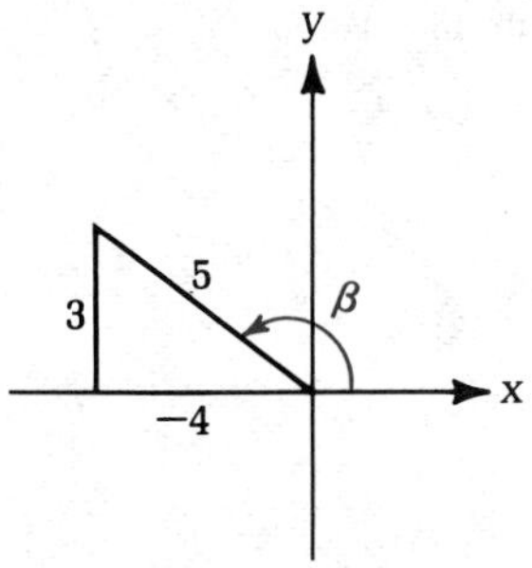

Figure 19-3

But β is in quadrant II where sine is positive (Fig. 19-3); therefore, $\sin \beta = +\dfrac{3}{5}$.

Step 3. Substituting in (19-9), we have

$$\cos (\alpha + \beta) = \cos \alpha \cos \beta - \sin \alpha \sin \beta$$

$$\cos (\alpha + \beta) = \left(\dfrac{12}{13}\right)\left(-\dfrac{4}{5}\right) - \left(\dfrac{5}{13}\right)\left(\dfrac{3}{5}\right)$$

$$= -\dfrac{48}{65} - \dfrac{15}{65} = -\dfrac{63}{65}$$

Tangent of a sum or difference

We may use the sum and difference formulas (19-9)–(19-12) to find a similar relationship for the tangent function:

$$\dfrac{\sin (\alpha \pm \beta)}{\cos (\alpha \pm \beta)} = \dfrac{\sin \alpha \cos \beta \pm \cos \alpha \sin \beta}{\cos \alpha \cos \beta \mp \sin \alpha \sin \beta}$$

Dividing each term of the fraction on the right by $\cos \alpha \cos \beta$, we have

$$= \dfrac{\dfrac{\sin \alpha}{\cos \alpha} \pm \dfrac{\sin \beta}{\cos \beta}}{1 \mp \dfrac{\sin \alpha}{\cos \alpha} \cdot \dfrac{\sin \beta}{\cos \beta}}$$

Using (19-4), we obtain the following formula:

$$\tan (\alpha \pm \beta) = \frac{\tan \alpha \pm \tan \beta}{1 \mp \tan \alpha \tan \beta} \qquad\qquad (19\text{-}13)$$

NOTE ▶ Identity (19-13) is equivalent to two formulas, one using the upper signs and one using the lower signs. ◀

EXAMPLE 12   Find tan 15° in simplest radical form by using 15° = 60° − 45°.

Solution $\tan 15° = \tan (60° - 45°) = \dfrac{\tan 60° - \tan 45°}{1 + \tan 60° \tan 45°}$

$$= \frac{\sqrt{3} - 1}{1 + \sqrt{3} \cdot 1}$$

$$= \frac{\sqrt{3} - 1}{1 + \sqrt{3}} \cdot \frac{1 - \sqrt{3}}{1 - \sqrt{3}}$$

$$= \frac{2\sqrt{3} - 4}{-2}$$

In lowest terms,

$$\tan 15° = 2 - \sqrt{3}$$

Proving identities   The sum and difference formulas can be used to prove many identities.

EXAMPLE 13   Prove that: sin (90° + x) = cos x

Solution   Using formula (19-10), we have

sin (90° + x) = sin 90° cos x + cos 90° sin x

Since sin 90° = 1 and cos 90° = 0, then

sin (90° + x) = 1 · cos x + 0 = cos x

EXAMPLE 14   Show that: $\cos x \cos y = \dfrac{1}{2}[\cos (x + y) + \cos (x - y)]$

Solution   Working on the right side, we use (19-9) and (19-11) to write

$$\frac{1}{2}[\cos (x + y) + \cos (x - y)]$$

$$= \frac{1}{2}[(\cos x \cos y - \sin x \sin y) + (\cos x \cos y + \sin x \sin y)]$$

$$= \frac{1}{2}[2 \cos x \cos y] = \cos x \cos y$$

$$\frac{1}{2}[\cos (x + y) + \cos (x - y)] = \cos x \cos y$$

EXAMPLE 15 Reduce the following expression to a single term:

$$\cos A \cos 3A - \sin A \sin 3A$$

Solution Careful inspection of the given expression reveals that it is in the form of (19-9). By replacing β with $3A$, we have

$$\cos A \cos 3A - \sin A \sin 3A = \cos (A + 3A) = \cos 4A$$

EXAMPLE 16 Prove that: $\tan x - \tan y = \dfrac{\sin (x - y)}{\cos x \cos y}$

Solution Working on the right side, we use the difference formula for sine (19-12) to write

$$\frac{\sin (x - y)}{\cos x \cos y} = \frac{\sin x \cos y - \cos x \sin y}{\cos x \cos y}$$

$$= \frac{\sin x \cos y}{\cos x \cos y} - \frac{\cos x \sin y}{\cos x \cos y}$$

$$= \frac{\sin x}{\cos x} - \frac{\sin y}{\cos y}$$

$$\frac{\sin (x - y)}{\cos x \cos y} = \tan x - \tan y$$

EXAMPLE 17 Show that: $\cos \left(\dfrac{\pi}{3} - x \right) = \dfrac{\cos x + \sqrt{3} \sin x}{2}$

Solution Using the difference formula for cosine (19-11) to transform the left side, we have

$$\cos \left(\frac{\pi}{3} - x \right) = \cos \frac{\pi}{3} \cos x + \sin \frac{\pi}{3} \sin x$$

$$= \frac{1}{2} \cdot \cos x + \frac{\sqrt{3}}{2} \cdot \sin x$$

$$\cos \left(\frac{\pi}{3} - x \right) = \frac{\cos x + \sqrt{3} \sin x}{2}$$

Exercises 19-2

Use the sum and difference formulas to find each of the following in simplest radical form:

1. $\cos 15°$ using $15° = 45° - 30°$ **2.** $\sin 75°$ using $75° = 45° + 30°$
3. $\tan 75°$ using $75° = 45° + 30°$ **4.** $\sin 105°$ using $105° = 60° + 45°$

Given that $\sin \alpha = \dfrac{3}{5}$ (α in quadrant I) and $\cos \beta = \dfrac{5}{13}$ (β in quadrant I), find the following without the use of a calculator:

5. $\sin (\alpha + \beta)$ **6.** $\cos (\alpha + \beta)$ **7.** $\cos (\alpha - \beta)$ **8.** $\sin (\beta - \alpha)$

Given that $\sin \alpha = \dfrac{24}{25}$ (α in quadrant I) and $\cos \beta = -\dfrac{3}{5}$ (β in quadrant III), find the following without the use of a calculator:

9. $\cos (\alpha + \beta)$ **10.** $\sin (\alpha + \beta)$ **11.** $\sin (\alpha - \beta)$ **12.** $\cos (\beta - \alpha)$

Reduce each expression to a single term.

13. $\sin A \cos 2A - \cos A \sin 2A$
15. $\cos (x + y) \cos (x - y) + \sin (x + y) \sin (x - y)$
17. $\dfrac{\tan m + \tan 2m}{1 - \tan m \tan 2m}$

14. $\cos 4B \cos B - \sin 4B \sin B$
16. $\sin 3x \cos (x - y) + \sin (x - y) \cos 3x$
18. $\dfrac{\tan (y - x) - \tan y}{1 + \tan (y - x) \tan y}$

Prove the following identities:

19. $\cos (90° + x) = -\sin x$

20. $\sin \left(\dfrac{\pi}{2} - x \right) = \cos x$

21. $\sin \left(\dfrac{3\pi}{2} + x \right) = -\cos x$

22. $\cos (60° - x) = \dfrac{\cos x + \sqrt{3} \sin x}{2}$

23. $\cos \left(x - \dfrac{\pi}{4} \right) = \dfrac{\sqrt{2}}{2} (\cos x + \sin x)$

24. $\cos (x + 180°) = -\cos x$

25. $\sin \left(\dfrac{5\pi}{6} - x \right) = \dfrac{\cos x + \sqrt{3} \sin x}{2}$

26. $\sin (x + 45°) = \dfrac{\sqrt{2}}{2} (\cos x + \sin x)$

27. $\tan (x + 45°) = \dfrac{1 + \tan x}{1 - \tan x}$

28. $\tan \left(\dfrac{\pi}{3} - x \right) = \dfrac{\sqrt{3} - \tan x}{1 + \sqrt{3} \tan x}$

29. $\dfrac{\cos (x + y)}{\cos x \sin y} = \cot y - \tan x$

30. $\sin x \sin y = \dfrac{1}{2} [\cos (x - y) - \cos (x + y)]$

31. $\sin x \cos y = \dfrac{1}{2} [\sin (x + y) + \sin (x - y)]$

32. $\tan (x + y) \tan (x - y) = \dfrac{\sec^2 x - \sec^2 y}{\sec^2 y - \sec^2 x (\sec^2 y - 1)}$

33. $\tan (90° - x) = \cot x$ [Hint: To avoid undefined values, use identity (19-4) to rewrite the left side.]
34. $\tan (-x) = -\tan x$ [Hint: $\tan (-x) = \tan (0 - x)$]
35. $\sin (-x) = -\sin x$ (See Problem 34.)
36. $\cos (-x) = \cos x$ (See Problem 34.)
37. *Electrical engineering*   The potential difference in a certain three-phase alternator is given by

$$V = -E \left[\sin \left(2\pi ft - \dfrac{5\pi}{6} \right) + \sin \left(2\pi ft - \dfrac{7\pi}{6} \right) \right]$$

Show that $V = E\sqrt{3} \sin (2\pi ft)$.

38. *Physics*   The formula below is used to describe an elliptically polarized light component. Express this equation in terms of sine functions only.

$$E = A \sin \gamma \cos \left(wt - \phi - \frac{\pi}{2} \right)$$

39. *Space technology*   In analyzing the geometric distortions caused by the earth's curvature as seen from an orbiting satellite, space scientists use the following equation:

$$\frac{a}{\sin \alpha} = \frac{a + h}{\sin (\pi - \alpha - \beta)}$$

Expand the denominator of the right side to eliminate π and solve the equation for a.

40. *Physics*   A sinusoidal wave moving on a water surface is described by the equation

$$y = 25 \sin \left(\frac{\pi t}{5} - 2 \right)$$

Write an equivalent expression by expanding the right member and then evaluating where possible.

41. Use $\dfrac{e^{j\alpha}}{e^{j\beta}}$ to derive the difference identities (19-11) and (19-12).

19-3 | Double-Angle and Half-Angle Formulas

Double-angle formulas   It is often necessary to express a function of twice an angle as a function of the angle itself. To derive expressions for $\sin 2\alpha$ and $\cos 2\alpha$, first note that $2\alpha = \alpha + \alpha$. Substituting α for β in the sum formulas (19-10) and (19-9), we obtain

$$\sin 2\alpha = \sin (\alpha + \alpha) = \sin \alpha \cos \alpha + \cos \alpha \sin \alpha = 2 \sin \alpha \cos \alpha$$

$$\cos 2\alpha = \cos (\alpha + \alpha) = \cos \alpha \cos \alpha - \sin \alpha \sin \alpha = \cos^2\alpha - \sin^2\alpha$$

Two additional expressions can be written for $\cos 2\alpha$ using variations of (19-6). First, substituting $1 - \sin^2\alpha$ for $\cos^2\alpha$:

$$\cos 2\alpha = 1 - \sin^2\alpha - \sin^2\alpha = 1 - 2 \sin^2\alpha$$

Substituting $1 - \cos^2\alpha$ for $\sin^2\alpha$:

$$\cos 2\alpha = \cos^2\alpha - (1 - \cos^2\alpha) = 2 \cos^2\alpha - 1$$

Summarizing these results, we have the following **double-angle formulas:**

$\sin 2\alpha = 2 \sin \alpha \cos \alpha$	(19-14)
$\cos 2\alpha = \cos^2\alpha - \sin^2\alpha$	(19-15)
$\cos 2\alpha = 1 - 2 \sin^2\alpha$	(19-16)
$\cos 2\alpha = 2 \cos^2\alpha - 1$	(19-17)

Finding function values   These identities can be used in many ways. For example, they can be used to find functions of an unknown angle from functions of a known angle.

EXAMPLE 18   Given that $\cos \theta = -\dfrac{4}{5}$ in the second quadrant, use the double-angle formulas to find:

(a) $\cos 2\theta$ (b) $\sin 2\theta$

Solution

(a)   Substituting directly into (19-17), we have

$$\cos 2\theta = 2\left(-\frac{4}{5}\right)^2 - 1 = 2\left(\frac{16}{25}\right) - 1 = \frac{7}{25}$$

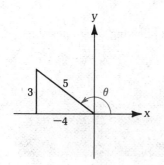

Figure 19-4

(b)   The formula for $\sin 2\theta$ requires $\sin \theta$ as well as $\cos \theta$. Using (19-6), we determine $\sin \theta$.

$$\sin^2 \theta = 1 - \cos^2 \theta$$
$$= 1 - \left(-\frac{4}{5}\right)^2 = \frac{9}{25}$$

Taking the square root of both sides:

$$\sin \theta = \pm \frac{3}{5}$$

But since θ is in quadrant II, $\sin \theta$ is positive (Fig. 19-4). Thus, $\sin \theta = \dfrac{3}{5}$. We can now substitute directly into (19-14):

$$\sin 2\theta = 2\left(\frac{3}{5}\right)\left(-\frac{4}{5}\right) = -\frac{24}{25}$$

Proving identities   The double-angle formulas may also be used to prove additional identities. In some cases, equivalent expressions can be found by simply recognizing a familiar form.

EXAMPLE 19   Write the expression $2 \cos^2 3x - 2 \sin^2 3x$ as a single term.

Solution   A close inspection reveals that the given expression has the form of (19-15).

Factoring 2 from each term gives us: $2(\cos^2 3x - \sin^2 3x)$

Substituting $3x$ for α in (19-15), we obtain: $= 2 \cos 6x$

EXAMPLE 20   Prove the identity: $\sin 2x = \dfrac{2 \cot x}{1 + \cot^2 x}$

Solution   We begin with the right side.

Use (19-5).

$$\frac{2 \cot x}{1 + \cot^2 x} = \frac{2\left(\dfrac{\cos x}{\sin x}\right)}{1 + \left(\dfrac{\cos x}{\sin x}\right)^2} = \frac{\dfrac{2 \cos x}{\sin x}}{\dfrac{\sin^2 x + \cos^2 x}{\sin^2 x}}$$

Use (19-6).

$$= \dfrac{\dfrac{2 \cos x}{\sin x}}{\dfrac{1}{\sin^2 x}}$$

Multiply the numerator and denominator by $\sin^2 x$.

$$= 2 \cos x \sin x$$

Use (19-14).

$$\dfrac{2 \cot x}{1 + \cot^2 x} = \sin 2x$$

EXAMPLE 21   Prove: $\cos 3x = \cos x - 4 \cos x \sin^2 x$

Solution   Since we have no identity for $\cos 3x$, we let $3x = 2x + x$ and attempt to transform the left side into the right.

Use (19-9).

$$\cos 3x = \cos (2x + x) = \cos 2x \cos x - \sin 2x \sin x$$

Use (19-16) and (19-14).

$$= (1 - 2 \sin^2 x) \cos x - (2 \cos x \sin x) \sin x$$

Simplify.

$$= \cos x - 2 \cos x \sin^2 x - 2 \cos x \sin^2 x$$

$$\cos 3x = \cos x - 4 \cos x \sin^2 x$$

Half-angle formulas   We will now develop and apply formulas for $\sin (\alpha/2)$ and $\cos (\alpha/2)$ in terms of $\sin \alpha$ and $\cos \alpha$. These are known as the **half-angle formulas.** First, write the identity (19-16) in terms of x:

$$\cos 2x = 1 - 2 \sin^2 x$$

Now, solve for $\sin x$:

$$2 \sin^2 x = 1 - \cos 2x$$

$$\sin^2 x = \dfrac{1 - \cos 2x}{2}$$

$$\sin x = \pm \sqrt{\dfrac{1 - \cos 2x}{2}}$$

Finally, replace x with $\dfrac{\alpha}{2}$.   Since $x = \dfrac{\alpha}{2}$,   then $2x = \alpha$.   Thus:

$$\sin \dfrac{\alpha}{2} = \pm \sqrt{\dfrac{1 - \cos \alpha}{2}} \tag{19-18}$$

In a similar way, we can derive the following formula for $\cos (\alpha/2)$ from (19-17):

$$\cos \frac{\alpha}{2} = \pm \sqrt{\frac{1 + \cos \alpha}{2}} \qquad\qquad (19\text{-}19)$$

In each case, the correct sign depends on the quadrant in which $\alpha/2$ lies.

REMEMBER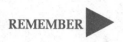

The sine function is negative in quadrants III and IV, and the cosine function is negative in quadrants II and III. ◄

Finding function values

The half-angle formulas can be used to obtain unknown function values in terms of known values.

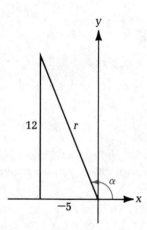

Figure 19-5

EXAMPLE 22 Given $\tan \alpha = -\dfrac{12}{5}$ and $90° < \alpha < 180°$, find the exact value of $\sin (\alpha/2)$.

Solution
Step 1. To use the half-angle formula for sine (19-18), we must find $\cos \alpha$. From Fig. 19-5 we see that

$$r^2 = (-5)^2 + 12^2 = 169$$

So

$$r = 13 \quad \text{and} \quad \cos \alpha = \frac{x}{r} = \frac{-5}{13}$$

Step 2. Because $90° < \alpha < 180°$, then $45° < \alpha/2 < 90°$. This means that $\alpha/2$ is in quadrant I where sine is positive. Therefore substitute $\cos \alpha = \dfrac{-5}{13}$ into the "+" version of (19-18):

$$\sin \frac{\alpha}{2} = \sqrt{\frac{1 - \cos \alpha}{2}} = \sqrt{\frac{1 - (-\frac{5}{13})}{2}} = \sqrt{\frac{9}{13}} = \frac{3\sqrt{13}}{13}$$

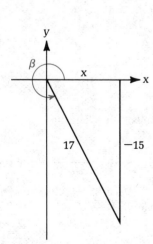

Figure 19-6

EXAMPLE 23 Given $\sin \beta = -\dfrac{15}{17}$ and $270° < \beta < 360°$, find the exact value of $\cos (\beta/2)$.

Solution
Step 1. To find $\cos (\beta/2)$ from (19-19), we must determine $\cos \beta$. From Fig. 19-6,

$$x^2 + (-15)^2 = 17^2$$

$$x^2 = 289 - 225 = 64$$

$$x = \pm 8$$

But β is in quadrant IV; therefore, $\cos \beta$ is positive and

$$\cos \beta = \frac{x}{r} = \frac{8}{17}$$

Step 2. Because $270° < \beta < 360°$, then $135° < \beta/2 < 180°$. This means that $\beta/2$ is in quadrant II where cosine is negative. Therefore substitute in the "$-$" version of (19-19):

$$\cos \frac{\beta}{2} = -\sqrt{\frac{1 + \cos \beta}{2}} = -\sqrt{\frac{1 + \frac{8}{17}}{2}} = -\sqrt{\frac{25}{34}} \quad \frac{-5\sqrt{34}}{34}$$

Proving identities   The half-angle formulas can also be used to prove additional identities.

EXAMPLE 24   Prove that: $2 \sin^2\left(\frac{x}{2}\right) = \frac{\sin^2 x}{1 + \cos x}$

Solution   Because the half-angle formulas are given in terms of $x/2$, we begin with the left side.

Use (19-18).

$$2 \sin^2\left(\frac{x}{2}\right) = 2\left(\sqrt{\frac{1 - \cos x}{2}}\right)^2$$

$$= 2\left(\frac{1 - \cos x}{2}\right) = 1 - \cos x$$

Multiply by $\dfrac{1 + \cos x}{1 + \cos x}$.

$$= \frac{(1 - \cos x)(1 + \cos x)}{1 + \cos x} = \frac{1 - \cos^2 x}{1 + \cos x}$$

Use a variation of (19-6).

$$= \frac{\sin^2 x}{1 + \cos x}$$

EXAMPLE 25   Prove the identity: $\tan \frac{x}{2} = \frac{\sin x}{1 + \cos x}$

Solution   Begin with the left side.

Use (19-4), (19-18), and (19-19).

$$\tan \frac{x}{2} = \frac{\sin \frac{x}{2}}{\cos \frac{x}{2}} = \frac{\sqrt{\dfrac{1 - \cos x}{2}}}{\sqrt{\dfrac{1 + \cos x}{2}}}$$

Simplify.

$$= \sqrt{\frac{1 - \cos x}{2} \cdot \frac{2}{1 + \cos x}} = \sqrt{\frac{1 - \cos x}{1 + \cos x}}$$

Rationalize the denominator.

$$= \frac{\sqrt{1 - \cos x}}{\sqrt{1 + \cos x}} \cdot \frac{\sqrt{1 + \cos x}}{\sqrt{1 + \cos x}} = \frac{\sqrt{1 - \cos^2 x}}{1 + \cos x}$$

Use a variation of (19-6) and simplify.

$$= \frac{\sqrt{\sin^2 x}}{1 + \cos x} = \frac{\sin x}{1 + \cos x}$$

Exercises 19-3

Use the double-angle formulas to find exact values for cos 2α and sin 2α from the given information.

1. $\cos \alpha = \dfrac{3}{5}$ (quadrant I)   **2.** $\sin \alpha = \dfrac{5}{13}$ (quadrant I)

3. $\sin \alpha = -\dfrac{4}{5}$ (quadrant III)   **4.** $\tan \alpha = -\dfrac{3}{4}$ (quadrant II)

Use the half-angle formulas to find exact values for cos (β/2) and sin (β/2) from the given information.

5. $\cos \beta = \dfrac{4}{5}$ $(0° < \beta < 90°)$ **6.** $\sin \beta = \dfrac{12}{13}$ $(0° < \beta < 90°)$

7. $\sin \beta = -\dfrac{8}{17}$ $(180° < \beta < 270°)$ **8.** $\tan \beta = -\dfrac{5}{4}$ $(270° < \beta < 360°)$

Write each expression as a single term.

9. $4 \sin 2x \cos 2x$ **10.** $3 \sin^2 4x - 3 \cos^2 4x$ **11.** $4 \sin^2 3x - 2$ **12.** $6 \cos^2 2x - 3$

Prove the following identities:

13. $\cos^2 x = \dfrac{1 + \cos 2x}{2}$ **14.** $\csc^2 x = \dfrac{2}{1 - \cos 2x}$

15. $\cos 2x = \dfrac{1 - \tan^2 x}{1 + \tan^2 x}$ **16.** $\tan 2x = \dfrac{2 \tan x}{1 - \tan^2 x}$

17. $\tan x + \cot x = \dfrac{2}{\sin 2x}$ **18.** $\sin 3x = 3 \sin x - 4 \sin^3 x$

19. $(\sin x + \cos x)^2 = 1 + \sin 2x$ **20.** $\cos x + \sin x = \dfrac{\cos 2x}{\cos x - \sin x}$

21. $2 \cos^2\left(\dfrac{x}{2}\right) = \dfrac{\sin^2 x}{1 - \cos x}$ **22.** $\cos^2\left(\dfrac{x}{2}\right) \tan \dfrac{x}{2} = \dfrac{\sin x}{2}$

23. $\dfrac{1 + \sin x}{\cos x} = \dfrac{\cos \dfrac{x}{2} + \sin \dfrac{x}{2}}{\cos \dfrac{x}{2} - \sin \dfrac{x}{2}}$ **24.** $\sin^2\left(\dfrac{x}{2}\right) \cos^2\left(\dfrac{x}{2}\right) = \dfrac{\sin^2 x}{4}$

25. $\tan \dfrac{x}{2} = \csc x - \cot x$ **26.** $\sec \dfrac{x}{2} = \pm\sqrt{\dfrac{2}{1 + \cos x}}$

27. $\csc \dfrac{x}{2} = \pm\sqrt{\dfrac{2}{1 - \cos x}}$ **28.** $\cot \dfrac{x}{2} = \dfrac{\sin x}{1 - \cos x}$

Solve.

29. *Physics* In the description of electromagnetic waves, physicists use the following equation:

$$R = \frac{\tan^2(\alpha - \beta)}{\tan^2(\alpha + \beta)}$$

Find an expression for R in terms of sin β when α = 2β. (Use the identity for sin 3x in Problem 18.)

30. *Physics* Another trigonometric equation that appears in electromagnetic theory is

$$T = \frac{4 \sin \alpha \cos^2\beta \sin \beta}{\sin^2(\alpha + \beta) \cos^2(\beta - \alpha)}$$

Simplify the right side for the case where $\alpha = \beta$.

31. *Transportation engineering*   In determining the width of the merging region of track for a rapid transit system, engineers use the equation below. Solve for x in terms of $\cos \theta$ only.

$$\frac{x}{2} = 2r \sin^2\left(\frac{\theta}{2}\right)$$

32. *Optics*   Under certain conditions, the transmission coefficient of light through a transport medium is given by the formula below.* Rewrite this in terms of $\cos \theta$ and simplify.

$$T = \frac{T_m}{1 + F \sin^2\left(\frac{\theta}{2}\right)}$$

19-4 | Solving Trigonometric Equations

So far in this chapter we have worked only with equations that are identities—that is, equations that are true for *all* values of the variable. However, many applications involve **conditional** trigonometric equations that are true for only *some* values of the variable.

Conditional trigonometric equations can be solved using the techniques used earlier to solve linear and quadratic equations. By substituting a variable for the trigonometric function itself, we can convert it to an algebraic equation that can be solved.

EXAMPLE 26   Solve for θ such that $0 \le \theta < 2\pi$: $2 \sin \theta - 1 = 0$

Solution   Let $x = \sin \theta$. Then the given equation becomes

$$2x - 1 = 0$$

Add 1. $2x = 1$

Divide by 2. $x = \dfrac{1}{2}$

Replace x with $\sin \theta$. $\sin \theta = \dfrac{1}{2}$

From the special angle values, we know that Ref $\theta = 30° = \pi/6$.[†] Because sine is positive in quadrants I and II,

---

* For further details, see Miles V. Klein, *Optics* (New York: Wiley, 1970), p. 207.
† Since these values will be used extensively in the remainder of this chapter, you might wish to review the exact values for functions of 30°, 45°, and 60° (Section 4-3) and the quadrantal angles (Section 8-3).

$$\theta_I = \text{Ref } \theta = \frac{\pi}{6} \quad \text{and} \quad \theta_{II} = \pi - \text{Ref } \theta = \pi - \frac{\pi}{6} = \frac{5\pi}{6}$$

Therefore, $\theta = \pi/6$ or $5\pi/6$. Check these values by substituting back into the original equation:

$$2 \sin\left(\frac{\pi}{6}\right) - 1 = 0 \qquad 2 \sin\left(\frac{5\pi}{6}\right) - 1 = 0$$

$$2\left(\frac{1}{2}\right) - 1 = 0 \qquad 2\left(\frac{1}{2}\right) - 1 = 0$$

NOTE ▶ Unless otherwise stated, assume that the solutions will be restricted to angles between 0 and 2π. All answers should be expressed in radians. ◀

Quadratic techniques can also be used to solve trigonometric equations.

 EXAMPLE 27 Solve: $3 \cos^2 y - \cos y = 0$

Solution Let $x = \cos y$. Then this trigonometric equation becomes the quadratic equation

$$3x^2 - x = 0$$

Factor. $\qquad\qquad\qquad\qquad\qquad\qquad\qquad x(3x - 1) = 0$

Set each factor equal to 0. $\qquad\qquad\quad x = 0 \quad \text{or} \quad 3x - 1 = 0$

Solve. $\qquad\qquad\qquad\qquad\qquad\quad x = 0 \quad \text{or} \qquad x = \frac{1}{3}$

Substitute $\cos y$ for x. $\qquad\qquad\quad \cos y = 0 \quad \text{or} \quad \cos y = \frac{1}{3}$

For $\cos y = 0$, $y = \frac{\pi}{2}$ or $y = \frac{3\pi}{2}$. For $\cos y = \frac{1}{3}$, find the reference angle:

$$\cos (\text{Ref } y) = \left|\frac{1}{3}\right| = \frac{1}{3}.$$

$\boxed{\text{RAD}}\ 1\ \boxed{\div}\ 3\ \boxed{=}\ \boxed{\text{INV}}\ \boxed{\text{COS}}\ \boxed{\text{STO}} \rightarrow$ ▨ *1.2309594* ← Ref y

Since $\cos y$ is positive, y is in quadrant I or quadrant IV.

$\qquad y_I = \text{Ref } y \approx 1.23$

$\qquad y_{IV} = 2\pi - \text{Ref } y \rightarrow 2\ \boxed{\times}\ \boxed{\pi}\ \boxed{-}\ \boxed{\text{RCL}}\ \boxed{=} \rightarrow$ ▨ *5.0522259*

We therefore have four possible solutions:

$$y \approx 1.23, 5.05 \quad \text{or} \quad y = \frac{\pi}{2}, \frac{3\pi}{2}$$

Check them.

NOTE ▶ For the special angle values of the function, such as $0, \frac{1}{2}, 1$, and so on, determine the reference angle as a rational multiple of π rad by inspection. Use a calculator only for function values that are not familiar. ◀

If an equation contains more than one trigonometric function, the equation can usually be solved more easily if it is rewritten in terms of a single function.

EXAMPLE 28 Solve: $-2 \sin^2\theta - 3 \cos\theta + 3 = 0$

Solution

First, use a variation of (19-6) to express the equation in terms of cosine only.

$$-2 \cdot (1 - \cos^2\theta) - 3 \cos\theta + 3 = 0$$

Second, simplify.

$$-2 + 2 \cos^2\theta - 3 \cos\theta + 3 = 0$$

$$2 \cos^2\theta - 3 \cos\theta + 1 = 0$$

Third, let $x = \cos\theta$.

$$2x^2 - 3x + 1 = 0$$

Finally, factor and solve.

$$(2x - 1)(x - 1) = 0$$

$$2x - 1 = 0 \quad \text{or} \quad x - 1 = 0$$

$$x = \frac{1}{2} \quad \text{or} \quad x = 1$$

Replace x with $\cos\theta$.

$$\cos\theta = \frac{1}{2} \quad \text{or} \quad \cos\theta = 1$$

Determine θ.

$$\theta = \frac{\pi}{3}, \frac{5\pi}{3} \quad \text{or} \quad \theta = 0$$

Check these three solutions in the original equation.

If a trigonometric equation in quadratic form is not factorable, use the quadratic formula to solve it.

EXAMPLE 29 Solve: $2 \sec^2 y = 3 \tan y + 5$

Solution **First,** use (19-7) to express the equation in terms of tangent only:

$$2 \cdot (1 + \tan^2 y) = 3 \tan y + 5$$

$$2 + 2 \tan^2 y = 3 \tan y + 5$$

$$2 \tan^2 y - 3 \tan y - 3 = 0$$

Now, let $x = \tan y$. Then this quadratic equation becomes

$$2x^2 - 3x - 3 = 0$$

From the quadratic formula,

$$x = \tan y = \frac{3 \pm \sqrt{9 - 4(2)(-3)}}{2(2)} = \frac{3 \pm \sqrt{33}}{4}$$

Calculate y for each value of $\tan y$ separately:

For $\dfrac{3 + \sqrt{33}}{4}$: $\tan y \rightarrow$ 3 $+$ 33 $\sqrt{}$ $=$ $\div$ 4 $=$ $\rightarrow$ 2.1861407

Because $\tan y$ is positive, y is in quadrant I or III.

Ref $y \rightarrow$ [RAD] [INV] [tan] [STO] $\rightarrow$ 1.1417832

Ref $y = y_I \approx 1.14$ rad

$y_{III} = \pi + $ Ref $y \rightarrow$ [π] $+$ [RCL] $=$ $\rightarrow$ 4.2833759

$y_{III} \approx 4.28$ rad

For $\dfrac{3 - \sqrt{33}}{4}$: $\tan y \rightarrow$ 3 $-$ 33 $\sqrt{}$ $=$ $\div$ 4 $=$ $\rightarrow$ −0.6861407

Since $\tan y$ is negative, y is in quadrant II or IV. Take the absolute value of $\tan y$ to find Ref y:

Ref $y \rightarrow$ [+/−] [INV] [tan] [STO] $\rightarrow$ 0.6013637

Then determine y_{II} and y_{IV}:

$y_{II} = \pi - $ Ref $y \rightarrow$ [π] $-$ [RCL] $=$ $\rightarrow$ 2.5402289

$y_{II} \approx 2.54$ rad

$y_{IV} = 2\pi - $ Ref $y \rightarrow$ 2 $\times$ [π] $-$ [RCL] $=$ $\rightarrow$ 5.6818216

$y_{IV} \approx 5.68$ rad

The four solutions are $y \approx 1.14$, 2.54, 4.28, or 5.68 rad. Check them.

   To solve an equation, it is not always necessary to rewrite it in terms of a single trigonometric function. But it is important to transform the equation so that all angles are the same.

EXAMPLE 30   Solve: $\sin 2x - \cos x = 0$

Solution   **First,** use the double-angle identity (19-14) to express $\sin 2x$ in terms of x:

$2 \sin x \cos x - \cos x = 0$

Second, solve. Although this equation contains two different functions, the left side may be factored.

$$\cos x (2 \sin x - 1) = 0$$

$\cos x = 0$ or $2 \sin x - 1 = 0$

$$\sin x = \frac{1}{2}$$

For cos x = 0: $x = \dfrac{\pi}{2}$ or $\dfrac{3\pi}{2}$

For sin x = $\dfrac{1}{2}$: $x = \dfrac{\pi}{6}$ or $\dfrac{5\pi}{6}$

The four solutions are $x = \pi/6,\ \pi/2,\ 5\pi/6,$ or $3\pi/2$. Check them.

Some trigonometric equations require that you square both sides or multiply both sides by the same function before they can be solved. However, this may introduce extraneous solutions, so be certain to check all solutions carefully.

EXAMPLE 31   Solve: $\sin x = \sin \dfrac{x}{2}$

Solution

First, use the half-angle identity (19-18) to replace $\sin (x/2)$ with an expression in terms of x.

$$\sin x = \pm \sqrt{\dfrac{1 - \cos x}{2}}$$

Second, to solve, square both sides.

$$\sin^2 x = \dfrac{1 - \cos x}{2}$$

Use a variation of (19-6) to replace $\sin^2 x$.

$$1 - \cos^2 x = \dfrac{1 - \cos x}{2}$$

Multiply by 2.

$$2 - 2\cos^2 x = 1 - \cos x$$

Move all terms to one side.

$$2\cos^2 x - \cos x - 1 = 0$$

Let $y = \cos x$.

$$2y^2 - y - 1 = 0$$

Factor.

$$(2y + 1)(y - 1) = 0$$

Set each factor equal to 0.

$$2y + 1 = 0 \quad \text{or} \quad y - 1 = 0$$

Solve for y.

$$y = -\dfrac{1}{2} \quad \text{or} \quad y = 1$$

For y = cos x = $-\dfrac{1}{2}$: $x = \dfrac{2\pi}{3}$ or $\dfrac{4\pi}{3}$

For y = cos x = 1: $x = 0$

The solutions appear to be $x = 0,\ 2\pi/3,$ or $4\pi/3$.

Third, check:

For x = 0: $\sin 0 \overset{?}{=} \sin \dfrac{0}{2}$

$$0 = 0$$

$$\text{For } x = \frac{2\pi}{3}: \quad \sin \frac{2\pi}{3} \overset{?}{=} \sin \frac{\pi}{3}$$

$$\frac{\sqrt{3}}{2} = \frac{\sqrt{3}}{2}$$

$$\text{For } x = \frac{4\pi}{3}: \quad \sin \frac{4\pi}{3} \overset{?}{=} \sin \frac{2\pi}{3}$$

$$-0.5 \neq 0.5 \qquad \text{This is not a valid solution.}$$

Squaring both sides introduced the extraneous solution $4\pi/3$. The correct solutions are $x = 0$ or $x = 2\pi/3$.

When the angle of the trigonometric function is a multiple of the variable, the full range of possible values must be tested so that all solutions are found.

EXAMPLE 32   Solve: $\cos x \cos 2x - \sin x \sin 2x = 0$

Solution   To solve this equation, first recognize that it has the form of the identity $\cos (\alpha + \beta)$. We can rewrite it as

$$\cos (x + 2x) = 0$$

$$\cos 3x = 0$$

To include all possible values of x between 0 and 2π, consider all possible values of $3x$ between 0 and $3 \cdot 2\pi = 6\pi$.

$$\cos 3x = 0 \quad \text{at } \frac{n\pi}{2}, \text{ where } n = 1, 3, 5, 7, \ldots$$

In other words,

$$\cos 3x = 0 \quad \text{at } 3x = \frac{\pi}{2}, \frac{3\pi}{2}, \frac{5\pi}{2}, \frac{7\pi}{2}, \frac{9\pi}{2}, \text{ or } \frac{11\pi}{2}$$

Dividing by 3, we obtain

$$x = \frac{\pi}{6}, \frac{\pi}{2}, \frac{5\pi}{6}, \frac{7\pi}{6}, \frac{3\pi}{2}, \text{ or } \frac{11\pi}{6}$$

Exercises 19-4

Solve for x such that $0 \leq x < 2\pi$.

1. $2 \sin x + 1 = 0$
3. $\cos x + 1 = 0$
5. $2 \sin x - \sqrt{3} = 0$
7. $\tan^2 x - 1 = 0$

2. $2 \cos x - 1 = 0$
4. $\tan x + 1 = 0$
6. $\tan x + \sqrt{3} = 0$
8. $4 \sin^2 x - 1 = 0$

9. $5 \cos^2 x - 2 = 0$
10. $6 \sin^2 x - 5 = 0$
11. $\sin^2 x - 2 \sin x = 0$
12. $4 \tan^2 x = \tan x$
13. $\tan x + \tan x \sin x = 0$
14. $\sec x \cos x + \cos x = 0$
15. $2 \cos^2 x + 3 \cos x + 1 = 0$
16. $4 \sin^2 x - 4 \sin x + 1 = 0$
17. $6 \sin^2 x + 7 \sin x + 1 = 0$
18. $4 \cos^2 x - 5 \cos x + 1 = 0$
19. $2 \cos^2 x - 3 \sin x = 0$
20. $4 \sin^2 x - \cos x - 2 = 0$
21. $\sin 2x - \sin x = 0$
22. $\sin 2x + 2 \cos x = 0$
23. $\cos 2x - \cos x = 0$
24. $\cos 2x - 9 \sin x + 4 = 0$
25. $\tan^2 x - \sec x - 1 = 0$
26. $2 \tan x + 3 \sec^2 x - 8 = 0$
27. $\sin 2x \cos x + \cos 2x \sin x = 0$
28. $\cos 5x \cos x + \sin 5x \sin x = 0$
29. $\dfrac{\cos x}{1 + \sin x} = 1$
30. $\cos x = \cos 2x - 2$
31. $3 \tan x - 1 = 4 \cot x$
32. $2 \csc x + 3 \sin x + 5 = 0$

Solve.

33. *Optics* The intensity I of a light beam transmitted by a pair of polarizing filters is given by the equation

$$I = I_m \cos^2 \theta$$

where I_m is the maximum intensity and θ is the angle between the filters. At what angles does the intensity drop by $\dfrac{1}{2}$?

$$\left[\text{Hint: } \dfrac{I}{I_m} = \dfrac{1}{2} \right]$$

34. *Surveying* The formula below is used to determine the new latitude L_n when the original latitude L_o, the distance d, and the angle θ are given. Solve for θ if $L_n = 5378$, $L_o = 5145$, and $d = 250.0$ ft.

$$L_n = d \cos \theta + L_o$$

35. *Energy technology* The height h (in feet) of a point on the rim of a certain waterwheel is given by

$$h = 5 + 6 \cos \frac{\pi}{6}(t - 2)$$

where h is measured from the water surface and t is the time in minutes. At what times t is the point at water level?

36. *Optics* The intensity pattern for a three-slit diffraction grating is given by the equation below.* Find x when $\dfrac{I_\theta}{I_m} = \dfrac{2}{3}$.

$$\dfrac{I_\theta}{I_m} = \dfrac{1}{9}(1 + 4 \cos x + 4 \cos^2 x).$$

19-5 | Inverse Trigonometric Functions

As we have seen in Chapter 3, a function by definition is a set of ordered pairs of numbers (x, y) such that every value of the x-component has associated with

* Based on David Halliday and Robert Resnick, *Fundamentals of Physics*, 2d ed. (New York: Wiley, 1981), p. 770.

it one and only one value of the y-component. The function $y = f(x)$ may be thought of as a rule that generates such ordered pairs, assigning a single unique value of y, the dependent variable, for every value of x, the independent variable. The set of all allowed values of x is called the **domain** of the function, and the set of resulting values of y is called the **range** of the function.

Finding inverse functions

When the x- and y-components in every pair defining the function $f(x)$ are interchanged, we create a new rule, called the **inverse** of $f(x)$, which may or may not itself be a function. If the new rule is a function, it is called the **inverse function** of $f(x)$ and is sometimes denoted by $f^{-1}(x)$.*

EXAMPLE 33   Find the inverse of the function: $y = 3x - 1$

Solution   **First,** interchange variables x and y, so that the equation becomes

$x = 3y - 1$

Second, solve for y:

$y = \dfrac{x}{3} + \dfrac{1}{3}$ This is the inverse of the function $y = 3x - 1$.

Third, check to determine whether the inverse is a function. Because each real number value of x produces a unique value of y, this inverse is a function. Notice that for the function $y = f(x) = 3x - 1$ and its inverse function $y = f^{-1}(x) = \dfrac{x}{3} + \dfrac{1}{3}$, the domain is the set of all real numbers.

For any function in which a particular value of y is associated with more than one value of x, the inverse will *not* be a function. In this case, finding an inverse that is itself a function will require that we redefine the original function by restricting its domain.

EXAMPLE 34   Find the inverse of $y = x^2 - 3$

Solution
First, interchange x and y. $x = y^2 - 3$

Second, solve for y. $y^2 = x + 3$

$$y = \pm\sqrt{x + 3}$$

Third, check to determine whether this inverse is a function. For each positive real number x, the equation $y = \pm\sqrt{x + 3}$ is satisfied by *two* real numbers y. For example, if $x = 1$, then $y = +2$ or $y = -2$ both satisfy the equation. This inverse is *not* a function.

---

* Be careful not to confuse $f^{-1}(x)$ with $\dfrac{1}{f(x)}$.

However, we can redefine the original function so that its inverse is a function. The original function was given with no restrictions on x, but if we restrict the domain of the original function so that $x \geq 0$, then the inverse of this new function is $y = \sqrt{x + 3}$, where $x \geq -3$. Here, the square root symbol is understood to mean that only the positive, or principal, root is used. This is shown graphically in Fig. 19-7.

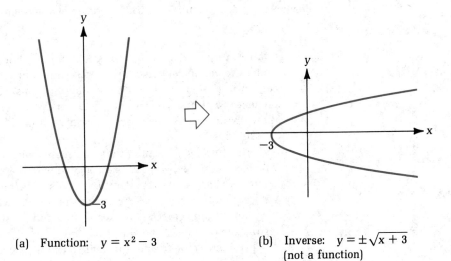

(a) Function: $y = x^2 - 3$

(b) Inverse: $y = \pm\sqrt{x + 3}$
(not a function)

Figure 19-7

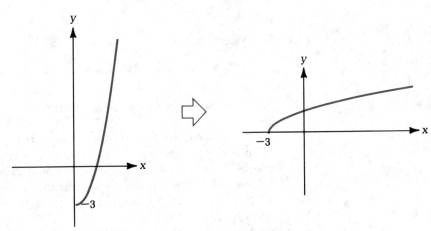

(c) New function:

$y = x^2 - 3 \quad (x \geq 0)$

(d) Inverse function:

$y = \sqrt{x + 3} \quad (x \geq -3)$

Defining inverses   The inverse of a function cannot always be found using algebraic methods. For example, to find the inverse of the exponential function $y = b^x$, we can interchange x and y to get

$x = b^y$

but it is not possible to solve this equation algebraically for y. Instead, we defined a new function in Chapter 13, the logarithm function:

$y = \log_b x$ is equivalent to $x = b^y$

and so, the functions $y = b^x$ and $y = \log_b x$ are inverse functions.

Inverse of the sine function

Similarly, if we want to find the inverse of a trigonometric function—for example, $y = \sin x$—we can interchange x and y to get

$x = \sin y$

but it is not possible to solve this equation algebraically for y. Previously, we found the value of y when $\sin y$ was known by entering on a calculator the sequence $\boxed{y}\,\boxed{\text{INV}}\,\boxed{\text{sin}}$. But for certain applications, such as solving literal trigonometric equations, it is necessary to work with an inverse function. Just as we did with the exponential function, we must now define a new function in order to write this.

Recall that for $y = \sin x$, there are many values of x that give the same y value. For example, $y = 1/2$ is associated with the x values $\pi/6$, $5\pi/6$, $13\pi/6$, $-7\pi/6$, $-11\pi/6$, and so on. Because there is more than one value of x associated with each value of y, we know that the inverse will not automatically be a function. In order to find an inverse function for the sine function, we must restrict the domain of the sine function.

The **inverse sine function** is defined as follows:*

$y = \text{Arcsin } x$ (19-20)

where $\text{Sin } y = x$ and $-\dfrac{\pi}{2} \le y \le \dfrac{\pi}{2}$

Notice the following regarding this definition:

- The Sine and Arcsine functions are written with initial capital letters to indicate that these are the functions restricted to the given values.
- The range of the Arcsine function includes all angles in quadrants I and IV as well as the quadrantal angles bordering these quadrants.
- Quadrant IV angles are expressed as *negative* angles, and all angles are expressed in radians.
- As usual with the sine function, $-1 \le x \le 1$. For positive values of x, y will be in quadrant I, and for negative values of x, y will be in quadrant IV.
- The function $y = \text{Arcsin } x$ is equivalent to having *both* $\text{Sin } y = x$ *and* $-\pi/2 \le y \le \pi/2$.

---

* You may also see $y = \text{Arcsin } x$ written as $y = \text{Sin}^{-1}x$. In this notation the "-1" is not an exponent, it simply indicates the inverse function.

In defining the Arcsine function, we have restricted the domain of the sine function to assure that the inverse is a function. Every value of x now corresponds to one and only one value of y, and every value of y corresponds to one and only one value of x. The graphs of y = Arcsin x and y = Sin x are shown in color in Fig. 19-8.

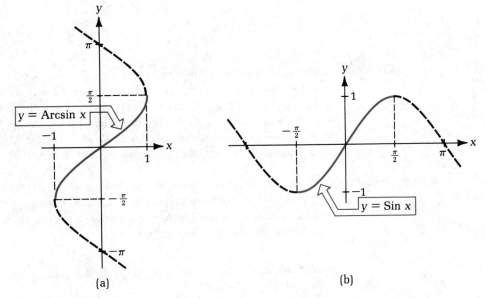

Figure 19-8

(a)

(b)

EXAMPLE 35 Find $y = \text{Arcsin} \frac{1}{2}$ without using a calculator.

Solution We must find a value of y such that $\text{Sin } y = \frac{1}{2}$ and $-\frac{\pi}{2} \le y \le \frac{\pi}{2}$.

Step 1. As usual, find the reference angle Ref y:

$$\text{Sin (Ref } y) = \left| \frac{1}{2} \right| = \frac{1}{2}$$

From our knowledge of special angle values, we know that

$$\text{Ref } y = \frac{\pi}{6}$$

Step 2. Use Ref y and the sign of the function to determine y within the limited range. For $-\pi/2 \le y \le \pi/2$, the sine function is positive only in quadrant I. Therefore,

$$y = \text{Ref } y = \frac{\pi}{6}$$

EXAMPLE 36   Find $y = \text{Arcsin}\,(-\sqrt{3}/2)$ without using a calculator.

Solution   We must find y such that $\text{Sin}\,y = -\sqrt{3}/2$ and $-\pi/2 \le y \le \pi/2$.

Step 1. $\text{Sin (Ref }y) = \left| -\dfrac{\sqrt{3}}{2} \right| = \dfrac{\sqrt{3}}{2}$ so $\text{Ref }y = \dfrac{\pi}{3}$

Step 2.   Within the restricted range, the sine function is negative only in quadrant IV. Therefore, expressing y as a negative angle,

$$y = 0 - \frac{\pi}{3} = -\frac{\pi}{3}$$

NOTE   To find the value of the inverse sine function $y = \text{Arcsin }x$ simply means to determine the angle y corresponding to a given value of the sine function x. We have performed this procedure throughout our study of trigonometry, but in the past we needed to find *all* corresponding values of the angle in the range of 0 to 2π. Now, because we are defining an inverse *function*, we are interested only in finding the *one* value of the angle within the restricted range.   ◄

Using a calculator   To find the value of the inverse sine function $y = \text{Arcsin }x$ using a calculator, first put the calculator in radian mode, and then enter $\boxed{x}\,\boxed{\text{INV}}\,\boxed{\sin}$. The calculator will automatically display the correct value of the function, and it will express quadrant IV angles as negative angles.

EXAMPLE 37   Find: (a) $y = \text{Arcsin }0.84$ (b) $y = \text{Arcsin }(-0.43)$

Solution

(a) $\boxed{\text{RAD}}\,.84\,\boxed{\text{INV}}\,\boxed{\sin} \rightarrow$ 0.9972832 $y = \text{Arcsin }0.84 \approx 1.0$ rad

(b) $.43\,\boxed{+/-}\,\boxed{\text{INV}}\,\boxed{\sin} \rightarrow$ -0.4444928 $y = \text{Arcsin }(-0.43) \approx -0.44$ rad

Inverse cosine   The definition of the inverse cosine function, like the inverse sine function, must also include restrictions on the range. To ensure that each value of x results in only one value of y, we define the **inverse cosine function** as follows:

$y = \text{Arccos }x$ $\hspace{6cm}$ **(19-21)**

where $\text{Cos }y = x$ and $0 \le y \le \pi$

Notice that the range is limited to quadrants I and II and the quadrantal angles 0 and π. When x is positive, y is in quadrant I, and when x is negative, y is in quadrant II. As before, $-1 \le x \le 1$. The graphs of $y = \text{Arccos }x$ and $y = \text{Cos }x$ are shown in color in Fig. 19-9.

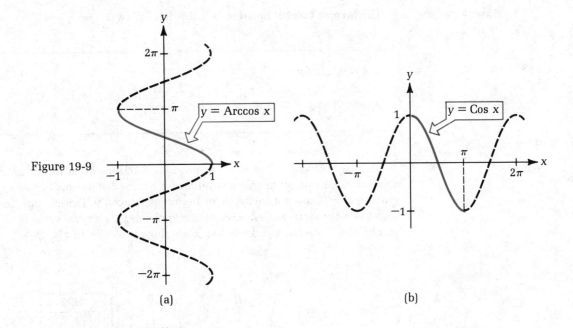

Figure 19-9

(a)   (b)

EXAMPLE 38 Find: (a) $y = \text{Arccos}\left(-\dfrac{\sqrt{2}}{2}\right)$ (b) $y = \text{Arccos } 1$

Solution
(a) The given expression means: Find a number y such that

$$\text{Cos } y = -\frac{\sqrt{2}}{2} \quad \text{and} \quad 0 \le y \le \pi$$

Step 1. $\text{Cos (Ref } y) = \left|-\dfrac{\sqrt{2}}{2}\right| = \dfrac{\sqrt{2}}{2}$ so $\text{Ref } y = \dfrac{\pi}{4}$

Step 2. Within its restricted range, cosine is negative only in quadrant II. Therefore,

$$y = \pi - \frac{\pi}{4} = \frac{3\pi}{4}$$

(b) Find y such that $\text{Cos } y = 1$. Since $\cos y = 1$ only at $y = 0, 2\pi$, and so on, then within the restricted range, $y = 0$.

EXAMPLE 39 Find: (a) $y = \text{Arccos } 0.385$ (b) $y = \text{Arccos } (-0.624)$

Solution
(a) [RAD] .385 [INV] [COS] → ▨ _1.1755885_ $y \approx 1.18$ rad
(b) .624 [+/−] [INV] [COS] → ▨ _2.2446475_ $y \approx 2.24$ rad

Inverse tangent   The **inverse tangent function** is defined as follows:

$$y = \text{Arctan } x \tag{19-22}$$

$$\text{where} \quad \text{Tan } y = x \quad \text{and} \quad -\frac{\pi}{2} < y < \frac{\pi}{2}$$

Because the range is restricted to quadrant I and quadrant IV angles, each value of x corresponds to only one value of y. Notice that the quadrantal angles $-\pi/2$ and $\pi/2$ are not included in the range, since the tangent function is undefined for these values. Also notice that x can be any real number. The graphs of $y = \text{Arctan } x$ and $y = \text{Tan } x$ are shown in color in Fig. 19-10.

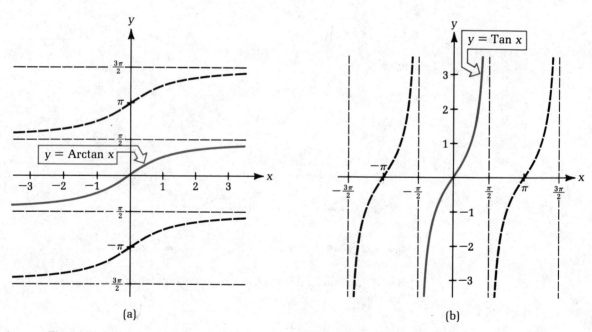

(a)   (b)

Figure 19-10

EXAMPLE 40   Find: (a) $y = \text{Arctan } \dfrac{\sqrt{3}}{3}$ ▦ (b) $y = \text{Arctan } (-2.65)$

Solution

(a)   This is equivalent to Tan $y = \sqrt{3}/3$ for $-\pi/2 < y < \pi/2$.

Step 1.   Tan (Ref y) = $\left| \dfrac{\sqrt{3}}{3} \right| = \dfrac{\sqrt{3}}{3}$ so   Ref y = $\dfrac{\pi}{6}$

Step 2.   Within the allowable range, y is positive only in quadrant I. Therefore,

$$y = \text{Ref } y = \frac{\pi}{6}$$

(b) [RAD] 2.65 [+/-] [INV] [tan] →   *-1.2099595* $y \approx -1.21$ rad

Notice that the calculator automatically displays the quadrant IV angle as a negative angle.

In some instances you may need to find the function of a function.

EXAMPLE 41 Evaluate: sin (Arctan 1)

Solution

Step 1. Find the angle. Arctan 1 is in quadrant I; therefore, Arctan 1 = $\pi/4$.

Step 2. Evaluate the function. From Step 1,

$$\sin (\text{Arctan } 1) = \sin \frac{\pi}{4} = \frac{\sqrt{2}}{2}$$

Expressions such as the one in Example 41 can be evaluated using a calculator if the function value is unfamiliar.

EXAMPLE 42 Evaluate: cos (2 Arcsin 0.589)

Solution [RAD] .589 [INV] [sin] [×] 2 [=] [cos] →   *0.3061580*

cos (2 Arcsin 0.589) ≈ 0.306 rad

Sometimes, the function of a function can be rewritten as an algebraic expression.

EXAMPLE 43 Find an algebraic expression for cos (Arctan x).

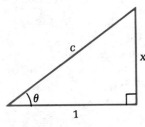

Figure 19-11

Solution

Step 1. Let θ = Arctan x. Then, x = tan θ. Draw a right triangle (Fig. 19-11) using

$$\tan \theta = \frac{x}{1} = \frac{\text{Opposite side}}{\text{Adjacent side}}$$

Step 2. Use the Pythagorean theorem to find the hypotenuse:

$$c = \sqrt{a^2 + b^2} = \sqrt{x^2 + 1}$$

Step 3. State the ratio for cos θ:

$$\cos \theta = \frac{\text{Adjacent side}}{\text{Hypotenuse}} = \frac{1}{\sqrt{x^2 + 1}}$$

Therefore, since θ = Arctan x,

$$\cos (\text{Arctan } x) = \frac{1}{\sqrt{x^2 + 1}} = \frac{\sqrt{x^2 + 1}}{x^2 + 1}$$

EXAMPLE 44 Find an algebraic expression for $\sin (2 \text{ Arcsin } x)$.

Solution

Step 1. Let $\theta = \text{Arcsin } x$. Then $x = \sin \theta$. Draw a right triangle (Fig. 19-12) using

$$\sin \theta = \frac{x}{1} = \frac{\text{Opposite side}}{\text{Hypotenuse}}$$

Step 2. Determine the missing leg using the Pythagorean theorem:

$$a = \sqrt{c^2 - b^2} = \sqrt{1 - x^2}$$

Step 3. The problem asks us to find $\sin (2 \text{ Arcsin } x)$, or $\sin 2\theta$. The double-angle identity (19-14) tells us that

$$\sin 2\theta = 2 \sin \theta \cos \theta$$

From Fig. 19-12,

$$\sin \theta = x \quad \text{and} \quad \cos \theta = \sqrt{1 - x^2}$$

Substituting, we get

$$2 \sin \theta \cos \theta = 2(x)\left(\sqrt{1 - x^2}\right)$$

Thus,

$$\sin (2 \text{ Arcsin } x) = 2x\sqrt{1 - x^2}$$

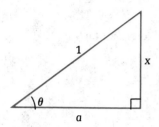

Figure 19-12

Exercises 19-5

Evaluate without using a calculator. Express all answers in radians.

1. $\text{Arcsin } \dfrac{\sqrt{3}}{2}$ **2.** $\text{Arcsin } \left(-\dfrac{1}{2}\right)$ **3.** $\text{Arccos } (-1)$ **4.** $\text{Arccos } \dfrac{1}{2}$

5. $\text{Arctan } 1$ **6.** $\text{Arctan } \left(-\sqrt{3}\right)$ **7.** $\text{Arcsin } (-1)$ **8.** $\text{Arcsin } \dfrac{\sqrt{2}}{2}$

9. $\text{Arccos } \left(-\dfrac{\sqrt{3}}{2}\right)$ **10.** $\text{Arccos } 0$ **11.** $\text{Arctan } \left(-\dfrac{\sqrt{3}}{3}\right)$ **12.** $\text{Arctan } (-1)$

Evaluate using a calculator. Express all answers in radians.

13. $\text{Arcsin } 0.74$ **14.** $\text{Arcsin } (-0.15)$ **15.** $\text{Arccos } 0.21$ **16.** $\text{Arccos } (-0.68)$
17. $\text{Arctan } 1.7$ **18.** $\text{Arctan } (-0.94)$ **19.** $\text{Arcsin } (-0.375)$ **20.** $\text{Arcsin } 0.543$
21. $\text{Arccos } (-0.875)$ **22.** $\text{Arccos } 0.226$ **23.** $\text{Arctan } (-3.16)$ **24.** $\text{Arctan } 2.59$

Evaluate each of the following. Use a calculator only for Problems 29–32.

25. $\sin \left[\text{Arccos } \left(-\dfrac{\sqrt{3}}{2}\right)\right]$ **26.** $\sin \left(\text{Arcsin } \dfrac{\sqrt{2}}{2}\right)$ **27.** $\cos \left[\text{Arctan } \left(-\dfrac{\sqrt{3}}{3}\right)\right]$

28. $\tan\left[\text{Arccos}\left(-\dfrac{1}{2}\right)\right]$ **29.** $\tan\,(\text{Arcsin } 0.427)$ **30.** $\cos[\text{Arcsin}\,(-0.127)]$

31. $\sin[\text{Arctan}\,(-4.16)]$ **32.** $\tan\,(2\,\text{Arctan } 0.754)$

Find an algebraic expression for each of the following:

33. $\sin\,(\text{Arccos } x)$ **34.** $\cos\,(\text{Arcsin } x)$ **35.** $\tan\,(\text{Arccos } x)$ **36.** $\tan\,(\text{Arctan } x)$

37. $\sin\,(\text{Arcsin } x)$ **38.** $\tan\,(\text{Arcsin } x)$ **39.** $\cos\,(2\,\text{Arccos } x)$ **40.** $\sin\,(2\,\text{Arctan } x)$

Solve.

41. *Machine technology*   The formula below applies to the tapered steel machine component shown in the figure. Solve for the angle α in terms of Arctan.

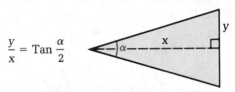

$$\frac{y}{x} = \text{Tan}\,\frac{\alpha}{2}$$

42. *Electronics*   The average power P dissipated in an ac circuit element is given by the equation $P = \dfrac{1}{2}V_m i_m \cos\,\phi$, where V_m and i_m are the maximum values of the voltage and current, respectively, and ϕ is the phase angle. Find ϕ if $V_m = 130$ V, $i_m = 12$ A, and $P = 110$ W.

43. *Optics*   Brewster's law states that a beam of light is reflected from a glass surface and will be polarized when it is reflected at some critical angle. The index of refraction n is given by $n = \text{Tan }\theta_p$, where θ_p is the critical angle. If the angle of refraction of the light is θ_r, then Snell's law gives $\text{Sin }\theta_p = n\,\text{Sin }\theta_r$. Use these two equations to find θ_r when $n = 1.60$.

44. *Optics*   To determine the length of the image arc on the surface of the retina, it is first necessary to find an expression for angle β in terms of α, y and R (see the figure). Find such an expression by solving for x and then θ, and using the fact that $\beta = \alpha + \theta$.*

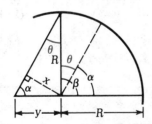

19-6 | Review of Chapter 19

Important Terms and Concepts

identity (p. 646)   inverse function (p. 672)   range (p. 672)

conditional trigonometric equation (p. 665)   domain (p. 672)

*Courtesy of John Merritt, Human Factors Research Inc., Santa Barbara, California.

Identities and Formulas

- basic trigonometric identities (p. 648):

$$\csc \theta = \frac{1}{\sin \theta} \quad \textbf{(19-1)} \qquad \sec \theta = \frac{1}{\cos \theta} \quad \textbf{(19-2)}$$

$$\cot \theta = \frac{1}{\tan \theta} \quad \textbf{(19-3)} \qquad \tan \theta = \frac{\sin \theta}{\cos \theta} \quad \textbf{(19-4)}$$

$$\cot \theta = \frac{\cos \theta}{\sin \theta} \quad \textbf{(19-5)} \qquad \sin^2 \theta + \cos^2 \theta = 1 \quad \textbf{(19-6)}$$

$$1 + \tan^2 \theta = \sec^2 \theta \quad \textbf{(19-7)} \qquad 1 + \cot^2 \theta = \csc^2 \theta \quad \textbf{(19-8)}$$

- sum and difference identities (Section 19-2):

$$\sin (\alpha \pm \beta) = \sin \alpha \cos \beta \pm \cos \alpha \sin \beta \quad \textbf{(19-10), (19-12)}$$

$$\cos (\alpha \pm \beta) = \cos \alpha \cos \beta \mp \sin \alpha \sin \beta \quad \textbf{(19-9), (19-11)}$$

$$\tan (\alpha \pm \beta) = \frac{\tan \alpha \pm \tan \beta}{1 \mp \tan \alpha \tan \beta} \quad \textbf{(19-13)}$$

- double-angle formulas (p. 659):

$$\sin 2\alpha = 2 \sin \alpha \cos \alpha \quad \textbf{(19-14)}$$

$$\cos 2\alpha = \cos^2 \alpha - \sin^2 \alpha \quad \textbf{(19-15)}$$

$$\cos 2\alpha = 1 - 2 \sin^2 \alpha \quad \textbf{(19-16)}$$

$$\cos 2\alpha = 2 \cos^2 \alpha - 1 \quad \textbf{(19-17)}$$

- half-angle formulas (p. 661):

$$\sin \frac{\alpha}{2} = \pm \sqrt{\frac{1 - \cos \alpha}{2}} \quad \textbf{(19-18)}$$

$$\cos \frac{\alpha}{2} = \pm \sqrt{\frac{1 + \cos \alpha}{2}} \quad \textbf{(19-19)}$$

- inverse trigonometric functions

  inverse sine function (p. 674): $y = \text{Arcsin } x$ where $\text{Sin } y = x$ and $-\frac{\pi}{2} \le y \le \frac{\pi}{2}$ **(19-20)**

  inverse cosine function (p. 676): $y = \text{Arccos } x$ where $\text{Cos } y = x$ and $0 \le y \le \pi$ **(19-21)**

  inverse tangent function (p. 678): $y = \text{Arctan } x$ where $\text{Tan } y = x$ and $-\frac{\pi}{2} < y < \frac{\pi}{2}$ **(19-22)**

Exercises 19-6

Given that $\sin \alpha = \frac{4}{5}$ $(0° < \alpha < 90°)$ and $\cos \beta = -\frac{7}{25}$ $(90° < \beta < 180°)$, find the exact value of each of the following:

1. $\sin (\alpha + \beta)$ 2. $\cos (\alpha + \beta)$ 3. $\cos (\beta - \alpha)$ 4. $\sin (\beta - \alpha)$

5. $\sin 2\alpha$ 6. $\cos 2\beta$ 7. $\sin \frac{\beta}{2}$ 8. $\cos \frac{\alpha}{2}$

Write each expression as a single term.

9. $6 \sin 4x \cos 4x$ 10. $5 \cos^2 2x - 5 \sin^2 2x$

11. $8 \cos^2 3x - 4$ 12. $3 - 6 \sin^2 4x$

13. $\sin x \cos 5x - \cos x \sin 5x$ 14. $\cos 6x \cos 2x + \sin 6x \sin 2x$

Prove the following identities:

15. $\dfrac{\csc x}{\sec x} = \cot x$

16. $\sin x = \cos x \tan x$

17. $\sin x \tan x \csc x \cot x = 1$

18. $\cos^4 x + 2 \cos^2 x \sin^2 x + \sin^4 x = 1$

19. $\sec \theta - \tan \theta = \dfrac{\cos \theta}{1 + \sin \theta}$

20. $\dfrac{\tan \theta \sin \theta}{\tan \theta - \sin \theta} = \dfrac{\sin \theta}{1 - \cos \theta}$

21. $\dfrac{\cos x}{1 - \sin x} + \dfrac{\cos x}{1 + \sin x} = 2 \sec x$

22. $\csc x = \dfrac{\csc x + \sec x}{1 + \tan x}$

23. $\tan x + \sin x = \dfrac{\tan^2 x \sin^2 x}{\tan x - \sin x}$

24. $\cot^2 x \sin x = (1 + \sin x)(\csc x - 1)$

25. $\cos \left(x - \dfrac{2\pi}{3} \right) = \dfrac{\sqrt{3} \sin x - \cos x}{2}$

26. $\sin \left(\dfrac{\pi}{4} - x \right) = \dfrac{\sqrt{2}}{2} (\cos x - \sin x)$

27. $\cos x \sin y = \dfrac{1}{2} [\sin (x + y) - \sin (x - y)]$

28. $\tan x + \tan y = \dfrac{\sin (x + y)}{\cos x \cos y}$

29. $\sin^2 \theta = \dfrac{1 - \cos 2\theta}{2}$

30. $\tan^2 \theta = \dfrac{1 - \cos 2\theta}{1 + \cos 2\theta}$

31. $\cos 3x = 4 \cos^3 x - 3 \cos x$

32. $\sin 4x = 4 \sin x \cos x - 8 \sin^3 x \cos x$

33. $\sec \dfrac{x}{2} \csc \dfrac{x}{2} = \dfrac{2}{\sin x}$

34. $\tan \theta = \dfrac{\tan 2\theta}{1 + \sec 2\theta}$

Solve for x such that $0 \le x < 2\pi$.

35. $2 \cos x + \sqrt{3} = 0$

36. $2 \sin x = \sqrt{2}$

37. $\cos^2 x - 1 = 0$

38. $\tan^2 x - \sqrt{3} \tan x = 0$

39. $6 \sin^2 x + \sin x = 2$

40. $2 \sin^2 x - 3 \cos^2 x + 2 \cos x = 1$

41. $\cos 2x - 4 \sin x = 0$

42. $\sin \dfrac{x}{2} = \cos x + 1$

Evaluate without the use of a calculator.

43. $\text{Arcsin } 1$

44. $\text{Arcsin} \left(-\dfrac{\sqrt{2}}{2} \right)$

45. $\text{Arccos} \left(-\dfrac{1}{2} \right)$

46. $\text{Arccos } \dfrac{\sqrt{3}}{2}$

47. $\text{Arctan } \sqrt{3}$

48. $\text{Arctan } 0$

49. $\sin \left(\text{Arccos } \dfrac{\sqrt{2}}{2} \right)$

50. $\cos [\text{Arctan } (-1)]$

Evaluate using a calculator.

51. $\text{Arcsin } (-0.34)$

52. $\text{Arcsin } 0.916$

53. $\text{Arccos } 0.195$

54. $\text{Arccos } (-0.585)$

55. $\text{Arctan } (-1.82)$

56. $\text{Arctan } 0.28$

57. $\tan (\text{Arccos } 0.727)$

58. $\sin [2 \text{ Arcsin } (-0.419)]$

Find an algebraic expression for each of the following:

59. $\cos (\text{Arccos } x)$

60. $\sin (\text{Arctan } x)$

61. $\sin (2 \text{ Arccos } x)$

62. $\cos (2 \text{ Arctan } x)$

Solve.

63. *Optics*   In studying the properties of light in a crystal, the equation below appears.* Use this equation to find expressions for $\sin^2 \alpha$ and $\tan^2 \alpha$ in terms of m, n, and p.

$$\cos^2 \alpha = \dfrac{n^{-2} - m^{-2}}{p^{-2} - m^{-2}}$$

* For further information, see Miles V. Klein, *Optics* (New York: Wiley, 1970), p. 608.

64. *Astrophysics*   The formula $\cos \theta = \sin x \cos \phi$ appears in studying the polarization of light scattered from particles in the atmosphere. If $\cos \phi = \dfrac{1}{2}$, find an expression for $\sin \theta$ in terms of $\cos x$.

65. *Cartography*   The first step in generalizing a line represented by two points $P_1(x_1, y_1)$ and $P_2(x_2, y_2)$ is to find θ such that

$$\frac{y_2 - y_1}{x_2 - x_1} = \text{Tan } \theta$$

Find θ for $P_1(3.0, 1.0)$ and $P_2(2.5, 2.2)$.

66. *Holography*   The radian spatial frequency w between a pair of reflected rays is given by

$$w = 2K \sin \frac{\alpha}{2} \cos \left(\theta - \frac{\alpha}{2} \right)$$

where $\alpha = 2\pi - \phi + \theta$. Show that $w = K(\sin \theta - \sin \phi)$.

67. *Optics*   The index of refraction n with respect to air of a glass of a triangular prism can be calculated from the formula

$$n = \frac{\sin \dfrac{1}{2}(\theta + \phi)}{\cos \dfrac{\theta}{2}}$$

Find an equivalent expression for n when $\phi = 3\theta$, and expand it using the double-angle and half-angle identities.

68. *Optics*   To determine the area of light transmitted through an opaque aperture, the following expression is used:

$$\frac{r^2}{2} \sin \left(2 \text{ Arccos } \frac{x}{r} \right)$$

Show that this is equal to $x\sqrt{r^2 - x^2}$.

69. *Energy technology*   At latitude 30°N, in order to find the tilt angle a of a solar panel that would provide 10% more energy in winter than in summer, the equation below must be solved.* Solve for a.

$$\cos (23.5° - a) = 1.10 \cos (-23.5° - a)$$

70. *Physics*   The vertical displacement in water level due to a sinusoidal wave is given by the equation below. Expand the right side using the difference formula, and then simplify it.

$$y = 6 \cos \left(\frac{\pi t}{12} - \frac{\pi}{2} \right)$$

71. *Physics*   The motion in a simple spring–mass system is given by the equation below. For what values of time t (in seconds) is the displacement $y = 0$?

$$y = 4.0 \sin (3.0t - 0.5)$$

72. *Electronics*   A sawtooth wave form can be represented by the Fourier series

$$y = \sin \pi x + \frac{\sin 2\pi x}{2} + \frac{\sin 3\pi x}{3} + \cdots + \frac{\sin N\pi x}{N}$$

For what values of x ($0 \leq x < 2$) do the first two terms sum to 0?

* Courtesy of Walter Hausz.

Topics in Analytic Geometry

20

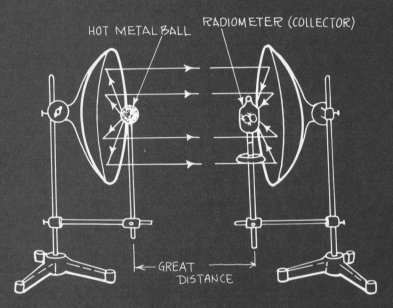

HOT METAL BALL

RADIOMETER (COLLECTOR)

GREAT DISTANCE

PARABOLIC RADIANT ENERGY REFLECTORS

For 25 centuries plane geometry has involved the study of figures such as lines, circles, and triangles by reasoning from definitions and basic assumptions. Analytic geometry began with the work of René Descartes in his famous book *La géométrie,* published in 1637. The rectangular, or cartesian, coordinate system he devised enables us to set up a correspondence between ordered pairs of real numbers and points in a plane. Analytic geometry allows us to set up a correspondence between an algebraic equation and the geometric properties of the curve associated with it.

Analytic geometry is not a branch of geometry. It is a powerful method of analysis that has extensive applications in science and technology. You will find it particularly valuable later in the study of calculus.

20-1 | Distance and Slope

Distance

To find the distance d between points P_1 and P_2 in Fig. 20-1, first locate point Q so that P_2QP_1 is a right triangle with hypotenuse d, horizontal leg P_1Q, and vertical leg P_2Q. According to the Pythagorean theorem,

$$d^2 = (x_2 - x_1)^2 + (y_2 - y_1)^2$$

Figure 20-1

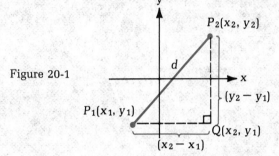

Taking the square root of both sides, we have the **distance formula:**

Distance Formula $\quad d = \sqrt{(x_2 - x_1)^2 + (y_2 - y_1)^2}$	**(20-1)**

EXAMPLE 1 Find the distance between the points $(-2, 5)$ and $(3, -1)$.

Solution Using $(-2, 5)$ as (x_1, y_1) and $(3, -1)$ as (x_2, y_2), substitute into the distance formula (20-1):

$$d = \sqrt{[3 - (-2)]^2 + (-1 - 5)^2} = \sqrt{5^2 + (-6)^2} = \sqrt{61} \approx 7.81$$

Notice that in using the distance formula, it does not matter which point is chosen as (x_1, y_1) and which is chosen as (x_2, y_2). In Example 1, if we had chosen $(-2, 5)$ as (x_2, y_2), we would have

$$d = \sqrt{(-2 - 3)^2 + [5 - (-1)]^2} = \sqrt{(-5)^2 + 6^2} = \sqrt{61}$$

LEARNING HINT ▶

If $y_1 = y_2$, the points P_1 and P_2 lie on the same horizontal line. Therefore,

$$d = \sqrt{(x_2 - x_1)^2} \qquad \text{and this may be written as} \qquad |x_2 - x_1|$$

Similarly, if $x_1 = x_2$, the points P_1 and P_2 lie on the same vertical line, and

$$d = \sqrt{(y_2 - y_1)^2} \qquad \text{or} \qquad |y_2 - y_1| \quad ◀$$

EXAMPLE 2   Find the distance between the indicated pairs of points.
(a) $A(4, -2)$ and $B(4, 6)$ (b) $C(-1, -3)$ and $D(-7, -3)$

Solution

(a) $d = \sqrt{(4 - 4)^2 + [6 - (-2)]^2} = \sqrt{8^2} = 8$

Notice that since A and B lie on the same vertical line, the distance can also be calculated as follows:

$$d = |6 - (-2)| = |8| = 8$$

(b) $d = \sqrt{[-7 - (-1)]^2 + [-3 - (-3)]^2} = \sqrt{(-6)^2} = 6$

Notice that since C and D lie on the same horizontal line, the distance can also be calculated as follows:

$$d = |-7 - (-1)| = |-6| = 6$$

IMPORTANT

The *distance* between two points is always a positive number. For two points aligned horizontally or vertically, a sign is sometimes used to indicate the **directed distance** from one point to the other. In Example 2(a), the directed distance from A to B is $6 - (-2) = +8$, while the directed distance from B to A is $-2 - 6 = -8$. ◀

The distance formula is an algebraic tool for solving a large variety of geometric problems.

EXAMPLE 3   Find k so that the distance between $(-1, k)$ and $(3, 7)$ is 5.

Solution

First, substitute into (20-1).

$$5 = \sqrt{[3 - (-1)]^2 + (7 - k)^2}$$

Then, solve for k.

$$5 = \sqrt{16 + (7 - k)^2}$$

Square both sides and expand the binomial.

$$25 = 16 + 49 - 14k + k^2$$

Combine all terms on one side.

$$k^2 - 14k + 40 = 0$$

Factor.

$$(k - 4)(k - 10) = 0$$

The solution is:

$$k = 4 \quad \text{or} \quad k = 10$$

Finally, check.
For $k = 4$:

$$d = \sqrt{16 + (7 - 4)^2} = \sqrt{25} = 5$$

For $k = 10$:

$$d = \sqrt{16 + (7 - 10)^2} = \sqrt{25} = 5$$

EXAMPLE 4   Show that $A(-3, -2)$, $B(16, 1)$, and $C(5, 9)$ are the vertices of an isosceles triangle.

Solution   An isosceles triangle has two sides that are the same length. The lengths of the three sides connecting the given points are

$$AB = \sqrt{[16 - (-3)]^2 + [1 - (-2)]^2} = \sqrt{361 + 9} = \sqrt{370}$$

$$BC = \sqrt{(5 - 16)^2 + (9 - 1)^2} = \sqrt{121 + 64} = \sqrt{185}$$

$$AC = \sqrt{[5 - (-3)]^2 + [9 - (-2)]^2} = \sqrt{64 + 121} = \sqrt{185}$$

Since two of the sides, BC and AC, have the same length, the triangle is isosceles.

Midpoint   The **midpoint** of the line segment connecting the points $P_1(x_1, y_1)$ and $P_2(x_2, y_2)$ may be found as shown in Fig. 20-2. The coordinates of the midpoint are given by the following formulas:

Midpoint of a Line Segment

$$x_m = \frac{x_1 + x_2}{2} \quad \text{and} \quad y_m = \frac{y_1 + y_2}{2} \qquad\qquad (20\text{-}2)$$

Figure 20-2

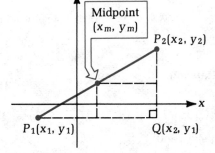

EXAMPLE 5   Find the midpoint of the line segment connecting the points $(-3, 2)$ and $(4, -2)$.

Solution   The coordinates of the midpoint are (x_m, y_m), where

$$x_m = \frac{x_1 + x_2}{2} \qquad\qquad y_m = \frac{y_1 + y_2}{2}$$

$$x_m = \frac{-3 + 4}{2} = \frac{1}{2} \qquad\qquad y_m = \frac{2 + (-2)}{2} = \frac{0}{2} = 0$$

The midpoint is located at $\left(\dfrac{1}{2}, 0\right)$.

Slope   The slope of a line, like the slope of a hillside or a road, refers to its steepness. As shown in Fig. 20-3, slope is often expressed as the ratio of the **rise** to the **run**. To find the slope of the line in Fig. 20-4, we determine this ratio for two points, $P_1(x_1, y_1)$ and $P_2(x_2, y_2)$, on the line. From P_1 to P_2, the rise is the vertical change $y_2 - y_1$, and the run is the horizontal change $x_2 - x_1$. The slope m of the line passing through P_1 and P_2 is therefore given by the following formula:*

Figure 20-3

Slope of a Line	$m = \dfrac{y_2 - y_1}{x_2 - x_1}$	(20-3)

Figure 20-4

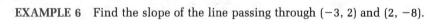

CAUTION ▶ Always check the slope calculated in this way to be certain you have placed $y_2 - y_1$ in the numerator and $x_2 - x_1$ in the denominator. Inexperienced students often mistakenly switch the two. ◀

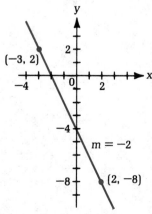

Figure 20-5

EXAMPLE 6   Find the slope of the line passing through $(-3, 2)$ and $(2, -8)$.

Solution   Using $(-3, 2)$ as (x_1, y_1) and $(2, -8)$ as (x_2, y_2), substitute into the slope formula (20-3) to obtain

$$m = \frac{-8 - 2}{2 - (-3)} = \frac{-10}{5} = -2$$

See Fig. 20-5.

As with the distance formula, it does not matter which point is chosen as (x_1, y_1) and which is chosen as (x_2, y_2) in calculating the slope of a line. For

* Using similar triangles it can be shown that the slope of a line is the same everywhere along the line and is independent of the points chosen to calculate it.

example, if we calculate the slope in Example 6 using $(2, -8)$ as (x_1, y_1), we have

$$m = \frac{2 - (-8)}{-3 - 2} = \frac{10}{-5} = -2$$

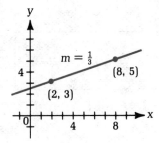

$m = \frac{1}{3}$

$(8, 5)$

$(2, 3)$

Figure 20-6

EXAMPLE 7   Find the slope of the line passing through the points $(2, 3)$ and $(8, 5)$.

Solution $m = \dfrac{5 - 3}{8 - 2} = \dfrac{2}{6} = \dfrac{1}{3}$

See Fig. 20-6.

The following statements are three basic properties of slope:

1. A line that rises to the right (increasing values of x) has a **positive** slope (see Fig. 20-6).
2. A line that rises to the left (decreasing values of x) has a **negative** slope (see Fig. 20-5).
3. The steeper the line, the larger will be the absolute value of the slope.
    $\left(\text{Compare Fig. 20-5 with Fig. 20-6; the first line is steeper than the second,}\right.$
    and the absolute value of its slope is also larger: $2 > \dfrac{1}{3}.\Big)$

Horizontal and vertical lines

Statement 3 might raise your curiosity about the slope of a vertical line (the steepest possible line) and the slope of a horizontal line (the flattest possible line).

EXAMPLE 8   (a) In Fig. 20-7, the line passing through points $(-3, 3)$ and $(-3, -2)$ is a vertical line. Calculating its slope, we have

$$m = \frac{-2 - 3}{-3 - (-3)} = \frac{-5}{0}$$

which is undefined.

Figure 20-7

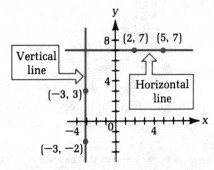

(b) The line passing through points (2, 7) and (5, 7) is a horizontal line. Calculating its slope, we have

$$m = \frac{7 - 7}{5 - 2} = \frac{0}{3} = 0$$

The results from Example 8 can be generalized as two more properties of slope:

4. The slope of a vertical line is undefined.
5. The slope of a horizontal line is 0.

Angle of inclination

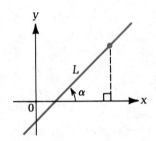

Figure 20-8

We may also use the slope to describe parallel and perpendicular lines. To do this, we must first introduce a new concept: the *angle of inclination*.

Any line L not parallel to the x-axis eventually intersects it. The **angle of inclination** α of L is the angle measured counterclockwise from the positive x-axis to L (see Fig. 20-8). By dropping a perpendicular from any point on L to the x-axis, we see from the resulting right triangle that

$$\tan \alpha = \frac{\text{Side opposite } \alpha}{\text{Side adjacent to } \alpha}$$

But the ratio on the right is also the slope of L. Therefore:

$$m = \tan \alpha \tag{20-4}$$

Slope and parallel lines

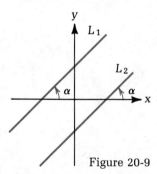

Figure 20-9

Now consider the two parallel lines L_1 and L_2 shown in Fig. 20-9. From geometry we know that their angles of inclination α must be equal; therefore, from equation (20-4), we know that their slopes are equal. This gives us another property of slope:

6. If two nonvertical lines are parallel, then their slopes are equal:

$$\text{Slopes of Parallel Lines} \qquad m_1 = m_2 \tag{20-5}$$

Slope and perpendicular lines

Figure 20-10 shows two perpendicular lines L_1 and L_2 with angles of inclination α_1 and α_2, respectively. Because L_1 and L_2 are perpendicular, we know that

$$\alpha_2 = \alpha_1 + 90°$$

$$-\alpha_1 = 90° - \alpha_2$$

Taking the tangent of both sides, we have

$$\tan(-\alpha_1) = \tan(90° - \alpha_2)$$

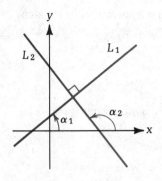

Figure 20-10

From trigonometry we know that this is equivalent to

$$-\tan \alpha_1 = \cot \alpha_2{}^*$$

which can also be written as

$$\tan \alpha_1 = -\frac{1}{\tan \alpha_2}$$

But $\tan \alpha_1 = m_1$ and $\tan \alpha_2 = m_2$, which leads to the following property of slope:

7. If two lines are perpendicular, and neither is horizontal, then their slopes satisfy the following equation:

Slopes of Perpendicular Lines

$$m_1 = -\frac{1}{m_2} \qquad \text{or} \qquad m_1 m_2 = -1 \qquad\qquad (20\text{-}6)$$

In other words, the slopes of two perpendicular lines are negative reciprocals of each other, and the product of their slopes is equal to -1.

These properties of the slopes of parallel and perpendicular lines enable us to solve a variety of problems in analytic geometry.

EXAMPLE 9   Determine whether the lines passing through the following pairs of points are parallel, perpendicular, or neither.
(a) L_1: $(-2, 6)$ and $(1, 12)$; L_2: $(1, 5)$ and $(-3, 7)$
(b) L_1: $(3, 5)$ and $(2, -1)$; L_2: $(2, 6)$ and $(4, 8)$
(c) L_1: $(5, -3)$ and $(1, 1)$; L_2: $(-6, -1)$ and $(1, -8)$

Solution

(a) The slope of L_1 is: $m_1 = \dfrac{12 - 6}{1 - (-2)} = \dfrac{6}{3} = 2$

The slope of L_2 is: $m_2 = \dfrac{7 - 5}{-3 - 1} = \dfrac{2}{-4} = -\dfrac{1}{2}$

Since

$$m_1 m_2 = 2\left(-\frac{1}{2}\right) = -1$$

the lines are perpendicular.

(b) $m_1 = \dfrac{-1 - 5}{2 - 3} = 6 \qquad m_2 = \dfrac{8 - 6}{4 - 2} = 1$

The slopes are not equal, and they are not negative reciprocals of each other. Therefore, the lines are neither parallel nor perpendicular.

---

* See Exercises 19-2, Problems 33 and 34.

(c) $m_1 = \dfrac{1 - (-3)}{1 - 5} = -1$ $m_2 = \dfrac{-8 - (-1)}{1 - (-6)} = -1$

Since $m_1 = m_2$, the lines are parallel.

EXAMPLE 10   Given the points $A(2, 1)$, $B(4, 7)$, and $C(7, 6)$:
(a)   Show that they are vertices of a right triangle.
(b)   Find the area of triangle ABC.

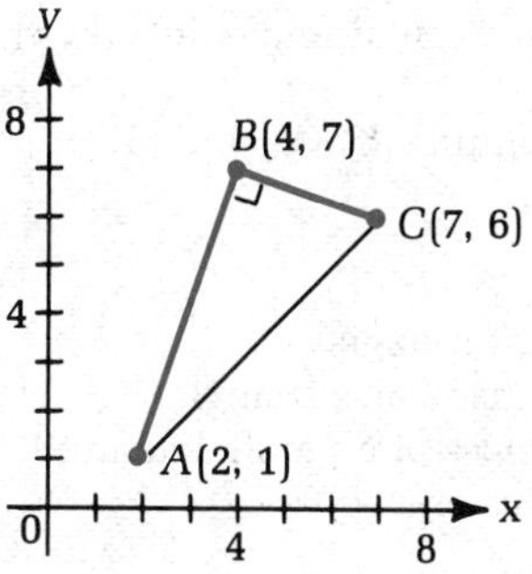

Solution
(a)   To prove that the points form a right triangle, we must show that two of the sides are perpendicular. The slopes of the three sides are

$$m_{AB} = \frac{7 - 1}{4 - 2} = 3 \qquad m_{BC} = \frac{6 - 7}{7 - 4} = -\frac{1}{3} \qquad m_{AC} = \frac{6 - 1}{7 - 2} = 1$$

Since

$$m_{AB} \cdot m_{BC} = 3\left(-\frac{1}{3}\right) = -1$$

Figure 20-11

then side AB is perpendicular to side BC, and ABC is a right triangle.

(b)   The area of a right triangle is half the product of the lengths of the legs. Use the distance formula to determine the lengths of AB and BC:

$$AB = \sqrt{(4 - 2)^2 + (7 - 1)^2} = \sqrt{40}$$
$$BC = \sqrt{(7 - 4)^2 + (6 - 7)^2} = \sqrt{10}$$

$$\text{Area of } ABC = \frac{1}{2}\left(\sqrt{10}\right)\left(\sqrt{40}\right) = 10 \text{ square units.}$$

Exercises 20-1

Find the distance between each pair of points. Round all irrational answers to the nearest hundredth.

1. $(3, -1)$ and $(4, 3)$ **2.** $(5, -2)$ and $(2, -5)$ **3.** $(-4, 8)$ and $(-1, 4)$
4. $(-2, 1)$ and $(6, -5)$ **5.** $(5, 7)$ and $(5, 2)$ **6.** $(-3, 6)$ and $(-3, -2)$
7. $(5, 3)$ and $(7, -1)$ **8.** $(5, -6)$ and $(-4, -9)$

9–16. *For Problems 9–16 find the slope of the line passing through each pair of points in Problems 1–8.*

Find the midpoint of the line segment joining each pair of points.

17. $(7, -2)$ and $(15, 6)$ **18.** $(6, -1)$ and $(-4, 4)$ **19.** $(14, 3)$ and $(7, 6)$ **20.** $(-5, -4)$ and $(0, -8)$

Determine whether the lines passing through the given pairs of points are parallel, perpendicular, or neither.

21. L_1: $(4, -2)$ and $(-2, 2)$; L_2: $(5, 6)$ and $(3, 3)$
22. L_1: $(6, -1)$ and $(4, 5)$; L_2: $(-1, 1)$ and $(0, -2)$
23. L_1: $(7, 5)$ and $(2, 4)$; L_2: $(-3, -6)$ and $(-1, 4)$
24. L_1: $(4, -3)$ and $(-6, 2)$; L_2: $(4, -5)$ and $(8, 3)$

25. L_1: $(-8, 1)$ and $(4, 7)$; L_2: $(-2, 11)$ and $(-6, 9)$
26. L_1: $(3, 9)$ and $(5, -5)$; L_2: $(-2, -4)$ and $(5, -3)$

Use the given information to solve for k.

27. The distance between $(-7, 2)$ and $(k, 5)$ is $\sqrt{13}$.
28. The distance between $(6, k)$ and $(-1, 3)$ is $\sqrt{74}$.
29. Points $(3, -7)$, $(-2, k)$, and $(-4, -14)$ are collinear—that is, they all lie along the same line.
   [*Hint:* The line segments joining all pairs of points must have equal slopes.]
30. Points $(4, 2)$, $(k, 4)$, and $(-8, 5)$ are collinear. (See Problem 29.)
31. The line passing through points $(-6, 7)$ and $(-2, 6)$ is perpendicular to the line containing points $(k, 5)$ and $(-1, 13)$.
32. The line passing through points $(-3, k)$ and $(3, 7)$ is parallel to the line containing $(5, -2)$ and $(8, -4)$.

Solve.

33. Show that the points $A(-4, 2)$, $B(0, 6)$, and $C(3, 3)$ are the vertices of a right triangle.
34. Show that the points $D(4, -7)$, $E(2, 6)$, and $F(-11, 8)$ are the vertices of an isosceles triangle.
35. Show that the points $G(2, 3)$, $H(-1, -1)$, $I(-8, 1)$, and $J(-5, 5)$ are the vertices of a parallelogram.
36. Show that the points $K(5, 3)$, $L(2, -1)$, $M(-2, 2)$, and $N(1, 6)$ are the vertices of a square.
37. Find the area of the triangle in Problem 33.
38. Find the perimeter of the square in Problem 36.

Use formula (20-4) to find the slope of a line with the given angle of inclination.

39. $60°$ (exact value) ▦ **40.** $80°$ (nearest hundredth)

▦ *Use formula (20-4) to find (to the nearest $0.1°$) the angle of inclination of a line with the given slope.*

41. -2.463 **42.** 5.729

20-2 | The Straight Line

In several earlier chapters we worked with equations of straight lines and graphed these equations. For example, we found the solutions to linear equations graphically by finding the zeros of the corresponding linear functions, and we found the solutions to systems of two linear equations in two unknowns by determining the coordinates of the intersection points of their graphs. In this section we shall continue to explore the relationship between the geometric properties of straight lines and their equations.

Point–slope form   A nonvertical straight line can be defined as a curve with a constant slope. This means that its equation can be written in terms of its slope. The slope between a fixed point (x_1, y_1) and any general point (x, y) is given by

$$m = \frac{y - y_1}{x - x_1}$$

Multiplying both sides by $(x - x_1)$, we obtain the **point–slope form** of the equation of a line:

Point–Slope Form	$y - y_1 = m(x - x_1)$	**(20-7)**

Given any point (x_1, y_1) on the line and the slope m of the line, we can immediately write its equation using the form (20-7).

EXAMPLE 11   Determine the equation of a line with slope of -2, passing through the point $(5, -4)$.

Solution   We are given $m = -2$, $x_1 = 5$, and $y_1 = -4$. Substitute these values into the point–slope form (20-7) to obtain

$$y - (-4) = -2(x - 5)$$

After simplifying and moving all terms to the left side, we have

$$y + 2x - 6 = 0$$

Check this answer by substituting the coordinates of the given point for x and y:

$$(-4) + 2(5) - 6 = 0$$
$$0 = 0$$

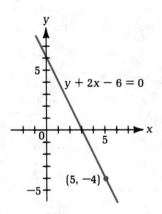

Figure 20-12

Sometimes, we must write the equation of a line when we know only the coordinates of two points on it. We use the point–slope equation (20-7) after first finding the slope using equation (20-3).

EXAMPLE 12   Find the equation of a line passing through the points $(-1, -5)$ and $(2, -3)$.

Solution   **First,** use (20-3) to calculate the slope of the line:

$$m = \frac{y_2 - y_1}{x_2 - x_1} = \frac{-3 - (-5)}{2 - (-1)} = \frac{2}{3}$$

Then, substitute the slope and *either* one of the given points into the point–slope equation (20-7). Using $(2, -3)$, we have

$$y - (-3) = \frac{2}{3}(x - 2)$$

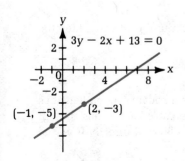

Figure 20-13

To simplify this equation, multiply each term by 3 to obtain

$$3y + 9 = 2x - 4$$
$$3y - 2x + 13 = 0$$

Finally, use the coordinates of the other point to check the answer:

$$3(-5) - 2(-1) + 13 = 0$$

$$-15 + 2 + 13 = 0$$

$$0 = 0$$

Slope–intercept form   Earlier, we defined the **intercepts** of a line as the points where the line intersects the axes. Specifically, the **x-intercept** is a point in the form **(a, 0)**, and the **y-intercept** is a point in the form **(0, b)**. Often, the equation of a line must be determined from its slope and the y-intercept. By substituting $(0, b)$ for (x_1, y_1) in the point–slope equation (20-7), we can derive another form of the equation of a line:

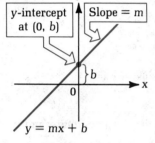

$$y - b = m(x - 0)$$

Simplifying and rearranging terms gives us the **slope–intercept form** of the equation of a line:

Figure 20-14

| Slope–Intercept Form | $y = mx + b$ | (20-8) |

Slope | y-intercept

In other words, the line having slope m and y-intercept at $(0, b)$ is represented by the equation $y = mx + b$ (Fig. 20-14).

NOTE ▶   If the y-intercept of a line is at $(0, b)$, we usually say "the line has a y-intercept of b." ◀

EXAMPLE 13   Find the equation of a line with a slope of 4 and y-intercept of -3.

Solution   We are given $m = 4$ and $b = -3$. Substituting directly into the slope–intercept equation (20-8), we obtain

$$y = 4x - 3$$

The slope of a line is a very useful concept in more advanced mathematics and its applications. If the equation of a line is solved for y and put in slope–intercept form, the slope and y-intercept can be read directly from the equation.

EXAMPLE 14
(a)   State the slope and y-intercept of the line given by the equation $3x - 5y = 10$.
(b)   Determine k so that $6y + kx - 9 = 0$ is perpendicular to the line given in part (a).

Solution

(a) **First,** solve for y. $-5y = -3x + 10$

$$y = \frac{3}{5}x - 2$$

Then, read m and b directly.

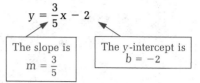

| The slope is $m = \dfrac{3}{5}$ | The y-intercept is $b = -2$ |

(b) **First,** solve for y. $6y = -kx + 9$

$$y = -\frac{k}{6}x + \frac{3}{2}$$

For this line to be perpendicular to the line given in part (a), the product of their slopes must be -1. Since the line in part (a) has a slope of $3/5$, and this line has a slope of $-k/6$, then

$$\left(\frac{3}{5}\right)\left(-\frac{k}{6}\right) = -1$$

$$\frac{-k}{10} = -1$$

$$k = 10$$

Horizontal lines   In Section 20-1 we showed that every horizontal line has a slope of 0. Substituting $m = 0$ into the slope–intercept equation (20-8), we have

$$y = 0 \cdot x + b$$

This tells us that every point on a horizontal line has the same y-coordinate, and that the equation of any line parallel to the horizontal or x-axis can be written in the following form:

Equation of a Horizontal Line $y = b$ **(20-9)**

---

where b is the y-coordinate of any point on the line. The equation of the x-axis is $y = 0$.

Vertical lines   Since the slope of a vertical line is undefined, we cannot use equation (20-8) to derive its equation. However, since the x-coordinate of every point on a vertical line is the same, then the form of equation (20-9) suggests that any line parallel to the vertical or y-axis can be written in the following form:

Equation of a Vertical Line $x = a$ **(20-10)**

---

where a is the x-coordinate of any point on the line. The equation of the y-axis is $x = 0$.

EXAMPLE 15   Write the equation of:
(a)   A horizontal line passing through $(3, -5)$
(b)   A line parallel to the y-axis passing through $(-2, -4)$

Solution
(a)   The y-coordinate of the given point is -5.
       Using equation (20-9), we have: $y = -5$
(b)   The x-coordinate of the given point is -2.
       Using equation (20-10), we obtain: $x = -2$

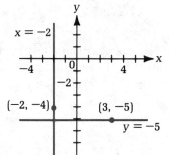

Figure 20-15

General form   Whether the equation of a line is in point–slope form or slope–intercept form, it normally contains an x-term, a y-term, and a constant term. Thus, any equation of a straight line can be put in the **general form** below.

General Form of the Equation of a Line

$$Ax + By + C = 0$$

(20-11)

---

where A, B, and C are real numbers, and A and B are not both 0. Any equation of this form is called a **linear equation.**

The general form of a linear equation (20-11) is helpful when we use the intercept method of graphing a line. We review this method from Chapter 3 in the next example.

EXAMPLE 16   Find the x- and y-intercepts of the line $3x - 8y - 12 = 0$ and sketch its graph.

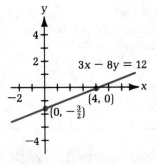

Figure 20-16

Solution

First, move the constant term to the right. $3x - 8y = 12$

Second, substitute $y = 0$ to find the $3x - 8(0) = 12$
x-intercept. $x = 4$

Third, substitute $x = 0$ to find the $3(0) - 8y = 12$
y-intercept.
                                                    $y = -\dfrac{3}{2}$

Since the x-intercept is 4, the line crosses the x-axis at (4, 0). Since the y-intercept is $-\dfrac{3}{2}$, the line crosses the y-axis at $\left(0, -\dfrac{3}{2}\right)$. Use these two points to sketch the graph, as shown in Fig. 20-16.

Applications   There are many technical applications in which physical quantities have a linear relationship. Knowing this, we can apply the techniques of this section to experimental data and formulate equations relating these quantities. The next examples illustrate two such applications.

EXAMPLE 17   The resistance of a circuit element varies directly with its temperature. The resistance at 0°C is found to be 5.00 Ω. Further tests show that the resistance increases by 0.02 Ω for every 1°C increase in temperature.
(a) Find the equation for the resistance R in terms of the temperature T.
(b) Use the equation from part (a) to find the resistance at 24°C.

Solution
(a) First, note that to find R in terms of T, R must be the dependent variable. Since resistance varies directly with temperature, the two quantities have a linear relationship. In deciding which form of the equation of a line to use, notice that we are given the R-intercept—that is, $R = 5.00$ when $T = 0$. We therefore use the slope–intercept form (20-8) to write

$$R = mT + b$$

$$R = mT + 5.00 \quad \longleftarrow \quad \boxed{\begin{array}{l} R = 5.00 \\ \text{at } T = 0 \end{array}}$$

To find m, recall that

$$m = \frac{\text{Change in } R}{\text{Change in } T}$$

Since R changes 0.02 Ω for every 1°C change in T, then

$$m = \frac{0.02\ \Omega}{1°C} = 0.02\ \Omega/°C$$

Substituting 0.02 for m in the slope–intercept equation above, we have the complete equation:

$$R = 0.02T + 5.00$$

(b) To find the resistance at 24°C, substitute 24°C for T in the equation from part (a) to obtain

$$R = 0.02(24) + 5.00 = 5.48\ \Omega$$

EXAMPLE 18   The pressure P in a body of wastewater varies directly with the depth D. If the pressure is measured to be 35.5 lb/in.^{2} at 40.0 ft and 45.9 lb/in.^{2} at 60.0 ft, then:
(a) Find the equation for P in terms of D.*
(b) Graph the equation.

* Remember, this means that P is the dependent variable.

Solution

(a)   We are given two points, but neither of them is the P-intercept. Therefore, we must use the technique of Example 12.

Step 1.   Find the slope.

$$m = \frac{P_2 - P_1}{D_2 - D_1} = \frac{45.9 \text{ lb/in.}^2 - 35.5 \text{ lb/in.}^2}{60.0 \text{ ft} - 40.0 \text{ ft}}$$

$$= 0.52 \text{ lb/in.}^2/\text{ft}$$

Step 2.   Write the equation in point–slope form (20-7), substituting 0.52 for m and one of the given pairs of values for (D_1, P_1). Using (40.0, 35.5), we obtain

$$P - 35.5 = 0.52(D - 40.0)$$

$$P = 0.52D + 14.7$$

(b)   To graph this equation, plot the two pairs of values given in the problem and connect them with a straight line, as shown in Fig. 20-17. Notice that a negative depth makes no physical sense in this case.

Figure 20-17

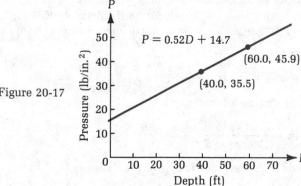

Exercises 20-2

Determine the equation of the line having the given properties. Express all answers in general form, $Ax + By + C = 0.$

1. Slope of 6, passing through $(-3, 5)$
2. Slope of -2, passing through $(5, -1)$
3. Slope of $-\dfrac{3}{4}$, passing through $(2, -6)$
4. Slope of $\dfrac{5}{2}$, passing through $(8, 3)$
5. Slope of 4, passing through $(0, -7)$
6. Slope of -5, y-intercept of 6
7. Slope of $-\dfrac{1}{3}$, x-intercept of 9
8. Slope of 1, passing through $(-2, 0)$
9. Passing through $(-5, 4)$ and $(-6, 2)$
10. Passing through $(3, -7)$ and $(-1, 9)$
11. Passing through $(3, -3)$ and $(-6, -9)$
12. Passing through $(10, 1)$ and $(8, -4)$
13. Horizontal line, passing through $(6, 5)$
14. Vertical line, passing through $(-2, -3)$

15. Parallel to the y-axis and 2 units to the left of it
16. Parallel to the x-axis and 4 units above it
17. Parallel to the line $y = 5x - 2$, passing through $(3, -1)$
18. Parallel to the line $6x - 4y = 3$, passing through $(-6, 3)$
19. Perpendicular to the line $6y - 2x - 7 = 0$, passing through $(5, 7)$
20. Perpendicular to the line $y = -x + 5$, passing through $(-8, 7)$

Find the slope m, y-intercept b, and x-intercept a, and sketch the graph of each of the following lines:

21. $3x - y - 3 = 0$ 22. $y - 4x + 2 = 0$ 23. $2y + 6x - 1 = 0$
24. $5x + 3y - 9 = 0$ 25. $7x - 3y = 0$ 26. $4y - 3x - 12 = 0$
27. $y = -2$ 28. $x = 4$

Determine whether the given lines are parallel, perpendicular, or neither.

29. $4x - 7y + 6 = 0$ and $4x - 5y + 2 = 0$
30. $6x - 2y + 1 = 0$ and $3x - y - 5 = 0$
31. $8x - 2y + 3 = 0$ and $y + 4x = 0$
32. $3x - 4y + 5 = 0$ and $4x + 3y + 1 = 0$
33. $2y = 3x$ and $6x - 4y + 7 = 0$
34. $2x - 5y + 7 = 0$ and $5x - 2y + 3 = 0$

Solve for k given the following conditions:

35. $2y - kx + 5 = 0$ is parallel to $3x + 6y + 1 = 0$
36. $3x + ky - 4 = 0$ is parallel to $12y + 9x - 7 = 0$
37. $5x - ky + 2 = 0$ is perpendicular to $10y = 4x - 3$
38. $kx - 6y + 5 = 0$ is perpendicular to $4x + 2y + 1 = 0$

Solve.

39. **Physics** The instantaneous velocity v of an object under constant acceleration a during an elapsed time t is given by $v = v_0 + at$, where v_0 is its initial velocity. If an object has an initial velocity of 8 ft/s and a velocity of 20 ft/s after 6 s of constant acceleration, write the equation relating velocity to time and sketch its graph for $0 \le t \le 8$ s.

40. **Physics** The displacement s of an object moving at a constant velocity v during an elapsed time t is given by $s = s_0 + vt$, where s_0 is its initial displacement. If $s = 1.0$ m at $t = 0$ and $s = 3.4$ m at $t = 4.0$ s, find the equation relating s to t and sketch the graph for $0 \le t \le 8.0$ s.

41. **Electronics** The resistance of a circuit element is found to increase by 0.008 Ω for every 1°C increase in temperature over a wide range of temperature. If the resistance is 4.000 Ω at 0°C, write the equation relating the resistance R to the temperature T, and find R when $T = 13$°C. (See Example 17.)

42. **Hydrology** The pressure P in a mountain lake is measured to be 24.0 lb/in.^{2} at a depth D of 30.0 ft and 43.5 lb/in.^{2} at 75.0 ft. Find the equation for P in terms of D and use it to calculate the pressure at a depth of 60.0 ft. (See Example 18.)

43. **Physics** Until it reaches its elastic limit, the length of a spring varies linearly with the amount of weight hung from it. If a spring has an initial length of 8.0 in. and stretches to 9.0 in. when a 2.0 lb weight is hung from it, find the equation relating the length L to the weight W. What is the spring length when a 5.0 lb weight is suspended from it?

44. **Physics** The length of a heated object varies linearly with temperature. A certain metal rod was found to be 120.00 cm long at 0°C and 120.10 cm long at 50°C. Write an equation relating its length L and temperature T. What will be its length at 20°C?

45. **Industrial management** In production of a certain type of electronic component, a manufacturer has fixed costs of $650 and additional costs of $1.50 per unit produced. Write an equation relating the total cost C to the number of units N, and use this equation to determine the cost of an order of 10,000 units.

46. *Electronics*   The potential drop V (in volts) across a circuit element is given by $V = IR$, where I is the current (in amperes) and R is the resistance (in ohms). Plot V as a function of I when $R = 1.4 \ \Omega$.

47. *Business*   In accounting, the depreciated value V of a deductible item using the linear method is given by

$$V = C\left(1 - \frac{n}{N}\right)$$

where C is the original cost, n is the number of years elapsed since the depreciation began, and N is the usable life of the item. Plot V as a function of C when $n = 4$ and $N = 7$.

48. *Industrial management*   A study performed on an assembly line showed that the ambient temperature increased linearly throughout the afternoon. If the temperature T was 74°F at 12:30 and 82°F at 4:30, write an equation for T in terms of the time t (in hours) elapsed since 12:30. What was the temperature at 2:00?

20-3 | The Circle

As we have seen, a first-degree equation in two variables graphs as a straight line in the coordinate plane. The graph of a second-degree equation in two variables is called a **conic section** because it appears as the intersection of a plane with a cone. Figure 20-18 illustrates the four types of conic sections—the circle, the parabola, the ellipse, and the hyperbola—which we have already discussed briefly in Chapter 14. Here, and in the next three sections, we shall examine the equations and graphs of these figures. We begin with the circle.

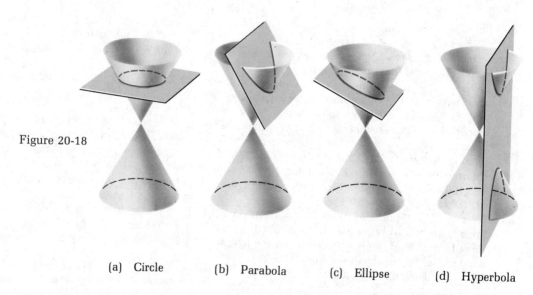

Figure 20-18

(a) Circle   (b) Parabola   (c) Ellipse   (d) Hyperbola

Definition and equation   A **circle** is defined as the set of all points equidistant from a fixed point. The fixed point is called the **center** of the circle, and the distance from the center to the circle is called the **radius**. If the point $C(h, k)$ represents the center, r is the

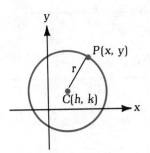

Figure 20-19

radius, and P(x, y) is a point on the circle (Fig. 20-19), then the definition can be stated mathematically using the distance formula:

$$\sqrt{(x - h)^2 + (y - k)^2} = r$$

Squaring both sides, we obtain the **standard form** of the equation of a circle with radius r and center at (h, k):

Standard Form of the Equation of a Circle

$$(x - h)^2 + (y - k)^2 = r^2 \qquad \textbf{(20-12)}$$

EXAMPLE 19 Determine the center and radius, and sketch the graph, of the circle $(x - 3)^2 + (y + 4)^2 = 25$.

Solution Rewriting the given equation in the form of (20-12), we have

$$(x - 3)^2 + [y - (-4)]^2 = 5^2$$

These must be subtraction signs. | Radius = 5

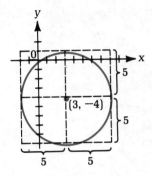

Figure 20-20

The center of the circle is at the point (3, −4), and its radius is 5. To sketch the graph, first plot the center point. Then either use a drawing compass or draw a square 2r on each side around the center point and sketch in the circle. The sketch is shown in Fig. 20-20. Every point on this circle satisfies the equation $(x - 3)^2 + (y + 4)^2 = 25$.

Center at the origin If the center of the circle is at the origin, then h and k in equation (20-12) are both 0, and the equation becomes:

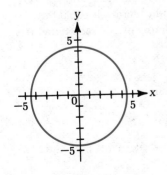

Figure 20-21

Circle Centered at the Origin $x^2 + y^2 = r^2$ \qquad \textbf{(20-13)}

EXAMPLE 20 Determine the center and radius of the circle $x^2 + y^2 = 20$, and sketch its graph.

Solution Since the equation is in the form of (20-13), the center is at the origin: (0, 0). The radius is $\sqrt{20} \approx 4.5$. The graph is shown in Fig. 20-21.

General equation of a circle When the equation of a circle is in the form of equation (20-12), we can determine its radius and center quickly. However, the equation of a circle may often appear in the form of the **general equation** of a circle:

> **General Equation of a Circle**
>
> $$Ax^2 + Ay^2 + Bx + Cy + D = 0 \qquad \text{where } A \neq 0 \qquad \text{(20-14)}$$

To work with an equation in this form, it is usually best to rewrite it in the standard form (20-12).

EXAMPLE 21   Write the following equation of a circle in standard form, and determine its center and radius: $x^2 + y^2 + 8x - 2y - 8 = 0$

Solution   **First,** rearrange terms as follows:

$$(x^2 + 8x \quad) + (y^2 - 2y \quad) = 8$$

Second, complete the trinomial squares by adding appropriate numbers within the parentheses:

$$(x^2 + 8x + 16) + (y^2 - 2y + 1) = 8 + 16 + 1$$

$$\left(\frac{8}{2}\right)^2 = 16 \qquad \left(-\frac{2}{2}\right)^2 = 1$$

Remember:
Add the same numbers to *both* sides.

Third, rewrite by factoring the trinomials on the left:

$$(x + 4)^2 + (y - 1)^2 = 25$$
$$[x - (-4)]^2 + [y - (+1)]^2 = 5^2$$

Center at $(-4, 1)$ Radius $= 5$

Comparing the equation given in Example 21 to the general equation (20-14), we see that A, the common coefficient of the x^2- and y^2-terms, is equal to 1. To analyze the equation of a circle in which $A \neq 1$, simply divide each term by A and then proceed as before.

EXAMPLE 22   Find the center and radius of the circle:

$$2x^2 + 2y^2 + 8y + 3 = 0$$

Solution   The coefficients of the squared terms are equal, so this is an equation of a circle. Dividing each term by 2, we obtain

$$x^2 + y^2 + 4y + \frac{3}{2} = 0$$

Continuing as in Example 21, we have

$$x^2 + (y^2 + 4y + 4) = -\frac{3}{2} + 4$$

This is already a perfect square.

$$\left(\frac{4}{2}\right)^2 = 4$$

$$x^2 + (y + 2)^2 = \frac{5}{2}$$

Thus, $h = 0, k = -2$, and $r = \sqrt{\frac{5}{2}} \approx 1.6$. The center is at $(0, -2)$, and the radius is approximately 1.6.

Finding the equation of a circle

So far, we have determined the geometric properties of a circle from its equation. We can also do the reverse; that is, given certain geometric properties, we can find the equation of a circle.

EXAMPLE 23 State the equation of a circle with its center at $(-4, 1)$ and a radius of 6.

Solution Substitute the given information into the standard form (20-12):

$$[x - (-4)]^2 + (y - 1)^2 = 6^2$$

$$(x + 4)^2 + (y - 1)^2 = 36$$

In many problems, the coordinates of the center or the radius may not be given directly. However, using other techniques of analytic geometry, we can determine the missing information and then write the equation.

EXAMPLE 24 Find the equation of a circle with center at $(2, 3)$ and passing through the point $(5, -6)$.

Solution We know h and k but not r. Using the standard form (20-12), we can write

$$(x - 2)^2 + (y - 3)^2 = r^2$$

Now, since the point $(5, -6)$ is on the circle, its coordinates must satisfy this equation. Thus,

$$(5 - 2)^2 + (-6 - 3)^2 = r^2$$

We can now solve for r^2:

$$9 + 81 = r^2$$

$$r^2 = 90$$

Substituting this into the original equation, we have

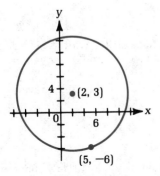

Figure 20-22

$$(x - 2)^2 + (y - 3)^2 = 90$$

The graph is shown in Fig. 20-22.

NOTE   Not every equation of the general form $Ax^2 + Ay^2 + Bx + Cy + D = 0$ produces a circle graph. When this equation is rewritten in standard form, $(x - h)^2 + (y - k)^2 = \ell$:

- If $\ell > 0$, the graph is a *circle* with radius $r = \sqrt{\ell}$.
- If $\ell = 0$, the graph is a single *point* at (h, k).
- If $\ell < 0$, there is no pair of real numbers (x, y) that satisfies the equation, and therefore there is *no graph*. ◀

Exercises 20-3

Determine the center and radius (to the nearest tenth, if necessary) of each circle, and sketch its graph.

1. $x^2 + y^2 = 81$
2. $x^2 + (y - 7)^2 = 16$
3. $(x - 4)^2 + y^2 = 64$
4. $(x - 2)^2 + (y + 6)^2 = 100$
5. $(x + 5)^2 + (y - 1)^2 = 18$
6. $(x + 2)^2 + (y + 8)^2 = 50$
7. $x^2 + y^2 - 6x - 10y + 25 = 0$
8. $x^2 + y^2 + 8x - 12y + 16 = 0$
9. $x^2 + y^2 - 4y - 4 = 0$
10. $x^2 + y^2 + 10x + 13 = 0$
11. $x^2 + y^2 + 2x + 6y - 39 = 0$
12. $x^2 + y^2 - 14x - 8y + 37 = 0$
13. $2x^2 + 2y^2 + 12x - 8y + 19 = 0$
14. $3x^2 + 3y^2 - 6x + 36y + 97 = 0$

Write an equation in standard form (20-12) for the circle with the given properties.

15. Center at $(0, 0)$, radius of 11
16. Center at the origin, radius of 15
17. Center at $(0, 5)$, radius of 4
18. Center at $(8, 0)$, radius of 9
19. Center at $(3, -8)$, radius of 1
20. Center at $(-6, 4)$, radius of 12
21. Center at $(-8, -9)$, radius of $\sqrt{11}$
22. Center at $(5, 8)$, radius of $\sqrt{17}$
23. Center at $(-3, -4)$, passing through $(2, -1)$
24. Center at $(-5, 5)$, passing through $(3, -2)$
25. Center at $(-8, 4)$, tangent to the x-axis
26. Center at $(6, -2)$, tangent to the y-axis
27. A diameter is the segment from $(6, 2)$ to $(12, 6)$
28. A diameter is the segment from $(-3, 5)$ to $(-7, 11)$

Solve.

29. *Industrial design*   In designing a surgical tool, an engineer calls for a hole to be drilled with its center 2.25 in. directly above a specified origin. If the hole must have a diameter of 0.40 in., find the equation of the hole.

30. *Physics*   When a radial force F acts on a particle of mass m moving with speed v, the particle travels in a circle whose radius r is given by the equation

$$F = \frac{mv^2}{r}$$

Find the equation of the circular path of a particle of mass 6.3×10^{-4} kg moving at 2.0×10^3 m/s under a radial force of 840 N, if the particle moves in the xy-plane and passes through both the origin and the point $(4, 0)$.

20-4 | The Parabola

The parabolic form, which is used in mirrors, solar ovens, sound and micro-wave detectors, and radar antennae, is an important curve in the hardware of technology. In addition, many technological and scientific formulas are parabolic. We already encountered the parabola in graphing certain quadratic functions in earlier chapters. In this section we shall systematically study the geometric properties and equations of the parabola.

Definitions   A **parabola** (Fig. 20-23) is defined as the set of points equidistant from both a given point and a given line. The given point is called the **focus** of the parabola, and the given line is called its **directrix.** The line through the focus and perpendicular to the directrix is called the **axis** of the parabola. The point on the parabola that intersects the axis is called the **vertex** of the parabola.

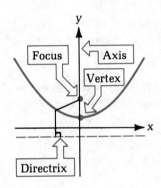

Figure 20-23

Vertex at origin   Figure 20-24 shows a parabola whose vertex is at the origin and whose axis is the x-axis. If we represent the focus by the point $F(p, 0)$, then the directrix is the line $x = -p$. If $P(x, y)$ represents a general point on the parabola, then the distance from point P to the directrix is $x + p$. Using the distance formula, the distance from P to F may be calculated as

$$PF = \sqrt{(x - p)^2 + (y - 0)^2}$$

Since the distance $x + p$ must be equal to the distance PF by definition, we can write

$$\sqrt{(x - p)^2 + (y - 0)^2} = x + p$$

Squaring both sides, we get

$$(x - p)^2 + (y - 0)^2 = (x + p)^2$$

And simplifying,

$$x^2 - 2px + p^2 + y^2 = x^2 + 2px + p^2$$

or

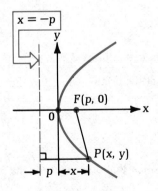

Figure 20-24

$$y^2 = 4px \qquad\qquad (20\text{-}15)$$

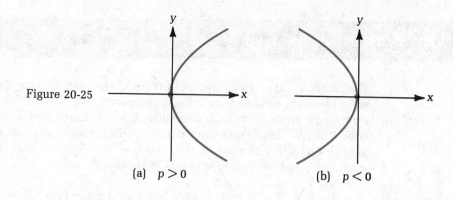

Figure 20-25

(a) $p > 0$ (b) $p < 0$

Equation (20-15) represents a parabola whose vertex is at the origin, and whose axis is the x-axis. If $p > 0$, the parabola opens to the right, as shown in Fig. 20-25(a). If $p < 0$, the parabola opens to the left, as shown in Fig. 20-25(b).

Sketching the graph   By writing the equation of a parabola in the form (20-15), we can determine quickly the coordinates of the focus $(p, 0)$ and the equation of the directrix $(x = -p)$. Then we can sketch the graph of the parabola.

EXAMPLE 25   For $y^2 = 8x$, find the coordinates of the focus and the equation of the directrix. Sketch the graph.

Solution
Step 1.   Solve for p. Comparing the given equation to (20-15), we have

$$4p = 8$$

$$p = 2$$

Step 2.   The focus is the point $(2, 0)$. The directrix is the line $x = -2$.

Step 3.   To sketch the graph, first plot the vertex at the origin. Then find the y-coordinates corresponding to at least one nonzero value of x. At $x = 2$,

$$y^2 = 8(2)$$

$$y = \pm\sqrt{16} = \pm 4$$

The completed sketch is shown in Fig. 20-26. Notice that since $p > 0$, the parabola opens to the right.

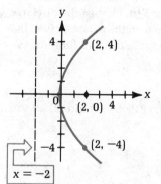

Figure 20-26

If we choose the point $F(0, p)$ as the focus and the line $y = -p$ as the directrix, the equation of the parabola takes the following form:

$$x^2 = 4py \qquad\qquad (20\text{-}16)$$

The vertex is still the origin, but the axis of the parabola is now the y-axis. If $p > 0$, the parabola opens upward; and if $p < 0$, the parabola opens downward.

EXAMPLE 26 For $x^2 = -6y$, find the coordinates of the focus and the equation of the directrix, and sketch the graph.

Solution

Step 1. Solve for p: $\qquad 4p = -6$

$$p = -\frac{3}{2}$$

Step 2. The parabola has an x^2-term, so the focus is on the y-axis at $\left(0, -\frac{3}{2}\right)$.

The directrix is the line $y = \frac{3}{2}$.

Step 3. To sketch the graph, plot the vertex at the origin and find two additional points. At $y = -2$,

$$x^2 = -6(-2)$$

$$x = \pm\sqrt{12} \approx \pm 3.5$$

As shown in Fig. 20-27, the parabola opens downward. This checks with the fact that p is negative.

Figure 20-27

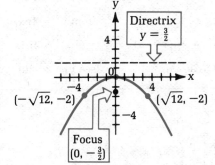

Translation of axes   Suppose we want to consider the equation of a parabola with vertex at the point (h, k) instead of at the origin. To find this new equation, we introduce a second coordinate system with axis x' parallel to the x-axis and h units away, and with axis y' parallel to the y-axis and k units away. As shown in Fig. 20-28, the point P with coordinates (x, y) in the first system now has coordinates (x', y') in the new system. We see that the coordinates are related as follows:

$$x = x' + h \qquad y = y' + k$$

$$x' = x - h \qquad y' = y - k$$

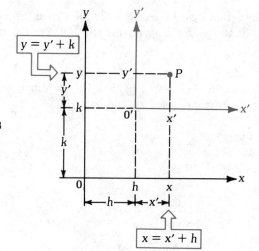

Figure 20-28

We have moved, or **translated,** the origin from $(0, 0)$ to (h, k). To **translate** in this case means to move without rotation, keeping the axes parallel. Every point $P(x, y)$ in the first coordinate system is now located at $(x - h, \; y - k)$ in the translated coordinate system.

The **standard equation** for a parabola with vertex at (h, k) can now be stated in the following two forms:

Standard Form of the Equation of a Parabola

From $x^2 = 4py$:   From $y^2 = 4px$:

$(x - h)^2 = 4p(y - k)$ $(y - k)^2 = 4p(x - h)$ **(20-17)**

(see Fig. 20-29)   (see Fig. 20-30)

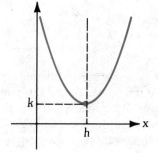

Figure 20-29

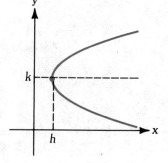

Figure 20-30

The geometric properties of these two parabolas are summarized in Table 20-1.

TABLE 20-1

Equation	Vertex	Axis	Focus	Directrix	Direction
$(x - h)^2 = 4p(y - k)$ (see Fig. 20-29)	(h, k)	$x = h$ Parallel to y-axis	$(h, \; k + p)$	$y = k - p$	$p > 0$, opens upward $p < 0$, opens downward
$(y - k)^2 = 4p(x - h)$ (see Fig. 20-30)	(h, k)	$y = k$ Parallel to x-axis	$(h + p, \; k)$	$x = h - p$	$p > 0$, opens right $p < 0$, opens left

EXAMPLE 27 For $(y - 5)^2 = 10(x + 1)$, find the coordinates of the vertex and focus, find the equation of the directrix, and sketch the graph.

Solution

Step 1. The equation is in standard form, so $h = -1$ and $k = 5$, and the vertex is at $(-1, 5)$.

Step 2. Note that the axis of the parabola is parallel to the x-axis and that $p = \dfrac{10}{4} = \dfrac{5}{2}$. Therefore, the focus is at $\left(-1 + \dfrac{5}{2}, 5\right)$ or $\left(\dfrac{3}{2}, 5\right)$ and the directrix is the line

$$x = -1 - \frac{5}{2} = -\frac{7}{2}$$

Step 3. To sketch the graph, plot the vertex and find two additional points. At $x = 3$,

$$(y - 5)^2 = 10(3 + 1)$$

$$y \approx 11.3 \quad \text{or} \quad y \approx -1.3$$

The graph is shown in Fig. 20-31. Notice that since $p > 0$, the graph opens to the right.

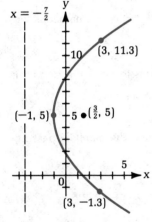

Figure 20-31

Equations not in standard form   If the equation of a parabola is not in either of the standard forms given in (20-17), it should be rewritten by completing the square.

EXAMPLE 28 Find the vertex of the given parabola and sketch its graph: $x^2 + 10x + 12y - 11 = 0$

Solution   First, to rewrite this equation in standard form, place the terms containing the squared variable on the left:

$$(x^2 + 10x \qquad) = -12y + 11$$

Complete the square:

$$(x^2 + 10x + 25) = -12y + 11 + 25$$

$$= -12y + 36$$

Write in standard form:

$$(x + 5)^2 = -12(y - 3)$$

$h = -5$	$4p = -12$	$k = 3$
	$p = -3$	

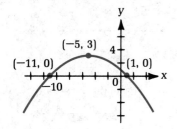

Figure 20-32

Next, analyze the equation. This is a parabola with axis parallel to the y-axis. The vertex is at $(-5, 3)$. Because $p < 0$, the parabola opens downward.

Finally, find a few points on the curve and sketch it (Fig. 20-32). At $y = 0$,

$$(x + 5)^2 = -12(0 - 3)$$

$$(x + 5)^2 = 36$$

$$x + 5 = \pm 6$$

$$x = 1 \text{ or } -11$$

Finding the equation of a parabola

As with the circle, if we are given the geometric properties of a parabola, we can find its equation.

EXAMPLE 29 Use the following geometric properties to write the standard equation of each parabola:
(a) Focus at $(2, 5)$ and vertex at $(2, -1)$
(b) Directrix $x = 7$ and vertex at $(0, 4)$
(c) Axis parallel to the x-axis, vertex at $(-2, -3)$, and the parabola passing through the point $(4, 3)$

Solution To write the equation of a parabola, we must find h, k, and p.
(a) **First,** from the vertex, note that $h = 2$ and $k = -1$.

Then, to find p, note that the focus and vertex are both on the line $x = 2$, so the axis of the parabola is parallel to the y-axis. Thus, the focus is at $(h, k + p) = (2, 5)$. Equating the y-coordinates, we have

$$k + p = 5$$

$$-1 + p = 5$$

$$p = 6$$

As shown in Fig. 20-33, we can also find p geometrically.

Now, since the axis is parallel to the y-axis, the equation is in the form $(x - h)^2 = 4p(y - k)$. Substituting, we have

$$(x - 2)^2 = 24(y + 1) \qquad \boxed{4p = 4 \cdot 6 = 24}$$

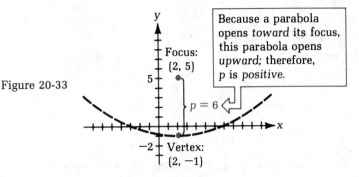

Figure 20-33

Because a parabola opens *toward* its focus, this parabola opens *upward*; therefore, p is *positive*.

Focus: (2, 5)

5

p = 6

−2 Vertex: (2, −1)

(b) **First,** from the vertex, h = 0 and k = 4.

Then, from the equation of the directrix, we know that the axis of the parabola is parallel to the x-axis, and

$$h - p = 7$$

$$0 - p = 7$$

$$p = -7$$

Using Fig. 20-34, we can also find p geometrically.

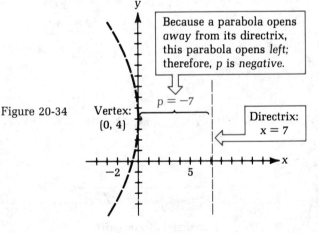

Figure 20-34

Because a parabola opens *away* from its directrix, this parabola opens *left*; therefore, p is *negative*.

p = −7

Vertex: (0, 4)

Directrix: x = 7

−2   5

Now, since the axis is parallel to the x-axis, the equation is in the form $(y - k)^2 = 4p(x - h)$. Substituting, we have

$$(y - 4)^2 = -28x$$

(c) **First,** from the vertex, h = −2 and k = −3.

Then, from the axis, we know the equation is in the form

$$(y + 3)^2 = 4p(x + 2)$$

Now, since the parabola passes through (4, 3), substitute 4 for x and 3 for y in the equation to solve for p:

$$(3 + 3)^2 = 4p(4 + 2)$$

$$36 = 24p$$

$$p = \frac{3}{2}$$

The equation is

$$(y + 3)^2 = 6(x + 2)$$

Applications   Applications of parabolas occur in two different ways. First, many physical objects are parabolic in shape. Writing their equations and graphing their shapes in the coordinate plane may be useful in designing the objects or in analyzing the way they operate. Second, many technical formulas are parabolic equations. Graphing these formulas is helpful in analyzing the relationship between the variables.

EXAMPLE 30   A cable hanging between two vertical supports approximates the shape of a parabola. The supports are 150 ft apart, and the points where the cable attaches to each support are vertically 25 ft above the lowest point of the cable. At a horizontal distance of 35 ft in from each support, how high above its lowest point is the cable?

Solution

Step 1.   Sketch the parabola choosing the y-axis for one cable connection (Fig. 20-35).*

Figure 20-35

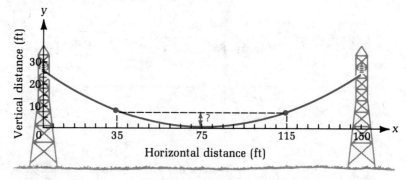

Step 2.   Find the equation of the parabola. The vertex is at $h = 75$, $k = 0$, and the axis is parallel to the y-axis. Therefore, the equation is in the form

$$(x - 75)^2 = 4py$$

The point $(0, 25)$ is on the curve, so we have

$$(0 - 75)^2 = 4p(25)$$

$$(-75)^2 = 100p$$

* It would be equally convenient to choose $(0, 0)$ as the vertex with cable connections at $(-75, 25)$ and $(75, 25)$.

$$75 \boxed{x^2} \boxed{\div} 100 \boxed{=} \rightarrow \quad \textit{56.25}$$

$$p = 56.25$$

The equation is

$$(x - 75)^2 = 4(56.25)y$$

$$(x - 75)^2 = 225y$$

Step 3. Now, substitute x = 35 ft into the equation and solve for y:

$$(35 - 75)^2 = 225y$$

$$1600 = 225y \qquad 1600 \boxed{\div} 225 \boxed{=} \rightarrow \quad \textit{7.1111111}$$

$$y \approx 7.1 \text{ ft}$$

The cable is approximately 7.1 ft above the lowest point at 35 ft from the support.

EXAMPLE 31 The height of a projectile is given by the equation below, assuming that air friction can be ignored:

$$y = v_0(\sin \theta_0)t - \frac{1}{2}gt^2$$

If $v_0 = 192$ ft/s, $\theta_0 = 30°$, and $g = 32$ ft/s², write the equation in standard form and plot the graph of this motion.

Solution

To write this equation in standard form, first substitute the given values of v_0, θ_0, and g:

$$y = (192)(\sin 30°)t - \frac{1}{2}(32)t^2$$

$$y = 96t - 16t^2$$

Then, proceed as follows: Move all terms involving t to the left.

$$16t^2 - 96t = -y$$

> Be sure to add 16·9 to the right side also.

Complete the square on the left as shown.

$$16(t^2 - 6t \quad) = -y$$

$$16(t^2 - 6t + 9) = -y + 16(9)$$

Divide both sides by 16 and factor.

$$(t - 3)^2 = -\frac{1}{16}(y - 144)$$

> In standard form the y-term must be positive.

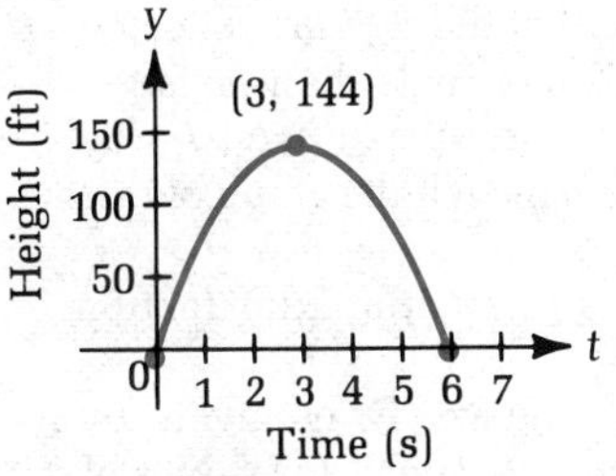

Figure 20-36

This is a parabola with axis parallel to the y-axis, vertex at t = 3 s, y = 144 ft, and opening downward. To sketch the graph, find the t-intercepts. At y = 0, t = 0 or t = 6. As shown in Fig. 20-36, negative values of t and y have no physical meaning.

Exercises 20-4

Find the coordinates of the vertex and focus, find the equation of the directrix, and sketch the graph of each parabola.

1. $y^2 = 12x$
2. $y^2 = -16x$
3. $x^2 = -12y$
4. $x^2 = 4y$
5. $y^2 = -2x$
6. $y^2 = x$
7. $x^2 = 3y$
8. $x^2 = -14y$
9. $(y - 5)^2 = -4(x - 3)$
10. $(y + 4)^2 = 8(x - 2)$
11. $(y + 3)^2 = -6x$
12. $y^2 = 10(x + 2)$
13. $x^2 = -8(y + 5)$
14. $(x - 1)^2 = 12(y - 4)$
15. $(x + 1)^2 = 2(y - 6)$
16. $(x + 4)^2 = 5y$

Find the coordinates of the vertex, and sketch the graph of each parabola.

17. $y^2 - 4y - 4x + 20 = 0$
18. $y^2 + 10y + 12x + 13 = 0$
19. $x^2 - 2y - 16 = 0$
20. $x^2 - 6x - 6y + 9 = 0$
21. $x^2 - 12x - y + 41 = 0$
22. $2x^2 + 16x - y + 27 = 0$
23. $4y^2 + 16y - x + 16 = 0$
24. $y^2 - 8x + 24 = 0$

Use the given geometric properties to write the standard equation of each parabola.

25. Focus (0, 5), vertex (0, 0)
26. Focus $\left(\dfrac{3}{2}, -7\right)$, vertex (2, −7)
27. Focus (−4, 4), directrix $x = 2$
28. Focus (2, 6), directrix $y = 8$
29. Vertex (−3, −8), directrix $y = -5$
30. Vertex (0, 0), directrix $x = -6$
31. Axis parallel to the x-axis, vertex (3, 5), passes through (5, 12)
32. Axis parallel to the y-axis, vertex (−1, −4), passes through (−5, −16)
33. **Physics**   Use the formula in Example 31 to plot the graph of the motion of a projectile given $v_0 = 56$ m/s, $\theta_0 = 45°$, and $g = 10$ m/s². [Round $v_0(\sin \theta)$ to the nearest whole number before completing the square.]
34. **Electronics**   For a simple linear resistance R (in ohms), the power P (in watts) dissipated in the circuit depends on the current I (in amperes) according to the equation $P = RI^2$. Plot the graph of P as a function of I for $0 \le I \le 4.0$ A when $R = 0.50\ \Omega$.
35. **Civil engineering**   A cable hangs in a parabolic curve between two vertical supports 88 ft apart. At a distance of 33 ft in from each support, the cable is 2.0 ft above its lowest point. How high up on each support is the cable attached?
36. **Civil engineering**   A parabolic archway is used as the entrance to a large exposition hall. If the maximum height of the arch is 48 ft and the width of the entrance along the ground is 36 ft, find the equation describing the archway. (Choose the left end of the entrance as the origin.)
37. **Optics**   If a light source is placed at the focus of a parabolic reflector, the light rays will be reflected on a line parallel to the axis of the parabola. If a certain parabolic reflector is 10 in. across at its widest point and 4 in. deep at its deepest point, where should the light source be placed so that the rays are reflected parallel to the axis? (Place the parabola so that its vertex is at the origin and its axis is the y-axis.)
38. **Fire science**   The velocity v (in feet per second) of water discharge from a hose nozzle is given by $v = 12.14\sqrt{P}$, where P is the pressure at the nozzle (in pounds per square inch). Sketch the graph of this equation using P as the independent variable for $0 \le P \le 100$ lb/in.².
39. **Police science**   The stopping distance y (in feet) of a car moving at v mi/h is given approximately by the equation $y = v + v^2/20$. Plot y as a function of v for $0 \le v \le 60$ mi/h.
40. **Electronics**   For a certain temperature-sensitive electronic device, the voltage V is given by $V = 4.0T - 0.016T^2$, where T is the ambient temperature (in °C). Plot V as a function of T for values of $V \ge 0$. At what temperature is $V = 0$?

20-5 | The Ellipse

Ancient astronomers, philosophers, and mathematicians saw the circle as the most perfect closed geometric figure, and to explain the movement of the planets they devised elaborate models using circular paths. But early in the 17th century, it was found that the astronomical measurements fit the ellipse much better. Today we know that a satellite orbiting the earth or a planet orbiting the sun moves in an elliptical orbit, and elliptical shapes are often used in technology, architecture, and industrial design.

Definitions   Most people think of an ellipse as an oval or flattened circle, but it is a very carefully defined mathematical curve. An **ellipse** is defined as the set of points in the plane, the sum of whose distances from two fixed points is a positive constant. The two fixed points are called the **foci** of the ellipse. The midpoint of the line segment connecting the foci is the **center** of the ellipse.

Ellipses centered at the origin   To derive the equation of an ellipse with center at the origin, first choose the foci to be at $(c, 0)$ and $(-c, 0)$. Then let the sum of the distances from the foci to any point $P(x, y)$ on the ellipse be $2a$; that is, $d_1 + d_2 = 2a$. From Fig. 20-37 we can see that $2a > 2c$, and thus, $a > c$. We will use this fact later.

Figure 20-37

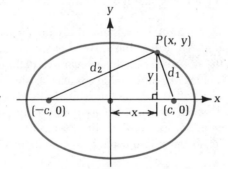

Using the distance formula to find d_1 and d_2, we can express the equation $d_1 + d_2 = 2a$ as

$$\sqrt{(x - c)^2 + y^2} + \sqrt{(x + c)^2 + y^2} = 2a$$

Now, move one of the radicals to the right,

$$\sqrt{(x - c)^2 + y^2} = 2a - \sqrt{(x + c)^2 + y^2}$$

square both sides, and simplify to obtain

$$x^2 - 2cx + c^2 + y^2 = 4a^2 - 4a\sqrt{(x + c)^2 + y^2}$$
$$+ x^2 + 2cx + c^2 + y^2$$

This simplifies further to

$$a\sqrt{(x + c)^2 + y^2} = a^2 + cx$$

Squaring both sides again, we get

$$a^2(x^2 + 2cx + c^2 + y^2) = a^4 + 2a^2cx + c^2x^2$$

$$a^2x^2 - c^2x^2 + a^2y^2 = a^4 - a^2c^2$$

$$x^2(a^2 - c^2) + a^2y^2 = a^2(a^2 - c^2)$$

Dividing both sides by $a^2(a^2 - c^2)$,

$$\frac{x^2}{a^2} + \frac{y^2}{a^2 - c^2} = 1$$

Earlier, we saw that $a > c$, which also means that $a^2 - c^2 > 0$, and we can replace $a^2 - c^2$ with another positive quantity b^2. Thus, the equation becomes:

$$\frac{x^2}{a^2} + \frac{y^2}{b^2} = 1 \qquad \text{where } b^2 = a^2 - c^2 \qquad \text{(20-18)}$$

Because $c > 0$, we also know that $a^2 > b^2$, or $a > b$.

This form of the equation of an ellipse leads to several further definitions:

- Replacing y with 0, we see that the x-intercepts are at $(-a, 0)$ and $(a, 0)$. These points are called the **vertices** of the ellipse. The line segment connecting the vertices is called the **major axis** of the ellipse. Because $a > b$, the ellipse is elongated along its major axis, as shown in Fig. 20-38(a).
- Replacing x with 0, we find the y-intercepts to be at $(0, b)$ and $(0, -b)$. Since the line segment connecting these points is shorter than the major axis, it is called the **minor axis** of the ellipse, as indicated in Fig. 20-38(b). We will often refer to the y-intercepts as the *endpoints* of the minor axis of this ellipse.
- The major and minor axes intersect at the center of the ellipse. In this case, the center is at the origin [Fig. 20-38(b)].

Figure 20-38

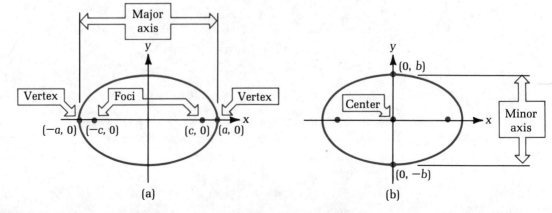

(a)   (b)

NOTE ▶ The major and minor axes of the ellipse are actually *lines* and have infinite length, but for analytic purposes these terms are commonly used to refer to the line *segments* as defined above. ◀

Equation (20-18) is the standard equation of an ellipse with center at the origin and foci and major axis on the x-axis. If we choose foci on the y-axis, a similar derivation results in another form of the standard equation:

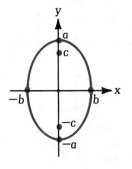

$$\frac{x^2}{b^2} + \frac{y^2}{a^2} = 1 \qquad \text{where } b^2 = a^2 - c^2 \text{ and } a > b \qquad \text{(20-19)}$$

Figure 20-39

In this case, the foci are at $(0, c)$ and $(0, -c)$, the vertices are the y-intercepts $(0, a)$ and $(0, -a)$, and the major axis is along the y-axis (Fig. 20-39). The minor axis coincides with the x-axis, and the endpoints are at $(-b, 0)$ and $(b, 0)$.

LEARNING HINT ▶ The x-axis is the major, or elongated, axis if the denominator of x^2 is the larger denominator, and the y-axis is the major axis if the denominator of y^2 is the larger denominator. ◀

EXAMPLE 32   Determine the coordinates of the vertices, the endpoints of the minor axis, and the foci, and sketch the graph of each ellipse:

(a) $\dfrac{x^2}{9} + \dfrac{y^2}{4} = 1$ (b) $\dfrac{x^2}{9} + \dfrac{y^2}{36} = 1$

Solution
(a) **First,** find the larger denominator. This is a^2.

$$\frac{x^2}{3^2} + \frac{y^2}{2^2} = 1$$

Major axis ▶

The ellipse is longer on the x-axis.

Then:

• Since $a^2 = 9$, then $a = 3$, and the vertices are at $(3, 0)$ and $(-3, 0)$.
• Since $b^2 = 4$, then $b = 2$, and the endpoints of the minor axis are at $(0, 2)$ and $(0, -2)$.
• Since $b^2 = a^2 - c^2$, then $c = \sqrt{a^2 - b^2} = \sqrt{9 - 4} = \sqrt{5} \approx 2.2$. The foci are at approximately $(2.2, 0)$ and $(-2.2, 0)$.
• To sketch the graph, first construct a rectangle using the lines $x = a$, $x = -a$, $y = b$, and $y = -b$. Then draw in the ellipse tangent to the sides of this rectangle (Fig. 20-40).

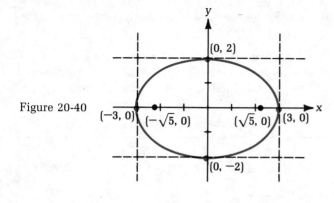

Figure 20-40

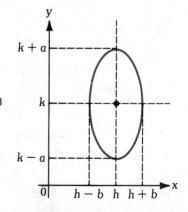

Figure 20-41

(b)   Here, $a^2 = 36$, and the major axis is along the y-axis.
   • Since $a = 6$, the vertices are at $(0, 6)$ and $(0, -6)$.
   • Since $b^2 = 9$, then $b = 3$, and the endpoints of the minor axis are at $(3, 0)$
     and $(-3, 0)$.
   • From $b^2 = a^2 - c^2$, we have $c = \sqrt{36 - 9} = 3\sqrt{3} \approx 5.2$. The foci are at
     approximately $(0, 5.2)$ and $(0, -5.2)$.
   • The graph is shown in Fig. 20-41.

Ellipses centered at   If we translate the center of the ellipse from the origin $(0, 0)$ to some point
(h, k) (h, k), then the equations for the ellipse become:

$$\frac{(x - h)^2}{a^2} + \frac{(y - k)^2}{b^2} = 1 \qquad \text{where } a > b \text{ (see Fig. 20-42)} \qquad \textbf{(20-20)}$$

$$\frac{(x - h)^2}{b^2} + \frac{(y - k)^2}{a^2} = 1 \qquad \text{where } a > b \text{ (see Fig. 20-43)} \qquad \textbf{(20-21)}$$

The geometric properties of these two ellipses are summarized in Table 20-2.

Figure 20-42

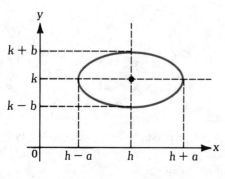

Figure 20-43

TABLE 20-2

Equation	Center	Vertices	Foci	Minor Axis Endpoints
$\dfrac{(x - h)^2}{a^2} + \dfrac{(y - k)^2}{b^2} = 1$ $a > b$ (see Fig. 20-42)	(h, k)	$(h \pm a,\ k)$ Major axis parallel to x-axis	$(h \pm c,\ k)$ $c = \sqrt{a^2 - b^2}$	$(h,\ k \pm b)$
$\dfrac{(x - h)^2}{b^2} + \dfrac{(y - k)^2}{a^2} = 1$ $a > b$ (see Fig. 20-43)	(h, k)	$(h,\ k \pm a)$ Major axis parallel to y-axis	$(h,\ k \pm c)$ $c = \sqrt{a^2 - b^2}$	$(h \pm b,\ k)$

LEARNING HINT ▶ You may find it easier to remember the information in Table 20-2 by interpreting the formulas geometrically as shown in Fig. 20-44. For example, since you already know that the quantity a is associated with the vertex, think of the vertices as being a units to the right and left of the center when the major axis is parallel to the x-axis, and a units above and below the center if the major axis is parallel to the y-axis. Similarly, the foci are aligned with the vertices c units from the center. The minor axis is perpendicular to the major axis, and its endpoints are b units from the center. Since a also represents the distance from a focus to the endpoints of the minor axis, the Pythagorean relationship among a, b, and c can be remembered by visualizing Fig. 20-44(c). ◀

Figure 20-44

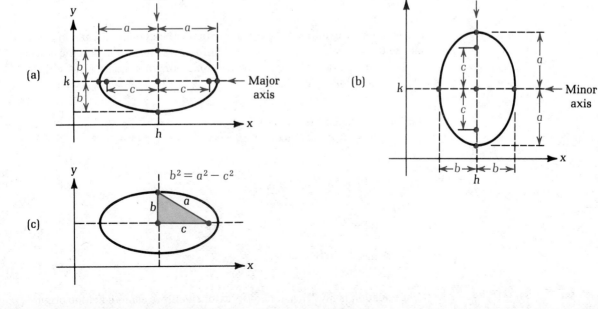

EXAMPLE 33 Determine the coordinates of the center, the vertices, and the foci, and sketch the graph of

$$\frac{(x + 2)^2}{16} + \frac{(y - 4)^2}{6} = 1$$

Solution   First, note that $h = -2$ and $k = 4$; therefore, the center is at $(-2, 4)$.

Second, since the major axis is parallel to the x-axis, $a^2 = 16$, so $a = 4$. The vertices are 4 units to the right and left of the center at $(2, 4)$ and $(-6, 4)$.

Third, $b^2 = 6$, so $c = \sqrt{a^2 - b^2} = \sqrt{16 - 6} = \sqrt{10} \approx 3.2$. Therefore, the foci are approximately 3.2 units to the right and left of the center at $(1.2, 4)$ and $(-5.2, 4)$.

Finally, notice that since $b = \sqrt{6} \approx 2.4$, the endpoints of the minor axis are approximately 2.4 units above and below the center at $(-2, 6.4)$ and $(-2, 1.6)$. The graph is shown in Fig. 20-45.

Figure 20-45

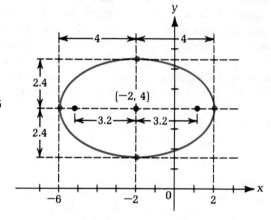

The equation of an ellipse may appear written without fractions, and in order to analyze the geometric properties of the ellipse, we must transform the equation into standard form. This may involve completing the square if the ellipse is not centered at the origin.

EXAMPLE 34 Sketch the graph of:
(a) $7x^2 + 2y^2 = 98$ (b) $9x^2 - 72x + 4y^2 - 24y + 144 = 0$

Solution
(a)   **First,** divide both sides by 98:

$$\frac{x^2}{14} + \frac{y^2}{49} = 1$$

This must be +

This number must be 1

The equation is now in standard form for an ellipse centered at the origin.

Then, proceed as before. Since $a^2 = 49$, $a = 7$, and the major axis is along the y-axis. The vertices are at $(0, 7)$ and $(0, -7)$. Also, $b = \sqrt{14} \approx 3.7$, so the endpoints of the minor axis are at $(3.7, 0)$ and $(-3.7, 0)$. The graph is shown in Fig. 20-46.

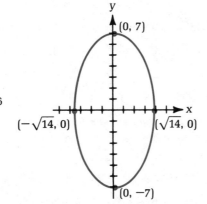

Figure 20-46

(b)   **First,** rewrite the equation in standard form. Because this equation contains first-degree terms, we must separate the x- and y-terms and complete the squares:

$$(9x^2 - 72x \qquad) + (4y^2 - 24y \qquad) = -144$$

$$9(x^2 - 8x \qquad) + 4(y^2 - 6y \qquad) = -144 \qquad \boxed{\text{Careful!}}$$

$$9(x^2 - 8x + 16) + 4(y^2 - 6y + 9) = -144 + 9(16) + 4(9)$$

$$9(x - 4)^2 + 4(y - 3)^2 \qquad = 36$$

Divide by 36 to obtain

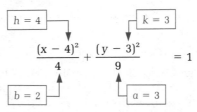

$$\boxed{h = 4} \qquad \boxed{k = 3}$$

$$\frac{(x - 4)^2}{4} + \frac{(y - 3)^2}{9} = 1$$

$$\boxed{b = 2} \qquad \boxed{a = 3}$$

Next, analyze the equation. It is an ellipse with major axis parallel to the y-axis and center at $(4, 3)$. The vertices are at $(4, 3 \pm 3)$, or $(4, 6)$ and $(4, 0)$. The endpoints of the minor axis are at $(4 \pm 2, 3)$, or $(6, 3)$ and $(2, 3)$.

Finally, sketch the graph, as shown in Fig. 20-47.

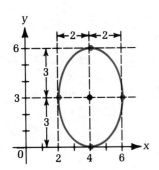

Figure 20-47

Determining the equation of an ellipse   Knowing the relationship between the equation of an ellipse and its geometric properties, we can determine the equation from the properties.

EXAMPLE 35 Find the equation of the ellipse with the given properties.
(a)   Center at $(-1, 1)$, vertex at $(-1, 6)$, one focus at $(-1, 4)$
(b)   Foci at $(5, -4)$ and $(-7, -4)$, length of minor axis is 6

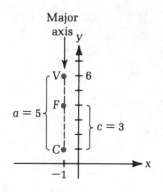

Major
axis

$a = 5$

$c = 3$

Figure 20-48

Solution   To find the equation of an ellipse, we must determine h, k, a^2, b^2, and the orientation of the major axis.

(a)   The coordinates of the center give us $h = -1$ and $k = 1$.

• We know that a is the distance from the center to each vertex. Therefore,

$$a = |6 - 1| = 5 \text{ (Fig. 20-48)} \qquad \text{and} \qquad a^2 = 25$$

• Since c is the distance from the center to each focus (Fig. 20-48), we have

$$c = |4 - 1| = 3 \qquad \text{and} \qquad c^2 = 9$$

Therefore,

$$b^2 = a^2 - c^2 = 25 - 9 = 16$$

• The given center, vertex, and focus are aligned vertically, so the major axis is parallel to the y-axis. The equation is

$$\frac{(x + 1)^2}{16} + \frac{(y - 1)^2}{25} = 1$$

(b)   • The distance between the foci is $2c = |-7 - 5| = 12$, so $c = 6$ and $c^2 = 36$.

• The center is halfway between the foci at $(-1, -4)$ (Fig. 20-49), so $h = -1$ and $k = -4$.

• The length of the minor axis is $2b = 6$, so $b = 3$ and $b^2 = 9$. Therefore,

$$a^2 = b^2 + c^2 = 9 + 36 = 45$$

• Since the foci are aligned horizontally, the major axis is parallel to the x-axis and the equation is

$$\frac{(x + 1)^2}{45} + \frac{(y + 4)^2}{9} = 1$$

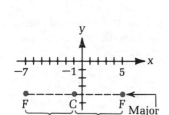

Figure 20-49

Applications

Here is one example of a practical application involving an ellipse.

EXAMPLE 36   A satellite orbits the earth in an elliptical path with focus at the center of the earth. The altitude of the satellite ranges from 525 to 1525 mi. Given that the radius of the earth is approximately 3960 mi, what is the equation of the path of the satellite?

Solution   The minimum distance from the earth's center is

$$525 + 3960 = 4485 \text{ mi}$$

and the maximum distance is

$$1525 + 3960 = 5485 \text{ mi}$$

If we place the major axis along the x-axis and the center of the earth at a focus, we can sketch the graph of the orbit as shown in Fig. 20-50. We can now see that

$$a - c = 4485 \text{ mi}$$

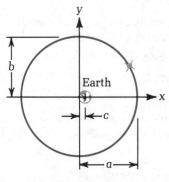

Figure 20-50

$$a + c = 5485 \text{ mi}$$

Adding these two equations, we get

$$2a = 9970$$

$$a = 4985 \quad \text{and} \quad c = 500$$

Therefore,

$$a^2 = 4985^2 \qquad 4985\,\boxed{x^2} \rightarrow \quad \boxed{24850225.}$$

$$b^2 = a^2 - c^2 \qquad \boxed{-}\;500\,\boxed{x^2}\,\boxed{=} \rightarrow \quad \boxed{24600225.}$$

The equation of the orbit is approximately

$$\frac{x^2}{2.49 \times 10^7} + \frac{y^2}{2.46 \times 10^7} = 1$$

Exercises 20-5

Determine the coordinates of the center, vertices, and foci, and sketch the graph of each ellipse. Round to the nearest tenth if necessary.

1. $\dfrac{x^2}{16} + \dfrac{y^2}{25} = 1$

2. $\dfrac{x^2}{49} + \dfrac{y^2}{64} = 1$

3. $\dfrac{x^2}{16} + \dfrac{y^2}{4} = 1$

4. $\dfrac{x^2}{100} + \dfrac{y^2}{36} = 1$

5. $\dfrac{x^2}{8} + \dfrac{y^2}{36} = 1$

6. $\dfrac{x^2}{20} + y^2 = 1$

7. $\dfrac{(x-3)^2}{4} + \dfrac{(y-2)^2}{9} = 1$

8. $\dfrac{(x-2)^2}{36} + \dfrac{(y+7)^2}{49} = 1$

9. $\dfrac{(x+4)^2}{64} + \dfrac{(y-1)^2}{16} = 1$

10. $\dfrac{(x+5)^2}{81} + \dfrac{(y+1)^2}{25} = 1$

11. $\dfrac{(x-1)^2}{9} + (y+5)^2 = 1$

12. $\dfrac{(x-6)^2}{25} + \dfrac{(y-3)^2}{49} = 1$

13. $\dfrac{(x+3)^2}{12} + \dfrac{y^2}{24} = 1$

14. $\dfrac{x^2}{40} + \dfrac{(y-4)^2}{18} = 1$

Determine the coordinates of the center and vertices, and sketch the graph of each ellipse.

15. $x^2 + 4y^2 = 4$

16. $36x^2 + 16y^2 = 576$

17. $5x^2 + 3y^2 = 15$

18. $x^2 + 2y^2 = 12$

19. $16x^2 - 96x + 9y^2 + 90y + 225 = 0$

20. $4x^2 - 8x + y^2 + 2y + 1 = 0$

21. $4x^2 + 32x + y^2 + 4y + 60 = 0$

22. $x^2 + 12y^2 - 120y + 288 = 0$

Use the given geometric properties to find the equation of each ellipse.

23. Center $(0, 0)$, vertex $(0, 4)$, focus $(0, 3)$

24. Center $(2, 5)$, vertex $(8, 5)$, focus $(6, 5)$

25. Vertices $(-3, 2)$ and $(-7, 2)$, length of minor axis is 2

26. Vertices $(6, 0)$ and $(6, 10)$, foci $(6, 2)$ and $(6, 8)$

27. Foci $(-8, 7)$ and $(2, 7)$, length of major axis is 14

28. Foci $(-1, 2\sqrt{2})$ and $(-1, 4\sqrt{2})$, length of minor axis is $4\sqrt{3}$

29. Center $(-4, -1)$, vertex $(-4, 8)$, passes through $(2, -1)$

30. Center $(0, 0)$, focus $(2, 0)$, passes through $(\sqrt{5}, \sqrt{12})$

31. *Space technology*   Solve Example 36 if the altitude of the satellite ranges from 475 to 1875 mi.

32. *Architecture*   A semielliptical archway is planned at the entrance to a courtyard. If the arch has a span (major axis) of 5 ft and a height of 1.5 ft (half the minor axis), where should the foci be placed in order to sketch the curve? (Place the origin at the center of the ellipse.)

33. *Industrial design*   A solar hot-tub blanket must be designed for an elliptical tub with major axis 6.0 ft long and minor axis 4.0 ft long. If the center of the ellipse is the origin of the coordinate system and the major axis is along the *x*-axis, find the coordinates of the foci.

34. *Mechanical engineering*   If a circular bit of diameter *d* is used to drill into a surface at an angle θ, an elliptical hole is formed with minor axis *d* units long and major axis $d/(\sin\theta)$ units long. Find the equation of such a hole if $d = 0.60$ in. and $\theta = 30°$. (Assume the hole is drilled so that the center of the hole is at the origin and the major axis is along the *x*-axis.)

35. *Transportation engineering*   A road passes through a tunnel whose cross section is a semiellipse 48 ft wide and 12 ft high at the center. How tall is the tallest vehicle that can pass under the tunnel at a point 15 ft from the center?

36. *Mechanical engineering*   Find the equation of the ellipse described in Problem 34 if the origin is placed at the left end of the major axis.

20-6 | The Hyperbola

When an electrically charged particle is repelled or scattered by a second similarly charged particle, it moves along a hyperbola. Using the difference in reception times of synchronized radio signals from two widely spaced transmitters, a ship can determine its position by the intersection of two hyperbolas. From nuclear physics to hyperbolic navigation, this two-part curve, first studied more than 21 centuries ago, appears in many scientific and technical applications.

Definitions   A **hyperbola** is the set of all points in the plane, the difference of whose distances from two fixed points is a positive constant. The two fixed points are called the **foci** of the hyperbola. The midpoint of the line segment connecting the foci is called the **center** of the hyperbola.

This definition is identical to that of the ellipse except for the word *difference*, which replaces the word *sum*. It follows that the derivation of the equation of the hyperbola will be similar to the derivation for the ellipse, and it will not be presented here.

Hyperbolas centered at the origin

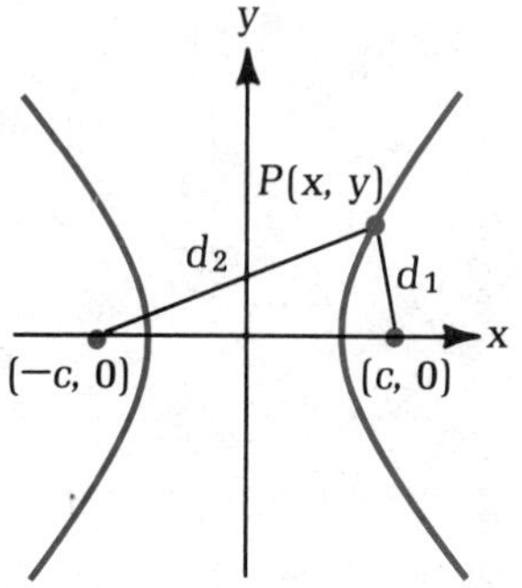

For the hyperbola centered at the origin with foci at $(c, 0)$ and $(-c, 0)$ and a constant difference $|d_2 - d_1| = 2a$ (Fig. 20-51), the resulting standard equation of the hyperbola is:

$$\frac{x^2}{a^2} - \frac{y^2}{b^2} = 1 \qquad \text{where } b^2 = c^2 - a^2 \qquad (20\text{-}22)$$

Figure 20-51

This formula leads to the following geometric properties of the hyperbola shown in Fig. 20-52.

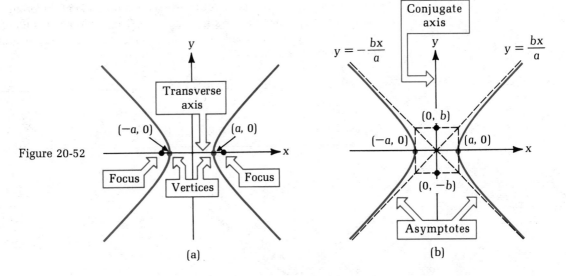

Figure 20-52

(a)   (b)

- Replacing y with 0, we find the x-intercepts to be at $(a, 0)$ and $(-a, 0)$. These are called the **vertices** of the hyperbola, and the line through them is called the **tranverse axis.** For analytic purposes, the transverse axis also refers to the line segment joining the vertices and has a length of $2a$. In this case, the transverse axis coincides with the x-axis, and therefore the vertices and the foci are also on the x-axis, as indicated in Fig. 20-52(a).
- The line through the points $(0, b)$ and $(0, -b)$ is called the **conjugate axis.** The conjugate axis also refers to the line segment joining these points and therefore has a length of $2b$. In this case, the conjugate axis coincides with the y-axis, as shown in Fig. 20-52(b).
- Every hyperbola is associated with a pair of **asymptotes.** The two parts, or **branches,** of the curve approach the asymptote lines but never quite reach them. The asymptotes of this hyperbola are the lines

$$y = \frac{bx}{a} \quad \text{and} \quad y = -\frac{bx}{a} \qquad (20\text{-}23)$$

(A derivation of these equations is given at the end of this section.) Notice in Fig. 20-52(b) that the asymptote lines are the diagonals of the rectangle formed by the points $(\pm a, 0)$ and $(0, \pm b)$. This fact will help you find the asymptotes and sketch the hyperbola very quickly.

- The axes and the asymptotes intersect at the center of the hyperbola. In this case, the center is at the origin.

If we choose the foci of the hyperbola to be the points $(0, c)$ and $(0, -c)$ on the y-axis, the graph is rotated 90°, and the resulting equation is

$$\frac{y^2}{a^2} - \frac{x^2}{b^2} = 1 \qquad \text{where } b^2 = c^2 - a^2 \qquad (20\text{-}24)$$

This is the standard equation of the hyperbola centered at the origin with transverse axis along the y-axis (Fig. 20-53). The vertices are at $(0, a)$ and $(0, -a)$, the endpoints of the conjugate axis are $(b, 0)$ and $(-b, 0)$, and the asymptotes are the lines

$$y = \frac{ax}{b} \quad \text{and} \quad y = -\frac{ax}{b} \tag{20-25}$$

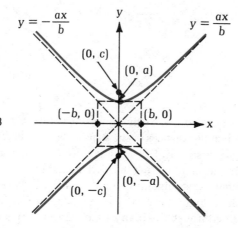

Figure 20-53

- Notice that unlike the ellipse, the hyperbola has an orientation determined not by the relative size of a and b, but by the position of the negative sign in the equation. The transverse axis (with the vertices and foci) is on the axis associated with the positive term. If the x^2-term is positive, the transverse axis coincides with the x-axis. If the y^2-term is positive, the transverse axis is along the y-axis.

CAUTION ▶ The formulas relating a, b, and c for the hyperbola are different from the corresponding formulas for the ellipse.

For the ellipse:

$$b^2 = a^2 - c^2 \quad \text{Fig. 20-54(a)} \qquad a > c \quad \text{Fig. 20-54(b)}$$

Figure 20-54

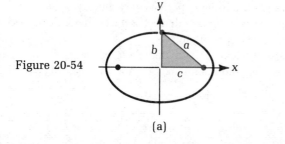

(a)

(b)

For the hyperbola:

$$b^2 = c^2 - a^2 \quad \text{Fig. 20-55(a)} \qquad c > a \quad \text{Fig. 20-55(b)} \blacktriangleleft$$

Figure 20-55

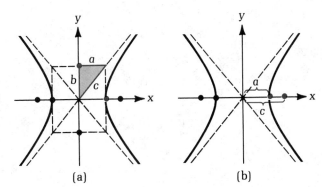

(a)   (b)

- Notice from the preceding formulas that for an ellipse, a cannot equal b (the result is a circle). However, for a hyperbola, a can equal b, and the result is a **rectangular** or **equilateral hyperbola**.

EXAMPLE 37 Find the coordinates of the vertices and foci, and sketch the graph of each hyperbola.

(a) $\dfrac{x^2}{25} - \dfrac{y^2}{36} = 1$ (b) $\dfrac{y^2}{16} - x^2 = 1$

Solution

(a) Since the x^2-term is positive, the transverse axis coincides with the x-axis, and therefore the vertices and foci are located on the x-axis. Thus,

$$a^2 = 25 \quad \text{and} \quad b^2 = 36$$
$$a = 5 \qquad\qquad\quad b = 6$$

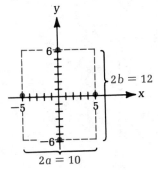

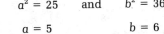

Figure 20-56

Vertices: $a = 5$, so the vertices are $(5, 0)$ and $(-5, 0)$.

Foci: $c = \sqrt{a^2 + b^2} = \sqrt{25 + 36} = \sqrt{61} \approx 7.8$, so the foci are at approximately $(7.8, 0)$ and $(-7.8, 0)$.

To graph: First, plot the vertices and draw a rectangle with sides $2a$ by $2b$ centered on the origin (Fig. 20-56).

Second, draw the diagonals of the rectangle. These are the asymptotes of the hyperbola (Fig. 20-57).

Finally, sketch the hyperbola through the vertices and approaching the asymptotes. (Remember, the asymptotes are merely graphing aids; they are *not* part of the graph. That is why dashed lines are used.)

(b) The y^2-term is positive, so the transverse axis is along the y-axis, and the vertices and foci are located on the y-axis. From the denominators, $a^2 = 16$ and $b^2 = 1$; therefore, $a = 4$ and $b = 1$.

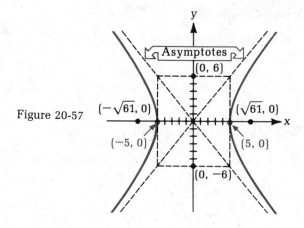

Figure 20-57

Vertices: $(0, \pm 4)$

Foci: $c = \sqrt{16 + 1} = \sqrt{17} \approx 4.1$, so the foci are at approximately $(0, \pm 4.1)$.

Graph: Draw a rectangle centered on the origin with sides $2a$ by $2b$—that is, 8 by 2. Using the diagonals of this rectangle as asymptotes, sketch the graph as shown in Fig. 20-58.

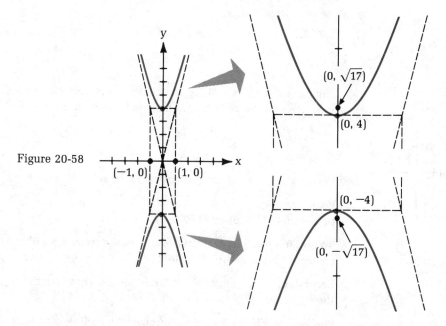

Figure 20-58

Hyperbolas of the form
$xy = k$

The hyperbolas determined by equations (20-22) and (20-24) are centered at the origin and have their vertices and foci on the coordinate axes. As we saw in Section 14-1, there is another class of hyperbolas centered at the origin with asymptotes on the coordinate axes, and vertices and foci on the line $y = x$ or $y = -x$. The equation of this type of hyperbola is in the following form:

$$xy = k \qquad \text{where } k \neq 0 \tag{20-26}$$

If k is positive, the vertices are at $\left(\sqrt{k}, \sqrt{k}\right)$ and $\left(-\sqrt{k}, -\sqrt{k}\right)$, and the foci are at $\left(\sqrt{2k}, \sqrt{2k}\right)$ and $\left(-\sqrt{2k}, -\sqrt{2k}\right)$. The transverse axis is $y = x$, and the asymptotes are the x- and y-axes [see Fig. 20-59(a)].

If k is negative [Fig. 20-59(b)], the vertices are at $\left(\sqrt{|k|}, -\sqrt{|k|}\right)$ and $\left(-\sqrt{|k|}, \sqrt{|k|}\right)$, and the foci are at $\left(\sqrt{|2k|}, -\sqrt{|2k|}\right)$ and $\left(-\sqrt{|2k|}, \sqrt{|2k|}\right)$. The transverse axis is $y = -x$, and the asymptotes are again the x- and y-axes.

Figure 20-59

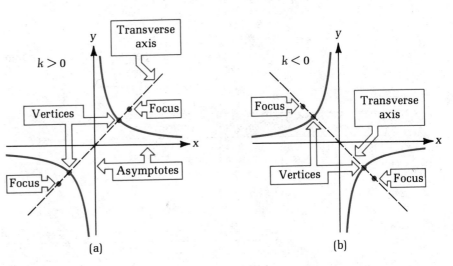

(a)   (b)

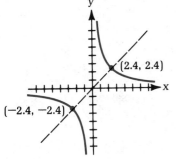

Figure 20-60

EXAMPLE 38　Sketch the graph of $xy = 6$.

Solution　**First,** find the vertices. Since $k = 6$, the vertices are at $\left(\sqrt{6}, \sqrt{6}\right)$ and $\left(-\sqrt{6}, -\sqrt{6}\right)$, or approximately $(2.4, 2.4)$ and $(-2.4, -2.4)$.

Second, draw the transverse axis through these points.

Finally, sketch the hyperbola using the x- and y-axes as asymptotes. The graph must pass through the vertices (Fig. 20-60).

Hyperbolas centered at (h, k)　　If we translate the centers of the hyperbolas described in (20-23) and (20-24) from the origin $(0, 0)$ to some point (h, k), the equations become:

$$\frac{(x - h)^2}{a^2} - \frac{(y - k)^2}{b^2} = 1 \qquad \text{(see Fig. 20-61)} \tag{20-27}$$

$$\frac{(y - k)^2}{a^2} - \frac{(x - h)^2}{b^2} = 1 \qquad \text{(see Fig. 20-62)} \tag{20-28}$$

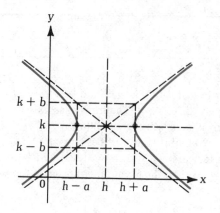

Figure 20-61

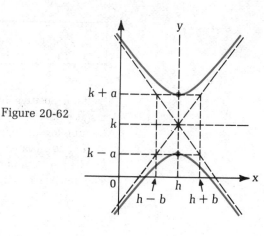

Figure 20-62

The geometric properties of these two hyperbolas are summarized in Table 20-3.

TABLE 20-3

Equation	Center	Vertices	Foci	Transverse Axis
$\dfrac{(x - h)^2}{a^2} - \dfrac{(y - k)^2}{b^2} = 1$ (see Fig. 20-61)	(h, k)	$(h \pm a, \ k)$	$(h \pm c, \ k)$ $c = \sqrt{a^2 + b^2}$	$y = k$ Parallel to x-axis
$\dfrac{(y - k)^2}{a^2} - \dfrac{(x - h)^2}{b^2} = 1$ (see Fig. 20-62)	(h, k)	$(h, \ k \pm a)$	$(h, \ k \pm c)$ $c = \sqrt{a^2 + b^2}$	$x = h$ Parallel to y-axis

LEARNING HINT    As with the ellipse, you may wish to determine the key points for the hyperbola geometrically. If the transverse axis is parallel to the x-axis, the vertices are a units to the right and left of the center, and the foci are c units to the right and left of the center. If the transverse axis is parallel to the y-axis, these points are above and below the center. ◀

EXAMPLE 39   Given the hyperbola

$$\frac{(y - 2)^2}{9} - \frac{(x + 4)^2}{16} = 1$$

find the coordinates of the center, the vertices, and the foci, and sketch the graph.

Solution   Center: Note that $h = -4$ and $k = 2$, so the center is at $(-4, 2)$.

Vertices: Because the $(y - k)^2$ term is positive, the transverse axis is parallel to the y-axis. Also, $a^2 = 9$, so $a = 3$. Therefore, the vertices are 3 units above and below the center at $(-4, 5)$ and $(-4, -1)$.

Foci: Since $c = \sqrt{a^2 + b^2} = \sqrt{9 + 16} = 5$, the foci are 5 units above and below the center at $(-4, 7)$ and $(-4, -3)$.

Graph: First, plot the center and vertices. Then note that $b^2 = 16$, so $b = 4$. Draw a rectangle with sides $2a$ by $2b$, or 6 by 8, about the center of the hyperbola and sketch the asymptotes and the curve as before (Fig. 20-63).

Figure 20-63

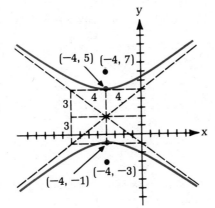

Putting the equation in standard form   If the equation of a hyperbola is not given in standard form, perform the algebraic operations necessary to rewrite it. This will involve completing the square if the hyperbola is not centered at the origin.

EXAMPLE 40   Sketch the graph of:

(a) $4y^2 - 9x^2 = 48$ (b) $2x^2 - 8x - 3y^2 - 18y - 31 = 0$

Solution

(a)   Since there is no x-term or y-term, this hyperbola is centered at the origin.

First, divide each side by 48 to obtain a constant term of 1:

$$\frac{y^2}{12} - \frac{3x^2}{16} = 1$$

Then, rewrite the second term with a numerator of x^2 and identify the important quantities:

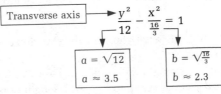

The vertices are at approximately $(0, 3.5)$ and $(0, -3.5)$. The graph is shown in Fig. 20-64.

(b)   Since this equation contains first-degree terms, the resulting hyperbola is *not* centered at the origin.

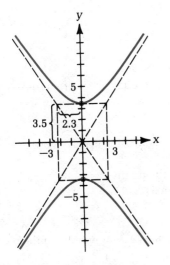

Figure 20-64

First, separate the x- and y-terms and complete the squares:

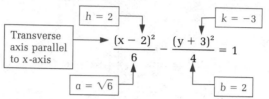

$$(2x^2 - 8x \qquad) + (-3y^2 - 18y \qquad) = 31$$

$$2(x^2 - 4x \qquad) - 3(y^2 + 6y \qquad) = 31 \quad \boxed{\text{Careful!}}$$

$$2(x^2 - 4x + 4) - 3(y^2 + 6y + 9) = 31 + 2(4) - 3(9)$$

$$2(x - 2)^2 - 3(y + 3)^2 = 12$$

Then, divide both sides by 12 and identify the important quantities from the resulting standard form:

$$\boxed{h = 2} \qquad \boxed{k = -3}$$

Transverse axis parallel to x-axis $\longrightarrow \dfrac{(x - 2)^2}{6} - \dfrac{(y + 3)^2}{4} = 1$

$$\boxed{a = \sqrt{6}} \qquad \boxed{b = 2}$$

The center is at $(2, -3)$. The vertices are at $(2 \pm \sqrt{6}, -3)$ or approximately $(4.4, -3)$ and $(-0.4, -3)$.

Finally, sketch the graph as shown in Fig. 20-65.

Figure 20-65

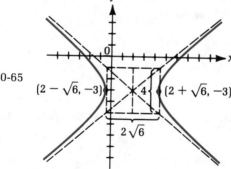

Finding the equation   As we did with the other conic sections, we may use the geometric properties of the hyperbola to determine its equation.

EXAMPLE 41   Determine the equation of each hyperbola using the given properties.

(a)   Vertex at (6, 0), one focus at (10, 0), center at the origin
(b)   Foci at (2, 5) and (2, −3), length of conjugate axis = 6
(c)   Vertices at (1, −8) and (1, 2), passing through (−5, 7)

Solution   For each problem we must find h, k, a^2, and b^2, and determine the transverse axis.

(a)   From the center we know that $h = 0$ and $k = 0$.

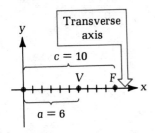

Figure 20-66

From the vertex and focus we know that $a = 6$ and $c = 10$. Therefore, $a^2 = 36$, $c^2 = 100$, and

$$b^2 = c^2 - a^2$$

$$= 100 - 36 = 64$$

Because the vertices and foci are on the x-axis (Fig. 20-66), the transverse axis is also on the x-axis, and the x^2-term is positive. The equation is

$$\frac{x^2}{36} - \frac{y^2}{64} = 1$$

(b) The center of the hyperbola is halfway between the foci at $(2, 1)$, and thus $h = 2$ and $k = 1$ (Fig. 20-67).

The distance between the foci is $2c = |-3 - 5| = 8$, so $c = 4$ and $c^2 = 16$.

The length of the conjugate axis is $2b = 6$, so $b = 3$ and $b^2 = 9$. Therefore,

$$a^2 = c^2 - b^2$$

$$= 16 - 9 = 7$$

The transverse axis is parallel to the y-axis, and the equation is

$$\frac{(y - 2)^2}{7} - \frac{(x - 1)^2}{9} = 1$$

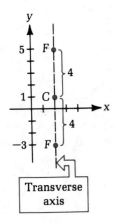

Figure 20-67

(c) The center is halfway between the vertices at $(1, -3)$, and thus $h = 1$ and $k = -3$ (Fig. 20-68).

The distance between the vertices is $2a = 10$, so $a = 5$ and $a^2 = 25$.

The transverse axis is parallel to the y-axis, so the equation is in the form

$$\frac{(y + 3)^2}{25} - \frac{(x - 1)^2}{b^2} = 1$$

The ordered pair $(-5, 7)$ must satisfy the equation, so substitute $x = -5$ and $y = 7$ to find b^2:

$$\frac{(7 + 3)^2}{25} - \frac{(-5 - 1)^2}{b^2} = 1$$

$$4 - \frac{36}{b^2} = 1$$

$$\frac{-36}{b^2} = -3$$

$$b^2 = 12$$

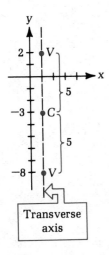

Figure 20-68

The equation is

$$\frac{(y + 3)^2}{25} - \frac{(x - 1)^2}{12} = 1$$

Derivation of the asymptote equations

Earlier, we stated without proof that the asymptotes of the hyperbola

$$\frac{x^2}{a^2} - \frac{y^2}{b^2} = 1$$

are the lines

$$y = \frac{bx}{a} \quad \text{and} \quad y = -\frac{bx}{a}$$

To prove this, solve the hyperbola equation for y:

$$\frac{y^2}{b^2} = \frac{x^2}{a^2} - 1$$

$$y^2 = \frac{x^2 b^2}{a^2} - b^2$$

$$y = \pm\sqrt{\frac{b^2(x^2 - a^2)}{a^2}} = \pm\frac{b}{a}\sqrt{x^2 - a^2}$$

The asymptotes of this type of hyperbola are lines that the curve approaches more and more closely as x becomes much larger than a. We can see that when x is much larger than a, the quantity $\sqrt{x^2 - a^2}$ is approximately equal to x, and thus the equation for y can be rewritten

$$y = \pm\frac{bx}{a} \quad \text{as } x \text{ becomes very large}$$

Similarly, if we solve for y in the equation

$$\frac{y^2}{a^2} - \frac{x^2}{b^2} = 1$$

we find the asymptotes to be

$$y = \pm\frac{ax}{b}$$

Also, note that if the center of the hyperbola is at (h, k), the corresponding asymptote equations are

$$(y - k) = \pm\frac{b}{a}(x - h) \quad \text{and} \quad (y - k) = \pm\frac{a}{b}(x - h)$$

Exercises 20-6

Determine the coordinates of the center, vertices, and foci, and sketch the graph of each hyperbola. Round to the nearest tenth if necessary.

1. $\dfrac{x^2}{16} - \dfrac{y^2}{9} = 1$

2. $\dfrac{x^2}{25} - \dfrac{y^2}{49} = 1$

3. $\dfrac{y^2}{64} - \dfrac{x^2}{40} = 1$

4. $y^2 - \dfrac{x^2}{4} = 1$

5. $x^2 - \dfrac{y^2}{9} = 1$

6. $\dfrac{x^2}{28} - \dfrac{y^2}{18} = 1$

7. $\dfrac{(x-2)^2}{49} - \dfrac{(y-5)^2}{100} = 1$ **8.** $\dfrac{(x+3)^2}{36} - \dfrac{(y+3)^2}{16} = 1$ **9.** $\dfrac{(y+1)^2}{9} - (x-4)^2 = 1$

10. $\dfrac{(y-6)^2}{121} - \dfrac{(x+4)^2}{81} = 1$ **11.** $\dfrac{(x-3)^2}{12} - \dfrac{y^2}{25} = 1$ **12.** $\dfrac{(x-1)^2}{3} - (y-2)^2 = 1$

13. $\dfrac{(y-7)^2}{6} - \dfrac{(x+1)^2}{6} = 1$ **14.** $\dfrac{y^2}{8} - \dfrac{(x+4)^2}{8} = 1$

Determine the coordinates of the center and vertices, and sketch the graph of each hyperbola.

15. $xy = 4$ **16.** $xy = -9$ **17.** $xy = -2$
18. $xy = 8$ **19.** $y^2 - 4x^2 = 16$ **20.** $9x^2 - y^2 = 81$
21. $72x^2 - 25y^2 = 200$ **22.** $4y^2 - 3x^2 = 18$
23. $25y^2 - 300y - 9x^2 + 36x + 639 = 0$ **24.** $4x^2 - y^2 - 8y - 32 = 0$
25. $2x^2 - 12x - 5y^2 - 20y - 22 = 0$ **26.** $2y^2 + 4y - x^2 + 10x - 25 = 0$
27. $2x^2 + 4x - 2y^2 + 10y + 2 = 0$ **28.** $3y^2 + 9y - 3x^2 - 18x - 27 = 0$

Determine the equation of each hyperbola using the given properties.

29. Center $(0, 0)$, vertex $(0, 5)$, focus $(0, 7)$
30. Center $(2, -6)$, vertex $(8, -6)$, focus $(11, -6)$
31. Vertices $(-9, 3)$ and $(1, 3)$, length of conjugate axis = 14
32. Foci $(-3, 7)$ and $(-3, -1)$, length of conjugate axis = $\sqrt{5}$
33. Foci $\left(-2, 3\sqrt{3}\right)$ and $\left(-2, -\sqrt{3}\right)$, length of transverse axis = 6
34. Center $(0, 0)$, vertex $\left(\sqrt{7}, 0\right)$, length of conjugate axis = 10
35. Center $(-4, 0)$, vertex $(-4, -4)$, passes through $\left(2, 3\sqrt{5}\right)$
36. Vertices $(5, 2)$ and $(-5, 2)$, passes through $\left(10\sqrt{3}, 12\right)$

Solve.

37. *Thermodynamics* The ideal gas law states that for a constant temperature, the pressure P and volume V of an ideal gas satisfy the equation $PV = K$, where K is a constant. At room temperature, a certain gas has a pressure of 250 kPa when the volume is 4.8 L. Sketch a graph of pressure as a function of volume for the constant temperature.

38. *Electronics* For any given voltage the product of the current and resistance in a simple circuit is constant. If the current in a circuit is 1.5 A when the resistance is 8.0 Ω, sketch the graph of current as a function of resistance for this value of the voltage.

39. The area of a rectangle is given by $A = LW$, where L is the length and W is the width. If a warehouse must contain 24,000 ft^2 of floor space, sketch the graph of possible values of L and W.

40. For a pipe of length L, outside diameter D, and inside diameter d, the interior volume V of the pipe is given by

$$V = \frac{\pi L}{4}(D^2 - d^2)$$

Sketch a graph of the possible values of D as a function of d for $V = 40.0$ ft^3 and $L = 20.0$ ft.

20-7 | Second-Degree Equations

In previous sections we discussed equations of the conic sections—circle, parabola, ellipse, and hyperbola—with axes parallel to the coordinate axes. Each type of curve has a standard form that provides readily accessible information useful in graphing. We have also seen that each curve can be written as follows:

General Form of a Second-Degree Equation

$$Ax^2 + Bxy + Cy^2 + Dx + Ey + F = 0 \qquad \text{(20-29)}$$

where A and C are not both 0.

In this section we will show how to use the coefficients of (20-29) to determine the type of curve represented by the equation. In addition, we will examine briefly the degenerate forms of the second-degree equation and the rotation of axes.

Identifying curves in general form

To understand the connection between the coefficients of (20-29) and the type of curve it represents, we must first transform an example of the standard form of each curve into general form.

EXAMPLE 42

(a) The equation

$$(x - 2)^2 + (y + 5)^2 = 9$$

represents a circle. Its equivalent general form is

$$x^2 + y^2 - 4x + 10y + 20 = 0$$

(b) The equations

$$(y - 3)^2 = 8(x + 2) \qquad \text{and} \qquad (x - 3)^2 = 8(y + 2)$$

represent parabolas. Their equivalent general forms are

$$y^2 - 6y - 8x - 7 = 0 \qquad \text{and} \qquad x^2 - 6x - 8y - 7 = 0$$

(c) The equation

$$\frac{(x + 2)^2}{4} + \frac{(y + 1)^2}{9} = 1$$

represents an ellipse. Its equivalent general form is

$$9x^2 + 4y^2 + 36x + 8y + 4 = 0$$

(d) The equation

$$\frac{(x + 2)^2}{4} - \frac{(y + 1)^2}{9} = 1$$

represents a hyperbola. Its equivalent general form is

$$9x^2 - 4y^2 + 36x - 8y - 4 = 0$$

Notice that all these equations are general second-degree equations in the form of (20-29) with $B = 0$. They contain no xy-term. If a conic section has axes parallel to the x- and y-axes, then its general equation will have $B = 0$ and the following statements are true with respect to equation (20-29):

- If $A = C$, ⟹ the graph is a **circle**.
- If either A or C is 0, but not both, ⟹ the graph is a **parabola**.
- If A and C are not equal but have the same sign, ⟹ the graph is an **ellipse**.
- If A and C have opposite signs, ⟹ the graph is a **hyperbola**.

EXAMPLE 43   Identify each of the following curves:

(a) $2x^2 - y^2 + 6x - 8 = 0$ (b) $5y^2 - 17y = 8 - 6x^2$

(c) $2x(x + 5) + y(2y - 3) = 7$ (d) $(y - 2)^2 = 5x - x^2 + y^2$

Solution

(a)   From equation (20-29), we note that $B = 0$ and that A and C have opposite signs. This is a hyperbola.

(b)   Moving all terms to the left, we have $5y^2 + 6x^2 - 17y - 8 = 0$. Since $A \neq C$ but they have the same sign, the curve is an ellipse.

(c)   Multiplying, we have

$$2x^2 + 10x + 2y^2 - 3y = 7$$

or, in general form,

$$2x^2 + 2y^2 + 10x - 3y - 7 = 0$$

Since A and C are equal, this is a circle.

(d)   Multiplying, we have

$$y^2 - 4y + 4 = 5x - x^2 + y^2$$

Moving all terms to the left,

$$x^2 - 5x - 4y + 4 = 0$$

Since $C = 0$, this is a parabola.

Degenerate forms   It is possible to write a general second-degree equation whose graph is not a conic section. For example,

$x^2 + y^2 + 4x - 6y + 12 = 0$ produces a circle $(x + 2)^2 + (y - 3)^2 = 1$

but

$x^2 + y^2 + 4x - 6y + 13 = 0$ gives a point $(x + 2)^2 + (y - 3)^2 = 0$

$x^2 + y^2 + 4x - 6y + 14 = 0$ has no graph at all $(x + 2)^2 + (y - 3)^2 = -1$

Second-degree equations that give a point, line, pair of lines, or no graph at all are said to be **degenerate forms.** Each degenerate form is easily recognizable when the equation is in standard form. Compare the equations of the following degenerate forms to the equations of the actual curves:

- $\dfrac{(x - h)^2}{a^2} + \dfrac{(y - k)^2}{b^2} = 1$ An ellipse that becomes a circle for $a = b$

  $\dfrac{(x - h)^2}{a^2} + \dfrac{(y - k)^2}{b^2} = -1$ An imaginary ellipse or imaginary circle—there is no graph

  $\dfrac{(x - h)^2}{a^2} + \dfrac{(y - k)^2}{b^2} = 0$ A point (h, k)

- $\dfrac{(x - h)^2}{a^2} - \dfrac{(y - k)^2}{b^2} = \pm 1$ A hyperbola

  $\dfrac{(x - h)^2}{a^2} - \dfrac{(y - k)^2}{b^2} = 0$ A point (h, k)

- $(y - k)^2 = 4p(x - h)$

  and

  $(x - h)^2 = 4p(y - k)$ Parabolas

  $(y - k)^2 = a^2$

  and

  $(x - h)^2 = a^2$ Each produces a pair of parallel lines, $y = k \pm a$ or $x = k \pm a$

  $(y - k)^2 = -a^2$

  and

  $(x - h)^2 = -a^2$ Each results in a pair of imaginary parallel lines—no graph

  $(y - k)^2 = 0$

  and

  $(x - h)^2 = 0$ Each produces a line, $y = k$ or $x = h$

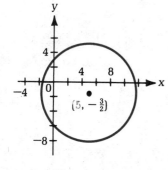

Figure 20-69

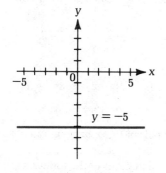

Figure 20-70

EXAMPLE 44   Identify each of the following curves and sketch their graphs if they exist:

(a) $(x - 1)^2 = 8x + 18 - y(y + 3)$ (b) $y^2 + 10y = -25$

(c) $3x^2 + 2y^2 - 6x - 8y + 23 = 0$ (d) $3x(x + 12) - 2y(y + 4) = -94$

Solution

(a)   In general form this becomes $x^2 + y^2 - 10x + 3y - 17 = 0$. Since $B = 0$ and A and C are equal and have the same sign, this appears to be a circle. Completing the squares, we obtain

$$(x - 5)^2 + \left(y + \frac{3}{2}\right)^2 = \frac{177}{4}$$

which is a circle with center at $\left(5, -\dfrac{3}{2}\right)$ and radius $r = \sqrt{177/4} \approx 6.7$. The graph is shown in Fig. 20-69.

(b)   Here $A = B = 0$ and $C \neq 0$, so this appears to be a parabola. However, if we complete the square, we obtain

$$(y + 5)^2 = 0$$

which is a degenerate form of a parabola, namely the line $y = -5$. The graph is shown in Fig. 20-70.

(c) Here, $B = 0$, $A \neq C$, and A and C have the same sign, so this appears to be an ellipse. Completing the square, the equation becomes

$$\frac{(x - 1)^2}{4} + \frac{(y - 2)^2}{6} = -1$$

Notice that both terms on the left are positive for all real values of x and y, but the right member is negative. There are no real numbers x and y that satisfy the equation. This is a degenerate form of the equation that has no graph.

(d) In general form this becomes $3x^2 - 2y^2 + 36x - 8y + 94 = 0$. Since $B = 0$ and A and C have opposite signs, this appears to be a hyperbola. Completing the squares, we obtain

$$\frac{(x + 6)^2}{2} - \frac{(y + 2)^2}{3} = 1$$

which is a hyperbola centered at $(-6, -2)$. The vertices are at $(-6 \pm \sqrt{2}, -2)$ or approximately $(-4.6, -2)$ and $(-7.4, -2)$. The graph is shown in Fig. 20-71.

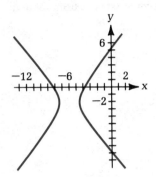

Figure 20-71

Rotation of axes   When $B \neq 0$, equation (20-29) will, in general, represent a parabola, ellipse, or hyperbola whose axis has been rotated. We can identify the graph and the angle of rotation from the coefficients.

- If $B^2 - 4AC = 0$,   it is a **parabola.**
- If $B^2 - 4AC < 0$,   it is an **ellipse.**
- If $B^2 - 4AC > 0$,   it is a **hyperbola.**

The angle of rotation θ can be found from the equation

$$\cot 2\theta = \frac{A - C}{B}$$

Again, the graph may be a degenerate form—that is, a point, line, pair of lines, or no graph at all.

EXAMPLE 45   Identify each of the following, assuming that the graph exists:
(a) $x^2 + 12xy - 6y^2 = 30$ (b) $x^2 + y^2 - 2xy - 2x - 2y = 0$
(c) $x^2 - xy + 3y^2 + 2x = 0$

Solution
(a)   We have $A = 1$, $B = 12$, and $C = -6$, so

$$B^2 - 4AC = 12^2 - 4(1)(-6)$$

$$= 144 + 24 = 168 > 0$$

$$\cot 2\theta = \frac{A - C}{B}$$

$$= \frac{1 + 6}{12} = \frac{7}{12}$$

so $\tan 2\theta = \dfrac{12}{7}$

$$2\theta \approx 59.74°$$

$$\theta \approx 29.87°$$

The equation is that of a hyperbola whose transverse axis has been rotated through an angle of approximately 30°.

(b)   We have $A = 1$, $B = -2$, and $C = 1$, so

$$B^2 - 4AC = 4 - 4 \cdot 1 \cdot 1 = 0$$

$$\cot 2\theta = \frac{A - C}{B}$$

$$= \frac{1 - 1}{-2} = 0 \qquad \text{so} \qquad 2\theta = 90°$$

$$\theta = 45°$$

The equation is that of a parabola rotated 45°.

(c)   Here, $A = 1$, $B = -1$, and $C = 3$, so

$$B^2 - 4AC = 1 - 4 \cdot 1 \cdot 3 = -11 < 0$$

$$\cot 2\theta = \frac{A - C}{B}$$

$$= \frac{1 - 3}{-1} = \frac{-2}{-1} = 2 \qquad \text{so} \qquad \tan 2\theta = \frac{1}{2}$$

$$2\theta \approx 26.56°$$

$$\theta \approx 13.28°$$

The equation is that of an ellipse whose major axis has been rotated about 13°.

The derivation of the rotation formula and the techniques for graphing rotated curves are beyond the scope of this text.

Exercises 20-7

Identify each equation as representing a circle, parabola, ellipse, or hyperbola. There are no degenerate forms represented, so you do not have to rewrite the equation in standard form.

1. $x^2 + 6x + y^2 - 5y + 12 = 0$
2. $4x^2 + 9y^2 - 18y + 4 = 0$
3. $2x^2 - 14x - y^2 - 6 = 0$
4. $3x^2 + 9x + 3y^2 - 4 = 0$
5. $x^2 - 4x + 2y^2 - 12y - 3 = 0$
6. $y^2 - 8y - 6x - 4 = 0$
7. $x^2 + 6x + 8y - 2 = 0$
8. $2x^2 - 4x - 2y^2 + 6x = 1$
9. $2x(x - 5) + y(2y + 7) = 2$
10. $3x(x + 2) - 2y(5 - y) = 8$
11. $x(2x - 1) - y(4y + 3) = 6$
12. $3y(y - 5) = 12x$

13. $4y^2 - 6x^2 - 12 = 2y(2y - 3)$ **14.** $(x + 4)^2 = 2y - y^2$
15. $(x - 5)^2 = 9 - (x^2 + y^2)$ **16.** $2(y - 3)^2 = (x + 4)^2$

Determine the coordinates of the center (or vertex if the curve is a parabola) and sketch the graph of each equation. Identify any degenerate forms.

17. $x^2 - 4y^2 + 24y - 40 = 0$ **18.** $x^2 - 4x - 4y + 8 = 0$
19. $9x^2 - 36x + y^2 + 6y + 126 = 0$ **20.** $x^2 + 10x + y^2 + 4y + 29 = 0$
21. $x^2 + 2x + 5 = 0$ **22.** $x^2 + 16y^2 - 160y + 384 = 0$
23. $2x^2 - 12x + 2y^2 + 24y + 71 = 0$ **24.** $3x^2 + 18x - 2y^2 + 32y - 101 = 0$

20-8 | Polar Coordinates

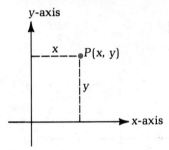

Figure 20-72

Up to now we have used the rectangular, or xy-coordinate, system for all graphing of real functions in a plane. However, other coordinate systems exist and are useful for certain applications. In this section and the next, we shall consider one such alternative system—the *polar coordinate system.*

As we have seen, a rectangular coordinate system is created by two fixed lines, the x- and y-axes, intersecting at right angles in a plane. The location of any point P in the plane is determined by an ordered pair (x, y) giving the directed distance from each axis to the point P (Fig. 20-72).

To create a **polar coordinate system** in a plane, we begin with a half-line called the **polar axis,** which has an endpoint O called the **pole** or **origin** [Fig. 20-73(a)]. In this system, the location of any point P is specified by the distance from O to P and by the angle formed by the line OP and the polar axis. If r represents the distance and θ represents the angle, then the **polar coordinates** of P can be stated as the ordered pair **(r, θ)** [Fig. 20-73(b)].

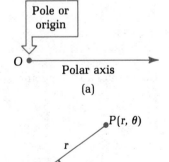

Figure 20-73

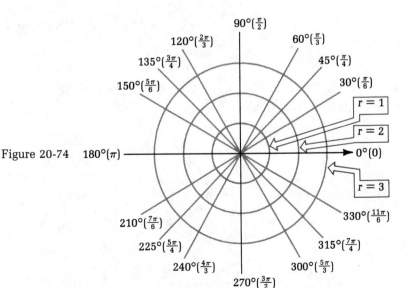

Figure 20-74

Plotting points   To facilitate plotting points in a polar coordinate system, it is convenient either to make or to obtain polar graph paper similar to that shown in Fig. 20-74. The lines extending from the origin represent the terminal sides of various values of θ, and the circles centered about the origin represent different values of r.

To plot the point $P(r, \theta)$, first measure the angle θ. Then measure the distance r along its terminal side. As in trigonometry, *positive* values of θ are measured *counterclockwise* from the polar axis, and *negative* values of θ are measured in a *clockwise* direction.

EXAMPLE 46   Plot the following points in the polar coordinate system:

(a) $(1, 45°)$ (b) $\left(2, \dfrac{5\pi}{6}\right)$ ▦ (c) $(2.8, -2.5)$

Solution

(a)
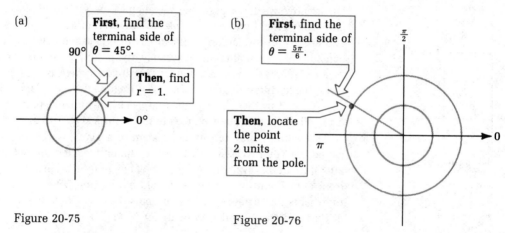

First, find the terminal side of $\theta = 45°$.

Then, find $r = 1$.

Figure 20-75

(b)

First, find the terminal side of $\theta = \frac{5\pi}{6}$.

Then, locate the point 2 units from the pole.

Figure 20-76

(c)   To measure $\theta = -2.5$, first convert -2.5 rad to degrees:

$$2.5 \;\boxed{+/-}\;\boxed{\times}\; 180 \;\boxed{\div}\;\boxed{\pi}\;\boxed{=}\; \rightarrow\quad \boxed{-143.23945}$$

Since θ is negative, we measure 143° *clockwise* from θ, and find the point at a radius of 2.8 on this axis. (Fig. 20-77).

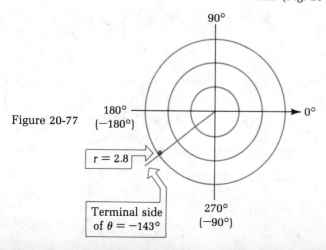

Figure 20-77

$r = 2.8$

Terminal side of $\theta = -143°$

Negative values of r   When r is *negative*, the point (r, θ) is located on the *opposite* side of the pole from θ on the extension of its terminal side.

EXAMPLE 47   Plot the following points in the polar coordinate system:

(a) $\left(-2, \dfrac{\pi}{3}\right)$ (b) $(-3, -250°)$

Solution

(a)

Figure 20-78

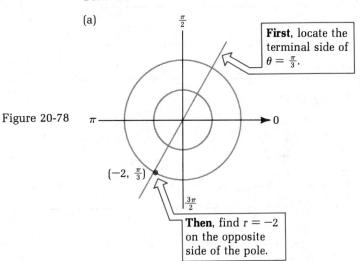

First, locate the terminal side of $\theta = \frac{\pi}{3}$.

$(-2, \frac{\pi}{3})$

Then, find $r = -2$ on the opposite side of the pole.

(b)

Figure 20-79

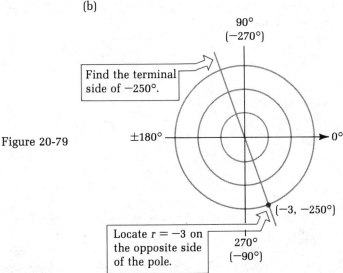

Find the terminal side of $-250°$.

Locate $r = -3$ on the opposite side of the pole.

$(-3, -250°)$

NOTE ▶   Since r may be negative, and since θ may be replaced by any angle coterminal to θ, there are an unlimited number of polar coordinates corresponding to each point in the coordinate system. For example, the point $(-2, \pi/3)$ in Fig. 20-78 could also be given as $(2, 4\pi/3)$, $(2, 10\pi/3)$, and so on. Similarly, the point $(-3, -250°)$ in Fig. 20-79 could be given as $(-3, 110°)$, $(3, 290°)$, and so on. ◀

Relationship between polar and rectangular coordinates

It is often useful to relate polar coordinates to rectangular coordinates. If we superimpose one system on the other so that the pole coincides with the origin and the polar axis coincides with the positive x-axis, we can use the trigonometric functions to state the following relationships (see Fig. 20-80):*

$$x = r \cos \theta \qquad y = r \sin \theta \qquad\qquad (20\text{-}30)$$

Or, solving for r and θ, we obtain:

$$r^2 = x^2 + y^2$$

$$r = \sqrt{x^2 + y^2}$$

$$\tan (\text{Ref } \theta) = \left| \frac{y}{x} \right| \qquad\qquad (20\text{-}31)$$

Determine θ using the signs of x and y.

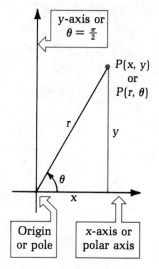

Figure 20-80

Converting: polar to rectangular

First we will use equations (20-30) to convert from polar coordinates to rectangular coordinates.

EXAMPLE 48 Find the rectangular coordinates corresponding to each set of polar coordinates:

(a) $\left(3, \dfrac{\pi}{3} \right)$ ▦ (b) $(-2.1, 5.4)$ ▦ (c) $(2, 235°)$

Solution
(a)   Using (20-30), we have

$$x = 3 \cos \frac{\pi}{3} = 3\left(\frac{1}{2} \right) = \frac{3}{2}$$

$$y = 3 \sin \frac{\pi}{3} = 3\left(\frac{\sqrt{3}}{2} \right) = \frac{3\sqrt{3}}{2}$$

The equivalent rectangular coordinates are $\left(\dfrac{3}{2}, \dfrac{3\sqrt{3}}{2} \right)$, as shown in Fig. 20-81.

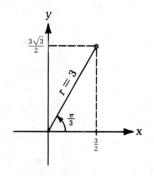

Figure 20-81

(b) $x = -2.1 \cos 5.4$ [RAD] 2.1 [+/-] [×] 5.4 [cos] [=] → *−1.3328550*

$y = -2.1 \sin 5.4$ [RAD] 2.1 [+/-] [×] 5.4 [sin] [=] → *1.6228054*

---

* These relationships are identical to those used in Section 12-4 to define the polar form of a complex number.

The equivalent rectangular coordinates are approximately $(-1.3, 1.6)$, as shown in Fig. 20-82.

(c) $x = 2 \cos 235°$ $\boxed{\text{DEG}}\ 2\ \boxed{\times}\ 235\ \boxed{\cos}\ \boxed{=} \rightarrow$ *−1.1471529*

   $y = 2 \sin 235°$ $\boxed{\text{DEG}}\ 2\ \boxed{\times}\ 235\ \boxed{\sin}\ \boxed{=} \rightarrow$ *−1.6383041*

The equivalent rectangular coordinates are approximately $(-1.1, -1.6)$, as shown in Fig. 20-83.

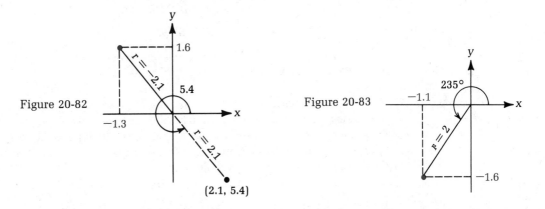

Figure 20-82

(2.1, 5.4)

Figure 20-83

Converting:
rectangular to polar   Now we will use equations (20-31) to convert from rectangular coordinates to polar coordinates.

EXAMPLE 49   Find a set of polar coordinates equivalent to each set of rectangular coordinates:

(a) $(\sqrt{3}, -1)$ (θ in radians, $0 \le \theta < 2\pi$)

(b) $(-3, -7)$ (θ in degrees, $0° \le \theta < 360°$)

Solution

(a) From (20-31), we have

$$r = \sqrt{(\sqrt{3})^2 + (-1)^2} = \sqrt{4} = 2$$

and

$$\tan (\text{Ref } \theta) = \left| \frac{-1}{\sqrt{3}} \right| = \frac{\sqrt{3}}{3} \qquad \text{so} \qquad \text{Ref } \theta = \frac{\pi}{6} \text{ rad}$$

Since x is positive and y is negative, θ is located in quadrant IV:

$$\theta = 2\pi - \frac{\pi}{6} = \frac{11\pi}{6} \text{ rad}$$

The equivalent polar coordinates for the point $(\sqrt{3}, -1)$ are $(2, 11\pi/6)$ as shown in Fig. 20-84.

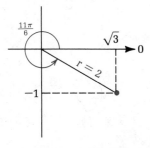

Figure 20-84

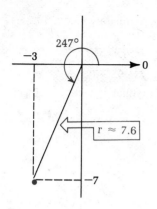

Figure 20-85

(b)   From (20-31), we obtain

$$r = \sqrt{(-3)^2 + (-7)^2} = \sqrt{58}$$

58 $\boxed{\sqrt{}}$ → ▨ 7.6157731 ▨ $r \approx 7.6$

Also,

$$\tan (\text{Ref } \theta) = \left| \frac{-7}{-3} \right| = \frac{7}{3}$$ $\boxed{\text{DEG}}$ 7 $\boxed{\div}$ 3 $\boxed{=}$ $\boxed{\text{INV}}$ $\boxed{\tan}$ $\boxed{\text{STO}}$ → ▨ 66.801409 ▨

Since x and y are both negative, θ is in quadrant III:

$\boxed{\text{RCL}}$ $\boxed{+}$ 180 $\boxed{=}$ → ▨ 246.80141 ▨

Thus, $\theta \approx 247°$ and the polar coordinates of the point $(-3, -7)$ are approximately $(7.6, 247°)$, as shown in Fig. 20-85.

Converting equations   Just as the coordinates of a point may be given in either system, equations also can be written and graphed in both rectangular and polar forms. To convert equations, we must use formulas (20-30) and (20-31) and the basic trigonometric identities. An equation in polar form usually is written explicitly with r or a power of r isolated on one side.

EXAMPLE 50   Find the polar equation of:
(a) $x^2 + y^2 = 16$ (b) $x^2 = 2y$

Solution

(a)   From (20-31), we know that: $x^2 + y^2 = r^2$

Substituting, we have: $r^2 = 16$

Notice that the polar form of this equation does not contain the variable θ.

(b)   From (20-30), we have $y = r \sin \theta$ and $x = r \cos \theta$.

Substituting, we see that the given equation: $x^2 = 2y$

is equivalent to: $r^2 \cos^2 \theta = 2r \sin \theta$

Dividing both sides by r: $r \cos^2 \theta = 2 \sin \theta$

Solving for r, we obtain: $r = \dfrac{2 \sin \theta}{\cos^2 \theta}$

EXAMPLE 51   Find the rectangular form of these equations:
(a) $r = 2a \sin \theta$ (b) $r^2 = 4 \sin 2\theta$

Solution

(a)   By definition, $\sin \theta = y/r$. Therefore: $r = 2a\left(\dfrac{y}{r}\right)$

$$r^2 = 2ay$$

Substituting for r^2, we obtain: $x^2 + y^2 = 2ay$

(b) **Step 1.** Equations (20-30) are given in terms of θ; therefore, we use the identity $\sin 2\theta = 2 \sin \theta \cos \theta$ to replace $\sin 2\theta$:

$$r^2 = 4(2 \sin \theta \cos \theta) = 8 \sin \theta \cos \theta$$

Step 2. Substitute for r^2, $\sin \theta$, and $\cos \theta$:

$$x^2 + y^2 = 8\left(\frac{y}{r}\right)\left(\frac{x}{r}\right) = \frac{8xy}{r^2} = \frac{8xy}{x^2 + y^2}$$

Step 3. Multiply by $(x^2 + y^2)$:

$$(x^2 + y^2)^2 = 8xy$$

Exercises 20-8

Plot each point on polar graph paper.

1. $(2, 30°)$ **2.** $(3, 135°)$ **3.** $(-1, 60°)$ **4.** $(-4, -15°)$

5. $\left(1, \dfrac{\pi}{2}\right)$ **6.** $\left(4, \dfrac{2\pi}{3}\right)$ **7.** $\left(-3, \dfrac{7\pi}{6}\right)$ **8.** $\left(-2, \dfrac{5\pi}{3}\right)$

9. $\left(-\dfrac{7}{2}, \dfrac{3\pi}{4}\right)$ **10.** $(2.5, \pi)$ **11.** $(3.5, 1.3)$ **12.** $(-2.6, 0.4)$

Find a pair of polar coordinates equivalent to each pair of rectangular coordinates. For Problems 13–16 express θ in radians, $0 \le \theta < 2\pi$, and for Problems 17–20 express θ in degrees, $0° \le \theta < 360°$.

13. $\left(\sqrt{3}, 1\right)$ **14.** $(1, -1)$ **15.** $(-3, 2)$ **16.** $(-2, -5)$
17. $(-2, -2)$ **18.** $(-4, 0)$ **19.** $(4, -3)$ **20.** $(4, 6)$

Find the rectangular coordinates corresponding to each set of polar coordinates.

21. $\left(2, \dfrac{\pi}{4}\right)$ **22.** $\left(-3, -\dfrac{\pi}{3}\right)$ **23.** $(4.3, -2.8)$ **24.** $(-1.7, 4.6)$

25. $\left(-2, -\dfrac{\pi}{2}\right)$ **26.** $(5, \pi)$ **27.** $(-3.5, 37°)$ **28.** $(6.1, 240°)$

Find the polar equation equivalent to each rectangular equation.

29. $x = 4$ **30.** $y = -3$ **31.** $x^2 + y^2 = 25$ **32.** $3x^2 + 3y^2 = 40$
33. $y^2 = 4x$ **34.** $y = 2x^2$ **35.** $x^2 + 2y^2 = 9$ **36.** $4x^2 - y^2 = 1$

Find the rectangular form of each polar equation.

37. $r = 3 \sin \theta$ **38.** $r = 2 \cos \theta$ **39.** $r \cos \theta = -6$

40. $r \sin \theta = 3$ **41.** $r = 3 + \sin \theta$ **42.** $r = \dfrac{1}{4 - \cos \theta}$

43. $r = \cos 2\theta$ **44.** $r^2 = 9 \sin 2\theta$

Solve.

45. *Architecture*   In the design of a geodesic dome, an architect uses the equation

$$x^2 + \frac{y^2}{E^2} = 1$$

where E is a constant. Write the polar form of this equation.*

46. *Physics*   The maximum speed of a pendulum bob is given by

$$v^2 = 64L(1 - \cos \theta)$$

This equation is now in polar form where v represents r. Express it in rectangular form for $L = 0.5$ ft.

47. *Electrical engineering*   The equation $r = 2 \cos^2 \theta$ describes the intensity of radiation from a radar antenna. Express this equation in rectangular form.

48. *Optics*   Express the equation of the parabolic reflector described in Problem 37, Exercises 20-4, in polar form.

20-9 | Graphs in Polar Coordinates

Many curves that are useful in technology are easier to plot using the polar form than the rectangular form of their equations. To graph an equation given in polar form, first prepare a table of ordered pairs (r, θ) by selecting values of θ and calculating the corresponding values of r. Then plot the points in the polar coordinate system and connect them with a smooth curve.

EXAMPLE 52   Plot the graph of $r = 4 \sin \theta$.

Solution   **First,** prepare a table of values by choosing values of θ and calculating the corresponding values of r:

θ	0°	30°	45°	60°	90°	120°	135°	150°	180°
r	0	2.0	2.8	3.5	4.0	3.5	2.8	2.0	0

Notice that values of θ, $180° < \theta \le 360°$, will simply duplicate these points. For example, for $\theta = 210°$, $r = -2$, and the point $(-2, 210°)$ is the same as the point $(2, 30°)$.

Then, plot these ordered pairs and connect them with a smooth curve (Fig. 20-86). The graph is a circle with radius 2 and center at $(2, 90°)$.

In general, polar equations of the form

$$r = \pm 2a \sin \theta \quad \text{or} \quad r = \pm 2a \cos \theta$$

represent circles of radius a with centers at $(a, \pm 90°)$ for the first equation, and

---

* For more information, see Hugh Kenner, *Geodesic Math and How to Use It* (Berkeley: University of California Press, 1976), p. 85.

at $(a, 0)$ or $(a, 180°)$ for the second equation. An equation of the form

$r = a$

represents a circle with center at the pole $(0, 0°)$.

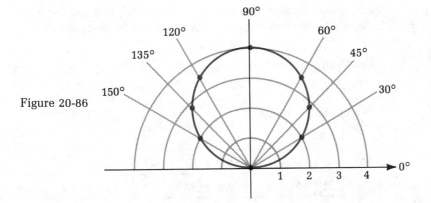

Figure 20-86

EXAMPLE 53 Plot the graph of $r = 2 + \sin \theta$.

Solution   **First,** prepare a table of values:

θ	0°	30°	45°	90°	135°	150°	180°	210°	225°	270°	315°	330°
r	2.0	2.5	2.7	3.0	2.7	2.5	2.0	1.5	1.3	1.0	1.3	1.5

Then, plot the ordered pairs and connect them with a smooth curve (Fig. 20-87). This type of curve is known as a *limaçon,* (from the Latin word for *snail*).

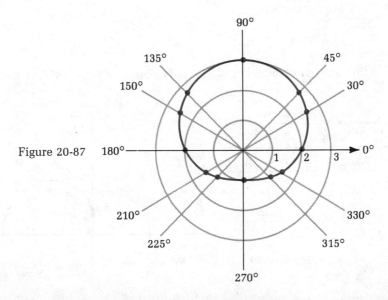

Figure 20-87

A **limaçon** is a polar curve whose equation has the form

$$r = \pm a \pm b \sin\theta \quad \text{or} \quad r = \pm a \pm b \cos\theta$$

where a and b are both greater than 0. When $a > b$, we have the smooth curve shown in Example 53. When $a = b$, the graph has a sharp point or cusp, and because of its heart shape, it is called a **cardioid**. When $a < b$, the graph has a loop. The last two cases are shown in the next example.

EXAMPLE 54   Plot the graph of:
(a) $r = 1 - \cos\theta$ (b) $r = 1 + 2\cos\theta$

Solution

(a)

θ	0°	30°	45°	60°	90°	120°	135°	150°	180°
r	0	0.1	0.3	0.5	1.0	1.5	1.7	1.9	2.0

The lower half of the curve can be found from symmetry or by plotting the points with $180° < \theta \leq 360°$ (see Fig. 20-88).

Figure 20-88

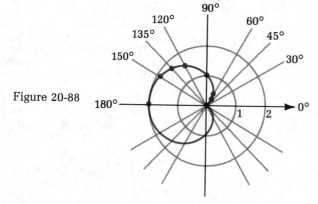

(b)   The graph should be plotted in four steps, as shown in Figs. 20-89–20-92.

For 0° ≤ θ ≤ 120°:

θ	0°	30°	45°	60°	90°	120°
r	3.0	2.7	2.4	2.0	1.0	0

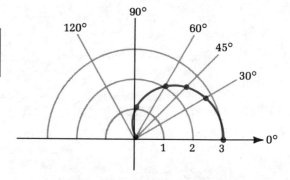

Figure 20-89

For 120° < θ ≤ 180°:

θ	135°	150°	180°
r	−0.4	−0.7	−1.0

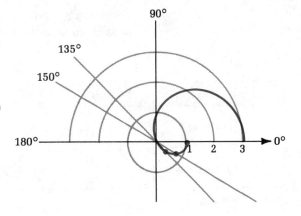

Figure 20-90

For 180° < θ < 270°:

θ	210°	225°	240°
r	−0.7	−0.4	0

Figure 20-91

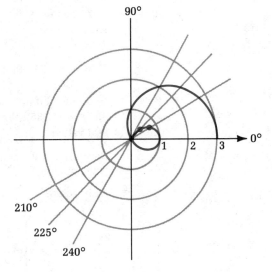

For 270° ≤ θ ≤ 360°:

θ	270°	300°	315°	330°
r	1.0	2.0	2.4	2.7

Figure 20-92

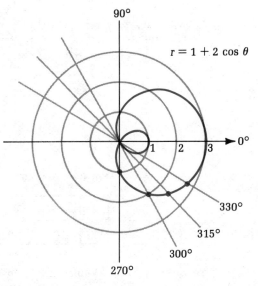

$r = 1 + 2\cos\theta$

The limaçon curves are related to the conic section curves we studied earlier, as outlined in Table 20-4.

TABLE 20-4

Limaçon Equation	Graph		Conic Equation	Graph	
$r = \pm a \pm b \sin \theta$	$a = b$	Cardioid	$r = \dfrac{1}{\pm a \pm b \sin \theta}$	$a = b$	Parabola
	$a > b$	No loop		$a > b$	Ellipse
$r = \pm a \pm b \cos \theta$	$a < b$	Loop	$r = \dfrac{1}{\pm a \pm b \cos \theta}$	$a < b$	Hyperbola

Note: $a > 0$ and $b > 0$.

Lemniscates   A second class of polar equations, the *lemniscates*, produces propeller-shaped curves.

EXAMPLE 55   Plot the graph of $r^2 = 4 \cos 2\theta$ in polar coordinates.

Solution   First, note that $\cos 2\theta$ is negative for $90° < 2\theta < 270°$ or $45° < \theta < 135°$. When $\cos 2\theta$ is negative, r is imaginary and the graph has no real values.

Second, plot the graph for $0° \le \theta \le 45°$ [Fig. 20-93(a)] and for $135° \le \theta \le 180°$ [Fig. 20-93(b)].

Figure 20-93

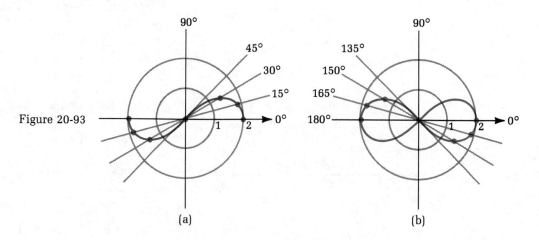

(a)   (b)

For $0° \le \theta \le 45°$:

θ	0°	15°	30°	45°
r	±2	±1.9	±1.4	0

For $135° \le \theta \le 180°$:

θ	135°	150°	165°	180°
r	0	±1.4	±1.9	±2

Values of θ in quadrants III and IV reproduce these points.

The general equation of a **lemniscate** is

$$r^2 = \pm a^2 \sin 2\theta \quad \text{or} \quad r^2 = \pm a^2 \cos 2\theta$$

where a is the maximum value of r. The graph of the reciprocal equation of a lemniscate,

$$r^2 = \frac{1}{\pm a^2 \sin 2\theta} \quad \text{or} \quad r^2 = \frac{1}{\pm a^2 \cos 2\theta}$$

is a hyperbola.

Roses   If you noticed that the circle and limaçon—both one-loop curves—involve $\sin \theta$ and $\cos \theta$, and that the lemniscate—a two-loop curve—involves $\sin 2\theta$ and $\cos 2\theta$, then you may guess that curves involving $\sin n\theta$ and $\cos n\theta$, where n is an integer greater than 1, produce multiloop curves.

Equations of the form

$$r = a \sin n\theta \quad \text{or} \quad r = a \cos n\theta$$

where a is any real number and $n > 1$ represent flower-shaped curves called **roses**. The rose has n equally spaced petals, or loops, if n is odd and $2n$ equally spaced petals if n is even.

EXAMPLE 56   Plot the graph of $r = \sin 3\theta$.

Solution   The sine function has zeros at $0°$ and $180°$; therefore, $\sin 3\theta = 0$ at $\theta = 0°$ and at $\theta = 180°/3 = 60°$. Plot points for $0° \le \theta \le 60°$ [Fig. 20-94(a)]. Because $n = 3$, we know there are three equally spaced petals. The others will be found by plotting the points where $60° \le \theta \le 120°$ and $120° \le \theta \le 180°$ [Fig. 20-94(b)].

Figure 20-94

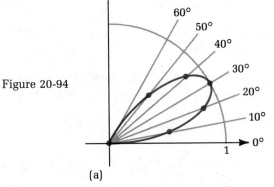

(a)

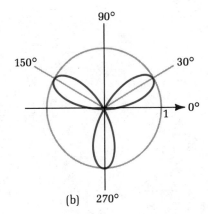

(b) $270°$

Spirals   A **spiral** is a curve that represents the path of a point moving continuously around the origin while steadily receding from it or approaching it. The equation of a spiral has the form $r = a\theta$, where a is a real number and θ is in radians.

EXAMPLE 57   Plot the spiral $r = 2\theta$ in polar coordinates.

Solution   The angle θ must be in radians.

θ	0	$\dfrac{\pi}{4}$	$\dfrac{\pi}{2}$	$\dfrac{3\pi}{4}$	π	$\dfrac{5\pi}{4}$	$\dfrac{3\pi}{2}$	$\dfrac{7\pi}{4}$	2π
r	0	1.6	3.1	4.7	6.3	7.9	9.4	11.0	12.6

The spiral shown in Fig. 20-95 is called the **spiral of Archimedes** (named after the 3rd century BC mathematician who first studied it).

Figure 20-95

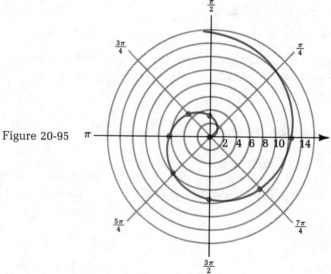

Straight line   A straight line through the polar origin (Fig. 20-96) has the equation

$\theta = a$ where a is any real number

More generally, the straight line $y = mx + b$ can be rewritten in polar form as follows:

$$y = mx + b$$

$$r \sin \theta = mr \cos \theta + b$$

Solving for r, we get the general formula:

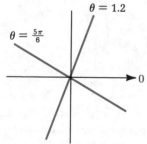

Figure 20-96

$$r = \frac{b}{\sin \theta - m \cos \theta} \qquad (20\text{-}32)$$

EXAMPLE 58 Plot the graph of $2r \cos \theta + 3r \sin \theta - 6 = 0$.

Solution **First,** solve for r:

$$r = \frac{6}{3 \sin \theta + 2 \cos \theta}$$

To graph it in polar form, construct a table of values and plot them as shown in Fig. 20-97.

θ	0°	45°	90°	135°	180°
r	3.0	1.7	2.0	8.5	−3.0

This is the same point as at $\theta = 0$.

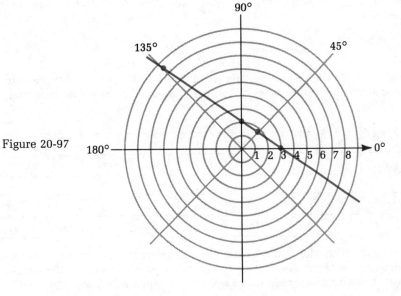

Figure 20-97

To check the graph, express the equation in the form of (20-32):

$$r = \frac{2 \quad \boxed{b}}{\sin \theta - (-\tfrac{2}{3}) \cos \theta}$$

$$\boxed{m}$$

As indicated by this form of the equation, the line shown does have a slope of $-\tfrac{2}{3}$ and a y-intercept of 2 in rectangular coordinates.

Exercises 20-9

Plot each equation on polar graph paper.

1. $r = 3$ **2.** $r = 5$ **3.** $\theta = \dfrac{2\pi}{3}$

4. $\theta = \dfrac{\pi}{4}$ **5.** $r = 6 \sin \theta$ **6.** $r = 2 \cos \theta$

7. $r = 8 \sec \theta$ **8.** $r = 4 \csc \theta$ **9.** $r = 1 + \sin \theta$

10. $r = \cos \theta - 1$ **11.** $r = 2 - \cos \theta$ **12.** $r = 3 + 4 \sin \theta$

13. $r = \dfrac{1}{1 + \sin \theta}$ **14.** $r = \dfrac{1}{2 - 2 \cos \theta}$ **15.** $r = \dfrac{1}{1 - 2 \cos \theta}$

16. $r = \dfrac{1}{2 + \sin \theta}$ **17.** $r^2 = 4 \sin 2\theta$ **18.** $r^2 = \cos 2\theta$

19. $r = -2 \sin 4\theta$ **20.** $r = \cos 3\theta$ **21.** $r = 3\theta$

22. $r = 2^\theta$ **23.** $4r \sin \theta - 3r \cos \theta - 2 = 0$ **24.** $r \sin \theta + r \cos \theta + 4 = 0$

Solve.

25. *Electrical engineering*   For two antennas in phase, the intensity distribution in their interference pattern is given by

$$I = 2I_0[1 + \cos (kd \sin \theta)]$$

Plot the curve of I/I_0 versus θ for $k = (\pi/150)$ m^{-1} and $d = 450$ m.

26. *Automotive engineering*   The theoretical data given in the table show the voltage output of an automobile alternator as the alternator rotates.* Plot these points on polar graph paper and connect them with a smooth curve.

Voltage (V)	0	10	18	23	25	23	18	10
Shaft Orientation	0°	$22\frac{1}{2}°$	45°	$67\frac{1}{2}°$	90°	$112\frac{1}{2}°$	135°	$157\frac{1}{2}°$

Voltage (V)	0	10	18	23	25	23	18	10	0
Shaft Orientation	180°	$202\frac{1}{2}°$	225°	$247\frac{1}{2}°$	270°	$292\frac{1}{2}°$	315°	$337\frac{1}{2}°$	360°

27. *Electrical engineering*   The data in the table represent the signal strength i at a constant radius surrounding a transmitting antenna. The angle θ represents the position receiving the signal relative to due north. Plot these points on polar graph paper and connect them with a smooth curve.

θ	0°	40°	45°	55°	70°	90°	100°	110°	120°	130°	180°
i (μA)	0	3	6	8	4	0	2	6	8	6	0

θ	220°	230°	235°	245°	250°	270°	290°	295°	305°	320°	360°
i (μA)	4	7	8	6	4	0	4	6	8	4	0

28. *Optics*   In wave theory, the Bragg relation for constructive interference of waves scattered at an angle θ from crystal planes spaced a distance d apart is

* Adapted from Ernest Zebrowski, Jr., *Fundamentals of Physical Measurement* (Boston: Duxbury, 1979), p. 158.

$$2d \sin \theta = n\lambda$$

Plot the curve of λ versus θ on polar graph paper for $n = 1$ and $d = 8.5 \times 10^{-11}$ m.

29. The graph of the equation

$$r = 2a(\sec \theta - \cos \theta)$$

is called the **cissoid of Diocles.** Plot this curve for $a = 1$.

30. The graph of the equation

$$r^2 \theta = a^2$$

is called a **lituus.** Plot this curve for $a = 2$.

20-10 | Review of Chapter 20

Important Terms and Concepts

distance (p. 686)

midpoint (p. 688)

slope (p. 689)

angle of inclination (p. 691)

x-intercept (p. 696)

y-intercept (p. 696)

circle (p. 702)
  center, radius (p. 702)

parabola (p. 707)
  focus, directrix, axis,
  vertex (p. 707)

translation of axes (p. 709)

ellipse (p. 717)
  foci, center, vertices, major
  axis, minor axis (p. 717–719)

hyperbola (p. 726)
  foci, center, vertices,
  transverse axis, conjugate
  axis, asymptotes (p. 726)

degenerate forms (p. 739)

polar coordinate system (p. 743)

polar axis (p. 743)

pole/origin (p. 743)

polar coordinates (p. 743)

Formulas and Equations

- for the general points $P_1(x_1, y_1)$ and $P_2(x_2, y_2)$:
  distance formula (p. 686): $d = \sqrt{(x_2 - x_1)^2 + (y_2 - y_1)^2}$ **(20-1)**
- for two points aligned horizontally $(y_1 = y_2)$ (p. 686): $d = |x_2 - x_1|$
- for two points aligned vertically $(x_1 = x_2)$ (p. 686): $d = |y_2 - y_1|$

- midpoint (p. 688): $x_m = \dfrac{x_1 + x_2}{2}$ and $y_m = \dfrac{y_1 + y_2}{2}$ **(20-2)**

- slope (p. 689): $m = \dfrac{y_2 - y_1}{x_2 - x_1}$ **(20-3)**

  $m = \tan \alpha$ **(20-4)**

  where α is the angle of inclination of a given line

- slopes of special lines and pairs of lines
  slope of a vertical line is undefined (p. 691)
  slope of a horizontal line is 0 (p. 691)
  for two nonvertical parallel lines with slopes m_1 and m_2 (p. 691): $m_1 = m_2$ **(20-5)**
  for two perpendicular lines with defined slopes $m_1 \neq 0$ and $m_2 \neq 0$ (p. 692):

  $m_1 = -\dfrac{1}{m_2}$ or $m_1 m_2 = -1$ **(20-6)**

- equations of a straight line
    point–slope form (p. 695): $y - y_1 = m(x - x_1)$ **(20-7)**
    slope–intercept form, where $(0, b)$ is the y-intercept (p. 696): $y = mx + b$ **(20-8)**
    horizontal line (p. 697): $y = b$ **(20-9)**
    vertical line (p. 697): $x = a$ **(20-10)**
    general form (p. 698): $Ax + By + C = 0$ **(20-11)**
- equations of the circle, where r represents the radius (Section 20-3):
    standard form, centered at (h, k) (p. 703): $(x - h)^2 + (y - k)^2 = r^2$ **(20-12)**
    standard form, centered at the origin (p. 703): $x^2 + y^2 = r^2$ **(20-13)**
    general equation (p. 704): $Ax^2 + Ay^2 + Bx + Cy + D = 0$, where $A \neq 0$ **(20-14)**
- equations and properties of the parabola (Section 20-4):

Equation	Vertex	Axis	Focus	Directrix	Direction
$y^2 = 4px$	$(0, 0)$	x-axis	$(p, 0)$	$x = -p$	$p > 0$, opens right
$(y - k)^2 = 4p(x - h)$	(h, k)	$y = k$	$(h + p,\ k)$	$x = h - p$	$p < 0$, opens left
$x^2 = 4py$	$(0, 0)$	y-axis	$(0, p)$	$y = -p$	$p > 0$, opens upward
$(x - h)^2 = 4p(y - k)$	(h, k)	$x = h$	$(h,\ k + p)$	$y = k - p$	$p < 0$, opens downward

- equations and properties of the ellipse, where $a > b$ and $c = \sqrt{a^2 - b^2}$ (Section 20-5):

Equation	Center	Vertices	Foci	Minor Axis Endpoints
$\dfrac{x^2}{a^2} + \dfrac{y^2}{b^2} = 1$	$(0, 0)$	$(\pm a, 0)$	$(\pm c, 0)$	$(0, \pm b)$
$\dfrac{(x - h)^2}{a^2} + \dfrac{(y - k)^2}{b^2} = 1$	(h, k)	$(h \pm a,\ k)$	$(h \pm c,\ k)$	$(h,\ k \pm b)$
$\dfrac{x^2}{b^2} + \dfrac{y^2}{a^2} = 1$	$(0, 0)$	$(0, \pm a)$	$(0, \pm c)$	$(\pm b, 0)$
$\dfrac{(x - h)^2}{b^2} + \dfrac{(y - k)^2}{a^2} = 1$	(h, k)	$(h,\ k \pm a)$	$(h,\ k \pm c)$	$(h \pm b,\ k)$

- equations and properties of the hyperbola, where $c = \sqrt{a^2 + b^2}$ (Section 20-6):

Equation	Center	Vertices	Foci	Transverse Axis
$\dfrac{x^2}{a^2} - \dfrac{y^2}{b^2} = 1$	$(0, 0)$	$(\pm a, 0)$	$(\pm c, 0)$	x-axis
$\dfrac{(x - h)^2}{a^2} - \dfrac{(y - k)^2}{b^2} = 1$	(h, k)	$(h \pm a, \cdot\ k)$	$(h \pm c,\ k)$	$y = k$
$\dfrac{y^2}{a^2} - \dfrac{x^2}{b^2} = 1$	$(0, 0)$	$(0,\ \pm a)$	$(0, \pm c)$	y-axis
$\dfrac{(y - k)^2}{a^2} - \dfrac{(x - h)^2}{b^2} = 1$	(h, k)	$(h,\ k \pm a)$	$(h,\ k \pm c)$	$x = h$
$xy = k \quad (k > 0)$	$(0, 0)$	$(\sqrt{k}, \sqrt{k}),$ $(-\sqrt{k}, -\sqrt{k})$	$(\sqrt{2k}, \sqrt{2k})$ $(-\sqrt{2k}, -\sqrt{2k})$	$y = x$
$xy = k \quad (k < 0)$	$(0, 0)$	$(\sqrt{|k|}, -\sqrt{|k|}),$ $(-\sqrt{|k|}, \sqrt{|k|})$	$(\sqrt{|2k|}, -\sqrt{|2k|}),$ $(-\sqrt{|2k|}, \sqrt{|2k|})$	$y = -x$

- for the general second-degree equation $Ax^2 + Cy^2 + Bxy + Dx + Ey + F = 0$, where $B = 0$ (p. 739):
  - if $A = C$, then the graph is a circle
  - if either A or C is 0 but not both, then the graph is a parabola
  - if A and C are not equal but have the same sign, then the graph is an ellipse
  - if A and C have opposite signs, then the graph is a hyperbola
- relationship between polar and rectangular coordinates (p. 746):

$$x = r \cos \theta, \qquad\qquad y = r \sin \theta \qquad\qquad\qquad \textbf{(20-30)}$$

$$r = \sqrt{x^2 + y^2}, \quad \tan (\text{Ref } \theta) = \left| \frac{y}{x} \right| \qquad\qquad \textbf{(20-31)}$$

Exercises 20-10

Find the distance d (to the nearest 0.01) between each pair of points, and find the midpoint M and slope m of the line segment joining them.

1. $(-2, 7)$ and $(-4, 13)$ **2.** $(6, -4)$ and $(18, -1)$ **3.** $(5, -3)$ and $(11, -5)$ **4.** $(3, 4)$ and $(-5, -2)$

Sketch the curve represented by each equation, and determine all the following that are applicable: slope, y-intercept, center, radius, directrix, vertices, and foci.

5. $x = -4$

6. $y = 1$

7. $2x + y - 5 = 0$

8. $4x - 3y + 9 = 0$

9. $x^2 = -8y$

10. $(x - 6)^2 + (y + 4)^2 = 36$

11. $\dfrac{(x + 1)^2}{4} + (y - 5)^2 = 1$

12. $\dfrac{y^2}{18} - \dfrac{x^2}{16} = 1$

13. $3x^2 + 24x - 4y^2 + 40y - 64 = 0$

14. $y^2 - 4y - 2x - 4 = 0$

15. $x^2 + 8x + y^2 + 4y - 5 = 0$

16. $25x^2 + 18y^2 = 150$

17. $x^2 + 10x + y + 24 = 0$

18. $2x^2 + 16x + y^2 - 6y + 37 = 0$

19. $9y^2 - 90y - x^2 + 216 = 0$

20. $x^2 + y^2 - 6y - 21 = 0$

Use the given geometric properties to find the equation of each curve.

21. Straight line: slope of -3, passes through $(-5, 6)$

22. Straight line: passes through $(-8, 4)$ and $(-5, 16)$

23. Straight line: parallel to the y-axis, passes through $(-5, 4)$

24. Straight line: perpendicular to the line $5x - 2y + 8 = 0$, passes through $(-10, 6)$

25. Circle: center $(6, -2)$, radius of 8

26. Circle: center $(-3, 3)$, passes through $(4, -1)$

27. Parabola: focus $(-2, 7)$, vertex $(-2, 5)$

28. Parabola: vertex $(3, -2)$, directrix $x = 4$

29. Ellipse: center $(0, 0)$, vertex $(6, 0)$, focus $(4, 0)$

30. Ellipse: foci $(4, -3)$ and $(4, 9)$, major axis of length $6\sqrt{6}$

31. Hyperbola: center $(0, 0)$, focus $(0, 7)$, conjugate axis of length 8

32. Hyperbola: vertices $(-8, 3)$ and $(6, 3)$, passes through $(10, 15)$

Find the polar equation equivalent to each rectangular equation.

33. $x^2 + y^2 = 81$ **34.** $x^2 = 6y$ **35.** $y^2 + 3x^2 = 12$ **36.** $x^2 - y^2 = 4$

Find the rectangular equation equivalent to each polar equation.

37. $r = 5 \sin \theta$ **38.** $r \cos \theta = 8$ **39.** $r = \dfrac{1}{1 + \cos \theta}$ **40.** $r^2 = 4 \cos 2\theta$

Plot each equation on polar graph paper.

41. $r = 16 \cos \theta$ **42.** $r = 1 + 2 \sin \theta$ **43.** $r = 1 - \sin \theta$ **44.** $r = \dfrac{1}{2 - \cos \theta}$

45. $r^2 = 9 \sin 2\theta$ **46.** $r = 3 \cos 4\theta$ **47.** $r = \dfrac{\theta}{2}$ **48.** $5r \sin \theta + 2r \cos \theta - 2 = 0$

Solve.

49. Find k such that the distance between $(4, -8)$ and $(-2, k)$ is $3\sqrt{5}$.
50. Find k such that the line $9x + ky + 5 = 0$ is parallel to the line $6y - 4x + 3 = 0$.
51. Is the line $4x - 12y - 6 = 0$ parallel or perpendicular to the line $3x + y - 5 = 0$?
52. Show that the points $A(-3, 7)$, $B(-1, 2)$, and $C(4, 4)$ are the vertices of a right triangle.
53. Find the area of the triangle given in Problem 52.
54. The points $P(-2, 7)$, $Q(5, 3)$, and $R(4, 0)$ are the vertices of a triangle. Show that the line through the midpoints of sides PR and RQ is parallel to side PQ.

Solve.

55. *Physics* If we can ignore air resistance, the velocity v (in feet per second) of an object launched vertically into the air is related to the elapsed time t (in seconds) by the formula $v = v_0 - 32t$, where v_0 is the initial velocity. For an object propelled upward at 180 ft/s, write the velocity equation and plot the graph for $0 \le t \le 10$ s.
56. *Material science* The density of mercury decreases linearly as the temperature increases from $-20°C$ to $100°C$. If the density of mercury is 13.644 g/cm^3 at $-20°C$ and 13.350 g/cm^3 at $100°C$, find the equation relating density to temperature. Using this equation, calculate the density of mercury at room temperature, $23°C$.
57. *Electronics* The current I flowing in a generator circuit changes linearly starting from an initial value of 110 mA and increasing by 15 mA/s for 10 s. Write an equation for I in terms of the time t (in seconds) and plot the graph for $0 \le t \le 10$ s.
58. *Environmental science* The pollutant standard index (PSI) in a particular metropolitan location was found to vary linearly with time during daylight hours. If the index reading was 27 at 6 AM and 52 at 4 PM, write an equation relating this index to the time (in hours) elapsed since 6 AM. From the equation determine the PSI at 11:30 AM.
59. *Police science* The formula $v = \sqrt{30kd}$ is used to estimate the speed v (mi/h) at which a car was moving if it skidded d ft after the brakes were applied. The number k is a constant that depends on the frictional characteristics of the road surface. If $k = 0.8$, plot v as a function of d for $0 \le d \le 100$ ft.
60. *Electronics* For a certain electrical circuit, the power P generated (in watts) is given by $P = RI^2 + VI$, where R is the resistance, V is the impressed voltage, and I is the current flowing in the circuit. Plot P as a function of I for $R = 2.0 \ \Omega$, $V = 6.0$ V, and $0 \le I \le 4.0$ A.
61. *Optics* Solve Problem 37 in Exercises 20-4 for a reflector 48 in. across and 16 in. deep.
62. A wire hanging between two supports 120 ft apart assumes a roughly parabolic shape. The ends of the wire are fastened 50 ft above the base on each support. If the lowest point of the wire is 20 ft above the base, find the equation describing the wire shape. (Place the leftmost support on the y-axis with its base at the origin.)
63. *Physics* The height h (in feet) of an object propelled vertically at 80 ft/s from a cliff 60 ft high is given by $h = 60 + 80t - 16t^2$, where t is the time in seconds. Sketch this curve for values of $h \ge 0$. After how many seconds (to the nearest tenth) does the object hit the ground below the cliff?
64. *Space technology* Solve Example 36 if the altitude of the satellite ranges from 625 to 2455 mi.
65. *Industrial design* Solve Problem 34 in Exercises 20-5 for $d = 4.50$ in. and $\theta = 60.0°$.
66. *Civil engineering* A bridge over a river is constructed in the form of a single elliptical arch. The width of the arch is 80.0 ft at water level, and the highest point is 24.0 ft above the water. Assuming that the river coincides with the x-axis and the origin is at the center of the river, find the equation of the ellipse.

67. *Civil engineering*   In Problem 66, how tall is the tallest sailboat that can pass under the bridge 15 ft from the center of the river?

68. *Thermodynamics*   At a certain temperature, a gas has a volume of 5.5 L when the pressure is 420 kPa. Sketch the graph of pressure as a function of volume at this temperature. (See Problem 37 in Exercises 20-6.)

69. *Mechanical engineering*   In the study of structure failures the formula below is known as the **Mises ellipse:***

$$\left(\frac{s_1}{s_3}\right)^2 + \left(\frac{s_2}{s_3}\right)^2 - \frac{s_1 s_2}{s_3 s_3} = 1 \qquad \text{If} \qquad \frac{s_1}{s_3} = \frac{1}{\sqrt{2}}(x - y) \qquad \text{and} \qquad \frac{s_2}{s_3} = \frac{1}{\sqrt{2}}(x + y)$$

what are the lengths of the major and minor axes of the ellipse?

70. *Wastewater technology*   The formula $A = 8.34FC$ is used to determine the amount A of chlorine (in pounds) to add to a basin with a flow rate of F million gallons per day to achieve a chlorine concentration of C parts per million. Sketch the values of F versus C that correspond to adding 1750 lb of chlorine to a certain reservoir.

71. *Optics*   The angular intensity distribution of a light source is measured by the equation $I = 5 + 5 \cos \theta$. Plot this curve on polar graph paper.

72. *Optics*   Express the equation in Problem 71 in rectangular form. [Hint: Use $I = r$.]

73. *Optics*   The optical path difference x between a typical pair of reflected rays is

$$x = 2d \cos \frac{\theta}{2}$$

Plot x versus θ for $d = 1.0$ mm.

74. *Hydraulics*   Poiseuille's law tells us that for a fluid flowing in a pipe, the speed v (in inches per second) at a distance d from the center of a pipe of radius r is $v = k(r^2 - d^2)$, where $d \le r$ and k is a positive constant. Plot v versus d for $k = 2$ and $r = 6$ in.

75. Given the ellipse

$$\frac{x^2}{169} + \frac{y^2}{144} = 1$$

find the equation of a hyperbola centered at the origin with the same foci and one vertex at (4, 0).

76. The equations of some conics can be written in the form

$$y^2 = Lx + kx^2 \qquad k \ge -1$$

where L, the *latus rectum*, is the "width" of the curve at a focus. For what values of k is this a circle? An ellipse? A hyperbola? A parabola?

*For additional background, refer to Nelson R. Bauld, Jr., *Mechanics of Materials* (Monterey, Ca.: Brooks/Cole Engineering Division, Wadsworth, 1982), p. 646.

Introduction to Probability and Statistics

21

DAISY WHEEL FOR WORD PROCESSOR PRINTER

In both our jobs and our personal lives, we must constantly make decisions. An industrial engineer must decide how many of each manufactured item to produce, a navigator must choose the best route to take, a baseball manager must select the most effective pinch-hitter, and an electronics engineer must decide which circuit components provide maximum reliability. Because none of us has a crystal ball, our decisions are normally based on past experience combined with a knowledge of the intrinsic characteristics of the alternatives. The science of recording, organizing, analyzing, and using past experiences in decision-making is known as *statistics*, and the mathematical estimate of the likelihood of some future outcome is called its *probability*.

Most of us have dealt with probability on an intuitive level when making such statements as "I have a good chance of passing this course" or "The Yankees are not likely to win today." In this section we shall examine how statements like these can be expressed quantitatively, and how the mathematical probability of certain events can be measured.

Definition

If some event can occur in s successful ways out of a total of n equally likely ways, the **probability** P of that event is defined as follows:

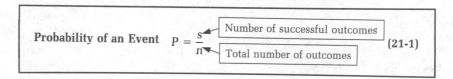

Probability of an Event $\quad P = \dfrac{s}{n}$ Number of successful outcomes   Total number of outcomes   **(21-1)**

EXAMPLE 1 When a coin is flipped, there are two equally likely outcomes: heads and tails. Since only one of these outcomes is heads, the probability that a single flip of the coin will result in heads is

$$P = \frac{s}{n} = \frac{1}{2} = 0.5$$

EXAMPLE 2 What is the probability that the number resulting from a single roll of a standard six-sided die will be 5 or more?

Solution There are two ways the roll can result in a 5 or greater: it can show a 5 or a 6. Since there are a total of six equally likely outcomes, the probability of rolling a 5 or more is

$$P = \frac{s}{n} = \frac{2}{6} = \frac{1}{3} \approx 0.33$$

Empirical probability

Examples 1 and 2 involve **a priori probability,** from the Latin phrase meaning "based on prior study." This kind of theoretical probability is found from logical conclusions and assumptions based on the event: the number of sides of a perfect coin, the number of faces on a balanced die, and so on.

Probability based on experimental data from past events is called **a posteriori,** or **empirical, probability,** from the Latin phrase meaning "based on experimental study." The empirical probability of an event can also be calculated using equation (21-1), but in this case we assume n to be the total number of trials in an experiment, and s is the number of times the event of interest actually occurred. Empirical probability is used in many technical applications.

EXAMPLE 3   In tooling a delicate instrument, a certain machine produced 105 instruments, 7 of which did not meet quality control specifications. Based on these results, the empirical probability of the machine producing a reject is

$$P = \frac{s}{n} = \frac{7}{105} \approx 0.067$$

The information in Example 3 can be very useful to the manufacturer in planning future production and estimating costs. However, the value and accuracy of empirical probability depends upon the number of trials on which it is based. If n is too small, the result may be unreliable, and if n is very large, getting the quality control information may be unneccessarily expensive and time-consuming.

Certainty   Since s can never be larger than n in equation (21-1), the probability of an event will always be a number between 0 and 1.

EXAMPLE 4   In one roll of a single die, what is the probability of each of the following?
(a)   Rolling a 10   (b)   Rolling less than an 8

Solution

(a) $P = \dfrac{s}{n} = \dfrac{0}{6}$ There is no 10 on a die.

$= 0$ A probability of 0 means that the desired event will *never* occur.

(b) $P = \dfrac{s}{n} = \dfrac{6}{6}$ There are six possible numbers, and they are all less than 8.

$= 1$ A probability of 1 means that the desired event will *always* occur.

Since a given event will either occur successfully or fail to occur, the **probability of failure F** is defined as follows:

Probability of Failure $F = 1 - P = 1 - \dfrac{s}{n}$ (21-2)

EXAMPLE 5   In Example 2, the probability of failure—that is, the probability of a roll of 4 or less—is

$$F = 1 - \frac{1}{3} = \frac{2}{3} \approx 0.67$$

We can verify this using (21-1) by noting that there are four ways in which a roll of 4 or less can result out of six possible outcomes, and therefore

$$P = \frac{4}{6} = \frac{2}{3}$$

EXAMPLE 6   In Example 3, what is the probability that an instrument will *not* be rejected?

Solution   Since P (the probability of the instrument being rejected) is 0.067, then F (the probability of it not being rejected) is

$$F = 1 - 0.067 \approx 0.933$$

Combined probability: A or B

So far, we have considered only the probability of a single event. It is also useful to determine the overall probability of occurrence of a combination of events.

EXAMPLE 7   A bag contains 2 red jellybeans, 5 yellow jellybeans, and 3 black jellybeans. Suppose a certain boy dislikes black jellybeans. If he randomly selects a jellybean from the bag, what is the probability that it will be either red or yellow?

Solution   The total number of outcomes n (jellybeans to be picked) is $2 + 5 + 3 = 10$. The number of successful outcomes s (red or yellow jelly-beans) is $2 + 5 = 7$. The probability of selecting a red or yellow jellybean is

$$P = \frac{s}{n} = \frac{7}{10} = 0.7$$

Although we used (21-1) to calculate P in Example 7, the probability of one or the other of two events occurring is often stated as follows:

Probability of A or B
If A and B are two mutually exclusive events with independent probabilities P_A and P_B, respectively, then the probability of A or B is given by

$$P_{A\,or\,B} = P_A + P_B \qquad\qquad (21\text{-}3)$$

By *mutually exclusive* we mean that events A and B cannot occur simultaneously. In Example 7,

$$P_{red\,or\,yellow} = P_{red} + P_{yellow}$$

$$= \frac{2}{10} + \frac{5}{10} = \frac{7}{10}$$

The events are mutually exclusive: You cannot pick one jellybean and get *both* red and yellow.

EXAMPLE 8 Out of 80 machined parts, 6 were found to be too long and 8 were found to be too short to pass quality control tests. What is the probability of a part being either too long or too short?

Solution Using L for "too long" and S for "too short," we have

$$P_{L \text{ or } S} = P_L + P_S = \frac{6}{80} + \frac{8}{80} = \frac{14}{80} = 0.175$$

Probability of A and B   Another type of combined probability involves the occurrence of two or more events.

EXAMPLE 9 If a single die is rolled twice, what is the probability that both tosses will show an even number?

Solution Note that each roll will be either even or odd, and that each result is equally likely. To use (21-1) we must first list all possible outcomes and then determine how many are successful. From the following *tree diagram*, we see that there are four equally likely possible outcomes:

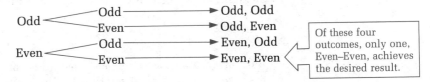

First Roll   **Second Roll**   **Overall Outcome of Two Rolls**

Odd — Odd → Odd, Odd
Odd — Even → Odd, Even
Even — Odd → Even, Odd
Even — Even → Even, Even

Of these four outcomes, only one, Even–Even, achieves the desired result.

Therefore,

$$P = \frac{s}{n} = \frac{1}{4}$$

One Even–Even outcome

Four possible outcomes

When the number of possibilities increases, listing all potential outcomes becomes more difficult. But notice that the result of this problem can be obtained also by finding the product of the probability of each event. That is, since the probability of rolling an even number each time is $\frac{1}{2}$, then the probability of rolling an even number twice in a row is $(\frac{1}{2})(\frac{1}{2}) = \frac{1}{4}$. It can be shown that the multiplication method is valid for all such cases, and the result can be stated as follows:

Probability of A and B

If A and B are two independent events with probabilities P_A and P_B, respectively, then the probability of A and B is

$$P_{A \text{ and } B} = P_A \cdot P_B \tag{21-4}$$

By *independent* we mean that the occurrence or nonoccurrence of one event does not influence the probability of the second event.

EXAMPLE 10 A study of wind patterns near Vandenburg Air Force Base showed that the harmful exhaust of the space shuttle vehicle would be carried away from a vulnerable environment 94% of the time. Assuming that the launch times are chosen at random, what is the probability that the exhaust from all six scheduled launches of the space shuttle will be dispersed away from this area?*

Solution We can express the probability of each exhaust dispersing success-fully as $\frac{94}{100} = 0.94$. The probability that the exhaust from all six launches will be safely dispersed is thus

$$P = (0.94)^6 \approx 0.69$$

In calculating the probability of some compound events, we must consider the fact that certain outcomes may be "used up."

EXAMPLE 11 If a card is drawn from a 52 card deck and discarded, and then a second card is drawn from the same deck, what is the probability that both cards will be aces?

Solution The probability of drawing an ace the first time is

$$P_1 = \frac{4}{52} = \frac{1}{13}$$

If the first draw is successful, and the card is not replaced, there will be only 3 aces remaining out of 51 total cards. The probability of drawing an ace the second time is therefore

$$P_2 = \frac{3}{51} = \frac{1}{17}$$

Using (21-4), we can compute the probability of drawing two consecutive aces as

$$P = P_1 \cdot P_2 = \frac{1}{13} \cdot \frac{1}{17} = \frac{1}{221} \approx 0.0045$$

In this section we have examined some basic kinds of probability computations. There are many additional and more complex types of situations that are beyond the scope of this text.

---

*For more information, see Department of the Air Force Space and Missile Systems Organization, *Candidate Environmental Statement. Space Shuttle Program* (Vandenburg Air Force Base, Ca. Nov. 22, 1976, Vol. 11.

Exercises 21-1

If a standard six-sided die is rolled, what is the probability of occurrence of each of the following? (Express your answers in fraction form.)

1. A 2
2. A 3 or a 5
3. A 3 or less

4. Anything but a 6
5. A 7
6. A 3 or more

7. Two consecutive 5's
8. Three consecutive rolls of 5 or more

A total of 18 unmarked identical boxes of computer chips are stored on the shelves of a warehouse. If 9 boxes contain chip A, 5 contain chip B, and 4 contain chip C, and if you randomly select boxes, what is the probability that you get each of the following? (Express your answers in decimal form rounded to the nearest hundredth.)

9. A box containing chip A
10. A box containing chip C
11. A box containing either chip A or B
12. A box containing either chip A or C
13. Two successive boxes containing chip A if the first box is not replaced
14. Three successive boxes containing chip B if the first two boxes are not replaced
15. Two successive boxes containing chip C if the first box is replaced before the second one is picked at random
16. A box containing chip B and then a box containing chip C if the first box is replaced before the second one is picked at random

In a test run of 60 machined parts, 4 were rejected because there were undersized and 8 were rejected because they were oversized. Based on these results, what is the probability of each of the following (to the nearest thousandth) in some future run?

17. A part will be undersized
18. A part will be oversized
19. A part will be rejected

20. Two successive undersized parts will be produced
21. Two successive oversized parts will be produced
22. Three successive parts will not be rejected

Using a standard deck of 52 cards, determine the probability of drawing each of the following. (Express your answers in fraction form.)

23. A club
24. A red card
25. A face card (K, Q, J)
26. A 7 or less

27. Two successive hearts if the first card is replaced before drawing the second
28. Two successive black cards if the first card is discarded before the second is drawn

A technical researcher must prepare a report for a congressional committee investigating drunk driving. Preliminary studies based on police data reveal the following probabilities:*

A driver being intoxicated: 0.020
An intoxicated driver having an accident: 0.00060
An intoxicated driver being arrested before having an accident: 0.00050
An intoxicated driver being convicted after arrest: 0.70

According to these records, what is the probability of each of the following for a driver?

*Courtesy of William Jago, Tecolote Research, Inc., Santa Barbara, California.

29. Being sober
30. Being intoxicated and having an accident
31. Being intoxicated and arrested before having an accident
32. Being intoxicated, arrested, and convicted
33. Being intoxicated, arrested, and not being convicted
34. Being intoxicated and either having an accident or being arrested

Solve.

35. *Environmental science*   In Example 10, what is the probability that the exhaust from four consecutive launches will be safely dispersed?
36. *Industrial engineering*   A die must be made for manufacturing a plastic part. There are three types available, and their costs are given in the table. Preliminary market research indicates a 0.2 probability that 3000 parts will be needed, a 0.5 probability for 12,000 parts, and a 0.3 probability for 20,000 parts. Based on this information, which die should be ordered?
[*Hint:* First, determine the cost of each amount with each die. Then, multiply each of the nine combinations by its probability. Finally, calculate the totals for each die, compare, and pick the least expensive one.]

Type	Cost of Die	Cost per Part for 10,000 or Fewer	Cost per Part for More than 10,000
A	$3000	$0.30	$0.25
B	$4650	$0.17	$0.13
C	$5800	$0.11	$0.09

21-2 | Frequency Distributions and Measures of Central Tendency

In most of the applied problems we have solved so far, physical quantities have been given as single values—for example, a radius of 2.50 in., a temperature of 8°C, and so on. However, most real-world applications involve a collection of numerical values, or **data,** such as a series of repeated measurements of the dimensions of an object or a list of temperatures taken at different times. In our discussion of statistics, we will show how to organize and analyze data so that meaningful conclusions can be made.

Frequency distributions

As the first step in organizing data, it is often helpful to construct a **frequency distribution.** This is a chart or graph showing the number of times each measurement value occurs in a series of measurements.

EXAMPLE 12   Twenty measurements taken on the speed of a computer printer resulted in the following data, where the numbers give printer speed in characters per second (cps):

156, 160, 158, 159, 161, 160, 157, 162, 159, 160,

157, 159, 161, 162, 159, 160, 163, 161, 159, 158

It is difficult to detect any clear pattern in the data, but the frequency distribution of these values shown below clearly reveals that the speeds vary in a predictable way:

Speed (cps)	156	157	158	159	160	161	162	163
Frequency, f	1	2	2	5	4	3	2	1

The speeds that occur most often are 159 and 160 cps. Printer speeds slower than 157 cps or higher than 162 cps seldom occur.

For sets of data involving very many numbers with a wider range of values, it may be more useful to organize the data within carefully chosen equal intervals.

EXAMPLE 13 As part of the Mellowville Electric Company's conservation efforts, data were collected on the use of electricity in dwellings of a similar size. In 40 homes with interior areas of 2500–3000 ft^2, the following amounts of monthly energy consumption (in kilowatt-hours) were recorded:

669, 684, 622, 675, 690, 671, 597, 654,

629, 647, 679, 638, 671, 666, 644, 682,

669, 612, 601, 659, 671, 699, 720, 654,

662, 631, 602, 677, 664, 718, 657, 641,

675, 662, 670, 622, 654, 676, 685, 666

Since there are many measurements and few repeated values, a frequency distribution based on single values would not be very useful. However, a frequency distribution based on intervals of 10 kWh simplifies the data and reveals a pattern:

Energy Usage (kWh)	590–599	600–609	610–619	620–629	630–639	640–649	650–659
Frequency, f	1	2	1	3	2	3	5

Energy Usage (kWh)	660–669	670–679	680–689	690–699	700–709	710–719	720–729
Frequency, f	7	9	3	2	0	1	1

Choosing an interval of 20 kWh or 30 kWh for organizing the set of measurements in Example 13 would produce a different, though similar, distribution. There is no single, "correct" rule for choosing the width of the intervals that can be used in all cases. The main guidelines are to reduce the number of categories but avoid intervals that are so large they obscure any trends or patterns in the data.

Graphical representation: histograms

Graphs can be useful in illustrating the frequency of values and revealing patterns in the data. A **histogram** is a bar graph in which the values or data intervals are represented on the horizontal axis, and the frequency of occurrence of each value is represented on the vertical axis.

EXAMPLE 14   Using the frequency distribution of printer speeds from Example 12, we can construct the histogram shown in Fig. 21-1. Notice that the widths of the bars are equal and centered on the data number. The height of each bar shows visually the frequency of occurrence for each printer speed.

Figure 21-1

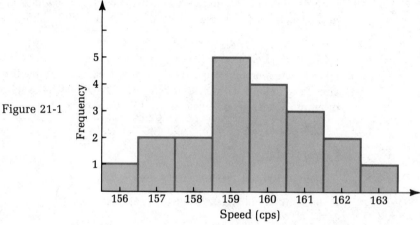

Frequency polygons

A similar type of graph, more commonly used in technical work, is the **frequency polygon.** This is a broken-line graph in which a single point represents the frequency of each value, and adjacent points are connected with line segments.

EXAMPLE 15   Figure 21-2 compares a frequency polygon (in color) with a histogram for the data from Fig. 21-1. Notice that the lines join the midpoints of the tops of the rectangles. This is especially important when the bars represent intervals. Also notice that line segments are added at each end of the graph so that the polygon is bounded by the horizontal axis.

EXAMPLE 16   Draw a frequency polygon from the frequency distribution given in Example 13.

Solution   Notice in Fig. 21-3 that the horizontal axis is marked with the initial values of the intervals, and the points are plotted at the *midpoint* of each interval.

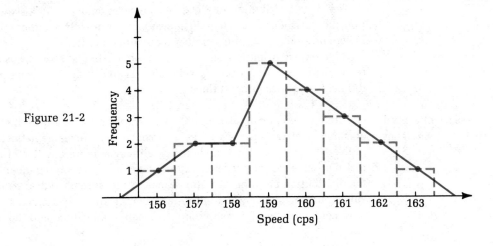

Figure 21-2

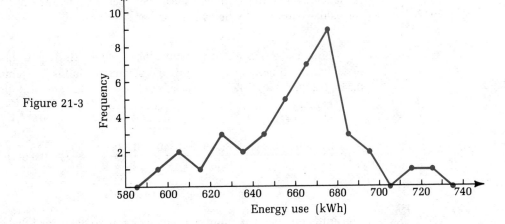

Figure 21-3

Types of frequency distributions

Frequency distributions appear in an unlimited variety of possible shapes, but a few appear regularly in scientific and technical work.

The symmetric, bell-shaped distribution shown in Fig. 21-4(a) is known as a **normal distribution.** Scientific and technical situations usually produce measurements that follow a normal or approximately normal distribution.

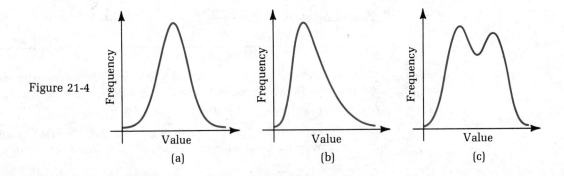

Figure 21-4

The **skew** distribution in Fig. 21-4(b) is not symmetric. In a skew distribution, measurement values tend to appear more frequently at either the upper or lower end of the distribution rather than at the middle.

A **bimodal** distribution, such as the one shown in Fig. 21-4(c), has two peaks rather than one. If the peaks are well-defined and distinct, it may mean that two different processes or conditions are involved.

Measures of central tendency

Although frequency distributions and graphs provide a meaningful way of organizing data, it is usually necessary to obtain a more mathematical summary of the data. We may want to choose a value most representative of the whole distribution, especially if certain values tend to appear more frequently so that the distribution has a peaked appearance. A **measure of central tendency** is a single number that can be used to represent all the measurements. The three most commonly used measures are the *mean*, the *median*, and the *mode*.

Mean

The **arithmetic mean** is the most widely used of all measures of central tendency. To find the arithmetic mean $\bar{x}$ of a set of values, first find the sum of the values and then divide by the number of values.

EXAMPLE 17 A machinist measures the thickness (in inches) of a uniform metal bar five times along its length and obtains the following values: 0.340, 0.341, 0.338, 0.340, 0.336. The mean thickness is calculated as follows:

Sum of values

$$\bar{x} = \frac{0.340 + 0.341 + 0.338 + 0.340 + 0.336}{5} = \frac{1.695}{5} = 0.339 \text{ in.}$$

Divided by number of values

NOTE ▶ The arithmetic mean is simply the commonly used *average* of the set of numbers. Because there are other kinds of averages, we need to be careful and use its correct name. ◀

Stated formally, the arithmetic mean $\bar{x}$ of the n numbers $x_1, x_2, \ldots, x_n$ is defined as follows:

| Arithmetic Mean | $\bar{x} = \dfrac{x_1 + x_2 + \cdots + x_n}{n} = \dfrac{\sum\limits_{i=1}^{n} x_i}{n}$ | (21-5) |

Summation notation

The Greek letter Σ ("sigma," a capital S for "sum") is traditionally used to indicate a sum. In this **sigma,** or **summation, notation,** i represents a positive integer and x_i stands for one of the values of x. The expression $\sum\limits_{i=1}^{n} x_i$ is a shorthand way of writing "the sum of all n numbers from the first through the nth." For convenience, we will often abbreviate this further by writing $\sum x_i$ or $\sum x$.

To calculate the arithmetic mean from a frequency distribution in which the values $x_1, x_2, \ldots, x_n$ are repeated $f_1, f_2, \ldots, f_n$ times, respectively, the following formula is used:

Arithmetic Mean from a Frequency Distribution

$$\overline{x} = \frac{f_1 x_1 + f_2 x_2 + \cdots + f_n x_n}{f_1 + f_2 + \cdots + f_n} = \frac{\sum\limits_{i=1}^{n} f_i x_i}{\sum\limits_{i=1}^{n} f_i} \qquad (21\text{-}6)$$

In other words, multiply each value by its frequency, add these products, and divide by the sum of the frequencies.

EXAMPLE 18   Using the frequency distribution in Example 12, we find the arithmetic mean of the printer speeds to be

Multiply each value by its frequency and add.

$$\overline{x} = \frac{1(156) + 2(157) + 2(158) + 5(159) + 4(160) + 3(161) + 2(162) + 1(163)}{1 + 2 + 2 + 5 + 4 + 3 + 2 + 1}$$

Divide by the sum of the frequencies.

$$= \frac{3191}{20} \approx 159.6 \text{ cps}$$

If a frequency distribution is stated in terms of intervals rather than exact values, use the midpoint of each interval for x_i and apply (21-6) as before.

EXAMPLE 19   From the frequency distribution in Example 13, we can calculate the mean electricity consumption. The midpoints of the intervals, 594.5, 604.5, and so on, are used for x_i to obtain

$$\overline{x} = \frac{1(594.5) + 2(604.5) + \cdots + 7(664.5) + \cdots + 1(724.5)}{40}$$

$$= \frac{26{,}370}{40} \approx 659.3 \text{ kWh}$$

If we calculate the actual mean from the individual values listed in the original example, we find that $\overline{x} \approx 659.1$ kWh. The estimate obtained using the interval method is very close to this value and was easier to calculate.

Median   The **median** is the value in a distribution that falls in the exact middle when the measurements are arranged in numerical order. It is a less commonly used measure of central tendency, but it is very easy to find.

EXAMPLE 20   The annual salaries paid to the seven staff members of the Synapse Research Company are as follows:

| $21,200 | $22,350 | $22,450 | $26,800 |
| $30,170 | $65,600 | $195,000 |

The median salary is $26,800. There are three salaries less than $26,800 and three salaries greater than this value. Notice that the arithmetic mean of these numbers, approximately $54,800, is very deceptive. If prospective employees were given this mean figure, they might assume that their starting salary would be quite high. The median of $26,800 is more representative of the overall pattern of numbers. The median is usually helpful with a skew distribution where a few very large or very small numbers distort the data.

The median of an *even* number of values is the arithmetic mean of the middle two values.

EXAMPLE 21   Find the median printer speed from Example 12.

Solution   Since there are 20 values, the median will be the arithmetic mean of the middle two—the 10th and 11th values. From the frequency distribution chart, we see that these values are 159 and 160. Therefore, the median is 159.5 cps. Notice that for this symmetric distribution the median is very close to the *mean* speed of 159.6 cps.

Mode   A third measure of central tendency—the *mode*—is rarely used in technical work. The **mode** of a collection of data is the value that occurs with the greatest frequency. There may be more than one mode for a given set of data.

EXAMPLE 22   The mode of the printer speeds in Example 12 is 159 cps, because it has the greatest frequency (5). Similarly, the most frequently found amount of energy consumption for the grouped data in Example 13 was in the interval 670–679 kWh. The mode is 674.5 kWh, the center of this interval.

Exercises 21-2

Use the following sets of data for Problems 1–26.

A. Number of defective light bulbs in lots of 500:   6, 4, 7, 5, 4, 8, 5, 6, 3, 5, 4, 3, 5

B. Measurements of the thickness (in centimeters) of a piece of steel: 0.654, 0.656, 0.653, 0.654, 0.657, 0.655, 0.656, 0.655, 0.656, 0.655

C. Weekly consumption of fuel (in gallons) for a delivery service: 225, 238, 233, 241, 229, 240, 236, 238, 251, 229, 240, 226, 235, 237, 246, 222, 231, 239, 238, 244

D. *Test scores for a Tech Math class:* 98, 74, 86, 88, 91, 66, 83, 70, 95, 68, 73, 78, 89, 88, 74, 96, 47, 72, 80, 77, 90, 74, 82, 59, 79

Construct a frequency distribution table for each set, as indicated.

1. Set A   2. Set B
3. Set C using intervals 220–224, 225–229, etc.
4. Set D using intervals 40–49, 50–59, etc.

Use the results of Problems 1–4 to draw histograms for each set of data.

5. Set A   6. Set B   7. Set C   8. Set D

Plot a frequency polygon for each of the indicated sets.

9. Set A   10. Set B   11. Set C   12. Set D

Use the frequency distributions from Problems 1–4 to compute the arithmetic mean of each set of data.

13. Set A   14. Set B   15. Set C   16. Set D

Use the original data to compute the arithmetic mean for the given sets of data.

17. Set C   18. Set D

Determine the median value for each set of data.

19. Set A   20. Set B   21. Set C   22. Set D

Determine the mode of each set of data.

23. Set A   24. Set B   25. Set C (Use grouped data.)   26. Set D (Use grouped data.)

27. *Energy technology*   A study of the monthly natural gas consumption in a number of similar dwellings produces the data in the table. Draw a histogram of the data.

Gas Consumption (therms)	75–79	80–84	85–89	90–94	95–99	100–104	105–109	110–114	115–119	120–124	125–129
Number of Dwellings	1	4	12	18	37	51	42	26	14	3	1

28. Plot a frequency polygon for the data in Problem 27.
29. Calculate the arithmetic mean of the data in Problem 27.
30. Determine the median of the data in Problem 27.
31. *Automotive engineering*   In testing the acceleration from 0 to 60 mi/h for a certain automobile, the following times (in seconds) resulted from a series of trials: 10.8, 10.9, 10.8, 10.7, 10.8, 10.6, 10.9, 11.0, 10.7, 10.8, 10.6, 10.7, 11.1, 10.9, 10.5. Construct a frequency distribution table for the data.
32. Plot a frequency polygon of the data in Problem 31.
33. Determine the median of the data in Problem 31.
34. Calculate the mean of the data in Problem 31.
35. *Auto mechanics*   A mechanic must estimate the labor cost of a certain maintenance job. Company files indicate that eight similar jobs in the past took the following times (in hours) to complete: 18.0, 16.5, 16.0, 19.5, 17.5, 19.0, 18.5, 16.5. Based on the arithmetic mean of the previous jobs, what is the best estimate of the labor cost at $32 per hour?

36. *Hydrology* A hydrologist estimates the velocity of a stream by timing the movement of an object over a distance of 200 ft. In six trials the object was timed at 21 s, 23 s, 20 s, 23 s, 22 s, and 24 s, rounded to the nearest second. Using the arithmetic mean of these trials, estimate the velocity of the stream in feet per second.*

37. *Graphics technology* During the past 7 months a print shop has used the following amounts of a certain type of paper: 6240 lb, 4750 lb, 5560 lb, 6080 lb, 5170 lb, 3860 lb, 4420 lb. Based on the arithmetic mean of the data, how many pounds of paper will be needed for the next 12 months?

38. *Wastewater technology* The water quality board in a community has ruled that the 7 day mean volume of settleable solids cannot exceed 0.15 mL/L. During the past 5 days, the concentrations have been measured at 0.18, 0.12, 0.21, 0.14, 0.16.
    (a) What is the mean concentration so far?
    (b) What is the maximum mean concentration allowable over the next 2 days in order to meet the board's requirement?

21-3 | Measures of Dispersion

In Section 21-2 we saw how measures of central tendency are used to obtain an estimate of the most representative value of a collection of data. It is equally important in technical work that we have an indication of the spread of the values. This characteristic of the distribution is known as a **measure of dispersion.**

Range; mean deviation

Several measures of dispersion are used by statisticians. The **range** is simply the difference between the two extreme values in a given set of data. For example, for the data in Example 20 (Section 21-2), the range of salaries is $195,000–$21,200 = $173,800.

The **mean deviation,** or **average deviation** (AD), is the mean distance of all values from the mean. It can be described by the following formula:

$$\text{Mean deviation} = \frac{\sum\limits_{i=1}^{n} |x_i - \overline{x}|}{n} \qquad (21\text{-}7)$$

Standard deviation

The measure of dispersion used most widely, and the one we shall emphasize in this section, is the standard deviation. The **standard deviation** s of a set of n values $x_1, x_2, \ldots, x_n$ is defined as follows:

$$\textbf{Standard Deviation} \qquad s = \sqrt{\frac{\sum\limits_{i=1}^{n} (x_i - \overline{x})^2}{n}} \qquad (21\text{-}8)$$

*Courtesy of Robert Blecker, U.S. Forest Service.

Stated in words, the standard deviation is the square root of the arithmetic mean of the squared deviations from the mean. We can best understand this formula by following the step-by-step procedure for finding the standard deviation of a set of values:

Step 1. Find the arithmetic mean of the values.
Step 2. Subtract the mean from each of the values.
Step 3. Square each of the differences from Step 2.
Step 4. Find the arithmetic mean of the squared differences.
Step 5. Take the principal square root of the result.

It is usually best to organize this procedure in a chart, as shown in the next example.

EXAMPLE 23 Suppose we are given the following set of numbers: 5, 4, 2, 4, 7, 5, 3, 4, 6, 5.
(a) Find the standard deviation.   (b) Find the mean deviation.

TABLE 21-1

x	$x - \bar{x}$	$(x - \bar{x})^2$
5	0.5	0.25
4	−0.5	0.25
2	−2.5	6.25
4	−0.5	0.25
7	2.5	6.25
5	0.5	0.25
3	−1.5	2.25
4	−0.5	0.25
6	1.5	2.25
5	0.5	0.25
45		18.50
$\sum x$		$\sum (x - \bar{x})^2$

Solution
(a) **Step 1.** Calculate the mean from the first column of Table 21-1:

$$\bar{x} = \frac{\sum x}{n} = \frac{45}{10} = 4.5$$

Step 2. Subtract the mean from each value and list this difference in the second column.

Step 3. Write the squares of the differences in the third column.

Step 4. Calculate the mean of the squares:

$$\frac{\sum (x - \bar{x})^2}{n} = \frac{18.50}{10} = 1.85$$

Step 5. Take the square root:

$$s = \sqrt{1.85} \approx 1.4$$

(b) The mean, or average, deviation as defined in (21-7) can be found by adding the absolute values of the numbers in the second column and dividing this sum by n:

$$\sum |x - \bar{x}| = 11 \quad \text{so} \quad AD = \frac{11}{10} = 1.1$$

Although the standard deviation is a bit more difficult to calculate than the AD, the standard deviation is more widely used in statistical calculations because it is easier to interpret mathematically.

Standard deviation from grouped data   To calculate the standard deviation from a frequency distribution, use the following variation of (21-8):

Standard Deviation from a Frequency Distribution

$$s = \sqrt{\dfrac{\sum\limits_{i=1}^{n} f_i(x_i - \overline{x})^2}{n}} \qquad \text{where } n = \sum\limits_{i=1}^{n} f_i \qquad (21\text{-}9)$$

EXAMPLE 24   The first two columns of Table 21-2 show the daily output of computers by an assembly group as measured over a 15 day period. To calculate the standard deviation from this frequency distribution, proceed as follows:

First, fill in the four additional column headings in Table 21-2.

TABLE 21-2

x	f	fx	$x - \overline{x}$	$(x - \overline{x})^2$	$f(x - \overline{x})^2$
24	1	24	−2.3	5.29	5.29
25	2	50	−1.3	1.69	3.38
26	6	156	−0.3	0.09	0.54
27	4	108	0.7	0.49	1.96
28	2	56	1.7	2.89	5.78
Σ: 15		394			16.95

Second, fill in the fx column and calculate the mean:

$$\overline{x} = \frac{\sum fx}{\sum f} = \frac{394}{15} \approx 26.3$$

Third, fill in the last three columns as shown.

Finally, use equation (21-9) to calculate the standard deviation:

$$s = \sqrt{\frac{\sum f(x - \overline{x})^2}{n}} = \sqrt{\frac{16.95}{15}} \approx 1.1$$

Calculating standard deviation from raw scores   Using formulas (21-8) and (21-9) can be very difficult when the data become more extensive or complex. With a bit of algebraic manipulation, we can derive alternative forms of the standard deviation formulas that are easier to use.

To help you understand this derivation, note the following:

1. If k represents a constant, then $\sum kx_i = k \sum x_i$ because

$$kx_1 + kx_2 + \cdots + kx_n = k(x_1 + x_2 + \cdots + x_n)$$

2. Since

$$(x_1 + y_1) + (x_2 + y_2) + \cdots + (x_n + y_n)$$
$$= (x_1 + x_2 + \cdots + x_n) + (y_1 + y_2 + \cdots + y_n)$$

by the associative law, we have

$$\sum (x_i + y_i) = \sum x_i + \sum y_i$$

3. For any constant k, $\sum\limits_{i=1}^{n} k$ means to add k to itself n times. Therefore, $\sum k = nk$.

We now alter (21-8) as follows:

Square both sides.

$$s^2 = \frac{\sum (x_i - \overline{x})^2}{n}$$

Expand the binomial.

$$s^2 = \frac{\sum (x_i^2 - 2x_i\overline{x} + \overline{x}^2)}{n}$$

Use statement 2 from above.

$$= \frac{\sum x_i^2 - \sum 2x_i\overline{x} + \sum \overline{x}^2}{n}$$

Since $2\overline{x}$ is a constant, apply statement 1 from above to obtain:

$$= \frac{\sum x_i^2 - 2\overline{x}\sum x_i + \sum \overline{x}^2}{n}$$

Since $\overline{x}$ is a constant, apply statement 3 to get:

$$= \frac{\sum x_i^2 - 2\overline{x}\sum x_i + n\overline{x}^2}{n}$$

Divide.

$$= \frac{\sum x_i^2}{n} - 2\overline{x}\frac{\sum x_i}{n} + \overline{x}^2$$

Since $\sum \overline{x}_i/n = \overline{x}$, the middle term becomes $-2\overline{x}^2$.

$$s^2 = \frac{\sum x_i^2}{n} - 2\overline{x}^2 + \overline{x}^2$$

$$= \frac{\sum x_i^2}{n} - \overline{x}^2$$

Replacing $\overline{x}$ with $\sum x_i/n$ and taking the square root of both sides, we have the following result:

Alternate Form of Standard Deviation Formula

$$s = \sqrt{\frac{\sum\limits_{i=1}^{n} (x_i^2)}{n} - \left(\frac{\sum\limits_{i=1}^{n} x_i}{n}\right)^2} \qquad (21\text{-}10)$$

Mean of the squares

Square of the mean

For calculations with a calculator, this formula is easier to use than (21-8) because it involves the values themselves rather than their individual differences from the mean.

EXAMPLE 25   Find the standard deviation of the following measurements (in inches):   1.25, 1.22, 1.24, 1.22, 1.26, 1.23, 1.27.

Solution
Step 1.   Calculate the square of the mean and store in memory:

$$\left(\frac{\sum x}{n}\right)^2:\quad 1.25\ \boxed{+}\ 1.22\ \boxed{+}\ \cdots\ \boxed{+}\ 1.27\ \boxed{=}\ \boxed{\div}\ 7\ \boxed{=}\ \boxed{x^2}\ \boxed{\text{STO}}\ \rightarrow\quad \boxed{1.5411449}$$

Step 2.   Calculate the mean of the squares:

$$\frac{\sum (x^2)}{n}:\quad 1.25\ \boxed{x^2}\ \boxed{+}\ 1.22\ \boxed{x^2}\ \boxed{+}\ \cdots\ \boxed{+}\ 1.27\ \boxed{x^2}\ \boxed{=}\ \boxed{\div}\ 7\ \boxed{=}\ \rightarrow$$

$$\boxed{1.5414714}$$

Step 3.   Subtract the square of the mean from this:

$$\boxed{-}\ \boxed{\text{RCL}}\ \boxed{=}\ \rightarrow\quad \boxed{0.0003265}$$

Step 4.   Take the square root:

$$\boxed{\sqrt{\ }}\ \rightarrow\quad \boxed{0.0180702}\qquad s \approx 0.018 \text{ in.}$$

NOTE   Many calculators have special keys for determining the mean, $\sum x$, $\sum x^2$, and standard deviation. Consult your instruction manual for details. ◀

To calculate the standard deviation from a frequency distribution, use the following modification of (21-10):

$$s = \sqrt{\frac{\sum\limits_{i=1}^{n} x_i^2 f_i}{n} - \left(\frac{\sum\limits_{i=1}^{n} x_i f_i}{n}\right)^2} \tag{21-11}$$

where f_i is the frequency value of x_i and $n = \sum f_i$.

EXAMPLE 26   Use (21-11) to calculate the standard deviation of the printer speeds from Example 12 of Section 21-2.

Solution   Using frequency distributions, we need four columns, as shown in Table 21-3 at the top of page 789.
Step 1.   From the second and third columns,

$$\left(\frac{\sum xf}{n}\right)^2 = \left(\frac{3191}{20}\right)^2 = (159.55)^2 = 25{,}456.203$$

TABLE 21-3

> To get this column, multiply each value in the third column by the corresponding value in the first column.

x	f	xf	x²f
156	1	156	24,336
157	2	314	49,298
158	2	316	49,928
159	5	795	126,405
160	4	640	102,400
161	3	483	77,763
162	2	324	52,488
163	1	163	26,569
Σ: 20		3191	509,187

Step 2. From the second and fourth columns,

$$\frac{\Sigma\, x^2 f}{n} = \frac{509{,}187}{20} = 25{,}459.35$$

Step 3. Subtract:

$$25{,}459.35 - 25{,}456.203 = 3.147$$

Step 4. Take the square root:

$$s = \sqrt{3.147} \approx 1.8 \text{ cps}$$

When frequency distributions involve intervals of values, the standard deviations, like the mean, should be calculated using the midpoints of the intervals.

EXAMPLE 27 Calculate the standard deviation for the distribution of electricity consumption from Example 13 of Section 21-2.

Solution As shown in Table 21-4, we use the midpoints of the intervals, 594.5, 604.5, and so on for x_i.

TABLE 21-4

x	594.5	604.5	614.5	624.5	634.5	644.5	654.5
f	1	2	1	3	2	3	5

x	664.5	674.5	684.5	694.5	704.5	714.5	724.5
f	7	9	3	2	0	1	1

Step 1. First, note that $n = \Sigma f = 40$. Then, find the square of the mean and store in memory:

$$\left(\frac{\sum xf}{n}\right)^2: \quad 594.5 \boxed{+} 604.5 \boxed{\times} 2 \boxed{+} \cdots \boxed{+} 724.5 \boxed{=} \boxed{\div} 40 \boxed{=} \boxed{x^2} \boxed{\text{STO}} \rightarrow$$

$$\boxed{434610.56}$$

Step 2. Find the mean of the squares:

$$\frac{\sum x^2 f}{n}: \quad 594.5 \boxed{x^2} \boxed{+} 604.5 \boxed{x^2} \boxed{\times} 2 \boxed{+} \cdots \boxed{+} 724.5 \boxed{x^2} \boxed{=} \boxed{\div} 40 \boxed{=} \rightarrow$$

$$\boxed{435405.5}$$

Step 3. Subtract. $\boxed{-}\boxed{\text{RCL}}\boxed{=} \rightarrow \quad \boxed{794.9375}$

Step 4. Take the square root: $\boxed{\sqrt{\;}} \rightarrow \quad \boxed{28.194636}$

$$s \approx 28.2 \text{ kWh}$$

Significance of standard deviation   In Section 21-2 we showed the graph of a symmetric distribution of data known as the *normal*, or *Gaussian*, **distribution.** This curve has the following equation:

$$y = \left(\frac{1}{s\sqrt{2\pi}}\right) e^{-(x-\bar{x})^2/2s^2}$$

where y is the height of the curve for a given value of x, $\bar{x}$ is the arithmetic mean, and s is the standard deviation. Statisticians have found that, given a large enough sample, the distributions of measurements of many natural objects or processes tend to fit the Gaussian curve. As shown in Fig. 21-5, this distribution has the following characteristics:

- The mean is at the peak of the curve.
- The distribution is symmetric.
- Roughly 68% of the measurements fall within 1 standard deviation of the mean—that is, between $\bar{x} - s$ and $\bar{x} + s$. About 95% of the data fall within 2 standard deviations of the mean, and approximately 99% of the values are between $\bar{x} - 3s$ and $\bar{x} + 3s$. These percentages are often referred to as **confidence levels.** When a specific percent of confidence is called for, the following list of corresponding standard deviations will be useful.

Figure 21-5

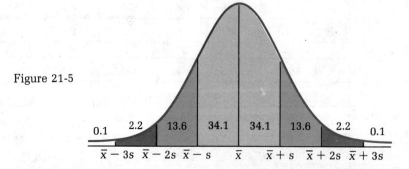

Confidence Level	Number of Standard Deviations
80%	1.28
90%	1.65
95%	1.96
96%	2.05
98%	2.33
99%	2.58

The standard deviation is an important measure of the spread of a distribution because it appears in the equation for a normal distribution. This allows statisticians to find the probability that any given measurement will fall within any specified interval of measurement.

EXAMPLE 28   State the range of values that are within 1 standard deviation of the mean for the data in Example 26. What actual percent of the data fell within this range? What percent of the measurements would you expect to find in that interval?

Solution $\bar{x} - s \approx 159.6 - 1.8 \approx 157.8$

$\bar{x} + s \approx 159.6 + 1.8 \approx 161.4$

We see that 14 out of 20 values, or 70% fell within 1 standard deviation of the mean. We would expect to find about 68% of the 20 values in this interval.

EXAMPLE 29   From a series of measurements, the mean voltage in a certain line is found to be $V = 22{,}680 \pm 128$ V, where 128 V is the standard deviation. The voltage varies randomly and normally about the mean with this standard deviation. A breaker must be installed to protect against surges in voltage. If we set the breaker so that it is activated only about 0.1% of the time, at what level is it set?

Solution   According to Fig. 21-5, the number of values in a normal distribution that exceeds $\bar{x} + 3s$ is approximately 0.1%. The equation given becomes $\bar{x} + 3s = 22{,}680 + 3(128) = 23{,}064$ V. The breaker is designed to activate when the voltage exceeds 23,064 V.

Exercises 21-3

Use data sets A–D for Problems 1–8.

A. *Yearly rainfall (in inches):*   17.52, 21.28, 18.33, 33.15, 16.51, 11.24, 25.96, 17.24, 15.25, 22.18, 14.36, 27.59, 19.73, 16.25, 20.48

B. *Length of steel pin (in millimeters):*   3.24, 3.25, 3.23, 3.25, 3.26, 3.24, 3.25, 3.25, 3.24, 3.26

C.   *Monthly sales of calculators:*   57, 48, 71, 55, 62, 29, 51, 44, 71, 60, 58, 41, 61, 59, 36, 44, 58, 52, 70, 62, 53, 66, 49, 61, 73, 56, 62, 49, 61, 50

D.   *Test scores:*   86, 94, 71, 82, 77, 65, 88, 91, 66, 78, 83, 54, 92, 75, 77, 72, 80, 49, 69, 73, 97, 84, 71, 77, 55, 68, 82, 91, 76, 38

Find the standard deviation of the numbers in the indicated set. Use the interval method (intervals of 5) for Problems 3 and 4.

1.   Set A   2.   Set B   3.   Set C   4.   Set D

In Problems 5–10, determine the percentage of values within 1 standard deviation of the mean for each set of data.

5.   Set A   6.   Set B   7.   Set C   8.   Set D

9.   The daily outputs of Example 24   10.   The measurements of Example 25

Calculate the standard deviation for each set of data.

11.   *Energy technology*   The gas consumption in Problem 27 of Exercises 21-2
12.   *Automotive engineering*   The acceleration data in Problem 31 of Exercises 21-2
13.   *Machine technology*   The measurements in Example 17 of Section 21-2
14.   *Business*   The salaries in Example 20 of Section 21-2

Solve.

15.   *Machine technology*   Measurements on a test batch of steel rods resulted in a mean length of 21.38 cm and a standard deviation of 0.09 cm. What tolerance about the mean should be specified so that only 5% of the rods are rejected?
16.   *Business*   In Problem 12, what range of 0–60 mi/h times should a car salesperson tell a customer so that the figures have a 90% certainty? (Assume the data will result in a normal distribution over a large number of measurements.)

21-4 | Empirical Curve Fitting

So far, we have analyzed measurement data involving only one measurable quantity at a time. However, our experience with functions and formulas has shown us that consistent relationships often may be found to exist between two measurable quantities.

Many of these relationships can be derived mathematically. For example, from geometry we know that the relationship between the circumference C and the diameter d of a circle is given by the equation $C = \pi d$. Relationships in technology often cannot be derived mathematically, but must be discovered empirically—that is, by measuring one variable while systematically varying the other. Because of measurement errors and other factors, these relationships rarely turn out to be exact; instead, we look for a pattern that can be approximated by the equation of some familiar curve. The process of finding such an equation is called **empirical curve fitting**.

Before attempting to derive an empirical relationship between the measured values of two variables, we must first decide on the type of curve that will fit the data best. This decision can be based on either some prior knowledge of the relationship or the pattern evident when the data values are plotted as ordered pairs.

EXAMPLE 30 The length L of a metal rod is measured repeatedly as its temperature T changes. The table shows the resulting data, and the graph of the data is shown in Fig. 21-6.

T(°C)	L(cm)
10	64.600
15	64.603
20	64.605
25	64.608
30	64.612
35	64.614

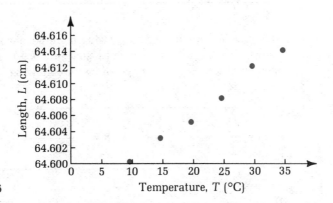

Figure 21-6

The data seem to follow an approximately straight line. A student of physics would know that the length of a rod is related to its temperature by the general linear equation

$$L = L_0 + L_0 \alpha T$$

where L_0 is the length of the rod at $T = 0°C$ and α is the coefficient of linear expansion. Thus, it seems clear from the graph and from prior knowledge that we should attempt to fit the data to a straight line.

The least-squares method   Because of errors of measurement, the points in Fig. 21-6 do not fall in a perfectly straight line. Any straight line drawn in the vicinity of the points will miss at least a few of them. The difference between the ordinate of a data point and the ordinate of a point on an empirical curve with the same abscissa is called the **deviation** of the data point (Fig. 21-7).

Figure 21-7

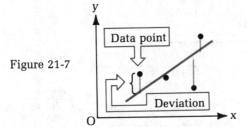

The most commonly used method for determining the equation of the best-fitting curve for a set of data is the **method of least squares.** This method uses the data points to find the equation of the curve for which the sum of the squares of the deviation of each point is a minimum.

Linear curve fitting   The equation of any straight line can be expressed in slope–intercept form:

$$y = mx + b \qquad\qquad (21\text{-}12)$$

where m and b are constants. Using more advanced mathematics, it can be shown that the values of m and b that produce the best fit according to the method of least squares are given by the following formulas:

$$m = \frac{n\sum xy - \sum x \sum y}{n \sum x^2 - (\sum x)^2} \qquad b = \frac{\sum x^2 \sum y - \sum x \sum xy}{n \sum x^2 - (\sum x)^2} \qquad (21\text{-}13)$$

Here, x and y represent the coordinates of the data points, and n is the number of points. Notice that the denominator for each formula is the same. This fact will save computational time.

EXAMPLE 31   Find the least-squares equation of the line that best fits the data in Example 30. Use the result to find L for a temperature of 50°C.

Solution   Once again it is best to organize the data as shown in Table 21-5. Notice that the columns represent the quantities in (21-13) that must be summed, where x represents the temperature T and y represents the length L.

TABLE 21-5

x	y	xy	x²
10	64.600	646	100
15	64.603	969.045	225
20	64.605	1292.1	400
25	64.608	1615.2	625
30	64.612	1938.36	900
35	64.614	2261.49	1225
$\sum$: 135	387.642	8722.195	3475

Using the sums at the foot of each column, and the fact that $n = 6$, we substitute into (21-13) to obtain

$$m = \frac{6(8722.195) - (135)(387.642)}{6(3475) - (135)^2}$$

6 $\boxed{\times}$ 8722.195 $\boxed{-}$ 135 $\boxed{\times}$ 387.642 $\boxed{=}$ $\boxed{\div}$ $\boxed{(}$ 6 $\boxed{\times}$ 3475 $\boxed{-}$ 135 $\boxed{x^2}$ $\boxed{)}$ $\boxed{STO}$

$\boxed{=}$ $\rightarrow$ *0.0005714*

(Notice that we stored the denominator in memory to avoid recomputing it for b.)

$$m \approx 0.00057 \text{ cm/}°\text{C}$$

$$b = \frac{(3475)(387.642) - (135)(8722.195)}{6(3475) - (135)^2}$$

3475 $\boxed{\times}$ 387.642 $\boxed{-}$ 135 $\boxed{\times}$ 8722.195 $\boxed{=}$ $\boxed{\div}$ $\boxed{RCL}$ $\boxed{=}$ $\rightarrow$ *64.594143*

$$b \approx 64.594 \text{ cm}$$

The least-squares equation is

$$y \approx 0.00057x + 64.594 \qquad \text{or} \qquad L \approx 64.594 + 0.00057T$$

At $T = 50°\text{C}$,

$$L \approx 64.594 \text{ cm} + (0.00057 \text{cm/}°\text{C})(50°\text{C}) \approx 64.623 \text{ cm}$$

The graph in Fig. 21-8 shows that this line fits the data points very closely.

Figure 21-8

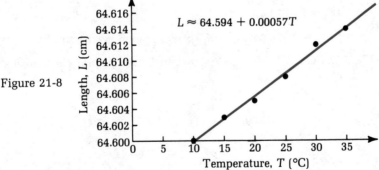

CAUTION ▶ Always plot both the least-squares line and the data points to be certain that the line does provide a good fit and that no great errors have been made. ◀

EXAMPLE 32 The following experimental results show the stretch y of a wire when a load L is hung from it:

y (cm)	4.20	4.25	4.42	4.47	4.55
L (kg)	0.00	1.00	2.00	3.00	4.00

Find the equation of the least-squares line that fits the data, and determine y when $L = 7.00$ kg.

Solution **First,** construct a table of values (Table 21-6).

TABLE 21-6

L (kg)	y (cm)	Ly	L²
0.00	4.20	0	0
1.00	4.25	4.25	1.00
2.00	4.42	8.84	4.00
3.00	4.47	13.41	9.00
4.00	4.55	18.20	16.00
$\sum$: 10.00	21.89	44.70	30.00

Then, substitute into the least-squares formulas (21-13) to find m and b:

$$m = \frac{5(44.70) - (10.00)(21.89)}{5(30.00) - (10.00)^2} = 0.092 \text{ cm/kg}$$

$$b = \frac{(30.00)(21.89) - (10.00)(44.70)}{5(30.00) - (10.00)^2} = 4.194 \text{ cm}$$

The least-squares equation is

$$y = 0.092L + 4.194$$

When $L = 7.00$ kg,

$$y = (0.092 \text{ cm/kg})(7.00 \text{ kg}) + 4.194 \text{ cm} \approx 4.84 \text{ cm}$$

The graph in Fig. 21-9 shows that the equation obtained fits the data very closely.

Figure 21-9

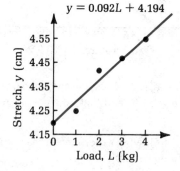

Nonlinear curve fitting   Of course, many pairs of physical quantities are related by nonlinear functions. For example, current in a simple dc circuit is inversely proportional to resistance, and for an object with a constant acceleration, displacement varies with the square of time. If the graph of the experimental data or prior knowledge suggests a nonlinear function $f(x)$, then the relationship between the variables can be expressed by rewriting the slope–intercept form (21-12) as

$$y = m[f(x)] + b \tag{21-14}$$

where $f(x)$ is some simple, usually algebraic, function. The values of m and b can then be found by replacing x with $f(x)$ in (21-13) as follows:

$$m = \frac{n\sum[f(x)]y - \sum f(x)\sum y}{n\sum[f(x)]^2 - [\sum f(x)]^2} \qquad b = \frac{\sum[f(x)]^2\sum y - \sum f(x)\sum[f(x)]y}{n\sum[f(x)]^2 - [\sum f(x)]^2} \tag{21-15}$$

EXAMPLE 33 Find the least-squares equation of the form $y = mt^2 + b$ that best fits the following data. Then find y at $t = 10.0$ s.

Distance, y (m)	0.8	1.4	2.4	3.3	4.7	6.2	7.1
Time, t (s)	0.0	1.0	2.0	3.0	4.0	5.0	6.0

Solution Using $f(t) = t^2$, we set up Table 21-7. Noting that $n = 7$, we substitute into (21-15) to obtain

$$m = \frac{7(526.5) - (91.0)(25.9)}{7(2275.0) - (91.0)^2} \approx 0.174 \text{ m/s}^2$$

$$b = \frac{(2275.0)(25.9) - (91.0)(526.5)}{7(2275.0) - (91.0)^2} \approx 1.44 \text{ m}$$

TABLE 21-7

t	y	$f(t) = t^2$	$f(t)y$	$[f(t)]^2$
0.0	0.8	0.0	0.0	0.0
1.0	1.4	1.0	1.4	1.0
2.0	2.4	4.0	9.6	16.0
3.0	3.3	9.0	29.7	81.0
4.0	4.7	16.0	75.2	256.0
5.0	6.2	25.0	155.0	625.0
6.0	7.1	36.0	255.6	1296.0
$\sum$: 25.9		91.0	526.5	2275.0

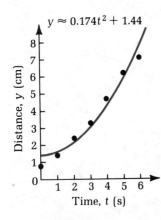

Figure 21-10

The least-squares equation is

$$y \approx 0.174t^2 + 1.44$$

For $t = 10.0$ s,

$$y \approx (0.174 \text{ m/s}^2)(10.0 \text{ s})^2 + 1.44 \text{ m} \approx 19 \text{ m}$$

The graph in Fig. 21-10 verifies that we have obtained the best fit.

▦ **EXAMPLE 34**   The temperature T of a heated object was measured at various times t as it cooled in a constant environment. Fit a curve to the resulting data using the least-squares method. Use $T = me^{-0.2t} + b$ and find T when $t = 7.00$ min.

Temperature, T (°F)	207	185	155	140	120
Time, t (min)	1.00	2.00	3.00	4.00	5.00

Solution   **First,** note that $f(t) = e^{-0.2t}$ and that $f(1.00) = e^{-0.2}$, $f(2.00) = e^{-0.4}$, and so on.

Next, calculate $\sum T$, $\sum f(t)$, $\sum f(t)T$, and $\sum [f(t)]^2$.

$\sum T$:   207 ⊕ 185 ⊕ ··· ⊕ 120 ⊜ →　　　　　　　*807.*

$\sum f(t)$:   .2 ⊬ eˣ ⊕ .4 ⊬ eˣ ⊕ ··· ⊕ 1 ⊬ eˣ ⊜ →　　*2.8550708*

$\sum f(t)T$:   .2 ⊬ eˣ ⊗ 207 ⊕ .4 ⊬ eˣ ⊗ 185 ⊕ ··· ⊕ 1 ⊬ eˣ ⊗ 120

　　　　　　　　　　　　　　　　　　　⊜ →　　*485.60387*

$\sum [f(t)]^2$:   .2 ⊬ eˣ x² ⊕ .4 ⊬ eˣ x² ⊕ ··· ⊕ 1 ⊬ eˣ x²

　　　　　　　　　　　　　　　　　　　⊜ →　　*1.7580750*

Finally, substitute into (21-15):

$$m = \frac{5(485.60387) - (2.8550708)(807)}{5(1.7580750) - (2.8550708)^2} \approx 194.1°F$$

$$b = \frac{(1.7580750)(807) - (2.8550708)(485.60387)}{5(1.7580750) - (2.8550708)^2} \approx 50.60°F$$

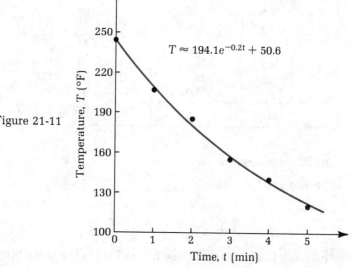

Figure 21-11

$T \approx 194.1e^{-0.2t} + 50.6$

The least-squares equation is

$$T \approx 194.1e^{-0.2t} + 50.60$$

At $t = 7.00$ min,

$$T \approx 194.1e^{-0.2(7.00)} + 50.60 \approx 98.5°F$$

Check your work by graphing, as in Fig. 21-11.

Exercises 21-4

Find the equation of the line that best fits each set of data using the least-squares method. Plot the data points and the best-fitting line on the same graph to check your result.

1.

x	1	2	3	4	5	6
y	5.6	6.1	7.1	10.5	11.1	12.3

2.

x	2	4	6	8	10	12
y	8	25	34	52	67	71

3.

x	5	10	15	20	25	30	35
y	32	24	21	19	11	8	3

4.

x	1	7	4	2	8	11	6	5
y	69	34	56	64	40	24	38	50

Find the best-fitting curve of the given form for each set of data using the least-squares method. Plot the data points and the best-fitting curve on the same graph to check your results.

5.

x	1	2	3	4	5
y	3.3	23.8	62.4	102.6	188.0

$y = mx^2 + b$

6.

x	10	20	30	40	50
y	19	25	26	29	32

$y = m\sqrt{x} + b$

7.

x	0.1	0.2	0.3	0.4	0.5	0.6
y	25	31	35	46	50	53

$y = m(10^x) + b$

8.

x	2	4	6	8
y	5	15	80	605

$y = m(e^x) + b$

Solve and graph as in Problems 1–8.

9. *Physics*   The following data were collected on the speed of sound in dry air at various temperatures:*

Temperature, T (°F)	9.0	32.0	48.0	68.0	91.0
Speed, v (ft/s)	1050	1080	1100	1110	1140

Assuming speed to be a function of temperature, find the least-squares line that best fits the data. What is the speed of sound at 0°F?

10. *Electronics*   The impedance Z in an RL circuit was measured at various frequencies f as follows:

f (Hz)	120	156	196	220	240
Z (Ω)	7.2	8.1	8.8	9.9	10.6

Fit the data to a straight line of the form $Z^2 = mf^2 + b$. What is the impedance at a frequency of 320 Hz?

11. *Physics*   The displacement s of an object moving at a constant velocity was measured at various elapsed times t as follows:

t (h)	1.0	2.0	3.0	4.0	5.0	6.0
s (mi)	77	132	174	220	298	360

Using the least-squares method, fit the data to a line of the form $s = mt + b$. Using this line, estimate the displacement after 2.5 h.

12. *Electrical engineering*   Measurement of the resistance R of a certain wire at various temperatures T resulted in the following data:

*Based on Ernest Zebrowski, Jr., *Fundamentals of Physical Measurement* (Boston: Duxbury, 1979), p. 121.

T (°C)	5.0	10.0	15.0	20.0	25.0	30.0
R (Ω)	8.5	12.1	16.4	18.6	22.0	25.3

Use the least-squares method to find the best-fitting line for R as a function of T. Use the resulting equation to determine the resistance at a temperature of 50.0°C.

13. *Automotive engineering*   The torque output T of an engine was measured at various frequencies of rotation f, resulting in the following data:

f (rev/min)	500	1000	1500	2000	2500	3000
T (lb·ft)	162	75	58	37	32	22

Determine the best-fitting curve of the form

$$T = m\left(\frac{1}{f}\right) + b$$

for the data. Use the best-fitting equation to find the torque at 1200 rev/min.

14. *Biology*   The following table lists data on the growth of bacteria in large containers over a period of time:

Time, t (h)	1.0	2.0	3.0	4.0
Estimated Number of Bacteria, n (thousands)	7	12	21	32

Using the least-squares method, fit the data to a curve of the form $n = m(2^t) + b$. Use the best-fitting equation to determine how many bacteria were present at the start of the experiment ($t = 0$ h).

15. *Acoustical engineering*   The sound level L at a given radial distance from a source was measured at various angles θ around the source as follows:

θ	0°	30°	90°	135°	180°	225°	270°	330°
L (dB)	82	76	63	41	38	48	61	81

Determine the best-fitting curve of the form $L = m \cos \theta + b$ using the least-squares method.

16. *Energy technology*   The power P generated by a windmill was measured for various wind velocities v as follows:

v (mi/h)	10	14	22	31	37	44
P (W)	88	220	870	2300	4000	6800

Fit the data to a curve of the form $P = mv^3 + b$. How much power would you expect to find generated by a wind of 25 mi/h?

21-5 | Review of Chapter 21

Important Terms and Concepts

probability (p. 770)
  a priori (p. 770)
  a posteriori (p. 770)

frequency distribution (p. 776)

histogram (p. 778)

frequency polygon (p. 778)

normal distribution (p. 779)

measure of central tendency (p. 780)
  arithmetic mean (p. 780)
  median (p. 781)
  mode (p. 782)

summation notation (p. 780)

measure of dispersion (p. 784)

standard deviation (p. 784)

empirical curve fitting (p. 792)

Formulas

- probability of an event (p. 770): $P = \dfrac{s}{n}$ **(21-1)**

    where s is the number of desired or successful outcomes, and n is the total number of outcomes

- probability of failure (p. 771): $F = 1 - P = 1 - \dfrac{s}{n}$ **(21-2)**

- probability of A or B (p. 772): $P_{A\,or\,B} = P_A + P_B$ **(21-3)**
- probability of A and B (p. 773): $P_{A\,and\,B} = P_A \cdot P_B$ **(21-4)**

- arithmetic mean (p. 780): $\bar{x} = \dfrac{x_1 + x_2 + \cdots + x_n}{n} = \dfrac{\displaystyle\sum_{i=1}^{n} x_i}{n}$ **(21-5)**

- arithmetic mean from a frequency distribution (p. 781):

    $$\bar{x} = \frac{f_1 x_1 + f_2 x_2 + \cdots + f_n x_n}{f_1 + f_2 + \cdots + f_n} = \frac{\displaystyle\sum_{i=1}^{n} f_i x_i}{\displaystyle\sum_{i=1}^{n} f_i}$$ **(21-6)**

- standard deviation (p. 784): $s = \sqrt{\dfrac{\displaystyle\sum_{i=1}^{n} (x_i - \bar{x})^2}{n}}$ **(21-8)**

- alternate form of (21-8) (p. 787): $s = \sqrt{\dfrac{\displaystyle\sum_{i=1}^{n} (x_i^2)}{n} - \left(\dfrac{\displaystyle\sum_{i=1}^{n} x_i}{n}\right)^2}$ **(21-10)**

- standard deviation from a frequency distribution (p. 786):

    $$s = \sqrt{\frac{\displaystyle\sum_{i=1}^{n} f_i (x_i - \bar{x})^2}{n}} \qquad \text{where } n = \sum_{i=1}^{n} f_i$$ **(21-9)**

- alternate form of (21-9) (p. 788): $s = \sqrt{\dfrac{\displaystyle\sum_{i=1}^{n} x_i^2 f_i}{n} - \left(\dfrac{\displaystyle\sum_{i=1}^{n} x_i f_i}{n}\right)^2}$ **(21-11)**

- least-squares method of curve-fitting: to fit a set of data to the line $y = mx + b$, calculate m and b as follows (p. 794):

$$m = \frac{n \sum xy - \sum x \sum y}{n \sum x^2 - (\sum x)^2} \qquad b = \frac{\sum x^2 \sum y - \sum x \sum xy}{n \sum x^2 - (\sum x)^2} \quad \textbf{(21-13)}$$

- to fit a set of data to the curve $y = m[f(x)] + b$, calculate m and b as follows (p. 797):

$$m = \frac{n \sum [f(x)]y - \sum f(x) \sum y}{n \sum [f(x)]^2 - [\sum f(x)]^2} \qquad b = \frac{\sum [f(x)]^2 \sum y - \sum f(x) \sum [f(x)]y}{n \sum [f(x)]^2 - [\sum f(x)]^2}$$

Exercises 21-5

A standard roulette wheel has 38 equally large slots marked at random with the numbers 1–36, 0, and 00. Half the numbers 1–36 are red and half are black; 0 and 00 are green. As the wheel spins, a ball is tossed in at random so that it will land in one of the slots. Find the probability to the nearest 0.01 of each of the following results:

1. A positive even number
2. Landing in slot 0 or 00
3. One of the numbers 1–12
4. A black slot
5. Two red numbers appearing on successive spins
6. The number 15

A jar contains 20 prize certificates for a drawing. There are 12 third-prize certificates for movie passes, 6 second-prize certificates for record albums, and 2 first-prize certificates for cassette decks. Determine the probability (to the nearest 0.01) of randomly drawing each of the following:

7. A movie pass
8. A record album
9. A movie pass or a cassette deck
10. A record album or a cassette deck
11. Two movie passes on successive choices if the first one is not replaced
12. A movie pass, followed by a record album, followed by a cassette deck if none of the picks is replaced

Use the following data for Problems 13–18:   87, 65, 94, 71, 81, 91, 77, 75, 62, 74, 83, 75, 98, 86, 72, 64, 76, 80, 79, 84.

13. Determine the median score.
14. Calculate the arithmetic mean.
15. Compute the standard deviation.
16. Set up a frequency distribution using the intervals 61–65, 66–70, etc.
17. Use the result from Problem 16 to make a histogram.
18. Use the result from Problem 17 to plot a frequency polygon.

Automotive engineering   A series of ten measurements of the torque (in newton–meters) produced by an engine rotating at a constant frequency yielded the following results: 385.6, 388.1, 384.8, 386.2, 385.7, 387.4, 386.0, 384.1, 388.8, 386.1. Use these data for Problems 19–22.
19. Determine the median of these measurements.
20. Calculate the arithmetic mean of the torque.
21. Find the standard deviation of the measurements.
22. What percent of the measurements are within 1 standard deviation of the mean?

Electronics   The following frequency distribution was the result of a series of measurements of the current in a certain circuit:

Current (A)	2.4	2.5	2.6	2.7	2.8	2.9
Frequency	1	5	16	18	6	2

Use these data for Problems 23–26.

23. Construct a frequency polygon for the data.
24. Determine the median value of the current.
25. Calculate the arithmetic mean of the current.
26. Compute the standard deviation of the current measurements.

Solve.

27. *Graphics technology*   A print shop ran 800 copies of a flyer, and 16 were damaged in printing. Based on this run, what is the probability that a flyer will be damaged in some future run? How many flyers should be printed in order to obtain 10,000 perfect ones?

28. *Electronics technology*   An electronics technician must estimate the cost of a certain repair job for a customer. The technician discovers that eight similar repairs in the past have required the following times (in hours) to complete: 7.5, 6.5, 6.0, 6.0, 8.5, 7.0, 8.0, 8.5. Based on the arithmetic mean of these times, what is the best estimate of the total cost of the repair at $36 per hour?

29. *Industrial management*   During the past 5 years, the annual increases in sales at the Tymrite Stopwatch Co. have been 4.7%, 5.4%, 3.8%, 5.6%, and 4.4%. Given that 68,200 stopwatches were sold by Tymrite last year, use the arithmetic mean of the annual increases to estimate sales for next year.

30. *Electronics*   For the circuit in Problem 26, what percent of the data falls within 1 standard deviation of the mean?

Using the least-squares method, determine the best-fitting curve of the form shown for each set of data.

31.

x	2	5	8	11	14
y	6	24	48	70	84

$y = mx + b$

32.

x	5	10	15	20	25
y	16	37	80	140	195

$y = mx^2 + b$

33. *Physics*   The following measurements were taken on the force p required to lift a weight w with a pulley block:

w (lb)	20	40	60	90	120
p (lb)	8	11	16	22	25

Fit the data to a linear function of the form $p = mw + b$ using the least-squares method. How much force would be needed to lift a weight of 250 lb in the same way?

34. *Physics* The pressure P at various depths d below the surface of a container of fluid was measured with the following results:

d (m)	2.0	10.0	14.0	20.0	26.0
P (kPa)	115	173	220	265	321

Determine the best-fitting line for P as a linear function of d. Use the equation to calculate the pressure at a depth of 55.0 m.

35. *Optics* The illuminance i of a light source was measured at various distances d from the source, and the results are shown in the following table:

d (ft)	1.0	2.0	4.0	6.0	8.0
i (lx)	498	138	39	16	13

Fit the data to a curve of the form

$$i = m\left(\frac{1}{d^2}\right) + b$$

using the least-squares method. Use this result to find the illuminance at a distance of 10.0 ft.

36. *Chemistry* The following table gives a series of measurements showing the boiling point T of water at various pressures P:

Pressure, P (kPa)	10.1	16.5	41.3	80.2	100.7	150.4
Boiling Point, T (°C)	44.8	55.2	75.7	94.0	94.4	111.6

Find the best-fitting curve of the form $T = m(\log P) + b$ using the least-squares method. Use this best-fitting equation to find the boiling point at a pressure of 900.0 kPa.

Introduction to Differentiation

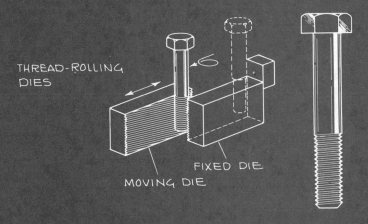

THREAD-ROLLING DIES

FIXED DIE

MOVING DIE

THREADING A BOLT

The idea of change and its mathematical description are fundamental to any understanding of scientific or technical phenomena. In the 17th century scientists became aware of the importance of the idea of change through their study of the motion of planets, projectiles, sound, and light. It soon became clear that many of the most important problems could not be solved—or solved very easily—using the traditional mathematics of algebra, geometry, and trigonometry. In order to study problems in which some quantity changed, scientists and mathematicians developed a package of new mathematical tools now known as *calculus*. This branch of mathematics has become of vital importance to all scientific and technical work.

The following problems show the need for some new mathematics beyond algebra and trigonometry.

PROBLEM 1 Figure 22-1 shows how fuel efficiency changes with speed for an automobile with an experimental Rankine engine. Fuel efficiency E is a function of the vehicle speed v. At what speed is the fuel efficiency a maximum?

Figure 22-1

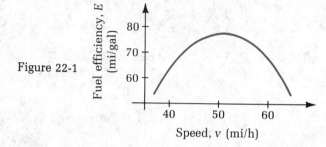

The maximum value of fuel efficiency is at the highest point on the curve. At the maximum, a horizontal line will touch the curve at only a single point (Fig. 22-2). A good estimate from the graph is that maximum fuel efficiency is roughly 75 mi/gal and occurs at the point where v is about 50 mi/h. A more careful measurement on the graph would show that the maximum is closer to $E \approx 77$ mi/gal at $v \approx 51$ mi/h, but this is not the *exact* answer.

To find an exact or analytical answer, we must use the equation of the curve rather than its graph. In general, given some function $f(x)$, it may be necessary to find the value of the variable x for which the function is either a maximum or a minimum. Many scientific and technical applications involve the solution of problems of this kind, and we will study them in detail in the next few chapters.

Figure 22-2

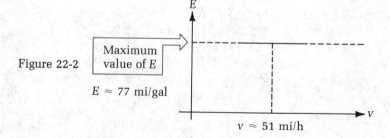

PROBLEM 2 Figure 22-3 shows the changing position of a rocket that was launched vertically and is accelerating upward. Here, the height h of the rocket is given by the function $h(t) = 40t^2$, where h is measured in meters and t is the time in seconds after the launch.

Time after Launch, t (s)	Height, h (m)
0	0
1	40
2	160
3	360
4	640
5	1000

Figure 22-3

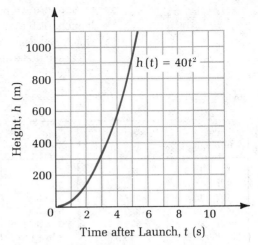

Because we have a mathematical description of the height of the rocket at any time, it is possible to calculate its average speed over any time interval. For example, the average speed of the rocket during the first 5 s is

$$\text{Average speed} = \frac{\text{Distance traveled}}{\text{Time of travel}} = \frac{\overbrace{h(5)}^{\text{Height at 5 s}} - \overbrace{h(0)}^{\text{Height at start}}}{5}$$

$$= \frac{1000 \text{ m}}{5 \text{ s}}$$

$$= 200 \text{ m/s}$$

You may recognize from your work in Chapter 20 that the average speed is the slope of the line segment joining the points at $t = 0$ and $t = 5$ (Fig. 22-4).

Figure 22-4

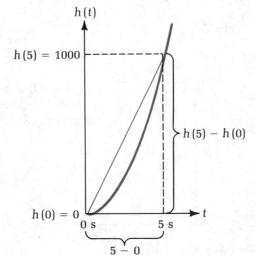

Suppose we need to determine the speed of the rocket at *exactly* 5 s after launch. Table 22-1 shows a series of calculations of the average speed for smaller and smaller intervals near 5 s after launch. The average speeds in Table 22-1 seem to be approaching a value of 400 m/s. We can estimate that at exactly 5 s after launch, the rocket is moving at a speed of 400 m/s.

TABLE 22-1

Time Interval (s)	Average Speed (m/s)	Time Interval (s)	Average Speed (m/s)	Speed at Exactly 5 s Is between:
4 to 5	360	5 to 6	440	360 and 440 m/s
4.5 to 5	380	5 to 5.5	420	380 and 420 m/s
4.8 to 5	392	5 to 5.2	408	392 and 408 m/s
4.9 to 5	396	5 to 5.1	404	396 and 404 m/s
4.99 to 5	399.6	5 to 5.01	400.4	399.6 and 400.4 m/s
4.999 to 5	399.96	5 to 5.001	400.04	399.96 and 400.04 m/s

Again, as with Problem 1, we need a mathematical procedure that will enable us to find the exact value of this speed analytically—that is, directly from the function $h(t)$ and without drawing a graph or making repeated calculations. In order to develop the calculus tools designed to solve these problems, we need to consider some important preliminary concepts: functions, limits, and continuity.

Functions

The concept of a *function* was introduced in Chapter 3, Section 3-1. You may want to review that material before you continue here.

A function can be described by a graph, by a list of data, by a formula, or by two or more formulas.

EXAMPLE 1   Graph the following functions:

(a) $f(x) = \begin{cases} x^2 & \text{for } 0 \le x < 2 \\ 4 & \text{for } x \ge 2 \end{cases}$ (b) $g(x) = x + 2$

(c) $h(x) = \dfrac{x^2 - 4}{x - 2}$

Solution

(a)

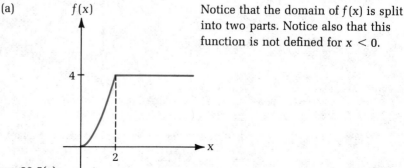

Notice that the domain of $f(x)$ is split into two parts. Notice also that this function is not defined for $x < 0$.

Figure 22-5(a)

(b)

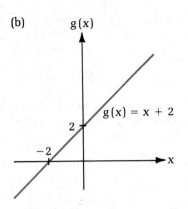

Figure 22-5(b)

(c)

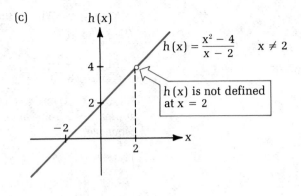

$$h(x) = \frac{x^2 - 4}{x - 2} \qquad x \neq 2$$

h(x) is not defined at x = 2

Figure 22-5(c)

Note that the functions g(x) and h(x) are not the same since their domains are different. The domain of g(x) is the set of all real numbers. However, because the value x = 2 produces a zero in the denominator, the domain of h(x) has the restriction x ≠ 2.

Limits   The idea of a **limit** is one of the most important concepts needed for the study of calculus. In everyday use, the word *limit* refers to a boundary or endpoint. In calculus, the limit of a function at a certain point is a number that is dependent on the behavior of the function near that point.

We have already seen one kind of limit in Chapter 20 in our discussion of the hyperbola. The graph of the function

$$y = 1 + \frac{1}{x}$$

is a hyperbola (Fig. 22-6). For both positive and negative values of x, as the absolute value of x becomes larger and larger, the value of the function y approaches 1. In other words, the graph gets closer and closer to the line y = 1, but never intersects it. Therefore, we say that for very large absolute values of x, *the limit of y is 1*.

Figure 22-6

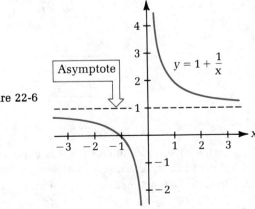

EXAMPLE 2   What is the limit of the function below as x approaches 2? [See Fig. 22-5(c).]

$$h(x) = \frac{x^2 - 4}{x - 2}$$

Solution 1   Using a calculator, we can prepare a table of values for $h(x)$ as x approaches 2 from values both less than 2 and greater than 2. Use the calculator sequence

x [STO] [x²] [−] 4 [=] [÷] [(] [RCL] [−] 2 [)] [=] →

For x < 2:

x	1.5	1.9	1.99	1.999
$h(x) = \dfrac{x^2 - 4}{x - 2}$	3.5	3.9	3.99	3.999

For x > 2:

x	2.5	2.1	2.01	2.001
$h(x) = \dfrac{x^2 - 4}{x - 2}$	4.5	4.1	4.01	4.001

As x approaches 2, $h(x)$ seems to approach 4. We say that *the limit of $h(x)$ as x approaches 2 is 4.*

Solution 2:   The Increment Method   Because the limit is determined only by the behavior of the function near x = 2, it is also possible to find the limit by examining the function at x = 2 + d, where the increment d is a very small positive or negative number.

Substitute (2 + d) for x. $h(2 + d) = \dfrac{(2 + d)^2 - 4}{(2 + d) - 2}$

Simplify. $= \dfrac{4 + 4d + d^2 - 4}{2 + d - 2}$

$$= \frac{d^2 + 4d}{d}$$

$$= \frac{d(d + 4)}{d}$$

$$= d + 4$$

For very small values of d, the number d + 4 is approximately equal to 4. Therefore, the value of $h(x)$ approaches 4 as x takes values near 2.

Notice that although the function $h(x)$ does not exist at x = 2, its limit does exist.

Notation   We will use the following notation:

- $\lim\limits_{x \to a} f(x)$ means the limit of a function $f(x)$ as x approaches a.
- $\lim\limits_{x \to a^+} f(x)$ means that we are computing the limit as x approaches a from the right side (values greater than a).
- $\lim\limits_{x \to a^-} f(x)$ means that we are computing the limit as x approaches a from the left side (values less than a).

Definition of a Limit

The limit of a function $f(x)$ is the number L that the function approaches as x approaches some value a.

$$\lim_{x \to a} f(x) = \lim_{x \to a^+} f(x) = \lim_{x \to a^-} f(x) = L$$

A function has a limit at a^* only if both the left and right limits exist and are equal.

The limit of $f(x)$ exists and is equal to L if the values of $f(x)$ approach, or cluster about, the number L as x assumes values near a.

EXAMPLE 3   The efficiency E of a heat engine is a function of the temperature T (in degrees Kelvin) at which energy is supplied:

$$E(T) = 1 - \frac{T_e}{T}$$

Find $\lim\limits_{T \to 400} E(T)$, where the exhaust temperature T_e is 200 K.

Solution 1   Using a calculator, we find the values listed in the table.

T	$E(T) = 1 - \dfrac{200}{T}$	T	$E(T) = 1 - \dfrac{200}{T}$
300	0.3333333	500	0.6
350	0.4285714	450	0.5555556
390	0.4871795	410	0.5121951
399	0.4987469	401	0.5012469
399.9	0.4998750	400.1	0.5001250
399.99	0.4999875	400.01	0.5000125

$\lim\limits_{T \to 400^-} E(T)$
From the left

$\lim\limits_{T \to 400^+} E(T)$
From the right

*The number a cannot be the endpoint of the domain of the function, since only one of the limits $\lim\limits_{x \to a^+} f(x)$ or $\lim\limits_{x \to a^-} f(x)$ can exist at the endpoint.

Because both

$$\lim_{T \to 400^-} \left(1 - \frac{200}{T}\right) \quad \text{and} \quad \lim_{T \to 400^+} \left(1 - \frac{200}{T}\right)$$

are approaching the number 0.5 as T approaches 400, we can say that

$$\lim_{T \to 400} \left(1 - \frac{200}{T}\right) = 0.5$$

Solution 2   Using the increment method, let $T = 400 + d$. Then

$$E(400 + d) = 1 - \frac{200}{400 + d}$$

$$= \frac{400 + d - 200}{400 + d}$$

$$= \frac{200 + d}{400 + d}$$

For very small values of d, the d term is negligible, and

$$E(400 + d) \approx \frac{200}{400} = \frac{1}{2}$$

Therefore, $\lim_{T \to 400} E(T) = 0.5$.

EXAMPLE 4   Find the limit of the function

$$f(x) = \frac{x}{\sqrt{x + 4} - 2}$$

as: (a) $x \to -5$ ▦ (b) $x \to 0$

Solution

(a)   The function $f(x)$ is not defined over the real numbers for $x < -4$, since these numbers lead to imaginary values for $\sqrt{x + 4}$. Because the function does not exist anywhere near the point $x = -5$, the limit of $f(x)$ as $x \to -5$ does not exist.

(b)   First, notice that for $x = 0$, $f(x)$ has the form $\frac{0}{0}$. Therefore, the function is not defined at the point $x = 0$. However, it is defined at all points nearby, so it is perfectly reasonable to seek the limit as $x \to 0$. Remember, the limit at any point depends on the behavior of the function in the neighborhood of that point (Fig. 22-7). Using the following calculator key sequence, we can construct the table shown.

$$x \boxed{+} 4 \boxed{=} \boxed{\sqrt{}} \boxed{-} 2 \boxed{=} \boxed{1/x} \boxed{\times} x \boxed{=} \to \boxed{}$$

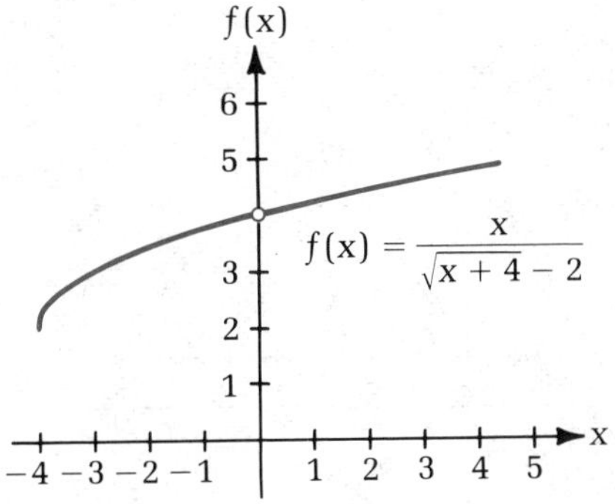

Figure 22-7

$$f(x) = \frac{x}{\sqrt{x+4} - 2}$$

x	0.1	0.01	0.001
$f(x)$	4.0248457	4.0024985	4.0002512

x	−0.1	−0.01	−0.001
$f(x)$	3.9748418	3.9974984	3.9997488

$\lim\limits_{x \to 0^+}$

$\lim\limits_{x \to 0^-}$

It seems reasonable that

$$\lim_{x \to 0^+} f(x) = \lim_{x \to 0^-} f(x) = \lim_{x \to 0} f(x) = 4$$

REMEMBER ▶ In general, the limit $\lim\limits_{x \to a} f(x)$ cannot be found by simply finding $f(a)$. ◀

EXAMPLE 5 Evaluate $\lim\limits_{x \to 1} f(x)$, where $f(x) = \begin{cases} x + 1 & \text{for } x \leq 1 \\ -1 & \text{for } x > 1 \end{cases}$

Solution From Fig. 22-8 we can see that

$\lim\limits_{x \to 1^-} f(x) = 2$ From the left, the function approaches 2.

$\lim\limits_{x \to 1^+} f(x) = -1$ From the right, the function approaches −1.

Both $\lim\limits_{x \to 1^-} f(x)$ and $\lim\limits_{x \to 1^+} f(x)$ exist, but they are not equal. Therefore, $\lim\limits_{x \to 1} f(x)$ does not exist.

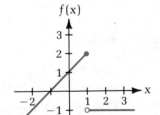

Figure 22-8

EXAMPLE 6 The coefficient of performance n of a certain refrigeration system is a function of its cold reservoir temperature T, as given below. Find $\lim\limits_{T \to 4} n(T)$.

$$n(T) = \frac{T}{4 - T}$$

Solution From Fig. 22-9 we can see that:

● The function is not defined for $T = 4$.

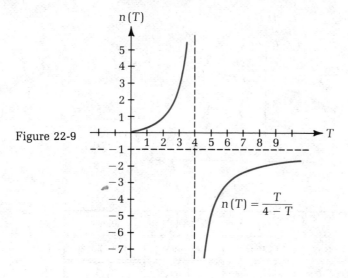

Figure 22-9

$$n(T) = \frac{T}{4 - T}$$

- The limit $\lim\limits_{T \to 4^-} n(T)$ does not exist. As T approaches the point $T = 4$ from the left, $n(T)$ increases without bound.
- The limit $\lim\limits_{T \to 4^+} n(T)$ does not exist. As T approaches 4 from the right, $n(T)$ becomes more negative without bound.

Because $\lim\limits_{T \to 4^+} n(T)$ and $\lim\limits_{T \to 4^-} n(T)$ do not exist, the limit $\lim\limits_{T \to 4} n(T)$ does not exist.

For a convenient notation and a simple way of talking about functions that increase without bound in this way, we can use the symbols ∞ or $+\infty$ (positive infinity) and $-\infty$ (negative infinity).

- $\lim\limits_{x \to a} f(x) = \infty$ means that the limit does not exist because $f(x)$ increases without bound as x approaches a.
- $\lim\limits_{x \to a} f(x) = -\infty$ means that the limit does not exist because $f(x)$ is negative and $|f(x)|$ increases without bound as x approaches a.

CAUTION   The symbols ∞ and $-\infty$ do not represent real numbers, and no arithmetic operations may be performed with them. They are simply a convenient way to indicate "increase without bound." ◀

Limits at infinity   As originally defined, a limit describes the behavior of a function $f(x)$ as the variable x approaches some real number a. But we can also use the limit concept to describe how a function behaves as x increases without bound in either the positive or negative direction.

- $\lim\limits_{x \to \infty} f(x)$ is read "limit of $f(x)$ as x approaches positive infinity" and is the limit of $f(x)$ as x increases without bound.
- $\lim\limits_{x \to -\infty} f(x)$ is read "limit of $f(x)$ as x approaches negative infinity" and is the limit of $f(x)$ as x decreases without bound.

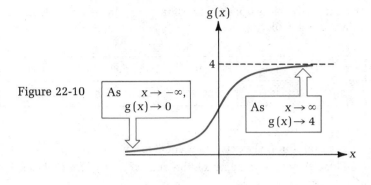

Figure 22-10

In Fig. 22-10, as $x \to \infty$, the graph approaches the line $g(x) = 4$, and we can write

$$\lim_{x \to \infty} g(x) = 4$$

As $x \to -\infty$, the graph approaches the x-axis, and we can write

$$\lim_{x \to -\infty} g(x) = 0$$

LEARNING HINT ▶ It is particularly helpful to note that the functions $1/x$, $1/x^2$, $1/x^3$, and so on, all approach zero as $x \to \infty$ or $x \to -\infty$. As x takes on larger and larger values, these functions become smaller and smaller, and thus, approach zero. ◀

EXAMPLE 7 Find $\lim_{x \to \infty} f(x)$, where $f(x) = \dfrac{4x^2 + 5x - 2}{2x^2 + 1}$

Solution As $x \to \infty$, both the numerator and denominator become very large. Because it is the x^2-term in the denominator that grows large as $x \to \infty$, we divide each term in the numerator and denominator by x^2, and the function becomes

$$f(x) = \frac{4 + \dfrac{5}{x} - \dfrac{2}{x^2}}{2 + \dfrac{1}{x^2}}$$

As $x \to \infty$, the terms $5/x$, $2/x^2$, and $1/x^2$ all approach zero, and we have

$$\boxed{\frac{5}{x} \to 0 \text{ as } x \to \infty} \qquad \boxed{\frac{2}{x^2} \to 0 \text{ as } x \to \infty}$$

$$\lim_{x \to \infty} \frac{4x^2 + 5x - 2}{2x^2 + 1} = \lim_{x \to \infty} \frac{4 + \dfrac{5}{x} - \dfrac{2}{x^2}}{2 + \dfrac{1}{x^2}} = \frac{4}{2} = 2$$

$$\boxed{\frac{1}{x^2} \to 0 \text{ as } x \to \infty}$$

We can verify the reasonableness of this limit by using a calculator to find the values in the table.

x	1	10	100	1000
$f(x)$	2.3333333	2.2288557	2.0247988	2.0024980

The limit $\lim_{x \to a} f(x)$ exists and is equal to L even if the values of $f(x)$ oscillate, provided they tend to approach the number L as x approaches a.

EXAMPLE 8

(a)   Figure 22-11 is a graph of the changing amplitude (in centimeters) of a mechanical system. As the time t increases, the amplitude of motion of the system oscillates about $A(t) = 4$ cm. For t sufficiently large, $A(t)$ approaches arbitrarily close to 4 cm. Therefore, $\lim_{t \to \infty} A(t) = 4$.

Figure 22-11

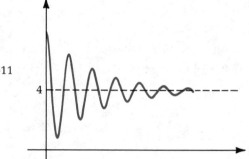

(b)   Figure 22-12 is a graph of the electrical current (in amperes) in a circuit. As the time t increases, $i(t)$ oscillates between $+2$ A and -2 A. The function does not approach any fixed value. Therefore, the limit $\lim_{t \to \infty} i(t)$ does not exist.

Figure 22-12

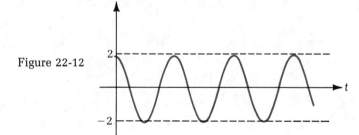

Because they will be useful in later work, the following theorems on limits are given without proof:

If $\lim\limits_{x \to a} f(x) = F$ and $\lim\limits_{x \to a} g(x) = G$, then

$$\lim_{x \to a} [f(x) \pm g(x)] = F \pm G \qquad\qquad (22\text{-}1)$$

$$\lim_{x \to a} f(x)g(x) = FG \qquad\qquad (22\text{-}2)$$

$$\lim_{x \to a} \frac{f(x)}{g(x)} = \frac{F}{G} \qquad G \neq 0 \qquad\qquad (22\text{-}3)$$

$$\lim_{x \to a} kf(x) = kF \qquad\qquad (22\text{-}4)$$

Continuity   Most functions describing physical phenomena are "well-behaved," that is, they seldom contain breaks (missing or displaced points) or make abrupt changes. A moving object, for example, cannot change speed or direction instantly, and it cannot disappear at one point in its path and reappear at another time or place. Physical change is usually gradual and smooth.

Continuity at a point   A function $f(x)$ is said to be **continuous at a point a** if all three of the following conditions are satisfied:

1. $f(x)$ is defined at $x = a$
2. $\lim\limits_{x \to a} f(x)$ exists and is a real number
3. $\lim\limits_{x \to a} f(x) = f(a)$

If the function fails to meet any one of these conditions, we say it is **discontinuous** at a, or that it has a **discontinuity** at a.

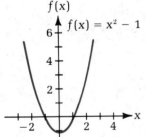

$f(x)$

$f(x) = x^2 - 1$

EXAMPLE 9   The function $f(x) = x^2 - 1$, shown in Fig. 22-13, is continuous at $x = 2$, because the following conditions are satisfied:

1. $f(x)$ is defined at $x = 2$: $f(2) = 2^2 - 1 = 3$
2. $\lim\limits_{x \to 2} f(x) = 3$
3. Therefore $\lim\limits_{x \to 2} f(x) = f(2)$

Figure 22-13

Since these conditions are satisfied for all points in the domain, the function is continuous, and there are no breaks in its graph (Fig. 22-13).

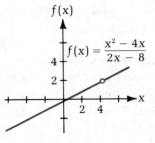

$f(x)$

$f(x) = \dfrac{x^2 - 4x}{2x - 8}$

EXAMPLE 10   The function

$$f(x) = \frac{x^2 - 4x}{2x - 8}$$

is defined for all values of x except $x = 4$. Because $f(4)$ does not exist, this function is not continuous at $x = 4$. Figure 22-14 shows that the graph of this function has a gap, or discontinuity at $x = 4$. Note that this function is continuous at all other points in its domain.

Figure 22-14

EXAMPLE 11   Show that the following function is continuous at $x = 4$.

$$g(x) = \begin{cases} \dfrac{x^2 - 4x}{2x - 8} & \text{for } x \neq 4 \\[2mm] 2 & \text{for } x = 4 \end{cases}$$

Solution   This function is identical to the function in Example 10, except that we have defined $g(4) = 2$. This satisfies the first condition of continuity.
   To find

$$\lim_{x \to 4} \frac{x^2 - 4x}{2x - 8}$$

use the increment method of Example 2. Let $x = 4 + d$, where d is a very small positive or negative number. Then

$$\frac{x^2 - 4x}{2x - 8} = \frac{(4 + d)^2 - 4(4 + d)}{2(4 + d) - 8}$$

$$= \frac{16 + 8d + d^2 - 16 - 4d}{8 + 2d - 8}$$

$$= \frac{d^2 + 4d}{2d}$$

$$= \frac{d + 4}{2}$$

As $d \to 0$, $g(x) \to \frac{4}{2} = 2$. Therefore, $\lim\limits_{x \to 4} g(x) = 2$. This satisfies the second condition of continuity.
   Finally, because $g(4) = 2$, this function satisfies all three conditions for continuity and is continuous at $x = 4$. By redefining the function, we have removed the discontinuity.

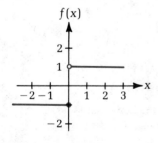

Figure 22-15

EXAMPLE 12   For the function

$$f(x) = \begin{cases} 1 & \text{for } x > 0 \\ -1 & \text{for } x \leq 0 \end{cases}$$

the left side limit, the limit as x approaches 0 from the left, is $\lim\limits_{x \to 0^-} f(x) = -1$, and the right side limit, the limit as x approaches 0 from the right, is $\lim\limits_{x \to 0^+} f(x) = +1$. Because these limits are not equal, $\lim\limits_{x \to 0} f(x)$ does not exist. Therefore, this function is discontinuous at $x = 0$ (Fig. 22-15).

Continuity over an interval

   A function is said to be **continuous over an interval** if for the open interval $p < x < q$ or the closed interval $p \leq x \leq q$ it is continuous at every point in the interval. For example, at every point a in the interval $-4 \leq x \leq 4$, the function $f(x) = \sqrt{16 - x^2}$ exists and $\lim\limits_{x \to a} f(x) = f(a)$. Therefore, $f(x)$ is continuous over the interval $-4 \leq x \leq 4$ (Fig. 22-16).

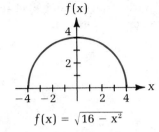

Figure 22-16

$$f(x) = \sqrt{16 - x^2}$$

LEARNING HINT ▶ If a function is continuous over some interval, its graph can be drawn without lifting the pencil from the paper. ◀

Here are some continuous functions:

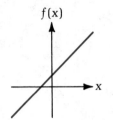 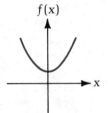 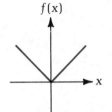 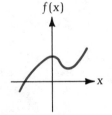 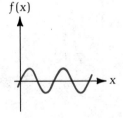

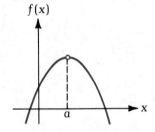

The following functions are *not* continuous:

This is a *missing point discontinuity*. The function is not defined at a. This is called a *removable* discontinuity, because we can define $f(a)$ to get a similar but continuous function.

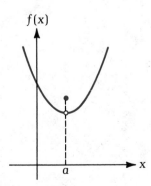

This is a *displaced point discontinuity*. The limit at $x = a$ exists, but it is not equal to $f(a)$.

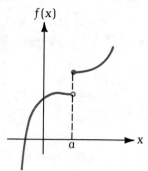

This is a *jump discontinuity*. The value of $f(x)$ changes abruptly at $x = a$. The limit of this function does not exist at a.

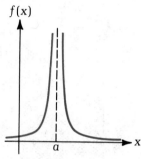

This is an *infinite discontinuity*. The value of $f(x)$ increases without bound as $x \to a$. The limit of this function does not exist at $x = a$.

Exercises 22-1

In Problems 1–6, use a calculator to evaluate each function for the values of x given. Use these values to find the limit indicated.

1. $\lim\limits_{x \to 1} 3x^2$ for $x = 0.5, 0.9, 0.99, 0.999, 1.1, 1.01, 1.001$

2. $\lim\limits_{x \to 2} \dfrac{1}{x - 2}$ for $x = 1.5, 1.9, 1.99, 1.999, 2.5, 2.1, 2.01, 2.001$

3. $\lim\limits_{x \to 3} f(x)$,   where $f(x) = \begin{cases} \dfrac{x^2}{2} & \text{for } x \le 3 \\ 6 & \text{for } x > 3 \end{cases}$
   for $x = 2.5, 2.9, 2.99, 2.999, 3.5, 3.1, 3.01, 3.001$

4. $\lim\limits_{x \to \infty} f(x)$,   where $f(x) = \dfrac{1 - 2x^2}{x^2 + 4}$
   for $x = 1, 5, 10, 100, 1000$

5. $\lim\limits_{x \to \infty} f(x)$,   where $f(x) = \dfrac{2x^2 - 8x + 3}{6x^2 + 5x - 1}$
   for $x = 1, 5, 10, 100, 1000$

6. $\lim\limits_{x \to 3} f(x)$,   where $f(x) = \dfrac{x^2 - 9}{2 - \sqrt{x^2 - 5}}$
   for $x = 2.5, 2.9, 2.99, 2.999, 3.5, 3.1, 3.01, 3.001$

For Problems 7–14, use the increment method to find the limits indicated.

7. $\lim\limits_{x \to 1} \dfrac{x^2 + 2}{x + 2}$
8. $\lim\limits_{x \to 4} \dfrac{x^2 + 4}{x^2 - 16}$
9. $\lim\limits_{x \to -3} \dfrac{x^2 + 2x - 1}{x + 3}$

10. $\lim\limits_{x \to -2} \dfrac{x^2 + x + 2}{x + 2}$
11. $\lim\limits_{x \to 2} (x^2 + 5)$
12. $\lim\limits_{x \to 5} (x - 5)\sqrt{x - 4}$

13. $\lim\limits_{x \to 2} \dfrac{2x^2 - 8}{x^2 - 3x + 2}$
14. $\lim\limits_{x \to -1} \dfrac{x^3 - x}{1 + x}$

In Problems 15–34, use any convenient method to find the limits indicated, if they exist.

15. $\lim\limits_{x \to 1} x^3$
16. $\lim\limits_{x \to 0} (1 - x^2)$
17. $\lim\limits_{x \to 1} (x^2 + 5x - 6)$

18. $\lim\limits_{x \to -2} \dfrac{x^2 - 4}{x + 2}$
19. $\lim\limits_{x \to \infty} \dfrac{x + 2}{x}$
20. $\lim\limits_{x \to 0} \dfrac{2x^2 + 3x}{x}$

21. $\lim\limits_{x \to \infty} \dfrac{x^2 - 3x + 2}{2x^2 + x - 1}$
22. $\lim\limits_{x \to 0} \dfrac{x^2 - 3x + 2}{2x^2 + x - 1}$
23. $\lim\limits_{x \to 1} \dfrac{2}{x - 1}$

24. $\lim\limits_{x \to 2} f(x)$, where
$f(x) = \begin{cases} x^2 & \text{for } x \neq 2 \\ 6 & \text{for } x = 2 \end{cases}$

25. $\lim\limits_{x \to 0} f(x)$, where
$f(x) = \begin{cases} \dfrac{|x|}{x} & \text{for } x \neq 0 \\ 0 & \text{for } x = 0 \end{cases}$

26. $\lim\limits_{x \to 0} f(x)$, where
$f(x) = \begin{cases} |x| & \text{for } x \neq 0 \\ 2 & \text{for } x = 0 \end{cases}$

27. $\lim\limits_{x \to -1} \dfrac{3x^2 - 2x}{2 + x}$
28. $\lim\limits_{x \to 0^+} \sqrt{x}$
29. $\lim\limits_{x \to 4} \sqrt{2 - x}$

30. $\lim\limits_{x \to 0} \dfrac{x^2}{\sqrt{9 - x} - 3}$
31. $\lim\limits_{x \to -3} (x^2 + 3x)$
32. $\lim\limits_{x \to -\infty} (1 + 2^x)$

33. For $g(x) = \begin{cases} x^2 & \text{for } x < 0 \\ x - 2 & \text{for } x \geq 0 \end{cases}$ find:
(a) $\lim\limits_{x \to 0^-} g(x)$
(b) $\lim\limits_{x \to 0^+} g(x)$
(c) $\lim\limits_{x \to 0} g(x)$

34. For $h(x) = \begin{cases} x - 2 & \text{for } x < 2 \\ 2x - 3 & \text{for } x \geq 2 \end{cases}$ find:
(a) $\lim\limits_{x \to 2^-} h(x)$
(b) $\lim\limits_{x \to 2^+} h(x)$
(c) $\lim\limits_{x \to 2} h(x)$

Determine whether each of the following functions is continuous at the given point. If the function is not continuous, state the reason.

35. $f(x) = 3x^2 + 2$ at $x = 2$

36. $g(x) = \dfrac{2x^2 - 7x}{3x - 9}$ at $x = 3$

37. $g(x) = \begin{cases} \dfrac{5x^2 - 6x}{2x - 4} & \text{for } x \neq 2 \\ 4 & \text{for } x = 2 \end{cases}$ at $x = 2$

38. $f(x) = \begin{cases} 2 & \text{for } x > 1 \\ 1 & \text{for } x \leq 1 \end{cases}$ at $x = 1$

39. $g(x) = \sqrt{16 - x^2}$ at $x = 5$

40. $f(x) = \begin{cases} \dfrac{x^2 - 3x}{4x - 12} & \text{for } x \neq 3 \\ \dfrac{3}{4} & \text{for } x = 3 \end{cases}$ at $x = 3$

For the functions whose graphs are shown, find the limits indicated if they exist, and determine if the function is continuous.

41. (a) $\lim\limits_{x \to -2} f(x)$ (b) $\lim\limits_{x \to 2} f(x)$

  (c) $\lim\limits_{x \to \infty} f(x)$

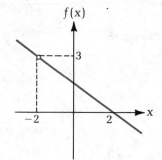

42. (a) $\lim\limits_{x \to 4^-} f(x)$ (b) $\lim\limits_{x \to 4^+} f(x)$

  (c) $\lim\limits_{x \to 0} f(x)$ (d) $\lim\limits_{x \to \infty} f(x)$

  (e) $\lim\limits_{x \to 4} f(x)$

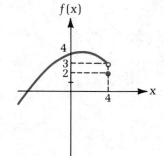

43. (a) $\lim\limits_{x \to \infty} f(x)$ (b) $\lim\limits_{x \to 0} f(x)$

  (c) $\lim\limits_{x \to -\infty} f(x)$

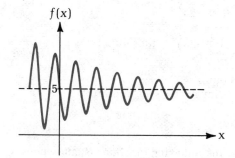

44. (a) $\lim\limits_{x \to 0} f(x)$ (b) $\lim\limits_{x \to \infty} f(x)$

  (c) $\lim\limits_{x \to -\infty} f(x)$

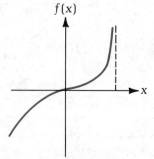

45. (a) $\lim\limits_{x \to 2^-} f(x)$ (b) $\lim\limits_{x \to 2^+} f(x)$

(c) $\lim\limits_{x \to +\infty} f(x)$ (d) $\lim\limits_{x \to -\infty} f(x)$

(e) $\lim\limits_{x \to 2} f(x)$

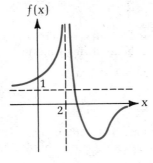

46. (a) $\lim\limits_{x \to 0^-} f(x)$ (b) $\lim\limits_{x \to 0^+} f(x)$

(c) $\lim\limits_{x \to 0} f(x)$ (d) $\lim\limits_{x \to +\infty} f(x)$

(e) $\lim\limits_{x \to -\infty} f(x)$

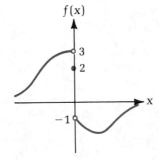

Solve.

47. *Electronics*   Two long parallel wires carrying equal currents in opposite directions will exhibit an inductance (in henries) given by the equation

$$L(d) = k \ln\left(\frac{d - a}{a}\right)$$

where $k = 4.0 \times 10^{-7}$ H. If the radius of the wire is $a = 0.100$ cm and d is the center-to-center separation of the wires, find $\lim\limits_{d \to 0.5} L(d)$ by evaluating $L(d)$ for the following values of d: 0.400, 0.490, 0.499, 0.4999, 0.600, 0.510, 0.501, 0.5001.

48. *Physics*   The displacement (in centimeters) of an object moving in a very viscous medium is described by the equation

$$y(t) = (6 - 2t)e^{-0.5t}$$

where t is time (in seconds). Find $\lim\limits_{t \to 3.0} y(t)$ by evaluating $y(t)$ for the following values of t: 2.00, 2.90, 2.99, 2.999, 4.00, 3.10, 3.01, 3.001.

49. *Electronics*   The magnetic field B in the shielding wall of a coaxial cable depends on the distance r from the center of the cable according to the equation

$$B(r) = 2\left(\frac{4 - r^2}{r}\right) \qquad \text{for } 0 < r \le 2$$

Find $\lim\limits_{r \to 1.5} B(r)$ by evaluating $B(r)$ for the following values of r: 1.400, 1.490, 1.499, 1.600, 1.510, 1.501.

50. *Civil engineering*   In a tension test, the stress–strain curve for structural steel is found to have the following shape:

$$C(q) = \begin{cases} (3.0 \times 10^4)q & \text{for } 0 \le q < 0.001 \\ 15 + (1.5 \times 10^4)q & \text{for } 0.001 \le q \le 0.003 \end{cases}$$

where C is the stress (ksi) and q is the strain (kip). Show that this function is continuous at $q = 0.001$ kip.*

51. *Civil engineering*   The critical stress C (ksi) at which a cylindrical steel column ruptures depends on its stenderness ratio s according to the following empirical equation:

$$C(s) = \begin{cases} 62.897 - \dfrac{s}{3.040} & \text{for } 20 \le s < 100 \\[2ex] \dfrac{30{,}397\,\pi^2}{s^2} & \text{for } 100 \le s \end{cases}$$

Show that this function is continuous at $s = 100$ in./in.

52. *Electronics*   The magnetic field B inside a coaxial cable depends on the radial distance r from its center according to the function

$$B(r) = \begin{cases} \dfrac{2r}{c^2} & \text{for } 0 \le r \le c \\[2ex] \dfrac{2}{r} & \text{for } c < r \le b \\[2ex] \dfrac{2(a^2 - r^2)}{r(a^2 - b^2)} & \text{for } b < r \le a \end{cases}$$

Show that this function is continuous at $r = c$ and $r = b$.

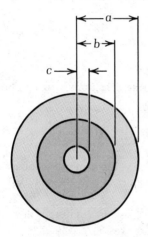

22-2 | The Derivative

Calculus was developed in the 17th century in an effort to solve a number of important scientific problems. In both the study of the motion of objects and the design of lenses in optics, it is often necessary to solve the following problem.

The tangent problem   Given a function $f(x)$ and a point $P(x_0, y_0)$ on its graph, find the equation of the line tangent to the graph at point P.

---

*For more information, see Nelson R. Bauld, Jr., *Mechanics of Materials* (Monterey, CA: Brooks/Cole, 1982), p. 42.

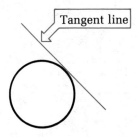

Tangent line

Figure 22-17

To solve the problem, we must first specify exactly what is meant by a tangent. In plane geometry, a line is said to be tangent to a circle if it meets the circle at only one point (Fig. 22-17).

This definition is not always adequate when it is applied to curves other than the circle, as the following graphs show:

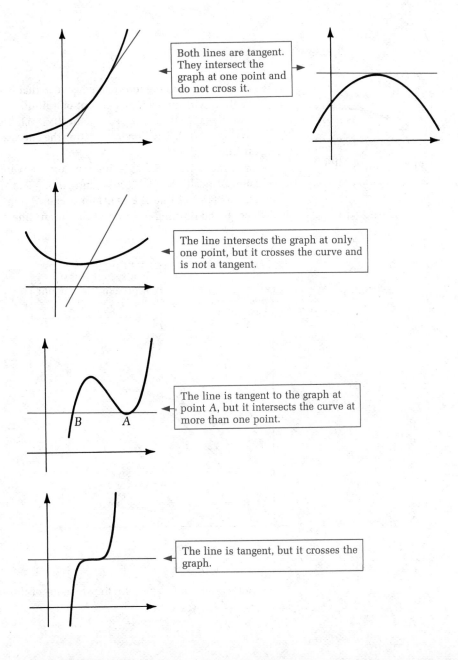

Both lines are tangent. They intersect the graph at one point and do not cross it.

The line intersects the graph at only one point, but it crosses the curve and is *not* a tangent.

The line is tangent to the graph at point A, but it intersects the curve at more than one point.

The line is tangent, but it crosses the graph.

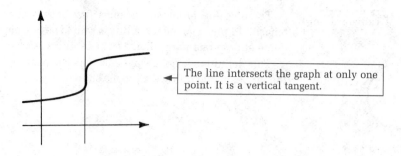

The line intersects the graph at only one point. It is a vertical tangent.

$y = f(x)$

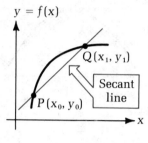

Figure 22-18

We need to redefine the idea of a tangent so that it applies to any curve, and this definition will involve the concept of a limit.

To find the tangent to a curve at point $P(x_0, y_0)$, first choose a second point $Q(x_1, y_1)$ (Fig. 22-18). The line intersecting the curve at points P and Q is called a **secant line** for the curve.

Second, move the point Q along the curve toward P so that it assumes a succession of positions Q_1, Q_2, Q_3, and so on. The secant lines thus formed approach the tangent line at P as Q approaches P (Fig. 22-19). The **tangent line** is defined as the limiting position of the secant line as $Q \to P$.

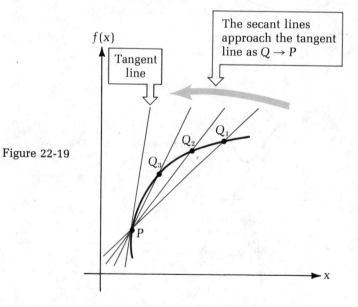

Figure 22-19

As you learned in Chapter 20, the slope m_S of the secant line in Fig. 22–18 is:

Slope of a Secant Line $m_S = $ Slope of PQ $= \dfrac{y_1 - y_0}{x_1 - x_0}$ (22-5)

↑

The subscript indicates that
this is the slope of a secant line.

Therefore, m_T, the slope of the tangent line at x_0 is:

Slope of a Tangent Line $m_T = $ Slope of tangent at (x_0, y_0)

$= \lim\limits_{x_1 \to x_0} \dfrac{y_1 - y_0}{x_1 - x_0}$ (22-6)

EXAMPLE 13   Find the slope of a line tangent to the graph of $y = x^2 + 2$ at the point $P(2, 6)$.

Solution   Calculate the slopes of secant lines through points $Q(x_1, y_1)$ for $x_1 = 1.5, 1.9, 1.99, 1.999$ and $2.5, 2.1, 2.01, 2.001$. Table 22-2 shows the result of these calculations.

TABLE 22-2
$y = x^2 + 2$ at $P(2, 6)$: $x_0 = 2$, $y_0 = 6$

x_1	$y_1 = (x_1)^2 + 2$	$x_1 - x_0$	$y_1 - y_0$	$m_S = \dfrac{y_1 - y_0}{x_1 - x_0}$	
2.5	8.25	0.5	2.25	4.5	
2.1	6.41	0.1	0.41	4.1	
2.01	6.0401	0.01	0.0401	4.01	
2.001	6.004001	0.001	0.004001	4.001	$\lim\limits_{x_1 \to 2^+} m_S = 4.000$
1.5	4.25	−0.5	−1.75	3.5	
1.9	5.61	−0.1	−0.39	3.9	
1.99	5.9601	−0.01	−0.0399	3.99	
1.999	5.996001	−0.001	−0.003999	3.999	$\lim\limits_{x_1 \to 2^-} m_S = 4.000$

As point Q approaches point P along the curve, $x_1 \to 2$, and the limiting value of the slope of the secant line is $\lim\limits_{x_1 \to 2} m_S = 4$. Therefore, by equation (22-6), the slope m_T of the tangent line at $x = 2$ is

$m_T = $ Slope of tangent $= \lim\limits_{x_1 \to 2} m_S = 4$ See Fig. 22-20 on p. 826.

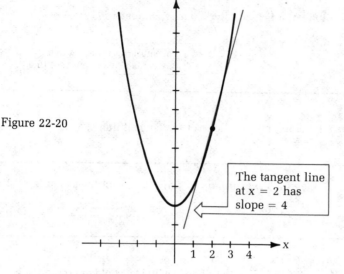

Figure 22-20

The tangent line at $x = 2$ has slope = 4

EXAMPLE 14   Find the slope of the tangent to $f(x) = y = 3x^2 - 2x$ at the point $(2, 8)$ using the increment method.

Solution

Step 1.   We have $x_0 = 2$ and $y_0 = 8$. Let $x_1 = 2 + d$, where d is some very small number.

Step 2.   Substitute $2 + d$ for x and find y_1.

$$y_1 = 3(2 + d)^2 - 2(2 + d)$$
$$= 3(4 + 4d + d^2) - 4 - 2d$$
$$= 12 + 12d + 3d^2 - 4 - 2d$$
$$= 3d^2 + 10d + 8$$

Step 3. m_T = Slope of tangent $= \lim_{x_1 \to 2} \dfrac{y_1 - y_0}{x_1 - x_0}$

If $x_1 = 2 + d$, then as $x_1 \to 2, d \to 0$. Therefore:

$$m_T = \lim_{d \to 0} \frac{(3d^2 + 10d + 8) - 8}{(2 + d) - 2}$$

$$= \lim_{d \to 0} \frac{3d^2 + 10d}{d}$$

$$= \lim_{d \to 0} (3d + 10) = 10$$

The slope at $x_0 = 2$ is 10.

This procedure leads to one of the most important concepts in calculus. But first, some new notation is needed.

Δ Notation   We define the quantity Δx (read "delta x") as an increment or small change in the variable x:

$$x_1 = x_0 + \Delta x \quad \text{or} \quad \Delta x = x_1 - x_0 \tag{22-7}$$

The resulting change in y is Δy, where

$$y_1 = y_0 + \Delta y \quad \text{or} \quad \Delta y = y_1 - y_0 \tag{22-8}$$

The symbol Δ is the Greek capital letter "delta," which is equivalent to the English letter d and stands for *difference*. Scientists and mathematicians find the Δ notation a convenient shorthand to denote the change in any quantity. For example, if V is the voltage across a dc circuit, ΔV is a small change in the voltage. If L is the length of a metal bar at some temperature T, then when the temperature increases to $T + \Delta T$, the length expands to $L + \Delta L$.

CAUTION ▶ Δx does *not* mean "Δ times x." ◀

Definition of the derivative

Note in equation (22-7) that as $x_1 \to x_0$, $\Delta x \to 0$ (Fig. 22-21), and (22-6) becomes

$$\text{Slope of tangent at } (x_0, y_0) = \lim_{\Delta x \to 0} \frac{\Delta y}{\Delta x}$$

In function notation, if $y = f(x)$, then

$$y_0 = f(x_0)$$

$$y_1 = f(x_1) = f(x_0 + \Delta x)$$

and from equation (22-8),

$$\Delta y = y_1 - y_0 = f(x_0 + \Delta x) - f(x_0)$$

Figure 22-21

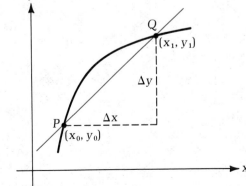

Then

$$\frac{\Delta y}{\Delta x} = \frac{y_1 - y_0}{x_1 - x_0} = \frac{f(x_0 + \Delta x) - f(x_0)}{\Delta x}$$

The **derivative** of a function $f(x)$ is a new function $f'(x)$ defined as follows:

The Derivative $f'(x) = \lim\limits_{\Delta x \to 0} \dfrac{\Delta y}{\Delta x} = \lim\limits_{\Delta x \to 0} \dfrac{f(x + \Delta x) - f(x)}{\Delta x}$	(22-9)

The notation $f'(x)$ is read "f prime of x."

CAUTION ▶ The derivative exists only if the limit exists. ◀

The derivative of some function $f(x)$ is a function $f'(x)$ whose value $f'(x_0)$ at point (x_0, y_0) is the slope of the line tangent to $f(x)$ at $x = x_0$.

Δ-limit process   To find the derivative, we will use a four-step procedure called the **Δ-limit process.**

Step 1.   Given $y = f(x)$, replace x in the function by $x + \Delta x$, and replace y by $y + \Delta y$ to get $y + \Delta y = f(x + \Delta x)$.

Step 2.   Solve for Δy. $\Delta y = f(x + \Delta x) - y$

$= f(x + \Delta x) - f(x)$

> This side of the equation contains only x and Δx —no y or Δy.

Simplify the expression on the right.

Step 3.   Divide each side of the equation from Step 2 by Δx to get

$$\frac{\Delta y}{\Delta x} = \frac{f(x + \Delta x) - f(x)}{\Delta x}$$

Step 4.   Find the limit of this expression as $\Delta x \to 0$:

$$f'(x) = \lim_{\Delta x \to 0} \frac{\Delta y}{\Delta x} = \lim_{\Delta x \to 0} \frac{f(x + \Delta x) - f(x)}{\Delta x}$$

EXAMPLE 15   Find the derivative of the function $y = x^2 - x + 2$ using the Δ-limit process.

Solution
Step 1.   Replace x with $x + \Delta x$, and y with $y + \Delta y$. $y + \Delta y = (x + \Delta x)^2 - (x + \Delta x) + 2$

Step 2.   Solve for Δy. $\Delta y = (x + \Delta x)^2 - (x + \Delta x) + 2 - y$

$= (x + \Delta x)^2 - (x + \Delta x) + 2 - (x^2 - x + 2)$

$$= x^2 + 2x\Delta x + (\Delta x)^2 - x - \Delta x + 2 - x^2 + x - 2$$

$$\Delta y = (\Delta x)^2 + 2x\Delta x - \Delta x$$

Step 3. Divide
by Δx.

$$\frac{\Delta y}{\Delta x} = \frac{(\Delta x)^2 + 2x\Delta x - \Delta x}{\Delta x} = \frac{(\Delta x)^2}{\Delta x} + \frac{2x\Delta x}{\Delta x} - \frac{\Delta x}{\Delta x}$$

$$= \Delta x + 2x - 1$$

Step 4. Find
the limit as
$\Delta x \to 0$.

$$f'(x) = \lim_{\Delta x \to 0} \frac{\Delta y}{\Delta x} = 2x - 1$$

The derivative of the function $y = x^2 - x + 2$ is the function $f'(x) = 2x - 1$.

Because the derivative is a function, we can evaluate it for any value of x. For the derivative found in Example 15, $f'(2) = 2(2) - 1 = 3$. Therefore, the slope of the tangent line to the graph of $y = x^2 - x + 2$ at $x = 2$ is 3.

$\dfrac{dy}{dx}$ **notation**   The derivative $f'(x)$ is often written dy/dx, where

$$f'(x) = \frac{dy}{dx} = \lim_{\Delta x \to 0} \frac{\Delta y}{\Delta x} \qquad\qquad (22\text{-}10)$$

The Greek letter Δ is replaced by its English equivalent d after the limit has been taken. Read dy/dx as "dee y dee x," or more formally, as "the derivative of y with respect to x."

The process of finding the derivative function is called **differentiation**.

The d notation was invented by the mathematician Gottfried Wilhelm von Leibniz in 1685, and several variations of it are used. For example, if $y = f(x)$, then the derivative might be written

$$f'(x) = \frac{dy}{dx} = \frac{df(x)}{dx} = \frac{d}{dx} f(x) = D_x f(x)$$

We will use the dy/dx notation in this textbook.

EXAMPLE 16 Find the derivative of $f(x) = x^3$ and use it to determine the slope of the tangent to the graph of this function at $x = -1$.

Solution

Step 1. $y + \Delta y = (x + \Delta x)^3$

Step 2. $\Delta y = (x + \Delta x)^3 - y$

$$= (x + \Delta x)^3 - x^3$$

$$= [x^3 + 3x^2\Delta x + 3x(\Delta x)^2 + (\Delta x)^3] - x^3$$

$$\Delta y = 3x^2\Delta x + 3x(\Delta x)^2 + (\Delta x)^3$$

Step 3.
$$\frac{\Delta y}{\Delta x} = \frac{3x^2 \Delta x + 3x(\Delta x)^2 + (\Delta x)^3}{\Delta x}$$

$$= 3x^2 + 3x\,\Delta x + (\Delta x)^2$$

Step 4.
$$\frac{dy}{dx} = \lim_{\Delta x \to 0} [3x^2 + 3x\,\Delta x + (\Delta x)^2]$$

$$\frac{dy}{dx} = 3x^2$$

> This derivative function gives the slope of the line tangent to $y = x^3$ for every value of x.

To denote the value of the derivative dy/dx at $x = a$, write

$$\left. \frac{dy}{dx} \right|_{x=a}$$

Thus, at $x = -1$, the value of the derivative, the slope of the tangent line, is

$$\left. \frac{dy}{dx} \right|_{x=-1} = 3(-1)^2 = 3$$

CAUTION ▶

The symbol $\Delta y/\Delta x$ is a fraction or ratio, and dy/dx looks like a fraction, but it is not. The symbol dy/dx is used to denote the function found by taking the limit of the ratio $\Delta y/\Delta x$ as $\Delta x \to 0$. ◀

EXAMPLE 17

(a) Find the derivative of $y = \dfrac{x-1}{x}$

(b) Evaluate the derivative at $x = 2$ and at $x = 0$.

Solution

(a) **Step 1.**
$$y + \Delta y = \frac{(x + \Delta x) - 1}{x + \Delta x}$$

Step 2.
$$\Delta y = \frac{(x + \Delta x) - 1}{x + \Delta x} - y$$

$$= \frac{(x + \Delta x) - 1}{x + \Delta x} - \frac{x - 1}{x}$$

Simplify this expression by combining the fractions.

$$= \frac{x(x + \Delta x) - x - (x - 1)(x + \Delta x)}{x(x + \Delta x)}$$

$$= \frac{x^2 + x\Delta x - x - x^2 - x\Delta x + x + \Delta x}{x^2 + x\Delta x}$$

$$\Delta y = \frac{\Delta x}{x^2 + x\Delta x}$$

Step 3. Divide both sides by Δx.
$$\frac{\Delta y}{\Delta x} = \frac{1}{x^2 + x\Delta x}$$

Step 4. Find the limit as $\Delta x \to 0$.

$$\frac{dy}{dx} = \lim_{\Delta x \to 0} \frac{\Delta y}{\Delta x} = \frac{1}{x^2 + 0}$$

$$\frac{dy}{dx} = \frac{1}{x^2}$$

(b) $\left.\dfrac{dy}{dx}\right|_{x=2} = \dfrac{1}{(2)^2} = \dfrac{1}{4}$

$\left.\dfrac{dy}{dx}\right|_{x=0} = \dfrac{1}{0^2} = \dfrac{1}{0}$ Not defined.

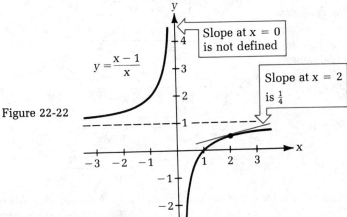

Figure 22-22

$y = \dfrac{x-1}{x}$

Slope at x = 0 is not defined

Slope at x = 2 is $\frac{1}{4}$

Differentiable function

A function $y = f(x)$ is said to be **differentiable** at some point $x = x_0$ if its derivative $f'(x) = dy/dx$ exists at $x = x_0$. The function discussed in Example 17 is differentiable everywhere except at $x = 0$.

A function $f(x)$ will not be differentiable at any point where its graph has a break or jump discontinuity, where it has a sharp point or a corner, or where the tangent of the graph is vertical. The graphs in Fig. 22-23 show some functions that are not differentiable at point a.

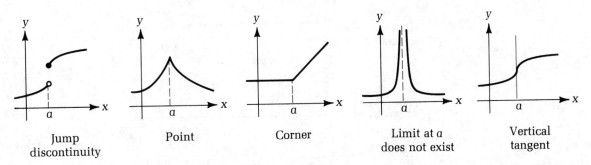

Jump discontinuity

Point

Corner

Limit at a does not exist

Vertical tangent

Figure 22-23

EXAMPLE 18   Find the derivative of $y = \sqrt{x - 2}$ and determine the values of x for which the function is differentiable.

Solution

Step 1. $y + \Delta y = \sqrt{(x + \Delta x) - 2}$

Step 2. $\Delta y = \sqrt{x + \Delta x - 2} - y$

$\qquad = \sqrt{x + \Delta x - 2} - \sqrt{x - 2}$

To simplify, multiply by a unity fraction consisting of the conjugate in the numerator and denominator.

$\Delta y = \dfrac{(\sqrt{x + \Delta x - 2} - \sqrt{x - 2})(\sqrt{x + \Delta x - 2} + \sqrt{x - 2})}{(\sqrt{x + \Delta x - 2} + \sqrt{x - 2})}$

$\qquad = \dfrac{(x + \Delta x - 2) - (x - 2)}{\sqrt{x + \Delta x - 2} + \sqrt{x - 2}}$

$\Delta y = \dfrac{\Delta x}{\sqrt{x + \Delta x - 2} + \sqrt{x - 2}}$

$$\sqrt{A} - \sqrt{B} = \frac{(\sqrt{A} - \sqrt{B})(\sqrt{A} + \sqrt{B})}{(\sqrt{A} + \sqrt{B})}$$

$$= \frac{A - B}{\sqrt{A} + \sqrt{B}}$$

Step 3. $\dfrac{\Delta y}{\Delta x} = \dfrac{1}{\sqrt{x + \Delta x - 2} + \sqrt{x - 2}}$

Step 4. $\dfrac{dy}{dx} = \lim\limits_{\Delta x \to 0} \dfrac{1}{\sqrt{x + \Delta x - 2} + \sqrt{x - 2}}$

As $\Delta x \to 0$, $\sqrt{x + \Delta x - 2} \to \sqrt{x - 2}$, so $\dfrac{dy}{dx} = \dfrac{1}{2\sqrt{x - 2}}$

Notice that dy/dx does not exist for $x \le 2$. At these points dy/dx is not defined. Therefore, the function $y = \sqrt{x - 2}$ is differentiable only for $x > 2$.

Derivative as a rate of change

Geometrically, the derivative dy/dx of a function $f(x)$ gives the slope of a tangent to the curve $y = f(x)$. However, the derivative may also be interpreted more generally as the rate of change of the function $f(x)$. For example, the curve in Fig. 22-24 shows the air temperature T as we move in a straight-line path through a small annealing furnace. For this graph, the derivative dT/dx has units of degrees Celsius per centimeter (°C/cm).

Figure 22-24

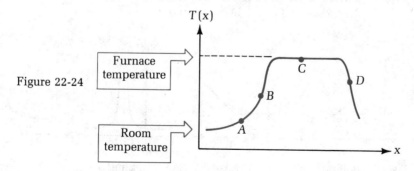

Notice the following:

• At point A, the temperature is increasing slowly; dT/dx at A is a small positive number. A positive derivative means that the original function $T(x)$ is increasing.

- At point B, the temperature is increasing rapidly; dT/dx at B is a large positive number. A large value of the derivative means that the function $T(x)$ is changing rapidly.
- At point C inside the furnace, the temperature is constant, and dT/dx at point C is equal to zero. A zero derivative means that the function $T(x)$ is not changing: The temperature remains at a constant value.
- At point D (exiting the furnace), the temperature is decreasing rapidly; dT/dx at D is a large negative number. A negative derivative means that the original function is decreasing.

This kind of interpretation of the derivative as a rate of change of a function is often very important when we use calculus to describe and solve scientific or technical problems.

EXAMPLE 19   The power P developed by a windmill depends on the cube of the wind velocity v, for a wide range of velocities. If $P = 5v^3$, find the rate of change of P with respect to v using the Δ-limit process.

Solution

Step 1. Substitute $P + \Delta P$ for P and $v + \Delta v$ for v.

$$P + \Delta P = 5(v + \Delta v)^3$$
$$= 5v^3 + 5 \cdot 3v^2(\Delta v) + 5 \cdot 3v(\Delta v)^2 + 5(\Delta v)^3$$

Step 2. Subtract $P = 5v^3$ from each side.

$$\Delta P = 15v^2(\Delta v) + 15v(\Delta v)^2 + 5(\Delta v)^3$$

Step 3. Divide both sides by Δv.

$$\frac{\Delta P}{\Delta v} = 15v^2 + 15v(\Delta v) + 5(\Delta v)^2$$

Step 4. Find the limit as $\Delta v \to 0$.

$$\frac{dP}{dV} = \lim_{\Delta v \to 0} \frac{\Delta P}{\Delta v}$$

As $\Delta v \to 0$,
$15v(\Delta v) \to 0$ and
$5(\Delta v)^2 \to 0$.

$$\frac{dP}{dv} = 15v^2$$

Exercises 22-2

Use the calculator method of Example 13 to find the slope of the line tangent to the graph of each function at the given point P. Use the values of x given.

1. $f(x) = x^3 - x^2 - 4$ at $P(2, 0)$;   for $x = 1.000,\ 1.900,\ 1.990,\ 1.999,\ 3.000,\ 2.100,\ 2.010,\ 2.001$

2. $f(x) = \dfrac{2}{x^2}$ at $P\left(4, \dfrac{1}{8}\right)$;   for $x = 3.000,\ 3.900,\ 3.990,\ 3.999,\ 5.000,\ 4.100,\ 4.010,\ 4.001$

3. $f(x) = 2 + \sqrt{x}$ at $P(4, 4)$, for $x = 3.000, 3.900, 3.990, 3.999, 5.000, 4.100, 4.010, 4.001$

4. $f(x) = \dfrac{x\sqrt{x} - 2}{x - 2}$ at $P(3.5, 3.03)$; for $x = 3.000, 3.400, 3.490, 3.499, 4.000, 3.600, 3.510, 3.501$

5. $f(x) = \dfrac{3x^2 - 7x}{x^2 + 5x - 1}$ at $P\left(-2, -\dfrac{26}{7}\right)$; for $x = -3.000, -2.100, -2.010, -2.001, -1.000, -1.900,$
$-1.990, -1.999$

6. $f(x) = (1 - 3x)^3$ at $P(1, -8)$; for $x = 0.000, 0.900, 0.990, 0.999, 2.000, 1.100, 1.010, 1.001$

Use the increment method (Example 14) to find the slope of the line tangent to the graph of each function at the given point P.

7. $f(x) = x^3 - x^2 - 4$ at $P(2, 0)$

8. $f(x) = \dfrac{2}{x^2}$ at $P\left(4, \dfrac{1}{8}\right)$

9. $f(x) = x + \dfrac{1}{x}$ at $P(1, 2)$

10. $f(x) = (1 - 3x)^3$ at $P(1, -8)$

Use the Δ-limit process to find the derivative of each function.

11. $f(x) = 4x + 5$

12. $v(t) = 30t - 10$

13. $f(x) = x^2 - 4$

14. $f(x) = 3x - 4x^2$

15. $f(x) = 4x^2$

16. $v(r) = 4r^3$

17. $f(x) = \dfrac{1}{x}$

18. $f(x) = x^2 + \dfrac{1}{x}$

19. $f(x) = \dfrac{3}{x} + 1$

20. $f(x) = \dfrac{2}{x + 2}$

21. $f(x) = \sqrt{x}$

22. $f(x) = \sqrt{x^2 + 2}$

23. $f(x) = 1 + \sqrt{2x}$

24. $f(x) = \sqrt{x - 4}$

25. $f(x) = \dfrac{1}{x^2 + 2}$

26. $f(x) = \dfrac{x + 1}{x^2}$

27. $f(x) = x^3 + x^2 - x + 2$

28. $s(t) = 10t^2 + 20t + 4$

Use the Δ-limit process to find the derivative of each function, evaluate the derivative at the given points, and determine whether the function is differentiable at those points.

29. $L(t) = 4 - 2t^2$ at $t = 0, 1, -1$

30. $V(t) = t - t^2$ at $t = 0, \dfrac{1}{2}, 1, -1$

31. $f(x) = \dfrac{1}{x - 1}$ at $x = 0, 1, 2, -1$

32. $g(x) = \dfrac{1}{x^2}$ at $x = 0, 1, 2$

33. $f(x) = 3x^2 - x^3$ at $x = 0, 1, 2$

34. $f(x) = x^3 - x$ at $x = 0, 1, -1$

35. $f(x) = x + \dfrac{1}{x}$ at $x = 0, 1.15, -1.25$

36. $f(x) = \sqrt{3x^2 + 4}$ at $x = -1, 0, 1.5$

Solve.

37. *Physics* At time t (in seconds), the position x (measured in centimeters) of a particle moving along the x-axis is given by the equation $x = 4t - 2t^2 + t^3$. Use the Δ-limit process to find the derivative dx/dt, the speed of the particle, and evaluate this derivative at $t = 2$ s.

38. *Mechanical engineering* The force F (in pounds) exerted by a nonlinear servomechanism depends on the displacement x (in inches) of the system according to the equation

$$F = 10.0\sqrt{2.0x - 6.0}$$

Use the Δ-limit process to find the derivative dF/dx, the rate at which the force changes with x, and evaluate this derivative at $x = 5.2$ in.

22-3 | Derivatives of Polynomials

The Δ-limit process enables us to find the derivative of a function, but the process is often long and tedious and may involve many ingenious mathematical manipulations. Rather than apply the Δ-limit process to every function individually, we can develop formulas that will enable us to find the derivative quickly and with very little paperwork.

Derivative of a constant

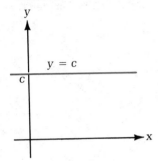

Figure 22-25

The constant function $y = c$, where c is a constant, has the graph shown in Fig. 22-25. Because the graph is a horizontal line, we know that its slope is zero. Therefore, $dc/dx = 0$. We can verify this using the Δ-limit process.

Given: $y = c$

Step 1. $y + \Delta y = c$

Step 2. $\Delta y = c - y = c - c = 0$

Step 3. $\dfrac{\Delta y}{\Delta x} = 0$

Step 4. $\dfrac{dy}{dx} = \lim_{\Delta x \to 0} 0 = 0$

Derivative of a Constant

If $y = c$, where c is a constant, then $\dfrac{dy}{dx} = 0$ (22-11)

EXAMPLE 20 Find the derivative of each function.
(a) $y = 8$ (b) $y = -4$

Solution
(a) Because 8 is a constant, we can use equation (22-11): $dy/dx = 0$.
(b) Again, -4 is a constant: $dy/dx = 0$.

Derivative of the identity function

The graph of the identity function $y = x$ is a straight line with a slope of 1, as shown in Fig. 22-26. Therefore:

If $y = x$, then $\dfrac{dy}{dx} = 1$ (22-12)

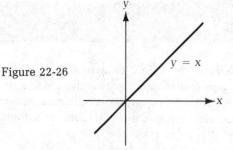

Figure 22-26

Derivative of a power

The identity function $y = x$ is a special case of the general power function $y = x^n$, where n is a positive integer. To find a general formula for the derivative of this function, we must use the binomial formula introduced in Chapter 18. For n a positive integer,

$$(a + b)^n = a^n + na^{n-1}b + \frac{n(n-1)}{2!}a^{n-2}b^2 + \cdots + b^n$$

Using the Δ-limit process on the function $y = x^n$, we have:

Step 1. $y + \Delta y = (x + \Delta x)^n$

Step 2. $\Delta y = (x + \Delta x)^n - y$

$$= (x + \Delta x)^n - x^n$$

Expand $(x + \Delta x)^n$ using the binomial formula:

$$(x + \Delta x)^n = x^n + nx^{n-1}\Delta x + \frac{n(n-1)}{2!}x^{n-2}(\Delta x)^2 + \cdots + (\Delta x)^n$$

Therefore, $\Delta y = nx^{n-1}\Delta x + \dfrac{n(n-1)}{2!}x^{n-2}(\Delta x)^2 + \cdots + (\Delta x)^n$

Step 3. $\dfrac{\Delta y}{\Delta x} = \dfrac{nx^{n-1}\Delta x + \dfrac{n(n-1)}{2!}x^{n-2}(\Delta x)^2 + \cdots + (\Delta x)^n}{\Delta x}$

$$\frac{\Delta y}{\Delta x} = nx^{n-1} + \underbrace{\frac{n(n-1)}{2!}x^{n-2}\Delta x + \cdots + (\Delta x)^{n-1}}$$

All terms have Δx as a factor.

Step 4. $\dfrac{dy}{dx} = \lim_{\Delta x \to 0}\left[nx^{n-1} + \underbrace{\dfrac{n(n-1)}{2!}x^{n-2}\Delta x + \cdots + (\Delta x)^{n-1}} \right]$

As $\Delta x \to 0$, each term approaches zero.

$$\frac{dy}{dx} = nx^{n-1}$$

For a positive integer n,* the following rule allows us to differentiate $y = x^n$:

*This differentiation formula is actually true for any rational number exponent n, positive or negative. We will show this for rational exponents later in this chapter.

The Power Formula

If $y = x^n$, then $\dfrac{dy}{dx} = nx^{n-1}$ 　　　　　　(22-13)

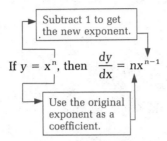

Notice that for $n = 1$, the power formula reduces to equation (22-12).

EXAMPLE 21 Find the derivative of each function:
(a) $y = x^2$ 　　(b) $y = x^{10}$ 　　(c) $y = x$
(d) If the charge Q flowing into a capacitor varies with time as $Q = t^3$, find the current $i = dQ/dt$.

Solution

(a) If $y = x^2$, then $\dfrac{dy}{dx} = 2x^{2-1} = 2x^1 = 2x$

(b) If $y = x^{10}$, then $\dfrac{dy}{dx} = 10x^{10-1} = 10x^9$

(c) If $y = x = x^1$, then $\dfrac{dy}{dx} = 1x^{1-1} = x^0 = 1$

(d) If $Q = t^3$, then $i = \dfrac{dQ}{dt} = 3t^{3-1} = 3t^2$

Derivative of cx^n 　　When the expression x^n in the power formula is multiplied by a constant, the constant can be moved outside the derivative symbol:

$$\frac{d}{dx}(2x^3) = 2 \cdot \frac{d}{dx}(x^3)$$

Move the constant outside the derivative symbol.

We can show this as follows: Let $y = c \cdot f(x)$, where $f(x)$ is a positive integer power of x. Then, using the Δ-limit process:

Step 1. 　　$y + \Delta y = c \cdot f(x + \Delta x)$

Step 2. 　　$\Delta y = c \cdot f(x + \Delta x) - y$

$$= c \cdot f(x + \Delta x) - c \cdot f(x)$$

$$\Delta y = c \cdot [f(x + \Delta x) - f(x)]$$

Step 3.
$$\frac{\Delta y}{\Delta x} = c \cdot \left[\underbrace{\frac{f(x + \Delta x) - f(x)}{\Delta x}} \right]$$

But the limit of this as $\Delta x \to 0$ is $\dfrac{df(x)}{dx}$.

Step 4.
$$\frac{dy}{dx} = c \frac{df(x)}{dx}$$

and we can state the following rule, where c is a constant and n is a positive integer:

The Constant Multiplier Rule

If $y = cx^n$, then $\quad \dfrac{dy}{dx} = c\dfrac{d(x^n)}{dx} = cnx^{n-1}$ ⁣ ⁣ ⁣ ⁣ ⁣ ⁣(22-14)

EXAMPLE 22   Find the derivative of each function:

(a) $y = 3x^5$ (b) $y = -x^4$ (c) $y = 8x$

(d)   The crushing load L for a square concrete pillar 2 ft tall depends on its thickness T according to the formula $L = 25T^4/4$. Find the rate of change of L with respect to T.

Solution

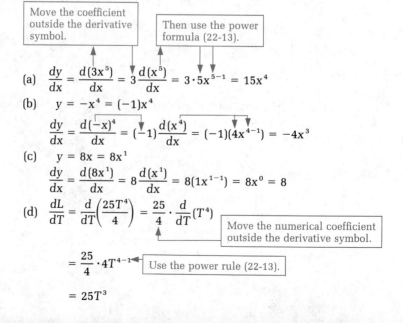

Move the coefficient outside the derivative symbol.

Then use the power formula (22-13).

(a) $\dfrac{dy}{dx} = \dfrac{d(3x^5)}{dx} = 3\dfrac{d(x^5)}{dx} = 3 \cdot 5x^{5-1} = 15x^4$

(b) $y = -x^4 = (-1)x^4$

$\dfrac{dy}{dx} = \dfrac{d(-x)^4}{dx} = (-1)\dfrac{d(x^4)}{dx} = (-1)(4x^{4-1}) = -4x^3$

(c) $y = 8x = 8x^1$

$\dfrac{dy}{dx} = \dfrac{d(8x^1)}{dx} = 8\dfrac{d(x^1)}{dx} = 8(1x^{1-1}) = 8x^0 = 8$

(d) $\dfrac{dL}{dT} = \dfrac{d}{dT}\left(\dfrac{25T^4}{4}\right) = \dfrac{25}{4} \cdot \dfrac{d}{dT}(T^4)$

Move the numerical coefficient outside the derivative symbol.

$= \dfrac{25}{4} \cdot 4T^{4-1}$

Use the power rule (22-13).

$= 25T^3$

CAUTION ▶

In Example 22(c), for $y = 8x$: $\dfrac{dy}{dx} = 8\dfrac{d(x)}{dx} = 8$

For the constant function $y = 8$: $\dfrac{dy}{dx} = 0$ ◀

In order to differentiate a polynomial expression of more than one term, an additional rule is needed. If $u(x)$ and $v(x)$ are two differentiable power functions, and if $y = u(x) + v(x)$,* then

$$\frac{dy}{dx} = \frac{du(x)}{dx} + \frac{dv(x)}{dx}$$

We can show that this is true using the Δ-limit process.

Given: $\qquad\qquad\qquad y = u(x) + v(x)$

Step 1. Replace x by $(x + \Delta x)$ and y by $(y + \Delta y)$: $\qquad y + \Delta y = u(x + \Delta x) + v(x + \Delta x)$

Step 2. Subtract y: $\qquad \Delta y = u(x + \Delta x) + v(x + \Delta x) - y$

$\qquad\qquad\qquad\qquad\qquad = u(x + \Delta x) + v(x + \Delta x) - u(x) - v(x)$

Rearrange terms: $\qquad \Delta y = u(x + \Delta x) - u(x) + v(x + \Delta x) - v(x)$

Step 3. $\qquad\qquad\qquad \dfrac{\Delta y}{\Delta x} = \dfrac{u(x + \Delta x) - u(x)}{\Delta x} + \dfrac{v(x + \Delta x) - v(x)}{\Delta x}$

Step 4. Take the limit as $\Delta x \to 0$ using equation (22-1):

$$\lim_{\Delta x \to 0} \frac{\Delta y}{\Delta x} = \lim_{\Delta x \to 0} \frac{u(x + \Delta x) - u(x)}{\Delta x} + \lim_{\Delta x \to 0} \frac{v(x + \Delta x) - v(x)}{\Delta x}$$

Therefore,

$$\frac{dy}{dx} = \frac{du}{dx} + \frac{dv}{dx}$$

If $u(x)$ and $v(x)$ are differentiable functions, then the following rule can be used to differentiate their sum:

The Sum Rule

If $y = u(x) + v(x)$, then $\dfrac{dy}{dx} = \dfrac{du}{dx} + \dfrac{dv}{dx}$ $\qquad\qquad$ (22-15)

In other words, the derivative of the sum of two functions is equal to the sum of the derivatives of the functions.

*The domain of $y(x)$ contains only values of x common to the domains of both $u(x)$ and $v(x)$.

The sum rule may be extended to include the difference of two functions, since $y = u(x) - v(x) = u(x) + [-v(x)]$.

The Difference Rule

If $y = u(x) - v(x)$, then $\dfrac{dy}{dx} = \dfrac{du}{dx} - \dfrac{dv}{dx}$ (22-16)

The sum and difference rules may be used to find the derivatives of the sums and differences of any number of differentiable functions.

EXAMPLE 23   Differentiate each function.
(a) $y = x^2 + x^3$ (b) $y = x^4 - x^7$ (c) $y = 5x^3 + 2$
(d)   Find the slope of the tangent line to the graph of
       $y = 2x^4 - 7x^3 + 4x^2 - x + 12$ at $x = 1$.

Solution
(a) $y = x^2 + x^3$:
       For this function, $u(x) = x^2 \qquad v(x) = x^3$

       Using the power rule (22-13), $\dfrac{du}{dx} = 2x \qquad \dfrac{dv}{dx} = 3x^2$

       Therefore, by the sum rule (22-15), $\dfrac{dy}{dx} = \dfrac{d(x^2)}{dx} + \dfrac{d(x^3)}{dx} = 2x + 3x^2$

(b) $y = x^4 - x^7$:
       For this function, $u(x) = x^4 \qquad v(x) = x^7$

       Using the power rule (22-13), $\dfrac{du}{dx} = 4x^3 \qquad \dfrac{dv}{dx} = 7x^6$

       Therefore, by the difference rule (22-16), $\dfrac{dy}{dx} = 4x^3 - 7x^6$

(c) $y = 5x^3 + 2$

       $u(x) = 5x^3 \qquad v(x) = 2$

       $\dfrac{du}{dx} = 15x^2 \qquad \dfrac{dv}{dx} = 0$

       $\dfrac{dy}{dx} = 15x^2$

(d)   **Step 1.**   Find the derivative of $y = 2x^4 - 7x^3 + 4x^2 - x + 12$:

$$\frac{dy}{dx} = 8x^3 - 21x^2 + 8x - 1 + 0 \longleftarrow \boxed{\frac{d(12)}{dx} = 0}$$

$$\boxed{\frac{d(2x^4)}{dx}} \quad \boxed{\frac{d(-7x^3)}{dx}} \quad \boxed{\frac{d(4x^2)}{dx}} \quad \boxed{\frac{d(-x)}{dx}}$$

Step 2. Evaluate dy/dx at $x = 1$.

$$\left.\frac{dy}{dx}\right|_{x=1} = 8(1)^3 - 21(1)^2 + 8(1) - 1 = -6$$

The slope of the tangent line is -6 at $x = 1$.

EXAMPLE 24 When a certain time-dependent force acts on a particle, its position s (in meters) varies with time (in seconds) according to the equation $s = 3t^4 - t^2 + 6t$. Find the instantaneous velocity $v = ds/dt$ at the given time.

(a) $\left.\dfrac{ds}{dt}\right|_{t=2\,s}$ ▦ (b) $\left.\dfrac{ds}{dt}\right|_{t=0.05\,s}$

Solution First, find the derivative ds/dt $\dfrac{ds}{dt} = 12t^3 - 2t + 6$
using the differentiation rules.

(a) Substitute $t = 2$ s to find

$$\left.\frac{ds}{dt}\right|_{t=2} = 12(2)^3 - 2(2) + 6$$

$$= 96 - 4 + 6 = 98 \text{ m/s} \longleftarrow \boxed{\begin{array}{l}\text{This is the slope of the}\\ \text{line tangent to the graph}\\ \text{of } s = 3t^4 - t^2 + 6t \text{ at}\\ t = 2 \text{ s.}\end{array}}$$

(b) Substitute $t = 0.05$ s to find

$$\left.\frac{ds}{dt}\right|_{t=0.05} \quad : \quad 12 \,\boxed{\times}\,.05\,\boxed{y^x}\,3\,\boxed{-}\,2\,\boxed{\times}\,.05\,\boxed{+}\,6\,\boxed{=} \rightarrow \qquad \textbf{5.9015}$$

EXAMPLE 25 For the function $y = x^2 - 6x + 2$, find the point where the slope of the tangent line to the curve is equal to 4.

Solution
Step 1. Find the derivative. If $y = x^2 - 6x + 2$, then $\dfrac{dy}{dx} = 2x - 6$.

Step 2. The derivative $\dfrac{dy}{dx}$ is the slope of the tangent line to the curve at each value of x. Therefore,

$$\text{Slope} = 4 = 2x - 6$$

$$10 = 2x$$

$$x = 5$$

Step 3.   Substitute this value of x into the original function and solve for y:

$$y = x^2 - 6x + 2 = (5)^2 - 6(5) + 2 = -3$$

The slope of the tangent line is equal to 4 at the point (5, −3).

EXAMPLE 26   Find the equation of the line tangent to the curve $y = 3x^2 + 2x - 1$ at the point where x = 1.

Solution
Step 1.   Find the coordinates of the tangent point. At x = 1,

$$y = 3(1)^2 + 2(1) - 1 = 4$$

Step 2.   The derivative of the function $y = 3x^2 + 2x - 1$ is

$$\frac{dy}{dx} = 6x + 2 \qquad \text{This represents the slope of the tangent line.}$$

At x = 1, this slope is

$$\frac{dy}{dx}\bigg|_{x=1} = 6(1) + 2 = 8$$

Step 3.   Now find the equation of the line with slope m = 8, passing through the point (1, 4). Use the point–slope equation of a line given in Chapter 20. The equation of the line is

$$y - 4 = 8(x - 1)$$

$$y = 8x - 4$$

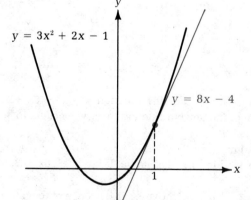

Figure 22-27

EXAMPLE 27   The voltage output V (in millivolts) of a certain electronic device varies with time t (in seconds) according to the formula $V = t^3 - 3t^2 - 3t + 4$. Find the value of t when $dV/dt = 0$.

Solution

Step 1. Find the derivative.

$$\frac{dV}{dt} = 3t^2 - 6t - 3$$

Step 2. Set the derivative equal to zero and solve. If

$$\frac{dV}{dt} = 0$$

then

$$3t^2 - 6t - 3 = 0$$

$$t^2 - 2t - 1 = 0$$

We can solve this equation using the quadratic formula.

$$t = \frac{2 \pm \sqrt{(-2)^2 - 4(-1)}}{2}$$

$$= \frac{2 \pm \sqrt{8}}{2} = \frac{2 \pm 2\sqrt{2}}{2}$$

$$t = 1 \pm \sqrt{2}$$

$$t \approx 2.414 \quad \text{or} \quad t \approx -0.414$$

The negative value of t is not physically meaningful, so $t \approx 2.414$ s.

Exercises 22-3

Find the derivative of each function.

1. $y = 4$ **2.** $y = -6$ **3.** $y = x^7$

4. $y = x^5$ **5.** $y = 5x^3$ **6.** $y = -4x^6$

7. $y = -5x^{11}$ **8.** $y = 10x^{20}$ **9.** $y = \frac{1}{2}x^4$

10. $y = \frac{2}{3}x^9$ **11.** $y = -\frac{x^2}{6}$ **12.** $y = \frac{x^{10}}{5}$

13. $y = 1 - 3x^2$ **14.** $y = 2 + 4x^5$ **15.** $y = x + x^4$

16. $y = x^2 - x^9$ **17.** $y = 5x^4 - 4x^5$ **18.** $y = x^2 - 4x^4$

19. $y = 3x^7 - x^{10}$ **20.** $y = 5x^2 - 2$

21. $y = x^6 - x^5 + x^4 - 2x^3 + 3$ **22.** $s = 10 + 20t - 40t^2 + 5t^3$

23. $L = 20 + 0.06T$ **24.** $Q = 2.5f - 1.2f^2$

Find the slope of the tangent line to the curve at the given value of the variable.

25. $y = 6x^2 - x$ at $x = -1$ **26.** $y = x^4 - 4x$ at $x = 1$

27. $y = 4x^4 + 2x^3 - 1$ at $x = 0.02$

28. $y = 2.4x^6 - 1.4x^5 + 0.7x^3 + 12$ at $x = 1.5$

29. $S = 16t^2 - 20t + 3.25$ at $t = 4.25$

30. $L = 0.00428T^3 - 0.0095T^2$ at $T = 3.5$

Evaluate.

31. $\left.\dfrac{dy}{dx}\right|_{x=1}$ for $y = 12x^3 - x$

32. $\dfrac{dy}{dx}\bigg|_{x=-2}$ for $y = 8x^4 + x^3 + x^2 - 1$ **33.** $\dfrac{dQ}{dt}\bigg|_{t=0.01}$ for $Q = 3t^2 + t + 1$

34. $\dfrac{dS}{dr}\bigg|_{r=20}$ for $S = 30r^2 - 10r$

Find the coordinates of the point or points where the graph has the given slope. Round irrational answers to the nearest hundredth.

35. $y = 5x^2 - 3x + 1$; Slope $= 7$ **36.** $y = 3x^2$; Slope $= -9$
37. $y = 4x^3$; Slope $= 24$ **38.** $y = x^2 + 8x + 40$; Slope $= -10$
39. $y = x^3 - x^2$; Slope $= 1$ **40.** $y = 2x^3 + \dfrac{9}{4}x^2 - x + 5$; Slope $= \dfrac{1}{2}$

For each function, find the value or values of x where the derivative is equal to zero. Round irrational answers to the nearest thousandth.

41. $y = x^2 - x$ **42.** $y = x^3 + x^2$ **43.** $y = 5x^2 - 3x + 4$
44. $y = 1 - x^4$ **45.** $y = x^3 - 2x^2 - 6x$ **46.** $y = x^3 + x^2 - x$
47. $y = 1 + x - 3x^2 + 2x^3$ **48.** $y = 1 + 2x - x^3$

Solve.

49. Find the equation of the line tangent to the graph of $y = 4x^2 + 16x + 9$ at the point where $x = -1$.
50. Find the equation of the line tangent to the graph of $y = 2x^3$ at the point where $x = \frac{1}{2}$.
51. Find an expression for the instantaneous rate of change of the volume of a sphere with respect to its radius.
52. *Electronics*   The output voltage V (in volts) of a circuit increases gradually with time t (in seconds) according to the equation $V(t) = 30.0 + 0.02t^2$. Find an expression for the instantaneous rate of change of the voltage with time.
53. The length L of the chord of a circle of radius r depends on the height h of the arc according to the equation $L = 2rh - h^2$.
(a) Find an expression for dL/dh, the rate of change of L with respect to h for a fixed value of r.
(b) What is the value of this rate when $h = \frac{1}{4}r$?
54. *Physics*   If the angular speed ω of a rotating flywheel varies with time according to the equation $\omega = 28 - 6t + 2t^4$, find $d\omega/dt$, the instantaneous angular acceleration of the flywheel.
55. For a cylinder with spherical caps, find an expression for the rate of change of the total surface area with respect to its radius (see the figure). Evaluate this rate if $h = 10$ cm when the radius $r = 4$ cm. (Round to the nearest tenth.)
56. If the width of a plate must always be $\frac{2}{3}$ of its length, find the rate of change of its perimeter with respect to its length.
57. *Material science*   The electrical resistivity ρ of a metal varies with temperature T according to the equation $\rho = \rho_0(1 + \alpha T + \beta T^2)$. Find the instantaneous rate of change of resistivity with temperature at room temperature $T = 23°C$ for Monel metal alloy if $\rho_0 = 42$ $\mu\Omega \cdot$cm, $\alpha = 0.0020$, and $\beta = 0.00012$. (Express in scientific notation and round to two decimal places.)
58. *Physics*   The energy E radiated by a hot object depends on its temperature T according to the Stefan–Boltzmann radiation law $E = k(T^4 - T_0^4)$, where k is a constant and T_0 is the temperature of the surroundings. Find an expression for the rate of energy radiation with temperature.
59. The altitude of a prism with a square base must be twice the side length of the base. Find the rate of change of the volume with respect to the side length.
60. A steel plate is in the shape of a square with a semicircle attached to one side. If the plate is heated, its area increases. Find the rate of change of area with respect to the side length of the square (see the figure).

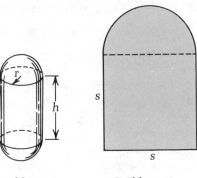

Problem 55   Problem 60

22-4 | Derivatives of Products and Quotients of Functions

The differentiation rules developed in Section 22-3 enable us to find the derivative for any polynomial function. We now need to develop rules for finding the derivatives of algebraic functions that are written as products, quotients, or powers of polynomial functions.

For example, to find the derivative of the function

$$y = (6x^4 - 15x^3 + 9x^2 - 17)(x^{12} - 6x^3 - 5)$$

we could multiply the factors to obtain a polynomial that we know how to differentiate. However, it is easier to think of this as the product $y = u(x) \cdot v(x)$, where $u(x)$ and $v(x)$ are the expressions in parentheses.

Let $u(x)$ and $v(x)$ be differentiable functions and

$$y = u(x) \cdot v(x)$$

Then as x changes to $x + \Delta x$,

y	becomes	$y + \Delta y$
$u(x)$	becomes	$u(x) + \Delta u$
$v(x)$	becomes	$v(x) + \Delta v$

Therefore, we can write

$$y + \Delta y = [u(x) + \Delta u][v(x) + \Delta v]$$

$$= u(x)v(x) + u(x)\Delta v + v(x)\Delta u + \Delta u \Delta v$$

$$\Delta y = u(x)v(x) + u(x)\Delta v + v(x)\Delta u + \Delta u\,\Delta v - u(x)v(x)$$

$$\Delta y = u(x)\Delta v + v(x)\Delta u + \Delta u\,\Delta v$$

Dividing by Δx,

$$\frac{\Delta y}{\Delta x} = u(x)\frac{\Delta v}{\Delta x} + v(x)\frac{\Delta u}{\Delta x} + \Delta u\frac{\Delta v}{\Delta x}$$

Now, take the limit as $\Delta x \to 0$ and use equation (22-1):

$$\lim_{\Delta x \to 0}\frac{\Delta y}{\Delta x} = u(x)\lim_{\Delta x \to 0}\frac{\Delta v}{\Delta x} + v(x)\lim_{\Delta x \to 0}\frac{\Delta u}{\Delta x} + \lim_{\Delta x \to 0}\Delta u\frac{\Delta v}{\Delta x}$$

$$\frac{dy}{dx} = u(x)\frac{dv}{dx} + v(x)\frac{du}{dx}$$

Notice that the last term is

$$\lim_{\Delta x \to 0}\Delta u\frac{\Delta v}{\Delta x}$$

As $\Delta x \to 0$, $\Delta v/\Delta x \to dv/dx$, but $\Delta u \to 0$, so this limit approaches zero.

If $u(x)$ and $v(x)$ are differentiable functions, then the following rule allows us to differentiate their product.

The Product Rule

If $y = u(x)\cdot v(x)$, then $\quad \dfrac{dy}{dx} = u(x)\dfrac{dv}{dx} + v(x)\dfrac{du}{dx}$ $\qquad$ (22-17)

In other words, the derivative of the product of two functions is equal to the first function multiplied by the derivative of the second plus the second function multiplied by the derivative of the first.

The following diagram may help you remember this rule:

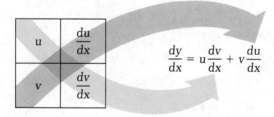

$$\frac{dy}{dx} = u\frac{dv}{dx} + v\frac{du}{dx}$$

EXAMPLE 28   Use the product rule to find $\dfrac{dy}{dx}$ for:
(a) $y = (2 + 3x)(x^2 - x)$
(b) $y = (x^3 - x^2 + 4x - 1)(x^2 + 2x - 1)$

Solution

(a) **First,** let

$$u(x) = 2 + 3x \qquad \text{so that} \qquad \frac{du}{dx} = 3$$

$$v(x) = x^2 - x \qquad \text{so that} \qquad \frac{dv}{dx} = 2x - 1$$

Then, using the product rule (22-17),

$$\frac{dy}{dx} = u\frac{dv}{dx} + v\frac{du}{dx}$$

$$= (2 + 3x)(2x - 1) + (x^2 - x)(3)$$

$$= 4x - 2 + 6x^2 - 3x + 3x^2 - 3x$$

$$= 9x^2 - 2x - 2$$

(b) **First,** let

$$u(x) = x^3 - x^2 + 4x - 1 \qquad \text{so} \qquad \frac{du}{dx} = 3x^2 - 2x + 4$$

$$v(x) = x^2 + 2x - 1 \qquad \text{so} \qquad \frac{dv}{dx} = 2x + 2$$

Then,

$$\frac{dy}{dx} = u\frac{dv}{dx} + v\frac{du}{dx}$$

$$= (x^3 - x^2 + 4x - 1)(2x + 2) + (x^2 + 2x - 1)(3x^2 - 2x + 4)$$

$$= (2x^4 - 2x^3 + 8x^2 - 2x + 2x^3 - 2x^2 + 8x - 2)$$

$$\qquad + (3x^4 + 6x^3 - 3x^2 - 2x^3 - 4x^2 + 2x + 4x^2 + 8x - 4)$$

$$= 5x^4 + 4x^3 + 3x^2 + 16x - 6$$

Notice that in both parts of Example 28 we could have expanded the original expression for y and calculated the derivative directly from the polynomial without using the product rule. In many cases, however, the product rule will greatly simplify the calculation.

If the function $u(x)$ is a constant, $du/dx = 0$, and the product rule reduces to the following differentiation rule:

Constant Product Rule

If $y = c \cdot v(x)$, then $\dfrac{dy}{dx} = c\dfrac{dv}{dx}$ (22-18)

EXAMPLE 29 Differentiate: $y = 4(2x^3 - x^2 + 5x + 2)$

Solution **First,** let $c = 4$ and

$$v(x) = 2x^3 - x^2 + 5x + 2$$

Then,

$$\frac{dv}{dx} = 6x^2 - 2x + 5$$

and using the constant product rule (22-18),

$$\frac{dy}{dx} = 4(6x^2 - 2x + 5) = 24x^2 - 8x + 20$$

Derivative of a quotient   When the function to be differentiated is written as the quotient of poly-nomial expressions, there is usually no simple way to find the derivative using the polynomial rules of Section 22-3. Instead, a special quotient rule must be developed.

If $y = \dfrac{u(x)}{v(x)}$

where $u(x)$ and $v(x)$ are differentiable functions, then we can write

$$y + \Delta y = \frac{u(x) + \Delta u}{v(x) + \Delta v}$$

$$\Delta y = \frac{u(x) + \Delta u}{v(x) + \Delta v} - y$$

$$= \frac{u(x) + \Delta u}{v(x) + \Delta v} - \frac{u(x)}{v(x)}$$

$$= \frac{[u(x) + \Delta u]v(x) - [v(x) + \Delta v]u(x)}{v(x)[v(x) + \Delta v]}$$

$$= \frac{u(x)v(x) + v(x)\Delta u - v(x)u(x) - u(x)\Delta v}{v(x)[v(x) + \Delta v]}$$

$$= \frac{v(x)\Delta u - u(x)\Delta v}{[v(x)]^2 + v(x)\Delta v}$$

Dividing by Δx,

$$\frac{\Delta y}{\Delta x} = \frac{v(x)\dfrac{\Delta u}{\Delta x} - u(x)\dfrac{\Delta v}{\Delta x}}{[v(x)]^2 + v(x)\Delta v}$$

Take the limit as $\Delta x \to 0$:

$$\lim_{\Delta x \to 0} \frac{\Delta y}{\Delta x} = \frac{v(x) \displaystyle\lim_{\Delta x \to 0} \frac{\Delta u}{\Delta x} - u(x) \displaystyle\lim_{\Delta x \to 0} \frac{\Delta v}{\Delta x}}{[v(x)]^2 + \underbrace{v(x) \displaystyle\lim_{\Delta x \to 0} \Delta v}}$$

This term approaches zero as $\Delta x \to 0$.

Notice that as $\Delta x \to 0$, $\Delta v \to 0$, so the second term in the denominator is equal to zero. Therefore,

$$\frac{dy}{dx} = \frac{v(x)\dfrac{du}{dx} - u(x)\dfrac{dv}{dx}}{[v(x)]^2}$$

If $u(x)$ and $v(x)$ are differentiable functions, then the following rule allows you to differentiate their quotient:

The Quotient Rule

If $y = \dfrac{u(x)}{v(x)}$, then $\dfrac{dy}{dx} = \dfrac{v(x)\dfrac{du}{dx} - u(x)\dfrac{dv}{dx}}{[v(x)]^2}$ (22-19)

NOTE ▶ Notice the order of the terms in the numerator. The $v(x)$ term appears first and the $u(x)$ term appears second. ◀

EXAMPLE 30 (a) Differentiate: $\dfrac{2 - 3x}{x^2 - 1}$

(b) The sensitivity S of a certain electrical circuit depends on the resistance R of a temperature sensor according to the equation below. Find $\dfrac{dS}{dR}$.

$$S = \frac{5R}{10 + R^2}$$

Solution

(a) **First,** let

$$u(x) = 2 - 3x \qquad \text{so} \qquad \frac{du}{dx} = -3$$

$$v(x) = x^2 - 1 \qquad \text{so} \qquad \frac{dv}{dx} = 2x$$

Second, substitute in the quotient rule (22-19).

$$\frac{dy}{dx} = \frac{(x^2 - 1)(-3) - (2 - 3x)(2x)}{(x^2 - 1)^2}$$

Third, simplify by multiplying in the numerator.

$$= \frac{-3x^2 + 3 - 4x + 6x^2}{(x^2 - 1)^2}$$

$$\frac{dy}{dx} = \frac{3x^2 - 4x + 3}{(x^2 - 1)^2}$$

(b) **First,** let

$$u(R) = 5R \qquad \text{so} \qquad \frac{du}{dR} = 5$$

$$v(R) = 10 + R^2 \qquad \text{so} \qquad \frac{dv}{dR} = 2R$$

Second, substitute in the quotient rule (22-19).

$$\frac{dS}{dR} = \frac{(10 + R^2)5 - (5R)2R}{(10 + R^2)^2}$$

$$= \frac{50 + 5R^2 - 10R^2}{(10 + R^2)^2}$$

$$\frac{dS}{dR} = \frac{50 - 5R^2}{(10 + R^2)^2}$$

LEARNING HINT ▶ If the quotient to be differentiated has a common factor in the numerator and denominator, simplify it before applying the quotient rule. For example, if

$$y = \frac{2x^2 - x}{x^2}$$

divide the numerator and denominator by x to get

$$y = \frac{2x - 1}{x}$$

and then continue as before. ◀

EXAMPLE 31   For $y = \dfrac{x^2 + 2}{x - 1}$ find:

(a) $\left.\dfrac{dy}{dx}\right|_{x=1}$ ▦ (b) $\left.\dfrac{dy}{dx}\right|_{x=0.0067}$

(c)   The values of x where $dy/dx = 0$

Solution   **First,** use the quotient rule (22-19) to find the derivative of the function.

$$\frac{dy}{dx} = \frac{x^2 - 2x - 2}{(x - 1)^2}$$

(a) $\left.\dfrac{dy}{dx}\right|_{x=1}$ is not defined, because when $x = 1$, the denominator is equal to zero.

(b) $\left.\dfrac{dy}{dx}\right|_{x=0.0067}$:

.0067 $\boxed{\text{STO}}$ $\boxed{x^2}$ $\boxed{-}$ 2 $\boxed{\times}$ $\boxed{\text{RCL}}$ $\boxed{-}$ 2 $\boxed{=}$ $\boxed{\div}$ $\boxed{(}$ $\boxed{\text{RCL}}$ $\boxed{-}$ 1 $\boxed{)}$ $\boxed{x^2}$ $\boxed{=}$ →

$$\boxed{-2.0406077}$$

(c) $\dfrac{dy}{dx} = 0$ if $x^2 - 2x - 2 = 0$, and we can solve this equation using the quadratic formula.

$$x = \frac{2 \pm \sqrt{4 + 8}}{2} = \frac{2 \pm 2\sqrt{3}}{2}$$

$$x = 1 \pm \sqrt{3}$$

$$x \approx 2.732 \quad \text{or} \quad x \approx -0.732$$

Negative exponents   The quotient rule (22-19) can be used to extend the power rule (22-17) so that we can differentiate functions where the exponent is a negative integer. If

$$y = x^{-n} = \frac{1}{x^n}$$

where n is a positive integer, then we can let

$$u(x) = 1 \qquad \text{so that} \qquad \frac{du}{dx} = 0$$

$$v(x) = x^n \qquad \text{so that} \qquad \frac{dv}{dx} = nx^{n-1}$$

Using the quotient rule,

$$\frac{dy}{dx} = \frac{(x^n)(0) - (1)(nx^{n-1})}{(x^n)^2}$$

$$= \frac{-nx^{n-1}}{x^{2n}}$$

$$= -nx^{n-1} \cdot x^{-2n}$$

$$= -nx^{n-1-2n}$$

$$\frac{dy}{dx} = -nx^{-n-1}$$

This last equation gives us a form of the power rule for negative exponents.

Power Rule for Negative Exponents

If $y = \dfrac{1}{x^n} = x^{-n}$, then $\dfrac{dy}{dx} = -nx^{-n-1} = \dfrac{-n}{x^{n+1}}$ \qquad (22-20)

for n a positive integer

Remember it this way:

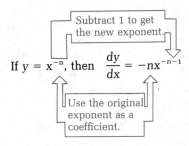

EXAMPLE 32   Find the derivative of each function:

(a) $y = \dfrac{1}{x}$ \qquad (b) $y = \dfrac{1}{x^4}$ \qquad (c) $y = -\dfrac{2}{x^{12}}$

(d)   The intensity I of a sound decreases with the distance r from its source according to the equation $I = k/r^2$, where k is a constant.

Solution

(a)   Use the power rule for negative exponents (22-20):

If $y = \dfrac{1}{x} = x^{-1}$, then: $\dfrac{dy}{dx} = (-1)x^{-1-1}$ ◄ Subtract 1 from the exponent.

$$\frac{dy}{dx} = -x^{-2} = -\frac{1}{x^2}$$

(b)   If $y = \dfrac{1}{x^4} = x^{-4}$, then: $\dfrac{dy}{dx} = (-4)x^{-4-1}$

$$\frac{dy}{dx} = -4x^{-5} = -\frac{4}{x^5}$$

(c)   If $y = -\dfrac{2}{x^{12}} = -2x^{-12}$, then: $\dfrac{dy}{dx} = (-12)(-2)x^{-12-1}$

$$\frac{dy}{dx} = 24x^{-13} = \frac{24}{x^{13}}$$

(d)   If $I = \dfrac{k}{r^2} = kr^{-2}$, then: $\dfrac{dI}{dr} = k(-2)r^{-2-1}$

$$\frac{dI}{dr} = -2kr^{-3} = -\frac{2k}{r^3}$$

Exercises 22-4

Find the derivative of each function.

1. $y = 4(x^3 - 2x^2)$

2. $y = -6(2x^2 - 3x)$

3. $y = x^2(x + x^4)$

4. $y = \dfrac{x^2}{2}(x^4 - 4x)$

5. $y = 3x(x^3 + 2)$

6. $y = -4x^2(3x^2 + 5x)$

7. $y = (1 + x^3)(x - x^2)$

8. $y = (x^2 + 1)(1 + x + x^2)$

9. $y = (-2x^4 + x^3)(4x^2 + 3x)$

10. $y = (10 - x^2)(2x^2 - x - 1)$

11. $y = (x - 4)(x^4 - 2x^3 + x^2)$

12. $y = (x^2 + 3x)(x^4 - 2x^2 - 1)$

13. $y = (x - x^2)(x + x^3)$

14. $y = (2x + x^2)(4 - x + 2x^2)$

15. $y = (2 - x - x^2)(1 + 2x - 2x^2)$

16. $y = (x^3 - 4x)(3 + 2x - x^2)$

17. $y = (10x - 5x^2)(x^5 - x^3 + x)$

18. $y = (x^6 + x^4)(x - x^3 + x^5)$

19. $y = \dfrac{x}{x - 3}$

20. $y = \dfrac{2x}{1 - x}$

21. $y = \dfrac{x^2 + 1}{2x}$

22. $y = \dfrac{1 - x^3}{2x}$

23. $y = \dfrac{2x - 3}{x}$

24. $y = \dfrac{4x + 3}{x^2}$

25. $y = \dfrac{3x^2 - x^3}{2x^4}$

26. $y = \dfrac{x^5 - 1}{x^4 - 1}$

27. $y = \dfrac{8x}{x^2 + 3}$

28. $y = \dfrac{x - 1}{x + 1}$

29. $y = \dfrac{5x - 2x^2}{2x + 3}$

30. $y = \dfrac{1 - 2x}{x^2 + 2}$

31. $y = \dfrac{x^2 - 2x + 2}{2x^2 + 1}$ **32.** $y = \dfrac{2 + x^2 - 3x^4}{x + x^3}$ **33.** $y = \dfrac{1}{x^5}$

34. $y = \dfrac{4}{x^7}$ **35.** $y = \dfrac{-3}{x^6}$ **36.** $y = \dfrac{5}{3x^2}$

Find the derivative of each function. Do not simplify the function before taking the derivative.

37. $y = \dfrac{x^2(2 - 3x)}{2x + 1}$ **38.** $y = \dfrac{(x^2 + 1)(2x + 5)}{x - 1}$

39. $y = \dfrac{1}{2x^2 + 1} + x^2$ **40.** $y = \dfrac{x}{x + 2} + 2x^3$

For the following functions, find:

(a) $\dfrac{dy}{dx}\bigg|_{x=0}$ (b) $\dfrac{dy}{dx}\bigg|_{x=1}$ (c)   The values of x where $\dfrac{dy}{dx} = 0$

41. $y = (2x - 1)(3x^2 - x - 1)$ **42.** $y = (x - 1)(x^2 - 2x + 1)$

43. $y = (4x + 3)(2x - 1)$ **44.** $y = \dfrac{x^2 + 1}{x + 1}$

45. $y = \dfrac{2x^2 - 1}{x - 1}$ **46.** $y = \dfrac{3x + 2}{x^2 + 1}$

47. Find the slope of the line tangent to $y = 2x(x^2 - 3)$ at $x = 3$.

48. Find the slope of the line tangent to $y = \dfrac{x^2 - 2x^3}{x^2 + 1}$ at $x = -1.25$.

49. Find the equation of the line tangent to the graph of $y = \dfrac{x^2 - 3x + 1}{x - 5}$ at $x = 2$.

50. Find the equation of the line tangent to the graph of $y = (x^2 + 2x)(2 - x^2 + x^4)$ at $x = -2$.

Solve.

51. Electronics   In a voltage divider circuit the output V depends on the value of a variable resistor R according to the equation below. Find $\dfrac{dV}{dR}$.

$$V = \dfrac{110R}{R + 60}$$

52. Electronics   The impedance Z of a Wein bridge circuit depends on the resistance R of one arm of the bridge so that

$$Z = \dfrac{(1 + 8R)^2 + 8R}{8(1 + 8R)}$$

Find $\dfrac{dZ}{dR}$.

53. Robotics   There exist certain self-sustaining physical systems such that for small oscillations energy is fed into the system, and for large oscillations energy is taken from the system.* These Van der Pol systems are important in the study of robotics and vacuum tube circuits and are described by equations of the form

*See Erwin Kreysig, *Advanced Engineering Mathematics*, 4th ed. (New York: Wiley, 1979), p. 139.

$$v = \frac{y}{a(1 - y^2) - b}$$

where a and b are constants. Find $\dfrac{dv}{dy}$.

54. *Physics*   For a rotating object the angular acceleration α (in rad/s²) is equal to the ratio of the torque T applied and the moment of inertia I—that is, $\alpha = T/I$. If the object changes shape as it rotates, then both T and I may change with time. Find $d\alpha/dt$ at $t = 4$ s if $T = 6(t^2 + 2t)$ and $I = 10t^2$.

55. *Thermodynamics*   The Van der Waal's model of a gas takes into account the finite size of gas molecules and their intermolecular forces. If P, V, and T are the pressure, volume, and temperature, respectively, of a sample of gas, then

$$\left(P + \frac{an^2}{V^2}\right)(V - nb) = nRT$$

where n is the number of moles of gas in the sample, and a, b, and R are constants. Find dP/dV for a sample of helium gas undergoing an isothermal expansion (that is, T is constant), where $n = 1$, $a = 3$, and $b = 24$. [*Hint:*   First solve for P; then differentiate with respect to V.]

56. *Electronics*   According to Ohm's law, the voltage V across a resistor R is related to the current i passing through the resistor by the equation $V = iR$. If the current varies with time t according to the equation

$$i = \frac{2.0t}{t + 2.0}$$

and if the resistance increases with time according to the equation $R = 20.0 + 0.04t^2$, find dV/dt at $t = 2.0$ s.

22-5 | Composite Functions

To find the derivative of the function $y = (3x^2 - 7x)^5$, we could expand the expression on the right using the binomial theorem from Chapter 18, and then differentiate the resulting polynomial. This is a lengthy procedure, especially for a function in which the exponent is very large. A much simpler approach involves the recognition of this function as a composite function.

Composite function   When a functional rule is applied to an expression that is itself a function, the result is called a **composite function**. A composite function can be thought of as a "function of a function." For example, the composite function $y = (3x^2 - 7x)^5$ can be written $y = u^5$, where u is the function $u(x) = 3x^2 - 7x$.

We can develop a rule for finding the derivative of a composite function as follows: Suppose $y = f(u)$ and $u = g(x)$ are both differentiable functions. When x changes by an amount Δx, $g(x)$ changes by an amount Δu and $f(u)$ changes by an amount Δy. As Δx approaches zero, Δu also approaches zero, since

$$\lim_{\Delta x \to 0} \Delta u = \lim_{\Delta x \to 0} \left(\Delta u \cdot \frac{\Delta x}{\Delta x}\right)$$

$$= \lim_{\Delta x \to 0} \left(\frac{\Delta u}{\Delta x} \cdot \Delta x \right) \qquad \boxed{\lim_{\Delta x \to 0} \Delta x = 0}$$

$$= 0 \qquad \boxed{\lim_{\Delta x \to 0} \frac{\Delta u}{\Delta x} = \frac{du}{dx}}$$

According to equation (22-2), the limit of the product of two functions equals the product of their limits; that is,

$$\lim_{x \to a} f(x)g(x) = \left[\lim_{x \to a} f(x) \right]\left[\lim_{x \to a} g(x) \right]$$

Therefore, we can write

$$\frac{\Delta y}{\Delta x} = \frac{\Delta y}{\Delta x} \cdot \frac{\Delta u}{\Delta u} = \frac{\Delta y}{\Delta u} \cdot \frac{\Delta u}{\Delta x}^{*}$$

$$\lim_{\Delta x \to 0} \frac{\Delta y}{\Delta x} = \lim_{\Delta u \to 0} \frac{\Delta y}{\Delta u} \cdot \lim_{\Delta x \to 0} \frac{\Delta u}{\Delta x}$$

Therefore, if $y = f(u)$, where $u = g(x)$, then the following rule enables us to find dy/dx:

The Chain Rule $\qquad \dfrac{dy}{dx} = \dfrac{dy}{du} \cdot \dfrac{du}{dx}$ $\qquad\qquad$ (22-21)

EXAMPLE 33 Differentiate the function $y = (3x^2 - 7x)^5$.

Solution The function $y = (3x^2 - 7x)^5$ can be thought of as

$$y = u^5 \qquad \text{where} \qquad u = 3x^2 - 7x$$

$$\frac{dy}{du} = 5u^4 \qquad\qquad \frac{du}{dx} = 6x - 7$$

Then,

$$\frac{dy}{dx} = \frac{dy}{du} \cdot \frac{du}{dx}$$

$$= (5u^4)(6x - 7)$$

$$= 5(3x^2 - 7x)^4(6x - 7)$$

Extended power rule $\qquad$ For functions like that of Example 33, where y has the form $y = [u(x)]^n$, the chain rule takes the following form:

---

*This step is only possible if $u(x)$ is a function such that Δu is never equal to zero when $\Delta x \neq 0$. Functions encountered in science and technology are usually of this kind.

The Extended Power Rule

If $y = [u(x)]^n$, then $\dfrac{dy}{dx} = n[u(x)]^{n-1} \cdot \dfrac{du}{dx}$ (22-22)

This result comes directly from the chain rule.

$$y = [u(x)]^n \implies \frac{dy}{dx} = \frac{dy}{du} \cdot \frac{du}{dx}$$

$$= \overbrace{n[u(x)]^{n-1}} \cdot \frac{du}{dx}$$

LEARNING HINT ▶

When the function to be differentiated has the form

$$y = \boxed{}^{\,n}$$

write

$$\frac{dy}{dx} = n\,\boxed{}^{\,n-1} \cdot \frac{d\,\boxed{}}{dx} \longleftarrow \boxed{\text{Be careful not to forget this part.}}$$

where any differentiable function can be written in the shaded boxes.   ◀

EXAMPLE 34   (a) Differentiate: $y = (x^2 - 3x + 2)^6$

(b) The impedance Z of a bridge circuit, written as a function of a variable resistance R, involves the expression shown.

$$Z = \frac{1}{(1 + R^2)^2} \qquad \text{Find } \frac{dZ}{dR}.$$

Solution

(a) **Step 1.**   Write $y = u^6$, where $u = x^2 - 3x + 2$

Then $\dfrac{du}{dx} = 2x - 3$

Step 2.   Use the extended power rule $\dfrac{dy}{dx} = 6u^{6-1} \cdot (2x - 3)$
(22-22).

$$= 6(x^2 - 3x + 2)^5(2x - 3)$$

(b) Let $u = 1 + R^2$, then $\dfrac{du}{dR} = 2R$

and $Z = \dfrac{1}{u^2} = u^{-2}$

Use the extended power rule (22-22). $\dfrac{dZ}{dR} = (-2)(u^{-2-1})\dfrac{du}{dR}$

$$= -2u^{-3} \cdot 2R$$

$$= -\frac{4R}{u^3}$$

$$= -\frac{4R}{(1 + R^2)^3}$$

CAUTION ▶ The most frequent student error is to omit the du/dx factor in the derivative. ◀

Finding the derivative of many composite functions involves using a combination of the differentiation rules we have developed.

EXAMPLE 35 Differentiate $y = (x^2 - 2)^3 + (2x^3 + x)^4$.

Solution
Step 1. Write this as $y = f(u) + g(v)$ and apply the sum rule (22-15),

$$\frac{dy}{dx} = \frac{df}{dx} + \frac{dg}{dx}$$

Step 2. Use the extended power rule (22-22) to find df/dx and dg/dx. Since $f(u) = u^3$, where $u = x^2 - 2$, then $du/dx = 2x$ and

$$\frac{df}{dx} = 3u^2 \cdot \frac{du}{dx} = 3u^2(2x) = 6x(x^2 - 2)^2$$

Since $g(v) = v^4$, where $v = 2x^3 + x$, then $dv/dx = 6x^2 + 1$ and

$$\frac{dg}{dx} = 4v^3 \cdot \frac{dv}{dx} = 4v^3(6x^2 + 1) = 4(2x^3 + x)^3(6x^2 + 1)$$

Step 3. Combine Steps 1 and 2:

$$\frac{dy}{dx} = \frac{df}{dx} + \frac{dg}{dx} = 6x(x^2 - 2)^2 + 4(2x^3 + x)^3(6x^2 + 1)$$

EXAMPLE 36 Differentiate $y = (2x - 3)^3(1 - x^2)^5$.

Solution
Step 1. Write this as $y = f(u) \cdot g(v)$ and apply the product rule (22-17),

$$\frac{dy}{dx} = f(u)\frac{dg}{dx} + g(v)\frac{df}{dx}$$

Step 2. Use the extended power rule (22-22) to find dg/dx and df/dx. Since $g(v) = v^5$, where $v = 1 - x^2$, then $dv/dx = -2x$ and

$$\frac{dg}{dx} = 5v^4 \cdot \frac{dv}{dx} = 5v^4(-2x) = -10x(1 - x^2)^4$$

Since $f(u) = u^3$, where $u = 2x - 3$, then $du/dx = 2$ and

$$\frac{df}{dx} = 3u^2 \cdot \frac{du}{dx} = 3u^2 \cdot 2 = 6(2x - 3)^2$$

Step 3.   Combine Steps 1 and 2:

$$\frac{dy}{dx} = (2x - 3)^3[-10x(1 - x^2)^4] + (1 - x^2)^5[6(2x - 3)^2]$$

$$= -10x(1 - x^2)^4(2x - 3)^3 + 6(1 - x^2)^5(2x - 3)^2$$

Note that this expression can be simplified by factoring the expression $(1 - x^2)^4(2x - 3)^2$ from each term, and combining the remaining expressions:

$$\frac{dy}{dx} = (1 - x^2)^4(2x - 3)^2[-10x(2x - 3) + 6(1 - x^2)]$$

$$= (1 - x^2)^4(2x - 3)^2(-26x^2 + 30x + 6)$$

EXAMPLE 37   Differentiate: $y = \left(\dfrac{2x + 1}{x^2 - 1}\right)^3$

Solution

Step 1.   Write this as $y = u^3$, where $u = \dfrac{2x + 1}{x^2 - 1}$ and apply the extended power rule (22-22):

$$\frac{dy}{dx} = 3u^2 \cdot \frac{du}{dx}$$

Step 2.   To find du/dx, use the quotient rule (22-19):

$$\frac{du}{dx} = \frac{g(x)\dfrac{df}{dx} - f(x)\dfrac{dg}{dx}}{[g(x)]^2} \qquad \text{for} \quad u = \frac{f(x)}{g(x)}$$

$$\frac{df}{dx} = \frac{d}{dx}(2x + 1) = 2 \qquad \frac{dg}{dx} = \frac{d}{dx}(x^2 - 1) = 2x$$

Therefore,

$$\frac{du}{dx} = \frac{(x^2 - 1)(2) - (2x + 1)(2x)}{(x^2 - 1)^2}$$

$$= \frac{2x^2 - 2 - 4x^2 - 2x}{(x^2 - 1)^2}$$

$$= \frac{(-2)(x^2 + x + 1)}{(x^2 - 1)^2}$$

Step 3.   Combine Steps 1 and 2 to find dy/dx:

$$\frac{dy}{dx} = 3u^2 \cdot \frac{du}{dx} = 3\left(\frac{2x+1}{x^2-1}\right)^2 \cdot \frac{(-2)(x^2+x+1)}{(x^2-1)^2}$$

$$= \frac{-6(2x+1)^2(x^2+x+1)}{(x^2-1)^4}$$

Exercises 22-5

Find the derivative of each function.

1. $y = (1 + 2x)^4$ **2.** $y = (1 - 4x)^3$

3. $y = (x^2 - 2)^5$ **4.** $y = (3 - 2x^2)^4$

5. $y = (4 + 3x^3)^3$ **6.** $y = (3x - 5x^2)^5$

7. $y = (1 + 2x - x^2)^5$ **8.** $y = (2x^2 + x - 2)^7$

9. $y = (x^2 - x^3)^{10}$ **10.** $y = 2(x^3 - 2x^2 - x)^6$

11. $y = -4(1 - x^4 + x^5)^4$ **12.** $y = (2x^3 - 3x^2 + 6x)^4$

13. $y = \dfrac{1}{x^2 - 1}$ **14.** $y = \dfrac{1}{x^3 + x}$

15. $y = \dfrac{1}{(2x + 3)^2}$ **16.** $y = \dfrac{1}{(5x^2 + 4)^2}$

17. $y = \dfrac{2}{(x^3 - x^2 + 1)^4}$ **18.** $y = \dfrac{4}{(1 + x - x^3)^5}$

19. $y = x(x^2 + 2)^4$ **20.** $y = 2x^2(3 - x^3)^5$

21. $y = 2x(x^2 - 2)^3$ **22.** $y = x^3(x^3 - x^2)^5$

23. $y = (1 - x)^2(2 + x^2)^4$ **24.** $y = (x^2 - 3)^4(x^3 - 2x)^3$

25. $y = x^2 + (1 + x^3)^5$ **26.** $y = 3x^3 - (2x + 3x^2)^4$

27. $y = (1 + x^2)^5 - (2x - x^2)^4$ **28.** $y = (3x - 5x^2)^5 + (2x - 1)^4$

29. $y = \dfrac{1}{x^3} - (x^3 - 2)^4$ **30.** $y = (4x^2 - 1)^7 - (2 - 3x)^6$

31. $y = \dfrac{4x}{(x^2 - 2)}$ **32.** $y = \dfrac{6x^2}{x^2 - 1}$

33. $y = \dfrac{7x^2}{(2x - 3)^2}$ **34.** $y = \dfrac{4x^3}{(3x^2 - 1)^4}$

35. $y = \left(\dfrac{1 + x^3}{2 - x^2}\right)^3$ **36.** $y = \left(\dfrac{2x - 1}{x^2 + 2}\right)^4$

37. $y = \left(\dfrac{5x^2 - 2}{2x + x^2}\right)^4$ **38.** $y = \left(\dfrac{1 + x - x^2}{1 - x - x^2}\right)^4$

Solve.

39. For what values of x is the derivative of $y = (x^2 - 4)^3$ equal to zero?

40. For what values of x is the derivative of $y = \left(\dfrac{x^2 - 9}{x}\right)^3$ equal to zero?

41. Find the slope of the line tangent to the curve of $y = (x^2 - 4x - 5)^3$ at $x = 4$.

42. Find the slope of the line tangent to the graph of $y = 2x(x^2 + 1)^4$ at $x = 0.2$.

43. *Electronics*   If two parallel wires a distance d apart carry equal currents i in opposite directions, then the magnetic field B (in webers) at a point between the wires at a distance x (in meters) from their midpoint is given by the equation

$$B = \frac{kid}{\pi(d^2 - 4x^2)}$$

where k is a constant. Find an expression for dB/dx when $x = \frac{1}{4}d$.* (*Hint:* Find dB/dx and evaluate it at $x = d/4$.)

44. *Electronics*   If the current i in a coil changes with time t, an electromotive force E (in volts) is induced in the coil and is proportional to the rate of change of the current. This self-induced electromotive force is given by the equation

$$E = -L\frac{di}{dt}$$

Find E when $t = 1.0$ s for a coil with inductance $L = 25 \times 10^{-3}$ H if

$$i = \frac{4.0(t + 2.0)^2}{2.0 - t + t^2}$$

45. *Energy Technology*   The energy output of an electrical heater varies with time t according to the equation $E = 6(1 + 2t^2)^3$. Find the power P (in kilowatts) generated by the heater at $t = 3.0$ s if $P = dE/dt$. (Round to the nearest kilowatt.)

46. *Physics*   The kinetic energy E (in joules) of a moving object depends on the square of its velocity v according to the equation $E = \frac{1}{2}mv^2$. If mass $m = 40$ kg and $v = 1 - 2t^2 + t^3$ m/s, find dE/dt when $t = 2.0$ s.

22-6 | Implicit Differentiation

Implicit function   An equation in x and y that has been solved for y and written in the form $y = f(x)$ is said to be expressed as an **explicit function** of y. If the equation has not been solved explicitly for y, then y is said to be an **implicit function** of x. For example, in the equations

$$xy = 1 \qquad 2x + 3y = 1 \qquad y^2 + 3xy + x - 1 = 0$$

y is an implicit function of x. In the equations

$$y = x^2 - 2x \qquad y = \frac{x - 1}{3x} \qquad y = x - 1$$

y is an explicit function of x.

An equation in x and y may define more than one explicit function. For example, $y^2 + x = 4$ can be solved explicitly for y:

$$y^2 = 4 - x$$

$$y = \pm\sqrt{4 - x}$$

*See David Halliday and Robert Resnick, *Fundamentals of Physics*, 2nd edition (New York: Wiley, 1981), p. 561

This equation is not a function as defined in Chapter 3. As Fig. 22-28(a) shows, two values of y are associated with each x < 4. However, we can define two functions: $y = \sqrt{4 - x}$ and $y = -\sqrt{4 - x}$.

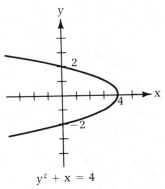

$y^2 + x = 4$

Figure 22-28(a)

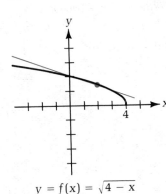

$y = f(x) = \sqrt{4 - x}$

Figure 22-28(b)

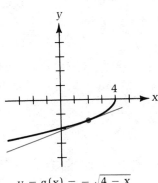

$y = g(x) = -\sqrt{4 - x}$

Figure 22-28(c)

The derivative dy/dx for the equation $y^2 + x = 4$ has two values at $x = 2$. There is one tangent line at $x = 2$ for each of the functions shown in Fig. 22-28(b) and (c).

To find the derivative $\dfrac{dy}{dx}$ when y is an implicit function of x, it may be possible to solve for y and use the differentiation rules. For example, for $xy = 1$,

$$y = \frac{1}{x} = x^{-1}$$

and using the power rule (22-20),

$$\frac{dy}{dx} = (-1)x^{-1-1} = -x^{-2} = -\frac{1}{x^2}$$

If solving explicitly for y is difficult, or leads to a function too complicated for easy differentiation, follow these steps.

Implicit differentiation **Step 1.** Take the derivative with respect to x of each side of the equation separately. Use the differentiation rules developed in this chapter.

Step 2. Solve the resulting equation for dy/dx.

EXAMPLE 38 Find dy/dx for $x^2 - xy = 2$.

Solution

Step 1. $\qquad\qquad\qquad\qquad\qquad\qquad\qquad \dfrac{d}{dx}(x^2 - xy) = \dfrac{d}{dx}(2)$

Using the sum rule (22-15), $\qquad\qquad\qquad \dfrac{d}{dx}(x^2) - \dfrac{d}{dx}(xy) = 0$

To find $\dfrac{d}{dx}(xy)$, use the product rule (22-17).

$$2x - \left[x\dfrac{dy}{dx} + y\dfrac{dx}{dx}\right] = 0$$

Because $dx/dx = 1$, this becomes

$$2x - x\dfrac{dy}{dx} - y = 0$$

Step 2. Solve for dy/dx.

$$x\dfrac{dy}{dx} = 2x - y$$

$$\dfrac{dy}{dx} = \dfrac{2x - y}{x}$$

Notice that

$$\dfrac{d(x^n)}{dx} = nx^{n-1}$$

but by the chain rule (22-21),

$$\dfrac{d(y^n)}{dx} = ny^{n-1} \cdot \dfrac{dy}{dx}$$

LEARNING HINT ▶

In other words, when any expression involving y is differentiated, dy/dx will be a factor in the result. When an expression involves only x, powers of x, and constants, there will be no dy/dx factor in the derivative. ◀

EXAMPLE 39 Find dy/dx for $xy^2 + xy = 2x - 2$ and evaluate it at the point $(1, 0)$.

Solution

Step 1.

$$\dfrac{d}{dx}(xy^2 + xy) = \dfrac{d}{dx}(2x - 2)$$

Using the sum rule (22-15),

$$\dfrac{d}{dx}(xy^2) + \dfrac{d}{dx}(xy) = 2$$

Using the product rule (22-17),

$$\dfrac{d}{dx}(xy^2) = \left[x\left(2y\dfrac{dy}{dx}\right) + y^2\dfrac{dx}{dx}\right]$$

$$\dfrac{d}{dx}(xy) = \left[x\dfrac{dy}{dx} + y\dfrac{dx}{dx}\right]$$

where $dx/dx = 1$. So the equation becomes

$$2xy\dfrac{dy}{dx} + y^2 + x\dfrac{dy}{dx} + y = 2$$

Step 2. Solve for dy/dx.

$$2xy\dfrac{dy}{dx} + x\dfrac{dy}{dx} = 2 - y - y^2$$

$$\frac{dy}{dx}(2xy + x) = 2 - y - y^2$$

$$\frac{dy}{dx} = \frac{2 - y - y^2}{2xy + x}$$

Before evaluating dy/dx at the point $(1, 0)$, check to be certain that $(1, 0)$ is a point on the graph.

$$(1)(0)^2 + (1)(0) = 2(1) - 2$$

$$0 + 0 = 0$$

Then,

$$\frac{dy}{dx}\bigg|_{(1,0)} = \frac{2 - (0) - (0)^2}{2(1)(0) + (1)} = \frac{2}{1} = 2$$

Rational exponents   Using implicit differentiation, we can now extend the power rule to include rational exponents. If $y = x^n$, where n is a rational number, we can write $n = p/q$, where p and q are integers, $q \neq 0$:

$$y = x^{p/q}$$

By raising both sides to the q power, this can be written as

$$y^q = (x^{p/q})^q = x^p$$

Using implicit differentiation,

$$\frac{d}{dx}(y^q) = \frac{d}{dx}(x^p)$$

$$q \cdot y^{q-1} \cdot \frac{dy}{dx} = p \cdot x^{p-1} \cdot \frac{dx}{dx}$$

$$\frac{dy}{dx} = \frac{px^{p-1}}{qy^{q-1}}$$

But

$$y^{q-1} = \frac{y^q}{y} = \frac{x^p}{x^{p/q}} = x^{p-(p/q)}$$

Therefore,

$$\frac{dy}{dx} = \left(\frac{p}{q}\right)\frac{x^{p-1}}{x^{p-(p/q)}} = \left(\frac{p}{q}\right)x^{p-1-p+(p/q)}$$

$$= \left(\frac{p}{q}\right)x^{(p/q)-1}$$

$$= nx^{n-1}$$

The power rules (22-13) and (22-18) for positive and negative integer exponents can now be extended to include rational exponents of the form a/b, where $b \neq 0$.

> **Rational Power Rule**
>
> If $y = x^n$, then $\dfrac{dy}{dx} = nx^{n-1}$ (22-23)

This rule includes equation (22-14) and equation (22-20). It holds for any rational number exponent n.

EXAMPLE 40 Differentiate the following functions:

(a) $y = \sqrt{x}$ (b) $y = \sqrt[3]{x^2 + 2}$

(c) $y = x^{2/3} + x^{1/3}$ (d) $y = \sqrt{\dfrac{x}{x + 1}}$

(e) The height h of the water stream from a fire hose is related to the pressure P by the equation $h = -15 + \sqrt{1900 + 240P - P^2}$. Find dh/dP.

Solution

(a) **First,** write this as $y = x^{1/2}$

   Then, $\dfrac{dy}{dx} = \dfrac{1}{2}x^{(1/2)-1} = \dfrac{1}{2}x^{-1/2}$

   or $\dfrac{dy}{dx} = \dfrac{1}{2\sqrt{x}} = \dfrac{\sqrt{x}}{2x}$

(b) **First,** write this as $y = (x^2 + 2)^{1/3}$. We can think of this as $y = u^{1/3}$, where $u = x^2 + 2$, $du/dx = 2x$.

   Then, use the rational power rule (22-23) to write

   $\dfrac{dy}{dx} = \dfrac{1}{3}u^{(1/3)-1} \cdot \dfrac{du}{dx}$

   $= \dfrac{1}{3} \cdot (x^2 + 2)^{-2/3} \cdot 2x$

   $= \dfrac{2x}{3}(x^2 + 2)^{-2/3} = \dfrac{2x}{3(x^2 + 2)^{2/3}}$

   Written using radicals, this becomes

   $\dfrac{dy}{dx} = \dfrac{2x}{3\sqrt[3]{(x^2 + 2)^2}}$

(c) This is already written with rational exponents: $y = x^{2/3} + x^{1/3}$

   Use the sum rule (22-15): $\dfrac{dy}{dx} = \dfrac{d}{dx}(x^{2/3}) + \dfrac{d}{dx}(x^{1/3})$

   Then use the rational power rule (22-23): $= \dfrac{2}{3}x^{(2/3)-1} + \dfrac{1}{3}x^{(1/3)-1}$

$$= \frac{2}{3}x^{-1/3} + \frac{1}{3}x^{-2/3}$$

$$= \frac{2}{3x^{1/3}} + \frac{1}{3x^{2/3}}$$

This can also be written using radicals: $\quad \dfrac{dy}{dx} = \dfrac{2}{3\sqrt[3]{x}} + \dfrac{1}{3\sqrt[3]{x^2}}$

(d) To differentiate

$$y = \sqrt{\frac{x}{x + 1}}$$

think of this as

$$y = \sqrt{u} = u^{1/2} \qquad \text{where} \quad u = \frac{x}{x + 1}$$

Then, use the rational power rule (22-23):

$$\frac{dy}{dx} = \frac{1}{2} \cdot u^{(1/2)-1} \cdot \frac{du}{dx}$$

For $u = x/(x + 1)$, use the quotient rule (22-19):

$$\frac{du}{dx} = \frac{(x + 1)\dfrac{dx}{dx} - x \cdot \dfrac{d}{dx}(x + 1)}{(x + 1)^2}$$

$$= \frac{x + 1 - x}{(x + 1)^2} = \frac{1}{(x + 1)^2}$$

So

$$\frac{dy}{dx} = \frac{1}{2} \cdot \left(\frac{x}{x + 1}\right)^{-1/2} \cdot \frac{1}{(x + 1)^2}$$

$$= \frac{1}{2(x + 1)^2} \cdot \frac{1}{\sqrt{\dfrac{x}{x + 1}}}$$

$$= \frac{\sqrt{x + 1}}{2(x + 1)^2\sqrt{x}}$$

(e) **First,** let $u = 1900 + 240P - P^2$; then $du/dP = 240 - 2P$ and $h = -15 + u^{1/2}$.

Then, by the rational power rule (22-23) and the sum rule (22-15),

$$\frac{dh}{dP} = \frac{1}{2}u^{(1/2)-1} \cdot \frac{du}{dP}$$

$$= \frac{1}{2} \cdot u^{-1/2} \cdot (240 - 2P)$$

$$= \frac{120 - P}{\sqrt{1900 + 240P - P^2}}$$

EXAMPLE 41   Find dy/dx for $\sqrt{x} + \sqrt{y} = 2$.

Solution   **First,** write this as

$$x^{1/2} + y^{1/2} = 2$$

Second, use implicit differentiation.

$$\frac{d}{dx}(x^{1/2} + y^{1/2}) = \frac{d}{dx}(2)$$

Third, apply the sum rule (22-15).

$$\frac{d}{dx}(x^{1/2}) + \frac{d}{dx}(y^{1/2}) = \frac{d}{dx}(2)$$

Fourth, use the rational power rule (22-23).

$$\frac{1}{2} \cdot x^{(1/2)-1} \cdot \frac{dx}{dx} + \frac{1}{2} y^{(1/2)-1} \cdot \frac{dy}{dx} = 0$$

Multiply by 2.

$$x^{-1/2} + y^{-1/2} \cdot \frac{dy}{dx} = 0$$

Write as radicals.

$$\frac{1}{\sqrt{x}} + \frac{1}{\sqrt{y}} \cdot \frac{dy}{dx} = 0$$

Finally, solve for dy/dx.

$$\frac{1}{\sqrt{y}} \cdot \frac{dy}{dx} = -\frac{1}{\sqrt{x}}$$

$$\frac{dy}{dx} = -\frac{\sqrt{y}}{\sqrt{x}}$$

Exercises 22-6

Find dy/dx for each function using implicit differentiation.

1. $4x + 5y = 4$
2. $2x^3 - 4y = 1$
3. $xy + x = 2$
4. $xy + y^2 = 1$
5. $2y^2 - xy = x$
6. $y^2 = x + 2$
7. $y^4 = x^3$
8. $2x^3 = y^2$
9. $x^2 - y^2 = 3$
10. $x^2 + y^2 = 2$
11. $x^2 + xy + y^2 = 0$
12. $2y + x^2 = y^2 + 1$
13. $2x^3 - y^2 = 1 - x^2y$
14. $x^3y + y^3x = 2$
15. $x^3 + y^3 = 5x$
16. $x^2 - y^3 = 2x - y$
17. $y = \dfrac{x^2}{x + y}$
18. $y^2 = \dfrac{x}{x + 1}$
19. $xy = 0$
20. $x + y + y^2 = 0$
21. $(2x + 3)(3y + 1) = 4$
22. $(x^2 - 1)(y^2 + 1) = 2$
23. $(x + 1)^2 + (2y - 3)^2 = 0$
24. $2(x^2 - 2)^2 - (y^2 - 1)^2 = 1$

Find dy/dx for each function.

25. $y = \sqrt[3]{x}$
26. $y = \sqrt{x - 2}$
27. $y = \dfrac{1}{\sqrt{x}}$
28. $y = \dfrac{1}{\sqrt{1 + x}}$
29. $y = \dfrac{2}{\sqrt[3]{x}}$
30. $y = \dfrac{1}{\sqrt{x^2 + 4}}$
31. $y = x\sqrt{x}$
32. $y = x\sqrt{x^2 + 1}$
33. $y = \dfrac{\sqrt{x}}{x - 1}$
34. $y = \sqrt{1 - x^3}$
35. $y = 1 - \sqrt{x^2 + 1}$
36. $y = \left(x - \dfrac{1}{x}\right)^{1/4}$

37. $y = x^{3/4} - x^{1/4}$ **38.** $y = 2x^{1/2} + x^{3/2}$ **39.** $y = (2x - 1)^{1/3}$

40. $y = (x^2 - 4x)^{3/4}$ **41.** $y = \sqrt[3]{2x^4 - 1}$ **42.** $y = \sqrt[4]{2x^3 + 3x}$

43. $y = \dfrac{2}{(x^2 + 3x)^{3/2}}$ **44.** $y = \dfrac{1}{\sqrt[3]{1 - 2x^2}}$ **45.** $\sqrt{x} - \sqrt{y} = 4$

46. $x^2 - \sqrt{2y} = 2$

Find dy/dx for each function and evaluate it at the point indicated.

47. $\dfrac{1}{x} + \dfrac{1}{y} = 1$ at (2, 2)   **48.** $x^2 y + xy^2 + 2 = 0$ at (−1, 2)

49. $\dfrac{1}{x^2} + \dfrac{1}{y^2} = 5$ at $\left(\dfrac{1}{2}, 1\right)$ **50.** $2x^2 + y^2 = 9$ at (2, 1)

Solve.

51. Find the slope of the tangent line to the curve of $x^3 - 2xy = 4$ at the point (2, 1).
52. Find the slope of the tangent line to the curve of $y = \sqrt[3]{2 - 3x}$ at the point (−2, 2)
53. Physics   A charged particle in an accelerating device travels along a curved path so that its displacement s (in meters) varies with time t (in seconds). If $s = (2t^3 + 3t)^{2/3}$, find the instantaneous speed of the particle ds/dt at t = 4.0 s.
54. Electronics   The magnetic field B (in gauss) at a point on the axis of a circular loop of wire carrying a current i is given by the equation*

$$B = \frac{kiR^2}{2(R^2 + x^2)^{3/2}}$$

where x is the distance along the axis from the loop, R is the radius, and k is a constant. If $k = 4\pi \times 10^{-7}$ and i = 80 A:
  (a)   For a loop of fixed radius R = 4.0 cm, find dB/dx at x = 6.0 cm.
  (b)   If the loop varies in size, find dB/dR when x = 6.0 cm and R = 4.0 cm.
55. Electronics   If the wire in Problem 54 is bent into a square loop of side d, then the magnetic field B (in gauss) on the axis a distance x from the center is given by the equation

$$B = \frac{4kid^2}{\pi(4x^2 + d^2)(4x^2 + 2d^2)^{1/2}}$$

Find dB/dx when x = 6.0 cm and d = 5.0 cm.
56. Physics   The electrical potential V of a charged circular disk at a point r from the disk along its axis is given by the equation

$$V = k\left(\sqrt{R^2 + r^2} - r\right)$$

where k is a constant and R is the radius of the disk. Find an expression for the electrical field E(r) = dV/dr.
57. Physics   The frequency f of a simple physical pendulum depends on its length L according to the equation

$$f = \frac{1}{2\pi}\sqrt{\frac{g}{L}}$$

where g is the acceleration due to gravity. Find an expression for df/dL.

*See David Halliday and Robert Resnick, *Fundamentals of Physics*, 2nd edition (New York: Wiley, 1981), p. 567.

58. *Physics*   A solid cylinder rolls without slipping down an inclined plane. If the cylinder starts from rest, its instantaneous speed v after it has moved a vertical distance h is given by the equation $v^2 = \frac{4}{3}gh$. Find dv/dh when $h = 4.0$ ft if $g = 32$ ft/s².

59. *Optics*   If an object is placed at a distance o from the vertex of a spherical mirror of radius r, then an image will be formed at a distance i from the mirror, where

$$\frac{1}{o} + \frac{1}{i} = \frac{2}{r}$$

Find an expression for di/do.

60. *Machine technology*   In the device shown, a spindle s slides along the rod. If the distance h is fixed at 4.5 cm, and if the displacement of the spindle changes so that $x = 2t + t^2$, find dL/dt, the rate at which thread is moving from the spindle when $t = 2.0$ s. [*Hint:* Use $L^2 = h^2 + x^2$.]

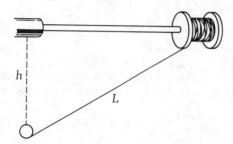

22-7 | Review of Chapter 22

Important Terms and Concepts

function (p. 806)

limit (p. 807)

increment method (p. 808)

limits at infinity (p. 812)

continuity/continuous (p. 815)

discontinuity/discontinuous (p. 815)

continuity at a point (p. 815)

continuity over an interval (p. 816)

secant line (p. 824)

tangent line (p. 824)

Δ-notation (p. 826)

the derivative (p. 827)

Δ-limit process (p. 828)

$\dfrac{dy}{dx}$ notation (p. 829)

differentiation (p. 829)

differentiable function (p. 831)

rate of change (p. 832)

composite function (p. 854)

explicit function (p. 860)

implicit function (p. 860)

implicit differentiation (p. 861)

Formulas and Rules

- the derivative $f'(x) = \lim\limits_{\Delta x \to 0} \dfrac{\Delta y}{\Delta x} = \lim\limits_{\Delta x \to 0} \dfrac{f(x + \Delta x) - f(x)}{\Delta x}$ **(22-9)**

- $f'(x) = \dfrac{dy}{dx} = \lim\limits_{\Delta x \to 0} \dfrac{\Delta y}{\Delta x}$ **(22-10)**

Summary of Differentiation Rules

Name of Rule	Function	Derivative	
Derivative of a constant	$y = c$	$\dfrac{dy}{dx} = 0$	(22-11)
Derivative of identity function	$y = x$	$\dfrac{dy}{dx} = 1$	(22-12)
Power formula	$y = x^n$ for n an integer	$\dfrac{dy}{dx} = nx^{n-1}$	(22-13)
Constant multiplier rule	$y = cx^n$	$\dfrac{dy}{dx} = cnx^{n-1}$	(22-14)
Sum/difference rule	$y = u(x) \pm v(x)$	$\dfrac{dy}{dx} = \dfrac{du}{dx} \pm \dfrac{dv}{dx}$	(22-15), (22-16)
Product rule	$y = u(x)v(x)$	$\dfrac{dy}{dx} = u\dfrac{dv}{dx} + v\dfrac{du}{dx}$	(22-17)
Quotient rule	$y = \dfrac{u(x)}{v(x)}$	$\dfrac{dy}{dx} = \dfrac{v\dfrac{du}{dx} - u\dfrac{dv}{dx}}{v^2}$	(22-19)
Chain rule	$y = u[v(x)]$	$\dfrac{dy}{dx} = \dfrac{dy}{du} \cdot \dfrac{du}{dx}$	(22-21)
Extended power rule	$y = [u(x)]^n$	$\dfrac{dy}{dx} = n[u(x)]^{n-1} \cdot \dfrac{du}{dx}$	(22-22)
Rational power rule	$y = x^n$ for n a rational number	$\dfrac{dy}{dx} = nx^{n-1}$	(22-23)

Exercises 22-7

Find the limit indicated for each function.

1. $\lim\limits_{x \to 3} (2x^2 - 5)$

2. $\lim\limits_{x \to -1} (3x + 2)$

3. $\lim\limits_{x \to 4} \dfrac{1}{x - 4}$

4. $\lim\limits_{x \to \infty} \dfrac{x + 3}{x}$

5. $\lim\limits_{x \to \infty} \dfrac{2 - x^2}{x^2 + 4}$

6. $\lim\limits_{x \to 4} \dfrac{x^2 - 16}{\sqrt{x^2 - 7} - 3}$

7. $\lim\limits_{x \to 0} \dfrac{3x^2 - 4x}{x}$

8. $\lim\limits_{x \to -5} \dfrac{x^2 - 25}{x + 5}$

9. $\lim\limits_{x \to \infty} \dfrac{x^2 - 6x + 8}{2x^2 - 5x - 12}$

10. $\lim\limits_{x \to \infty} \dfrac{3x + 1}{2x^3 - x}$

11. $\lim\limits_{x \to -\infty} (2 + 3^x)$

12. $\lim\limits_{x \to 2} f(x)$, where
$$f(x) = \begin{cases} 2x^3 & \text{for } x \neq 2 \\ 4 & \text{for } x = 2 \end{cases}$$

Use the Δ-limit process to find the derivative of each function.

13. $f(x) = 6x + 2$ **14.** $f(x) = x^2 - 1$ **15.** $f(x) = 2x^3$

16. $f(x) = \dfrac{1}{x + 4}$ **17.** $f(x) = \sqrt{x^2 + 4}$ **18.** $f(x) = 2x^2 - 7x + 6$

Find the derivative of each function.

19. $y = -3$ **20.** $y = -6x^3$ **21.** $y = \dfrac{x^8}{2}$

22. $y = 3x^2 - 5x + 4$ **23.** $y = x^4 - 2x^3 + 7x^2 - 4x + 1$

24. $y = 5(3x^2 - 7x)$ **25.** $y = -2x^2(x^3 - 4)$

26. $y = (x^2 + 2)(3 + 2x - x^2)$ **27.** $y = \dfrac{3x}{x - 5}$ **28.** $y = \dfrac{x^3 - 1}{4x}$

29. $y = \dfrac{x^2 - 5x + 8}{3x^2 - 2}$ **30.** $y = \dfrac{2}{x^3}$ **31.** $y = (4 - 3x^2)^3$

32. $y = (2x^2 + 7x - 3)^4$ **33.** $y = \dfrac{1}{(3x^2 - 2)^2}$ **34.** $y = (x^3 - 1)^3 + (3x^2 + 5)^5$

35. $y = (x^2 - 2)^4(4x + 2x^2)^3$ **36.** $y = \left(\dfrac{3x - 1}{x^2 - 5}\right)^4$ **37.** $4x - 3y = 7$

38. $xy - x^2 = y^2 - 1$ **39.** $x^2 + y^2 = (x - y)^3$ **40.** $y = \sqrt{2x + 3}$

41. $y = \dfrac{x}{\sqrt{x^2 + 4}}$ **42.** $y = 2x^{5/2} - 5x^{3/2}$

Solve.

43. Find the slope of the line tangent to the curve of $y = 2x^3 - 6x^2 + 5$ at $(2, -3)$.

44. Find the equation of the line tangent to the graph of $y = 3x^2 - 5x + 2$ at the point where $x = -1$.

45. Find the equation of the line tangent to $y = (2x - 3)(x^2 - 5x + 4)$ at $x = 2$.

46. Find the values of x where $\dfrac{dy}{dx} = 0$ for $y = \dfrac{2x - 3}{x^2 - 1}$.

47. Find dy/dx for $x + y^2 = xy + 5$ and evaluate it at the point $(2, -1)$.

48. Find the slope of the line tangent to $y = (2x - 7)^3$ at the point where $x = 3$.

49. *Mechanical engineering*   In a tension test a sample of structural steel is found to exhibit the following stress–strain curve:

$$c = \begin{cases} (7 \times 10^4)q & \text{for } 0 \le q \le 0.002 \\ 140 & \text{for } 0.002 < q \le 0.006 \end{cases}$$

where c is the stress (in millipascals) and q is the strain (measured in inches per inch). Show that this function is continuous at $q = 0.002$ in./in.

50. *Electronics*   When a parallel plate capacitor with circular plates of radius R is being charged, a magnetic field B is induced in the capacitor. If i_0 is the maximum value of the displacement current, and r is the radial distance from the center of the plates, then

$$B(r) = \begin{cases} \dfrac{\mu_0 i_0}{2\pi r} & \text{for } r \ge R \\[2ex] \dfrac{\mu_0 i_0}{2\pi} \cdot \dfrac{r}{R^2} & \text{for } r < R \end{cases}$$

Show that the field is continuous at $r = R$.

51. *Electrical engineering*   If a uniform magnetic field in a circular region of radius R is increasing in magnitude at a constant rate Q, then an electric field E will be induced in the region. The magnitude of E at any radius r is

$$E(r) = \begin{cases} -\dfrac{1}{2}rQ & \text{for } r \leq R \\[2ex] -\dfrac{1}{2}\dfrac{R^2Q}{r} & \text{for } r > R \end{cases}$$

When R = 0.100 m and Q = 0.20 T/s, show that $\lim\limits_{r \to R^-} E(r) = \lim\limits_{r \to R^+} E(r)$ by evaluating E(r) for the following values of r: 0.000, 0.090, 0.099, 0.0999, 0.200, 0.190, 0.109, 0.1009.

52. *Material science*   A certain alloy undergoes a crystalline phase change at 400°C, and as a result its electrical resistance changes abruptly according to the equation

$$R(T) = \begin{cases} 2.00 + \dfrac{T}{200} & \text{for } T \leq 400°C \\[2ex] 2.00\left(\dfrac{T}{400}\right)^2 & \text{for } T > 400°C \end{cases}$$

Show that the resistance is not continuous at T = 400°C.

53. *Physics*   When an oscillating system with natural frequency f_0 undergoes forced oscillation at frequency f, its amplitude A is described by a function of the form

$$A(f) = \frac{A_0}{f_0^2 - f^2}$$

If $A_0 = 4.0$ and $f_0 = 10.0$, show that $\lim\limits_{f \to f_0^-} A \neq \lim\limits_{f \to f_0^+} A$ by evaluating A(f) for the following values of f: 9.00, 9.90, 9.99, 11.00, 10.10, 10.01

54. *Physics*   The speed v (in centimeters per second) of a particle moving along a curved path depends on its travel distance x along the path according to the equation

$$v = \sqrt{4x - 9}$$

Use the Δ-limit process to find the derivative dv/dx, and evaluate this derivative at x = 3 cm.

55. *Physics*   The acceleration g due to the gravitational attraction of the earth on an object a distance r from the center of the earth is given by the equation

$$g = \frac{GM_e}{r^2}$$

where G is a constant and M_e is the mass of the earth. If $g = 9.8$ m/s² and $r = 6.4 \times 10^6$ m, find dg/dr at the surface of the earth.

56. *Electronics*   An electron, accelerated between the charged plates of a cathode ray oscilloscope, moves so that its deflection y is related to its horizontal position x by the equation

$$y = \frac{eEx^2}{4K_0}$$

where e (the charge on the electron) is 1.6×10^{-19} C, the electric field $E = 1.2 \times 10^4$ N/C, and K_0 (the initial kinetic energy) is 3.2×10^{-16} J. Find dy/dx, the slope of the path of this particle when x = 0.012 m.

57. *Electronics*   The current in a circuit varies with time according to the equation $i = 4.0/t$, and the resistance increases with time according to $R = 10.0 + 0.2t + 0.05t^2$. Using Ohm's law, $V = iR$, find dV/dt at $t = 3.0$ s.

58. *Electronics* For a current-carrying loop of wire bent into the shape of a rectangle of length L and width W, the magnetic field at the center of the rectangle is given by the equation

$$B = \frac{2ki(L^2 + W^2)^{1/2}}{\pi LW}$$

If the length is fixed at 6.0 cm and the width is allowed to vary, find dB/dW when $W = 2.0$ cm. Use $k = 4\pi \times 10^{-7}$ and $i = 80$ A.

59. *Physics* The torsion pendulum shown consists of a cylindrical rod of radius r suspended on a stiff wire. When the rod is twisted about the wire axis, it oscillates with period P given by the equation

$$P^2 = \frac{2\pi^3 \rho L r^4}{k}$$

where ρ is the density of the rod, L is its length, and k is a constant related to the stiffness of the wire. Find an expression for dP/dr in terms of r.*

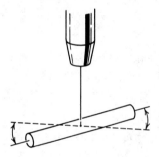

60. *Electronics* If the voltage output of a certain electronic device varies with time according to the equation

$$V(t) = 2.45(1 + 2t^2 - t) \qquad 0 < t < 2.0 \text{ s}$$

find dV/dt when $t = 1.5$ s.

61. *Electronics* The electrical field E on the axis of a simple electric quadrupole has the form

$$E = \frac{Q}{4\pi\epsilon_0 r^4}$$

where Q is the quadrupole moment of the charge distribution, ϵ_0 is a constant, and r is the distance from the center of the charge distribution. Find an expression for dE/dr, the variation in the field along the axis.

62. *Electronics* The capacitance of a capacitor made of two concentric shells of radii a and b $(a < b)$ is

$$C = \frac{kab}{b - a}$$

If k is a constant and if radius a is held fixed while radius b changes, find the rate of change of capacitance dC/db when $b = 2.4$ cm if $k = 1.1 \times 10^{-10}$ F and $a = 1.0$ cm.

*See Frederich Bueche, *Technical Physics* (New York: Harper & Row, 1981), p. 264

63. *Thermodynamics*   If a liquid is heated, its mass will remain constant while its volume increases, causing the density of the liquid to decrease according to the equation

$$D = \frac{D_0}{1 + kT}$$

where D_0 is the density at $T = 0$ and k is the coefficient of expansion. Find dD/dT at $T = 100°C$ for the liquid in a thermometer if $D_0 = 0.916$ g/cm³ and $k = 51.4 \times 10^{-5} °C^{-1}$.

64. *Electronics*   The voltage output V of a thermister depends on its temperature T according to the equation $V = aT + bT^2$. Find dV/dT, the rate at which voltage changes with temperature, at $T = 35°C$ if $a = 0.067$ and $b = 0.000014$.

Applications of Differentiation

23

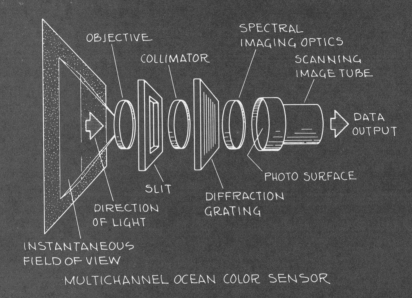

OBJECTIVE

COLLIMATOR

SPECTRAL
IMAGING OPTICS

SCANNING
IMAGE TUBE

DATA
OUTPUT

PHOTO SURFACE

SLIT

DIFFRACTION
GRATING

DIRECTION
OF LIGHT

INSTANTANEOUS
FIELD OF VIEW

MULTICHANNEL OCEAN COLOR SENSOR

Using the differentiation techniques developed in the previous chapter, you should be able to find the derivative of an algebraic function. In this chapter we will explore the mathematical and physical meaning of the derivative as a rate of change. We will also develop applications in the analysis and plotting of graphs; in physics, electronics, and other technical areas; and in the solution of optimization problems.

Tangent   As explained in Section 22-2, the derivative of a function gives the slope of a line tangent to that function at any point where the derivative exists. Using the point–slope equation of a straight line presented in Chapter 20, we can write the equation of the line tangent to the graph of the function at a given point.

If $y = f(x)$ is a differentiable function and (x_1, y_1) is a point on its graph, then the equation of the tangent line at (x_1, y_1) is given by the following:

Tangent Line Equation

$$y - y_1 = m_T(x - x_1) \qquad \text{where } m_T = \left.\frac{dy}{dx}\right|_{x=x_1} \qquad \text{(23-1)}$$

EXAMPLE 1   Find the equation of the line tangent to $y = x^2 - 4x$:
(a)   At the point $(3, -3)$ (b)   At the point where $x = 2$

Solution   The derivative of this function is

$$\frac{dy}{dx} = 2x - 4$$

(a)   At the point $(3, -3)$,

$$\left.\frac{dy}{dx}\right|_{x=3} = 2(3) - 4 = 2$$

Therefore, the slope of the tangent line is $m_T = 2$, and the tangent line equation (23-1) gives

$$y - (-3) = 2(x - 3)$$
$$y + 3 = 2x - 6$$
$$y = 2x - 9$$

This is the equation of the tangent line at $(3, -3)$, as shown in Fig. 23-1.

(b)   When $x = 2$, $y = (2)^2 - 4(2) = -4$. Also,

$$\left.\frac{dy}{dx}\right|_{x=2} = 2(2) - 4 = 0$$

The slope of the tangent line is $m_T = 0$, and equation (23-1) gives

$$y - (-4) = 0(x - 2)$$
$$y + 4 = 0$$
$$y = -4$$

This is the equation of the tangent line at $(2, -4)$ (Fig. 23-1).

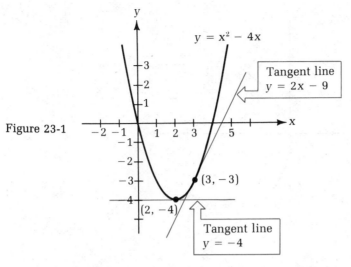

Figure 23-1

EXAMPLE 2 Find the equation of the line tangent to $y^2 = 6x - y$ at the point (1, 2).

Solution Use implicit differentiation to find the derivative of $y^2 = 6x - y$.

First, take the derivative of each side.

$$\frac{d}{dx}(y^2) = \frac{d}{dx}(6x - y)$$

Use the difference rule (22-16) on the right.

$$\frac{d}{dx}(y^2) = \frac{d}{dx}(6x) - \frac{d}{dx}(y)$$

$$2y\frac{dy}{dx} = 6 - \frac{dy}{dx}$$

Move all terms containing dy/dx to the left.

$$2y\frac{dy}{dx} + \frac{dy}{dx} = 6$$

Factor dy/dx from each term.

$$\frac{dy}{dx}(2y + 1) = 6$$

Divide by $2y + 1$.

$$\frac{dy}{dx} = \frac{6}{2y + 1}$$

At (1, 2),

$$\frac{dy}{dx}\bigg|_{y=2} = \frac{6}{2(2) + 1}$$

$$= \frac{6}{5}$$

From (23-1), the equation of the tangent line is

$$y - 2 = \frac{6}{5}(x - 1)$$

Multiplying by 5, we get or, in standard form:

$$5y - 10 = 6x - 6$$
$$6x - 5y + 4 = 0$$

The graph of $y^2 = 6x - y$ and the tangent line at (1, 2) are shown in Fig. 23-2 on page 878.

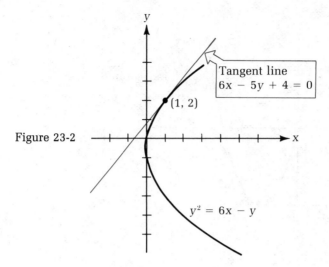

Figure 23-2

In Chapter 20 we showed that if two straight nonvertical lines are perpendicular, then the slope of one line is the negative reciprocal of the other. For slopes m_1 and m_2, $m_1 = -\dfrac{1}{m_2}$.

Normal line   A line perpendicular to the tangent line to a curve at the point where the tangent line intersects the curve is called a **normal line** to the curve (Fig. 23-3). If $y = f(x)$ is a differentiable function, then the equation of the normal line at (x_1, y_1) is given by the following formula:

Normal Line Equation

$$y - y_1 = m_N(x - x_1) \qquad \text{where } m_N = -\frac{1}{m_T}$$

$$\text{and } m_T = \left.\frac{dy}{dx}\right|_{x=x_1}, \; m_T \neq 0 \qquad (23\text{-}2)$$

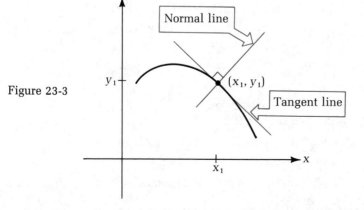

Figure 23-3

EXAMPLE 3   Find the equation of the normal line to the curve $y = \sqrt{x^2 + 5}$ at the point (2, 3).

Solution   We can use the chain rule (22-21) to find the derivative.

First, identify the function $u(x)$, and write y as a function of u.

$$u(x) = x^2 + 5$$
$$y = u^{1/2}$$

Second, use the extended power rule (22-22) to find $\dfrac{dy}{dx}$ and $\dfrac{du}{dx}$.

$$\frac{du}{dx} = 2x$$

$$\frac{dy}{dx} = \frac{1}{2} u^{-1/2} \cdot \frac{du}{dx}$$

Next, substitute $u(x)$ and $\dfrac{du}{dx}$ into the equation for $\dfrac{dy}{dx}$.

$$\frac{dy}{dx} = \frac{1}{2} \cdot \frac{1}{\sqrt{x^2 + 5}} \cdot 2x$$

$$\frac{dy}{dx} = \frac{x}{\sqrt{x^2 + 5}}$$

Finally, evaluate this function at (2, 3).

$$m_T = \frac{dy}{dx}\bigg|_{x=2} = \frac{2}{\sqrt{2^2 + 5}} = \frac{2}{3}$$

Therefore, the slope of the normal line is

$$m_N = -\frac{1}{m_T} = -\frac{3}{2}$$

Now the equation of the normal line can be found from (23-2).

Substituting,

$$y - 3 = -\frac{3}{2}(x - 2)$$

Multiplying by 2,

$$2y - 6 = -3x + 6$$

or

$$3x + 2y - 12 = 0$$

The graph of $y = \sqrt{x^2 + 5}$ and the normal line at (2, 3) are shown in Fig. 23-4.

Figure 23-4

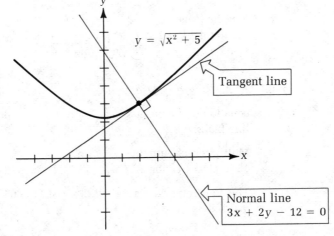

$$y = \sqrt{x^2 + 5}$$

Tangent line

Normal line
$3x + 2y - 12 = 0$

▦ **EXAMPLE 4**   The curve on an automobile test track is banked according to the equation $y = 4/x$, where x is the horizontal distance in feet from the outer edge, $1 \leq x \leq 10$, and y is the vertical height of the track (Fig. 23-5).

(a)   Find the slope of the normal line to the track at $x = 3$ ft.

(b)   Find the equation of the normal line at this point.

(c)   Find the angle the normal line makes with the horizontal.

Figure 23-5

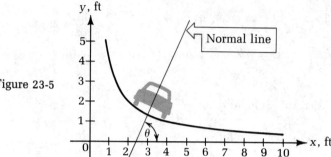

Solution

(a)   Using the extended power rule (22-22), if

$$y = \frac{4}{x} = 4x^{-1}$$

then

$$\frac{dy}{dx} = (-1)4x^{-1-1}$$

$$= -4x^{-2} = -\frac{4}{x^2}$$

The slope of the tangent line is

$$m_T = \frac{dy}{dx}\bigg|_{x=3}$$

$$= -\frac{4}{3^2} = -\frac{4}{9}$$

and the slope of the normal line is

$$m_N = -\frac{1}{m_T} = \frac{9}{4}$$

(b)   At $x = 3$, $y = \frac{4}{3}$. Use the equation of the normal line (23-2).

$$y - \frac{4}{3} = \frac{9}{4}(x - 3)$$

Multiply each term by 12 to clear fractions.

$$12y - 16 = 27(x - 3)$$

The equation of the normal line is

$$27x - 12y - 65 = 0$$

(c)   As you learned in Chapter 20, equation (20-4), if a straight line makes an angle θ with the x-axis, then

Slope $= \tan \theta$ for $0° \leq \theta < 180°$

For the normal line, if θ is measured as shown in Fig. 23-5,

$$\tan \theta = \frac{9}{4}$$ 9 $\boxed{\div}$ 4 $\boxed{=}$ $\boxed{\text{INV}}$ $\boxed{\tan}$ $\rightarrow$ **66.037511**

$$\theta \approx 66°$$

Higher derivatives   Because the derivative of a function is also a function, it is possible to find the derivative of this new function. If $y = f(x)$ is a differentiable function, then $dy/dx = f'(x)$ is called the **first derivative** of the function $y = f(x)$. If the function $f'(x)$ has a derivative, it is called the **second derivative** of the function $y = f(x)$ and is written

$$\frac{d^2y}{dx^2} = f''(x)$$

Similarly, third and higher-order derivatives may be found:

$$y = f(x)$$

$$\frac{dy}{dx} = f'(x)$$ **First derivative**

$$\frac{d^2y}{dx^2} = \frac{d}{dx}\left(\frac{dy}{dx}\right) = f''(x)$$ **Second derivative**

$$\frac{d^3y}{dx^3} = \frac{d}{dx}\left(\frac{d^2y}{dx^2}\right) = f'''(x)$$ **Third derivative**

and so on to

$$\frac{d^ny}{dx^n} = f^{(n)}(x)$$ **nth derivative**

The integer n in $\dfrac{d^ny}{dx^n}$ denotes the **order** of the derivative.

CAUTION ▶   The raised 2's in $\dfrac{d^2y}{dx^2}$ are symbols that tell you this is the derivative of a derivative. They have nothing to do with squaring.   ◀

EXAMPLE 5   If $y = x^4 + 3x^3 - x^2 + 5x - 3$, find $\dfrac{dy}{dx}$, $\dfrac{d^2y}{dx^2}$, and $\dfrac{d^3y}{dx^3}$.

Solution

First, find the derivative. $$\frac{dy}{dx} = 4x^3 + 9x^2 - 2x + 5$$

Second, take the derivative of this new function to get the second derivative.

$$\frac{d^2y}{dx^2} = \frac{d}{dx}(4x^3 + 9x^2 - 2x + 5)$$

$$= 12x^2 + 18x - 2$$

Next, take the derivative of the second derivative to find the third derivative.

$$\frac{d^3y}{dx^3} = \frac{d}{dx}(12x^2 + 18x - 2)$$

$$= 24x + 18$$

EXAMPLE 6   Find the second derivative d^2y/dx^2 of the equation $y^2 - x^2 = 5$. Evaluate this at the point (2, 3).

Solution
First, use implicit differentiation to find the first derivative. Take the derivative of each side.

$$2y\frac{dy}{dx} - 2x = 0$$

Solve for dy/dx.

$$\frac{dy}{dx} = \frac{2x}{2y} = \frac{x}{y}$$

Next, take the derivative of this function to get the second derivative.

$$\frac{d^2y}{dx^2} = \frac{d}{dx}\left(\frac{dy}{dx}\right)$$

$$= \frac{d}{dx}\left(\frac{x}{y}\right)$$

Use the quotient rule (22-19).

$$= \frac{y\dfrac{dx}{dx} - x\dfrac{dy}{dx}}{y^2}$$

Substitute x/y for dy/dx.

$$= \frac{y - x\left(\dfrac{x}{y}\right)}{y^2}$$

Simplify.

$$\frac{d^2y}{dx^2} = \frac{y^2 - x^2}{y^3}$$

Finally, evaluate this function at (2, 3).

$$\frac{d^2y}{dx^2}\bigg|_{(2,\,3)} = \frac{3^2 - 2^2}{3^3} = \frac{5}{27}$$

As you will see in Section 23-2, the first and second derivatives have many scientific and technical applications.

Exercises 23-1

Find the equations of the tangent and normal lines to each curve at the given point. Sketch each curve showing the tangent and normal lines.

1. $y = 3x^2$ at $(-1, 3)$

2. $y = x^2 + 4x$ at $(1, 5)$

3. $y = \dfrac{x}{x + 1}$ at $\left(1, \dfrac{1}{2}\right)$

4. $y = x^3 - 4x + 2$ at $(0, 2)$

5. $y = x\sqrt{x^2 + 1}$ at $\left(\sqrt{3}, 2\sqrt{3}\right)$

6. $y = \dfrac{2x^2}{3x + 1}$ at $(-1, -1)$

7. $x + 4y - y^2 = 2$ at $(-1, 1)$ **8.** $xy - y^2 = -2$ at $(1, 2)$
9. The circle $x^2 + y^2 = 4$ at $\left(1, \sqrt{3}\right)$ **10.** The hyperbola $2x^2 - y^2 = 4$ at $(-2, 2)$
11. The hyperbola $y = \dfrac{1}{x - 2}$ at $\left(0, -\dfrac{1}{2}\right)$ **12.** $y = \dfrac{x^2}{x + 1}$ at $\left(1, \dfrac{1}{2}\right)$
13. $y^3 = 3x + 5$ at $(1, 2)$ **14.** $y^2 - x^2 = 1 - x$ at $(0, 1)$

For each equation, find the derivatives d^2y/dx^2 and d^3y/dx^3.

15. $y = 3x^5$ **16.** $y = 1 - x^3$ **17.** $y = x^2 - 1/x^2$ **18.** $y = 1/x^3$
19. $y = 2x^4 - x^2$ **20.** $y = 4 - x^2 - 3x^4$ **21.** $y = x^2 - 6x + 4$
22. $y = x^2 - x^4$ **23.** $y = \sqrt{x}$ **24.** $y = \sqrt[3]{x}$

For each equation, find the second derivative d^2y/dx^2.

25. $y = \sqrt{x^2 - 1}$ **26.** $y = (x^2 + 1)^{2/3}$ **27.** $y^3 = x^5$
28. $x^2 - y^2 = 1$ **29.** $y = (x^3 - 2x)^4$ **30.** $y = (1 + x^3)^5$
31. $y = \dfrac{2x}{x + 1}$ at $(0, 0)$ **32.** $y = \dfrac{2x - 1}{x^2 + 2}$ at $\left(2, \dfrac{1}{2}\right)$ **33.** $y = \dfrac{1}{\sqrt{x}}$ at $(1, 1)$
34. $y = \dfrac{x}{\sqrt{x}}$ at $\left(\dfrac{1}{4}, \dfrac{1}{2}\right)$ **35.** $2x - y^2 = x^2$ at $(1, 1)$ **36.** $y^2 = x^2 - 3$ at $(2, 1)$

37. *Physics*   According to Coulomb's law of electrostatic attraction and repulsion, the magnitude of the force on a point charge q_1, due to a second point charge q_2 a distance r away is $F = \dfrac{kq_1 q_2}{r^2}$, where k is a constant. Find $\dfrac{d^2F}{dr^2}$.

38. *Physics*   At audiofrequencies, the speed v of sound in a gas depends on the gas pressure P according to the equation $v = \sqrt{\dfrac{\gamma P}{D}}$, where γ is a constant that depends on the molecular nature of the gas and D is the gas density. Find an expression for $\dfrac{d^2v}{dP^2}$.

39. *Thermodynamics*   The equation of state of a certain gaseous material is

$$P = \frac{AT - BT^2}{V}$$

where A and B are constants, T is the temperature of the gas, V is its volume, and P is the gas pressure. If P is held constant, find an expression for $\dfrac{d^2V}{dT^2}$.

40. *Physics*   The equation of the trajectory of a projectile is

$$y = (\tan \theta)x - \left(\frac{g}{2v_0^2 \cos^2 \theta}\right)x^2$$

where θ is the launch angle, v_0 is the initial speed, and g is the acceleration due to gravity. Find the direction of motion of the projectile—that is, the slope of the tangent line—for $\theta = 42°$, $v_0 = 150$ ft/s, and $g = 32.2$ ft/s² when $x = 525$ ft.

41. *Physics*   The lines of force of an electric field are defined so that at any point the direction of a tangent line to a line of force gives the direction of the field. The equipotential lines of the field are perpendicular to the lines of force. If the equation $y = 8/x^{3/2}$ defines an equipotential line, find the equation of the line of force intersecting this curve at a point where $x = 4$.

42. A bicycle racing track is banked according to the equation $y = \dfrac{4}{x + 1}$, where x is the horizontal distance in feet from the outer edge, $x \geq 0$, and y is the vertical height of the track.
(a)   Find the slope of the normal line to the track at $x = 2$ ft.
(b)   Find the equation of the normal line to the track at this point.
(c)   Find the angle made by the normal line with the horizontal at this point.

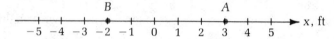

23-2 | Motion in a Plane

One of the great scientific achievements of the 17th century was the development of a mathematical description of motion. Calculus was invented for this purpose, and its successful union with physics led to a better understanding of many areas of science and to many technological applications.

Particle

An actual moving object can be any size or shape and can rotate or vibrate as well as move along a path in space, so we will simplify our description of motion by treating an object as if it were a particle. Mathematically, a **particle** is a geometric point with no extent in space, and we can ignore all characteristics of the object except its motion along a path. Many real objects, from planets and stars to rockets, baseballs, and atoms, can be treated in this way.

Rectilinear motion
Displacement

The simplest kind of motion we can consider is **rectilinear** motion, or motion along a straight line. The **displacement** of a particle, as defined in Chapter 9, Section 9-3, is a vector giving its position in space. For rectilinear motion, the displacement is its distance from some starting point or origin. In Fig. 23-6 the line of motion is the x-axis. Point A has displacement $x_A = 3$ ft from the origin, and point B has displacement $x_B = -2$ ft from the origin.

Figure 23-6

$$B \qquad\qquad A$$
$$\xrightarrow{\hspace{2cm}} x, \text{ ft}$$
$$-5 \quad -4 \quad -3 \quad -2 \quad -1 \quad 0 \quad 1 \quad 2 \quad 3 \quad 4 \quad 5$$

CAUTION ▶

The negative sign on $x_B = -2$ ft is an important part of the displacement value. ◀

The velocity of a particle was defined in Chapter 9 as the rate at which the displacement changes with time. As the particle moves along the x-axis, if its displacement x is specified at each instant of time t, we may be able to write the displacement as a differentiable function of time,

$$x = x(t) \qquad \text{where } t \geq 0$$

Instantaneous velocity

Then the **instantaneous velocity** of the particle is defined as the derivative dx/dt.*

Instantaneous Velocity	$v = \dfrac{dx}{dt}$	where x is the displacement and t is the time (23-3)

The instantaneous velocity is the velocity of the particle at any instant of time, and the sign of the instantaneous velocity shows its direction of motion

---

*Isaac Newton, one of the inventors of calculus, used the notation $\dot{x}$ to denote the derivative with respect to time. Many physicists and physics textbooks still use this dot notation.

along a straight-line path (Fig. 23-7). Motion to the right is positive and motion to the left is negative.*

Figure 23-7

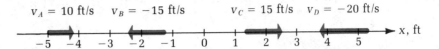

$v_A = 10$ ft/s $v_B = -15$ ft/s $v_C = 15$ ft/s $v_D = -20$ ft/s

EXAMPLE 7 Suppose a particle moves along the x-axis according to the equation $x = t^2 - 4t$, where x is in feet, and t is in seconds. Find the displacement and instantaneous velocity for each of the first 5 s, and describe the motion.

Solution Using equation (23-3), the instantaneous velocity is

$$v = \frac{dx}{dt} = 2t - 4 \text{ ft/s}$$

To describe the motion of the particle, prepare the following table:

Time, t (s)	0	1	2	3	4	5
Displacement, x (ft)	0	−3	−4	−3	0	5
Velocity, v (ft/s)	−4	−2	0	2	4	6

- At the start of the motion, when $t = 0$ s, the particle is at the origin and is moving to the left at 4 ft/s.
- At $t = 1$ s, the particle is at $x = -3$ ft. It is still moving left, but has slowed to 2 ft/s.
- At $t = 2$ s, the particle is at $x = -4$ ft, and has come to a stop. It reached the limit of its motion to the left and is ready to start back to the right.
- At $t = 3$ s, the particle is at $x = -3$ ft, and is now moving to the right at 2 ft/s.
- At $t = 4$ s, the particle has returned to the origin, $x = 0$ ft, and is moving to the right at 4 ft/s.
- At $t = 5$ s, the particle has moved to $x = 5$ ft and is moving to the right at 6 ft/s.

As t increases, the particle continues to move to the right, increasing its velocity by 2 ft/s in each successive second. This motion is shown in Fig. 23-8 on p. 886.

For the motion described in Example 7, the velocity is increasing at a constant rate. In each second the velocity increases by 2 ft/s. For example,

$$v(4) = v(3) + 2 \text{ ft/s} = 2 \text{ ft/s} + 2 \text{ ft/s} = 4 \text{ ft/s}$$

$$v(1) = v(0) + 2 \text{ ft/s} = -4 \text{ ft/s} + 2 \text{ ft/s} = -2 \text{ ft/s}$$

*Do not confuse the words *speed* and *velocity*. Velocity is a vector quantity having both size and direction. For motion in a straight line, velocity may be negative. Speed is the magnitude of the velocity vector and is always a positive quantity.

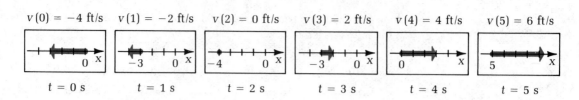

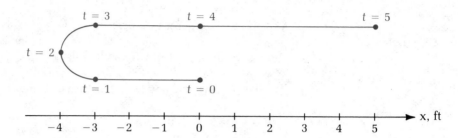

Figure 23-8

Acceleration

When the velocity of a moving object changes in either magnitude or direction or both, the object is said to be **accelerating.** The **acceleration** of a particle is defined as the derivative with respect to time of the velocity.

Acceleration	$a = \dfrac{dv}{dt}$	where v is the velocity and t is time	(23-4)

For motion in a straight line, $v = dx/dt$; therefore, the acceleration is the second derivative of the displacement with respect to t:

$$a = \frac{d}{dt}\left(\frac{dx}{dt}\right) = \frac{d^2x}{dt^2}$$

Units of acceleration

The derivative dv/dt has units of

$$\frac{\text{Velocity}}{\text{Time}} \quad \text{or} \quad \frac{\text{ft/s}}{\text{s}} \text{ in English units}$$

$$\text{and} \quad \frac{\text{m/s}}{\text{s}} \text{ in metric units}$$

These are usually written as ft/s² or ft·s⁻² and m/s² or m·s⁻².

EXAMPLE 8   The particle in Example 7 moves along the x-axis according to the equation $x = t^2 - 4t$ ft. Find its acceleration.

Solution
First, calculate the velocity equation.

$$v = \frac{dx}{dt} = 2t - 4 \text{ ft/s}$$

Then, calculate the acceleration.

$$a = \frac{dv}{dt} = \frac{d}{dt}(2t - 4)$$

$$a = 2 \text{ ft/s}^2$$

This agrees with Fig. 23-8, where we saw that the velocity increased 2 ft/s in each second.

EXAMPLE 9 A spacecraft is moving along a straight line in space so that its position s in time is described by the equation $s = t^3 - 3t^2 + 6$, where s is in kilometers and t is in seconds. Find the velocity and the acceleration at $t = 0$, 1, 2, and 3 s and describe the motion.

Solution

From equation (23-3), the velocity is

$$v = \frac{ds}{dt} = 3t^2 - 6t \text{ km/s}$$

From equation (23-4), the acceleration is

$$a = \frac{dv}{dt} = \frac{d}{dt}(3t^2 - 6t)$$

$$= 6t - 6 \text{ km/s}^2$$

The table lists the position, velocity, and acceleration each second for the first 3 s.

t (s)	s (km)	v (km/s)	a (km/s²)
0	6	0	−6
1	4	−3	0
2	2	0	6
3	6	9	12

During the motion, when $v = 0$, the object is momentarily at rest. It has paused to reverse the direction of its motion. To find these turning points, we must set $v = 0$ and solve:

$$v = 3t^2 - 6t = 3t(t - 2)$$

Figure 23-9

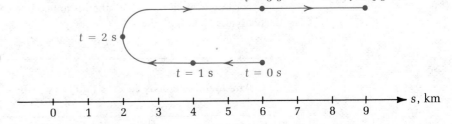

Therefore $v = 0$ at $t = 0$ or $t = 2$ s. We can show the motion using the diagram in Fig. 23-9.

The spacecraft is at rest at $s = 6$ km when we begin to time the motion at $t = 0$ s. It moves left, increasing speed for 1 s, then slows to a stop at $t = 2$ s and $s = 2$ km. Here, it turns and begins to move to the right with increasing speed.

Jerk   Just as the second derivative, or acceleration, measures the change in velocity, the third derivative, or **jerk**,

$$j = \frac{d^3x}{dt^3}$$

is a measure of change in the acceleration. This quantity has applications in engineering in the design of intermittent motion mechanisms such as cams and in the specification of transition curves from straight lines to circular arcs in railways and highways. We will not use the concept of jerk in this book.

Constant acceleration   A very important kind of motion is that of a particle moving in a straight line with **constant acceleration.** The most common example of this type of motion occurs when an object falls near the surface of the earth. More than 300 years ago the Italian mathematician–physicist Galileo found that if there is no air friction, and if we can ignore the rotation of the earth and any rotation or tumbling of the falling object, then all objects, regardless of their size, weight, or composition, fall with constant acceleration. The acceleration varies slightly at different geographical points near the earth's surface, but if the distance of the fall is not too great, the acceleration is constant throughout the fall. This ideal-**Free fall**   ized motion is called **free fall.**

If we set up a coordinate system with the origin at the earth's surface, then free fall is described by the following equation:

Free Fall Motion $y = -\dfrac{1}{2}gt^2 + v_0t + y_0$ (23-5)

In this equation, y is the height of the object above the earth at some time t seconds after it is released; y_0 and v_0 are the initial position and initial velocity, respectively, at time $t = 0$ s. The acceleration g is called the *acceleration due to gravity.* Near the earth's surface, g is approximately 32 ft/s² or 9.8 m/s².

According to equation (23-3), the instantaneous velocity is given by the first derivative of the displacement y.

Instantaneous Velocity in Free Fall $v = \dfrac{dy}{dt} = -gt + v_0$ (23-6)

The velocity is positive when the object is moving upward and negative when it is moving downward.

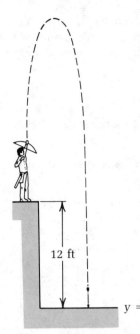

Figure 23-10

EXAMPLE 10   An arrow is shot vertically upward from a point 12 ft above the earth with an initial velocity of 88 ft/s (Fig. 23-10).

(a)   Find the time needed for the arrow to reach the top of its path.
(b)   Find the maximum height attained.
(c)   Find the time when the arrow hits the ground.

Solution   Substituting the values of g, v_0, and y_0 into equation (23-5), we have

$$y = -\frac{1}{2}(32)t^2 + 88t + 12$$

$$y = -16t^2 + 88t + 12$$

Differentiating,

$$v = \frac{dy}{dt} = -32t + 88$$

(a)   At the top of the path, $v = 0$; therefore

$$-32t + 88 = 0$$

$$t = \frac{88}{32} = 2.75 \approx 2.8 \text{ s}$$

(b)   The maximum height is reached in 2.75 s, so

$$y = -16(2.75)^2 + 88(2.75) + 12$$

$$= 133 \approx 130 \text{ ft}$$

(c)   At the ground, $y = 0$:

$$0 = -16t^2 + 88t + 12$$

$$16t^2 - 88t - 12 = 0$$

$$4t^2 - 22t - 3 = 0$$

$$t = \frac{22 \pm \sqrt{(-22)^2 - 4(4)(-3)}}{2(4)}$$

$$t = 22 \boxed{x^2} \boxed{+} 4 \boxed{\times} 4 \boxed{\times} 3 \boxed{=} \boxed{\sqrt{}} \boxed{+} 22 \boxed{=} \boxed{\div} 8 \boxed{=} \rightarrow \quad \boxed{5.6331406}$$

$$t \approx 5.6 \text{ s}$$

NOTE   Only the positive value of t has any physical meaning.   ◄

Curvilinear motion

Rectilinear motion is movement along a straight-line path. **Curvilinear motion** is the more general case of movement along a curve in a plane or in space.

As we have seen, a plane curve may be represented mathematically by an equation $y = f(x)$. The description of motion along a curve is greatly simplified if we can specify the coordinates x and y of each point (x, y) on the curve as a function of time.

$$x = x(t) \qquad y = y(t)$$

Parametric equations

Equations describing a curve in this way are called **parametric equations,** and the variable t is called a **parameter.**

LEARNING HINT

Imagine the curve to be traced by a moving pencil whose changing coordinates are given at each instant by the parametric equations. ◄

When the parametric equations of the path are given, the velocity of the moving particle can be found by taking derivatives.

x-Component of Velocity	$v_x = \dfrac{dx}{dt}$	**(23-7)**
y-Component of Velocity	$v_y = \dfrac{dy}{dt}$	**(23-8)**

As you learned in Chapter 9, the magnitude of the velocity can be found using the Pythagorean theorem. The direction of the velocity can be found using the tangent function, since the direction of motion is along the tangent line to the path (Fig. 23-11).

Figure 23-11

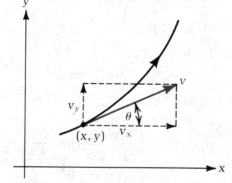

Magnitude of the Velocity	$v = \sqrt{v_x^2 + v_y^2}$	**(23-9)** .
Direction of the Velocity	$\text{Ref } \theta = \text{Arctan} \left| \dfrac{v_y}{v_x} \right|$	**(23-10)**

In equation (23-10), use the signs of v_y and v_x to find θ from Ref θ.

▦ **EXAMPLE 11**   A projectile moves in a path described by the following parametric equations:

$$x = 58t + 3$$

$$y = 5 + 110t - 16t^2$$

where x and y are measured in feet and t is in seconds. Find the magnitude and direction of its velocity at:
(a) $t = 1.0$ s   (b) $t = 4.0$ s

Solution   First, using equations (23-7) and (23-8), find expressions for the velocity components:

$$v_x = \frac{d}{dt}(58t + 3) = 58 \text{ ft/s}$$

$$v_y = \frac{d}{dt}(5 + 110t - 16t^2) = 110 - 32t \text{ ft/s}$$

Next, evaluate each component at the given time and find v and θ from equations (23-9) and (23-10).
(a)   At $t = 1$ s, $v_x = 58$ ft/s and $v_y = 110 - 32(1) = 78$ ft/s. Then

$$v(1) = \sqrt{58^2 + 78^2} \qquad 58\,\boxed{x^2}\,\boxed{+}\,78\,\boxed{x^2}\,\boxed{=}\,\boxed{\sqrt{}} \rightarrow \quad \boxed{\mathit{97.200823}}$$

$$v(1) \approx 97 \text{ ft/s}$$

and

$$\text{Ref } \theta(1) = \text{Arctan} \left| \frac{78}{58} \right| \qquad 78\,\boxed{\div}\,58\,\boxed{=}\,\boxed{\text{INV}}\,\boxed{\text{tan}} \rightarrow \quad \boxed{\mathit{53.365886}}$$

Then from Fig. 23-12, $\theta(1) \approx 53°$.

Figure 23-12

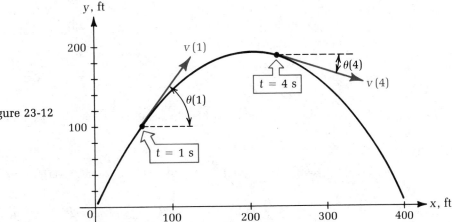

(b)   At $t = 4$ s, $v_x = 58$ ft/s and $v_y = 110 - 32(4) = 110 - 128 = -18$ ft/s. Then

$$v(4) = \sqrt{58^2 + (-18)^2}$$

$$v(4) \approx 61 \text{ ft/s}$$

and

$$\text{Ref } \theta(4) = \text{Arctan} \left| \frac{-18}{58} \right| \approx 17°$$

Then $\theta(4) \approx -17°$

See Fig. 23-12.

Notice that for a projectile moving with negligible air resistance, the x-component of velocity is constant, and the y-component of velocity is a linear function of time. The velocity component in the vertical direction changes because of the downward acceleration due to gravity.

If the path of motion is defined using parametric equations, the acceleration can be found using second derivatives.

x-Component of Acceleration	$a_x = \dfrac{d^2x}{dt^2}$	(23-11)
y-Component of Acceleration	$a_y = \dfrac{d^2y}{dt^2}$	(23-12)
Magnitude of Acceleration	$a = \sqrt{a_x^2 + a_y^2}$	(23-13)
Direction of Acceleration	$\text{Ref } \theta = \text{Arctan} \left\| \dfrac{a_y}{a_x} \right\|$	(23-14)

EXAMPLE 12   A particle moves along a curvilinear path described by the parametric equations

$$x = 3t^2 + t$$

$$y = t^3 - 2t^2 + 4$$

where x and y are measured in meters and t is in seconds. Find the acceleration at $t = 2$ s.

Solution $\dfrac{dx}{dt} = 6t + 1$ and using equation (23-11), $a_x = \dfrac{d^2x}{dt^2} = 6$

$\dfrac{dy}{dt} = 3t^2 - 4t$ and using equation (23-12), $a_y = \dfrac{d^2y}{dt^2} = 6t - 4$

At t = 2 s,

$$a_x(2) = 6 \text{ m/s}^2$$

$$a_y(2) = 6(2) - 4 = 8 \text{ m/s}^2$$

Therefore, from equation (23-13), the magnitude of the acceleration is

$$a(2) = \sqrt{6^2 + 8^2} = \sqrt{100} = 10 \text{ m/s}^2$$

and, from equation (23-14),

$$\text{Ref } \theta(2) = \text{Arctan} \left| \frac{8}{6} \right| \approx 53°$$

Exercises 23-2

Each of the following equations describes the motion of a particle moving along the x-axis. Find the velocity and acceleration of the particle at the time indicated. (x is in meters)

1. $x = t^2 - 10t + 1$; $t = 4$ s

2. $x = 7 - 2t^2$; $t = 0$ s

3. $x = 5 + t^2 - 2t^3$; $t = 2$ s

4. $x = 3t^3 - 2t^2 + 5t - 1$; $t = 1.5$ s

5. $x = 5t^3 - 6t + 1$; $t = 0.3$ s

6. $x = t^3 - 6t^2 + 4$; $t = 4$ s

7. $x = t^4 - 4t^3 + 6t^2 - 7t + 8$; $t = 1$ s

8. $x = 0.6t^3 + t^2 - 3$; $t = 0.2$ s

9. $x = \dfrac{t-1}{t} - t^2$; $t = 2$ s

10. $x = \dfrac{t+2}{t}$; $t = 1$ s

11. *Electronics* An electron moves in a straight-line path in an electric field according to the equation

$$s = \frac{10t^2}{t+1}$$

where s is in meters and t is in seconds. Find the velocity and acceleration of the particle when t = 5.0 s.

12. *Physics* A particle moves so that $s = t^2 + 2\sqrt{t}$, where s is in centimeters and t is in seconds. Find the velocity and acceleration of the particle when t = 4.0 s.

13. *Physics* If an object moves in a path so that $s = t^3 + t\sqrt{t}$, where s is in feet and t is in seconds, find the velocity and acceleration of the particle at t = 1.0 s.

14. *Hydrodynamics* A marker in a fluid flow test moves with variable speed so that its position along its path depends on the time according to the equation $s = 12 + 0.75t + 0.12t^3$, for s in centimeters and t in seconds. Find its velocity and acceleration at time t = 2.0 s.

Each of the following pairs of parametric equations describes the motion of an object along a curved path. For each, find the magnitude and direction of the velocity and acceleration at the indicated time (x and y in feet).

15. $x = 20t$
$y = -16t^2$ at t = 5.0 s

16. $x = 10t - \dfrac{1}{t}$
$y = 10 + t - t^2$ at t = 4.0 s

17. $x = \dfrac{t}{t+1}$
$y = \dfrac{t^2}{t+1}$ at t = 2.0 s

18. $x = 2\sqrt{t}$
$y = \dfrac{1}{t^2}$ at t = 4.0 s

19. *Physics*   A particle launched horizontally with initial velocity $v_0 = 300$ m/s falls in a gravitational field according to the parametric equations $x = v_0 t$ and $y = -9.8t^2$.
  (a)   How far will it fall vertically when it moves 10.0 m horizontally?
  (b)   What will be the magnitude and direction of the velocity of the object after $t = 3.0$ s?

20. *Physics*   Find the magnitude and direction of the acceleration for the particle in Problem 19 at $t = 3.0$ s.

21. *Electronics*   An electron in an electric field moves in a path described by the parametric equations

$$x = \frac{10}{\sqrt{t^2 + 1}} \quad \text{and} \quad y = \frac{10t}{\sqrt{t^2 + 1}}$$

where x and y are in meters $\times 10^4$ and t is in seconds. Find the magnitude and direction of the velocity of this electron when $t = 2.0$ s.

22. *Physics*   Find the magnitude and direction of the acceleration of a particle moving according to the parametric equations $x = t^2 + 2t$ and $y = t\sqrt{t}$ when $t = 4.0$ s.

23. *Physics*   The projectile shown in the figure is launched vertically with an initial velocity of 125 ft/s.
  (a)   How high will it rise?
  (b)   When will it reach point A?
  (c)   What will its velocity be at point A?
  (d)   What velocity will it have at a height B that is 20 ft above its starting point?

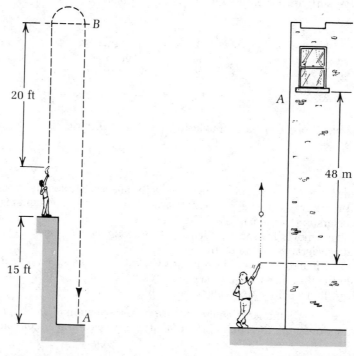

Problem 23   Problem 24

24. *Physics*   In the figure the ball is thrown upward with velocity 32 m/s.
  (a)   When will it reach point A, 48 m above the release point?
  (b)   What will be the velocity of the ball at point A?
  (c)   What will be the maximum height of the ball?

23-3 | **Related Rates**

The concepts of change and rate of change are very important in science and technology, not only in specifying the motion of an object (as we showed in Section 23-2), but also in describing such diverse concepts as the flow of electric charge in a circuit, pressure change, diffusion rates, growth and decay rates, and so on. In this section, we will be particularly interested in situations in which two or more related quantities change simultaneously. In a typical problem, the rate of change of one quantity must be found in terms of the related rate of change of one or more other quantities.

Use the following six-step method for solving related rate problems:

Step 1.   Determine the quantities that vary with time and **label** them with letters. Identify the quantities that will remain fixed.

Step 2.   Identify the rate of change to be found and the rate of change that is known.

Step 3.   Find an equation relating the quantity whose rate of change is to be found to the quantity whose rate of change is known.

Step 4.   Differentiate both sides of this equation with respect to time.

Step 5.   Solve for the derivative that gives the unknown rate of change.

Step 6.   Evaluate this derivative according to the conditions given in the problem.

EXAMPLE 13   The pressure P, volume V, and temperature T of an ideal gas are related by the equation $PV = RT$, where R is the constant 0.082 L·atm/K. If the volume of a sample of gas is held constant at 12 L while the gas is warmed gradually at the rate of 2.5 K/s, find the rate at which the pressure changes.

Solution

Step 1.   The pressure P and temperature T vary with time, while the volume V does not change and R is a constant.

Step 2.   We are asked to find dP/dt, the rate at which the pressure changes with time, and we know that the rate of temperature change dT/dt is 2.5 K/s.

Step 3. $PV = RT$ is the equation relating P and T.

Step 4.   Differentiate this equation using the product rule (22-17):

$$P\frac{dV}{dt} + V\frac{dP}{dt} = R\frac{dT}{dt}$$

But since V does not change,

$$\frac{dV}{dt} = 0$$

Therefore, we are left with

$$V\frac{dP}{dt} = R\frac{dT}{dt}$$

Step 5. Solve for $\dfrac{dP}{dt}$.
$$\frac{dP}{dt} = \frac{R}{V}\frac{dT}{dt}$$

Step 6. Evaluate.
$$\left.\frac{dP}{dt}\right|_{V=12\,L} = \frac{0.082\ \text{L}\cdot\text{atm/K}}{12\ \text{L}} \cdot (2.5\ \text{K/s})$$

$$\approx 0.017\ \text{atm/s}$$

Notice that the units for each quantity are included in the calculation in Step 6. This acts as a check on your work.

If the equation relating the variable quantities is not given or previously known, you may need to develop it using the information given in the problem. It may help to sketch a diagram, or even several diagrams showing the problem situation.

EXAMPLE 14   A police helicopter, hovering 0.75 mi above the highway, detects an automobile approaching along the road (Fig. 23-13). The radar display indicates that the car is moving at 58 mi/h and is 2.3 mi away. What is the actual speed of the car?

Figure 23-13

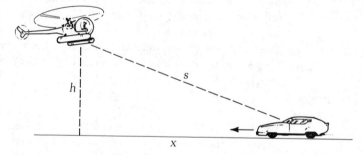

Solution

Step 1. The horizontal distance x and the diagonal distance s both vary with time. The height h is fixed.

Step 2. The speed of the car is dx/dt, and the speed recorded by the radar is ds/dt. We need to find dx/dt. Note that $ds/dt = -58$ mi/h. This rate is negative since s is decreasing as the car approaches. We expect that dx/dt will also be negative since x also decreases.

Step 3. In the right triangle shown in Fig. 23-13, $s^2 = h^2 + x^2$.

Step 4. Taking the time derivative of each side of this equation, we obtain
$$2s\frac{ds}{dt} = 2h\frac{dh}{dt} + 2x\frac{dx}{dt}$$

But $dh/dt = 0$ since the height is constant, so we have
$$2s\frac{ds}{dt} = 2x\frac{dx}{dt}$$

Step 5. Solve for dx/dt.
$$\frac{dx}{dt} = \frac{s}{x}\frac{ds}{dt}$$

To eliminate x from the
right side of the equation,
we use the equation from
Step 3 to find x in terms of
the variable s:

$$s^2 = h^2 + x^2$$

$$x = \sqrt{s^2 - h^2}$$

Then:

$$\frac{dx}{dt} = \frac{s}{\sqrt{s^2 - h^2}} \frac{ds}{dt}$$

Step 6. Evaluate.

$$\left.\frac{dx}{dt}\right|_{s=2.3 \text{ mi}} = \frac{2.3 \text{ mi}}{\sqrt{2.3^2 - 0.75^2} \text{ mi}} \cdot (-58 \text{ mi/h})$$

$$2.3 \boxed{\times} 58 \boxed{+/-} \boxed{\div} \boxed{(} 2.3 \boxed{x^2} \boxed{-} .75 \boxed{x^2} \boxed{)} \boxed{\sqrt{}} \boxed{=} \rightarrow \quad \boxed{-61.353601}$$

$$\left.\frac{dx}{dt}\right|_{s=2.3 \text{ mi}} \approx -61 \text{ mi/h}$$

IMPORTANT ▶ Because time can only increase, any rate of change dQ/dt will be **positive** when Q is increasing, and will be **negative** when Q is decreasing. ◀

EXAMPLE 15 The conical water tank shown in Fig. 23-14(a) is being filled by an inlet pipe at the rate of 12 ft³/s. At what rate is the water surface rising when the water is:

(a)   6 ft deep in the tank?   (b)   0.8 ft deep?

(The volume of a cone is $V = \frac{1}{3}\pi r^2 h$.)

Figure 23-14

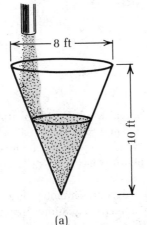

(a)

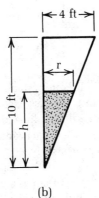

(b)

Solution

Step 1. The volume V, the depth h, and the radius r of the water surface are all changing as the tank fills.

Step 2. We want to find dh/dt, the rate at which the surface of the water is rising. The rate of volume increase dV/dt is 12 ft³/s.

Step 3. We must express the volume V as a function of the height h alone. Using Fig. 23-14(b), we can write an equation relating r and h. From the similar triangles, we can see that

$$\frac{r}{4} = \frac{h}{10}$$

$$r = \frac{4h}{10} = \frac{2h}{5}$$

Substituting this into the equation given for the volume of a cone, we obtain

$$V = \frac{1}{3}\pi\left(\frac{2h}{5}\right)^2 h$$

$$= \frac{4\pi h^3}{75}$$

Step 4. Differentiate using the power rule (22-13).

$$\frac{dV}{dt} = \frac{4\pi}{75} \cdot 3h^2 \frac{dh}{dt}$$

$$= \frac{4\pi h^2}{25}\frac{dh}{dt}$$

Step 5. Solve for $\dfrac{dh}{dt}$.

$$\frac{dh}{dt} = \frac{25}{4\pi h^2}\frac{dV}{dt}$$

Step 6. Evaluate.

(a) $\left.\dfrac{dh}{dt}\right|_{h=6\text{ ft}} = \dfrac{25}{4\pi(6\text{ ft})^2}\cdot(12\text{ ft}^3/\text{s})$

$$\approx 0.7\text{ ft/s}$$

(b) $\left.\dfrac{dh}{dt}\right|_{h=0.8\text{ ft}} = \dfrac{25}{4\pi(0.8\text{ ft})^2}\cdot(12\text{ ft}^3\text{ s})$

$$\approx 40\text{ ft/s}$$

It is important to notice that we did not substitute a numerical value for h until the very last step of Example 15. The equation in Step 5 tells us that the rate dh/dt depends on h. When h is small, dh/dt is very large, and the water level is rising rapidly. When h is large, dh/dt is small, and the water level is rising slowly.

If the path of a particle in curvilinear motion is given as an equation in x and y rather than as a pair of parametric equations, treat it as a related rate problem to find the velocity components.

EXAMPLE 16   A particle moves along the upper half of an elliptical path described by the equation $x^2 + 4y^2 = 4$. Find the magnitude and direction of the velocity at the point $\left(1, \sqrt{3}/2\right)$ cm if the x-component of the velocity is constant at $v_x = 2.0$ cm/s.

Solution **First,** using implicit differentiation and the power rule (22-13), differentiate each term of the equation with respect to time.

$$\frac{d}{dt}(x^2 + 4y^2) = \frac{d}{dt}(4)$$

$$2x\frac{dx}{dt} + 8y\frac{dy}{dt} = 0$$

Second, we are given v_x and we need to find $v_y = \dfrac{dy}{dt}$ in order to find the magnitude and direction of the velocity.

Solve for $\dfrac{dy}{dt}$.

$$\frac{dy}{dt} = -\frac{2x}{8y}\frac{dx}{dt}$$

$$= -\frac{x}{4y}\frac{dx}{dt}$$

Third, substitute $v_x = \dfrac{dx}{dt} = 2.0$ cm/s, $x = 1$ cm, and $y = \sqrt{3}/2$ cm into this equation.

$$\frac{dy}{dt} = \frac{-(1\ cm)}{4(\sqrt{3}/2\ cm)}(2.0\ cm/s)$$

$$= -\frac{1}{\sqrt{3}}\ cm/s = v_y$$

Next, substitute in equation (23-9) to find the magnitude of the velocity.

$$v = \sqrt{v_x^2 + v_y^2}$$

$$= \sqrt{2.0^2 + \left(-\frac{1}{\sqrt{3}}\right)^2}$$

$$= \sqrt{4.0 + \frac{1}{3}}$$

$$v \approx 2.1\ cm/s$$

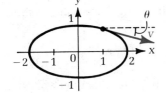

Figure 23-15

Finally, use equation (23-10) to find the direction.

$$Ref\ \theta = Arctan\left|\frac{-1/\sqrt{3}}{2}\right|$$

$$3\ \boxed{\sqrt{}}\ \boxed{1/x}\ \boxed{\div}\ 2\ \boxed{=}\ \boxed{INV}\ \boxed{tan}\ \rightarrow\ \boxed{16.102114}$$

From Fig. 23-15, we see that $\theta \approx -16°$.

Exercises 23-3

1. *Space technology* A rocket is launched so that it travels in a path described by the equation

$$y = x\left(20 - \frac{x^2}{60}\right)$$

where x and y are in kilometers. If the horizontal component of velocity is $v_x = 2x + 1$ (t in minutes) throughout the flight, find:

(a) The vertical component of velocity at $x = 10$ km

(b) The magnitude and direction of the velocity when the rocket hits the earth.

2. *Physics*   A particle moves clockwise along the upper half of the ellipse $x^2 + 16y^2 = 40$ with vertical velocity $v_y = y + 1$. Find v_x at the point $(2\sqrt{2}, \sqrt{2})$.

3. *Physics*   An object moves along the curve $y = \sqrt{x - 2}$, $x \geq 2$, with horizontal velocity component $v_x = 2x$. Find v_y at the point $(6, 2)$.

4. *Physics*   A particle moves in a magnetic field along the curve $y = 2x^3 + x$ with horizontal velocity component $v_x = x + 1$. Find the magnitude (in 10^4 m/s) and direction of its velocity at the point $(1, 3)$.

5. *Space technology*   The acceleration g due to the gravitational attraction of the earth on an object a distance r from the earth's center is $g = \dfrac{GM_e}{r^2}$, where M_e is the mass of the earth, 5.98×10^{24} kg, and G is a constant, 6.67×10^{-11} m³/kg·s². At what rate is g changing for an object 1000 m above the earth's surface ($r = 6.37 \times 10^6$ m), moving upward at 2000 m/s?

6. *Thermodynamics*   The equation of state for a certain gas is

$$P = \frac{0.2T - 0.00015T^2}{V}$$

where P is the pressure of the gas (in atmospheres), T is its temperature, and V is its volume. If the volume is kept fixed at 3.0 L, and the temperature is increased at the rate of 1.0 K/s, at what rate does P change when $T = 300$ K?

7. *Electronics*   The electric field E of an electric dipole at a distance r along the perpendicular bisector of the line joining the charges is

$$E = \frac{kp}{(a^2 + r^2)^{3/2}}$$

where p is the dipole moment, a is the charge separation distance, and k is a constant. If $p = 2 \times 10^{-6}$ C·m, $a = 10^{-2}$ m, and $k = 9 \times 10^9$ Nm²·C⁻², find the rate at which E changes at $r = 0.50$ m if the point r is moving away from the dipole at 1.5 m/s.

8. *Electronics*   The resistance of a strip of metal wire depends upon its temperature T according to the equation $R = 24.8 + 1.26T + 0.54T^2$, where R is in ohms. Find the rate at which the resistance changes if $T = 23°$C and is changing at 0.20°C/s.

9. *Electronics*   The energy E (in joules) stored in a capacitor is given by the equation $E = \dfrac{1}{2}CV^2$, where C is the capacitance and V is the voltage across the capacitor. At what rate is E changing if the voltage is increasing at 2.0 V/s when $V = 1000$ V and $C = 2 \times 10^{-6}$ F?

10. *Electronics*   The capacitance of a spherical capacitor is

$$C = \frac{kab}{b - a}$$

where a is the inner radius and b is the outer radius of the sphere. At what rate is the capacitance changing if the inner radius is increasing at a rate of 0.20 cm/s, when $a = 2.0$ cm, $b = 3.0$ cm, and $k = 2.5 \times 10^{-3}$ F?

11. Sand leaks out of a container forming a conical pile whose altitude is always equal to its radius. If the height of the pile is increasing at the rate of 3.0 in./min, find the rate at which sand is being added to the pile when the height is 2.0 ft.

12. The figure shows a ladder that is 14 ft long leaning against a wall. If the ladder begins to slip so that the bottom is moving at 4.5 ft/s along the floor, at what speed is the top of the ladder moving down the wall when the bottom edge is 4.0 ft from the wall?

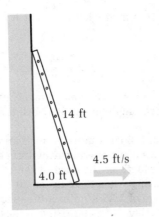

14 ft

4.5 ft/s

4.0 ft

13. If a soap bubble is inflated at 0.45 cm³/s, how fast is its diameter increasing when it is 3.0 cm across?

14. *Physics* The block shown is being dragged up the inclined plane at the rate of 0.5 m/s. At what rate is the block being moved vertically when it has been lifted 3 m?

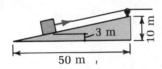

3 m

10 m

50 m

15. *Petroleum engineering* An oil spill on the open sea spreads in a circular pattern. If the radius of the oil slick increases at the constant rate of 1.2 m/s, how fast is the area of the spill increasing when its radius is 150 m?

16. *Physics* The moment of inertia of a thin rod about an axis through its center and perpendicular to its length is $I = ML^2/12$, where M is the mass of the rod and L is its length. If the length of the rod expands at 0.60 cm/min as it is heated, at what rate does I change when $M = 250$ g and $L = 35$ cm?

17. *Physics* In Problem 16, if the rod is suspended at its center to form a torsion pendulum, the frequency of oscillation of the rod is given by the equation

$$f = \frac{1}{2\pi}\sqrt{\frac{k}{I}}$$

where k is a constant equal to 9.0×10^6. At what rate is the frequency changing when $L = 35$ cm?

18. Two cyclists start from an intersection and travel on perpendicular roads. Cyclist A travels east at 10 mi/h and cyclist B leaves the intersection 12 min later and travels north at 15 mi/h. Find the rate at which they are moving apart 36 min after B leaves the intersection.

19. *Thermodynamics* When a diatomic gas undergoes adiabatic expansion, its pressure P and volume V are related by the equation $PV^{1.41} = k$, where k is a constant. If the pressure of this gas changes at 6.0 lb/in.²·s, when $P = 2$ lb/in.² and $V = 4$ ft³, find the rate at which its volume changes.

20. A spherical mothball sublimes (changes directly from the solid to the vapor state) so that its radius changes at −0.005 cm/day.
   (a) Find the rate at which the volume is decreasing when the mothball has a diameter of 1 cm.
   (b) Show that the rate of decrease of the volume is proportional to the surface area of the ball.

21. *Electronics* When resistors R_1 and R_2 are connected in parallel, the total resistance of the combination is

$$R = \frac{R_1 R_2}{R_1 + R_2}$$

If R_2 is fixed at 30 Ω, and R_1 decreases at 0.5 Ω/min, at what rate does R change when R_1 is 45 Ω?

22. *Optics*   The thin lens equation

$$\frac{1}{o} + \frac{1}{i} = \frac{1}{f}$$

relates the object distance o, the image distance i, and the focal length f of a thin lens. If an object is moving at 1.0 cm/s away from a lens with focal length 35 cm, how fast is its image moving when the object is 45 cm from the lens?*

23. Two sailboats leave the dock at the same instant. Boat A moves directly north at 6.5 mi/h, and boat B moves east at 4.5 mi/h. At what rate are the two boats moving away from each other?

24. A metal cube expands as it is heated. If the length of the edge of the cube increases at the rate of 0.20 cm/min, at what rate is its volume increasing when the edge is 3.0 cm long?

25. Water is added to a cone-shaped tank at the rate of 2.25 ft³/min. If the diameter of the top of the cone is 32 ft, and its height is 24 ft, at what rate is the level of water rising in the tank at the instant its depth is 10 ft?

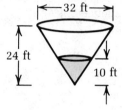

26. The hot-air balloon shown in the figure is over a point 85 ft away from the observer and is rising at 2.5 ft/s. At what rate will the balloon be receding from the observer when it is at an altitude of 60 ft?

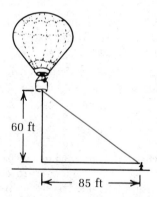

27. In the triangle shown, side a is increasing at 1.5 cm/s and side b is decreasing at 2.5 cm/s. At what rate is the area changing when a = 6.0 cm and b = 4.0 cm?

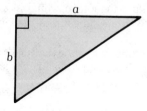

*See Arthur Beiser, *Physics*, 2nd edition (Menlo Park, CA: Benjamin/Cummings, 1983), p. 654

28. A liquid storage tank has the form of a trough 30 ft long with cross section in the shape of an equilateral triangle. If the tank is leaking at the rate of 1.5 ft³/min, at what rate (in inches per hour) is the surface level falling when the depth of liquid is 6.0 ft?

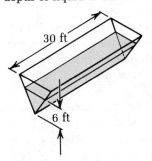

29. A rectangular swimming pool 50 ft long and 20 ft wide is 4 ft deep at one end and 12 ft deep at the other end. If water is pumped into the pool at the rate of 200 gal/min:
   (a)   At what rate is the water level rising when it is 2 ft from the top (1 gal ≈ 0.134 ft³)?
   (b)   At what rate is the water level rising when it is 5 ft from the top?

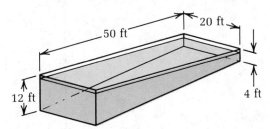

30. *Electronics*   In a mass spectrometer the mass m of an ion is measured by accelerating it through a potential difference V and allowing it to enter a magnetic field where it moves in a semicircle. The position of the ion on the photographic plate is related to the magnetic field strength B by the equation $x^2 = \dfrac{8mV}{B^2q}$, where q is the charge on the ion and x is in meters. If the magnetic field changes at the rate of 0.10 T/s, find the rate at which x changes for a singly charged chlorine ion when B = 0.70 T, V = 7.5 × 10³ V, m = 6.0 × 10⁻²⁶ kg, and q = 1.6 × 10⁻¹⁹ C.*

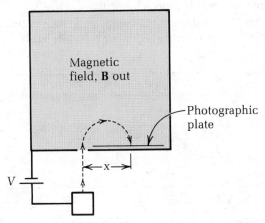

*For more information see Francis W. Sears, Mark W. Zemansky, and Hugh D. Young, *University Physics*, 6th edition (Reading, MA: Addison-Wesley, 1982), p. 591

23-4 | Curve Sketching

The shape of the graph of a function is related to its algebraic equation, and from an analysis of the equation we can determine many of the properties of the graph. The information obtained will enable us to sketch the graph much more quickly and easily than if we plotted it point-by-point.

Increasing and decreasing functions   A function $f(x)$ is said to be **increasing** if its graph rises as it is traced from left to right, and **decreasing** if its graph falls as it is traced from left to right. For any two values of x, say x_1 and x_2 where $x_1 < x_2$ (Fig. 23-16),

$f(x)$ is increasing if $f(x_2) > f(x_1)$

$f(x)$ is decreasing if $f(x_2) < f(x_1)$

Figure 23-16

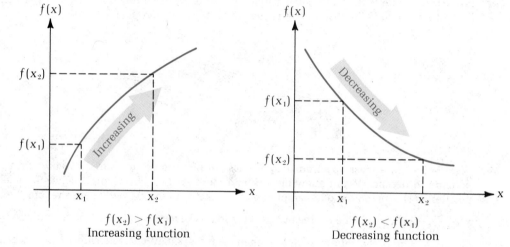

$f(x_2) > f(x_1)$
Increasing function

$f(x_2) < f(x_1)$
Decreasing function

It is possible to test a function to determine where the function is increasing and where it is decreasing without actually plotting its graph. If $y = f(x)$ is a function that is continuous and differentiable on some open interval $a < x < b$, then the following rule will hold:

Test for Increasing or Decreasing Function

- If $\dfrac{dy}{dx} > 0$, then $y = f(x)$ is increasing on the interval $a < x < b$.

- If $\dfrac{dy}{dx} < 0$, then $y = f(x)$ is decreasing on the interval $a < x < b$.

If a function is increasing on an interval, then a line tangent to the graph has positive slope everywhere on that interval [Fig. 23-17(a)]. If the function is

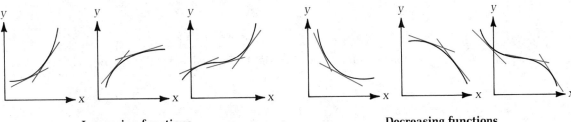

Increasing functions
Slope is positive everywhere

Decreasing functions
Slope is negative everywhere

(a)   (b)

Figure 23-17

decreasing on an interval, then a line tangent to the graph has negative slope everywhere on that interval [Fig. 23-17(b)].

It is possible for a function to increase on one interval of its domain and decrease on another interval, as shown in Fig. 23-18.

Figure 23-18

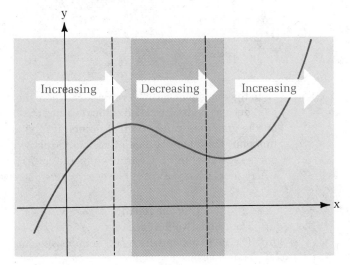

EXAMPLE 17 Use the test for increasing and decreasing functions to determine the intervals on which the function $y = 4x - x^2$ is increasing or decreasing.

Solution **First,** find the derivative. $\dfrac{dy}{dx} = 4 - 2x$

Second, apply the test for an increasing function. $\dfrac{dy}{dx} > 0$ gives $4 - 2x > 0$

Solve. $4 > 2x$

$2 > x$ or $x < 2$

The function is increasing on the interval $x < 2$ (Fig. 23-19).

Finally, apply the test for a decreasing function. $\dfrac{dy}{dx} < 0$ gives $4 - 2x < 0$

Solve.

$$4 < 2x$$

$$2 < x \quad \text{or} \quad x > 2$$

The function is decreasing on the interval $x > 2$ (Fig. 23-19).

Figure 23-19

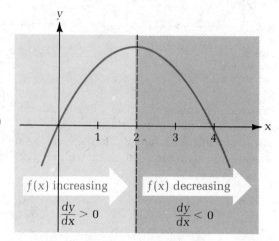

Critical values In the domain of $f(x)$, the values of x where the derivative $f'(x) = 0$ or where $f'(x)$ does not exist are called **critical values** of x, and the corresponding points on the graph are called **critical points.** If $f'(x)$ is a quadratic or higher-order function, there may be more than one critical value. The critical points of a curve separate it into regions in which the corresponding function $f(x)$ is everywhere increasing [$f'(x) > 0$] or everywhere decreasing [$f'(x) < 0$].

EXAMPLE 18 For the function $f(x) = x^3 - 4x^2 - 16x + 8$ find:
(a) The critical values of x
(b) The intervals on which $f(x)$ is increasing and the intervals on which $f(x)$ is decreasing.

Solution **First,** find the derivative: $f'(x) = 3x^2 - 8x - 16$
(a) To locate the critical values, set $f'(x) = 0$ and solve the resulting equation:

$$3x^2 - 8x - 16 = 0$$

$$(3x + 4)(x - 4) = 0$$

$$3x + 4 = 0 \qquad x - 4 = 0$$

$$x = -\frac{4}{3} \qquad x = 4$$

The critical values of x are $-\dfrac{4}{3}$ and 4.

(b) The critical values divide the x-axis into three open intervals. In each interval the function $f(x)$ will be everywhere increasing or everywhere decreasing, depending on the sign of $f'(x)$. Choose a point in each interval and use it to determine the sign of each factor $(3x + 4)$ and $(x - 4)$. Multiply the signs of these factors to find the sign of $f'(x)$ for each interval. The table gives a summary of the work.

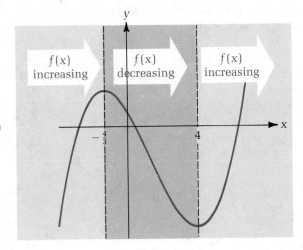

Figure 23-20

Interval	Test Point	3x + 4	Sign	x − 4	Sign	Sign of $f'(x) = (3x + 4)(x − 4)$
$x < -\dfrac{4}{3}$	−2	3(−2) + 4 = −2	−	(−2) − 4 = −6	−	+ $f'(x) > 0$
$-\dfrac{4}{3} < x < 4$	0	3(0) + 4 = 4	+	(0) − 4 = −4	−	− $f'(x) < 0$
$x > 4$	5	3(5) + 4 = 19	+	(5) − 4 = 1	+	+ $f'(x) > 0$

Therefore, using the test for increasing and decreasing functions,

$f(x)$ is increasing when $x < -\dfrac{4}{3}$

$f(x)$ is decreasing when $-\dfrac{4}{3} < x < 4$

$f(x)$ is increasing when $x > 4$

See Fig. 23-20.

Information that is very useful in sketching the graph of a function can be obtained by studying the behavior of the graph at the critical points. At these points the function may have a maximum or a minimum value so that the graph shows a corresponding peak or valley.

Absolute maximum and minimum

The **absolute maximum** value of a function is the point where it attains its largest value over the entire domain of the function. Similarly, the **absolute minimum** value of a function is the point where the function attains its smallest value over the domain of the function.

Local maximum and minimum

A **relative maximum**, or **local maximum**, value of a function occurs where the function has a larger value than at any point nearby. A **relative minimum**, or **local minimum**, value of a function occurs where the function has a smaller value than at any point nearby. Maxima and minima are often called the **extreme values** of the function.

Extreme values

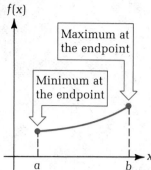

The extreme values of a function defined on some closed interval $a \leq x \leq b$ may occur only at:

- An endpoint of the interval (Fig. 23-21)
- A critical point inside the interval (Fig. 23-22)

Figure 23-21

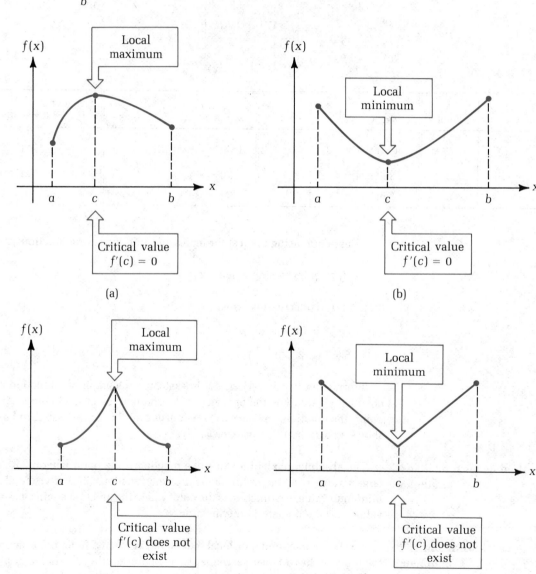

Figure 23-22

After the critical points of a function have been found, it is necessary to determine the nature of the function at the critical points. A critical point may signal a relative maximum [Fig. 23-23(a); see p. 910], a relative minimum [Fig. 23-23(b)], or a change in slope not associated with an extreme value (Fig. 23-24; see p. 910). The most direct way to check for an extreme value of a function is to use the following test:

First Derivative Test
If c is a critical value of the function $f(x)$, choose two values of x, one slightly less than c and the other slightly greater than c. Evaluate the derivative $f'(x)$ at each of these points to determine its sign.

- If the sign of $f'(x)$ changes from **positive to negative** as x increases in the neighborhood of c, then $f(c)$ is a local **maximum** [Fig. 23-23(a)].

- If the sign of $f'(x)$ changes from **negative to positive** as x increases in the neighborhood of c, then $f(c)$ is a local **minimum** [Fig. 23-23(b)].

- If the sign of $f'(x)$ does not change, then $f(c)$ is neither a maximum nor a minimum (Fig. 23-24 on page 910).

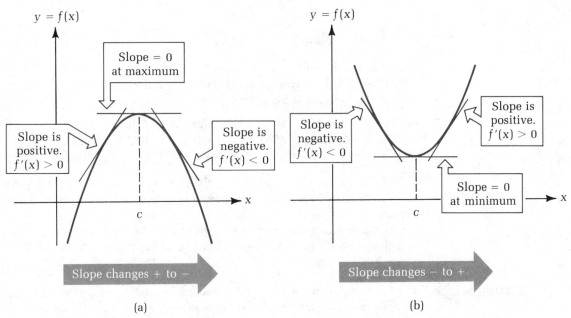

Figure 23-23

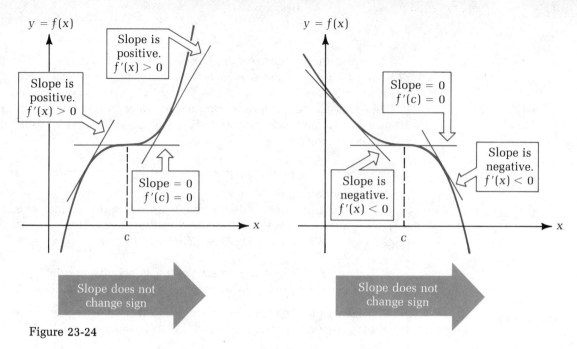

Figure 23-24

EXAMPLE 19 For the equation $y = x^3 - 6x^2 + 9x + 1$, find the values of x, if any, where relative maxima or relative minima occur.

Solution **First,** find the derivative. $f'(x) = \dfrac{dy}{dx} = 3x^2 - 12x + 9$

Second, set $f'(x) = 0$ and solve the resulting equation to find the critical values:

$$3x^2 - 12x + 9 = 0$$

Factoring, $3(x - 1)(x - 3) = 0$

$$x - 1 = 0 \qquad x - 3 = 0$$

$$x = 1 \qquad\qquad x = 3$$

The critical values are $x = 1$ and $x = 3$.

Third, use the first derivative test to determine if the function has a maximum or a minimum value at the critical points:

- Test $x = 0$: $f'(0) = 3(0)^2 - 12(0) + 9 = 9$ $f'(0)$ is positive
- Test $x = 2$: $f'(2) = 3(2)^2 - 12(2) + 9 = -3$ $f'(2)$ is negative
- Test $x = 4$: $f'(4) = 3(4)^2 - 12(4) + 9 = 9$ $f'(4)$ is positive

For the critical point where $x = 1$, the sign of the derivative changes from positive to negative as x increases from 0 to 2. The function has a maximum value at $x = 1$.

For the critical point where $x = 3$, the sign of the derivative changes from negative to positive as x increases from 2 to 4. The function has a minimum value at $x = 3$ (Fig. 23-25).

CAUTION ▶ If a function $f(x)$ has a relative maximum or a relative minimum value, this extreme point will always occur at a critical point where either $f'(c) = 0$ or $f'(c)$ does not exist. However, a critical point can exist and *not* lead to a relative

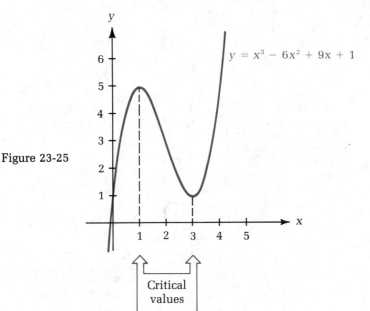

Figure 23-25

$$y = x^3 - 6x^2 + 9x + 1$$

Critical
values

maximum or minimum value, as shown in Fig. 23-24. The next example illustrates this. ◄

EXAMPLE 20   Test the critical points of the function $y = 3x^4 - 4x^3 + 1$ for a
maximum or a minimum.

Solution   **First,** find the derivative. $f'(x) = \dfrac{dy}{dx} = 12x^3 - 12x^2$

Second, find the critical values by setting $f'(x) = 0$:

$$12x^3 - 12x^2 = 0$$

$$12x^2(x - 1) = 0$$

The critical values are $x = 0$ and $x = 1$.
Third, apply the first derivative test:

- Test $x = -1$: $f'(-1) = 12(-1)^2(-1 - 1) = -24$ $f'(-1)$ is negative
- Test $x = \dfrac{1}{2}$: $f'\left(\dfrac{1}{2}\right) = 12\left(\dfrac{1}{2}\right)^2\left(\dfrac{1}{2} - 1\right) = -\dfrac{3}{2}$ $f'\left(\dfrac{1}{2}\right)$ is negative
- Test $x = 2$: $f'(2) = 12(2)^2(2 - 1) = 48$ $f'(2)$ is positive

For the critical point where $x = 0$, the sign of the derivative does not change
as x increases from -1 to $+\dfrac{1}{2}$. The function has neither a maximum nor a
minimum at $x = 0$. (This critical point is a "dud.")
For the critical point where $x = 1$, the sign of the derivative changes from
negative to positive as x increases from $\dfrac{1}{2}$ to 2. The function has a minimum
value at $x = 1$ (Fig. 23-26).

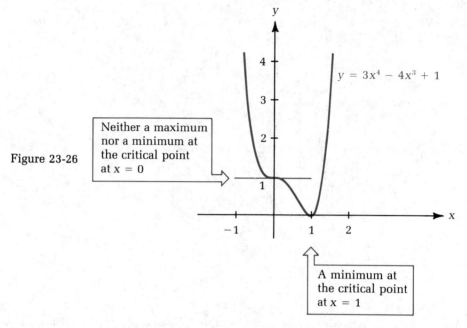

Figure 23-26

Neither a maximum
nor a minimum at
the critical point
at $x = 0$

$y = 3x^4 - 4x^3 + 1$

A minimum at
the critical point
at $x = 1$

Concavity   Another concept that is useful in sketching the graph of a function involves
Concave up   the **concavity** of a curve. A curve or portion of a curve is said to be **concave up**
if the slope of its tangent line is steadily increasing (Fig. 23-27).

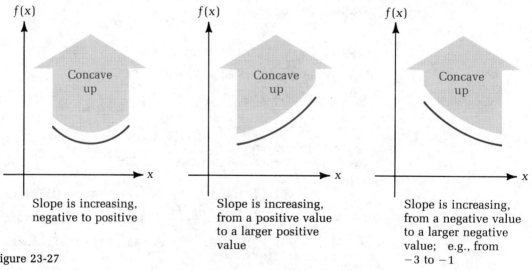

Slope is increasing,
negative to positive

Slope is increasing,
from a positive value
to a larger positive
value

Slope is increasing,
from a negative value
to a larger negative
value;   e.g., from
-3 to -1

Figure 23-27

Concave down   A curve is said to be **concave down** if the slope of its tangent line is steadily
decreasing (Fig. 23-28).

LEARNING   Concave up is a *cup.*
HINT   Concave *down* is a *cap.*

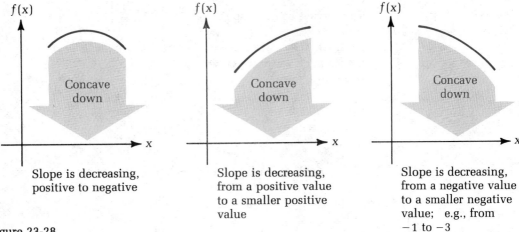

Slope is decreasing, positive to negative	Slope is decreasing, from a positive value to a smaller positive value	Slope is decreasing, from a negative value to a smaller negative value; e.g., from -1 to -3

Figure 23-28

Because the second derivative gives the rate of change of the slope of the tangent line, we can use the second derivative to identify regions of different concavity on the graph. If $f(x)$ is a differentiable function, and both $f'(x)$ and $f''(x)$ exist, then the following test may be used to find the concavity of $f(x)$:

Concavity Test
For any open interval $a < x < b$:

- The curve is **concave up** where $f''(x) > 0$.
- The curve is **concave down** where $f''(x) < 0$.

EXAMPLE 21 For the function $f(x) = x^3 - 2x^2 - 3x + 20$, find the regions where the curve is concave up and the regions where it is concave down.

Solution
First, find the derivatives $f'(x)$ and $f''(x)$.

$$f'(x) = 3x^2 - 4x - 3$$

$$f''(x) = \frac{d}{dx}[f'(x)] = 6x - 4$$

Second, set $f''(x) > 0$ and solve to determine where the curve is concave up.

$$6x - 4 > 0$$

$$6x > 4$$

$$x > \frac{4}{6}$$

$$x > \frac{2}{3}$$

The curve is concave up where $x > \dfrac{2}{3}$.

Next, set $f''(x) < 0$ and solve to determine where the curve is concave down.

$$6x - 4 < 0$$
$$6x < 4$$
$$x < \frac{4}{6}$$
$$x < \frac{2}{3}$$

The curve is concave down where $x < \dfrac{2}{3}$ (Fig. 23-29).

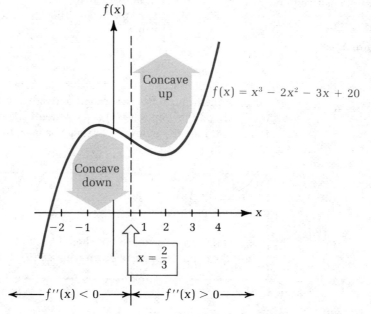

Figure 23-29

$f(x)$

Concave up

$f(x) = x^3 - 2x^2 - 3x + 20$

Concave down

$x = \dfrac{2}{3}$

$\longleftarrow f''(x) < 0 \longrightarrow \longleftarrow f''(x) > 0 \longrightarrow$

Inflection point   A point on the graph where the concavity changes from concave up to concave down or from concave down to concave up is called a **point of inflection** (Fig. 23-30). The point where $x = x_0$ is an inflection point of the curve $f(x)$ if the following conditions are satisfied:

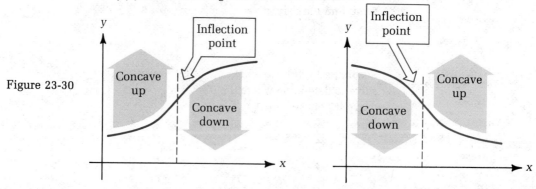

Figure 23-30

y

Inflection point

Concave up

Concave down

y

Inflection point

Concave down

Concave up

1. $f(x)$ exists and is continuous at the point where $x = x_0$.
2. The concavity changes at $x = x_0$.
3. The second derivative $f''(x)$ either is equal to zero or is undefined at $x = x_0$.

Figure 23-31 shows three types of inflection point.

Figure 23-31

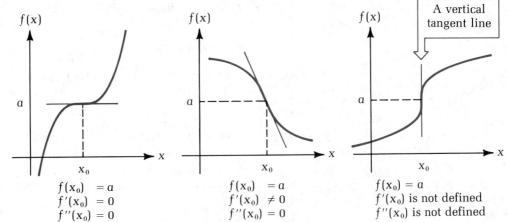

$f(x_0) = a$
$f'(x_0) = 0$
$f''(x_0) = 0$

$f(x_0) = a$
$f'(x_0) \neq 0$
$f''(x_0) = 0$

$f(x_0) = a$
$f'(x_0)$ is not defined
$f''(x_0)$ is not defined

In Example 21, the point on the curve where $x = \dfrac{2}{3}$ is an inflection point of the curve and can be shown to satisfy these three conditions.

1. At $x = \dfrac{2}{3}$, the function $f(x) = x^3 - 2x^2 - 3x + 20$ exists and is continuous (Fig. 23-29).
2. The concavity changes at this point.
3. At $x = \dfrac{2}{3}$, the second derivative $f''(x)$ is equal to zero.

$$6\left(\frac{2}{3}\right) - 4 = 0$$

$$4 - 4 = 0$$

The inflection points of a curve separate it into regions, and in each of these regions the concavity is everywhere the same.

EXAMPLE 22 Locate the inflection points for the function
$f(x) = x^4 - 6x^2 + 2$.

Solution **First,** calculate the derivatives $f'(x)$ and $f''(x)$. $\quad f'(x) = 4x^3 - 12x$
$$f''(x) = 12x^2 - 12$$

Second, solve $f''(x) = 0$ to find $12x^2 - 12 = 0$
possible inflection points.
$$x^2 - 1 = 0$$

$$x = \pm 1$$

Third, check to be certain that $f(x)$ exists at these points:

$$f(1) = (1)^4 - 6(1)^2 + 2 = -3$$
$$f(-1) = (-1)^4 - 6(-1)^2 + 2 = -3$$

Fourth, the possible inflection points divide the x-axis into three regions. Determine the concavity of the graph in each region by finding the sign of $f''(x)$ at a point in each region.

- For $x < -1$: $f''(-2) = 12(-2)^2 - 12 = 36$ Concave up since $f''(x)$ is positive.

- For $-1 < x < 1$: $f''(0) = 12(0)^2 - 12 = -12$ Concave down since $f''(x)$ is negative.

- For $x > 1$: $f''(2) = 12(2)^2 - 12 = 36$ Concave up since $f''(x)$ is positive.

The concavity changes at both $x = 1$ and $x = -1$; therefore, both are inflection points.

Figure 23-32

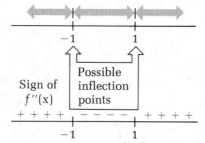

NOTE ▶ It is possible to have a function $f(x)$ for which $f''(x) = 0$ at a point that is *not* an inflection point. For example, the function $f(x) = x^4$ has $f'(x) = 4x^3$ and $f''(x) = 12x^2$. Therefore, $f''(x) = 0$ at $x = 0$, but because the curve is concave up on both sides of this point, it cannot be an inflection point. ◀

A corner point need not always be an inflection point (Fig. 23-33).

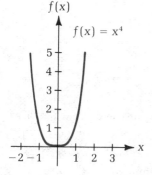

Figure 23-33

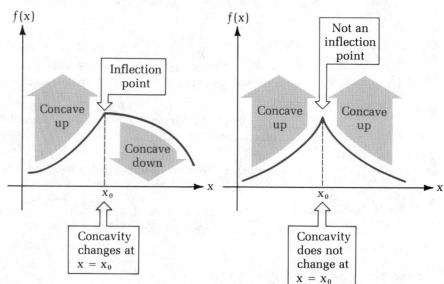

Second derivative test   If a function $f(x)$ is found to be concave down on a certain interval, and if the derivative $f'(x)$ is equal to zero at some point inside that interval, then the function has a **local maximum** at that point. If the function is concave up on the interval and the derivative $f'(x)$ is equal to zero at some point inside the interval, then the function has a **local minimum** at that point. Although the first derivative test will always work, the following second derivative test is usually a simpler, quicker test for extreme values.

For a function $f(x)$, if $f'(c) = 0$, then c is a critical value and the following test can be used to determine whether the function has either a maximum or a minimum at c.

Second Derivative Test

- If $f''(c) > 0$, then $f(c)$ is a minimum value.
- If $f''(c) < 0$, then $f(c)$ is a maximum value.
- If $f''(c) = 0$ or does not exist, the second derivative test fails, and the first derivative test must be used to determine the nature of the graph at $x = c$.

EXAMPLE 23   Locate any local maximum and minimum points of the function

$f(x) = \dfrac{1}{3}x^3 - 3x^2 + 8x + 1$ using the second derivative test.

Solution   **First,** find the first and second derivatives. $f'(x) = x^2 - 6x + 8$

$$f''(x) = 2x - 6$$

Second, find the critical values $x^2 - 6x + 8 = 0$
by solving $f'(x) = 0$.

$$(x - 2)(x - 4) = 0$$

The critical values are $x = 2$ and $x = 4$.

Third, evaluate the second de- $f''(2) = 2(2) - 6 = -2$
rivative at these values.

$$f''(4) = 2(4) - 6 = 2$$

Finally, apply the second derivative test:

- Since $f''(2)$ is negative, the function has a local maximum at $x = 2$.
- Since $f''(4)$ is positive, the function has a local minimum at $x = 4$.

Figure 23-34 on p. 918 shows the graph of this function.

**LEARNING
HINT** ▶   Both the second derivative test and the first derivative test allow you to determine whether a critical point is a local maximum or a local minimum. Use the second derivative test if the second derivative is easy to find. If the second

derivative is difficult to find, or if $d^2y/dx^2 = 0$, then use the first derivative test. ◄

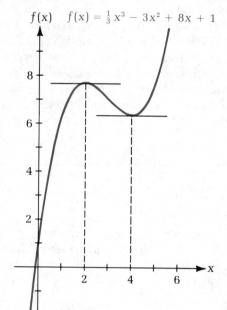

$f(x)$ $f(x) = \frac{1}{3}x^3 - 3x^2 + 8x + 1$

Figure 23-34

TABLE 23-1

$f(x) = x^5 - 2x^2 + 10$

x	$f(x)$		$f'(x)$	$f''(x)$
0	10		0	-4
0.1	9.98001		-0.3995	-3.98
0.2	9.92032		-0.792	-3.84
0.3	9.82243		-1.1595	-3.46
0.4	9.69024	Function is	-1.472	-2.72
0.5	9.53125	decreasing	-1.6875	-1.5
0.6	9.35776		-1.752	0.320000002
0.7	9.18807		-1.5995	2.86000001
0.8	9.04768		-1.152	6.24000002
0.9	8.97049		-0.319499996	10.58
1.0	9		1.00000001	16
1.1	9.19051		2.92050001	22.62
1.2	9.60832001		5.56800004	30.5600001
1.3	10.33293		9.08050005	39.9400001
1.4	11.45824		13.6080001	50.8800002
1.5	13.09375	Function is	19.3125	63.5000001
1.6	15.3657601	increasing.	26.3680001	77.9200003
1.7	18.4185701		34.9605002	94.2600003
1.8	22.4156801		45.2880002	112.64
1.9	27.5409901		57.5605002	133.18

There is a local minimum between x = 0.9 and x = 1.0

First derivative test:
The sign of $f'(x)$ changes from − to +. There is a local minimum between x = 0.9 and x = 1.0

Second derivative test:
$f''(x) > 0$ means that there is a local minimum. The curve is concave up.

Table 23-1 shows how the first and second derivative tests can be used to find a local maximum or minimum for some function. In this case, the x-coordinate of the inflection point is not a rational number and must be approximated. A calculator or computer is used to calculate values of the first and second derivative.

Procedure for Graphing a Polynomial Function

Each of the techniques illustrated reveals an important characteristic of the graph of a function. We will now combine these into the following step-by-step procedure for graphing any polynomial function:

Step 1.   Find $f'(x)$ and $f''(x)$.

Step 2.   Let $f'(x) = 0$ and solve the resulting equation to find the critical values of x. Find $f(x)$ at each critical value and plot the critical points.

Step 3.   Use the second derivative test to find any local maxima or minima.

- For a maximum, $f'(x) = 0$, $f''(x) < 0$.
- For a minimum, $f'(x) = 0$, $f''(x) > 0$
- If $f''(x) = 0$, use the first derivative test to check for a maximum or minimum.

Step 4.   Determine the values of x for which $f''(x) = 0$. Use these values of x to divide the x-axis into intervals. Test a point in each interval to find the concavity of the graph there.

- If $f''(x) > 0$, the graph is concave up.
- If $f''(x) < 0$, the graph is concave down.

Step 5.   Use the information from Step 4 to find any inflection points and mark these points.

Step 6.   List in a table all the points found so far and use this information to determine a reasonable scale for the graph. Draw a horizontal tangent line at each local maximum or minimum. Sketch the graph starting at an inflection point and draw to the next inflection point, maintaining the proper concavity. Plot any additional points needed.

EXAMPLE 24   Sketch the graph of the function $y = f(x) = x^3 - 12x + 3$.

Solution

Step 1.   Find the derivatives.

$$f'(x) = \frac{dy}{dx} = 3x^2 - 12$$

$$f''(x) = \frac{d^2y}{dx^2} = 6x$$

Step 2.   To find the critical values, set $f'(x) = 0$.

$$3x^2 - 12 = 0$$

$$3(x^2 - 4) = 0$$

Factoring,

$$3(x - 2)(x + 2) = 0$$

Therefore, the critical values are $x = 2$ and $x = -2$
The critical points are:

- For $x = 2$, $y = (2)^3 - 12(2) + 3 = -13$
- For $x = -2$, $y = (-2)^3 - 12(-2) + 3 = 19$

Plot these points [Fig. 23-35(a)].

Figure 23-35 (a)

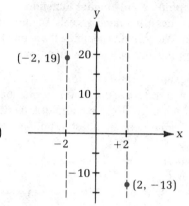

Step 3. Apply the second derivative test to determine whether the critical points are at a maximum or minimum:

$$f''(-2) = 6(-2) = -12 \qquad \text{Negative}$$

There is a local maximum at $x = -2$.

$$f''(2) = 6(2) = 12 \qquad \text{Positive}$$

There is a local minimum at $x = 2$ [Fig. 23-35(b)].

Figure 23-35 (b)

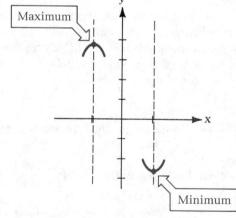

Step 4. Determine where $f''(x) = 0$:

$$f''(x) = 6x = 0$$
$$x = 0$$

There is a possible inflection point.

- Test a point where x < 0:

$$f''(-1) = 6(-1) = -6 \qquad \text{Negative}$$

The graph is concave down for x < 0.
- Test a point where x > 0:

$$f''(1) = 6(1) = 6 \qquad \text{Positive}$$

The graph is concave up for x > 0 [Fig. 23-35(c)].

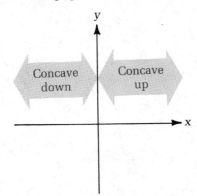

Figure 23-35 (c)

Step 5. At x = 0, the graph has an inflection point since $f(0)$ exists, $f''(0) = 0$, and the concavity changes [Fig. 23-35(d)].

$$f(0) = (0)^3 - 12(0) + 3 = 3$$

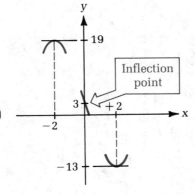

Figure 23-35 (d)

Step 6.

x	−2	0	2
y	19	3	−13

Set up the x-axis extending from x = −4 to x = 4. The y-axis will extend roughly from -13 to $+19$. Plot a few additional points to help determine the shape of the graph [Fig. 23-35(e)].

x	−1	−3	1	3
y	14	12	−8	−6

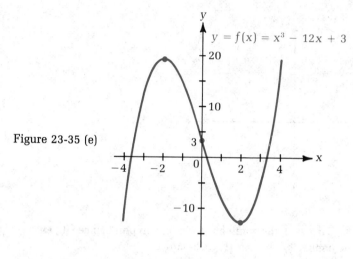

Figure 23-35 (e)

Exercises 23-4

For each function determine the values of x for which the graph:

(a) *Is increasing or decreasing*   (b) *Is concave up or concave down*
(c) *Has a local maximum or minimum*   (d) *Has an inflection point*

Sketch the graph of each function.

1. $y = x^2 - 4x$ 　　　　　　　 2. $y = 8 - x^2$ 　　　　　　　 3. $y = x^2 + x - 2$
4. $y = x^2 + 3x + 4$ 　　　　　 5. $y = x^2 - 4x + 3$ 　　　　 6. $y = 2 - 3x - 2x^2$
7. $y = 3x^3$ 　　　　　　　　　 8. $y = 1 + x^3$ 　　　　　　　 9. $y = x^3 - 4x$
10. $y = x - 4x^3$ 　　　　　　 11. $y = x^3 - 6x^2 + 4$ 　　 12. $y = x^3 - 4x^2$
13. $y = 3x^4 - 4x^3$ 　　　　　 14. $y = 2x^5 - 8x^3$ 　　　　 15. $y = x^3 - 2x^2 + 4x + 2$
16. $y = 2x^3 + 4x^2 - 8x$ 　　 17. $y = x^4 - 4x^2 + 2$ 　　 18. $y = x^5 - x^4 - 2x^3$

Solve.

19. *Business*   For a certain manufacturing process, the cost per unit C of producing N units is given by the function

$$C = \frac{N^3}{72,000} - \frac{N^2}{4000} - \frac{N}{2} + 400$$

Sketch the graph of C as a function of N.

20. *Energy technology* The energy output of a solar power generating system depends on time according to an equation of the form $E(t) = t^3 - 6t^2 + 9t$. Sketch the graph of this function.

21. *Electrical engineering* The family of polynomials called the *Legendre polynomials* have many important technical applications, especially in the study of electrical transmission lines. Sketch the graph of each of the following Legendre polynomials:

$$P_0(x) = 1 \qquad P_1(x) = x \qquad P_2(x) = \frac{1}{2}(3x^2 - 1)$$

$$P_3(x) = \frac{1}{2}(5x^3 - 3x) \qquad P_4(x) = \frac{1}{8}(35x^4 - 30x^2 + 3)$$

22. In many engineering applications it is useful to work with a polynomial approximation to a given function. The function $f(x) = x^4$ in the interval $-1 \le x \le 1$ can be closely approximated by the function

$$g(x) = \begin{cases} -2x^3 - x^2 & \text{for } -1 \le x < 0 \\ 2x^3 - x^2 & \text{for } 0 \le x \le 1 \end{cases}$$

Sketch the graphs of $f(x)$ and $g(x)$ on the same coordinate axes.*

23-5 | Graphing Rational Functions

The techniques outlined in Section 23-4 are sufficient for graphing most polynomial functions. However, there are certain rational functions, useful in technical work, whose graphs have features that are not revealed by these techniques. In this section we will develop some additional procedures that will enable us to graph these functions.

Vertical asymptote If the value of $|f(x)|$ becomes infinitely large as x approaches some value x_0, and if

$$\lim_{x \to x_0^-} f(x) = \pm\infty \qquad \text{or} \qquad \lim_{x \to x_0^+} f(x) = \pm\infty$$

then the line $x = x_0$ is called a **vertical asymptote** of the curve $f(x)$ (Fig. 23-36). The curve can never intersect a vertical asymptote.

Horizontal asymptote If the limit of $f(x)$ approaches some value a as $x \to +\infty$ or $x \to -\infty$, then the line $f(x) = a$ is called a **horizontal asymptote** of the curve $f(x)$. In Fig. 23-37, on the left, as $x \to -\infty, f(x) \to a$; on the right, as $x \to +\infty, f(x) \to a$. Notice that the curve may intersect a horizontal asymptote.

*For more information, see Erwin Kreyszig, *Advanced Engineering Mathematics*, 4th edition (New York: Wiley, 1979), p. 781

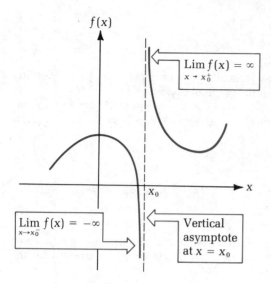

Figure 23-36

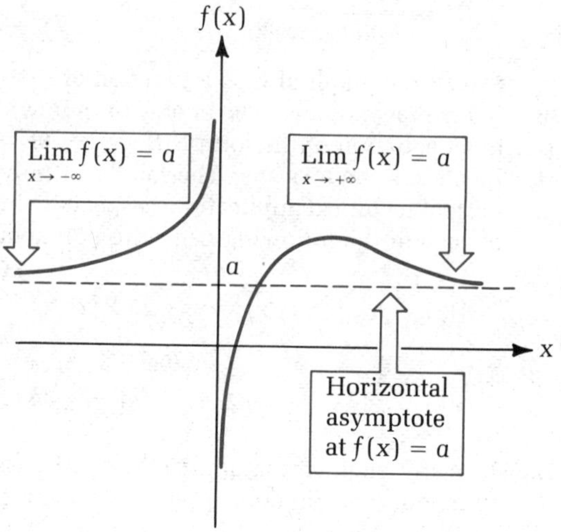

$$\text{Lim } f(x) = \infty$$
$$x \to x_0^+$$

$$\text{Lim } f(x) = -\infty$$
$$x \to x_0^-$$

Vertical asymptote at $x = x_0$

$f(x)$

$$\text{Lim } f(x) = a$$
$$x \to -\infty$$

$$\text{Lim } f(x) = a$$
$$x \to +\infty$$

a

Horizontal asymptote at $f(x) = a$

Figure 23-37

EXAMPLE 25   Find the vertical and horizontal asymptotes of the graph of the function $f(x) = \dfrac{x - 1}{x - 2}$

Solution   **First,** to find any vertical asymptotes, set the denominator equal to zero and solve:

$$x - 2 = 0$$

$$x = 2$$

The function is discontinuous at $x = 2$; therefore, we must check $\lim\limits_{x \to 2} f(x)$ for a vertical asymptote.

To check $\lim\limits_{x \to 2} \dfrac{x - 1}{x - 2}$, for some positive number d, let $x = 2 + d$. Then

$$\frac{x - 1}{x - 2} = \frac{(2 + d) - 1}{(2 + d) - 2}$$

$$= \frac{d + 1}{d} = 1 + \frac{1}{d}$$

As $x \to 2$, $d \to 0$, and $1/d \to \infty$; therefore, $\lim\limits_{x \to 2} \dfrac{x - 1}{x - 2} = \infty$, and the graph has a vertical asymptote at $x = 2$.

Second, to find any horizontal asymptotes, find $\lim\limits_{x \to \infty} \dfrac{x - 1}{x - 2}$. Divide the numerator and denominator of the fraction by x to get

$$\lim_{x \to \infty} \frac{1 - \dfrac{1}{x}}{1 - \dfrac{2}{x}}$$

But $1/x \rightarrow 0$ as $x \rightarrow \infty$ and $2/x \rightarrow 0$ as $x \rightarrow \infty$. Therefore,

$$\lim_{x \to \infty} \frac{1 - \dfrac{1}{x}}{1 - \dfrac{2}{x}} = \lim_{x \to \infty} \frac{1}{1} = 1$$

Similarly,

$$\lim_{x \to -\infty} \frac{x - 1}{x - 2} = 1$$

The graph of the function shows how the curve approaches the asymptotes (Fig. 23-38).

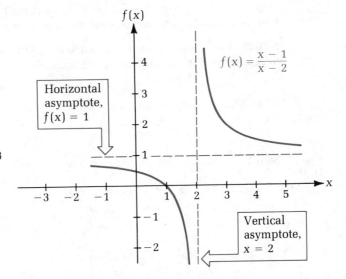

Figure 23-38

NOTE ▶ A zero denominator at a point is not itself sufficient to assure that a vertical asymptote exists at that point. For example, the function

$$g(x) = \frac{(x + 1)(x - 2)}{(x - 2)}$$

does not exist at $x = 2$, but there is no asymptote at this point, since $\lim_{x \to 2} g(x) = 3$. ◀

To sketch the graph of a rational function $f(x) = P(x)/Q(x)$, follow these steps.

Procedure for Graphing a Rational Function

Step 1. Reduce the given function to lowest terms if necessary. Solve the equation $P(x) = 0$, formed from the numerator function, to find the x-intercepts of the graph. If $P(x)$ is a constant or if the equation $P(x) = 0$ has no real roots, the graph does not intersect the x-axis.

Step 2.   Solve the equation $Q(x) = 0$, formed from the denominator function, to find the location of any possible vertical asymptotes. If $Q(x) = 0$ at $x = x_0$, and $f(x)$ approaches $+\infty$ or $-\infty$, then the graph has a vertical asymptote at $x = x_0$.

Step 3.   Find $\lim\limits_{x \to \infty} f(x)$ and $\lim\limits_{x \to -\infty} f(x)$. If either limit is a finite number a, then the line $y = a$ is a horizontal asymptote.

Step 4.   The function $f(x)$ can change sign only at the points found in Steps 1 and 2. These points divide the x-axis into two or more regions, and $f(x)$ must have the same sign in each region. Select a value of x in each region, calculate the corresponding value of $f(x)$, and use this information to determine whether the graph is above or below the x-axis in each region.

Step 5.   Use the procedure for sketching the graph of a polynomial function to find the critical points, maxima and minima, concavity, and inflection points.

Step 6.   Plot any other points needed to sketch the graph. Be especially careful to determine if the graph crosses the horizontal asymptote.

EXAMPLE 26   Sketch the graph of the rational function $f(x) = \dfrac{x - 2}{x^2}$

Solution
Step 1.   Find and plot the x-intercepts. $P(x) = x - 2 = 0$

$$x = 2 \quad \text{[Fig. 23-39(a)]}$$

Figure 23-39(a)

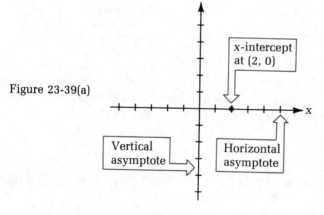

Step 2.   Find the vertical asymptotes. $Q(x) = x^2 = 0$

$$x = 0$$

Checking,

$$\lim_{x \to 0} \left(\frac{x - 2}{x^2} \right) = -\infty$$

Therefore at $x = 0$ the graph has a vertical asymptote [Fig. 23-39(a)].

Step 3. Find the horizontal asymptotes.

$$\lim_{x \to +\infty} \frac{x - 2}{x^2} = \lim_{x \to +\infty} \frac{\dfrac{x}{x^2} - \dfrac{2}{x^2}}{\dfrac{x^2}{x^2}}$$

$$= \lim_{x \to +\infty} \frac{\dfrac{1}{x} - \dfrac{2}{x^2}}{1} = 0$$

Similarly,

$$\lim_{x \to -\infty} \frac{x - 2}{x^2} = 0$$

The line $y = 0$ (the x-axis) is the horizontal asymptote [Fig. 23-39(a)].

Step 4. Steps 1 and 2 define the following intervals on the x-axis: $x < 0$, $0 < x < 2$, $x > 2$. Test points in each of these intervals to determine the sign of $f(x)$ in each interval:

- At $x = -1$: $f(-1) = \dfrac{(-1) - 2}{(-1)^2} = -3$

  The graph is below the x-axis for $x < 0$.

- At $x = 1$: $f(1) = \dfrac{(1) - 2}{(1)^2} = -1$

  The graph is below the x-axis for $0 < x < 2$.

- At $x = 3$: $f(3) = \dfrac{(3) - 2}{(3)^2} = \dfrac{1}{9}$

  The graph is above the x-axis for $x > 2$. [Fig. 23-39(b)]

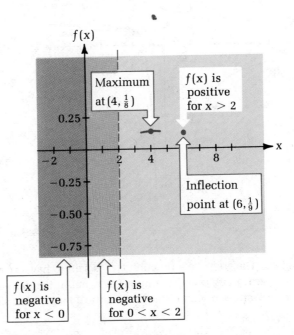

Figure 23-39(b)

Step 5.   To find the inflection points, find the first and second derivatives using the quotient rule (22-19).

$$f'(x) = \frac{x^2 - 2x(x - 2)}{x^4} = \frac{4 - x}{x^3}$$

$$f''(x) = \frac{-x^3 - (4 - x)(3x^2)}{x^6} = \frac{2x - 12}{x^4}$$

- $f'(x) = 0$ at $4 - x = 0$ or $x = 4$. Since $f''(4)$ is negative, the function has a maximum value at $x = 4$.

$$f(4) = \frac{1}{8} \quad \text{[Fig. 23-39(b)]}$$

- $f''(x) = 0$ at $2x - 12 = 0$ or $x = 6$. Because $f''(5)$ is negative and $f''(7)$ is positive, the graph is concave down for $x < 6$ and concave up for $x > 6$. The graph has an inflection point at $x = 6$, $f(x) = \frac{1}{9}$. [Fig. 23-39(b)]

Step 6.   At $x = -2$, $f(-2) = -1$; and at $x = -3$, $f(-3) = -\frac{5}{9}$.

The completed sketch is shown in Fig. 23-39(c).

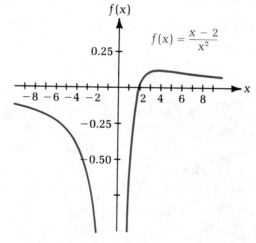

Figure 23-39(c)

Vertical tangent   The graph of a function $f(x)$ is said to have a **vertical tangent line** at some point x_0 if $f(x)$ is continuous at x_0 and if the derivative $f'(x) \rightarrow \pm\infty$ as $x \rightarrow x_0$. Examples of curves having a vertical tangent are shown in Fig. 23-40.

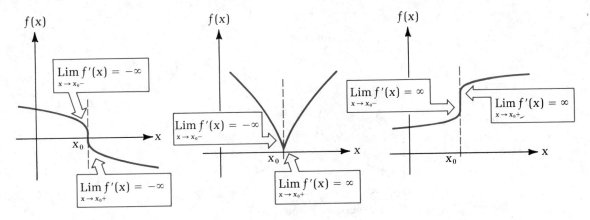

Figure 23-40

EXAMPLE 27   Find the vertical tangent line and sketch the graph of $f(x) = (x - 1)^{1/3}$.

Solution   Find the first derivative:

$$f'(x) = \frac{1}{3}(x - 1)^{-2/3} = \frac{1}{3(x - 1)^{2/3}}$$

As $x \to 1$, $f'(x) \to \infty$; therefore, the graph has a vertical tangent line at $x = 1$. Complete the other steps for graphing a rational function to get Fig. 23-41.

Figure 23-41

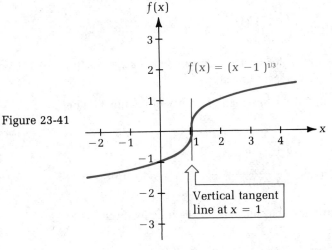

Cusp   The graph of a function $f(x)$ is said to have a **cusp** at some point $x = x_0$ if:

1. It has a vertical tangent line at x_0.
2. The concavity is the same on both sides of x_0.
3. The function is increasing in direction on one side of x_0 and decreasing on the other side (Fig. 23-42).

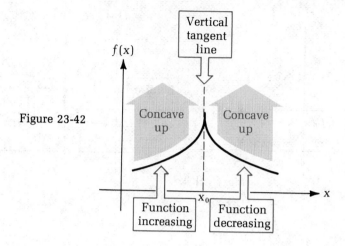

Figure 23-42

EXAMPLE 28 Locate the vertical tangent line and cusp, and sketch the graph of the function $f(x) = (x - 2)^{2/3}$.

Solution First, find the derivatives:

$$f'(x) = \frac{2}{3}(x - 2)^{-1/3} = \frac{2}{3(x - 2)^{1/3}}$$

$$f''(x) = \left(-\frac{1}{3}\right)\left(\frac{2}{3}\right)(x - 2)^{(-1/3)-1} = -\frac{2}{9(x - 2)^{4/3}}$$

Second, test for a vertical tangent line. As $x \to 2$, $f'(x) \to -\infty$. The function has a vertical tangent line at $x = 2$.

Third, test the direction of change of the function near $x = 2$:

- $f'(1) = \dfrac{2}{3(1 - 2)^{1/3}} = -\dfrac{2}{3}$ $f(x)$ is decreasing for $x < 2$.

- $f'(3) = \dfrac{2}{3(3 - 2)^{1/3}} = \dfrac{2}{3}$ $f(x)$ is increasing for $x > 2$.

Fourth, test the concavity of the function near $x = 2$:

- $f''(1) = -\dfrac{2}{9(1 - 2)^{4/3}} = -\dfrac{2}{9}$ $f(x)$ is concave down for $x < 2$.

- $f''(3) = -\dfrac{2}{9(3 - 2)^{4/3}} = -\dfrac{2}{9}$ $f(x)$ is concave down for $x > 2$.

The graph of $f(x)$ has a cusp at $x = 2$ (Fig. 23-43).

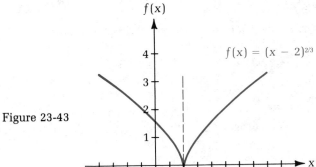

Figure 23-43

Vertical tangent line at x = 2

Summary

	Negative	Zero	Positive	Not Defined
$f(x)$	Curve is below the x-axis	x-intercept	Curve is above the x-axis	Discontinuous; possible vertical asymptote
$f'(x)$	Decreasing function	Critical value; possible maximum, minimum, or inflection point	Increasing function	Possible vertical tangent line, cusp, or asymptote
$f''(x)$	Concave down	Possible inflection point	Concave up	Possible vertical tangent line, inflection point, or asymptote

Exercises 23-5

Sketch the graph of each function.

1. $y = \dfrac{3}{x}$

2. $y = -\dfrac{1}{x^2}$

3. $y = \dfrac{x - 3}{x - 2}$

4. $y = \dfrac{x + 4}{x - 1}$

5. $y = \dfrac{x^2}{x^2 - 4}$

6. $y = \dfrac{x - 1}{x^2}$

7. $y = (x - 1)^{2/3}$

8. $y = (x - 3)^{1/3}$

9. $y = \dfrac{x - 1}{x^2 + 2}$

10. $y = \dfrac{6}{x^2} - \dfrac{6}{x}$

11. $P = \dfrac{3t^3 - t^2 + 5t}{t^3 + 10}$

12. $Q = \dfrac{t - 4}{t^2 - 10t + 27}$

Solve.

13. *Physics* The potential energy of a spherically symmetric nuclear system has the form

$$U(r) = \frac{1}{R} - \frac{1}{R^3} \qquad r > 0$$

Sketch the graph of this function.

14. *Physics* The amplification A of a vibrating mechanical system depends on the frequency ω of the external driving force and ω_0, the natural frequency of the system, according to the formula

$$A(\omega) = \frac{1}{\sqrt{m^2(\omega_0^2 - \omega^2)^2 + c^2\omega^2}}$$

where m is the virtual mass of the vibrating system and c is the damping constant. Sketch the shape of this curve with $m = 1$ and $\omega_0 = 1$ for:

(a) $c = \frac{1}{2}$ (b) $c = 1$ (c) $c = 2$

15. *Electronics* The power P (in watts) dissipated in a small electric motor is related to the current i (in amperes) drawn by the motor according to the equation

$$P(i) = \frac{2i^2 + 4}{0.5i^2 + i + 2}$$

Sketch the graph of this function if $i \geq 0$.

16. *Electronics* The magnetic field B due to a certain configuration of current loops is given by the equation

$$B(x) = \frac{2x}{(1 + x^2)^{3/2}}$$

Sketch the graph of this function.

23-6 | Applied Maximum/Minimum Problems

An important application of calculus to science and technology involves finding the optimal value of a quantity, and this is often equivalent to determining where the maximum or minimum value of a function occurs. For example, in physics:

- A light ray traveling through an optical medium takes the path that requires minimum time compared to neighboring paths.
- Any surface such as a soap bubble will contract under tension until it assumes the minimum area consistent with its boundary and the pressure difference across the surface.
- A collection of electrically charged particles will arrange themselves on a surface so as to minimize the electrical potential energy of the system.

If the quantity being optimized can be stated as a polynomial or rational function whose maximum or minimum value is to be found, we can use the techniques of Section 23-4 to find the optimal value.

EXAMPLE 29 The information transmission rate I (in bits per second) for a communication link is found to be a function of the signal speed v (in kilometers per second). If $I = 2000v - 400v^2$, where $v < 5$ km/s, at what speed should signals be sent to achieve maximum information throughput?

Solution **First,** find the critical value for the function.
$$\frac{dI}{dv} = 2000 - 800v$$

At $\frac{dI}{dv} = 0$,
$$2000 - 800v = 0$$
$$v = 2.5 \text{ km/s}$$

Second, use the first or second derivative test to determine whether the function has a maximum or minimum value at this critical point.

$$\frac{d^2I}{dv^2} = -800$$

Since the second derivative is negative, the critical point is at a maximum of the function. Send signals at 2.5 km/s for maximum information flow.

When finding a maximum or minimum, we must be careful to distinguish between local and absolute maxima or minima. In Example 29, the value of I at the endpoints $v = 0$ and $v = 5$ is less than the value at $v = 2.5$: $I(0) = 0$ and $I(5) = 0$. Therefore, the value of I at $v = 2.5$ is an absolute maximum.

EXAMPLE 30 The voltage drop V (in millivolts) across a circuit changes according to the equation

$$V = t^3 - 9t^2 + 24t$$

where t is time (in seconds) and $0 \le t \le 6$ s. Find the maximum value of the voltage.

Solution **First,** find the critical values.
$$\frac{dV}{dt} = 3t^2 - 18t + 24$$

At $\frac{dV}{dt} = 0$
$$3t^2 - 18t + 24 = 0$$
$$t^2 - 6t + 8 = 0$$
$$(t - 2)(t - 4) = 0$$
$$t = 2 \qquad t = 4$$

The critical values are $t = 2$ s and $t = 4$ s.

Second, use the second derivative test to check both critical points:

$$\frac{d^2V}{dt^2} = 6t - 18$$

At $t = 2$ s,

$$\frac{d^2V}{dt^2} = 6(2) - 18 = -6$$

The second derivative is negative; hence, the function has a local maximum at $t = 2$ s. The maximum voltage at this point is

$$V(2) = (2)^3 - 9(2)^2 + 24(2) = 20 \text{ mV}$$

At $t = 4$ s,

$$\frac{d^2V}{dt^2} = 6(4) - 18 = 6$$

The second derivative is positive; hence, the function has a local minimum at $t = 4$ s.

Third, check the value of the function at the endpoints of its domain interval to see if the local maximum at $t = 2$ s is also an absolute maximum:

$$V(0) = 0$$

$$V(6) = (6)^3 - 9(6)^2 + 24(6)$$

$$= 36 \text{ mV} \qquad \text{The maximum voltage is 36 mV.}$$

The absolute maximum value of this function over the given interval occurs at an endpoint of the interval rather than at a critical point (Fig. 23-44).

Figure 23-44

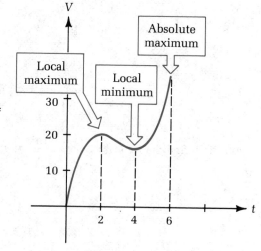

In most technical maximum/minimum problems, the function whose maximum or minimum is to be found is not given, but must be obtained from the information in the problem.

■ **EXAMPLE 31** A cylindrical container must have a volume of 60.0 ft³. What must be its height and the radius of its base if the amount of sheet metal used to make it is to be a minimum?

Solution The surface area of the container, including a flat circular top and bottom, is

$$A = 2\pi rh + 2\pi r^2$$

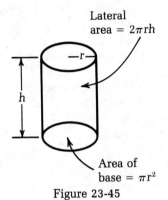

Lateral area = $2\pi rh$

Area of base = πr^2

Figure 23-45

The volume of the cylinder is

$$V = \pi r^2 h = 60.0 \text{ ft}^3$$

First, to obtain an equation for A as a function of r only, solve the volume equation for h.

$$h = \frac{60.0}{\pi r^2}$$

Now, substitute this expression for h into the area equation.

$$A = 2\pi r\left(\frac{60.0}{\pi r^2}\right) + 2\pi r^2$$

$$A = \frac{120.0}{r} + 2\pi r^2$$

Next, find the first derivative, and use it to locate the minimum.

$$\frac{dA}{dr} = -\frac{120.0}{r^2} + 4\pi r$$

For a minimum, $\dfrac{dA}{dr} = 0$:

$$-\frac{120.0}{r^2} + 4\pi r = 0$$

$$4\pi r = \frac{120.0}{r^2}$$

$$r^3 = \frac{120.0}{4\pi} = \frac{30.0}{\pi}$$

$$30 \boxed{\div} \boxed{\pi} \boxed{=} \boxed{\sqrt[x]{y}} 3 \boxed{=} \boxed{\text{STO}} \rightarrow \boxed{2.1215688}$$

$$r \approx 2.12 \text{ ft}$$

At r ≈ 2.12 ft.

$$h = \frac{60.0}{\pi r^2}$$

$$60 \boxed{\div} \boxed{\pi} \boxed{\div} \boxed{\text{RCL}} \boxed{x^2} \boxed{=} \rightarrow \boxed{4.2431377}$$

$$h \approx 4.24 \text{ ft}$$

Then, test this critical point using the second derivative.

$$\frac{d^2A}{dr^2} = \frac{240}{r^3} + 4\pi$$

At r ≈ 2.12

$$\frac{d^2A}{dr^2} \approx \frac{240}{(2.12)^3} + 4\pi$$

Positive

Since the second derivative is positive, we know that the function has a local minimum at r ≈ 2.12 ft. To minimize the amount of sheet metal used, r ≈ 2.12 ft and h ≈ 4.24 ft.

General Procedure for Solving Applications

To solve applied maximum/minimum problems, follow these steps.

Step 1.   Read the problem carefully. Identify the quantity to be optimized, and determine whether it is to be a maximum or a minimum.

Step 2.   Use the information given in the problem to express the variable to be optimized as a function of one other variable. If this quantity is given as a function of two other variables, then you must find another relationship that can be used to eliminate one of the variables.

Step 3.   Find the critical values by setting $df/dx = 0$.

Step 4.   Test the critical values for a maximum or a minimum using the second derivative test, if possible.

Step 5.   From the problem statement, find the interval over which the variable has physically meaningful values, and test the endpoints of this interval to determine if an absolute maximum or minimum occurs there.

Step 6.   Check your answer by rereading the original problem to be certain your answer is reasonable and answers the question posed.

EXAMPLE 32   The strength S of a rectangular beam depends on its width w and thickness t according to the equation

$$S = kwt^2$$

where k is a positive constant dependent on the material of the beam. What would be the dimensions of the strongest beam that could be cut from a cylindrical log of diameter 10 cm?

Solution

Step 1.   We want to find values of w and t that give a maximum for S.

Step 2.   Using the Pythagorean theorem (see Fig. 23-46),

$$t^2 + w^2 = D^2 = 100$$

$$t^2 = 100 - w^2$$

Therefore,

$$S = kw(100 - w^2)$$

$$S = 100kw - kw^3$$

Step 3.   Find dS/dw and set it equal to zero to find the critical values.

$$\frac{dS}{dw} = 100k - 3kw^2 = 0$$

$$w^2 = \frac{100}{3}$$

$$w = \sqrt{\frac{100}{3}} = \frac{10}{\sqrt{3}}$$

10 ⊡ 3 √ ⊟ STO → **5.7735027**

$$w \approx 5.77 \text{ cm}$$

At $w \approx 5.77$ cm,

$$t = \sqrt{100 - w^2}$$

100 ⊟ RCL x² ⊟ √ → **8.1649658**

$$t \approx 8.16 \text{ cm}$$

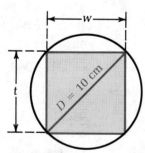

Figure 23-46

Step 4. Since k and w are both positive, the second derivative is negative, and we know that the critical value gives a maximum. $\qquad \dfrac{d^2S}{dw^2} = -6kw$

Step 5. The width of the beam must be within the limits $0 < w < 10$. At both $w = 0$ and $w = 10$, the value of S is zero, so the maximum value of S must occur inside the interval.

Step 6. The optimal dimensions of the beam are approximately 5.8 cm by 8.2 cm.

EXAMPLE 33   An industrial designer wants to construct an open-top box from a single 16 in. by 20 in. sheet of cardboard by cutting squares from the corners, folding as shown in Fig. 23-47, and taping the corners. What length of corner cut will give the maximum volume for the box? (Round to the nearest 0.1 in.)

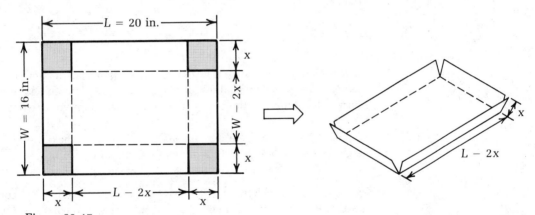

Figure 23-47

Solution

Step 1.   We need to write the volume of the completed box as a function of the length of the cut-out corner square, and then find the maximum volume.

Step 2.   Express the volume V in terms of the variable x.

$$V = (x)(L - 2x)(w - 2x)$$

Height | Length of base | Width of base

$$V = x(20 - 2x)(16 - 2x)$$

Multiply.

$$V = 4x^3 - 72x^2 + 320x$$

Step 3.

$$\frac{dV}{dx} = 12x^2 - 144x + 320$$

$$= 4(3x^2 - 36x + 80)$$

For a maximum value, $dV/dx = 0$.
Therefore, solve

$$3x^2 - 36x + 80 = 0$$

Use the quadratic formula.

$$x = \frac{36 \pm \sqrt{(-36)^2 - 4 \cdot 3 \cdot 80}}{2 \cdot 3}$$

$$x = \frac{36 \pm \sqrt{336}}{6} = \frac{36 \pm 4\sqrt{21}}{6}$$

$$x = \frac{18 \pm 2\sqrt{21}}{3}$$

$$x \approx 9.1 \text{ in.} \quad \text{or} \quad x \approx 2.9 \text{ in.}$$

The first value of x is *not* an acceptable solution since it makes the width of the base of the box a negative number.

Step 4. Use the second derivative test to determine if $x \approx 2.9$ in. will produce a maximum value for the volume.

$$\frac{d^2V}{dx^2} = 4(6x - 36)$$

At $x \approx 2.9$,

$$\frac{d^2V}{dx^2} \approx 4[6(2.9) - 36] \approx -74.4$$

Since the second derivative is negative, the function has a maximum at $x \approx 2.9$.
Step 5. The acceptable values for x are $0 < x < 8$. At both endpoints of this interval the volume of the box is zero, so the maximum volume occurs within the interval.
Step 6. The approximate dimensions of the box with maximum volume are: height ≈ 2.9 in., length ≈ 14.2 in., width ≈ 10.2 in.

EXAMPLE 34 As shown in Fig. 23-48, an electrical cable must connect the tower at P with a second tower Q across the river, then connect points Q and R over land. If it costs $20 per linear foot for cable over the river and $10 per foot over land, where should tower Q be placed in order to minimize the cost?

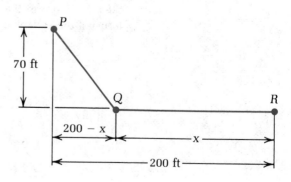

Figure 23-48

Solution

Step 1. If x is the distance from tower Q to point R, the problem is to find the value of x that makes cost a minimum.

Step 2. Write the cost as a function of x. From the Pythagorean theorem (see Fig. 23-48),

Distance $PQ = \sqrt{70^2 + (200 - x)^2}$

Then,

Cost of cable from P to $Q = 20\sqrt{70^2 + (200 - x)^2}$

Cost of cable from Q to $R = 10x$

Total cost = $\qquad C = 10x + 20\sqrt{70^2 + (200 - x)^2}$

$$= 10x + 20[70^2 + (200 - x)^2]^{1/2}$$

Step 3. Find the critical value for this function. Differentiate using the sum rule (22-15) and the extended power rule (22-22):

$$\frac{dC}{dx} = 10 + 20\left(\frac{1}{2}\right)[70^2 + (200 - x)^2]^{-1/2} \cdot \frac{d}{dx}(200 - x)^2$$

$$= 10 + \frac{10 \cdot 2(200 - x)(-1)}{\sqrt{70^2 + (200 - x)^2}}$$

$$= 10 - \frac{20(200 - x)}{\sqrt{70^2 + (200 - x)^2}}$$

$$= \frac{10\sqrt{70^2 + (200 - x)^2} - 20(200 - x)}{\sqrt{70^2 + (200 - x)^2}}$$

Set $\dfrac{dC}{dx} = 0$. Then,

$$10\sqrt{70^2 + (200 - x)^2} - 20(200 - x) = 0$$

$$\sqrt{70^2 + (200 - x)^2} = 2(200 - x)$$

Square both sides. $\qquad\qquad 70^2 + (200 - x)^2 = 4(200 - x)^2$

$$70^2 = 3(200 - x)^2$$

Take the square root of both sides of the equation. $\qquad 70 = \sqrt{3}(200 - x)$

Solve for x. $\qquad\qquad \dfrac{70}{\sqrt{3}} = 200 - x$

$$x = 200 - \frac{70}{\sqrt{3}}$$

$$200 \;\boxed{-}\; 70 \;\boxed{\div}\; 3 \;\boxed{\sqrt{}}\;\boxed{=} \rightarrow \quad \boxed{159.58548}$$

$$x \approx 160 \text{ ft}$$

Step 4. Because finding the second derivative d^2C/dx^2 is a bit involved, we can use the first derivative test to check that this critical point is a minimum. Test dC/dx at points greater than and less than $x \approx 160$ ft:

- At $x = 150$ ft, $\quad \dfrac{dC}{dx} = 10 - \dfrac{20(200 - 150)}{\sqrt{70^2 + (200 - 150)^2}}$

$$= 10 - \dfrac{1000}{\sqrt{4900 + 2500}}$$

$$10 \;\boxed{-}\; 1000 \;\boxed{\div}\;\boxed{(}\; 4900 \;\boxed{+}\; 2500 \;\boxed{)}\;\boxed{\sqrt{}}\;\boxed{=}\; \rightarrow \quad \boxed{-1.6247639}$$

- At $x = 170$ ft, $\quad \dfrac{dC}{dx} = 10 - \dfrac{20(200 - 170)}{\sqrt{70^2 + (200 - 170)^2}}$

$$= 10 - \dfrac{600}{\sqrt{4900 + 900}}$$

$$10 \;\boxed{-}\; 600 \;\boxed{\div}\;\boxed{(}\; 4900 \;\boxed{+}\; 900 \;\boxed{)}\;\boxed{\sqrt{}}\;\boxed{=}\; \rightarrow \quad \boxed{2.1216140}$$

The sign of the first derivative changes from negative to positive; therefore, the critical point is a minimum.

Step 5.   The distance x must satisfy the restriction $0 \le x \le 200$ ft. Test the endpoints of this interval, and find C at the critical point:

- At $x = 0$ ft, $\qquad C = 20\sqrt{70^2 + 200^2} \approx \$4200$
- At $x = 200$ ft, $\quad C = 10(200) + 20\sqrt{70^2} = \$3400$
- At $x \approx 160$ ft, $\quad C = 10(160) + 20\sqrt{70^2 + (200 - 160)^2} \approx \$3200$

At the critical point, the function has its minimum value over the interval $0 \le x \le 200$ ft.

Step 6.   Place tower Q about 160 ft from point R.

Exercises 23-6

1. *Electronics*   The vibration frequency f (in kilohertz) of a quartz crystal oscillator depends upon its thickness t (in micrometers) according to the equation $f = 5t^2 - 4t + 384$. For what crystal thickness will the vibration frequency be a minimum?
2. *Electronics*   According to the manufacturer's specifications, a power supply produces a heat output (in kilowatt-hours) of $H = 3V - 2V^2 + 416$, where V is the voltage. For what voltage will the heat output be a maximum?
3. What positive number exceeds its square by the greatest amount?
4. Find two positive numbers whose sum is 30 such that:
   (a)   Their product is a maximum.
   (b)   The sum of their squares is a maximum.
   (c)   The sum of their cubes is a maximum.
   (d)   The sum of their nth powers is a maximum, for n an integer > 1.
5. *Chemistry*   In an autocatalytic chemical reaction a substance is formed that causes an increase in its rate of formation. The reaction rate R is related to the amount x of the substance present at any time according to the equation $R = kx(x_0 - x)$, where k is a constant and x_0 is the initial value of x. If $k = 6.2$ and $x_0 = 25.2$ g, for what value of x will the reaction rate R be a maximum?

6. *Physics* The specific density σ of water at temperature T is given by the equation

$$\sigma = 1 + aT + bT^2 + cT^3 \qquad 0 < T < 100°C$$

where $a = 5.30 \times 10^{-5}$, $b = -6.53 \times 10^{-6}$, and $c = 1.40 \times 10^{-8}$. At what temperature does water have the maximum specific density?

7. *Industrial engineering* The cost of producing n computer components per day is $C = 65 + 3.5n + 0.08n^2$, where \$65 is the daily start-up cost, \$3.50 is the cost of materials and labor for each unit, and \$0.08 is the cost of equipment deterioration. If each unit sells for \$49.95, find the value of n that will maximize profits. [*Hint:* Profit $= P = 49.95n - C$.]

8. *Electronics* The electromotive force V induced in a current loop through which a magnetic flux Φ flows is given by the equation $V = -d\Phi/dt$. If the field varies with time so that $\Phi = 8t^3 - 20t^2 - t$, at what time will the electromotive force be a minimum?

9. At what point is the slope of the tangent to the curve $y = x^3 + x^2 - 5x$ a minimum (for $x \geq 0$)?

10. *Industrial design* An open-top box must be constructed from a single 20 in. by 32 in. sheet of cardboard by cutting squares from the corners, folding the sides up, and taping the corners. What length of corner cut will give the maximum volume? (Round to the nearest 0.1 in.)

11. *Industrial design* A utility box must be constructed from a single 24 in. by 40 in. piece of sheet metal by cutting squares from the corners, bending up the sides, and spot welding the edges. What length of corner cut will give the maximum volume? (Round to the nearest 0.1 in.)

12. *Industrial design* Determine the value of x that gives the maximum volume for the self-holding open box shown.

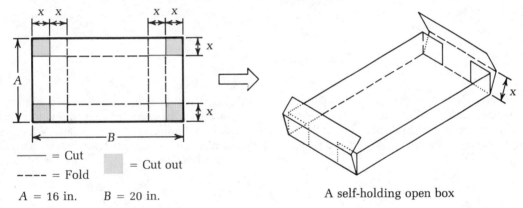

——— = Cut
---- = Fold
▦ = Cut out

$A = 16$ in. $B = 20$ in.

A self-holding open box

13. *Industrial design* What value of x will produce maximum volume for the self-holding box with a folding lid shown?

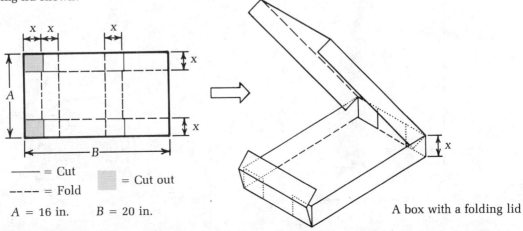

——— = Cut
---- = Fold
▦ = Cut out

$A = 16$ in. $B = 20$ in.

A box with a folding lid

14. *Construction technology* A construction worker wishes to fence a rectangular equipment yard border-ing along a river bank. Assuming that she does not fence along the river, and that 480 yd of fencing is available, what is the largest area that can be enclosed?

15. *Energy technology* A rectangular solar collector plate with area 63 cm² must be constructed with a 2.0 cm margin at the top, 1.5 cm at the bottom, and 1.0 cm on each side. What should be the overall dimensions of the rectangle in order to maximize the central area?

16. *Industrial design* A cylindrical tank with capacity 5250 ft³ must be constructed with flat circular ends. If the side material costs \$2.29/ft², and the ends are made of a heavier material costing \$3.74/ft², what dimensions should the tank have to minimize the cost of materials?

17. *Industrial design* In Problem 16, what dimensions for the tank would minimize the cost of materials if the ends are hemispherical caps?

18. *Industrial design* What are the dimensions of the right circular cone of maximum volume that can be inscribed in a sphere of radius 6 cm? [*Hint:* In triangle ABC, $AB = h - r$, $AC = r$, $BC = b/2$, and angle ABC is a right angle. The volume of the cone is $\frac{1}{3} \cdot$ Area of base $\cdot$ Height.]

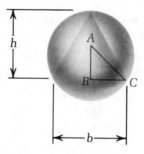

19. *Industrial design* A right circular cylinder is inscribed in a cone as shown. Find the max-imum volume this cylinder can have if the cone has altitude 12 in. and base diameter 6 in.

20. *Construction technology* The bending strength of a wooden beam is proportional to its width and to the square of its thickness. What are the dimensions of the strongest beam that can be cut from a circular log of diameter 16 in.?

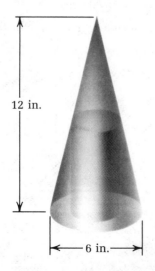

12 in.

6 in.

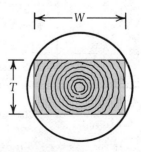

W

T

21. *Construction technology*   If the beam in Problem 20 must be cut from a log whose cross section is the ellipse

$$\frac{x^2}{64} + \frac{y^2}{36} = 1$$

what are the dimensions of the strongest beam that can be cut? [*Hint:* For a coordinate system with origin at the center of the rectangle, $x = w/2$ and $y = T/2$ are the coordinates of point A.]

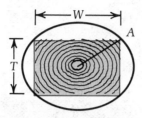

22. Find the coordinates of the point on the curve $xy^2 = 16$ that is closest to the origin. [*Hint:* If D is the distance from the origin to the point (x, y), find the minimum value of D^2.]

23. What point on the curve $y = \sqrt{x + 1}$ is closest to the point $(3.5, 0)$?

24. *Electronics*   If two identical electrolytic cells, each with internal resistance r and each producing electromotive force V, are connected in series with a resistance R, the power dissipated in the resistance R is

$$P = \frac{4V^2R}{(2r + R)^2}$$

For what value of R will the power dissipated be a maximum?

25. *Electronics*   If the two cells in Problem 24 are connected in parallel across the resistance R, the power dissipated in R is

$$P = \frac{V^2(r + 2R)^2}{Rr^2}$$

For what value of R will the power dissipated be a maximum?

26. What is the shortest distance from the point $(-2, 3)$ to the line $3x + 4y + 4 = 0$?

27. Find the shortest distance from the point $(1, -2)$ to the line $3x - y + 5 = 0$.

28. *Electronics*   The electric field E due to a certain charge configuration depends on the radial distance from the center of charge according to the equation

$$E(r) = r^2 + \frac{2r^2}{(r - 2)} \qquad r > 2$$

Where will the field have a minimum value?

29. *Construction technology*   A cable must be strung from point A to some point C on the wall and then back to point B. Where should point C be located in order to minimize the length of cable?

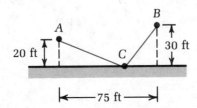

30. *Physics*   A particle travels from point A to point B, touching the surface at C en route, as shown. What value of x will result in the minimum distance of travel from A to B?

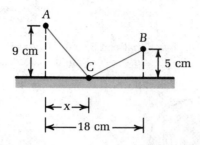

23-7 | Review of Chapter 23

Important Terms and Concepts

tangent line (p. 876)

normal line (p. 878)

first derivative (p. 881)

second derivative (p. 881)

order of a derivative (p. 881)

particle (p. 884)

rectilinear motion (p. 884)

displacement (p. 884)

instantaneous velocity (p. 884)

acceleration (p. 886)

constant acceleration (p. 888)

free fall (p. 888)

curvilinear motion (p. 889)

parametric equations/parameter (p. 890)

related rates (p. 895)

increasing and decreasing functions (p. 904)

critical values (p. 906)

critical points (p. 906)

absolute maximum/minimum (p. 907)

relative or local maximum/ minimum (p. 907)

extreme values (p. 907)

first derivative test (p. 908)

concavity (p. 912)

concavity test (p. 913)

point of inflection (p. 914)

second derivative test (p. 917)

vertical asymptote (p. 923)

horizontal asymptote (p. 923)

vertical tangent (p. 928)

cusp (p. 929)

Formulas and Rules

- tangent line equation (p. 876)

$$y - y_1 = m_T(x - x_1) \qquad \text{where } m_T = \left.\frac{dy}{dx}\right|_{x=x_1} \qquad \textbf{(23-1)}$$

- normal line equation (p. 878)

$$y - y_1 = m_N(x - x_1) \qquad \text{where } m_N = -\frac{1}{m_T} \quad (m_T \neq 0) \quad \textbf{(23-2)}$$

- instantaneous velocity (p. 884)

$$v = \frac{dx}{dt} \qquad \text{where } x \text{ is the displacement and } t \text{ is time} \qquad \textbf{(23-3)}$$

- acceleration (p. 886)

$$a = \frac{dv}{dt} \qquad \text{where } v \text{ is the velocity and } t \text{ is time} \qquad \textbf{(23-4)}$$

- free fall motion (p. 888)

$$y = -\frac{1}{2}gt^2 + v_0 t + y_0 \qquad \textbf{(23-5)}$$

- instantaneous velocity in free fall (p. 888)

$$v = \frac{dy}{dt} = -gt + v_0 \qquad \textbf{(23-6)}$$

- curvilinear motion equations
  components of velocity (p. 890)

$$v_x = \frac{dx}{dt} \qquad \textbf{(23-7)}$$

$$v_y = \frac{dy}{dt} \qquad \textbf{(23-8)}$$

magnitude of the velocity (p. 890)

$$v = \sqrt{v_x^2 + v_y^2} \qquad \textbf{(23-9)}$$

direction of the velocity (p. 890)

$$\text{Ref } \theta = \text{Arctan} \left| \frac{v_y}{v_x} \right| \qquad \textbf{(23-10)}$$

- procedure for graphing a polynomial function: see p. 919
- procedure for graphing a rational function: see p. 925
- summary of curve-sketching tests: see p. 931
- procedure for solving applied maximum/minimum problems: see p. 936

Exercises 23-7

Find the equations of the tangent and normal lines to each curve at the given point:

1. $y = 2x - x^2$ at (2, 0) **2.** $y = x\sqrt{x}$ at (4, 8)

3. $x^2 + y^2 = 10$ at (1, 3) **4.** $y = \dfrac{x}{x - 1}$ at (2, 2)

5. Physics If an isolated conducting sphere of radius R carries a charge q, then the energy density in the space surrounding the sphere at any radius r (r > R) is

$$u = \frac{q^2}{32\pi^2 \varepsilon_0 r^4}$$

where ε_0 is a constant. Find an expression for $\dfrac{d^2u}{dr^2}$.

Each of the following equations describes the motion of a particle moving along the x-axis. Find the velocity and acceleration of the particle at the time indicated (x is in meters).

6. $x = 3t^2 - 4t + 2$; $t = 2$ s **7.** $x = 4t^3 + t^2 - 6t - 5$; $t = 0$ s

8. $x = 3t + \dfrac{4}{t}; \quad t = 3$ s

9. $x = \dfrac{t - 2}{t + 3}; \quad t = 1$ s

10. *Electronics*   An electron moves in a straight-line path in an electric field according to the equation

$$s = \frac{2t^3}{t^2 + 1}$$

where s is in meters and t is in seconds. Find the velocity and acceleration of the particle when $t = 2.0$ s.

Each of the following pairs of parametric equations describes the motion of an object along a curved path. For each, find the magnitude and direction of the velocity and acceleration at the indicated time.

11. $x = 45t$
  $y = -16t^2 + 12t$ at $t = 5.0$ s

12. $x = \dfrac{10}{t + 2}$
  $y = 4\sqrt{t}$ at $t = 1.0$ s

13. *Physics*   An object moves according to the parametric equations $x = 190t$ and $y = 192t - 16t^2$, where x and y are in feet and t is in seconds.
  (a)   Find the magnitude and direction of the velocity at $t = 5.0$ s.
  (b)   Find the magnitude and direction of the acceleration at $t = 5.0$ s.

14. *Electronics*   In a cathode ray tube, a beam of electrons is projected horizontally with a speed of 1.5×10^9 cm/s into the electrical field between two charged plates. If the horizontal velocity is kept constant and the vertical motion satisfies the equation $y = \sqrt{1 + 2t} \times 10^{10}$ cm, find the magnitude and direction of the velocity of the beam at $t = 1.2$ s.

15. *Physics*   On the opening kickoff of a football game, the ball travels on a path given by the equation

$$y = x - \frac{x^2}{82}$$

where x and y are in yards.
  (a)   At what value of x will its vertical component of velocity be zero?
  (b)   What will be the height of the ball at this point?
  (c)   If the horizontal component of velocity is always constant at $v_x = 2.0$ yd/s, what will be the magnitude and direction of the velocity of the ball when it lands?

16. *Physics*   A projectile is launched vertically from a point 35 ft above ground level with initial velocity 75 ft/s.
  (a)   How high will it rise above ground level?
  (b)   When will it reach a point 115 ft above ground level?
  (c)   What velocity will it have when it passes a point 50 ft above its starting point?

17. *Physics*   A particle moves along the path $y = 1 + 3x + x^3$ so that its horizontal component of velocity is 2 cm/s. Find the magnitude and direction of the velocity of the object when it is at the point (1, 5).

18. *Physics*   An object moves along the path $y^2 = 2x + 4$ with a vertical component of velocity of constant magnitude 16 ft/s. Find the magnitude and direction of the velocity when the object reaches the point (6, 4).

19. *Physics*   A pebble dropped in a pond forms circular ripples. If the radius of a ripple is increasing at 4.5 cm/s, how fast is the area of the ripple increasing when the ripple has a radius of 35 cm?

20. *Petroleum engineering*   A petroleum geologist testing rock strata sets off an underground explosion. The shock wave spreads out spherically at about 11,440 ft/s.
  (a)   At what rate is the volume covered by this wave increasing when $r = 25,000$ ft?
  (b)   If the energy density is $E = k/V$, where $k = 10^6$ and V is the volume, at what rate is E changing?

21. *Physics*   An airplane is flying horizontally at an elevation of 14,000 ft with a speed of 180 mi/h. At what speed is it approaching the observer shown in the figure when it passes over a point exactly 3 mi from the observer?

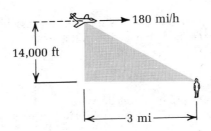

180 mi/h

14,000 ft

3 mi

22. A nocturnal jogger 5.0 ft tall is running away from a lamppost 16 ft tall. If she runs at 12 ft/s, at what rate is the tip of her shadow moving when she is 30 ft from the lamppost?

23. *Navigation*   The radar scanner shown indicates that the plane is moving at 325 mi/h. If the plane is 6.45 mi from the antenna and 2.15 mi above the ground (assumed level), what is its actual speed?

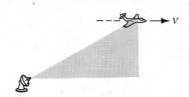

v

24. *Electronics*   The resistance of a certain conducting rod depends on its temperature T according to the equation $R = 18.4 + 1.08T + 0.24T^2$, where R is in ohms. Find the rate at which the resistance is changing when $T = 23°C$ if the temperature is increasing at 0.25°C/s.

25. *Meteorology*   If a spherical weather balloon is being inflated at 0.24 ft³/s, how fast is its diameter increasing when it is 8.0 ft across?

26. Two runners start simultaneously at an intersection and run on perpendicular roads. If one travels west at 8.4 mi/h, and the other travels south at 6.5 mi/h, at what rate are they moving apart after 20 min?

27. In a right triangle with legs a and b, side a is decreasing at 2.4 in./s and side b is increasing at 3.2 in./s. At what rate is the area changing when $a = 6.5$ in. and $b = 4.5$ in?

28. Water is drained from an inverted cone-shaped tank at the rate of 4.5 ft³/min. If the diameter of the top of the cone is 24 ft and its height is 20 ft, at what rate is the level of water falling at the instant its depth is 15 ft?

For each function determine the values of x for which the graph:

(a) Is increasing or decreasing   (b) Is concave up or concave down
(c) Has a local maximum or minimum   (d) Has an inflection point

Sketch the graph of each function.

29. $y = 6x - x^2$ **30.** $y = x^2 + 2x - 1$
31. $y = 1 - x^3$ **32.** $y = 2x^3 - 4x^2$
33. $y = 20x^3 + 36x^2 + 7x$ **34.** $y = 2x^3 - 3x^2 - 72x$

Sketch the graph of each function.

35. $y = \dfrac{2}{x + 1}$ **36.** $y = \dfrac{x^2 + 2}{x}$ **37.** $y = \dfrac{2}{x} - \dfrac{4}{x^2}$

38. $y = (x + 2)^{1/3}$ **39.** $y = \dfrac{2x + 1}{x - 1}$ **40.** $y = (x + 3)^2(x - 2)^2$

41. $y = (x - 4)^{2/3}$ **42.** $y = 2x^5 - 10x^3$

43. *Mechanical engineering*   The total potential energy V (in joules) of a mechanical system is related to its system displacement x (in centimeters) by an equation of the form

$$V(x) = \frac{1}{4}x^4 - \frac{a}{3}x^3 - \frac{2}{3}a^2x^2$$

where a is a system constant. Sketch the graph of this curve for $a = 1$.*

44. *Hydraulics*   The hydrostatic potential V near an object with circular cross section is

$$V(r) = 2r + \frac{2}{r} - 1$$

Sketch the graph of this potential function for $r > 0$.

45. Find two positive numbers whose sum is 40 and whose product is a maximum.

46. What positive number exceeds its cube by the greatest amount?

47. What point on the curve $y^2 = 4x$ is closest to the point $(2, 0)$?

48. Show that the minimum value of the potential function

$$V = ar + \frac{b}{r}$$

for a and b positive real numbers and $r > 0$, occurs at $r = \sqrt{b/a}$.

49. *Electronics*   If the current in a certain circuit varies with time (in seconds) according to the equation

$$i = \frac{8(t - 2)}{t^2 + 4}$$

at what time is the current a maximum?

50. *Electronics*   An electrolytic cell has internal resistance r and produces an electromotive force V. When it is placed in series with a resistance R, the power dissipated in the resistance R is

$$P = \frac{V^2R}{(r + R)^2}$$

For what value of R will the power dissipated be a maximum?

51. *Physics*   If a ball is thrown vertically upward with initial speed 45 ft/s, then, neglecting air friction, its height (in feet) after time t is $h = 45t - 16t^2$. Find the time at which the ball reaches its maximum height.

52. What are the dimensions of the rectangle of maximum area that can be inscribed in a semicircle of radius 8 cm?

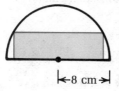

$\leftarrow$8 cm$\rightarrow$

53. What are the dimensions of the cylinder of maximum volume that can be inscribed in a sphere of radius 12 cm?

*See Nelson R. Bauld, Jr., *Mechanics of Materials*, Monterey, CA: Brooks/Cole, 1982), p. 493

54. Find the dimensions of the rectangle in the first quadrant with the largest possible area such that two sides are along the axes and one vertex is on the curve $y^2 = 4 - x$.

55. In order to travel from starting point A to the lighthouse at point C, an athlete jogs from A to some point B at 6 mi/h, then rows from B to C at 4 mi/h. Where should point B be located in order to make the trip in minimum time?

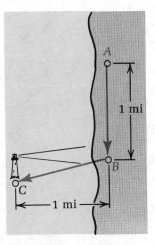

56. *Industrial design*   For the cereal box shown, what value of x will produce the maximum volume?

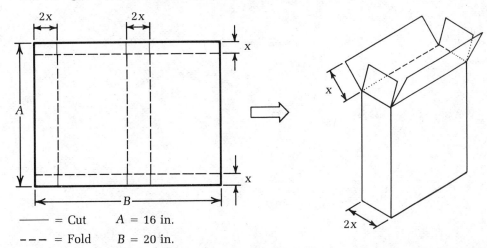

———— = Cut A = 16 in.

– – – = Fold B = 20 in.

57. The U.S. Postal Service requires that a package must satisfy the specifications Girth + Length ≤ 108 in. What are the dimensions of the rectangular box of largest volume that can be mailed if its length must be double its width?

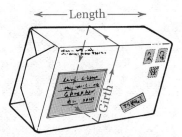

Integration

24

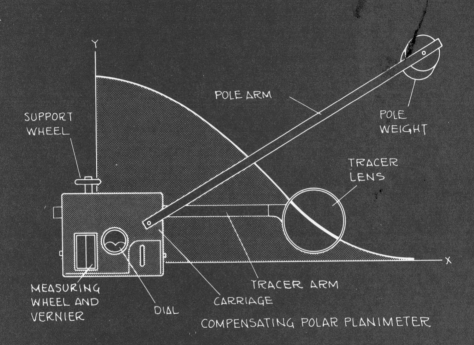

SUPPORT
WHEEL

POLE ARM

POLE
WEIGHT

TRACER
LENS

MEASURING
WHEEL AND
VERNIER

DIAL

CARRIAGE

TRACER ARM

COMPENSATING POLAR PLANIMETER

In the preceding chapters we were concerned with differential calculus, in which we found the derivative of a function and used it to solve applied problems. In this chapter we will examine integral calculus and the inverse problem: Given the derivative of a function, find the function itself. The concepts of integral calculus were developed historically before those of differential calculus and are extremely important in the study of physics and engineering.

First, we must develop some useful notation. When the derivative of a function $y = f(x)$ is written as dy/dx, there is a natural inclination to think of this symbol as a fraction or quotient.* Because it is sometimes useful in integral calculus to think of the derivative in this way, it is necessary to define the separate quantities dx and dy so that their quotient dy/dx is equal to $f'(x)$.

Differential   If $y = f(x)$ is a differentiable function, and if the independent variable x is changed an amount Δx, then the **differentials** dx and dy are defined as follows:

Definition of Differentials	$dx = \Delta x$
	$dy = f'(x)\, dx \qquad\qquad$ (24-1)

Notice that with this definition, if $\Delta x = dx \neq 0$, then the quotient

$$\frac{dy}{dx} = \frac{f'(x)\, dx}{dx} = f'(x)$$

We have chosen the definition of the differentials so that their quotient is equal to the derivative.

Graphically, Δx and Δy are increments defined for points P and Q on the curve [Fig. 24-1(a)]. The differential dy is defined with respect to the tangent line, with $dx = \Delta x$ [Fig. 24-1(b)].

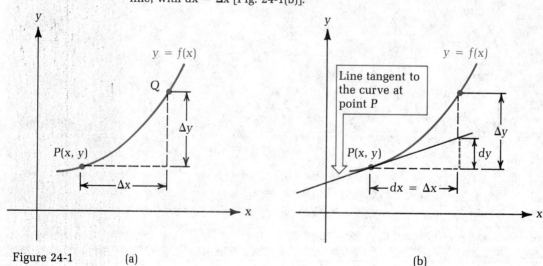

Figure 24-1   (a)   (b)

*In fact, Leibniz, who originally developed many of the concepts of calculus, used the symbols dy and dx and the word *differential* to refer to "infinitesimally small quantities." His critics called these quantities "ghosts."

For small values of Δx, the tangent line is close to the curve at point P. The increment Δy is the actual change in the value of y resulting from an increment Δx in variable x. The differential dy is the change in the height of the tangent line resulting from an increment Δx in variable x. Notice that when Δx is small, $\Delta y \approx dy$.

The slope of the line tangent to $y = f(x)$ at piont P is $f'(x)$, and from Fig. 24-1(b) we see that this is

$$\text{Slope} = \frac{dy}{\Delta x} = \frac{dy}{dx} = f'(x)$$

so that $dy = f'(x)\, dx$, which is the definition of the differential dy.

EXAMPLE 1   Find the differential of each of the following.

(a) $y = x^3 + 2x$ (b) $y = (3x^4 - 2)^3$ (c) $y = \dfrac{x^2}{2x + 1}$

Solution

(a)   For $f(x) = x^3 + 2x$, differentiation rules (22-9) and (22-11) give $f'(x) = 3x^2 + 2$ and definition (24-1) gives

$$dy = f'(x)\, dx$$
$$= (3x^2 + 2)\, dx$$

(b)   For $f(x) = (3x^4 - 2)^3$, the power rule (22-18) gives

$$f'(x) = 3(3x^4 - 2)^2(12x^2)$$
$$= 36x^2(3x^4 - 2)^2$$

and definition (24-1) gives

$$dy = 36x^2(3x^4 - 2)^2\, dx$$

(c)   For $f(x) = \dfrac{x^2}{2x + 1}$, the quotient rule (22-15) gives

$$f'(x) = \frac{(2x + 1)2x - x^2(2)}{(2x + 1)^2} = \frac{2x(x + 1)}{(2x + 1)^2}$$

so that the differential is

$$dy = \frac{2x(x + 1)}{(2x + 1)^2}\, dx$$

EXAMPLE 2   A projectile is launched so that it follows a path given by the equation

$$y = 20x - x^2$$

where $x > 0$ and x and y are measured in kilometers. Find:
(a) The increment in height Δy at $x = 3.0$ km for $\Delta x = 0.1$ km
(b) The differential dy at $x = 3.0$ km for $\Delta x = 0.1$ km

Solution

(a) Substituting $y + \Delta y$ for y and $x + \Delta x$ for x, we obtain

$$y + \Delta y = 20(x + \Delta x) - (x + \Delta x)^2$$

Subtract $y = 20x - x^2$:

$$\Delta y = 20x + 20\Delta x - x^2 - 2x\Delta x - (\Delta x)^2 - 20x + x^2$$

$$= 20\Delta x - 2x\Delta x - (\Delta x)^2$$

Substituting $x = 3.0$ and $\Delta x = 0.1$ into this equation, we get

$$\Delta y = 20(0.1) - 2(3.0)(0.1) - (0.1)^2$$

$$= 1.39 \text{ km}$$

This is the actual change in height as x increases from 3.0 km to 3.1 km.

(b) According to the definition (24-1),

$$dx = \Delta x = 0.1 \text{ km}$$

$$dy = f'(x)\, dx$$

But $f'(x) = 20 - 2x$, so

$$dy = (20 - 2x)\Delta x$$

Evaluating this expression,

$$dy = [20 - 2(3.0)](0.1)$$

$$= 1.4 \text{ km}$$

This is the estimated change in height as x increases from 3.0 km to 3.1 km (Fig. 24-2).

Figure 24-2

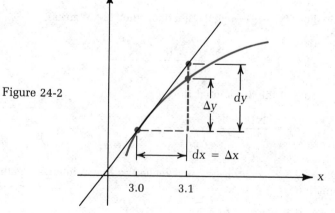

Table 24-1 shows that for Example 2, $dy \approx \Delta y$ for small values of the increment Δx.

TABLE 24-1 $y = 20x - x^2$ at x = 3.0	Δx	$dy = (20 - 2x)\Delta x$	$\Delta y = 20\Delta x - 2x\Delta x - (\Delta x)^2$	$dy - \Delta y$
	0.1	1.4	1.39	0.01
	0.01	0.14	0.1399	0.0001
	0.001	0.014	0.013999	0.000001
	0.0001	0.0014	0.00139999	0.00000001

Error　　In scientific and technical work, when the variable x is a measured quantity, the increment Δx may be interpreted as the **error** of measurement of x. If the value of y is calculated from $y = f(x)$, then Δy is the resulting error in the calculated value of y, and dy is an approximation to this calculated error.

Relative error　　Given the function $y = f(x)$, the **relative error** in y is defined as the ratio of the magnitude of the measurement error $|\Delta y|$ to the exact value of the measurement, y.

Relative Error

$$\text{Relative error} = \frac{|\Delta y|}{y} = \frac{|\text{Error in measurement}|}{\text{Exact value}} \qquad \textbf{(24-2)}$$

Because the exact value Δy is usually not known, the calculated value dy is used instead to find the relative error.

Percent error　　The **percent error** is the relative error expressed as a percent.

Percent Error　　$\text{Percent error} = (\text{Relative error} \times 100)\%$ 　　**(24-3)**

EXAMPLE 3　　The speed of an automobile is measured manually over a calibrated 400 m test track. The time is measured as 12.4 ± 0.1 s. Find:
(a) The average speed　　(b) The error in the average speed
(c) The relative error　　(d) The percent error.

Solution

(a) Average speed v is　　$v = \dfrac{\text{Distance}}{\text{Time}} = \dfrac{d}{t}$

Substituting:　　$v = \dfrac{400 \text{ m}}{12.4 \text{ s}} \approx 32.3 \text{ m/s}$

(b) The error in the measured time, $\Delta t = 0.1$ s, produces an error Δv in the average speed v. But for small values of Δt, $\Delta v \approx dv$. So,

$$\frac{dv}{dt} = -\frac{d}{t^2}$$

From the definition of the differential (24-1),

$$dv = \left(-\frac{d}{t^2}\right) dt$$

$$= \left(-\frac{d}{t^2}\right)\Delta t$$

Substituting, we obtain

$$dv = -\frac{(400 \text{ m})}{(12.4 \text{ s})^2} \cdot (0.1 \text{ s})$$

$$400 \;\boxed{+/-}\boxed{\times}\; .1 \;\boxed{\div}\; 12.4 \;\boxed{x^2}\;\boxed{=} \rightarrow \quad \boxed{-0.2601457}$$

$$dv \approx -0.3 \text{ m/s}$$

The negative sign indicates that the average speed decreases as the time increases.

(c)   From the definition of relative error (24-2),

$$\text{Relative error} = \frac{|\Delta v|}{v} = \frac{|-0.26 \text{ m/s}|}{32.3 \text{ m/s}} \approx 0.008$$

(d)   Percent error $= (0.008 \times 100)\%$

$$= 0.8\%$$

Round the percent error to the nearest tenth.
The speed of the automobile should be expressed as 32.3 ± 0.3 m/s or 32.3 m/s $\pm 0.8\%$.

NOTE ▶

- In the calculation of relative error, the units are the same in the numerator and denominator, so the relative error is a dimensionless ratio.
- Although we round dv to the same decimal place as v, we need to use the unrounded form in calculating the relative error.   ◀

▦ **EXAMPLE 4**   The frequency f of the tuning fork in an electronic timer is related to the length L of the fork by the equation

$$f = \frac{240}{\sqrt{L}}$$

where f is in millihertz and L is in millimeters. If the steel fork is exactly 5 mm long when calibrated at room temperature, and if a change in temperature of 20°C causes the fork to expand 0.002 mm, by how much will the frequency change? Write the answer as a percent error.

Solution **First,** find the value of the frequency f for $L = 5$ mm.

$$f = \frac{240}{\sqrt{5}} = 107.33126\ldots \text{ mHz}$$

Next, calculate df using the definition of a differential (24-1) with $\Delta L = 0.002$ mm.

$$f = 240L^{-1/2}$$

$$\frac{df}{dL} = -\frac{1}{2} \cdot 240 \cdot L^{-3/2}$$

$$= -\frac{120}{L^{3/2}}$$

Then,

$$df = -\frac{120}{L^{3/2}} \cdot dL$$

Substituting,

$$df = -\frac{120(0.002)}{(5)^{3/2}}$$

$$120 \; \boxed{+/-} \; \boxed{\times} \; .002 \; \boxed{\div} \; 5 \; \boxed{y^x} \; 1.5 \; \boxed{=} \rightarrow \boxed{-0.0214663}$$

$$df \approx -0.02 \text{ mHz}$$

The negative sign indicates that the frequency decreases as the length L increases.

Finally, calculate the relative error and use it to find the percent error:

$$\text{Relative error} = \frac{|df|}{f} = \frac{0.0214663}{107.33126} \approx 0.0002$$

$$\text{Percent error} = (\text{Relative error} \times 100)\%$$

$$= (0.0002 \times 100)\%$$

$$= 0.02\%$$

The frequency changes by 0.02% due to the temperature change.

NOTE ▶ The word error usually carries the connotation of a mistake or blunder, but in a scientific or technical sense the word error refers to the inevitable uncertainty that is a part of the measurement process. ◀

EXAMPLE 5 The diameter of a ball bearing is measured repeatedly using a micrometer, and from these measurements its radius is found to be 0.315 ± 0.002 cm. Find its volume, and the error, relative error, and percent error in the volume.

Solution **First,** calculate the volume V. For a sphere of radius r,

$$V = \frac{4}{3}\pi r^3$$

$$= \frac{4\pi(0.315)^3}{3}$$

$$4 \; \boxed{\times} \; \boxed{\pi} \; \boxed{\times} \; .315 \; \boxed{y^x} \; 3 \; \boxed{\div} \; 3 \; \boxed{=} \rightarrow \boxed{0.1309243}$$

$$V \approx 0.131 \text{ cm}^3$$

Second, calculate the differential dV.

$$\frac{dV}{dr} = 4\pi r^2$$

Then, use the definition (24-1) to find the error.

$$dV = 4\pi r^2\, dr = 4\pi r^2 \Delta r$$
$$= 4\pi (0.315)^2 (0.002)$$
$$= 0.0024937\ldots$$
$$\approx 0.002 \text{ cm}^3$$

Finally, calculate the relative error and percent error. Using the definition (24-2),

$$\text{Relative error} = \frac{|dV|}{V} = \frac{0.0024937}{0.1309243}$$
$$= 0.019047\ldots$$
$$\approx 0.019$$

Using definition (24-3),

$$\text{Percent error} = (0.019047 \times 100)\%$$
$$\approx 1.9\%$$

LEARNING HINT

The following procedure allows you to calculate the relative error more directly. Since

$$dV = 4\pi r^2\, dr \quad \text{and} \quad V = \frac{4}{3}\pi r^3$$

Dividing,

$$\frac{dV}{V} = \frac{4\pi r^2\, dr}{\frac{4}{3}\pi r^3}$$

$$\frac{dV}{V} = 3\left(\frac{dr}{r}\right)$$

| Relative error in V | Relative error in r |

$$= 3\left(\frac{0.002}{0.315}\right) \approx 0.019 \text{ or } 1.9\% \quad \blacktriangleleft$$

Exercises 24-1

Find the differential dy for each function.

1. $y = x^3 - 2x^2$
2. $y = 4x^5 - 3$
3. $y = (2 - x^2)^3$
4. $y = (x^3 - 7)^5$
5. $y = x^2(x^2 - 1)^3$
6. $y = x(2x - 1)^{1/3}$

7. $y = \dfrac{x^2}{x - 1}$ **8.** $y = \dfrac{2x}{x + 5}$ **9.** $y = x\sqrt{x}$

10. $y = \dfrac{1}{\sqrt{x^2 + 1}}$ **11.** $xy + x^3 = x$ **12.** $y = x^2 - xy + 2x$

Calculate both dy and Δy for each function at the values of x and Δx given.

13. $y = 9x - x^2$; $x = 5$, $\Delta x = 0.1$ **14.** $y = x^2 - 5x + 1$; $x = 3.5$, $\Delta x = 0.2$

15. $y = 2x^3 - 3x^2$; $x = 2.5$, $\Delta x = 0.06$ **16.** $y = 3x^4 - 1$; $x = 4$, $\Delta x = 0.05$

17. $y = (2x + 3)^3$; $x = 2$, $\Delta x = 0.02$ **18.** $y = (1 - 4x)^4$; $x = 0.5$, $\Delta x = 0.01$

19. $y = \dfrac{x}{1 + x}$; $x = -0.2$, $\Delta x = 0.04$ **20.** $y = \dfrac{x}{\sqrt{4x + 7}}$; $x = -1.5$, $\Delta x = 0.08$

Solve.

21. *Machine technology* A metal cylinder 10.0 cm in height has an almost circular but irregular cross section. The diameter of the cylinder is measured repeatedly using a micrometer and is found to be 3.42 ± 0.20 cm. Find its volume and the error, relative error, and percent error in the volume.

22. *Physics* The acceleration due to gravitational attraction can be measured using a simple pendulum:

$$g = \frac{4\pi^2 L}{P^2}$$

where L is the length of the pendulum (in meters) and P is the period of its oscillation (in seconds). Calculate g and the percent error in g if L is 0.80 m and P is measured to be 1.80 ± 0.10 s.

23. Find the error in the calculated area of a square 5.50 cm on each side if the error in measuring the side length is 0.04 cm.

24. *Electronics* In a simple series circuit, the voltage V, current i, and resistance R are related by Ohm's law: $V = iR$. If V is 1.54 V and i is measured as 0.31 A with an error of 0.02 A, find the value of R (in ohms) and the error in R.

25. *Physics* A wooden cylindrical storage tank is secured by a continuous steel band around its circumference. If the temperature of the band increases 10°C during the course of a day, the band will expand its length by 0.014 ft.
(a) By how much does the radius of the band increase?
(b) If the same metal band were placed around the earth, a 10°C increase in temperature would cause it to expand its length by 40 ft. By what amount would the radius of this band increase?

26. *Machine technology* A circular plate 2.4 cm thick was originally made 25.0 cm in diameter, but 0.2 cm was trimmed from its edge. Estimate the volume of metal wasted.

27. *Thermodynamics* The pressure P, volume V, and temperature T of a gas in thermal equilibrium are related by the equation $PV = nRT$, where n and R are constants. Show that if the volume of a gas sample is held fixed, the relative error in P caused by a change in T is equal to the relative error in T.

28. *Acoustical engineering* The frequency f of the fundamental note on a violin string is related to the tension T in the string, the length L of the string, and the linear density μ of the string according to the equation

$$f = \frac{1}{2L} \sqrt{\frac{T}{\mu}}$$

Estimate the amount by which the frequency will change if the tension is adjusted 1% during tuning. Use $L = 0.22$ m, $\mu = 0.0036$ kg/m, $T = 590$ N (f is in hertz, H).

29. *Physics* The acceleration g due to gravity at some point a distance r from the center of the earth is related to M_e, the mass of the earth, by the equation $g = GM_e/r^2$, where G is a constant.

Find $|dg|/g$, the relative change in g, corresponding to a change in position $dr = 1.0$ m near the earth if $g = 9.806$ m/s² and $r = 6.41 \times 10^6$ m at the earth's surface.

30.   *Electronics*   Two long wires a distance a apart carry equal currents i in opposite directions. At a distance R from the center line of the pair of wires a magnetic field B is set up, which obeys the following equation:*

$$B = \frac{kia}{4R^2 + a^2}$$

where $k = 8 \times 10^{-7}$. If the spacing a is allowed to vary slightly, by what amount will B vary? Use $R = 6.0$ cm, $i = 50$ A, $a = 4.0$ cm, $da = 0.10$ cm, and find the percent change in B.

24-2 | The Indefinite Integral

In Chapter 22 we developed rules that enabled us to find the derivative dy/dx of a polynomial or rational function $y = f(x)$ when the function was given. In this chapter we will examine the reverse process: Given the derivative dy/dx of a function, find the corresponding function $y = f(x)$. Many applications of calculus in science and technology require that we be able to perform this inverse operation; some of these applications are listed in Table 24-2.

TABLE 24-2

Function	Derivative
Position	Velocity
Velocity	Acceleration
Amount of a quantity	Rate of increase or decrease
Work done	Power
Potential energy	Force

EXAMPLE 6   If the velocity v of an object moving along a straight-line path is given by the function $v(t) = 10t$, and if $v = ds/dt$, where s is the position of the object at time t, then find an equation for $s(t)$.

Solution   To find the function $s(t)$, we need to work backward from the rules for differentiation. **First,** from the power rule for differentiation (22-13), we

---

* David Halliday and Robert Resnick, *Fundamentals of Physics*, 2nd edition (New York: Wiley, 1981), p. 561

know that differentiation reduces the power of the variable by 1. If $s = t^2$, then $ds/dt = 2t$. Therefore, the function $s(t)$ must contain t^2, and some experimentation reveals that the given derivative must have come from a function of the form $s = 5t^2$.

Second, because the derivative of a constant is equal to zero (22-11), the desired function could also be $s = 5t^2 + C$, where C is any constant.

Third, check the function by differentiating.

$$\frac{ds}{dt} = \frac{d}{dt}(5t^2 + C) = 10t$$

Antiderivative   Given the derivative function $f(x)$, we can find a function $g(x)$ such that $g'(x) = f(x)$. The function $g(x)$ is called the **antiderivative** of $f(x)$.

EXAMPLE 7   Find the antiderivative of the function $f(x) = 6x^2$.

Solution

Step 1.   From (22-13), we know that for $y = x^n$, $dy/dx = nx^{n-1}$. We can guess from this that the antiderivative is of the form $g(x) = ax^3$, where a is some constant.

Step 2.   Find the derivative of this possible antiderivative and compare it with the given function:

$$\frac{d}{dx}(ax^3) = 3ax^2 = 6x^2$$

Therefore,

$$3a = 6$$

$$a = 2$$

The antiderivative is of the form $2x^3$.

Step 3.   To get the antiderivative, add a constant to this function. The antiderivative of $f(x) = 6x^2$ is $g(x) = 2x^3 + C$, where C is any constant.

Step 4.   Check this by finding the derivative:

$$\frac{d}{dx}g(x) = \frac{d}{dx}(2x^3 + C) = 6x^2$$

Note that any of the functions $2x^3$, $2x^3 + 1$, $2x^3 - 10$, and so on, can be the antiderivative of $6x^2$, since each of these functions has $6x^2$ as its derivative. For any given function $f(x)$, the antiderivative $g(x)$ is never a single unique function. The antiderivative gives a family of functions, each differing only by an additive constant, and each having $f(x)$ as its derivative (see Fig. 24-3 on p. 962).

Notice that at a given value of x, each curve in Fig. 24-3 has the same slope. For example, at $x = 1$, the slope of the line tangent to each curve is 6.

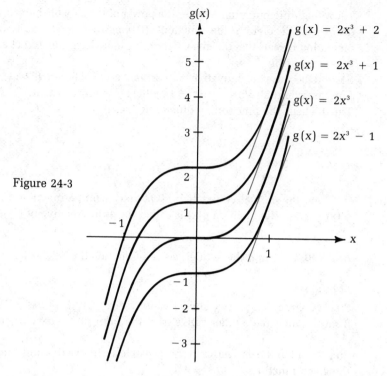

Figure 24-3

Integration   This reverse process to differentiation is called **antidifferentiation** or **integration**. The antiderivative of some function $f(x)$ can be written using the following notation:

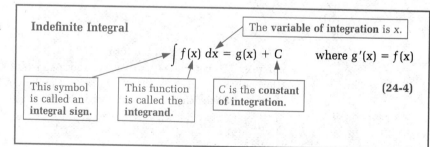

Variable of integration

Indefinite Integral

The **variable of integration** is x.

$$\int f(x)\,dx = g(x) + C \qquad \text{where } g'(x) = f(x)$$

Integrand

This symbol is called an **integral sign**.

This function is called the **integrand**.

C is the **constant of integration**.

(24-4)

Indefinite integral   The quantity $\int f(x)\,dx$ is called the **indefinite integral** of the function $f(x)$, and the process of finding the function $g(x)$ from $f(x)$ is called **integrating** $f(x)$. The differential dx indicates that x is the variable of integration.*

---

* Historically, the process of integration was developed before differentiation, and the integral symbol (an elongated S) arose from these early applications. We shall see later how this notation originated.

Using this notation, the antiderivative problem in Example 6 may be written

$$\int 6x^2\ dx = 2x^3 + C$$

The differentiation rules presented in Chapter 22 allow us to develop formulas that can be used to evaluate some indefinite integrals.

Integration of a Constant	$\int k\ dx = k \int dx$	(24-5)
	$\int k\ dx = kx + C$	(24-6)

where k is a constant.

These integration formulas follow from the differentiation rules

$$\frac{d}{dx}(kx) = k\frac{dx}{dx} = k$$

The first formula (24-5) tells us that a constant coefficient can be written as a multiplier in the front of the integral sign:

$$\int k{\cdot}f(x)\ dx = k \int f(x)\ dx$$

Note that the integral $\int dx$ is a shorthand way of writing $\int 1\ dx$, which is equal to $x + C$.

EXAMPLE 8 If a rocket is fired vertically upward with acceleration $a = 4$ m/s^2, find an expression for the velocity v, where $v = \int a\ dt$. Assume that $v = 0$ at $t = 0$.

Solution
Use (24-5).

$$v = \int 4\ dt = 4 \int dt$$

Then, from (24-6):

$$= 4t + C$$

If $v = 0$ at $t = 0$, then:

$$0 = 4(0) + C$$

Therefore:

$$C = 0$$

The velocity is $v = 4t$.

A second integration formula follows directly from the power rule (22-22):

$$\frac{d}{dx}(x^n) = nx^{n-1}$$

$$\text{Power Rule of Integration} \qquad \int x^n\, dx = \frac{x^{n+1}}{n+1} + C \qquad n \neq -1$$

(24-7)

The case where $n = -1$ is important in calculus, but the antiderivative of $1/x$ is not a polynomial function. We will study this special integral in a later chapter.

Notice that for $n = 0$ the power rule (24-7) reduces to equation (24-6) with $k = 1$.

EXAMPLE 9   Integrate the following functions:
(a) $\int (-12x^5)\, dx$ (b) $\int \sqrt{x}\, dx$
(c) The total electrostatic energy V stored in the field surrounding a charged conducting sphere is given by the integral

$$V = \frac{Q^2}{k} \int \frac{dr}{r^2}$$

where r is the distance from the center of the sphere, Q is the total charge, and k is a constant. Integrate to find an expression for V.

Solution

(a) **First,** using (24-5), move the coefficient outside the integral sign. $\int (-12x^5)\, dx = -12 \int x^5\, dx$

Next, use the power rule (24-7) to integrate.

$$-12 \int x^5\, dx = -12\,\frac{x^{5+1}}{5+1} + C = -12 \cdot \frac{x^6}{6} + C$$

$$= -2x^6 + C$$

Notice that we add the constant of integration *after* finding the antiderivative $-2x^6$.

Finally, check the integration by differentiating.

$$\frac{d}{dx}(-2x^6 + C) = -12x^5$$

(b) **First,** write the integrand as a power of x. $\sqrt{x} = x^{1/2}$

Next, integrate using the power rule (24-7).

$$\int \sqrt{x}\, dx = \int x^{1/2}\, dx$$

$$= \frac{x^{(1/2)+1}}{\frac{1}{2}+1} + C$$

$$= \frac{x^{3/2}}{\frac{3}{2}} + C = \frac{2x^{3/2}}{3} + C = \frac{2x\sqrt{x}}{3} + C$$

The final result should be written as a radical so that it agrees in form with the original integrand.

Finally, check by differentiating.

$$\frac{d}{dx}\left(\frac{2x^{3/2}}{3} + C\right) = \frac{2}{3}\frac{d}{dx}(x^{3/2})$$

$$= \frac{2}{3} \cdot \frac{3}{2}x^{(3/2)-1}$$

$$= x^{1/2} = \sqrt{x}$$

(c) **First,** write the integrand as a power.

$$V = \frac{Q^2}{k}\int r^{-2}\, dr$$

Next, integrate using the power rule (24-7).

$$V = \frac{Q^2}{k}\int r^{-2}\, dr = \frac{Q^2}{k}\left(\frac{r^{-2+1}}{-2+1}\right) + C$$

$$= \frac{Q^2}{k}\left(\frac{r^{-1}}{-1}\right) + C$$

$$= -\frac{Q^2}{kr} + C$$

Finally, check by differentiating.

$$\frac{d}{dr}\left(-\frac{Q^2}{kr} + C\right) = -\frac{Q^2}{k}\cdot\frac{d}{dr}\left(\frac{1}{r}\right)$$

$$= -\frac{Q^2}{k}\cdot\frac{d}{dr}(r^{-1})$$

$$= -\frac{Q^2}{k}(-1)r^{-2}$$

$$= \frac{Q^2}{k}\cdot\frac{1}{r^2}$$

The sum and difference rules for differentiation (22-15, 22-16) can be used to develop a formula that allows us to integrate when the integrand is the sum or difference of several functions.

Sum Rule for Integration

$$\int [f(x) \pm h(x)]\, dx = \int f(x)\, dx \pm \int h(x)\, dx \qquad \text{(24-8)}$$

EXAMPLE 10 Integrate:

(a) $\displaystyle\int (x - 1)\, dx$

(b) $\displaystyle\int \left(2x^3 - 3x^2 - \frac{1}{x^3}\right) dx$

(c) $\displaystyle\int \left(\frac{1}{\sqrt{x}} + \sqrt[3]{x}\right) dx$

Solution

(a) **First,** use the sum rule (24-8) to write this as the difference of two integrals.

$$\int (x - 1)\, dx = \int x\, dx - \int dx$$

Next, integrate using the power rule (24-7).

$$\int x\ dx = \frac{x^2}{2} + C_1$$

$$\int dx = x + C_2$$

Therefore:

$$\int (x - 1)\ dx = \frac{x^2}{2} - x + C$$

Notice that we have combined the constants C_1 and C_2 into one arbitrary constant C.

Finally, check by differentiating.

$$\frac{d}{dx}\left(\frac{x^2}{2} - x + C\right) = x - 1$$

(b) **First,** use the sum rule (24-8).

$$\int \left(2x^3 - 3x^2 - \frac{1}{x^3}\right)\ dx = \int 2x^3\ dx - \int 3x^2\ dx - \int x^{-3}\ dx$$

Next, integrate using the power rule (24-7).

$$= 2 \cdot \frac{x^4}{4} - 3 \cdot \frac{x^3}{3} - \frac{x^{-3+1}}{-3+1} + C$$

$$= \frac{x^4}{2} - x^3 + \frac{1}{2x^2} + C$$

Finally, check by differentiating.

(c) Use the sum rule (24-8).

$$\int \left(\frac{1}{\sqrt{x}} + \sqrt[3]{x}\right)\ dx = \int \frac{dx}{\sqrt{x}} + \int \sqrt[3]{x}\ dx$$

Then, use the power rule (24-7).

$$= \int x^{-1/2}\ dx + \int x^{1/3}\ dx$$

$$= \frac{x^{(-1/2)+1}}{-\frac{1}{2}+1} + \frac{x^{(1/3)+1}}{\frac{1}{3}+1} + C$$

$$= \frac{x^{1/2}}{\frac{1}{2}} + \frac{x^{4/3}}{\frac{4}{3}} + C$$

$$= 2x^{1/2} + \frac{3}{4}x^{4/3} + C$$

$$= 2\sqrt{x} + \frac{3x\ \sqrt[3]{x}}{4} + C$$

Check by differentiating.

CAUTION ▶ The sum and difference rules for differentiation have a corresponding integration rule (24-8), but the product and quotient rules for differentiation do *not*. That is,

$$\int f(x) \cdot h(x)\ dx \qquad \text{is **not** equal to} \qquad \int f(x)\ dx \cdot \int h(x)\ dx \quad \blacktriangleleft$$

EXAMPLE 11   Find the equation of the curve passing through the point $(1, 2)$ and having slope given by the function $f'(x) = 8x^3 - 2x$.

Solution

The equation of the curve will be

$$f(x) = \int (8x^3 - 2x)\, dx$$

Use the sum rule (24-8).

$$= \int 8x^3\, dx - \int 2x\, dx$$

Then, use equation (24-5).

$$= 8 \int x^3\, dx - 2 \int x\, dx$$

Integrate using the power rule (24-7).

$$= 8 \cdot \frac{x^4}{4} - 2 \cdot \frac{x^2}{2} + C$$

$$f(x) = 2x^4 - x^2 + C$$

But the problem states that $(1, 2)$ is a point on this curve. Therefore:

$$2 = 2(1)^4 - (1)^2 + C$$

$$2 = 2 - 1 + C$$

$$C = 1$$

The equation of the curve is $f(x) = 2x^4 - x^2 + 1$

Notice that integrating gives a family of possible functions. Additional information must be used to determine the value of the constant of integration.

u Substitution   The integration formulas (24-7) and (24-8) may often be used to perform integrations in which the variable appears in a form more complicated than a power of x. In the method of **u substitution,** the integrand is transformed so that the existing formulas apply. Use u substitution if the integral can be rewritten in terms of some function u(x), as shown.

---

u Substitution $\displaystyle \int f(x)\, dx = \int u(x) \cdot \frac{du}{dx} \cdot dx = \int u(x)\, du$ (24-9)

u(x) is some function that can be integrated using existing formulas.

Derivative of u(x)

In particular, if the integrand can be written as a power of u, then we can integrate using a formula similar to the power rule (24-7).

Generalized Power Rule of Integration

$$\int u^n\, du = \frac{u^{n+1}}{n+1} + C \qquad n \neq -1$$ (24-10)

EXAMPLE 12 Integrate: $\displaystyle \int 2x(x^2 + 6)^3\, dx$

Solution This integral could be evaluated by expanding the integrand and integrating term-by-term using the sum rule (24-8). However, it is much simpler to use u substitution. Follow these steps.

Step 1. Choose as $u(x)$ some expression that appears in the integrand. This integral may be written as

$$\int (x^2 + 6)^3 (2x)\, dx \qquad \text{Try } u(x) = (x^2 + 6).$$

Step 2. Find du/dx. Differentiating,

$$\frac{du}{dx} = \frac{d}{dx}(x^2 + 6) = 2x$$

Step 3. Substitute these values of $u(x)$ and du/dx into the original integral:

$$\int 2x(x^2 + 6)^3\, dx = \int \overbrace{(x^2 + 6)^3}^{u(x)}\,\overbrace{(2x)\, dx}^{\frac{du}{dx}}$$

$$= \int u^3\, du$$

Step 4. Integrate. Use the generalized power rule (24-10):

$$\int u^3\, du = \frac{u^4}{4} + C$$

> Do not forget the constant of integration.

Step 5. Replace u by the expression chosen in Step 1:

$$\int 2x(x^2 + 6)^3\, dx = \frac{(x^2 + 6)^4}{4} + C$$

Step 6. Check the answer by differentiating:

$$\frac{d}{dx}\left[\frac{(x^2 + 6)^4}{4} + C\right] = \frac{4(x^2 + 6)^3}{4} \cdot \frac{d}{dx}(x^2 + 6)$$

$$= (x^2 + 6)^3 \cdot 2x \qquad \text{This } \textit{is} \text{ the original integrand.}$$

LEARNING HINT

- Step 1 may require some trial and error to find the correct expression for $u(x)$.
- The result of Step 3 should be an integral written completely in terms of u. No x's should appear in the integral after the substitution is made. If the variable x is not eliminated, choose a different expression for $u(x)$ and begin again. ◀

The following steps show why u substitution (24-9) is a valid procedure: **First,** from the definition of the indefinite integral,

$$\int f(x)\, dx = g(x) + C \qquad \text{where } g'(x) = f(x)$$

Then, using the chain rule (22-21),

$$\frac{d}{dx}\, g(x) = \frac{d}{du}\, g(x) \cdot \frac{du}{dx}$$

If we write

$$f(x) = u(x)\frac{du}{dx} \qquad \text{where } u(x) = \frac{d}{du}\, g(x)$$

then,

$$\int f(x)\, dx = \int u(x) \cdot \frac{du}{dx} \cdot dx$$

which is the u substitution formula (24-9).

EXAMPLE 13 Integrate: $\displaystyle\int (2x^3 - 1)^4 x^2\, dx$

Solution

Step 1. Let $u(x) = 2x^3 - 1$.

Step 2. Then $\dfrac{du}{dx} = 6x^2$, so that $x^2 = \dfrac{1}{6}\dfrac{du}{dx}$

Step 3. Substituting, the integral becomes

$$\int (2x^3 - 1)^4 x^2\, dx = \int u^4 \cdot \frac{1}{6}\frac{du}{dx}\, dx = \frac{1}{6}\int u^4 \cdot \frac{du}{dx}\, dx$$

$$= \frac{1}{6}\int u^4\, du \quad \longleftarrow \boxed{\begin{array}{l}\text{The integral is now} \\ \text{written in terms of } u \\ \text{only. No } x\text{'s appear.}\end{array}}$$

Step 4. Using the power rule (24-10),

$$= \frac{1}{6}\cdot\frac{u^5}{5} + C = \frac{u^5}{30} + C$$

Step 5. Substitute $2x^3 - 1 = u$, so that the integral becomes

$$\int (2x^3 - 1)^4 x^2\, dx = \frac{(2x^3 - 1)^5}{30} + C$$

Step 6. Check this by differentiating.

CAUTION ▶ The following integrals cannot be evaluated using equations (24-9) and (24-10):

$$\boxed{\begin{array}{l}\text{If } u = 2x^3 - 1, \text{ then } du/dx = 6x^2. \text{ But this} \\ \text{integrand has a factor of } x \text{ rather than } x^2.\end{array}}$$

$$\int (2x^3 - 1)^4 x\, dx$$

$$\boxed{\text{The integrand has a factor of } x^3 \text{ rather than } x^2.}$$

$$\int (2x^3 - 1)^4 x^3\, dx \quad ◀$$

EXAMPLE 14 The axial electrostatic potential difference V at a distance r from a uniformly charged disk can be found using the integral

$$V = k \int \frac{r\, dr}{\sqrt{1 + r^2}} \qquad \text{where } k \text{ is a constant}$$

Integrate to obtain an algebraic expression for V.

Solution

Step 1. Let $u = 1 + r^2$

Step 2. Then $\dfrac{du}{dr} = 2r$, so that $r = \dfrac{1}{2}\dfrac{du}{dr}$

Step 3. The integral becomes

$$V = k \int (1 + r^2)^{-1/2}\left(\frac{1}{2}\frac{du}{dr}\right) dr$$

$$= \frac{k}{2}\int u^{-1/2}\, du$$

Step 4. Integrate using equation (24-10).

$$V = \frac{k}{2}\cdot\frac{u^{(-1/2)+1}}{-\frac{1}{2}+1} + C$$

$$V = ku^{1/2} + C = k\sqrt{u} + C$$

Step 5. Substitute $1 + r^2 = u$. $V = k\sqrt{1 + r^2} + C$

Step 6. Check by differentiating.

EXAMPLE 15   Find the equation of the curve passing through the point $(0, 1)$ and having slope given by the function $f'(x) = x\sqrt{2x^2 + 4}$.

Solution $f(x) = \displaystyle\int f'(x)\, dx = \int x\sqrt{2x^2 + 4}\, dx$

Step 1. Let $u(x) = 2x^2 + 4$

Step 2. Then $\dfrac{du}{dx} = 4x$, so $x = \dfrac{1}{4}\dfrac{du}{dx}$

Step 3. The integral becomes

$$f(x) = \int \sqrt{u}\cdot\frac{1}{4}\frac{du}{dx}\cdot dx$$

$$= \frac{1}{4}\int \sqrt{u}\, du = \frac{1}{4}\int u^{1/2}\, du$$

Step 4. Integrate using equation (24-10).

$$f(x) = \frac{1}{4}\cdot\frac{u^{(1/2)+1}}{\frac{1}{2}+1} + C$$

$$= \frac{1}{4}\cdot\frac{u^{3/2}}{\frac{3}{2}} + C = \frac{u^{3/2}}{6} + C$$

Step 5. Substitute.

$$f(x) = \frac{(2x^2 + 4)^{3/2}}{6} + C$$

The problem states that when $x = 0$, $f(0) = 1$, so

$$1 = \frac{[2(0)^2 + 4]^{3/2}}{6} + C$$

$$1 = \frac{4^{3/2}}{6} + C$$

$$1 = \frac{8}{6} + C$$

Therefore: $$C = -\frac{1}{3}$$

and $$f(x) = \frac{(2x^2 + 4)^{3/2}}{6} - \frac{1}{3}$$

Step 6. Check by differentiating.

REMEMBER ▶

1. Check your answer to be certain you have included a constant of integration.
2. Double-check by differentiating. ◀

Exercises 24-2

Integrate.

1. $\int 3\, dx$

2. $\int 5\, dr$

3. $\int 3x\, dx$

4. $\int 4x^2\, dx$

5. $\int t^{-4}\, dt$

6. $\int x^6\, dx$

7. $\int x^{5/2}\, dx$

8. $\int \sqrt[4]{x}\, dx$

9. $\int (2m - 5)\, dm$

10. $\int (x^2 - 4x + 3)\, dx$

11. $\int \left(4x^3 - \frac{3}{x^2}\right) dx$

12. $\int \left(2\sqrt{x} - \frac{1}{x^3}\right) dx$

13. $\int \sqrt{x}(1 - x + x^2)\, dx$

14. $\int (x^{-1/4} + x^{3/4})\, dx$

15. $\int (3x + 2)^2\, dx$

16. $\int (t - 3)^3\, dt$

17. $\int 2x(x^2 + 5)^3\, dx$

18. $\int 3x^2(x^3 - 4)^4\, dx$

19. $\int 2x^2(2x^3 + 1)^5\, dx$

20. $\int x(2x^2 - 3)^3\, dx$

21. $\int 2x\sqrt{3x^2 + 2}\, dx$

22. $\int 4x(2x^2 - 1)^{1/3}\, dx$

23. $\int \sqrt[3]{6x - 2}\, dx$

24. $\int \sqrt{5x + 9}\, dx$

25. $\int \frac{2x}{(x^2 + 4)^3}\, dx$

26. $\int \frac{x + 2}{(x^2 + 4x)^2}\, dx$

27. $\int \frac{x^2 - 4x}{\sqrt{x^3 - 6x^2}}\, dx$

28. $\int \frac{x^2}{\sqrt{x^3 + 2}}\, dx$

29. $\int (x^4 - 2x^2)^4(2x^3 - 2x)\, dx$

30. $\int (3x^2 - 1)(x^3 - x)^5\, dx$

Find the equation of the curve passing through the given point P and having slope f'(x).

31. $P(-1, -2)$; $f'(x) = 4x^2$

32. $P(2, 2)$; $f'(x) = x(x^2 - 6)^3$

33. $P(4, 7)$; $f'(x) = \sqrt{2x + 1}$

34. $P(0, 2)$; $f'(x) = -2x\sqrt{1 - 2x^2}$

Solve.

35. *Industrial engineering* A stamping machine produces computer chassis cases at a rate that varies over the working day according to the equation $dN/dt = 150 + 2t - 0.1t^2$, where t is time (in hours). Integrate to find an expression for N and calculate N for an 8-hour day if $N = 0$ at $t = 0$.

36. *Thermodynamics*   The temperature T (in °C) in an industrial furnace varies from its center to a point outside so that the rate of temperature change is given by the equation

$$\frac{dT}{dx} = -\frac{5000}{(x + 1)^3}$$

Integrate to find an expression for T at a distance x (in meters) from the center if $T = 2700$°C at $x = 0$.

37. *Physics*   If a charged particle moves in an electrical field with acceleration

$$a = 2 + t^2 + \frac{2}{(t + 1)^2}$$

find an expression for its velocity $v = \int a\, dt$. Assume that $v = 0$ when $t = 0$.

38. *Physics*   The energy output E of a certain device varies with time t according to the equation

$$\frac{dE}{dt} = \frac{20t}{(t^2 + 4)^{3/2}}$$

Integrate to find an expression for E. Assume $E = 0$ at $t = 0$.

39. *Physics*   The power output P of a mechanical system varies with time t according to the equation

$$P = \frac{2t}{\sqrt{t^2 + 1}}$$

Given that power is defined as the rate of doing work, $P = dW/dt$, find an expression for W as a function of t. Assume $W = 0$ at $t = 0$.

24-3 | Area under a Curve

In geometry, formulas have been developed for finding the areas of plane figures such as regular polygons and circles. One of the earliest problems to be solved using calculus involved finding the area of much more general plane figures formed by the intersection of curves and straight lines.

For example, we can find the area under the curve $y = f(x)$ between $x = a$ and $x = b*$ by following these steps (Fig. 24-4).

Rectangle Sum Approximation
Step 1.   Divide the interval $a \le x \le b$ into n subintervals by choosing points $x_1, x_2, x_3, \ldots, x_n = b$. Any points will do, but for convenience we can make all the intervals the same width, Δx (Fig. 24-5).

$$\Delta x = \frac{b - a}{n} \tag{24-11}$$

*The phrase "under the curve" means bounded by the curve, the x-axis, and the vertical lines $x = a$ and $x = b$.

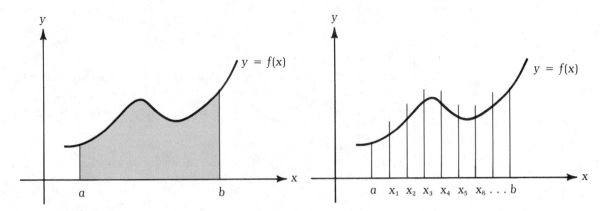

Figure 24-4   Figure 24-5

To avoid complications with negative quantities, we have assumed in this development that $f(x) > 0$ on the interval $a \le x \le b$. This will ensure that we get a nonnegative value for the area under the curve.

Step 2. If $f(x)$ is continuous over the interval $a \le x \le b$, it is continuous on each subinterval, and therefore it has a minimum value and a maximum value on each subinterval. Within each subinterval we may select a point x_i and construct a rectangle of width Δx and height $y_i = f(x_i)$. In Fig. 24-6, each x_i was chosen so that the height h_i is the minimum value of y in that subinterval. In Fig. 24-7 on p. 974, each x_i was chosen so that the height H_i is the maximum value of y in that subinterval.

Figure 24-6

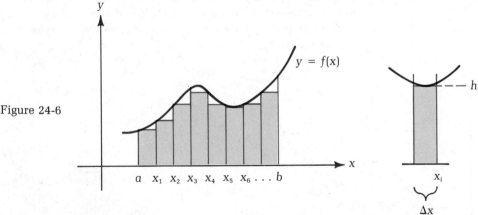

The area A_i of each rectangle can be calculated by multiplying its length and width:

Lower area $= h_i \Delta x$ Upper area $= H_i \Delta x$

Step 3. The sum of the n shaded areas gives an approximation to the actual area A under the curve.

Figure 24-7

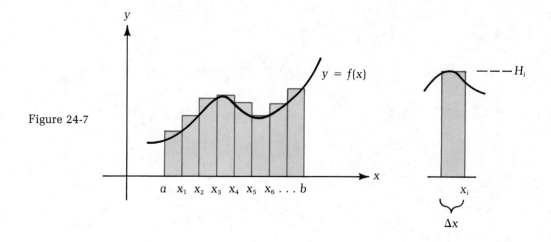

$$\text{Lower sum} = \sum_{i=1}^{n} h_i \Delta x \qquad \text{Upper sum} = \sum_{i=1}^{n} H_i \Delta x \qquad \text{(24-12)}$$

Notice that the lower sum is an underestimate of the area under the curve, and the upper sum is an overestimate of this area. Therefore,

Lower sum $\leq A \leq$ Upper sum

Step 4. As the number of intervals n increases, the interval width Δx approaches zero and the approximation improves (Fig. 24-8). The lower and upper sums approach A from above and below; therefore, the sums of the rectangular areas approach the area under the curve more and more closely. We can express this as a limit.

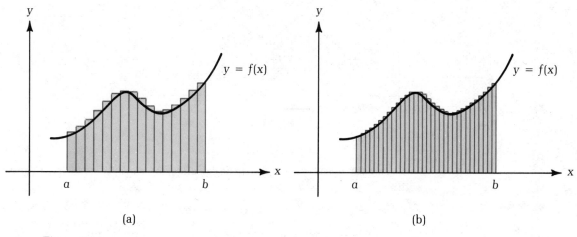

(a)   (b)

Figure 24-8

$$\text{Area under the Curve} \qquad A = \lim_{n \to \infty} \sum_{i=1}^{n} f(x_i) \Delta x \qquad \text{(24-13)}$$

The rectangle sum, often called a **Riemann sum** after the German mathematician Georg Riemann (1826–1866), gives a procedure for calculating the area under a curve that is ideally suited for use with a calculator. Organize your work as shown in Example 16.

EXAMPLE 16 Find the area under the curve $y = x^2 + 2$ from $x = 1$ to $x = 2$ for:
(a) $n = 2$ (b) $n = 10$

Solution

(a) **Step 1.** Sketch the curve [Fig. 24-9(a)] and find Δx. From equation (24-11), for $n = 2$, $a = 1$, $b = 2$,

$$\Delta x = \frac{b - a}{n} = \frac{2 - 1}{2} = \frac{1}{2} = 0.5$$

Step 2. Calculate the endpoint of each interval:

$$x_1 = a + \Delta x = 1 + 0.5 = 1.5$$

$$x_2 = a + 2\Delta x = 1 + 2(0.5) = 2$$

Using an "upper sum," we can approximate the area under the curve by constructing rectangles with heights y_1 and y_2, where $y_1 = f(x_1)$ and $y_2 = f(x_2)$. The area under the curve is approximated by two rectangles [Fig. 24-9(a)]. The heights of the rectangles are

$$y_1 = x_1^2 + 2 = (1.5)^2 + 2 = 2.25 + 2 = 4.25$$

$$y_2 = x_2^2 + 2 = (2)^2 + 2 = 6$$

Figure 24-9

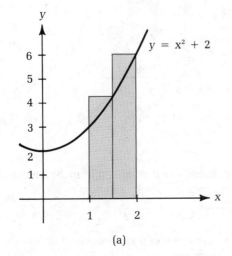

(a)

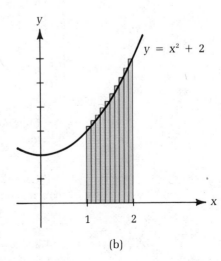

(b)

Step 3. The area of each rectangle can be calculated by multiplying its height and width:

$$A_1 = y_1 \Delta x = 4.25(0.5) = 2.125$$

$$A_2 = y_2 \Delta x = 6(0.5) = 3.000$$

Step 4. Total area of rectangles = 2.125 + 3.000 = 5.125 square units

(b) **Step 1.** For $n = 10$ [Fig. 24-9(b)],

$$\Delta x = \frac{2 - 1}{10} = 0.1$$

Step 2. $x_1 = 1 + 0.1 = 1.1$ $\qquad$ $y_1 = (1.1)^2 + 2 = 3.21$

$\qquad\qquad$ $x_2 = 1 + 2(0.1) = 1.2$ $\qquad$ $y_2 = (1.2)^2 + 2 = 3.44$

$\qquad\qquad$ $x_3 = 1 + 3(0.1) = 1.3$ $\qquad$ $y_3 = (1.3)^2 + 2 = 3.69$

$\qquad\qquad$ . . . and so on $\qquad\qquad\qquad$ . . . and so on

Step 3. $A_1 = y_1 \Delta x = 3.21(0.1) = 0.321$

$\qquad\qquad$ $A_2 = y_2 \Delta x = 3.44(0.1) = 0.344$

$\qquad\qquad$ $A_3 = y_3 \Delta x = 3.69(0.1) = 0.369$

$\qquad\qquad$ . . . and so on

The completed calculations are shown in Table 24-3.

TABLE 24-3
For $y = x^2 + 2$ between $x = 1$ and $x = 2$, and $n = 10$

x_i	$y_i = (x_i)^2 + 2$	$A_i = y_i \cdot \Delta x$
1.1	3.21	0.321
1.2	3.44	0.344
1.3	3.69	0.369
1.4	3.96	0.396
1.5	4.25	0.425
1.6	4.56	0.456
1.7	4.89	0.489
1.8	5.24	0.524
1.9	5.61	0.561
2.0	6.00	0.600

Step 4. Sum the last column to find the total area of the rectangles: 4.485 square units.

Because Δx is a constant, we can factor it from the sum and use the following sequence on a calculator:

$$1.1 \; \boxed{x^2} \; \boxed{+} \; 2 \; \boxed{+} \; 1.2 \; \boxed{x^2} \; \boxed{+} \; 2 \; \boxed{+} \cdots \boxed{+} \; 2 \; \boxed{x^2} \; \boxed{+} \; 2 \; \boxed{=} \; \boxed{\times} \; .1 \; \boxed{=} \rightarrow \qquad \textsf{4.485}$$

Table 24-4 shows the result of this approximation as calculated for various values of n using a computer.

TABLE 24-4

For $y = x^2 + 2$ between $x = 1$ and $x = 2$

Number of Intervals, n	Width of Interval, Δx	Total Area of Rectangles, A_n (square units)
2	0.5	5.125
5	0.2	4.640
10	0.1	4.485
100	0.01	4.34835
1,000	0.001	4.3348335
10,000	0.0001	4.33348335

This approximation appears to be approaching 4.333 . . . , or $4\frac{1}{3}$, as the number of intervals increases. To find the exact value of the area under the curve using (24-13), we must develop an expression for the total area of n rectangles, written as a function of n and then find the limit of this expression as n increases without bound. This procedure is shown in Example 17, but to use it we need the following formulas:

$$\sum_{i=1}^{n} c = nc \qquad \text{where } c \text{ is a constant} \qquad (24\text{-}14)$$

$$\sum_{i=1}^{n} i = \frac{1}{2}n(n+1) \qquad (24\text{-}15)$$

$$\sum_{i=1}^{n} i^2 = \frac{1}{6}n(n+1)(2n+1) \qquad (24\text{-}16)$$

$$\sum_{i=1}^{n} i^3 = \frac{1}{4}n^2(n+1)^2 \qquad (24\text{-}17)$$

Formulas (24-14) and (24-15) were proved in Chapter 18. The proofs of formulas (24-16) and (24-17) are beyond the scope of this book.

EXAMPLE 17   Use the definition in equation (24-13) to find the exact value of the area under the curve $y = x^2 + 2$ between $x = 1$ and $x = 2$.

Solution   Let
$$\Delta x = \frac{2 - 1}{n} = \frac{1}{n}$$

Then
$$x_1 = 1 + \Delta x = 1 + \frac{1}{n}$$

$$x_2 = 1 + 2\Delta x = 1 + \frac{2}{n}$$

$$\vdots$$

$$x_i = 1 + i\Delta x = 1 + \frac{i}{n}$$

The height of the ith rectangle at $x = x_i$ is
$$y_i = (x_i)^2 + 2 = \left(1 + \frac{i}{n}\right)^2 + 2$$

$$= 3 + \frac{2i}{n} + \frac{i^2}{n^2}$$

and the area of this rectangle is

$$A_i = y_i \cdot \Delta x = \left(3 + \frac{2i}{n} + \frac{i^2}{n^2}\right)\left(\frac{1}{n}\right)$$

$$= \frac{3}{n} + \frac{2i}{n^2} + \frac{i^2}{n^3}$$

The total area of these n rectangles is

$$A_n = \sum_{i=1}^{n} A_i = \sum_{i=1}^{n} \left(\frac{3}{n} + \frac{2i}{n^2} + \frac{i^2}{n^3}\right)$$

Now, using (24-14),

$$\sum_{i=1}^{n} \frac{3}{n} = \frac{1}{n} \sum_{i=1}^{n} 3 = \frac{1}{n} \cdot 3n = 3$$

and from (24-15),

$$\sum_{i=1}^{n} \frac{2i}{n^2} = \frac{2}{n^2} \sum_{i=1}^{n} i = \frac{2}{n^2} \cdot \frac{1}{2} n(n+1)$$

$$= \frac{n+1}{n} = 1 + \frac{1}{n}$$

and from (24-16),

$$\sum_{i=1}^{n} \frac{i^2}{n^3} = \frac{1}{n^3} \sum_{i=1}^{n} i^2 = \frac{1}{n^3} \cdot \frac{1}{6} n(n+1)(2n+1)$$

$$= \frac{2n^2 + 3n + 1}{6n^2}$$

$$= \frac{1}{3} + \frac{1}{2n} + \frac{1}{6n^2}$$

Adding these sums, we get

$$A_n = 4\tfrac{1}{3} + \frac{1}{n} + \frac{1}{2n} + \frac{1}{6n^2}$$

The area A under the curve now can be found using (24-13):

$$A = \lim_{n \to \infty} \left(4\tfrac{1}{3} + \frac{1}{n} + \frac{1}{2n} + \frac{1}{6n^2}\right)$$

$$= 4\tfrac{1}{3}$$

Since $\lim_{n \to \infty} \dfrac{1}{n} = 0$, $\lim_{n \to \infty} \dfrac{1}{2n} = 0$,

and $\lim_{n \to \infty} \dfrac{1}{6n^2} = 0$.

As n increases without bound, the quantity $(1/n) + (1/2n) + (1/6n^2)$ approaches zero. Therefore, the area under the curve is $4\tfrac{1}{3}$ square units.

EXAMPLE 18

(a)   Use the rectangle sum method with $n = 5$ to calculate the area under the curve $y = x^3$ from $x = 0$ to $x = 1$.

(b)   Use the limit method to calculate the same area.

Solution

(a)   **First,** sketch the curve, as shown in Fig. 24-10.

Second, from equation (24-11),

$$\Delta x = \frac{1-0}{5} = 0.2$$

Third, calculate the values for x_i and y_i as in Table 24-5:

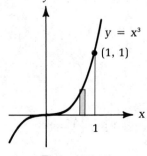

y

$y = x^3$

(1, 1)

x

1

Figure 24-10

TABLE 24-5

n	$x_i = 0 + n(\Delta x)$	$y_i = f(x_i) = (x_i)^3$
1	0.2	0.008
2	0.4	0.064
3	0.6	0.216
4	0.8	0.512
5	1.0	1.000

Finally, calculate the approximate area from equation (24-12):

$$\text{Area of rectangles } A_n = \Delta x \sum_{i=1}^{n} f(x_i)$$

$$= (0.2)(1.800)$$

$$= 0.360 \text{ square unit}$$

Table 24-6 shows how the approximation improves as n increases.

TABLE 24-6
For $y = x^3$ between
$x = 0$ and $x = 1$

Number of Intervals, n	Width of Interval, Δx	Total Area of Rectangles, A_n (square units)
5	0.2	0.360000
10	0.1	0.302500
100	0.01	0.255025
1000	0.001	0.250500

(b) If $\qquad \Delta x = \dfrac{1 - 0}{n} = \dfrac{1}{n}$

then $\qquad\qquad\qquad\qquad\qquad x_1 = 0 + \Delta x = 0 + \dfrac{1}{n} = \dfrac{1}{n}$

$$x_2 = 0 + 2\Delta x = 0 + \dfrac{2}{n} = \dfrac{2}{n}$$

$$\vdots$$

$$x_i = 0 + i\Delta x = 0 + \dfrac{i}{n} = \dfrac{i}{n}$$

and $\qquad\qquad\qquad\qquad\qquad y_i = (x_i)^3 = \dfrac{i^3}{n^3}$

The area of the ith rectangle is $\qquad A_i = y_i \Delta x$

The total area of the n rectangles is $\qquad A_n = \sum_{i=1}^{n} A_i = \sum_{i=1}^{n} y_i \Delta x$

$$= \sum_{i=1}^{n} \dfrac{i^3}{n^3} \cdot \dfrac{1}{n} = \dfrac{1}{n^4} \sum_{i=1}^{n} i^3$$

Now, using (24-17), $A_n = \dfrac{n^2(n + 1)^2}{4n^4}$

$$= \dfrac{n^4 + 2n^3 + n^2}{4n^4}$$

$$= \dfrac{1}{4} + \dfrac{1}{2n} + \dfrac{1}{4n^2}$$

The area A under the curve now can be found using (24-13):

$$A = \lim_{n \to \infty} \left(\dfrac{1}{4} + \dfrac{1}{2n} + \dfrac{1}{4n^2} \right) = \dfrac{1}{4} \text{ square unit}$$

As $n \to \infty$, these terms approach zero.

Notice that the calculated areas shown in Table 24-6 seem to approach this exact value for A.

Exercises 24-3

Use the rectangle sum method (upper sum) to find the approximate area under each curve for the given values of n. If necessary, round to two decimal places.

1. $y = 4x$ from $x = 0$ to $x = 2$ for: (a) $n = 4$ (b) $n = 10$

2. $y = \dfrac{1}{2}x$ from $x = 0$ to $x = 3$ for: (a) $n = 3$ (b) $n = 9$

3. $y = 2x^2$ from $x = 0$ to $x = 1$ for: (a) $n = 2$ (b) $n = 8$

4. $y = x^2 + 4$ from $x = 0$ to $x = 3$ for: (a) $n = 3$ (b) $n = 10$

5. $y = x^2 - 2x$ from $x = 2$ to $x = 4$ for: (a) $n = 4$ (b) $n = 10$

6. $y = x - x^2$ from $x = 0$ to $x = 1$ for: (a) $n = 4$ (b) $n = 10$

7. $y = \dfrac{1}{x^2}$ from $x = 0.5$ to $x = 2$ for: (a) $n = 5$ (b) $n = 10$

8. $y = \dfrac{x^3}{2}$ from $x = 0$ to $x = 2$ for: (a) $n = 4$ (b) $n = 12$

9. $y = x^2 + x + 2$ from $x = 1$ to $x = 4$ for: (a) $n = 3$ (b) $n = 12$

10. $y = \sqrt{x}$ from $x = 1$ to $x = 4$ for: (a) $n = 3$ (b) $n = 9$

11. $y = \dfrac{1}{\sqrt{x}}$ from $x = 1$ to $x = 4$ for: (a) $n = 3$ (b) $n = 9$

12. $y = x + \sqrt{x}$ from $x = 1$ to $x = 4$ for: (a) $n = 3$ (b) $n = 12$

13. $y = \dfrac{2x}{(1 + x^2)^{3/2}}$ from $x = 1$ to $x = 2$ for: (a) $n = 4$ (b) $n = 8$

14. $y = \sqrt[3]{\dfrac{2x + 3}{2x + 1}}$ from $x = 0$ to $x = 3$ for: (a) $n = 6$ (b) $n = 12$

Find the exact area under each curve using the limit method.

15. $y = 3x$ from $x = 0$ to $x = 2$

16. $y = \dfrac{x}{3}$ from $x = 0$ to $x = 3$

17. $y = x^2$ from $x = 1$ to $x = 5$

18. $y = x^2 - 3$ from $x = 2$ to $x = 5$

19. $y = x^2 + 3$ from $x = 1$ to $x = 3$ **20.** $y = 4x - x^2$ from $x = 0$ to $x = 4$
21. $y = x^3 + 2$ from $x = 0$ to $x = 2$ **22.** $y = 2x^3$ from $x = 0$ to $x = 1$

24-4 | The Definite Integral

The limit process (24-13) discussed in Section 24-3 enables us to find the exact area under a curve, but it usually involves a great many algebraic manipulations, and for many functions the limit may be difficult to find. In this section we will show that for most functions the process of integration introduced in Section 24-2 provides a simpler way of finding the area under a curve.

Consider the curve shown in Fig. 24-11. Let $A(x)$ be the area under the curve between $x = a$ and some general point x. We need to find a general expression for the function $A(x)$ in terms of the equation of the curve $f(x)$.

Figure 24-11

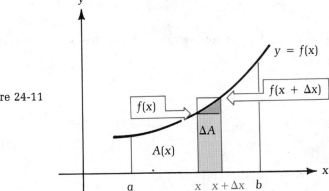

When x increases by an amount Δx, the area $A(x)$ increases by ΔA, where

$$\Delta A = A(x + \Delta x) - A(x)$$

- The area of the shaded rectangle is $f(x)\Delta x$.
- The area of the larger rectangle is $f(x + \Delta x)\Delta x$.
- The area under the curve between x and $(x + \Delta x)$ is ΔA.

Therefore, we can write

$$f(x)\Delta x \leq A(x + \Delta x) - A(x) \leq f(x + \Delta x)\Delta x$$

Divide each term of this inequality by Δx:

$$f(x) \leq \frac{A(x + \Delta x) - A(x)}{\Delta x} \leq f(x + \Delta x)$$

For each member of this inequality, take the limit as Δx approaches zero:

- $\lim\limits_{\Delta x \to 0} f(x) = f(x)$

- $\lim\limits_{\Delta x \to 0} f(x + \Delta x) = f(x)$

- $\lim\limits_{\Delta x \to 0} \dfrac{A(x + \Delta x) - A(x)}{\Delta x} = \dfrac{dA}{dx}$ From the definition of a derivative

Then: $f(x) \le \dfrac{dA}{dx} \le f(x)$

But this must mean that $\dfrac{dA}{dx} = f(x)$

or $dA = f(x)\,dx$

To find the function $A(x)$, take the indefinite integral of both sides of this last equation. $\displaystyle\int dA = \int f(x)\,dx$

$A(x) = \displaystyle\int f(x)\,dx$

This tells us that we can use the integration process of Section 24-2 to find the area under the curve.* From the definition of the indefinite integral (24-4),

$A(x) = g(x) + C$

where $g(x)$ is the antiderivative of $f(x)$; that is, $g'(x) = f(x)$. Therefore, when $x = a$,

$A(a) = g(a) + C$

$0 = g(a) + C$

$C = -g(a)$

In general,

$A(x) = g(x) - g(a)$

and for the area under the curve between $x = a$ and $x = b$,

$A(b) = g(b) - g(a)$

Therefore, the area under the curve is given by equation (24-18).

Area under a Curve between a and b

$$A = \lim_{n \to \infty} \sum_{i=1}^{n} f(x_i)\Delta x = g(b) - g(a)$$ (24-18)

where $g'(x) = f(x)$ and $\Delta x = \dfrac{b - a}{n}$.

---

*This equivalence of integration and the limit of a sum is the reason that Leibniz in 1675 used an elongated S-shaped symbol for the integral sign. The symbol stands for the Latin word *summa*, or sum.

We have reduced the problem of finding the area under the curve $f(x)$ to that of evaluating $g(x)$, the antiderivative of $f(x)$.

Because the limit process (24-13) is important in mathematical, scientific, and technical applications that do not directly involve the area under a curve, a more general notation for this process has been developed. The **definite integral** of a function $f(x)$ is defined as shown in (24-19).

Definite Integral $\displaystyle\int_a^b f(x)\ dx = \lim_{n\to\infty} \sum_{i=1}^n f(x_i)\Delta x$ (24-19)

where $f(x)$ is defined on the interval $a \le x \le b$, and provided that the limit exists.

Limits of integration

The real numbers a and b are called the **limits of integration.** The number a is the **lower limit** and b is the **upper limit** of the integral. Read the quantity on the left as "the integral of $f(x)$ with respect to x from a to b."

Combining equations (24-18) and (24-19), the **fundamental theorem of calculus** states that we may evaluate the definite integral of a function $f(x)$ directly using its antiderivative, without any need to interpret it as an area.

Fundamental Theorem of Calculus $\displaystyle\int_a^b f(x)\ dx = g(b) - g(a)$

(24-20)

where $f(x)$ is a continuous function on $a \le x \le b$, and $g'(x) = f(x)$ for x in the interval.

The difference $g(b) - g(a)$ is usually denoted as

$$g(x)\ \Big]_a^b$$

and equation (24-20) can be written as

$$\int_a^b f(x)\ dx = g(x)\ \Big]_a^b = g(b) - g(a)$$

EXAMPLE 19 Find the area under the curve $y = x^2$:
(a) From $x = 1$ to $x = 4$ (b) From $x = -2$ to $x = 2$

Solution

(a) Area $= \displaystyle\int_1^4 x^2\ dx = \dfrac{x^3}{3}\ \Big]_1^4$

> The antiderivative of x^2 is $x^3/3$.

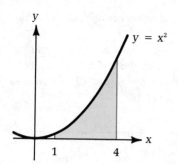

Figure 24-12

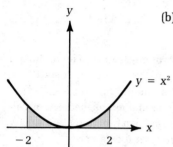

Figure 24-13

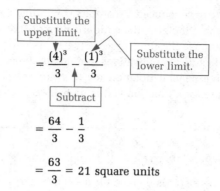

$$= \frac{(4)^3}{3} - \frac{(1)^3}{3}$$

Substitute the upper limit.

Substitute the lower limit.

Subtract

$$= \frac{64}{3} - \frac{1}{3}$$

$$= \frac{63}{3} = 21 \text{ square units}$$

(b)   Area $= \displaystyle\int_{-2}^{2} x^2\, dx = \left.\frac{x^3}{3}\right]_{-2}^{2}$

Substitute the upper limit.

$$= \frac{(2)^3}{3} - \frac{(-2)^3}{3}$$

Substitute the lower limit

$$= \frac{8}{3} - \left(-\frac{8}{3}\right)$$

$$= \frac{16}{3} = 5\frac{1}{3} \text{ square units}$$

Notice that the resulting value of the area is a number that depends only on a and b, the limits of integration. Because the variable x does not appear in the result, it is called a *dummy variable,* and any letter will do.

$$\int_a^b f(x)\, dx = \int_a^b f(z)\, dz = \int_a^b f(q)\, dq$$

IMPORTANT ▶ The definite integral $\int_a^b f(x)\, dx$ and the indefinite integral $\int f(x)\, dx$ are quite different mathematical quantities. The indefinite integral is the antiderivative of $f(x)$, a family of functions. The definite integral is a real number. ◀

EXAMPLE 20   Integrate: (a) $\displaystyle\int_{-2}^{0} 3x^5\, dx$ (b) $\displaystyle\int_{-1}^{3} (3x^2 - 4x)\, dx$

Solution

(a)   Equation (24-5) holds for both definite and indefinite integrals. Thus,

$$\int_a^b k\cdot f(x)\, dx = k \int_a^b f(x)\, dx \qquad \text{where } k \text{ is a constant}$$

Using this fact, integrate using (24-20).

$$\int_{-2}^{0} 3x^5\, dx = 3 \int_{-2}^{0} x^5\, dx = 3 \cdot \left.\frac{x^6}{6}\right]_{-2}^{0}$$

Simplify, then substitute the limits.

$$= \frac{x^6}{2}\bigg]_{-2}^{0}$$

Evaluate each expression separately.

$$= \left(\frac{0^6}{2}\right) - \left[\frac{(-2)^6}{2}\right]$$

Subtract.

$$= 0 - 32$$

$$= -32$$

When $f(x) < 0$ on the interval $a \le x \le b$, the integral $\int_a^b f(x)\,dx$ will be negative.

(b) The sum rule for integration (24-8) also holds for both definite and indefinite integrals. Thus,

$$\int_a^b [f(x) \pm h(x)]\,dx = \int_a^b f(x)\,dx \pm \int_a^b h(x)\,dx$$

Then:

$$\int_{-1}^{3} (3x^2 - 4x)\,dx = \left(3 \cdot \frac{x^3}{3} - 4 \cdot \frac{x^2}{2}\right)\bigg]_{-1}^{3}$$

Using equation (24-20),

$$= (x^3 - 2x^2)\bigg]_{-1}^{3}$$

Substitute the upper and lower limits.

Substitute the upper limit.

Substitute the lower limit.

$$= [(3)^3 - 2(3)^2] - [(-1)^3 - 2(-1)^2]$$

$$= [27 - 18] - [-1 - 2]$$

$$= 9 + 3$$

$$= 12$$

Evaluate each quantity before subtracting.

The definite integral of a function is important in many scientific and technical applications where area is not involved. The integral $\int_a^b f(x)\,dx$ defines the area between the curve $y = f(x)$ and the x-axis only when $f(x)$ is a nonnegative continuous function for all x in the interval $a \le x \le b$ (as in Example 19). If $f(x)$ is continuous but $f(x) < 0$ for all x in the interval $a \le x \le b$ [as in Example 20(a)], then $\int_a^b f(x)\,dx < 0$, and we define the area to be the absolute value of the integral:

$$\text{Area} = \left| \int_a^b f(x)\,dx \right|$$

If $f(x)$ is continuous and assumes both positive and negative values on the interval $a \le x \le b$, then the area bounded by the curve, the x-axis, and the lines $x = a$ and $x = b$ can be found by separating the integral into two or more parts:

$$\text{Area} = \left| \int_a^c f(x)\, dx \right| + \left| \int_c^b f(x)\, dx \right| \qquad \text{(24-21)}$$

where $f(x) < 0$ for $a \le x < c$
$\quad\quad f(x) = 0$ at $x = c$
$\quad\quad f(x) > 0$ for $c < x \le b$

EXAMPLE 21

(a) Integrate $\displaystyle \int_{-1}^{2} (x^2 - 1)\, dx$.

(b) Find the area under the curve $y = x^2 - 1$ between $x = -1$ and $x = 2$.

Solution

(a) Since the problem statement makes no mention of area, we need not worry about any change in the sign of $f(x)$. Determine the value of the integral using the sum rule (24-8) and the power rule (24-7).

First, integrate.
$$\int_{-1}^{2} (x^2 - 1)\, dx = \left(\frac{x^3}{3} - x \right) \Bigg]_{-1}^{2}$$

Next, substitute each limit separately.
$$= \left[\frac{(2)^3}{3} - (2) \right] - \left[\frac{(-1)^3}{3} - (-1) \right]$$

$$= \left[\frac{8}{3} - 2 \right] - \left[-\frac{1}{3} + 1 \right]$$

$$= \frac{2}{3} - \frac{2}{3} = 0$$

(b) In this case, because an area is to be found, we must examine the curve to determine the position of areas above and below the x-axis. **First,** sketch the graph, as shown in Fig. 24-14. Note that $f(x) = 0$ at $x = 1$. The area

Figure 24-14

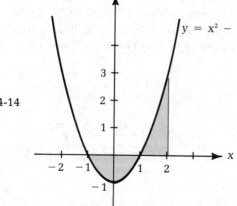

bounded by the curve between $x = -1$ and $x = 1$ is below the x-axis, and the area between $x = 1$ and $x = 2$ is above the x-axis. Using (24-21), we can express the area as a sum of two integrals:

$$\text{Area} = \left| \int_{-1}^{1} (x^2 - 1)\, dx \right| + \left| \int_{1}^{2} (x^2 - 1)\, dx \right|$$

Next, integrate using the sum and power rules, as in part (a). For the first integral,

$$\int_{-1}^{1} (x^2 - 1)\, dx = \left(\frac{x^3}{3} - x \right) \Big]_{-1}^{1}$$

$$= \left[\frac{(1)^3}{3} - (1) \right] - \left[\frac{(-1)^3}{3} - (-1) \right]$$

$$= \left(-\frac{2}{3} \right) - \left(\frac{2}{3} \right) = -\frac{4}{3}$$

For the second integral,

$$\int_{1}^{2} (x^2 - 1)\, dx = \left(\frac{x^3}{3} - x \right) \Big]_{1}^{2}$$

$$= \left[\frac{(2)^3}{3} - (2) \right] - \left[\frac{(1)^3}{3} - (1) \right]$$

$$= \left(\frac{2}{3} \right) - \left(-\frac{2}{3} \right) = \frac{4}{3}$$

Finally, calculate the total area by adding the absolute values:

$$\text{Area} = \left| -\frac{4}{3} \right| + \left| \frac{4}{3} \right| = \frac{8}{3}$$

$$= 2\frac{2}{3} \text{ square units}$$

Notice that for $-1 < x < 1$ where $f(x) < 0$, the integral has a negative value.

EXAMPLE 22   (a) $\displaystyle \int_{0}^{2} \frac{2x\, dx}{(x^2 + 2)^2}$

(b) The energy U (in joules) stored in a certain radially symmetric magnetic field is given by an equation of the form

$$U = \int_{a}^{b} 6kr\sqrt{r^2 + 1}\, dr$$

where r is the distance from the center of the field and k is a constant. Integrate to find the energy stored in the field between radial distances $b = 0.10$ cm and $a = 0.01$ cm. Round to two decimal places.

Solution
(a) **First,** use u substitution. Let $u = x^2 + 2$; then $du = 2x\, dx$

Next, find the new limits of integration. When $x = 0$, $u = 2$ and when $x = 2$, $u = 6$.

Finally, rewrite the integral as

$$\int_2^6 \frac{du}{u^2} = \int_2^6 u^{-2} \, du \quad \text{and integrate.}$$

$$= \frac{u^{-1}}{-1}\Bigg]_2^6 = -\frac{1}{u}\Bigg]_2^6$$

$$= \left(-\frac{1}{6}\right) - \left(-\frac{1}{2}\right) = \frac{1}{3}$$

(b) We want to integrate

$$U = \int_{0.01}^{0.1} 6kr\sqrt{r^2 + 1} \, dr$$

First, use u substitution. Let $u = r^2 + 1$; then $du = 2r \, dr$, and the integral becomes

$$U = 3k \int_{r=0.01}^{r=0.1} u^{1/2} \, du = 3k\left[\frac{u^{3/2}}{\frac{3}{2}}\right]_{r=0.01}^{r=0.1}$$

We have labeled the upper and lower limits as $r = 0.1$ and $r = 0.01$ to remind us that they are limits on r and not on u.

$$= 2k\left[u^{3/2}\right]_{r=0.01}^{r=0.1}$$

Second, use the substitutuion rule $u = r^2 + 1$ to change the limits to values of u:

- At $r = 0.1$, $u = (0.1)^2 + 1 = 1.01$
- At $r = 0.01$, $u = (0.01)^2 + 1 = 1.0001$

$$U = 2k\left[u^{3/2}\right]_{1.0001}^{1.01}$$

Finally, evaluate the integral:

$$= 2k[(1.01)^{3/2} - (1.0001)^{3/2}]$$

$$1.01 \; \boxed{y^x} \; 1.5 \; \boxed{-} \; 1.0001 \; \boxed{y^x} \; 1.5 \; \boxed{=} \boxed{\times} \; 2 \boxed{=} \rightarrow \boxed{\textit{0.0297749}}$$

$$U \approx 0.03k \text{ J}, \quad \text{rounded}$$

Exercises 24-4

Integrate.

1. $\displaystyle\int_0^2 2x \, dx$ 　　　　**2.** $\displaystyle\int_0^1 4x \, dx$ 　　　　**3.** $\displaystyle\int_1^3 4x^3 \, dx$

4. $\displaystyle\int_1^4 2x^2 \, dx$ 　　　　**5.** $\displaystyle\int_1^4 x^{3/2} \, dx$ 　　　　**6.** $\displaystyle\int_{-1}^8 x^{1/3} \, dx$

7. $\int_0^2 (2x^2 - 3)\, dx$ **8.** $\int_{-1}^2 (x^3 + 5x^2 - 1)\, dx$ **9.** $\int_1^2 \sqrt{5x - 1}\, dx$

10. $\int_0^1 \sqrt{4 - 3x}\, dx$ **11.** $\int_1^9 \dfrac{dx}{x\sqrt{x}}$ **12.** $\int_1^4 \left(\sqrt{x} - \dfrac{1}{\sqrt{x}}\right) dx$

13. $\int_1^2 \dfrac{2x\, dx}{(4x^2 - 1)^2}$ **14.** $\int_0^1 x(2 - x^2)^3\, dx$ **15.** $\int_2^4 x(2x^2 - 7)^{3/2}\, dx$

16. $\int_0^2 \dfrac{3x^2}{\sqrt{x^3 + 1}}\, dx$ **17.** $\int_{0.5}^1 \dfrac{1 - x}{(2x - x^2)^3}\, dx$ **18.** $\int_0^2 (x^2 - 2x)(2x^3 - 6x^2)^2\, dx$

Find the area under each curve between the given values of x.

19. $y = 5x$ from $x = 1$ to $x = 4$

20. $y = 3x^2$ from $x = 0$ to $x = 3$

21. $y = x^2 - 5x$ from $x = -3$ to $x = 3$

22. $y = x^3 + 8$ from $x = -2$ to $x = 2$

23. $y = \sqrt{2x + 3}$ from $x = -1$ to $x = 3$

24. $y = \dfrac{x}{\sqrt{x^2 + 1}}$ from $x = 0$ to $x = 4$

25. $y = x(2 - x^2)^2$ from $x = -1$ to $x = 2$

26. $y = \dfrac{x - 2}{(x^2 - 4x)^2}$ from $x = 1$ to $x = 3$

Solve.

27. Petroleum engineering   In the study of petroleum geology, the following integral appears:

$$Q(a) = \int_0^{\sqrt{a}} \dfrac{f(x)}{\sqrt{a - x^2}}\, dx$$

Integrate to find $Q(a)$ when $f(x) = 2x$.

28. Hydraulics   The following integral arises from Poiseuille's law, which describes viscous fluid flow in a pipe:

$$F(R) = \int_0^R \dfrac{\pi P}{2kL}(R^2 - r^2)r\, dr$$

Integrate to find the fluid flow rate $F(R)$.

29. Electronics   The electrical potential V at a point P on the axis of a uniformly charged irregular disk can be found from the integral

$$V = k \int_0^a \dfrac{x\, dx}{\sqrt{x^2 + r^2}}$$

where r is the distance of the point P from some reference point on the disk, a is a constant related to the size of the disk, and k is a constant. Integrate to find $V(a)$.

24-5 | Numerical Integration

If a continuous function $f(x)$ has an antiderivative $g(x)$, then according to the fundamental theorem of calculus (24-20),

$$\int_a^b f(x)\, dx = g(b) - g(a)$$

But not every function has an antiderivative that is itself a simple function or a combination of simple functions. In fact, in many technical applications there occur integrals whose integrand is an empirical function given by a table of data or a graph plotted from measured values. For such functions, integration requires methods other than the fundamental theorem.

EXAMPLE 23   A recording tachymeter plots on a stripchart the rotation rate R of a flywheel as a function of time t (Fig. 24-15). How many revolutions did this flywheel make in the first 8 min?

Figure 24-15

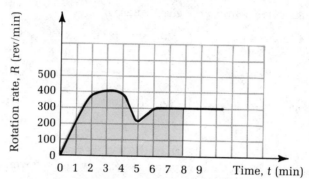

Solution   To calculate the number of revolutions, we need to determine the value of the integral $\int_0^8 R(t)\, dt$. The function $R(t)$ gives the value of the rotation rate R as a function of time t. The limits on the integral indicate that we are counting revolutions from the start of motion at $t = 0$ min to some later time $t = 8$ min.

Because an equation for $R(t)$ is not available, we can evaluate the integral directly from the graph by estimating the area beneath the curve.

The area of one square on the graph is

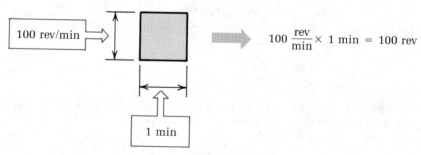

Interval	0–1	1–2	2–3	3–4	4–5	5–6	6–7	7–8	
Estimated Number of Squares	1	3	4	4	3	$2\frac{1}{2}$	3	3	Total number of squares = $23\frac{1}{2}$

Number of revolutions = $23\frac{1}{2} \times 100 = 2350$ revolutions

More accurate results can be obtained using an area-measuring device called a **planimeter**.

Numerical integration involves a set of techniques that allow us to calculate an approximate value for a definite integral. The rectangle, or Riemann, sum given by equation (24-13) is one example of numerical integration. A much better approximation may be obtained by summing the areas of trapezoids inscribed in the curve (Fig. 24-16).

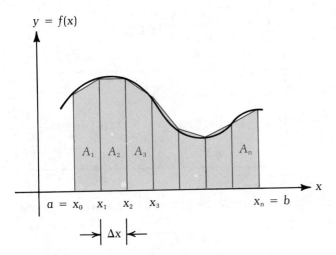

Figure 24-16

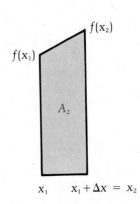

Figure 24-17

As before, we choose points $x_1, x_2, x_3, \ldots, x_n$ along the x-axis so that the distance ab is divided into n equal intervals. Then the width of an interval is

$$\Delta x = \frac{b - a}{n}$$

The area of a trapezoid is equal to its altitude, h, multiplied by the average length of its bases, b_1 and b_2. For example, in Fig. 24-17,

$$A_2 = \Delta x \cdot \frac{1}{2} \cdot [f(x_1) + f(x_2)] = \Delta x \left[\frac{1}{2} f(x_1) + \frac{1}{2} f(x_2) \right]$$

The area of each trapezoid can be calculated in this way:

$$A_1 = \Delta x \left[\frac{1}{2} f(x_0) + \frac{1}{2} f(x_1) \right] = \Delta x \left[\frac{1}{2} f(a) + \frac{1}{2} f(x_1) \right]$$

$$A_2 = \Delta x \left[\frac{1}{2} f(x_1) + \frac{1}{2} f(x_2) \right]$$

$$A_3 = \Delta x \left[\frac{1}{2} f(x_2) + \frac{1}{2} f(x_3) \right]$$

$$\vdots$$

$$A_n = \Delta x \left[\frac{1}{2} f(x_{n-1}) + \frac{1}{2} f(x_n) \right] = \Delta x \left[\frac{1}{2} f(x_{n-1}) + \frac{1}{2} f(b) \right]$$

$$A = h \cdot \frac{b_1 + b_2}{2}$$

Total area $= A_1 + A_2 + A_3 + \cdots + A_n$

$$= \Delta x \left[\frac{1}{2}f(a) + \frac{1}{2}f(x_1) + \frac{1}{2}f(x_1) + \frac{1}{2}f(x_2) + \frac{1}{2}f(x_2) \right.$$

$$\left. + \frac{1}{2}f(x_3) + \cdots + \frac{1}{2}f(x_{n-1}) + \frac{1}{2}f(b) \right]$$

$$= \Delta x \left[\frac{1}{2}f(a) + f(x_1) + f(x_2) + \cdots + \frac{1}{2}f(b) \right]$$

Trapezoidal Rule

$$\int_a^b f(x)\, dx \approx \frac{b-a}{n} \left[\frac{1}{2}f(a) + f(x_1) + f(x_2) + \cdots + f(x_{n-1}) + \frac{1}{2}f(b) \right]$$

$$(24\text{-}22)$$

▦ **EXAMPLE 24**   Use the trapezoidal rule (24-22) to find the area of the quarter circle shown in Fig. 24-18. Use $n = 5$.

Figure 24-18

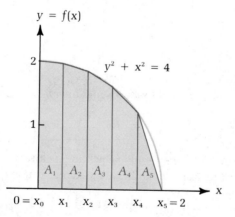

Solution   For $n = 5$, $\qquad \Delta x = \dfrac{b-a}{n} = \dfrac{2-0}{5}$

$$= 0.4$$

The equation of the circle is $\qquad y^2 + x^2 = 2^2$

Therefore $\qquad f(x) = y = \sqrt{4 - x^2}$

The work is shown in Table 24-7.
From the trapezoidal rule (24-22),

$$\text{Area} \approx (0.4)\left[\frac{1}{2}(2.0) + (1.9595918) + (1.8330303) + (1.6) + (1.2) + \frac{1}{2}(0) \right]$$

$$\approx 3.0370488 \quad \text{or about 3.04 square units}$$

TABLE 24-7

i	$x_i = 0 + i(\Delta x)$	$f(x_i) = \sqrt{4 - (x_i)^2}$
0	$x_0 = 0$	$f(x_0) = \sqrt{4 - (0)^2} = 2.0$
1	$x_1 = 0.4$	$f(x_1) = \sqrt{4 - (0.4)^2} \approx 1.9595918$
2	$x_2 = 0.8$	$f(x_2) = \sqrt{4 - (0.8)^2} \approx 1.8330303$
3	$x_3 = 1.2$	$f(x_3) = \sqrt{4 - (1.2)^2} = 1.6$
4	$x_4 = 1.6$	$f(x_4) = \sqrt{4 - (1.6)^2} = 1.2$
5	$x_5 = 2.0$	$f(x_5) = \sqrt{4 - (2.0)^2} = 0$

The trapezoidal rule calculation above can be performed on a calculator as follows:

4 ⊟ 0 x^2 ⊟ √ ✕ .5 ⊟ STO   4 ⊟ 0.4 x^2 ⊟ √ SUM   4 ⊟ 0.8 x^2 ⊟ √ SUM ...

$$\underbrace{}_{\frac{1}{2} \cdot f(x_0)} \qquad \underbrace{}_{f(x_1)} \qquad \underbrace{}_{f(x_2)}$$

4 ⊟ 2 x^2 ⊟ √ ✕ .5 ⊟ SUM RCL ✕ .4 ⊟ → **3.0370488**

$$\underbrace{}_{\frac{1}{2} \cdot f(x_5)}$$

Table 24-8 shows that the approximation improves as the number of intervals increases.

TABLE 24-8
For $f(x) = \sqrt{4 - x^2}$
between $x = 0$ and $x = 2$

Number of Intervals, n	$\int_a^b f(x)\, dx$
5	3.0370488
10	3.1045183
100	3.1304179
1,000	3.1405555
10,000	3.1414915

The area of a quarter circle can be calculated directly using the geometric formula to confirm the approximation in Table 24-8:

$$\text{Area} = \frac{1}{4}(\pi r^2) = \frac{1}{4}(\pi)(2^2) = \pi$$

$$\approx 3.1415927 \text{ square units}$$

EXAMPLE 25   The inductance L (in millihenries) of a toroidal coil with rectangular cross section, inner radius 0.05 m, and outer radius 0.10 m, is given by the equation

$$L = k\int_{0.05}^{0.10} \frac{dr}{r}$$

where the constant $k = 2$ mH. Evaluate this integral using the trapezoidal rule (24-22) with $n = 5$. Round to the nearest tenth.

Solution   For $n = 5$, $\Delta r = \dfrac{0.10 - 0.05}{5} = 0.01$

The calculation is summarized in Table 24-9.

TABLE 24-9

i	$r_i = 0.05 + i(\Delta r)$	$f(r_i) = \dfrac{1}{r_i}$
0	$r_0 = 0.05$	$f(r_0) = \dfrac{1}{0.05} = 20.0$
1	$r_1 = 0.06$	$f(r_1) = \dfrac{1}{0.06} = 16.666667$
2	$r_2 = 0.07$	$f(r_2) = \dfrac{1}{0.07} = 14.285714$
3	$r_3 = 0.08$	$f(r_3) = \dfrac{1}{0.08} = 12.5$
4	$r_4 = 0.09$	$f(r_4) = \dfrac{1}{0.09} = 11.111111$
5	$r_5 = 0.10$	$f(r_5) = \dfrac{1}{0.10} = 10.0$

From the trapezoidal rule (24-22),

$$L \approx 2(0.01)\left[\frac{1}{2}(20) + (16.666667) + (14.285714) + (12.5) + (11.111111) + \frac{1}{2}(10)\right]$$

(where the box labeled k points to the factor 2)

$$\approx (0.02)(69.563492)$$

$$\approx 1.3912698 \approx 1.4 \text{ mH}$$

A more exact value ($L = 1.3862943\ldots$) can be found using an integration formula we will study in Chapter 27.

The calculation of L using the trapezoidal rule can be performed directly using a calculator. The sequence of operations looks like this:

.05 $[1/x]$ $[\times]$ .5 $[=]$ $[\text{STO}]$ .05 $[+]$ .01 $[=]$ $[1/x]$ $[\text{SUM}]$ .05 $[+]$ .02 $[=]$ $[1/x]$ $[\text{SUM}]$ . . .

$\underbrace{\hspace{2cm}}$ $\underbrace{\hspace{2cm}}$ $\underbrace{\hspace{2cm}}$

$\frac{1}{2}f(r_0)$ $f(r_1)$ $f(r_2)$

.05 $[+]$ .05 $[=]$ $[1/x]$ $[\times]$ .5 $[=]$ $[\text{SUM}]$ $[\text{RCL}]$ $[\times]$ .02 $[=]$ → ▓ *1.3912698*

$\underbrace{\hspace{3cm}}$

$\frac{1}{2}f(b)$ $\boxed{k\Delta r}$

EXAMPLE 26   Calculate the value of the following integral using the trapezoidal rule (24-22) with $n = 10$. Round to three decimal places.

$$\int_0^1 \frac{dx}{\sqrt{x^2 + 1}}$$

Solution   For $n = 10$, $\Delta x = \frac{1 - 0}{10} = 0.1$

Here is the sequence of calculator displays for this calculation:

1 $[\times]$ .5 $[=]$ $[\text{STO}]$ →   ▓ *0.5* $\frac{1}{2}f(x_0)$

.1 $[x^2]$ $[+]$ 1 $[=]$ $[\sqrt{\ }]$ $[1/x]$ $[\text{SUM}]$ →   ▓ *0.9950372* $f(x_1)$

.2 $[x^2]$ $[+]$ 1 $[=]$ $[\sqrt{\ }]$ $[1/x]$ $[\text{SUM}]$ →   ▓ *0.9805807* $f(x_2)$

.3 $[x^2]$ $[+]$ 1 $[=]$ $[\sqrt{\ }]$ $[1/x]$ $[\text{SUM}]$ →   ▓ *0.9578263* $f(x_3)$

.4 $[x^2]$ $[+]$ 1 $[=]$ $[\sqrt{\ }]$ $[1/x]$ $[\text{SUM}]$ →   ▓ *0.9284767* $f(x_4)$

.5 $[x^2]$ $[+]$ 1 $[=]$ $[\sqrt{\ }]$ $[1/x]$ $[\text{SUM}]$ →   ▓ *0.8944272* $f(x_5)$

.6 $[x^2]$ $[+]$ 1 $[=]$ $[\sqrt{\ }]$ $[1/x]$ $[\text{SUM}]$ →   ▓ *0.8574929* $f(x_6)$

.7 $[x^2]$ $[+]$ 1 $[=]$ $[\sqrt{\ }]$ $[1/x]$ $[\text{SUM}]$ →   ▓ *0.8192319* $f(x_7)$

.8 $[x^2]$ $[+]$ 1 $[=]$ $[\sqrt{\ }]$ $[1/x]$ $[\text{SUM}]$ →   ▓ *0.7808688* $f(x_8)$

.9 $[x^2]$ $[+]$ 1 $[=]$ $[\sqrt{\ }]$ $[1/x]$ $[\text{SUM}]$ →   ▓ *0.7432941* $f(x_9)$

1 $[+]$ 1 $[=]$ $[\sqrt{\ }]$ $[1/x]$ $[\times]$ .5 $[=]$ $[\text{SUM}]$ →   ▓ *0.3535534* $\frac{1}{2}f(x_{10})$

$[\text{RCL}]$ $[\times]$ .1 $[=]$ →   ▓ *0.8810789*

Therefore,

$$\int_0^1 \frac{dx}{\sqrt{x^2 + 1}} \approx 0.881, \quad \text{rounded}$$

EXAMPLE 27   In the development of a new internal combustion engine, measurements are made of the pressure of the gas in the cylinder for various volumes, as shown in Table 24-10. Calculate the work W done by the gas on the expansion part of the cycle from the integral

$$W = \int_{0.10}^{0.35} P \, dV$$

TABLE 24-10

Volume, V (L)	Pressure, P (atm)
0.10	1.5
0.15	2.2
0.20	3.1
0.25	4.2
0.30	5.5
0.35	7.8

Solution $\Delta V = 0.05$

$V_0 = 0.10$ $P(V_0) = 1.5$

$V_1 = 0.15$ $P(V_1) = 2.2$

$V_2 = 0.20$ $P(V_2) = 3.1$

$V_3 = 0.25$ $P(V_3) = 4.2$

$V_4 = 0.30$ $P(V_4) = 5.5$

$V_5 = 0.35$ $P(V_5) = 7.8$

Using the trapezoidal rule (24-22),

$$W \approx (0.05)\left[\frac{1}{2}(1.5) + (2.2) + (3.1) + (4.2) + (5.5) + \frac{1}{2}(7.8)\right]$$

$$\approx (0.05)(19.65)$$

$$\approx 0.9825 \quad \text{or about } 0.98 \text{ L·atm}$$

Notice that W has the units of the product of P and V.

Exercises 24-5

Use the trapezoidal rule to find the approximate value of each integral for the given value of n. Check your answers to Problems 1–4 by direct integration. Round to two decimal places.

1. $\int_0^3 (x^2 + 1) \, dx; \quad n = 6$

2. $\int_0^2 \frac{x^3}{2} \, dx; \quad n = 4$

3. $\int_1^2 \frac{dx}{x^2}; \quad n = 3$

4. $\int_1^4 \sqrt{2x + 1} \, dx; \quad n = 6$

5. $\int_2^6 \sqrt{x^2 + 2} \, dx; \quad n = 8$

6. $\int_0^2 \sqrt{9 - x^3} \, dx; \quad n = 4$

7. $\int_1^4 x\sqrt{x + 1} \, dx; \quad n = 9$

8. $\int_2^4 \frac{x \, dx}{\sqrt{x + 1}}; \quad n = 8$

9. $\int_0^3 2^x \, dx; \quad n = 10$

10. $\int_1^5 e^x \, dx; \quad n = 8$

11. $\int_1^2 \frac{dx}{x^2 + 2x}; \quad n = 10$

12. $\int_0^2 \frac{dx}{1 + x^2}; \quad n = 10$

Use the given set of points to find the approximate value of each integral. Round to two decimal places.

13. Find $\int_{0.1}^{0.6} y \, dx$ for:

x	0.1	0.2	0.3	0.4	0.5	0.6
y	5.8	4.7	3.7	2.8	3.9	5.9

14. Find $\int_1^5 y\,dx$ for:

x	1	2	3	4	5
y	14.7	15.5	16.2	18.8	21.9

15. Find $\int_2^{10} y\,dx$ for:

x	2	4	6	8	10
y	0.48	0.31	0.15	0.18	0.26

16. Find $\int_{0.25}^{1.50} y\,dx$ for:

x	0.25	0.50	0.75	1.00	1.25	1.50
y	18.5	40.2	44.7	48.6	46.9	27.3

Solve.

17. *Mechanical engineering*   The graph gives the rotation rate R of a flywheel as a function of time t. How many revolutions did the flywheel make in the first 6 min?

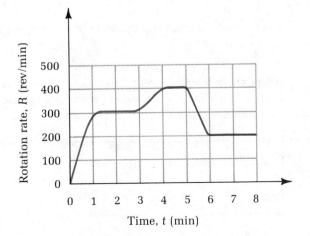

18. *Physics* The force F exerted by an explosion is measured using a pressure sensor and is shown in the graph. Integrate using the trapezoidal rule to find the impulse $\int_0^{12} F\, dt$.

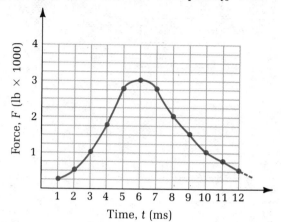

Time, t (ms)

19. *Electronics* The electric potential V at a point b on the axis of a uniformly charged disk of radius a is given by an equation of the form

$$V = \frac{\sigma}{2\epsilon_0} \int_0^a \frac{dx}{\sqrt{x^2 + r^2}}$$

Integrate to find V (in volts) using the trapezoidal rule with $n = 10$ for $\sigma/2\epsilon_0 = 600$ V, $r = 1.0$ cm, and $a = 2.0$ cm.

20. Use the trapezoidal rule with $n = 15$ to find the approximate value of the area bounded by the curve $y = x(1 - x^2)^2$, the x-axis, and the lines $x = -1$ and $x = 2$. Round to the nearest tenth.

21. Use the trapezoidal rule with $n = 12$ to find the approximate value of the area bounded by the curve $y = 4x - x^2$, the x-axis, and the lines $x = -2$ and $x = 4$. Round to the nearest tenth.

24-6 | Applications in Physics

In Chapter 23 we used the derivative to describe the rectilinear or straight-line motion of a particle. The instantaneous velocity $v(t)$ of a particle moving along the x-axis was defined as

$$v(t) = \frac{dx}{dt} \tag{23-3}$$

and the instantaneous acceleration $a(t)$ was defined as

$$a(t) = \frac{dv}{dt} \tag{23-4}$$

where $x(t)$ is the position of the particle at time t.

Because differentiation and integration are inverse operations, each of these formulas may also be written as an integral.

Rectilinear Motion	$x(t) = \int v(t)\, dt$	(24-23)
	$v(t) = \int a(t)\, dt$	(24-24)

These integrals can be used to develop equations that describe the straight-line motion of many objects: a rocket launching, electrons moving in a conductor, falling objects, a vibrating spring, the motion of a piston in a cylinder, and so on.

▦ **EXAMPLE 28**   For an object in free fall, the acceleration due to gravitational attraction is $a(t) = -g$, where $g \approx 32.2$ ft/s^2 or 9.8 m/s^2.

(a)   Integrate to find equations for the velocity $v(t)$ and displacement $x(t)$ of an object that is at $x = 0$ and is moving with velocity v_0 at time $t = 0$.

(b)   How fast is this object moving at $t = 4.0$ s if $v_0 = 25$ m/s?

(c)   What is the location of the object at $t = 4.0$ s?

(d)   When does the object pass a point 12 m above the starting point?

Solution

(a)   From (24-24):
$$v(t) = \int a(t)\, dt = \int -g\, dt$$
$$= -g \int dt$$
$$= -gt + C$$

At $t = 0$, $v(0) = v_0$, so
$$v_0 = -g(0) + C$$
$$C = v_0$$

Therefore:
$$v(t) = -gt + v_0 \tag{24-25}$$

From (24-23):
$$x(t) = \int v(t)\, dt = \int (-gt + v_0)\, dt$$
$$x(t) = -\frac{1}{2}gt^2 + v_0 t + C$$

At $t = 0$, $x(0) = 0$, so
$$0 = -\frac{1}{2}g(0)^2 + v_0(0) + C$$
$$C = 0$$

Therefore:
$$x(t) = -\frac{1}{2}gt^2 + v_0 t \tag{24-26}$$

(b)   Use equation (24-25) from part (a) and substitute $g = 9.8$ m/s^2, $v_0 = 25$ m/s, $t = 4.0$ s:

$$v(4.0) = -9.8(4.0) + 25 = -39.2 + 25$$
$$= -14.2 \text{ m/s}$$
$$\approx -14 \text{ m/s}$$

> The negative sign means that the object is moving downward.

(c)   Use equation (24-26) from part (a) and substitute $g = 9.8$ m/s^2, $v_0 = 25$ m/s, $t = 4.0$ s:

$$x(4.0) = -\frac{1}{2}(9.8)(4.0)^2 + 25(4.0) = -78.4 + 100$$

$$= 21.6 \text{ m}$$

$$\approx 22 \text{ m}$$

$t = 4$ s

At $t = 4.0$ s, the object is 22 m above its starting point and is moving downward at 14 m/s (Fig. 24-19).

(d) Use equation (24-26) from part (a) and substitute $g = 9.8 \text{ m/s}^2$, $v_0 = 25$ m/s, $x = 12$ m:

$$12 = -\frac{1}{2}(9.8)t^2 + 25t$$

$$4.9t^2 - 25t + 12 = 0$$

Solve using the quadratic formula:

$$t = \frac{25 \pm \sqrt{(-25)^2 - 4(4.9)(12)}}{2(4.9)}$$

25 $\boxed{x^2}$ $\boxed{-}$ 4 $\boxed{\times}$ 4.9 $\boxed{\times}$ 12 $\boxed{=}$ $\boxed{\sqrt{}}$ $\boxed{\text{STO}}$

t_1: $\boxed{+}$ 25 $\boxed{=}$ $\boxed{\div}$ 2 $\boxed{\div}$ 4.9 $\boxed{=}$ → *4.5656483*

t_2: 25 $\boxed{-}$ $\boxed{\text{RCL}}$ $\boxed{=}$ $\boxed{\div}$ 2 $\boxed{\div}$ 4.9 $\boxed{=}$ → *0.5363925*

The object is 12 m from the start at $t \approx 0.5$ s on the way up, and $t \approx 4.6$ s on the way down.

Figure 24-19

CAUTION ▶

It is important to realize that $x(t)$ is **not** necessarily the distance the object travels. It is the position of the object at a given time.   ◀

Equations (24-25) and (24-26) describe motion under constant acceleration, but the general equations for rectilinear motion (24-23) and (24-24) may be used to develop equations describing situations in which the acceleration is not constant.

EXAMPLE 29   If the acceleration (in m/s²) of an ion engine varies with time according to the equation

$$a = \frac{4t}{\sqrt{1 + t^2}}$$

find its velocity v (in meters per second) at $t = 2.0$ s after ignition. Assume $v(0) = 0$.

Solution   From equation (24-24):

$$v(t) = \int a \, dt = \int \frac{4t \, dt}{\sqrt{1 + t^2}}$$

Use u substitution with $u = 1 + t^2$ and $du = 2t \, dt$.

$$v(t) = \int \frac{2 \, du}{u} = 2 \int u^{-1/2} \, du$$

$$= 2\left(\frac{u^{1/2}}{\frac{1}{2}}\right) + C$$

$$= 4\sqrt{u} + C$$

$$= 4\sqrt{1 + t^2} + C$$

At $t = 0$ s, $v(0) = 0$:
$$0 = 4\sqrt{1 + (0)^2} + C$$

$$C = -4$$

Thus, the velocity varies with time according to the equation
$$v(t) = 4\sqrt{1 + t^2} - 4$$

At $t = 2.0$ s,
$$v(2.0) = 4\sqrt{1 + (2.0)^2} - 4$$

$$\approx 4.9 \text{ m/s}$$

EXAMPLE 30 In a study of automobile safety, a collision between a freely moving object and a padded surface produces an impulse force on the object. The resulting acceleration can be described by the equation $a(t) = t - t^2$, where $t \le 1$ s.

(a) Find the velocity v (in meters per second) of the object at $t = 0.80$ s, assuming it is at rest at $x = 0$ m when $t = 0$ s.

(b) Find the displacement of the object from $t = 0.2$ s to $t = 1.0$ s. Round to the nearest hundredth.

Solution

(a) From (24-24),
$$v(t) = \int a(t)\, dt$$

$$= \int (t - t^2)\, dt$$

$$= \frac{t^2}{2} - \frac{t^3}{3} + C$$

At $t = 0$,
$$v(0) = 0 = \frac{0^2}{2} - \frac{0^3}{3} + C$$

Therefore, $C = 0$, and the velocity at any time t is
$$v(t) = \frac{1}{2}t^2 - \frac{1}{3}t^3$$

At $t = 0.80$ s,
$$v(0.8) = \frac{1}{2}(0.8)^2 - \frac{1}{3}(0.8)^3$$

$$\approx 0.15 \text{ m/s, rounded}$$

(b) From (24-23), the displacement function is
$$x(t) = \int v(t)\, dt$$

$$= \int \left(\frac{t^2}{2} - \frac{t^3}{3}\right) dt$$

$$= \frac{t^3}{6} - \frac{t^4}{12} + C$$

The change in position from $t = 0.2$ s to $t = 1.0$ s is

$$x = \int_{0.2}^{1.0} \left(\frac{t^2}{2} - \frac{t^3}{3} \right) dt$$

$$= \frac{t^3}{6} - \frac{t^4}{12} \Bigg]_{0.2}^{1.0}$$

$$= \left[\frac{(1.0)^3}{6} - \frac{(1.0)^4}{12} \right] - \left[\frac{(0.2)^3}{6} - \frac{(0.2)^4}{12} \right]$$

$$\approx 0.833\ldots - 0.0012$$

$$\approx 0.08 \text{ m}$$

Work   In ordinary use, the word *work* is used as a synonym for terms such as power, energy, or force. In physics, work is a very important concept with a precise meaning. If an object is acted upon by a constant force F and moves in a straight line in the direction of the force, then the **work** done by the force is defined as follows:

Work $W = Fd$	(24-27)

where F is the magnitude of the force and d is the distance through which the object moves (Fig. 24-20).

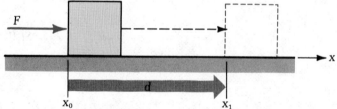

Figure 24-20

The metric and English units for work are shown in Table 24-11. The newton is a metric unit of force, equivalent to about 0.225 lb. A 1 kg mass weighs about 0.1 N.

TABLE 24-11

Force	× Distance	= Work
Pound, lb	Foot, ft	ft·lb
Newton, N	Meter, m	N·m (or joule, J)

If the constant force does not act in the direction in which the object moves, then the work done on the object depends on the component of the force along the direction of motion.

> **Work Done by a Constant Force** $W = Fd \cos \theta$ (24-28)
>
> where θ is the angle between the force and the direction of motion
> (Fig. 24-21).

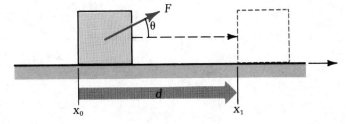

Figure 24-21

Other forces, such as friction or weight, may be acting on the object at the same time F is exerted, but the work defined is that due only to force F. The work done by the other forces must be considered separately.

NOTE ▶   If the object does not move, $d = 0$, and no work is done.   ◀

Variable force   If the force F is not constant, but varies in magnitude with the position of the object, then the work done in moving the object through a small displacement Δx is

$$\Delta W = F \cdot \Delta x$$

where F is the average value of the force component along the x-axis, the direction of motion.

The total work done in moving the object from position x_0 to x_1 is therefore given by equation (24-29):

> **Work Done by a Variable Force** $W = \displaystyle\int_{x_0}^{x_1} F(x)\, dx$ (24-29)

EXAMPLE 31   If a spring is stretched so that its endpoint moves a distance x, then the spring will exert an opposing force tending to restore the spring to its normal unstretched state (Fig. 24-22). For small extensions, the force F needed

Figure 24-22

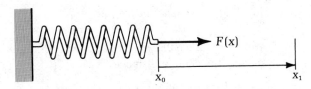

to stretch the spring is proportional to the amount of stretch; that is,

$$F(x) = kx$$

where k is a constant depending on the stiffness of the spring. If the spring has a natural length of 2 in. and requires a 48 lb force to stretch it to 4 in., then find the work done in stretching the spring from 2.0 in. to 6.0 in..

Solution   Calculate k: $k = \dfrac{48 \text{ lb}}{4 \text{ in.} - 2 \text{ in.}} = 24 \text{ lb/in.}$

Then, from (24-29), the work done is

$$W = \int_2^6 kx \, dx = 24 \int_2^6 x \, dx$$

$$= 24 \left[\frac{x^2}{2} \right]_2^6$$

$$= 24 \left[\frac{(6)^2}{2} - \frac{(2)^2}{2} \right]$$

$$= 24(18 - 2)$$

$$= 384 \text{ lb} \cdot \text{in.} = 32 \text{ ft} \cdot \text{lb}$$

Notice that because both k and x are given in inch units, the final calculation has units of lb·in. and must be converted to the standard English units, ft·lb.

EXAMPLE 32   Calculate the work that must be done in order to pull the chain in Fig. 24-23 onto the table top. The chain is exactly 3 ft long and weighs 12 lb.

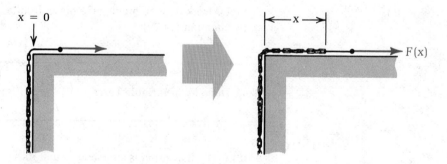

Figure 24-23

Solution   When x ft of chain is on the table, $(3 - x)$ ft remains hanging over the edge. This amount hanging over the edge weighs

$$\left(\frac{12 \text{ lb}}{3 \text{ ft}} \right)(3 - x) \text{ ft} = 4(3 - x) \text{ lb}$$

Therefore, the force that must be applied is $F(x) = 4(3 - x)$, and the work done in moving the chain a distance Δx is

$$\Delta W = F(x)\Delta x = 4(3 - x)\Delta x$$

The work done in moving the entire length of chain onto the table top is found using equation (24-29):

$$W = \int_0^3 4(3 - x)\ dx$$

$$= \int_0^3 (12 - 4x)\ dx$$

$$= \left(12x - \frac{4x^2}{2} \right) \Bigg]_0^3$$

$$= [12(3) - 2(3)^2] - [12(0) - 2(0)^2]$$

$$= 36 - 18 = 18\ \text{ft} \cdot \text{lb}$$

Improper integrals   If the function $f(x)$ in the equation for the definite integral (24-19) is continuous on the interval $a \le x \le b$ and is bounded so that a and b are finite limits of integration, the integral $\int_a^b f(x)\ dx$ is said to be a **proper integral.** An **improper integral** is one for which at least one of the following is true:

1. One or both of the limits of integration is infinite.
2. The integrand function becomes infinite for some value of x in the interval of integration $a \le x \le b$.

If one or both of the limits of integration is infinite, the improper integral is defined as follows:

Improper Integral $\displaystyle \int_a^\infty f(x)\ dx = \lim_{b \to \infty} \int_a^b f(x)\ dx$ (24-30)

$\displaystyle \int_{-\infty}^b f(x)\ dx = \lim_{a \to -\infty} \int_a^b f(x)\ dx$ (24-31)

If the limit on the right exists, the improper integral is said to **converge,** and the value of the integral is equal to the limit. If the limit does not exist, the integral is said to **diverge,** and it is not given a value.

EXAMPLE 33   The gravitational force of attraction of a planet on a nearby object has the form

$$F(r) = \frac{k}{r^2}$$

where r is the distance from the object to the center of the planet and k is a constant that depends on the mass of the planet and the mass of the object. Calculate the work that must be done to toss a 16 lb (7.3 kg) shotput completely free of a very small planet or asteroid. The planet has radius 5200 m, and k for this situation is approximately 4.4×10^7 N·m².

Solution   According to equation (24-29),

$$W = \int_{5200}^{\infty} \frac{k}{r^2}\, dr$$

The object moves from the surface of the planet, 5200 m from its center, to an infinite distance from the center of the planet. Use (24-30) to evaluate the improper integral:

$$W = \lim_{b \to \infty} \int_{5200}^{b} kr^{-2}\, dr$$

$$= k \cdot \lim_{b \to \infty} \left[-\frac{1}{r} \right]_{5200}^{b}$$

$$= k \cdot \lim_{b \to \infty} \left[\left(-\frac{1}{b} \right) - \left(-\frac{1}{5200} \right) \right]$$

But $\lim\limits_{b \to \infty} (-1/b) = 0$. Therefore,

$$W = \frac{k}{5200} = \frac{4.4 \times 10^7}{5200}$$

$$\approx 8500 \text{ J}$$

An integral is also said to be improper if the integrand function $f(x)$ approaches $+\infty$ or $-\infty$ somewhere in the interval $a \le x \le b$. For example, if the function has an infinite discontinuity at the upper limit $x = b$ (Fig. 24-24), we evaluate the integral using equation (24-32).

$$\int_a^b f(x)\, dx = \lim_{c \to b^-} \int_a^c f(x)\, dx \tag{24-32}$$

where $f(x) \to \infty$ or $f(x) \to -\infty$ as $x \to b$

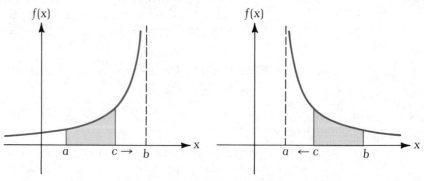

Figure 24-24   Figure 24-25

The limit c approaches point b from within the interval of integration. If the integrand function $f(x)$ has an infinite discontinuity at the lower limit $x = a$ (Fig. 24-25), evaluate the integral using equation (24-33).

$$\int_a^b f(x)\,dx = \lim_{c \to a^+} \int_c^b f(x)\,dx \qquad (24\text{-}33)$$

where $f(x) \to \infty$ or $f(x) \to -\infty$ as $x \to a$

The limit c approaches point a from within the interval of integration.

EXAMPLE 34   Evaluate: $\displaystyle\int_0^3 \frac{dx}{\sqrt{x}}$

Solution   **First,** note that the function $f(x) = 1/\sqrt{x}$ has an infinite discontinuity at $x = 0$ (Fig. 24-26). Using (24-33), evaluate this integral from the limit:

$$\int_0^3 \frac{dx}{\sqrt{x}} = \lim_{c \to 0} \int_c^3 \frac{dx}{\sqrt{x}} = \lim_{c \to 0} \left[2\sqrt{x}\right]_c^3$$

$$= \lim_{c \to 0} \left[(2\sqrt{3}) - (2\sqrt{c})\right]$$

Second, determine the limit. As $c \to 0$, $\sqrt{c} \to 0$. Therefore,

$$\int_0^3 \frac{dx}{\sqrt{x}} = 2\sqrt{3}$$

If the integrand function $f(x)$ is continuous in the interval $a \le x \le b$ except for a point c, $a < c < b$, where $f(x)$ tends to $+\infty$ or $-\infty$ as x approaches c from either the left or the right, we can write the integral as a sum of two improper integrals (Fig. 24-27).

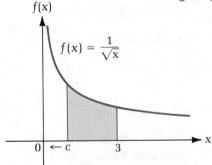

Figure 24-26

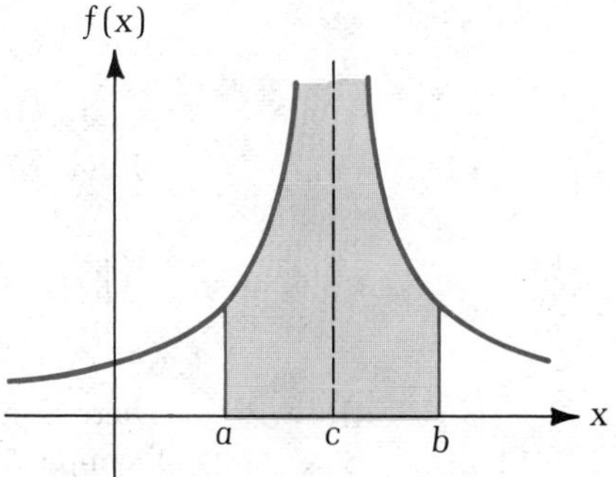

$f(x)$

$f(x) = \dfrac{1}{\sqrt{x}}$

$0 \quad \leftarrow c \qquad 3$

$f(x)$

$a \quad c \quad b$

Figure 24-27

$$\int_a^b f(x)\,dx = \int_a^c f(x)\,dx + \int_c^b f(x)\,dx \qquad (24\text{-}34)$$

where $f(x) \to \infty$ or $f(x) \to -\infty$ as $x \to c$

We can evaluate the improper integrals on the right using equations (24-32) and (24-33). If *either* of the improper integrals on the right of (24-34) diverges, then the original integral diverges.

EXAMPLE 35   Evaluate: $\displaystyle\int_1^4 \frac{dx}{(x-2)^{2/3}}$

Solution   **First,** note that the integrand function $f(x) = 1/(x-2)^{2/3}$ approaches ∞ as $x \to 2$. Use (24-34) to separate this integral into two improper integrals:

$$\int_1^4 \frac{dx}{(x-2)^{2/3}} = \int_1^2 \frac{dx}{(x-2)^{2/3}} + \int_2^4 \frac{dx}{(x-2)^{2/3}}$$

Next, evaluate these integrals using (24-32) and (24-33):

$$\int_1^2 \frac{dx}{(x-2)^{2/3}} = \lim_{c \to 2^-} \int_1^c \frac{dx}{(x-2)^{2/3}}$$

$$= \lim_{c \to 2^-} [3(c-2)^{1/3} - 3(1-2)^{1/3}]$$

$$= 3$$

> As $c \to 2^-$, this term $\to 0$.

$$\int_2^4 \frac{dx}{(x-2)^{2/3}} = \lim_{c \to 2^+} \int_c^4 \frac{dx}{(x-2)^{2/3}}$$

$$= \lim_{c \to 2^+} [3(4-2)^{1/3} - 3(c-2)^{1/3}]$$

$$= 3\sqrt[3]{2}$$

> As $c \to 2^+$, this term $\to 0$.

Finally, since both improper integrals converge, we can add to find the value of the original integral:

$$\int_1^4 \frac{dx}{(x-2)^{2/3}} = 3 + 3\sqrt[3]{2}$$

CAUTION ▶

There is no immediately obvious way to tell that an integral such as that in Example 35 is an improper integral. You must therefore test values of x within the limits of integration to determine if an infinite discontinuity exists. Ignoring this necessary test may lead to an incorrect result. For example,

$$\int_{-2}^2 \frac{dx}{x^2} = \left[-\frac{1}{x}\right]_{-2}^2 = \left(-\frac{1}{2}\right) - \left(-\frac{1}{-2}\right) = -\frac{1}{2} - \frac{1}{2} = -1$$

Since the integrand $1/x^2$ is always positive, the value of the integral cannot be negative. Using (24-34),

$$\int_{-2}^{2} \frac{dx}{x^2} = \int_{-2}^{0} \frac{dx}{x^2} + \int_{0}^{2} \frac{dx}{x^2}$$

$$= \lim_{c \to 0^-} \int_{-2}^{c} \frac{dx}{x^2} + \lim_{c \to 0^+} \int_{c}^{2} \frac{dx}{x^2}$$

$$= \lim_{c \to 0^-} \left[-\frac{1}{x} \right]_{-2}^{c} + \lim_{c \to 0^+} \left[-\frac{1}{x} \right]_{c}^{2}$$

$$= \lim_{c \to 0^-} \left[\left(-\frac{1}{c} \right) - \left(-\frac{1}{-2} \right) \right] + \lim_{c \to 0^+} \left[\left(-\frac{1}{2} \right) - \left(-\frac{1}{c} \right) \right]$$

But $1/c \to \infty$ as $c \to 0^-$ or $c \to 0^+$, so the limits do not exist, and therefore the original improper integral diverges. ◀

Exercises 24-6

1. The acceleration of an object is given by $a = 4t$ ft/s^2. Find its velocity at $t = 2.5$ s if $v = 4.0$ ft/s when $t = 1.0$ s.

2. Find the velocity (in meters per second) of a ball, falling freely under the influence of gravity, 4.0 s after it is dropped.

3. A rock is thrown vertically downward from a rooftop with an initial velocity of 15 m/s. How fast is it moving after 3.0 s?

4. The acceleration of a sliding object is given by the equation $a = 2t - \frac{1}{2}t^2$, where $t \le 4$ s. Find the displacement (in meters) of the object from $t = 1.0$ to $t = 3.0$ s. (Assume that $v = 0$ and $x = 0$ when $t = 0$.)

5. If the velocity (in meters per second) of a space vehicle powered by an ion engine is given by $v = 2t^2 + 3t + 1$, find the displacement 3.0 s after ignition if $x = 0$ m at $t = 0$ s.

6. The acceleration of a skier sliding down a hillside varies with time according to the equation $a = \sqrt{2t + 3}$. If its initial velocity is $v = 0$ at $t = 0$, find its velocity (in meters per second) after 4.0 s.

7. If the acceleration of an object subject to a steadily increasing force is given by the equation $a = 2t$ m/s^2, find the displacement after 4.0 s if $v = 8.0$ m/s and $x = 2.0$ m at $t = 0$.

8. A baseball is hit vertically upward and leaves the bat with an initial velocity of 41 m/s. What is the location of the object with respect to the batter at $t = 3.0$ s after it is hit?

9. If a spring gun shoots a pellet vertically upward with an initial velocity of 18 m/s, when will it pass a point 15 m above its starting point?

10. What must be the initial vertical velocity of a baseball if it just reaches the ceiling of an indoor stadium 63 m high?

11. Find the work done in stretching a spring from 8 in. to 12 in. if $k = 20$ lb/in.

12. Calculate the work done in stretching a spring from 6 in. to 8 in. if $k = 16$ lb/in.

13. A spring has a natural length of 12 in. When a 15 lb weight is applied, the spring stretches to 14 in. Determine k for this spring, and calculate the amount of work done in stretching the spring from 14 in. to 21 in.

14. When a 12 N force is applied to a spring with a 16 cm natural length, it stretches an additional 2 cm. Calculate k in N/cm and then find the amount of work done in stretching the spring from 16 cm to 20 cm.

15. A cable 12 ft long and weighing 24 lb hangs vertically from a table edge. Find the amount of work done in pulling the cable up onto the level surface of the table.

16. A bucket of water weighing 16 lb is suspended 48 ft down a well on a rope weighing 0.30 lb/ft. How much work is done in winding the bucket of water up to the surface?

17. The gravitational force between two objects whose centers are a distance r apart is given by Newton's equation for gravitational attraction, $F(r) = k/r^2$, where k is a constant depending on the masses of the objects (see Example 33). Find the work done in separating two objects initially a distance 0.5 m apart to a distance of 2.0 m. Express your answer as a multiple of k.

18. Find the work done against their gravitational attraction in separating two objects from a distance of 0.2 m to 1.5 m.

19. *Physics* According to Coulomb's law for electrostatic attraction and repulsion, two charges q_1 and q_2 a distance r apart each experience a force $F = kq_1q_2/r^2$, where $k = 9.0 \times 10^9$ N·m²/C². What work is done in separating a pair of oppositely charged objects if $q_1 = 1.0 \times 10^{-6}$ C, $q_2 = -3.0 \times 10^{-6}$ C, and the charges are initially 0.05 m apart and are separated to $r = 0.15$ m?*

20. *Physics* What work must be done to bring to a separation of $r = 0.01$ m two protons $(q = 1.6 \times 10^{-19}$ C) initially 0.10 m apart? (See Problem 19.)

21. *Mechanical engineering* A flywheel initially at rest rotates with angular acceleration $\alpha = t^2 + t$ s⁻².
    (a) Find an expression for its angular velocity ω at any time t if $\alpha = d\omega/dt$.
    (b) If $\omega = d\theta/dt$, where θ is the angle of rotation of the flywheel, find the total angle through which it rotates from $t = 0$ to $t = 10$ s.

22. *Mechanical engineering* A heavy flywheel rotating on its axis is slowing down because of friction in its bearings. If $\alpha = \dfrac{d\omega}{dt} = \dfrac{200}{\sqrt{t+1}}$, find ω at time $t = 5.0$ s, if $\omega = 0$ at $t = 0$.

Evaluate each improper integral.

23. $y = \displaystyle\int_0^1 \frac{dx}{\sqrt{1-x}}$

24. $y = \displaystyle\int_2^4 \frac{dx}{(x-2)^{3/4}}$

25. $y = \displaystyle\int_2^\infty \frac{dx}{(x-1)^2}$

26. $y = \displaystyle\int_{-\infty}^0 \frac{dx}{(2x-1)^3}$

27. $y = \displaystyle\int_1^\infty \frac{dx}{\sqrt{x}}$

28. $y = \displaystyle\int_0^8 x^{-1/3}\, dx$

29. $y = \displaystyle\int_{-2}^7 \frac{dx}{(x+1)^{2/3}}$

30. $y = \displaystyle\int_{-\infty}^\infty x^3\, dx$

31. Find the area under the curve $y = 1/\sqrt{x}$ from $x = 0$ to $x = 1$.

32. *Electronics* The magnetic potential V at a point on the axis of a certain coil is given by an equation of the form

$$V = k \int_a^\infty \frac{x\, dx}{(R^2 + x^2)^{3/2}}$$

Integrate to find an expression for V.

33. *Physics* In Problem 19, how much work must be done to move one object infinitely far from the other?

24-7 | Review of Chapter 24

Important Terms and Concepts

differential calculus (p. 951)

integral calculus (p. 951)

differentials (p. 952)

integrand (p. 962)

variable of integration (p. 962)

limits of integration: lower limit/upper limit (p. 983)

numerical integration (p. 989)

*William M. Hayt, Jr., *Engineering Electromagnetics*, 4th edition (New York: McGraw-Hill, 1981), p. 30

Formulas and Rules

- definition of differentials (p. 952)

$$dx = \Delta x$$
$$dy = f'(x)\, dx \qquad \text{for } y = f(x) \tag{24-1}$$

- relative error (p. 955)

$$\text{Relative error} = \frac{|\Delta y|}{y} = \frac{|\text{Error in measurement}|}{\text{Exact value}} \tag{24-2}$$

- percent error (p. 955)

$$\text{Percent error} = (\text{Relative error} \times 100)\% \tag{24-3}$$

- indefinite integral (p. 962)

$$\int f(x)\, dx = g(x) + C \qquad \text{where } g'(x) = f(x) \tag{24-4}$$

- integration of a constant (p. 963)

$$\int k\, dx = k \int dx \tag{24-5}$$

$$\text{where } k \text{ is a constant}$$

$$\int k\, dx = kx + C \tag{24-6}$$

- power rule of integration (p. 964)

$$\int x^n\, dx = \frac{x^{n+1}}{n+1} + C \qquad n \neq -1 \tag{24-7}$$

- sum rule for integration (p. 965)

$$\int [f(x) \pm h(x)]\, dx = \int f(x)\, dx \pm \int h(x)\, dx \tag{24-8}$$

- u substitution (p. 967)

$$\int f(x)\, dx = \int u(x) \cdot \frac{du}{dx} \cdot dx = \int u(x)\, du \tag{24-9}$$

- generalized power rule of integration (p. 967)

$$\int u^n\, du = \frac{u^{n+1}}{n+1} + C \qquad n \neq -1 \tag{24-10}$$

- area under a curve: rectangle sum approximation (p. 975)

$$A = \lim_{n \to \infty} \sum_{i=1}^{n} f(x_i)\Delta x \tag{24-13}$$

where

$$\Delta x = \frac{b-a}{n} \quad \text{for the interval } a \leq x \leq b \tag{24-11}$$

For a more complete description of this procedure, see p. 972.

- area under a curve between a and b (p. 982)

$$A = \lim_{n \to \infty} \sum_{i=1}^{n} f(x_i)\Delta x = g(b) - g(a) \tag{24-18}$$

where $g'(x) = f(x)$ and $\Delta x = \dfrac{b-a}{n}$

- definite integral (p. 983)

$$\int_a^b f(x)\, dx = \lim_{n \to \infty} \sum_{i=1}^{n} f(x_i)\Delta x \tag{24-19}$$

where $f(x)$ is defined on the interval $a \le x \le b$, and provided that the limit exists.

- fundamental theorem of calculus (p. 983)

$$\int_a^b f(x)\, dx = g(b) - g(a) \tag{24-20}$$

where $f(x)$ is a continuous function on $a \le x \le b$, and $g'(x) = f(x)$ for x in the interval.

- area under a curve where $f(x) < 0$ in the interval (p. 986)

$$A = \left| \int_a^c f(x)\, dx \right| + \left| \int_c^b f(x)\, dx \right| \tag{24-21}$$

where $f(x) < 0$ for $a \le x < c$
$\qquad f(x) = 0$ at $x = c$
$\qquad f(x) > 0$ for $c < x \le b$

- trapezoidal rule (p. 992)

$$\int_a^b f(x)\, dx \approx \frac{b-a}{n}\left[\frac{1}{2}f(a) + f(x_1) + f(x_2) + \cdots + f(x_{n-1}) + \frac{1}{2}f(b)\right] \tag{24-22}$$

- rectilinear motion (p. 998)

$$x(t) = \int v(t)\, dt \tag{24-23}$$

$$v(t) = \int a(t)\, dt \tag{24-24}$$

where $x(t)$ is the position of the particle at time t, $v(t)$ is its instantaneous velocity, and $a(t)$ is its instantaneous acceleration.

- work (p. 1002)

$$W = Fd \tag{24-27}$$

where F is the magnitude of the force and d is the distance through which the object moves.

- work done by a constant force (p. 1003)

$$W = Fd \cos \theta \tag{24-28}$$

where θ is the angle between the force and the direction of motion.

- work done by a variable force (p. 1003)

$$W = \int_{x_0}^{x_1} F(x)\, dx \tag{24-29}$$

- improper integrals (p. 1005)

$$\int_a^{\infty} f(x)\, dx = \lim_{b \to \infty} \int_a^b f(x)\, dx \tag{24-30}$$

$$\int_{-\infty}^{b} f(x)\, dx = \lim_{a \to -\infty} \int_a^b f(x)\, dx \tag{24-31}$$

If $f(x)$ approaches $+\infty$ or $-\infty$ somewhere within the interval $a \le x \le b$, then one of the following formulas applies:

$$\int_a^b f(x)\, dx = \lim_{c \to b^-} \int_a^c f(x)\, dx \tag{24-32}$$

where $f(x) \to \infty$ or $f(x) \to -\infty$ as $x \to b$.

$$\int_a^b f(x)\ dx = \lim_{c \to a^+} \int_c^b f(x)\ dx \tag{24-33}$$

where $f(x) \to \infty$ or $f(x) \to -\infty$ as $x \to a$.

If $f(x)$ is continuous in the interval $a \le x \le b$ except for point c, then the following formula applies:

$$\int_a^b f(x)\ dx = \int_a^c f(x)\ dx + \int_c^b f(x)\ dx \tag{24-34}$$

where $f(x) \to \infty$ or $f(x) \to -\infty$ as $x \to c$.

Exercises 24-7

Find the differential dy for each function.

1. $y = 3x^4 - 5x$ **2.** $y = 2x(x^2 - 2x)^3$

3. $y = \dfrac{1}{\sqrt{4 - x^2}}$ **4.** $y^2 - x^2 = xy - 3x$

Calculate both dy and Δy for each function for the given values of x and Δx.

5. $y = 4x^2$; $x = 3$, $\Delta x = 0.2$ **6.** $y = (3x - 1)^3$; $x = 2$, $\Delta x = 0.1$

Integrate.

7. $\displaystyle\int 2\ dx$ **8.** $\displaystyle\int_0^3 7x\ dx$ **9.** $\displaystyle\int_1^2 5x^2\ dx$

10. $\displaystyle\int 4x^{-3}\ dx$ **11.** $\displaystyle\int x^{7/2}\ dx$ **12.** $\displaystyle\int \sqrt[3]{x}\ dx$

13. $\displaystyle\int \sqrt{6x - 5}\ dx$ **14.** $\displaystyle\int_{-1}^{12} (2x + 3)^{2/3}\ dx$ **15.** $\displaystyle\int (3x^2 - x + 4)\ dx$

16. $\displaystyle\int \left(\sqrt{x} + \dfrac{1}{x^2}\right) dx$ **17.** $\displaystyle\int_1^3 2x(x - 3)\ dx$ **18.** $\displaystyle\int_0^1 \dfrac{dx}{(4x + 1)^3}$

19. $\displaystyle\int (5x - 2)^3\ dx$ **20.** $\displaystyle\int 6x(x^2 + 4)^4\ dx$ **21.** $\displaystyle\int \sqrt{x}(x^3 - 4x)\ dx$

22. $\displaystyle\int_0^2 2x^2\sqrt{2x^3 + 1}\ dx$ **23.** $\displaystyle\int_1^2 \dfrac{4x^3}{\sqrt{2x^4 - 1}}\ dx$ **24.** $\displaystyle\int \dfrac{2x - 1}{(3x^2 - 3x)^{4/3}}\ dx$

Find the equation of the curve passing through the given point P and having slope f'(x).

25. $P(3, -5)$; $f'(x) = 2x^2 - 6x$ **26.** $P\left(1, \dfrac{5}{2}\right)$; $f'(x) = \dfrac{x}{\sqrt{2x^2 - 1}}$

For Problems 27–30, use the method indicated to find the area under the curve $y = 3 + 2x - x^2$ between $x = 1$ and $x = 3$.

27. Rectangle sum method for n = 4 **28.** Limit method
29. Trapezoidal rule for n = 4 **30.** Direct integration

For Problems 31–34 use the method indicated to find the area under the curve $y = \sqrt{3x + 1}$ between $x = 1$ and $x = 5$.

31. Limit method **32.** Rectangle sum method for n = 8
33. Direct integration **34.** Trapezoidal rule for n = 8

Use the trapezoidal rule to solve Problems 35 and 36.

35. For $n = 8$, find the approximate value of $\displaystyle\int_1^3 \frac{dx}{x^2 + 3}$

36. Approximate the value of the integral given by the following set of points:

x	2	4	6	8	10	12
y	4.6	8.7	7.2	6.5	7.5	5.9

Solve.

37. *Physics* If a stone is dropped in a well, the depth h (in meters) of the well and the time of fall t (in seconds) of the stone are related by the equation $h = 4.9t^2$. If t is measured as 3.0 ± 0.5 s, calculate the depth of the well and the error in this calculation.

38. *Electronics*   If an inductor coil is connected across a charged capacitor, current oscillations will occur, and the maximum current is given by the equation

$$i_m = V\sqrt{\frac{C}{L}}$$

where V is the original voltage across the capacitor, C is the capacitance, and L is the coil inductance. If $V = 50$ V, $C = 1.0 \times 10^{-6}$ F, and $L = 0.010$ H, estimate the change in i_m if L changes by 0.001 H.

39. *Chemistry*   In a certain manufacturing process, the main ingredient M is used up at a rate given by the equation

$$\frac{dM}{dt} = -\frac{10}{(t + 1)^{3/2}}$$

Integrate to find the amount remaining at any time t if $M = 5$ g at $t = 0$ s.

40. *Physics*   If a collision force F varies with time t according to the equation

$$F = 1 + \frac{1}{(t - 2)^2}$$

find the impulse of this force $\int F\, dt$.

41. *Industrial engineering*   In a continuous manufacturing process, the output rate is sampled every hour, and the results are plotted as shown. Use numerical integration to determine the total number of units produced in this 7-hour period.

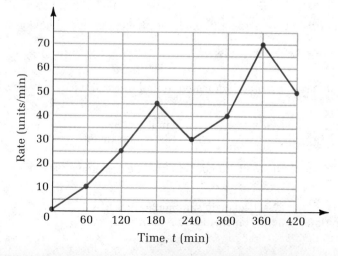

42. *Physics*   Find the work done in stretching a rubber band from length 3.0 cm to length 7.0 cm if the tension in the band varies with length according to the equation $F = 4.2x - 0.2x^2$, where $x \leq 20$ cm.

43. *Physics*   How much work is required to lift a 30 ft cable hanging down an elevator shaft if it weighs 45 lb?

44. *Physics*   Find the work done against their gravitational attraction in separating two objects from a distance of 0.3 m to 1.2 m. (See Problem 17, Exercises 24-6.)

45. *Physics*   What work must be done to separate a pair of oppositely charged objects if $q_1 = 2.0 \times 10^{-6}$ C, $q_2 = -2.0 \times 10^{-6}$ C, and the charges are initially 0.04 in. apart and are separated to $r = 0.12$ in.? (See Problem 19, Exercises 24-6.)

46. *Physics*   A charged particle in an accelerating device has an acceleration a that varies with time t according to the equation $a = t + \sqrt{t} + 2t^2$.
   (a)   Integrate to find an expression for the velocity of the particle.
   (b)   Find the displacement (in meters) of the particle from $t = 1$ to $t = 1.5$ s.

47. *Physics*   What must have been the initial velocity of a projectile if it was launched vertically upward and reached a maximum height of 145 ft?

48. *Mechanical engineering*   The angular speed of rotation (in s^{-1}) of the shaft of an experimental engine varies with time t as given by the equation $\omega = 20 + 6t + 0.1t^2$. What total angular rotation does the shaft go through between $t = 0$ and $t = 10$ s?

49. *Physics*   The range R of a projectile launched with initial speed v_0 at an angle of 45° with the horizontal is $R = v_0^2/g$.
   (a)   If g changes from 9.806 to 9.795 m/s^2 at the altitude of Mexico City, by what amount should a shot-putter competing in Mexico City improve her sea level record of 18.4 m?
   (b)   If $v = 13.4$ m/s, by what percent must the launch speed v_0 at sea level be increased to get the improvement calculated in part (a)?

50. *Electronics*   The magnetic field in a betatron* is measured at different radii, and the data are recorded in the table. The average value of the field can be found using the integral

$$B = \frac{1}{\pi R^2} \int_0^R B(r)(2\pi r)\, dr$$

where R = 80 cm for this machine. Evaluate this integral numerically using the trapezoidal rule.

r (cm)	0	10	20	30	40	50	60	70	80
B (T)	0.350	0.740	0.950	0.940	0.542	0.411	0.385	0.352	0.310

Evaluate each improper integral.

51. $y = \displaystyle\int_1^\infty \frac{dx}{x^{4/3}}$ **52.** $y = \displaystyle\int_{-\infty}^\infty \frac{x\,dx}{(x^2 + 3)^2}$ **53.** $y = \displaystyle\int_0^5 \frac{dx}{(x - 2)^2}$ **54.** $y = \displaystyle\int_3^\infty \frac{10x\,dx}{(x^2 - 4)^2}$

*See W. J. Duffin, *Electricity and Magnetism* (New York: McGraw-Hill, 1980), p. 254

Applications of Integration

25

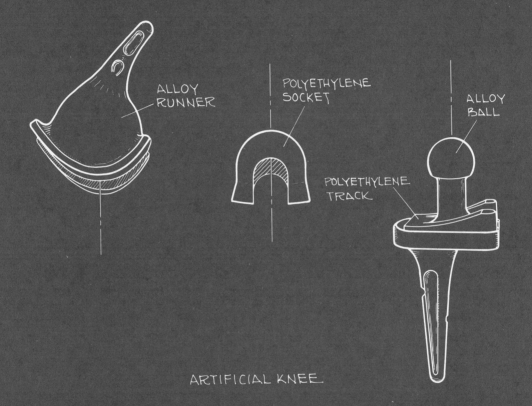

ALLOY
RUNNER

POLYETHYLENE
SOCKET

ALLOY
BALL

POLYETHYLENE
TRACK

ARTIFICIAL KNEE

The mathematical techniques of integration are very useful in the development of concepts and applications in science, engineering, and technology. In this chapter we will use the definite integral to calculate the area of a region bounded by coplanar curves, the volume of a solid of revolution, and the center of mass and moment of inertia of solid objects. We will also show how integration is used in the study of hydrostatic forces.

In Chapter 24 we used integration to find the area under a curve—that is, the area of a region bounded by the curve, the horizontal axis, and two vertical lines. We can use similar integration techniques to solve the more general problem of calculating the area between two curves, when one is not necessarily the horizontal axis.

Vertical area elements   In Fig. 25-1 the area between the curves is divided into vertical rectangular strips, or **elements of area.** The top of each element is on $y = f(x)$, and the bottom of each element is on $y = g(x)$. The area of the element ΔA is

$$\Delta A = \text{Length} \cdot \text{Width}$$

$$= [f(x) - g(x)]\Delta x$$

Figure 25-1

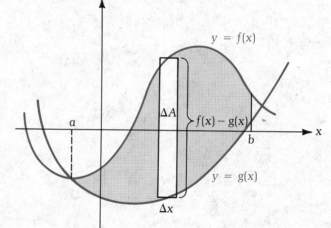

If $f(x)$ and $g(x)$ are continuous on the interval $a \leq x \leq b$, and $f(x) \geq g(x)$ for each value of x in this interval, then the area elements can be summed using a definite integral.

Area between Two Curves: Vertical Elements

$$\text{Area} = \int_{a}^{b} [f(x) - g(x)] \, dx \qquad \text{where } a \leq x \leq b \qquad (25\text{-}1)$$

NOTE

1. Since $f(x)$ is the upper curve throughout the interval from a to b, the height $f(x) - g(x)$ of the area element will be nonnegative, and the definite integral will give a nonnegative value for the area.

2. The height of the area element is found by subtracting the function describing the lower curve from the function describing the upper curve.
3. The left and right boundaries of the area may be intersection points (the point when $x = a$ in Fig. 25-1) or vertical lines (the line $x = b$ in Fig. 25-1).
4. The left boundary $x = a$ gives the lower limit of the integral in (25-1), and the right boundary $x = b$ gives the upper limit of the integral. The area elements are summed from left to right.

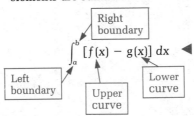

We can use this method in the simple case where one curve is the horizontal axis.

EXAMPLE 1   Find the area bounded by the curve $y = 16 - x^2$, the x-axis, and the lines $x = 1$ and $x = 3$.

Solution   First, sketch the curve $y = 16 - x^2$ and draw in a sample vertical area element (Fig. 25-2).
Second, the limits on x are given as $x = 1$ and $x = 3$.
Third, set up the integral (25-1):

$$\text{Area} = \int_1^3 [(16 - x^2) - 0] \, dx$$

Notice that $y = 16 - x^2$ is the upper curve throughout and that $y = 0$, the x-axis, is the lower curve.

$$= \left(16x - \frac{x^3}{3} \right) \Bigg]_1^3$$

$$= \left(48 - \frac{27}{3} \right) - \left(16 - \frac{1}{3} \right)$$

$$= 23 \tfrac{1}{3} \text{ square units}$$

Figure 25-2

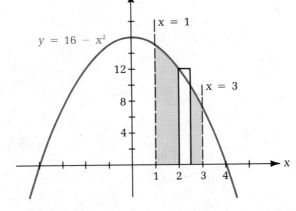

EXAMPLE 2   Figure 25-3 shows the theoretical shape of a cam, similar to those used in Nautilus exercise machines and robots, that produces an automatically variable resistance throughout its range of motion. Find the area of the cam.

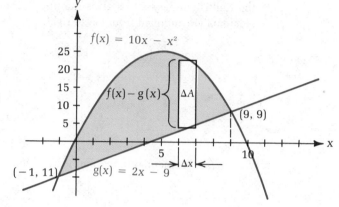

Figure 25-3

Solution   **First,** sketch the curves $y = f(x)$ and $y = g(x)$. Draw in a sample area element strip. Note which curve is the upper one.

Second, find the x-coordinates of the points of intersection of the two curves:

$$2x - 9 = 10x - x^2$$

$$x^2 - 8x - 9 = 0$$

$$(x + 1)(x - 9) = 0$$

The x-coordinates of the intersection points are -1 and 9. These will be the limits of the integral: $a = -1, b = 9$. Check the graph to be certain that all of the area falls between these two points of intersection, and that $f(x)$ is the upper curve throughout.

Third, set up the integral (25-1), and integrate using the power rule (22-13) to find the area:

$$\text{Area} = \int_{-1}^{9} [(10x - x^2) - (2x - 9)]\, dx$$

$$= \int_{-1}^{9} [8x - x^2 + 9]\, dx$$

$$= \left(4x^2 - \frac{x^3}{3} + 9x\right)\Bigg]_{-1}^{9}$$

$$= (324 - 243 + 81) - \left(4 + \frac{1}{3} - 9\right)$$

$$= 162 - \left(-4\tfrac{2}{3}\right)$$

$$= 166\tfrac{2}{3} \text{ square units}$$

Horizontal area elements   To find the area of the region between the curves in Fig. 25-4, it is much simpler to use a horizontal strip as an area element rather than attempt to apply equation (25-1). If the curves are described by the functions $x = p(y)$ and $x = q(y)$, then the area of the element ΔA is

$$\Delta A = \text{Length} \cdot \text{Width}$$

$$= [p(y) - q(y)]\Delta y$$

and the area can be found using a definite integral. If the functions are continuous on the interval $c \le y \le d$, and $p(y) \ge q(y)$ for all y in that interval, then the area can be calculated using equation (25-2).

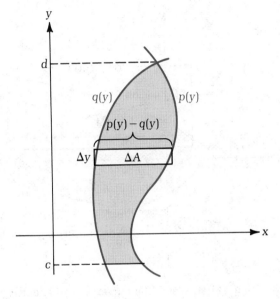

Figure 25-4

Area between Two Curves: Horizontal Elements

$$\text{Area} = \int_c^d [p(y) - q(y)] \, dy \qquad \text{where } c \le y \le d \qquad (25\text{-}2)$$

NOTE ▶

1. The curve $p(y)$ is on the right throughout the interval $c \le y \le d$, and $q(y)$ is always on the left.
2. The length of the area element is found by subtracting the function describing the left curve from the function describing the right curve.
3. The upper and lower boundaries of the area may be intersection points (the point at $y = d$ in Fig. 25-4) or horizontal lines (the line $y = c$ in Fig. 25-4).
4. The lower boundary c gives the lower limit of the integral in (25-2), and the upper boundary d gives the upper limit of the integral. Horizontal area elements are summed from bottom to top.

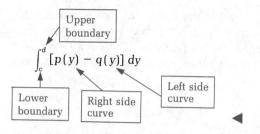

EXAMPLE 3 Find the area of the region enclosed by the curves $x = y^2 - 1$ and $x = 5 - 2y^2$ (Fig. 25-5).

Figure 25-5

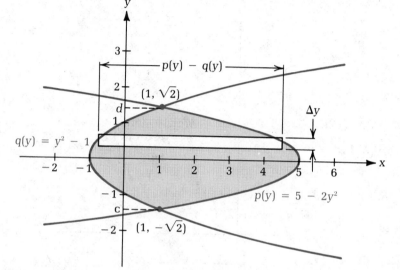

Solution **First,** sketch the curves $x = p(y)$ and $x = q(y)$. Draw in a sample area element. Note which curve is on the right.

Second, find the y-coordinates of the points of intersection of the curves:

$$y^2 - 1 = 5 - 2y^2$$

$$3y^2 = 6$$

$$y^2 = 2$$

$$y = \pm\sqrt{2}$$

The intersection points are at $y = \sqrt{2}$ and $y = -\sqrt{2}$. These will be the limits of the integral: $c = -\sqrt{2}$, $d = \sqrt{2}$.

Third, set up the integral, and integrate to find the area:

$$\text{Area} = \int_{-\sqrt{2}}^{\sqrt{2}} [(5 - 2y^2) - (y^2 - 1)] \, dy$$

$$= \int_{-\sqrt{2}}^{\sqrt{2}} [6 - 3y^2] \, dy$$

$$= [6y - y^3]\Big]_{-\sqrt{2}}^{\sqrt{2}}$$

$$= [6(\sqrt{2}) - (\sqrt{2})^3] - [6(-\sqrt{2}) - (-\sqrt{2})^3]$$

$$= 6\sqrt{2} - 2\sqrt{2} + 6\sqrt{2} - 2\sqrt{2}$$

$$= 8\sqrt{2} \text{ square units}$$

EXAMPLE 4   Find the area of the region bounded by the curves $y = x^3 - 3x + 2$ and $y = 2 - x^2/2$. (Round to the nearest hundredth.)

Solution   First, sketch the curves $f(x) = x^3 - 3x + 2$ and $g(x) = 2 - x^2/2$, as shown in Fig. 25-6. Notice that the region bounded by the curves has two distinct parts. For the region on the left, $f(x)$ is the upper curve; and for the region on the right, $g(x)$ is the upper curve. To find the total area of the shaded region, we must find the area of each part separately and then add them together.

Figure 25-6

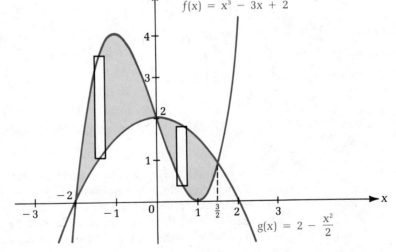

Second, find the x-coordinates of the points of intersection of the curves:

$$x^3 - 3x + 2 = 2 - \frac{x^2}{2}$$

$$x^3 + \frac{x^2}{2} - 3x = 0$$

$$2x^3 + x^2 - 6x = 0$$

$$x(x + 2)(2x - 3) = 0$$

The x-coordinates of the intersection points are $x = 0$, $x = -2$, and $x = \dfrac{3}{2}$.

Third, use (25-1) to set up two integrals:

$$A_1 = \int_{-2}^{0} \left[(x^3 - 3x + 2) - \left(2 - \frac{x^2}{2} \right) \right] dx \longleftarrow$$

For the left region,
$y = x^3 - 3x + 2$
is the upper curve.

$$= \int_{-2}^{0} \left[x^3 + \frac{x^2}{2} - 3x \right] dx$$

$$= \left[\frac{x^4}{4} + \frac{x^3}{6} - \frac{3x^2}{2} \right] \Bigg]_{-2}^{0}$$

$$= \left[\left(\frac{(0)^4}{4} + \frac{(0)^3}{6} - \frac{3(0)^2}{2} \right) \right] - \left[\frac{(-2)^4}{4} + \frac{(-2)^3}{6} - \frac{3(-2)^2}{2} \right]$$

$$= 0 + 3\tfrac{1}{3} = 3\tfrac{1}{3} \text{ square units}$$

$$A_2 = \int_{0}^{3/2} \left[\left(2 - \frac{x^2}{2} \right) - (x^3 - 3x + 2) \right] dx \longleftarrow$$

For the right region,
$y = 2 - \dfrac{x^2}{2}$
is the upper curve.

$$= \int_{0}^{3/2} \left[-x^3 - \frac{x^2}{2} + 3x \right] dx$$

$$= \left[-\frac{x^4}{4} - \frac{x^3}{6} + \frac{3x^2}{2} \right] \Bigg]_{0}^{3/2}$$

$$= \left[-\frac{1}{4}\left(\frac{3}{2}\right)^4 - \frac{1}{6}\left(\frac{3}{2}\right)^3 + \frac{3}{2}\left(\frac{3}{2}\right)^2 \right] - \left[-\frac{(0)^4}{4} - \frac{(0)^3}{6} + \frac{3(0)^2}{2} \right]$$

$$= 1\tfrac{35}{64} \text{ square units}$$

The total area is

$$A_1 + A_2 = 3\tfrac{1}{3} + 1\tfrac{35}{64} = 4\tfrac{169}{192} \approx 4.88 \text{ square units}$$

EXAMPLE 5   Figure 25-7(a) shows experimentally obtained curves describing the rate of energy release during the burning of two industrial fuel samples. The area under each curve is proportional to the energy released in the combustion. The area between the curves tells us the difference in energy output of the samples. Calculate the area of the region between the curves.

Solution   **First,** examine the curves. We can find the area between the curves using either equation (25-1) or equation (25-2). If we use equation (25-1) with a vertical strip area element [Fig. 25-7(b)], we must split the region into two parts. From $t = 0$ s to $t = 2$ s, the lower boundary is the curve $E = 4 - t^2$; and from $t = 2$ s to $t = 16$ s, the lower boundary is the horizontal axis. A simpler procedure would be to use equation (25-2) with a horizontal strip area element [Fig. 25-7(c)].

Second, find the intersection point. Notice that at $t = 0$ s both curves have $E = 4$.

Third, integrate to find the area (energy difference) using equation (25-2). The right curve is $t = 16 - E^2$, and the left curve is $t = \sqrt{4 - E}$. Then,

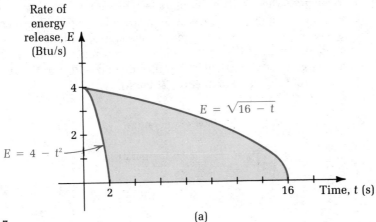

Figure 25-7

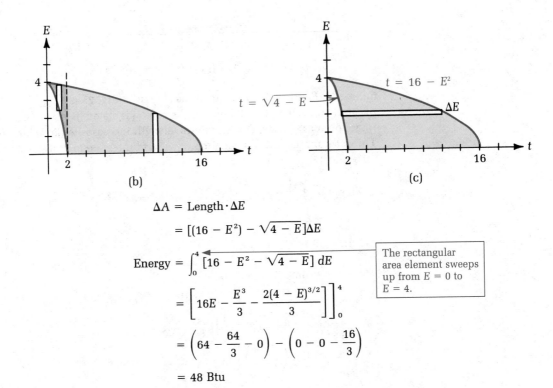

$$\Delta A = \text{Length} \cdot \Delta E$$

$$= [(16 - E^2) - \sqrt{4 - E}]\Delta E$$

$$\text{Energy} = \int_0^4 [16 - E^2 - \sqrt{4 - E}] \, dE$$

> The rectangular area element sweeps up from $E = 0$ to $E = 4$.

$$= \left[16E - \frac{E^3}{3} - \frac{2(4 - E)^{3/2}}{3} \right]_0^4$$

$$= \left(64 - \frac{64}{3} - 0 \right) - \left(0 - 0 - \frac{16}{3} \right)$$

$$= 48 \text{ Btu}$$

EXAMPLE 6   The Rayleigh–Jeans radiation law,

$$R = \frac{2\pi kT}{x^4}$$

describes the spectrum of infrared radiation R emitted by a hot solid at long wavelengths, where T is the absolute temperature, x is the wavelength, and k is a constant. Figure 25-8 shows an approximation to this wavelength distribution for two hot objects at different temperatures, with $2\pi kT = 50$ for a low temperature and $2\pi kT = 100$ for a higher temperature. Find the difference in total energy emitted (in joules) for all radiation with wavelength $x > 2 \mu m$ ($1 \mu m = 10^{-6}$ m; round to the nearest tenth).

Figure 25-8

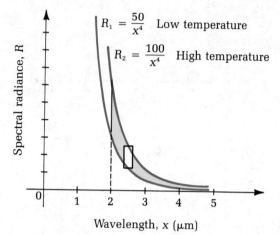

Solution   **First,** sketch the curves, as shown in Fig. 25-8.

Second, notice that the left boundary is $x = 2$ (as stated in the problem) and the right boundary is $+\infty$; that is, the wavelength increases without bound.

Third, set up the integral equation (25-1) and integrate to find the area (energy difference):

$$\text{Energy} = \int_2^\infty \left[\frac{100}{x^4} - \frac{50}{x^4} \right] dx$$

$$= \int_2^\infty \frac{50}{x^4} dx = 50 \int_2^\infty x^{-4} dx$$

$$= \lim_{a \to \infty} 50 \int_2^a x^{-4} dx$$

$$= \lim_{a \to \infty} \left(50 \cdot \frac{x^{-4+1}}{-4+1} \right) \Big]_2^a$$

$$= \lim_{a \to \infty} \left(-\frac{50}{3x^3} \right) \Big]_2^a$$

$$= \lim_{a \to \infty} \left(-\frac{50}{3(a)^3} \right) - \left(-\frac{50}{3(2)^3} \right)$$

$$= 0 + \frac{50}{24}$$

The limit of this term approaches 0 as a increases without bound

$$= 2.1 \text{ J}$$

Exercises 25-1

Find the areas of the regions bounded by the given curves. For problems marked with the calculator symbol, give answers rounded to two decimal places; otherwise, give exact answers.

1. $y = 8x - x^2$, $y = 0$, $x = 2$, $x = 6$

2. $y = x^2 + 2$, $y = 0$, $x = -2$, $x = 2$

3. $y = 8 - x^2$, $y = 0$

4. $y = 16 - x^4$, $y = 0$

5. $y = 2x$, $y = x^2$

6. $y = 3x$, $y = \frac{1}{2}x^2$

7. $x = y$, $x = y^2$

8. $x = 4y$, $x = y^2$

9. $y = x^3$, $y = 9x$

10. $y = \frac{1}{2}x^3$, $y = 2x$

11. $x = y$, $x = y^3$

12. $x = 4y$, $x = \frac{1}{4}y^3$

13. $y = 4x - x^2 + 12$, $y = 3x + 6$

14. $y = x^2 - 3x$, $y = 3 - x$, $x = 0$, $x = 2$

15. $x = y^2 - 5$, $x = -3y + 5$, $y = 1$, $y = -4$

16. $x = 6y - y^2$, $x = 2y - 5$

17. $y = 2x^2$, $y = x^2 + 9$

18. $y = 2 - x^2$, $y = x^2 - 2$

19. $y = \sqrt{x + 1}$, $y = \frac{1}{2}x$, $x = 0$, $x = 4$

20. $y = \sqrt{x}$, $y = x - 4$, $x = 0$, $x = 6$

21. $x^2 = y$, $x = y^3$

22. $x^2 = y$, $x = y^2$

23. $y = \frac{1}{x^2}$, $y = -x + 6$, $x = 1$, $x = 3$

24. $y = \frac{2}{x^2}$, $y = -2x + 5$, $x = 1$, $x = 2$

25. $y = \sqrt{9 - x}$, $y = 3 - x$, x-axis

26. $y = \sqrt{4 - x}$, $y = -6x + 2$, x-axis

27. $y = 2x - x^2$, $y = x^2 - 2x$

28. $y = 4x - x^2$, $y = x^2$

29. $y = x^3 - 2x$, $y = -x^2$, $x = 0$, $x = 1$

30. $y = 4 - x^2$, $y = x^3 - 2x + 4$

31. $y = x^3 - 4x + 3$, $y = x^2 + 2x + 3$

32. $y = 3x - x^3$, $y = 2x^2$

33. $y = x^2 + x + 1$, $y = (x - 2)^2$, $x = 1$, $x = 4$

34. $y = x^3 - 2x^2 + 2$, $4y = x^2$, $x = 0$, $x = 4$

Solve.

35. *Automotive engineering* The work cycle of an internal combustion engine may be approximated by the paths of the pressure–volume graphs shown in the figure. The area of the region between the curves is proportional to the work done by the engine during one cycle. The curves represent adiabatic or zero heat input/output processes, where PV^γ is constant and $\gamma \approx 1.33$ for a polyatomic gas such as gasoline

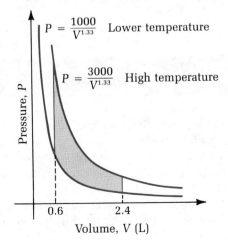

$P = \dfrac{1000}{V^{1.33}}$ Lower temperature

$P = \dfrac{3000}{V^{1.33}}$ High temperature

Pressure, P

0.6   2.4

Volume, V (L)

vapor. Find the work $\int P\,dV$ done by the cycle shown. (The constant has been chosen so that the work will be given in British thermal units, Btu.)

36. *Thermodynamics*   An isothermal process is one in which a gas changes volume V and pressure P while its temperature remains constant. This kind of process is defined by the equation $PV = k$, where k is a constant. Find the work $\int P\,dV$ done by the gas as it expands from $V = 1.0$ L to $V = 3.0$ L if $k = 42$ L·atm.

37. *Machine technology*   The gusset outlined in the figure is defined by the region in the first quadrant bounded by the curve $y = x^3 + 1$, the y-axis, the x-axis, and the line $x = 2$ in. Find the area of this part and the cost of materials for producing 100 gussets at \$1.75 per square inch.

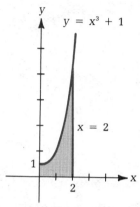

38. *Industrial design*   A double-channel trough used on an assembly line is constructed with end plates defined to be the area bounded by the curve $y = x^4 - 2x^2 + 1$ and the lines $x = -1.7$, $x = 1.7$, and $y = 5$, where x and y are measured in inches. What is the volume of a trough that is 20 ft long and has these end plates?

39. *Industrial design*   A CAM (computer-aided manufacture) system uses a computer-controlled cutting device to produce sheet metal parts having the shape of the area bounded by the curve $y = x^3 - 5x^2 + 4x + 8$, the y-axis, the x-axis, and the line $x = 4$. If all of these dimensions are in centimeters, find the area of the part to the nearest 0.01 cm^2.

40. *Machine technology*   Find the area of the cam outlined by the curves $y = 8x - x^2$ and $4y = x^2$. Round to the nearest 0.1 cm^2.

25-2 | Volume by Integration

Disk method

The volume of many solid objects can be calculated using integration. In order to perform such a calculation, we must first imagine the object cut into a series of similar infinitesimal elements of volume. The **disk method** involves representing the volume as a set of thin circular disks. Figure 25-9 shows some familiar solid objects divided in this way.

Once the object has been sliced into disks, the volume of each disk element must be expressed as a function of a single variable and the sequence of elements summed using integration.

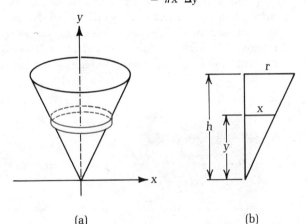

Figure 25-9

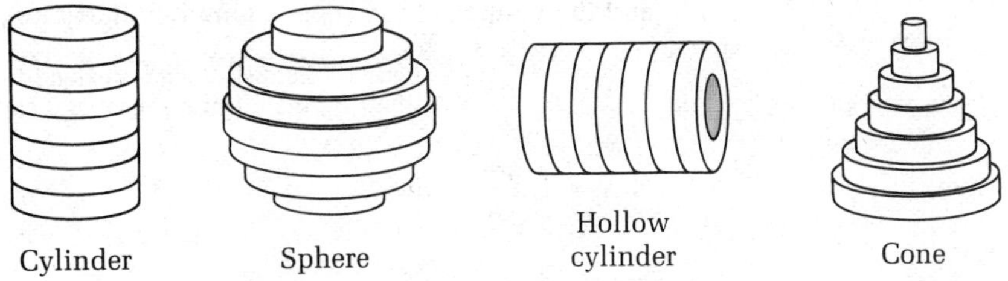

Cylinder   Sphere   Hollow cylinder   Cone

EXAMPLE 7   Find the volume of a right circular cone with base radius r and height h.

Solution   Set up the cone axis along the y-axis, as shown in Fig. 25-10(a). The volume element is a circular disk perpendicular to the y-axis and parallel to the base. Consider a disk element situated at a distance y from the origin. The radius of this disk is x and its thickness is Δy. Each disk is a thin right circular cylinder whose volume is given by Volume $= \pi \cdot (\text{Radius})^2 \cdot \text{Height}$. Therefore,

$$\text{Volume of disk} = \Delta V = \pi \cdot (\text{Radius})^2 \cdot \text{Thickness}$$

$$= \pi x^2 \Delta y$$

Figure 25-10

(a)   (b)

Because we are going to integrate to sum these disk elements, we need to replace x in this formula with an expression in y.

From the similar triangles shown in Fig. 25-10(b), we may form the proportion

$$\frac{x}{r} = \frac{y}{h} \qquad \text{solving for } x,$$

$$x = \frac{ry}{h}$$

Therefore, the volume of a typical disk element at a distance y from the origin is

$$\Delta V = \frac{\pi r^2 y^2}{h^2} \Delta y$$

and the volume of the cone is found by integrating:

$$V = \int_0^h dV$$

Because the disk elements extend from $y = 0$ to $y = h$, the limits on the integral are set at 0 and h.

$$V = \int_0^h \frac{\pi r^2 y^2}{h^2} \, dy$$

$$= \frac{\pi r^2}{h^2} \int_0^h y^2 \, dy$$

$$= \frac{\pi r^2}{h^2} \cdot \frac{y^3}{3} \Big]_0^h$$

$$= \left(\frac{\pi r^2}{h^2} \cdot \frac{(h)^3}{3} \right) - \left(\frac{\pi r^2}{h^2} \cdot \frac{(0)^3}{3} \right)$$

$$= \frac{1}{3} \pi r^2 h \text{ cubic units}$$

Solid of revolution   When a plane figure is rotated about an axis in its plane, the three-dimensional figure generated is called a **solid of revolution.** Notice that each of the figures shown in Fig. 25-11 has a cross section perpendicular to the axis of rotation. This cross section is a circular disk. The volume of each of these solids of revolution can be calculated using the disk method of Example 7.

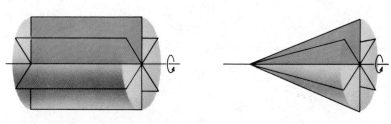

A rectangle rotated about one side forms a cylinder.

A triangle rotated about one leg forms a cone.

Figure 25-11

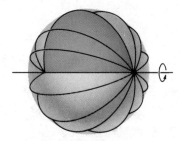

A semicircle rotated about its diameter forms a sphere.

In general, any nonnegative function $y = f(x)$ that is continuous on some interval $a \leq x \leq b$ can be used to generate a solid of revolution (Fig. 25-12). If the area bounded by the function $y = f(x)$, the x-axis, and the lines $x = a$ and $x = b$ is rotated about the x-axis, the resulting volume can be calculated using the following formula:

Volume of a Solid of Revolution: Area under a Curve, Rotated About the x-axis

$$V = \pi \int_a^b [f(x)]^2 \, dx \qquad\qquad (25\text{-}3)$$

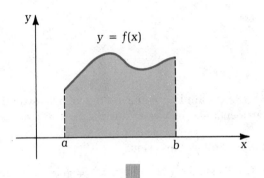

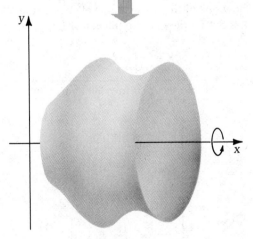

Figure 25-12

This definite integral can be obtained directly from the disk method (Fig. 25-13 on p. 1032):

ΔV = Volume of disk

$\quad = \pi(\text{Radius})^2(\text{Thickness})$

$\quad = \pi[f(x)]^2\Delta x$

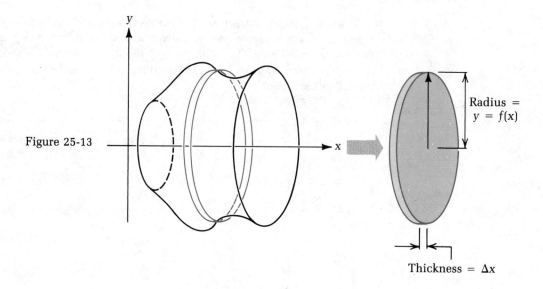

Figure 25-13

EXAMPLE 8   Find the volume of the solid of revolution generated by revolving the area enclosed by $y = x^2$, the x-axis, $x = 0$, and $x = 4$ about the x-axis.

Solution   Follow these steps.

Step 1.   Sketch the graph of $y = x^2$. Sketch the shape of the solid of revolution and show one of the disk volume elements (Fig. 25-14).

Step 2.   Set up the definite integral (25-3):

$$V = \pi \int_0^4 [x^2]^2 \, dx = \pi \int_0^4 x^4 \, dx$$

Sum the disks from left to right.

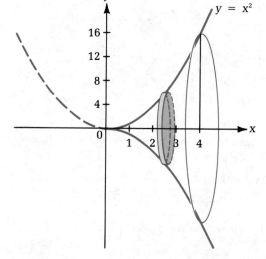

Figure 25-14

Step 3. Integrate using the power rule (22-13):

$$V = \frac{\pi x^5}{5}\Big]_0^4 = \left(\frac{\pi \cdot (4)^5}{5}\right) - \left(\frac{\pi \cdot (0)^5}{5}\right)$$

$$= \frac{1024\pi}{5} \text{ cubic units}$$

LEARNING HINT ▶ If the area is rotated about the x-axis, the disk will be perpendicular to the x-axis. ◀

EXAMPLE 9 A dirigible is a gas-filled, lighter-than-air craft with a rigid frame and a propelling and steering system. If the gas envelope of a certain industrial dirigible has the form of an ellipsoid 148 ft long and 48 ft in diameter, what volume of helium gas does it contain?

Solution

Step 1. Sketch the ellipse centered at the origin, draw in the disk volume element, and identify the limits a and b. Notice that the disks in Fig. 25-15 extend from $x = -a$ to $x = +a$, where $a = 74$ ft.

Figure 25-15

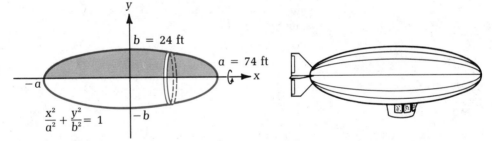

Step 2. Set up the volume integral (25-3):

$$\frac{x^2}{a^2} + \frac{y^2}{b^2} = 1 \qquad \text{gives} \qquad y^2 = b^2 - \frac{b^2 x^2}{a^2}$$

Therefore,

$$V = \pi \int_{-a}^{a} y^2 \, dx = \int_{-a}^{a} \left(b^2 - \frac{b^2 x^2}{a^2}\right) dx$$

Because the solid is symmetric, we can simplify this by integrating to find half the volume from $x = 0$ to $x = a$ ft and multiplying the integral by 2:

$$V = 2\pi \int_0^a \left(b^2 - \frac{b^2 x^2}{a^2}\right) dx$$

Step 3. Integrate. The values of a and b are large numbers, so we will evaluate the integral in terms of a and b, and then substitute $a = 74$ ft and $b = 24$ ft at the last stage of the calculation. Using the power rule (22-13),

$$V = 2\pi\left(b^2 x - \frac{b^2 x^3}{3a^2}\right)\Big]_0^a$$

$$= 2\pi\left(b^2(a) - \frac{b^2(a)^3}{3a^2}\right) - 2\pi\left(b^2(0) - \frac{b^2(0)^3}{3a^2}\right)$$

$$= 2\pi\left(b^2a - \frac{1}{3}b^2a\right) = \frac{4\pi b^2 a}{3}$$

Substituting $a = 74$ ft and $b = 24$ ft into this last expression, we obtain

$$V = \frac{4\pi(24)^2(74)}{3} \approx 180{,}000 \text{ ft}^3$$

Washer method   In the disk method of finding the volume of a solid of revolution, slicing the object perpendicular to its axis of rotation produces disk-shaped volume elements. However, if for example, the area to be rotated about the x-axis is that enclosed between the graphs of the continuous functions $f(x)$ and $g(x)$ and bounded by $x = a$ and $x = b$ (Fig. 25-16), then the volume elements produced by slicing may be "washers," or circular rings, and the following formula can be used in finding the volume of the solid of revolution:

Volume of a Solid of Revolution: Area between Two Curves, Rotated about the x-axis

$$V = \pi \int_a^b \{[f(x)]^2 - [g(x)]^2\} \, dx \qquad\qquad (25\text{-}4)$$

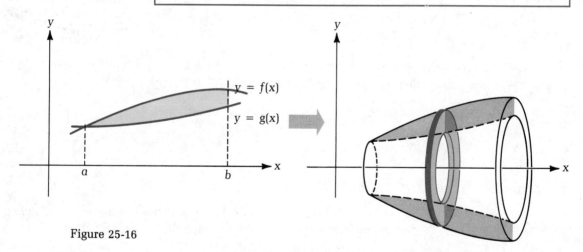

Figure 25-16

The volume of a typical washer element located at position x may be calculated as indicated in Fig. 25-17:

$\Delta V =$ Volume of washer

$\quad =$ (Area of disk $-$ Area of hole)·Thickness

$\quad = \{\pi[f(x)]^2 - \pi[g(x)]^2\}\Delta x$

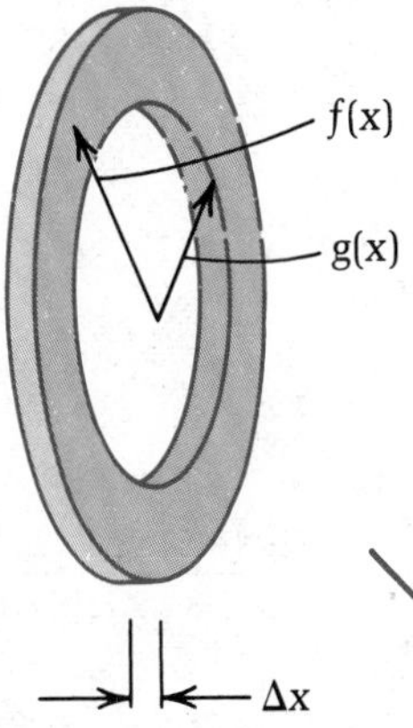

EXAMPLE 10 Find the approximate volume of the material forming the Venturi nozzle shown in Fig. 25-18. It is generated by revolving the area between $y = 1/x$ and $y = 3 - x$ about the x-axis, and is bounded by $a = 0.5$ on the left. (Round the answer to the nearest tenth.)

Solution

Step 1. From the curves $y = 1/x$ and $y = 3 - x$, sketch the solid of revolution and draw in a washer volume element perpendicular to the x-axis (Fig. 25-18).

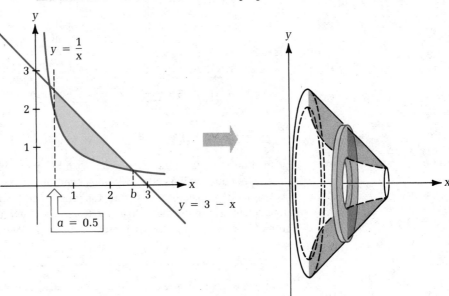

Figure 25-17

Figure 25-18

Step 2. Find the x-coordinate of the intersection point, $x = b$, and set up the definite integral using equation (25-4):

$$\frac{1}{x} = 3 - x$$

$$1 = 3x - x^2$$

$$x^2 - 3x + 1 = 0$$

Using the quadratic formula, the solution is

$$x = \frac{3 \pm \sqrt{5}}{2} \approx 0.38, \ 2.62$$

Use $a = 0.5$ as the lower limit (as given in the problem), and since $x = 2.62$ is to the right of a, use $b = 2.62$ as the upper limit for an approximate value of the volume. The volume integral (25-4) is therefore

$$V \approx \pi \int_{0.5}^{2.62} \left\{ [3 - x]^2 - \left[\frac{1}{x} \right]^2 \right\} dx$$

Sum the washer volume elements from left to right.

Step 3. Integrate:

$$V \approx \pi \int_{0.5}^{2.62} \left(9 - 6x + x^2 - \frac{1}{x^2} \right) dx$$

$$\approx \pi \left(9x - 3x^2 + \frac{x^3}{3} + \frac{1}{x} \right) \Bigg]_{0.5}^{2.62}$$

$$\approx \pi \left[9(2.62) - 3(2.62)^2 + \frac{(2.62)^3}{3} + \frac{1}{(2.62)} \right]$$

$$- \pi \left[9(0.5)^2 - 3(0.5)^2 + \frac{(0.5)^3}{3} + \frac{1}{(0.5)} \right]$$

$$\pi \boxed{\times} \boxed{(} \boxed{9} \boxed{\times} \boxed{2.62} \boxed{\text{STO}} \boxed{-} \boxed{3} \boxed{\times} \boxed{\text{RCL}} \boxed{x^2} \boxed{+} \boxed{\text{RCL}} \boxed{y^x} \boxed{3} \boxed{\div} \boxed{3} \boxed{+} \boxed{\text{RCL}} \boxed{\frac{1}{x}} \boxed{)}$$

$$\boxed{-} \boxed{\pi} \boxed{\times} \boxed{(} \boxed{9} \boxed{\times} \boxed{.5} \boxed{\text{STO}} \boxed{-} \boxed{3} \boxed{\times} \boxed{\text{RCL}} \boxed{x^2} \boxed{+} \boxed{\text{RCL}} \boxed{y^x} \boxed{3} \boxed{\div} \boxed{3} \boxed{+} \boxed{\text{RCL}} \boxed{\frac{1}{x}} \boxed{)}$$

$$\boxed{=} \rightarrow \quad \textit{11.220896}$$

$$\approx 11.2 \text{ cubic units}$$

Rotation about the y-axis   Any line may serve as an axis for a solid of revolution. If the graph of $x = p(y)$ is nonnegative and continuous over the interval $c \leq y \leq d$, then the area bounded by the curve $x = p(y)$, the y-axis, and the lines $y = c$ and $y = d$ can be rotated about the y-axis to generate a solid of revolution (Fig. 25-19). The volume of this solid can be calculated using the following formula:

Volume of a Solid of Revolution: Area under a Curve, Rotated about the y-axis

$$V = \pi \int_c^d [p(y)]^2 \, dy \qquad\qquad (25\text{-}5)$$

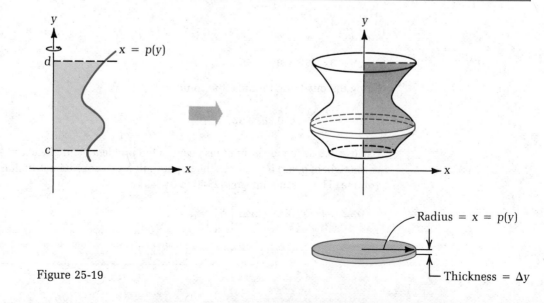

Figure 25-19

Notice in Fig. 25-19 that the volume of the disk element is

ΔV = Volume of disk

$= \pi (\text{Radius})^2 (\text{Thickness})$

$= \pi [p(y)]^2 \Delta y$

EXAMPLE 11 Find the volume of the solid generated when the region enclosed by $x = y^2 - 1$, $y = 1$, $y = 3$, and the y-axis is revolved about the y-axis.

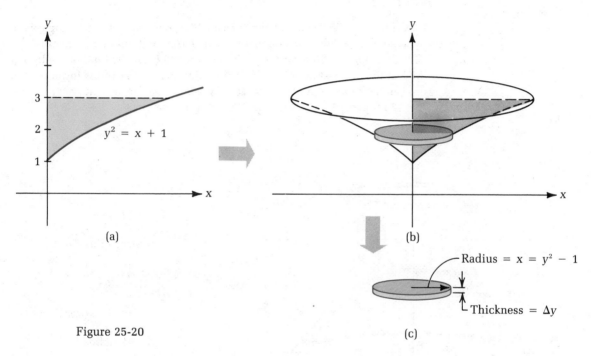

(a)

$y^2 = x + 1$

(b)

Radius = $x = y^2 - 1$

Thickness = Δy

Figure 25-20

(c)

Solution

Step 1. Sketch the curve [Fig. 25-20(a)], and draw a disk volume element perpendicular to the y-axis [Fig. 25-20(b)].

Step 2. Set up the integral (25-5). The radius of the disk is $x = y^2 - 1$ and the thickness is Δy:

$$V = \pi \int_1^3 [y^2 - 1]^2 \, dy$$

Sum the horizontal disk volume elements from bottom to top.

Step 3. Integrate by expanding the integrand and using the power rule (22-13):

$$V = \pi \int_1^3 [y^4 - 2y^2 + 1] \, dy$$

$$= \pi \left[\frac{y^5}{5} - \frac{2y^3}{3} + y \right] \Bigg]_1^3$$

$$= \pi \left[\frac{(3)^5}{5} - \frac{2(3)^3}{3} + (3) \right] - \left[\frac{(1)^5}{5} - \frac{2(1)^3}{3} + (1) \right]$$

$$= \pi \left[\frac{243}{5} - 18 + 3 \right] - \left[\frac{1}{5} - \frac{2}{3} + 1 \right]$$

$$= \pi \left(33\frac{3}{5} - \frac{8}{15} \right)$$

$$= 33\frac{1}{15}\pi \approx 103.9 \text{ cubic units}$$

EXAMPLE 12   The cooling tower for a nuclear power plant has the shape of the hyperbola $x^2 - by^2 = a^2$ on both the exterior and interior surfaces [Fig. 25-21(a)]. If $a_1 = 150$ ft for the exterior, $a_2 = 149$ ft for the interior (walls 1 ft thick at the thinnest point), and $b = 0.15$, find the volume of the tower wall to determine the amount of reinforced concrete needed in its construction. Use $y = 120$ ft and $y = -440$ ft as the limits of integration.

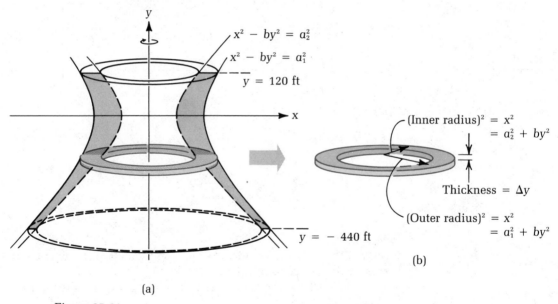

Figure 25-21

Solution
Step 1.   Sketch the hyperbolas and the solid of revolution obtained by rotating them about the y-axis. Show a "washer" element of volume [Fig. 25-21(b)].
Step 2.   Set up a definite integral to find the volume:

$$\Delta V = \text{Volume of the washer}$$

$$= \pi[(\text{Outer radius})^2 - (\text{Inner radius})^2](\text{Thickness})$$

$$= \pi[(a_1^2 + by^2) - (a_2^2 + by^2)]\Delta y$$

$$= \pi(a_1^2 - a_2^2)\Delta y$$

$$\text{Volume} = \int_{-440}^{120} \pi(a_1^2 - a_2^2)\, dy$$

Step 3. Integrate:

$$\text{Volume} = \pi(a_1^2 - a_2^2) \int_{-440}^{120} dy$$

$$= \pi(a_1^2 - a_2^2)[120 - (-440)]$$

$$= \pi(150^2 - 149^2)(120 + 440)$$

$$\approx 530{,}000 \text{ ft}^3$$

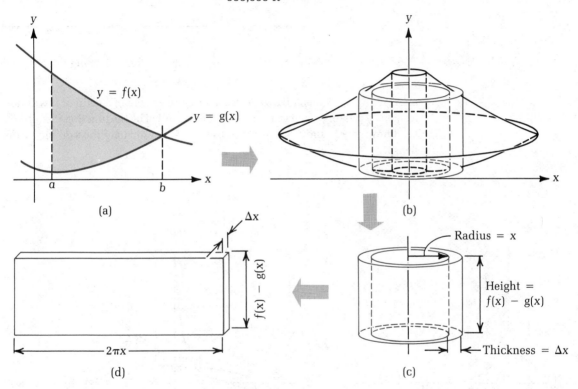

Figure 25-22

Shell method

Rotation about the y-axis

For some solids of revolution, calculating the volume using either disk or washer volume elements can be very difficult. In the **shell method,** the solid is cut into a series of concentric cylindrical shells. If the area bounded by the curves $y = f(x)$ and $y = g(x)$ between the lines $x = a$ and $x = b$ is rotated about the y-axis, the resulting volume can be calculated by summing concentric cylindrical shells (Fig. 25-22). If the shell is cut vertically and flattened, it is a thin strip with thickness Δx, height $f(x) - g(x)$, and length equal to the circumference of the shell $= 2\pi(\text{Radius})$ [Fig. 25-22(d)]. Therefore,

$$\Delta V = \text{Volume of shell}$$

$$\Delta V = 2\pi(\text{Radius})(\text{Height})(\text{Thickness})$$

$$= 2\pi x[f(x) - g(x)]\Delta x$$

The volume of the solid of revolution can be found by summing these nested cylindrical shells from $x = a$ to $x = b$.

Volume of a Solid of Revolution: Shell Method, Rotated about the y-axis

$$V = 2\pi \int_a^b x[f(x) - g(x)]\, dx \qquad\qquad \textbf{(25-6)}$$

CAUTION   The function $f(x)$ is the upper curve everywhere in the area being rotated.   ◀

EXAMPLE 13   A metal flange is designed in the shape of a solid of revolution formed by rotating about the y-axis the area bounded by the function $y = 10/x^3$, the line $y = -1$, and the lines $x = 1$ and $x = 5$. Calculate the volume of the flange.

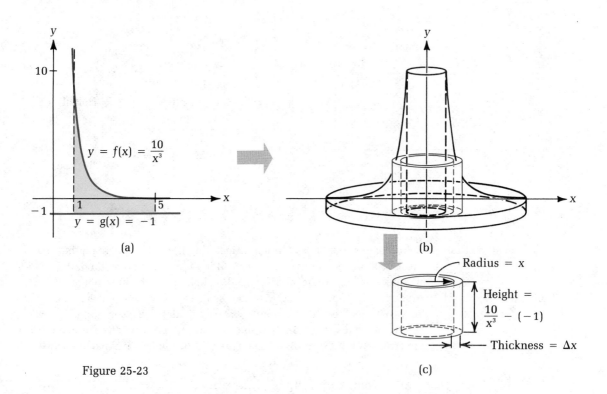

(a)

(b)

(c)

Figure 25-23

Solution

Step 1.   Sketch the function [Fig. 25-23(a)] and the resulting solid of revolution. Show a cylindrical shell volume element [Fig. 25-23(c)].

Step 2.   Set up the definite integral (25-6) with $f(x) = 10/x^3$ as the upper curve and $g(x) = -1$ as the lower curve:

$$V = 2\pi \int_1^5 x\left[\frac{10}{x^3} - (-1)\right] dx$$

Shell radius varies from x = 1 to x = 5.

Step 3.   Simplify the integrand and perform the integration using the rational power rule (22-23):

$$V = 2\pi \int_1^5 \left[\frac{10}{x^2} + x\right] dx$$

$$= 2\pi\left[-\frac{10}{x} + \frac{x^2}{2}\right]_1^5$$

$$= 2\pi\left\{\left[-\frac{10}{(5)} + \frac{(5)^2}{2}\right] - \left[-\frac{10}{(1)} + \frac{(1)^2}{2}\right]\right\}$$

$$= 2\pi\{10.5 + 9.5\}$$

$$= 40\pi \text{ cubic units}$$

Rotation about the x-axis   If the area bounded by the curves $x = p(y)$ and $x = q(y)$ between the lines $y = c$ and $y = d$ is rotated about the x-axis, the volume of the resulting solid of revolution can be calculated by summing cylindrical shells (Fig. 25-24 on p. 1042). Notice that the shell has radius y, thickness Δy, and height $p(y) - q(y)$. Therefore, the volume of the shell element is

$$\Delta V = \text{Volume of shell}$$

$$= 2\pi(\text{Radius})(\text{Height})(\text{Thickness})$$

$$= 2\pi y[p(y) - q(y)]\Delta y$$

The volume of the solid of revolution can be found by summing these shells with the definite integral (25-7).

Volume of a Solid of Revolution: Shell Method, Rotated about the x-axis

$$V = 2\pi \int_c^d y[p(y) - q(y)] \, dy \tag{25-7}$$

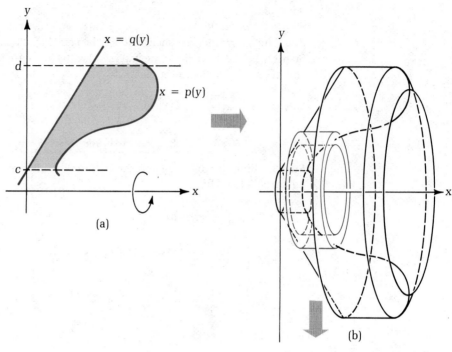

Figure 25-24

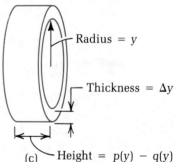

CAUTION ▶   In this formula the graph of the function $p(y)$ is to the right of the graph of $q(y)$ everywhere in the interval $c \leq y \leq d$. ◀

EXAMPLE 14 Calculate the volume of the solid generated by rotating about the x-axis the area bounded by the curve $y = x^2/4$, the line $y = 2x$, and the lines $y = 0$ and $y = 6$. (Round to the nearest whole number.)

Solution

Step 1. Sketch the functions [Fig. 25-25(a)] and the solid of revolution [Fig. 25-25(b)]. Draw a typical cylindrical shell volume element [Fig. 25-25(c)].

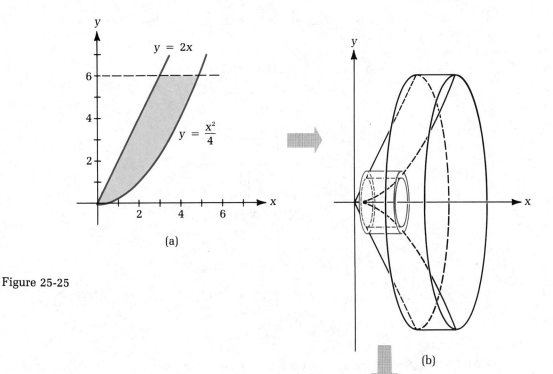

Figure 25-25

(a)

(b)

(c)

Radius = y
Thickness = Δy
Height = $2\sqrt{y} - \dfrac{y}{2}$

Step 2. Find $p(y)$ and $q(y)$ by solving the equations $y = x^2/4$ and $y = 2x$ for x:

$$p(y) = \sqrt{4y} = 2\sqrt{y} \qquad \text{and} \qquad q(y) = \frac{y}{2}$$

The limits are given as $c = 0$ and $d = 6$.

Step 3. Set up the definite integral (25-7):

$$V = 2\pi \int_0^6 y\left[2\sqrt{y} - \frac{y}{2}\right] dy$$

Shell radius varies from $y = 0$ to $y = 6$.

Step 4. Simplify the integrand and use the extended power rule (22-22) to integrate:

$$V = 2\pi \int_0^6 \left[2y\sqrt{y} - \frac{y^2}{2} \right] dy$$

$$2y\sqrt{y} = 2 \cdot y \cdot y^{1/2} = 2y^{3/2}$$

$$= 2\pi \int_0^6 \left[2y^{3/2} - \frac{y^2}{2} \right] dy$$

$$= 2\pi \left[2\left(\frac{y^{5/2}}{5/2} \right) - \frac{y^3}{6} \right]_0^6$$

$$= 2\pi \left\{ \left[\frac{4(6)^{5/2}}{5} - \frac{(6)^3}{6} \right] - \left[\frac{4(0)^{5/2}}{5} - \frac{(0)^3}{6} \right] \right\}$$

$$= 2\pi \left\{ \frac{4 \cdot 6^{5/2}}{5} - 36 \right\}$$

$$4 \boxed{\times} 6 \boxed{y^x} 2.5 \boxed{\div} 5 \boxed{-} 36 \boxed{=} \boxed{\times} 2 \boxed{\times} \boxed{\pi} \boxed{=} \rightarrow \boxed{217.05455}$$

$$\approx 217 \text{ cubic units}$$

CAUTION ▶

It is very important that you do not attempt to use the volume formulas (25-3) through (25-7) by simply "plugging" equations in and integrating. **First,** sketch the solid and analyze it to determine which type of volume element should be used. Choose the element that produces the simplest integral. **Second,** think about the volume integrals as described in Table 25-1. ◀

TABLE 25-1

Disk

$$V = \pi \int_a^b [R(x)]^2 \, dx$$

For an area rotated about the x-axis, where $R(x)$ is an expression giving the radius of the vertical disk located at position x

$$V = \pi \int_c^d [R(y)]^2 \, dy$$

For an area rotated about the y-axis, where $R(y)$ is an expression giving the radius of the horizontal disk located at position y

Washer

$$V = \pi \int_a^b \{[R_o(x)]^2 - [R_i(x)]^2\} \, dx$$

For an area rotated about the x-axis, where $R_o(x)$ and $R_i(x)$ are expressions giving the outer and inner radii, respectively, of the vertical washer located at position x

$$V = \pi \int_c^d \{[R_o(y)]^2 - [R_i(y)]^2\} \, dy$$

For an area rotated about the y-axis, where $R_o(y)$ and $R_i(y)$ are expressions giving the outer and inner radii, respectively, of the horizontal washer located at position y

TABLE 25-1, cont.

Shell

$$V = 2\pi \int_c^d y[H(y)] \, dy$$

For an area rotated about the x-axis, where $H(y)$ is an expression giving the (horizontal) height of the shell at position y, and y is the radius of the shell

$$V = 2\pi \int_a^b x[H(x)] \, dx$$

For an area rotated about the y-axis, where $H(x)$ is an expression giving the (vertical) height of the shell at position x, and x is the radius of the shell.

Exercises 25-2

For each of Problems 1–16, find the volume of the solid generated by rotating about the x-axis the region bounded by the given curves. Use the method indicated for each problem. Round answers to calculator problems to one decimal place.

1. $y = x$, $x = 4$, $x = 0$; disks

2. $y = 2x$, $y = 5$, $x = 0$; shells

3. $y = 3 - 2x$, $x = 0$, $y = 0$; shells

4. $y = 2 - x$, $x = 0$, $y = 0$; disks

5. $y = x^2 + 2$, $x = 0$, $x = 2$, $y = 0$; disks

6. $y = 2x^2$, $x = 0$, $y = 8$; shells

7. $y = x^3$, $y = 6$, $x = 0$; shells

8. $y = 4x - x^2$, $y = 0$; disks

9. $y = \sqrt{x}$, $y = 0$, $x = 4$; disks

10. $y = 2\sqrt{x}$, $y = 1$, $x = 4$; shells

11. $y = \dfrac{4}{x^3}$, $x = 0$, $y = 1$, $y = 4$; shells

12. $y = \dfrac{2}{x}$, $x = 1$, $x = 4$, $y = 0$; disks

13. $y = \dfrac{3}{x}$, $y = 4 - x$; washers

14. $y = x^2 + 3$, $y = 1$, $x = 0$, $x = 4$; washers

15. $y = 3\sqrt{2x}$, $x = 2$, $x = 8$, $y = 2$; washers

16. $8y + 3x = 24$, $y = \dfrac{1}{x}$, $x = 1$, $x = 6$; washers

For each of Problems 17–32, find the volume of the solid generated by rotating about the y-axis the region bounded by the given curves. Use the method indicated for each problem. Round answers to calculator problems to one decimal place.

17. $y = 2x$, $y = 5$, $x = 0$; disks

18. $y = \dfrac{x}{2}$, $x = 4$, $y = 0$; shells

19. $x + y^2 - 2y = 0$, $x = 0$; disks

20. $y = x^2$, $x = 4$, $y = 0$; shells

21. $x = 2y^2 + 1$, $y = 1$, $y = 4$, $x = 0$; disks

22. $y = x^3$, $y = 0$, $x = 2$; shells

23. $y = x^2$, $y = 2x$; shells

24. $y = x^3$, $y = 6$, $x = 0$; disks

25. $y = \sqrt{4 - x}$, $x = 0$, $y = 0$; disks

26. $y = \dfrac{2}{x^3}$, $y = 0$, $x = 1$, $x = 4$; shells

27. $y = \sqrt{x}$, $x = 1$, $x = 4$, $y = 0$; shells

28. $9x + 4y^2 = 36$, in quadrant I; disks

29. $x = y^2 + 4$, $x = 2$, $y = 0$, $y = 2$; washers

30. $y = 2\sqrt{x}$, $x = 1$, $x = 4$, $y = 2$; washers

31. $y = \dfrac{2}{x}$, $y = 4 - x$; washers

32. $x^2 - 4y^2 = 4$, $x = 1$, $y = 2$, $y = -2$; washers

Solve. Round to one decimal place if necessary.

33. Use any convenient method to find the volume of the solid generated by rotating about the x-axis the region bounded by $y = \sqrt{x - 2}$, $x = 6$, and $y = 0$.

34. Use any convenient method to find the volume of the solid generated by rotating about the y-axis the region bounded by $y = 0.2(x^2 - 1)^{3/2}$, the y-axis, $x = 0$, and $x = 3$.

35. *Optics*   An 8-in. diameter parabolic mirror has the shape of the solid of revolution formed by rotating about the y-axis the region bounded by the parabola $y = (x^2/16) + 3$, the x-axis, the y-axis, and the line $x = 4$. Find the volume of glass used in this mirror.

36. Find the volume of the paraboloid formed by rotating about the x-axis the region bounded by the parabola $y^2 = x - 3$, the x-axis, and the line $x = 6$.

37. *Industrial design*   Find the volume of the acoustic horn formed by rotating about the x-axis the region bounded by the curve $y = \sqrt{x^2 - 1}$, the x-axis, and the line $x = 4$.

38. *Civil engineering*   A hill, formed with a core of gypsum and a covering layer of soil, is estimated from photographs to have approximately the shape of the solid of revolution generated by rotating about the y-axis the region in the first quadrant under the curve $y = 30 - (x^2/120)$, where x and y are in meters.*
(a) What volume of material must be moved if the top 15 m of the hill is removed?
(b) If the top 2 m is soil overburden, what volume of gypsum would be obtained from the top 15 m of the hill? [*Hint*: Assume the gypsum core is the solid obtained by rotating the curve $y = 28 - (x^2/120)$.]

39. *Machine technology*   A flange is formed in the shape of a solid of revolution generated by rotating about the x-axis the region bounded by the curve $y = x^2 + 3$ (x and y in centimeters), the y-axis, and the lines $y = 1$ and $x = 3$. What is the mass of this flange if it is made of a material of density 0.018 kg/cm³? [*Hint*:   Mass = Density·Volume.]

40. *Mechanical engineering*   A float valve is constructed in the shape of a solid of rotation formed by rotating about the y-axis the region bounded by the curve $y = x^2$, the line $y = 20 - x$, and the line $x = 1$. Find the volume (in in.³) of this float.

25-3 | Centroids

Center of mass   A uniform circular disk can be made to balance if it is supported at its center point. At what point will a semicircular disk balance (Fig. 25-26)? This "balance point" is known in physics as the **center of gravity** or **center of mass** of an object. Under the influence of external forces, any object behaves as if all of its mass is concentrated at its center of mass. In this section we will show how the center of mass of a flat plate or solid of revolution may be calculated using integration.

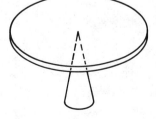
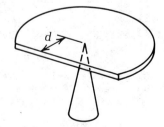

Figure 25-26

*Application suggested by Edward B. Graper of Lebow Co., Santa Barbara, CA.

Moment   If a particle of mass m is located on the x-axis at a distance x from the origin, then the **moment** M_x of the particle with respect to the origin is defined as the product of its mass and the distance x.*

Moment of a Particle	$M_x = mx$	(25-8)

The distance x from the mass to the origin is called the **moment arm** for the mass.

If an array of n particles is constructed with masses $m_1, m_2, m_3, \ldots, m_n$ located along the x-axis at points $x_1, x_2, x_3, \ldots, x_n$, then the moment of this array is the sum of the individual moments.

Moment of a Linear Array of n Particles with Respect to the Origin $$M_x = m_1x_1 + m_2x_2 + \cdots + m_nx_n = \sum_{i=1}^{n} m_ix_i$$	(25-9)

The center of mass of this array is located at the point x_{cm}:

$$x_{cm} = \frac{\text{Total moment of the array}}{\text{Total mass of the array}}$$

Center of Mass of n Particles	$$x_{cm} = \frac{\sum\limits_{i=1}^{n} m_ix_i}{\sum\limits_{i=1}^{n} m_i}$$	(25-10)

EXAMPLE 15   Figure 25-27 shows three particles fastened to a rod of negligible mass. Find the location of the center of mass of this system.

Figure 25-27

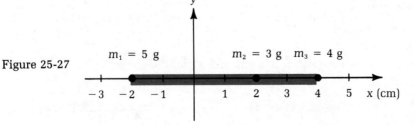

*More accurately, this is the *first* moment of the mass, since x appears to the first power. The second moment will be discussed in the next section.

Solution   **First,** the total mass of the system is

$$M = \sum m_i = 5 + 3 + 4 = 12 \text{ g}$$

Second, the total moment of the system with respect to the origin can be calculated using (25-9):

$$M_x = \sum m_i x_i = 5(-2) + 3(2) + 4(4)$$
$$= 12 \text{ g·cm}$$

Finally, the location of the center of mass can be calculated using (25-10):

$$x_{cm} = \frac{12 \text{ g·cm}}{12 \text{ g}} = 1.0 \text{ cm}$$

This system will respond to external forces in the same way as a single 12 g mass located at $x = 1$ cm. The rod and attached masses will balance at the center of mass.

NOTE ▶   The location of the center of mass of this array does not depend on where the coordinate system is placed. The center of mass will be 1 cm left of m_2 no matter where the rod is positioned on the x-axis, providing that the relative positions of the masses do not change. ◀

If the mass particles are distributed on the xy-plane so that mass m_i is located at the point (x_i, y_i), then the coordinates of the center of mass are (x_{cm}, y_{cm}), as defined in (25-11):

Center of Mass of a Planar Array of Particles

$$x_{cm} = \frac{M_x}{M} = \frac{\sum m_i x_i}{\sum m_i} \qquad y_{cm} = \frac{M_y}{M} = \frac{\sum m_i y_i}{\sum m_i} \qquad (25\text{-}11)$$

EXAMPLE 16   Figure 25-28 shows an array of five particles fastened to the corners of a light frame. Calculate the center of mass of this system.

Solution   **First,** calculate the total mass of the system:

$$M = \sum m_i = 5 + 6 + 4 + 2 + 3 = 20 \text{ g}$$

Second, calculate M_x and M_y for the system.

| Mass of m_5 | x-coordinate for particle m_5 |

$$M_x = 5(2) + 6(4) + 4(2) + 2(-1) + 3(-2)$$
$$= 34 \text{ g·cm}$$

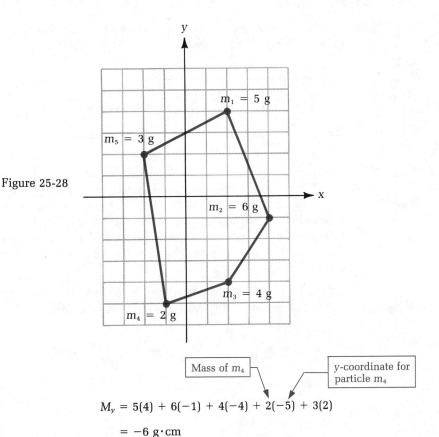

Figure 25-28

$$M_y = 5(4) + 6(-1) + 4(-4) + 2(-5) + 3(2)$$

$$= -6 \text{ g} \cdot \text{cm}$$

Finally, use equation (25-11) to calculate the coordinates of the center of mass:

$$x_{cm} = \frac{34 \text{ g} \cdot \text{cm}}{20 \text{ g}} = 1.7 \text{ cm} \qquad y_{cm} = \frac{-6 \text{ g} \cdot \text{cm}}{20 \text{ g}} = -0.3 \text{ cm}$$

The center of mass of this system of particles is at the point $(1.7, -0.3)$ cm. Note that in this case the center of mass is not a point on the frame.

Center of mass of a plate

Density

It is not usually possible to represent a physical object as a rigid arrangement of massive particles. Instead, we assume that the material of which any object is made is continuously and evenly distributed. The **mass density** of a three-dimensional solid object is defined as follows:

Mass Density $\qquad \rho = \dfrac{\text{Mass}}{\text{Volume}} \qquad$ or $\qquad$ Mass $= \rho \cdot$ Volume $\qquad$ **(25-12)**

For example, $\rho \approx 2.7 \text{ g/cm}^3$ for aluminum and $\rho \approx 11.4 \text{ g/cm}^3$ for lead. If the density of the material is the same throughout, the object is said to be **Homogeneous** **homogeneous.**

Lamina

If the object is a homogeneous, thin rigid sheet, or **lamina**, it is useful to define its density as follows:

Area Density	$\delta = \dfrac{\text{Mass}}{\text{Area}}$ or Mass $= \delta \cdot \text{Area}$	**(25-13)**

For example, a sheet of 26 gauge galvanized steel has an area density of $\delta \approx 0.44$ g/cm^2 and ordinary copy paper has $\delta \approx 0.0075$ g/cm^2.

For some symmetrical, homogeneous objects, the center of mass can be determined by inspection. The center of mass of a uniform circular disk is located at its center, the center of mass of a thin rectangular plate is located at the intersection of its diagonals, and so on.

EXAMPLE 17 Find the center of mass of the thin L-shaped plate shown in Fig. 25-29. The density is $\delta \approx 2.0$ g/cm^2. (Round to the nearest tenth.)

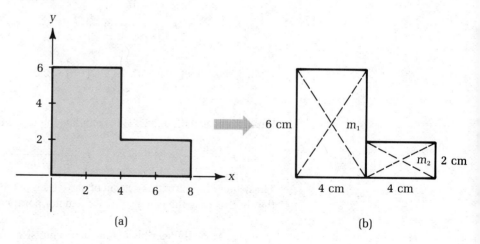

(a)   (b)

Figure 25-29

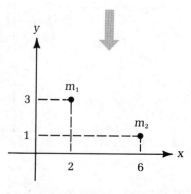

(c)

Solution   First, calculate the mass of the object. Notice that it can be separated into two rectangles:

$$\text{Mass of larger rectangle} = m_1 = \delta \cdot \text{Area}$$

$$= (2.0 \text{ g/cm}^2)(6 \text{ cm} \times 4 \text{ cm})$$

$$= 48 \text{ g}$$

$$\text{Mass of smaller rectangle} = m_2 = \delta \cdot \text{Area}$$

$$= (2.0 \text{ g/cm}^2)(4 \text{ cm} \times 2 \text{ cm})$$

$$= 16 \text{ g}$$

Second, locate the center of mass for each rectangle. Because these are symmetric plates, the center of mass is at the intersection of the diagonals. For m_1,

$$x_{cm} = 2 \text{ cm} \qquad y_{cm} = 3 \text{ cm}$$

For m_2,

$$x_{cm} = 6 \text{ cm}, \qquad y_{cm} = 1 \text{ cm}$$

Third, the mass of each plate can be thought of as being concentrated at its center [Fig. 25-29(c)]. We can find the center of mass of this two-particle system using equations (25-11):

$$x_{cm} = \frac{(48 \text{ g})(2 \text{ cm}) + (16 \text{ g})(6 \text{ cm})}{48 \text{ g} + 16 \text{ g}} = \frac{192 \text{ g} \cdot \text{cm}}{64 \text{ g}} = 3.0 \text{ cm}$$

$$y_{cm} = \frac{(48 \text{ g})(3 \text{ cm}) + (16 \text{ g})(1 \text{ cm})}{48 \text{ g} + 16 \text{ g}} = \frac{160 \text{ g} \cdot \text{cm}}{64 \text{ g}} = 2.5 \text{ cm}$$

The center of mass of the L-shaped plate is at (3.0, 2.5) cm.

In general, the center of mass of a homogeneous nonsymmetrical lamina must be found by integration. If the lamina is formed by the region bounded by the curve $y = f(x)$, the x-axis, and the lines $x = a$ and $x = b$, we can draw a vertical area element (Fig. 25-30).

Figure 25-30

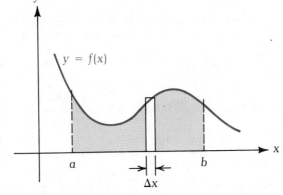

- The area of this element is $\Delta A = f(x)\Delta x$.
- The mass of this element is $\Delta m = \delta \Delta A = \delta f(x)\Delta x$.
- The moment arm parallel to the x-axis of this element is x (Fig. 25-31).

Figure 25-31

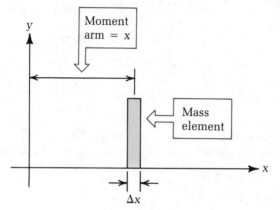

- The moment of this element with respect to the x-axis is $x\Delta m = \delta x\, f(x)\Delta x$.
- The x-coordinate of the center of mass is therefore

$$X_{cm} = \frac{\int_a^b x\, dm}{\int_a^b dm} = \frac{\delta \int_a^b x\, f(x)\, dx}{\delta \int_a^b f(x)\, dx}$$

We may divide the numerator and denominator of the above fraction by the constant density δ to get the following:

---

x-Coordinate of the Center of Mass of a Lamina

$$X_{cm} = \frac{\int_a^b x\, f(x)\, dx}{\int_a^b f(x)\, dx}$$

(25-14)

To find the y-coordinate of the center of mass, first note that the mass Δm of the element in Fig. 25-30 can be thought of as concentrated at its center. Therefore, as shown in Fig. 25-32, the moment arm parallel to the y-axis is

$$\frac{1}{2}y = \frac{1}{2}f(x)$$

The moment of this element with respect to the y-axis is

- $$\frac{1}{2}y\Delta m = \delta\left[\frac{1}{2}f(x)\right]f(x)\Delta x$$

$$= \frac{1}{2}\delta[f(x)]^2\Delta x$$

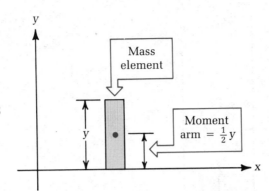

Figure 25-32

● The y-coordinate of the center of mass is

$$y_{cm} = \frac{\int_a^b \frac{1}{2}y \; dm}{\int_a^b dm} = \frac{\frac{1}{2}\delta \int_a^b [f(x)]^2 \; dx}{\delta \int_a^b f(x) \; dx}$$

Again, since δ is a constant, we may divide the numerator and denominator by δ.

---

y-Coordinate of the Center of Mass of a Lamina

$$y_{cm} = \frac{\frac{1}{2} \int_a^b [f(x)]^2 \; dx}{\int_a^b f(x) \; dx} \qquad (25\text{-}15)$$

CAUTION  The variable $f(x)$ cannot be "canceled" from this equation, since it is not a constant. ◀

Centroid Notice in equations (25-14) and (25-15) that only the shape of the lamina $y = f(x)$ and the limits $x = a$ and $x = b$ appear. The mass and density are not involved. For a homogeneous lamina or solid, the center of mass is a "center" point, a purely geometric property of the object, called the **centroid** of the solid.

EXAMPLE 18 Find the coordinates of the centroid of the triangular lamina bounded by the lines $y = 3x$, $y = 0$, and $x = 3$ (Fig. 25-33 on p. 1054), where x and y are measured in inches.

Solution
Step 1. Calculate the denominator of the fractions (25-14) and (25-15). This is the area of the lamina:

$$\int_0^3 f(x) \; dx = \int_0^3 3x \; dx$$

$$= \frac{3x^2}{2}\Bigg]_0^3 = \frac{27}{2} \text{ in.}^2$$

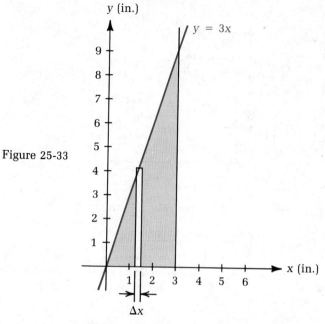

Figure 25-33

Step 2. Calculate the numerator of the fraction (25-14):

$$\int_0^3 x\,f(x)\,dx = \int_0^3 x(3x)\,dx$$

$$= \int_0^3 3x^2\,dx$$

$$= x^3\Big]_0^3 = 27 \text{ in.}^3$$

This final expression contains x^3, so the answer will have units of in.^{3}.

Step 3. Calculate the x-coordinate of the centroid using equation (25-14) and the numerator and denominator found in Steps 1 and 2:

$$x_{cm} = \frac{27 \text{ in.}^3}{\frac{27}{2} \text{ in.}^2} = 2 \text{ in.}$$

Step 4. Calculate the numerator of the fraction in equation (25-15):

$$\frac{1}{2}\int_0^3 [f(x)]^2\,dx = \frac{1}{2}\int_0^3 (3x)^2\,dx$$

$$= \frac{1}{2}\int_0^3 9x^2\,dx$$

$$= \frac{3x^3}{2}\Big]_0^3 = \frac{81}{2} \text{ in.}^3$$

Step 5. Use equation (25-15), the numerator from Step 4, and the denominator from Step 1 to calculate the y-coordinate of the centroid:

$$y_{cm} = \frac{\frac{81}{2} \text{ in.}^3}{\frac{27}{2} \text{ in.}^2} = 3 \text{ in.}$$

The coordinates of the centroid are (2, 3) in.

Equations (25-14) and (25-15) are valid only for a lamina formed by the region bounded by the curve $y = f(x)$, the x-axis, and the lines $x = a$ and $x = b$, using a vertical area element. If the lamina is formed by a region bounded by two curves, with either a vertical or horizontal rectangular area element ΔA, its centroid can be found using the more general equations given in (25-16).

Coordinates of the Centroid of a Lamina

$$x_{cm} = \frac{\int L_x \, dA}{\int dA} \qquad y_{cm} = \frac{\int L_y \, dA}{\int dA} \qquad (25\text{-}16)$$

Here, L_x is the moment arm of the area element with respect to the x-axis; it is the perpendicular distance from the y-axis to the center of the area element. The moment arm of the area element with respect to the y-axis is L_y; it is the perpendicular distance from the x-axis to the center of the area element (Fig. 25-34).

Figure 25-34

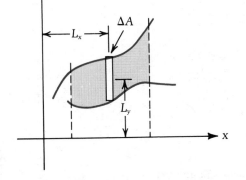
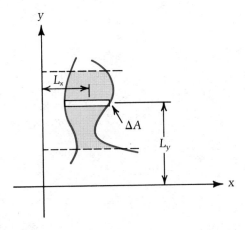

EXAMPLE 19 Find the centroid of the area bounded by the curve $y = \sqrt{x}$ and the lines $y = x/10$ and $x = 4$ [Fig. 25-35(a)], where x and y are measured in feet.

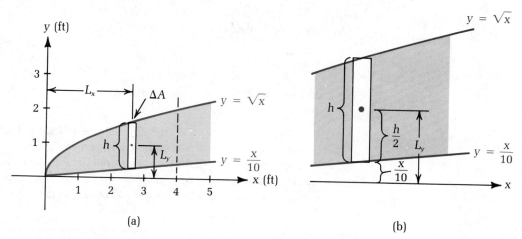

Figure 25-35

Solution

Step 1. Find the height h of the area element:

$$h = \left(\sqrt{x} - \frac{x}{10} \right)$$

$$\text{Area of element} = \Delta A = h\Delta x = \left(\sqrt{x} - \frac{x}{10} \right) \Delta x$$

Then the area of the lamina is

$$A = \int_0^4 dA = \int_0^4 \left[\sqrt{x} - \frac{x}{10} \right] dx$$

$$= \left[\frac{x^{3/2}}{\frac{3}{2}} - \frac{x^2}{20} \right]_0^4$$

$$= \frac{16}{3} - \frac{16}{20} = \frac{68}{15} \text{ ft}^2$$

Step 2. Find an expression for the moment arm L_x:

$$L_x = x$$

Then

$$\int_0^4 L_x \, dA = \int_0^4 x \left[\sqrt{x} - \frac{x}{10} \right] dx$$

$$= \int_0^4 \left[x^{3/2} - \frac{x^2}{10} \right] dx$$

Use the rational power rule (22-23):

$$= \left[\frac{x^{5/2}}{\frac{5}{2}} - \frac{x^3}{30} \right]_0^4$$

$$= \frac{64}{5} - \frac{64}{30} = \frac{32}{3} \text{ ft}^3$$

Step 3. Use equation (25-16) to find the x-coordinate of the centroid:

$$X_{cm} = \frac{\frac{32}{3} \text{ ft}^3}{\frac{68}{15} \text{ ft}^2}$$

$$= 2\frac{6}{17} \text{ ft}$$

$$\approx 2.35 \text{ ft}$$

Step 4. Find an expression for the moment arm L_y [Fig. 25-35(b)]:

$$L_y = \frac{h}{2} + \frac{x}{10}$$

$$= \frac{1}{2}\left(\sqrt{x} - \frac{x}{10}\right) + \frac{x}{10}$$

$$= \frac{1}{2}\left(\sqrt{x} + \frac{x}{10}\right)$$

Then:

$$\int_0^4 L_y \, dA = \int_0^4 \frac{1}{2}\left(\sqrt{x} + \frac{x}{10}\right)\left(\sqrt{x} - \frac{x}{10}\right) dx$$

$$= \frac{1}{2}\int_0^4\left[x - \frac{x^2}{100}\right] dx$$

$$= \frac{1}{2}\left[\frac{x^2}{2} - \frac{x^3}{300}\right]\Bigg]_0^4$$

$$= 4 - \frac{8}{75} = \frac{292}{75} \text{ ft}^3$$

Step 5. Use equation (25-16) to find the y-coordinate of the centroid:

$$y_{cm} = \frac{\frac{292}{75} \text{ ft}^3}{\frac{68}{15} \text{ ft}^2}$$

$$= \frac{73}{85} \text{ ft} \approx 0.86 \text{ ft}$$

The centroid of the area is at about (2.35, 0.86) ft.

Centroid of a solid of revolution

This procedure for finding the centroid of a lamina can be extended to allow us to find the centroid of a solid of revolution. If ΔV is a disk or washer volume element for a solid of revolution formed by rotating an area about the x-axis, the coordinates of the centroid can be found from equation (25-17).

Centroid of a Solid of Revolution: Rotation about the x-Axis

$$X_{cm} = \frac{\int L_x \, dV}{\int dV} \qquad y_{cm} = 0 \qquad \text{(25-17)}$$

As before, L_x is the moment arm of the volume element with respect to the x-axis; it is the perpendicular distance from the y-axis to the center of the volume element (Fig. 25-36). Because the solid object has been formed by

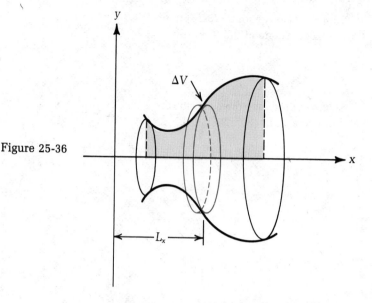

Figure 25-36

rotation about the x-axis, it is symmetric, and the centroid must be on the x-axis. Therefore, $y_{cm} = 0$.

For a solid of revolution formed by rotating an area about the y-axis, the coordinates of the centroid can be found from equation (25-18).

Centroid of a Solid of Revolution: Rotation about the y-axis

$$x_{cm} = 0 \qquad y_{cm} = \frac{\int L_y \, dv}{\int dV}$$

(25-18)

Here, L_y is the moment arm of the volume element with respect to the y-axis; it is the perpendicular distance from the x-axis to the center of the volume element (Fig. 25-37). Because the solid is symmetric about the y-axis, the centroid is on the y-axis, so that $x_{cm} = 0$.

EXAMPLE 20   Find the coordinates of the centroid of the solid of revolution formed by rotating the curve $y = 4x - x^2$ between $x = 0$ and $x = 3$ about the x-axis (Fig. 25-38), where x is measured in feet.

Solution

Step 1.   The radius of the disk volume element is

$$r = y = 4x - x^2$$

The volume of this element is

$$\Delta V = \pi r^2 \Delta x$$

$$= \pi (4x - x^2)^2 \Delta x$$

$$= \pi (16x^2 - 8x^3 + x^4) \Delta x$$

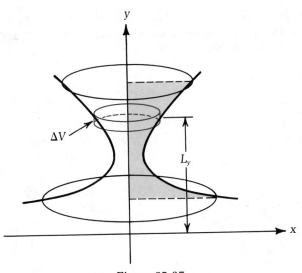

Figure 25-37

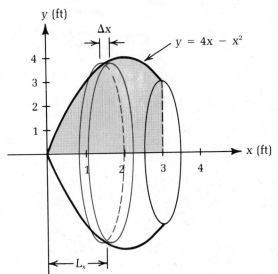

Figure 25-38

The volume of the solid of revolution is

$$\int_0^3 dV = \pi \int_0^3 (16x^2 - 8x^3 + x^4)\, dx$$

$$= \pi\left(\frac{16x^3}{3} - 2x^4 + \frac{x^5}{5}\right)\Bigg]_0^3$$

$$= \frac{153\pi}{5}\ \text{ft}^3$$

Step 2. Find an expression for the moment arm:

$$L_x = x$$

Then:

$$\int_0^3 L_x\, dV = \int_0^3 x \cdot \pi(16x^2 - 8x^3 + x^4)\, dx$$

$$= \pi \int_0^3 (16x^3 - 8x^4 + x^5)\, dx$$

$$= \pi\left(4x^4 - \frac{8x^5}{5} + \frac{x^6}{6}\right)\Bigg]_0^3$$

$$= \frac{567\pi}{10}\ \text{ft}^4$$

Step 3. Use equation (25-17) to find the x-coordinate of the centroid.

$$x_{cm} = \frac{567\pi/10\ \text{ft}^4}{153\pi/5\ \text{ft}^3}$$

$$= \frac{567}{306}\ \text{ft} \approx 1.85\ \text{ft}$$

Step 4. By symmetry, the centroid must be on the axis of revolution. Therefore, $y_{cm} = 0$.

The centroid is located at roughly (1.85, 0) ft.

EXAMPLE 21   Find the centroid of the solid of revolution obtained by rotating about the y-axis the area in the first quadrant bounded by the curve $y = x^3$ and the line $y = x$ (Fig. 25-39), where x and y are measured in centimeters.

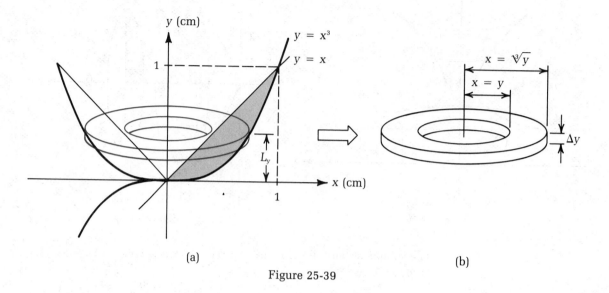

(a)   (b)

Figure 25-39

Solution

Step 1.   The outer radius of the washer volume element is $x = \sqrt[3]{y}$ and the inner radius of the washer is $x = y$, as shown in Fig. 25-39(b).

The volume of this washer element is

$$\Delta V = [\pi(\text{Outer radius})^2 - \pi(\text{Inner radius})^2]\Delta y$$
$$= \pi[(\sqrt[3]{y})^2 - (y)^2]\Delta y$$
$$= \pi[y^{2/3} - y^2]\Delta y$$

The volume of the solid of revolution is

$$\int_0^1 dV = \pi \int_0^1 (y^{2/3} - y^2)\, dy$$

> The limits of integration $x = 0$ and $x = 1$ are found by solving the system of equations $y = x$ and $y = x^3$.

$$= \pi\left(\frac{y^{5/3}}{\frac{5}{3}} - \frac{y^3}{3}\right)\Bigg]_0^1$$
$$= \pi\left(\frac{3}{5} - \frac{1}{3}\right) = \frac{4\pi}{15}\ \text{cm}^3$$

Step 2.   The moment arm is the perpendicular distance from the x-axis to the volume element:

$$L_y = y$$

Then:

$$\int_0^1 L_y\, dV = \int_0^1 \pi y(y^{2/3} - y^2)\, dy$$
$$= \pi \int_0^1 (y^{5/3} - y^3)\, dy$$

$$= \pi\left(\frac{y^{8/3}}{\frac{8}{3}} - \frac{y^4}{4}\right)\Big]_0^1$$

$$= \pi\left(\frac{3}{8} - \frac{1}{4}\right) = \frac{\pi}{8} \text{ cm}^4$$

Step 3. Use equation (25-18) to find the y-coordinate of the centroid:

$$y_{cm} = \frac{\pi/8 \text{ cm}^4}{4\pi/15 \text{ cm}^3}$$

$$= \frac{15}{32} \text{ cm} \approx 0.47 \text{ cm}$$

Step 4. By symmetry, the centroid must be on the axis of revolution. Therefore, $x_{cm} = 0$.

The centroid is located at approximately (0, 0.47) cm. Notice that this is not a point on the solid itself.

Theorem of Pappus

In the Egyptian city of Alexandria in the late 4th century, the Greek mathematician Pappus wrote a textbook reviewing all existing mathematical knowledge. This book, titled *Mathematical Collection*, still exists, and to it we owe much of our knowledge of ancient mathematics. It includes the following very useful theorem:

Theorem of Pappus   If a region in a plane is revolved about a line that is in that plane but does not intersect the region, then a doughnut-shaped solid called a **torus** is produced.

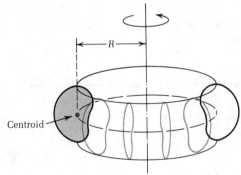

Centroid

The volume of this solid is

$$\text{Volume} = \begin{bmatrix}\text{Area of the} \\ \text{rotated region}\end{bmatrix} \times \begin{bmatrix}\text{Distance traveled} \\ \text{by the centroid} \\ \text{of the region}\end{bmatrix}$$

$$= \text{Area} \cdot 2\pi R$$

where R is the perpendicular distance from the axis to the centroid of the area.

EXAMPLE 22 Find the volume of the torus formed by rotating a circle of radius r about a line a distance R from the center of the circle, where $R > r$ (Fig. 25-40).

Figure 25-40

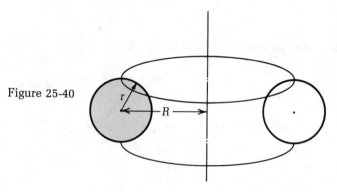

Solution   Volume = Area $\cdot 2\pi R$

$$= (\pi r^2)(2\pi R)$$

$$= 2\pi^2 R r^2$$

Exercises 25-3

Find the center of mass x_{cm} for each linear array of particles. Round to one decimal place if necessary.

1.   14 g   16 g 20 g

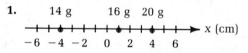

2.   2.1 kg   4.2 kg   3.7 kg

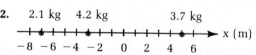

3.   8 kg 4 kg   7 kg   6 kg

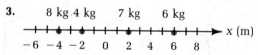

4.   14 g   15 g 18 g 13 g

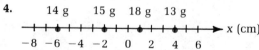

Find the center of mass (x_{cm}, y_{cm}) for each planar array of particles. Round to one decimal place if necessary.

5.

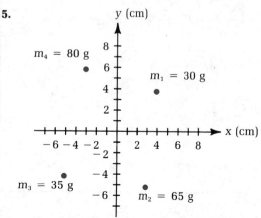

6.

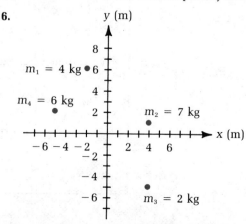

7.

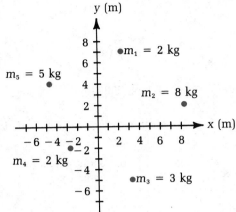

8.

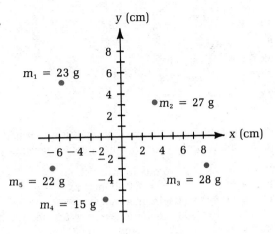

📖 Find the center of mass (x_{cm}, y_{cm}) of each lamina. Use $\delta = 2.0$ g/cm². Round to one decimal place if necessary.

9.

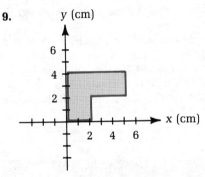

10.

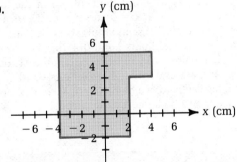

11.

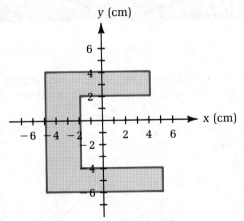

12.

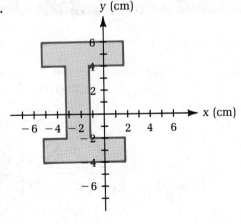

📖 Find the coordinates of the centroid of each region bounded as indicated. Round to one decimal place if necessary.

13. $y = 2x$, $y = 0$, $x = 4$

14. $y = \dfrac{x}{2}$, $y = 0$, $x = 3$

15. $y = 6 - x$, both axes

16. $y = 3 - 2x$, both axes

17. $y = x^2$, $y = 3$

18. $y = \sqrt{x}$, $y = 0$, $x = 4$

19. $y = 2\sqrt{x}, \quad y = \dfrac{x}{2}, \quad x = 4$

20. $y = 4 - x^2, \quad y = 0$

21. $y = 6x - x^2, \quad y = 5$

22. $y = 3x, \quad y = x, \quad x = 3$

23. $y = x^2, \quad y = 3x$

24. $y = x^3, \quad y = 4x, \quad$ quadrant I

25. $y = x^{2/3}, \quad x = 8, \quad y = 0$

26. $y = x^2, \quad y = x$

27. $y = \dfrac{x^2}{2} + 2, \quad y = 0, \quad x = -2, \quad x = 2$

28. $y = x^3 - 2x, \quad y = 0, \quad$ quadrant IV

Find the centroid of the solid of revolution formed by rotating about the given axis the region bounded as indicated. Round to one decimal place if necessary.

29. $y = \sqrt{9 - x^2}$ in quadrant I; about x-axis

30. $y = x^2, \quad x = 3, \quad y = 0;$ about x-axis

31. $x = 2y - y^2, \quad x = 0;$ about y-axis

32. $y^2 = 9x, \quad y = 0, \quad x = 4;$ about y-axis

33. $x^2 + 4y^2 = 4,$ in quadrant I; about x-axis

34. $9x^2 - y^2 = 9,$ quadrant I; $y = 5;$ about y-axis

35. $y = x^2, \quad y = 2x;$ about y-axis

36. $y = x^{2/3}, \quad y = \dfrac{x}{2};$ about x-axis

Solve.

37. *Industrial design* Find the centroid of the foundry pattern fillet formed by the region bounded by $y = x^2/4 - 2x + 4, x = 0$ and $y = 0,$ where all dimensions are in inches.

38. At what point will a semicircular disk balance? Find the centroid of the region bounded by $x^2 + y^2 = 1$ and $y = 0$ in quadrants I and II. [*Hint:* Use area of disk $= \pi/2.$]

39. At what point will an elliptical half-disk balance? Find the centroid of the region bounded by $x^2 + 4y^2 = 4$ and $y = 0$ in quadrants I and II. [*Hint:* The area of this disk is π square units.]

40. Find the centroid of a hemispherical solid of radius 1 cm formed by rotating about the y-axis the region in quadrant I bounded by the curve $x^2 + y^2 = 1.$

25-4 | Moment of Inertia

Consider the following closely related questions:

1. For a given size, mass, and type of material, why is the flywheel shown in Fig. 25-41(a) more effective than the design shown in Fig. 25-41(b)?

Figure 25-41

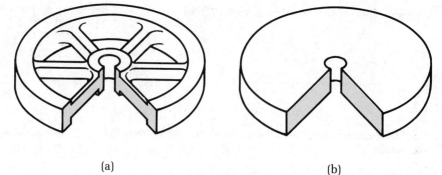

(a)   (b)

2. Why does a diver doing a somersault or multiple turn tuck in his or her arms and legs?

3. Why is an I-beam able to support a larger load than a similar strut with a uniform cross section?

The answer to each of these questions involves an important concept from physics known as the *moment of inertia*.

Moment of inertia

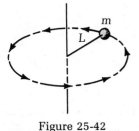

Figure 25-42

If a point mass m is attached to a light rigid rod of length L, it may be rotated using the rod as a pivot arm, as shown in Fig. 25-42. The **moment of inertia I** of the particle is defined by equation (25-19).

Moment of Inertia of a Point Mass $I = L^2m$	(25-19)

The moment of inertia is often called the *second moment of mass*, because the distance L appears to the second power. Notice that L is the perpendicular distance from the mass to the axis of rotation.

EXAMPLE 23   Calculate the moment of inertia of a particle of mass 50 g rotating in a circle of radius 40 cm.

Solution $I = L^2m$

$$= (40 \text{ cm})^2(50 \text{ g})$$

$$= 80{,}000 \text{ g} \cdot \text{cm}^2$$

When a particle moves in a straight line, its resistance to any change in velocity is measured by its mass. When a particle rotates about an axis, its resistance to any change in either the direction or magnitude of its rotational motion is measured by its moment of inertia.

For an array of n particles, the moment of inertia about some axis is the sum of the individual moments about that axis. For each mass m_i, the distance L_i is the perpendicular distance from the mass to the axis of rotation.

Moment of Inertia for an Array of Particles $I = \sum\limits_{i=1}^{n} L_i^2 m_i$	(25-20)

EXAMPLE 24   Calculate the moment of inertia of the array of masses shown in Fig. 25-43:
(a) About the x-axis   (b) About the y-axis
(c) About an axis that is perpendicular to the xy-plane at the origin

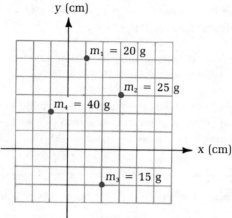

Figure 25-43

Solution

(a) The perpendicular distance L is equal to the y-coordinate when the rotation is about the x-axis. Arrange your work in a table like this:

	y (cm)	y^2 (cm^2)	m (g)	y^2m (g·cm^2)
m_1	5	25	20	500
m_2	3	9	25	225
m_3	-2	4	15	60
m_4	2	4	40	160

Then

$$I_x = \sum y^2m = 945 \text{ g·cm}^2$$

Using a calculator,

$$5 \;\boxed{x^2}\;\boxed{\times}\; 20 \;\boxed{+}\; 3 \;\boxed{x^2}\;\boxed{\times}\; 25 \;\boxed{+}\; 2 \;\boxed{x^2}\;\boxed{\times}\; 15 \;\boxed{+}\; 2 \;\boxed{x^2}\;\boxed{\times}\; 40 \;\boxed{=}$$

$$\rightarrow \qquad \boxed{945.}$$

(b) The perpendicular distance L to be used in (25-20) is the x-coordinate when the rotation is about the y-axis.

	x (cm)	x^2	m (g)	x^2m (g·cm^2)
m_1	1	1	20	20
m_2	3	9	25	225
m_3	2	4	15	60
m_4	-1	1	40	40

Then,

$$I_y = \sum x^2m = 345 \text{ g·cm}^2$$

Using a calculator,

$$1 \;\boxed{\times}\; 20 \;\boxed{+}\; 3 \;\boxed{x^2}\;\boxed{\times}\; 25 \;\boxed{+}\; 2 \;\boxed{x^2}\;\boxed{\times}\; 15 \;\boxed{+}\; 1 \;\boxed{\times}\; 40 \;\boxed{=}\; \rightarrow \qquad 345.$$

(c) For an axis perpendicular to the xy-plane and located at the origin, the distance L is the radial distance from the origin to each mass. Therefore,

$$L^2 = x^2 + y^2$$

	x	x^2	y	y^2	$L^2 = x^2 + y^2$	m	$L^2 m$
m_1	1	1	5	25	26	20	520
m_2	3	9	3	9	18	25	450
m_3	2	4	-2	4	8	15	120
m_4	-1	1	2	4	5	40	200

Then,

$$I_o = \sum L^2 m = 1290 \text{ g} \cdot \text{cm}^2$$

Using a calculator,

$$1 \;\boxed{x^2}\;\boxed{+}\; 5 \;\boxed{x^2}\;\boxed{=}\;\boxed{\times}\; 20 \;\boxed{=}\;\boxed{\text{STO}}\; 3 \;\boxed{x^2}\;\boxed{+}\; 3 \;\boxed{x^2}\;\boxed{=}\;\boxed{\times}\; 25 \;\boxed{=}\;\boxed{\text{SUM}}\; 2 \;\boxed{x^2}\;\boxed{+}$$
$$2 \;\boxed{x^2}\;\boxed{=}\;\boxed{\times}\; 15 \;\boxed{=}\;\boxed{\text{SUM}}\; 1 \;\boxed{x^2}\;\boxed{+}\; 2 \;\boxed{x^2}\;\boxed{=}\;\boxed{\times}\; 40 \;\boxed{=}\;\boxed{\text{SUM}}\;\boxed{\text{RCL}}\; \rightarrow \qquad 1290.$$

For any physical object, the moment of inertia I depends on the total mass, the shape or mass distribution, and the location of the axis of rotation. The moment of inertia is larger for objects whose mass is concentrated farther from the axis of rotation. The flywheel shown in Fig. 25-41(a) has a larger moment of inertia than the flywheel shown in Fig. 25-41(b), because most of its mass is concentrated in its heavy rim. This produces a highly stable rotation with a strong resistance to the small forces and fluctuations in motion that would tend to destabilize it.

A diver can change his or her moment of inertia about an axis of rotation or somersault axis by tucking in or straightening out arms and legs. When the body mass is extended as in Fig. 25-44(a) on p. 1068, I is large—approximately 20 kg·m^2 for a diver 1.8 m tall with a mass of 65 kg. In the pike position shown in Fig. 25-44(b), with legs straight and folded, I drops to about 6 kg·m^2. For the full tuck position shown in Fig. 25-44(c), I is about 4 kg·m^2. The smaller the moment of inertia, the faster the diver will rotate once the motion has begun.

Moment of inertia of a lamina

For a continuous distribution of mass in a thin lamina, the moment of inertia about a given axis may be found using integration. In Fig. 25-45, the moment arm L is the perpendicular distance from the mass element Δm to the axis of rotation. If the area density δ of the lamina is constant, equation (25-21) allows us to find I.

Figure 25-44

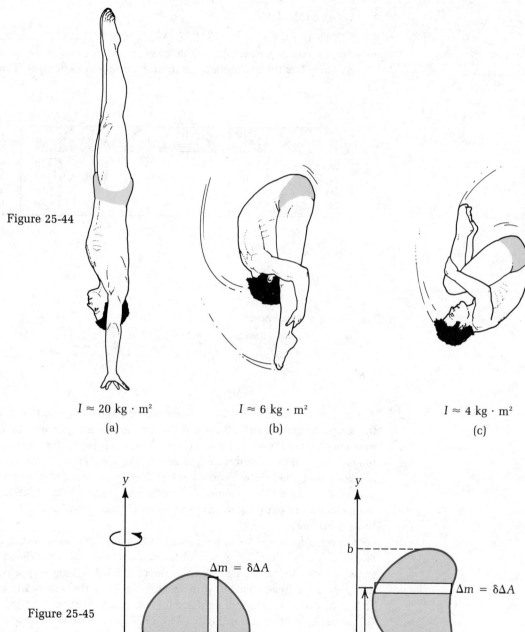

$I \approx 20$ kg $\cdot$ m^2 $I \approx 6$ kg $\cdot$ m^2 $I \approx 4$ kg $\cdot$ m^2

(a)   (b)   (c)

Figure 25-45

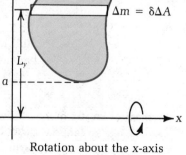

Rotation about the y-axis

$$I_y = \delta \int_a^b (L_x)^2 \, dA$$

Rotation about the x-axis

$$I_x = \delta \int_a^b (L_y)^2 \, dA$$

| Moment of Inertia of a Lamina | $I = \int_a^b L^2 \, dm$ | (25-21) |

EXAMPLE 25 Find the moment of inertia for the rectangle shown in Fig. 25-46.

(a) About the y-axis   (b) About the x-axis

(c) About the line $x = b/2$

Solution

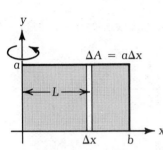

Figure 25-46(a)

(a) **Step 1.** Choose a rectangular area element parallel to the axis of rotation, the y-axis [Fig. 25-46(a)]. Then $\Delta m = \delta \Delta A$, but the area of the rectangular element is $a\Delta x$, so $\Delta m = \delta a \Delta x$.

Step 2. The moment arm is $L = x$.

Step 3. From equation (25-21), the moment of inertia about the y-axis is

$$I_y = \int_0^b L^2 \, dm$$

$$= \delta a \int_0^b x^2 \, dx \qquad \boxed{\text{The limits on x range from } x = 0 \text{ to } x = b.}$$

$$= \delta a \left(\frac{x^3}{3} \right) \Bigg]_0^b$$

$$= \frac{1}{3} a b^3 \delta$$

Step 4. In technical work, the moment of inertia is usually written in terms of the mass m of the object rather than its area density δ. By definition,

$$\delta = \frac{\text{Mass}}{\text{Area}} = \frac{m}{ab}$$

Therefore,

$$I_y = \frac{1}{3} a b^3 \left(\frac{m}{ab} \right)$$

$$= \frac{1}{3} m b^2 \qquad \text{Notice that in this form, it is easy to see that } I \text{ has the dimension of Mass} \times (\text{Distance})^2.$$

(b) **Step 1.** Choose a rectangular area element parallel to the axis of rotation, the x-axis [Fig. 25-46(b)]. Then $\Delta m = \delta \Delta A = \delta b \Delta y$.

Step 2. The moment arm is $L = y$.

Step 3. From equation (25-21),

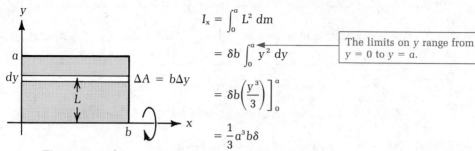

$$I_x = \int_0^a L^2 \, dm$$

$$= \delta b \int_0^a y^2 \, dy \qquad \boxed{\text{The limits on } y \text{ range from } y = 0 \text{ to } y = a.}$$

$$= \delta b \left(\frac{y^3}{3} \right) \Bigg]_0^a$$

$$= \frac{1}{3} a^3 b \delta$$

Step 4. Since $\delta = m/ab$,

$$I_x = \frac{1}{3} a^3 b \left(\frac{m}{ab} \right) = \frac{1}{3} m a^2$$

Figure 25-46(b)

(c)   **Step 1.** Choose ΔA parallel to the axis of rotation [Fig. 25-46(c)]. Then $\Delta m = \delta \Delta A = \delta a \Delta x$.

Step 2. The moment arm is $L = x - (b/2)$.

Step 3. From equation (25-21),

$$I_{b/2} = \int_0^b L^2 \, dm \qquad \boxed{\text{The limits on } x \text{ are from } x = 0 \text{ to } x = b.}$$

$$= \delta a \int_0^b \left[x - \frac{b}{2} \right]^2 dx = \delta a \int_0^b \left[x^2 - bx + \frac{b^2}{4} \right] dx$$

$$= \delta a \left[\frac{x^3}{3} - \frac{bx^2}{2} + \frac{b^2 x}{4} \right] \Bigg]_0^b = \delta a \left[\frac{b^3}{3} - \frac{b^3}{2} + \frac{b^3}{4} \right]$$

$$= \frac{1}{12} a b^3 \delta$$

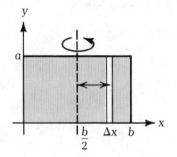

Figure 25-46(c)

Step 4. Since $\delta = \dfrac{m}{ab}$,

$$I_{b/2} = \frac{1}{12} a b^3 \left(\frac{m}{ab} \right) = \frac{1}{12} m b^2$$

Notice in Example 25 that since $b > a$, then $I_y > I_x$. The moment of inertia is smaller for objects where the mass is concentrated nearer the axis of rotation. The concept of moment of inertia enables us to relate the deflection property of a beam to its cross section. When a beam is bent, the cross section rotates about an axis perpendicular to the beam (Fig. 25-47). The upper edge rotates forward and the lower edge rotates backward. In response to a given force, the beam deflection will depend on the moment of inertia of its cross section. The mass of an I-beam is concentrated farther from the rotation axis than that of a uniform beam of the same mass; therefore, the moment of inertia of the I-beam is larger. The I-beam cross section resists rotation more and thus deflects less than a beam of uniform cross section.

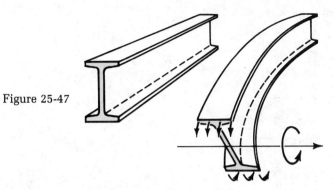

Figure 25-47

Radius of gyration   It is often helpful to be able to relate the rotational behavior of a complex object to the rotation of the simplest object, a particle. The **radius of gyration** R_G for any object is the rotation radius of a particle with equal mass having the same moment of inertia. If the object has moment of inertia I and mass m, then by equation (25-19), $I = mR_G^2$ and we can solve for R_G.

Radius of Gyration $R_G = \sqrt{\dfrac{I}{m}}$ (25-22)

If all the mass of the object were concentrated in a particle at this distance from the axis, or in a thin ring or hoop with radius R_G, the particle or hoop would have the same moment of inertia as the original object.

In Example 24, the radius of gyration with respect to the x-axis is

$$R_{Gx} = \sqrt{\frac{945}{100}} \approx 3.1 \text{ cm}$$

and the radius of gyration with respect to the y-axis is

$$R_{Gy} = \sqrt{\frac{345}{100}} \approx 1.9 \text{ cm}$$

EXAMPLE 26   Find (a) the moment of inertia and (b) the radius of gyration of the lamina formed by the region in the first quadrant bounded by the curve $y = x^2$, the y-axis, and the line $y = 4$ for rotation about the x-axis (Fig. 25-48 on p. 1072). The mass is in grams and the distances are in centimeters.

Solution

(a) **Step 1.** Choose rectangular area element ΔA parallel to the axis of rotation. Then $\Delta m = \delta \Delta A = \delta x \Delta y$. Write Δm in terms of the integration variable y: $\Delta m = \delta \sqrt{y} \Delta y$.

   Step 2. The moment arm is $L = y$.

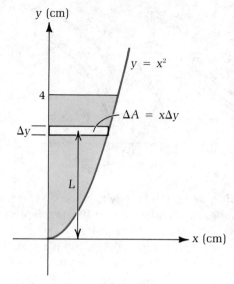

y (cm)

$y = x^2$

4

$\Delta A = x\Delta y$

Δy

L

x (cm)

Figure 25-48

Step 3. From equation (25-21),

$$I_x = \int_0^4 L^2 \, dm$$

$$= \delta \int_0^4 y^2 \sqrt{y} \, dy = \delta \int_0^4 y^{5/2} \, dy$$

Using the rational power rule (22-23),

$$I_x = \delta \left(\frac{y^{7/2}}{\frac{7}{2}} \right) \Bigg]_0^4$$

$$= \frac{2\delta}{7} (y^{7/2}) \Bigg]_0^4$$

$$= \frac{256\delta}{7} \text{ g cm}^2$$

Step 4. To express I_x in terms of the total mass m rather than the area density δ, we must calculate the area of the lamina:

$$A = \int dA = \int_0^4 x \, dy$$

$$= \int_0^4 y^{1/2} \, dy$$

$$= \frac{y^{3/2}}{\frac{3}{2}} \Bigg]_0^4$$

$$= \frac{16}{3} \text{ cm}^2$$

Then $\delta = m/A = 3m/16$, and

$$I_x = \frac{256}{7}\left(\frac{3m}{16}\right) = \frac{48m}{7} \text{ g cm}^2$$

(b) Using equation (25-22), the radius of gyration is

$$R_{Gx} = \sqrt{\frac{I}{m}}$$

$$= \sqrt{\frac{48m}{7m}}$$

$$= \sqrt{\frac{48}{7}} = \frac{4\sqrt{21}}{7} \approx 2.6 \text{ cm}$$

Moment of inertia for a solid of revolution   The shell method (25-6) may be used to calculate the moment of inertia of a solid of revolution for rotation of the solid about its axis.

EXAMPLE 27   For rotation about the y-axis, find (a) the moment of inertia and (b) the radius of gyration of the cone formed by rotating the area in quadrant I under the line $y = 4 - 2x$ about the y-axis (Fig. 25-49).

Solution

(a) **Step 1.**   The volume of the shell element is

$$\Delta V = 2\pi(\text{Radius})(\text{Height})(\text{Thickness})$$

$$= 2\pi xy\Delta x$$

Then the mass of this shell element is $\Delta m = \rho\Delta V$, where ρ is the mass density of the cone, and thus,

$$\Delta m = 2\pi\rho xy\Delta x$$

$$= 2\pi\rho x(4 - 2x)\Delta x$$

Figure 25-49

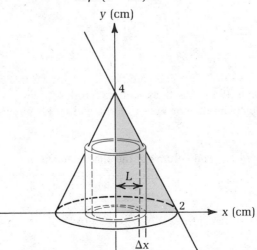

Step 2. The moment arm of the shell element is $L = x$.

Step 3. The moment of inertia can be found using equation (25-21):

$$I_y = \int_0^2 L^2 \, dm$$

> The limits of integration $x = 0$ and $x = 2$ are found by noting where the line $y = 4 - 2x$ intersects the x and y axes.

$$= \int_0^2 x^2 \cdot 2\pi\rho x \,(4 - 2x) \, dx$$

$$= 2\pi\rho \int_0^2 (4x^3 - 2x^4) \, dx$$

$$= 2\pi\rho \left[x^4 - \frac{2x^5}{5} \right]\Bigg]_0^2$$

$$= 2\pi\rho \left[(2)^4 - \frac{(2)^6}{5} \right]$$

$$= \frac{32\pi\rho}{5}$$

Step 4. To write I_y in terms of the mass of the cone m rather than its mass density ρ, note that $\rho = m/V$. The volume of a cone (from Example 7) is $V = \frac{1}{3}\pi r^2 h$, and for $r = 2$, $h = 4$, $V = 16\pi/3$. Therefore, $\rho = 3m/16\pi$, and

$$I_y = \frac{32\pi}{5}\left(\frac{3m}{16\pi}\right) = \frac{6m}{5}$$

(b) The radius of gyration of this cone for rotation about the y-axis can be found from equation (25-22):

$$R_{Gy} = \sqrt{\frac{I}{m}}$$

$$= \sqrt{\frac{6m}{5m}}$$

$$= \sqrt{\frac{6}{5}} = \frac{\sqrt{30}}{5} \approx 1.1$$

EXAMPLE 28 The spindle weight shown in Fig. 25-50 is found by rotating about the y-axis the area between the curves $y = 4 - x^2$ and $y = (x^2/2) - 2$. Calculate its moment of inertia and radius of gyration about the y-axis.

Solution

Step 1. The volume of the shell element is

> Height of the shell

$$\Delta V = 2\pi x\left[(4 - x^2) - \left(\frac{x^2}{2} - 2\right) \right]\Delta x$$

$$= 2\pi\left(6x - \frac{3x^3}{2} \right)\Delta x$$

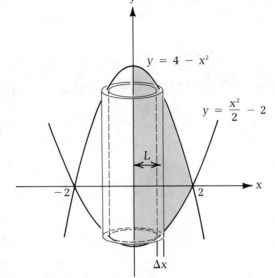

Figure 25-50

The mass of the shell element is

$$\Delta m = 2\pi\rho\left(6x - \frac{3x^3}{2}\right)\Delta x$$

Step 2. The moment arm of the shell element is $L = x$.

Step 3. The moment of inertia can be found from equation (25-21):

$$I_y = \int_0^2 L^2 \, dm$$

> The limit of integration $x = 0$ is found by noting where the curve $y = 4 - x^2$ intersects the y-axis. The limit $x = 2$ is found by solving the system of equations $y = 4 - x^2$ and $y = \frac{x^2}{2} - 2$.

$$= 2\pi\rho \int_0^2 x^2\left(6x - \frac{3x^3}{2}\right) dx$$

$$= 2\pi\rho\left(\frac{3x^4}{2} - \frac{x^6}{4}\right)\Bigg]_0^2$$

$$= 16\pi\rho$$

Step 4. $V = \int dV = \int_0^2 2\pi\left(6x - \frac{3x^3}{2}\right) dx$

$$= 2\pi\left(3x^2 - \frac{3x^4}{8}\right)\Bigg]_0^2$$

$$= 12\pi$$

Then $\rho = m/V = m/12\pi$.

Step 5. Now we can calculate the moment of inertia I_y and radius of gyration R_{Gy}:

$$I_y = 16\pi\left(\frac{m}{12\pi}\right) \qquad R_{Gy} = \sqrt{\frac{I}{m}} = \sqrt{\frac{4m}{3m}}$$

$$= \frac{4m}{3} \qquad\qquad\qquad = \sqrt{\frac{4}{3}} = \frac{2\sqrt{3}}{3} \approx 1.2$$

NOTE ▶   This method of finding the moment of inertia can be used only where the volume element is a cylindrical shell centered on the axis of rotation.   ◀

Exercises 25-4

▦ Find the moment of inertia and radius of gyration for each array of masses about (a) the x-axis, (b) the y-axis. Round to one decimal place if necessary.

1.
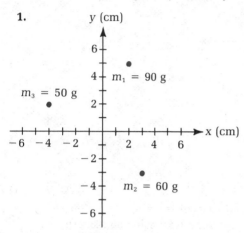

2.
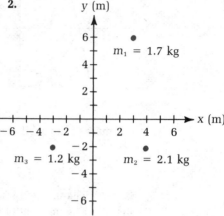

3.
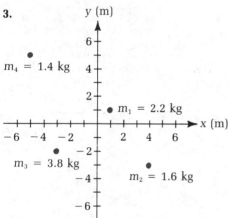

4.
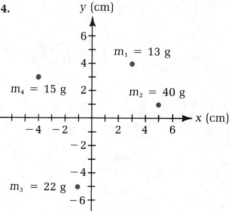

▦ For the axis given, calculate (a) the moment of inertia in terms of the mass m and (b) the radius of gyration for the lamina bounded by the given curve. Round to one decimal place if necessary.

5. $y = 3x$, x-axis, $x = 3$;   about the x-axis
6. $y = 2x$, $y = 2$, $x = 1$, $x = 4$;   about the y-axis
7. $y = 2x^2$, y-axis, $y = 2$;   about the y-axis
8. $y = x^3$, y-axis, $y = 8$;   about the x-axis
9. $y = \sqrt{x}$, x-axis, $x = 4$;   about the x-axis
10. $y^2 = 2x$, x-axis, $x = 2$;   about the y-axis

11. $y = 6 - x$, in the first quadrant; about the y-axis
12. $y = 3x - x^2$, in the first quadrant; about the y-axis

Find (a) the moment of inertia and (b) the radius of gyration with respect to the axis of rotation for each solid of revolution described below. Round to one decimal place if necessary.

13. The solid generated by rotating about the x-axis the region bounded by the line $y = 3 - x$ and the x- and y-axes.
14. The solid generated by rotating about the x-axis the region bounded by $y = 2x$, $y = 3$, and the y-axis.
15. The solid generated by rotating about the y-axis the region bounded by $y = x^2$, $x = 3$, and the x-axis.
16. The solid generated by rotating about the y-axis the region bounded by $y^2 = x$, $y = 2$, and the y-axis.
17. The solid generated by rotating about the x-axis the region bounded by $y = x^3$, $x = 2$, and the x-axis.
18. The solid generated by rotating about the y-axis the region in the first quadrant bounded by $y = 9 - x^2$.
19. The solid generated by rotating about the y-axis the region bounded by $y = 4x$ and $y = x^2$.
20. The solid generated by rotating about the y-axis the region bounded by $y = 6x - x^2$, the x-axis, and $x = 3$.

Solve.

21. Find the moment of inertia of a rectangle with width a and length b with respect to rotation about side b.

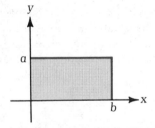

22. Find the moment of inertia, for rotation about the y-axis, of the plate formed by the region bounded by the curve $y = 4/x^2$ and the lines $x = 1$, $x = 2$, and $y = 1$.
23. Find the moment of inertia for rotation about the y-axis of the parabolic lamina bounded by $y = 1 - x^2$ and the x- and y-axis.
24. Find the moment of inertia of a right circular cone of radius r and height h with respect to rotation about its axis.
25. *Marine technology* In the calculation of the stability of a ship's hull, it is necessary to know the moment of inertia about the center line of the ship of the cross section of the hull at the water line. The hull cross section of a certain ship has the shape of the lamina formed by the region bounded by

$y = 4 - 0.01x^2$ in quadrant I   and $y = -4 + 0.01x^2$ in quadrant IV

Find the moment of inertia of this lamina about the y-axis.

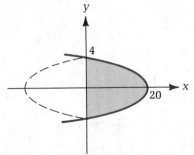

26. *Physics*   Find the moment of inertia of a lawn roller about its central axis. Consider the roller to be a thin-spoked hollow cylinder of outer radius 0.40 m, inner radius 0.30 m, and mass 60 kg. (*Hint:* the roller is formed by rotating the rectangle of length L, from $y = 0.3$ to $y = 0.4$ about the x-axis.)

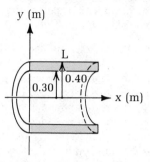

25-5 | More Applications

Average value of a function

In a certain metallurgical process, the temperature T varies with time as shown in Fig. 25-51. What should you report as the average temperature during the 4-hour process? One average you might use is the arithmetic average, or *mean*, of the temperatures at the beginning and end of the 4-hour period.

$$\text{Average}_1 = \frac{1400 + 2360}{2} = 1880°F$$

Time, t (h)	Temperature, T (°F)
0	1400
1	1460
2	1640
3	1940
4	2360

Figure 25-51

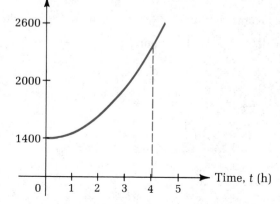

A more representative average could be found by using the mean temperature at the end of each hour.

$$\text{Average}_2 = \frac{1460 + 1640 + 1940 + 2360}{4} = 1850°F$$

We could continue this process, averaging the temperature at the end of each 10 min or 1 min interval, but if we have a continuous, analytical function describing the temperature–time relationship, it is possible to use integration to find an exact value for the average.

If $f(x)$ is a continuous function defined on the interval $a \le x \le b$, then for some integer n we can divide this interval into subintervals of width

$$\Delta x = \frac{b - a}{n}$$

The endpoints of these subintervals are at

$$x_1 = a + \Delta x$$

$$x_2 = a + 2\Delta x$$

$$\vdots$$

$$x_n = a + n\Delta x$$

Then the average of the function values $f(x_1), f(x_2), f(x_3), \ldots, f(x_n)$ is

$$\text{Average} = \frac{f(x_1) + f(x_2) + f(x_3) + \cdots + f(x_n)}{n}$$

And since

$$\frac{1}{n} = \frac{\Delta x}{b - a}$$

we have

$$\text{Average} = \frac{f(x_1)\Delta x}{b - a} + \frac{f(x_2)\Delta x}{b - a} + \cdots + \frac{f(x_n)\Delta x}{b - a}$$

$$= \frac{1}{b - a} \sum_{i=1}^{n} f(x_i)\Delta x$$

The average value of $f(x)$ is defined as

$$f_{\text{AVE}} = \lim_{n \to \infty} \left(\frac{1}{b - a} \right) \sum_{i=1}^{n} f(x_i)\Delta x$$

The quantity on the right is a definite integral.

Average Value of a Function	$f_{\text{AVE}} = \dfrac{1}{b - a} \displaystyle\int_a^b f(x)\, dx$ (25-23)

EXAMPLE 29   Find the average temperature from $t = 0$ to $t = 4$ h for the process shown in Fig. 25-51 if the function is $T(t) = 1400 + 60t^2$.

Solution   Using the equation (25-23),

$$T_{AVE} = \frac{1}{4 - 0} \int_0^4 (1400 + 60t^2) \, dt$$

$$= \frac{1}{4}(1400t + 20t^3) \Big]_0^4$$

$$= 1720°F$$

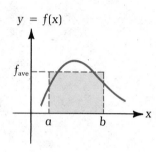

$y = f(x)$

f_{ave}

a b

Figure 25-52

If we multiply both sides of (25-23) by $(b - a)$, then the average value of a function f_{AVE} may be thought of graphically as the height of a rectangle whose base is $(b - a)$ and whose area is equal to the area of the region under the graph of $f(x)$ between $x = a$ and $x = b$ (Fig. 25-52).

$$\int_a^b f(x) \, dx \quad = \quad (b - a)f_{AVE}$$

Area under $f(x)$

Rectangle area:
Base $= b - a$
Height $= f_{AVE}$

LEARNING HINT   This graphic interpretation of the average value of a function allows you to make a quick calculation of, or check on, the value of a definite integral. From a sketch of the function graph, estimate f_{AVE} and multiply it by $(b - a)$. The product is approximately equal to the value of the indefinite integral.   ◀

EXAMPLE 30   The current in a circuit varies according to the equation $i = 8t - 3t^2$, where i is in milliamps and t is in seconds. Find the average value of the current from 0.1 s to 0.5 s. (Round to the nearest tenth.)

Solution

$$i_{AVE} = \frac{1}{0.5 - 0.1} \int_{0.1}^{0.5} (8t - 3t^2) \, dt$$

$$= \frac{1}{0.4}(4t^2 - t^3) \Big]_{0.1}^{0.5}$$

$$= \frac{1}{0.4}[(1 - 0.125) - (0.04 - 0.001)]$$

$$1 \boxed{-} .125 \boxed{-} \boxed{(} \boxed{)} .04 \boxed{-} .001 \boxed{)} \boxed{=} \boxed{÷} .4 \boxed{=} \rightarrow \quad \boxed{\textit{2.09}}$$

$$\approx 2.1 \text{ A}$$

Root mean square average   When $y = f(x)$ is negative in the interval $a \le x \le b$, then the definite integral $\int_a^b f(x) \, dx$ will have a negative value. In scientific and technical work, the root mean square (RMS) average is often used when the function fluctuates or takes on negative values.

Root Mean Square Average $f_{RMS} = \left\{ \dfrac{1}{b-a} \displaystyle\int_a^b [f(x)]^2 \, dx \right\}^{1/2}$

(25-24)

EXAMPLE 31   Calculate the RMS average value of the function $f(x) = x^3 + 1$ from $x = -2$ to $x = 2$.

Solution   Using (25-24),

$$\frac{1}{b-a} \int_a^b [f(x)]^2 \, dx = \frac{1}{2 - (-2)} \int_{-2}^2 (x^3 + 1)^2 \, dx$$

$$= \frac{1}{4} \int_{-2}^2 (x^6 + 2x^3 + 1) \, dx$$

$$= \frac{1}{4} \left(\frac{x^7}{7} + \frac{x^4}{2} + x \right) \Bigg]_{-2}^2 = \frac{71}{7}$$

Then,

$$f_{RMS} = \left(\frac{71}{7} \right)^{1/2} \approx 3.18$$

Because the function $f(x) = x^3 + 1$ is negative for $x < -1$, the average calculated using (25-23), $f_{AVE} = 1.00$, is not equal to f_{RMS}.

Work   In Chapter 24 we defined the work done by a variable force $F(x)$ moving an object in a straight line from x_0 to x_1 as

$$W = \int_{x_0}^{x_1} F(x) \, dx$$

(24-29)

When a problem involves calculating the work required to lift a volume of fluid, it is often necessary to know the weight density of the fluid. For water, the weight density is

$$\rho g \approx 62.4 \text{ lb/ft}^3 \approx 9800 \text{ N/m}^3$$

EXAMPLE 32   A cylindrical water tank 30 ft high and 10 ft in radius is filled to a height of 25 ft. How much work must be done to pump water from the tank so that the level drops to 5 ft?

Solution
Step 1.   Sketch the cylinder and set up a convenient coordinate system, as shown in Fig. 25-53 on p. 1082.

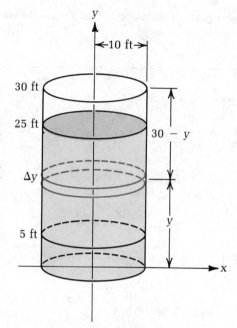

Figure 25-53

Step 2.   For the circular disk volume element shown, the radius is 10 ft. Therefore,

$$\text{Volume of element} = \pi \cdot 10^2 \cdot \Delta y$$

$$\text{Weight of element} = \pi \rho g \, 10^2 \Delta y = 6240 \pi \Delta y$$

Step 3.   As we discussed in Chapter 24, the work done in lifting this element to the top of the tank is

$$\text{Work} = (\text{Force})(\text{Distance})$$

$$\Delta W = (6240 \pi \Delta y)(30 - y)$$

| Force needed to lift the disk | Distance the element is lifted |

Step 4.   Integrate:

> The limits indicate that the elements sweep out a volume from $y = 5$ ft to $y = 25$ ft.

$$\text{Work done} = W = \int_5^{25} 6240 \pi (30 - y) \, dy$$

$$W = 6240 \pi \left(30y - \frac{y^2}{2} \right) \Bigg]_5^{25}$$

$$= 1{,}872{,}000 \pi \ \text{ft} \cdot \text{lb}$$

$$\approx 5.9 \times 10^6 \ \text{ft} \cdot \text{lb}$$

NOTE It is possible to show that the work done in moving a mass in this way is

$$\text{Work} = \begin{pmatrix} \text{Total weight} \\ \text{moved} \end{pmatrix} \cdot \begin{pmatrix} \text{Change in position} \\ \text{of the center of mass} \end{pmatrix}$$ ◄

EXAMPLE 33 The water in a reservoir with parabolic cross section $y = x^2/40$ must be pumped out by raising the water to a height of 5 m above water level. How much work is done in pumping out the reservoir if the original depth of the water is 10 m?

Solution
Step 1. Sketch the reservoir and set up a coordinate system, as shown in Fig. 25-54.

Figure 25-54

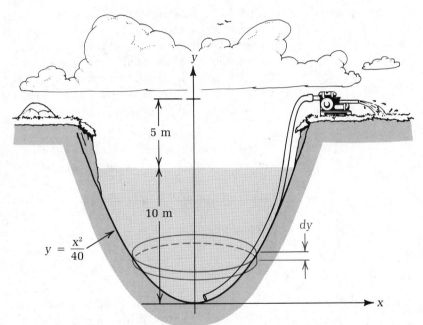

Step 2. For the disk element shown, we see that the radius is x. From $y = x^2/40$, we have $x = \sqrt{40y}$, and thus

$$\text{Volume of element} = \pi x^2 \Delta y = 40\pi y \Delta y$$

$$\text{Mass of element} = 40\pi \rho g y \Delta y$$

Step 3. The work done in lifting this element to a point 15 m above the bottom of the reservoir is

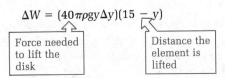

$$\Delta W = (40\pi \rho g y \Delta y)(15 - y)$$

Force needed to lift the disk

Distance the element is lifted

Step 4. Integrate (use $\rho g = 9800$ N/m³, the metric weight density of water):

> The disk elements extend from $y = 0$ to $y = 10$ m.

$$W = 392{,}000\pi \int_0^{10} y(15 - y)\, dy$$

$$= 392{,}000\pi \left(\frac{15y^2}{2} - \frac{y^3}{3} \right) \Big]_0^{10}$$

$$\approx 392{,}000\pi (416.7)$$

$$\approx 5.1 \times 10^8 \text{ N}\cdot\text{m}$$

Fluid pressure and force

A **fluid** is a substance characterized by its ability to flow under the action of relatively small forces and to assume the shape of its container. Both liquids (water, oil, molten metals) and gases (air, steam) are fluids. When a fluid is in contact with a surface, it will exert a force on that surface, and if the fluid is at rest, this contact force will be perpendicular to the surface. If the force were not perpendicular, it would have a component along the surface—a tangential force—and the fluid would begin to flow.

Pressure

The **pressure** exerted by a fluid on a surface is defined as the magnitude of the normal force per unit of surface area.

Fluid Pressure $P = \dfrac{F}{A}$ (25-25)

Pressure units

A great many pressure units have been devised to serve the different needs of physicists, chemists, meteorologists, engineers, and others. The standard metric unit of pressure is the pascal, abbreviated Pa:

1 Pa = 1 N/m²

Because the pascal is a very small pressure unit, weather reports of air pressure usually are given in bar units, where 1 bar = 10^5 Pa. In engineering and technology, pressure is often expressed in English units of pounds per square inch (lb/in.²) or atmospheres. Normal atmospheric pressure at sea level is about 14.7 lb/in.² or 1.013×10^5 Pa or 1 atmosphere.

Pressure–depth equation

The pressure on a surface immersed in a liquid depends on the depth of the surface. In Fig. 25-55, the pressure P_0 on the upper surface of the rectangular volume element is due to the atmosphere above the liquid. The pressure P on the bottom surface is due to the weight of the column of liquid.

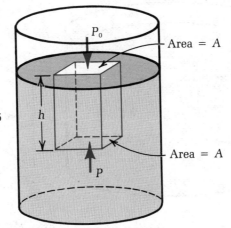

Figure 25-55

Using equation (25-25),

Force on top = $P_0 A$

Force on bottom = PA

Weight of liquid in the volume element = $\rho g \cdot$ Volume = $\rho g A h$

Because the fluid is at rest, these forces must balance:

$PA - P_0 A = \rho g A h$

Dividing by A,

$P - P_0 = \rho g h$

Pressure–Depth Equation	$P = P_0 + \rho g h$	(25-26)

In this equation, P is the **absolute pressure** at a depth h in the liquid and P_0 is the air pressure at the outer surface of the liquid. The term $\rho g h$ is called the **gauge pressure** at a depth h in the liquid. It is the pressure exerted by the liquid at that depth.

Because most liquids such as water or oil are essentially incompressible, the density ρ does not change appreciably with depth, and the pressure is directly proportional to depth. Because a gas is a compressible fluid, its pressure is not proportional to depth.

For any surface, the pressure is perpendicular to the surface and has the same magnitude for any orientation of the surface (Fig. 25-56). If the surface is horizontal, the pressure is the same all along it, and the total force can be found from (25-25). If the surface is not horizontal, the pressure varies with depth, and we must integrate the changing pressure to find the total force on the surface.

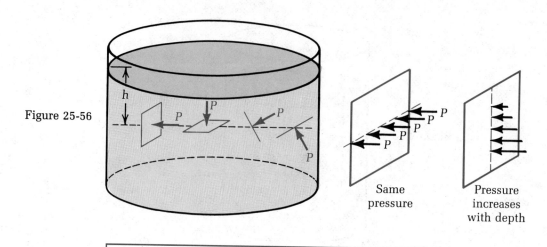

Figure 25-56

Same pressure

Pressure increases with depth

Total Force Due to Fluid Pressure $F = \int_a^b P\,dA$ (25-27)

EXAMPLE 34   Figure 25-57 shows a 2 ft high by 3 ft wide rectangular observation window in a marine technology equipment testing tank. Find the total force on the window if its upper edge is 4 ft below the water level. (Round to the nearest hundred pounds.)

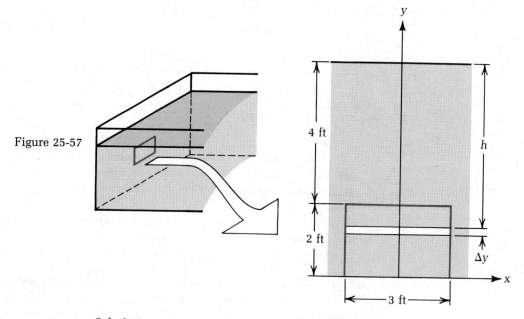

Figure 25-57

Solution

Step 1.   Sketch the area and draw a horizontal area element. Set up a coordinate system.

Step 2. The area of the rectangular element is

$$\Delta A = 3\Delta y$$

Width
Length

The depth of this element below the water level is $h = 6 - y$.

Step 3. Using equation (25-27) and the pressure–depth equation (25-26), we have

$$F = \int_0^2 \rho g (6 - y)(3\ dy)$$

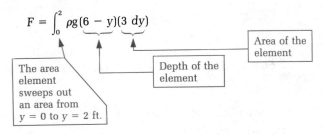

Area of the element

The area element sweeps out an area from $y = 0$ to $y = 2$ ft.

Depth of the element

Step 4. Integrate:

$$F = 3\rho g \int_0^2 (6 - y)\ dy$$

$$= 3\rho g \left(6y - \frac{y^2}{2} \right) \Bigg]_0^2 \qquad \text{Use } \rho g = 62.4\ \text{lb/ft}^3.$$

$$= 1872\ \text{lb} \approx 1900\ \text{lb}$$

This is the total force on the observation window due to the water in the tank.

NOTE ▶ It is possible to show that the total force on the surface can be calculated by multiplying the area of the surface by the pressure at the centroid of the area. In Example 34, the centroid of the rectangular window is at its midpoint 5 ft below the water level. Then

$$P = \rho g h_{cm} \cdot A = (62.4\ \text{lb/ft}^3)(5\ \text{ft})(2\ \text{ft} \times 3\ \text{ft})$$

$$= 1872\ \text{lb}$$

This is especially useful when the area under pressure is symmetric so that its centroid can be found by inspection. ◀

With some problems it is more effective to set up the coordinate axes with the origin at the surface of the liquid and the y-axis directed downward.

EXAMPLE 35 A 10 m diameter cylindrical oil tank, positioned horizontally, is half-filled with oil of weight density 7500 N/m³. Calculate the total force on the circular end of the tank (Fig. 25-58 on p. 1088).

Solution

Step 1. Set up the y-axis directed downward. Sketch a rectangular area element parallel to the oil surface.

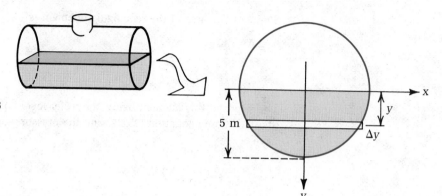

Figure 25-58

Step 2.   The area of the rectangular element is

$$\Delta A = 2x\Delta y$$

The equation of the circle is $x^2 + y^2 = 5^2$. Therefore, $x = \sqrt{25 - y^2}$ and

$$\Delta A = 2\sqrt{25 - y^2}\Delta y$$

The depth of the area element is y.

Step 3.   Using equation (25-27), we obtain

$$F = \int_0^{-5} \rho g \cdot y \cdot 2\sqrt{25 - y^2}\ dy$$

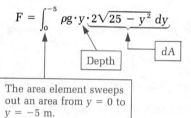

Depth

dA

The area element sweeps
out an area from $y = 0$ to
$y = -5$ m.

Step 4.   Integrate:

$$F = 2\rho g \int_0^{-5} y\sqrt{25 - y^2}\ dy$$

Use u substitution with $u = 25 - y^2$. Then at $y = 0$, $u = 25$, and at $y = -5$, $u = 0$. These are the new limits.

$$du = -2y\ dy$$

Therefore, $y\ dy = -\frac{1}{2}\ du$ and

$$2\rho g \int y\sqrt{u}\ dy = -\rho g \int \sqrt{u}\ du$$

Then the integral is

$$F = -\rho g \int_{25}^0 \sqrt{u}\ du$$

$$= (-\rho g)\left(\frac{u^{3/2}}{\frac{3}{2}}\right)\Bigg]_{25}^0$$

$$= (-7500)\left(-\frac{(25)^{3/2}}{\frac{3}{2}}\right)$$

$$= 625,000 \text{ N} \approx 6.3 \times 10^5 \text{ N}$$

EXAMPLE 36 Calculate the total force exerted by the water on a triangular hatch cover on the vertical side of a dam. The triangle is 8 ft wide at the base, 4 ft high, and positioned with the base 12 ft below water level (Fig. 25-59.)

Solution
Step 1. Sketch the triangle, setting up a coordinate system and a horizontal area element [Fig. 25-59(b)].
Step 2. Knowing that (0, 4) and (4, 0) are the endpoints of one side, we can find the equation of the line forming a side to be $y = 4 - x$ [Fig. 25-59(c)].

$$\text{Area of rectangular element} = \Delta A = \text{Length} \cdot \text{Width}$$

$$= (2x)\Delta y$$

$$= 2(4 - y)\Delta y$$

Depth of the element $= 12 - y$
Step 3. Using (25-27), the total force on the triangular cover is

$$F = \int_0^4 \rho g \underbrace{(12 - y)}_{\text{Depth}} \underbrace{2(4 - y) \, dy}_{dA}$$

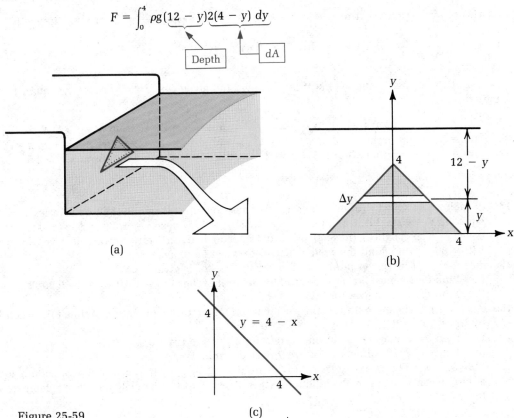

(a)

(b)

(c)

Figure 25-59

Step 4. Integrate:

$$F = 2\rho g \int_0^4 (12 - y)(4 - y)\, dy$$

$$= 2\rho g \int_0^4 (48 - 16y + y^2)\, dy$$

$$= 2\rho g \left(48y - 8y^2 + \frac{y^3}{3} \right) \Bigg]_0^4$$

$$= \frac{512\rho g}{3}$$ Use g = 62.4 lb/ft³.

$$= 10{,}649.6 \text{ lb} \approx 11{,}000 \text{ lb}$$

Exercises 25-5

Find the average value of each function over the given interval. If necessary, round to the nearest hundredth.

1. $f(x) = x^2$; from x = 0 to x = 4

2. $f(x) = x^3 - 3x + 3$; from x = 0 to x = 3

3. $f(x) = \sqrt{x}$; from x = 1 to x = 9

4. $f(x) = \dfrac{x}{\sqrt{2 + x^2}}$; from x = 2 to x = 5

5. $f(x) = \dfrac{x}{(x^2 + 3)^2}$; from x = 1 to x = 3

6. $f(x) = \sqrt{4 - x}$; from x = 0 to x = 4

Find the RMS average value of each function over the given interval. If necessary, round to the nearest hundredth.

7. $f(x) = x^2 - 4$; from x = -2 to x = 2

8. $f(x) = 2x^3$; from x = -3 to x = 3

9. $f(x) = \sqrt{4 - x}$; from x = 0 to x = 4

10. $f(x) = 4x - x^2$; from x = 0 to x = 4

11. $f(x) = x^3 - 6x$; from x = -2.5 to x = 2.5

12. $f(x) = x^5 - 5x^3$; from x = -2.5 to x = 2.5

Solve.

13. *Physics*   The speed v (in miles per hour) of a car moving along a straight path varies according to the equation $v = 40 - 10t + 6t^2$. Find the average speed between t = 1.0 s and t = 3.0 s.

14. *Physics*   The height h (in feet) of an object launched vertically upward is given by the equation $h(t) = 80t - 16t^2$. Find the average height between t = 0 and t = 5.0 s.

15. *Electronics*   The power produced in a certain resistor is given by $P = 1.2i^3$.
    (a)   Find the average power produced for a current range of i = 0.10 A to i = 0.60 A.
    (b)   Find the RMS power for this change in current.

16. *Electronics*   The current i (in amperes) in a certain circuit as a function of time t (in seconds) is given by $i(t) = 10t - 2t^2$.
    (a)   Find the average value of the current from t = 0 to t = 5.0 s.
    (b)   Find the RMS value of the current over this interval.

17. *Meteorology*   The temperature T (in °F) from 7 AM to 5 PM (t = 0 to 10) at a certain location was approximated one day by the function $T(t) = 45 + 12t - t^2$. Calculate the average temperature during the 10 h time period.

18. *Environmental technology* The rate of production of a particulate pollutant (in pounds per day) is given by the equation $P(t) = 1.26\sqrt{t/32.4}$. The rate P increases as time goes on and the scrubbers and filters in a smokestack allow more particles to be released. Find the average value of $P(t)$ over the first week of use, from $t = 0$ to $t = 7$ days.

19. *Electronics* If the power output in a signal pulse varies with time according to the equation $P = 3.05t^2 - 2.1t^3$, where P is in watts and t is in seconds, find the RMS value of the power from $t = 0$ to $t = 1.5$ s.

20. *Hydraulics* The flow rate R of a certain automatic valve decreases with time according to the equation $R = 2.50 + (0.95/t^2)$, where t is in seconds and R is in cubic centimeters per second. What is the average value of the flow rate between $t = 1.0$ s and $t = 3.0$ s?

21. *Chemical engineering* A cylindrical tank 12 m high has a radius of 6 m and is filled with water to a depth of 8 m. Calculate the work done in pumping the water out of the top of the tank so that the level drops to 2 m.

22. *Chemical engineering* A reservoir is in the shape of a hemisphere with a radius of 35 ft. If it is filled with water to a depth of 25 ft, find the work done in pumping the water to the top of the tank so that the level drops to 12 ft.

23. *Wastewater technology* A rectangular septic tank is 22 ft long, 16 ft wide, and 14 ft deep. If it is filled to a depth of 12 ft, calculate the amount of work done in pumping out the entire tank to a height 4 ft above the original level. (Use $\rho g = 66$ lb/ft³.)

24. A swimming pool filled with water has a trapezoidal cross section 3 ft deep at the shallow end and 9 ft deep at the deep end. If the pool is 15 ft wide and 25 ft long, calculate the work done in pumping water to a height of 1 ft above the original level so that the level of the pool drops 2 ft.

25. A chemical holding tank filled with water has a cross section in the shape of an isosceles trapezoid 12 ft wide across the top, 8 ft wide at the bottom, and 6 ft deep. If the tank is 15 ft long, calculate the work done in pumping out all the water to a height of 2 ft above the original level. [Hint: Using similar triangles, we can show that the width W of the tank varies with the depth y above the bottom according to the equation $W = 8 + 2y/3$.]

26. A tank is in the form of an inverted cone with a base diameter of 16 m. If the altitude of the tank is 8 m and it is filled with water to a depth of 5 m, calculate the amount of work done in pumping water out of the top of the tank until the level drops to 2 m.

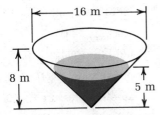

27. A reservoir has the shape of a solid formed by rotating the curve $y = x^3$ about the y-axis. If the reservoir originally is filled with water to a depth of 22 m, calculate the amount of work done to pump water to a height 6 m above the original water level until the water level drops to 10 m.

28. A cylindrical drinking glass 3 in. in diameter and 8 in. high is filled to the top with orange juice. Find the amount of work done (in foot-pounds) in drinking all the juice through a 10 in. straw. (Assume that the straw is positioned perpendicular to the base of the glass cylinder, and assume that $\rho g = 62.4$ lb/ft³.)

29. A rectangular water tank has an access plate on one end. This plate has the shape of an isosceles triangle with a base of 2 ft and a height of 1 ft. If the upper vertex of the triangle is 3 ft below the water level, find the total hydraulic force on the plate. (Assume that the base of the triangle is parallel to the base of the tank.) [Hint: Using similar triangles, we can show that the width W of the triangle a distance y up from its lower edge is $W = 2 - 2y$.]

30. An aquarium has a square observation window with 8 in. sides. Find the total force on the window if its upper edge is 6 in. below water level.

31. A metal cube with edge length 9 in. is suspended in water so that its top surface is 4 in. below water level.
  (a) Find the total force on the top of the cube.
  (b) Find the total force on the bottom of the cube.
  (c) Find the **buoyant force** on the cube by subtracting result (a) from result (b).

32. The vertical face of a dam is parabolic in shape and described by the equation $y = x^2/10$. Find the total force on the face when the water has a depth of 22.5 ft.

33. The vertical face of a dam contains a rectangular gate 12 ft high and 5 ft wide. Find the total force on the gate if the lower edge is 18 ft below water level.

34. *Marine technology*   In the ADS-4 (Advanced Diving System) deep submersible vehicle, the square observation port is 12 in. on each side. What is the total force acting on the window when its top edge is 1000 ft below the surface of the sea? (Use $\rho g \approx 64$ lb/ft³.)

35. The side wall of a water trough is in the shape of an isosceles trapezoid with an upper base of 4 ft, a lower base of 2 ft, and a height of 5 ft. What is the total hydraulic force on the wall when the trough is full? [*Hint:* The width W of the wall is related to the altitude y of the trapezoid by the equation $W = 2 + 2y/5$.]

36. *Marine technology*   Part of the undersurface of a barge is a flat rectangular face with width 8.0 m and height 3.5 m. Its upper edge is 1.5 m below the surface of the water. Find the total force on the surface.

25-6 | Review of Chapter 25

Important Terms and Concepts

vertical area elements (p. 1018)

horizontal area elements (p. 1021)

disk method of finding volume (p. 1028)

solid of revolution (p. 1030)

washer method of finding volume (p. 1034)

shell method of finding volume (p. 1039)

center of gravity/center of mass (p. 1046)

moment/moment arm (p. 1047)

mass density (p. 1049)

homogeneous (p. 1049)

lamina (p. 1050)

centroid (p. 1053)

moment of inertia (p. 1065)

radius of gyration (p. 1071)

work (p. 1081)

fluid (p. 1084)

pressure (p. 1084)

absolute pressure (p. 1085)

gauge pressure (p. 1085)

Formulas and Rules

- area between two curves
    vertical elements (p. 1018)

$$\text{Area} = \int_a^b [f(x) - g(x)]\, dx \qquad a \le x \le b \qquad \text{(25-1)}$$

    horizontal elements (p. 1021)

$$\text{Area} = \int_c^d [p(y) - q(y)]\, dy \qquad c \le y \le d \qquad \text{(25-2)}$$

- volume of a solid of revolution

  area under a curve, rotated about the x-axis (p. 1031)

  $$V = \pi \int_a^b [f(x)]^2 \, dx \qquad\qquad (25\text{-}3)$$

  area between two curves, rotated about the x-axis (p. 1034)

  $$V = \pi \int_a^b \{[f(x)]^2 - [g(x)]^2\} \, dx \qquad\qquad (25\text{-}4)$$

  area under a curve, rotated about the y-axis (p. 1036)

  $$V = \pi \int_c^d [p(y)]^2 \, dy \qquad\qquad (25\text{-}5)$$

  shell method, rotated about the y-axis (p. 1040)

  $$V = 2\pi \int_a^b x[f(x) - g(x)] \, dx \qquad\qquad (25\text{-}6)$$

  shell method, rotated about the x-axis (p. 1041)

  $$V = 2\pi \int_c^d y[p(y) - q(y)] \, dy \qquad\qquad (25\text{-}7)$$

- moment of a particle (p. 1047)

  $$M_x = mx \qquad\qquad (25\text{-}8)$$

- moment of a linear array of n particles with respect to the origin (p. 1047)

  $$M_x = m_1 x_1 + m_2 x_2 + \cdots + m_n x_n = \sum_{i=1}^n m_i x_i \qquad (25\text{-}9)$$

- center of mass of n particles (p. 1047)

  $$x_{cm} = \frac{\displaystyle\sum_{i=1}^n m_i x_i}{\displaystyle\sum_{i=1}^n m_i} \qquad\qquad (25\text{-}10)$$

- center of mass of a planar array of particles (p. 1048)

  $$x_{cm} = \frac{M_x}{M} = \frac{\sum m_i x_i}{\sum m_i} \qquad y_{cm} = \frac{M_y}{M} = \frac{\sum m_i y_i}{\sum m_i} \qquad (25\text{-}11)$$

- mass density (p. 1049)

  $$\rho = \frac{\text{Mass}}{\text{Volume}} \qquad \text{or} \qquad \text{Mass} = \rho \cdot \text{Volume} \qquad (25\text{-}12)$$

- area density (p. 1050)

  $$\delta = \frac{\text{Mass}}{\text{Area}} \qquad \text{or} \qquad \text{Mass} = \delta \cdot \text{Area} \qquad (25\text{-}13)$$

- coordinates of the center of mass of a lamina (pp. 1052, 1053)

  $$x_{cm} = \frac{\int_a^b x f(x) \, dx}{\int_a^b f(x) \, dx} \qquad y_{cm} = \frac{\frac{1}{2} \int_a^b [f(x)]^2 \, dx}{\int_a^b f(x) \, dx} \qquad (25\text{-}14,\ 25\text{-}15)$$

- coordinates of the centroid of a lamina (p. 1055)

  $$x_{cm} = \frac{\int L_x \, dA}{\int dA} \qquad y_{cm} = \frac{\int L_y \, dA}{\int dA} \qquad\qquad (25\text{-}16)$$

- centroid of a solid of revolution

  rotation about the x-axis (p. 1057)

  $$x_{cm} = \frac{\int L_x \, dV}{\int dV} \qquad y_{cm} = 0 \qquad\qquad (25\text{-}17)$$

  rotation about the y-axis (p. 1058)

  $$x_{cm} = 0 \qquad y_{cm} = \frac{\int L_y \, dV}{\int dV} \qquad\qquad (25\text{-}18)$$

- moment of inertia of a point mass (p. 1065)

$$I = L^2 m \qquad\qquad (25\text{-}19)$$

- moment of inertia for an array of particles (p. 1065)

$$I = \sum_{i=1}^{n} L_i^2 m_i \qquad\qquad (25\text{-}20)$$

- moment of inertia of a lamina (p. 1069)

$$I = \int_a^b L^2 \, dm \qquad\qquad (25\text{-}21)$$

- radius of gyration (p. 1071) $\quad R_G = \sqrt{\dfrac{I}{M}} \qquad\qquad (25\text{-}22)$

- average value of a function (p. 1079)

$$f_{\text{AVE}} = \frac{1}{b-a} \int_a^b f(x) \, dx \qquad\qquad (25\text{-}23)$$

- root mean square (RMS) average (p. 1081)

$$f_{\text{RMS}} = \left\{ \frac{1}{b-a} \int_a^b [f(x)]^2 \, dx \right\}^{1/2} \qquad\qquad (25\text{-}24)$$

- pressure–depth equation (p. 1085)

$$P = P_0 + \rho g h \qquad\qquad (25\text{-}26)$$

- total force due to fluid pressure (p. 1086)

$$F = \int_a^b P \, dA \qquad\qquad (25\text{-}27)$$

Exercises 25-6

1. Find the area of the region bounded by the line $y = x$ and the curve $y = x^2/4$.
2. Find the area of the region bounded by the curves $y^2 = 2x$ and $y = 2x - 20$.
3. Find the area bounded by $y = x^4$, $y = 4 - x^4$, and the y-axis. Round to the nearest hundredth.
4. Find the area bounded by the parabolas $x = y^2$ and $x = 2y^2 - 4$.
5. Find the area bounded by the curves $y = 1 + x^2$ and $y = 3 - x^2$.
6. Find the area bounded by $y = x$ and $y = x^3 + 1$ between $x = 0$ and $x = 2$.
7. Find the volume of the solid generated by rotating about the x-axis the region bounded by $y = 5 - 2x$, $x = 0$, and $y = 0$.
8. Find the volume of the solid generated by rotating about the y-axis the region bounded by the curve $y = x^2 + 4$, the line $y = 8$, and the y-axis.
9. Find the volume of the solid generated by rotating about the y-axis the region bounded by $y^2 = 9x$ and the line $x = 4$.
10. Find the volume of the solid generated by rotating about the x-axis the area bounded by $y = x^3/2$, the line $y = 4$, and the y-axis.
11. Find the volume of the solid generated by rotating about the x-axis the region bounded by $y = 3/x^2$ and $y = 4 - x$ between $x = 1$ and $x = 3$. Round to two decimal places.
12. Find the volume of the solid ring generated by rotating about the y-axis the region bounded by $y = 1 + \dfrac{1}{x}$, $y = 1 - \dfrac{1}{x}$, $x = 1$, and $x = 3$. Round to two decimal places.
13. Find (to the nearest tenth) the center of mass x_{cm} for the following linear array of particles:

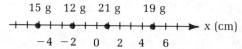

- volume of a solid of revolution
  area under a curve, rotated about the x-axis (p. 1031)

$$V = \pi \int_a^b [f(x)]^2 \, dx \qquad (25\text{-}3)$$

  area between two curves, rotated about the x-axis (p. 1034)

$$V = \pi \int_a^b \{[f(x)]^2 - [g(x)]^2\} \, dx \qquad (25\text{-}4)$$

  area under a curve, rotated about the y-axis (p. 1036)

$$V = \pi \int_c^d [p(y)]^2 \, dy \qquad (25\text{-}5)$$

  shell method, rotated about the y-axis (p. 1040)

$$V = 2\pi \int_a^b x[f(x) - g(x)] \, dx \qquad (25\text{-}6)$$

  shell method, rotated about the x-axis (p. 1041)

$$V = 2\pi \int_c^d y[p(y) - q(y)] \, dy \qquad (25\text{-}7)$$

- moment of a particle (p. 1047)

$$M_x = mx \qquad (25\text{-}8)$$

- moment of a linear array of n particles with respect to the origin (p. 1047)

$$M_x = m_1 x_1 + m_2 x_2 + \cdots + m_n x_n = \sum_{i=1}^n m_i x_i \qquad (25\text{-}9)$$

- center of mass of n particles (p. 1047)

$$x_{cm} = \frac{\displaystyle\sum_{i=1}^n m_i x_i}{\displaystyle\sum_{i=1}^n m_i} \qquad (25\text{-}10)$$

- center of mass of a planar array of particles (p. 1048)

$$x_{cm} = \frac{M_y}{M} = \frac{\sum m_i x_i}{\sum m_i} \qquad y_{cm} = \frac{M_y}{M} = \frac{\sum m_i y_i}{\sum m_i} \qquad (25\text{-}11)$$

- mass density (p. 1049)

$$\rho = \frac{\text{Mass}}{\text{Volume}} \qquad \text{or} \qquad \text{Mass} = \rho \cdot \text{Volume} \qquad (25\text{-}12)$$

- area density (p. 1050)

$$\delta = \frac{\text{Mass}}{\text{Area}} \qquad \text{or} \qquad \text{Mass} = \delta \cdot \text{Area} \qquad (25\text{-}13)$$

- coordinates of the center of mass of a lamina (pp. 1052, 1053)

$$x_{cm} = \frac{\int_a^b x f(x) \, dx}{\int_a^b f(x) \, dx} \qquad y_{cm} = \frac{\frac{1}{2} \int_a^b [f(x)]^2 \, dx}{\int_a^b f(x) \, dx} \qquad (25\text{-}14, \, 25\text{-}15)$$

- coordinates of the centroid of a lamina (p. 1055)

$$x_{cm} = \frac{\int L_x \, dA}{\int dA} \qquad y_{cm} = \frac{\int L_y \, dA}{\int dA} \qquad (25\text{-}16)$$

- centroid of a solid of revolution
  rotation about the x-axis (p. 1057)

$$x_{cm} = \frac{\int L_x \, dV}{\int dV} \qquad y_{cm} = 0 \qquad (25\text{-}17)$$

  rotation about the y-axis (p. 1058)

$$x_{cm} = 0 \qquad y_{cm} = \frac{\int L_y \, dV}{\int dV} \qquad (25\text{-}18)$$

- moment of inertia of a point mass (p. 1065)

$$I = L^2 m \qquad \text{(25-19)}$$

- moment of inertia for an array of particles (p. 1065)

$$I = \sum_{i=1}^{n} L_i^2 m_i \qquad \text{(25-20)}$$

- moment of inertia of a lamina (p. 1069)

$$I = \int_a^b L^2 \, dm \qquad \text{(25-21)}$$

- radius of gyration (p. 1071) $R_G = \sqrt{\dfrac{I}{M}}$ \qquad (25-22)

- average value of a function (p. 1079)

$$f_{\text{AVE}} = \frac{1}{b-a} \int_a^b f(x) \, dx \qquad \text{(25-23)}$$

- root mean square (RMS) average (p. 1081)

$$f_{\text{RMS}} = \left\{ \frac{1}{b-a} \int_a^b [f(x)]^2 \, dx \right\}^{1/2} \qquad \text{(25-24)}$$

- pressure–depth equation (p. 1085)

$$P = P_0 + \rho g h \qquad \text{(25-26)}$$

- total force due to fluid pressure (p. 1086)

$$F = \int_a^b P \, dA \qquad \text{(25-27)}$$

Exercises 25-6

1. Find the area of the region bounded by the line $y = x$ and the curve $y = x^2/4$.
2. Find the area of the region bounded by the curves $y^2 = 2x$ and $y = 2x - 20$.
3. Find the area bounded by $y = x^4$, $y = 4 - x^4$, and the y-axis. Round to the nearest hundredth.
4. Find the area bounded by the parabolas $x = y^2$ and $x = 2y^2 - 4$.
5. Find the area bounded by the curves $y = 1 + x^2$ and $y = 3 - x^2$.
6. Find the area bounded by $y = x$ and $y = x^3 + 1$ between $x = 0$ and $x = 2$.
7. Find the volume of the solid generated by rotating about the x-axis the region bounded by $y = 5 - 2x$, $x = 0$, and $y = 0$.
8. Find the volume of the solid generated by rotating about the y-axis the region bounded by the curve $y = x^2 + 4$, the line $y = 8$, and the y-axis.
9. Find the volume of the solid generated by rotating about the y-axis the region bounded by $y^2 = 9x$ and the line $x = 4$.
10. Find the volume of the solid generated by rotating about the x-axis the area bounded by $y = x^3/2$, the line $y = 4$, and the y-axis.
11. Find the volume of the solid generated by rotating about the x-axis the region bounded by $y = 3/x^2$ and $y = 4 - x$ between $x = 1$ and $x = 3$. Round to two decimal places.
12. Find the volume of the solid ring generated by rotating about the y-axis the region bounded by $y = 1 + \dfrac{1}{x}$, $y = 1 - \dfrac{1}{x}$, $x = 1$, and $x = 3$. Round to two decimal places.
13. Find (to the nearest tenth) the center of mass x_{cm} for the following linear array of particles:

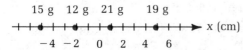

14. Find the center of mass (x_{cm}, y_{cm}) for the following array of particles:

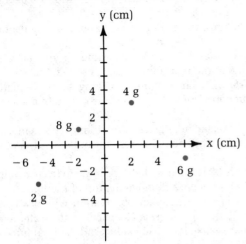

Find the center of mass (x_{cm}, y_{cm}) of each lamina. Round to the nearest tenth.

15.

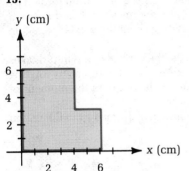

16.

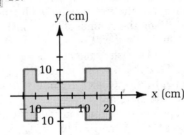

17. Find the centroid of the triangular region bounded by $y = 2 - x$ and both axes.

18. Find the centroid of the region bounded by $y = 4 - (x^2/4)$, $x = 4$, and $y = 4$.

19. Find the centroid of the region bounded by $x^2 + y^2 = 16$ and $y = 4 - x$. Round to the nearest hundredth. [*Hint:* Use area of a quarter circle $= \pi r^2/4$ and area of a triangle $= ab/2$.]

20. Find the centroid of the volume generated by rotating about the x-axis the region bounded by $y = 6x - x^2$ between $x = 0$ and $x = 3$.

21. Find the centroid of the volume generated by rotating about the y-axis the area bounded by $y^2 = 4x$, $y = 0$, and $x = 4$.

22. Find the centroid of the volume generated by rotating about the x-axis the region in quadrant I bounded by the curve $y = x^3/2$ and the line $y = 2x$. Round to the nearest hundredth.

23. Calculate (a) the moment of inertia and (b) the radius of gyration for the lamina bounded by $y = x^2/2$, the y-axis, and the line $y = 2$ for rotation about the x-axis.

24. Calculate (a) the moment of inertia and (b) the radius of gyration for the lamina bounded by $y = 3\sqrt{x}$, $y = 0$, and $x = 4$ for rotation about the y-axis. Round to the nearest hundredth.

25. Calculate (a) the moment of inertia and (b) the radius of gyration with respect to the axis of rotation for the solid generated by rotating about the y-axis the area bounded by $y = 3x$, $y = 6$, and the y-axis.

26. Calculate (a) the moment of inertia and (b) the radius of gyration with respect to the axis of rotation for the solid generated by rotating about the x-axis the area bounded by $y = x^3$, $y = 2$, and the y-axis. Round to the nearest hundredth.

27. Find the average value of the function $f(x) = \dfrac{x}{(x^2 + 3)^{3/2}}$ from $x = 1$ to $x = 3$. Round to one decimal place.

28. Find the RMS average value of the function $f(x) = x^3 - 1$ from $x = -1$ to $x = 2$. Round to one decimal place.

29. *Astronomy*   The light output from a star system is found to vary according to the equation $y = 1.5 + 4x + 3x^2 - 2x^3 - x^4$. Find the average value of the light output from $x = -2.7$ to $x = 1.7$.

30. *Electronics*   The power output in a signal pulse varies with current according to the equation $P = 1.2i^2 - 0.9i$, where P is in watts and i is in milliamperes. Find the RMS value of P with respect to i from $i = -1.5$ mA to $i = 1.0$ mA.

31. *Petroleum engineering*   An oil reservoir in the shape of a rectangular prism is 34 ft long, 26 ft wide, and 16 ft deep. If it is filled with oil to a depth of 10 ft, find the work done in pumping out of the top of the reservoir all but 2 ft of the oil. (Use $\rho g = 60$ lb/ft³.)

32. *Petroleum engineering*   If the cylindrical tank of Example 32 is filled with oil ($\rho g = 60$ lb/ft³) to a height of 25 ft, find the work done in pumping all of the oil to a height 2 ft above the top of the tank.

33. *Marine technology*   A marine technology testing tank has a rectangular observation window 12 cm high by 18 cm wide. Find the total force on the window if its upper edge is 15 cm below the water level.

34. *Civil engineering*   The vertical face of a dam has a square gate 6 ft on each side. Find the total force on the gate when the upper edge is 4 ft below water level.

35. *Machine technology*   Find the area of the cam formed by the region between the curve $y = 4x - x^2$ and the straight line $2y = 4 - x$.

36. *Machine technology*   Find the volume of the flange made in the shape of a solid of rotation formed by rotating about the x-axis the region bounded by the curve $y = x^3 + 2$ and the lines $x = 2$ and $x = -1$.

37. *Industrial design*   Find the volume of the acoustic horn formed by rotating about the y-axis the region bounded by the curve $x = y^{3/2}$, the y-axis, and the line $y = 4$.

38. *Optics*   When a circular dish of mercury is rotated about its center axis, the surface of the liquid takes a parabolic shape. Find the moment of inertia for rotation about the y-axis of the solid formed by rotating the region bounded by $y = 0.04x^2 + 1.0$, $y = 0$, and $x = 5$ about the y-axis. (Express the moment of inertia in terms of the mass m.)

39. Find the moment of inertia for rotation about the x-axis of a semiparabolic disk formed from the region bounded by $2y = x^2$, $y = 8$, and $x = 0$.

Derivatives of Transcendental Functions

26

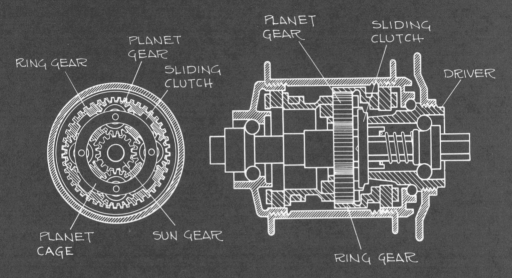

RING GEAR

PLANET GEAR

SLIDING CLUTCH

PLANET GEAR

SLIDING CLUTCH

DRIVER

PLANET CAGE

SUN GEAR

RING GEAR

BICYCLE PLANETARY GEARING SYSTEM

In Chapter 22 we found rules that allow us to calculate the derivatives of algebraic functions, including polynomials and algebraic fractions. Any equation that is not algebraic is said to be transcendental. For an algebraic equation $y = f(x)$, values of the dependent variable y will be either rational (integers, fractions) or irrational (for example, $\sqrt{2}$, $\sqrt[3]{4}$, $5^{7/8}$, and so on). For a transcendental function, the dependent variable is generally irrational, but its values usually cannot be expressed as fractions or as rational numbers raised to a rational power. Numbers such as π, e, log 2 and sin 1.5 are transcendental numbers.

In this chapter we will develop rules for finding the derivatives of three very useful classes of transcendental functions: the trigonometric functions and their inverses, the logarithmic function $y = \log_b x$, and the exponential function $y = e^x$.

The derivative of the sine function may be found using the definition of a derivative (22-9) and the Δ-limit process used in Chapter 22. However, we will need two preliminary results in order to use the Δ-limit process.

Preliminary Result 1 $\lim\limits_{\theta \to 0} \dfrac{\sin \theta}{\theta} = 1$ for θ in radians

Proof **First,** use a calculator to check a few values of $(\sin \theta)/\theta$, as listed in Table 26-1.

TABLE 26-1

θ	$\sin \theta$	$\dfrac{\sin \theta}{\theta}$
0.10	0.0998334	0.9983342
0.05	0.0499792	0.9995834
0.01	0.0099998	0.9999833
0.001	0.0009999	0.9999998

Notice that as $\theta \to 0$, $\sin \theta \to 0$, and the value of $(\sin \theta)/\theta$ has the form $0/0$. From the table, the limit of $(\sin \theta)/\theta$ seems to approach 1 as $\theta \to 0$. To prove this, consider Fig. 26-1. If the radius of the circle is r, then $OA = OP = r$. We assume that angle θ is expressed in radians and $0 < \theta < \pi/2$, since we are interested in small positive values of θ. Then we have the following:

- $A_1 = $ Area of triangle $AOP = \dfrac{1}{2} \cdot \text{Base} \cdot \text{Altitude}$

  $= \dfrac{1}{2} \cdot OA \cdot BP$

  But $OA = r$ and $\sin \theta = BP/OP = BP/r$, so $BP = r \sin \theta$; therefore,

  $A_1 = \dfrac{1}{2} r^2 \sin \theta$

- $A_2 = $ Area of sector $AOP = \dfrac{1}{2} r^2 \theta$ From equation (8-6)

- $A_3 = $ Area of triangle $AOQ = \dfrac{1}{2} \cdot OA \cdot AQ$

  But $OA = r$ and $\tan \theta = AQ/OA = AQ/r$, so $AQ = r \tan \theta$; therefore,

  $A_3 = \dfrac{1}{2} r^2 \tan \theta = \dfrac{1}{2} r^2 \dfrac{\sin \theta}{\cos \theta}$ From identity (19-4)

  As you can see in Fig. 26-1, $A_1 < A_2 < A_3$, so

  $$\dfrac{1}{2} r^2 \sin \theta < \dfrac{1}{2} r^2 \theta < \dfrac{1}{2} r^2 \dfrac{\sin \theta}{\cos \theta}$$

Figure 26-1

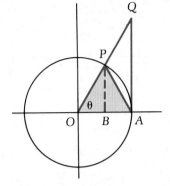

Multiplying each term by $2/(r^2 \sin \theta)$, we obtain

$$1 < \frac{\theta}{\sin \theta} < \frac{1}{\cos \theta} \qquad \text{for } 0 < \theta < \frac{\pi}{2}$$

After taking the reciprocal of each term and reversing the inequality symbols, the result is*

$$1 > \frac{\sin \theta}{\theta} > \cos \theta$$

Since $0 < \theta < \pi/2$, we can take the limit of each term as θ approaches 0 through positive values:

$$\lim_{\theta \to 0^+} 1 > \lim_{\theta \to 0^+} \frac{\sin \theta}{\theta} > \lim_{\theta \to 0^+} \cos \theta$$

By definition, $\lim_{\theta \to 0^+} 1 = 1$. Also because $\cos \theta$ is continuous at $\theta = 0$, $\lim_{\theta \to 0^+} \cos \theta = 1$. Therefore,

$$1 > \lim_{\theta \to 0^+} \frac{\sin \theta}{\theta} > 1$$

and since the remaining limit is bounded above and below by 1, by a theorem on limits called the *pinching theorem* we know that

$$\lim_{\theta \to 0^+} \frac{\sin \theta}{\theta} = 1$$

To find the limit as $\theta \to 0^-$, note that

$$\lim_{\theta \to 0^-} \frac{\sin \theta}{\theta} \quad \text{is equivalent to} \quad \lim_{\theta \to 0^+} \frac{\sin (-\theta)}{-\theta} = \lim_{\theta \to 0^+} \frac{-\sin \theta}{-\theta}$$

$$= \lim_{\theta \to 0^+} \frac{\sin \theta}{\theta}$$

$$= 1$$

*Using the multiplication rule for inequalities, it is possible to show that if $a < b < c$, then $1/a > 1/b > 1/c$ for positive values of a, b, and c. For example, multiplying each side of $a < b$ by $1/ab$ gives $1/b < 1/a$.

Therefore: $\lim\limits_{\theta \to 0^+} \dfrac{\sin \theta}{\theta} = \lim\limits_{\theta \to 0^-} \dfrac{\sin \theta}{\theta} = 1$

Preliminary Result 1 $\lim\limits_{\theta \to 0} \dfrac{\sin \theta}{\theta} = 1$ **(26-1)**

CAUTION ▶

This result is true only if θ is expressed in radians. If α is the same angle expressed in degrees, then

$$\alpha = \frac{180\theta}{\pi}, \text{ and } \lim\limits_{\alpha \to 0^\circ} \frac{\sin \alpha}{\alpha} = \frac{\sin \alpha}{180\theta/\pi}$$

But $\sin \alpha = \sin \theta$, since α and θ are the same angle. Therefore,

$$\lim\limits_{\alpha \to 0^\circ} \frac{\sin \alpha}{\alpha} = \frac{\pi}{180} \cdot \lim\limits_{\theta \to 0} \frac{\sin \theta}{\theta} = \frac{\pi}{180} \quad \blacktriangleleft$$

Preliminary Result 2 $\lim\limits_{\theta \to 0} \dfrac{\cos \theta - 1}{\theta} = 0$

Proof   First, check a few values on a calculator, as shown in Table 26-2.

TABLE 26-2

θ	$\cos \theta$	$\cos \theta - 1$	$\dfrac{\cos \theta - 1}{\theta}$
0.10	0.9950042	−0.0049958	−0.0499583
0.05	0.9987503	−0.0012497	−0.0249948
0.01	0.9999500	−0.0000499	−0.0049999
0.001	0.9999995	−0.0000005	−0.0004999

Then, proceed as follows:

$$\frac{\cos \theta - 1}{\theta} = \frac{\cos \theta - 1}{\theta} \cdot \frac{\cos \theta + 1}{\cos \theta + 1}$$

$$= \frac{\cos^2 \theta - 1}{\theta(\cos \theta + 1)} = \frac{-\sin^2 \theta}{\theta(\cos \theta + 1)} \qquad \text{From identity (19-6)}$$

$$= -\frac{\sin \theta}{\theta} \cdot \frac{\sin \theta}{\cos \theta + 1}$$

Taking limits as $\theta \to 0^+$,

$$\lim\limits_{\theta \to 0^+} \frac{\cos \theta - 1}{\theta} = \left(-\lim\limits_{\theta \to 0^+} \frac{\sin \theta}{\theta} \right) \cdot \left(\lim\limits_{\theta \to 0^+} \frac{\sin \theta}{\cos \theta + 1} \right)$$

$$= (-1) \cdot \left(\frac{0}{1 + 1} \right) = 0$$

Again, because $\sin \theta$ and $\cos \theta$ are continuous at $\theta = 0$, we know that $\lim\limits_{\theta \to 0} \sin \theta = 0$ and $\lim\limits_{\theta \to 0} \cos \theta = 1$. The result is exactly the same for the limit as $\theta \to 0^-$; therefore, we have the result shown in (26-2).

Preliminary Result 2 $\lim\limits_{\theta \to 0} \dfrac{\cos \theta - 1}{\theta} = 0$ (26-2)

Derivative of the sine function

Armed with these two preliminary results, we can now find the derivative of the sine function using the Δ-limit process of Chapter 22. If $f(u) = \sin u$, for u in radians, then

$$f(u + \Delta u) = \sin(u + \Delta u)$$

$$\Delta f = f(u + \Delta u) - f(u)$$

$$= \sin(u + \Delta u) - \sin u$$

and

$$\frac{\Delta f}{\Delta u} = \frac{\sin(u + \Delta u) - \sin u}{\Delta u}$$

But, from Chapter 19 we have the trigonometric identity $\sin(A + B) = \sin A \cos B + \cos A \sin B$. Therefore,

$$\frac{\Delta f}{\Delta u} = \frac{\sin u \cos \Delta u + \cos u \sin \Delta u - \sin u}{\Delta u}$$

Factoring, we obtain

$$= \frac{\cos u \sin \Delta u + \sin u (\cos \Delta u - 1)}{\Delta u}$$

$$= \cos u \left(\frac{\sin \Delta u}{\Delta u} \right) + \sin u \left(\frac{\cos \Delta u - 1}{\Delta u} \right)$$

From the definition of a derivative (22-9),

$$\frac{df}{du} = \lim_{\Delta u \to 0} \left[\cos u \left(\frac{\sin \Delta u}{\Delta u} \right) + \sin u \left(\frac{\cos \Delta u - 1}{\Delta u} \right) \right]$$

$$= \cos u \lim_{\Delta u \to 0} \left(\frac{\sin \Delta u}{\Delta u} \right) + \sin u \lim_{\Delta u \to 0} \left(\frac{\cos \Delta u - 1}{\Delta u} \right)$$

By (26-1), this limit equals 1.   By (26-2), this limit equals 0.

$$= \cos u$$

Finally, if u is some function of x, $u = u(x)$, we can find df/dx using the chain rule (22-21):

$$\frac{df}{dx} = \frac{df}{du} \cdot \frac{du}{dx}$$

$$= \cos u \cdot \frac{du}{dx}$$

Derivative of sin u	$\dfrac{d}{dx}(\sin u) = \cos u \cdot \dfrac{du}{dx}$	(26-3)

EXAMPLE 1 Find dy/dx if:
(a) $y = \sin 4x$
(b) The range R of a projectile, launched with initial speed v_0 at an angle θ $(0 \le \theta \le \pi/2)$ with the horizontal, is given by the equation

$$R = \frac{v_0^2}{g} \sin 2\theta$$

Find $dR/d\theta$, the rate at which the range changes with θ, and use it to determine the angle at which R is a maximum.

Solution
(a) **First,** let $u = 4x$, so that $du/dx = 4$.

   Then, using (26-3). $\dfrac{dy}{dx} = \cos u \cdot 4 = 4 \cos 4x$

(b) $\dfrac{dR}{d\theta} = \dfrac{v_0^2}{g} \cdot \dfrac{d}{d\theta}(\sin 2\theta) = \dfrac{v_0^2}{g} \cdot 2 \cos 2\theta$

   The range R is a maximum at $dR/d\theta = 0$; therefore

$$\frac{2v_0^2}{g} \cos 2\theta = 0 \qquad \text{when } 2\theta = \frac{\pi}{2}$$

$$\theta = \frac{\pi}{4} = 45°$$

   Check: Use the first derivative test of Chapter 23 to check that R is a maximum rather than a minimum at $\theta = \pi/4$. Using a calculator, $\cos 2(40°) \approx 0.1736$ and $\cos 2(50°) \approx -0.1736$.

CAUTION ▶ A common error is to omit the factor du/dx. Be careful. ◀

EXAMPLE 2 Find dy/dx if:
(a) $y = \sin t^2$ (b) $y = \sin^2 t$ (c) $y = \sin^3(4t^2)$

Solution

(a) **Step 1.** Recall that $y = \sin t^2$ means $y = \sin (t^2)$.

  Step 2. Let $u = t^2$; then $du/dt = 2t$.

  Step 3. Use (26-3). $\dfrac{dy}{dt} = (\cos t^2)(2t)$

$$= 2t \cos t^2$$

(b) **Step 1.** Recall that $y = \sin^2 t$ means $y = (\sin t)^2$.

  Step 2. Let $u = \sin t$; then $du/dt = \cos t$ and $y = u^2$.

  Step 3. Use the chain rule (22-21). $\dfrac{dy}{dt} = \dfrac{dy}{du} \cdot \dfrac{du}{dt}$

$$= 2u \cdot \cos t$$

$$= 2 \sin t \cos t = \sin 2t \quad \text{By equation (19-14)}$$

(c) **Step 1.** Let $u = \sin (4t^2)$; then $y = u^3$ and

$$\frac{dy}{dt} = 3u^2 \cdot \frac{du}{dt}$$

$$= 3 \sin^2(4t^2) \frac{du}{dt}$$

  Step 2. Use (26-3). $\dfrac{du}{dt} = \cos (4t^2) \cdot \dfrac{d}{dt}(4t^2)$

$$= 8t \cos (4t^2)$$

  Step 3. Combine the results of Steps 1 and 2. $\dfrac{dy}{dt} = 24t \sin^2(4t^2) \cos (4t^2)$

CAUTION ▶   Be careful to distinguish $\sin^2 x$ from $\sin 2x$ or $\sin x^2$:

$\sin^2 x = (\sin x)^2 \qquad \sin 2x = \sin (2x) \qquad \sin x^2 = \sin (x^2)$ ◀

EXAMPLE 3   Find dy/dx if:

(a) $y = x^3 \sin x^2$ (b) $y = \dfrac{\sin^2 x}{x^2}$

Solution

(a) **Step 1.** Use the product rule (22-17). $\dfrac{dy}{dx} = x^3 \cdot \dfrac{d}{dx}(\sin x^2) + \sin x^2 \cdot \dfrac{d}{dx}(x^3)$

  Step 2. As in Example 2(a), $\dfrac{d}{dx}(\sin x^2) = 2x \cos x^2$.

  Step 3. Therefore $\dfrac{dy}{dx} = 2x^4 \cos x^2 + 3x^2 \sin x^2$

(b) **Step 1.** Use the quotient rule (22-19). $\dfrac{dy}{dx} = \dfrac{x^2 \cdot \dfrac{d}{dx}(\sin^2 x) - \sin^2 x \cdot \dfrac{d}{dx}(x^2)}{(x^2)^2}$

Step 2. As in Example

2(b), $\dfrac{d}{dx}(\sin^2 x) = 2\sin x \cdot \cos x.$

Step 3. Therefore:

$$\frac{dy}{dx} = \frac{x^2(2\sin x \cos x) - \sin^2 x\,(2x)}{x^4}$$

$$= \frac{2x\sin x \cos x - 2\sin^2 x}{x^3}$$

EXAMPLE 4 Find $dy/d\theta$ if:

(a) $y = \sin(4\theta - 1)^2$ (b) $y = \sin\sqrt{\theta^2 + 2\theta}$

(c) The acceleration g due to gravity is not constant at the surface of the earth but varies slightly with latitude.* If ϕ is the latitude in radians, then at sea level $g \approx 9.78049(1 + 0.005288\ \sin^2\phi - 0.000006\ \sin^2 2\phi)$ m/s². Find $dg/d\phi$, the rate at which g changes with latitude.

Solution

(a) **Step 1.** Let $u = (4\theta - 1)^2$; $\dfrac{du}{d\theta} = 2(4\theta - 1)\cdot 4 = 8(4\theta - 1)$
then $y = \sin u$, and

Step 2. Use the sine de- $\dfrac{dy}{d\theta} = \cos u \cdot \dfrac{du}{d\theta} = \cos u \cdot 8(4\theta - 1)$
rivative rule (26-3).

$$= 8(4\theta - 1)\cos(4\theta - 1)^2$$

Notice that the coefficient $8(4\theta - 1)$ is placed *before* the trigonometric function so that it will not be confused with the argument of the sine function.

(b) **Step 1.** Let $u = \sqrt{\theta^2 + 2\theta} = (\theta^2 + 2\theta)^{1/2}$.

Step 2. Use the rational $\dfrac{du}{d\theta} = \dfrac{1}{2}(\theta^2 + 2\theta)^{-1/2}\cdot(2\theta + 2)$
power rule (22-23).

$$= \frac{\theta + 1}{\sqrt{\theta^2 + 2\theta}}$$

Step 3. From the sine de- $\dfrac{dy}{d\theta} = \cos\sqrt{\theta^2 + 2\theta}\cdot\dfrac{du}{d\theta}$
rivative rule (26-3):

$$= \cos\sqrt{\theta^2 + 2\theta}\cdot\frac{\theta + 1}{\sqrt{\theta^2 + 2\theta}}$$

$$= \frac{\theta + 1}{\sqrt{\theta^2 + 2\theta}}\cdot\cos\sqrt{\theta^2 + 2\theta}$$

(c) **First,** note that this function has the form $g = a + b\,\sin^2\phi + c\,\sin^2 2\phi$,

*See G. Kuiper, ed., *Earth as a Planet* (Chicago: University of Chicago Press, 1954), *The Solar System*, Vol. 2, p. 16.

where $a = 9.78049$, $b = 9.78049(0.005288) \approx 0.051719$, and
$c = 9.78049(-0.000006) \approx -0.0000587$.
Then,

$$\frac{dg}{d\phi} = b\frac{d}{d\phi}(\sin^2\phi) + c\frac{d}{d\phi}(\sin^2 2\phi)$$

$$= b \cdot 2 \sin\phi \cdot \cos\phi + c \cdot 2 \sin 2\phi \cdot \cos 2\phi \cdot 2$$

$$= 2b \sin\phi \cos\phi + 4c \sin 2\phi \cos 2\phi$$

Finally, using the trigonometric identity $2 \sin A \cos A = \sin 2A$, we can simplify this as

$$\frac{dg}{d\phi} = \sin 2\phi(b + 4c \cos 2\phi)$$

$$\approx \sin 2\phi(0.051719 - 0.000235 \cos 2\phi)$$

EXAMPLE 5 Find dy/dx if $y^2 = x^2 + \sin 3x^2$

Solution Using the technique of implicit differentiation from Chapter 22, we first take the derivative of both members of the equation:

$$2y\frac{dy}{dx} = 2x + 6x \cos 3x^2$$

$$\boxed{\frac{d}{dx}(\sin 3x^2) = \cos 3x^2 \cdot \frac{d}{dx}(3x^2)}$$

Now we solve for dy/dx:

$$\frac{dy}{dx} = \frac{x + 3x \cos 3x^2}{y}$$

$$= \frac{x + 3x \cos 3x^2}{\pm\sqrt{x^2 + \sin 3x^2}} = \pm\frac{x + 3x \cos 3x^2}{\sqrt{x^2 + \sin 3x^2}}$$

Derivative of the cosine function To obtain a formula for the derivative of $f(u) = \cos u$, use the Δ-limit process:

$$\frac{\Delta f}{\Delta u} = \frac{\cos(u + \Delta u) - \cos u}{\Delta u}$$

From Chapter 19 we have the trigonometric identity

$$\cos(A + B) = \cos A \cos B - \sin A \sin B$$

Then,

$$\frac{\Delta f}{\Delta u} = \frac{\cos u \cos \Delta u - \sin u \sin \Delta u - \cos u}{\Delta u}$$

$$= \cos u\left(\frac{\cos \Delta u - 1}{\Delta u}\right) - \sin u\left(\frac{\sin \Delta u}{\Delta u}\right)$$

By the definition of the derivative (22-9),

$$\frac{df}{du} = \lim_{\Delta u \to 0} \frac{\Delta f}{\Delta u} = \cos u \cdot \lim_{\Delta u \to 0} \left(\frac{\cos \Delta u - 1}{\Delta u} \right) - \sin u \cdot \lim_{\Delta u \to 0} \left(\frac{\sin \Delta u}{\Delta u} \right)$$

We can evaluate the limits using (26-1) and (26-2):

$$\frac{df}{du} = \cos u \cdot 0 - \sin u \cdot 1 = -\sin u$$

If u is some function of x, u = u(x), then from the chain rule (22-21),

$$\frac{df}{dx} = \frac{df}{du} \cdot \frac{du}{dx} = -\sin u \cdot \frac{du}{dx}$$

Derivative of cos u	$\dfrac{d}{dx}(\cos u) = -\sin u \cdot \dfrac{du}{dx}$	(26-4)

EXAMPLE 6   Find dy/dx for:

(a) $y = \cos 6x$ (b) $y = \cos(3x^2 - x)$ (c) $y = \cos^2 4x$

(d)   The voltage drop V_L across an inductor L in an RL circuit is given by the equation

$$V_L = L \frac{di}{dt}$$

Find V_L if $i = 25 \cos[60\pi t - (\pi/4)]$ and L = 5.0 H.

Solution

(a)   **Step 1.**   Let u = 6x; then du/dx = 6.

Step 2.   Use (26-4). $\dfrac{dy}{dx} = (-\sin u)(6) = -6 \sin 6x$

(b)   **Step 1.**   If we let u = $3x^2 - x$, then du/dx = 6x − 1.

Step 2.   Use (26-4). $\dfrac{dy}{dx} = (-\sin u)(6x - 1) = (1 - 6x) \sin (3x^2 - x)$

(c)   **Step 1.**   Recall that $y = \cos^2 4x$ means $y = (\cos 4x)^2$. If we let u = cos 4x, then $y = u^2$ and du/dx = −4 sin 4x.

Step 2.   Use the chain rule (22-21). $\dfrac{dy}{dx} = \dfrac{dy}{du} \cdot \dfrac{du}{dx} = 2u(-4 \sin 4x)$

$$= -8 \sin 4x \cos 4x$$

Using the double-angle identity (19-14) we can also write this as $\dfrac{dy}{dx} = -4 \sin 8x$

(d)   Let $u = 60\pi t - \left(\dfrac{\pi}{4}\right)$; then:

$$\frac{di}{dt} = 25\Big[\underbrace{-\sin\left(60\pi t - \frac{\pi}{4}\right)}_{-\sin u}\Big]\underbrace{(60\pi)}_{\frac{du}{dt}}$$

$$= -1500\pi \sin\left(60\pi t - \frac{\pi}{4}\right)$$

$$V_L = 5.0\,\frac{di}{dt} = -7500\pi \sin\left(60\pi t - \frac{\pi}{4}\right) \text{ V}$$

EXAMPLE 7   Find dy/dx for:

(a) $y = x^2 \cos x$ (b) $y = \cos 2x + 3x^4$

Solution

(a)   **Step 1.**   Use the product rule (22-17).

$$\frac{dy}{dx} = x^2 \cdot \frac{d}{dx}(\cos x) + \cos x \cdot \frac{d}{dx}(x^2)$$

Step 2.   From (26-4):

$$= -x^2 \sin x + 2x \cos x$$

(b)   **Step 1.**   Use the sum rule (22-15).

$$\frac{dy}{dx} = \frac{d}{dx}(\cos 2x) + \frac{d}{dx}(3x^4)$$

Step 2.   From (26-4):

$$= -2 \sin 2x + 12x^3$$

EXAMPLE 8   Find dy/dx for:

(a) $y = \sin x^2 \cos 2x$ (b) $y = \dfrac{\sin 2x}{\cos x^2}$

Solution

(a)   Use the product rule (22-17).

$$\frac{dy}{dx} = \sin x^2 \frac{d}{dx}(\cos 2x) + \cos 2x \frac{d}{dx}(\sin x^2)$$

From (26-3) and (26-4):

$$= \sin x^2 \cdot (-2 \sin 2x) + \cos 2x \cdot (2x \cos x^2)$$

$$= -2 \sin x^2 \sin 2x + 2x \cos x^2 \cos 2x$$

(b)   Use the quotient rule (22-19).

$$\frac{dy}{dx} = \frac{\cos x^2 \cdot \dfrac{d}{dx}(\sin 2x) - \sin 2x \cdot \dfrac{d}{dx}(\cos x^2)}{\cos^2 x^2}$$

$$= \frac{2 \cos x^2 \cos 2x + 2x \sin 2x \sin x^2}{\cos^2 x^2}$$

EXAMPLE 9

(a)   Find the value(s) of x where the graph of the function $y = 3 \sin 2x^2 - 4x^2$ has a horizontal tangent line.

(b)   Find the slope of this function at $x = 0.25$ rad.

Solution

(a)   Use (26-4).

$$\frac{dy}{dx} = 3(\cos 2x^2)(4x) - 8x$$

$$= 12x \cos 2x^2 - 8x$$

The tangent to the curve is horizontal where $dy/dx = 0$:

$$12x \cos 2x^2 - 8x = 0$$

$$4x(3 \cos 2x^2 - 2) = 0$$

Therefore: $x = 0$ or $3 \cos 2x^2 - 2 = 0$

$$\cos 2x^2 = \frac{2}{3}$$

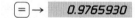 $\boxed{\text{RAD}}\ 2 \left(\div\right) 3 \left(=\right) \boxed{\text{INV}} \boxed{\text{COS}} \left(\div\right) 2 \left(=\right) \boxed{\sqrt{\ }} \rightarrow$ `0.6484862`

$$x \approx \pm 0.65$$

The graph has horizontal tangents at $x = 0$ and $x \approx \pm 0.65$.

(b)   Using a calculator, we can substitute $x = 0.25$ into the derivative found in part (a):

$$\text{Slope} = \frac{dy}{dx} = 12x \cos 2x^2 - 8x$$

$\boxed{\text{RAD}}\ .25\ \boxed{\text{STO}} \boxed{x^2} \left(\times\right) 2 \left(=\right) \boxed{\text{COS}} \left(\times\right) \boxed{\text{RCL}} \left(\times\right) 12 \left(-\right) 8 \left(\times\right) \boxed{\text{RCL}}$

$\left(=\right) \rightarrow$ `0.9765930`

Slope ≈ 0.98

EXAMPLE 10   Figure 26-2 shows a crate of weight W being pulled at constant speed up a ramp by a force P. Using the laws of physics that describe this kind of motion, it is possible to show that $P = W(\mu \cos \theta + \sin \theta)$, where μ is a constant related to the sliding friction between the crate and the ramp. If $\mu = 0.72$, for what value of θ (in degrees) is force P a maximum?

Solution
Use the sum rule (22-15).

$$\frac{dP}{d\theta} = W\left[\frac{d}{d\theta}(\mu \cos \theta) + \frac{d}{d\theta}(\sin \theta)\right]$$

Now, use (26-3) and (26-4).

$$\frac{dP}{d\theta} = W[-\mu \sin \theta + \cos \theta]$$

For a maximum, $dP/d\theta = 0$, so

$$0 = -\mu \sin \theta + \cos \theta$$

or

$$\sin \theta = \frac{1}{\mu} \cos \theta$$

$$\tan \theta = \frac{1}{\mu}$$

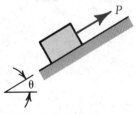

Figure 26-2

For $\mu = 0.72$, $\qquad\qquad\qquad$ $\tan \theta = \dfrac{1}{0.72}$

$\boxed{\text{RAD}}$ .72 $\boxed{\frac{1}{x}}$ $\boxed{\text{INV}}$ $\boxed{\tan}$ $\rightarrow$ $\qquad$ $\boxed{0.9467733}$

In degrees: $\boxed{\times}$ 180 $\boxed{\div}$ $\boxed{\pi}$ $\boxed{=}$ $\rightarrow$ $\qquad$ $\boxed{54.246113}$

Force P is a maximum at $\theta \approx 54°$.

Check that P is a maximum for this value of θ by using the second derivative test:

$$\frac{d^2P}{d\theta^2} = -\mu \cos \theta - \sin \theta \text{ and } \frac{d^2P}{d\theta^2} < 0 \text{ when } \theta \approx 54°.$$

Exercises 26-1

Find the derivative of each function.

1. $y = \sin 2x$ $\qquad\qquad$ **2.** $y = 3 \sin 5x$ $\qquad\qquad$ **3.** $y = 2 \sin (4x - 3)$
4. $y = \sin t^3$ $\qquad\qquad$ **5.** $y = 4 \cos 3t$ $\qquad\qquad$ **6.** $y = \cos (7x^2 + 4)$

7. $y = \cos (5x - 2)$ $\qquad$ **8.** $y = 3 \cos 4x^2$ $\qquad$ **9.** $y = \sin \left(\dfrac{\pi}{2} - x\right)$

10. $y = \cos \left(x + \dfrac{\pi}{4}\right)$ $\qquad$ **11.** $y = \sin (x^2 + 2x)$ $\qquad$ **12.** $y = \cos (2x + 3)^2$

13. $y = \sin^2 3x$ $\qquad\qquad$ **14.** $y = 6 \sin^3 2x^4$ $\qquad$ **15.** $y = 2 \cos^3 (3\theta - 1)$
16. $y = \cos^2 8x$ $\qquad\qquad$ **17.** $y = \sqrt{\sin 2x}$ $\qquad$ **18.** $y = (1 + \cos 3x)^2$
19. $y = x \sin 2x$ $\qquad\qquad$ **20.** $y = 2x^2 \sin x^2$ $\qquad$ **21.** $y = x^3 \cos 4x$
22. $y = 5x \cos x$ $\qquad\qquad$ **23.** $y = \sin\sqrt{x^2 + 1}$ $\qquad$ **24.** $y = 3 \cos\sqrt{2\theta}$
25. $y = x^2 + \sin 5x$ $\qquad$ **26.** $y = \cos x^2 - 4x^3$ $\qquad$ **27.** $y = \cos 2x + 2 \sin x$
28. $y = \sin^2 x - 4 \cos 3x$ $\qquad$ **29.** $y = x \cos x - 2x^2$ $\qquad$ **30.** $y = 3x \sin x + x^2 \cos x$

31. $y = \dfrac{\sin 2x}{x}$ $\qquad\qquad$ **32.** $y = \dfrac{4x}{\sin^2 4x}$ $\qquad$ **33.** $y = \dfrac{x^3}{\cos^2 x}$

34. $y = \dfrac{3 \cos x^2}{5x}$ $\qquad$ **35.** $y = \dfrac{\cos 3x}{\sin x^2}$ $\qquad$ **36.** $y = \dfrac{\cos^2 x}{1 + \cos x}$

37. $y = \sin 3x \cos x^2$ $\qquad$ **38.** $y = \cos 4x \sin x^3$ $\qquad$ **39.** $y = 4 \sin x \cos 2x$
40. $y = 2 \sin x^2 \cos x$ $\qquad$ **41.** $y^2 = 3x - \cos x^3$ $\qquad$ **42.** $y^3 = \sin 5x + 2x^3$

Solve.

43. Find the slope of a line tangent to the curve $y = 2 \cos 4x$ at the point where $x = \pi/4$.
44. Find the values of x where the curve $y = 2x^2 + 3 \cos x^2$ has a horizontal tangent line. ($0 \le x \le \pi/2$)
45. Find the value of x where the function $y = 3 \sin 2x$ has a maximum. ($0 \le x \le 1$)
46. Find the slope of a line tangent to the curve $y = 2 \sin x^2$ at $x = 0.75$.
47. *Physics* The displacement of a longitudinal wave in a string is given by $y = A \sin (kx - \omega t)$. Find dy/dt, the transverse speed of the string.
48. *Physics* The displacement of an object moving in a straight line with simple harmonic motion is given

by the equation $y = y_0 \cos(\omega t + \phi)$, where y_0, ω, and ϕ are constants. Find the speed dy/dt of the object when $t = 4.0$ s if $y_0 = 8.5$ cm, $\omega = 0.15$ s^{-1}, and $\phi = 1.8$.

49. *Optics*   The following formula appears in the study of optical transmission. If T_M and F are constants, find $dT/d\theta$.

$$T = \frac{T_M}{1 + F \sin^2(\theta/2)}$$

50. *Industrial design*   A conical container is constructed with a slant height of 4.0 ft. Find the radius and height of the cone having the maximum volume. [*Hint:* Express both r and h in terms of θ and substitute these values into the formula $V = \frac{1}{3}\pi r^2 h$.]

51. *Optics*   The intensity I of the light reaching the screen in a double-slit diffraction system depends on the angle θ a ray of light makes with the perpendicular:

$$I = \frac{I_m}{9}\left[1 + 8\cos^2\left(\frac{\pi d \sin \theta}{\lambda}\right)\right]$$

where I_m is the maximum intensity, d is the slit separation, and λ is the wavelength of the light.* Find $dI/d\theta$.

52. *Physics*   A wave travels out uniformly in all directions from a point source. At a distance r from the source, the displacement y (in meters) of the medium varies with time t (in seconds) according to the equation

$$y = \frac{25}{r}\sin 4.0(r - 15t) \qquad \text{for } r = 25 \text{ m}, t = 1.6 \text{ s}$$

Find: (a) $\dfrac{dy}{dt}$ (b) $\dfrac{dy}{dr}$

26-2 | Derivatives of Other Trigonometric Functions

The derivatives of the four remaining trigonometric functions may be found using (26-3) and (26-4) with trigonometric identities involving the sine and cosine functions.

Derivative of tan u   From (19-4):

$$\tan u = \frac{\sin u}{\cos u}$$

Then, use the quotient rule (22-19).

$$\frac{d}{dx}(\tan u) = \frac{\cos u \cdot \dfrac{d}{dx}(\sin u) - \sin u \cdot \dfrac{d}{dx}(\cos u)}{\cos^2 u}$$

*David Halliday and Robert Resnick, *Fundamentals of Physics* (New York: Wiley, 1981), p. 753.

From (26-3) and (26-4):

$$= \frac{\cos^2 u \cdot \dfrac{du}{dx} + \sin^2 u \cdot \dfrac{du}{dx}}{\cos^2 u}$$

$$= \frac{\cos^2 u + \sin^2 u}{\cos^2 u} \frac{du}{dx}$$

By trigonometric identity (19-6):

$$= \frac{1}{\cos^2 u} \frac{du}{dx}$$

From the definition of the secant function (19-2):

$$= \sec^2 u \frac{du}{dx}$$

Derivative of tan u $\dfrac{d}{dx}(\tan u) = \sec^2 u \dfrac{du}{dx}$ (26-5)

EXAMPLE 11   Find the derivative dy/dx for each of the following:
(a) $y = \tan 8x$ (b) $y = 4 \tan x^2$ (c) $y = x^2 \tan (x^2 + 2)$

Solution
(a)   Use (26-5).

$$\frac{dy}{dx} = \sec^2 8x \cdot \frac{d}{dx}(8x)$$

$$= (\sec^2 8x)(8)$$

$$= 8 \sec^2 8x$$

(b) **First:**

$$\frac{dy}{dx} = 4 \frac{d}{dx}(\tan x^2)$$

Then, from (26-5):

$$= 4(\sec^2 x^2)(2x) \qquad \boxed{\dfrac{d}{dx}(x^2) = 2x}$$

$$= 8x \sec^2 x^2$$

(c) **Step 1.** Use the product rule (22-17).

$$\frac{dy}{dx} = x^2 \frac{d}{dx}[\tan (x^2 + 2)]$$

$$+ \tan (x^2 + 2) \cdot \frac{d}{dx}(x^2)$$

Step 2. From (26-5):

$$\boxed{\dfrac{d}{dx}(x^2 + 2) = 2x}$$

$$= x^2 \cdot \sec^2 (x^2 + 2) \cdot 2x + \tan (x^2 + 2) \cdot 2x$$

$$= 2x^3 \sec^2 (x^2 + 2) + 2x \tan (x^2 + 2)$$

Derivative of cot u   The cotangent function is defined as the reciprocal of the tangent.

From equation (19-5): $\cot u = \dfrac{\cos u}{\sin u}$

Use the quotient rule (22-19).

$$\frac{d}{dx}(\cot u) = \frac{\sin u \cdot \frac{d}{dx}(\cos u) - \cos u \cdot \frac{d}{dx}(\sin u)}{\sin^2 u}$$

From (26-3) and (26-4):

$$= \frac{-\sin^2 u - \cos^2 u}{\sin^2 u}\frac{du}{dx}$$

$$= -\frac{1}{\sin^2 u}\frac{du}{dx}$$

From the definition of the cosecant function (19-1),

$$= -\csc^2 u \frac{du}{dx}$$

Derivative of cot u $\dfrac{d}{dx}(\cot u) = -\csc^2 u \dfrac{du}{dx}$ $\qquad$ (26-6)

EXAMPLE 12   Find dy/dx when:
(a) $y = \cot 4x^2$ $\qquad$ (b) $y = \cot^4 2x$
(c) $y = \sin(x^2 + 1) + \cot(x^2 - 1)$ $\qquad$ (d) $y^2 = \cot\sqrt{x}$

Solution

(a)   **Step 1.**   Let $u = 4x^2$; then $du/dx = 8x$.

   Step 2.   Use (26-6). $\qquad\qquad \dfrac{dy}{dx} = (-\csc^2 4x^2)(8x)$

$$= -8x \csc^2 4x^2$$

(b)   **Step 1.**   Use the power rule (22-22). $\qquad \dfrac{dy}{dx} = 4\cot^3 2x \cdot \dfrac{d}{dx}(\cot 2x)$

   Step 2.   From (26-6): $\qquad\qquad\qquad\qquad \boxed{\dfrac{d}{dx}(2x) = 2}$

$$= (4\cot^3 2x)(-\csc^2 2x)(2)$$

$$= -8\cot^3 2x \csc^2 2x$$

(c)   **Step 1.**   Use the sum rule (22-15). $\qquad \dfrac{dy}{dx} = \dfrac{d}{dx}[\sin(x^2 + 1)] + \dfrac{d}{dx}[\cot(x^2 - 1)]$

   Step 2.   Use (26-3) and (26-6). $\qquad\qquad = 2x\cos(x^2 + 1) - 2x\csc^2(x^2 - 1)$

(d)   **Step 1.**   Use implicit differentiation. $\qquad 2y\dfrac{dy}{dx} = \dfrac{d}{dx}(\cot\sqrt{x})$

   Step 2.   Let $u = \sqrt{x}$; then $du/dx = \dfrac{1}{2\sqrt{x}}$.

Next, use (26-6).

$$2y \frac{dy}{dx} = -\csc^2\sqrt{x} \cdot \frac{1}{2\sqrt{x}}$$

Step 3. Solve for dy/dx. Note that $y = \pm\sqrt{\cot\sqrt{x}}$.

$$\frac{dy}{dx} = \frac{-\csc^2\sqrt{x}}{\pm 4\sqrt{x}\sqrt{\cot\sqrt{x}}}$$

$$= \pm \frac{\csc^2\sqrt{x}}{4\sqrt{x}\sqrt{\cot\sqrt{x}}}$$

Derivative of csc *u*   By definition (19-1),

$$\csc u = \frac{1}{\sin u} = (\sin u)^{-1}$$

Therefore, using the power rule (22-22), we have:

$$\frac{d}{dx}(\csc u) = (-1)(\sin u)^{-2}\frac{d}{dx}(\sin u)$$

By (26-3),

$$= -\frac{1}{\sin^2 u}\cdot\cos u\cdot\frac{du}{dx}$$

We can rewrite this as

$$= -\frac{1}{\sin u}\cdot\frac{\cos u}{\sin u}\cdot\frac{du}{dx}$$

$$= -\csc u\cot u\frac{du}{dx}$$

Derivative of csc *u* $\dfrac{d}{dx}(\csc u) = -\csc u \cot u \dfrac{du}{dx}$ (26-7)

EXAMPLE 13 Find dy/dx if:
(a) $y = \csc 6x$ (b) $y = x^3 \csc^2 2x$

Solution
(a) Use (26-7).

$$\frac{dy}{dx} = (-\csc 6x)(\cot 6x)(6) \qquad \boxed{\frac{d}{dx}(6x) = 6}$$

$$= -6 \csc 6x \cot 6x$$

(b) **First,** use the product rule (22-17).

$$\frac{dy}{dx} = x^3 \cdot \frac{d}{dx}(\csc^2 2x) + \csc^2 2x \cdot \frac{d}{dx}(x^3)$$

Next, from the power rule (22-22),

$$= x^3 \cdot 2 \csc 2x \cdot \frac{d}{dx}(\csc 2x) + (\csc^2 2x)(3x^2)$$

Finally, use (26-7).

$$= 2x^3 \csc 2x \, (-\csc 2x \cot 2x)\cdot 2 + 3x^2 \csc^2 2x$$

$$= -4x^3 \csc^2 2x \cot 2x + 3x^2 \csc^2 2x$$

Derivative of sec u   By definition (19-2):

$$\sec u = \frac{1}{\cos u} = (\cos u)^{-1}$$

Therefore, using the power rule (22-22), we have:

$$\frac{d}{dx}(\sec u) = (-1)(\cos u)^{-2}\frac{d}{dx}(\cos u)$$

From (26-4),

$$\frac{d}{dx}(\sec u) = -\frac{1}{\cos^2 u}\cdot(-\sin u)\frac{du}{dx} = \frac{\sin u}{\cos^2 u}\cdot\frac{du}{dx}$$

Write this as:

$$= \frac{1}{\cos u}\cdot\frac{\sin u}{\cos u}\frac{du}{dx}$$

$$= \sec u \tan u \frac{du}{dx}$$

Derivative of sec u $\dfrac{d}{dx}(\sec u) = \sec u \tan u \dfrac{du}{dx}$ (26-8)

EXAMPLE 14   Find dy/dx if:

(a) $y = 4 \sec 3x$ (b) $y = \dfrac{\sec x^3}{x^2}$

Solution

(a)   From (26-8),

$$\frac{dy}{dx} = 4\frac{d}{dx}(\sec 3x)$$

$$\boxed{\frac{d}{dx}(3x) = 3}$$

$$= 4(\sec 3x)(\tan 3x)(3)$$

$$= 12\sec 3x \tan 3x$$

(b)   **Step 1.** Use the quotient rule (22-19).

$$\frac{dy}{dx} = \frac{x^2\dfrac{d}{dx}(\sec x^3) - (\sec x^3)(2x)}{(x^2)^2}$$

$$\boxed{\frac{d}{dx}(x^2) = 2x}$$

Step 2. Use (26-8).

$$\frac{dy}{dx} = \frac{x^2(\sec x^3)(\tan x^3)(3x^2) - 2x\sec x^3}{x^4}$$

Step 3. Simplify.

$$= \frac{3x^3\sec x^3\tan x^3 - 2\sec x^3}{x^3}$$

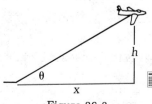

Figure 26-3

EXAMPLE 15   The airplane shown in Fig. 26-3 is flying horizontally at a constant speed of 240 mi/h at a height of 5500 ft. At what rate is the angle of elevation changing when $\theta = 35°$?

Solution   First, from Fig. 26-3: $\tan \theta = \dfrac{h}{x}$

$$x = \frac{h}{\tan \theta} = h \cot \theta$$

The speed of the plane is the rate dx/dt at which the horizontal distance x is changing, where t is time.

Second, find dx/dt using (26-6).

$$\frac{dx}{dt} = h\frac{d}{dt}(\cot \theta)$$

$$= -h \csc^2 \theta \frac{d\theta}{dt}$$

But $\csc \theta = \dfrac{1}{\sin \theta}$;

therefore:

$$\frac{dx}{dt} = \frac{-h}{\sin^2 \theta}\frac{d\theta}{dt}$$

Figure 26-3

Third, solve for $d\theta/dt$
and evaluate.

$$\frac{d\theta}{dt} = -\frac{\sin^2 \theta}{h}\cdot\frac{dx}{dt}$$

Convert the given quan-
tities to the same units.

$$\frac{dx}{dt} = (240 \text{ mi/h})\cdot\left(\frac{5280 \text{ ft}}{1 \text{ mi}}\right)\cdot\left(\frac{1 \text{ h}}{3600 \text{ s}}\right) = 352 \text{ ft/s}$$

$$h = 5500 \text{ ft} \qquad \text{and} \qquad \theta = 35°$$

Then substitute.

$$\frac{d\theta}{dt} = -\frac{\sin^2(35°)}{5500 \text{ ft}}\cdot 352 \text{ ft/s}$$

DEG 35 sin x² +/- ÷ 5500 × 352 = → $\boxed{-0.0210554}$

$$\frac{d\theta}{dt} = -0.021 \text{ s}^{-1} \longleftarrow \boxed{\text{Unit is radian per second.}}$$

$\boxed{\begin{array}{c}\text{The negative sign means}\\ \text{that } \theta \text{ is decreasing.}\end{array}}$

▦ **EXAMPLE 16**   The honeycomb structure in a beehive is a prism with hexagonal cross section, closed at one end with three quadrilaterals (Fig. 26-4 on
p. 1116). The bee is a skillful structural engineer who chooses θ, the angle of
closure at the base of the cell, so as to minimize the cell surface area A, and
therefore the amount of wax needed in its construction. If

$$A = 6sh - \frac{3s^2}{2}\cot \theta + \frac{3s^2\sqrt{3}}{2}\csc \theta$$

find the value of θ that makes A a minimum.*

Solution   Using equations (26-6) and (26-7), we obtain

---

* See D'Arcy Thompson, *On Growth and Form* (New York: Macmillan, 1945),
pp. 525–544.

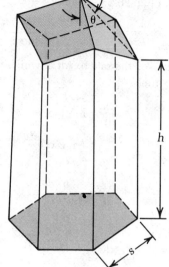

Figure 26-4

$$\frac{dA}{d\theta} = -\frac{3s^2}{2}(-\csc^2\theta) + \frac{3s^2\sqrt{3}}{2}(-\csc\theta\cot\theta)$$

$$= \frac{3s^2}{2}\csc\theta[\csc\theta - \sqrt{3}\cot\theta]$$

Area A will be a minimum when $dA/d\theta = 0$. But since $\csc\theta \neq 0$, then

$$\csc\theta - \sqrt{3}\cot\theta = 0$$

$$\csc\theta = \sqrt{3}\cot\theta$$

By definition, $\csc\theta = 1/(\sin\theta)$, and from identity (19-5),
$\cot\theta = (\cos\theta)/(\sin\theta)$. Then, substituting into the equation for $\csc\theta$, we have

$$\frac{1}{\sin\theta} = \sqrt{3}\,\frac{\cos\theta}{\sin\theta}$$

$$\cos\theta = \frac{1}{\sqrt{3}}$$

[DEG] 3 [√] [1/x] [INV] [COS] → ⬛ 54.735610 ⬛

$$\theta \approx 55°$$

Check that A is a minimum at this value of θ by using the second derivative test.

Summary $\dfrac{d}{dx}(\sin u) = \cos u\,\dfrac{du}{dx}$ **(26-3)** $\dfrac{d}{dx}(\csc u) = -\csc u\cot u\,\dfrac{du}{dx}$ **(26-7)**

$$\frac{d}{dx}(\cos u) = -\sin u \frac{du}{dx} \quad \textbf{(26-4)} \qquad \frac{d}{dx}(\sec u) = \sec u \tan u \frac{du}{dx} \quad \textbf{(26-8)}$$

$$\frac{d}{dx}(\tan u) = \sec^2 u \frac{du}{dx} \quad \textbf{(26-5)} \qquad \frac{d}{dx}(\cot u) = -\csc^2 u \frac{du}{dx} \quad \textbf{(26-6)}$$

Exercises 26-2

Find the derivative of each function.

1. $y = \tan 4x$

2. $y = 2 \tan (x^3)$

3. $y = 3 \cot 2x^4$

4. $y = \cot (2x - 1)$

5. $y = \csc \sqrt{x}$

6. $y = 4 \csc 5x$

7. $y = 5 \sec \left(x + \dfrac{\pi}{4} \right)$

8. $y = \sec (x^2 - 1)$

9. $y = \tan^2 3\theta$

10. $y = \sqrt{\tan \theta}$

11. $y = 3 \cot^3 (x^2)$

12. $y = \cot^2 5x$

13. $y = \sqrt{\csc 2x}$

14. $s = 2 \csc^4 3t$

15. $s = \sec^2 (3t)^2$

16. $y = 6 \sec^3 (3x + 2)$

17. $y = 2x \tan (x^3)$

18. $y = x^2 \cot 5x$

19. $s = t^2 \csc^3 t$

20. $s = 3t \sec^2 2t$

21. $y = \cos 2x \tan x^2$

22. $y = 2 \sin \theta \sec 3\theta$

23. $y = 3 \cot \theta \sin 2\theta$

24. $y = \csc 2x \cos x^3$

25. $y = \dfrac{\tan x^4}{2x}$

26. $y = \dfrac{x^2}{\cot 2x}$

27. $y = \dfrac{\csc 3x}{1 + \cos x^2}$

28. $y = \dfrac{\sec^2 x}{2 - \sin 3x}$

29. $y = \tan x^2 + \sec 2x$

30. $y = \csc (1 + x) - \cos (1 - x)$

31. $y = 2 \cot 3x - \cot^2 3x$

32. $y = 2x + \tan x + \sin^2 x$

33. $y^2 = x + \cot x$

34. $y^3 = \tan x^2$

Solve.

35. *Navigation* A freighter is moving directly toward a lighthouse at a speed of 15 mi/h. If the observation deck of the lighthouse is 85 ft above the water, at what rate is the angle of elevation θ changing when $\theta = 2.5°$?

36. *Electronics* An electronics engineer wants to use the least possible amount of wire in connecting four contact points located at the corners of a square. Contrary to your first guess, the shortest route has the form shown in the figure.
(a) What value of θ gives the minimum path length?
(b) Find the minimum path length L if $s = 1.5$ cm.
[Hint: Find expressions for both a and b in terms of θ and the side length s. Then substitute in $L = 4a + b$ and find $dL/d\theta$.]

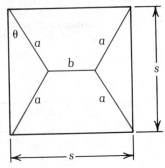

37. Find the slope of a line tangent to $y = 3 \tan 2x$ at the point where $x = \pi/6$.

38. Find the value of x where $y = 2 \csc [2x - (\pi/4)]$ has a minimum.

39. *Physics* If the displacement in meters of an object is given by the equation $s = 4t^2 \csc (t^2)$, find its velocity when $t = 1.2$ s.

40. *Electrical engineering* The strength of a radar signal in millivolts varies with the orientation of the antenna according to the formula $I = 16 \cot^2 2\theta$. Find the rate of change of signal strength with angle when $\theta = 5\pi/8$.

26-3 | Derivatives of the Inverse Trigonometric Functions

In Example 15 we found the derivative $d\theta/dt$ from the equation $x = h \cot \theta$ by taking the derivative with respect to time of both members of the equation. However, we could have solved the equation explicitly for angle θ using the inverse trigonometric functions defined in Chapter 19 and then calculated the derivative directly. In this section we will develop the rules for differentiating the inverse trigonometric functions.

Derivative of Arcsin u

From the definition of the Arcsine function (19-20), we know that $y = $ Arcsin u implies

$$u = \text{Sin } y \qquad \text{for } -\frac{\pi}{2} \le y \le \frac{\pi}{2}$$

Use (26-3) to find the derivative:

$$\frac{du}{dx} = \cos y \frac{dy}{dx} \qquad \cos y > 0 \text{ for } -\frac{\pi}{2} < y < \frac{\pi}{2}$$

$$\frac{dy}{dx} = \frac{1}{\cos y} \cdot \frac{du}{dx}$$

We can express the right side of this equation in terms of u, since $\cos^2 y + \sin^2 y = 1$ or

$$\cos y = \sqrt{1 - \sin^2 y} \qquad \sin^2 y < 1 \text{ for } -\frac{\pi}{2} < y < \frac{\pi}{2}$$

$$= \sqrt{1 - u^2}$$

For $-1 < u < 1$, $\sqrt{1 - u^2}$ is a nonzero real number; therefore,

$$\frac{dy}{dx} = \frac{1}{\sqrt{1 - u^2}} \cdot \frac{du}{dx}$$

Derivative of Arcsin u	$\dfrac{d}{dx}(\text{Arcsin } u) = \dfrac{1}{\sqrt{1 - u^2}} \dfrac{du}{dx}$
where $-1 < u < 1$	(26-9)

EXAMPLE 17 Find dy/dx if:

(a) $y = \text{Arcsin } 3x$ (b) $y = \text{Arcsin } 2x^3$

Solution

(a) **First,** note that $u = 3x$. Therefore, $du/dx = 3$ and $u^2 = 9x^2$, so that $1 - u^2 = 1 - 9x^2$.

Then, from (26-9),

$$\frac{dy}{dx} = \frac{1}{\sqrt{1 - 9x^2}} \cdot 3 = \frac{3}{\sqrt{1 - 9x^2}}$$

(b) **First,** note that $u = 2x^3$. Therefore, $du/dx = 6x^2$ and $u^2 = 4x^6$, so that $1 - u^2 = 1 - 4x^6$.

Then, from (26-9),

$$\frac{dy}{dx} = \frac{1}{\sqrt{1 - 4x^6}} \cdot 6x^2 = \frac{6x^2}{\sqrt{1 - 4x^6}}$$

IMPORTANT

Both derivatives in this example exist only for nonzero real number values of the denominator. ◄

Derivative of Arccos u

Similarly, we can find the derivative of the inverse cosine function. If $y = \text{Arccos } u$, then from (19-21), we have

$$u = \text{Cos } y \qquad \text{for } 0 \le y \le \pi$$

Now use (26-4):

$$\frac{du}{dx} = -\sin y \frac{dy}{dx} \qquad \sin y > 0 \text{ for } 0 \le y \le \pi$$

$$\frac{dy}{dx} = -\frac{1}{\sin y} \cdot \frac{du}{dx}$$

$$\sin y = \sqrt{1 - \cos^2 y} \qquad \cos^2 y < 1 \text{ for } 0 \le y \le \pi$$

$$\qquad\quad = \sqrt{1 - u^2}$$

For $0 \le y \le \pi$, $-1 < u < 1$, and $\sqrt{1 - u^2}$ is a nonzero real number; therefore,

$$\frac{dy}{dx} = -\frac{1}{\sqrt{1 - u^2}} \cdot \frac{du}{dx}$$

Derivative of Arccos u $\dfrac{d}{dx}(\text{Arccos } u) = -\dfrac{1}{\sqrt{1 - u^2}}\dfrac{du}{dx}$

where $-1 < u < 1$ (26-10)

EXAMPLE 18 Find dy/dx if:

(a) $y = 3x \text{ Arccos } 2x$ (b) $y = 2 \text{ Arccos } (x^2)$ (c) $y = 3 \text{ Arccos}^2 4x$

Solution

(a) **First,** use the product rule (22-17).

$$\frac{dy}{dx} = 3x \cdot \frac{d}{dx}(\text{Arccos } 2x) + \text{Arccos } 2x \cdot \frac{d}{dx}(3x)$$

Second, in the Arccosine function, $u = 2x$; therefore, $1 - u^2 = 1 - 4x^2$ and $du/dx = 2$. Now, use (26-10):

$$\frac{d}{dx}(\text{Arccos } 2x) = \frac{-2}{\sqrt{1 - 4x^2}}$$

Finally, substitute in the expression for dy/dx.

$$\frac{dy}{dx} = \frac{-6x}{\sqrt{1 - 4x^2}} + 3 \text{ Arccos } 2x$$

(b) **First,** note that $u = x^2$; therefore, $1 - u^2 = 1 - x^4$ and $du/dx = 2x$. **Then,** use (26-10).

$$\frac{dy}{dx} = -\frac{2}{\sqrt{1 - x^4}} \cdot 2x = \frac{-4x}{\sqrt{1 - x^4}}$$

(c) **First,** let $u = \text{Arccos } 4x$, and, by (26-10), we have

$$\frac{du}{dx} = -\frac{1}{\sqrt{1 - (4x)^2}} \frac{d(4x)}{dx} = -\frac{4}{\sqrt{1 - 16x^2}}$$

Then, we have $y = 3u^2$ and $\dfrac{dy}{dx} = 6u \dfrac{du}{dx}$

$$= 6(3 \text{ Arccos } 4x)\left(-\frac{4}{\sqrt{1 - 16x^2}}\right)$$

$$= -\frac{72 \text{ Arccos } 4x}{\sqrt{1 - 16x^2}}$$

NOTE The derivative of the Arccosine function is the negative of the derivative of the Arcsine function. ◀

Derivative of Arctan u   From the definition of the Arctangent function (19-22), if $y = \text{Arctan } u$, then

$$u = \tan y \qquad \text{for } -\frac{\pi}{2} < y < \frac{\pi}{2}$$

Using (26-5), we have

$$\frac{du}{dx} = \sec^2 y \cdot \frac{dy}{dx}$$

$$\sec y > 0 \text{ for } -\frac{\pi}{2} < y < \frac{\pi}{2}$$

therefore,

$$\frac{dy}{dx} = \frac{1}{\sec^2 y} \cdot \frac{du}{dx}$$

From identity (19-7),

$$\sec^2 y = 1 + \tan^2 y$$

$$= 1 + u^2$$

Therefore,

$$\frac{dy}{dx} = \frac{1}{1 + u^2} \cdot \frac{du}{dx}$$

Derivative of Arctan u $\dfrac{d}{du}(\text{Arctan } u) = \dfrac{1}{1 + u^2}\dfrac{du}{dx}$	(26-11)

Notice that this derivative exists for every real number value of x or u.

It is interesting that, even though the Arcsine, Arccosine, and Arctangent are transcendental functions, their derivatives are algebraic functions.

EXAMPLE 19 Find dy/dx if:

(a) $y = \dfrac{1}{1 + 9x^2} + \text{Arctan } 3x$ (b) $y = \text{Arctan}\left(\dfrac{x}{x + 1}\right)$

Solution

(a) **First,** from the sum rule (22-15), we have

$$\frac{dy}{dx} = \frac{d}{dx}\left(\frac{1}{1 + 9x^2}\right) + \frac{d}{dx}(\text{Arctan } 3x)$$

Second, differentiate the first term on the right:

$$\frac{d}{dx}(1 + 9x^2)^{-1} = -1(1 + 9x^2)^{-2}\frac{d}{dx}(9x^2)$$

$$= -\frac{18x}{(1 + 9x^2)^2}$$

Next, differentiate the second term on the right using (26-11):

$$\frac{d}{dx}(\text{Arctan } 3x) = \frac{1}{1 + (3x)^2}\frac{d}{dx}(3x)$$

$$= \frac{3}{1 + 9x^2}$$

Finally, combine the results of these differentiations:

$$\frac{dy}{dx} = -\frac{18x}{(1 + 9x^2)^2} + \frac{3}{1 + 9x^2}$$

$$= \frac{-18x + 3(1 + 9x^2)}{(1 + 9x^2)^2}$$

$$= -\frac{3(1 - 6x + 9x^2)}{(1 + 9x^2)^2}$$

$$= -\frac{3(1 - 3x)^2}{(1 + 9x^2)^2}$$

(b) **First,** since $u = \dfrac{x}{x + 1}$, $u^2 = \dfrac{x^2}{(x + 1)^2}$

and $1 + u^2 = 1 + \dfrac{x^2}{(x + 1)^2}$

$$= \frac{2x^2 + 2x + 1}{(x + 1)^2}$$

Second, use the quotient rule (22-19). $\dfrac{du}{dx} = \dfrac{(x + 1) \cdot 1 - x(1)}{(x + 1)^2} = \dfrac{1}{(x + 1)^2}$

Finally, use (26-11). $\dfrac{dy}{dx} = \dfrac{(x + 1)^2}{2x^2 + 2x + 1} \cdot \dfrac{1}{(x + 1)^2}$

$$= \frac{1}{2x^2 + 2x + 1}$$

EXAMPLE 20 In Fig. 26-5, a winch 12 ft above the bow of the boat is pulling cable in at 2.5 ft/s. At what rate is the angle θ changing when the boat is 45 ft from the dock?

Figure 26-5

Solution

Step 1. From the drawing, we can see that $\sin\theta = h/x$. Then, $\theta = \text{Arcsin}\,(h/x)$. We use the sine function because it can be written in terms of the constant h and the variable x whose rate of change is given.

Step 2. Differentiate using (26-9):

$$\frac{d\theta}{dt} = \frac{1}{\sqrt{1-\left(\dfrac{h}{x}\right)^2}} \cdot \frac{d}{dt}\left(\frac{h}{x}\right)$$

$$\boxed{\frac{d}{dt}\left(\frac{h}{x}\right) = -\frac{h}{x^2}\cdot\frac{dx}{dt}}$$

$$= -\frac{h}{x^2\sqrt{1-\left(\dfrac{h}{x}\right)^2}} \cdot \frac{dx}{dt}$$

Step 3. Evaluate this expression for $h = 12$ ft, $L = 45$ ft, and $dx/dt = -2.5$ ft/s. From the Pythagorean theorem,

$$x^2 = L^2 + h^2$$

$$= 45^2 + 12^2 = 2169$$

Then,

$$\frac{d\theta}{dt} = \frac{12(2.5)}{2169\sqrt{1-\dfrac{12^2}{2169}}}$$

$1\;\boxed{-}\;12\;\boxed{x^2}\;\boxed{\div}\;2169\;\boxed{=}\;\boxed{\sqrt{\ }}\;\boxed{\times}\;2169\;\boxed{=}\;\boxed{1/x}\;\boxed{\times}\;12\;\boxed{\times}\;2.5$

$\boxed{=}\rightarrow$ ▓ *0.0143146*

$$\frac{d\theta}{dt} \approx 0.014 \text{ rad/s}$$

■ **EXAMPLE 21** For the tapered pin shown in Fig. 26-6, the size of angle θ can be estimated by measuring the dimension x. If the length of the pin is exactly 12 cm, and if $x = 4.35$ cm measured to the nearest half-millimeter (0.05 cm), find the value of θ and the percent error in this value.

Solution **First,** from the figure, $\tan\theta = \dfrac{x}{12}$. Then, $\theta = \text{Arctan}\left(\dfrac{x}{12}\right)$

$$= \text{Arctan}\left(\frac{4.35}{12}\right).$$

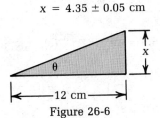

$x = 4.35 \pm 0.05$ cm

$\boxed{\text{RAD}}\;4.35\;\boxed{\div}\;12\;\boxed{=}\;\boxed{\text{INV}}\;\boxed{\tan}\rightarrow$ ▓ *0.3477670*

Next, using (26-11), calculate $d\theta/dx$.

$$\frac{d\theta}{dx} = \frac{1}{1+\left(\dfrac{x}{12}\right)^2} \cdot \frac{d}{dx}\left(\frac{x}{12}\right)$$

—12 cm—

Figure 26-6

$$= \frac{1}{12\left[1 + \left(\dfrac{4.35}{12}\right)^2\right]}$$

$$4.35 \,\boxed{\div}\, 12 \,\boxed{=}\, \boxed{x^2} \,\boxed{+}\, 1 \,\boxed{=}\, \boxed{\times}\, 12 \,\boxed{=}\, \boxed{1/x} \to \quad \boxed{0.0736547}$$

Finally, from the definition of the differential (24-1), the error in θ is

$$d\theta = \frac{d\theta}{dx} \cdot \Delta x$$

$$\approx (0.0736547)(0.05)$$

$$\approx 0.0036827$$

By (24-2) and (24-3), the relative error is $\dfrac{d\theta}{\theta} \approx \dfrac{0.0036827}{0.3477670} \approx 0.0105897$ and the percent error is approximately 1.05897%. Therefore, the angle is $\theta \approx 0.35$ rad $\pm$ 1.1%.

EXAMPLE 22   A billboard 8 ft high is located so that its lower edge is 4 ft above eye level for the observer in Fig. 26-7(a). How far from the billboard should the observer stand so that the viewing angle θ is a maximum?

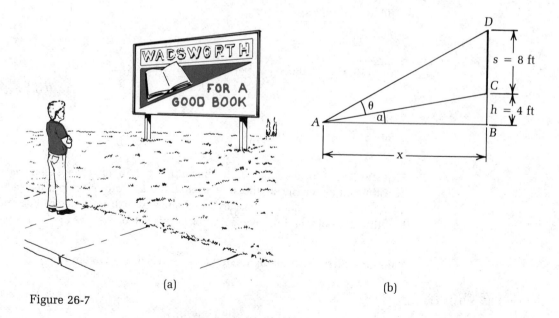

(a)   (b)

Figure 26-7

Solution **First,** find an expression for θ in terms of x. In Fig. 26-7(b), from triangle ABD, we have

$$\tan(\theta + a) = \frac{h + s}{x} = \frac{12}{x} \qquad \text{or} \qquad \theta + a = \text{Arctan}\left(\frac{12}{x}\right)$$

Also, from triangle ABC,

$$\tan a = \frac{h}{x} = \frac{4}{x} \qquad \text{or} \qquad a = \text{Arctan}\left(\frac{4}{x}\right)$$

Then,

$$\theta = (\theta + a) - a$$

$$= \text{Arctan}\left(\frac{12}{x}\right) - \text{Arctan}\left(\frac{4}{x}\right)$$

Next, differentiate using (26-11) and the difference rule (22-16):

$$\frac{d\theta}{dx} = \left(\frac{1}{1 + \frac{144}{x^2}}\right)\left(-\frac{12}{x^2}\right) - \left(\frac{1}{1 + \frac{16}{x^2}}\right)\left(-\frac{4}{x^2}\right)$$

Finally, solve $d\theta/dx = 0$ to find the value of x that makes the viewing angle θ a maximum. This gives the equation

$$\left(\frac{1}{1 + \frac{144}{x^2}}\right)\left(-\frac{12}{x^2}\right) = \left(\frac{1}{1 + \frac{16}{x^2}}\right)\left(-\frac{4}{x^2}\right)$$

Multiply each side by $-x^2/4$ to get

$$3\left(\frac{1}{1 + \frac{144}{x^2}}\right) = \left(\frac{1}{1 + \frac{16}{x^2}}\right)$$

Then, take reciprocals.

$$1 + \frac{144}{x^2} = 3\left(1 + \frac{16}{x^2}\right)$$

$$x^2 + 144 = 3x^2 + 48$$

$$2x^2 = 96$$

$$x^2 = 48$$

$$x = 4\sqrt{3} \approx 7 \text{ ft}$$

The second derivative test can be used to show that this value of x does indeed give a maximum and not a minimum value of θ. The angle θ has no local minimum; θ decreases as x increases.

Exercises 26-3

Find the derivative of each function.

1. $y = 3 \text{ Arcsin } 2x$

2. $y = \text{Arcsin } x^2$

3. $y = \text{Arcsin}\left(\frac{1}{x + 1}\right)$

4. $y = 2 \text{ Arcsin } \sqrt{x}$

5. $y = \text{Arccos } x^3$

6. $y = 5 \text{ Arccos } 4x$

7. $y = 4 \text{ Arccos } \sqrt{1 - x}$

8. $y = \text{Arccos } \left(\dfrac{1}{x}\right)$

9. $y = \text{Arctan } (3x^4)$

10. $y = 2 \text{ Arctan } (2x + 5)$

11. $y = 6 \text{ Arctan } \sqrt{2x}$

12. $y = \text{Arctan } \left(\dfrac{1}{x - 1}\right)$

13. $y = x \text{ Arccos } x^2$

14. $y = x^2 \text{ Arcsin } 4x$

15. $y = (x^2 - 1) \text{ Arctan } 3x$

16. $y = 3x \text{ Arctan } x^3$

17. $y = \text{Arcsin}^2 \, 2x$

18. $y = 2 \text{ Arccos}^2 \, x$

19. $y = 2 \text{ Arctan}^3 \, x$

20. $y = \sqrt{\text{Arctan } x}$

21. $y = \dfrac{x}{\text{Arcsin } x}$

22. $y = \dfrac{1}{\text{Arccos } 2x}$

23. $y = \dfrac{\text{Arctan } x}{2x}$

24. $y = \dfrac{x}{\text{Arctan } 2x}$

25. $y = x + \text{Arctan } 2x^2$

26. $y = \dfrac{1}{1 + x^2} + \text{Arctan } x$

27. $y = \text{Arcsin } 3x + \sqrt{1 - 9x^2}$

28. $y = \text{Arccos } 2x - \sqrt{1 - 4x^2}$

Solve.

29. *Machine technology*   For the tapered pin in the figure find the value of θ and the percent error in this value.

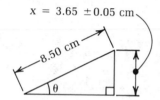

$x = 3.65 \pm 0.05$ cm

8.50 cm

θ

30. *Physics*   An observer is watching a rocket being launched. If the rocket has a length of 65 ft, at what horizontal distance x from the launch pad should the observer stand so that the viewing angle θ is a maximum when the base of the rocket has reached an altitude of 540 ft?

31. Find the slope of the tangent line to the curve $y = x \text{ Arctan } 2x$ at $x = 1.5$.

32. Find the equation of the normal line to the curve $y = 2 \text{ Arcsin } 5x$ at $x = 0.12$.

33. *Electronics technology*   A race car is traveling at a speed of 210 ft/s as it moves past a television camera at trackside 35 ft away. Find $d\theta/dt$, the rate at which the camera must turn to follow the car when $\theta = 20°$ (see the figure).

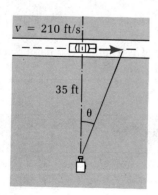

$v = 210$ ft/s

35 ft

θ

34. The figure shows a 12-ft ladder leaning against a wall. If point A moves horizontally away from the wall at 2.5 m/s, how fast is θ changing when $h = 7.5$ m?

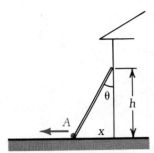

35. Physics   A radar device at point A tracks the moving object shown. If distance a is 650 m, find the rate at which distance L is changing when $L = 1540$ m and θ is increasing at 1.0×10^{-3} rad/s.

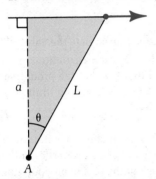

26-4 | Derivative of the Logarithmic Function

The logarithmic function $y = \log_b x$, defined in Chapter 13, is a second class of transcendental functions. The Δ-limit process can be used to find the derivative of $y = \log_b u$, where u is some differentiable function of x. If $y = \log_b u$, then

$$y + \Delta y = \log_b(u + \Delta u)$$

$$\Delta y = \log_b(u + \Delta u) - \log_b u$$

By logarithmic property (13-8), we have

$$\Delta y = \log_b\left(\frac{u + \Delta u}{u}\right)$$

$$= \log_b\left(1 + \frac{\Delta u}{u}\right)$$

Divide by Δu.

$$\frac{\Delta y}{\Delta u} = \frac{1}{\Delta u} \log_b\left(1 + \frac{\Delta u}{u}\right)$$

We can write this as

$$\frac{\Delta y}{\Delta u} = \frac{1}{u}\left(\frac{u}{\Delta u}\right) \log_b\left(1 + \frac{\Delta u}{u}\right)$$

and by logarithmic property (13-9),

$$= \frac{1}{u} \log_b \left(1 + \frac{\Delta u}{u} \right)^{u/\Delta u}$$

Then, by definition of the derivative,

$$\frac{dy}{du} = \lim_{\Delta u \to 0} \frac{\Delta y}{\Delta u} = \lim_{\Delta u \to 0} \frac{1}{u} \log_b \left(1 + \frac{\Delta u}{u} \right)^{u/\Delta u}$$

But $\lim\limits_{\Delta u \to 0} (1/u) = 1/u$, since u does not depend on Δu, and because the logarithm function is continuous we can write this as,

$$\frac{dy}{du} = \frac{1}{u} \log_b \left[\lim_{\Delta u \to 0} \left(1 + \frac{\Delta u}{u} \right)^{u/\Delta u} \right]$$

This limit is of the form $\lim\limits_{h \to 0} (1 + h)^{1/h}$. We can approximate the value of this limit using a calculator, as shown in Table 26-3. These calculations indicate that the limit exists and is approximately equal to 2.718, but the proof that the limit exists is difficult and we will not show it in this text. The limit is

$$e = \lim_{h \to 0} (1 + h)^{1/h} \approx 2.718281828459045\ldots$$

The transcendental number e is sometimes called *Euler's number* after the 18th century Swiss mathematician who discovered many of its properties and gave it the symbol "e," for "exponential." Like π, it is not the root of any polynomial or other algebraic equation with rational coefficients.

TABLE 26-3

	As $h \to 0^-$		As $h \to 0^+$	
h	$(1 + h)^{1/h}$		h	$(1 + h)^{1/h}$
−0.1	2.8679720		0.1	2.5937425
−0.01	2.7319990		0.01	2.7048138
−0.001	2.7196422		0.001	2.7169239
−0.0001	2.7184178		0.0001	2.7181459

Derivative of log u   Using this limit in the above expression for dy/du, we have

$$\frac{dy}{du} = \frac{1}{u} \log_b e$$

The chain rule (22-21) allows us to write this derivative with respect to x.

Derivative of the Logarithmic Function

$$\frac{d}{dx}(\log_b u) = \frac{1}{u} \cdot \log_b e \cdot \frac{du}{dx} \qquad (26\text{-}12)$$

When $b = 10$, this is written

$$\frac{d}{dx}(\log u) = \frac{1}{u} \cdot \log e \cdot \frac{du}{dx} \qquad \text{where } \log e \approx 0.4342945$$

Derivative of ln u If $b = e$, then $\log_b e = \log_e e = \ln e = 1$ by (13-10), and we can rewrite (26-12) for natural logarithms.

Derivative of the Natural Logarithmic Function

$$\frac{d}{dx}(\ln u) = \frac{1}{u} \cdot \frac{du}{dx} \qquad\qquad\qquad (26\text{-}13)$$

CAUTION ▶ The value of u in (26-12) and (26-13) must always be *positive* since $\log_b u$ is defined only for positive values of u. The function u(x) cannot be equal to zero. ◀

Logarithms using the transcendental number e as base may not seem "natural," but (26-13) reveals that choosing this base gives a very simple form for the derivative.

EXAMPLE 23 Find dy/dx if:
(a) $y = \log 3x$ (b) $y = \ln (x^2 + 1)$ (c) $y = [\log (4x^2)]^3$

Solution

(a) If $u = 3x$, then $du/dx = 3$, and by (26-12), $\dfrac{dy}{dx} = \dfrac{1}{3x} \cdot \log e \cdot 3$

$$= \frac{\log e}{x}$$

(b) If $u = x^2 + 1$, then $du/dx = 2x$, and by (26-13), $\dfrac{dy}{dx} = \dfrac{1}{x^2 + 1} \cdot 2x$

$$= \frac{2x}{x^2 + 1}$$

(c) For $y = [\log (4x^2)]^3$, use the power rule (22-22) to find $\dfrac{dy}{dx} = 3 \log^2(4x^2) \cdot \dfrac{d}{dx}[\log (4x^2)]$

By (26-12), $= 3 \log^2(4x^2) \cdot \dfrac{1}{4x^2} \cdot \log e \cdot 8x$

$$= \frac{24x \log e}{4x^2} \log^2(4x^2)$$

$$= \frac{6 \log e}{x} \log^2(4x^2)$$

EXAMPLE 24   Find dy/dx if:

(a) $y = \ln \sqrt[3]{x^2 + 2}$ (b) $y = \ln [x^3(x^2 + 1)^5]$

(c) $y = \ln \left(\dfrac{x - 2}{x + 2} \right)$ (d) $y = \log (\sin 3x)$

Solution

(a)   **First,** rewrite the function as: $y = \ln (x^2 + 2)^{1/3}$

Use (13-9). $= \dfrac{1}{3} \ln (x^2 + 2)$

Then: $\dfrac{dy}{dx} = \dfrac{1}{3} \dfrac{d}{dx} [\ln (x^2 + 2)]$

Finally, by (26-13), $= \dfrac{1}{3} \cdot \dfrac{1}{x^2 + 2} \cdot 2x \longleftarrow \boxed{\dfrac{d}{dx}(x^2 + 2)}$

$= \dfrac{2x}{3x^2 + 6}$

(b)   Using the multiplication property of logarithms (13-7) and
the power property (13-9), we
can write this function as

$y = \ln x^3 + \ln (x^2 + 1)^5$

$= 3 \ln x + 5 \ln (x^2 + 1)$

Then, using the sum rule
(22-15), we obtain:

$\dfrac{dy}{dx} = 3 \dfrac{d}{dx} (\ln x) + 5 \dfrac{d}{dx} [\ln (x^2 + 1)]$

By (26-13),

$= \dfrac{3}{x} + \dfrac{5}{x^2 + 1} \cdot 2x$

$= \dfrac{3}{x} + \dfrac{10x}{x^2 + 1}$

This can be written as

$= \dfrac{13x^2 + 3}{x^3 + x}$

(c)   **Step 1.**   The division property of logarithms (13-8) allows us to write this
function as $y = \ln (x - 2) - \ln (x + 2)$.

Step 2.   Then, using the difference rule (22-16), we have:

$\dfrac{dy}{dx} = \dfrac{d}{dx} [\ln (x - 2)] - \dfrac{d}{dx} [\ln (x + 2)]$

Step 3.   By (26-13),

$= \dfrac{1}{x - 2} \cdot 1 - \dfrac{1}{x + 2} \cdot 1$

$= \dfrac{1}{x - 2} - \dfrac{1}{x + 2}$

$= \dfrac{(x + 2) - (x - 2)}{(x - 2)(x + 2)}$

$= \dfrac{4}{x^2 - 4}$

(d) **First,** use (26-12) to find the derivative of the log function with $u = \sin 3x$.

$$\frac{dy}{dx} = \frac{1}{\sin 3x} \cdot \log e \cdot \frac{d}{dx}(\sin 3x)$$

Next, use (26-3).

$$\frac{d}{dx}(\sin 3x) = 3\cos 3x$$

Then:

$$\frac{dy}{dx} = \frac{1}{\sin 3x} \cdot \log e \cdot 3\cos 3x$$

Finally, simplify using $(\cos 3x)/(\sin 3x) = \cot 3x$.

$$\frac{dy}{dx} = 3\log e \cdot \cot 3x$$

EXAMPLE 25 A coaxial cable is a pair of long concentric cylinders. The inductance of a coaxial cable of length s is given by the equation $L = ks \ln (a/r)$, where a is the radius of the outer cylinder, r is the radius of the inner cylinder, and k is a constant dependent on the insulating material between the cylinders. Find dL/dr, the rate of change of inductance as the inner radius is varied.

Solution Using (26-13), with $u = a/r$, we have:

$$\frac{dL}{dr} = ks \cdot \frac{1}{a/r} \cdot \frac{d}{dr}\left(\frac{a}{r}\right)$$

$$= \frac{ksr}{a} \cdot \left(-\frac{a}{r^2}\right)$$

$$= -\frac{ks}{r}$$

Exercises 26-4

Find the derivative of each function.

1. $y = \log x^3$
2. $y = 2 \log 4x$
3. $y = 3 \log (5x - 1)$
4. $y = \log (\tan 2x)$
5. $y = \ln (x^2 - 3)$
6. $y = \ln 3x^4$

7. $y = \ln (\cos 3x)$
8. $y = \ln \sqrt[3]{2x^2 + 1}$
9. $y = \log \dfrac{x}{x + 1}$

10. $y = \ln \dfrac{2x + 1}{2x - 1}$
11. $y = 3 \ln \dfrac{2x^3}{3x^2 + 2}$
12. $y = \log \dfrac{x - 5}{x}$

13. $y = \ln \sqrt{5 + x^3}$
14. $y = \ln (x \sin x)$
15. $y = \ln (\sec^2 x + x)$
16. $y = \ln x\sqrt{1 + x^2}$
17. $y = \ln (x + 1)^2$
18. $y = \ln (3 - 2x)^3$
19. $y = [\ln 2x]^2$
20. $y = [\ln (x + 5)]^3$
21. $y = x \ln x^2$

22. $y = x^2 \ln (x - 1)$
23. $y = \dfrac{4x}{\ln 3x}$
24. $y = \dfrac{\ln (x + 1)}{x}$

Solve.

25. Find the slope of a line tangent to the curve $y = \ln (x^3 + 4)$ at $x = 1$.

26. Find the slope of a line tangent to the curve $y = \ln (\sin 2x)$ at $x = \pi/8$.
27. Find the value of x where the slope of $y = \ln \sqrt{3x} - x$ is equal to zero.
28. Find the value of x where the slope of $y = 2x \ln 2x$ is zero.
29. *Physics* The loudness L (in decibels) of a sound is measured by the logarithmic equation

$$L = 10 \log \frac{I}{I_0}$$

where $I_0 = 10^{-12}$ W/m² for a standard 1 kHz sound. Find dL/dI when $I = 80$ dB.

30. *Electronics* A cylindrical capacitor consists of a pair of oppositely charged coaxial cylinders. Its capacitance is given by the equation

$$C = \frac{kL}{\ln (b/r)}$$

where b is the outer radius, r is the inner radius, L is the length, and k is a constant. Find dC/dr when $b = 1.5$ cm, $r = 1.0$ cm, $L = 3.2$ cm, and $k = 5.6 \times 10^{-11}$ C/cm.

31. *Thermodynamics* In a reversible isothermal expansion from volume V_0 to volume V, the entropy change S can be calculated using the equation

$$S = nR \ln \frac{V}{V_0}$$

If $V_0 = 2.1$ liter, $n = 0.10$ mole, $R \approx 8.314$ liter·atm/mole·K, find $\dfrac{dS}{dV}$ when $V = 3.5$ liter.

32. *Physics* If a rod of length L is uniformly charged, then the electrostatic potential V at a point along the axis of the rod and a distance x from it is given by the equation below. Find dV/dx.

$$V = k \ln \frac{L + x}{x}$$

26-5 | Derivative of the Exponential Function

As we discussed in Chapter 13, an exponential function has the form $y = a^u$, where a is a positive constant not equal to 1, and u is a function of x. It is defined for all real number values of u. This class of transcendental functions is very important in science and technology.

Derivative of a^u

To find the derivative of the exponential function, first take the natural logarithm of both sides of the equation; then differentiate using implicit differentiation.

For

$$y = a^u$$

$$\ln y = \ln a^u$$

$$= u \ln a$$

Now, use (26-13).

$$\frac{d}{dx}(\ln y) = \frac{d}{dx}(u \ln a)$$

$$\frac{1}{y}\frac{dy}{dx} = \ln a \frac{du}{dx} \qquad \text{Since } \ln a \text{ is a constant}$$

Solve for $\dfrac{dy}{dx}$.

$$\frac{dy}{dx} = y \ln a \frac{du}{dx}$$

Substituting a^u for y gives the result shown in (26-14).

Derivative of a^u $\quad \dfrac{d}{dx}(a^u) = a^u \cdot \ln a \cdot \dfrac{du}{dx}$	**(26-14)**

CAUTION ▶ Be very careful to distinguish between

Constant exponent

$$\frac{d}{dx}(x^n) = nx^{n-1} \qquad \text{for any real number } n$$

and

Variable exponent

$$\frac{d}{dx}(a^u) = a^u \ln a \frac{du}{dx} \qquad \text{for } u \text{ a function of } x \quad ◀$$

Derivative of e^u   For the special case where the base a is equal to e, the base of natural logarithms, we have

$$\log_a e = \log_e e$$
$$= \ln e$$
$$= 1$$

Derivative of e^u $\quad \dfrac{d}{dx}(e^u) = e^u \cdot \dfrac{du}{dx}$	**(26-15)**

NOTE ▶ The function $y = e^x$ is called *the* exponential function to distinguish it from other exponential functions such as $y = 2^x$ or $y = e^{3x}$. ◀

Figure 26-8 on p. 1134 shows a remarkable property of the exponential

Figure 26-8

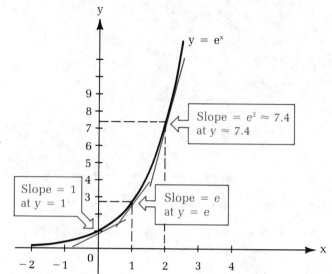

function. For u = x, equation (26-15) gives

$$\frac{d}{dx}(e^x) = e^x$$

The slope of the function $y = e^x$ is everywhere equal to the function itself. The exponential function is its own derivative.

EXAMPLE 26   Find dy/dx if:
(a) $y = 2^{3x}$ (b) $y = e^{4x^2}$ (c) $y = 6^{2x^3+1}$

Solution
(a)   Using (26-14) with $a = 2$, $u = 3x$, and $du/dx = 3$, we have

$$\frac{dy}{dx} = 2^{3x}(\ln 2)(3) = 3(2^{3x} \ln 2)$$

$$\boxed{\ln a} \qquad \boxed{\frac{du}{dx}}$$

(b)   Using (26-15) with $u = 4x^2$ and $du/dx = 8x$, we obtain

$$\frac{dy}{dx} = e^{4x^2} \cdot 8x = 8xe^{4x^2}$$

(c)   Using (26-14) with $a = 6$, $u = 2x^3 + 1$, and $du/dx = 6x^2$, we find

$$\frac{dy}{dx} = 6^{2x^3+1}(\ln 6)(6x^2)$$

$$= 6x^2(6^{2x^3+1})(\ln 6)$$

$$= x^2(6^1)(6^{2x^3+1})(\ln 6)$$

$$= x^2(6^{2x^3+2})(\ln 6)$$

EXAMPLE 27 Find dy/dx if:

(a) $y = x^2 e^{\tan x}$

(b) $y = x^2 - e^{-2x^2}$

(c) $y = \ln(\sin e^{3x})$

(d) $y = (4e^{3x})^2$

Solution

(a) **First**, use the product rule (22-13).

$$\frac{dy}{dx} = x^2 \frac{d}{dx}(e^{\tan x}) + e^{\tan x}\frac{d}{dx}(x^2)$$

Then, use (26-15).

$$= x^2 e^{\tan x}\frac{d}{dx}(\tan x) + e^{\tan x}(2x)$$

Finally, use (26-5) to find the derivative of the tangent function.

$$= x^2 e^{\tan x}(\sec^2 x) + 2xe^{\tan x}$$

$$= x^2 \sec^2 x e^{\tan x} + 2xe^{\tan x}$$

(b) From the difference rule (22-16):

$$\frac{dy}{dx} = \frac{d}{dx}(x^2) - \frac{d}{dx}(e^{-2x^2})$$

Use (26-15).

$$= 2x + 4xe^{-2x^2}$$

(c) **First**, use (26-13) with $u = \sin e^{3x}$.

$$\frac{dy}{dx} = \frac{1}{\sin e^{3x}} \cdot \frac{d}{dx}(\sin e^{3x})$$

Next, use (26-3) with $u = e^{3x}$.

$$= \frac{1}{\sin e^{3x}}(\cos e^{3x}) \cdot \frac{d}{dx}(e^{3x})$$

Finally, use (26-15).

$$= \frac{\cos e^{3x}}{\sin e^{3x}} \cdot 3e^{3x}$$

$$= 3e^{3x} \cot e^{3x}$$

(d) **First**, use the power rule (22-22) to write:

$$\frac{dy}{dx} = 2(4e^{3x}) \cdot \frac{d}{dx}(4e^{3x})$$

Then, use (26-15) to find the derivative of e^{3x}.

$$= 2(4e^{3x})12e^{3x}$$

Finally, simplify.

$$= 96e^{6x} \qquad \text{Since } e^u \cdot e^u = e^{2u}$$

We also could have found this derivative by first writing the original function as $y = 16e^{6x}$, and then using (26-15) to find the derivative.

EXAMPLE 28 For what value of t is the function $y = te^{-4t}$ a maximum?

Solution Using (26-15) and the product rule, we find that

$$\frac{dy}{dt} = -4te^{-4t} + e^{-4t}$$

$$\frac{dy}{dt} = 0 \quad \text{at} \quad -4te^{-4t} + e^{-4t} = 0$$

$$e^{-4t}(-4t + 1) = 0$$

Since the exponential function is always positive, $e^{-4t} \neq 0$, and we must have

$$1 - 4t = 0$$

$$t = 0.25$$

Check: Use the second derivative test from Chapter 23 to verify that the function is a maximum at $t = 0.25$. Again, using (26-15) and the product and sum rules, we have

$$\frac{d^2y}{dt^2} = (-4t)(-4)e^{-4t} - 4e^{-4t} - 4e^{-4t}$$

$$= 16te^{-4t} - 8e^{-4t}$$

For $t = 0.25$, this expression is

$$\left.\frac{d^2y}{dt^2}\right|_{t=0.25} = 4e^{-1} - 8e^{-1}$$

$$= -4e^{-1}$$

and, since $e^{-1} = 1/e \approx 1/2.7$, the second derivative is negative. Therefore, the function is a maximum at $t = 0.25$.

Applications

Exponential functions are very useful in describing mathematically the process of change, particularly growth and decay, in a great many physical quantities in technology, physics, and other sciences. We will describe a few of the most important applications.

Charge and discharge of a capacitor

A **capacitor** is an electronic circuit element used in devices to generate, detect, and transmit electromagnetic signals. It consists of two insulated conducting surfaces or plates isolated from their surroundings and carrying a positive charge on one plate and an equal negative charge on the other. Because work must be done to move charge onto the capacitor plates, electrical energy will be stored in an electric field between the plates.

In Fig. 26-9, connecting the switch to A completes the RC circuit, and current begins to flow, building up charge on capacitor C. Because the resistor R is in the circuit, the charge q on the capacitor rises slowly to some maximum value q_m that depends on the voltage V and the capacitance C. The rate dq/dt at which charge builds up is directly proportional to $q - q_m$.

If the switch is connected to B after the capacitor is fully charged, the capacitor will begin to discharge as charge leaves the capacitor and flows through the resistor R. The current dq/dt is the rate of decay of the charge on the capacitor and is directly proportional to q, the charge remaining.

These proportions between the rate of charge flow and the amount of charge lead to the following equations:

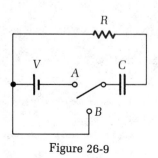

Figure 26-9

Charging a Capacitor

$$q = q_m(1 - e^{-t/RC}) \qquad \text{where } q_m = VC \qquad\qquad (26\text{-}16)$$

Discharging a Capacitor

$$q = q_m e^{-t/RC} \qquad \text{where } q_m = VC \qquad\qquad (26\text{-}17)$$

The constant RC has the dimension of time, and is called the *capacitance time constant* of the circuit. It is the time needed for the charge on the capacitor to increase to within $1 - (1/e) \approx 63\%$ of its maximum value.

EXAMPLE 29   In the circuit shown in Fig. 26-9, find the current $i = dq/dt$ flowing 2.0 ms after the switch is first moved to position A. Use $V = 240$ V, $R = 2500 \ \Omega$, and $C = 1.0 \ \mu$F $(=1.0 \times 10^{-6}$ F).

Solution   **First,** the time constant is: $\quad RC = (2500 \ \Omega)(1.0 \times 10^{-6}$ F)

$$= 2.5 \times 10^{-3} \text{ s} = 2.5 \text{ ms}$$

Next, from equation (26-16): $\quad q = q_m - q_m e^{-t/RC}$

Use (26-15).
$$i = \frac{dq}{dt} = -q_m\left(-\frac{1}{RC}\right)e^{-t/RC}$$

$$= \frac{q_m}{RC}e^{-t/RC}$$

Substitute $q_m = VC$.
$$i = \frac{dq}{dt} = \frac{V}{R}e^{-t/RC}$$

Finally, use a calculator to evaluate this expression for i.
$$i = \frac{240}{2500} \cdot e^{-2.0\,\text{ms}/2.5\,\text{ms}}$$

$$2 \;\boxed{\div}\; 2.5 \;\boxed{=}\;\boxed{+/-}\;\boxed{e^x}\;\boxed{\times}\; 240 \;\boxed{\div}\; 2500 \;\boxed{=} \rightarrow \quad \boxed{0.0431356}$$

$$i \approx 0.043 \text{ A}$$

Damped harmonic oscillations   In any oscillating mechanical system such as a spring or pendulum, the motion gradually decreases to zero as energy is dissipated by friction forces in the system. In the most interesting situation, the magnitude of the friction force is proportional to the speed of motion of the moving component of the system and opposes its motion (Fig. 26-10). The oscillations of the system are said to be **damped** by the friction force.

If the undamped system oscillates in harmonic motion, its amplitude y will vary according to the equation $y = A \cos(\omega t + \phi)$. For damped harmonic motion, the amplitude of the motion decreases exponentially according to the equation $y = Ae^{-bt} \cos(kt + \phi)$, where A is the maximum amplitude, t is time, b, k, and ϕ are constants (Fig. 26-11). In damped harmonic motion, the total

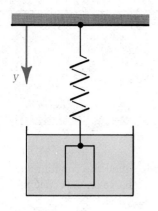

Figure 26-10

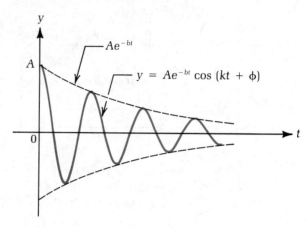

Figure 26-11

mechanical energy of the oscillating system decreases to zero and is trans-formed into heat by the damping mechanism.

EXAMPLE 30   Find the vertical speed dy/dt at $t = 1.2$ s for the damped oscil-lating weight shown in Fig. 26-10 if

$$y = 10e^{-2t} \cos\left(3t + \frac{\pi}{4}\right)$$

for y in centimeters and t in seconds.

Solution   First, use the product rule (22-17):

$$\frac{dy}{dt} = 10e^{-2t} \frac{d}{dt}\left[\cos\left(3t + \frac{\pi}{4}\right)\right] + 10 \cos\left(3t + \frac{\pi}{4}\right)\frac{d}{dt}(e^{-2t})$$

Next, use (26-4) for the derivative of cos u and (26-15) for the derivative of the exponential:

$$\frac{dy}{dt} = -30e^{-2t} \sin\left(3t + \frac{\pi}{4}\right) - 20e^{-2t} \cos\left(3t + \frac{\pi}{4}\right)$$

$$= -10e^{-2t}\left[3 \sin\left(3t + \frac{\pi}{4}\right) + 2 \cos\left(3t + \frac{\pi}{4}\right)\right]$$

Finally, evaluate this expression for $t = 1.2$ s:

3 ⊗ 1.2 ⊕ π ÷ 4 ⊜ STO RAD sin ⊗ 3 ⊕ 2 ⊗ RCL COS ⊜ ⊗ 2.4 +/− eˣ ⊗

10 +/− ⊜ →   **3.1600955**

$$\frac{dy}{dt} \approx 3.2 \text{ cm/s}$$

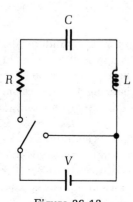

Figure 26-12

Damped oscillations may also occur in an *LC* circuit (Fig. 26-12) when the capacitor is fully charged and is allowed to discharge. If resistance R is present in the circuit, the oscillations are damped so that at any instant the charge on the capacitor is

$$q = q_m e^{-Rt/2L} \cos (kt + \phi)$$

The total electromagnetic energy stored in the circuit shifts back and forth from the capacitor to the inductor, gradually decreasing as it is transferred to heat in the resistor.

Exponential growth and decay

The charging of a capacitor is one example of the **exponential growth** of a quantity. Equations similar to (26-16) can be used to describe the growth of many other kinds of systems. For example, if N is the number of bacteria in a sample, then N increases as the bacteria divide, and under conditions of unlimited resources and a hospitable environment, the rate of change dN/dt of the sample population will depend on N. When the population size is small, it will increase slowly, but as the number increases, the rate of increase also increases. This description of the population growth can be expressed by the equation*

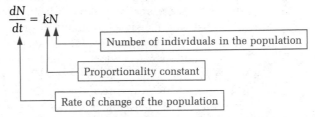

The constant k is the net rate of increase in each time interval. For population growth, k = Birth rate − Death rate. The model of population growth described by this equation leads to the exponential growth equation (26-18) and the growth pattern shown in Fig. 26-13.

Figure 26-13

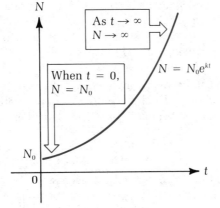

*This is an example of a feedback mechanism where the growth rate determines the future population, which in turn determines the growth rate.

Exponential Growth $\qquad N = N_0 e^{kt}$ $\qquad\qquad\qquad$ (26-18)

The exponential growth model is a good approximation for the description of very many situations involving growth and decay. Here are a few examples:

- In the open air, moisture will evaporate from a wet cloth at a rate roughly proportional to its moisture content. The moisture remaining at any time follows an exponential decay curve.
- In a photomultiplier tube, an incoming photon frees an electron from the first electrode, and this electron is accelerated by an electric field. It collides with a second electrode, producing two or three more electrons by a secondary emission process. These collisions continue with further electrodes, and the number of electrons in the resulting cascade increases exponentially.
- When a microphone is placed near a public address system speaker, faint sounds picked up by the microphone are amplified and broadcast by the speaker. If the rebroadcast sound is louder than the original, then the cycle continues, and the signal is amplified each time around the loop. The resulting exponential growth of the signal can double the sound amplitude in about a millisecond, producing a loud squawk.
- The growth in demand for highways may in some cases be proportional to the miles of highway already built. If there are no financial or political constraints, the miles of highway constructed will grow exponentially, especially where the total mileage is still low.

EXAMPLE 31 In the early 19th century, the English economist T. R. Malthus proposed that, since population seems to increase faster than the resources necessary to support it, the population of the earth would grow exponentially according to (26-18) unless it is checked by war, famine, disease, or changes in reproductive rate. The population of the world was 4.09 billion in 1975. If the worldwide average yearly birth rate is 30.4 per 1000, or 0.0304, and the death rate is 12.3 per 1000, or 0.0123, then find:

(a) k, the growth rate
(b) The size of the world population in the year 2000
(c) The yearly rate dN/dt by which world population is increasing in 1985
(d) The time in which the population will double if it continues to increase at this rate

Solution
(a) $k = 0.0304 - 0.0123 = 0.0181$ per year
(b) $N_0 = 4.09$ billion; therefore, in the year 2000 at $t = 25$ years,

$$N(25) = 4.09e^{0.0181(25)}$$

$$.0181 \boxed{\times} 25 \boxed{=} \boxed{e^x} \boxed{\times} 4.09 \boxed{=} \rightarrow \boxed{6.4304529}$$

$$\approx 6.43 \text{ billion}$$

(c)   Using (26-15), $dN/dt = N_0 k e^{kt}$; and in the year 1985 at $t = 10$ years,

$$\left.\frac{dN}{dt}\right|_{t=10} = (4.09)(0.0181)e^{0.0181(10)}$$

$$\approx 0.0887175 \text{ billion/year}$$

This is about one-third the population of the United States.

$$\approx 89 \text{ million/year}$$

(d)   The population will be doubled when $N = 2N_0$. If this happens at time t_d, then

$$2N_0 = N_0 e^{kt_d}$$

$$2 = e^{kt_d}$$

Taking the natural logarithm of both sides, we have

$$\ln 2 = kt_d$$

Remember: $\ln x = \log_e x$, so $\log_e e^u = u \log_e e = u$.

$$t_d = \frac{\ln 2}{k}$$

$2 \boxed{\ln x} \boxed{\div} .0181 \boxed{=} \rightarrow$ **38.295424**

$$t_d \approx 38 \text{ years}$$ This will be the year 2013.

Radioactive decay   In a large sample of a radioactive element, nuclei decay spontaneously and randomly into a new element. Because each nucleus in the sample has the same probability of disintegrating at any time, the number of disintegrations per second is proportional to the number of nuclei of the original element remaining. If A is a measure of the radioactivity of the sample at time t, then

$$-\frac{dA}{dt} = kA$$

where k is called the **disintegration constant.** Notice that the decay rate is negative. The rate of decay is decreasing as the number of original nuclei declines. The level of activity $A(t)$ at any time t is given by the following exponential decay equation:

Radioactive Decay $A(t) = A_0 e^{-kt}$ (26-19)

where $A_0 = A(0)$ is the activity level at $t = 0$.

The activity level A can be the actual number of radioactive nuclei remaining in the sample, or it can be some equivalent measure of the radioactivity level.

EXAMPLE 32   Every nuclear reactor produces the highly toxic radioactive element plutonium-239 (Pu-239) as a by-product. For this isotope, k is about 2.85×10^{-5}/year.

(a) How much time is required for the radioactivity level of a sample of Pu-239 to decrease to 10% of its starting level?

(b) What is the half-life of Pu-239?

(c) If a 100 g sample of Pu-239 contains about 2.52×10^{23} radioactive nuclei at time $t = 0$, how many disintegrations are taking place per second 10 years later?

Solution

(a) If $A = 0.10A_0$, then

$$0.10A_0 = A_0e^{-kt}$$

$$0.10 = e^{-(2.85 \times 10^{-5})t}$$

Take the natural log of both sides.

$$\ln 0.10 = (-2.85 \times 10^{-5})t$$

$$t = \frac{\ln 0.10}{-2.85 \times 10^{-5}}$$

$$.1 \boxed{\ln x} \boxed{\div} 2.85 \boxed{+/-} \boxed{\text{EE}} 5 \boxed{+/-} \boxed{=} \rightarrow \boxed{80792.459}$$

$$t \approx 80{,}800 \text{ years}$$

(b) The half-life of a radioactive decay process is defined as the time at which one-half of the original sample remains. Therefore, if t_H is the half-life,

$$0.5A_0 = A_0e^{-kt_H}$$

$$0.5 = e^{-kt_H}$$

Take the natural logarithm of both sides.

$$\ln 0.5 = -kt_H$$

$$t_H = \frac{\ln 0.5}{-k} = \frac{\ln 2}{k}$$

$$\approx 24{,}320.954$$

$$\approx 24{,}300 \text{ years}$$

(c) Use (26-15).

$$\frac{dA}{dt} = -kA_0e^{-kt}$$

Substitute the given values for k, A_0, and t:

$$= -(2.85 \times 10^{-5})(2.52 \times 10^{23})e^{-(2.85 \times 10^{-5})10}$$

$$\approx -7.1799534 \times 10^{18} \text{ decays/year}$$

$$\approx -7.18 \times 10^{18} \text{ decays/year} \times \frac{1 \text{ year}}{3.15 \times 10^7 \text{ s}}$$

$$\approx -2.3 \times 10^{11} \text{ decays/s} \qquad \text{The negative sign tells us that the amount of Pu-239 is decreasing.}$$

Newton's law of cooling   Hot objects tend to cool down and cool objects tend to warm up to the temperature of their surroundings. In 1701, the physicist Isaac Newton suggested that the rate of temperature change of an object is proportional to the difference in temperature between the object and its surroundings. That is,

$$\frac{d}{dt}(T - T_s) = -k(T - T_s)$$

where T is the temperature at the time t, T_s is the temperature of the surroundings, and k is a constant that depends on the shape and composition of the object. This leads to the following exponential function:

Newton's Law of Cooling	$T = T_s + (T_0 - T_s)e^{-kt}$	**(26-20)**

In this equation, T_0 is the temperature of the object at time $t = 0$. This function is usually called Newton's law of cooling, even though it applies to both the cooling and the warming of an object. For most materials, this relationship holds approximately over some range of temperature, although not for temperature extremes.

In Fig. 26-14 we can see that for both cooling and warming, at $t = 0$, the original temperature of the object is $T = T_0$. Also, after a long time t, $\lim T = T_s$; that is, the temperature of the object approaches the temperature of the surroundings.

Figure 26-14

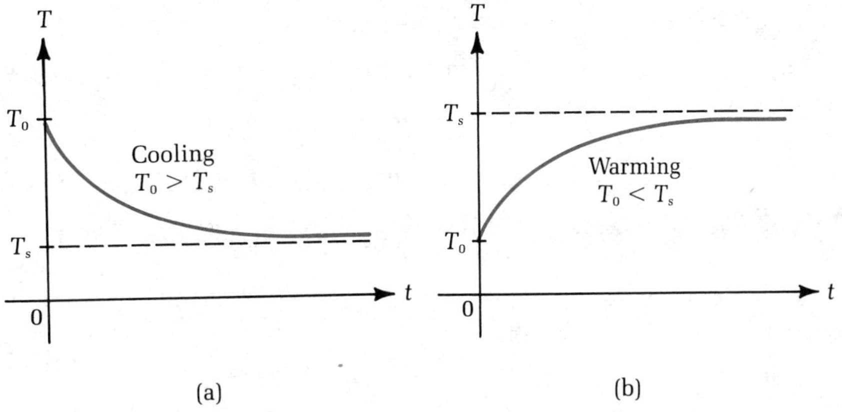

(a)

(b)

▦ **EXAMPLE 33**   A cast steel fitting initially at 450°F cools to 400°F in 60 s when the surrounding air temperature is 60°F.

(a) How much time is needed for its temperature to drop to 100°F? (Round to the nearest second.)

(b) At what rate is its temperature falling after 2 min?

Solution

(a) **First,** use (26-20) to find the value of constant k. After 60 s,

$$400°F = 60°F + (450°F - 60°F)e^{-60k}$$

$$400 - 60 = (450 - 60)e^{-60k}$$

$$340 = 390e^{-60k}$$

Divide by 390.

$$\frac{340}{390} = e^{-60k}$$

Take the natural logarithm of both sides.

$$\ln\left(\frac{340}{390}\right) = -60k$$

$$k = \frac{\ln\left(\frac{340}{390}\right)}{-60}$$

$$340 \;\boxed{\div}\; 390 \;\boxed{=}\;\boxed{\ln x}\;\boxed{\div}\; 60 \;\boxed{+/-}\;\boxed{=}\;\boxed{\text{STO}} \rightarrow \quad \boxed{2.2867 \quad -03}$$

$$k \approx 0.0022867$$

Next, use this value of k to find the time when $T = 100°F$ in (26-20).

$$100 = 60 + (450 - 60)e^{-kt}$$

$$\frac{40}{390} = e^{-kt}$$

Take the natural logarithm of both sides and solve for t.

$$t = \frac{\ln\left(\frac{40}{390}\right)}{-k}$$

$$40 \;\boxed{\div}\; 390 \;\boxed{=}\;\boxed{\ln x}\;\boxed{\div}\;\boxed{\text{RCL}}\;\boxed{+/-}\;\boxed{=} \rightarrow \quad \boxed{995.88134}$$

$$t \approx 996 \text{ s or about 17 min}$$

(b)   Using (26-15), the derivative of (26-20) is

$$\frac{dT}{dt} = -k(T_0 - T_s)e^{-kt}$$

At $t = 120$ s,

$$\frac{dT}{dt} = -k(450 - 60)e^{-k(120)}$$

$$\boxed{\text{RCL}}\;\boxed{\times}\; 120 \;\boxed{=}\;\boxed{+/-}\;\boxed{e^x}\;\boxed{\times}\;\boxed{(}\; 450 \;\boxed{-}\; 60 \;\boxed{)}\;\boxed{\times}\;\boxed{\text{RCL}}\;\boxed{+/-}\;\boxed{=}$$

$$\rightarrow \quad \boxed{-0.6778001}$$

$$\frac{dT}{dt} \approx -0.7°F/s$$

The negative sign on dT/dt indicates that the temperature is decreasing—the object is cooling.

Compound interest   If an amount of money P is deposited for 1 year in an account earning interest, then at the end of the first year the total amount of money in the account will be

$$A_1 = P(1 + r)$$

where r is the annual interest (expressed as a decimal number). If both the original deposit and the earned interest remain in the account, then at the end of the second year the total in the account will be

$$A_2 = P(1 + r)(1 + r) = P(1 + r)^2$$

After t years the amount in the account is given by

$$A_n = P(1 + r)^t$$

Interest earned on deposited interest is called **compound interest.** If the interest earned by the account is calculated or compounded n times each year, then the total amount increases more rapidly and is described by the following exponential function:

Compound Interest, Calculated n Times per Year

$$A_n = P\left(1 + \frac{r}{n}\right)^{nt} \tag{26-21}$$

Interest can be compounded over any time interval, and it is useful to examine the limit of A_n as $n \to \infty$ and interest is compounded continuously. A comparison of different compounding intervals is given in Table 26-4 for $P = \$100$ and an interest rate of 10%—that is, $r = 0.10$.

TABLE 26-4
Compounding $P = \$100$
at $r = 0.10$

Year	Annually $(n = 1)$	Quarterly $(n = 4)$	Monthly $(n = 12)$	Daily $(n = 365)$	Continuously $(n \to \infty)$
1	$110.00	$110.38	$110.47	$110.52	$110.52
2	$121.00	$121.84	$122.04	$122.14	$122.14
5	$161.05	$163.86	$164.53	$164.86	$164.87
10	$259.37	$268.51	$270.70	$271.79	$271.83

If we take the limit of (26-21) as $n \to \infty$, we obtain

$$A = \lim_{n \to \infty} A_n(t) = \lim_{n \to \infty} P\left(1 + \frac{r}{n}\right)^{nt}$$

Then, take the constant P outside the limit and rewrite the exponent:

$$A = P \lim_{n \to \infty} \left(1 + \frac{r}{n}\right)^{(n/r)rt}$$

Now, move the exponent rt outside the limit:

$$A = P\left[\lim_{n\to\infty}\left(1 + \frac{r}{n}\right)^{n/r}\right]^{rt}$$

But if we let $h = r/n$, then as $n \to \infty$, $h \to 0$, the limit inside the brackets becomes $\lim_{h\to 0}(1 + h)^{1/h} = e$, as was shown in Section 26-4. Therefore, for continuous compounding, (26-21) becomes the following exponential function:

| Continuous Compounding | $A = Pe^{rt}$ | (26-22) |

EXAMPLE 34 A bank offers the following possibilities for 3 year, $1000 fixed-rate savings certificates: X, 10.12% compounded annually; Y, 9.8% compounded quarterly; Z, $8\frac{7}{8}$% compounded continuously.
(a) Which represents the highest interest earnings?
(b) At what rate is the amount of interest increasing on each of these certificates after 1 year?

Solution
(a) For X: $A = \$1000(1 + 0.1012)^3 \approx \$1335.36$

For Y: $A = \$1000\left(1 + \frac{0.098}{4}\right)^{12} \approx \$1337.04$

$\boxed{r = 8\frac{7}{8} \div 100 = 0.08875}$

For Z: $A = \$1000e^{(0.08875)3} \approx \$1305.06$

Certificate Y earns the most interest over the 3-year period.
(b) The rate of increase for X and Y can be found by differentiating (26-21) using (26-14).

$$\frac{dA}{dt} = P\left(1 + \frac{r}{n}\right)^{nt}\cdot\ln\left(1 + \frac{r}{n}\right)\cdot n$$

For X ($n = 1$, $t = 1$, $r = 0.1012$):

$$\frac{dA}{dt} = 1000(1 + 0.1012)^1\cdot\ln(1 + 0.1012)\cdot 1$$

$$\approx \$106.16 \text{ per year}$$

For Y ($n = 4$, $t = 1$, $r = 0.098$):

$$\frac{dA}{dt} = 1000\left(1 + \frac{0.098}{4}\right)^4\cdot\ln\left(1 + \frac{0.098}{4}\right)\cdot 4$$

$$\approx \$106.66 \text{ per year}$$

The rate of increase for Z can be found by differentiating (26-22) using (26-15).

$$\frac{dA}{dt} = Pre^{rt}$$

Then, for Z ($r = 0.08875$, $t = 1$):

$$\frac{dA}{dt} = 1000(0.08875)e^{(0.08875)1}$$

$$\approx \$96.99 \text{ per year}$$

Exercises 26-5

Find the derivative of each function.

1. $y = 3^{4x}$
2. $y = 2^{x/4}$
3. $y = 4^{x^2}$
4. $y = 5^{x^3 + x}$
5. $y = e^{x^2}$
6. $y = 2e^{3x}$
7. $y = 4e^{x + x^2}$
8. $y = e^{-2x}$
9. $y = 3^{-x}$
10. $y = 6^{\sqrt{x}}$
11. $y = xe^{\sqrt{x}}$
12. $y = xe^{\tan x}$
13. $y = e^{\sin x}$
14. $y = e^{\text{Arcsin} \, x}$
15. $y = \dfrac{e^{2x}}{2x}$
16. $y = \dfrac{2e^{x^2}}{x + 2}$
17. $y = 2^x + 2^{3x}$
18. $y = e^x - e^{2x}$
19. $y = x^3 + e^{3x}$
20. $y = 3(x + 2^x)$
21. $y = e^{\ln 2x}$
22. $y = (3e^{2x})^2$
23. $y = (e^{1/x})^2$
24. $y = e^x \ln x$
25. $y = \ln (\cos e^{2x})$
26. $y = \ln (\tan e^{4x})$
27. $y = \text{Arcsin} \, e^{3x}$
28. $y = 2^{\cos x}$
29. $y = \sqrt{1 + e^x}$
30. $y = 3 \, \text{Arctan} \, e^{2x}$
31. $y = \sin e^{x^2}$
32. $y = \cos e^{3x}$
33. $y = (e^{1+x})^2 + 2$
34. $y = 1 - (1 - e^x)^6$
35. $y = \dfrac{a}{1 + bc^{-kx}}$
36. $y = \dfrac{ae^{kx}}{1 + e^{-kx}}$

Solve.

37. Find the value of x where $y = xe^x$ has a minimum.
38. Find the value of x where $y = xe^{-\sqrt{x}}$ has a maximum.
39. Find the value of x where $xe^{1/x}$ has a minimum ($x > 0$).
40. Find the value of x where $y = x^2 e^{-x}$ has a maximum.
41. *Biology*   The Gompertz growth function describing the growth of certain biological systems is given by

$$N(t) = ae^{-be^{-kt}}$$

where a, b, and k are constants greater than 0. Find dN/dt.

42. *Chemistry*   Fick's law for the diffusion of molecules through an organic membrane is given by

$$R = R_0(1 + ae^{-kt})$$

where R is the concentration of molecules at time t and R_0 is the initial concentration.* Find dR/dt at $t = 1.4$ s for $R_0 = 2.0$ mg/cm^3, $a = 3.5$, and $k = 1.15$ s^{-1}.

43. *Biology*   The logistic growth equation

$$N = \frac{N_m}{1 + ae^{-kt}}$$

*See Desmond M. Burns and Simon G. G. MacDonald, *Physics for Biology and Premedical Students* (Reading, Mass: Addison-Wesley, 1980), p. 201.

describes the spread of a contagious disease in a certain group of people, where N is the number of cases at the end of week t, and $a = 50$, $k = 0.28$, and the total number of susceptible people in the group is $N_m = 100$. Find dN/dt, the rate of spread of the disease at $t = 5$ weeks.

44. *Physics*   If air resistance for a skydiver is proportional to the square of the skydiver's speed v, then

$$v(t) = v_t\left(\frac{1 + ce^{-pt}}{1 - ce^{-pt}}\right)$$

Find dv/dt, the acceleration of the skydiver, at $t = 1.0$ s, if $v_t = 5.0$ m/s, $c = 0.35$, and $p = 4.0$ s^{-1}.

45. *Meteorology*   If we assume that the temperature of the atmosphere is constant, then for $h < 50$ mi above sea level, atmospheric pressure P will change with altitude h according to the equation

$$P = P_0 e^{-1.93h}$$

For $P_0 \approx 14.7$ lb/in.^{2}, find dP/dh at the top of Mt. Everest, where $h \approx 5.5$ mi.

46.   Certain dynamic processes are given by the growth curve defined by the equation below. Find $F'(t)$.

$$F(t) = \frac{be^{kt}}{a + e^{kt}}$$

47. *Civil engineering*   The hanging cable on a suspension bridge forms a curve called a **catenary**, which is given by the equation

$$y = \frac{a}{2}(e^{bx} + e^{-bx}) \qquad a > 0, b \neq 0$$

Find the slope of this curve at $x = 45$ ft if $a = 120$ ft and $b = 0.016$ ft^{-1}.

48. *Nuclear physics*   The radioactive element iodine-128 is used as a biological tracer in measurements of the activity of the thyroid gland. The decay constant for this isotope is $k = 0.0279$ min^{-1}.
   (a)   If the initial level of radioactivity is 2.64 counts/s, how much time elapses before the count drops to 0.50 count/s?
   (b)   What is the half-life of I-128?
   (c)   At what rate is the level of radioactivity changing at $t = 20$ min?

49. *Nuclear physics*   A sample of carbon taken from a wooden beam found at an archaeological site has a carbon-14 activity of 63.0 disintegrations per minute. If the half-life of C-14 is 5730 years, find:
   (a)   The decay constant k
   (b)   The time required for the radioactivity level of this sample to have decayed to this level if it started at 76.5 disintegrations per minute when the beam was a living tree
   (c)   The rate at which the level of radioactivity is changing when $t = 5000$ years.

50. *Machine technology*   A heated steel part initially at 380°F cools to 320°F in 30 s when the surrounding air temperature is 45°F.
   (a)   How much additional time is needed for its temperature to drop to 150°F?
   (b)   At what rate is its temperature falling after 90 s?

51. *Chemistry*   A chemical solution initially at 100°C cools to 70°C in 75 s when the surrounding air temperature is 20°C.
   (a)   At the same ambient temperature, how much additional time is needed for its temperature to drop to 40°C?
   (b)   At what rate is its temperature falling after 2 min?

52. *Finance*   Which represents a higher effective interest rate, 11.25% compounded annually or 10.85% compounded continuously?

53. *Finance*   If $1000 invested in a Keogh account earns 9.75% compounded quarterly, at what rate is the amount in the account increasing after 2 years?

54. *Economics*   When an electric utility company changes over to a new type of energy generator, the cost (in thousands of dollars) of the changeover is given by

$$C = 180(1 - e^{-15x})$$

where x is the fraction of market penetration. Find dC/dx when $x = 0.20$.

55. *Electronics*   In an RC series circuit where $R = 2000 \ \Omega$ and $C = 10 \ \mu F$, find the current flowing 15 ms after a 140-V power source is connected across the circuit.

56. *Electronics*   If the voltage source in Problem 55 is removed after the capacitor is fully charged, what current will be flowing 5 ms later?

57. *Chemistry*   In the early stages, growth of a crystal is exponential in nature. If an alkali halide crystal increases in weight from 5 g to 10 g in the first 60 s of growth, what will be its rate of growth at 120 s?

58. *Mechanical engineering*   A mechanical system oscillates with damped harmonic motion described by the equation

$$y = Ae^{-bt} \cos (kt + \phi)$$

where $A = 25$ cm, $b = 1.5 \ s^{-1}$, $k = 4.0 \ s^{-1}$, $\phi = \pi/6$, and y is the amplitude of the motion in centimeters. Find the speed of oscillation of the system at $t = 1.0$ s.

26-6 | Review of Chapter 26

Important Terms and Concepts

transcendental function (p. 1097)

capacitor (p. 1136)

damped harmonic oscillations (p. 1137)

exponential growth/decay (p. 1139)

compound interest (p. 1145)

radioactive decay (p. 1141)

Formulas and Rules

- derivative of sin u (p. 1102)

$$\frac{d}{dx}(\sin u) = \cos u \cdot \frac{du}{dx} \qquad (26\text{-}3)$$

- derivative of cos u (p. 1106)

$$\frac{d}{dx}(\cos u) = -\sin u \cdot \frac{du}{dx} \qquad (26\text{-}4)$$

- derivative of tan u (p. 1111)

$$\frac{d}{dx}(\tan u) = \sec^2 u \frac{du}{dx} \qquad (26\text{-}5)$$

- derivative of cot u (p. 1112)

$$\frac{d}{dx}(\cot u) = -\csc^2 u \frac{du}{dx} \qquad (26\text{-}6)$$

- derivative of csc u (p. 1113)

$$\frac{d}{dx}(\csc u) = -\csc u \cot \frac{du}{dx} \qquad (26\text{-}7)$$

- derivative of sec u (p. 1114)

$$\frac{d}{dx}(\sec u) = \sec u \tan u \frac{du}{dx} \qquad (26\text{-}8)$$

- derivative of Arcsin u (p. 1118)

$$\frac{d}{dx}(\text{Arcsin } u) = \frac{1}{\sqrt{1-u^2}}\frac{du}{dx} \qquad \text{where } -1 < u < 1 \qquad \textbf{(26-9)}$$

- derivative of Arccos u (p. 1119)

$$\frac{d}{dx}(\text{Arccos } u) = -\frac{1}{\sqrt{1-u^2}}\frac{du}{dx} \qquad \text{where } -1 < u < 1 \qquad \textbf{(26-10)}$$

- derivative of Arctan u (p. 1121)

$$\frac{d}{du}(\text{Arctan } u) = \frac{1}{1+u^2}\frac{du}{dx} \qquad \textbf{(26-11)}$$

- derivative of the logarithmic function (p. 1128)

$$\frac{d}{dx}(\log_b u) = \frac{1}{u}\cdot\log_b e \cdot\frac{du}{dx} \qquad \textbf{(26-12)}$$

- derivative of the natural logarithmic function (p. 1129)

$$\frac{d}{dx}(\ln u) = \frac{1}{u}\cdot\frac{du}{dx} \qquad \textbf{(26-13)}$$

- derivative of a^u (p. 1133)

$$\frac{d}{dx}(a^u) = a^u\cdot\ln a\cdot\frac{du}{dx} \qquad \textbf{(26-14)}$$

- derivative of e^u (p. 1133)

$$\frac{d}{dx}(e^u) = e^u\cdot\frac{du}{dx} \qquad \textbf{(26-15)}$$

- charging a capacitor (p. 1137)

$$q = q_m(1 - e^{-t/RC}) \qquad \text{where } q_m = VC \qquad \textbf{(26-16)}$$

- discharging a capacitor (p. 1137)

$$q = q_m e^{-t/RC} \qquad \text{where } q_m = VC \qquad \textbf{(26-17)}$$

- exponential growth (p. 1139)

$$N = N_0 e^{kt} \qquad \textbf{(26-18)}$$

- radioactive decay (p. 1141)

$$A(t) = A_0 e^{-kt} \qquad \text{where } A_0 = A(0) \qquad \textbf{(26-19)}$$

- Newton's law of cooling (p. 1143)

$$T = T_s + (T_0 - T_s)e^{-kt} \qquad \textbf{(26-20)}$$

- compound interest, calculated n times per year (p. 1145)

$$A_n = P\left(1 + \frac{r}{n}\right)^{nt} \qquad \textbf{(26-21)}$$

- continuous compounding (p. 1146)

$$A = Pe^{rt} \qquad \textbf{(26-22)}$$

Exercises 26-6

Find the derivative of each function.

1. $y = \sin 6x$
4. $y = \cot(5x - 2)$

2. $y = 2\cos(x^2 - 4)$
5. $y = \sec^2(x^4)$

3. $y = 6\tan\sqrt{x}$
6. $y = x\csc 2x$

7. $y = 4 \cos^2 x^3$

8. $y = \tan^3 \sqrt{x}$

9. $y = \sin 3x \sec x^2$

10. $y = \cot (x + 1) \cos (x - 1)$

11. $y = 2x^4 - \csc (2x + 5)$

12. $y = 2 \sin x^2 + 4 \tan 2x$

13. $y = \dfrac{\cos x^2}{1 + \cot x}$

14. $y = \dfrac{2x}{\sin^3 2x}$

15. $y = 2 \operatorname{Arcsin} x^3$

16. $y = \operatorname{Arccos} \dfrac{1}{x}$

17. $y = \operatorname{Arctan} \sqrt{x}$

18. $y = \sqrt{\operatorname{Arcsin} 3x}$

19. $y = \dfrac{x}{\operatorname{Arccos} x^2}$

20. $y = \cos (\operatorname{Arcsin} x)$

21. $y = \log 5x^2$

22. $y = \ln \sqrt{2x - 1}$

23. $y = x^3 \ln x^2$

24. $y = \ln (\cos x)$

25. $y = \ln (\operatorname{Arctan} 2x)$

26. $y = [\ln (x^2 - 1)]^2$

27. $y = 5^{x^3}$

28. $y = e^{x+1}$

29. $y = e^{\cos 3x}$

30. $y = \sqrt{1 - e^{2x}}$

31. $y = xe^{\sqrt{x+1}}$

32. $y = e^{2x} \cos 2x$

33. $y = e^{x^2} \ln x$

34. $y = \ln (\sin e^{3x})$

35. $y = (2e^{4x})^2$

36. $y = \sqrt{1 - x^2} + \operatorname{Arcsin} x$

37. $y^2 = 2x - \tan 2x$

38. $x \cos y = x + y$

Solve.

39. Find the equation of the line tangent to $y = 2 \sin 3x$ at $x = \pi/6$.

40. Find the equation of the line tangent to $y = \operatorname{Arcsin} x$ at $x = \frac{1}{2}$.

41. Find the equation of the line tangent to $y = \ln (\tan x)$ at $x = \pi/4$.

42. Find the equation of the line tangent to $y = 2e^{3x}$ at $x = 0.25$.

43. Physics   The displacement of a standing wave in a stretched string is given by the equation $y = 2y_m \sin kx \cos \omega t$, where y_m, k, and ω are constants. Find an expression for the transverse velocity dy/dt of the string.

44. Physics   In Problem 43 find an expression for the slope dy/dx of the string.

45. Electronics   The magnetic field B produced by a rotating armature varies with the angle of rotation according to the equation

$$B = \frac{2 \sin \theta}{1 + \cos^2 (2\theta)}$$

Find the rate of change of B with θ when $\theta = 2\pi/3$ (B is measured in gauss).

46. Physics   If a jogger 5 ft tall moves horizontally at 11 ft/s toward a building 65 ft tall, at what rate is the angle of elevation θ changing when she is 540 ft from the building?

47. Machine technology   For the tapered pin shown in the figure, find the value of θ and the percent error in this value.

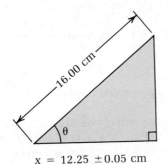

$x = 12.25 \pm 0.05$ cm

48. A balloon is rising vertically at 25 m/s. At what rate is θ changing when $h = 75$ m?

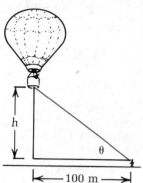

h

θ

$\leftarrow$ 100 m $\rightarrow$

49. A police officer in a helicopter hovering 150 ft over a highway watches a car speeding away (along its line of sight) from the helicopter at 85 mi/h. Find $d\theta/dt$ when the car is 450 ft away from the helicopter. [*Hint:* convert 85 mi/h to ft/s to get an answer in rad/s.]

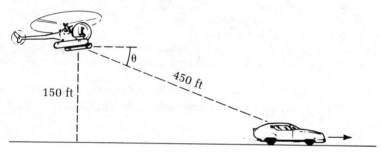

θ

150 ft

450 ft

50. *Electronics*   The electrical potential at any point between two infinitely long parallel charged wires is given by the equation below. Find dV/dR_1.

$$V = -\frac{\lambda}{2}(\ln R_1 - \ln R_2)$$

51. *Electronics*   The impressed voltage in an electrical circuit is given by the equation $V(t) = 6.0 \ln (1 + 0.25t^2)$ volts. Find the rate of change of $V(t)$ when $t = 1.6$ s.

52. *Electronics*   The power dissipated in an electrical circuit is given by the formula $P(t) = 1250\ e^{0.45t}$ watts. Find the rate of power dissipation when $t = 0.40$ s.

53. *Electronics*   A 2.2 μF capacitor is placed in series with a 5000 Ω resistor, and a constant 220 V potential is applied to the combination, charging the capacitor. Find dq/dt, the rate of buildup of charge at $t = 15$ ms.

54. *Machine technology*   A heated steel rod initially at 230°C cools to 150°C in 60 s when the surrounding air temperature is 28°C.
 (a) How much time is needed for its temperature to fall to 80°C?
 (b) At what rate is its temperature falling after 30 s?

55. *Finance*   If $100 had been invested in 1940 at 6% annual interest compounded continuously, at what rate would the investment be growing 50 years later in 1990?

56. *Nuclear physics*   The radioactive isotope phosphorus-32, used as a tracer in chemical reactions, has a half-life of 14.28 days. Find (a) The decay constant k, (b) the time required for a level of 3500 counts/s to drop to 35 counts/s, and (c) the rate at which the radioactivity level is changing after 1 week.

Techniques of Integration

27

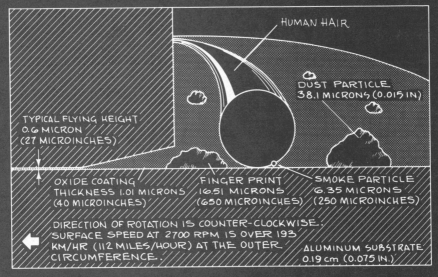

HUMAN HAIR

DUST PARTICLE
38.1 MICRONS (0.015 IN)

TYPICAL FLYING HEIGHT
0.6 MICRON
(27 MICROINCHES)

OXIDE COATING
THICKNESS 1.01 MICRONS
(40 MICROINCHES)

FINGER PRINT
16.51 MICRONS
(650 MICROINCHES)

SMOKE PARTICLE
6.35 MICRONS
(250 MICROINCHES)

DIRECTION OF ROTATION IS COUNTER-CLOCKWISE.
SURFACE SPEED AT 2700 RPM IS OVER 193
KM/HR (112 MILES/HOUR) AT THE OUTER
CIRCUMFERENCE.

ALUMINUM SUBSTRATE
0.19 cm (0.075 IN.)

RIGID DISC DRIVE OPERATING ENVIRONMENT

In earlier chapters the concept of integration was introduced, and integrals involving algebraic functions were solved using the generalized power rule (24-10). However, a very wide variety of integrals appear in applications of science and technology, including many based on nonalgebraic functions. In this chapter we will use the concept of the antiderivative to integrate basic trigonometric, logarithmic, and exponential functions, and we will develop techniques that allow us to increase greatly the variety and complexity of integrals that can be solved.

One of the most powerful techniques of integration is the process of substitution shown in Chapter 24. This procedure often may be used to transform an integral into a simpler form that can be solved quickly. The generalized power rule (24-10),

$$\int u^n \, du = \frac{u^{n+1}}{n+1} + c \qquad \text{for } n \neq -1$$

holds where $u(x)$ is any algebraic or transcendental function, including the trigonometric, logarithmic, or exponential functions.

EXAMPLE 1   Integrate: $\int \sin^2 x \cos x \, dx$

Solution   From (26-3) we know that if we let $u = \sin x$, then $du = \cos x \, dx$, and we can simplify the integral using substitution.

Replace $\sin^2 x$ with u^2 and $\cos x \, dx$ with du.

$$\int \underbrace{\sin^2 x}_{u^2} \underbrace{\cos x \, dx}_{du} = \int u^2 \, du$$

Use the power rule (25-10).

$$= \frac{u^3}{3} + C$$

$$= \frac{\sin^3 x}{3} + C$$

Check this result by differentiating. From (22-22),

$$\frac{d}{dx}\left(\frac{\sin^3 x}{3} + C\right) = \frac{3 \sin^2 x}{3} \cdot \frac{d(\sin x)}{dx}$$

$$= \sin^2 x \cos x \, dx$$

The key to successful use of the substitution technique is quick and correct recognition of the function $u(x)$ to be used in the integration. For practice in finding u and du, complete the table in Example 2 and perform the integrations.

EXAMPLE 2

Integral:	Let $u(x) =$	Then $du =$	Integral becomes:
$\int \cos^3 x \sin x \, dx$			
$\int \sqrt{\sin x} \cos x \, dx$			
$\int \dfrac{e^x \, dx}{\sqrt{e^x - 1}}$			

Integral:	Let $u(x)$ =	Then du =	Integral becomes:
$\int \dfrac{\text{Arctan } 3x}{1 + 9x^2}\, dx$			
$\int \dfrac{\ln (x + 1)}{x + 1}\, dx$			

Solution

(a) If $u = \cos x$, then by (26-4),
$du = -\sin x\, dx$, and the original integral becomes:

$$\int \cos^3 x \sin x\, dx = -\int u^3\, du$$

Use the power rule.

$$= -\frac{u^4}{4} + C$$

$$= -\frac{\cos^4 x}{4} + C$$

(b) If $u = \sin x$, then by (26-3),
$du = \cos x\, dx$. Substitute $\sqrt{u}$ for $\sqrt{\sin x}$ and du for $\cos x\, dx$, and the integral becomes:

$$\int \sqrt{\sin x}\, \cos x\, dx = \int \sqrt{u}\, du = \int u^{1/2}\, du$$

Use the power rule.

$$= \frac{u^{3/2}}{\frac{3}{2}} + C = \frac{2u^{3/2}}{3} + C$$

$$= \frac{2(\sin x)^{3/2}}{3} + C = \frac{2\sqrt{\sin^3 x}}{3} + C$$

$$= \frac{2 \sin x \sqrt{\sin x}}{3} + C$$

(c) Recall from (26-15) that

$$\frac{d}{dx}(e^x) = e^x$$

Therefore, if $u = e^x - 1$, then
$du = e^x\, dx$. Substitute $\sqrt{u}$ for $\sqrt{e^x - 1}$ and du for $e^x\, dx$, and the integral may be written as:

$$\int \frac{e^x\, dx}{\sqrt{e^x - 1}} = \int \frac{du}{\sqrt{u}} = \int u^{-1/2}\, du$$

Use the power rule.

$$= \frac{u^{1/2}}{\frac{1}{2}} + C = 2u^{1/2} + C$$

$$= 2\sqrt{e^x - 1} + C$$

(d) Let $u = $ Arctan $3x$; then by (26-11): $du = \dfrac{1}{1 + (3x)^2} \cdot d(3x)$

$$= \dfrac{3\,dx}{1 + 9x^2}$$

There-fore, the original integral becomes:

$$\int \dfrac{\text{Arctan } 3x}{1 + 9x^2}\,dx = \int u\!\left(\dfrac{du}{3}\right) \quad \boxed{\text{Substitute } \dfrac{du}{3} \text{ for } \dfrac{dx}{1 + 9x^2}.}$$

$$= \dfrac{1}{3}\int u\,du$$

Use the power rule.

$$= \dfrac{1}{3}\cdot\dfrac{u^2}{2} + C$$

$$= \dfrac{(\text{Arctan } 3x)^2}{6} + C$$

(e) Let $u = \ln(x + 1)$; then by (26-13),
$$du = \dfrac{1}{x + 1}\cdot dx$$

The integral becomes:
$$\int \dfrac{\ln(x + 1)}{x + 1}\,dx = \int \ln(x + 1)\cdot\dfrac{dx}{x + 1}$$

$$= \int u\,du$$

Use the power rule.

$$= \dfrac{u^2}{2} + C$$

$$= \dfrac{1}{2}[\ln(x + 1)]^2 + C$$

LEARNING HINTS ▶

1. If the integrand contains the pairs of functions (sin x, cos x), as in Examples 2(a) and (b), (sec x, tan x), or (csc x, cot x), seek a substitution that will enable you to write it as

$$\int f(\sin x)[\cos x\,dx] \quad \text{or} \quad \int f(\cos x)[\sin x\,dx]$$

$$\text{or} \quad \int f(\tan x)[\sec^2 x\,dx]$$

and so on.

2. If the integrand is a function of e^x as in Example 2(c), try substituting $u = e^x$. If the integrand contains a radical expression, try setting u equal to the radicand.

3. If the integrand has the form

$$\dfrac{\text{Arcsin } a}{\sqrt{1 - a^2}} \quad \text{or} \quad \dfrac{\text{Arccos } a}{\sqrt{1 - a^2}} \quad \text{or} \quad \dfrac{\text{Arctan } a}{1 + a^2}$$

as in Example 2(d), try substituting u = Arcsin a, u = Arccos a, or u = Arctan a, respectively. ◀

Many integrals may be simplified using the sum rule (24-8).

EXAMPLE 3 Integrate: $\int (\cos^3 x - 3 \cos^2 x) \sin x \, dx$

Solution **First,** use (24-8) to write this as a difference of two integrals:

$$\int (\cos^3 x - 3 \cos^2 x) \sin x \, dx = \int \cos^3 x \sin x \, dx - 3 \int \cos^2 x \sin x \, dx$$

Second, let u = cos x; then du = −sin x dx, and these integrals can be written:

$$= -\int u^3 \, du + 3 \int u^2 \, du$$

Substitute −du for sin x dx.

Next, integrate using the power rule (24-10).

$$= -\frac{u^4}{4} + \frac{3u^3}{3} + C$$

Finally, rewrite in terms of the original functions.

$$= \cos^3 x - \frac{\cos^4 x}{4} + C$$

Notice that each integral in the second step has a constant of integration, but these are combined into a single constant in the next step.

EXAMPLE 4 The transient current in a certain circuit immediately after a power failure can be described by the equation

$$i = \frac{ae^{-kt}}{(1 - e^{-kt})^2}$$

where k = 2.0 s^{-1} and a = 250 A. Find the total charge q (in coulombs) moving past a point in the circuit between 0.10 s and 0.20 s, given that i = dq/dt.

Solution Since i = dq/dt, then: $q = \int_{0.10}^{0.20} i \, dt = \int_{0.10}^{0.20} \frac{250 e^{-2.0t} \, dt}{(1 - e^{-2.0t})^2}$

Step 1. Let u = 1 − e$^{-2.0t}$, then du = 2e$^{-2.0t}$ dt. Substitute u for (1 − e$^{-2.0t}$) and du/2 for e$^{-2.0t}$ dt in the integral.

$$q = 250 \int_{t=0.10}^{t=0.20} \frac{1}{u^2} \cdot \frac{du}{2}$$

Step 2. Change the limits. At $t = 0.10$, $u \approx 0.1813$, and at $t = 0.20$, $u \approx 0.3297$.

$$\approx 125 \int_{0.1813}^{0.3297} \frac{du}{u^2}$$

Step 3. Use the power rule (24-10) to integrate.

$$\approx 125 \left[-\frac{1}{u} \right]_{0.1813}^{0.3297} = 125 \left[\left(-\frac{1}{0.3297} \right) - \left(-\frac{1}{0.1813} \right) \right]$$

$$125 \;\boxed{\times}\;\boxed{(}\;.3297\;\boxed{1/x}\;\boxed{+/-}\;\boxed{+}\;.1813\;\boxed{1/x}\;\boxed{)}\;\boxed{=} \rightarrow \quad \boxed{310.33243}$$

$$\approx 310 \text{ C}$$

Note that this last step could also be written in terms of the original variable t. Since $u = 1 - e^{-2.0t}$,

$$125 \left[-\frac{1}{u} \right]_{0.1813}^{0.3297} = 125 \left(-\frac{1}{1 - e^{-2.0t}} \right) \Bigg]_{0.10}^{0.20}$$

$$= 125 \left[\left(-\frac{1}{1 - e^{-0.4}} \right) - \left(-\frac{1}{1 - e^{-0.2}} \right) \right]$$

$$\approx 125 \left[-\frac{1}{1 - 0.6703} + \frac{1}{1 - 0.8187} \right] \approx 310 \text{ C}$$

Exercises 27-1

Integrate. Round answers to calculator problems to three decimal places.

1. $\displaystyle \int \frac{dx}{(2x - 1)^2}$

2. $\displaystyle \int \frac{x \, dx}{\sqrt{1 - 3x^2}}$

3. $\displaystyle \int 2x\sqrt{x^2 + 2} \, dx$

4. $\displaystyle \int \frac{dx}{\sqrt{x - 1}}$

5. $\displaystyle \int \cos x \sin^3 x \, dx$

6. $\displaystyle \int \cos^2 x \sin x \, dx$

7. $\displaystyle \int_0^{\pi/4} \sqrt{\tan x} \sec^2 x \, dx$

8. $\displaystyle \int_{\pi/4}^{\pi/2} (\sin^2 x - 2 \sin^3 x) \cos x \, dx$

9. $\displaystyle \int (4 \cos x + \cos^4 x)\sin x \, dx$

10. $\displaystyle \int \frac{\sin x \, dx}{\cos^2 x}$

11. $\displaystyle \int \frac{\cos x \, dx}{(1 - \sin x)^3}$

12. $\displaystyle \int \frac{\sec^2 x \, dx}{(1 + \tan x)^2}$

13. $\displaystyle \int \frac{e^x \, dx}{(e^x + 2)^{3/2}}$

14. $\displaystyle \int e^x(e^x + 1)^2 \, dx$

15. $\displaystyle \int \frac{(\text{Arcsin } x)^2}{\sqrt{1 - x^2}} \, dx$

16. $\displaystyle \int \frac{\text{Arccos } 2x}{\sqrt{1 - 4x^2}} \, dx$

17. $\displaystyle \int \sqrt{\sin 2x} \cos 2x \, dx$

18. $\displaystyle \int \sin 3x \sqrt{\cos 3x} \, dx$

19. $\displaystyle \int \frac{\ln x \, dx}{x}$

20. $\displaystyle \int \frac{2 - \ln x}{2x} \, dx$

21. $\displaystyle \int \frac{e^{2t}}{(1 - e^{2t})^2} \, dt$

22. $\displaystyle \int \frac{1 - e^x}{\sqrt{x - e^x}} \, dx$

23. $\displaystyle \int_1^e \frac{(\ln x)^2}{x} \, dx$

24. $\displaystyle \int_{-3}^{-2} \frac{2e^t}{\sqrt{1 - 4e^t}} \, dt$

25. $\int_0^1 \dfrac{(1 - 2e^{-x})^2}{e^x}\, dx$ **26.** $\int_{\pi/8}^{3\pi/8} \dfrac{\csc^2 2\theta}{(4 + \cot 2\theta)^3}\, d\theta$ **27.** $\int \cos\theta(\sin^2\theta + \sin^{-2}\theta)\, d\theta$

28. $\int \dfrac{t - \sin 2t}{(t^2 + \cos 2t)^4}\, dt$

Solve.

29. *Electronics* If the charge stored in a capacitive circuit decreases at the rate

$$\frac{dq}{dt} = e^{-4t}(e^{-4t} - 1)^2$$

(in coulombs), what amount of charge is dissipated from $t = 0.50$ s to $t = 1.50$ s?

30. Determine the average value of the function $y = \sin\theta \cos^2\theta$ over the interval $\theta = 0$ to $\theta = \pi$.

31. Find the area of the region bounded by the curve $y = \dfrac{(\ln x)^3}{x}$ the x-axis, and the lines $x = 3$ and $x = 1$.

32. *Electrical engineering* If a circular disk of radius r carries a uniform electrical charge, then the electric potential on the axis of the disk at a point a from its center is given by the equation

$$V = k \int_0^r (x^2 + a^2)^{-1/2} x\, dx$$

where k is a constant depending on the charge density. Integrate to find V as a function of r and a.

33. *Physics* If an object moves in a straight line with acceleration $a = 10e^{-t}(1 + e^{-t})$, find an expression for the velocity of the object. [*Hint:* Use $v = \int a\, dt$.]

34. Find the RMS value of the function $y = e^{2x} - e^x$ from $x = 0$ to $x = 1$. Round to two decimal places.

27-2 | Integrating Exponential and Logarithmic Functions

Integrating the reciprocal function To every differentiation formula there corresponds a related integration formula. According to equation (26-13),

$$\frac{d}{dx}(\ln u) = \frac{1}{u} \cdot \frac{du}{dx} \qquad \text{for } u > 0$$

For $u < 0$, we may write

$$\frac{d}{dx}[\ln(-u)] = \frac{1}{-u} \cdot \frac{d(-u)}{dx} = \frac{1}{u} \cdot \frac{du}{dx} \qquad \text{for } u < 0$$

Then these two formulas may be combined as

$$\frac{d}{dx}(\ln|u|) = \frac{1}{u} \cdot \frac{du}{dx} \qquad \text{for } u \ne 0$$

This leads to the related integration formula (27-1):

Integration of the Reciprocal Function

$$\int \frac{du}{u} = \ln |u| + C \qquad u \neq 0 \tag{27-1}$$

Notice that this formula completes the power rule (24-10), which states

$$\int u^n \, du = \frac{u^{n+1}}{n + 1} + C \qquad n \neq -1$$

Formula (27-1) is used when $n = -1$.

EXAMPLE 5   Integrate: $\displaystyle\int \frac{dx}{2x + 1}$

Solution
Step 1.   Rewrite the integrand. Let $u = 2x + 1$; then $du = 2 \, dx$ or $dx = \dfrac{1}{2} \cdot du$.

Substitute. $$\int \frac{dx}{2x + 1} = \frac{1}{2} \int \frac{du}{u}$$

Step 2.   Use equation (27-1) to integrate. $$= \frac{1}{2} \ln |u| + C$$

$$= \frac{1}{2} \ln |2x + 1| + C$$

IMPORTANT ▶ The absolute value sign is important, since the logarithmic function is defined only for positive values of $2x + 1$. ◀

EXAMPLE 6   If a gas expands isothermally—that is, at constant temperature— from an initial volume V_1 to some final volume V_2, the work done by the expanding gas is

$$W = \int_{V_1}^{V_2} \frac{nRT}{V} \, dV$$

where n, R, and T are constants. For $nRT = 3200$ J, find the work done in expanding this gas from 14 L to 28 L.

Solution
Step 1.   Simplify the integral by bringing the constant nRT outside the integral sign. $$W = nRt \int_{V_1}^{V_2} \frac{dV}{V} = 3200 \int_{14}^{28} \frac{dV}{V}$$

Step 2. Integrate using equation (27-1).

$$= 3200(\ln |V|)\Big]_{14}^{28}$$

$$= 3200(\ln 28 - \ln 14)$$

The absolute value sign may be omitted when the variable is positive.

$$= 3200\left(\ln \frac{28}{14}\right) = 3200(\ln 2)$$

$$\approx 2200 \text{ J}$$

EXAMPLE 7 Integrate:

(a) $\displaystyle\int \frac{x^2 \, dx}{1 - x^3}$ (b) $\displaystyle\int_0^{\pi/6} \frac{\cos \theta \, d\theta}{1 - \sin \theta}$ (c) $\displaystyle\int \frac{e^{2x} \, dx}{1 - e^{2x}}$

Solution

(a) **Step 1.** Let $u = 1 - x^3$, then $du = -3x^2 \, dx$ or $x^2 \, dx = -\dfrac{1}{3} du$.

    Step 2. Substitute. $\displaystyle\int \frac{x^2 \, dx}{1 - x^3} = -\frac{1}{3} \int \frac{du}{u}$

    Use equation (27-1). $\displaystyle = -\frac{1}{3} \ln |u| + C$

Do not omit the constant of integration for an indefinite integral.

$$= -\frac{1}{3} \ln |1 - x^3| + C$$

(b) **First,** use substitution to rewrite the integrand in a form that allows us to integrate using a basic integration formula. Let $u = 1 - \sin \theta$; then by equation (26-3), $du = -\cos \theta \, d\theta$. The integral can be rewritten in the form $-\int du/u$.

Second, rewrite the limits of integration so that they are values of u. At $\theta = 0$, $u = 1 - \sin 0 = 1$, and at $\theta = \pi/6$, $u = 1 - \sin (\pi/6) = 1 - \dfrac{1}{2} = \dfrac{1}{2}$.

Next, substitute and integrate using equation (27-1):

$$\int_0^{\pi/6} \frac{\cos \theta \, d\theta}{1 - \sin \theta} = -\int_1^{1/2} \frac{du}{u}$$

$$= -\Big[\ln |u|\Big]_1^{1/2}$$

$$= -\left[\ln \frac{1}{2} - \ln 1\right] = -\ln \frac{1}{2}$$

$$= \ln 2 \approx 0.693$$

Note that we could also write the final result of this integration in terms of the original variable θ:

$$-(\ln |u|) \Big]_1^{1/2} = -(\ln |1 - \sin \theta|) \Big]_0^{\pi/6}$$

$$= -\left[\ln \left| 1 - \sin \frac{\pi}{6} \right| - \ln |1 - \sin 0| \right]$$

$$= -\ln |1 - 0.5| + \ln |1 - 0|$$

$$= -\ln 0.5 = \ln 2 \approx 0.693$$

(c)   **Step 1.**   Let $u = 1 - e^{2x}$, then by equation (26-15), $du = -2e^{2x} dx$ or $e^{2x} dx = -\dfrac{1}{2} du$.

Step 2.   Substitute. $\displaystyle\int \frac{e^{2x} dx}{1 - e^{2x}} = -\frac{1}{2} \int \frac{du}{u}$

Use equation (27-1). $= -\dfrac{1}{2} \ln |u| + C$

$$= -\frac{1}{2} \ln |1 - e^{2x}| + C$$

LEARNING HINT

If the integrand of an indefinite integral is always positive, or positive within the interval of integration for a definite integral, then the value of the integral will be positive. Use this fact to check your answer wherever possible. For example, in Example 7(b), the integrand

$$\frac{\cos \theta}{1 - \sin \theta} > 0 \qquad \text{for } 0 \le \theta \le \frac{\pi}{6}$$

Therefore, we know that the value of the integral will be positive. The answer obtained, $\ln 2$, is positive, and is therefore a reasonable answer.   ◀

Integrating the exponential function

If $u(x)$ is a differentiable function of x, then by equation (26-15),

$$\frac{d}{dx}(e^u) = e^u \frac{du}{dx}$$

This result allows us to find the integral of the exponential function.

Integral of the Exponential Function

$$\int e^u \, du = e^u + C \tag{27-2}$$

EXAMPLE 8   Integrate:

(a) $\int e^{4x} \, dx$ (b) $\int x e^{x^2} \, dx$ (c) $\displaystyle\int_0^2 \frac{e^{\sqrt{x}}}{\sqrt{x}} \, dx$

Solution

(a) **Step 1.** Let $u = 4x$; then $du = 4\,dx$ or $dx = \dfrac{1}{4}\,du$.

Step 2. Substitute in the original integral.　$\displaystyle\int e^{4x}\,dx = \frac{1}{4}\int e^{u}\,du$

Integrate using equation (27-2).　$= \dfrac{1}{4}e^{u} + C$

$= \dfrac{1}{4}e^{4x} + C$

(b) **Step 1.** Use substitution to rewrite the integrand as a basic integral form. Let $u = x^2$, then $du = 2x\,dx$, so that $x\,dx = \dfrac{1}{2}\,du$.

Substitute in the original integral.　$\displaystyle\int xe^{x^2}\,dx = \int e^{x^2}\cdot x\,dx$

$= \dfrac{1}{2}\int e^{u}\,du$

Step 2. Integrate using equation (27-2).　$= \dfrac{1}{2}e^{u} + C$

$= \dfrac{1}{2}e^{x^2} + C$

(c) **Step 1.** Use substitution to rewrite the integral as a basic form. Let $u = \sqrt{x} = x^{1/2}$, then

$$du = \frac{1}{2}x^{-1/2}\,dx = \frac{1}{2}\cdot\frac{1}{\sqrt{x}}\,dx$$

Therefore, $dx/\sqrt{x} = 2\,du$, and the integral has the form $\int e^{u}\,du$.

Step 2. Change the limits of integration from values of x to corresponding values of u. At $x = 0$, $u = \sqrt{0} = 0$, and at $x = 2$, $u = \sqrt{2}$.

Step 3. Substitute into the original integral and integrate using equation (27-2).

$$\int_{0}^{2}\frac{e^{\sqrt{x}}}{\sqrt{x}}\,dx = 2\int_{0}^{\sqrt{2}}e^{u}\,du$$

$$= 2(e^{u})\Big]_{0}^{\sqrt{2}}$$

$$= 2\left(e^{\sqrt{2}} - e^{0}\right)$$

$$= 2\left(e^{\sqrt{2}} - 1\right) \approx 6.227$$

Again, in terms of the original variable,

$$2(e^u)\Big]_0^{\sqrt{2}} = 2(e^{\sqrt{x}})\Big]_0^2$$

$$= 2(e^{\sqrt{2}} - e^0) = 2(e^{\sqrt{2}} - 1) \approx 6.227$$

EXAMPLE 9   Find the area bounded by the curve $y = e^x$, the straight line $y = x$, and the vertical lines $x = 0$ and $x = 3$. (Round to the nearest tenth of a square unit.)

Solution   **First,** sketch the graph (Fig. 27-1).

Figure 27-1

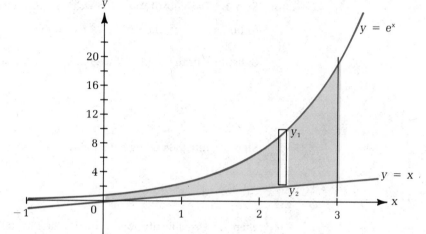

Second, set up an integral representing the given area:

$$A = \int_0^3 (y_1 - y_2)\, dx$$

$$= \int_0^3 (e^x - x)\, dx$$

Third, integrate. By equation (27-2),

$$\int_0^3 e^x\, dx = e^x\Big]_0^3$$

and by the power rule,

$$\int_0^3 x\, dx = \frac{x^2}{2}\Big]_0^3$$

Therefore,

$$A = \left(e^x - \frac{x^2}{2}\right)\Big]_0^3$$

$$= \left(e^3 - \frac{9}{2} \right) - (e^0 - 0)$$

$$= e^3 - \frac{9}{2} - 1$$

$3 \boxed{e^x} \boxed{-} 4.5 \boxed{-} 1 \boxed{=} \rightarrow \quad \boxed{14.585537}$

≈ 14.6 square units

Integrating a^u   For any positive real number a, if $u(x)$ is a differentiable function of x, then by equation (26-14),

$$\frac{d}{dx}(a^u) = a^u \ln a \, \frac{du}{dx}$$

and a corresponding integration formula can be obtained.

$$\int a^u \, du = \frac{a^u}{\ln a} + C \qquad a > 0 \qquad\qquad (27\text{-}3)$$

EXAMPLE 10   Integrate: (a) $\displaystyle\int 10^{x-1} \, dx$ (b) $\displaystyle\int_1^2 2^{3x} \, dx$

Solution
(a) **Step 1.**   Rewrite the integral as a standard form. This integral has the form of the left side of equation (27-3), where $a = 10$ and $u = x - 1$. Then $du = dx$, and the integral becomes

$$\int 10^{x-1} \, dx = \int 10^u \, du$$

Step 2.   Integrate using equation (27-3).

$$\int 10^u \, du = \frac{10^u}{\ln 10} + C$$

$$= \frac{10^{x-1}}{\ln 10} + C$$

(b) **Step 1.**   Use substitution to rewrite this integral as a standard form. Let $u = 3x$, then $du = 3 \, dx$ or $dx = \dfrac{1}{3} \, du$. Therefore, the integral has the form $\int a^u \, du$ with $a = 2$.

Step 2.   Change the limits of integration from values of x to values of u. At $x = 1$, $u = 3$, and at $x = 2$, $u = 6$. Substitute.

$$\int_1^2 2^{3x} \, dx = \frac{1}{3} \int_3^6 2^u \, du$$

Step 3. Integrate using equation (27-3):

$$\frac{1}{3}\int_3^6 2^u\,du = \frac{1}{3}\left(\frac{2^u}{\ln 2}\right)\Bigg]_3^6$$

$$= \frac{1}{3}\left(\frac{2^6}{\ln 2} - \frac{2^3}{\ln 2}\right)$$

$$= \frac{1}{3}\left(\frac{64 - 8}{\ln 2}\right)$$

$$\approx \frac{56}{3\ln 2} \approx 26.9$$

CAUTION ▶   Never confuse the exponential function a^x with the power function x^a.

$$a^x \qquad \neq \qquad x^a$$

Variable exponent: Integrate using equation (27-2) or equation (27-3).

Constant real number exponent: Integrate using the power rule (24-10).

◀

Exercises 27-2

Integrate. Round answers to calculator problems to two decimal places.

1. $\displaystyle\int \frac{dx}{1-x}$

2. $\displaystyle\int \frac{d\theta}{2+3\theta}$

3. $\displaystyle\int \frac{x\,dx}{1-2x^2}$

4. $\displaystyle\int \frac{x^2\,dx}{2x^3+1}$

5. $\displaystyle\int \frac{\sec^2\theta\,d\theta}{\tan\theta}$

6. $\displaystyle\int \frac{\sin 2t\,dt}{\cos 2t + 2}$

7. $\displaystyle\int \frac{dx}{x\ln x}$

8. $\displaystyle\int \frac{e^{4x}\,dx}{1+e^{4x}}$

9. $\displaystyle 3\int e^{3x}\,dx$

10. $\displaystyle 2\int xe^{3x^2}\,dx$

11. $\displaystyle\int e^{x/2}\,dx$

12. $\displaystyle\int e^{-x/3}\,dx$

13. $\displaystyle\int \frac{dx}{e^{2x}}$

14. $\displaystyle\int e^{x-1}\,dx$

15. $\displaystyle\int e^{2x-3}\,dx$

16. $\displaystyle\int \frac{e^{\sqrt[3]{x}}\,dx}{3\sqrt[3]{x^2}}$

 17. $\displaystyle\int_0^1 \frac{e^x\,dx}{\sqrt{1+e^x}}$

18. $\displaystyle\int (2-e^{2x})\,dx$

19. $\displaystyle\int (x+e^{3x})\,dx$

20. $\displaystyle\int (1-e^x)^2\,dx$

21. $\displaystyle\int (1+e^{-x})^2\,dx$

22. $\displaystyle\int e^{\cos 2t}\sin 2t\,dt$

 23. $\displaystyle\int_0^{\pi/3} e^{\tan x}\sec^2 x\,dx$

24. $\displaystyle\int_1^4 \frac{e^{2x}}{e^{2x}+1}\,dx$

25. $\displaystyle\int_0^2 2x^2 e^{x^3+1}\,dx$

26. $\displaystyle\int_0^{\pi/4} \frac{\sin\theta\,d\theta}{\cos\theta}$

27. $\displaystyle\int_1^2 \frac{e^{1/t}}{t^2}\,dt$

28. $\displaystyle\int_0^a \frac{e^{\text{Arctan}\,x}}{1+x^2}\,dx$

29. $\displaystyle\int 2^{4x}\,dx$

30. $\displaystyle\int 10^{2x+1}\,dx$

31. $\displaystyle\int_0^1 3^{2x}\, dx$ **32.** $\displaystyle\int_{-1}^1 4^{t+2}\, dt$ **33.** $\displaystyle\int \frac{4x+1}{x+2x^2}\, dx$

34. $\displaystyle\int_{0.1}^{0.2} \frac{x+3x^2}{x^2+2x^3}\, dx$ **35.** $\displaystyle\int_{\pi/4}^{\pi/2} \frac{1+2\sin 2x}{x-\cos 2x}\, dx$ **36.** $\displaystyle\int \frac{e^{3x}\, dx}{1+2e^{3x}}$

37. $\displaystyle\int_1^2 \frac{x}{1+3x^2}\, dx$ **38.** $\displaystyle\int_2^4 \frac{dx}{1+2x}$

Solve.

39. *Electronics* If the electric field E (in newtons per coulomb) set up in a capacitive control device increases so that $dE/dt = 1.5e^{0.3t}$, find the average rate of growth of the field strength from $t = 0$ s to $t = 3.0$ s.

40. Determine the volume of the solid generated by rotating about the x-axis the region bounded by $y = 2\sqrt{x}\, e^{x^2}$, the x-axis, and the line $x = 1$. Round to the nearest tenth.

41. Calculate the area under the curve $y = 4.0e^{-0.01x}$ from $x = 0$ to $x = 12$. Round to the nearest hundredth.

42. *Physics* A falling parachutist will experience a resisting force a due to air friction that is approximately proportional to the speed v. Therefore, $a = dv/dt = g - kv$, where g and k are constants. Find v as a function of t (time). [*Hint:* Integrate $t = \int dv/(g - kv)$ and solve for v.]

43. *Nuclear engineering* Integrals of the following kind appear in the analysis of failure rates and reliability in nuclear reactor design. Integrate to find $F(x)$.

$$F(x) = \int_0^x 12te^{-3t^2}\, dt$$

44. *Electronics* If the current in a circuit varies with time according to the equation $i = 10e^{-t/2}$, find the charge flowing through the circuit from $t = 0$ s to $t = 0.2$ s.

45. Determine the area between the curves $y = 1.5^x$ and $y = \sqrt{x}$ from $x = 0$ to $x = 2$. Round to the nearest hundredth.

46. Determine the area enclosed by the curves $y = 2x + 1$, $x = 1$, and $y = e^x$. Round to the nearest hundredth.

47. Calculate the volume of the solid of rotation formed by rotating about the x-axis the region bounded by the curve $y = \sqrt{x}\, 2^{x^2+1}$ and the lines $x = 0$ and $x = 1$. Round to the nearest hundredth.

48. *Physics* Newton's law of cooling states that the rate dT/dt at which an object changes temperature is proportional to the difference between its temperature T and that of its surroundings T_o. $dT/dt = -k(T - T_o)$. (Note that the negative sign is needed since T decreases with time if $T > T_o$ and increases with time if $T < T_o$.) Find the time t, in minutes, required for the temperature of an object to drop from 45°C to 25°C, for $k = 0.025$ and $T_o = 20$°C.

$$\left[\text{Hint:}\quad t = -\frac{1}{k}\int_{45}^{25} \frac{dT}{T - T_o} \right]$$

27-3 | Integrating Basic Trigonometric Functions

Because equation (24-4) defines the indefinite integral in terms of an antiderivative, every derivative formula will have a corresponding integration formula. In particular, we can immediately write integration formulas corresponding to differentiation of each of the basic trigonometric functions (26-3) through (26-8). In each of the following formulas, u is some differentiable function of x.

Differentiation Formula	Integration Formula	
$\dfrac{d}{du}(\sin u) = \cos u$	$\displaystyle\int \cos u \; du = \sin u + C$	(27-4)
$\dfrac{d}{du}(\cos u) = -\sin u$	$\displaystyle\int \sin u \; du = -\cos u + C$	(27-5)
$\dfrac{d}{du}(\tan u) = \sec^2 u$	$\displaystyle\int \sec^2 u \; du = \tan u + C$	(27-6)
$\dfrac{d}{du}(\cot u) = -\csc^2 u$	$\displaystyle\int \csc^2 u \; du = -\cot u + C$	(27-7)
$\dfrac{d}{du}(\csc u) = -\csc u \cot u$	$\displaystyle\int \csc u \cot u \; du = -\csc u + C$	(27-8)
$\dfrac{d}{du}(\sec u) = \sec u \tan u$	$\displaystyle\int \sec u \tan u \; du = \sec u + C$	(27-9)

EXAMPLE 11 Integrate: $\displaystyle\int 4 \cos x \; dx$

Solution **First,** use equation (24-5) to write: $\displaystyle\int 4 \cos x \; dx = 4 \int \cos x \; dx$

Then, integrate using equation (27-4). $= 4 \sin x + C$

Notice that we do not write the result as $4(\sin x + C) = 4 \sin x + 4C$. The factor 4 is included in the integration constant C.

EXAMPLE 12 In a resistanceless *LC* circuit, the current oscillates so that between time $t = 0$ and $t = a$, the charge accumulating on the capacitor is given by the integral

$$q = q_m \int_0^a \omega \sin (\omega t + \phi) \; dt$$

Integrate to find the total charge accumulating on the capacitor between $t = 0$ s and $t = 1.0$ s if $\omega = \pi/2$ s^{-1} and $\phi = \pi/4$.

Solution
Step 1. Use substitution to rewrite the integral in a basic form. Let $u = \omega t + \phi$, then $du = \omega \; dt$. The integral now has the form $\int \sin u \; du$.

Step 2. Change the limits of integration from values of t to corresponding values of u. At $t = 0$, $u = \phi = \pi/4$, and at $t = 1$, $u = \omega + \phi = (\pi/2) + (\pi/4) = 3\pi/4$.

Step 3. Substitute to rewrite the integral; then integrate using equation (27-5):

$$q = q_m \int_0^1 \omega \sin(\omega t + \phi) \, dt$$

$$= q_m \int_{\pi/4}^{3\pi/4} \sin u \, du$$

$$= q_m(-\cos u) \Big]_{\pi/4}^{3\pi/4}$$

$$= q_m\left[\left(-\cos\frac{3\pi}{4}\right) - \left(-\cos\frac{\pi}{4}\right)\right] = q_m\left[-\left(-\frac{\sqrt{2}}{2}\right) + \left(\frac{\sqrt{2}}{2}\right)\right]$$

$$= q_m\sqrt{2}$$

$$\boxed{\frac{\sqrt{2}}{2} + \frac{\sqrt{2}}{2} = \frac{2\sqrt{2}}{2} = \sqrt{2}}$$

As a check on the integration, note that sin u is positive for $\pi/4 \leq u \leq 3\pi/4$. Therefore, we expect the integral to be positive, and $\sqrt{2}$ is indeed positive.

EXAMPLE 13   Integrate: $\int x \sec^2(x^2) \, dx$

Solution   **First,** let $u = x^2$, then $du = 2x \, dx$ or $x \, dx = \dfrac{1}{2} du$.

Next, substitute to rewrite the integral in a basic form.

$$\int \sec^2(x^2)x \, dx = \frac{1}{2}\int \sec^2 u \, du$$

Finally, use integration formula (27-6).

$$= \frac{1}{2}\tan u + C$$

$$= \frac{1}{2}\tan x^2 + C$$

EXAMPLE 14   Integrate: $\int \dfrac{\tan(1-x)}{\cos(1-x)} \, dx$

Solution   **First,** recall that

$$\sec(1-x) = \frac{1}{\cos(1-x)}$$

Therefore, this integral can be rewritten as

$$\int \sec(1-x)\tan(1-x) \, dx$$

Next, Let $u = 1 - x$, then $du = -dx$, and the integral becomes:

$$= -\int \sec u \tan u \, du$$

Finally, use (27-9) to integrate.

$$= -\sec u + C$$

$$= -\sec(1-x) + C$$

Equations (27-4) and (27-5) allow us to integrate two of the six basic trigonometric functions. We can obtain integration formulas for the remaining trigonometric functions by using these formulas and a few trigonometric identities.

Integrating tan x

To evaluate $\int \tan x \, dx$, first recall that by equation (19-4),

$$\tan x = \frac{\sin x}{\cos x}$$

so that

$$\int \tan x \, dx = \int \frac{\sin x \, dx}{\cos x}$$

Using the substitution method, if $v = \cos x$, then $dv = -\sin x \, dx$, and the integral becomes

$$\int \frac{\sin x \, dx}{\cos x} = -\int \frac{dv}{v} = -\ln |v| + C$$

$$= -\ln |\cos x| + C$$

$$\int \tan u \, du = -\ln |\cos u| + C \qquad\qquad (27\text{-}10)$$

where $u = u(x)$ is a differentiable function of x.

Integrating cot x

Evaluate $\int \cot x \, dx$ in the same way using equation (19-5):

$$\cot x = \frac{\cos x}{\sin x}$$

$$\int \cot x \, dx = \int \frac{\cos x \, dx}{\sin x}$$

For $v = \sin x$, $dv = \cos x \, dx$, and the integral becomes

$$\int \frac{\cos x \, dx}{\sin x} = \int \frac{dv}{v} = \ln |v| + C$$

$$= \ln |\sin x| + C$$

$$\int \cot u \, du = \ln |\sin u| + C \qquad\qquad (27\text{-}11)$$

Integrating sec x

To evaluate $\int \sec x \, dx$, we must first rewrite the integrand as a fraction whose numerator is the derivative of the denominator. The trigonometric identity (19-7) can be used to rewrite sec x.

From (19-7), $$\sec^2 x - \tan^2 x = 1$$

Factor. $$(\sec x - \tan x)(\sec x + \tan x) = 1$$

Then, $$\sec x - \tan x = \frac{1}{\sec x + \tan x}$$

$$\sec x = \tan x + \frac{1}{\sec x + \tan x}$$

$$= \frac{\sec x \tan x + \tan^2 x + 1}{\sec x + \tan x}$$

$$= \frac{\sec x \tan x + \sec^2 x}{\sec x + \tan x}$$

But, if $v = \sec x + \tan x$, then $dv = \sec x \tan x + \sec^2 x$. Therefore, the integral $\int \sec x \, dx$ has the form

$$\int \frac{dv}{v} = \ln |v| + C$$

If u is a differentiable function of x and $\sec u + \tan u \neq 0$, this leads to the following integration formula:

$$\int \sec u \, dx = \ln |\sec u + \tan u| + C \qquad \text{(27-12)}$$

Integrating csc x   The integral $\int \csc x \, dx$ can be evaluated in exactly the same way using trigonometric identity (19-8):

$$\int \csc x \, dx = \int \frac{(\csc^2 x - \csc x \cot x) \, dx}{\csc x - \cot x}$$

$$= \int \frac{dv}{v} \qquad \text{Where } v = \csc x - \cot x$$

$$= \ln |v| + C$$

If u is a differentiable function of x and $\csc u - \cot u \neq 0$, this gives the following integration formula:

$$\int \csc u \, du = \ln |\csc u - \cot u| + C \qquad \text{(27-13)}$$

EXAMPLE 15 Integrate:

(a) $\displaystyle\int \sec(4x + 1)\, dx$ (b) $\displaystyle\int e^{3x} \tan(e^{3x})\, dx$ (c) $\displaystyle\int_{\pi/2}^{\pi} \cot \frac{x}{2}\, dx$

(d) Find $A(\theta) = \displaystyle\int \csc(2\theta)\, d\theta$ if $A(\theta) = 1$ at $\theta = \pi/4$.

Solution

(a) **First,** use substitution to rewrite the integral in a basic trigonometric form.

If $u = 4x + 1$, then $du = 4\, dx$ or $dx = \dfrac{1}{4} \cdot du$.

Substitute.

$$\int \sec(4x + 1)\, dx = \frac{1}{4} \int \sec u\, du$$

Then, use equation (27-12) to integrate.

$$= \frac{1}{4} \ln |\sec u + \tan u| + C$$

$$= \frac{1}{4} \ln |\sec(4x + 1) + \tan(4x + 1)| + C$$

(b) **First,** note that if $u = e^{3x}$, then $du = 3e^{3x}\, dx$.

Substitute.

$$\int e^{3x} \tan(e^{3x})\, dx = \frac{1}{3} \int \tan u\, du$$

Then, integrate using equation (27-10).

$$= -\frac{1}{3} \ln |\cos u| + C$$

$$= -\frac{1}{3} \ln |\cos(e^{3x})| + C$$

(c) **Step 1.** Use substitution to rewrite the integrand. If $u = x/2$, then $du = \dfrac{1}{2}\, dx$ or $dx = 2\, du$, and the integral has the form $\displaystyle\int \cot u\, du$.

Step 2. Change the limits of integration. At $x = \pi/2$, $u = \pi/4$, and at $x = \pi$, $u = \pi/2$.

Substitute.

$$\int_{\pi/2}^{\pi} \cot \frac{x}{2}\, dx = 2 \int_{\pi/4}^{\pi/2} \cot u\, du$$

Step 3. Use equation (27-11) to integrate.

$$= 2(\ln |\sin u|) \Big]_{\pi/4}^{\pi/2}$$

$$= 2 \left[\ln\left(\sin \frac{\pi}{2}\right) - \ln\left(\sin \frac{\pi}{4}\right) \right]$$

> Omit the absolute value symbols since $\sin(\pi/2)$ and $\sin(\pi/4)$ are both positive.

$$= 2 \left[\ln 1 - \ln \frac{\sqrt{2}}{2} \right] = -2 \ln \frac{\sqrt{2}}{2}$$

$$= -\ln\left(\frac{\sqrt{2}}{2}\right)^2 = -\ln\frac{1}{2}$$

$$= \ln 2$$

$$\approx 0.693$$

(d) Let $u = 2\theta$, then $du = 2\,d\theta$.
Substitute.

$$A(\theta) = \int \csc(2\theta)\,d\theta$$

$$= \frac{1}{2}\int \csc u\,du$$

Use equation (27-13).

$$= \frac{1}{2}\ln|\csc u - \cot u| + C$$

$$= \frac{1}{2}\ln|\csc(2\theta) - \cot(2\theta)| + C$$

At $\theta = \pi/4$, we have:

$$1 = \frac{1}{2}\ln\left|\csc\frac{\pi}{2} - \cot\frac{\pi}{2}\right| + C$$

$$1 = \frac{1}{2}\ln|1 - 0| + C$$

$$1 = \frac{1}{2}\ln 1 + C$$

But $\ln 1 = 0$, so:

$$1 = 0 + C$$

$$1 = C$$

Therefore:

$$A(\theta) = \frac{1}{2}\ln|\csc(2\theta) - \cot(2\theta)| + 1$$

Exercises 27-3

Integrate. Round answers to calculator problems to the nearest hundredth.

1. $\displaystyle\int \sin 3x\,dx$

2. $\displaystyle 4\int \tan \pi x\,dx$

3. $\displaystyle\int \sec(-2t)\,dt$

4. $\displaystyle\int \cot 2x\,dx$

5. $\displaystyle\int \sec^2\left(\frac{x}{2}\right)dx$

6. $\displaystyle\int \csc^2 5\theta\,d\theta$

7. $\displaystyle\int_0^{\pi/3} \cos 4x\,dx$

8. $\displaystyle\int x\cos(\pi x^2)\,dx$

9. $\displaystyle\int_0^{3\pi/16} \tan 2\theta\,d\theta$

10. $\displaystyle\int \sec(3x - 2)\,dx$

11. $\displaystyle\int \cos(x + 2)\,dx$

12. $\displaystyle\int \csc\theta\cot\theta\,d\theta$

13. $\displaystyle\int \sec 2t\tan 2t\,dt$

14. $\displaystyle\int x^2\sin 2x^3\,dx$

15. $\displaystyle\int x\tan x^2\,dx$

16. $\displaystyle\int \frac{\sin\sqrt{2x}}{\sqrt{x}}\,dx$

17. $\displaystyle\int \frac{\cos x}{\sin^2 x}\,dx$

18. $\displaystyle\int e^x\sin e^x\,dx$

19. $\displaystyle\int \frac{\sin x}{\cos^3 x}\,dx$

20. $\displaystyle\int xe^{x^2}\cos e^{x^2}\,dx$

21. $\displaystyle\int \frac{\sin(\ln x)}{x}\,dx$

22. $\int \dfrac{\cot(\ln x)}{x}\,dx$ ▦ **23.** $\int_0^{\pi/4} x\sec x^2 \tan x^2\,dx$ ▦ **24.** $\int_{0.5}^{1.5} \dfrac{\sin(1/t)}{t^2}\,dt$

Solve.

▦ **25.** *Physics* The work done in changing the orientation of an electric dipole from θ_0 to θ_1 in an external electric field is $W = \int_{\theta_0}^{\theta_1} \tau\,d\theta$, where $\tau(\theta)$ is the torque exerted by an outside force. Find W (in Newton meters) if $\tau = k\sin 2\theta$ and $k = 0.002$ N·m for $\theta_0 = 0$, $\theta_1 = 1.0$.

26. *Planetary science* In 1609, in his book *Astronomia Nova*, the astronomer Johannes Kepler calculated the dimensions of the orbit of Mars. His lengthy calculations were equivalent to finding the value of the integral shown. Integrate to find d.

$$d = \frac{\pi}{180} \int_{0°}^{15°} \sin x\,dx$$

▦ **27.** Find the average value of $y = \tan x$ on the interval $x = 0$ to $x = \pi/4$. Round to the nearest hundredth.

▦ **28.** *Electronics* If the current in an RC circuit varies according to the equation $i = 2.50\sin 60\pi t$, find the voltage V_c across a 500 μF ($= 5.00 \times 10^{-4}$ F) capacitor after 0.150 s, if $V_c = 0$ at $t = 0$.

$$\left[\text{Hint: } V_c = \frac{1}{C}\int_0^{0.15} i\,dt\right]$$

▦ **29.** Find the volume of the solid of revolution obtained by revolving about the x-axis the region under the graph of $y = \sec x$ from $x = 0$ to $x = \pi/4$. Round to the nearest hundredth.

▦ **30.** Find the volume of the solid of revolution obtained by revolving about the x-axis the region bounded by $y = 2\sec x$, $x = 0$, $x = 1$, and the x-axis. Round to the nearest tenth.

▦ **31.** Find the area of the region bounded by the curve

$$y = \csc\left(1 + \frac{x}{2}\right)$$

the x-axis, and the lines $x = 1$ and $x = 4$. Round to the nearest hundredth.

▦ **32.** Find the area of the region bounded by the curve $y = e^{x/4}\tan(e^{x/4})$, the x-axis, the y-axis, and the line $x = 1$. Round to the nearest hundredth.

▦ **33.** *Physics* If the acceleration of a moving particle is given by the equation $a = \cos(1 - 2t)$, find the velocity of the particle if it is at rest at time $t = 0$.

▦ **34.** *Electronics* The current in an RL circuit varies with the impressed voltage according to the integral

$$i(t) = \frac{1}{L}\int V(t)\,dt$$

If $i = 0$ at $t = 0$ and $L = 2.0$ H, find $i(t)$ for

$$V(t) = 10.0\cos\left(120\pi t + \frac{\pi}{6}\right).$$

27-4 | Integrating Products of Trigonometric Functions

The method of substitution and the trigonometric identities of Chapter 19 may be used to evaluate many integrals involving products of trigonometric functions. The following identities are most useful:

Trigonometric Identities

$$\cos^2 x + \sin^2 x = 1 \qquad\qquad\qquad (27\text{-}14)$$

$$\tan^2 x + 1 = \sec^2 x \qquad\qquad\qquad (27\text{-}15)$$

$$\cot^2 x + 1 = \csc^2 x \qquad\qquad\qquad (27\text{-}16)$$

$$\cos^2 x = \frac{1}{2}(1 + \cos 2x) \qquad\qquad\qquad (27\text{-}17)$$

$$\sin^2 x = \frac{1}{2}(1 - \cos 2x) \qquad\qquad\qquad (27\text{-}18)$$

Using these identities, we will examine the following types of integrals:

$$\int \sin^m u \cos^n u \, du \qquad \int \tan^m u \sec^n u \, du \qquad \int \cot^m u \csc^n u \, du$$

The exponents m and n are nonnegative integers, and, in each case, one of them may equal zero. For each type of integral the process of integration will depend on whether the exponents m and n are odd, even, or zero. But in each situation a trigonometric identity can be used to transform the integral into a form that can be integrated using the techniques already developed.

Type A: **Case 1.** If the exponent n is an **odd** positive integer, use identity (27-14) to
$\int \sin^n u \, du$ rewrite the integrand, and use substitution with $u = \cos x$ for the first integral
or $\int \cos^n u \, du$ or $u = \sin x$ for the second integral.

EXAMPLE 16 Integrate: (a) $\displaystyle\int \sin^5 x \, dx$ (b) $\displaystyle\int \cos^3 4x \, dx$

Solution

(a) **First,** since the exponent is odd, factor
out sin x and group it
with the differential
dx.

$$\int \sin^5 x \, dx = \int (\sin^4 x)(\sin x \, dx)$$

Second, use (27-14) to
rewrite $(\sin^4 x)$ in
terms of cos x.

$$= \int (\sin^2 x)^2 (\sin x \, dx)$$

$$= \int (1 - \cos^2 x)^2 (\sin x \, dx)$$

Third, use substi-
tution with $u = \cos x$;
then by (26-4),
$du = -\sin x \, dx$.

$$= -\int (1 - u^2)^2 \, du$$

Finally, expand the integrand, and integrate using the power rule (24-10).

$$= -\int (1 - 2u^2 + u^4)\, du$$

$$= -u + \frac{2u^3}{3} - \frac{u^5}{5} + C$$

$$= -\cos x + \frac{2}{3} \cos^3 x - \frac{1}{5} \cos^5 x + C$$

(b)   Follow the same steps used in part (a).

$$\int \cos^3 4x\, dx = \int (\cos^2 4x)(\cos 4x\, dx)$$

$$= \int (1 - \sin^2 4x)(\cos 4x\, dx)$$

Let $u = \sin 4x$, then by (26-3), $du = 4 \cos 4x\, dx$ or $\cos 4x\, dx = \frac{1}{4} du$.

$$= \int (1 - u^2)\left(\frac{1}{4}\, du\right)$$

$$= \frac{1}{4} \int (1 - u^2)\, du$$

$$= \frac{u}{4} - \frac{u^3}{12} + C$$

$$= \frac{1}{4} \sin 4x - \frac{1}{12} \sin^3 4x + C$$

Case 2.   If the exponent n is an **even** positive integer, use either identity (27-17) or (27-18) to rewrite the integrand; then integrate using the formulas (27-4) or (27-5) and substitution.

EXAMPLE 17   Integrate: $\int \sin^4 x\, dx$

Solution   First, write the integrand in terms of $\sin^2 x$ and use identity (27-18) to transform it into a sum of cosine functions.

$$\int \sin^4 x\, dx = \int (\sin^2 x)^2\, dx$$

$$= \int \left[\frac{1}{2}(1 - \cos 2x)\right]^2 dx$$

Expand.

$$= \frac{1}{4} \int (1 - 2\cos 2x + \cos^2 2x)\, dx$$

Second, if any $\cos^2$ terms appear, use identity (27-17) to rewrite them. Here,

$$\cos^2 2x = \frac{1}{2} \cdot (1 + \cos 2 \cdot 2x).$$

$$= \frac{1}{4} \int \left[1 - 2\cos 2x + \frac{1}{2}(1 + \cos 4x)\right] dx$$

Finally, write as a sum of integrals and integrate each using (27-4) and substitution if necessary.

$$= \frac{1}{4} \int dx - \frac{1}{2} \int \cos 2x \, dx + \frac{1}{8} \int dx$$
$$+ \frac{1}{8} \int \cos 4x \, dx$$

$$= \frac{x}{4} - \frac{1}{4} \int \cos u \, du + \frac{x}{8} + \frac{1}{32} \int \cos v \, dv$$

$$\boxed{\begin{array}{l} u = 2x \\ du = 2 \, dx \end{array}} \qquad \boxed{\begin{array}{l} v = 4x \\ dv = 4 \, dx \end{array}}$$

Use (27-4) to evaluate the integrals.

$$= \frac{3x}{8} - \frac{1}{4} \sin 2x + \frac{1}{32} \sin 4x + C$$

IMPORTANT ▶

Because some integrals can be evaluated using more than one of the techniques shown, it is important that you check your answer by differentiating. For example, the integral $I = \int \cos x \sin x \, dx$ can be evaluated in several different ways:

- Using (26-3), $I = \int \sin x \, d(\sin x) = \frac{1}{2} \sin^2 x + C_1$
- Using (26-4), $I = -\int \cos x \, d(\cos x) = -\frac{1}{2} \cos^2 x + C_2$
- Using (19-14), $I = \frac{1}{2} \int \sin 2x \, dx = -\frac{1}{4} \cos 2x + C_3$

These answers appear to be different, but because the constants are different, the answers are equivalent. Differentiating will show that $dI/dx = \cos x \sin x$ for all three answers. Always **check** your answers. ◀

EXAMPLE 18   If the current in an RCL circuit varies so that $i = i_m \sin \omega t$, where i_m is the maximum value of the current and ω is a constant, find the root mean square (RMS) value of i over one cycle of the current—that is, from $t_0 = 0$ s to $t_1 = 2\pi/\omega$ s.

Solution   As shown in Section 25-5, the RMS value of a function $F(x)$ is defined as

$$F_{RMS} = \left\{ \frac{1}{b-a} \int_a^b [F(x)]^2 \, dx \right\}^{1/2}$$

Therefore,

$$i_{RMS} = \left\{ \frac{1}{\dfrac{2\pi}{\omega} - 0} \int_0^{t_1} [i_m \sin \omega t]^2 \, dt \right\}^{1/2}$$

$$= \left\{ \frac{i_m^2 \omega}{2\pi} \int_0^{t_1} \sin^2 \omega t \, dt \right\}^{1/2}$$

This expression contains a type A integral with an even exponent. Substitute into the integral using equation (27-18):

$$\int_0^{t_1} \sin^2 \omega t \, dt = \frac{1}{2} \int_0^{t_1} (1 - \cos 2\omega t) \, dt$$

Let $u = 2\omega t$, then $du = 2\omega \, dt$ or $dt = (1/2\omega) \, du$. Change the limits: At $t_0 = 0$, $u = 0$, and at $t_1 = 2\pi/\omega$, $u = 4\pi$. Substitute:

$$\frac{1}{2} \int_0^{t_1} (1 - \cos 2\omega t) \, dt = \frac{1}{2} \int_0^{2\pi/\omega} dt - \frac{1}{4\omega} \int_0^{4\pi} \cos u \, du$$

$$= \frac{1}{2}(t) \Big]_0^{2\pi/\omega} - \frac{1}{4\omega}(\sin u) \Big]_0^{4\pi}$$

$$= \frac{1}{2}\left(\frac{2\pi}{\omega} - 0\right) - \frac{1}{4\pi}(\sin 4\pi - \sin 0)$$

$$= \frac{\pi}{\omega}$$

Therefore,

$$i_{RMS} = \left\{ \frac{i_m^2 \omega}{2\pi} \cdot \frac{\pi}{\omega} \right\}^{1/2} = \sqrt{\frac{i_m^2}{2}}$$

$$= \frac{i_m}{\sqrt{2}} \approx 0.707 i_m$$

The RMS value of the current is often called the "effective" current. Alternating current instruments such as ammeters and voltmeters are calibrated to read i_{RMS} and V_{RMS}. The RMS voltage at a household electrical outlet is $V_{RMS} \approx 120$ V, and this corresponds to a maximum voltage of $V_m = V_{RMS}\sqrt{2} \approx 170$ V.

Type B: **Case 1.**   If at least one of the exponents m or n is an **odd** positive integer, use
$\int \sin^m u \cos^n u \, du$ the trigonometric identity (27-14) to rewrite the integrand in a form that allows substitution to be used.

EXAMPLE 19   Integrate: (a) $\displaystyle\int \sin^3 x \cos^2 x \, dx$ (b) $\displaystyle\int \sin^5 2x \cos^3 2x \, dx$

Solution

(a)   **First,** factor out sin x from the odd exponent and group it with the differential dx.

$$\int \sin^3 x \cos^2 x \, dx = \int (\sin^2 x)(\cos^2 x)(\sin x \, dx)$$

Second, use (27-14) to rewrite the factor remaining from the original odd-exponent term.

$$= \int (1 - \cos^2 x)(\cos^2 x)(\sin x \, dx)$$

Third, use substitution to simplify this integrand, and then integrate using the power rule (24-10). Let $u = \cos x$, then $du = -\sin x\, dx$ or $\sin x\, dx = -du$, and the integral becomes:

$$= -\int (1 - u^2)u^2\, du$$

$$= -\int (u^2 - u^4)\, du$$

$$= -\frac{u^3}{3} + \frac{u^5}{5} + C$$

$$= -\frac{1}{3}\cos^3 x + \frac{1}{5}\cos^5 x + C$$

(b) If both exponents are odd positive integers, use the factor with the smaller exponent to form the new differential. Then proceed as in part (a).

$$\int \sin^5 2x \cos^3 2x\, dx = \int (\sin^5 2x)(\cos^2 2x)(\cos 2x\, dx)$$

Use (27-14).

$$= \int (\sin^5 2x)(1 - \sin^2 2x)(\cos 2x\, dx)$$

Let $u = \sin 2x$; then $du = 2\cos 2x\, dx$ or $\frac{1}{2}du = \cos 2x\, dx$.

$$= \frac{1}{2}\int u^5(1 - u^2)\, du$$

$$= \frac{1}{2}\int (u^5 - u^7)\, du$$

$$= \frac{u^6}{12} - \frac{u^8}{16} + C$$

$$= \frac{1}{12}\sin^6 2x - \frac{1}{16}\sin^8 2x + C$$

CAUTION ▶ Notice that these methods of integration are valid only when each function has the same argument.

$$\int \sin^m u \cos^n u\, du$$

Must be the same. ◀

The method of integration used in Example 19 is valid if either one of the exponents is an odd positive integer. The other exponent may be any real number.

EXAMPLE 20   Integrate: $\int \sin^{1/2}x \cos^3 x \, dx$

Solution

$$\int \sin^{1/2}x \cos^3 x \, dx = \int (\sin^{1/2}x)(\cos^2 x)(\cos x \, dx)$$

Use (27-14).

$$= \int (\sin^{1/2}x)(1 - \sin^2 x)(\cos x \, dx)$$

Let u = sin x, then
du = cos x dx.

$$= \int u^{1/2}(1 - u^2) \, du$$

$$= \int (u^{1/2} - u^{5/2}) \, du$$

$$= \frac{u^{3/2}}{\frac{3}{2}} - \frac{u^{7/2}}{\frac{7}{2}} + C$$

$$= \frac{2}{3} \sin^{3/2}x - \frac{2}{7} \sin^{7/2}x + C$$

Case 2. If both exponents m and n are **even** positive integers, use identity (27-14) to rewrite the factor with the smaller exponent. This will enable you to express the integrand as a sum of even powers that can be integrated as in Example 17.

EXAMPLE 21   Integrate: $\int \sin^2 x \cos^2 x \, dx$

Solution

$$\int \sin^2 x \cos^2 x \, dx = \int (\sin^2 x)(1 - \sin^2 x) \, dx$$

$$= \int (\sin^2 x - \sin^4 x) \, dx$$

$$= \int \sin^2 x \, dx - \int \sin^4 x \, dx$$

These are both type A,
case 2 integrals and can be
integrated using the techniques already shown.

$$= \frac{x}{8} - \frac{1}{32} \sin 4x + C$$

Type C:
$\int \tan^m u \sec^n u \, du$
or $\int \cot^m u \csc^n u \, du$

Not all integrals of this form can be found using the methods shown in this section, but the following cases cover most integrals of this kind.

Case 1A. If the exponent m of the tangent or cotangent function is an **odd** positive integer, use identity (27-15) or (27-16) to rewrite the integrand. The first step is to separate out a factor of the form (sec u tan u du) or (csc u cot u du).

EXAMPLE 22 Integrate: $\int \tan^3 4x \sec^2 4x \, dx$

Solution **First,** factor out sec 4x tan 4x dx, and use identity (27-15) to write the remaining factors in terms of sec 4x.

$$\int \tan^3 4x \sec^2 4x \, dx = \int (\tan^2 4x)(\sec 4x) \cdot (\sec 4x \tan 4x \, dx)$$

$$= \int (\sec^2 4x - 1)(\sec 4x) \cdot (\sec 4x \tan 4x \, dx)$$

Next, let u = sec 4x, then by (26-8), du = 4 sec 4x tan 4x dx, and the integral becomes:

$$= \frac{1}{4} \int (u^2 - 1)u \, du$$

$$= \frac{1}{4} \int (u^3 - u) \, du$$

$$= \frac{u^4}{16} - \frac{u^2}{8} + C$$

$$= \frac{1}{16} \sec^4 4x - \frac{1}{8} \sec^2 4x + C$$

Case 1B. If the exponent n of the secant or cosecant function is an **even** positive integer, use identity (27-15) or (27-16) to rewrite the integrand. The first step is to separate out a factor of the form $\sec^2 u \, du$ or $\csc^2 u \, du$.

EXAMPLE 23 Integrate: $\int \cot^2 x \csc^4 x \, dx$

Solution **First,** factor out csc²x dx, and use identity (27-16) to write the remaining factors in terms of cot x.

$$\int \cot^2 x \csc^4 x \, dx = \int (\cot^2 x)(\csc^2 x)(\csc^2 x \, dx)$$

$$= \int (\cot^2 x)(\cot^2 x + 1)(\csc^2 x \, dx)$$

Next, let u = cot x, then by (26-6), du = −csc²x dx, and the integral becomes:

$$= -\int u^2(u^2 + 1) \, du$$

$$= -\int (u^4 + u^2) \, du$$

$$= -\frac{u^5}{5} - \frac{u^3}{3} + C$$

$$= -\frac{1}{5} \cot^5 x - \frac{1}{3} \cot^3 x + C$$

If m is an odd positive integer and n is an even positive integer, then treat the integral as an example of case 1B.

Case 2. If the exponent n of the secant or cosecant factor is equal to zero, the integrals become

$$\int \tan^m u \; du \quad \text{and} \quad \int \cot^m u \; du$$

- If m is an **odd** positive integer, factor the integrand as

$$\tan^m u = \tan u \cdot \tan^{m-1} u \quad \text{or} \quad \cot^m u = \cot u \cdot \cot^{m-1} u$$

Then use (27-15) or (27-16) to rewrite the factor with the even exponent.
- If m is an **even** positive integer, factor the integrand as

$$\tan^m u = \tan^2 u \cdot \tan^{m-2} u \quad \text{or} \quad \cot^m u = \cot^2 u \cdot \cot^{m-2} u$$

Then use (27-15) or (27-16) to rewrite the factor with the smaller exponent.

EXAMPLE 24 Integrate: (a) $\displaystyle\int \cot^3 2x \; dx$ (b) $\displaystyle\int \tan^4 x \; dx$

Solution

(a) **First,** since the exponent is odd, factor out cot 2x.

$$\int \cot^3 2x \; dx = \int (\cot 2x)(\cot^2 2x) \; dx$$

Next, rewrite $\cot^2 2x$ using (27-16).

$$= \int (\cot 2x)(\csc^2 2x - 1) \; dx$$

$$= \int (\cot 2x \csc^2 2x - \cot 2x) \; dx$$

$$= \int \cot 2x (\csc^2 2x \; dx) - \int \cot 2x \; dx$$

If $u = \cot 2x$ in the first integral, then by (26-6), $du = -2 \csc^2 2x \; dx$, and

$$\int \cot 2x (\csc^2 2x \; dx) = -\frac{1}{2} \int u \; du$$

Let $v = 2x$ in the second integral; then $dv = 2 \; dx$, so

$$\int \cot 2x \; dx = \frac{1}{2} \int \cot v \; dv$$

The original integral is therefore

$$\int \cot^3 2x \; dx = -\frac{1}{2} \int u \; du - \frac{1}{2} \int \cot v \; dv \qquad \boxed{\begin{array}{l}\text{Integrate}\\ \text{using (27-11).}\end{array}}$$

$$= -\frac{u^2}{4} - \frac{1}{2} \ln|\sin v| + C$$

$$= -\frac{1}{4} \cot^2 2x - \frac{1}{2} \ln|\sin 2x| + C$$

(b) **First,** since the
exponent is
even, factor out
$\tan^2 x$.

$$\int \tan^4 x \; dx = \int (\tan^2 x)(\tan^2 x) \; dx$$

Next, rewrite
one of the $\tan^2 x$
factors using
identity (27-15).

$$= \int (\tan^2 x)(\sec^2 x - 1) \; dx$$

$$= \int (\tan^2 x \sec^2 x - \tan^2 x) \; dx$$

Again, replace
$\tan^2 x$ with
$(\sec^2 x - 1)$.

$$= \int (\tan^2 x \sec^2 x - \sec^2 x + 1) \; dx$$

Finally, use the
sum rule (24-8)
to write this as
three separate
integrals.

$$= \int \tan^2 x \sec^2 x \; dx - \int \sec^2 x \; dx + \int dx$$

If we let
$u = \tan x$, then
by (26-5),
$du = \sec^2 x \; dx$.

$$= \int u^2 \; du - \int du + \int dx$$

$$= \frac{u^3}{3} - u + x$$

$$= \frac{1}{3} \tan^3 x - \tan x + x + C$$

Case 3.   If the exponent m of the tangent or cotangent factor is equal to zero,
the integrals become

$$\int \sec^n u \; du \qquad \text{or} \qquad \int \csc^n u \; du$$

- If n is an **even** positive integer, use (27-15) or (27-16) to rewrite the integrand
  as a sum of tangent or cotangent terms.
- If n is an **odd** positive integer, we must use techniques of integration that
  will be explained in Section 27-6.

EXAMPLE 25   Integrate: $\int \sec^4 x \; dx$

Solution **First,** use
(27-15) to rewrite the
integral in terms of
$\tan x$.

$$\int \sec^4 x \; dx = \int (\sec^2 x)^2 \; dx$$

$$= \int (\tan^2 x + 1)^2 \; dx$$

$$= \int (\tan^4 x + 2 \tan^2 x + 1) \; dx$$

Write as a sum of integrals using the sum rule (24-8).

$$= \int \tan^4 x \, dx + 2 \int \tan^2 x \, dx + \int dx$$

Next, evaluate each of the first two integrals as in Example 24(b).

$$= \frac{1}{3} \tan^3 x + \tan x + C$$

Note that if this integral is rewritten, it can also be evaluated using u substitution:

$$\int \sec^4 x \, dx = \int \sec^2 x \sec^2 x \, dx = \int (1 + \tan^2 x)\sec^2 x \, dx$$

Let $u = \tan x$; then $du = \sec^2 x \, dx$, and the integral becomes

$$\int (1 + u^2) \, du = u + \frac{u^3}{3} + C = \tan x + \frac{1}{3} \tan^3 x + C$$

EXAMPLE 26   Find the volume of the solid generated by rotating about the x-axis the region bounded by $y = \sin^{3/2} x$ and the x-axis between $x = 0$ and $x = \pi$.

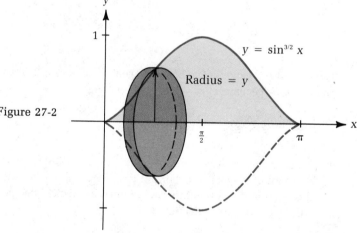

Figure 27-2

Solution   From Fig. 27-2:   Volume of element $= \pi y^2 \Delta x$

$$V = \int_0^\pi \pi (\sin^{3/2} x)^2 \, dx$$

$$= \pi \int_0^\pi \sin^3 x \, dx$$

Follow the procedure for a type A integral.

$$= \pi \int_0^\pi (\sin^2 x)(\sin x \, dx)$$

Use (27-14).

$$= \pi \int_0^\pi (1 - \cos^2 x)(\sin x \; dx)$$

If we let u = cos x, then by
(26-4), du = −sin x dx.
Change the limits of integration: At x = 0, u = cos 0 = 1,
and at x = π,
u = cos π = −1.
Substitute.

$$V = -\pi \int_1^{-1} (1 - u^2) \; du$$

$$= (-\pi)\left(u - \frac{u^3}{3}\right)\Bigg]_1^{-1}$$

$$= (-\pi)\left[\left(-1 - \frac{(-1)^3}{3}\right) - \left(1 - \frac{1^3}{3}\right)\right]$$

$$= (-\pi)\left[-\frac{2}{3} - \frac{2}{3}\right] = \frac{4\pi}{3} \text{ cubic units}$$

Exercises 27-4

Integrate.

1. $\displaystyle\int \sin^2 x \; dx$

2. $\displaystyle\int \cos^2 x \; dx$

3. $\displaystyle\int \cos^5\left(\frac{x}{2}\right) dx$

4. $\displaystyle\int \sin^3 4x \; dx$

5. $\displaystyle\int \cos^4(x + 1) \; dx$

6. $\displaystyle\int \cos^3(2x - 1) \; dx$

7. $\displaystyle\int \frac{\sin x \; dx}{\cos^2 x}$

8. $\displaystyle\int \frac{\cos x \; dx}{\sin^2 x}$

9. $\displaystyle\int \cos^{1/3} t \sin^3 t \; dt$

10. $\displaystyle\int \sin^{1/2} y \cos^5 y \; dy$

11. $\displaystyle\int \sin^4 x \cos^3 x \; dx$

12. $\displaystyle\int \cos^3 2x \sin^5 2x \; dx$

13. $\displaystyle\int \cos^4 2x \sin^3 2x \; dx$

14. $\displaystyle\int \sin^2 t \cos^4 t \; dt$

15. $\displaystyle\int \sin^3 \theta \cos^3 \theta \; d\theta$

16. $\displaystyle\int \sin^3 z \cos^2 z \; dz$

17. $\displaystyle\int \tan^3 x \; dx$

18. $\displaystyle\int \tan^2(4x + 3) \; dx$

19. $\displaystyle\int \cot^2 \pi x \; dx$

20. $\displaystyle\int \cot^3(2 + x) \; dx$

21. $\displaystyle\int \sec^6 t \; dt$

22. $\displaystyle\int \csc^2 x \; dx$

23. $\displaystyle\int \tan^3 x \sec x \; dx$

24. $\displaystyle\int \tan^2 3x \sec^2 3x \; dx$

25. $\displaystyle\int \tan^4 \pi\theta \sec^2 \pi\theta \; d\theta$

26. $\displaystyle\int \tan^2\left(\frac{t}{2}\right) \sec^4\left(\frac{t}{2}\right) dt$

27. $\displaystyle\int x \cot^5(x^2) \csc^3(x^2) \; dx$

28. $\displaystyle\int \cot z \csc^4 z \; dz$

29. $\displaystyle\int_0^{\pi/2} \cos^2 \theta \; d\theta$

30. $\displaystyle\int_0^\pi (2 + \sin x)^2 \; dx$

31. $\displaystyle\int_0^{\pi/4} \sqrt{\tan t} \sec^4 t \; dt$

32. $\displaystyle\int_0^{\pi/6} \tan^3(x - 1) \; dx$

Solve.

33. *Physics* The moment of inertia about a diameter of a circular disk of radius r and mass m is

$$I = \frac{mr^2}{4} \int_0^{2\pi} \sin^2 \theta \; d\theta$$

Calculate I for a disk of mass 10 kg and radius 0.05 m.

34. Calculate the volume of the solid formed by rotating about the x-axis the region under the curve $y = \sin x$ between $x = 0$ and $x = \pi$.

35. Determine the area under the curve $y = \tan^2 x$ between $x = 0$ and $x = \pi/4$.

36. *Electronics*   Find the RMS average current in a circuit from $t = 0$ s to $t = 0.5$ s if $i = i_0 \sin t \sqrt{\cos t}$.

37. *Electrical engineering*   If the power (in watts) dissipated in an alternating current circuit is

$$P = 1200 \cos \left(60\pi t + \frac{\pi}{4} \right)$$

calculate the RMS average power over the interval from $t = 0$ s to $t = \dfrac{1}{60}$ s. Round to the nearest tenth.

38. Find the area bounded by the curve $y = \sin^2 x$, $y = x/10$, $x = 1$, and $x = 0$. Round to the nearest hundredth.

27-5 | Trigonometric Substitutions

Several integrals containing expressions of the form $a^2 + u^2$, where $a > 0$ and $u = u(x)$, may be obtained directly from the derivatives of inverse trigonometric functions. From equation (26-9),

$$\frac{d}{dx}(\text{Arcsin } u) = \frac{1}{\sqrt{1 - u^2}} \frac{du}{dx} \qquad \text{for } -1 < u < 1$$

and we can write this in the more general form

$$\frac{d}{dx}\left[\text{Arcsin} \left(\frac{u}{a} \right) \right] = \frac{1}{\sqrt{1 - \left(\dfrac{u}{a} \right)^2}} \frac{d}{dx} \left(\frac{u}{a} \right)$$

$$= \frac{1}{a} \cdot \frac{1}{\sqrt{1 - \left(\dfrac{u}{a} \right)^2}} \cdot \frac{du}{dx}$$

$$= \frac{1}{\sqrt{a^2 - u^2}} \frac{du}{dx}$$

This leads to the following integration formula:

$$\int \frac{du}{\sqrt{a^2 - u^2}} = \text{Arcsin} \left(\frac{u}{a} \right) + C \qquad \text{where } -1 < \frac{u}{a} < 1 \qquad (27\text{-}19)$$

Similarly, the differentiation formula (26-11)

$$\frac{d}{dx}(\text{Arctan } u) = \frac{1}{1 + u^2}\frac{du}{dx}$$

leads to the following integration formula:

$$\int \frac{du}{a^2 + u^2} = \frac{1}{a} \text{ Arctan}\left(\frac{u}{a}\right) + C \qquad \text{(27-20)}$$

EXAMPLE 27 Integrate: (a) $\displaystyle\int \frac{dx}{\sqrt{16 - x^2}}$ (b) $\displaystyle\int \frac{2\,dx}{x^2 + 8}$

Solution

(a) This integral has the form of integration formula (27-19) with $a^2 = 16$ or $a = 4$ and $u = x$. Therefore,

$$\int \frac{dx}{\sqrt{16 - x^2}} = \int \frac{dx}{\sqrt{4^2 - x^2}}$$

$$= \text{Arcsin}\left(\frac{x}{4}\right) + C$$

(b) This integral has the form of integration formula (27-20) with $a^2 = 8$ or $a = \sqrt{8} = 2\sqrt{2}$ and $u = x$. Therefore:

Rationalize the fractions.

$$\int \frac{2\,dx}{x^2 + 8} = \int \frac{2\,dx}{x^2 + \left(2\sqrt{2}\right)^2}$$

$$= 2\cdot\frac{1}{2\sqrt{2}} \text{ Arctan}\left(\frac{x}{2\sqrt{2}}\right) + C$$

$$= \frac{\sqrt{2}}{2} \text{ Arctan}\left(\frac{x\sqrt{2}}{4}\right) + C$$

CAUTION ▶ If Example 27(a) had a factor of x in the numerator—for example,

$$\int \frac{x\,dx}{\sqrt{16 - x^2}}$$

it would have to be integrated using the power rule rather than (27-19). Similarly, if Example 27(b) had a factor of x in the numerator—for example,

$$\int \frac{2x\,dx}{x^2 + 8}$$

it would have to be integrated as a reciprocal function using (27-1) rather than (27-20). ◀

EXAMPLE 28 The strength and duration of a collision force F (in Newtons) can be specified by calculating the **impulse** J as the force acts from time t_1 to t_2.

The impulse is defined as

$$J = \int_{t_1}^{t_2} F \, dt$$

Calculate the impulse for the force acting from $t_1 = 0$ ms to $t_2 = 6$ ms that is described by the equation (see Fig. 27-3)

$$F = \frac{1}{0.2(t - 3)^2 + 0.03}$$

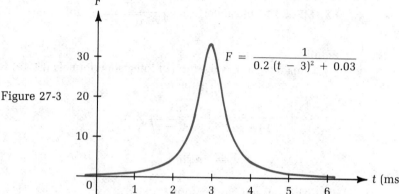

Figure 27-3

Solution   From the definition, the impulse is:

$$J = \int_0^6 \frac{dt}{0.2(t - 3)^2 + 0.03}$$

Multiply the numerator and denominator of the integrand by 5.

$$= 5 \int_0^6 \frac{dt}{(t - 3)^2 + 0.15}$$

This has the form of (27-20) with $a^2 = 0.15$ or $a = \sqrt{0.15}$ and $u = t - 3$, $du = dt$. Substitute.

$$= \frac{5}{\sqrt{0.15}} \text{Arctan} \left(\frac{t - 3}{\sqrt{0.15}}\right) \Bigg]_0^6$$

$$= \frac{5}{\sqrt{0.15}} \left[\text{Arctan} \left(\frac{3}{\sqrt{0.15}}\right) - \text{Arctan} \left(\frac{-3}{\sqrt{0.15}}\right) \right]$$

Using a calculator in radian mode, we find

3 ÷ .15 √ = STO INV tan − RCL +/− INV tan = × 5 ÷ .15 √ =

→   37.242789

$$J \approx 37 \text{ N·ms}$$

$$\approx 0.037 \text{ N·s}$$

Notice that the units are N·ms, and we must divide by 1000 to convert to standard SI units.

If the expression in the denominator of the integrand is not a sum or differ-ence of squares, as shown in equations (27-19) and (27-20), it may be possible to rewrite it in this form by completing the square.

EXAMPLE 29   Integrate: $\displaystyle\int_{-2}^{3} \frac{dx}{x^2 + 4x + 9}$

Solution   First, rewrite the denominator. $x^2 + 4x + 9 = x^2 + 4x + 4 + 5$

$$= (x + 2)^2 + 5$$

The integral be-comes:

$$\int_{-2}^{3} \frac{dx}{x^2 + 4x + 9} = \int_{-2}^{3} \frac{dx}{(x + 2)^2 + 5}$$

$$= \int_{-2}^{3} \frac{dx}{5 + (x + 2)^2}$$

Next, this inte-gral has the form of (27-20) with $a^2 = 5$ or $a = \sqrt{5}$ and $u^2 = (x + 2)^2$ or $u = x + 2$, $du = dx$. Substi-tute.

$$= \frac{1}{\sqrt{5}} \text{Arctan} \left(\frac{x + 2}{\sqrt{5}}\right)\Bigg]_{-2}^{3}$$

$$= \frac{1}{\sqrt{5}} \left[\text{Arctan} \left(\frac{5}{\sqrt{5}}\right) - \text{Arctan} \left(\frac{0}{\sqrt{5}}\right)\right]$$

$$= \frac{1}{\sqrt{5}} \text{Arctan} \left(\frac{5}{\sqrt{5}}\right)$$

Rationalize the fractions.

$$= \frac{\sqrt{5}}{5} \text{Arctan} \sqrt{5}$$

$$\approx 0.514$$

Trigonometric substitution   In general, if an integrand contains an expression of the form $\sqrt{a^2 + u^2}$ or an integer power of such an expression, where $a > 0$ and u is some differentiable function of the variable of integration, then the integral often can be integrated by making a substitution involving a trigonometric function. Table 27-1 on p. 1190 summarizes this process of **trigonometric substitution.** In each case, the indicated substitution transforms the integral from $\int f(u)\, du$ to an equivalent integral $\int g(\theta)\, d\theta$ that can be solved using one of the basic integration formulas already presented. The variable of integration has been changed from u to θ, where $-\pi/2 \le \theta \le \pi/2$. This kind of substitution is often called an *inverse trigonometric substitution*. The substitutions listed in the second column of Table 27-1 may be memorized, but when you need them, you may find it easier to derive them by drawing a reference triangle, as shown in the last column.

TABLE 27-1

For integrands containing the expression:	Substitute:	The integrand expression becomes:	Reference Triangle
I. $\sqrt{a^2 - u^2}$ $\quad a > 0$	$u = a \sin \theta$ $du = a \cos \theta \, d\theta$	$\sqrt{a^2 - u^2} = a \cos \theta$	
II. $\sqrt{a^2 + u^2}$ $\quad a > 0$	$u = a \tan \theta$ $du = a \sec^2 \theta \, d\theta$	$\sqrt{a^2 + u^2} = a \sec \theta$	
III. $\sqrt{u^2 - a^2}$ $\quad a > 0$	$u = a \sec \theta$ $du = a \sec \theta \tan \theta \, d\theta$	$\sqrt{u^2 - a^2} = a\|\tan \theta\|$	

EXAMPLE 30   Integrate: $\displaystyle\int \frac{dx}{x^2\sqrt{9 - x^2}}$

Solution   **First,** notice that the integrand contains an expression of the form $\sqrt{a^2 - u^2}$, where $a^2 = 9$ or $a = 3$ and $u = x$.* Therefore, as shown in the first row of Table 27-1, use the substitution $u = a \sin \theta$ or $x = 3 \sin \theta$ with $-\pi/2 \le \theta \le \pi/2$. Then $dx = 3 \cos \theta \, d\theta$ and

$$\sqrt{9 - x^2} = \sqrt{9 - (3 \sin \theta)^2}$$

$$= \sqrt{9 - 9 \sin^2 \theta}$$

$$= \sqrt{9(1 - \sin^2 \theta)} = \sqrt{9 \cos^2 \theta}$$

$$= 3 \cos \theta$$

This is the result shown in the third column of Table 27-1.

*Notice that integration formula (27-19) gives a special case that can be solved with this substitution.

Next, rewrite the integral using these substitutions.

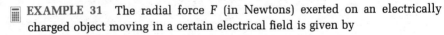

$$\int \frac{dx}{x^2\sqrt{9-x^2}} = \int \frac{(3 \cos \theta \, d\theta)}{(3 \sin \theta)^2(3 \cos \theta)}$$

$$= \frac{1}{9} \int \frac{d\theta}{\sin^2\theta}$$

$$= \frac{1}{9} \int \csc^2\theta \, d\theta$$

Integrate using equation (27-7).

$$= -\frac{1}{9} \cot \theta + C$$

Finally, this result is expressed in terms of the variable θ, so we must rewrite it in terms of the original variable x. Draw a right triangle (Fig. 27-4) with an angle θ and sides 3 and x such that $x = 3 \sin \theta$ or $\sin \theta = x/3$. From the Pythagorean theorem, the remaining side is $\sqrt{3^2 - x^2} = \sqrt{9 - x^2}$. Using this triangle, all the basic trigonometric functions may be written in terms of x. For $\cot \theta$,

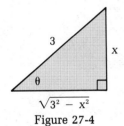

Figure 27-4

$$\cot \theta = \frac{\sqrt{9-x^2}}{x}$$

Therefore, the result of the integration is

$$\int \frac{dx}{x^2\sqrt{9-x^2}} = -\frac{\sqrt{9-x^2}}{9x} + C$$

EXAMPLE 31   The radial force F (in Newtons) exerted on an electrically charged object moving in a certain electrical field is given by

$$F(r) = \frac{1}{(\sqrt{r^2+4})^3}$$

Find the work done on this object by the force as the object moves from $r = 0.10$ m to $r = 1.00$ m.

Solution   Work $= \displaystyle\int_{r_1}^{r_2} F(r) \, dr = \int_{0.10}^{1.00} \frac{dr}{(\sqrt{r^2+4})^3}$

First, notice that the integrand contains an expression of the form $\sqrt{u^2 + a^2}$, where $a^2 = 4$ or $a = 2$ and $u = r$. As shown in the second row of Table 27-1, we can simplify this integral by using the substitution $r = 2 \tan \theta$, $-\pi/2 \le \theta \le \pi/2$, so that $dr = 2 \sec^2\theta \, d\theta$ and $\sqrt{r^2 + 4} = 2 \sec \theta$.

Next, use these substitutions to rewrite the integral:

$$\int \frac{dr}{(\sqrt{r^2+4})^3} = \int \frac{2 \sec^2\theta \, d\theta}{(2 \sec \theta)^3}$$

Notice that we integrate without regard to the original limits of integration. We will return to these limits after the integration is complete.

$$= \frac{1}{4} \int \frac{d\theta}{\sec \theta}$$

$$= \frac{1}{4} \int \cos \theta \, d\theta$$

Finally, integrate using (27-4):

$$\frac{1}{4} \int \cos \theta \, d\theta = \frac{1}{4} \sin \theta + C$$

Draw a reference triangle (Fig. 27-5) and write $\sin \theta$ in terms of the original variable of integration r:

$$\sin \theta = \frac{r}{\sqrt{r^2 + 4}}$$

Therefore,

$$W = \int_{0.10}^{1.00} \frac{dr}{(\sqrt{r^2 + 4})^3} = \frac{r}{4\sqrt{r^2 + 4}} \Bigg]_{0.10}^{1.00}$$

$$= \left(\frac{1}{4\sqrt{5}} - \frac{0.10}{4\sqrt{4.01}} \right)$$

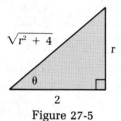

Figure 27-5

$4 \, \boxed{\times} \, 5 \, \boxed{\checkmark} \, \boxed{=} \, \boxed{1/x} \, \boxed{-} \, .1 \, \boxed{\div} \, 4 \, \boxed{\div} \, 4.01 \, \boxed{\checkmark} \, \boxed{=} \rightarrow \quad \boxed{\textit{0.0993190}}$

$$\approx 0.099 \text{ J}$$

EXAMPLE 32 Integrate: $\displaystyle\int \frac{\sqrt{4x^2 - 1}}{x} \, dx$

Solution **First,** notice that the integrand contains an expression of the form $\sqrt{u^2 - a^2}$, where $a^2 = 1$ or $a = 1$ and $u^2 = 4x^2$ or $u = 2x$. As shown in the third row of Table 27-1, we can simplify this integral by using the substitution

$2x = \sec \theta$, $-\pi/2 \le \theta \le \pi/2$, so that $dx = \dfrac{1}{2} \sec \theta \tan \theta \, d\theta$ and

$$\sqrt{4x^2 - 1} = |\tan \theta|$$

Next, rewrite the integral. $\quad \displaystyle\int \frac{\sqrt{4x^2 - 1}}{x} \, dx = \int \frac{|\tan \theta|}{\frac{1}{2} \sec \theta} \cdot \frac{1}{2} \sec \theta \tan \theta \, d\theta$

$$= \int \tan^2 \theta \, d\theta$$

Finally, integrate using equation (27-15).

$$= \int (\sec^2 \theta - 1) \, d\theta$$

Use equation (27-6).

$$= \tan \theta - \theta + C$$

Draw a reference triangle (Fig. 27-6) and write $\tan \theta$ in terms of x:

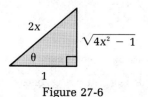

Figure 27-6

$$\tan \theta = \frac{\sqrt{4x^2 - 1}}{1}$$

Since $2x = \sec \theta$, $\theta = \text{Arcsec } 2x$, and the integral becomes

$$\int \frac{\sqrt{4x^2 - 1}}{x} \, dx = \sqrt{4x^2 - 1} - \text{Arcsec } 2x + C$$

Using the trigonometric substitutions $u = a \tan \theta$ or $u = a \sec \theta$, we can show the following:*

$$\int \frac{du}{\sqrt{u^2 \pm a^2}} = \ln \left| u + \sqrt{u^2 \pm a^2} \right| + C \qquad \text{(27-21)}$$

For example, if we let $u = a \tan \theta$, then from the second row of Table 27-1,

$$\int \frac{du}{\sqrt{u^2 + a^2}} = \int \frac{a \sec^2 \theta \, d\theta}{a \sec \theta}$$

$$= \int \sec \theta \, d\theta$$

Use equation (27-12). $\qquad = \ln |\sec \theta + \tan \theta| + C$

Use the reference triangle for the second row of Table 27-1 to show that

$$\sec \theta = \frac{\sqrt{u^2 + a^2}}{a} \qquad \text{and} \qquad \tan \theta = \frac{u}{a}$$

Then the integral becomes: $\displaystyle\int \frac{du}{\sqrt{u^2 + a^2}} = \ln \left| \frac{\sqrt{u^2 + a^2}}{a} + \frac{u}{a} \right| + C$

$$= \ln \left| \frac{\sqrt{u^2 + a^2} + u}{a} \right| + C$$

$$= \ln \left| \sqrt{u^2 + a^2} + u \right| - \ln a + C$$

$$= \ln \left| u + \sqrt{u^2 + a^2} \right| + C'$$

Note that we have replaced the constant $C - \ln a$ with a new constant of integration C'.

EXAMPLE 33 If an electrically charged surface loses charge at a rate determined by the equation

$$\frac{dq}{dt} = \frac{10}{\sqrt{0.01 + t^2}}$$

*Do not confuse this integral with (27-20), where the integrand does *not* contain a square root.

where charge q is in coulombs and time t is in seconds, find the net charge leaving the surface between $t = 0$ s and $t = 2.0$ s.

Solution

$$q = \int_0^{2.0} \frac{10\ dt}{\sqrt{0.01 + t^2}}$$

The integrand has the form of equation (27-21) with $a = 0.1$.

$$= 10 \int_0^{2.0} \frac{dt}{\sqrt{0.01 + t^2}}$$

Substitute.

$$= 10 \ln \left| t + \sqrt{0.01 + t^2} \right| \Big]_0^2$$

$$= 10 \left(\ln \left| 2 + \sqrt{4.01} \right| - \ln \left| \sqrt{0.01} \right| \right)$$

$2\ \boxed{+}\ 4.01\ \boxed{\sqrt{\ }}\ \boxed{=}\ \boxed{\ln x}\ \boxed{-}\ .01\ \boxed{\sqrt{\ }}\ \boxed{\ln x}\ \boxed{=}\ \boxed{\times}\ 10\ \boxed{=}\ \rightarrow$ *36.895039*

$$\approx 37 \text{ C}$$

Some integrands containing expressions not already in the form $\sqrt{u^2 \pm a^2}$ may often be put into this form by completing the square.

EXAMPLE 34 Find the area bounded by the curve

$$y = \frac{1}{\sqrt{x^2 + 4x - 5}}$$

the x-axis, and the lines $x = 2.0$ and $x = 5.0$.

Solution The area is given by the integral:

$$A = \int_{2.0}^{5.0} \frac{dx}{\sqrt{x^2 + 4x - 5}}$$

Rewrite the expression under the radical sign.

$$x^2 + 4x - 5 = x^2 + 4x + (4 - 9)$$

$$= (x + 2)^2 - 3^2$$

Therefore:

$$A = \int_{2.0}^{5.0} \frac{dx}{\sqrt{(x + 2)^2 - 3^2}}$$

The integrand now contains an expression of the form $\sqrt{u^2 - a^2}$, where $a = 3$ and $u = x + 2$. Use equation (27-21).

$$= \ln \left| (x + 2) + \sqrt{(x + 2)^2 - 3^2} \right| \Big]_{2.0}^{5.0}$$

$$= \ln \left| 7 + \sqrt{40} \right| - \ln \left| 4 + \sqrt{7} \right|$$

$40\ \boxed{\sqrt{\ }}\ \boxed{+}\ 7\ \boxed{=}\ \boxed{\ln x}\ \boxed{-}\ \boxed{(}\ 7\ \boxed{\sqrt{\ }}\ \boxed{+}\ 4\ \boxed{)}\ \boxed{\ln x}\ \boxed{=}\ \rightarrow$ *0.6956308*

$$= 0.70 \text{ square unit}$$

Exercises 27-5

Integrate. Round all answers to calculator problems to the nearest hundredth.

1. $\displaystyle\int \frac{dx}{\sqrt{9 - x^2}}$

2. $\displaystyle\int_0^2 \frac{dy}{\sqrt{25 - y^2}}$

3. $\displaystyle\int_{-1}^2 \frac{dx}{1 + x^2}$

4. $\displaystyle\int \frac{dx}{x^2 + 4}$

5. $\displaystyle\int \frac{dq}{3 + 4q^2}$

6. $\displaystyle\int \frac{dt}{\sqrt{5 - 9t^2}}$

7. $\displaystyle\int \frac{2\,dx}{\sqrt{16 - 9x^2}}$

8. $\displaystyle\int \frac{3\,d\theta}{12 + 16\theta^2}$

9. $\displaystyle\int x\sqrt{x^2 - 1}\,dx$

10. $\displaystyle\int 2x\sqrt{4x^2 - 3}\,dx$

11. $\displaystyle\int \frac{x^2\,dx}{\sqrt{4 - x^2}}$

12. $\displaystyle\int \frac{dx}{\sqrt{9x^2 + 4}}$

13. $\displaystyle\int \frac{dx}{\sqrt{1 - 8x^2}}$

14. $\displaystyle\int_0^2 \sqrt{16 - 4x^2}\,dx$

15. $\displaystyle\int_0^4 t^3\sqrt{9 + t^2}\,dt$

16. $\displaystyle\int x^{-2}\sqrt{4 - x^2}\,dx$

17. $\displaystyle\int \frac{2t\,dt}{\sqrt{9 - t^4}}$

18. $\displaystyle\int \frac{4x\,dx}{\sqrt{25 - 16x^4}}$

19. $\displaystyle\int \frac{1 + x}{1 + x^2}\,dx$

20. $\displaystyle\int \frac{2 - x}{\sqrt{9 - 4x^2}}\,dx$

21. $\displaystyle\int \frac{dx}{(4 - x^2)^{3/2}}$

[Hint: For Problems 19 and 20, use the sum rule (24-8) first.]

22. $\displaystyle\int \frac{dx}{(9x^2 - 1)^{3/2}}$

23. $\displaystyle\int \frac{dx}{x^2\sqrt{x^2 - 16}}$

24. $\displaystyle\int_1^2 \frac{x^3\,dx}{\sqrt{x^2 + 4}}$

25. $\displaystyle\int \sqrt{9 - (x + 1)^2}\,dx$

26. $\displaystyle\int \frac{dx}{\sqrt{(x - 1)^2 - 25}}$

27. $\displaystyle\int \frac{dx}{x^2 + 6x + 12}$

28. $\displaystyle\int \frac{2\,dx}{x^2 - 4x + 13}$

29. $\displaystyle\int_1^3 \frac{x\,dx}{\sqrt{x^2 + 6x + 5}}$

30. $\displaystyle\int \frac{x\,dx}{\sqrt{x^2 - 2x + 5}}$

31. $\displaystyle\int e^x\sqrt{1 - e^{2x}}\,dx$

32. $\displaystyle\int \frac{\cos\theta\,d\theta}{\sqrt{4 - \sin^2\theta}}$

Solve.

33. *Robotics* The acceleration of a robotic grasping device is given by the equation

$$a = \frac{2}{0.01 + t^2}$$

Find an equation for the velocity of this device if $v = 0$ cm/s at $t = 0$ s.

34. Calculate the area under the curve $y = \sqrt{1 - x^2}$ between $x = 0$ and $x = 1$.

35. Find the volume of the solid produced by rotating about the x-axis the region bounded by the graph of

$y = \dfrac{1}{\sqrt{5 + x^2}}$ and the x-axis, between $x = -1$ and $x = 2$. Round to the nearest hundredth.

36. *Chemistry* When the chemical material in a retinal cell is exposed to light, it decomposes according to the equation

$$\frac{dA}{dt} = k(A - A_0) - pA^2$$

where k, A_0, and p are constants, and A is the amount of chemical present. Then the time t (in milliseconds) for the amount of material to decrease from $A = a$ to $A = b$ is

$$t = \int_a^b \frac{dA}{k(A - A_0) - pA^2}$$

Solve if $k = 20$, $A_0 = 5.0$, $p = 4.0$, $a = 10$, and $b = 2$.

37. *Physics*   Find the average force (in newtons) from $t = 0.1$ s to $t = 0.2$ s if the force F is given by

$$F = \frac{1}{\left(\sqrt{4 - t^2}\right)^3}$$ Round to the nearest hundredth.

38. Find the volume of the solid produced by rotating about the x-axis the region under the graph

$$y = \frac{1}{\sqrt{x^2 + 2}}$$

from $x = 0$ to $x = 1$. Round to two decimal places.

39. Find the area bounded by the curve

$$y = \frac{1}{\sqrt{16 - x^2}}$$

the x-axis, the y-axis, and the line $x = 3$. Round to two decimal places.

40. *Electrical engineering*   A current surge in an electrical circuit varies with the time according to the equation

$$i = \frac{100}{4(t - 3)^2 + 10}$$

where i is in amperes and t is in milliseconds. What total charge (in coulombs) passes through the circuit in the surge from $t = 0$ ms to $t = 6$ ms? Round to three decimal places.

$$\left[\text{Hint:} \quad q = \frac{1}{1000} \int_0^6 i(t) \, dt \right]$$

27-6 | Integration by Parts

The substitution method is a general technique of integration in which we rewrite an integral so that it can be solved using one of the basic integration formulas. A second, and very powerful, general technique of integration separates the original integral into two parts, one of which is another—hopefully simpler—integral. This method of **integration by parts** depends on the formula for the derivative of a product (22-17). If u and v are differentiable functions of x, then

$$\frac{d}{dx}(uv) = u\frac{dv}{dx} + v\frac{du}{dx}$$

or, writing this in differential form,

$$d(uv) = u \, dv + v \, du$$

We can integrate both members of this equation with respect to x:

$$\int d(uv) = \int u \, dv + \int v \, du$$

Then, rearranging this equation, we obtain

$$\int u \, dv = \int d(uv) - \int v \, du$$

Integrating the first term on the right leads to the basic formula for integration by parts.

Integration by Parts $\displaystyle\int u \, dv = uv - \int v \, du$ (27-22)

Note that $\int d(uv) = uv + C$. We have omitted the constant of integration in (27-22), since a constant will appear in the integration of $\int v \, du$.

To apply the integration by parts technique, separate the integrand of the original integral into two factors or "parts," u and dv, where the factor dv contains the differential dx. This separation should be made so that the new integral $\int v \, du$ is easier to integrate than the original integral $\int u \, dv$. Using integration by parts, it is possible to reduce a difficult integration to an easier one.

EXAMPLE 35   Integrate: $\displaystyle\int xe^{2x} \, dx$

Solution   First, notice that this integral does not have the form of any of the basic integration formulas and therefore cannot be simplified by the substitution technique. To use integration by parts, separate the integrand into factors u and dv so that $u \, dv = xe^{2x} \, dx$. If we choose $u = x$ and $dv = e^{2x} \, dx$, then $du = dx$, and from equation (27-2), $v = \int dv = \int e^{2x} \, dx = \dfrac{1}{2}e^{2x} + C$. Use $v = \dfrac{1}{2}e^{2x}$.

Second, rewrite the integral using (27-22).

$$\int \overset{u}{x}e^{2x} \, \overset{dv}{dx} = \overset{u}{x} \cdot \overset{v}{\frac{1}{2}e^{2x}} - \int \overset{v}{\frac{1}{2}e^{2x}} \, \overset{du}{dx}$$

Integrate the integral on the right using (27-2).

$$= \frac{1}{2}xe^{2x} - \frac{1}{4}e^{2x} + C$$

$$= \frac{1}{4}e^{2x}(2x - 1) + C$$

The expression $xe^{2x} \, dx$ can be separated into factors u and dv in more than one way. For example,

u	dv
x	e^{2x} dx
e^{2x}	x dx
xe^{2x}	dx
xe^x	e^x dx

⟵ u dv = xe^{2x} dx

In this example we chose the first factorization. Choosing one of the other combinations may lead to an integral that is more difficult to integrate than the original. For example, choosing u = e^{2x}, dv = x dx gives

$$\int xe^{2x}\ dx = \frac{1}{2}x^2 e^{2x} - \int x^2 e^{2x}\ dx$$

and the integral on the right cannot be solved using a basic integration formula.

LEARNING HINT ▶ There is no foolproof method or fixed rule for choosing u and dv, and the process is usually one of trial and error, but the following hints may help:

1. Be certain to keep dx as a part of the factor dv.
2. It is often helpful to choose as dv the most complicated part of the integrand that can be integrated exactly using a basic integration. The integral ∫ dv must be easy to find.
3. Choose u so that du/dx is a simpler expression than u. For example, it may be helpful to choose as u a polynomial expression whose successive derivatives tend to zero. ◀

EXAMPLE 36   Integrate: $\int x \ln x\ dx$

Solution   Choose u = ln x and dv = x dx; then du = $\dfrac{dx}{x}$ and v = ∫ x dx or v = $\dfrac{1}{2}x^2$ + C. Substituting in the integration by parts formula (27-22), we have

$$\int x \ln x\ dx = \int (\ln x)(x\ dx) = \ln x \cdot \frac{1}{2}x^2 - \int \left(\frac{1}{2}x^2\right)\frac{dx}{x}$$

$$= \frac{1}{2}x^2 \ln x - \frac{1}{2}\int x\ dx$$

$$= \frac{1}{2}x^2 \ln x - \frac{1}{4}x^2 + C$$

LEARNING HINT ▶ When the factor ln x is present in the integrand, letting $u = \ln x$ will eliminate ln x from the integral on the right in integration by parts. ◀

For a definite integral, equation (27-22) can be written as follows:

$$\int_a^b u \; dv = uv \Big]_a^b - \int_a^b v \; du \qquad (27\text{-}23)$$

EXAMPLE 37   If the voltage output of an electronic device varies according to the function

$$V(t) = V_0 t \cos t$$

where V_0 is a constant, find the average value of the voltage from $t = 0$ s to $t = \pi/2$ s.

Solution   The average value of the voltage is

$$V_{\text{AVE}} = \frac{1}{(\pi/2) - 0} \int_0^{\pi/2} V_0 t \cos t \; dt = \frac{2V_0}{\pi} \int_0^{\pi/2} t \cos t \; dt$$

Choose $u = t$ and $dv = \cos t \; dt$; then $du = dt$ and $v = \sin t$.

Substitute in equation (27-23).

$$\int_0^{\pi/2} t \cos t \; dt = t \sin t \Big]_0^{\pi/2} - \int_0^{\pi/2} \sin t \; dt$$

Evaluate the integral on the right using equation (27-5).

$$= \left(\frac{\pi}{2} \sin \frac{\pi}{2} - 0 \cdot \sin 0 \right) + (\cos t) \Big]_0^{\pi/2}$$

$$= \frac{\pi}{2} + \left(\cos \frac{\pi}{2} - \cos 0 \right)$$

$$= \frac{\pi}{2} - 1$$

Therefore, the average value of the voltage is

$$V_{\text{AVE}} = \frac{2V_0}{\pi} \left(\frac{\pi}{2} - 1 \right)$$

$$= V_0 \left(1 - \frac{2}{\pi} \right) \approx 0.363 V_0 \text{ volts}$$

EXAMPLE 38   Integrate: $\int \text{Arctan } x \; dx$

Solution   Choose $u = \text{Arctan } x$ and $dv = dx$. Then from (26-11), $du = dx/(1 + x^2)$ and $v = x$. Substitute these expressions in the integration by parts formula (27-22):

$$\int \text{Arctan } x \, dx = (\text{Arctan } x)x - \int x \cdot \frac{dx}{1 + x^2}$$

$$= x \text{ Arctan } x - \int \frac{x \, dx}{1 + x^2}$$

The integral on the right can be solved using the substitution $w = 1 + x^2$, $dw = 2x \, dx$, and

$$\int \frac{x \, dx}{1 + x^2} = \frac{1}{2} \int \frac{dw}{w} = \frac{1}{2} \ln |w| + C$$

Therefore, the original integral is

$$\int \text{Arctan } x \, dx = x \text{ Arctan } x - \frac{1}{2} \ln (1 + x^2) + C$$

For many integrals, integration by parts may lead to a new integral $\int v \, du$ that is related closely to the original integral $\int u \, dv$, so that the formula must be applied again.

EXAMPLE 39 Calculation of the current $i(t)$ in a critically damped RL circuit involves an integral of the form $i = \int e^{-t} \cos \omega t \, dt$. Complete the integration.

Solution **First,** choose $u = e^{-t}$ and $dv = \cos \omega t \, dt$; then $du = -e^{-t} \, dt$ and $v = (1/\omega) \sin \omega t$. Substitute these expressions in the integration by parts formula (27-22):

$$i = (e^{-t})\left(\frac{1}{\omega} \sin \omega t\right) - \int \left(\frac{1}{\omega} \sin \omega t\right)(-e^{-t} \, dt)$$

$$= \frac{1}{\omega} e^{-t} \sin \omega t + \frac{1}{\omega} \int e^{-t} \sin \omega t \, dt$$

The integral on the right is very similar to the original integral, with $\sin \omega t$ replacing $\cos \omega t$. Applying integration by parts again will produce an integral with the original integrand.

Second, choose $u = e^{-t}$ and $dv = \sin \omega t \, dt$; then $du = -e^{-t} \, dt$ and $v = -(1/\omega) \cos \omega t$. Substituting into (27-22), gives

$$\int e^{-t} \sin \omega t \, dt = (e^{-t})\left(-\frac{1}{\omega} \cos \omega t\right) - \int \left(-\frac{1}{\omega} \cos \omega t\right)(-e^{-t} \, dt)$$

$$= -\frac{1}{\omega} e^{-t} \cos \omega t - \frac{1}{\omega} \int e^{-t} \cos \omega t \, dt$$

Then, the original integral is

$$\int e^{-t} \cos \omega t \, dt = \frac{1}{\omega} e^{-t} \sin \omega t + \frac{1}{\omega} \left[-\frac{1}{\omega} e^{-t} \cos \omega t - \frac{1}{\omega} \int e^{-t} \cos \omega t \, dt \right]$$

or

$$i = \frac{1}{\omega} e^{-t} \sin \omega t - \frac{1}{\omega^2} e^{-t} \cos \omega t - \frac{1}{\omega^2} \cdot i$$

$$i + \frac{1}{\omega^2} \cdot i = \frac{1}{\omega^2} e^{-t} (\omega \sin \omega t - \cos \omega t)$$

$$i = \left(\frac{1}{1 + \frac{1}{\omega^2}} \right) \left(\frac{1}{\omega^2} \right) e^{-t} (\omega \sin \omega t - \cos \omega t)$$

$$i = \frac{1}{1 + \omega^2} e^{-t} (\omega \sin \omega t - \cos \omega t)$$

NOTE ▶ Integrals of the form $\int e^{ax} \sin bx \, dx$ appear often in the study of Fourier series, an important application of mathematics to electronics, which we will examine in detail in Chapter 28. ◀

EXAMPLE 40 Integrate: $\int \sec^3 x \, dx$

Solution The technique of integration by parts can be used with this integral. Choose $u = \sec x$ and $dv = \sec^2 x \, dx$; then $du = \sec x \tan x \, dx$ from equation (26-8) and $v = \int \sec^2 x \, dx = \tan x + C$ from equation (27-6).

Substitute in the integration by parts formula (27-22).

$$\int \sec^3 x \, dx = \sec x \tan x - \int \tan x (\sec x \tan x \, dx)$$

$$= \sec x \tan x - \int \tan^2 x \sec x \, dx$$

But, by (27-15), $\tan^2 x = \sec^2 x - 1$; therefore:

$$= \sec x \tan x - \int (\sec^2 x - 1) \sec x \, dx$$

$$= \sec x \tan x - \int \sec^3 x \, dx + \int \sec x \, dx$$

Add $\int \sec^3 x \, dx$ to both sides of the equation.

$$2 \int \sec^3 x \, dx = \sec x \tan x + \int \sec x \, dx$$

The integral on the right can be integrated using equation (27-12).

$$= \sec x \tan x + \ln |\sec x + \tan x| + C$$

Therefore:

$$\int \sec^3 x \, dx = \frac{1}{2} \sec x \tan x + \frac{1}{2} \ln |\sec x + \tan x| + C$$

Exercises 27-6

Integrate.

1. $\int x \sin x \, dx$ **2.** $\int 2x \cos 4x \, dx$ **3.** $\int x e^{-2x} \, dx$

4. $\int \sqrt{x} \ln x \, dx$ **5.** $\int x (\ln x)^2 \, dx$ **6.** $\int \frac{\ln x \, dx}{x^2}$

7. $\int (\ln y)^3 \, dy$ **8.** $\int_0^{\pi/3} x \cos 4x \, dx$ **9.** $\int_{\pi/4}^{\pi/2} x \sin 2x \, dx$

10. $\int e^x \cos x \, dx$ **11.** $\int e^{2t} \sin 2t \, dt$ **12.** $\int x^2 \sin x \, dx$

13. $\int x^2 \cos 2x \, dx$ **14.** $\int x \sin^2 (3x) \, dx$ **15.** $\int y \tan^2 y \, dy$

16. $\int_0^\pi z \cos z \, dz$ **17.** $\int q \sqrt{q + 2} \, dq$ **18.** $\int \frac{x^3 \, dx}{\sqrt{x^2 + 1}}$

19. $\int_0^\pi x \cos (x - 1) \, dx$ **20.** $\int x(x + 1)^5 \, dx$ **21.** $\int y(y - 2)^4 \, dy$

22. $\int x^3 \sqrt{1 - x^2} \, dx$ **23.** $\int \sec^5 \theta \, d\theta$ **24.** $\int x^2 e^{2x} \, dx$

25. $\int x^3 e^{x^2} \, dx$

Solve.

26. Find the volume of the solid obtained by rotating about the x-axis the region bounded by the curve $y = \ln x$, the x-axis, and the line $x = 2$. Round to the nearest hundredth.

27. Find the area of the region bounded by the curve $y = \text{Arcsin } 2x$, the x-axis, and the line $x = 0.5$. Round to the nearest hundredth.

28. *Physics* Calculate the moment of inertia, with respect to the y-axis, of the solid obtained by rotating about the y-axis the region bounded by the curve $y = e^x$, the x-axis, and the line $x = 1$.

29. *Electronics* When a periodic electromotive force is applied to an *RL* circuit, the current can be calculated using the integral

$$i(t) = \frac{V}{L} \int e^{kt} \sin \omega t \, dt$$

Integrate to find $i(t)$.

30. *Electronics* The following integral appears in the study of critical damping in electrical oscillators:

$$A(t) = \int (c_1 t + c_2)e^{-\alpha t}\, dt$$

where t is the time. Find $A(t)$ for $c_1 = 2$, $c_2 = 3$, and $\alpha = \dfrac{1}{2}$.

31. *Physics* A charged particle in an accelerating device moves with velocity (in km/s) given by the equation $v = ds/dt = 2t\sqrt{1 + t}$. If $s = 0$ at $t = 0$, find $s(t)$.

27-7 | Integration Using Tables

Using algebraic manipulation, trigonometric identities, and the techniques of substitution and integration by parts presented in this chapter, we can rewrite many integrals so that they can be solved using the basic integration formulas. Integrals that are difficult to rewrite, or whose transformation to a basic form requires more advanced techniques or tricky substitutions, may be integrated using either computer programs designed to perform mathematical operations symbolically (see box on p. 1204) or tables of integrals.

A typical table of integrals used in scientific or engineering work may contain hundreds of integrals grouped by type of integrand. (A brief table is given as Appendix D.) For any specific application, it is usually necessary to rewrite the given integral so that it is in a form exactly matching some entry in the table. Rewriting the integral requires that you be able to recognize the basic integral forms and use the integration techniques already discussed. The following examples are designed to provide practice in the use of a table of integrals.

EXAMPLE 41 Verhurst's law of logistic growth states that under certain conditions of growth, the time for a quantity N to grow from size $N = N_1$ to size $N = N_2$ is

$$t = \int_{N_1}^{N_2} \frac{dN}{pN - qN^2}$$

Find t (in seconds) for $N_1 = 10$, $N_2 = 100$, $p = 20$, and $q = 0.1$.

Solution **First,** notice that this integral can be written as

$$t = \int \frac{dN}{N(p - qN)}$$

Search for a section in the table of integrals (Appendix D) listing integrals containing expressions of the form $p - qN$. Section B contains integrals involving expressions of the form $a + bu$, and integral B.22 matches the given form exactly.

Next, substitute the given values into the solution shown in B.22:

Computer Integration

The process of numerical approximation of a definite integral (discussed in Chapter 24) can be performed with a relatively simple computer program. Integration of an indefinite integral has very recently become possible using high-level computer programs arising out of work in artificial intelligence. These programs enable scientists and engineers to perform mathematical operations involving extensive algebraic manipulation symbolically on a computer, replacing the paper and pencil methods normally used.

In a typical system, if you wanted to solve the equation

$$x(3 + x^2) = 4x(1 + a^2) - x$$

you would enter into the computer the command

SOLVE [X*(3 + X∧2) = = 4*X*(1 + a∧2) − X·X];

and get the response

{X = = −2*C,   X = = 2*C,   X = = 0}

or   X = {±2C,0}

To find the value of the integral

$$\int (ax^2 + x \sin x^2)\, dx$$

you would enter

INT[a*X∧2 + X*SIN(X∧2),X];

and get the response

a*X∧3/3 − cos(X∧2)/2   or $\dfrac{1}{3}ax^3 - \dfrac{1}{2}\cos x^2 + c.$

Software systems such as MACSYMA (Symbolics, Inc. Cambridge, Massachusetts) and SMP (Inference Corp., Los Angeles, California) are designed for large, expensive mainframe computers. The muMATH system (Microsoft Corp., Bellevue, Washington) is designed for use with relatively inexpensive microcomputers, and was used to solve many of the exercises and examples in this textbook. The existence of mathematical symbol manipulation systems of this kind will place a high premium on *understanding* mathematical operations and their limitations, and on the ability to translate real-world problems into mathematical form. (For further information, see R. Pavelle, M. Rothstein, and J. Fitch, "Computer Algebra," *Scientific American*, Vol. 245, Dec. 1981, pp. 136–152.)

$$\int \frac{du}{u(a + bu)} = -\frac{1}{a} \ln \left| \frac{a + bu}{u} \right| + C$$

Therefore,

$$\int \frac{dN}{N(p - qN)} = -\frac{1}{p} \ln \left| \frac{p - qN}{N} \right| + C$$

and, substituting the given numerical values, we have

$$t = -\frac{1}{20} \left(\ln \left| \frac{20 - 0.1N}{N} \right| \right) \Bigg]_{10}^{100}$$

$$= -\frac{1}{20} \left[\ln \left| \frac{20 - 10}{100} \right| - \ln \left| \frac{20 - 1}{10} \right| \right]$$

Simplify the fractions and evaluate using a calculator:

$$t = -\frac{1}{20} \left[\ln \frac{1}{10} - \ln \frac{19}{10} \right] = -\frac{1}{20} \left[\ln \frac{\frac{1}{10}}{\frac{19}{10}} \right]$$

$$= -\frac{1}{20} \ln \frac{1}{19} = \frac{1}{20} \ln 19$$

$$\approx 0.15 \text{ s}$$

EXAMPLE 42 Integrate: $\displaystyle\int \frac{x^2 \, dx}{\sqrt{4x^2 - 3}}$

Solution **First,** notice the expression $\sqrt{4x^2 - 3}$ in the denominator of the integrand. Search the table in Appendix D for a list of integrals containing expressions of this form. Section E lists integrals containing $\sqrt{u^2 - a^2}$, which is of this form where $u = 2x$ and $a = \sqrt{3}$. Then $x = u/2$ and $dx = \frac{1}{2} du$, and we can rewrite the given integral as

$$\int \frac{\left(\frac{u}{2}\right)^2 \left(\frac{1}{2} du\right)}{\sqrt{u^2 - a^2}} = \frac{1}{8} \int \frac{u^2 \, du}{\sqrt{u^2 - a^2}} \quad \longleftarrow \quad \boxed{\text{This is } \textit{exactly} \text{ the form of integral E. 64.}}$$

Next, substitute the values for u and a into the solution given in the table of integrals:

$$\int \frac{u^2 \, du}{\sqrt{u^2 - a^2}} = \frac{u}{2} \sqrt{u^2 - a^2} + \frac{a^2}{2} \ln |u + \sqrt{u^2 - a^2}| + C$$

$$\boxed{\text{The + sign here goes with the} \\ - \text{ sign in the denominator.}}$$

Therefore,

$$\int \frac{x^2 \, dx}{\sqrt{4x^2 - 3}} = \frac{1}{8} \left[\frac{2x}{2} \cdot \sqrt{4x^2 - 3} + \frac{3}{2} \ln |2x + \sqrt{4x^2 - 3}| + C \right]$$

$$= \frac{x}{8} \sqrt{4x^2 - 3} + \frac{3}{16} \ln |2x + \sqrt{4x^2 - 3}| + C$$

A **reduction formula** is an integration formula that enables us to write an integral containing some power of an expression in terms of an integral containing a lower power of that expression.

EXAMPLE 43   Integrate: $\int x^2 \cos x \, dx$

Solution   **First,** notice that this integral has the form of F.82 in Appendix D, where $u = x$, $n = 2$, and $a = 1$. Therefore, we can write

$$\int x^2 \cos x \, dx = \frac{x^2}{1} \sin x - \frac{2}{1} \int x^{2-1} \sin x \, dx$$

$$= x^2 \sin x - 2 \int x \sin x \, dx$$

The original integral contained the factor x^2 and this reduction formula has reduced the integration to an integrand with the factor x.

Next, use formula F.79 to integrate $\int x \sin x \, dx$:

$$\int x \sin x \, dx = \sin x - x \cos x + C$$

Finally, substitute this result into the first integration step:

$$\int x^2 \cos x \, dx = x^2 \sin x + 2x \cos x - 2 \sin x - 2C$$

$$= x^2 \sin x + 2x \cos x - 2 \sin x + C$$

> Write the constant of integration as C.

EXAMPLE 44   In a hydrogen atom, the probability that a ground-state electron will be found inside a sphere of radius *s* centered on the nucleus is given by the following integral. Integrate to find *P*.

$$P = \frac{4}{b^3} \int_0^s e^{2r/b} \, r^2 \, dr$$

Solution   **First,** notice that this integral has the form of G.91 in Appendix D, with $r = u$ and $a = 2/b$. Therefore,

$$\int e^{2r/b} r^2 \, dr = \frac{e^{2r/b}}{(2/b)^3}\left[\left(\frac{2}{b}\right)^2 r^2 - 2\left(\frac{2}{b}\right)r + 2\right]$$

$$= \frac{b^3}{8} e^{2r/b}\left(\frac{4r^2}{b^2} - \frac{4r}{b} + 2\right)$$

Then, the original integral becomes

$$P = \frac{4}{b^3} \cdot \frac{b^3}{8}\left[e^{2r/b}\left(\frac{4r^2}{b^2} - \frac{4r}{b} + 2\right)\right]\Bigg|_0^s$$

$$= e^{2r/b}\left(\frac{2r^2}{b^2} - \frac{2r}{b} + 1\right)\Bigg|_0^s$$

$$= e^{2s/b}\left(\frac{2s^2}{b^2} - \frac{2s}{b} + 1\right) - e^0(0 - 0 + 1)$$

$$= e^{2s/b}\left(\frac{2s^2}{b^2} - \frac{2s}{b} + 1\right) - 1$$

Exercises 27-7

Integrate using the table of integrals in Appendix D. Round answers to calculator problems to two decimal places.

1. $\int (2 + 3x)^3 \, dx$

2. $\int 2x(1 + 3x)^4 \, dx$

3. $\int \dfrac{dt}{t(1 - 2t)}$

4. $\int \dfrac{z \, dz}{(2z + 3)^2}$

5. $\int \dfrac{x \, dx}{\sqrt{2 + 5x}}$

6. $\int \sqrt{1 + 3x} \, x^2 \, dx$

7. $\int \dfrac{d\theta}{\sin \theta \cos \theta}$

8. $\int \dfrac{dt}{1 + 2e^t}$

9. $\int_0^1 \dfrac{dq}{(4 - q^2)^{3/2}}$

10. $\int_3^5 \dfrac{dx}{(x^2 - 5)^{3/2}}$

11. $\int \dfrac{dx}{x\sqrt{2 + x}}$

12. $\int x^5 \ln x \, dx$

13. $\int_{0.5}^{1.0} \dfrac{dy}{\sqrt{4y - y^2}}$

14.. $\int_0^{1.5} \dfrac{dx}{3x^2 + 2x + 1}$

15. $\int \dfrac{dx}{x(3 + 5x)}$

16. $\int \dfrac{\sqrt{1 - x^2}}{x^2} \, dx$

17. $\int (x^2 + 5)^{3/2} \, dx$

18. $\int e^{2x} \sin x \, dx$

19. $\int_0^{0.5} 10x^2 e^{2x} \, dx$

20. $\int_0^{\pi/4} e^{4x} \sin 2x \, dx$

21. $\int \dfrac{x^2 \, dx}{\sqrt{2 + x}}$

22. $\int \sin^5 x \, dx$

23. $\int \cot^4 2x \, dx$

24. $\int e^x \sqrt{1 + 2e^x} \, dx$

Solve.

25. *Physics* The following equation appears in the analysis of the forced vibrations of a damped mechanical oscillator subjected to a sinusoidal driving force:

$$v(t) = \int e^{kt} \sin \omega t \, dt$$

Integrate to find $v(t)$ if $v = 0$ at $t = 0$.

26. Calculate the area of the region bounded by the curve $y = x^2 e^{2x}$, the x-axis, and the lines $x = 0$ and $x = 1$. Round to the nearest hundredth.

27. Find the volume of the solid produced by rotating about the x-axis, the region bounded by the curve $y = 3 \cot^2 x$, the x-axis, and the lines $x = 1$ and $x = 2$. Round to the nearest hundredth.

28. *Electronics* Find the average value of the voltage function $V(t) = t^2 e^{10t}$ from $t = 0$ s to $t = 2$ s.

29. *Electronics* Find the RMS value of the voltage function given in Problem 28.

30. *Electronics* If the electric current in a control circuit is described approximately by the equation

$$i(t) = \frac{1}{t\sqrt{t^2 + 100}}$$

find an expression for the charge that passes a given point in the circuit if $q = 0$ when $t = 1$ s.

31. *Robotics* The angular velocity $\omega = d\theta/dt$ of a certain rotating system depends on the time according to the equation $\omega(t) = te^{0.2t}$ (t in seconds). Calculate the total number of revolutions that occur in the first 10 s. $\left[\textit{Hint:}\ \ \text{Number of revolutions} = \frac{1}{2\pi} \int_0^{10} \omega(t) \, dt \right]$

32. *Electronics* In an RL circuit containing a 2.0 H inductor, the induced voltage varies with time according to the equation

$$V(t) = \frac{t}{\sqrt{2.5 + 6.5t}}$$

Find the current flowing after $t = 3.0$ s if the current is 1.2 A at $t = 1.0$ s. $\left[\textit{Hint:}\ \ i(t) = \frac{1}{L} \int V(t) \, dt \right]$

27-8 | Review of Chapter 27

Formulas and Rules

- integral of the reciprocal function (p. 1160)

$$\int \frac{du}{u} = \ln |u| + C \qquad u \neq 0 \qquad \textbf{(27-1)}$$

- integral of the exponential function (p. 1162)

$$\int e^u \, du = e^u + C \qquad \textbf{(27-2)}$$

- integral of a^u (p. 1165)

$$\int a^u \, du = \frac{a^u}{\ln a} + C \qquad a > 0 \qquad \textbf{(27-3)}$$

- integrals of the basic trigonometric functions (p. 1168)

$$\int \cos u \, du = \sin u + C \quad \textbf{(27-4)} \qquad \int \sec u \tan u \, du = \sec u + C \qquad \textbf{(27-9)}$$

$$\int \sin u \, du = -\cos u + C \quad \textbf{(27-5)} \qquad \int \tan u \, du = -\ln |\cos u| + C \qquad \textbf{(27-10)}$$

$$\int \sec^2 u \ du = \tan u + C \quad \textbf{(27-6)}$$

$$\int \csc^2 u \ du = -\cot u + C \quad \textbf{(27-7)}$$

$$\int \csc u \cot u \ du = -\csc u + C \quad \textbf{(27-8)}$$

$$\int \cot u \ du = \ln |\sin u| + C \quad \textbf{(27-11)}$$

$$\int \sec u \ du = \ln |\sec u + \tan u| + C \quad \textbf{(27-12)}$$

$$\int \csc u \ du = \ln |\csc u - \cot u| + C \quad \textbf{(27-13)}$$

- trigonometric identities (p. 1175)

$$\cos^2 x + \sin^2 x = 1 \quad \textbf{(27-14)}$$
$$\tan^2 x + 1 = \sec^2 x \quad \textbf{(27-15)}$$
$$\cot^2 x + 1 = \csc^2 x \quad \textbf{(27-16)}$$

$$\cos^2 x = \frac{1}{2}(1 + \cos 2x) \quad \textbf{(27-17)}$$

$$\sin^2 x = \frac{1}{2}(1 - \cos 2x) \quad \textbf{(27-18)}$$

- integrating products of trigonometric functions (pp. 1175–1185)

Type A: $\displaystyle\int \sin^n u \ du$ or $\displaystyle\int \cos^n u \ du$

Case 1. If n is odd, use (27-14) to rewrite the integrand, and substitute u = cos x in the first integral and u = sin x in the second integral.

Case 2. If n is even, use (27-17) or (27-18) to rewrite the integrand; then integrate using (27-4) or (27-5) and substitution.

Type B: $\displaystyle\int \sin^m u \cos^n u \ du$

Case 1. If at least one of the exponents m or n is an odd positive integer, use (27-14) to rewrite the integrand so substitution can be used. If both are odd, use the factor with the smaller exponent to form the new differential.

Case 2. If m and n are both even positive integers, use (27-14) to rewrite the factor with the smaller exponent. This will result in a type A, case 2 integral.

Type C: $\displaystyle\int \tan^m u \sec^n u \ du$ or $\displaystyle\int \cot^m u \csc^n u \ du$

Case 1A. If m is odd, factor out (sec u tan u du) or (csc u cot u du) and use (27-15) or (27-16) to rewrite the integrand.

Case 1B. If n is even, factor out (sec²u du) or (csc²u du) and use (27-15) or (27-16) to rewrite the integrand.

Case 2. If n is zero, the integrals become

$$\int \tan^m u \ du \quad \text{and} \quad \int \cot^m u \ du$$

If m is odd, factor the integrand as

$$\tan^m u = \tan u \cdot \tan^{m-1} u \quad \text{or} \quad \cot^m u = \cot u \cdot \cot^{m-1} u$$

and use (27-15) or (27-16) to rewrite the factor with the even exponent.
If m is even, factor the integrand as

$$\tan^m u = \tan^2 u \cdot \tan^{m-2} u \quad \text{or} \quad \cot^m u = \cot^2 u \cdot \cot^{m-2} u$$

and use (27-15) or (27-16) to rewrite the factor with the smaller exponent.

Case 3. If m is zero, the integrals become

$$\int \sec^n u \ du \quad \text{or} \quad \int \csc^n u \ du$$

If n is even, use (27-15) or (27-16) to rewrite the integrand as a sum of tangent or cotangent terms.
If n is odd, use integration by parts (Section 27-6).

- trigonometric substitutions (pp. 1189–1194)

$$\int \frac{du}{\sqrt{a^2 - u^2}} = \text{Arcsin}\left(\frac{u}{a}\right) + C \qquad \textbf{(27-19)}$$

$$\int \frac{du}{a^2 + u^2} = \frac{1}{a}\,\text{Arctan}\left(\frac{u}{a}\right) + C \qquad \textbf{(27-20)}$$

$$\int \frac{du}{\sqrt{u^2 \pm a^2}} = \ln\left|u + \sqrt{u^2 \pm a^2}\right| + C \qquad \textbf{(27-21)}$$

See also Table 27-1 (p. 1190).
- integration by parts (p. 1196)

$$\int u\,dv = uv - \int v\,du \qquad \textbf{(27-22)}$$

Exercises 27-8

Integrate. Express definite integrals rounded to three decimal places.

1. $\displaystyle\int \sin^4 x \cos x\,dx$

2. $\displaystyle\int \sin 2x \sqrt{\cos 2x}\,dx$

3. $\displaystyle\int \tan\theta \sec^2\theta\,d\theta$

4. $\displaystyle\int e^{2x}(3 + 2e^{2x})\,dx$

5. $\displaystyle\int \frac{e^x\,dx}{(e^x - 1)^2}$

6. $\displaystyle\int \frac{\text{Arctan } 2t}{1 + 4t^2}\,dt$

7. $\displaystyle\int \frac{(\text{Arccos } x)^3}{\sqrt{1 - x^2}}\,dx$

8. $\displaystyle\int \frac{\cos x\,dx}{\sin^3 x}$

9. $\displaystyle\int \frac{\ln 3t\,dt}{t}$

10. $\displaystyle\int \frac{[\ln (x + 1)]^3}{x + 1}\,dx$

11. $\displaystyle\int \frac{dx}{3 - 2x}$

12. $\displaystyle\int_0^3 \frac{x^2\,dx}{2 + x^3}$

13. $\displaystyle\int_0^{\pi/6} \frac{\cos 3\theta\,d\theta}{1 + \sin 3\theta}$

14. $\displaystyle\int \frac{e^{-2x}\,dx}{1 - e^{-2x}}$

15. $\displaystyle\int (\cos x)\,e^{\sin x}\,dx$

16. $\displaystyle\int e^{-2x}\,dx$

17. $\displaystyle\int_{1.0}^{1.5} x^2 e^{x^3}\,dx$

18. $\displaystyle\int \frac{e^{\text{Arcsin } x}}{\sqrt{1 - x^2}}\,dx$

19. $\displaystyle\int 4^{2x}\,dx$

20. $\displaystyle\int \sec 3\theta\,d\theta$

21. $\displaystyle\int x \cos x^2\,dx$

22. $\displaystyle\int e^{2x}\cot e^{2x}\,dx$

23. $\displaystyle\int \frac{\tan (\ln 2x)}{x}\,dx$

24. $\displaystyle\int \csc^3 3\theta\,d\theta$

25. $\displaystyle\int x^2 \sec x^3 \tan x^3\,dx$

26. $\displaystyle\int_{\pi/6}^{\pi/3} \frac{\sin x}{\cos^2 x}\,dx$

27. $\displaystyle\int \cos^3 3x\,dx$

28. $\displaystyle\int \sin^2(2x + 1)\,dx$

29. $\displaystyle\int \cos^5 x \sin^2 x\,dx$

30. $\displaystyle\int \sin^{1/3} x \cos^3 x\,dx$

31. $\displaystyle\int \sin^3 2x \cos^3 2x\,dx$

32. $\displaystyle\int \cot^3 x \csc^2 x\,dx$

33. $\displaystyle\int \tan^2 3x \sec^4 3x\,dx$

34. $\displaystyle\int \tan^5 x\,dx$

35. $\displaystyle\int \cot^4 6x\,dx$

36. $\displaystyle\int \csc^4 x\,dx$

37. $\displaystyle\int \frac{dt}{\sqrt{36 - t^2}}$

38. $\displaystyle\int \frac{dx}{x^2 + 9}$

39. $\displaystyle\int_0^{4.0} \frac{dx}{x^2 + 6x + 14}$

40. $\displaystyle\int \frac{dx}{\sqrt{x^2 - 6}}$

41. $\displaystyle\int \frac{x^2\,dx}{\sqrt{x^2 + 6}}$

42. $\displaystyle\int \frac{\sqrt{16 - x^2}}{x}\,dx$

43. $\displaystyle\int \frac{e^x\,dx}{\sqrt{1 - e^{2x}}}$

44. $\displaystyle\int_{1.0}^{2.5} te^t\,dt$

45. $\displaystyle\int x \cos 2x\,dx$

46. $\displaystyle\int e^{3x}\cos 3x\,dx$

47. $\displaystyle\int x^2 e^{-x}\,dx$

48. $\displaystyle\int x \sec^2 x\,dx$

Integrate using the table of integrals (Appendix D).

49. $\int x(2 + 3x)^2 \, dx$

50. $\int \dfrac{\theta^2 \, d\theta}{(4\theta^2 - 1)^{3/2}}$

51. $\int \dfrac{x \, dx}{\sqrt{1 + x}}$

52. $\int x^2 \sin 2x \, dx$

Solve.

53. Find the area bounded by the curve $y = \sin^4 x \cos x$, the x-axis, and the line $x = \pi/2$.

54. Determine the equation of the curve whose slope at any point is

$$\frac{dy}{dx} = \frac{1}{x^2 + 2x + 2}$$

if the curve passes through the point $\left(0, \dfrac{1}{2}\right)$

55. Find the volume generated by rotating about the x-axis the area bounded by the curve

$$y = \frac{1}{\sqrt{x^2 + 16}}$$

the x-axis, the y-axis, and the line $x = 2$.

56. Find the moment of inertia (in terms of the denisity δ) with respect to the y-axis of the area bounded by the curve

$$y = \frac{1}{1 + x^3}$$

the x-axis, the y-axis, and the line $x = 3$.

57. Calculate the centroid of the area bounded by the curve $y = \ln x$, the line $x = 3$, and the x-axis.

58. Find the volume generated by rotating about the x-axis the area in the first quadrant bounded by the curve

$$y = \frac{\sec 2x}{\sqrt{1 + \tan 2x}}$$

and the line $x = \pi/6$.

59. *Electronics* If the charge stored in a capacitative circuit increases at the rate $dq/dt = te^{t^2}$ (where q is in coulombs), what amount of charge is added from $t = 0.02$ s to $t = 0.08$ s?

60. *Physics* A proton in an accelerating device moves in a straight line with acceleration

$$s = \frac{10e^t}{e^t - 1}$$

Find an expression for the velocity of this particle ($v = \int a \, dt$).

61. *Electronics* Find the RMS value of the voltage output of a circuit from $t = 0$ s to $t = 1.50$ s if the output varies according to the equation $V = 2.00 \sec^2 t$.

62. *Electronics* If the current in an RC circuit varies according to the equation $i = 4.20 \cos 30\pi t$, find the voltage V_c across a 200 μF $(= 2.00 \times 10^{-4}$ F) capacitor after 0.125 s, if $V_c = 0$ at $t = 0$ s.

$$\left[\text{Hint:}\quad V_c = \frac{1}{C} \int i \, dt \right]$$

63. *Physics* If the acceleration of a vibrating system is given by the equation

$$a = 4.0 \sin \left(3t + \frac{\pi}{2} \right)$$

find its velocity if it is at rest at $t = 0$.

64. *Electronics*   If an electrically charged object is leaking charge at a rate given by the equation

$$\frac{dq}{dt} = \frac{2 \times 10^3}{\sqrt{0.20 - t^2}}$$

where q is in coulombs and t is in seconds, find the net charge lost between $t = 0.10$ s and $t = 0.20$ s.

65. *Physics*   Find the average force exerted from $t = 0$ s to $t = 5.0$ s if the force F (in newtons) varies according to the equation

$$F = \frac{40.0}{t^2 - 4t + 5}$$

66. *Electrical engineering*   If the current in a certain electrical circuit varies with the time according to the equation

$$i(t) = -0.2t^3\sqrt{t^2 + 4.0}$$

where i is in amperes and t is in seconds, what total charge passes through the circuit in the first 2 s?

Series Expansion of Functions | 28

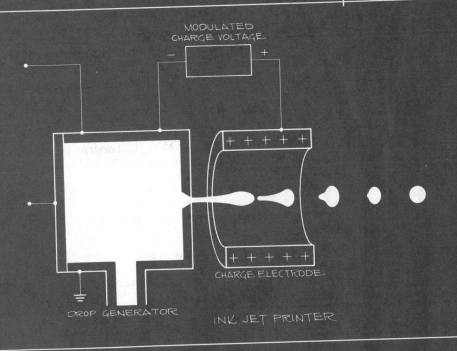

MODULATED
CHARGE VOLTAGE

CHARGE ELECTRODE

DROP GENERATOR

INK JET PRINTER

A glance at a table of trigonometric, exponential, or logarithmic functions, or a few minutes play with a scientific calculator, will show that for most values of the variable x the values of the functions $\sin x$, $\cos x$, e^x, $\ln x$, and so on, are not rational numbers or recognizable irrational numbers expressable as roots. If the internal logic of a calculator or computer is limited to the basic arithmetic operations of addition, subtraction, multiplication, and division, how does it calculate the value of these and other such functions?

In this chapter we will learn how to represent a nonalgebraic function by polynomial series, and this representation will allow us to find approximate values, to any desired accuracy, for many such functions. Furthermore, despite the many techniques of integration discussed so far, there are many functions whose integrals cannot be written as finite algebraic expressions or in terms of any of the functions we have seen. This new way of representing functions by polynomial series will allow us to find the values of such integrals.

If we use algebraic long division to divide $1 - x$ into 1, we get

$$\frac{1}{1 - x} = 1 + x + x^2 + x^3 + \cdots + x^n + \cdots$$

Since x represents a real number, the infinite series on the right is a geometric progression with first term $a = 1$ and common ratio $r = x$. As we discussed in Chapter 18, for $|r| < 1$, the sum of such a geometric progression is

$$S = \frac{a}{1 - r} = \frac{1}{1 - x}$$

which is exactly the expression on the left. Therefore, the infinite series on the right represents the function $f(x) = 1/(1 - x)$ for real values of x in the interval $-1 < x < 1$.

Power series   In general, if $c_0, c_1, c_2, \ldots$ are constants and x is a variable, then a series of the form

$$c_0 + c_1 x + c_2 x^2 + c_3 x^3 + \cdots + c_n x^n + \cdots$$

is called a **power series** in x. The real numbers c_n are called the *coefficients* of the series.

Convergence   An infinite series $a_1 + a_2 + a_3 + \cdots + a_n + \cdots$ has associated with it the sequence of partial sums

$$S_1 = a_1$$

$$S_2 = a_1 + a_2$$

$$S_3 = a_1 + a_2 + a_3$$

$$\vdots$$

$$S_n = a_1 + a_2 + a_3 + \cdots + a_n$$

The series is said to be **convergent** or to **converge** if $\lim_{n \to \infty} S_n = S$, where S, if it exists, is some real number called the **sum of the series**.* The series is said to be **divergent** or to **diverge** if this limit does not exist. A series that diverges has no sum.

*In order for the series $a_1 + a_2 + a_3 + \cdots + a_n$ to converge it must be true that $\lim_{n \to \infty} a_n = 0$, but many series whose general term approaches zero do *not* converge. Just because the terms get smaller does not guarantee that the series converges. For example, the series $1 + \frac{1}{2} + \frac{1}{3} + \cdots + \frac{1}{n} + \cdots$ diverges.

Integrate using the table of integrals (Appendix D).

49. $\displaystyle\int x(2 + 3x)^2 \, dx$

50. $\displaystyle\int \frac{\theta^2 \, d\theta}{(4\theta^2 - 1)^{3/2}}$

51. $\displaystyle\int \frac{x \, dx}{\sqrt{1 + x}}$

52. $\displaystyle\int x^2 \sin 2x \, dx$

Solve.

53. Find the area bounded by the curve $y = \sin^4 x \cos x$, the x-axis, and the line $x = \pi/2$.

54. Determine the equation of the curve whose slope at any point is

$$\frac{dy}{dx} = \frac{1}{x^2 + 2x + 2}$$

if the curve passes through the point $\left(0, \dfrac{1}{2}\right)$

55. Find the volume generated by rotating about the x-axis the area bounded by the curve

$$y = \frac{1}{\sqrt{x^2 + 16}}$$

the x-axis, the y-axis, and the line $x = 2$.

56. Find the moment of inertia (in terms of the denisity δ) with respect to the y-axis of the area bounded by the curve

$$y = \frac{1}{1 + x^3}$$

the x-axis, the y-axis, and the line $x = 3$.

57. Calculate the centroid of the area bounded by the curve $y = \ln x$, the line $x = 3$, and the x-axis.

58. Find the volume generated by rotating about the x-axis the area in the first quadrant bounded by the curve

$$y = \frac{\sec 2x}{\sqrt{1 + \tan 2x}}$$

and the line $x = \pi/6$.

59. *Electronics*   If the charge stored in a capacitative circuit increases at the rate $dq/dt = te^{t^2}$ (where q is in coulombs), what amount of charge is added from $t = 0.02$ s to $t = 0.08$ s?

60. *Physics*   A proton in an accelerating device moves in a straight line with acceleration

$$s = \frac{10e^t}{e^t - 1}$$

Find an expression for the velocity of this particle ($v = \int a \, dt$).

61. *Electronics*   Find the RMS value of the voltage output of a circuit from $t = 0$ s to $t = 1.50$ s if the output varies according to the equation $V = 2.00 \sec^2 t$.

62. *Electronics*   If the current in an RC circuit varies according to the equation $i = 4.20 \cos 30\pi t$, find the voltage V_c across a $200 \; \mu F$ ($= 2.00 \times 10^{-4}$ F) capacitor after 0.125 s, if $V_c = 0$ at $t = 0$ s.

$$\left[\text{Hint: } \; V_c = \frac{1}{C} \int i \, dt\right]$$

63. *Physics*   If the acceleration of a vibrating system is given by the equation

$$a = 4.0 \sin\left(3t + \frac{\pi}{2}\right)$$

find its velocity if it is at rest at $t = 0$.

64. Electronics   If an electrically charged object is leaking charge at a rate given by the equation

$$\frac{dq}{dt} = \frac{2 \times 10^3}{\sqrt{0.20 - t^2}}$$

where q is in coulombs and t is in seconds, find the net charge lost between $t = 0.10$ s and $t = 0.20$ s.

65. Physics   Find the average force exerted from $t = 0$ s to $t = 5.0$ s if the force F (in newtons) varies according to the equation

$$F = \frac{40.0}{t^2 - 4t + 5}$$

66. Electrical engineering   If the current in a certain electrical circuit varies with the time according to the equation

$$i(t) = -0.2t^3\sqrt{t^2 + 4.0}$$

where i is in amperes and t is in seconds, what total charge passes through the circuit in the first 2 s?

With every power series there is associated an **interval of convergence** such that the series converges if x is in the interval and diverges if x is outside the interval. For example, the power series $1 + x + x^2 + x^3 + \cdots$ converges on the interval $-1 < x < 1$.

Many functions, but not all functions, can be expressed as a power series in x:

$$f(x) = c_0 + c_1 x + c_2 x^2 + c_3 x^3 + \cdots + c_n x^n + \cdots$$

If the series converges, then we know that for a larger and larger number of terms, the difference between the value of the function and the sum on the right tends to zero. We say that the function $f(x)$ is represented by the power series expansion on the given interval of convergence.

This is not true for all functions or for all values of x for a given function, but it is true for values of x within an interval for a broad class of functions called **analytic functions.** Many mathematical descriptions of scientific and technical phenomena can be expressed in terms of analytic functions.

To specify the power series exactly and assure that it fits the given function, we must evaluate the coefficients $c_0, c_1, c_2, \ldots, c_n$. One way to do this is to determine the successive derivatives of the function and the series, and equate them at $x = 0$:

$$f(x) = c_0 + c_1 x + c_2 x^2 + c_3 x^3 + c_4 x^4 + \cdots \qquad\qquad f(0) = c_0$$

$$f'(x) = c_1 + 2c_2 x + 3c_3 x^2 + 4c_4 x^3 + \cdots \qquad\qquad f'(0) = c_1$$

$$f''(x) = 2c_2 + 2\cdot 3c_3 x + 3\cdot 4c_4 x^2 + \cdots \qquad\qquad f''(0) = 2c_2$$

$$f'''(x) = 2\cdot 3c_3 + 2\cdot 3\cdot 4c_4 x + \cdots \qquad\qquad f'''(0) = 2\cdot 3c_3$$

and so on. In each expression on the right, when $x = 0$, all terms except the first are equal to zero. For the nth term,

$$f^{(n)}(x) = 2\cdot 3\cdot 4 \cdots\cdot nc_n + \cdots \qquad\qquad f^{(n)}(0) = 2\cdot 3\cdot 4 \cdots\cdot nc_n$$

$$= n!\, c_n$$

When we solve for the coefficients, we obtain

$$c_0 = f(0) \qquad c_1 = f'(0) \qquad c_2 = \frac{1}{2}f''(0) = \frac{1}{2!}f''(0)$$

$$c_3 = \frac{1}{2\cdot 3}f'''(0) = \frac{1}{3!}f'''(0)$$

and so on. In general,

$$c_n = \frac{f^{(n)}(0)}{n!}$$

Maclaurin series   The coefficients determined in this way assure that the power series fits the function $f(x)$ over some interval centered at $x = 0$. The series determined in this way for some function $f(x)$ that can be differentiated n times at $x = 0$ is called the **Maclaurin series** or Maclaurin expansion of $f(x)$.

Maclaurin Series

$$f(x) = f(0) + f'(0)x + \frac{f''(0)x^2}{2!} + \frac{f'''(0)x^3}{3!} + \cdots$$

$$+ \frac{f^{(n)}(0)x^n}{n!} + \cdots \qquad (28\text{-}1)$$

This series is named after Colin Maclaurin (1698–1746), the Scottish mathematician who studied it. The series actually appeared earlier in the work of the Swiss physician and mathematician Johann Bernoulli (1667–1748), but his book on differential calculus, which was written in 1691, was not published until 1922, long after Maclaurin's name had become associated with the series.

EXAMPLE 1   Find the Maclaurin series expansion of $f(x) = e^x$.

Solution   **First,** differentiate the function using equation (26-15); then evaluate each derivative at $x = 0$:

$$f(x) = e^x \longrightarrow f(0) = e^0 = 1$$
$$f'(x) = e^x \longrightarrow f'(0) = e^0 = 1$$
$$f''(x) = e^x \longrightarrow f''(0) = e^0 = 1$$

and so on.

Next, substitute these values into (28-1) to find the Maclaurin series corresponding to $f(x) = e^x$:

$$e^x = 1 + 1 \cdot x + \frac{1 \cdot x^2}{2!} + \frac{1 \cdot x^3}{3!} + \cdots$$

$$= 1 + x + \frac{x^2}{2} + \frac{x^3}{6} + \cdots$$

The series on the right represents e^x for all x in the interval $-\infty < x < \infty$, the interval over which the series converges.

In Fig. 28-1, we have sketched the graph of $f(x) = e^x$ and the approximations to it provided by including successively more terms of its Maclaurin expansion.

EXAMPLE 2
(a)   Find the Maclaurin series expansion of $f(x) = \sin x$.
(b)   An electronics engineer designing the chip for a calculating device needs a set of instructions (an *algorithm*) that will enable him to program the function $f(x) = \cos x$ using only the arithmetic operations addition, subtraction, multiplication, and division. Find such a formula from the Maclaurin expansion of $f(x) = \cos x$.

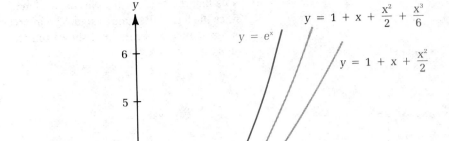

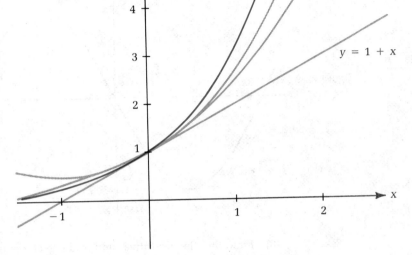

Figure 28-1

Solution

(a) **First,** find the derivatives and evaluate at x = 0:

$$f(x) = \sin x \longrightarrow f(0) = \sin 0 = 0$$

$$f'(x) = \cos x \longrightarrow f'(0) = \cos 0 = 1$$

$$f''(x) = -\sin x \longrightarrow f''(0) = -\sin 0 = 0$$

$$f'''(x) = -\cos x \longrightarrow f'''(0) = -\cos 0 = -1$$

Because $f^{(4)}(x) = \sin x$, this pattern of values for the derivatives at x = 0 will repeat endlessly.

Next, substitute these values into (28-1) to find the Maclaurin series corresponding to $f(x) = \sin x$:

$$\sin x = 0 + 1 \cdot x + \frac{0 \cdot x^2}{2!} + \frac{(-1)x^3}{3!} + \frac{0 \cdot x^4}{4!} + \frac{1 \cdot x^5}{5!} + \frac{0 \cdot x^6}{6!}$$

$$+ \frac{(-1)x^7}{7!} + \cdots$$

$$= x - \frac{x^3}{3!} + \frac{x^5}{5!} - \frac{x^7}{7!} + \cdots$$

Note that the Maclaurin series expansion of sin x is an *alternating* series, since the terms alternate in sign. The interval of convergence of this series is $-\infty < x < \infty$. Figure 28-2 shows the close fit of this expansion in the interval around $x = 0$.

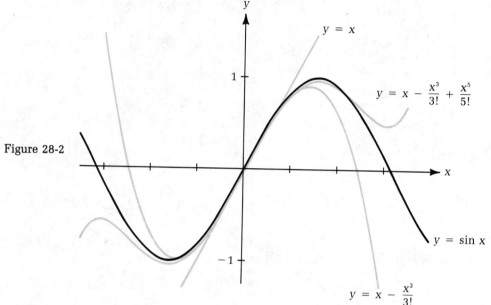

Figure 28-2

(b) **First,** find the derivatives and evaluate at $x = 0$:

$$f(x) = \cos x \qquad f(0) = \cos 0 = 1$$

$$f'(x) = -\sin x \qquad f'(0) = -\sin 0 = 0$$

$$f''(x) = -\cos x \qquad f''(0) = -\cos 0 = -1$$

$$f'''(x) = \sin x \qquad f'''(0) = \sin 0 = 0$$

As with $f(x) = \sin x$, this pattern of derivatives repeats endlessly. **Next,** substitute in (28-1) to get

$$\cos x = 1 + 0 \cdot x + \frac{(-1)x^2}{2!} + \frac{0 \cdot x^3}{3!} + \frac{1 \cdot x^4}{4!} + \frac{0 \cdot x^5}{5!} + \frac{(-1)x^6}{6!} + \cdots$$

$$= 1 - \frac{x^2}{2!} + \frac{x^4}{4!} - \frac{x^6}{6!} + \cdots$$

The value of the series on the right can be calculated using simple arithmetic operations for any real number x.

Once the Maclaurin series for $f(x)$ is known, the series for $f(ax)$ can be found by substitution.

EXAMPLE 3 Find the Maclaurin series expansion for:
(a) $f(x) = \ln(1 + x)$ (b) $f(x) = \text{Arcsin } x$ (c) $f(x) = \text{Arcsin } 2x$

Solution

(a) **First,** find the successive derivatives and evaluate each at $x = 0$:

$$f(x) = \ln (1 + x) \longrightarrow f(0) = \ln (1 + 0) = \ln 1 = 0$$

$$f'(x) = \frac{1}{1 + x} \longrightarrow f'(0) = \frac{1}{1 + 0} = 1$$

$$f''(x) = -\frac{1}{(1 + x)^2} \longrightarrow f''(0) = -\frac{1}{(1 + 0)^2} = -1$$

$$f'''(x) = \frac{2}{(1 + x)^3} \longrightarrow f'''(0) = \frac{2}{(1 + 0)^3} = 2$$

$$f^{(4)}(x) = -\frac{2 \cdot 3}{(1 + x)^4} \longrightarrow f^{(4)}(0) = -\frac{2 \cdot 3}{(1 + 0)^4} = -6$$

Next, substitute into (28-1) to find the Maclaurin expansion of $f(x)$:

$$\ln (1 + x) = 0 + 1 \cdot x + \frac{(-1)x^2}{2!} + \frac{2x^3}{3!} + \frac{(-6)x^4}{4!} + \cdots$$

$$= x - \frac{x^2}{2} + \frac{x^3}{3} - \frac{x^4}{4} + \cdots$$

If you are clever at finding patterns, you may have noticed that the successive values of the derivatives at $x = 0$ alternate in sign and have magnitude $n!$ for the $(n + 1)$th derivative. This series represents the function $\ln (1 + x)$ in the interval $-1 < x \leq 1$.

(b) $\quad f(x) = \text{Arcsin } x \longrightarrow f(0) = \text{Arcsin } 0 = 0$

$$f'(x) = \frac{1}{\sqrt{1 - x^2}}$$

$$= (1 - x^2)^{-1/2} \longrightarrow f'(0) = \frac{1}{\sqrt{1 - 0}} = 1 \quad \text{From (26-9)}$$

$$f''(x) = x(1 - x^2)^{-3/2} \longrightarrow f''(0) = 0(1 - 0)^{-3/2} = 0$$

$$f'''(x) = 3x^2(1 - x^2)^{-5/2} + (1 - x^2)^{-3/2}$$

$$\longrightarrow f'''(0) = 3 \cdot 0(1 - 0)^{-5/2} + (1 - 0)^{-3/2} = 1$$

$$f^{(4)}(x) = 15x^3(1 - x^2)^{-7/2} + 9x(1 - x^2)^{-5/2} \longrightarrow f^{(4)}(0) = 0$$

$$f^{(5)}(x) = 105x^4(1 - x^2)^{-9/2} + 90x^2(1 - x^2)^{-7/2} + 9(1 - x^2)^{-5/2}$$

$$\longrightarrow f^{(5)}(0) = 9$$

Substituting into (28-1), we have

$$\text{Arcsin } x = 0 + 1 \cdot x + 0 \cdot \frac{x^2}{2!} + 1 \cdot \frac{x^3}{3!} + 0 \cdot \frac{x^4}{4!} + 9 \cdot \frac{x^5}{5!} + \cdots$$

$$= x + \frac{x^3}{3!} + \frac{9x^5}{5!} + \cdots$$

(c)   Substitute 2x for x in the equation for Arcsin x from part (b):

$$\text{Arcsin } 2x = (2x) + \frac{(2x)^3}{3!} + \frac{9(2x)^5}{5!} + \cdots$$

$$= 2x + \frac{4x^3}{3} + \frac{12x^5}{5} + \cdots$$

If the function $f(x)$ is a polynomial, then its Maclaurin expansion is the finite series $f(x)$ itself.

EXAMPLE 4   Find the Maclaurin series expansion of the function $f(x) = 4x^3 - x^2 + 2x - 3$.

Solution $f(x) = 4x^3 - x^2 + 2x - 3 \longrightarrow f(0) = -3$

$f'(x) = 12x^2 - 2x + 2 \longrightarrow f'(0) = 2$

$f''(x) = 24x - 2 \longrightarrow f''(0) = -2$

$f'''(x) = 24 \longrightarrow f'''(0) = 24$

All further derivatives are equal to zero. Then, by (28-1),

$$f(x) = (-3) + 2 \cdot x + \frac{(-2)x^2}{2!} + \frac{24x^3}{3!}$$

$$= -3 + 2x - x^2 + 4x^3$$

$$= 4x^3 - x^2 + 2x - 3$$

and this is the original polynomial.

EXAMPLE 5   Find the first four terms of the Maclaurin series expansion of the function $f(x) = \sqrt{1 + x}$.

Solution $f(x) = \sqrt{1 + x} = (1 + x)^{1/2} \longrightarrow f(0) = (1 + 0)^{1/2} = 1$

$f'(x) = \frac{1}{2}(1 + x)^{-1/2} \longrightarrow f'(0) = \frac{1}{2}(1 + 0)^{-1/2} = \frac{1}{2}$

$f''(x) = -\frac{1}{4}(1 + x)^{-3/2} \longrightarrow f''(0) = -\frac{1}{4}(1 + 0)^{-3/2} = -\frac{1}{4}$

$f'''(x) = \frac{3}{8}(1 + x)^{-5/2} \longrightarrow f'''(0) = \frac{3}{8}(1 + 0)^{-5/2} = \frac{3}{8}$

Substitute in (28-1) to find the first four terms of the Maclaurin series expansion for $f(x)$:

$$\sqrt{1 + x} = 1 + \left(\frac{1}{2}\right)x + \left(-\frac{1}{4}\right)\frac{x^2}{2!} + \left(\frac{3}{8}\right)\frac{x^3}{3!} + \cdots$$

$$= 1 + \frac{x}{2} - \frac{x^2}{8} + \frac{x^3}{16} - \cdots$$

Notice that this is the same series we obtain if we expand the expression $(1 + x)^{1/2}$ using the binomial series introduced in Chapter 18.

In general, the power series

$$(1 + x)^n = 1 + nx + \frac{n(n - 1)x^2}{2!} + \frac{n(n - 1)(n - 2)x^3}{3!} + \cdots$$

where n is any real number, is a Maclaurin series that converges for all x in the interval $-1 < x < 1$.

Exercises 28-1

Find the first three nonzero terms of the Maclaurin series expansion of each function.

1. $f(x) = 4 \cos x$ **2.** $f(x) = x \sin x$ **3.** $f(x) = xe^{-x}$

4. $f(x) = \dfrac{e^x}{1 + 2x}$ **5.** $f(x) = \ln (1 - x)$ **6.** $f(x) = \ln (1 - 2x)$

7. $f(x) = \sqrt{1 + 2x}$ **8.** $f(x) = \dfrac{1}{\sqrt{x + 1}}$ **9.** $f(x) = \dfrac{1}{1 + x^2}$

10. $f(t) = \dfrac{1}{2t + 1}$ **11.** $f(x) = (x^2 - 1)^2$ **12.** $f(x) = x^3 - x^2 + 1$

13. $f(\theta) = \cos^2 \theta$ **14.** $f(t) = t^2 \cos t$ **15.** $f(x) = xe^{3x}$

16. $f(x) = x^2 e^{-x}$ **17.** $f(x) = \sqrt[3]{1 + x}$ **18.** $f(x) = \dfrac{1}{(2 + x)^{1/3}}$

Find the first two nonzero terms of the Maclaurin series expansion of each function.

19. $f(\theta) = \tan \theta$ **20.** $f(\theta) = \ln (1 + \cos \theta)$ **21.** $f(x) = \text{Arctan } x \quad x + 1$

22. $f(x) = \text{Arccos } x$ **23.** $f(t) = e^t \cos t$ **24.** $f(x) = e^{\sin x}$

25. $f(\theta) = \sin^2 \theta$ **26.** $f(\theta) = \dfrac{1 + 2 \sin \theta}{1 + \theta}$ **27.** $f(x) = \ln (1 + 2x)$

28. $f(x) = \sin (e^x)$

Solve.

29. Electronics The hyperbolic cosine function $\cosh x$, defined as

$$\cosh x = \frac{1}{2}(e^x + e^{-x})$$

is often used in engineering work. Find the first four terms of the Maclaurin series expansion of this function so that an electronics engineer could program it into a computer chip using only the basic arithmetic operations.

30. *Hydraulics*   The function

$$E(x) = \sec x$$

defines the Euler numbers used in the study of fluid flow. Find the first three nonzero terms of the Maclaurin expansion of this function.*

31. *Physics*   In the study of heat flow along a uniform bar, the function $F(x) = \cos ax + \sin ax$, where a is a constant, appears in the description of the temperature distribution along the bar. Find the first four terms of the Maclaurin expansion of this function.

32. *Physics*   When a small mass is hung on a vertically fixed spring and allowed to oscillate with damping proportional to its speed, its displacement can be described by an equation of the form $y(t) = e^{-t}(\cos t - \sin t)$. Find the first three terms of the Maclaurin expansion of this function.

28-2 | Operating with Series

Table 28-1 lists the Maclaurin series expansions of a few important functions. We can use these basic series to find the series expansions of many related functions without the necessity of using definition (28-1).

TABLE 28-1

Function	Maclaurin Series Expansion	Interval of Convergence
$e^x = 1 + x + \dfrac{x^2}{2!} + \dfrac{x^3}{3!} + \cdots$		$-\infty < x < \infty$
$\sin x = x - \dfrac{x^3}{3!} + \dfrac{x^5}{5!} - \dfrac{x^7}{7!} + \cdots$		$-\infty < x < \infty$
$\cos x = 1 - \dfrac{x^2}{2!} + \dfrac{x^4}{4!} - \dfrac{x^6}{6!} + \cdots$		$-\infty < x < \infty$
$\ln(1 + x) = x - \dfrac{x^2}{2} + \dfrac{x^3}{3} - \dfrac{x^4}{4} + \cdots$		$-1 < x \leq 1$
$\dfrac{1}{1 + x} = 1 - x + x^2 - x^3 + \cdots$		$-1 < x < 1$
$(1 + x)^n = 1 + nx + \dfrac{n(n-1)x^2}{2!}$		
$\qquad\qquad + \dfrac{n(n-1)(n-2)x^3}{3!} + \cdots$		$-1 < x < 1$ for any real number n

One very valuable property of power series is that many of the operations that are valid for finite polynomials may be applied to them. The operations of substitution of a variable, addition, subtraction, multiplication, division, and term-by-term differentiation and integration may be performed with these series.

*Erwin Kreyszig, *Advanced Engineering Mathematics*, 4th edition (New York: Wiley, 1979), p. 703

EXAMPLE 6 Find the first four terms of the Maclaurin series expansion of:

(a) e^{-x^2} (b) $\sin(2x^2)$ (c) $\dfrac{1}{2 - 4x^3}$

Solution

(a)   In the series for e^x in Table 28-1, substitute $-x^2$ for x:

$$e^{-x^2} = 1 + (-x^2) + \frac{(-x^2)^2}{2!} + \frac{(-x^2)^3}{3!} + \cdots$$

$$= 1 - x^2 + \frac{x^4}{2!} - \frac{x^6}{3!} + \cdots$$

(b)   In the series for $\sin x$ in Table 28-1, substitute $2x^2$ for x:

$$\sin(2x^2) = (2x^2) - \frac{(2x^2)^3}{3!} + \frac{(2x^2)^5}{5!} - \frac{(2x^2)^7}{7!} + \cdots$$

$$= 2x^2 - \frac{8x^6}{3!} + \frac{32x^{10}}{5!} - \frac{128x^{14}}{7!} + \cdots$$

(c)   **First,** rewrite as

$$\frac{1}{2 - 4x^3} = \frac{1}{2} \cdot \frac{1}{1 - 2x^3}$$

Then, in the series for $1/(1 + x)$ in Table 28-1, substitute $-2x^3$ for x:

$$\frac{1}{2 - 4x^3} = \frac{1}{2}[1 - (-2x^3) + (-2x^3)^2 - (-2x^3)^3 + \cdots]$$

$$= \frac{1}{2}[1 + 2x^3 + 4x^6 + 8x^9 + \cdots]$$

$$= \frac{1}{2} + x^3 + 2x^6 + 4x^9 + \cdots$$

Multiplication of series   **EXAMPLE 7**   Find the first three terms of the Maclaurin series expansion for $e^{-x}\sin x$.

Solution   We may multiply the series for e^{-x} and $\sin x$ to obtain a series expansion for the function $e^{-x}\sin x$. From Table 28-1, substituting $-x$ for x in the series for e^x, we have

$$e^{-x} = 1 - x + \frac{x^2}{2!} - \frac{x^3}{3!} + \cdots = 1 - x + \frac{x^2}{2} - \frac{x^3}{6} + \cdots$$

and

$$\sin x = x - \frac{x^3}{6} + \frac{x^5}{120} - \cdots$$

Multiply term-by-term:

$$1 - x + \frac{x^2}{2} - \frac{x^3}{6} + \cdots$$

$$x - \frac{x^3}{6} + \frac{x^5}{120} - \cdots$$

$$x - x^2 + \frac{x^3}{2} - \frac{x^4}{6} + \cdots$$

$$- \frac{x^3}{6} + \frac{x^4}{6} - \frac{x^5}{12} + \cdots$$

$$+ \frac{x^5}{120} + \cdots$$

$$x - x^2 + \frac{x^3}{3} + 0 + \cdots \quad \longleftarrow \quad \boxed{\begin{array}{l}\text{Add like terms}\\\text{of the product.}\end{array}}$$

Therefore,

$$e^{-x} \sin x = x - x^2 + \frac{x^3}{3} - \cdots \qquad \text{for all } x.$$

CAUTION ▶ Be careful to include enough terms in each series so that the product has the desired number of terms. In this example, finding the fourth term of the series would require that the fifth term of e^{-x} be used in the multiplication. ◀

Division of series

EXAMPLE 8 Find the first three terms of the Maclaurin series expansion of $\tan x$.

Solution Rather than determine the series from the definition (28-1), we can write

$$\tan x = \frac{\sin x}{\cos x}$$

and divide the series given in Table 28-1 for $\sin x$ and $\cos x$. We will use the "long-division" process for polynomials:

$$
\begin{array}{r}
x + \dfrac{x^3}{3} + \dfrac{2x^5}{15} + \cdots \\[2mm]
1 - \dfrac{x^2}{2} + \dfrac{x^4}{24} - \cdots \overline{\big)\, x - \dfrac{x^3}{6} + \dfrac{x^5}{120} - \cdots} \\[2mm]
x - \dfrac{x^3}{2} + \dfrac{x^5}{24} + \cdots \\[2mm]
\hline
\dfrac{x^3}{3} - \dfrac{x^5}{30} + \cdots \\[2mm]
\dfrac{x^3}{3} - \dfrac{x^5}{6} + \cdots
\end{array}
$$

$$\boxed{-\frac{x^3}{6} - \left(-\frac{x^3}{2}\right) = -\frac{x^3}{6} + \frac{3x^3}{6} = \frac{2x^3}{6} = \frac{x^3}{3}}$$

$$\dfrac{2x^5}{15} + \cdots$$

$$\dfrac{2x^5}{15} + \cdots$$

$$-\dfrac{x^5}{30} - \left(-\dfrac{x^5}{6}\right) = -\dfrac{x^5}{30} + \dfrac{5x^5}{30}$$

$$= \dfrac{4x^5}{30} = \dfrac{2x^5}{15}$$

Therefore,

$$\tan x = x + \dfrac{x^3}{3} + \dfrac{2x^5}{15} + \cdots \qquad \text{for } -\dfrac{\pi}{2} < x < \dfrac{\pi}{2}$$

Differentiation   A power series can be differentiated and integrated term-by-term within its interval of convergence. This is not true for all series, and it is true for a power series only *within* the interval of convergence and not at its endpoints. This fact enables us to differentiate and integrate functions that are difficult to work with in any other way.

EXAMPLE 9 If the electric potential $V(x)$ along the center line of an aperture in a charged surface is given by the equation

$$V(x) = \ln\left(\dfrac{1+x}{1-x}\right)$$

find a three-term polynomial approximation to the electric field $E(x)$ if

$$E(x) = \dfrac{d}{dx}V(x)$$

Solution **First,** find the Maclaurin series expansion of $V(x)$. Note that

$$\ln\left(\dfrac{1+x}{1-x}\right) = \ln(1+x) - \ln(1-x)$$

From Table 28-1, substitute $-x$ for x in the series for $\ln(1+x)$:

$$\ln(1-x) = (-x) - \dfrac{(-x)^2}{2} + \dfrac{(-x)^3}{3} - \dfrac{(-x)^4}{4} + \cdots$$

$$= -x - \dfrac{x^2}{2} - \dfrac{x^3}{3} - \dfrac{x^4}{4} - \cdots$$

But

$$\ln(1+x) = x - \dfrac{x^2}{2} + \dfrac{x^3}{3} - \dfrac{x^4}{4} + \cdots$$

so

$$V(x) = \ln\left(\dfrac{1+x}{1-x}\right) = \ln(1+x) - \ln(1-x)$$

$$= \left(x - \frac{x^2}{2} + \frac{x^3}{3} - \frac{x^4}{4} + \frac{x^5}{5} + \cdots \right)$$

$$- \left(-x - \frac{x^2}{2} - \frac{x^3}{3} - \frac{x^4}{4} - \frac{x^5}{5} - \cdots \right)$$

$$= 2x + \frac{2x^3}{3} + \frac{2x^5}{5} + \cdots$$

Next, find the series representation of $E(x)$ by differentiating $V(x)$ term-by-term:

$$E(x) = \frac{d}{dx} V(x) = \frac{d}{dx}(2x) + \frac{d}{dx}\left(\frac{2x^3}{3}\right) + \frac{d}{dx}\left(\frac{2x^5}{5}\right) + \cdots$$

$$= 2 + 2x^2 + 2x^4 + \cdots$$

This series expansion is valid for all x in the interval $-1 < x < 1$.

NOTE   We could have solved this problem by first taking the derivative of $V(x)$ and then finding the Maclaurin series expansion of this derivative function. The resulting series would be identical to that obtained in the solution shown. It is possible to prove that the Maclaurin series representation of a function is unique. ◄

Integration   Many integrals that occur in scientific or engineering work involve integrands for which antiderivatives do not exist. That is, there is no expression involving any finite combination of the algebraic or transcendental functions we have seen so far that has the integrand for its derivative. For example, the integrals

$$\int \frac{e^{ax}}{x} dx \qquad \int \frac{dx}{\ln x} \qquad \int \frac{\sin x}{x} dx \qquad \int \frac{\text{Arctan } x}{x} dx \qquad \int \frac{dx}{\sqrt{1 + x^3}}$$

are all of this kind. None of the techniques of integration discussed up to this point will enable you to evaluate these integrals.* But if we expand the integrand in a power series, it may be integrated term-by-term to obtain a series representation of the integral.

EXAMPLE 10   Use a Maclaurin expansion to find the value of the integral

$$\int \frac{e^{2x}}{x} dx$$

Solution   **First,** find the Maclaurin series expansion of e^{2x} by substituting $2x$ for x in the series for e^x given in Table 28-1:

---

*Numerical integration will give an approximate value for the definite integral of any continuous function, but it will not help us evaluate the indefinite integral.

$$e^{2x} = 1 + (2x) + \frac{(2x)^2}{2!} + \frac{(2x)^3}{3!} + \frac{(2x)^4}{4!} + \cdots$$

$$= 1 + 2x + 2x^2 + \frac{4x^3}{3} + \frac{2x^4}{3} + \cdots$$

Then,

$$\frac{e^{2x}}{x} = \frac{1}{x} + \frac{2x}{x} + \frac{2x^2}{x} + \frac{4x^3}{3x} + \frac{2x^4}{3x} + \cdots$$

$$= \frac{1}{x} + 2 + 2x + \frac{4x^2}{3} + \frac{2x^3}{3} + \cdots$$

Finally, the integral can be found by integrating this series term-by-term:

$$\int \frac{e^{2x}}{x} \, dx = \int \frac{dx}{x} + \int 2 \, dx + \int 2x \, dx + \int \frac{4x^2}{3} \, dx + \int \frac{2x^3}{3} \, dx + \cdots$$

$$= C + \ln |x| + 2x + x^2 + \frac{4x^3}{9} + \frac{x^4}{6} + \cdots \qquad \text{where } x \neq 0$$

Notice that we have combined the constants of integration into a single constant written at the start of the infinite sum.

EXAMPLE 11 If the current i (in amperes) in a damped oscillator decreases according to the function

$$i(t) = \frac{i_0 \sin t}{t}$$

where t is time in seconds and i_0 is the initial current, find the average value of the current from $t = 0$ s to $t = 1.2$ s.

Solution The average value of $i(t)$ is

$$i_{\text{AVE}} = \frac{i_0}{1.2 - 0} \int_0^{1.2} \frac{\sin t}{t} \, dt$$

Since this function is not integrable using the standard techniques, we will evaluate it by expanding as a Maclaurin series. Divide each term of the series given for $\sin t$ in Table 28-1 by t:

$$\frac{\sin t}{t} = \frac{t}{t} - \frac{t^3}{3! \, t} + \frac{t^5}{5! \, t} - \frac{t^7}{7! \, t} + \cdots$$

$$= 1 - \frac{t^2}{6} + \frac{t^4}{120} - \frac{t^6}{5040} + \cdots$$

Integrate term-by-term:

$$i_{\text{AVE}} = \frac{i_0}{1.2} \int_0^{1.2} \frac{\sin t}{t} \, dt \approx \frac{i_0}{1.2} \left(\int_0^{1.2} dt - \int_0^{1.2} \frac{t^2 \, dt}{6} + \int_0^{1.2} \frac{t^4 \, dt}{120} - \int_0^{1.2} \frac{t^6 \, dt}{5040} \right)$$

$$\approx \frac{i_0}{1.2}\left(t - \frac{t^3}{18} + \frac{t^5}{600} - \frac{t^7}{35{,}280}\right)\Bigg]_0^{1.2}$$

$$\approx \frac{i_0}{1.2}\left[\left(1.2 - \frac{1.2^3}{18} + \frac{1.2^5}{600} - \frac{1.2^7}{35{,}280}\right) - 0\right]$$

$$\approx i_0\left(1 - \frac{1.2^2}{18} + \frac{1.2^4}{600} - \frac{1.2^6}{35{,}280}\right)$$

$1\ \boxed{-}\ 1.2\ \boxed{\text{STO}}\ \boxed{x^2}\ \boxed{\div}\ 18\ \boxed{+}\ \boxed{\text{RCL}}\ \boxed{y^x}\ 4\ \boxed{\div}\ 600\ \boxed{-}\ \boxed{\text{RCL}}\ \boxed{y^x}\ 6\ \boxed{\div}\ 35280\ \boxed{=}$

$$\rightarrow \boxed{0.9233714}$$

$$i_{\text{AVE}} \approx 0.92 i_0$$

Euler's Identity

In Chapter 12 we defined the exponential form of a complex number in terms of the relationship

$$e^{j\theta} = \cos\theta + j\sin\theta \qquad \text{where } j^2 = -1$$

We can now show that this equation follows logically from the Maclaurin series expansions of the functions e^x, $\sin x$, and $\cos x$. If we assume that the Maclaurin series for these functions are valid for complex numbers, we may write

$$e^{j\theta} = 1 + j\theta + \frac{(j\theta)^2}{2!} + \frac{(j\theta)^3}{3!} + \frac{(j\theta)^4}{4!} + \frac{(j\theta)^5}{5!} + \cdots$$

$$= 1 + j\theta - \frac{\theta^2}{2!} - \frac{j\theta^3}{3!} + \frac{\theta^4}{4!} + \frac{j\theta^5}{5!} + \cdots$$

Separating the real and imaginary terms, we have

$$e^{j\theta} = \left(1 - \frac{\theta^2}{2!} + \frac{\theta^4}{4!} - \cdots\right) + j\left(\theta - \frac{\theta^3}{3!} + \frac{\theta^5}{5!} - \cdots\right)$$

But

$$\cos\theta = 1 - \frac{\theta^2}{2!} + \frac{\theta^4}{4!} - \cdots \qquad \text{and} \qquad \sin\theta = \theta - \frac{\theta^3}{3!} + \frac{\theta^5}{5!} - \cdots$$

Therefore,

$$e^{j\theta} = \cos\theta + j\sin\theta$$

The identity obtained when $\theta = \pi$ is particularly interesting.

$$e^{j\pi} = \cos\pi + j\sin\pi$$

$$e^{j\pi} = -1$$

This remarkable equation connects the constants e and π with -1 and $\sqrt{-1}$. It is called **Euler's identity** after the Swiss mathematician who first made extensive use of the symbols e, π, and j.

Exercises 28-2

Find the first four nonzero terms of the Maclaurin series expansion of each function using the series in Table 28-1.

1. $f(x) = e^{2x}$
2. $f(x) = e^{x^2}$
3. $f(x) = \ln (1 - 2x)$
4. $f(x) = \ln (1 - x^2)$
5. $f(\theta) = \cos 3\theta$
6. $f(t) = \sin \pi t$
7. $f(x) = \sin x^2$
8. $f(x) = \cos \sqrt{x}$
9. $f(t) = \dfrac{1}{3t + 1}$
10. $f(x) = \dfrac{2}{1 - x^2}$
11. $f(x) = \dfrac{1}{2 - 3x}$
12. $f(x) = \dfrac{1}{1 - x^3}$
13. $f(x) = \sin \dfrac{x}{2}$
14. $f(\theta) = \cos (2\theta^4)$
15. $f(t) = e^{-t} \cos t$
16. $f(t) = \sin t \cos t$
17. $f(x) = x^2 \cos x$
18. $f(x) = \dfrac{\sin x - x}{x^2}$
19. $f(x) = \dfrac{e^x}{x}$
20. $f(\theta) = \dfrac{\sin \theta}{\theta}$
21. $f(x) = \dfrac{\cos x}{1 + x}$
22. $f(x) = \dfrac{\ln (1 + x)}{1 - x}$

Integrate each of the following using the first three nonzero terms of the appropriate Maclaurin series expansion:

23. $\displaystyle\int e^{-x^2} \, dx$
24. $\displaystyle\int \dfrac{e^x}{x^2} \, dx$
25. $\displaystyle\int \dfrac{\ln (1 + x)}{x} \, dx$
26. $\displaystyle\int \dfrac{x \, dx}{e^x}$
27. $\displaystyle\int_0^{0.20} x \cos x \, dx$
28. $\displaystyle\int_{0.10}^{0.30} e^\theta \cos \theta \, d\theta$

Differentiate each function using the first three nonzero terms of the appropriate Maclaurin series.

29. $f(x) = (1 + 2x)^5$
30. $f(x) = x \cos x$
31. $f(x) = \dfrac{\sin x}{x^2}$
32. $f(x) = \dfrac{\cos x^2}{x}$

Solve.

33. *Environmental technology*   After a spill of a heavy-metal pollutant at a waste disposal site, the amount of pollutant in local groundwater decreases according to the equation

$$P = P_0(e^{-2t} + e^{-t})$$

where t is time in weeks and P is measured in parts per million (ppm). Use the first four terms of the Maclaurin expansion of this function to find dP/dt, the rate of pollution decrease, at $t = 1.1$ weeks.

34. *Electronics*   The electrical potential function near an array of charged plates is

$$V(r) = V_0 + \frac{ae^r}{r}$$

Use the first four terms of the Maclaurin series expansion to approximate $V(r)$, and differentiate to find an expression for the electrical field

$$E(r) = \frac{d}{dr} V(r)$$

35. Find the average value of the function given from x = 0.5 to x = 1.0. Use the first three terms of a Maclaurin series, and round to three decimal places.

$$f(x) = \frac{\cos x}{x}$$

36. Using the first three terms of a Maclaurin series, calculate the approximate area bounded by the curve

$$y = \frac{\ln (1 + x)}{x^2}$$

the x-axis, and the lines x = 0.1 and x = 0.5.

37. Find the volume generated by rotating about the x-axis the region bounded by the curve $y = e^{-x}$ and the lines y = 0, x = 0, and x = 0.5. Use the first three terms of a Maclaurin series, and round to two decimal places.

38. *Physics* Calculate the moment of inertia with respect to the y-axis of the area bounded by the curve

$$y = \cos^2 x$$

the x-axis, the y-axis, and the line x = 1. Use the first three terms of a Maclaurin series, express the answer in terms of the density δ, and round to two decimal places.

28-3 | Computing with Series

The Maclaurin series representation of a function may be used to obtain an approximate value for that function in some interval centered on x = 0. For any function whose Maclaurin series can be found, this allows us to prepare a table of values similar to the tables available for the trigonometric, exponential, and logarithmic functions. The Maclaurin series also enables us to calculate, or to write a computer program for calculating, the value of such a function for any value of the variable in its interval of convergence.

In Example 11, if the terms of the sum are calculated separately, the approximation can be written as

$$\bar{i} = 1.0 - 0.08 + 0.003456 - 0.00008464$$

Notice the following about this kind of series approximation:

1. The approximation improves as the number of terms included increases.
2. Because the Maclaurin series is designed to fit some function $f(x)$ best at x = 0, the terms of such an approximation approach zero more quickly for values of x near x = 0 than for values farther from x = 0. (You can check this for Example 11 by calculating the size of each term at x = 10.)

These basic properties are true for every Maclaurin series, and they are helpful when we compute the approximate value of a function using the first few terms of its series representation.

Error of a Maclaurin series   The absolute value of the error that results when a function $f(x)$ is approximated by the truncated Maclaurin series

$$f(x) = f(0) + f'(0)x + \frac{f''(0)x^2}{2!} + \cdots + \frac{f^{(n)}(0)x^n}{n!}$$

can be found using equation (28-2):

Error in a Truncated Maclaurin Series

$$E_n(x) = \left| \frac{f^{(n+1)}(c)x^{n+1}}{(n+1)!} \right| \qquad\qquad (28\text{-}2)$$

where c is some number between 0 and x, and the series approximation ends with the x^n term.

It is also possible to show that for an alternating series, if the function is approximated by the first n terms of the series, then the error of the approximation is less than the absolute value of the first term omitted.

EXAMPLE 12   Calculate the approximate value of $e^{0.2}$ using the first four terms of a Maclaurin series, and find the maximum value of the error in this approximation.

Solution   **First,** find the approximate value of $e^{0.2}$. From Table 28-1,

$$e^x = 1 + x + \frac{x^2}{2!} + \frac{x^3}{3!} + \frac{x^4}{4!} + \cdots$$

Substitute 0.2 for x:

$$e^{0.2} = 1 + 0.2 + \frac{(0.2)^2}{2!} + \frac{(0.2)^3}{3!} + \frac{(0.2)^4}{4!} + \cdots$$

$$\approx 1.0 + 0.2 + 0.02 + 0.00133333 + 0.00006667 + \cdots$$

Summing the first four terms, we have

$$e^{0.2} \approx 1.2213333$$

Next, use equation (28-2) to find the upper bound of the error in this approximation.

Since $f(x)$ is an increasing function and $c \le 0.2$:

$$E_{max} = \left| \frac{f^{(4)}(0.2)(0.2)^4}{4!} \right|$$

But $f^{(4)}(x) = e^x$, and this error becomes:

$$E_{max} = \frac{e^{0.2}(0.2)^4}{4!}$$

The quantity $e^{0.2}$ is the number we are calculating. Therefore, we can write $e \approx 3$.

$$E_{max} = \frac{3^{0.2}(0.2)^4}{4!} \approx 0.00008305$$

Therefore, $e^{0.2}$ is less than $1.2213333 + 0.00008305 = 1.22141635$ or $e^{0.2} \approx 1.2214$, accurate to four decimal places. Checking with a calculator, we find $e^{0.2} \approx 1.2214028$.

For a decreasing function, use $c = 0$ in (28-2) to find the maximum error.

EXAMPLE 13 Calculate the value of sin 10° using the first two terms of a Maclaurin series, and find the maximum error of this approximation.

Solution **First,** because the variable in the series expansion must be a real number, write 10° in radians:

$$10° = 10°\left(\frac{\pi}{180°}\right) = \frac{\pi}{18}\text{rad}$$

Then, from Table 28-1:

$$\sin x = x - \frac{x^3}{3!} + \frac{x^5}{5!} - \cdots$$

Use only the first two terms of the series and substitute $\pi/18$ for x.

$$\approx \frac{\pi}{18} - \frac{1}{6}\left(\frac{\pi}{18}\right)^3$$

$$\approx 0.1745329 - 0.0008861$$

$$\approx 0.1736468$$

Finally, this is an alternating series, so we know that the error in this approximation is less than the absolute value of the first term not included:

$$E_{max} = \frac{x^5}{5!} = \frac{1}{120}\left(\frac{\pi}{18}\right)^5$$

$$\approx 0.0000013$$

Therefore, we expect that the value of sin 10° calculated using the series is accurate to five decimal places; that is, sin 10° ≈ 0.17365. Checking with a calculator, we find sin 10° ≈ 0.1736482.

Using the series expansions of trigonometric and exponential functions, it is possible to calculate the values of π and e to any desired accuracy.

EXAMPLE 14 Use the first ten terms of a Maclaurin series to calculate:
(a) e (b) π

Solution
(a) From the series expansion for e^x in Table 28-1,

$$e^x = 1 + x + \frac{x^2}{2!} + \frac{x^3}{3!} + \frac{x^4}{4!} + \cdots$$

so that for $x = 1$,

$$e^1 = e = 1 + 1 + \frac{1}{2!} + \frac{1}{3!} + \frac{1}{4!} + \frac{1}{5!} + \cdots$$

Summing the first ten terms, we have

e ≈ 2.7182788

The actual value of e is e ≈ 2.7182818, so our ten-term sum is accurate to five decimal places.

(b) Applying the definition of a Maclaurin series,

$$\text{Arctan } x = x - \frac{x^3}{3} + \frac{x^5}{5} - \frac{x^7}{7} + \cdots$$

At x = 1, Arctan 1 = π/4, therefore,

$$\frac{\pi}{4} = 1 - \frac{1}{3} + \frac{1}{5} - \frac{1}{7} + \cdots$$

Summing the first ten terms of this series, we have

$$\frac{\pi}{4} \approx 0.7604599 \text{ or } \pi \approx 3.0418396$$

This ten-term series approximation for π/4 is accurate to only one decimal place.

This series was first discovered by the mathematician James Gregory in 1671 and was first used in 1699 by the astronomer Edmund Halley to calculate π to 72 decimal places.

EXAMPLE 15 The Fresnel integral,*

$$c(x) = \int_0^x \cos t^2 \, dt$$

is used to describe the longitudinal displacement of a beam subject to a periodic force. Use the first two terms of a Maclaurin series to evaluate c(0.4), and find the maximum error involved in this calculation.

Solution This integral cannot be found directly, but we can approximate it using a Maclaurin series. Write a series representing cos t² by substituting t² for x in the series for cos x in Table 28-1:

$$\cos t^2 = 1 - \frac{(t^2)^2}{2!} + \frac{(t^2)^4}{4!} - \cdots$$

The first two terms are

$$\cos t^2 \approx 1 - \frac{t^4}{2}$$

---

*R. Rothe, *Theory of Functions as Applied to Engineering Problems* (New York: Dover, 1961).

Therefore,

$$c(x) \approx \int_0^x \left(1 - \frac{t^4}{2}\right) dt$$

$$c(0.4) \approx \int_0^{0.4} \left(1 - \frac{t^4}{2}\right) dt$$

$$\approx \left(t - \frac{t^5}{10}\right)\Big]_0^{0.4}$$

$$\approx 0.4 - \frac{(0.4)^5}{10} \approx 0.398976$$

The error in this approximation is less than the next term in the sum. The third term in the series for $\cos t^2$ is $t^8/4!$; therefore, the third term in the integral is

$$\frac{t^9}{9 \cdot 4!} = \frac{t^9}{216}$$

and the error is

$$E_{max} = \frac{(0.4)^9}{216} \approx 0.0000012$$

Therefore, the answer, accurate to five decimal places, is $c(0.4) \approx 0.398998$.

Exercises 28-3

Calculate the value of each function using the first three nonzero terms of a Maclaurin series.

1. $e^{-0.75}$
2. $e^{0.08}$
3. $\sin 5°$
4. $\sin(-0.5)$
5. $\cos 0.8$
6. $\cos(-12°)$
7. $\ln 1.3$
8. $\ln 0.75$
9. $\sqrt{e}$
10. $\sqrt[3]{e}$
11. $\dfrac{1}{e}$
12. $\ln \sqrt{\pi}$
13. $\sqrt[3]{1.75}$
    [Hint: Use the series for $(1 + x)^n$.]
14. $\sqrt{0.885}$
    Hint: $\ln \sqrt{\pi} = \ln[1 + (\sqrt{\pi} - 1)]$

Calculate the value of each function using the first three nonzero terms of a Maclaurin series, and estimate the maximum error of the approximation.

15. $e^{1.1}$
16. $\sqrt[3]{0.950}$
17. $\sin 20°$
18. $\cos 0.65$
19. $\ln 1.65$
20. $\ln 0.44$

Solve.

21. Use the first three nonzero terms of the Maclaurin series expansion for Arcsin x to calculate the value for π. [Hint: Arcsin $0.5 = \pi/6$]

22. Use the first three nonzero terms of the Maclaurin series expansion for

$$\frac{\pi}{4} = \text{Arctan } \frac{1}{2} + \text{Arctan } \frac{1}{3}$$

to calculate a value for π.

23. *Electrical engineering* A very long transmission line of negligible inductance and capacitance per unit length has a voltage V_0 applied to its transmitting end. To calculate the voltage $V(x)$ along the line, we must evaluate the integral

$$V(x) = V_0 \int_0^x e^{-u^2} \, du$$

Find $V(1.2)$ using the first three nonzero terms of a Maclaurin series, and estimate the maximum error of the approximation.

24. *Thermodynamics* When heat flows along a very long insulated metal bar, the temperature difference between two points on the bar can be found by evaluating an integral of the form

$$T = \int_a^b e^{-x^2} \cos 2x \, dx$$

Calculate T using the first three terms of a Maclaurin series if $a = 0.1$ and $b = 1.1$.

25. *Electronics* The charge on the surface of a capacitative device decreases with time according to the equation

$$q(t) = \frac{\ln (1 + t^2)}{t^2}$$

Use the first four terms of a Maclaurin series expansion of this function to find the approximate value of the current $i = dq/dt$ flowing at time $t = 0.45$ s.

26. Use the first three terms of a Maclaurin series to calculate the area of the region bounded by the curve

$$y = e^x \ln (1 + x)$$

the x-axis, and the line $x = 1$.

28-4 | Taylor Series

The Maclaurin series (28-1) was defined so that it converges rapidly for values of x close to $x = 0$. If x is within the interval of convergence of the series, but not close to zero, a great many terms may be needed to provide a good approximation to $f(x)$. For example, although the series converges for all values of x, four terms of the Maclaurin series for e^x give an approximation accurate to better than 2% for $x = 1$, but accurate only to 35% for $x = 3$. For large values of x, the Maclaurin series may not be very useful.

The idea of a power series expansion can be generalized so that it provides a good approximation to $f(x)$ on an interval centered at $x = a$ rather than $x = 0$.

To assure that the power series

$$f(x) = c_0 + c_1(x - a) + c_2(x - a)^2 + c_3(x - a)^3 + \cdots + c_n(x - a)^n + \cdots$$

fits the function $f(x)$ exactly at $x = a$, we define the coefficients as we did for the Maclaurin series.

Determine the successive derivatives of the function and the series and equate them at $x = a$:

$$f(x) = c_0 + c_1(x - a) + c_2(x - a)^2 + c_3(x - a)^3 + c_4(x - a)^4 + \cdots$$

$$f(a) = c_0$$

$$f'(x) = c_1 + 2c_2(x - a) + 3c_3(x - a)^2 + 4c_4(x - a)^3 + \cdots \qquad f'(a) = c_1$$

$$f''(x) = 2c_2 + 2\cdot 3c_3(x - a) + 3\cdot 4c_4(x - a)^2 + \cdots \qquad f''(a) = 2c_2$$

and so on. In general,

$$f^{(n)}(a) = n!\, c_n$$

In each expression on the right, when $x = a$, the factor $(x - a) = 0$, and all terms except the first are equal to zero. If we solve for the coefficients c_n and substitute into the original power series, the resulting series expansion is called the **Taylor series** about a for $f(x)$.*

Taylor Series about a

$$f(x) = f(a) + f'(a)(x - a) + \frac{f''(a)(x - a)^2}{2!} + \frac{f'''(a)(x - a)^3}{3!} + \cdots$$

$$+ \frac{f^{(n)}(a)(x - a)^n}{n!} + \cdots \qquad (28\text{-}3)$$

If the derivatives of $f(x)$ exist at $x = 0$, then the series on the right converges to $f(x)$ in some interval about $x = a$.

NOTE ▶ The Taylor series includes the Maclaurin series as a special case where $a = 0$. ◀

EXAMPLE 16   Find the Taylor series expansion of e^x about $x = 4$.

Solution   **First,** find the successive derivatives of $f(x) = e^x$, and evaluate them at $x = 4$:

---

* The Taylor series was named after Brook Taylor (1685–1731), an English mathematician and physicist, who applied these ideas to problems in astronomy, thermodynamics, mechanics, and many other technical areas.

$$f(x) = e^x \longrightarrow f(4) = e^4$$

$$f'(x) = e^x \longrightarrow f'(4) = e^4$$

$$f''(x) = e^x \longrightarrow f''(4) = e^4$$

and so on.

Then, substitute these values in the definition (28-3):

$$e^x = e^4 + e^4(x - 4) + \frac{e^4(x - 4)^2}{2!} + \frac{e^4(x - 4)^3}{3!} + \cdots$$

$$= e^4 \left[1 + (x - 4) + \frac{(x - 4)^2}{2!} + \frac{(x - 4)^3}{3!} + \cdots \right]$$

For values of x near x = 4, this series converges much more rapidly than the Maclaurin series for the same function.

EXAMPLE 17 Find the Taylor series expansion for $\sqrt[3]{x}$ about x = 1. Use the first four terms of this series to calculate the approximate value of $\sqrt[3]{1.25}$.

Solution First, find the successive derivatives of $f(x) = \sqrt[3]{x}$, and evaluate them at x = 1:

$$f(x) = \sqrt[3]{x} = x^{1/3} \longrightarrow f(1) = 1$$

$$f'(x) = \frac{1}{3}x^{-2/3} \longrightarrow f'(1) = \frac{1}{3}$$

$$f''(x) = -\frac{2}{9}x^{-5/3} \longrightarrow f''(1) = -\frac{2}{9}$$

$$f'''(x) = \frac{10}{27}x^{-8/3} \longrightarrow f'''(1) = \frac{10}{27}$$

and so on.

Next, substitute these quantities into equation (28-3) with $a = 1$:

$$\sqrt[3]{x} = 1 + \frac{1}{3}(x - 1) - \frac{2}{9}\frac{(x - 1)^2}{2!} + \frac{10}{27}\frac{(x - 1)^3}{3!} - \cdots$$

$$= 1 + \frac{1}{3}(x - 1) - \frac{1}{9}(x - 1)^2 + \frac{5}{81}(x - 1)^3 - \cdots$$

Then, for x = 1.25, (x − 1) = 0.25. Therefore, using the first four terms of this Taylor series expansion, we have

$$\sqrt[3]{1.25} \approx 1 + \frac{1}{3}(0.25) - \frac{1}{9}(0.25)^2 + \frac{5}{81}(0.25)^3$$

$$\approx 1.0773534$$

The correct value to seven decimal places is 1.0772173; therefore, our approximation is accurate to three decimal places.

EXAMPLE 18   Expand ln x in a Taylor series about x = 1, and use the first four terms of this series to find the approximate value of ln 0.95.

Solution   **First,** find the successive derivatives of $f(x) = \ln x$, and evaluate them at x = 1:

$$f(x) = \ln x \qquad \longrightarrow \qquad f(1) = \ln 1 = 0$$

$$f'(x) = \frac{1}{x} \qquad \longrightarrow \qquad f'(1) = \frac{1}{1} = 1$$

$$f''(x) = -\frac{1}{x^2} \qquad \longrightarrow \qquad f''(1) = -\frac{1}{1} = -1$$

$$f'''(x) = \frac{2}{x^3} = \frac{2!}{x^3} \qquad \longrightarrow \qquad f'''(1) = 2!$$

$$f^{(4)}(x) = -\frac{3 \cdot 2}{x^4} = -\frac{3!}{x^4} \longrightarrow f^{(4)}(1) = -3!$$

and so on.

Next, using equation (28-3),

$$\ln x = 0 + 1(x - 1) - \frac{1(x-1)^2}{2!} + 2!\frac{(x-1)^3}{3!} - 3!\frac{(x-1)^4}{4!} + \cdots$$

$$= (x - 1) - \frac{1}{2}(x-1)^2 + \frac{1}{3}(x-1)^3 - \frac{1}{4}(x-1)^4 + \cdots$$

Then, for x = 0.95, (x − 1) = −0.05, and if we use the first four terms of this series to approximate ln 0.95, we have

$$\ln 0.95 \approx (-0.05) - \frac{1}{2}(-0.05)^2 - \frac{1}{3}(-0.05)^3 - \frac{1}{4}(-0.05)^4$$

$$\approx -0.0512099$$

which is accurate to four decimal places.

When you use a Taylor series expansion to find the approximate value of a function, the first step is to choose an appropriate value for the constant a. Use the following two rules to guide your choice:

1.   Choose a close to the value of x for which the function is to be evaluated. This assures that $|x - a|$ will be small, and therefore that successive powers of $(x - a)$ in equation (28-3) will be small so that the series will converge rapidly.
2.   Choose a so that the coefficients $f(a)$, $f'(a)$, $f''(a)$, and so on, are easy to calculate and to work with.

▦ **EXAMPLE 19**   Use a Taylor series expansion to find the value of sin 58°.

Solution **First,** choose a. Note that $58° = 58(\pi/180) \approx 1.0122910$ rad. Therefore, we could find the Taylor series about $a = 1$, but $\sin 1 \approx 0.8414710$ and $\cos 1 \approx 0.5403023$, and these decimal approximations are cumbersome. The calculation will be simplified if we choose $a = 60° = \pi/3$. For this value of a,

$$\sin \frac{\pi}{3} = \frac{\sqrt{3}}{2} \quad \text{and} \quad \cos \frac{\pi}{3} = \frac{1}{2}$$

Second, determine a numerical value for the factor $(x - a)$:

$$x - a = 58° - 60° = -2°$$

$$\approx -0.0349066 \text{ rad}$$

Third, find the successive derivatives of $f(x) = \sin x$, and evaluate them at $x = a = \pi/3$:

$$f(x) = \sin x \longrightarrow f(a) = \sin a = \sin \frac{\pi}{3} = \frac{\sqrt{3}}{2}$$

$$f'(x) = \cos x \longrightarrow f'(a) = \cos a = \cos \frac{\pi}{3} = \frac{1}{2}$$

$$f''(x) = -\sin x \longrightarrow f''(a) = -\sin a = -\sin \frac{\pi}{3} = -\frac{\sqrt{3}}{2}$$

$$f'''(x) = -\cos x \longrightarrow f'''(a) = -\cos a = -\cos \frac{\pi}{3} = -\frac{1}{2}$$

and so on.

Finally, substitute these values into (28-3) to find the Taylor series expansion of $\sin x$ about $a = \pi/3$:

$$\sin x = \frac{\sqrt{3}}{2} + \frac{1}{2} \cdot \left(x - \frac{\pi}{3} \right) + \left(-\frac{\sqrt{3}}{2} \right) \cdot \frac{(x - \pi/3)^2}{2!} + \left(-\frac{1}{2} \right) \cdot \frac{(x - \pi/3)^3}{3!} + \cdots$$

$$= \frac{\sqrt{3}}{2} + \frac{1}{2}\left(x - \frac{\pi}{3} \right) - \frac{\sqrt{3}}{4}\left(x - \frac{\pi}{3} \right)^2 - \frac{1}{12}\left(x - \frac{\pi}{3} \right)^3 + \cdots$$

Therefore, using the first four terms of the series, we have

$$\sin 58° \approx \frac{\sqrt{3}}{2} + \frac{1}{2}(-0.0349066) - \frac{\sqrt{3}}{4}(-0.0349066)^2 - \frac{1}{12}(-0.0349066)^3$$

3 $\boxed{\sqrt{}}$ $\boxed{\div}$ 2 $\boxed{-}$ .0349066 $\boxed{\text{STO}}$ $\boxed{\div}$ 2 $\boxed{-}$ 3 $\boxed{\sqrt{}}$ $\boxed{\div}$ 4 $\boxed{\times}$ $\boxed{\text{RCL}}$ $\boxed{x^2}$ $\boxed{+}$ $\boxed{\text{RCL}}$ $\boxed{y^x}$ 3 $\boxed{\div}$ 12 $\boxed{=}$

$$\rightarrow \quad \boxed{0.8480480}$$

$$\approx 0.8480480$$

This is very close to the calculator value of $\sin 58° \approx 0.8480481$.

Although this may seem an unnecessarily difficult way of arriving at a result that could be obtained with a few key strokes on a calculator, the Taylor series is very useful. It provides a way of evaluating functions for which tables or calculator buttons are not available.

Exercises 28-4

Expand each function in a Taylor series about the given value. Find the first three nonzero terms.

1. $\cos x$; $a = \dfrac{\pi}{2}$ **2.** $\sin 2x$; $a = \dfrac{\pi}{4}$

3. e^x; $a = 3$ **4.** e^{-x}; $a = 2$

5. $\dfrac{1}{\sqrt{x}}$; $a = 4$ **6.** $\sqrt{x + 3}$; $a = 1$

7. $\sqrt[3]{x}$; $a = 8$ **8.** $\sqrt[4]{x}$; $a = 1$

9. $\ln x$; $a = 2$ **10.** $\dfrac{1}{x}$; $a = 4$

Evaluate each function using the first three nonzero terms of an appropriate Taylor series.

11. $\tan 25°$; use $a = \dfrac{\pi}{6}$ **12.** e^π; use $a = 3$

13. $\sqrt[3]{7.55}$; use $a = 8$ **14.** $\sin 87°$; use $a = \dfrac{\pi}{2}$

15. $\ln 3.14$; use $a = 3$ **16.** $e^{-1.2}$; use $a = 1$

17. $\cos 41°$ **18.** $\sin 2.1\pi$

Solve.

19. Calculate a value for π using the first three terms of a Taylor series for Arctan x about $x = 0.5$. [*Hint:* Use the fact that $\tan \pi/6 \approx 0.57735$.]

20. *Electronics*   The damped current in a certain electronic circuit has the form

$$i(t) = \frac{\cos \pi t}{t}$$

Expand this function as a Taylor series about $t = 1$, and write the first three terms.

28-5 | Fourier Series

The Taylor and Maclaurin series provide polynomial approximations to a function, and these polynomials can be made to fit the function very well at some point $x = a$. However, a polynomial sum is not well-suited to represent all types of functions. For example, as shown in Fig. 28-2, the first three approximations of the Maclaurin series for $f(x) = \sin x$ fit the function very well near $x = 0$, but the absolute value of each polynomial grows quite large as x increases or decreases, while sin x is never greater than 1. This series representation gets worse as x increases.

Periodic functions

A function $f(x)$ is said to be **periodic** if, for all real values of x, there is some number P that is characteristic of the function such that $f(x + P) = f(x)$. The number P is called the **period** of the function (see Fig. 28-3).

For example, the period of $\sin x$ is 2π, since $\sin(x + 2\pi) = \sin x$. If n is any integer, and $f(x)$ is a periodic function with period P, then for all x, $f(x + nP) = f(x)$. That is, $2P$, $3P$, $4P$, and so on, are also periods of $f(x)$. For example, since the period of $\sin x$ and $\cos x$ is 2π, then

$$\sin(x + 2\pi) = \sin(x - 4\pi) = \sin x$$

$$\cos(x + 4\pi) = \cos(x - 6\pi) = \cos x$$

If both functions $f(x)$ and $g(x)$ are periodic with period P, then it is possible to show that, for constants a and b, the function $ag(x) + bf(x)$ is also a periodic function with period P. For example, $\sin x + \cos x$ is periodic with period 2π, $2\sin x - 3\cos x$ is periodic with period 2π, and so on.

Figure 28-3

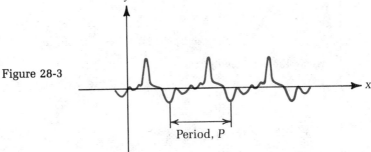

Period, P

The sine and cosine functions are the simplest, most natural examples of periodic functions, but many scientific and technical phenomena are described by periodic functions that may be very complicated in form. Periodic functions appear extensively in electronics; in the mechanics of vibrating or oscillating systems; and in the study of light, sound, and electromagnetic waves.

Rather than approximate a periodic function by a power series (a sum of polynomial terms), it is much more effective to represent a periodic function as a sum of terms that are themselves based on periodic functions.

Trigonometric series

If a_0, a_1, a_2, ... and b_1, b_2, b_3, ... are constants and x is a variable, then a series of the form

$$a_0 + a_1 \cos x + a_2 \cos 2x + a_3 \cos 3x + \cdots + a_n \cos nx + \cdots$$

$$+ b_1 \sin x + b_2 \sin 2x + b_3 \sin 3x + \cdots + b_n \sin nx + \cdots$$

is called a **trigonometric series**.* The real numbers a_n and b_n are the coefficients of the series. Because the functions $\sin x$ and $\cos x$ both have period 2π, if the

---

* All the basic trigonometric functions are periodic, but only the sine and cosine functions are used in trigonometric series, because only they do not have points of discontinuity.

trigonometric series converges, its sum will be a periodic function with period 2π.

If this trigonometric series is to represent some function $f(x)$, then the coefficients a_n and b_n must be determined from $f(x)$. For the Taylor and Maclaurin series, the coefficients of the series were determined by matching the derivatives of the function and its series. But, because many periodic functions of interest in scientific and technological applications have sharp corners or finite discontinuities, it is more reasonable to evaluate the coefficients by integration over one period of the function.

Fourier series   If a function $f(x)$ is defined over the interval $-\pi \leq x \leq \pi$, then the **Fourier series** is a periodic approximation to this function in that interval and is defined as follows:

Fourier Series for $f(x)$

$$f(x) = a_0 + a_1 \cos x + a_2 \cos 2x + \cdots + a_n \cos nx + \cdots$$
$$+ b_1 \sin x + b_2 \sin 2x + \cdots + b_n \sin nx + \cdots \qquad \text{(28-4)}$$

where the coefficients are

$$a_0 = \frac{1}{2\pi} \int_{-\pi}^{\pi} f(x)\, dx \qquad\qquad\qquad\qquad \text{(28-5)}$$

$$a_n = \frac{1}{\pi} \int_{-\pi}^{\pi} f(x) \cos nx\, dx \qquad n = 1, 2, 3, \ldots \qquad \text{(28-6)}$$

$$b_n = \frac{1}{\pi} \int_{-\pi}^{\pi} f(x) \sin nx\, dx \qquad n = 1, 2, 3, \ldots \qquad \text{(28-7)}$$

The series gives $f(x)$ in the interval $-\pi \leq x \leq \pi$, and gives a periodic repetition of it outside this interval. The original function repeats with a period of 2π.

The trigonometric series were first studied by Joseph Louis Lagrange (1736–1813), a French–Italian mathematician and astronomer, who applied them to the propagation of sound waves. Jean Joseph Fourier (1768–1830) developed the Fourier series as a tool in the mathematical theory of heat conduction, and his work led to many of our modern concepts of applied mathematics.

Finding Fourier coefficients   In order to perform the integrations (28-5)–(28-7) that define the coefficients of a Fourier series, you must be able to evaluate integrals of the following forms:

$$\int_{-\pi}^{\pi} \cos nx\, dx = \int_{-\pi}^{\pi} \sin nx\, dx = 0 \qquad n \neq 0 \qquad \text{(28-8)}$$

$$\int_{-\pi}^{\pi} \cos^2 nx\, dx = \int_{-\pi}^{\pi} \sin^2 nx\, dx = \pi \qquad n \neq 0 \qquad \text{(28-9)}$$

$$\int_{-\pi}^{\pi} \cos mx \cos nx \, dx = 0 \qquad m \neq n \tag{28-10}$$

$$\int_{-\pi}^{\pi} \sin mx \sin nx \, dx = 0 \qquad m \neq n \tag{28-11}$$

$$\int_{-\pi}^{\pi} \sin mx \cos nx \, dx = 0 \tag{28-12}$$

- Integrals such as (28-8) can be found using integration formulas (27-4) and (27-5).
- Integrals such as (28-9) have already been found using the techniques explained in Chapter 27.
- Integral (28-10) can be found by using the cosine sum and difference trigonometric identities given in Chapter 19:

$$\cos (\alpha + \beta) = \cos \alpha \cos \beta - \sin \alpha \sin \beta \tag{19-9}$$

$$\cos (\alpha - \beta) = \cos \alpha \cos \beta + \sin \alpha \sin \beta \tag{19-11}$$

Adding:

$$\cos (\alpha + \beta) + \cos (\alpha - \beta) = 2 \cos \alpha \cos \beta$$

so that

$$\cos mx \cos nx = \frac{1}{2} \left[\cos (mx + nx) + \cos (mx - nx) \right]$$

$$= \frac{1}{2} \left[\cos (m + n)x + \cos (m - n)x \right]$$

and

$$\int_{-\pi}^{\pi} \cos mx \cos nx \, dx = \frac{1}{2} \int_{-\pi}^{\pi} \cos (m + n)x \, dx + \frac{1}{2} \int_{-\pi}^{\pi} \cos (m - n)x \, dx$$

But both integrals on the right have the form of (28-8), and are therefore equal to zero. Thus,

$$\int_{-\pi}^{\pi} \cos mx \cos nx \, dx = 0$$

- Integral (28-11) can be found by subtracting the sum and difference trigonometric identities to get

$$\cos (\alpha - \beta) - \cos (\alpha + \beta) = 2 \sin \alpha \sin \beta$$

so that

$$\sin mx \sin nx = \frac{1}{2} \left[\cos (mx - nx) - \cos (mx + nx) \right]$$

$$= \frac{1}{2} \left[\cos (m - n)x - \cos (m + n)x \right]$$

and

$$\int_{-\pi}^{\pi} \sin mx \sin nx \, dx = \frac{1}{2} \int_{-\pi}^{\pi} \cos (m - n)x \, dx - \frac{1}{2} \int_{-\pi}^{\pi} \cos (m + n)x \, dx$$

From (28-8), the integrals on the right are both equal to zero. Therefore,

$$\int_{-\pi}^{\pi} \sin mx \sin nx \, dx = 0$$

- Integral (28-12) can be found by using the sine sum and difference trigonometric identities:

$$\sin (\alpha + \beta) = \sin \alpha \cos \beta + \cos \alpha \sin \beta \qquad \textbf{(19-10)}$$

$$\sin (\alpha - \beta) = \sin \alpha \cos \beta - \cos \alpha \sin \beta \qquad \textbf{(19-12)}$$

Adding:

$$\sin (\alpha + \beta) + \sin (\alpha - \beta) = 2 \sin \alpha \cos \beta$$

so that

$$\sin mx \cos nx = \frac{1}{2} \left[\sin (mx + nx) + \sin (mx - nx) \right]$$

$$= \frac{1}{2} \left[\sin (m + n)x + \sin (m - n)x \right]$$

and

$$\int_{-\pi}^{\pi} \sin mx \cos nx \, dx = \frac{1}{2} \int_{-\pi}^{\pi} \sin (m + n)x \, dx + \frac{1}{2} \int_{-\pi}^{\pi} \sin (m - n)x \, dx$$

By (28-8), both integrals on the right are equal to zero. Therefore,

$$\int_{-\pi}^{\pi} \sin mx \cos nx \, dx = 0$$

EXAMPLE 20 In the design of many electronic devices, including digital circuits, a **squarewave** (or pulsed rectangular waveform, Fig. 28-4) is used as a timing signal.

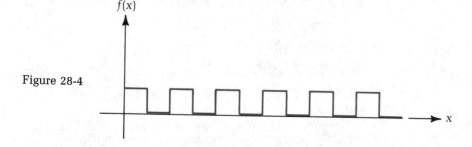

Figure 28-4

If the form of this signal is defined as

$$f(x) = \begin{cases} 0 & \text{for } -\pi \le x < 0 \\ 1 & \text{for } 0 \le x < \pi \end{cases}$$

find the Fourier series representing this function.

Solution   Figure 28-5 shows the squarewave function for $-\pi \le x < \pi$. The coefficients of the Fourier series for this function can be found by using equations (28-5), (28-6), and (28-7).

Figure 28-5

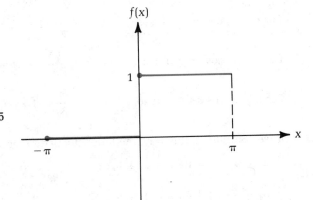

Use (28-5). $a_0 = \dfrac{1}{2\pi} \displaystyle\int_{-\pi}^{\pi} f(x)\, dx$

$f(x) = 0$ for $-\pi \le x < 0$

$$= \frac{1}{2\pi} \int_{-\pi}^{0} (0)\, dx + \frac{1}{2\pi} \int_{0}^{\pi} (1)\, dx$$

$f(x) = 1$ for $0 \le x < \pi$

$$= 0 + \frac{x}{2\pi} \Big]_{0}^{\pi} = \frac{1}{2}$$

Use (28-6). $a_n = \dfrac{1}{\pi} \displaystyle\int_{-\pi}^{\pi} f(x) \cos nx\, dx$

$$= \frac{1}{\pi} \int_{-\pi}^{0} (0) \cos nx\, dx + \frac{1}{\pi} \int_{0}^{\pi} (1) \cos nx\, dx = \frac{1}{\pi} \int_{0}^{\pi} \cos nx\, dx$$

$$= \frac{1}{n\pi} (\sin nx) \Big]_{0}^{\pi}$$

$$= \frac{1}{n\pi} \sin n\pi - \frac{1}{n\pi} \sin 0$$

$$= 0 \qquad \text{Since } \sin n\pi = 0 \text{ for any integer } n$$

Use (28-7). $b_n = \dfrac{1}{\pi} \displaystyle\int_{-\pi}^{\pi} f(x) \sin nx\, dx$

$$= \frac{1}{\pi} \int_{-\pi}^{0} (0) \sin nx \, dx + \frac{1}{\pi} \int_{0}^{\pi} (1) \sin nx \, dx$$

$$= \frac{1}{\pi} \int_{0}^{\pi} \sin nx \, dx$$

$$= -\frac{1}{n\pi} (\cos nx) \Big]_{0}^{\pi} = -\frac{1}{n\pi} (\cos n\pi - \cos 0)$$

$$= -\frac{1}{n\pi} (\cos n\pi - 1)$$

For $n = 1$, $\cos \pi = -1$; so: $\qquad b_1 = -\frac{1}{\pi}(-1 - 1) = \frac{2}{\pi}$

For $n = 2$, $\cos 2\pi = 1$; so: $\qquad b_2 = -\frac{1}{2\pi}(1 - 1) = 0$

For $n = 3$, $\cos 3\pi = -1$; so: $\qquad b_3 = -\frac{1}{3\pi}(-1 - 1) = \frac{2}{3\pi}$

If we continue in this way, the following pattern becomes apparent:

$$b_n = 0 \qquad \text{for } n = 2, 4, 6, \ldots$$

$$b_n = \frac{2}{n\pi} \qquad \text{for } n = 1, 3, 5, \ldots$$

Therefore, substituting in (28-4), the Fourier series representation of the square-wave is

$$f(x) = \frac{1}{2} + \frac{2}{\pi} \sin x + \frac{2}{3\pi} \sin 3x + \frac{2}{5\pi} \sin 5x + \cdots$$

$$= \frac{1}{2} + \frac{2}{\pi} \left(\sin x + \frac{\sin 3x}{3} + \frac{\sin 5x}{5} + \cdots \right)$$

Figure 28-6 on p. 1247 shows how the successive approximations obtained from this series approach the original squarewave function.

NOTE ▶ Notice in Fig. 28-6 that the value of the Fourier series representation of the function at the discontinuities at $x = \pi, 2\pi, 3\pi, 4\pi$, and so on, is 0.5. The series value is the average of the actual function values at a jump discontinuity. ◀

Functions with period other than 2π

In many technical applications a pulse $f(t)$ is specified over some time interval $0 \le t < 2P$, and the Fourier series can be used to give the mathematical equation of a time-dependent periodic function (or wave). If the period of the wave is $2P$, then substitute $\pi t/P$ for x in (28-4), so that

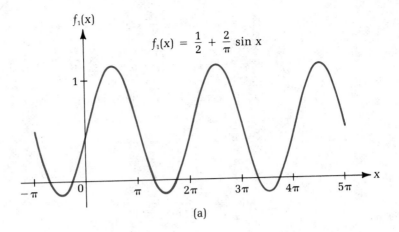

$$f_1(x) = \frac{1}{2} + \frac{2}{\pi} \sin x$$

(a)

Figure 28-6

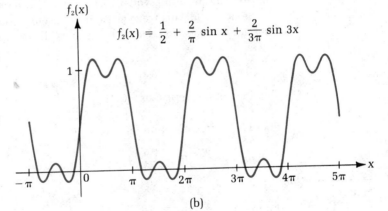

$$f_2(x) = \frac{1}{2} + \frac{2}{\pi} \sin x + \frac{2}{3\pi} \sin 3x$$

(b)

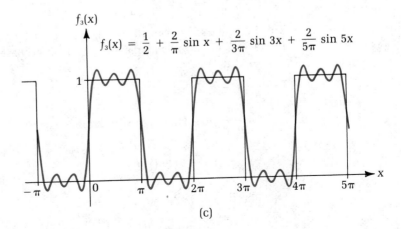

$$f_3(x) = \frac{1}{2} + \frac{2}{\pi} \sin x + \frac{2}{3\pi} \sin 3x + \frac{2}{5\pi} \sin 5x$$

(c)

$$f(t) = a_0 + a_1 \cos \frac{\pi t}{P} + a_2 \cos \frac{2\pi t}{P} + \cdots + a_n \cos \frac{n\pi t}{P} + \cdots$$

$$+ b_1 \sin \frac{\pi t}{P} + b_2 \sin \frac{2\pi t}{P} + \cdots + b_n \sin \frac{n\pi t}{P} + \cdots \qquad \textbf{(28-13)}$$

and the integrals defining the coefficients are

$$a_0 = \frac{1}{2P} \int_0^{2P} f(t)\, dt \qquad\qquad\qquad\qquad \textbf{(28-14)}$$

$$a_n = \frac{1}{P} \int_0^{2P} f(t) \cos \frac{n\pi t}{P}\, dt \qquad n = 1, 2, 3, \ldots \qquad \textbf{(28-15)}$$

$$b_n = \frac{1}{P} \int_0^{2P} f(t) \sin \frac{n\pi t}{P}\, dt \qquad n = 1, 2, 3, \ldots \qquad \textbf{(28-16)}$$

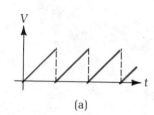

(a)

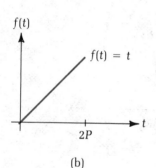

(b)

Figure 28-7

EXAMPLE 21   If the rotating shaft of a radar antenna is coupled to a potentiometer, the resulting voltage signal is a **sawtooth wave** [Fig. 28-7(a)]. The Fourier series for this wave gives the sinusoidal component signal that can be used for feedback to control the antenna. If the equation of one cycle of the sawtooth wave is

$$f(t) = t \qquad \text{for } 0 \le t < 2P$$

find the Fourier series for $f(t)$.

Solution   From equation (28-14),

$$a_0 = \frac{1}{2P} \int_0^{2P} t\, dt = \frac{1}{2P} \left(\frac{1}{2} t^2 \right) \Big]_0^{2P} = P$$

From equation (28-15),

$$a_n = \frac{1}{P} \int_0^{2P} f(t) \cos \frac{n\pi t}{P}\, dt = \frac{1}{P} \int_0^{2P} t \cos \frac{n\pi t}{P}\, dt$$

This integral can be evaluated using integration by parts or from integral F.80 in the table of integrals (Appendix D). Either of these methods gives

$$a_n = \frac{P}{n^2 \pi^2} \left(\cos \frac{n\pi t}{P} + \frac{n\pi t}{P} \sin \frac{n\pi t}{P} \right) \Big]_0^{2P}$$

$$= \frac{P}{n^2 \pi^2} [(\cos 2n\pi + 2n\pi \sin 2n\pi) - (\cos 0 + 0 \cdot \sin 0)]$$

$$= \frac{P}{n^2 \pi^2} [(1 + 0) - (1 + 0)] = 0$$

From equation (28-16),

$$b_n = \frac{1}{P} \int_0^{2P} f(t) \sin \frac{n\pi t}{P}\, dt = \frac{1}{P} \int_0^{2P} t \sin \frac{n\pi t}{P}\, dt$$

This integral can be evaluated using integration by parts or from integral F.79 in the table of integrals (Appendix D). Either of these methods gives

$$b_n = \frac{P}{n^2\pi^2}\left(\sin\frac{n\pi t}{P} - \frac{n\pi t}{P}\cos\frac{n\pi t}{P}\right)\Bigg]_0^{2P}$$

$$= \frac{P}{n^2\pi^2}[(\sin 2n\pi - 2n\pi\cos 2n\pi) - (\sin 0 - 0\cdot\cos 0)]$$

$$= \frac{P}{n^2\pi^2}[(0 - 2n\pi) - (0 - 0)] = -\frac{2P}{n\pi}$$

For $n = 1$:　　$b_1 = -\dfrac{2P}{\pi}$

For $n = 2$:　　$b_2 = -\dfrac{2P}{2\pi}$

For $n = 3$:　　$b_3 = -\dfrac{2P}{3\pi}$

And so on. Therefore, the Fourier series representation of $f(t)$ is

$$f(t) = P - \frac{2P}{\pi}\left(\sin\frac{\pi t}{P} + \frac{1}{2}\sin\frac{2\pi t}{P} + \frac{1}{3}\sin\frac{3\pi t}{P} + \cdots\right)$$

Figure 28-8 is a graph of the approximation obtained using the first four terms of this series.

Figure 28-8

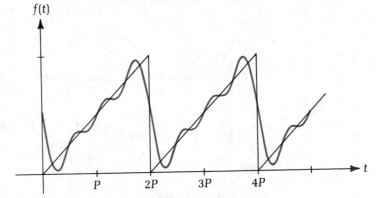

EXAMPLE 22　In the **half-wave rectifier** circuit shown in Fig. 28-9(a) on p. 1250, the diode D conducts only in one direction; therefore, all negative portions of the input signal V_s will be deleted. If the input is a sinusoidal voltage $V_s = V_0 \sin \omega t$ [Fig. 28-9(b)], then the output voltage $V(t)$ is the rectified (or "clipped") signal shown in Fig. 28-9(c). If

$$V(t) = \begin{cases} V_0 \sin t & \text{for } 0 \le t < \pi \\ 0 & \text{for } \pi \le t < 2\pi \end{cases}$$

find the Fourier series representation of the function $V(t)$.

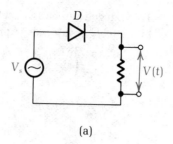

(a)

Figure 28-9

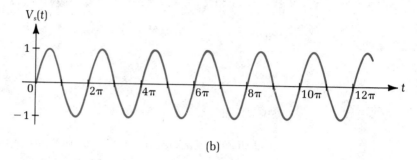

(b)

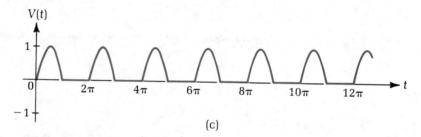

(c)

Solution The period of the periodic function is 2π; therefore, $P = \pi$. From equation (28-14),

$$a_0 = \frac{1}{2P} \int_0^{2P} V(t)\ dt = \frac{V_0}{2\pi} \int_0^\pi \sin t\ dt + \frac{V_0}{2\pi} \int_\pi^{2\pi} 0\ dt$$

$$= \frac{V_0}{2\pi}(-\cos t)\Big]_0^\pi = \frac{V_0}{\pi}$$

From equation (28-15),

$$a_n = \frac{1}{P} \int_0^{2P} V(t) \cos \frac{n\pi t}{P}\ dt = \frac{V_0}{\pi} \int_0^\pi \sin t \cos nt\ dt + \frac{V_0}{\pi} \int_\pi^{2\pi} 0 \cdot \cos nt\ dt$$

For $n = 1$,

$$a_1 = \frac{V_0}{\pi} \int_0^\pi \sin t \cos t\ dt = \frac{V_0}{2\pi}(\sin^2 t)\Big]_0^\pi = 0$$

For $n \neq 1$, recall that from the sum and difference equations

$$2 \sin \alpha \cos \beta = \sin (\alpha + \beta) + \sin (\alpha - \beta)$$

so that

$$\sin t \cos nt = \frac{1}{2}[\sin (t + nt) + \sin (t - nt)]$$

$$= \frac{1}{2}[\sin (1 + n)t + \sin (1 - n)t]$$

and thus, for $n \neq 1$,

$$a_n = \frac{V_0}{2\pi} \int_0^\pi [\sin (1 + n)t + \sin (1 - n)t]\, dt$$

$$= \frac{V_0}{2\pi} \left(-\frac{\cos (1 + n)t}{1 + n} - \frac{\cos (1 - n)t}{1 - n} \right) \Big]_0^\pi$$

$$= -\frac{V_0}{2\pi} \left[\frac{\cos (1 + n)\pi}{1 + n} + \frac{\cos (1 - n)\pi}{1 - n} - \frac{1}{1 + n} - \frac{1}{1 - n} \right]$$

For $n = 2$,

$$a_2 = -\frac{V_0}{2\pi} \left[\frac{\cos 3\pi}{3} + \frac{\cos (-\pi)}{-1} - \frac{1}{3} + 1 \right]$$

$$= -\frac{V_0}{2\pi} \left[-\frac{1}{3} + 1 - \frac{1}{3} + 1 \right] = -\frac{2V_0}{3\pi}$$

For $n = 3$,

$$a_3 = -\frac{V_0}{2\pi} \left[\frac{\cos 4\pi}{4} + \frac{\cos (-2\pi)}{-2} - \frac{1}{4} + \frac{1}{2} \right]$$

$$= -\frac{V_0}{2\pi} \left[\frac{1}{4} - \frac{1}{2} - \frac{1}{4} + \frac{1}{2} \right] = 0$$

In general,

$$a_n = 0 \qquad \text{for } n = 3, 5, 7, \ldots$$

$$a_n = -\frac{2V_0}{(n^2 - 1)\pi} \qquad \text{for } n = 2, 4, 6, \ldots$$

From equation (28-16),

$$b_n = \frac{1}{P} \int_0^{2P} V(t) \sin \frac{n\pi t}{P}\, dt = \frac{V_0}{\pi} \int_0^\pi \sin t \sin nt\, dt + \frac{V_0}{\pi} \int_\pi^{2\pi} 0 \cdot \sin nt\, dt$$

For $n = 1$,

$$b_1 = \frac{V_0}{\pi} \int_0^\pi \sin t \sin t\, dt$$

$$= \frac{V_0}{\pi} \int_0^\pi \sin^2 t\, dt = \frac{V_0}{2\pi} \left(t - \frac{1}{2} \sin 2t \right) \Big]_0^\pi$$

$$= \frac{V_0}{2\pi}\left[\left(\pi - \frac{1}{2}\sin 2\pi\right) - \left(0 - \frac{1}{2}\sin 0\right)\right]$$

$$= \frac{V_0}{2}$$

Recall that from the sum and difference equations,

$$2\sin\alpha\sin\beta = \cos(\alpha - \beta) - \cos(\alpha + \beta)$$

Therefore,

$$\sin t \sin nt = \frac{1}{2}[\cos(t - nt) - \cos(t + nt)]$$

and so, for $n \ne 1$,

$$b_n = \frac{V_0}{2\pi}\int_0^\pi [\cos(1 - n)t - \cos(1 + n)t]\, dt$$

$$= \frac{V_0}{2\pi}\left(\frac{\sin(1 - n)t}{1 - n} - \frac{\sin(1 + n)t}{1 + n}\right)\Big]_0^\pi$$

$$= \frac{V_0}{2\pi}\left[\frac{\sin(1 - n)\pi}{1 - n} - \frac{\sin(1 + n)\pi}{1 + n}\right] - \left[\frac{\sin 0}{1 - n} - \frac{\sin 0}{1 + n}\right]$$

$$= 0$$

$$\boxed{\begin{array}{l}\sin k\pi = 0 \\ \text{for } k \text{ an integer}\end{array}}$$

Then the Fourier series representation for the half-wave rectifier function is

$$V(t) = \frac{V_0}{\pi} + \frac{V_0}{2}\sin t - \frac{2V_0}{3\pi}\cos 2t - \frac{2V_0}{15\pi}\cos 4t - \frac{2V_0}{35\pi}\cos 6t - \cdots$$

A graph of the approximation to $V(t)$ obtained using the first four terms of this series is shown in Fig. 28-10.

Figure 28-10

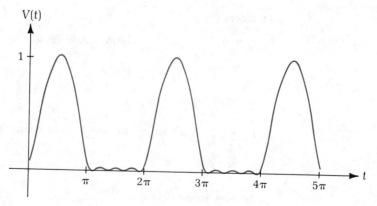

REMEMBER ▶ The Fourier series represents a function over an entire interval. The Taylor series represents a function in the neighborhood of a point where the derivatives of the function exist. ◀

Exercises 28-5

Find the Fourier series representation of each function, and include the first three coefficients for a_n and b_n.

1. $f(x) = \begin{cases} -1 & \text{for } -\pi \leq x < 0 \\ 1 & \text{for } 0 \leq x < \pi \end{cases}$

2. $f(x) = \begin{cases} 1 & \text{for } -\pi \leq x < 0 \\ 2 & \text{for } 0 \leq x < \pi \end{cases}$

3. $f(x) = \begin{cases} 0 & \text{for } -1 \leq x < 0 \\ 1 & \text{for } 0 \leq x < 1 \end{cases}$

4. $f(x) = \begin{cases} -2 & \text{for } -1 \leq x < 0 \\ 2 & \text{for } 0 \leq x < 1 \end{cases}$

5. $f(t) = \begin{cases} t & \text{for } -1 \leq t < 0 \\ -t & \text{for } 0 \leq t < 1 \end{cases}$

6. $f(t) = \begin{cases} t & \text{for } 0 \leq t < 1 \\ 1 - t & \text{for } 1 \leq t < 2 \end{cases}$

7. $f(x) = \begin{cases} 0 & \text{for } -\pi \leq x < 0 \\ x & \text{for } 0 \leq x < \pi \end{cases}$

8. $f(x) = \begin{cases} x & \text{for } 0 \leq x < \pi \\ 0 & \text{for } \pi \leq x < 2\pi \end{cases}$

9. $f(x) = \begin{cases} 0 & \text{for } -\pi \leq x < 0 \\ \dfrac{\pi}{2} - x & \text{for } 0 \leq x < \pi \end{cases}$

10. $f(x) = x - \pi \quad \text{for } -\pi \leq x < \pi$

11. $f(t) = \begin{cases} 0 & \text{for } -2 \leq t < -1 \\ 2 & \text{for } -1 \leq t < 1 \\ 0 & \text{for } 1 \leq t < 2 \end{cases}$

12. $f(t) = \begin{cases} 0 & \text{for } 0 \leq t < \pi \\ \cos t & \text{for } \pi \leq t < 2\pi \end{cases}$

Solve.

13. **Electronics** The circuit shown in the figure provides **full-wave rectification** of a sinusoidal voltage input V_s. Diode D_1 conducts on the positive half-cycle, and diode D_2 conducts on the negative half-cycle. The result, shown in the graph, is an output in which the original ac input has been converted to dc.* Find the Fourier series expansion of this output as given by

$$V(t) = \begin{cases} \sin t & \text{for } 0 \leq t < \pi \\ -\sin t & \text{for } \pi \leq t < 2\pi \end{cases}$$

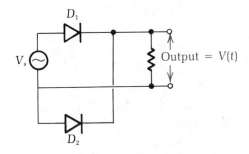

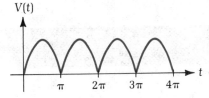

*Edward Pasahow, *Principles of Integrated Electronics* (Belmont, CA: Wadsworth, 1982), p. 281

14. *Physics* When an elastic system is forced to oscillate under a nonsinusoidal periodic driving force $F(t)$, the Fourier series expansion of $F(t)$ can be used to represent this force as a sum of sinusoidal component forces. Find the Fourier series expansion of the force function

$$F(t) = \begin{cases} \dfrac{\pi}{2} + t & \text{for } -\pi \leq t < 0 \\[2mm] \dfrac{\pi}{2} - t & \text{for } 0 \leq t < \pi \end{cases}$$

15. *Electronics* A relaxation oscillator circuit is shown in the figure. When the capacitor C is charged to some critical level, it discharges, and the neon lamp N lights briefly. If the voltage buildup on the lamp follows the equation (see the graph)

$$V(t) = 1 - e^{-t} \qquad \text{for } 0 \leq t < 2P$$

find the Fourier series expansion of this voltage signal.

$$\left[\text{Hint:} \quad \int e^{-x} \sin ax \, dx = -\frac{e^{-x}}{1 + a^2}(\sin ax + a \cos ax) + C \right.$$

$$\left. \text{and} \quad \int e^{-x} \cos ax \, dx = -\frac{e^{-x}}{1 + a^2}(\cos ax - a \sin ax) + C \right]$$

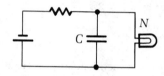

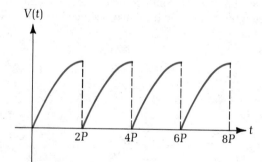

16. Find the Fourier series for the function

$$f(x) = \begin{cases} e^x & \text{for } 0 \leq x < \pi \\ e^\pi & \text{for } \pi \leq x < 2\pi \end{cases}$$

28-6 | Review of Chapter 28

Important Terms and Concepts

power series (p. 1214)

convergence (p. 1214)

analytic functions (p. 1215)

Maclaurin series (p. 1215)

periodic function (p. 1241)

trigonometric series (p. 1241)

Exercises 28-5

Find the Fourier series representation of each function, and include the first three coefficients for a_n and b_n.

1. $f(x) = \begin{cases} -1 & \text{for } -\pi \le x < 0 \\ 1 & \text{for } 0 \le x < \pi \end{cases}$

2. $f(x) = \begin{cases} 1 & \text{for } -\pi \le x < 0 \\ 2 & \text{for } 0 \le x < \pi \end{cases}$

3. $f(x) = \begin{cases} 0 & \text{for } -1 \le x < 0 \\ 1 & \text{for } 0 \le x < 1 \end{cases}$

4. $f(x) = \begin{cases} -2 & \text{for } -1 \le x < 0 \\ 2 & \text{for } 0 \le x < 1 \end{cases}$

5. $f(t) = \begin{cases} t & \text{for } -1 \le t < 0 \\ -t & \text{for } 0 \le t < 1 \end{cases}$

6. $f(t) = \begin{cases} t & \text{for } 0 \le t < 1 \\ 1 - t & \text{for } 1 \le t < 2 \end{cases}$

7. $f(x) = \begin{cases} 0 & \text{for } -\pi \le x < 0 \\ x & \text{for } 0 \le x < \pi \end{cases}$

8. $f(x) = \begin{cases} x & \text{for } 0 \le x < \pi \\ 0 & \text{for } \pi \le x < 2\pi \end{cases}$

9. $f(x) = \begin{cases} 0 & \text{for } -\pi \le x < 0 \\ \dfrac{\pi}{2} - x & \text{for } 0 \le x < \pi \end{cases}$

10. $f(x) = x - \pi$ for $-\pi \le x < \pi$

11. $f(t) = \begin{cases} 0 & \text{for } -2 \le t < -1 \\ 2 & \text{for } -1 \le t < 1 \\ 0 & \text{for } 1 \le t < 2 \end{cases}$

12. $f(t) = \begin{cases} 0 & \text{for } 0 \le t < \pi \\ \cos t & \text{for } \pi \le t < 2\pi \end{cases}$

Solve.

13. *Electronics*   The circuit shown in the figure provides **full-wave rectification** of a sinusoidal voltage input V_s. Diode D_1 conducts on the positive half-cycle, and diode D_2 conducts on the negative half-cycle. The result, shown in the graph, is an output in which the original ac input has been converted to dc.* Find the Fourier series expansion of this output as given by

$$V(t) = \begin{cases} \sin t & \text{for } 0 \le t < \pi \\ -\sin t & \text{for } \pi \le t < 2\pi \end{cases}$$

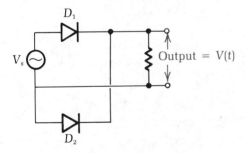

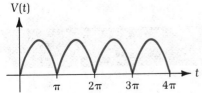

*Edward Pasahow, *Principles of Integrated Electronics* (Belmont, CA: Wadsworth, 1982), p. 281

14. *Physics*   When an elastic system is forced to oscillate under a nonsinusoidal periodic driving force $F(t)$, the Fourier series expansion of $F(t)$ can be used to represent this force as a sum of sinusoidal component forces. Find the Fourier series expansion of the force function

$$
F(t) = \begin{cases} \dfrac{\pi}{2} + t & \text{for } -\pi \leq t < 0 \\[2mm] \dfrac{\pi}{2} - t & \text{for } 0 \leq t < \pi \end{cases}
$$

15. *Electronics*   A relaxation oscillator circuit is shown in the figure. When the capacitor C is charged to some critical level, it discharges, and the neon lamp N lights briefly. If the voltage buildup on the lamp follows the equation (see the graph)

$$V(t) = 1 - e^{-t} \qquad \text{for } 0 \leq t < 2P \qquad \text{find the Fourier series expansion of this voltage signal.}$$

$$
\left[\text{Hint:} \quad \int e^{-x} \sin ax \, dx = -\frac{e^{-x}}{1 + a^2}(\sin ax + a \cos ax) + C \right.
$$

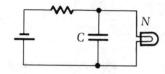

$$
\left. \text{and} \quad \int e^{-x} \cos ax \, dx = -\frac{e^{-x}}{1 + a^2}(\cos ax - a \sin ax) + C \right]
$$

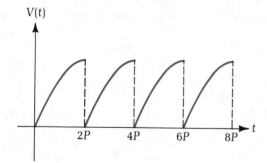

16. Find the Fourier series for the function

$$
f(x) = \begin{cases} e^x & \text{for } 0 \leq x < \pi \\ e^\pi & \text{for } \pi \leq x < 2\pi \end{cases}
$$

28-6 | Review of Chapter 28

Important Terms and Concepts

power series (p. 1214)

convergence (p. 1214)

analytic functions (p. 1215)

Maclaurin series (p. 1215)

periodic function (p. 1241)

trigonometric series (p. 1241)

Formulas and Rules

- Maclaurin series (p. 1216)

$$f(x) = f(0) + f'(0)x + \frac{f''(0)x^2}{2!} + \frac{f'''(0)x^3}{3!} + \cdots + \frac{f^{(n)}(0)x^n}{n!} + \cdots \tag{28-1}$$

- see Table 28-1 (p. 1222) for the Maclaurin series expansions of e^x, sin x, cos x, ln $(1 + x)$, $\dfrac{1}{1+x}$, and $(1 + x)^n$

- error in a truncated Maclaurin series (p. 1231)

$$E_n(x) = \left| \frac{f^{(n+1)}(c)x^{n+1}}{(n+1)!} \right| \tag{28-2}$$

where c is some number between 0 and x

- Taylor series about a (p. 1236)

$$f(x) = f(a) + f'(a)(x - a) + \frac{f''(a)(x-a)^2}{2!} + \frac{f'''(a)(x-a)^3}{3!} + \cdots + \frac{f^{(n)}(a)(x-a)^n}{n!} + \cdots \tag{28-3}$$

- Fourier series for $f(x)$ (p. 1242)

$$\begin{aligned}f(x) = a_0 + a_1 \cos x + a_2 \cos 2x + \cdots + a_n \cos nx + \cdots \\ + b_1 \sin x + b_2 \sin 2x + \cdots + b_n \sin nx + \cdots\end{aligned} \tag{28-4}$$

where the coefficients are

$$a_0 = \frac{1}{2\pi} \int_{-\pi}^{\pi} f(x)\, dx \tag{28-5}$$

$$a_n = \frac{1}{\pi} \int_{-\pi}^{\pi} f(x) \cos nx\, dx \qquad n = 1, 2, 3, \ldots \tag{28-6}$$

$$b_n = \frac{1}{\pi} \int_{-\pi}^{\pi} f(x) \sin nx\, dx \qquad n = 1, 2, 3, \ldots \tag{28-7}$$

- integrals used to determine the coefficients of a Fourier series (pp. 1242–1243)

$$\int_{-\pi}^{\pi} \cos nx\, dx = \int_{-\pi}^{\pi} \sin nx\, dx = 0 \qquad n \neq 0 \tag{28-8}$$

$$\int_{-\pi}^{\pi} \cos^2 nx\, dx = \int_{-\pi}^{\pi} \sin^2 nx\, dx = \pi \qquad n \neq 0 \tag{28-9}$$

$$\int_{-\pi}^{\pi} \cos mx \cos nx\, dx = 0 \qquad m \neq n \tag{28-10}$$

$$\int_{-\pi}^{\pi} \sin mx \sin nx\, dx = 0 \qquad m \neq n \tag{28-11}$$

$$\int_{-\pi}^{\pi} \sin mx \cos nx\, dx = 0 \tag{28-12}$$

- Fourier series for a function with period 2P (p. 1248)

$$\begin{aligned}F(t) = a_0 + a_1 \cos \frac{\pi t}{P} + a_2 \cos \frac{2\pi t}{P} + \cdots + a_n \cos \frac{n\pi t}{P} + \cdots \\ + b_1 \sin \frac{\pi t}{P} + b_2 \sin \frac{2\pi t}{P} + \cdots + b_n \sin \frac{n\pi t}{P} + \cdots\end{aligned} \tag{28-13}$$

$$a_0 = \frac{1}{2P} \int_0^{2P} f(t)\, dt \tag{28-14}$$

$$a_n = \frac{1}{P} \int_0^{2P} f(t) \cos \frac{n\pi t}{P} \, dt \qquad n = 1, 2, 3, \ldots \tag{28-15}$$

$$b_n = \frac{1}{P} \int_0^{2P} f(t) \sin \frac{n\pi t}{P} \, dt \qquad n = 1, 2, 3, \ldots \tag{28-16}$$

Exercises 28-6

Find the first three nonzero terms of the Maclaurin series expansion for each function.

1. $f(x) = e^{-3x}$
2. $f(x) = e^{x^2}$
3. $f(x) = 2x \cos x$
4. $f(x) = e^{2x} + e^{-2x}$
5. $f(x) = \ln(2x + 1)$
6. $f(x) = \sqrt{x + 1}$
7. $f(x) = 2x^3 - 5x + 2$
8. $f(x) = \dfrac{1}{(2x + 1)^2}$
9. $f(x) = e^x \sin x$
10. $f(x) = \dfrac{\cos x}{x + 1}$

Integrate each of the following using the first three nonzero terms of the appropriate Maclaurin series expansion:

11. $\displaystyle\int \frac{e^{x^2}}{x} \, dx$
12. $\displaystyle\int_0^{0.5} \cos \sqrt{x} \, dx$

Differentiate each function using the first three nonzero terms of the appropriate Maclaurin series.

13. $f(x) = x^2 \sin x$
14. $f(x) = \ln(1 - x^2)$

Calculate the value of each function using the first three nonzero terms of a Maclaurin series.

15. $e^{0.4}$
16. $\sin 0.6$
17. $\cos 7°$
18. $\ln 0.9$
19. $\operatorname{Arcsin} 0.2$
20. $\sqrt[3]{1.2}$

Use the results obtained in the indicated problem to estimate the maximum possible error of the approximation.

21. Problem 17
22. Problem 18

Expand each function in a Taylor series about the given value. Find the first three nonzero terms.

23. e^{-x}; $a = 4$
24. $\cos 2x$; $a = \dfrac{\pi}{3}$
25. $\ln(\sin x)$; $a = \dfrac{\pi}{6}$
26. $\dfrac{1}{x^2}$; $a = 1$

Evaluate each function using the first three nonzero terms of an appropriate Taylor series.

27. $\tan 15°$; $a = \dfrac{\pi}{10}$
28. $\sqrt[3]{6.8}$; $a = 8$
29. $e^{1.5}$; $a = 1$, use $e^1 \approx 2.71828$
30. $\ln 2.25$; $a = 2$, use $\ln 2 \approx 0.693147$

Find the Fourier series representation of each function, and include the first three coefficients for a_n and b_n.

31. $f(x) = \begin{cases} 1 & \text{for } -\pi \le x < 0 \\ -1 & \text{for } 0 \le x < \pi \end{cases}$

32. $f(x) = \begin{cases} 0 & \text{for } 0 \le x < \pi \\ x & \text{for } \pi \le x < 2\pi \end{cases}$

Solve.

33. Use the first three nonzero terms of a Maclaurin series to find the area between $x = 0$ and $x = \sqrt{\pi}$ bounded by the curve $y = \sin x^2$ and the x-axis. Round to two decimal places.

34. Find the Maclaurin series expansion for Arctan x by first finding a Maclaurin series expansion for

$$f(x) = \frac{1}{1 + x^2}$$

and integrating term-by-term.

35. *Statistics* Use the first three terms of a Maclaurin series to evaluate the error distribution

$$P = \frac{1}{\sqrt{2\pi}} \int_0^1 e^{-x^2/2} \, dx$$

$$\left[\text{Hint:}\quad \text{Expand } e^u \text{ as a Maclaurin series; then substitute } u = -\frac{1}{2}x^2. \right]$$

36. *Thermodynamics* The integral

$$v(t) = \int_0^t x^2 e^{-kx^2} \, dx$$

arises in the calculation of the mean speed of molecules in a gas. Use the first three terms of a Maclaurin series to evaluate this integral.

37. Find the Maclaurin series expansion for $1/(1 - x)^2$ by first finding a Maclaurin series expansion for $f(x) = 1/(1 - x)$ and differentiating term-by-term.

38. Using the first three terms of a Maclaurin series expansion, find the area bounded by the curve

$$y = \frac{x + \cos x^2}{x}$$

the x-axis, and the lines $x = 0.50$ and $x = 3.50$. Round to two decimal places.

39. *Physics* The observed, or *doppler-shifted*, frequency f of a light source of frequency f_0 moving away from the observer with relative velocity v is given by the equation

$$f = f_0 \frac{1 - \beta}{\sqrt{1 - \beta^2}}$$

where $\beta = v/c$ and c is the speed of light ($v < c$). Expand this in a Maclaurin series, and show the first three terms.
[Hint: Expand $(1 - \beta^2)^{-1/2}$ in a Maclaurin series and multiply by $(1 - \beta)$.]

40. *Physics* If a fixed amount of charge is distributed along a straight line of length L, the electrical potential at a distance x from one end is given by the equation

$$V(x) = k \ln \frac{L + x}{x}$$

Expand this expression as a Taylor series about $x = L$ and give the first three terms.

41.   *Atomic physics*   The expression

$$\cos \theta = \left(1 + \frac{1}{L}\right)^{-1/2}$$

relates the quantum number L to the angle θ that the angular momentum vector makes with some reference direction. By expanding each side of the equation $y = (1 + x)^{-1/2}$, where $x = 1/L$, in a Maclaurin series, show that $\theta \approx 1/\sqrt{L}$ for large L.

42.   Expand $\ln x$ in a Taylor series expansion about $x = e$.

43.   Use the first three terms of a Maclaurin series expansion to evaluate the integral

$$\int_0^{\pi/2} e^{\sin x} \, dx$$

44.   Use a Taylor series expansion to rewrite the function $f(x) = 4x^5 + 3x^4 - 2x^3 + x^2 + 5$ as a series in $(x - 1)$.

Introduction to Differential Equations

29

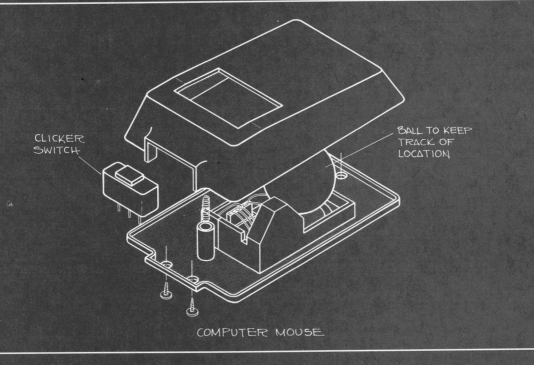

CLICKER SWITCH

BALL TO KEEP TRACK OF LOCATION

COMPUTER MOUSE

In physics and technology, the most useful mathematical descriptions of dynamic phenomena are not usually simple algebraic equations, but are equations in which derivatives of unknown functions appear. These differential equations are used to describe processes of continuous, gradual change, and to relate these changes to the various aspects of the physical situation in which they occur.

A very wide variety of differential equations appear in technical applications, and, like the process of integration, the solving of differential equations involves procedures and techniques that cannot be summarized by a few simple rules. In this chapter we will look at a few types of differential equations that appear often in scientific and technical work, and we will examine some of the most general techniques for solving differential equations.

A **differential equation** is an equation that includes some unknown function and one or more of its derivatives. For example, the equations

$$\frac{dy}{dx} + xy = 2 \qquad x\frac{dy}{dx} - \left(\frac{dy}{dx}\right)^2 = y \qquad \frac{d^2y}{dx^2} - 2\frac{dy}{dx} = x$$

are all differential equations. These equations may also be written using the notation $y' = dy/dx$, $y'' = d^2y/dx^2$, and so on, so that the examples given above become

$$y' + xy = 2 \qquad xy' - (y')^2 = y \qquad y'' - 2y' = x$$

The first example may also be written using differentials:

$$dy + xy\,dx = 2\,dx$$

Because the derivative dy/dx can be interpreted as a rate of change with respect to x of some function $y(x)$, many verbal descriptions of physical phenomena can be translated into differential equations.

EXAMPLE 1 Write a differential equation corresponding to each of the following statements:
(a) The angular velocity of the rotating shaft of a control device is proportional to the cosine of the angle through which it has rotated.
(b) The rate of change of the speed of a car moving in a line of traffic is proportional to the negative of the difference between the speed of the car and the speed of the truck it is following.
(c) A certain graph has a slope at each point that is equal to the reciprocal of the product of its x and y coordinates at that point.
(d) The acceleration of an object moving along the horizontal axis is proportional to the sum of its horizontal displacement and its velocity at any time t.

Solution
(a) If $\theta(t)$ is the angle of rotation at time t, then $d\theta/dt$ is the angular velocity of the shaft, and the differential equation describing the motion is $d\theta/dt = k\cos\theta$, where k is a proportionality constant.
(b) Let v be the speed of the car, and let v_L be the speed of the truck it is following. Then dv/dt is the rate of change of speed of the car. The differential equation is therefore

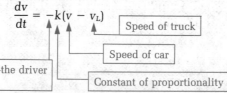

$$\frac{dv}{dt} = -k(v - v_L)$$

Speed of truck

Speed of car

Constant of proportionality

If $v > v_L$, then dv/dt is negative—the driver of the car must apply the brakes!

(c) If the function describing the graph is $y = y(x)$, then its slope at point (x, y) is dy/dx, and the differential equation describing it is

$$\frac{dy}{dx} = \frac{1}{xy}$$

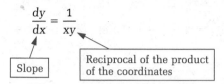

Slope

Reciprocal of the product of the coordinates

(d) If x is the displacement of the object at time t, then its velocity is dx/dt and its acceleration is d^2x/dt^2. Therefore, the differential equation describing its motion is

$$\frac{d^2x}{dt^2} = k\left(x + \frac{dx}{dt}\right)$$

Proportionality constant

Order   Before we go any further, some new terminology is needed. The **order** of a differential equation is the order of the highest derivative in the equation. If a differential equation contains derivatives of first order only, it is said to be a **first-order equation.** If a differential equation contains derivatives of second order, and possibly first order, but no third- or higher-order derivatives, it is said to be a **second-order equation.** For example:

- $y' + x^4 = 2$ is a first-order differential equation, since the highest-order derivative it contains is $y' = dy/dx$, a first derivative.
- $xy'' + y' = y$ is a second-order differential equation, since the highest-order derivative it contains is $y'' = d^2y/dx^2$, a second derivative.

Degree   The **degree** of a differential equation is the power of the highest-order derivative in the equation. For example:

- $x\left(\dfrac{dy}{dx}\right)^2 = 1 - y$ is a first-order differential equation of degree 2.

Exponent 2 means degree 2

First derivative, so first-order differential equation

The highest derivative has exponent 1, so the degree is 1.

- $\dfrac{d^2y}{dx^2} + x\left(\dfrac{dy}{dx}\right)^4 + y = 0$ is a second-order differential equation of degree 1.

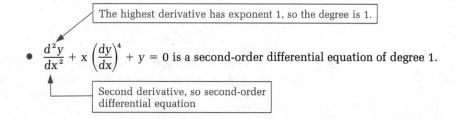

Second derivative, so second-order differential equation

EXAMPLE 2   Determine the order and degree of each differential equation.

(a) $y' = 2y$

(b) $x\dfrac{d^2y}{dx^2} = 4\dfrac{dy}{dx}$

(c) $\left(\dfrac{dQ}{dt}\right)^2 = \dfrac{dQ}{dt} + Q$

(d) $\dfrac{d^2p}{dt^2} = 1 - \left(\dfrac{dp}{dt}\right)^3$

(e) $(y'')^3 + xy' + y = 0$

(f) $(y')^4 - 2y'' = y$

Solution

(a)   First order, degree 1

(b)   Second order, degree 1

(c)   First order, degree 2

(d)   Second order, degree 1

(e)   Second order, degree 3

(f)   Second order, degree 1

Solution   The **solution** of a differential equation is some relation among variables in the differential equation that satisfies the equation. That is, given a differential equation in variables y and x, if the expression $f(x)$ is substituted for y in this equation and the result is an identity that is true for all x in some interval, then $y = f(x)$ is a solution of the differential equation.

EXAMPLE 3   Show that each equation is a solution of the given differential equation.

Differential Equation	**Solution**
(a) $y' = 4x$	$y = 2x^2 + 1$
(b) $\dfrac{dy}{dx} = 4y$	$y = e^{4x}$
(c) $\dfrac{d^2y}{dx^2} + x\dfrac{dy}{dx} - 2 = 2y - x$	$y = x^2 + x$
(d) $V'' + V = 0$	$V(x) = 3 \sin x$

Solution

(a)   If $y = 2x^2 + 1$, then $y' = dy/dx = 4x$, and substituting into the differential equation, we have $4x = 4x$. The resulting equation is true for all values of x, so we know that $y = 2x^2 + 1$ is a solution of the original differential equation.

(b)   If $y = e^{4x}$, then $dy/dx = 4 \cdot e^{4x}$. Substituting into the differential equation, we have $4e^{4x} = 4y = 4e^{4x}$.

(c)   If $y = x^2 + x$, then $dy/dx = 2x + 1$ and $d^2y/dx^2 = 2$. Substituting into the differential equation, we obtain

$$2 + x(2x + 1) - 2 = 2(x^2 + x) - x$$

$$2x^2 + x = 2x^2 + x \qquad \text{An identity}$$

(d)   If $V(x) = 3 \sin x$, then using derivative formulas (26-3) and (26-4), $V' = 3 \cos x$, and $V'' = -3 \sin x$. Substituting into the differential equation, we have $-3 \sin x + 3 \sin x = 0$, which is true for all x.

Each of the solutions in Example 3 is only one of an infinite number of solutions of the given differential equation. In Example 3(a), the functions $y = 2x^2$, $y = 2x^2 + 2$, $y = 2x^2 - 3, \ldots, y = 2x^2 + c$ (for some constant c) are also solutions of the differential equation $y' = 4x$. In Example 3(b), the functions $y = 2e^{4x}$, $y = 3e^{4x}$, $y = 4e^{4x}, \ldots, y = ce^{4x}$ (where c is some constant) are all solutions of $y' = 4y$. In part (d) of Example 3, the functions $y = c_1 \sin x$, $y = c_2 \cos x$, and $y = c_1 \sin x + c_2 \cos x$ (where c_1 and c_2 are constants) are all solutions of the given differential equation. (Check them.)

In general, a differential equation does not have only a single solution; it is satisfied by an infinite family of solutions.

General solution

Particular solution

The **general solution** of an nth-order differential equation will contain n arbitrary constants. If one or more of these constants are specified by giving it a numerical value, the solution is said to be a **particular solution.** The particular solution to a differential equation is obtained from the general solution and additional information or conditions that are given with the differential equation. These conditions that the general solution must satisfy are often called **initial,** or **boundary, conditions.** For example, the general solution of the first-order differential equation in Example 3(a) is $y = 2x^2 + c$, an equation with one arbitrary constant c. This equation defines the family of parabolas shown in Fig. 29-1. Each value of the arbitrary constant gives a particular solution of the differential equation.

Figure 29-1

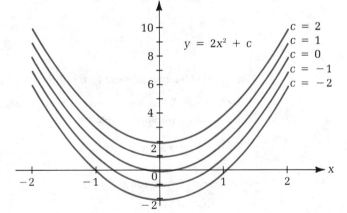

Solving differential equations

Solving many first-order, first-degree differential equations is a straightforward process of integration. If a first-order differential equation can be written in the form $y' = f(x)$, where $f(x)$ is some integrable function of x, then the solution can be found by integrating both sides of the equation with respect to x:

$$\int y'\, dx = \int f(x)\, dx$$

But

$$\int y'\, dx = \int \frac{dy}{dx}\, dx = \int dy = y$$

Therefore, the solution is: $y = \int f(x)\, dx$

Direct Integration

If $dy/dx = f(x)$ or $dy = f(x)\, dx$, then

$$y = \int f(x)\, dx \qquad\qquad (29\text{-}1)$$

EXAMPLE 4

(a) Solve the differential equation $y' = 6x^2$.

(b) A particle moves along a straight line under an applied force that is the product of a linearly increasing force and a periodic force. This motion is described by the differential equation $dv/dt = t \sin t$, where v is the velocity of the particle, and the initial speed at $t = 0$ s is $v = 1$ m/s. Find the particular solution of this equation under the given condition.

Solution

(a) Find the general solution by taking the integral of both sides of the equation:

$$y' = 6x^2$$

$$\int dy = \int 6x^2\, dx$$

$$y = 2x^3 + C$$

> The first-order equation has a solution with one arbitrary constant.

This is the general solution of the differential equation.

Check: Check this solution by differentiating and substituting back into the original differential equation.

$$y' = \frac{d}{dx}(2x^3 + C) = 6x^2 \qquad \text{The original differential equation}$$

(b) In differential notation we have

$$dv = t \sin t\, dt$$

or $\int dv = \int t \sin t\, dt$

First, the integral on the right can be evaluated using integration by parts with $u = t$, $dv = \sin t\, dt$. Then by (27-22),

$$v = -t \cos t + \int \cos t\, dt$$

$$v = -t \cos t + \sin t + C$$

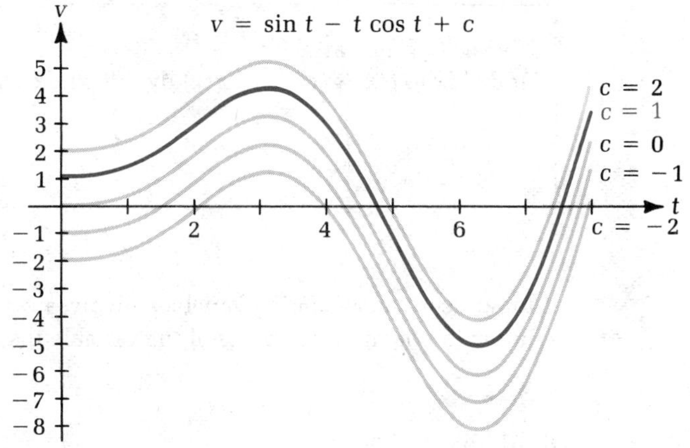

$$v = \sin t - t \cos t + c$$

Figure 29-2

This general solution of the differential equation defines the family of curves shown in Fig. 29-2.

Second, from the initial condition, $v = 1$ at $t = 0$. Therefore, substituting these values into the general solution, we find

$$(1) = -(0) \cdot \cos (0) + \sin (0) + C$$

$$1 = C$$

Then, the particular solution is

$$v = -t \cos t + \sin t + 1$$

Check: Check this solution by differentiating:

$$v' = (-t)(-\sin t) - \cos t + \cos t$$

$$= t \sin t \qquad \text{The original differential equation}$$

Verify that $v = 1$ when $t = 0$ in the particular solution:

$$1 = -0 + 0 + 1$$

IMPORTANT ▶

Always check the solution by substituting it and its derivatives into the original differential equation, and by verifying that the initial conditions hold for the particular solution. ◀

Separation of variables

A first-order, first-degree differential equation $dy/dx = P(x, y)$ is said to be **separable** if it can be written in the form $dy/dx = f(x)/g(y)$ or, in differential notation, $g(y) \, dy = f(x) \, dx$. In this equation, $f(x)$ is an integrable function of x only, and $g(y)$ is an integrable function of y only. The variables x and y have been "separated" from each other. The solution of the separated equation may be obtained by integrating both sides.

Separation of Variables
If $dy/dx = f(x)/g(y)$ or $g(y)\,dy = f(x)\,dx$, then

$$\int g(y)\,dy = \int f(x)\,dx \qquad\qquad (29\text{-}2)$$

CAUTION

If the process of separating variables involves division by an expression, the solution is not valid for values of the variable that make the divisor zero. ◄

EXAMPLE 5 Solve:

(a) $y' = \dfrac{3x^2}{y}$

(b) $y(4 + x^2)\dfrac{dy}{dx} = 1;$ for $y = 1$ when $x = 0$

(c) $(1 + x^2)\,dy + 2xy\,dx = 0;$ for $y = 1$ when $x = 1$

Solution

(a) **First,** write the equation using differential notation; then separate the variables:

$$\frac{dy}{dx} = \frac{3x^2}{y}$$

$$y\,dy = 3x^2\,dx$$

Second, integrate both sides of this equation:

$$\int y\,dy = \int 3x^2\,dx$$

$$\frac{1}{2}y^2 = x^3 + C_1 \quad\longleftarrow\quad \boxed{\text{Rather than include a constant of integration for each integral, we combine them into one constant on the right.}}$$

$$y^2 = 2x^3 + 2C_1$$

$$y^2 = 2x^3 + C \quad\longleftarrow\quad \boxed{\text{Notice that it is convenient to rewrite the constant } 2C_1 \text{ as } C.}$$

Check: Check the solution using implicit differentiation:

$$2y\frac{dy}{dx} = 3\cdot 2x^2 + 0$$

$$\frac{dy}{dx} = \frac{3x^2}{y}$$

(b) **First,** separate the variables by rewriting the given equation as

$$y\,dy = \frac{dx}{4 + x^2}$$

Second, integrate both sides of this equation:

$$\int y\ dy = \int \frac{dx}{4 + x^2}$$

From integration formula (27-20)

$$\frac{1}{2}y^2 = \frac{1}{2} \text{Arctan} \left(\frac{x}{2}\right) + C$$

Third, substitute the initial condition to obtain the particular solution:

$$\frac{1}{2} \cdot (1)^2 = \frac{1}{2} \text{Arctan} \frac{(0)}{2} + C$$

$$\frac{1}{2} = C$$

Therefore,

$$\frac{1}{2}y^2 = \frac{1}{2} \text{Arctan} \left(\frac{x}{2}\right) + \frac{1}{2}$$

$$y^2 = \text{Arctan} \left(\frac{x}{2}\right) + 1$$

Check: Check the solution by substituting it into the original differential equation. Using implicit differentiation,

$$2y \frac{dy}{dx} = \frac{1}{1 + \left(\frac{x}{2}\right)^2} \cdot \frac{d}{dx} \left(\frac{x}{2}\right) + 0$$

$$= \frac{\frac{1}{2}}{1 + \frac{x^2}{4}}$$

$$= \frac{2}{4 + x^2}$$

$$y(4 + x^2) \frac{dy}{dx} = 1$$

(c) **First,** separate variables.

$$\frac{dy}{y} + \frac{2x\ dx}{1 + x^2} = 0$$

Second, integrate.

$$\int \frac{dy}{y} + \int \frac{2x\ dx}{1 + x^2} = 0$$

Note that both integrands have the form du/u, therefore:

$$\ln |y| + \ln (1 + x^2) = C_1$$

Using the properties of logarithms, we can write this as:

$$\ln [y(1 + x^2)] = C_1$$

Now, since C_1 is an arbitrary constant, we can write it as $\ln C$, where C is a constant.

$$\ln [y(1 + x^2)] = \ln C$$

$$y(1 + x^2) = C$$

Therefore, the general solution of the differential equation is:

$$y = \frac{C}{1 + x^2}$$

Third, according to the condition given, $y = 1$ when $x = 1$, so that:

$$(1) = \frac{C}{1 + (1)^2}$$

$$C = 2$$

The particular solution is:

$$y = \frac{2}{1 + x^2}$$

Check: Check the solution by differentiating and substituting back into the original differential equation.

EXAMPLE 6 Expressed as a differential equation, Newton's law of cooling states that the rate of temperature change of an object is proportional to the difference between its temperature T and the temperature of its surroundings T_s.

(a) Write the differential equation to express Newton's law of cooling.
(b) Find the general solution using separation of variables.
(c) Find the particular solution corresponding to the initial condition $T = T_0$ at $t = 0$ s.
▦ (d) A cup of coffee initially at $T_0 = 34.0°C$ is in a room kept at a constant temperature of $23.0°C$. If it cools to $30.0°C$ in 2 min, what will be its temperature after 5 min?

Solution

T_s is a constant

(a) $\dfrac{dT}{dt} = -k(T - T_s)$ $k > 0$

> The negative sign is important. For $T > T_s$, the object is warmer than its surroundings and is cooling; therefore, the rate of temperature change dT/dt must be negative. Similarly, if $T < T_s$, dT/dt must be positive.

First, separate variables and integrate.

$$\frac{dT}{T - T_s} = -k\, dt$$

$$\int \frac{dT}{T - T_s} = -k \int dt$$

$$\ln (T - T_s) = -kt + C_1$$

Second, solve for T. Since the constant C_1 is arbitrary, we can write it as $C_1 = \ln C$.

$$\ln (T - T_s) - \ln C = -kt$$

$$\ln \left(\frac{T - T_s}{C}\right) = -kt$$

But from the definition of a logarithm (13-2), we know that if $\log_b y = x$, then $y = b^x$. Therefore:

$$\frac{T - T_s}{C} = e^{-kt}$$

The general solution is:

$$T = T_s + Ce^{-kt}$$

(c) Substitute the initial condition $T = T_0$ when $t = 0$, into the general solution.

$$T_0 = T_s + Ce^{-0}$$
$$= T_s + C$$
$$C = T_0 - T_s$$

Thus, the particular solution is:

$$T = T_s + (T_0 - T_s)e^{-kt}$$

This is the equation given in Chapter 26 when Newton's law of cooling was first discussed.

(d) **First,** determine the value of the constant k by solving the particular solution:

$$e^{-kt} = \frac{T - T_s}{T_0 - T_s}$$

$$k = -\frac{1}{t} \ln \left(\frac{T - T_s}{T_0 - T_s} \right) = -\frac{1}{2} \ln \left(\frac{30.0 - 23.0}{34.0 - 23.0} \right)$$

7 $\boxed{\div}$ 11 $\boxed{=}$ $\boxed{\text{lnx}}$ $\boxed{+/-}$ $\boxed{\div}$ 2 $\boxed{=}$ $\boxed{\text{STO}}$ → $\boxed{\;0.2259926\;}$

Next, use this value of k to find T at $t = 5$ min:

$$T(5) = 23.0 + (34.0 - 23.0)e^{-k(5)}$$

$\boxed{\text{RCL}}$ $\boxed{\times}$ 5 $\boxed{+/-}$ $\boxed{=}$ $\boxed{e^x}$ $\boxed{\times}$ 11 $\boxed{+}$ 23 $\boxed{=}$ → $\boxed{\;26.553498\;}$

$$T(5) \approx 26.6°C$$

Exercises 29-1

Find the (a) order and (b) degree of each differential equation.

1. $y' = 2x$ 　　　　　**2.** $y'' = 3x$ 　　　　　**3.** $y'' = 2x + (y')^3$

4. $(y')^3 = 4(y'')^2$ 　　**5.** $x\,dy = dx - y\,dx$ 　　**6.** $\cos x\,dy = dy - x\,dx$

7. $x\dfrac{d^2y}{dx^2} + \dfrac{dy}{dx} + y = 0$ 　**8.** $k\dfrac{d^3x}{dt^3} + px^4 = 0$ 　**9.** $\dfrac{1}{c}q + k\left(\dfrac{dq}{dt}\right)^3 = 0$

10. $L\dfrac{di}{dt} + Ri = E\cos \omega t$

Determine if the given equation is a solution of the accompanying differential equation.

11. $y = 5 - 2x$; $(y - 1)\,dx + (2 - x)\,dy = 0$ 　　**12.** $y = \dfrac{x^2}{4}$; $dy = \sqrt{y}\,dx$

13. $y = x^2 \ln x$; $x\dfrac{dy}{dx} - 2y = -x^2$ 　　　　**14.** $y = \dfrac{1}{x} + x$; $\dfrac{dy}{dx} = 2 - \dfrac{y}{x}$

15. $y = e^{2x}$; $\dfrac{d^2y}{dx^2} - \dfrac{dy}{dx} - 2y = 0$ 　　　　**16.** $y = 1 + e^{-x}$; $\dfrac{d^2y}{dx^2} = \dfrac{dy}{dx}\left(y + \dfrac{dy}{dx}\right)$

17. $y = \cos \omega x$; $y'' - \omega^2 y = 0$ 　　　　　　**18.** $y = ae^x + be^{-x}$; $y'' - y = 0$

Solve.

19. $y' = 2x^2 + 1$ 　　　　**20.** $y\dfrac{dy}{dx} = 4e^{2x}$

21. $dy - dx + \sin x \, dx = 0$ **22.** $2 \, dy = 3^x \, dx$

23. $\dfrac{dy}{dx} = x\sqrt{4 - x^2}$ **24.** $y' - \tan x = x$

25. $xy \, dy - x^2 \, dx = 0$ **26.** $x^2 \, dx + 4y^2 \, dy = 0$

27. $x^2 y' + y = 0$ **28.** $(1 + x)y' = 2 + y$

29. $y' + xy = x$ **30.** $2xy^2 + y^2 = y'$

31. $e^x \, dy = (y - 1)^2 \, dx$ **32.** $2y\sqrt{1 + x^2} \, dy = x \, dx$

33. $(x^2 \cos y)y' + 2 = 0$ **34.** $1 + y + (\cot x)y' = 0$

35. $e^{x+2} \, dx + dy = 0$ **36.** $x^2 \cos y \, dy + (x + x \sin y) \, dx = 0$

37. $xy + y = xy(y' - x)$ **38.** $y^2 - y' = xy' - 1$

Find the particular solution of each differential equation for the given condition.

39. $y^2 \dfrac{dy}{dx} = x$; $y = 1$ when $x = 0$ **40.** $\sqrt{x^2 - 1} \, \dfrac{dy}{dx} = xy$; $y = 4$ when $x = 1$

41. $\cot y + xy' = 0$; $y = 0$ when $x = 1$ **42.** $1 - xy' = y + y'$; $y = 0$ when $x = 0$

Solve.

43. **Electronics**   The charge q on a charging capacitor in an RC circuit with constant impressed voltage V is given by the differential equation

$$R\dfrac{dq}{dt} + \dfrac{1}{C}q = V$$

where $q = 0$ when $t = 0$. Solve for q.

44. **Chemistry**   If Q is the concentration of one component of a certain chemical reaction, then at fixed temperature, $dQ/dt = -10Q^2$. Solve this differential equation for Q if $Q = 0.5$ mole/L at $t = 0$ s.

45. **Hydrology**   If the rate of growth of the thickness of a layer of ice on a lake is determined by the differential equation $dx/dt = 4/x$, and if $x = 2$ cm at $t = 0$ h, when will the ice be 6 cm thick?

46. Find the equation of the curve whose slope at a point (x, y) is equal to twice the product of its coordinates, and for which $y = 2$ when $x = 0$.

29-2 | First-Order Linear Differential Equations

Integrable forms   For many first-order differential equations, the process of separating variables may be difficult or impossible. However, if some combination of terms in the equation can be written as the differential of an expression, then the differential equation can be integrated directly. For example, in the differential equation

$$x \, dy + y \, dx = 2 \, dy$$

you may recognize the expression on the left from the product rule (22-17):

$$d(xy) = x \, dy + y \, dx$$

Therefore, after substituting $u = xy$, the differential equation may be written as

$$du = 2 \, dy$$

Integrating both sides gives

$$u = 2y + C$$

Then,

$$xy = 2y + C$$

$$y = \frac{C}{x - 2}$$

The following differentials suggest possible integrable forms:

$$x \, dy + y \, dx = d(xy) \tag{29-3}$$

$$x \, dx + y \, dy = \frac{1}{2} d(x^2 + y^2) \tag{29-4}$$

$$x \, dx - y \, dy = \frac{1}{2} d(x^2 - y^2) \tag{29-5}$$

$$\frac{x \, dy - y \, dx}{x^2} = d\left(\frac{y}{x}\right) \tag{29-6}$$

$$\frac{y \, dx - x \, dy}{y^2} = d\left(\frac{x}{y}\right) \tag{29-7}$$

EXAMPLE 7   Solve the differential equation: $xy \, dy + y^2 \, dx = \dfrac{dx}{x^2}$

Solution   Factoring the left side of the equation, we can write

$$y(x \, dy + y \, dx) = \frac{dx}{x^2}$$

The expression in parentheses is the differential (29-3). Multiply both sides of the equation by x to get

$$(xy)(x \, dy + y \, dx) = \frac{dx}{x}$$

If we let $u = xy$, then the product on the left has the integrable form $u \, du$ and the differential equation becomes

$$u \, du = \frac{dx}{x}$$

Integrating, we have

$$\frac{1}{2} u^2 = \ln x + C$$

$$\frac{1}{2} x^2 y^2 = \ln x + C$$

Always check the solution by substituting back into the original differential equation.

EXAMPLE 8   Solve the differential equation $x \, dy - x^3 \, dx - y \, dx = 0$, subject to the condition $y = 0$ when $x = 2$.

Solution   Rearranging terms,

$$x \, dy - y \, dx = x^3 \, dx$$

The expression on the left appears in differential forms (29-6) and (29-7). To rewrite this in an integrable form, divide each side of the equation by x^2.

$$\frac{x \, dy - y \, dx}{x^2} = x \, dx$$

Therefore, by (29-6), we may write this as an integrable form.

$$d\left(\frac{y}{x}\right) = x \, dx$$

If we let $y/x = u$, then,

$$du = x \, dx$$

Integrate.

$$u = \frac{1}{2}x^2 + C$$

$$\frac{y}{x} = \frac{1}{2}x^2 + C$$

$$y = \frac{1}{2}x^3 + Cx$$

Substitute the condition $y = 0$ when $x = 2$.

$$(0) = \frac{1}{2}\cdot(2)^3 + C\cdot(2)$$

$$C = -2$$

The particular solution is: $y = \frac{1}{2}x^3 - 2x$
Check it.

The first approach to solving a first-order, first-degree differential equation is to attempt to separate the variables. If it is not possible to separate the variables, then try to rewrite the equation or some part of it as one of the integrable forms (29-3)–(29-7), or as an integrable form based on these.

EXAMPLE 9   Solve the differential equation $2xy^3 + 3x^2y^2y' = 1$, subject to the condition $y = 0$ when $x = 1$.

Solution   It is not possible to separate variables for this differential equation. Writing it in differential form, we have

$$2xy^3 \, dx + 3x^2y^2 \, dy = dx$$

This does not correspond directly to any of the given integrable forms (29-3)–(29-7). However, if we rewrite the expression on the left as

$$\underbrace{y^3}_{u}\cdot\underbrace{2x \, dx}_{dv} + \underbrace{x^2}_{v}\cdot\underbrace{3y^2 \, dy}_{du} = dx$$

then the equation can be written as

$$u \, dv + v \, du = dx$$

or, from (29-3),

$$d(uv) = dx$$

Integrating, we have

$$uv = x + C$$

$$x^2 y^3 = x + C$$

Substituting the given boundary condition $y = 0$ when $x = 1$, we find

$$(1)^2 \cdot (0)^3 = (1) + C$$

$$C = -1$$

The particular solution is $x^2 y^3 = x - 1$. Check it.

EXAMPLE 10 Solve $x^2 \cos y \, dy - 2x \sin y \, dx = x^4 y \, dy$, subject to the boundary condition $y = 0$ when $x = 1$.

Solution This differential equation is not separable, but notice that the expression on the left has the form $u \, dv - v \, du$, where $u = x^2$ and $v = \sin y$:

$$\underbrace{x^2}_{u} \cdot \underbrace{\cos y \, dy}_{dv} - \underbrace{\sin y}_{v} \cdot \underbrace{2x \, dx}_{du} = \underbrace{x^4}_{u^2} \cdot y \, dy$$

Therefore, we can rewrite this equation as

$$\frac{u \, dv - v \, du}{u^2} = y \, dy$$

and by (29-6), this is the integrable form

$$d\left(\frac{v}{u}\right) = y \, dy$$

Integrating,

$$\frac{v}{u} = \frac{1}{2} y^2 + C$$

$$\frac{\sin y}{x^2} = \frac{1}{2} y^2 + C$$

Substituting the boundary condition $y = 0$ when $x = 1$, we find

$$\frac{\sin (0)}{(1)^2} = \frac{1}{2} \cdot (0)^2 + C$$

$$C = 0$$

The particular solution is therefore $\sin y = \frac{1}{2} x^2 y^2$. Check it.

As Example 10 shows, it may take considerable ingenuity to rewrite a differential equation in an integrable form. In the remainder of this section we will show how this general procedure can be used to solve any first-order differential equation of a particular type.

Linear differential equations

A first-order differential equation is said to be **linear** if it can be written in the form

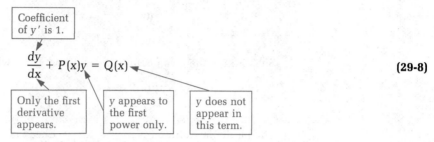

Coefficient of y' is 1.

$$\frac{dy}{dx} + P(x)y = Q(x) \qquad\qquad (29\text{-}8)$$

Only the first derivative appears.

y appears to the first power only.

y does not appear in this term.

In differential form, the first-order linear differential equation is written as

$$dy + P(x)y \, dx = Q(x) \, dx \qquad\qquad (29\text{-}9)$$

The functions P and Q may be constants or functions of x only. Equation (29-8) is the standard form for a first-order linear differential equation.*

EXAMPLE 11   Determine which of the following are first-order linear differential equations. For those that are, write the equation in standard form (29-8) and identify $P(x)$ and $Q(x)$.

(a) $dy + 2x^2y \, dx = x \, dx$ (b) $2y' - 4ye^x + 2 \sin x = 0$

(c) $dy + x^2y^2 \, dx = 2x \, dx$ (d) $xy' + y = x^2y - y' + x$

(e) $2 - \dfrac{dy}{dx} = 3y$ (f) $y'' - y' + xy = x^2$

Solution

(a)   Linear; $\dfrac{dy}{dx} + 2x^2y = x$; $P(x) = 2x^2$; $Q(x) = x$

(b)   Linear; $\dfrac{dy}{dx} - 2e^xy = -\sin x$; $P(x) = -2e^x$; $Q(x) = -\sin x$

(c)   Not linear; $\dfrac{dy}{dx} + x^2y^2 = 2x$

A linear equation contains y to the first power in this term.

(d)   Linear; $y' + xy' + y - x^2y = x$
$$(1 + x)y' + (1 - x^2)y = x$$
$$y' + \left(\frac{1 - x^2}{1 + x}\right)y = \frac{x}{1 + x}$$

---

*It may help you to identify a linear differential equation if you write it as $dy/dx = Q(x) - P(x)y = a + by$ and note that the expression on the right is linear in y.

$$\frac{dy}{dx} + (1 - x)y = \frac{x}{1 + x}$$

$$P(x) = 1 - x; \quad Q(x) = \frac{x}{1 + x}$$

(e)   Linear; $\dfrac{dy}{dx} + 3y = 2$; $P(x) = 3$; $Q(x) = 2$

(f)   Not first-order; $y'' - y' + xy = x^2$

> A first-order differential equation cannot contain a second derivative.

Integrating factor

In differential form, a first-order linear equation may be written as

$$dy + P(x)y \, dx = Q(x) \, dx$$

The left member of this equation will be an integrable form if we multiply each term of the equation by e^u, where $u = u(x)$ is some function of x:

$$e^u \, dy + ye^u \, P(x) \, dx = Q(x)e^u \, dx$$

But

$$d(ye^u) = e^u \, dy + ye^u \, du$$

and this is an integrable form similar to equation (29-3). Therefore, the left member of this modified differential equation will be an integrable form if $du = P(x) \, dx$, or, after integrating, $u = \int P(x) \, dx$.

The multiplier $F(x) = e^u = e^{\int P(x)\,dx}$ is called the **integrating factor** of the given differential equation. Multiplying by the integrating factor enables us to rewrite the original differential equation as

$$d[yF(x)] = Q(x)F(x) \, dx$$

with the left member in the form of a perfect differential. Integrating, we obtain

$$yF(x) = \int Q(x)F(x) \, dx$$

and we have the following general result:

The solution to the first-order linear differential equation

$$\frac{dy}{dx} + P(x)y = Q(x)$$

is

$$y = \frac{1}{F(x)}\left[\int Q(x)F(x) \, dx\right] \qquad (29\text{-}10)$$

where

$$F(x) = e^{\int P(x)\,dx} \qquad (29\text{-}11)$$

NOTE ▶

This method will produce a solution only if the following conditions are met:

1. The function $P(x)$ must be integrable; that is, the integral $\int P(x)\, dx$ must be one we can integrate.
2. The function $Q(x)F(x) = Q(x)e^{\int P(x)\, dx}$ must be integrable. ◀

EXAMPLE 12 Solve the linear differential equation: $\dfrac{dy}{dx} = y + e^x$

Solution

Step 1. Check to see if the equation is separable or an obvious integrable form. If it is not, rewrite it in standard linear form (29-8):

$$\frac{dy}{dx} - y = e^x$$

Step 2. Identify $P(x)$ and $Q(x)$, and calculate the integrating factor from (29-11):

$$P(x) = -1 \qquad Q(x) = e^x$$

$$\int P(x)\, dx = \int (-1)\, dx = -\int dx = -x$$

Therefore, the integrating factor is

$$F(x) = e^{-x}$$

Step 3. Use the integrating factor to obtain a solution in the form of equation (29-10):

$$y = \frac{1}{F(x)}\left[\int Q(x)F(x)\, dx\right]$$

$$= \frac{1}{e^{-x}}\left[\int e^x e^{-x}\, dx\right] = e^x\left[\int dx\right]$$

$$= e^x[x + C]$$

$$y = xe^x + Ce^x$$

Step 4. Check the solution by differentiating and substituting back into the original differential equation:

$$\frac{dy}{dx} = xe^x + e^x + Ce^x$$

Substituting into the original equation, we have

$$xe^x + e^x + Ce^x = (xe^x + Ce^x) + e^x \qquad \text{The solution satisfies the equation.}$$

LEARNING
HINT ▶

Finding the integrating factor may involve an expression of the form $e^{\ln u}$, where u is some function of x. From the definition of the logarithm (13-2), if

$$y = e^a \qquad \text{then} \qquad a = \log_e y = \ln y$$

Therefore, if

$$F(x) = e^{\ln u} \quad \text{then} \quad \ln u = \ln F(x)$$

$$F(x) = u$$

That is, $e^{\ln u} = u$ and, similarly, $e^{-\ln u} = 1/u$. ◀

EXAMPLE 13 Find the general solution of the differential equation:
$$x \, dy = x^2 \, dx - y \, dx$$

Solution
Step 1. The equation is not separable. Write it in standard form:

$$\frac{dy}{dx} + \frac{y}{x} = x$$

Step 2. Here, $P(x) = 1/x$ and $Q(x) = x$, so

$$\int P(x) \, dx = \int \frac{dx}{x} = \ln x$$

Therefore, the integrating factor is

$$F(x) = e^{\ln x} = x$$

Step 3. Now, the solution to the differential equation may be obtained using equation (29-10):

$$y = \frac{1}{F(x)} \left[\int Q(x)F(x) \, dx \right]$$

$$= \frac{1}{x} \left[\int x \cdot x \, dx \right] = \frac{1}{x} \left[\int x^2 \, dx \right]$$

$$= \frac{1}{x} \left[\frac{1}{3} x^3 + C \right]$$

$$y = \frac{x^2}{3} + \frac{C}{x}$$

Step 4. Check this solution by substituting back into the original equation.

EXAMPLE 14 Find the particular solution of the following differential equation, where $y(\pi/2) = 0$.

$$y \cos x = 2 \cos x \sin x - (\sin x)y' \qquad \text{for } 0 < x \leq \pi$$

Solution
Step 1. Rewriting this equation in standard form gives

$$y' + y \frac{\cos x}{\sin x} = 2 \cos x$$

$$y' + y \cot x = 2 \cos x$$

Step 2.   Identify $P(x)$ and $Q(x)$, and find the integrating factor:

$$P(x) = \cot x \qquad Q(x) = 2 \cos x$$

$$\int P(x)\, dx = \int \cot x\, dx$$

$$= \ln|\sin x| \qquad \text{From integration formula (27-11)}$$

Then, the integrating factor is

$$F(x) = e^{\ln|\sin x|}$$

$$= |\sin x| = \sin x \qquad \text{for } 0 < x \le \pi$$

Step 3.   From equation (29-10), the solution to the differential equation is

$$y = \frac{1}{F(x)}\left[\int Q(x)F(x)\, dx\right]$$

$$= \frac{1}{\sin x}\left[2\int \cos x \sin x\, dx\right]$$

Integrate using the methods of Section 27-1:

$$y = \frac{1}{\sin x}[\sin^2 x + C] \qquad \text{The general solution}$$

Substitute $y = 0$ when $x = \pi/2$ to find the particular solution:

$$0 = \frac{1}{1}[1^2 + C]$$

$$C = -1$$

$$y = \frac{1}{\sin x}[\sin^2 x - 1]$$

$$= \sin x - \frac{1}{\sin x}$$

$$= \sin x - \csc x$$

Step 4.   Check the solution.

EXAMPLE 15   Solve the following differential equation:

$$\frac{dy}{dx} = \frac{e^x - 2y}{1 + 2x} \quad \longleftarrow \boxed{\text{Where } y = 0 \text{ when } x = 0}$$

Solution

Step 1.   Rewrite the equation in standard form.

$$\frac{dy}{dx} = \frac{e^x}{1 + 2x} - \frac{2y}{1 + 2x}$$

$$\frac{dy}{dx} + \frac{2y}{1 + 2x} = \frac{e^x}{1 + 2x}$$

Step 2.   Identify P(x) and Q(x), and find the integrating factor.

$$P(x) = \frac{2}{1 + 2x} \qquad Q(x) = \frac{e^x}{1 + 2x}$$

$$\int P(x) \, dx = \int \frac{2}{1 + 2x} \, dx$$

$$= \int \frac{du}{u} \qquad \text{Where } u = 1 + 2x$$

$$= \ln u = \ln (1 + 2x)$$

Then, the integrating factor is

$$F(x) = e^{\ln (1+2x)} = 1 + 2x$$

Step 3.   Use equation (29-10) to find the general solution:

$$y = \frac{1}{F(x)} \left[\int Q(x)F(x) \, dx \right]$$

$$= \frac{1}{1 + 2x} \left[\int \frac{e^x}{1 + 2x} \cdot (1 + 2x) \, dx \right]$$

$$= \frac{1}{1 + 2x} \left[\int e^x \, dx \right]$$

$$y = \frac{1}{1 + 2x} (e^x + C) \qquad \boxed{\text{Do not forget the constant of integration.}}$$

Substitute the initial condition, y = 0 when x = 0:

$$0 = \frac{1}{1 + 0} (e^0 + C)$$

$$C = -1$$

The particular solution is

$$y = \frac{e^x - 1}{1 + 2x}$$

Step 4.   Check this solution by differentiating and substituting into the original equation given in the problem statement.

Exercises 29-2

Solve by finding integrable forms.

1. x dx + y dy = dx
2. x dy + y dx − y dy = 0
3. x dy − y dx − 2x dx = 0
4. x dx − y dy = 2 dy
5. y dx − 2y³ dy = x dy
6. x dx − 3x² dx = −y dy

7. $x^3\,dx - y^2x\,dx + y^3\,dy - x^2y\,dy = x\,dx + y\,dy$

8. $xy^2\,dx + y^3\,dy - x^2\,dy - y^2\,dy = 0$

9. $y\,dx - x\,dy = y^2\cos x\,dx$ **10.** $x^2e^x\,dx = x\,dy - y\,dx$

11. $y\,dy + x^2y^2\,dy + x\,dx + y^4\,dy = 0$ **12.** $y\,dx = x\,dy + 2x^2y^2\,dx + xy^2\,dx$

13. $(x\,dy - y\,dx)\sqrt{\dfrac{y}{x}} = x^3\,dx$ **14.** $(x\,dy + y\,dx)\sqrt{xy} = x\,dx$

15. $e^{xy}(x\,dy + y\,dx) = dx$ **16.** $(x\,dy - y\,dx) = x^2e^{y/x}\,dx$

17. $x\,dy - y\,dx = (x^2 + 4)x\,dx$ **18.** $(x\,dy - y\,dx) = x^3e^x\,dx$

Solve by using an integrating factor.

19. $y' + \dfrac{y}{x} = e^x$ **20.** $y' - 2y = xe^{2x}$

21. $dy = 3x^2e^{-4x}\,dx - 4y\,dx$ **22.** $dy - x^3e^x\,dx = y\,dx$

23. $2y\,dx + dy = e^{-2x}\ln x\,dx$ **24.** $e^{-x}\sin x\,dx = y\,dx + dy$

25. $dx - e^{3x}\,dy = 2ye^{3x}\,dx$ **26.** $dy - 3y\,dx = e^{5x}\,dx$

27. $xy' = 2x - y$ **28.** $2x^4 + y - xy' = 0$

29. $4x^4 + x\dfrac{dy}{dx} = 2y$ **30.** $\dfrac{y}{3x} = 1 - \dfrac{dy}{dx}$

31. $x(y' + 1) + y = x^2(x + 2)$ **32.** $y + 2x(x - 1)^2 = 2xy'$

33. $y' + \dfrac{2xy}{x^2 - 1} = \dfrac{x}{x - 1}$ **34.** $\dfrac{y'}{x} = \dfrac{y}{x + x^2} + \dfrac{1 + x}{x}$

Find the particular solution for each differential equation.

35. $y' + y = e^{-x}\sin x$; $y = 0$ when $x = 0$

36. $y' - y = e^{-x} - e^x$; $y = 2$ when $x = 0$

37. $y' + y\tan x = \sin x$; $y = 1$ when $x = 0$

38. $y' - \cos x + \dfrac{y}{\cos x} = 0$; $y = -1$ when $x = 0$

39. $y' + \dfrac{y}{x} = \sin 2x$; $y = 0$ when $x = \dfrac{\pi}{2}$

40. $y' + 2y\cot 2x = \sin 4x$; $y = \dfrac{1}{4}$ when $x = \dfrac{\pi}{4}$

29-3 | Applications of First-Order Differential Equations

Differential equations can be used to solve many problems in geometry. The solutions of the differential equation $y' = f(x, y)$ are curves whose slope at any point (x, y) is given by the function $f(x, y)$. If you are given $f(x, y)$, it may be possible to solve the differential equation of the curve.

EXAMPLE 16   Find the equation of the curve in the xy-plane that includes the point $(0, 3)$ and whose tangent line at any point (x, y) has the slope $3x^2/y$.

Solution   **First,** write a differential equation describing the curve.

$$\text{Slope} = y' = \frac{3x^2}{y}$$

Second, solve this equation. Separate variables.

$$y \, dy = 3x^2 \, dx$$

Integrate.

$$\frac{1}{2}y^2 = x^3 + C$$

$$y^2 = 2x^3 + 2C$$

Third, substitute the boundary condition $y = 3$ when $x = 0$.

$$9 = 2 \cdot 0^3 + 2C$$

Therefore:

$$y^2 = 2x^3 + 9$$

Finally, check to be certain this equation satisfies the original problem. At $(0, 3)$, we have $(3)^2 = 2 \cdot (0)^3 + 9$, which is correct. Differentiating,

$$2y \, \frac{dy}{dx} = 6x^2$$

$$\text{Slope} = \frac{dy}{dx} = \frac{3x^2}{y} \qquad \text{which is also correct.}$$

EXAMPLE 17 Find the equation of the family of curves whose slope at any point (x, y) is equal to 1 plus the square of the y-coordinate.

Solution

$$\text{Slope} = y' = 1 + y^2$$

Separate variables.

$$\frac{dy}{1 + y^2} = dx$$

Integrate using (27-20).

$$\text{Arctan } y = x + C$$

$$y = \tan (x + C)$$

Check:

$$y' = \sec^2 (x + C)$$

But $\sec^2 u = 1 + \tan^2 u$, so:

$$y' = 1 + \tan^2 (x + C) = 1 + y^2$$

Orthogonal trajectory   For each real value of the parameter c, the equation $F(x, y, c) = 0$ produces a curve in the xy-plane. If each curve in such a one-parameter family of curves intersects at right angles every curve of another, coplanar, family of curves, then each curve of one family is said to be an **orthogonal trajectory** of the curves in the other set.*

For example, Fig. 29-3 shows the electric field inside a cylindrical capacitor. The field lines $y = cx$ (in black) are the orthogonal trajectories of the concentric circles $x^2 + y^2 = k^2$ (in color) representing lines of constant electrical potential.

Systems of curves related in this way are of great importance in the study of gravitational, electrostatic, and magnetic fields; in the study of heat conduction and fluid flow; and in many other scientific and technical areas. For example, for the pair of oppositely charged electrical point charges shown in Fig. 29-4,

---

* The angle of intersection of two curves is, by definition, the angle between the tangent lines to the curves at their point of intersection.

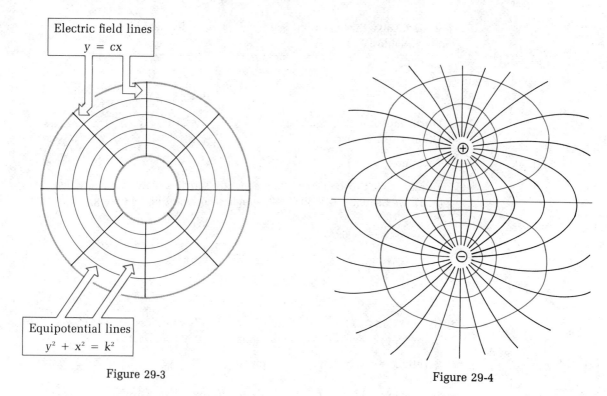

Electric field lines
$y = cx$

Equipotential lines
$y^2 + x^2 = k^2$

Figure 29-3

Figure 29-4

the lines of electrical force are the orthogonal trajectories of the field contour or equipotential lines. On the earth's surface, the latitude lines parallel to the equator are orthogonal trajectories of the meridian circles passing through the poles. On a contour map (Fig. 29-5), the curves of steepest descent are the orthogonal trajectories of the contour lines.

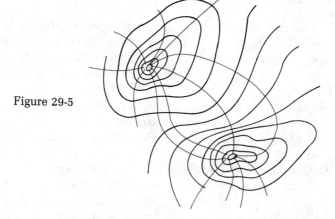

Figure 29-5

In Section 20-1 we showed that the slopes m_1 and m_2 of perpendicular lines are related by the equation $m_1 = -1/m_2$. Therefore, if the differential equation describing a family of curves can be written as $y' = F(x, y)$, then its orthogonal trajectories are described by the equation formed when y' is replaced by $-1/y'$:

$$-\frac{1}{y'} = F(x, y) \quad \text{or} \quad y' = -\frac{1}{F(x, y)}$$

Orthogonal Trajectories
The differential equations

$$y' = F(x, y) \quad \text{and} \quad y' = -\frac{1}{F(x, y)} \tag{29-12}$$

describe a system of orthogonal trajectories.

EXAMPLE 18 The **streamlines** of a fluid in motion are curves whose tangent lines give the direction of motion of the fluid. The orthogonal trajectories of the streamline curves are the **velocity equipotential** curves. For a fluid flowing around a corner, the streamlines are the family of curves $xy = k$. Find the velocity equipotential curves that are the orthogonal trajectories of the streamlines.

Solution
Step 1. Find an expression for the slope of the streamlines. Differentiate. Since $y = \dfrac{k}{x}$,

$$y' = -\frac{k}{x^2}$$

But $k = xy$

$$= -\frac{xy}{x^2}$$

$$y' = -\frac{y}{x}$$

Note that this equation must **not** contain the original constant k.

Step 2. Use (29-12) to find an expression for the slope of the orthogonal set of curves.

$$y' = \frac{x}{y}$$

Step 3. Solve this equation. Separate the variables.

$$y \, dy = x \, dx$$

Integrate.

$$\frac{1}{2}y^2 = \frac{1}{2}x^2 + c_1$$

$$y^2 = x^2 + 2c_1$$

$$y^2 - x^2 = c^2$$

The streamlines are a family of hyperbolas and their orthogonal trajectories are a second family of hyperbolas. Each set of curves involves a different constant or parameter, c or k. Figure 29-6 shows this system of curves plotted for a "corner" in the first quadrant.

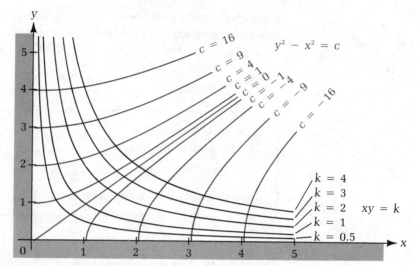

Figure 29-6

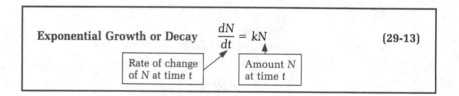

Growth and decay   A quantity $N(t)$ is said to undergo **exponential growth** or **decay** if at each instant of time its rate of increase or decrease is directly proportional to the amount present. We can express this as the following differential equation:

| **Exponential Growth or Decay** $\dfrac{dN}{dt} = kN$ | (29-13) |

Rate of change of N at time t → Amount N at time t

- If $k > 0$, the process represents unlimited growth, and k is called the **growth constant** of the process. This leads to the equation describing exponential growth given in Chapter 26:

    $N = N_0 e^{kt}$

- If $k < 0$, the process represents exponential decay, and k is called the **decay constant** of the process.

Radioactive decay   The process of **radioactive decay** occurs when an isotope spontaneously emits particles from its nucleus. For example, a 10 g sample of uranium metal contains about 2.5×10^{22} radioactive U-238 nuclei. During any given second, about 12,000 of these nuclei will disintegrate, emitting an alpha particle to form a thorium nucleus and releasing energy. There is no way to predict exactly which individual nuclei will decay in any given time interval, but this decay process may be described by the exponential decay formula, $dN/dt = -kN$, $k > 0$.

The **half-life** of a radioactive decay process is defined as the time t_H needed for the number of nuclei N to decrease to half of its original value. We can show that t_H and k are related as follows:

Half-Life for Exponential Decay $t_H = \dfrac{\ln 2}{k}$ (29-14)

EXAMPLE 19

(a) Solve the differential equation describing radioactive decay under the condition $N = N_0$ when $t = 0$.

(b) If $k = 2.9 \times 10^{-7} \text{ s}^{-1}$ for a sample of radioactive cesium-140 released from a nuclear reactor, what fraction of this sample is still present in the environment after 100 days?

(c) Find the half-life of Cs-140.

Solution

(a) **First,** separate variables and integrate.

$$\frac{dN}{N} = -k \, dt$$

$$\ln N = -kt + C_1$$

If we write the constant of integration as $\ln C$ this becomes

$$\ln N = -kt + \ln C$$

$$\ln N - \ln C = -kt$$

$$\ln \frac{N}{C} = -kt$$

$$N = Ce^{-kt}$$

Next, substitute the initial condition.

$$N_0 = Ce^0$$

$$C = N_0$$

The particular solution of the exponential decay equation is therefore:

$$N = N_0 e^{-kt}$$

This is the exponential decay formula given in Chapter 26.

(b) If $k = 2.9 \times 10^{-7} \text{ s}^{-1}$ and $t = 100$ days $= 8.64 \times 10^6$ s, then

$$\frac{N}{N_0} = e^{-kt}$$

$$= e^{-(2.9 \times 10^{-7})(8.64 \times 10^6)}$$

2.9 [EE] 7 [+/-] [×] 8.64 [EE] 6 [=] [+/-] [e^x] → `0.0816266`

$$\frac{N}{N_0} \approx 0.082$$

About 8% remains after 100 days.

(c) $t_H = \dfrac{\ln 2}{k}$

2 $\boxed{\ln x}$ $\boxed{\div}$ 2.9 $\boxed{\text{EE}}$ 7 $\boxed{+/-}$ $\boxed{=}$ $\rightarrow$ $\boxed{2390162.7}$

$t_H \approx 2.4 \times 10^6$ s $\approx$ 28 days

The amount of radioactivity of a Cs-140 sample decreases by one-half approximately every 28 days.

Restricted growth   The differential equation for exponential growth (29-13) assumes that the quantity N continues to grow without any limitation due to diminishing resources or other changes in the environment caused by the growth itself. A more realistic model is one of **restricted growth,** in which the changing quantity approaches some maximum value. In equation (29-15), M is the maximum or equilibrium size of N.

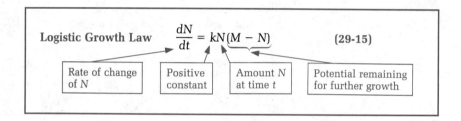

The solution of differential equations of this kind involves the following relationship:

$$\frac{1}{N(M-N)} = \frac{1}{M}\left(\frac{1}{N} + \frac{1}{M-N}\right) \tag{29-16}$$

Check this by simplifying the expression on the right.

EXAMPLE 20   In some chemical processes, certain reactants accelerate, or **catalyze,** their own formation. For these **autocatalytic** reactions, if $A(t)$ is the amount of the reactant present at time t, then the differential equation describing the process is

$$\frac{dA}{dt} = kA(P - A)$$

where P is the value of A at which one of the materials is used up and the reaction stops.

(a)   Solve this logistic growth equation subject to the condition $A = A_0$ at $t = 0$.

(b)   If $A_0 = 125$ g, $P = 350$ g, and $k = 0.0012$ h^{-1}, find A when $t = 4.0$ h.

(c)   Show that the rate of production dA/dt for this process is a maximum when $A = \dfrac{1}{2}P$.

Solution

(a) **Step 1.** Separate variables.

$$\frac{dA}{A(P-A)} = k\,dt$$

Use equation (29-16) to rewrite this as:

$$\frac{dA}{PA} + \frac{dA}{P(P-A)} = k\,dt$$

$$\frac{dA}{A} + \frac{dA}{P-A} = Pk\,dt$$

Step 2. Integrate.

$$\ln A - \ln(P-A) = Pkt + C_1$$

Write this as:

$$-\ln A + \ln(P-A) = -Pkt + \ln C$$

$$\ln\left(\frac{P-A}{CA}\right) = -Pkt$$

$$\frac{P-A}{CA} = e^{-Pkt}$$

$$\frac{P-A}{A} = Ce^{-Pkt}$$

Step 3. Evaluate C when $A = A_0$ and $t = 0$.

$$\frac{P-A_0}{A_0} = Ce^0 = C$$

Therefore:

$$\frac{P-A}{A} = \left(\frac{P-A_0}{A_0}\right)e^{-Pkt}$$

$$\frac{P}{A} - 1 = \left(\frac{P-A_0}{A_0}\right)e^{-Pkt}$$

$$\frac{P}{A} = 1 + \left(\frac{P-A_0}{A_0}\right)e^{-Pkt}$$

$$= \frac{A_0 + (P-A_0)e^{-Pkt}}{A_0}$$

$$\frac{A}{P} = \frac{A_0}{A_0 + (P-A_0)e^{-Pkt}}$$

The particular solution is:

$$A = \frac{A_0 P}{A_0 + (P-A_0)e^{-Pkt}}$$

Check: At $t = 0$:

$$A = \frac{A_0 P}{A_0 + (P-A_0)} = A_0$$

As t grows very large, $e^{-Pkt} \to 0$ and $A \to P$. Figure 29-7 shows a graph of this logistic growth function.

(b) For $A_0 = 125$ g, $P = 350$ g, $k = 0.0012$ h^{-1}, and $t = 4.0$ h, we calculate

$$A = \frac{125(350)}{125 + (350 - 125)e^{-350(0.0012)4.0}}$$

350 $\boxed{\times}$ .0012 $\boxed{\times}$ 4 $\boxed{=}$ $\boxed{+/-}$ $\boxed{e^x}$ $\boxed{\times}$ 225 $\boxed{+}$ 125 $\boxed{=}$ $\boxed{1/x}$ $\boxed{\times}$ 125 $\boxed{\times}$ 350 $\boxed{=}$

$$\rightarrow \boxed{262.07940}$$

$$A \approx 260 \text{ g}$$

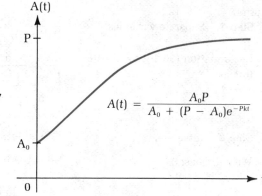

Figure 29-7

$$A(t) = \frac{A_0 P}{A_0 + (P - A_0)e^{-Pkt}}$$

(c)   If R is the rate of production dA/dt, then the original equation becomes

$$R = kA(P - A) = kPA - kA^2$$

The rate R is a maximum when $dR/dA = 0$, or

$$kP - 2kA = 0$$

$$A = \frac{1}{2}P$$

Use the second derivative test to check that it is a maximum. If the chemical process is regulated by removing the catalytic material as it is formed so that $A = \frac{1}{2}P$, then the production will be maximized.

LR circuit   In Chapter 26 we showed that in a series circuit containing resistance R and capacitance C, if a constant voltage V_0 is applied, the charge on the capacitor builds up gradually to some equilibrium value CV_0 according to the equation

$$Q = CV_0(1 - e^{-t/RC})$$

An analogous situation occurs in a circuit containing resistance R and inductance L (Fig. 29-8). If the switch is connected to A, the current in R starts to increase, and this rise is opposed by an opposite induced voltage in the inductor coil L. If the applied voltage is $V(t)$, then the current in the circuit varies according to the following differential equation:

$$L\frac{di}{dt} + Ri = V(t) \tag{29-17}$$

The solution of this equation depends on the nature of the function $V(t)$.

Figure 29-8

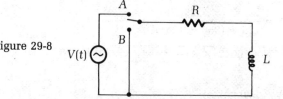

EXAMPLE 21

(a)   Solve equation (29-17) if $V(t) = V_0$, a constant voltage.

(b)   Find the value of i after 1.5 s if $R = 80\ \Omega$, $L = 100$ H, and $V_0 = 240$ V.

Solution

(a)   The equation may be solved using separation of variables.

Step 1.   Rewrite it as:

$$\frac{di}{dt} = \frac{V_0}{L} - \frac{R}{L}i$$

$$= \frac{R}{L}\left(\frac{V_0}{R} - i\right)$$

Separate the variables.

$$\frac{di}{\dfrac{V_0}{R} - i} = \frac{R}{L}\,dt$$

Step 2.   Integrate.

$$-\ln\left(\frac{V_0}{R} - i\right) = \frac{Rt}{L} + C_1$$

Step 3.   Simplify and solve for i.

$$\ln\left(\frac{V_0}{R} - i\right) = -\frac{Rt}{L} + \ln C \longleftarrow \boxed{\ln C = -C_1}$$

$$\ln\left(\frac{\dfrac{V_0}{R} - i}{C}\right) = -\frac{Rt}{L}$$

$$\frac{V_0}{R} - i = Ce^{-Rt/L}$$

Substituting $i = 0$ when $t = 0$, we find $C = V_0/R$. Thus:

$$i = \frac{V_0}{R} - \frac{V_0}{R}e^{-Rt/L}$$

$$i = \frac{V_0}{R}(1 - e^{-Rt/L})$$

The constant L/R is called the **inductive time constant** t_L for this circuit. Figure 29-9 shows the current increase described by this equation.

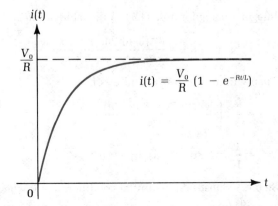

Figure 29-9

(b) $i = \dfrac{240}{80}(1 - e^{-(80)(1.5)/100})$

$80 \boxed{\times} 1.5 \boxed{\div} 100 \boxed{=} \boxed{+/-} \boxed{e^x} \boxed{+/-} \boxed{+} 1 \boxed{=} \boxed{\times} 240 \boxed{\div} 80 \rightarrow$ $\boxed{2.0964174}$

$i \approx 2 \text{ A}$

In an ac circuit, the applied voltage $V(t)$ will usually be periodic.

EXAMPLE 22   Solve equation (29-17) if $L = 100$ H, $R = 100$ Ω, and $V(t) = 200 \sin 3t$ V, where $i(0) = 0$.

Solution
Step 1.   The equation is not separable, but it can be written in the form of a linear first-order differential equation:

$$100\,\frac{di}{dt} + 100(i) = 200 \sin 3t$$

$$\frac{di}{dt} + i = 2 \sin 3t$$

Here, $P(t) = 1$ and $Q(t) = 2 \sin 3t$.
Step 2.   Find the integrating factor $F(t)$:

$$\int P(t)\, dt = \int dt = t$$

so that $F(t) = e^t$.
Step 3.   From equation (29-10),

$$i = \frac{1}{F(t)}\left[\int Q(t)F(t)\, dt\right]$$

$$= \frac{1}{e^t}\left[\int (2 \sin 3t)e^t\, dt\right]$$

Integrate using formula G.94 in the table of integrals (Appendix D):

$$\int e^{au} \sin bu \, du = \frac{e^{au}[a \sin bu - b \cos bu]}{a^2 + b^2} + C$$

Then with $a = 1$, $b = 3$, and $u = t$.

$$i = e^{-t}\left[\frac{2e^t(\sin 3t - 3 \cos 3t)}{1^2 + 3^2} + C\right]$$

$$= \frac{2}{10}(\sin 3t - 3 \cos 3t) + Ce^{-t}$$

Step 4.   Substitute the initial condition, $i = 0$ when $t = 0$:

$$0 = \frac{1}{5}(0 - 3) + C \cdot 1$$

$$C = 0.6$$

The particular solution is therefore:

$$i = \frac{1}{5}(\sin 3t - 3 \cos 3t) + \frac{3}{5}e^{-t}$$

$$= 0.2 \sin 3t - 0.6 \cos 3t + 0.6e^{-t}$$

A graph of this solution is shown in Fig. 29-10.

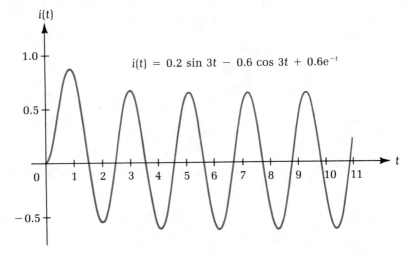

Figure 29-10

Motion with a resisting force

If an object of mass m is acted upon by a force $F(t)$, it will move with acceleration dv/dt according to the differential equation

$$m \frac{dv}{dt} = F(t)$$

For example, in the absence of air resistance, objects near the earth's surface fall freely under the influence of gravity with a constant acceleration $-g$ so that, if v is measured positively in the downward direction, $F(t) = mg$.

An object moving through a resisting medium experiences an additional friction, or **drag force,** that varies with the shape and size of the object and the nature of the medium through which it moves. This drag friction force depends on whether the fluid medium moving past the object flows smoothly or forms turbulent eddies.

Motion in a Resisting Medium $\qquad m \dfrac{dv}{dt} = F(t) - F_{\text{Drag}}$ $\qquad\qquad$ (29-18)

where $F_{\text{Drag}} = k_1 v$ for smooth flow, and $F_{\text{Drag}} = k_2 v^2$ for turbulent flow.

NOTE ▶   If v is measured positively in the downward direction, F_{Drag} opposes any increase in v and has a negative sign. ◀

EXAMPLE 23   A parachutist experiences a drag force $F_{Drag} \approx 27.3v$. The mass of the parachutist is 67 kg and $g = 9.8$ m/s².

(a)   Solve the differential equation describing this motion,

$$m\frac{dv}{dt} = mg - kv$$

for the condition $v = 0$ when $t = 0$.

(b)   What speed will the parachutist have reached when $t = 5.0$ s?

(c)   What is the maximum speed?

Solution

(a)   **First,** rewrite the equation as

$$\frac{dv}{dt} = g - \frac{kv}{m} \longleftarrow \boxed{\text{This equation is separable.}}$$

$$= \frac{k}{m}\left(\frac{mg}{k} - v\right)$$

Next, write it in differential form with variables separated, and then integrate:

$$\frac{dv}{\dfrac{mg}{k} - v} = \frac{k}{m}\,dt$$

$$-\ln\left(\frac{mg}{k} - v\right) = \frac{kt}{m} + C_1$$

Finally, solve for v as in Example 21:

$$v = \frac{mg}{k}(1 - e^{-kt/m})$$

(b)   Substituting $m = 67$ kg, $g = 9.8$ m/s², $k = 27.3$ kg/s and $t = 5.0$ s, we calculate

$$v(5.0) = \frac{67(9.8)}{27.3}(1 - e^{-(27.3)(5.0)/67})$$

$$27.3\;\boxed{\times}\;5\;\boxed{\div}\;67\;\boxed{=}\;\boxed{+/-}\;\boxed{e^x}\;\boxed{+/-}\;\boxed{+}\;1\;\boxed{=}\;\boxed{\times}\;67\;\boxed{\times}\;9.8\;\boxed{\div}\;27.3\;\boxed{=}$$

$$\rightarrow \quad \textit{20.915512}$$

$$v(5) \approx 21 \text{ m/s}$$

(c)   As $t \to \infty$, $e^{-kt/m} \to 0$, and $v \to mg/k$. Therefore, the maximum speed will be

$$v = \frac{67(9.8)}{27.3} \approx 24 \text{ m/s}$$

This speed is often called the **terminal** speed of a falling object.

Rate problems   Differential equations are particularly useful in the analysis of dynamic situations in which several related rates of change appear.

EXAMPLE 24   A tank contains a solution of 25 lb of salt dissolved in 100 gal of water. A more concentrated solution of 2 lb of salt per gallon of water flows continuously into the tank at 4 gal/min. The solution in the tank is kept uniform by stirring and is pumped out continuously at 4 gal/min.
(a)   Find an equation giving the amount of salt in the tank at any time t.
(b)   What amount of salt will be in the tank at $t = 30$ min?
(c)   How much salt will be in the tank at some very long time after the process is begun?

Solution
(a)   Let y represent the amount of salt in the tank at time t. Then dy/dt is the rate of change in the amount y (in pounds per minute), and

$$\frac{dy}{dt} = \begin{bmatrix} \text{Rate at which} \\ \text{salt is added} \end{bmatrix} - \begin{bmatrix} \text{Rate at which} \\ \text{salt is removed} \end{bmatrix}$$

$$\text{Rate in} = \left(2\,\frac{\text{lb}}{\text{gal}}\right)\left(4\,\frac{\text{gal}}{\text{min}}\right) = 8\,\frac{\text{lb}}{\text{min}}$$

$$\text{Rate out} = \left(\frac{y}{100}\,\frac{\text{lb}}{\text{gal}}\right)\left(4\,\frac{\text{gal}}{\text{min}}\right) = \frac{y}{25}\,\frac{\text{lb}}{\text{min}}$$

| Total amount of salt |
| Total volume of water |

Then:

$$\frac{dy}{dt} = 8 - \frac{y}{25} = \frac{200 - y}{25}$$

Solve by separation of variables.

$$\frac{dy}{200 - y} = \frac{dt}{25}$$

$$\ln(200 - y) = -0.04t + C_1$$

Write the constant of integration as $\ln C$.

$$\ln\left(\frac{200 - y}{C}\right) = -0.04t$$

$$\frac{200 - y}{C} = e^{-0.04t}$$

$$y = 200 - Ce^{-0.04t}$$

The problem states that $y = 25$ lb when $t = 0$, therefore:

$$25 = 200 - C$$

$$C = 175$$

and

$$y = 200 - 175e^{-0.04t}$$

(b)   When $t = 30$ min,

$$y = 200 - 175e^{-0.04(30)}$$

30 [×] .04 [=] [⁺⁄₋] [e^x] [×] 175 [⁺⁄₋] [+] 200 [=] → ▓▓▓ *147.29101*

$$y \approx 150 \text{ lb}$$

(c)   As $t \to \infty$, $e^{-0.04t} \to 0$, and the equilibrium value of y is 200 lb.

EXAMPLE 25   A small industrial lab stores and dilutes the pollutants produced in its work in a large holding tank. At time $t = 0$ the tank holds 4 lb of pollutant in 500 gal of water. Waste solution containing 0.2 lb of pollutant per gallon of water is pumped into the tank at the rate of 30 gal per day. The concentration of pollutant is kept uniform by stirring. If the solution in the tank is released at the rate of 20 gal per day:
(a)   Find an expression for the amount of pollutant in the tank at time t.
(b)   Find the amount of pollutant in the tank after a week.

Solution
(a)   Let y represent the amount of pollutant in the tank at time t. Then dy/dt is the rate of change of the amount of pollutant in the tank at time t (in pounds per day), and

$$\frac{dy}{dt} = \begin{bmatrix} \text{Rate at which} \\ \text{pollutant is added} \end{bmatrix} - \begin{bmatrix} \text{Rate at which} \\ \text{pollutant is removed} \end{bmatrix}$$

$$\text{Rate in} = \left(0.2 \frac{\text{lb}}{\text{gal}}\right)\left(30 \frac{\text{gal}}{\text{day}}\right) = 6 \frac{\text{lb}}{\text{day}}$$

To find the rate at which pollutant is removed, first find the concentration of pollutant in the tank at time t:

$$\frac{\text{Concentration}}{\text{of pollutant}} = \frac{\text{Amount of pollutant (lb)}}{\text{Amount of water (gal)}}$$

$$= \frac{y}{500 + (30 - 20)t}$$

| Original amount of water | Rate at which water is added | Rate at which water is removed |

$$= \frac{y}{500 + 10t}$$

Then the rate at which pollutant is removed is

$$\text{Rate out} = [\text{Concentration}] \cdot [\text{Release rate}]$$

$$= \frac{y}{500 + 10t} \cdot 20$$

$$= \frac{2y}{50 + t} \frac{\text{lb}}{\text{day}}$$

Therefore, the differential equation is

$$\frac{dy}{dt} = 6 - \frac{2y}{50 + t}$$

Or, written in standard form for a first-order linear differential equation,

$$\frac{dy}{dt} + \left(\frac{2}{50 + t}\right)y = 6$$

Solve using the integrating factor method:

$$P(t) = \frac{2}{50 + t} \qquad Q(t) = 6$$

$$\int P(t)\, dt = \int \frac{2\, dt}{50 + t}$$

$$= 2 \ln (50 + t)$$

$$= \ln (50 + t)^2$$

Then the integrating factor is $e^{\ln (50+t)^2} = (50 + t)^2$, and the solution may be found from equation (29-10):

$$y = \frac{1}{(50 + t)^2}\left[\int 6(50 + t)^2\, dt \right]$$

This is $6u^2\, du$ for $u = 50 + t$.

$$= \frac{1}{(50 + t)^2}\left[2(50 + t)^3 + C \right]$$

$$= 2(50 + t) + \frac{C}{(50 + t)^2}$$

The initial condition is $y = 4$ at $t = 0$, so

$$4 = 2(50 + 0) + \frac{C}{(50 + 0)^2}$$

$$C = -240,000$$

Therefore,

$$y = 100 + 2t - \frac{240,000}{(50 + t)^2}$$

(b) At $t = 7$ days,

$$y(7) = 100 + 14 - \frac{240,000}{(57)^2} = 40.1311 \ldots$$

$$\approx 40 \text{ lb}$$

Exercises 29-3

1. Find the equation of the curve that passes through the point $(0, 2)$ and whose slope is everywhere equal to $(1 + y)/(1 + x)$.

2. Find the equation of the curve that passes through the point $(1, 1)$ and whose slope at the point (x, y) is equal to y^2.

3. Find the equation of the curve that passes through the point $(0, 1)$ and whose tangent line at a point (x, y) has slope $2x(y + 1)$.

4. Find the equation of the curve that passes through the point $(1, 0)$ and whose normal line at the point (x, y) has slope $3yx^2$.

5. A function $y(x)$ has the property that the slope of the tangent line at every point is equal to $x - y$. Find the function if its graph passes through the origin.

6. Find the equation of the curve that passes through the point $(2, 1)$ and whose slope at every point (x, y) is equal to $1 - (y/x)$.

7. Find the equation of the curve passing through the origin whose slope at every point (x, y) is equal to $\sin x$.

8. Find the equation of the curve passing through the origin whose slope at any point (x, y) is equal to the square of its x-coordinate.

9. A curve in the xy-plane passing through the point $(1, 1)$ has the property that the slope at any point (x, y) is equal to the product of its coordinates. Find the equation of this curve.

10. A curve in the xy-plane passing through the point $(0, 1)$ has the property that its slope at point (x, y) is equal to the product of the squares of its coordinates. Find the equation of this curve.

11. Find the equation of the orthogonal trajectories of the family of parabolas $y = kx^2$.

12. Find the equation of the family of curves that intersect each of the curves $y = ke^x$ at right angles.

13. Find the equation of the orthogonal trajectories of the family of curves $y = 1 - kx^3$.

14. Find the equation of the orthogonal trajectories of the family of curves $x + ky^2 = 1$.

15. *Urban planning*   Studies in urban planning have shown that the rate of increase in highway construction is directly proportional to the total miles of highway already built.[*] Express this as a differential equation in terms of L, the miles of highway existing at time t, and solve for L as a function of t, where $L = L_0$ when $t = 0$.

16. *Planetary science*   In a planetary atmosphere where temperature is constant, the atmospheric pressure P varies with altitude above the planet's surface such that $dP/dh = kP$.
    (a) Solve for P if $P = P_0$ when $h = 0$.
    (b) Find P at $h = 8000$ ft if $P_0 = 15$ lb/in.^{2} and $k = -0.00005$.

17. *Biology*   A culture of bacteria grows at a rate proportional to the number of bacteria present. If 10,000 bacteria are present at the start of an experiment, and if the number doubles in 5 h, how many bacteria will be present 7 h after the start?

18. *Business*   The average starting annual salary of an engineer is now \$31,000, and it was \$15,000 only 10 years ago. If the rate of growth is proportional to the salary itself, what will be the average starting salary in 5 years?

19. *Statistics*   The growth of the population N of a country with constant immigration rate a is described by the differential equation $dN/dt = kN + a$. Solve for N as a function of t if $N = N_0$ when $t = 0$.

20. *Chemistry*   In the chemical process for producing the substrate material for circuit boards, each batch is seeded at $t = 0$ with an amount of material A_0 and produces a maximum amount A_m. If the rate of production satisfies the equation $dA/dt = k(A_m - A)$:
    (a) Find an equation giving A as a function of t.
    (b) Find A after 30 min if $k = 0.02$, $A_m = 85$ kg, and $A_0 = 5$ kg.

[*] See Clive L. Dym and Elizabeth S. Ivey, *Principles of Mathematical Modeling* (New York: Academic Press, 1980), p. 175–176.

21. If a growth process is described by the differential equation $dy/dt = k\sqrt{y}$:
(a) Find an equation giving y as a function of t for the initial condition $y = 4$ when $t = 0$.
(b) Find y when $t = 10$ s if $y = 25$ when $t = 2$ s.

22. *Material science* In its early stages of development a crystal grows so that its rate of increase of mass M is proportional to its surface area. This leads to the differential equation $dM/dt = kM^{2/3}$. If a potassium bromide crystal has mass 10 g at $t = 0$ hours, then for $k = 4.5$ find the time needed for it to grow to a mass of 150 g.

23. *Medical technology* Public health scientists have found that epidemics in the early stages spread according to a differential equation of the form $P' = kP(1 - 2P)$, where P is the proportion of the population that is infected at time t and $0 < P < 0.5$. Find the particular solution to this equation if $P = P_0$ when $t = 0$.

$$\left[\text{Hint:} \quad \frac{1}{P(1 - 2P)} = \frac{1}{P} + \frac{2}{1 - 2P} \right]$$

24. *Statistics* Rumors spread through a population at a rate proportional to the frequency of contact between those who have heard the rumor and those who have not. If N is the number of people who have heard the rumor at time t (in days) and T is the size of the population, then $T - N$ is the number of people who have not heard it. The differential equation describing the change in N is $N' = kN(T - N)$.*
(a) Find the general solution to this equation.
(b) If $T = 2000$, $k = 0.0001$, and $N_0 = 2$, that is, two people have heard the rumor at $t = 0$, then how many people will have heard it by $t = 10$ days?

25. *Nuclear physics* If a certain radioactive substance decays to 20% of its initial level in 5 years, what is its half-life?

26. *Nuclear physics* A radioactive tracer element with a half-life of 45 min is used in the detection of leaks in a hydraulic system. How much time is needed for the tracer to decay to 5% of its initial level?

27. *Nuclear physics* If 350 Ci of radioactive iodine-131 is used in a medical research study, how much time is needed for the source strength to decrease to 50 Ci? (The half-life of iodine-131 is approximately 8.1 days.)

28. *Nuclear physics* A long-life radioactive element used in dating archaeological artifacts decays 4.5% in 1200 years. What is the age of an object containing this element if the radioactive material has decayed 12% over the lifetime of the object?

29. *Chemistry* In a *unimolecular* chemical reaction, a single molecule changes into one or more molecules. If x is the amount (in moles) of the original material, then the rate at which this material reacts is given by the equation $dx/dt = k(A - x)$, where $k > 0$. If $x(0) = 1$, find the particular solution to this equation.

30. *Chemistry* In a *biomolecular* chemical reaction, two molecules of the same kind react to produce one or more new molecules. If x is the amount (in moles) of the original material, then the rate of reaction is given by the equation $dx/dt = k(A - x)^2$, where $k > 0$. If $x(0) = 0$, find the particular solution to this equation.

31. *Electronics* An LR series circuit with $R = 25\ \Omega$ and $L = 0.5$ H has a constant applied voltage of $V(t) = 60$ V. Find an equation giving the current as a function of time if $i = 0$ when $t = 0$.

32. *Electronics* In an LR series circuit with $R = 150\ \Omega$, $L = 30$ H, and $V(t) = 2 \cos 5t$, find an expression for the current if $i = 0$ when $t = 0$.

33. *Electronics* In an LR series circuit with $R = 10\ \Omega$, $L = 2$ H, and $V(t) = 110e^{-2t}$, find an expression for the current if $i = 0$ when $t = 0$.

34. *Electronics* In an RC circuit, the following differential equation describes the rate at which the charge on the capacitor changes with time:

---

*Laurence D. Hoffmann, *Calculus for the Social, Managerial, and Life Sciences*, 2nd edition (New York: McGraw-Hill, 1980), p. 240.

$$R\frac{dq}{dt} + \frac{q}{C} = V(t)$$

If $R = 25\ \Omega$, $C = 0.008$ F, and $V(t) = 50$ V:

(a) Find an equation giving the charge on the capacitor as a function of time if $q = 100$ C when $t = 0$.

(b) Find q at $t = 0.4$ s.

35. *Physics* The drag on a golf ball is proportional to its speed because of the boundary layer effect created by its "dimples." What would be the terminal speed (in feet per second) of a golf ball dropped from an airplane if $k = 0.0082$, $m = 0.12$ lb, and $m\frac{dv}{dt} = mg - kv$?

36. *Physics* For a bicyclist coasting on a level road, acceleration is proportional to the drag force $F = 4.8 + 0.024v$ lb so that $m\frac{dv}{dt} = -F$, where $m = 5$ and v is in feet per second.

(a) Find an expression for v as a function of t.

(b) If $v = 30$ ft/s when $t = 0$, after what time will the bicycle have slowed to 5 ft/s?

37. *Physics* When a speed boat moves through the water at speed v, the friction force is proportional to v^2 so that the motion of the boat when it is coasting to a stop is given by the equation $m\frac{dv}{dt} = -kv^2$.

If a 1200-lb boat (mass $= 36$) is going 40 ft/s and its power is cut, it slows to 10 ft/s in 10 s. What is the value of the drag coefficient k?

38. *Physics* For a skydiver, before the parachute opens the retarding force is $F_{Drag} = F_0 + kv$, where k and F_0 are constants. Solve the differential equation describing this motion for the condition $v = 0$ when $t = 0$.

39. *Industrial engineering* A settling tank contains 10 lb of solid industrial waste dissolved in 200 gal of water. A solution of 1 lb of solid per 5 gal of water is added at a rate of 10 gal/min and mixed to keep the tank contents at a uniform concentration. If the contents of the tank are being drained off at 10 gal/min, how much dissolved solid will it contain after 30 min?

40. *Industrial engineering* A tank contains 100 gal of pure water. Starting at time $t = 0$, brine containing 1.5 lb of dissolved salts per gal is pumped into the tank at a rate of 5 gal/min. The mixture is stirred and pumped out continuously at 5 gal/min.

(a) Find an equation giving the amount of salt in the tank at time t.

(b) How much salt will be present at $t = 15$ min?

(c) What will be the equilibrium amount of salt in the tank?

41. *Industrial engineering* A water tank initially contains 400 gal of pure water, and a salt solution is added at a rate of 10 gal/min. If the solution added contains 1 lb of salt for every 5 gal, and if the contents of the tank are being drained off at 6 gal/min, how much salt will the tank contain after 20 min?

42. *Industrial engineering* A large tank contains 300 gal of brine in which 10 lb of salt is dissolved. At time $t = 0$, a solution containing 2 lb of salt per gallon is added at a rate of 5 gal/min. The mixture is kept uniform by stirring and is pumped out of the tank at 4 gal/min. How much salt is in the tank at the end of 25 min?

29-4 | Higher-Order Differential Equations

The first-order linear differential equations we have worked with so far are a special case of the linear differential equation of order n:

$$P_0(x)\frac{d^n y}{dx^n} + P_1(x)\frac{d^{n-1}y}{dx^{n-1}} + \cdots + P_{n-1}(x)\frac{dy}{dx} + P_n(x)y = Q(x)$$

In general, the coefficients $P_i(x)$ are constants or continuous functions of x. This equation is linear, since y and all of its derivatives appear to the first power only.

In this brief introduction to higher-order differential equations, we will consider only the second-order differential equation

$$P_0(x) \frac{d^2y}{dx^2} + P_1(x) \frac{dy}{dx} + P_2y = Q(x)$$

The following three forms of the linear second-order differential equation can be solved directly.

Direct integration

Case 1. Any differential equation of the form

$$P_0(x) \frac{d^ny}{dx^n} = Q(x)$$

can be solved by writing the equation as

$$\frac{d^ny}{dx^n} = \frac{Q(x)}{P_0(x)}$$

$$\frac{d}{dx}\left[\frac{d^{n-1}y}{dx^{n-1}}\right] = \frac{Q(x)}{P_0(x)}$$

Multiplying by dx,

$$d\left[\frac{d^{n-1}y}{dx^{n-1}}\right] = \frac{Q(x)}{P_0(x)} dx$$

and integrating,

$$\frac{d^{n-1}y}{dx^{n-1}} = \int \frac{Q(x)}{P_0(x)} dx$$

This process can be repeated until y appears on the left and a solution is obtained.

EXAMPLE 26 Solve: $\dfrac{d^2y}{dx^2} = x + 2$

Solution

Write the left side as $\dfrac{d}{dx}\left(\dfrac{dy}{dx}\right)$, then: $\qquad d\left(\dfrac{dy}{dx}\right) = (x + 2)\, dx$

Integrate. $\qquad \dfrac{dy}{dx} = \dfrac{x^2}{2} + 2x + c_1$

Solve this equation by separation of variables. $\qquad dy = \left(\dfrac{x^2}{2} + 2x + c_1\right) dx$

$$y = \int \left(\frac{x^2}{2} + 2x + c_1\right) dx$$

$$y = \frac{x^3}{6} + x^2 + c_1x + c_2$$

Check this solution by differentiating and substituting into the original equation.

NOTE   This solution contains two arbitrary constants c_1 and c_2. The general solution of an nth-order differential equation contains n arbitrary constants. ◀

Reduction of order   **Case 2.**   If the variable y does not appear explicitly in the differential equation, then the substitution $u = dy/dx$ will produce a differential equation in u and x that is one order lower than the original.

EXAMPLE 27   Solve: $x\dfrac{d^2y}{dx^2} + \dfrac{dy}{dx} = x$

Solution   **First,** let $dy/dx = u$, then differentiate. $d^2y/dx^2 = du/dx$, and the differential equation can be written as

$$x\frac{du}{dx} + u = x$$

$$\frac{du}{dx} + \frac{u}{x} = 1$$

Second, since this new equation is a linear first-order differential equation, we can solve it using the integrating factor method explained in Section 29-2. The integrating factor is $F(x) = e^{\int P(x)\,dx} = e^{\ln x} = x$, and the solution is

$$u = \frac{1}{x}\left[\int x\,dx\right] = \frac{1}{x}\left[\frac{x^2}{2} + c_1\right]$$

$$= \frac{x}{2} + \frac{c_1}{x}$$

Third, substituting $u = dy/dx$, we have

$$\frac{dy}{dx} = \frac{x}{2} + \frac{c_1}{x}$$

and this equation can be solved by multiplying by dx and integrating:

$$y = \int\left(\frac{x}{2} + \frac{c_1}{x}\right)dx$$

$$y = \frac{x^2}{4} + c_1\ln x + c_2$$

Check it.

Case 3.   If the variable x does not appear explicitly in the given differential equation, then the substitution $u = dy/dx$ can be used to produce a differential equation in u and y. Note that for this substitution,

$$\frac{d}{dx}\left(\frac{dy}{dx}\right) = \frac{du}{dx}$$

$$\frac{d^2y}{dx^2} = \frac{du}{dy} \cdot \frac{dy}{dx}$$

$$= \frac{du}{dy} \cdot u$$

EXAMPLE 28   Solve: $y\dfrac{d^2y}{dx^2} = \left(\dfrac{dy}{dx}\right)^2$

Solution

First, let $dy/dx = u$, then

$$\frac{d^2y}{dx^2} = u\frac{du}{dy}$$

The differential equation can be written as

$$yu\frac{du}{dy} = u^2$$

Second, solve this first-order equation by separation of variables.

$$\frac{du}{u} = \frac{dy}{y}$$

$$\ln u = \ln y + c \longleftarrow \boxed{c = \ln c_1}$$

$$u = c_1 y$$

Third, substitute $u = dy/dx$ and solve the resulting first-order differential equation:

$$\frac{dy}{dx} = c_1 y$$

$$\frac{dy}{y} = c_1 \, dx$$

$$\ln y = c_1 x + c$$

$$\ln y = c_1 x + \ln c_2 \longleftarrow \boxed{c = \ln c_2}$$

$$y = c_2 e^{c_1 x}$$

Check it.

Homogeneous equations   If $Q(x) = 0$, the equation is said to be **homogeneous.** If $Q(x) \neq 0$, the equation is said to be **nonhomogeneous.** Both homogeneous and nonhomogeneous differential equations have important applications in physics and engineering.

The simplest form of the higher-order linear differential equation is one in which the functions $P_i(x)$ are all constants. In the remainder of this section we will consider only homogeneous linear differential equations with constant coefficients.

Operator notation   To solve more general forms of the linear homogeneous differential equation, it is very helpful to use **operator notation.** If we write dy/dx as Dy, then

$$\frac{d^2y}{dx^2} = \frac{d}{dx}\left(\frac{dy}{dx}\right)$$

$$= D(Dy) = D^2y$$

In general,

$$\frac{d^ny}{dx^n} = D^ny$$

For example, in differential operator notation, the differential equation

$$2\frac{d^2y}{dx^2} - \frac{dy}{dx} + 3y = 0 \qquad \text{is written} \qquad 2D^2y - Dy + 3y = 0$$

$$\text{or} \qquad (2D^2 - D + 3)y = 0$$

EXAMPLE 29 Write each equation using differential operator notation.

(a) $y' + 2y = 1$ (b) $\dfrac{dy}{dx} - 2x = y$

(c) $y'' - y' - y = 0$ (d) $\dfrac{d^3y}{dx^3} - 2\dfrac{d^2y}{dx^2} - 4\dfrac{dy}{dx} + 3y = 0$

Solution
(a) $Dy + 2y = 1$ or $(D + 2)y = 1$
(b) $Dy - 2x = y$, $Dy - y = 2x$,   or $(D - 1)y = 2x$
(c) $D^2y - Dy - y = 0$ or $(D^2 - D - 1)y = 0$
(d) $D^3y - 2D^2y - 4Dy + 3y = 0$ or $(D^3 - 2D^2 - 4D + 3)y = 0$

The symbol D used in this way is not an algebraic quantity, but it does have many algebraic properties and may be manipulated algebraically. Using this notation, the general form of the linear homogeneous differential equation with constant coefficients a_0, a_1, and a_2,

$$a_0\frac{d^2y}{dx^2} + a_1\frac{dy}{dx} + a_2y = 0$$

can be written as $(a_0D^2 + a_1D + a_2)y = 0$. This differential equation can be solved by working directly with the operator $(a_0D^2 + a_1D + a_2)$ as if it were an algebraic expression and solving the quadratic equation

$$a_0D^2 + a_1D + a_2 = 0$$

If the roots of this equation are m_1 and m_2, then it may be factored as

$$(D - m_1)(D - m_2) = 0$$

and we may write this as the pair of equations

$$D - m_1 = 0 \qquad \text{and} \qquad D - m_2 = 0$$

The first of these equations is equivalent to the first-order differential equation

$$(D - m_1)y = 0 \quad \text{or} \quad \frac{dy}{dx} - m_1 y = 0$$

Solve this equation using separation of variables or an integrating factor. Its solution is

$$y = c_1 e^{m_1 x}$$

Similarly, the second equation is equivalent to the differential equation

$$(D - m_2)y = 0 \quad \text{or} \quad \frac{dy}{dx} - m_2 y = 0$$

with solution

$$y = c_2 e^{m_2 x}$$

The sum rule (22-15) tells us that the derivative of a sum is equal to the sum of the derivatives, so we know that these two solutions can be combined to form the general solution of the original differential equation:

$$y = c_1 e^{m_1 x} + c_2 e^{m_2 x}$$

CAUTION This process of adding solutions to find the general solution is valid *only* for a *linear* differential equation. ◄

In general, a homogeneous linear differential equation with constant coefficients can be solved as follows: If

$$a_0 \frac{d^n y}{dx^n} + a_1 \frac{d^{n-1} y}{dx^{n-1}} + \cdots + a_{n-1} \frac{dy}{dx} + a_n y = 0$$

is a homogeneous linear differential equation with constant coefficients a_i, then the equation

$$a_0 m^n + a_1 m^{n-1} + \cdots + a_{n-1} m + a_n = 0 \tag{29-19}$$

is called the **auxiliary equation**. If $m_1, m_2, \ldots, m_n$ are distinct real roots of this equation, then the general solution of the differential equation is

$$y = c_1 e^{m_1 x} + c_2 e^{m_2 x} + \cdots + c_n e^{m_n x} \tag{29-20}$$

EXAMPLE 30 Solve: $2 \dfrac{d^2 y}{dx^2} + 3 \dfrac{dy}{dx} - 2y = 0$

Solution

Step 1. Check to be certain the given differential equation is linear (first power in y and its derivatives), homogeneous (no constant term or x-term), and has only constant coefficients. Then write the equation using D operator notation:

$$(2D^2 + 3D - 2)y = 0$$

Step 2. Write the auxiliary equation by substituting m for D.

$$2m^2 + 3m - 2 = 0$$

Step 3. Solve the auxiliary equation. Since we are working with second-order equations only, the auxiliary equation will be a quadratic. Factor it if possible or solve using the quadratic formula:

$$(2m - 1)(m + 2) = 0$$

The roots are $m_1 = \dfrac{1}{2}$ and $m_2 = -2$.

Step 4. Write the general solution using equation (29-20).

$$y = c_1 e^{x/2} + c_2 e^{-2x}$$

Step 5. Check to be certain the solution has two arbitrary constants (for a second-degree equation) and then check the solution by substituting it into the original differential equation.

EXAMPLE 31   Solve: $y'' - 2y' - y = 0$

Solution
Step 1. In operator notation this equation may be written as $(D^2 - 2D - 1)y = 0$.

Step 2. The auxiliary equation is $m^2 - 2m - 1 = 0$.

Step 3. Solve the auxiliary equation using the quadratic formula. We find $m = 1 \pm \sqrt{2}$, so the roots are $m_1 = 1 + \sqrt{2}$ and $m_2 = 1 - \sqrt{2}$.

Step 4. The solution of the differential equation is

$$\begin{aligned}
y &= c_1 e^{(1+\sqrt{2})x} + c_2 e^{(1-\sqrt{2})x} \\
&= c_1 e^x e^{x\sqrt{2}} + c_2 e^x e^{-x\sqrt{2}} \\
&= e^x(c_1 e^{x\sqrt{2}} + c_2 e^{-x\sqrt{2}})
\end{aligned}$$

Step 5. Check the solution by differentiating and substituting.

EXAMPLE 32   Solve $y'' - 3y' + 2y = 0$ subject to the initial conditions $y(0) = 0$ and $y'(0) = -1$.

Solution
Step 1. In operator notation, $(D^2 - 3D + 2)y = 0$.

Step 2. The auxiliary equation is $m^2 - 3m + 2 = 0$.

Step 3. Factoring, this is $(m - 1)(m - 2) = 0$, so the roots are $m_1 = 1$ and $m_2 = 2$.

Step 4. The general solution is $y = c_1 e^x + c_2 e^{2x}$.

Step 5. Substitute the initial conditions:

$$0 = c_1 e^0 + c_2 e^{2(0)} \longrightarrow c_1 + c_2 = 0$$

$$y' = c_1 e^x + 2c_2 e^{2x}$$

$$-1 = c_1 e^0 + 2c_2 e^{2(0)} \longrightarrow c_1 + 2c_2 = -1$$

Solving these two equations in c_1 and c_2, we find $c_1 = 1$ and $c_2 = -1$. Therefore, the particular solution is $y = e^x - e^{2x}$.

Step 6. Check this solution.

Equal roots   The solution in equation (29-20) holds only if the roots of the auxiliary equation are real and unequal. However, one of the following may happen:

1. The roots are real and equal.
2. The roots are complex.

If the **roots are equal**—that is, $m_1 = m_2$—then the differential operator form of $(a_0 D^2 + a_1 D + a_2)y = 0$ may be written as $(D - m_1)(D - m_1) y = 0$. If we let $(D - m_1)y = u$, then the auxiliary equation is $(D - m_1)u = 0$, and, as we have seen, the solution to this first-order differential equation is

$$u = c_1 e^{m_1 x}$$

Therefore, the auxiliary equation is

$$(D - m_1)y = c_1 e^{m_1 x}$$

or

$$\frac{dy}{dx} - m_1 y = c_1 e^{m_1 x}$$

But this is a linear first-order equation that may be solved using an integrating factor $F(x)$, as follows:

$$F(x) = e^{-\int m_1 \, dx} = e^{-m_1 x}$$

$$y = e^{m_1 x} \left[\int e^{-m_1 x} c_1 e^{m_1 x} \, dx \right]$$

$$y = e^{m_1 x}(c_1 + c_2 x)$$

In general, if the auxiliary equation has n real and equal roots m, the general solution is given by equation (29-21).

Repeating Roots

If the auxiliary equation has n real roots m, then the general solution of the differential equation is

$$y = c_1 e^{mx} + c_2 x e^{mx} + c_3 x^2 e^{mx} + \cdots + c_n x^{n-1} e^{mx} \tag{29-21}$$

EXAMPLE 33 Solve: $y'' + 4y' + 4y = 0$

Solution

Step 1. $(D^2 + 4D + 4)y = 0$

Step 2. $m^2 + 4m + 4 = 0$

Step 3. $(m + 2)^2 = 0$, so the roots are $m_1 = -2$ and $m_2 = -2$.

Step 4. From equation (29-21), the general solution is $y = c_1 e^{-2x} + c_2 x e^{-2x}$. Check it.

Complex roots   If the auxiliary equation has a pair of **complex conjugate roots**—that is, $m = a \pm jb$ or $m_1 = a + jb$, $m_2 = a - jb$ then the solution to the differential equation is, from equation (29-20),

$$y = c_1 e^{(a+jb)x} + c_2 e^{(a-jb)x}$$

$$= c_1 e^{ax} e^{jbx} + c_2 e^{ax} e^{-jbx}$$

$$= e^{ax} \left[c_1 e^{jbx} + c_2 e^{-jbx} \right]$$

But this solution may be simplified using equation (12-7):

$$e^{j\theta} = \cos \theta + j \sin \theta$$

$$e^{-j\theta} = \cos \theta - j \sin \theta$$

$$y = e^{ax} [c_1 (\cos bx + j \sin bx) + c_2 (\cos bx - j \sin bx)]$$

$$= e^{ax} [(c_1 + c_2) \cos bx + j(c_1 - c_2) \sin bx]$$

$$= e^{ax} [c_3 \cos bx + c_4 \sin bx]$$

where c_3 and c_4 are new arbitrary constants.

Complex Roots

If the auxiliary equation has complex roots $m = a \pm bj$, then the general solution of the differential equation is

$$y = e^{ax} (c_1 \cos bx + c_2 \sin bx) \qquad (29\text{-}22)$$

EXAMPLE 34 Solve: $\dfrac{d^2y}{dx^2} - 2\dfrac{dy}{dx} + 5y = 0$

Solution

Step 1. $(D^2 - 2D + 5)y = 0$

Step 2. $m^2 - 2m + 5 = 0$

Step 3. Solve using the quadratic formula.

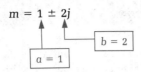

$$m = 1 \pm 2j$$

Step 4. From equation (29-22), the general solution is

$$y = e^x(c_1 \cos 2x + c_2 \sin 2x)$$

Exercises 29-4

Solve by direct integration.

1. $y'' = x^2$ **2.** $y''' = 0$ **3.** $\dfrac{d^2y}{dx^2} = \sin x$

4. $\dfrac{d^2y}{dx^2} = e^x$ **5.** $y'' = \dfrac{1}{x}$ **6.** $y'' - x = 0$

Solve by reduction of order.

7. $\dfrac{d^2y}{dx^2} = \dfrac{dy}{dx} + x$ **8.** $\dfrac{d^2y}{dx^2} = 2 - \dfrac{dy}{dx}$ **9.** $xy'' = y'$

10. $xy'' = x + y'$ **11.** $y'' = -\dfrac{x}{y'}$ **12.** $(1 + x^2)y'' - 2xy' = 0$

13. $y'' - y' = 0$ **14.** $xy'' - y' = x^2$ **15.** $yy'' + (y')^2 = 0$

16. $y'' - 2yy' = 0$

Solve each homogeneous linear differential equation.

17. $\dfrac{d^2y}{dx^2} + \dfrac{dy}{dx} - 2y = 0$ **18.** $\dfrac{d^2y}{dx^2} - 5\dfrac{dy}{dx} + 6y = 0$ **19.** $\dfrac{d^2y}{dx^2} + 3\dfrac{dy}{dx} + 2y = 0$

20. $\dfrac{d^2y}{dx^2} - \dfrac{dy}{dx} - 2y = 0$ **21.** $2y'' - y' = 3y$ **22.** $2y'' + y' = y$

23. $y'' - 9y = 0$ **24.** $y'' = y$ **25.** $y'' - y' = 0$

26. $y'' + 2y' = 0$ **27.** $y'' + 2y' + y = 0$ **28.** $y'' - 4y' + 4y = 0$

29. $4y'' - 4y' + y = 0$ **30.** $9y'' + 12y' + 4y = 0$ **31.** $\dfrac{d^2y}{dx^2} - 2\dfrac{dy}{dx} - 2y = 0$

32. $\dfrac{d^2y}{dx^2} - 2\dfrac{dy}{dx} = 7y$ **33.** $y = y'' + y'$ **34.** $2y' = 2y'' - y$

35. $y'' + 2y = 2y'$ **36.** $y'' - 4y' + 8y = 0$ **37.** $y'' + 4y' + 5y = 0$

38. $y'' + 2y' + 3y = 0$ **39.** $y'' + 4y' + 6y = 0$ **40.** $y'' + 2y' + 4y = 0$

41. $\dfrac{d^2y}{dx^2} + 2\dfrac{dy}{dx} = 8y$;   where $y = 0$ when $x = 0$, and $y' = 6$ when $x = 0$

42. $y'' + 4y' + 4 = 0$;   where $y(0) = 1$ and $y'(0) = 0$

43. $y'' - 2y' = y$;   where $y = 0$ when $x = 0$, and $y' = \sqrt{2}$ when $x = 0$

44. $y'' + 4y' + 5y = 0$;   where $y = 1$ when $x = 0$, and $y' = 0$ when $x = \pi/2$

29-5 | Nonhomogeneous Differential Equations

Using differential operator notation and the auxiliary equation as explained in Section 29-4, we can solve any second-order homogeneous linear differential equation with constant coefficients. In this section we will show how to solve linear **nonhomogeneous equations** of the form

$$a_0 \frac{d^2y}{dx^2} + a_1 \frac{dy}{dx} + a_2 y = Q(x)$$

or, in operator notation,

$$(a_0 D^2 + a_1 D + a_2)y = Q(x)$$

It is possible to show that the general solution y of this nonhomogeneous equation can be expressed in the form

$$y = y_c + y_p \tag{29-23}$$

where y_c, the **complementary solution** of the nonhomogeneous equation, is the general solution of the corresponding homogeneous equation,

$$(a_0 D^2 + a_1 D + a_2)y = 0$$

and y_p is a **particular solution** of the nonhomogeneous equation. Note that for a second-order differential equation, y_c will contain two arbitrary constants, and y_p will contain no arbitrary constants. Therefore, to find the general solution of a given nonhomogeneous linear differential equation with constant coefficients:

First, find the general solution y_c of the corresponding homogeneous equation $(a_0 D^2 + a_1 D + a_2)y = 0$.

Second, find a particular solution y_p of the given nonhomogeneous equation.

Method of undetermined coefficients

The methods of Section 29-4 can be used to find y_c. The simplest method for finding y_p is the **method of undetermined coefficients.** This involves assuming a general form for y_p and substituting it into the original differential equation. The resulting equations can be solved to provide the exact solution. Although the method of undetermined coefficients is useful only for a few functions $Q(x)$, these few functions happen to be those that are most important in applications of nonhomogeneous equations to physics and engineering.

We expect that the particular solution y_p will involve an expression very similar to $Q(x)$. If $y = y_c + y_p$ is a solution of $(a_0 D^2 + a_1 D + a_2)y = Q(x)$, then

$$(a_0 D^2 + a_1 D + a_2)y = (a_0 D^2 + a_1 D + a_2)y_c + (a_0 D^2 + a_1 D + a_2)y_p$$

But $(a_0 D^2 + a_1 D + a_2)y_c = 0$, since y_c is a solution of the corresponding homogeneous equation. Therefore, $(a_0 D^2 + a_1 D + a_2)y_p = Q(x)$. Because the derivatives of y_p appear on the left, we will assume that the particular solution y_p involves a sum of terms containing $Q(x)$ and its derivatives.

EXAMPLE 35   Solve the equation: $\dfrac{d^2y}{dx^2} - 3\dfrac{dy}{dx} + 2y = 4x$

Solution

Step 1.   Solve the corresponding homogeneous equation,

$$\frac{d^2y}{dx^2} - 3\frac{dy}{dx} + 2y = 0 \quad \text{or} \quad (D^2 - 3D + 2)y = 0$$

using the method of Section 29-4, equations (29-19) and (29-20). From the auxiliary equation $m^2 - 3m + 2 = 0$, we find that $m_1 = 1$ and $m_2 = 2$, so the complementary solution is

$$y_c = c_1e^x + c_2e^{2x}$$

Step 2.   Now, assume that $y_p = A + Bx$, where A and B are unspecified constants. Then, $y_p' = B$ and $y_p'' = 0$, and substituting in the original differential equation, we have

$$0 - 3B + 2(A + Bx) = 4x$$

$$-3B + 2A + 2Bx = 4x$$

Step 3.   Solve this equation for the undetermined coefficients A and B by equating like powers of x on both sides of the equation.

For x^0: $-3B + 2A = 0$
For x^1: $2Bx = 4x$, or $2B = 4$ for all real values of x

The second equation gives $B = 2$. Substituting this value of B into the first equation gives $A = 3$. Therefore, the particular solution is $y_p = 3 + 2x$, and the general solution of the original equation is

$$y = y_c + y_p$$

$$y = c_1e^x + c_2e^{2x} + 3 + 2x$$

Step 4.   Check it. First, notice that it contains two arbitrary constants c_1 and c_2, as required for the general solution of a second-order equation. Second, verify that the solution satisfies the equation by taking derivatives and substituting them back into the original nonhomogeneous differential equation.

NOTE ▶   If we had assumed a form for y_p that did not include both expressions like $Q(x)$ and the derivatives of $Q(x)$, then Step 3 would have produced an equation that holds for *no* real values of x. This would tell us that there is no particular solution of the form assumed. If we had included in our assumed expression for y_p terms that are neither in $Q(x)$ nor derivatives of $Q(x)$, then Step 3 would have produced zero coefficients, indicating that these terms are not needed. To see this "self-selecting" aspect of the method of undetermined coefficients, repeat Example 35 using the assumed particular solutions $y_p = Ax$ and $y_p = A + Bx + Cx^2 + Ee^x$. ◀

Table 29-1 will help you to make reasonable assumptions in the choice of a particular solution y_p.

TABLE 29-1

$Q(x)$	Assumed Form for y_p
x^n ax^n $ax^n + bx^{n-1} + \cdots$	$A + Bx + Cx^2 + \cdots + Mx^n$
ae^{bx}	Ae^{bx}
axe^{bx}	$Ae^{bx} + Bxe^{bx}$
ax^2e^{bx}	$Ae^{bx} + Bxe^{bx} + Cx^2e^{bx}$
$a \cos bx$ $a \sin bx$	$A \sin bx + B \cos bx$
$ax \cos bx$ $ax \sin bx$	$A \sin bx + B \cos bx + Cx \cos bx + Ex \sin bx$

EXAMPLE 36   Solve the differential equation: $y'' - 2y' + 2y = 2e^{2x}$

Solution

Step 1.   Solve the corresponding homogeneous equation:

$$y'' - 2y' + 2y = 0 \qquad \text{or} \qquad (D^2 - 2D + 2)y = 0$$

From the auxiliary equation, $m^2 - 2m + 2 = 0$, we find $m_1 = 1 + j$ and $m_2 = 1 - j$, and the complementary solution is

$$y_c = e^x(c_1 \cos x + c_2 \sin x)$$

for arbitrary constants c_1 and c_2.

Step 2.   Since $Q(x) = 2e^{2x}$, assume (as shown in Table 29-1) that the particular solution y_p has the form $y_p = Ae^{2x}$, where A is some unspecified coefficient. Then, differentiating, $y_p' = 2Ae^{2x}$ and $y_p'' = 4Ae^{2x}$. Substitute these expressions for y_p, y_p', and y_p'' into the original differential equation to get

$$4Ae^{2x} - 4Ae^{2x} + 2Ae^{2x} = 2e^{2x}$$

Step 3.   Solve this equation for the undetermined coefficient A by dividing each term by e^{2x}:

$$4A - 4A + 2A = 2$$

$$A = 1$$

Therefore, the particular solution is $y_p = e^{2x}$, and the general solution of the original nonhomogeneous equation is

$$y = y_c + y_p$$

$$y = e^x(c_1 \cos x + c_2 \sin x) + e^{2x}$$

Step 4.   Check this solution to be certain it satisfies the original differential equation.

EXAMPLE 37 Solve the differential equation $y'' - y' - 2y = 20 \cos 2x$, subject to the initial conditions $y(0) = 0$ and $y'(0) = 1$.

Solution

Step 1. For the corresponding homogeneous equation $y'' - y' - 2y = 0$, the auxiliary equation is $m^2 - m - 2 = 0$ or $(m + 1)(m - 2) = 0$, so $m_1 = -1$ and $m_2 = 2$. Therefore, by equation (29-20), the complementary solution is

$$y_c = c_1 e^{-x} + c_2 e^{2x}$$

Step 2. From Table 29-1, the particular solution may be assumed to have the form $y_p = A \sin 2x + B \cos 2x$, where A and B are the coefficients to be determined. Then,

$$y'_p = 2A \cos 2x - 2B \sin 2x$$

$$y''_p = -4A \sin 2x - 4B \cos 2x$$

and, substituting into the original nonhomogeneous equation,

$$(-4A \sin 2x - 4B \cos 2x) - (2A \cos 2x - 2B \sin 2x)$$

$$- 2(A \sin 2x + B \cos 2x) = 20 \cos 2x$$

Step 3. Solve this equation for the coefficients A and B by writing it as

$$(-4A + 2B - 2A) \sin 2x + (-4B - 2A - 2B) \cos 2x = 20 \cos 2x$$

$$(-6A + 2B) \sin 2x + (-2A - 6B) \cos 2x = 20 \cos 2x$$

Then, if this equation is to hold for all values of x, we have

$$-6A + 2B = 0$$

$$-2A - 6B = 20$$

Solving this system of equations for A and B, we find $A = -1$ and $B = -3$. The particular solution is $y_p = -\sin 2x - 3 \cos 2x$, and the general solution of the original equation is

$$y = c_1 e^{-x} + c_2 e^{2x} - \sin 2x - 3 \cos 2x$$

Step 4. Substitute the initial conditions $y(0) = 0$ and $y'(0) = 1$.

For y:
$$0 = c_1 e^0 + c_2 e^0 - \sin 0 - 3 \cos 0$$

$$c_1 + c_2 = 3$$

For y':
$$y' = -c_1 e^{-x} + 2c_2 e^{2x} - 2 \cos 2x + 6 \sin 2x$$

$$1 = -c_1 e^0 + 2c_2 e^0 - 2 \cos 0 + 6 \sin 0$$

$$-c_1 + 2c_2 = 3$$

Solving these two equations for c_1 and c_2, we get $c_1 = 1$ and $c_2 = 2$. Therefore, the particular solution of the nonhomogeneous differential equation is

$$y = e^{-x} + 2e^{2x} - \sin 2x - 3 \cos 2x$$

Check it.

If $Q(x)$ is a sum of two or more of the expressions in the first column of Table 29-1, use the sum of the corresponding terms in the second column as the assumed form for y_p.

EXAMPLE 38   Solve the differential equation: $y'' - 4y' + 4y = 4x + 2xe^x$

Solution
Step 1.   For the corresponding homogeneous equation $y'' - 4y' + 4y = 0$, the auxiliary equation is $m^2 - 4m + 4 = 0$ or $(m - 2)^2 = 0$. Therefore, the roots are $m_1 = m_2 = 2$ and, by equation (29-21), the complementary solution is

$$y_c = c_1 e^{2x} + c_2 x e^{2x}$$

Step 2.   Using Table 29-1, we assume that the particular solution has the form

$$y_p = A + Bx + Ce^x + Gxe^x$$

where A, B, C, and G are the coefficients to be determined. Differentiating, we have

$$y_p' = B + Ce^x + Ge^x + Gxe^x = B + (C + G)e^x + Gxe^x$$

$$y_p'' = Ce^x + Ge^x + Ge^x + Gxe^x = (C + 2G)e^x + Gxe^x$$

and the original nonhomogeneous equation becomes

$$(C + 2G)e^x + Gxe^x - 4B - 4(C + G)e^x - 4Gxe^x + 4A + 4Bx + 4Ce^x$$

$$+ 4Gxe^x = 4x + 2xe^x$$

$$(4A - 4B) + x(4B) + e^x(C + 2G - 4C - 4G + 4C)$$

$$+ xe^x(G - 4G + 4G) = 4x + 2xe^x$$

Step 3.   Equating the coefficients of like expressions in x, we obtain the equations

$$4A - 4B = 0$$

$$4B = 4$$

$$C - 2G = 0$$

$$G = 2$$

and solving this system of equations for the constants A, B, C, and G gives

$$A = B = 1 \qquad C = 4 \qquad G = 2$$

Therefore, the particular solution is

$$y_p = 1 + x + 4e^x + 2xe^x$$

and the general solution of the original equation is

$$y = y_c + y_p = c_1 e^{2x} + c_2 x e^{2x} + 1 + x + 4e^x + 2xe^x$$

Step 4.   Check it.

CAUTION ▶ 1. If Step 2 produces an equation that holds for *no* real values of x, then there is no particular solution of the form you have assumed for y_p. Try a different form.

2. If any of the terms you need to include in y_p are also terms of y_c, then you should rewrite y_p with those terms multiplied by some power of x. Problems of this kind are not included in this chapter, but may arise in other work. ◀

Applications

Higher-order differential equations arise often in the study of many phenomena in physics and engineering. In the remainder of this section we will examine the differential equations describing simple harmonic motion, damped oscillatory motion, and the flow of charge in an RCL circuit.

Simple harmonic motion

Any motion that repeats itself in equal time intervals is said to be *periodic* or *harmonic*. Motion of this kind is found in a great variety of physical situations: movement in mechanical systems, oscillation of charge in a circuit, motion of atoms in a molecule or in a crystal, vibration of air molecules due to a passing sound wave, and so on. If a mass m oscillates under the action of a force F whose magnitude is proportional to the displacement of the mass and whose direction is such that it opposes the displacement at any time, then the motion of the mass is a special kind of periodic motion called **simple harmonic motion (SHM).** Mechanical vibrations that are not SHM often can be approximated mathematically as SHM for small amplitudes of motion.

The mass shown in Fig. 29-11(a) on p. 1314 is free to slide on a horizontal friction-free surface under the action of a spring. When it is at rest, the block sits at its equilibrium position $x = 0$. If it is moved to the right to position $x = A$ and then released, as shown in Fig. 29-11(b), the mass will oscillate about the equilibrium position. If the spring is not overstretched, the displacement x of the mass at any time t, [Fig. 29-11(c)], is given by the differential equation

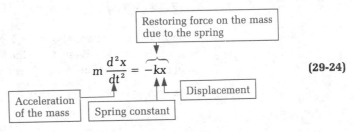

$$m \frac{d^2x}{dt^2} = -kx \qquad (29\text{-}24)$$

In standard form, this is the second-order homogeneous linear differential equation

$$m \frac{d^2x}{dt^2} + kx = 0$$

or, in operator form,

$$(mD^2 + k)x = 0 \qquad \text{or} \qquad \left(D^2 + \frac{k}{m}\right)x = 0$$

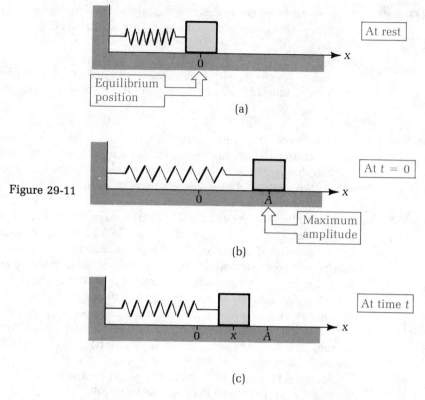

Figure 29-11

The auxiliary equation has complex conjugate roots $j\sqrt{k/m}$ and $-j\sqrt{k/m}$. Therefore, by equation (29-22), the solution to the original differential equation is

$$x = c_1 \cos \sqrt{\frac{k}{m}}\, t + c_2 \sin \sqrt{\frac{k}{m}}\, t$$

The initial condition, shown in Fig. 29-11(b), can be used to determine the constants c_1 and c_2. When $t = 0$ the object has a maximum displacement $x = A$, and it is held motionless, so $dx/dt = 0$. Therefore,

$$x = A = c_1 \cos 0 + c_2 \sin 0$$

$$c_1 = A$$

$$\frac{dx}{dt} = -c_1 \sqrt{\frac{k}{m}} \sin \sqrt{\frac{k}{m}}\, t + c_2 \sqrt{\frac{k}{m}} \cos \sqrt{\frac{k}{m}}\, t$$

When $t = 0$,

$$\frac{dx}{dt} = 0 = -c_1 \sqrt{\frac{k}{m}} \sin 0 + c_2 \sqrt{\frac{k}{m}} \cos 0$$

$$0 = c_2 \sqrt{\frac{k}{m}}$$

Since $\sqrt{k/m} \neq 0$, $c_2 = 0$, and the solution is

$$x = A \cos \sqrt{\frac{k}{m}} t \tag{29-25}$$

The period of the cosine function $\cos t$ is 2π; therefore, the period P of this motion is

$$P = 2\pi \sqrt{\frac{m}{k}}$$

and the frequency of motion is

$$f = \frac{1}{P} = \frac{1}{2\pi} \sqrt{\frac{k}{m}} \tag{29-26}$$

EXAMPLE 39   When a 1.0 kg block is fastened vertically to a certain spring, the spring stretches 0.098 m. The block and spring are then arranged horizontally as shown in Fig. 29-11, and the spring is stretched a distance of 0.12 m and released. Find:

(a)   The spring constant
(b)   An equation giving the displacement as a function of time
(c)   The frequency of oscillation of the block
(d)   The maximum speed of the block
(e)   The time elapsed as the block moves from the start at $x = 0.12$ m to the equilibrium point at $x = 0$.

Solution

(a)   Spring constant $= k = \dfrac{\text{Mass} \cdot 9.8}{\text{Displacement}} = \dfrac{1.0 \cdot 9.8}{0.098} = 100$ N/m

(b)   The block moves with simple harmonic motion, so its motion must be described by equation (29-25). Substituting the given initial conditions $x = 0.12$ and $dx/dt = 0$ when $t = 0$, we have

$$x = 0.12 \cos \sqrt{\frac{100}{0.5}} t$$

$$\approx 0.12 \cos 10t;$$

(c)   From equation (29-26),

$$f = \frac{1}{2\pi} \sqrt{\frac{100}{1.0}} \approx \frac{5}{\pi} \approx 1.6 \text{ Hz}$$

(d)   From part (b), $x = 0.12 \cos 10t$, so the speed at any time is

$$\frac{dx}{dt} = -(0.12)(10) \sin 10t$$

$$= -1.2 \sin 10t$$

Therefore, the maximum value of the speed is approximately 1.2 m/s.

(e)   At $x = 0$, from part (b), $0 = 0.12 \cos 10t$; but $\cos 10t = 0$ at $10t = \pi/2$, or

$$t = \frac{\pi}{20} \approx 0.16 \text{ s}$$

Damped oscillatory motion   In addition to the spring force in Fig. 29-11(c), if the mass m is also subject to a *damping* or *friction* force proportional to the speed of the object, then the differential equation of motion becomes:

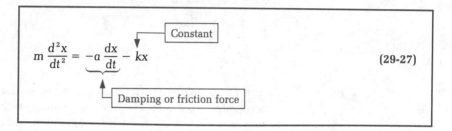

$$m \frac{d^2x}{dt^2} = -a \frac{dx}{dt} - kx \qquad (29\text{-}27)$$

This is a second-order homogeneous linear differential equation with constant coefficients. The standard form is

$$m \frac{d^2x}{dt^2} + a \frac{dx}{dt} + kx = 0$$

or, in operator notation,

$$(mD^2 + aD + k)x = 0$$

The roots of the auxiliary equation are

$$r_1 = \frac{-a + \sqrt{a^2 - 4mk}}{2m} \qquad \text{and} \qquad r_2 = \frac{-a - \sqrt{a^2 - 4mk}}{2m}$$

The nature of these roots, and therefore the form of the solution, will depend on the coefficients a and k. Three possible cases exist.

Damped oscillations   **Case 1.**   If $a^2 - 4mk < 0$ or $a < \sqrt{4mk}$, then the roots r_1 and r_2 are the complex conjugate numbers

$$r_1 = -\frac{a}{2m} + jb \qquad \text{and} \qquad r_2 = -\frac{a}{2m} - jb \qquad \text{where } b = \frac{\sqrt{4mk - a^2}}{2m}$$

and from equation (29-22), the solution is

$$x = e^{-at/2m}(c_1 \cos bt + c_2 \sin bt)$$

The object moves with an oscillatory motion in which the amplitude of the oscillations becomes successively smaller (Fig. 29-12). The periodic motion is said to be **damped,** and the factor $e^{-at/2m}$ is called the **damping factor.**

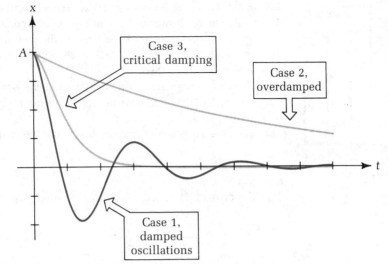

Figure 29-12

Overdamping   **Case 2.**   If $a^2 - 4mk > 0$ or $a > \sqrt{4mk}$, then the roots r_1 and r_2 are unequal negative real numbers $-\alpha$ and $-\beta$, where

$$\alpha = -r_1 = \frac{a}{2m} - \frac{\sqrt{a^2 - 4mk}}{2m} \quad \text{and} \quad \beta = -r_2 = \frac{a}{2m} + \frac{\sqrt{a^2 - 4mk}}{2m}$$

From equation (29-20), the solution has the form

$$x = c_1 e^{-\alpha t} + c_2 e^{-\beta t}$$

The block does not oscillate, but slides very gradually back to its equilibrium position at $x = 0$ (Fig. 29-12). The damping effect of the friction force is so great that no oscillations can occur, and the motion is said to be **overdamped.**

Critical damping   **Case 3.**   If $a^2 - 4mk = 0$ or $a = \sqrt{4mk}$, then the roots r_1 and r_2 are equal negative real numbers:

$$r_1 = r_2 = -\frac{a}{2m}$$

From equation (29-21), the solution has the form

$$x = e^{-at/2m}(c_1 + c_2 t)$$

The block does not oscillate but slides quickly back to its equilibrium position (Fig. 29-12). The damping effect of the friction force is just great enough to prevent oscillations. If the coefficient a were slightly smaller, oscillations would occur. The motion is said to be **critically damped.**

In each of these cases the arbitrary coefficients c_1 and c_2 can be found from the initial conditions that describe the given situation.

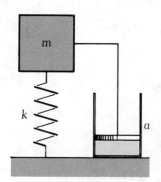

Figure 29-13

EXAMPLE 40   In order to analyze the vertical oscillation of an automobile with hydraulic shock absorbers, the car may be treated as a mass m connected to a spring with spring constant k and a hydraulic damping mechanism with damping constant a (see Fig. 29-13). If the mass of the car is 1500 kg and $k = 4.86 \times 10^5$ N/m, then:

(a)   To what value of a should the hydraulic damper be adjusted to give critical damping? (This will give no oscillation and optimal recovery from a single pulse shock.)

(b)   Find the equation of motion describing the critically damped situation if

Solution

(a)   For critical damping, case 3, $a = \sqrt{4mk}$. Thus,

$$a = \sqrt{4mk}$$

$$= \sqrt{4(1500)(4.86 \times 10^5)}$$

$$= 5.4 \times 10^4 \text{ kg/s}$$

(b)   The solution to equation (29-27), case 3, is

$$x = e^{-at/2m}(c_1 + c_2 t) \qquad \text{where } \frac{a}{2m} = \frac{5.4 \times 10^4}{2(1500)} = 18$$

When $t = 0$, $x = -12$ cm $= -0.12$ m, so

$$x = -0.12 = e^0(c_1 + c_2 \cdot 0)$$

$$c_1 = -0.12$$

and

$$\frac{dx}{dt} = -\left(\frac{a}{2m}\right)e^{-at/2m}(c_1 + c_2 t) + e^{-at/2m}(c_2)$$

When $t = 0$, $dx/dt = 0$, so

$$0 = -\left(\frac{a}{2m}\right)e^0(-0.12 + c_2 \cdot 0) + e^0(c_2)$$

$$c_2 = -0.12\left(\frac{a}{2m}\right)$$

$$= -2.16$$

The equation of motion is

$$x = -e^{-18t}(0.12 + 2.16t)$$

RCL circuit   The RCL series circuit shown in Fig. 29-14 contains a resistance R (in ohms), capacitance C (in farads), inductance L (in henries), and an impressed voltage $V(t)$ (in volts). The voltage drop across L is $L\dfrac{di}{dt}$, across R is iR, and across C is q/C, where q is the charge on the capacitor. Because i is defined as $i = dq/dt$,

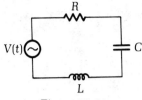

Figure 29-14

we can describe the charge on the capacitor at any time t by the following second-order nonhomogeneous linear differential equation with constant coefficients:

$$L \frac{d^2q}{dt^2} + R \frac{dq}{dt} + \frac{q}{C} = V(t) \qquad \text{(29-28)}$$

Notice that if $V(t) = 0$, this equation is exactly analogous to equation (29-27) describing the damped oscillatory motion of a vibrating object.

EXAMPLE 41   In the RCL circuit of Fig. 29-14, if $L = 1$ H, $R = 6$ Ω, $C = 0.1$ F, and $V(t) = 120 \sin 60t$, find an equation giving the charge on the capacitor at any time t.

Solution
Step 1.   The differential equation is

$$\frac{d^2q}{dt^2} + 6 \frac{dq}{dt} + 10q = 120 \sin 60t$$

The corresponding homogeneous equation is

$$\frac{d^2q}{dt^2} + 6 \frac{dq}{dt} + 10q = 0$$

and the auxiliary equation is

$$m^2 + 6m + 10 = 0$$

Step 2.   Use the quadratic formula to solve the auxiliary equation. Its roots are the complex conjugates

$$r_1 = -3 + j \qquad \text{and} \qquad r_2 = -3 - j$$

Therefore, the complementary solution to the differential equation is

$$q_c = e^{-3t}(c_1 \cos t + c_2 \sin t)$$

Step 3.   To find the particular solution, assume that

$$q_p = A \sin 60t + B \cos 60t$$

Then,

$$q_p' = 60A \cos 60t - 60B \sin 60t$$

$$q_p'' = -3600A \sin 60t - 3600B \cos 60t$$

Substitute into the original differential equation:

$$(-3600A \sin 60t - 3600B \cos 60t) + 6(60A \cos 60t - 60B \sin 60t)$$
$$+ 10(A \sin 60t + B \cos 60t) = 120 \sin 60t$$

$$\sin 60t(-3600A - 360B + 10A) + \cos 60t(-3600B + 360A + 10B)$$
$$= 120 \sin 60t$$

Therefore,

$$-3590B + 360A = 0$$

$$-3590A - 360B = 120$$

Solve this system of equations for A and B:

$$A \approx -0.033 \qquad B \approx -0.003$$

The general solution is $q(t) = q_c + q_p$

$$q(t) \approx e^{-3t}(c_1 \cos t + c_2 \sin t) - 0.033 \sin 60t - 0.003 \cos 60t$$

Notice that the solution for $q(t)$ involves an exponential term q_c and a sinusoidal term q_p. Because of the factor e^{-3t}, as $t \to \infty$, $q_c \to 0$. Since this term becomes negligible with time, it is called the **transient term.** For times much greater than $t = 0$, only the q_p or **steady-state term** remains.

Exercises 29-5

Solve each differential equation.

1. $y'' + y' - 6y = 4x$
2. $y'' + y' - 20y = 2x$
3. $y'' - 2y' = 3$
4. $y'' + y' = \dfrac{1}{2}$
5. $y'' - 2y' - 8y = \frac{1}{2} - 8x^2$
6. $y'' - 2y' - 3y = 2x^3$
7. $y'' - y' - 2y = \sin x$
8. $2y'' + 5y' - 3y = e^{2x}$
9. $y'' - 2y' + y = 2e^{3x}$
10. $y'' + 6y' - 9y = 4 \cos x$
11. $4y'' - 4y' + y = x + 2 \sin x$
12. $4y'' + 12y' + 9y = 4 \cos x - x^2$
13. $y'' - 8y' + 16y = 1 + 2x + e^{-x}$
14. $y'' + 10y' + 25y = x^2 - e^{-2x}$
15. $y'' + 2y' + 2y = \sin 2x$
16. $y'' - 4y' + 5y = x + e^x$
17. $y'' + 4y = x^2$
18. $y'' + 3y = 2x - x^3$
19. $y'' - 2y' + 5y = xe^{3x}$
20. $y'' - 6y' + 10y = x + 2xe^{-x}$
21. $y'' - 2y' + y = x \cos x$
22. $y'' - y' + y = 2x \sin x$
23. $y'' - 4y' - 5y = 2 - e^x$
24. $y'' - 2y' + 10y = xe^{2x} - e^{2x}$

Find the particular solution of each equation for the given conditions.

25. $4y'' + 4y' + y = 9e^x$;
    $y(0) = 1, \quad y'(0) = 0$
26. $2y'' - y' - 6y = 9xe^{-x}$;
    $y(0) = 0, \quad y'(0) = 3$
27. $y'' + 4y = 4x + \sin x$;
    $y(0) = 2, \quad y'(0) = 2$
28. $y'' - 2y' + 2y = 2x^2 - \cos x$;
    $y(0) = 0, \quad y'(0) = 0$

Solve.

29. **Physics** If an object moves with simple harmonic motion according to the equation

$$\frac{d^2x}{dt^2} + 0.25x = 0$$

find the displacement as a function of time subject to the conditions x = −4.0 cm and dx/dt = 3 cm/s when t = 0.

30. *Physics* A 2.0 kg mass is attached to the lower end of a spring with spring constant k = 450 N/m. This object is then pulled to a point 10 cm below its resting position and released at t = 0.

   (a) Assuming that the object moves with simple harmonic motion, find an expression giving its displacement as a function of time.
   (b) Where is the object at t = 3.0 s?
   (c) At what speed is the object moving at t = 3.0 s?

31. *Physics* The block shown slides horizontally on a friction-free surface under the control of two identical springs, each with k = 96 N/m. The motion is simple harmonic and is described by the differential equation

$$m \frac{d^2x}{dt^2} + 2kx = 0$$

If the mass of the block is 4.0 kg, and if it is at rest at x = 2 cm when t = 0, find an equation giving its displacement as a function of time.

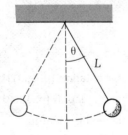

32. *Physics* The differential equation describing the motion of a simple pendulum is

$$\frac{d^2\theta}{dt^2} + \frac{g}{L} \sin \theta = 0$$

where θ is the angle (in radians) that the pendulum arm makes with the vertical, L is the length of the pendulum arm, and g is the acceleration due to gravity. For angles less than 15°, $\sin \theta \approx \theta$ to about 1% accuracy.

   (a) Show that the pendulum motion is approximately simple harmonic for such small angles and solve the equation to find $\theta(t)$.
   (b) Show that the frequency of the pendulum is given by the following equation for small values of θ.

$$f = \frac{1}{2\pi} \sqrt{\frac{g}{L}}$$

33. *Physics* A 1 kg mass is attached to the lower end of a coil spring which is suspended under water. The spring constant is k = 16 N/m, and the water exerts a damping force equal to four times the velocity of the spring. The spring is lifted 20 cm from its rest position and released with velocity v = 0 when t = 0. Find the equation describing the motion of the spring.

34. *Physics*   For the spring system in Problem 33, assume that the damping force is ten times the velocity of the spring. Find the equation of motion of the spring.

35. *Physics*   For the spring system in Problem 33, suppose the damping force is eight times the velocity of the spring. Find the equation of motion of the spring.

36. *Physics*   For the oscillating system shown, the block has mass 3.0 kg, and the spring constant is 15 N/m. The block is pulled down a distance of 10 cm, held motionless, and then released. If a friction force $a\dfrac{dx}{dt}$ is exerted by the water, where $a = 18$ kg/s, find the equation of motion of the block.

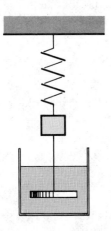

37. *Electronics*   Find the equation for $q(t)$ in an *RCL* series circuit with $L = 1.2$ H, $R = 5.0\ \Omega$, $C = 0.025$ F, and $V(t) = 15$ V.

38. *Electronics*   Find the equation for $q(t)$ in an *RCL* series circuit with $L = 0.23$ H, $R = 16\ \Omega$, $C = 0.0015$ F if $V(t) = 110$ V.

39. *Electronics*   Find the equation for $q(t)$ for the circuit of Problem 37 if the impressed voltage varies with time according to the equation $V(t) = 150 \cos t$ V.

40. *Electronics*   Find the equation for $q(t)$ for the circuit of Problem 38 if the impressed voltage varies with time according to the equation $V(t) = 1200 \sin 2t$ V.

29-6 | Review of Chapter 29

Important Terms and Concepts

differential equation (p. 1260)

order of a differential equation (p. 1261)

first-order equation (p. 1261)

second-order equation (p. 1261)

degree of a differential equation (p. 1261)

orthogonal trajectory (p. 1281)

exponential growth and decay (p. 1284)

  growth constant (p. 1284)

  decay constant (p. 1284)

  radioactive decay (p. 1284)

  half-life (p. 1284)

operator notation (p. 1301)

auxiliary equation (p. 1303)

complementary solution (p. 1308)

method of undetermined coefficients (p. 1308)

simple harmonic motion (p. 1313)

Formulas and Rules

- solution by direct integration (p. 1264)

  If $\dfrac{dy}{dx} = f(x)$ or $dy = f(x)\,dx$, then $y = \displaystyle\int f(x)$ **(29-1)**

- separation of variables (p. 1266)

  If $\dfrac{dy}{dx} = \dfrac{f(x)}{g(y)}$ or $g(y)\,dy = f(x)\,dx$, then $\displaystyle\int g(y)\,dy = \int f(x)\,dx$ **(29-2)**

- integrable forms (p. 1270)

$$x\,dy + y\,dx = d(xy) \qquad\qquad\qquad \textbf{(29-3)}$$

$$x\,dx + y\,dy = \frac{1}{2}d(x^2 + y^2) \qquad\qquad \textbf{(29-4)}$$

$$x\,dx - y\,dy = \frac{1}{2}d(x^2 - y^2) \qquad\qquad \textbf{(29-5)}$$

$$\frac{x\,dy - y\,dx}{x^2} = d\left(\frac{y}{x}\right) \qquad\qquad \textbf{(29-6)}$$

$$\frac{y\,dx - x\,dy}{y^2} = d\left(\frac{x}{y}\right) \qquad\qquad \textbf{(29-7)}$$

- linear differential equation (p. 1274)

$$\frac{dy}{dx} + P(x)y = Q(x) \qquad\qquad\qquad \textbf{(29-8)}$$

  or

$$dy + P(x)y\,dx = Q(x)\,dx \qquad\qquad\quad \textbf{(29-9)}$$

- solution using an integrating factor (p. 1275)

  If $\dfrac{dy}{dx} + P(x)y = Q(x)$, then

$$y = \frac{1}{F(x)}\left[\int Q(x)F(x)\,dx\right] \qquad\quad \textbf{(29-10)}$$

  where

$$F(x) = e^{\int P(x)\,dx} \qquad\qquad\qquad\quad \textbf{(29-11)}$$

- system of orthogonal trajectories (p. 1283)

$$y' = F(x, y) \qquad \text{and} \qquad y' = -\frac{1}{F(x, y)} \qquad \textbf{(29-12)}$$

- exponential growth or decay (p. 1284)

$$\frac{dN}{dt} = kN \qquad \text{(29-13)}$$

- half-life for exponential decay (p. 1285)

$$t_H = \frac{\ln 2}{k} \qquad \text{(29-14)}$$

- logistic growth law (p. 1286)

$$\frac{dN}{dt} = kN(M - N) \qquad \text{(29-15)}$$

When solving this type of differential equation, note that

$$\frac{1}{N(M - N)} = \frac{1}{M}\left(\frac{1}{N} + \frac{1}{M - N}\right) \qquad \text{(29-16)}$$

- current in an LR circuit (p. 1288)

$$L\frac{di}{dt} + Ri = V(t) \qquad \text{(29-17)}$$

- motion in a resisting medium (p. 1291)

$$m\frac{dv}{dt} = F(t) - F_{\text{Drag}} \qquad \text{(29-18)}$$

where $F_{\text{Drag}} = k_1 v$ for smooth flow, and $F_{\text{Drag}} = k_2 v^2$ for turbulent flow
- solving second-order equations by direct integration (see p. 1299)
- solving second-order equations by reduction of order (see p. 1300)
- general solution of homogeneous linear differential equations (p. 1303)
  For a homogeneous linear differential equation with constant coefficients of the form

$$a_0\frac{d^n y}{dx^n} + a_1\frac{d^{n-1}y}{dx^{n-1}} + \cdots + a_{n-1}\frac{dy}{dx} + a_n y = 0$$

  the auxiliary equation is (p. 1303)

$$a_0 m^n + a_1 m^{n-1} + \cdots + a_{n-1}m + a_n = 0 \qquad \text{(29-19)}$$

  If $m_1, m_2, \ldots, m_n$ are distinct real roots of (29-19), then the general solution of the differential equation is

$$y = c_1 e^{m_1 x} + c_2 e^{m_2 x} + \cdots + c_n e^{m_n x} \qquad \text{(29-20)}$$

- repeating roots (p. 1305)
  If the auxiliary equation (29-19) has n real roots m, the general solution of the differential equation is

$$y = c_1 e^{mx} + c_2 x e^{mx} + c_3 x^2 e^{mx} + \cdots + c_n x^{n-1} e^{mx} \qquad \text{(29-21)}$$

- complex roots (p. 1306)
  If the auxiliary equation (29-19) has complex roots $m = a + bj$, then the general solution of the differential equation is

$$y = e^{ax}(c_1 \cos bx + c_2 \sin bx) \qquad \text{(29-22)}$$

- general solution of nonhomogeneous linear differential equations (p. 1308)
  For a nonhomogeneous linear differential equation of the form

$$a_0\frac{d^2 y}{dx^2} + a_1\frac{dy}{dx} + a_2 y = Q(x) \qquad \text{or} \qquad (a_0 D^2 + a_1 D + a_2)y = Q(x)$$

  the general solution y has the form (p. 1308)

$$y = y_c + y_p \qquad \text{(29-23)}$$

  where y_c, the complementary solution, is the general solution of

$$(a_0 D^2 + a_1 D + a_2)y = 0$$

  and y_p is a particular solution of the nonhomogeneous equation.

Exercises 29-6

Solve using separation of variables.

1. $y' = -5x$ **2.** $y' = 3y$ **3.** $dy - dx = e^x \, dx$

4. $dy - x \, dx = \cos x \, dx$ **5.** $\dfrac{dy}{dx} = 1 + \sqrt{x}$ **6.** $y \dfrac{dy}{dx} = x^2$

7. $(1 - x)y' = 1 + y$ **8.** $e^x y' - \sqrt{y + 1} = 0$

9. $y'\sqrt{y} = x + 1$; if $y = 1$ when $x = 0$ **10.** $x + xy' = x^2 + y'$; if $y = 0$ when $x = 2$

Solve using an integrable form.

11. $x \, dy + y \, dx = 4 \, dx$ **12.** $2x \, dx + 4y \, dy = (x^2 - x) \, dx$

13. $x \, dy - y \, dx = x^2 e^x \, dx$ **14.** $y \, dx - x \, dy = y^2 \sin x \, dx$; if $y = 1$ when $x = 0$

Solve using an integrating factor.

15. $y' + \dfrac{y}{x} = \sqrt{x^2 + 1}$ **16.** $y' = e^x + 2y$

17. $dy + x \, dx + y \cot x \, dx = 0$ **18.** $x^2 \sin x^2 + \dfrac{y}{x} = \dfrac{dy}{dx}$

19. $\dfrac{y'}{x} = \dfrac{2y}{x^2 + 1} + 1$; if $y = 1$ when $x = 0$ **20.** $y' - \sin x = y \sin x$; if $y = 0$ when $x = \pi/2$

21. Find the equation of a curve that passes through the point $(1, 1)$ and whose tangent line at a point (x, y) has slope $dy/dx = e^x - (y/x)$.

22. Find the equation of a curve that passes through the origin and whose slope at any point is equal to 1 plus the square of its x-coordinate.

23. Find the orthogonal trajectories of the family of curves $k + y^2 = 2x$.

Solve.

24. $\dfrac{d^2 y}{dx^2} = e^{-x}$ **25.** $y'' - x^2 = 2x$

26. $y'' = y' + 2x$ **27.** $y \dfrac{d^2 y}{dx^2} = \left(\dfrac{dy}{dx}\right)^2 + \dfrac{dy}{dx}$

28. $y'' + 10y' + 25y = 0$ **29.** $y'' + 2y = 4y'$

30. $y'' = 8y - 2y'$ **31.** $y'' = 4(y' - 2y)$

32. $y'' = 2y$; where $y = 0$ and $y' = 2$ when $x = 0$

33. $y'' + 13y = 6y'$; where $y = 2$ and $y' = 4$ when $x = 0$

34. $y'' - 3y' + 2y = 2x^2$ **35.** $y'' + y' = 4 \sin x$

36. $y'' = 4y + 9xe^x$

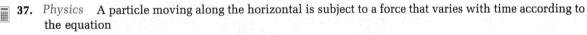 **37.** Physics   A particle moving along the horizontal is subject to a force that varies with time according to the equation

$$F = m \frac{dv}{dt} = 24me^{-2t}$$

(F in pounds, t in seconds). If the motion is opposed by a drag force $F_{Drag} = 4v$, and if $m = 1$ and $v = 16$ ft/s when $t = 0$, find an equation giving v at any time t.

38. *Industrial engineering*   A tank initially contains 30 lb of salt dissolved in 120 gal of water. Brine containing 5 lb of salt per gallon of water is allowed to enter the tank continuously at the rate of 2.5 gal/min, and the mixture is stirred and drained from the tank at 2.5 gal/min.
  (a) How much salt is in the tank at a time t?
  (b) How much salt is in the tank at $t = 20$ min?

39. *Industrial engineering*   In Problem 38, if the mixture is drained from the tank at 2.0 gal/min, how much salt is in the tank at time t?

40. *Electronics*   The charge q on a charging capacitor in an RC circuit with impressed voltage $V(t) = 2t$ is given by the differential equation

$$R \frac{dq}{dt} + \frac{q}{C} = V(t)$$

Solve for q if $q = 0$ when $t = 0$.

41. *Aeronautical engineering*   The rate of growth of the thickness T of a layer of ice on the control surfaces of an airplane is given by the differential equation $dT/dt = 4T^2$. If $T = 0.1$ cm when $t = 0$ s, when will the ice layer be 4.0 cm thick?

42. *Electronics*   The current i in an RL series circuit varies according to the differential equation

$$L \frac{di}{dt} + Ri = V(t)$$

If the impressed voltage $V(t)$ is $V(t) = 120$ volts, $R = 80$ Ω, and $L = 100$ H, and $i = 0$ at $t = 0$, find:
  (a) An expression for the current as a function of time
  (b) The value of i when $t = 2.50$ s.

43. *Nuclear physics*   Radiocarbon dating techniques are used to determine the age of organic materials. The radioisotope carbon-14, which is created in the upper atmosphere, is absorbed continuously by living cells, and when the cell dies, the C-14 present in the material decreases in an exponential decay process. If C-14 has a half-life of 5750 years, determine the age of a wooden archaeological artifact whose C-14 content has decreased to 0.460 of its level in living wood.

44. *Biology*   If the number of bacteria in a culture sample grows exponentially at 2% per hour, and 500 bacteria are present initially, find:
  (a) The number of bacteria present at any time t
  (b) The number of bacteria present after 6 h
  (c) The time needed for the number of bacteria to double

45. *Industrial engineering*   Production of a certain computer chip on an automated assembly line follows the restricted growth model $dN/dt = 0.0125N(52.5 - N)$, where $N = 10$ at $t = 0$. Find an expression for N as a function of time t.

46. *Physics*   When an object with mass 1.5 kg is suspended vertically from a spring, it stretches 0.35 m from its equilibrium position. If the object is then pulled 0.10 m below the equilibrium position and released, it will oscillate with simple harmonic motion. Assuming that no damping forces are acting, find:
  (a) The spring constant
  (b) An equation describing the displacement of the object as a function of time
  (c) The frequency of oscillation of the object
  (d) The speed of the object when it passes the equilibrium position

47. *Physics*   If the object in Problem 46 is also subject to a damping force proportional to its speed of motion, $F_{Damping} = -20v$, find the equation describing the damped motion.

48. *Physics*   In Problem 47, if an external driving force $F = 362 \sin 2t$ acts on the object in addition to the spring force and the damping force, find an equation describing the steady-state motion of the object. [Hint: Find the particular solution of the form $A \sin 2t + B \cos 2t$.]

49. *Electronics*   For an RCL series circuit in which $L = 2.5$ H, $R = 20$ Ω, $C = 0.025$ F, and $V(t) = 75 \cos 5t$, find an equation describing the charge on the capacitor at time t.

Appendix A: Use of the Scientific Calculator

Introduction   Since its introduction in the early 1970's, the hand-held electronic calculator has quickly become an indispensable tool in modern society. From clerks, carpenters, and shoppers to technicians, engineers, and scientists, people in virtually every occupation now use calculators to perform mathematical tasks. With this in mind, we have included in this text special instructions, examples, and exercises demonstrating the use of a calculator wherever it is appropriate.

To use a calculator intelligently and effectively in your work, you should remember the following:

1. Whenever possible, make an estimate of your answer before entering a calculation. Then check this estimate against your calculator result. You may get an incorrect answer by accidently pressing the wrong keys or by entering numbers in an incorrect sequence.
2. Always try to check your answers to equations and word problems by substituting them back into the original problem statement.
3. Organize your work on paper before using a calculator, and record any intermediate results that you may need later.
4. Calculators do not understand the meaning of significant digits. Remember to round your answer whenever necessary (see Appendix B).

Choosing a calculator   There are available many brands and models of electronic calculators with different capabilities, internal logic, and key symbols. For technical work you will need a calculator with a full array of scientific functions. Scientific calculators vary widely in sophistication and price, but even the most basic will have the functions required by this course. Before making your choice of model, check with your instructor and consider your future requirements for a calculator.

Among the various brands of scientific calculators, two basic types of internal logic are used: algebraic and reverse Polish. Calculators with algebraic notation have an $=$ key, while those with reverse Polish notation have an ENTER key instead. The two types differ slightly in the way calculations must be entered. Although either calculator will suit the needs of this course, the displays and instructions in this text have been geared toward the more widely used algebraic calculators.

There are also slight variations among calculators in the symbols used for various keys. Table A-1 is provided for reference. It shows the names and sequences used in this text along with equivalent names and sequences that may be used by your calculator. As you will see, these variations are minor and should not cause any difficulty in using this text. To become completely familiar with your calculator, be certain to read the instruction manual provided by the manufacturer.

NOTE ▶ Many of the keys on your calculator perform a second function if the [INV] or [2nd F] key is pressed first. Most models have the name of the second function written on or above the appropriate key, although these second functions are not always paired with the same primary function. Therefore, when listing the sequences for such functions, we will not usually mention the [2nd F] or [INV] key, and you must recognize when to press this key first. If your calculator does not indicate the names of the second functions above the keys, see the column headed "Other Symbols" in Table A-1 or consult your instruction manual for the proper sequence. ◀

TABLE A-1

Function	Example	Our Sequence and Symbols	Other Symbols	Display
Clear display and register		[C]	[ON/C]	0.
Clear last entry only		[CE]		0.

NOTE ▶ Once a calculation has been completed, the next calculation can be entered without clearing the display. ◀

Function	Example	Our Sequence and Symbols	Other Symbols	Display
Add	$5 + 3$	5 [+] 3 [=]		8.
Subtract	$5 - 3$	5 [−] 3 [=]		2.
Multiply	$5 \cdot 3$	5 [×] 3 [=]		15.
Divide	$15 \div 3$	15 [÷] 3 [=]		5.
Change sign	-5	5 [+/−]	5 [CS]	−5.
Reciprocal	$\dfrac{1}{3}$	3 [1/x]		0.3333333
Square	6^2	6 [x²]		36.
Square root	$\sqrt{25}$	25 [√]	25 [√x]	5.
Pi	π	[π]		3.1415927
Find a power	7^4	7 [yˣ] 4 [=]	7 [xʸ] 4 [=]	2401.
Extract a root	$\sqrt[3]{64}$	64 [ˣ√y] 3 [=]	64 [x^{1/y}] 3 [=] 64 [³√], 64 [INV] [yˣ] 3 [=]	4.
Common logarithm	$\log 26$	26 [log]		1.4149733
Natural logarithm	$\ln 0.84$	.84 [ln x]	.84 [ln]	−0.1743534
Power of 10	10^5	5 [10ˣ]	5 [INV] [log], 10 [yˣ] 5 [=]	100000.
Power of e	$e^{1.7}$	1.7 [eˣ]	1.7 [INV] [ln x]	5.4739474
Degree/Radian/Gradient modes		[DEG] [RAD] [GRAD] Press the [DRG] key until the display indicates the desired mode. Some calculators will display indication of only [RAD] and [GRAD] modes; no indication means it is in [DEG] mode.		DEG 0. RAD 0. GRAD 0.

NOTE ▶ Some calculators will automatically revert to [DEG] mode when turned on, while others will retain the mode being used when last turned off. Always check before beginning a calculation. ◀

Function	Example	Our Sequence and Symbols	Other Symbols	Display
Sine	sin 132°	132 [sin]		−0.7431448
Cosine	cos 27°15′	27 [+] 15 [÷] 60 [=] [cos]		0.8890171
Tangent	$\tan \dfrac{5\pi}{12}$	[RAD] 5 [×] [π] [÷] 12 [=] [tan]		3.7320508
Inverse sine	Arcsin 0.62*	.62 [INV] [sin]	.62 [sin⁻¹]	38.316135
Inverse cosine	Arccos −0.146	.146 [+/−] [INV] [cos]	.146 [+/−] [cos⁻¹]	98.395191
Inverse tangent	Arctan −2.78	2.78 [+/−] [INV] [tan]	2.78 [tan⁻¹]	−70.215721

NOTE ▶ To obtain angles measured in radians, put the calculator in radian mode before entering the function value. ◀

Factorial	7!	7 [x!]	7 [n!]	5040.
Parentheses	2(5 + 11)	2 [×] [(] 5 [+] 11 [)] [=]		32.
	$\dfrac{17}{3 + 4 \cdot 9}$	17 [÷] [(] 3 [+] 4 [×] 9 [)] [=]		0.4358974
Entering scientific notation	5×10^6	5 [EE] 6	5 [EXP] 6	5. 06
Changing display to scientific notation	7000.	[×] 1 [EE] [=]	[ENG] or [F↔E]	7. 03
Changing display to decimal notation	7. 03	[INV] [EE] [=]	[F↔E]	7000.

NOTE ▶ This will not work if the number exceeds the display length of the calculator. ◀

Storing in memory	Place 16 in memory	16 [STO]	16 [x→M] 16 [Min]	16.
Add to memory	Add 24	24 [SUM]	24 [M+]	24.
Recall memory		[RCL]	[MR] or [RM]	40.

*Prior to discussing the inverse trigonometric functions in Chapter 19, we used these same sequences of keys to solve problems such as "Find θ if sin θ = 0.62." Notice that the calculator displays the value of θ only within the range of the inverse function. To find *all* values of θ between 0° and 360° (2π), follow the procedures outlined in Sections 8-2 and 8-3.

Combined operations   Calculators with algebraic logic automatically perform the order of operations as defined in Chapter 1. Consequently, a calculation involving two or more operations can be entered in the same order as it is written algebraically. Proper calculator sequences for such problems are demonstrated in calculator examples throughout the text.

Error messages   Whenever a calculation results in an undefined value, a calculator will display an error message such as

Error ,   E   0. ,   0.0.0.0.0. ,   E.

and so on. For example, attempting to divide by 0, take the square root or logarithm of a negative number, or find tan 90° will all result in an error message. In addition, an error message will occur for certain operations even when a defined value results. Attempting to raise a negative number to a power is the most common example of this.

Accuracy and precision   Calculators that round differently or have different degrees of internal precision may also differ in the display of the final digit. Some calculators round the final digit, while others simply truncate the answer. For example, the answer 3.12345678. . . will be displayed as   *3.1234568* by a rounding calculator and as   *3.1234567* by a truncating calculator.

In addition, many calculators do not display a rounded 0 as the final digit of a display, incorrectly implying that the displayed answer is exact. These calculators will display the answer 4.56345796 . . . as   *4.563458* . Our displays will always show the eighth digit as a 0 in such cases.

Finally, the length of the display itself varies for different models. The displays shown in this text are all 8 digits long, but some models display 10 digits.

Because we will rarely require an answer accurate even to 8 digits, none of these differences should cause difficulty.

A related problem does cause difficulty for some students. Again, due to internal precision, some calculators may display a very small answer such as   *2   −10* (2×10^{-10}) when an answer is clearly 0. You must learn to recognize such instances if your calculator does this.

It would be impossible to cover every characteristic of every calculator in this appendix, so it is important that you read carefully the instruction manual accompanying your calculator.

Appendix B: Units of Measurement and Approximate Numbers

B-1 | Units and Unit Conversion

Denominate numbers

In technical applications of mathematics, numbers are used to describe physical quantities. For example, an electrician might measure the current in a certain circuit as 0.8 ampere, a meteorologist might record the outside temperature as 21 degrees Celsius, and a mechanical engineer might find the handbook value of the shear modulus of copper to be 6.09×10^6 pounds per square inch. Notice that to describe these quantities fully, each number must include both a measure of its magnitude and a statement of the units of measurement used. A number containing both a magnitude and a unit of measurement is called a **denominate number.**

Units

A **unit of measurement** may be defined as a selected magnitude of a physical quantity in terms of which other magnitudes of the same quantity may be expressed. To help clarify this definition, think of a denominate number as a product or quotient. For example, a length of 6 meters can be interpreted as 6×1 meter—that is, 6 times the length of an accepted standard meter length. Similarly, a speed of 12 ft/s (feet per second) can be thought of as $12 \times \dfrac{1 \text{ ft}}{1 \text{ s}}$, or 12 times a basic unit of speed. In this example, both the foot and the second are the fundamental units in terms of which the speed unit is defined.

If a measurement is to be understood or useful in technical work, its units must be precisely defined and standardized. Over the centuries the accepted units of measurement have evolved in a variety of ways, ranging from the ancient *digit*, defined as the width of a man's finger, to the modern *kilogram*, based on the mass of a standard metal cylinder stored in a vault in France. Two basic systems of measurement are used in technical work: the British, or engineering, system and the International (metric) System. To work effectively in modern technology, you must be familiar with both systems and be able to convert between them.

The International System of Units

Because different countries had developed their own variations of the metric system, a General Conference of Weights and Measures was held in 1960 to standardize the system. The result was the establishment of the **International**

System (SI) of units. Seven fundamental physical quantities form the basis of this system, and base units for each are defined in terms of a measurable standard. Table B-1 lists these fundamental quantities along with their base units and abbreviations.

TABLE B-1
Fundamental Quantities and Their Base Units

Quantity	SI Base Unit	Abbreviation	British Base Unit	Abbreviation
Length, L	meter	m	foot	ft
Mass, m	kilogram	kg	slug	
Time, t	second	s	second	s
Electric current, i, I	ampere	A	ampere	A
Temperature, T	Kelvin	K	Fahrenheit degree	°F
Amount of a substance, n	mole	mol	mole	mol
Luminous intensity, I	candela	cd	candlepower	cp

In addition to these base units, supplementary units are defined for the plane angle (the radian, rad) and the solid angle (the steradian, sr). The significance of the base and supplementary units is that none of them can be defined in terms of the others, yet all other units for physical quantities can be derived from various combinations of the base units.

Table B-2 lists the derived units most commonly used in the technologies. Notice that some of these units are expressed directly as combinations of the

TABLE B-2
SI-Derived Quantities

Quantity	Symbol	Derivation	Name	Abbreviation	British Unit	Abbreviation
Area	A	m^2				ft^2
Volume	V	m^3				ft^3
Capacity	V	$10^{-3} \cdot m^3$	liter	L	gallon	gal
Velocity	v	m/s				ft/s
Acceleration	a	m/s^2				ft/s^2
Mass density	d, ρ	kg/m^3				lb/ft^3
Force	F	$m \cdot kg \cdot s^{-2}$	newton	N	pound	lb
Pressure	P	$m^{-1} \cdot kg \cdot s^{-2}$ or N/m^2	pascal	Pa		lb/ft^2
Energy (work)	W	$m^2 \cdot kg \cdot s^{-2}$ or $N \cdot m$	joule	J	foot-pound	$ft \cdot lb$
Energy (heat)	E		joule	J	British thermal unit	Btu
Power	P	$m^2 \cdot kg \cdot s^{-3}$ or J/s	watt	W	horsepower	hp
Electric charge	q	$s \cdot A$	coulomb	C	coulomb	C
Electric potential	V, E	$m^2 \cdot kg \cdot s^{-3} \cdot A^{-1}$ or W/A	volt	V	volt	V
Capacitance	C	$m^{-2} \cdot kg^{-1} \cdot s^4 \cdot A^2$ or C/V	farad	F	farad	F
Electric resistance	R	$m^2 \cdot kg \cdot s^{-3} \cdot A^{-2}$ or V/A	ohm	Ω	ohm	Ω
Magnetic flux	ϕ	$m^2 \cdot kg \cdot s^{-2} \cdot A^{-1}$ or $V \cdot s$	weber	Wb	weber	Wb
Inductance	L	$m^2 \cdot kg \cdot s^{-2} \cdot A^{-2}$ or Wb/A	henry	H	henry	H
Frequency	f	s^{-1}	hertz	Hz	hertz	Hz
Luminous flux	F	$cd \cdot sr$	lumen	lm	lumen	lm
Illumination	E	$m^{-2} \cdot cd \cdot sr$	lux	lx	footcandle	fc

base units (for example, velocity as m/s), while others have been given their own distinctive names (for example, *pascal* is used to mean $m^{-1} \cdot kg \cdot s^{-2}$).

SI prefixes

Because the magnitudes of physical quantities can vary over a wide range of values, prefixes representing powers of 10 are often combined with SI units. These prefixes, their symbols, and their mathematical meaning are listed in Table B-3. The most commonly used prefixes are shown in color and illustrated by examples.

**TABLE B-3
Metric Prefixes**

Prefix	Symbol	Factor	Example
exa-	E	10^{18}	
peta-	P	10^{15}	
tera-	T	10^{12}	
giga-	G	10^{9}	
mega-	M	10^{6}	$1 \text{ MW} = 10^{6} \text{ W}$
kilo-	k	10^{3}	$1 \text{ kL} = 10^{3} \text{ L}$
hecto-	h	10^{2}	
deca-	da	10^{1}	
deci-	d	10^{-1}	
centi-	c	10^{-2}	$1 \text{ cm} = 10^{-2} \text{ m}$
milli-	m	10^{-3}	$1 \text{ mA} = 10^{-3} \text{ A}$
micro-	μ	10^{-6}	$1 \text{ } \mu\text{H} = 10^{-6} \text{ H}$
nano-	n	10^{-9}	$1 \text{ ns} = 10^{-9} \text{ s}$
pico-	p	10^{-12}	$1 \text{ pF} = 10^{-12} \text{ F}$
femto-	f	10^{-15}	
atto-	a	10^{-18}	

British system

Since they are based on powers of 10, the metric prefixes allow us to convert easily within the metric system. However, units describing the same physical quantities in the British system follow no such systematic pattern. Table B-4 summarizes the most common conversion factors within the British system. Other relationships can be derived from these as needed.

TABLE B-4
Common Conversion Factors within the British System

Length	1 mile (mi) = 5280 ft	1 mi = 1760 yards (yd)
	1 yd = 3 ft	1 ft = 12 inches (in.)
Weight/Force	1 ton = 2000 lb	1 lb = 16 ounces (oz)
Power	1 hp = 550 ft·lb/s	
Area	$1 \text{ yd}^2 = 9 \text{ ft}^2$	$1 \text{ ft}^2 = 144 \text{ in.}^2$
	$1 \text{ acre} = 43{,}560 \text{ ft}^2$	$1 \text{ mi}^2 = 640 \text{ acres}$
Volume/Capacity	$1 \text{ yd}^3 = 27 \text{ ft}^3$	$1 \text{ ft}^3 = 1728 \text{ in.}^3$
	1 (U.S.) gal = 4 quarts (qt)	
	1 qt = 2 pints (pt)	1 pt = 16 oz
	$1 \text{ (U.S.) gal} \approx 231 \text{ in.}^3$	$1 \text{ Imperial gal} \approx 277.4 \text{ in.}^3$
	$1 \text{ ft}^3 \approx 7.481 \text{ gal}$	
Pressure	$1 \text{ atmosphere (atm)} \approx 14.70 \text{ lb/in.}^2$	
Speed	1 knot ≈ 1.151 mi/h	
Energy	1 Btu ≈ 777.9 ft·lb	

NOTE Conversions written with "=" are considered exact, while those written with "≈" are approximate. ◀

Conversion factors: Both the British and metric systems are currently in use in the United States.
British–SI   Because of this, and because engineering units are now legally defined in terms of SI units, it is important that you be able to convert measurements from one system to the other. Table B-5 lists the basic conversion factors with which all other common SI–British unit conversions can be performed. All conversion factors are approximate except the first one, which is the legal definition of the inch.

TABLE B-5
SI–British Conversions

Length	1 in. = 2.54 cm
Weight/Force*	1 kg weighs 2.205 lb
Mass	1 slug = 14.59 kg
Force/Weight	1 lb = 4.448 N
Capacity	1 L = 1.057 qt
Area	1 hectare = 2.471 acres
Pressure	1 lb/in.^{2} = 6895 Pa
Work energy	1 ft·lb = 1.356 J
Heat energy	1 Btu = 1055 J
Power	1 hp = 745.7 W
Illumination	1 fc = 10.76 lx

*The *weight* of an object on the earth is a measure of the earth's gravitational pull on it, measured in force units (pounds or newtons). The *mass* of an object is a measure of its resistance to change of state of motion measured in mass units (kilograms or slugs). A mass of 1 kg weighs approximately 2.2 lb.

Time and angle measure   Units of angle measure and time are the same in both the metric and British systems. The following relationships are useful in technological work:

Time: 1 h = 60 min   1 min = 60 s

Angle measure: 1 rev = 2π rad   1 rad ≈ 57.3 degrees (°)

         1° = 60 minutes (')   1' = 60 seconds (")

Conversion of units   To convert from one unit of measurement to another, we use the unity fraction method. A **unity fraction** is simply a fraction equivalent to 1. Follow these steps:

Step 1. Find a conversion equation relating the two quantities and use it to form two unity fractions.

Step 2. Choose the unity fraction that eliminates the unwanted units and leaves the desired units.

Step 3. Multiply and round as needed.

EXAMPLE 1 Convert 6.0 lb to metric units:

6.0 lb = _____ N

Solution

Step 1. From Table B-5, 1 lb ≈ 4.448 N. The two possible unity fractions are

$$\frac{1 \text{ lb}}{4.448 \text{ N}} \quad \text{and} \quad \frac{4.448 \text{ N}}{1 \text{ lb}}$$

Step 2. Multiplying 6.0 lb by the second fraction eliminates the pound units so that our answer is expressed in newtons:

$$6.0 \text{ lb} = 6.0 \cancel{\text{ lb}} \left(\frac{4.448 \text{ N}}{1 \cancel{\text{ lb}}} \right) = \underline{\quad\quad} \text{ N}$$

Step 3. Multiplying and rounding from Step 2, we have the following:
6.0 lb ≈ 26.688 N ≈ 27 N

To convert within the metric system, we must use the appropriate factor of 10 from Table B-3.

EXAMPLE 2   Convert 0.15 L to milliliters.

Solution

Step 1.   From Table B-3, we have

1 mL = 10^{-3} L

The two possible unity fractions are

$$\frac{1 \text{ mL}}{10^{-3} \text{ L}} \quad \text{and} \quad \frac{10^{-3} \text{ L}}{1 \text{ mL}}$$

Step 2.   Multiply by the first unity fraction to eliminate the liter units and leave the milliliter units:

$$0.15 \text{ L} = 0.15 \cancel{\text{ L}} \left(\frac{1 \text{ mL}}{10^{-3} \cancel{\text{ L}}} \right) = \underline{\quad\quad} \text{ mL}$$

Step 3.   Multiply: 0.15 L = 150 mL

At times, it is necessary to multiply by more than one unity fraction to complete the conversion.

EXAMPLE 3   Convert 0.275 km to feet.

Solution

Step 1.   The only SI–British length conversion listed in Table B-5 is

1 in. = 2.54 cm

Therefore, we also need 1 km = 10^3 m, 1 cm = 10^{-2} m, and 1 ft = 12 in.

Step 2. Using the appropriate unity fractions, we have

$$0.275 \text{ km} = 0.275 \text{ km}\left(\frac{10^3 \text{ m}}{1 \text{ km}}\right)\left(\frac{1 \text{ cm}}{10^{-2} \text{ m}}\right)\left(\frac{1 \text{ in.}}{2.54 \text{ cm}}\right)\left(\frac{1 \text{ ft}}{12 \text{ in.}}\right) = \underline{\quad} \text{ ft}$$

Step 3. Performing the indicated arithmetic, we have

$$.275 \boxed{\times} 3 \boxed{10^x} \boxed{\div} 2 \boxed{+/-}\boxed{10^x} \boxed{\div} 2.54 \boxed{\div} 12 \boxed{=} \rightarrow \quad \boxed{902.23097}$$

Therefore, 0.275 km $\approx$ 902 ft.

Conversion of compound units is done in a similar manner.

EXAMPLE 4   Convert 6.42 rad/s to revolutions per minute.

Solution
Step 1. The necessary relationships are

$$1 \text{ rev} = 2\pi \text{ rad} \qquad \text{and} \qquad 1 \text{ min} = 60 \text{ s}$$

Step 2. $6.42 \text{ rad/s} = 6.42 \dfrac{\text{rad}}{\text{s}}\left(\dfrac{1 \text{ rev}}{2\pi \text{ rad}}\right)\left(\dfrac{60 \text{ s}}{1 \text{ min}}\right) = \underline{\quad} \dfrac{\text{rev}}{\text{min}}$

Step 3. $6.42 \boxed{\div} 2 \boxed{\div} \boxed{\pi} \boxed{\times} 60 \boxed{=} \rightarrow \quad \boxed{61.306484}$

$$6.42 \text{ rad/s} \approx 61.3 \text{ rev/min}$$

When dealing with measurements of area or volume, be certain to raise the unity fraction to the appropriate power.

EXAMPLE 5   Convert 51.5 m^3 to cubic feet.

Solution
Step 1. We need the following conversion equations: 1 cm = 10^{-2} m, 1 in. = 2.54 cm, and 1 ft^3 = 1728 in.^{3}.

Step 2. To achieve the desired conversion, the following calculation must be performed:

$$51.5 \text{ m}^3 = 51.5 \text{ m}^3\left(\frac{1 \text{ cm}}{10^{-2} \text{ m}}\right)^3\left(\frac{1 \text{ in.}}{2.54 \text{ cm}}\right)^3\left(\frac{1 \text{ ft}^3}{1728 \text{ in.}^3}\right)$$

$$= 51.5 \text{ m}^3\left(\frac{1 \text{ cm}^3}{10^{-6} \text{ m}^3}\right)\left(\frac{1 \text{ in.}^3}{2.54^3 \text{ cm}^3}\right)\left(\frac{1 \text{ ft}^3}{1728 \text{ in.}^3}\right) = \underline{\quad} \text{ ft}^3$$

Notice that to obtain volume units we had to cube the two unity fractions representing length. The third relationship, 1 ft^3 = 1728 in.^{3}, was already given in units of volume.

Step 3. $51.5 \boxed{\div} 6 \boxed{+/-}\boxed{10^x} \boxed{\div} 2.54 \boxed{y^x} 3 \boxed{\div} 1728 \boxed{=} \rightarrow \quad \boxed{1818.7053}$

$$51.5 \text{ m}^3 \approx 1820 \text{ ft}^3$$

Temperature conversion   Although the degree Kelvin is the SI base unit of temperature, the degree Celsius (°C) is a related metric unit of temperature that is widely used. The relationships among the Kelvin, Celsius, and British Fahrenheit scales are given by the following formulas:

$$K = C + 273.16 \qquad F = \frac{9}{5}C + 32$$

$$C = K - 273.16 \qquad C = \frac{5}{9}(F - 32)$$

EXAMPLE 6   Convert: (a) 27°C to K and °F   (b) 12°F to °C and K

Solution
(a) $K = C + 273.16 = 27 + 273.16 \approx 300 \text{ K}$

$$F = \frac{9}{5}C + 32 = \frac{9}{5}(27) + 32 \approx 81°F$$

(b) $C = \frac{5}{9}(F - 32) = \frac{5}{9}(12 - 32) = \frac{5}{9}(-20) \approx -11°C$

$$K = C + 273.16 = -11 + 273.16 \approx 262 \text{ K}$$

Exercises B-1

Perform the following conversions.

1. 4.5 kg to grams
2. 380 cm to meters
3. 0.120 mi to yards
4. 14 gal to cubic feet
5. 430,000 Hz to kilohertz
6. 0.025 MW to watts
7. 2.3 atm to pounds per square inch
8. 12,000 ft to miles
9. 1450 in.^{3} to gallons
10. 0.60 acre to square feet
11. 0.125 L to milliliters
12. 217° to radians
13. 9200 lb to tons
14. 0.75 km to meters
15. 2,450,000 ft·lb to British thermal units
16. 6.5 knots to miles per hour
17. 28.5 cm to inches
18. 48 kg to slugs
19. 12.5 L to quarts
20. 8500 J to British thermal units
21. 5800 W to horsepower
22. 88 J to foot-pounds
23. 14.5 fc to lux
24. 680.0 lb to newtons
25. 48 mi/h to feet per second
26. 2.7 yd^3 to cubic feet
27. 4.5 m/s to miles per hour
28. 2450 cm^2 to square meters
29. 18 m^3 to cubic yards
30. 1.7 ft/s^2 to meters per second per second
31. 3.25 ft^2 to square meters
32. 0.42 lb/ft^2 to pounds per square inch

33. The density of methyl alcohol is 792 kg/m^3. Convert this to grams per cubic centimeter.
34. A steel pin has a mass of 97.4 g. Find its weight in ounces.
35. A hot water tank has a capacity of 50.0 gal. What is the equivalent volume in liters?
36. An airplane is flying at 610 mi/h. Change this to kilometers per hour.
37. A concrete foundation of 125 ft^3 must be poured. How many cubic yards of concrete are needed?

38. An automobile accelerates at 6.1 mi/h/s. Convert this to feet per second per second.
39. The Russian meteorite fall of 1947 had a mass of 7.0×10^7 g. Express this in slugs.
40. A solar device can generate 0.015 MWh per day. How many horsepower-hours is this?
41. The melting point of tin is 505 K. Convert this to °C.
42. The boiling point of mercury is 357°C. Express this in °F.

B-2 | Approximate Numbers, Significant Digits, and Rounding

Exact and approximate numbers

An electronics technician receives a shipment of 25 resistors rated at 8 Ω each. How reliable are these numbers? Does the technician interpret them as being exact or approximate?

In science and technology denominate numbers are a major form of communication, and it is important that all engineers and scientists write and interpret these numbers in the same way. Because special rules apply to the use of approximate numbers, we must be able to distinguish them from exact numbers. In general, any number obtained by counting or given by definition is considered an **exact number,** while any number resulting from the use of a measuring device is considered an **approximate number.** The reliability of the latter is limited by the precision of the instrument and the skill of the person using it.

EXAMPLE 7

(a) The electronics technician mentioned above can interpret 25 as being an exact number (obtained by counting the number of resistors) and the number 8 as being an approximate number (obtained by measuring the resistance with an ohmmeter).

(b) Given the length of a steel rod as 14.6 in., we would consider this number to be approximate because it can be obtained only by means of a measuring device. However, if we want to convert this length to feet using the fact that there are 12 in. in a foot, we would consider the number 12 to be exact because it is given by definition.

Significant digits

In working with approximate numbers, we often need to characterize them in terms of the number of significant digits they contain. A **significant digit** is any digit in a measurement that denotes an actual physical amount. The rules for determining whether a digit is significant are listed below:

1. All nonzero digits are significant. For example, the number 3.75 contains three significant digits.
2. All 0's in between significant digits are significant. Therefore, the number 3.705 contains four significant digits.

3. Left-end 0's are *never* significant. The number 0.00 53 contains only two significant digits, the 5 and the 3. The leftmost three 0's are used only as place-holders to locate the decimal point.

4. Final (right-end) 0's are significant only when they lie to the right of the decimal point. For example, the numbers 2.70 0. 500 and 0.000 630 each have three significant digits.

5. Final (right-end) 0's are not significant on integers. Therefore, the numbers 20, 300, and 4000 each contain only one significant digit. To indicate that any of the final 0's of an integer are significant, write the number with an overbar or in scientific notation. For example, writing the number 4000 as $40\overline{0}0$ or as 4.00×10^3 means that it has three significant digits.

EXAMPLE 8   Determine the number of significant digits in each of the following approximate numbers:

(a) 2070   (b) 46.334   (c) 0.009   (d) 50.0

(e) 500   (f) 4.000   (g) 8.020   (h) 5.0×10^4

Solution (a) 3   (b) 5   (c) 1   (d) 3   (e) 1   (f) 4
(g) 4   (h) 2

Accuracy and precision   The **accuracy** of a number is defined as the number of significant digits it contains. The **precision** of a number is the decimal position of the rightmost significant digit.

EXAMPLE 9

(a) The number 45.3 is **accurate** to three significant digits, but it is **precise** to only one decimal place (tenths).

(b) The number 0.008 is **accurate** to only one significant digit, but it is **precise** to three decimal places (thousandths).

Rounding   When we write an approximate number it is understood that the rightmost significant digit contains a degree of uncertainty. For example, if a length is given as 5.0 m, the true length lies somewhere between 4.95 and 5.05 m and has been estimated or rounded to the nearest 0.1 m.

   In working with approximate numbers we will often need to round these numbers to a certain degree of accuracy or precision. To round a number, use the following procedure:

1. Locate the least accurate or precise digit required (referred to hereafter as L) and the digit to its immediate right (referred to hereafter as R).

2. If R is less than 5, leave L as is. For all digits to the right of L, discard any that follow the decimal point and change to 0 any that precede the decimal point.

3. If R is greater than or equal to 5, increase L by 1 and treat all numbers to the right of L as in Step 2.

EXAMPLE 10   Round each of the following as indicated:
(a)   87.266 to three significant digits
(b)   6,723,000 to nearest hundred thousand
(c)   1.536 to nearest tenth
(d)   0.05040 to two significant digits
(e)   4.895 to three significant digits
(f)   0.008654 to four decimal places
(g)   269,438,736.8 to five significant digits

Solution
(a)   **First,** identify L and R:

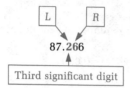

Second, because $R \geq 5$, increase L by 1. Discard all other digits since they lie to the right of the decimal point:

87.3, rounded

(b)   **First,** identify L and R:

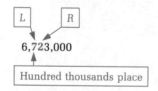

Second, because $R < 5$, leave L as is. Change all other digits to 0 since they precede the decimal point:

6,700,000, rounded

(c)   1.5   (d)   0.050   (e)   4.90   (f)   0.0087   (g)   269,440,000

Exercises B-2

Determine whether each of the following numbers is exact (E) or approximate (A):

1.   The voltage in a circuit is 108 V.
2.   The temperature at 8 AM was 46°F.
3.   There are 2.54 cm in an inch.
4.   There are 6 solar panels on the roof of a house.
5.   Two beams are joined at an angle of 115°.
6.   An aluminum casting weighs 0.74 kg.
7.   A quart contains 32 oz.
8.   The velocity of a projectile is 8.5 ft/s.

9. There are 1000 g in a kilogram.
10. An hour consists of 3600 s.
11. A furnace is rated at 150,000 Btu.
12. A certain automobile engine contains 6 cylinders.

State (a) the accuracy and (b) the precision of each of the following numbers:

13. 4.68   14. 2500   15. 0.0008   16. 38.9   17. 456   18. 0.257
19. 17.00   20. 0.0807

Round each of the following to the number of significant digits indicated in parentheses:

21. 568 (2)   22. 0.0405 (2)   23. 7.941 (3)   24. 15,743 (2)
25. 16.438 (3)   26. 0.8572 (1)   27. 3.195 (3)   28. 1,659,923 (4)

Round each of the following to the degree of precision indicated in parentheses:

29. 6.472 (tenths)   30. 268.27 (units)   31. 0.082 (two decimal places)
32. 15.5625 (nearest 0.001) 33. 1,572,600 (thousands)   34. 0.26854 (thousandths)

B-3 | Calculations with Approximate Numbers

Suppose a runner completes a 26.3 mi race in 3.4 h. To determine the runner's average speed we enter 26.3 $\div$ 3.4 on a calculator, and it quickly displays the answer *7.7352941* . However, if we express the result as 7.7352941 mi/h, we are implying that we know the runner's average speed to eight significant digits, even though his distance and time were accurate only to three and two significant digits, respectively. The precision or accuracy of the result of any calculation with approximate numbers must be related to the original numbers—not to the length of the calculator display.

For problems involving approximate numbers in this text, the rounding conventions described below are followed.

Addition and subtraction

For any calculation involving the addition or subtraction of approximate numbers, the result should be rounded to the same degree of **precision** as the *least* precise number involved in the calculation.

EXAMPLE 11 Calculate: 2.41 in. − 0.085 in. + 56.9 in.

Solution Combining the numbers as given, we obtain 59.225 in. However, the least precise of the original measurements, 56.9 in., is given only to the nearest tenth, and thus our result must also be rounded to the nearest tenth. Correctly stated, the answer is 59.2 in.

Multiplication and division

For any calculation involving multiplication or division of approximate numbers, the result should be rounded to the same degree of **accuracy** as the *least* accurate number involved in the calculation.

■ **EXAMPLE 12** Calculate: (a) (3.748 m)(0.0820 m)   (b) $\dfrac{6.4 \text{ ft}}{2.719 \text{ s}}$

Solution

(a) 3.748 ⊗ .082 ⊜ →   *0.307336*

To round properly, notice that the least accurate of the original factors, 0.0820 m, contains three significant digits. Therefore, the product must also contain three significant digits, and the correct answer is 0.307 m².

(b) 6.4 ⊘ 2.719 ⊜ →   *2.3538065*

The least accurate of the original numbers, 6.4 ft, contains two significant digits. The correctly rounded quotient is 2.4 ft/s.

Powers and roots   Any power or root of an approximate number should be rounded to the same degree of **accuracy** as the number itself.

■ **EXAMPLE 13** Calculate: (a) (2.7 cm)³   (b) $\sqrt{5.62 \text{ yd}^2}$

Solution

(a) 2.7 ⊡ 3 ⊜ →   *19.683*

Because 2.7 cm contains two significant digits, we round the result to 20 cm³.

(b) 5.62 ⊡ →   *2.3706539*

Since 5.62 yd² contains three significant digits, its square root is 2.37 yd, properly rounded.

CAUTION   Notice that for addition and subtraction we round according to the **precision** of the numbers involved, whereas for multiplication, division, powers, and roots, we round according to the **accuracy** of the numbers. ◀

Trigonometric calculations   In applications involving trigonometric functions, the measures of both sides and angles are considered approximate numbers. The result of any trigonometric calculation must be rounded to reflect the accuracy of side lengths and the precision of angle measurements. The correlation between side length and angle measure is given in Table B-6.

TABLE B-6

A Side Length Measured to:	Corresponds to an Angle Measured to the Nearest:
1 or 2 significant digits	1°
3 significant digits	0.1° or 10′
4 significant digits	0.01° or 1′

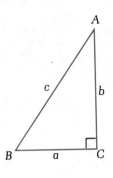

A

c

b

B   a   C

EXAMPLE 14 Find the indicated quantity. (Refer to the figure in the margin.)
(a) $a = 6.4$ in., $B = 27.5°$, $c = ?$
(b) $a = 26.5$ m, $b = 34.25$ m, $B = ?$

Solution
(a) Using the techniques explained in Chapter 4, we have

$$\frac{6.4}{c} = \cos 27.5°$$

$$c = \frac{6.4}{\cos 27.5°}$$

6.4 $\boxed{÷}$ 27.5 $\boxed{\cos}$ $\boxed{=}$ → **7.2152445**

To round properly, notice that the angle measure of 27.5° corresponds to a side length of three significant digits and that the given side length of 6.4 in. is accurate to two significant digits. Because we are dividing the two quantities, our answer must be expressed according to the accuracy of the least accurate number. Rounding to two significant digits, $c ≈ 7.2$ in.

(b) $\tan B = \dfrac{34.25}{26.5}$ 34.25 $\boxed{÷}$ 26.5 $\boxed{=}$ $\boxed{\text{INV}}$ $\boxed{\tan}$ → **52.270069**

The least accurate side, 26.5 m, contains three significant digits. This corresponds to an angle measure precise to the nearest 0.1°. The correctly rounded answer is 52.3°. Notice that in degrees and minutes, we round to the nearest 10′ to obtain 52°20′.

Premature rounding   When a calculation involves several steps, rounding should be avoided until the final step. This can usually be accomplished using a calculator with a memory. However, if an intermediate result must be recorded, always retain *at least* one more digit than the required precision or accuracy of the answer. Failure to do so may cause your final answer to be incorrect.

Exact numbers   The rules for rounding as outlined in this section apply only to approximate numbers. If both exact and approximate numbers are involved in a calculation, only the approximate numbers should be used to determine the precision or accuracy of the answer.

Exceptions   Most of the exercise sets in this text contain a number of drill problems in which the numbers given are not associated with units of measure and can therefore be considered exact. If no specific rounding instructions are given for such problems, round only if the answers are nonterminating decimals.

Finally, in certain applied problems the denominate numbers may contain more accuracy than implied by the number of digits displayed and may even be considered exact. Rounding procedures for such cases will be clarified either by a rounding instruction or by the precedent set in an example.

Exercises B-3

Perform the indicated operations. Assume all given numbers are approximate.

1. $6.7 + 0.074$

2. $54.83 - 7.992$

3. $13.3 - 8$

4. $0.874 + 0.04932$

5. $(11.4)(3.2)$

6. $(0.915)(43.63)$

7. $\dfrac{43.8}{7.2}$

8. $\dfrac{0.6784}{83}$

9. $\dfrac{(4.93)(275.9)}{0.56}$

10. $24.67 + 181.5 - 7.394$

11. $86.2 + (19.41)(7.2)$

12. $3.7(14.608 - 91)$

13. $\dfrac{24.62}{4.47} - 3.7$

14. 7.9^2

15. 13.0^3

16. 280^4

17. 0.8904^2

18. $\sqrt{3.4}$

19. $\sqrt{12.4}$

20. $\sqrt[3]{9180}$

21. $\sqrt[4]{0.0409}$

Use the given parts of the triangle shown here to calculate the indicated missing parts. Express all angles in both decimal degrees and, where appropriate, degrees and minutes.

22. $a = 96.25$ cm, $b = 37.5$ cm, $A = ?$

23. $B = 34.8°$, $c = 3.422$ in., $a = ?$

24. $a = 0.8$ ft, $c = 0.95$ ft, $B = ?$

25. $A = 23°40'$, $c = 76$ m, $b = ?$

26. $B = 67°22'$, $b = 3.9877$ in., $c = ?$

27. $a = 254.5$ km, $b = 176.7$ km, $B = ?$

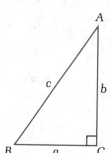

Solve.

28. A pipe has an outside radius of 26.8 in. and an inside radius of 22.75 in. What is the thickness of the wall of the pipe?

29. Three resistors are rated at 12 Ω, 9 Ω, and 4.5 Ω, respectively. When the resistors are connected in series, the total resistance is given by their sum. Find this sum.

30. The side length of a square piece of sheet steel is measured to be 14.8 cm. Find the area of the piece.

31. The illumination E (in lux) on a surface is found by dividing the luminous flux F (in lumens) of a light source by the area A (in square meters) of the surface ($E = F/A$). Find E when $F = 155$ lm and $A = 28$ m^2.

32. A manufacturer has 20 cube-shaped boxes with edge lengths of 7.5 in. each. How much total storage space do the boxes contain?

33. The power (in watts) in a certain dc circuit is given by the product of the resistance R and the square of the current I. Find the power in a circuit where $R = 285$ Ω and $I = 0.65$ A.

34. The period (in seconds) of a simple pendulum 1.75 ft long is determined by the equation

$$P = 2\pi\sqrt{\dfrac{1.75}{32.2}}$$

where the number 3.14159 can be used for π. Find the period.

Appendix C: Basic Concepts of Geometry

Many geometric figures, facts, and formulas are important for the study of technical math. This appendix provides a brief review of basic applied geometry. In this discussion, the meaning of the terms *point*, *line*, and *plane* are accepted without definition.

Lines   A **straight line** is assumed to extend infinitely in both directions (Fig. C-1). Two lines in a plane are said to be **intersecting lines** if they share a single common point (Fig. C-2).

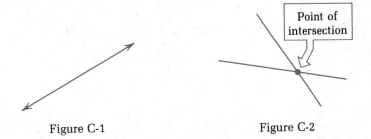

Figure C-1   Figure C-2

They are said to be **parallel lines** if they have no points in common (Fig. C-3) and **coincident lines** if they share all points in common (Fig. C-4).

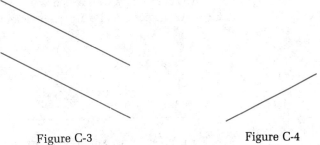

Figure C-3   Figure C-4

A **half-line** is the portion of a line extending infinitely to one side of a given endpoint (Fig. C-5). If the endpoint A itself is included, the figure is called a **ray**.

A **line segment** is the finite portion of a line bounded by and including two endpoints. Figure C-6 shows line segment AB.

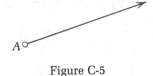

Figure C-5

Figure C-6

Angles   An **angle** is generated by rotating a half-line about its endpoint (Fig. C-7). Thus, an angle can also be defined as the union of two rays sharing a common endpoint. The endpoint O is called the **vertex** of the angle. The **magnitude** of an angle is measured in either degrees or radians. Angle measurement is discussed in Chapters 4 and 8.

Figure C-7

The following angles and classes of angles are often used in technical fields:

- An **acute angle** measures between 0° and 90° (Fig. C-8).
- An **obtuse angle** measures between 90° and 180° (Fig. C-9).
- A **straight angle** measures exactly 180° (Fig. C-10).
- A **right angle** measures exactly 90° (Fig. C-11).

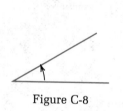

Figure C-8

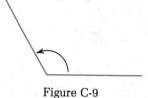

Figure C-9

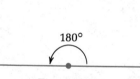

180°

Figure C-10

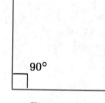

90°

Figure C-11

The sides of a right angle are said to be **perpendicular,** and the square symbol shown in Fig. C-11 is often used to denote perpendicular lines and right angles.

Certain pairs of angles have important relationships, as described below.

Two acute angles whose sum is 90° are said to be **complementary angles.** In Fig. C-12, angles a and b are complements of each other.

Figure C-12

40°

50°

Two angles whose sum is 180° are said to be **supplementary angles.** In Fig. C-13, angles c and d are supplements of each other.

Figure C-13

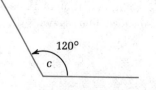

120°

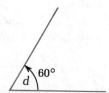

60°

Two angles that share a common vertex and one common side, and have no interior points in common, are called **adjacent angles.** In Fig. C-14, angles e and f are adjacent angles.

Figure C-14

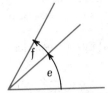

The two nonadjacent angles formed by two intersecting lines are known as **vertical angles.** Angles g and i in Fig. C-15 are vertical angles, as are angles j and h. Vertical angles are always equal in size: $\angle g = \angle i$ and $\angle j = \angle h$.

Figure C-15

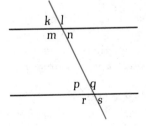

When two parallel lines are intersected by a third line (called a **transversal**), several pairs of equal angles are formed. In Fig. C-16 the two pairs of **alternate interior angles** are equal to each other. Thus, $\angle m = \angle q$ and $\angle n = \angle p$. Simi-

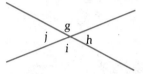

Figure C-16

larly, the two pairs of **alternate exterior angles** are equal. Therefore, $\angle k = \angle s$ and $\angle l = \angle r$. Finally, all four pairs of **corresponding angles** are equal. Thus, $\angle k = \angle p$, $\angle m = \angle r$, $\angle l = \angle q$, and $\angle n = \angle s$.

Polygon   A **polygon** (Fig. C-17) is a closed plane figure bounded by three or more line segments called **sides.** The **angles** of a polygon are formed by the intersection

Figure C-17

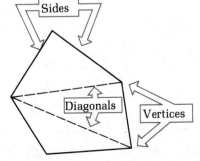

of its sides, and the points of intersection are called the **vertices** of the polygon. The line segments joining pairs of nonconsecutive vertices are the **diagonals** of the polygon.

A polygon is **equilateral** if all its sides are of equal length. It is **equiangular** if all its angles are equal. It is said to be **regular** if it is both equilateral and equiangular.

Polygons of particular importance in technical math are the triangle and the quadrilateral.

Triangle   A **triangle** is a polygon composed of three sides and three angles (Fig. C-18). As shown in Fig. C-19, the three angles of a triangle always sum to 180°.

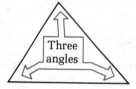

Figure C-18

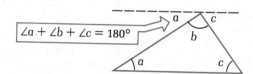

$$\angle a + \angle b + \angle c = 180°$$

Figure C-19

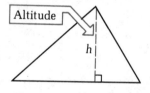

Figure C-20

An **altitude** of a triangle is the perpendicular line segment drawn from any vertex to the opposite side (Fig. C-20). Every triangle has three altitudes. The length of the altitude is often referred to as the **height**, h.

Triangles are classified according to either their sides or their angles.

Classification According to Sides:
- **Equilateral triangle:** All three sides are equal. This also means that all three angles are equal (Fig. C-21).
- **Isosceles triangle:** Two sides are equal. This also means that the two angles at the base of these sides are equal (Fig. C-22).
- **Scalene triangle:** No sides are equal (Fig. C-23).

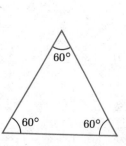

Figure C-21

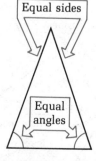

Figure C-22

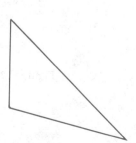

Figure C-23

Classification According to Angles:
- **Acute triangle:**   All three angles are acute (Fig. C-24).
- **Obtuse triangle:**   One of the angles is obtuse (Fig. C-25).
- **Right triangle:**   One of the angles is a right angle (Fig. C-26).
- **Oblique triangle:**   Any triangle without a right angle is said to be oblique.

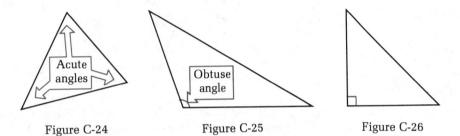

Figure C-24   Figure C-25   Figure C-26

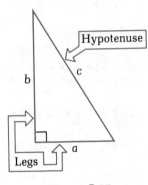

Figure C-27

Right triangles are particularly significant because they are the basis of trigonometry. The two perpendicular sides of a right triangle are called the **legs,** and the side opposite the right angle is called the **hypotenuse** (Fig. C-27). An important relationship involving right triangles is the **Pythagorean theorem:**

For any right triangle, the square of the hypotenuse is equal to the sum of the squares of the legs.

Referring to Fig. C-27, we have

$$c^2 = a^2 + b^2$$

Solving for c,

$$c = \sqrt{a^2 + b^2}$$

Similarly, solving for a or b,

$$a = \sqrt{c^2 - b^2}$$
$$b = \sqrt{c^2 - a^2}$$

See Section 2-5 for more on the Pythagorean theorem.

Certain pairs of triangles have special relationships.

- Two triangles in which all three pairs of corresponding sides and all three pairs of corresponding angles are equal are called **congruent triangles.** Congruent triangles have the same size and shape (Fig. C-28).

Figure C-28

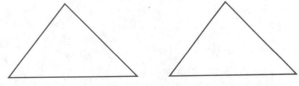

- Two triangles in which all three pairs of corresponding angles are equal are called **similar triangles** (Fig. C-29). Similar triangles have the same shape, and all pairs of corresponding sides are proportional in length. See Section 2-7 for more on similar triangles.

Figure C-29

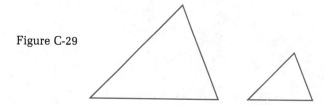

Quadrilaterals   A **quadrilateral** is any four-sided polygon. The following quadrilaterals are especially useful in various technical fields:

- **Parallelogram:**   Opposite sides of a parallelogram are parallel and equal in length (Fig. C-30).
- **Rhombus:**   A rhombus is a parallelogram in which all four sides are equal in length (Fig. C-31).
- **Rectangle:**   A rectangle is a parallelogram in which all pairs of adjacent sides are perpendicular (Fig. C-32). All four angles in a rectangle are right angles.
- **Square:**   A square is a rectangle in which all four sides are equal in length (Fig. C-33).
- **Trapezoid:**   A trapezoid has one pair of parallel sides and one pair of nonparallel sides (Fig. C-34).

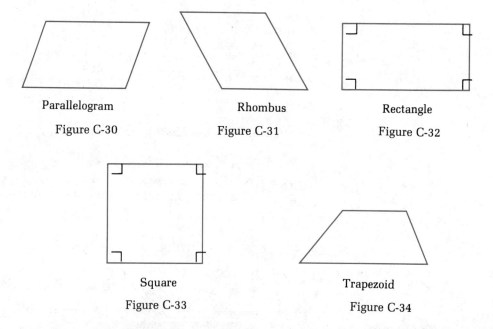

Parallelogram

Figure C-30

Rhombus

Figure C-31

Rectangle

Figure C-32

Square

Figure C-33

Trapezoid

Figure C-34

Circles   A **circle** is a closed plane figure in which each point is equidistant from a fixed point called its **center** (Fig. C-35).

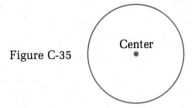

Figure C-35

The **radius** r is a line segment joining the center to any point on the circle (Fig. C-36). The **diameter** d is a line segment passing through the center of the circle and having its endpoints on the circle. As shown in Fig. C-36, the diameter d is twice the length of the radius r.

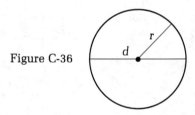

Figure C-36

Other terms of importance relating to circles are:

- A **chord** is a line segment joining any two points on a circle (Fig. C-37). The diameter is a chord containing the center of the circle.
- A **secant line** is a line containing the chord of a circle (Fig. C-37).
- An **arc** is that portion of a circle between any two points on the circle (Fig. C-37).
- A **tangent** is a line that touches the circle in exactly one point (Fig. C-38).
- A **central angle** is any angle formed by two radii (Fig. C-38).
- A **sector** is a region bounded by two radii and an arc of the circle (Fig. C-39).
- **Concentric circles** are two or more circles that have the same center (Fig. C-40).

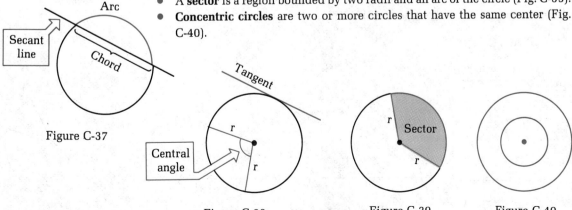

Figure C-37

Figure C-38

Figure C-39

Figure C-40

Two facts of importance in working with circles are:

- A line through the center of a circle and perpendicular to a chord bisects both the chord and the arc (Fig. C-41).
- A **radius** is perpendicular to the tangent at its endpoint (Fig. C-42).

Figure C-41

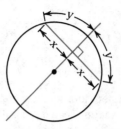

Figure C-42

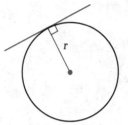

Circles are analyzed in more detail in Section 20-3.

Formulas for area and perimeter

The following list summarizes the formulas for finding the area and perimeter of the plane figures already described:

Figure	Perimeter, P	Area, A
Triangle (Fig. C-43)	$P = a + b + c$	$A = \frac{1}{2}bh$
Rectangle (Fig. C-44)	$P = 2l + 2w$	$A = lw$
Square (Fig. C-45)	$P = 4s$	$A = s^2$
Parallelogram (Fig. C-46)	$P = 2a + 2b$	$A = bh$
Trapezoid (Fig. C-47)	$P = a + b + c + d$	$A = \left(\dfrac{a + b}{2}\right)h$
Circle (Fig. C-48)	Circumference, $C = \pi d = 2\pi r$	$A = \pi r^2$

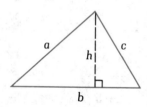

Figure C-43

Figure C-44

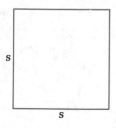

Figure C-45

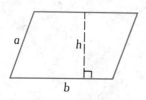

Figure C-46

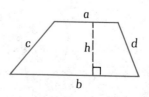

Figure C-47

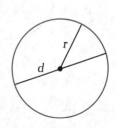

Figure C-48

Solid figures   In addition to these two-dimensional, or plane, figures, technical work also involves three-dimensional, or **solid, figures.** The following list gives the formulas for the surface area S and volume V of the most common solid figures. Other symbols not evident in the figures themselves are: B = area of the base; L = lateral surface area [area of all surfaces other than the base(s)]; P = perimeter of the base; C = circumference of the base.

Solid	Surface Area, S	Volume, V
General prism (Fig. C-49)	$S = Ph + 2B$	$V = Bh$
Rectangular prism (Fig. C-50)	$S = 2(lw + wh + lh)$	$V = lwh$
Cube (Fig. C-51)	$S = 6e^2$	$V = e^3$
General cylinder (Fig. C-52)	$S = Ch + 2B$	$V = Bh$
Right circular cylinder (Fig. C-53)	$S = 2\pi rh + 2\pi r^2$	$V = \pi r^2 h$
General pyramid or cone (Fig. C-54)	$S = L + B$	$V = \dfrac{1}{3}Bh$
Right circular cone (Fig. C-55)	$S = \pi rs + \pi r^2$	$V = \dfrac{1}{3}\pi r^2 h$
Sphere (Fig. C-56)	$S = 4\pi r^2$	$V = \dfrac{4}{3}\pi r^3$

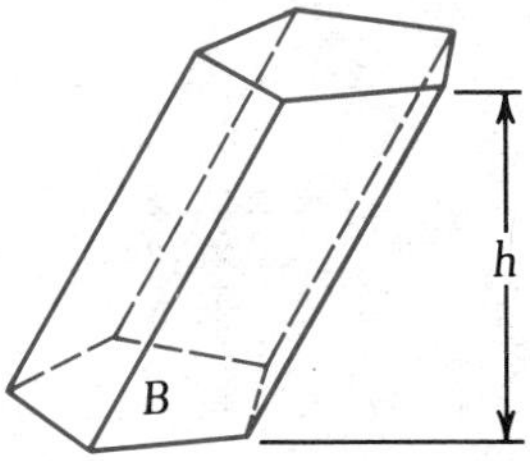

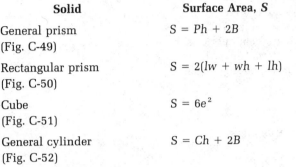

Figure C-49

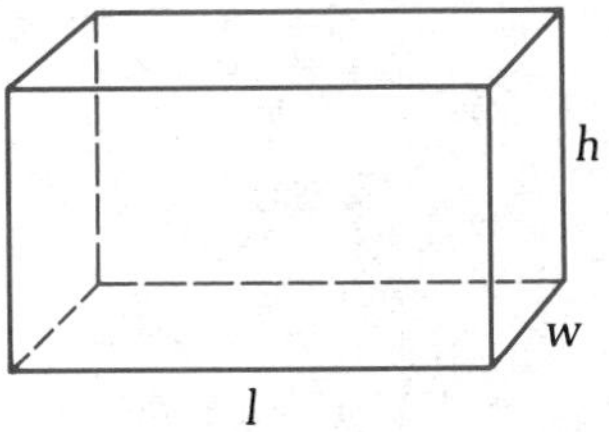

Figure C-50

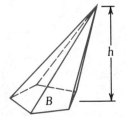

Figure C-51

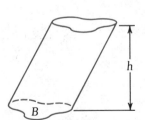

Figure C-52

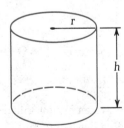

Figure C-53

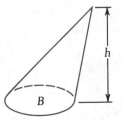

Figure C-54

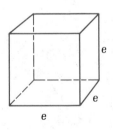

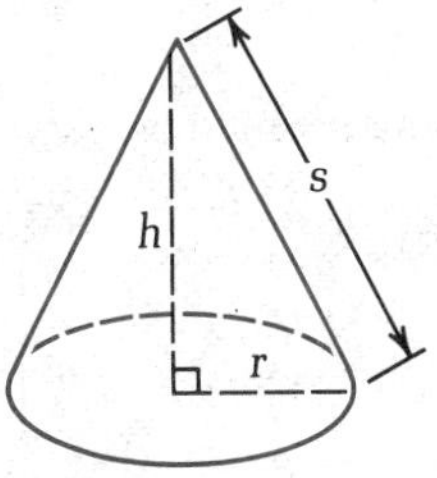

Figure C-55

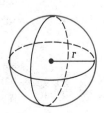

Figure C-56

Exercises

For Problems 1–12 refer to the figure shown below. If $\angle s = 65°$, and lines a and b are parallel, find the measure of each angle.

1. $\angle t$ 2. $\angle u$ 3. $\angle v$ 4. $\angle w$ 5. $\angle x$ 6. $\angle y$ 7. $\angle z$
8. Which of the angles are acute?
9. Which of the angles are obtuse?
10. Name both pairs of alternate interior angles.
11. Name both pairs of alternate exterior angles.
12. Name all pairs of corresponding angles.

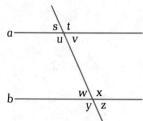

Find the complement of each angle.

13. 27°   14. 72°   15. 41.6°   16. 64°15′

Find the supplement of each angle.

17. 58°   18. 141°   19. 118°40′   20. 76.3°

Problems 21–23 refer to the triangle below.

21. Give two names for this triangle.
22. What is the size of $\angle A$?
23. What is the length of side BC? Of side AC?

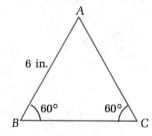

Problems 24–26 refer to the triangle below.

24. Give two names for this triangle.
25. What is the size of $\angle E$?
26. What is the size of $\angle F$?

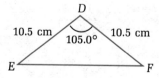

Problems 27 and 28 refer to the triangle below.

27. Give two names for this triangle.
28. What is the length of side GI?

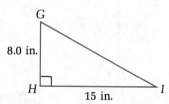

For Problems 29–32 find the indicated missing side of the right triangle shown in the figure below.

29. $a = 3.0$ in., $b = 5.0$ in., $c = ?$
30. $a = 26.5$ cm, $b = 19.3$ cm, $c = ?$
31. $a = 18$ m, $c = 61$ m, $b = ?$
32. $b = 0.250$ in., $c = 0.475$ in., $a = ?$

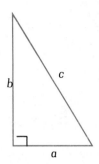

For each of the figures in Problems 33–40, (a) name the figure, (b) find its perimeter (or circumference), and (c) find its area.

33.

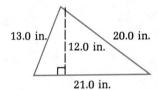

13.0 in.   12.0 in.   20.0 in.   21.0 in.

34.

13.70 cm   7.50 cm   8.50 cm

35.

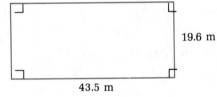

19.6 m   43.5 m

36.

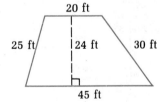

20 ft   25 ft   24 ft   30 ft   45 ft

37.

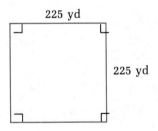

225 yd   225 yd

38.

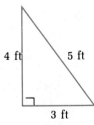

4 ft   5 ft   3 ft

39.

2.60 in.

40.

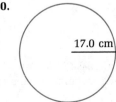

17.0 cm

For each of the figures in Problems 41–46, (a) name the figure, (b) find its total surface area, and (c) find its volume.

41.

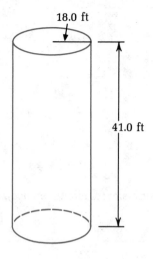

18.0 ft

41.0 ft

42.

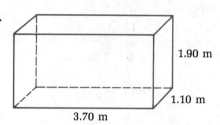

1.90 m

1.10 m

3.70 m

43.

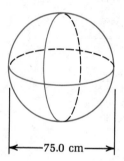

—75.0 cm—→

44.

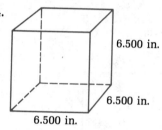

6.500 in.

6.500 in.

6.500 in.

45.

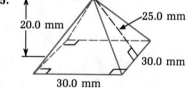

20.0 mm

25.0 mm

30.0 mm

30.0 mm

46.

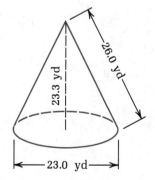

23.3 yd

26.0 yd

←— 23.0 yd —→

Find each of the following:

47. The area of a triangle with a base of 6.4 in. and a height of 3.5 in.

48. The area and perimeter of a rectangle with a length of 86.0 ft and a width of 55.0 ft.

49. The area and hypotenuse of a right triangle with legs measuring 21.0 m and 37.0 m.
50. The area of a parallelogram with a base of 4.25 cm and a height of 7.75 cm.
51. The area and circumference of a circle with a radius of 44.0 yd.
52. The area of a semicircle with a diameter of 13.0 m.
53. The area and perimeter of a square with a side length of 6.50 ft.
54. The volume and surface area of a rectangular prism with a length of 16.0 in., a width of 12.0 in., and a height of 10.0 in.
55. The volume and surface area of a sphere with a diameter of 11.50 cm.
56. The volume and surface area of a right circular cylinder with a diameter of 7.50 in. and a height of 23.5 in.
57. The surface area and volume of a cube with an edge length of 150.0 mm.
58. The volume and surface area of a cone with a radius of 6.0 yd, an altitude of 7.5 yd, and a slant height of 9.6 yd.
59. The volume of a hexagonal right pyramid with a height of 1.250 in. and whose base has an area of 0.785 in.
60. The volume of a pentagonal right prism with an altitude of 46.0 ft, and whose base has an area of 3650 ft.
61. The total surface area of a triangular prism with an altitude of 28.0 cm. The base is an equilateral triangle with sides of 14.0 cm and a height of 12.1 cm.
62. The surface area of a pyramid whose square base has a side length of 1.40 ft and whose four congruent triangular faces each have a height of 3.20 ft.

Solve.

63. A fog seal must be applied to a 24 ft wide roadway at the rate of 0.00055 gal/ft². How many gallons are needed for 4.7 mi of road?
64. What size diameter circular stock is needed to mill a square end 4.0 cm on a side? (See the figure.)

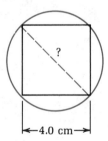

←4.0 cm→

65. The *Landsat* satellite circles the earth once every 100 min. Assuming that the earth moves in a circle of radius 6370 km, over what earth distance will the satellite fly in exactly 30 s?
66. A rectangular vent 12.0 in. wide must have the same total area as two smaller vents measuring 6.0 in. by 4.0 in. and 8.0 in. by 5.0 in. What must be the length of the vent?
67. A tank must be designed to hold 5750 gal of water. Given that there are 7.48 gal in a cubic foot:
(a) What is the diameter of a spherical tank that will hold this amount?
(b) What side length should a cube-shaped tank have to hold this amount?
68. A rectangular settling basin is 68.0 ft long, 25.0 ft wide, and 10.0 ft deep. How many gallons of water will it hold?
69. The *bore* of a cylinder in a piston engine is its diameter, the *stroke* is its height, and the *displacement* is its volume. What is the displacement of a cylinder with a bore of 58.0 mm and a stroke of 46.0 mm? Give the answer in cubic centimeters.

70. A manufacturer prepares an order of 50,000 conical paper cups with a base diameter of 3.20 in. and a slant height of 3.70 in. (see the figure). Adding 15% for overlap and waste, how many square *feet* of paper will be used?

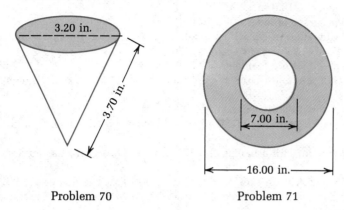

Problem 70   Problem 71

71. A ring is the area between two concentric circles. Find the area of the ring shown in the figure.

72. Find the shaded area in the figure.

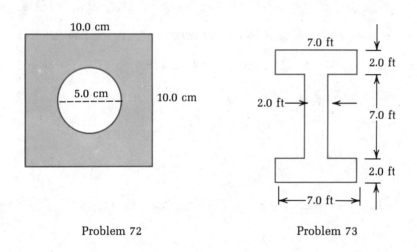

Problem 72   Problem 73

73. Find the cross-sectional area of the I-beam in the figure.

74. Which of properties 1–5 listed below are true for each of the following quadrilaterals?
   (a) Rectangle   (b) Parallelogram   (c) Square   (d) Trapezoid   (e) Rhombus
   1. All sides are equal.
   2. All angles are equal.
   3. Opposite sides are equal.
   4. Opposite angles are equal.
   5. Diagonals are equal.

Appendix D:
Table of Integrals

A. Basic Integrals

1. $\displaystyle\int u^n \, du = \frac{u^{n+1}}{n+1} + C \qquad n \neq -1$

2. $\displaystyle\int \frac{du}{u} = \ln|u| + C$

3. $\displaystyle\int e^u \, du = e^u + C$

4. $\displaystyle\int a^u \, du = \frac{a^u}{\ln a} + C \qquad a > 0$

5. $\displaystyle\int \sin u \, du = -\cos u + C$

6. $\displaystyle\int \cos u \, du = \sin u + C$

7. $\displaystyle\int \tan u \, du = -\ln|\cos u| + C$

8. $\displaystyle\int \cot u \, du = \ln|\sin u| + C$

9. $\displaystyle\int \sec u \, du = \ln|\sec u + \tan u| + C$

10. $\displaystyle\int \csc u \, du = \ln|\csc u - \cot u| + C$

11. $\displaystyle\int \sec^2 u \, du = \tan u + C$

12. $\displaystyle\int \csc^2 u \, du = -\cot u + C$

13. $\displaystyle\int \sec u \tan u \, du = \sec u + C$

14. $\displaystyle\int \csc u \cot u \, du = -\csc u + C$

15. $\displaystyle\int \frac{du}{\sqrt{a^2 - u^2}} = \text{Arcsin}\left(\frac{u}{a}\right) + C$

16. $\displaystyle\int \frac{du}{a^2 + u^2} = \frac{1}{a} \text{Arctan}\left(\frac{u}{a}\right) + C$

17. $\displaystyle\int \frac{du}{\sqrt{u^2 \pm a^2}} = \ln\left|u + \sqrt{u^2 \pm a^2}\right| + C$

18. $\displaystyle\int \frac{du}{u^2 - a^2} = \frac{1}{2a} \ln\left|\frac{u - a}{u + a}\right| + C$

B.   Integrals Involving $a + bu$

19. $\displaystyle \int (a + bu)^n \, du = \frac{(a + bu)^{n+1}}{b(n + 1)} + C \qquad n \neq -1$

20. $\displaystyle \int u(a + bu)^n \, du = \frac{(a + bu)^{n+1}}{b^2} \left(\frac{a + bu}{n + 2} - \frac{a}{n + 1} \right) + C \qquad n \neq -1, -2$

21. $\displaystyle \int \frac{du}{a + bu} = \frac{1}{b} \ln |a + bu| + C$

22. $\displaystyle \int \frac{du}{u(a + bu)} = -\frac{1}{a} \ln \left| \frac{a + bu}{u} \right| + C$

23. $\displaystyle \int \frac{du}{u^2(a + bu)} = -\frac{1}{au} + \frac{b}{a^2} \ln \left| \frac{a + bu}{u} \right| + C$

24. $\displaystyle \int \frac{u \, du}{a + bu} = \frac{1}{b^2}[a + bu - a \ln |a + bu|] + C$

25. $\displaystyle \int \frac{u^2 \, du}{a + bu} = \frac{1}{b^3}\left[\frac{1}{2}(a + bu)^2 - 2a(a + bu) + a^2 \ln |a + bu| \right] + C$

26. $\displaystyle \int \frac{du}{u(a + bu)^2} = \frac{1}{a(a + bu)} - \frac{1}{a^2} \ln \left| \frac{a + bu}{u} \right| + C$

27. $\displaystyle \int \frac{du}{u^2(a + bu)^2} = -\frac{a + 2bu}{a^2 u(a + bu)} + \frac{2b}{a^3} \ln \left| \frac{a + bu}{u} \right| + C$

28. $\displaystyle \int \frac{u \, du}{(a + bu)^2} = \frac{1}{b^2}\left[\ln |a + bu| + \frac{a}{a + bu} \right] + C$

C.   Integrals Involving $\sqrt{a + bu}$

29. $\displaystyle \int \sqrt{a + bu} \, du = \frac{2}{3b}(a + bu)^{3/2} + C$

30. $\displaystyle \int \left(\sqrt{a + bu} \right)^n du = \frac{2}{b} \frac{\left(\sqrt{a + bu} \right)^{n+2}}{n + 2} + C \qquad n \neq -2$

31. $\displaystyle \int u\sqrt{a + bu} \, du = \frac{2(3bu - 2a)}{15b^2}(a + bu)^{3/2} + C$

32. $\displaystyle \int u^2\sqrt{a + bu} \, du = \frac{2(15b^2 u^2 - 12abu + 8a^2)}{105b^3}(a + bu)^{3/2} + C$

33. $\displaystyle \int u^n\sqrt{a + bu} \, du = \frac{2u^n(a + bu)^{3/2}}{(2n + 3)b} - \frac{2an}{(2n + 3)b} \int u^{n-1}\sqrt{a + bu} \, du \qquad 2n \neq -3$

34. $\displaystyle \int \frac{\sqrt{a + bu}}{u} \, du = 2\sqrt{a + bu} + a \int \frac{du}{u\sqrt{a + bu}}$

35. $\displaystyle \int \frac{du}{\sqrt{a + bu}} = \frac{2}{b}\sqrt{a + bu} + C$

36. $\displaystyle \int \frac{u \, du}{\sqrt{a + bu}} = \frac{2(bu - 2a)}{3b^2}\sqrt{a + bu} + C$

37. $\displaystyle\int \frac{u^n \, du}{\sqrt{a + bu}} = \frac{2u^n\sqrt{a + bu}}{(2n + 1)b} - \frac{2an}{(2n + 1)b}\int \frac{u^{n-1}\, du}{\sqrt{a + bu}}$ $2n \neq -1$

38. $\displaystyle\int \frac{u \, du}{\sqrt{a + bu}} = \frac{2(bu - 2a)}{3b^2}\sqrt{a + bu} + C$

39. $\displaystyle\int \frac{du}{u\sqrt{a + bu}} = \frac{1}{\sqrt{a}} \ln \left| \frac{\sqrt{a + bu} - \sqrt{a}}{\sqrt{a + bu} + \sqrt{a}} \right| + C$ $a > 0$

40. $\displaystyle\int \frac{du}{u\sqrt{a + bu}} = \frac{2}{\sqrt{-a}} \operatorname{Arctan} \sqrt{\frac{a + bu}{-a}} + C$ $a < 0$

41. $\displaystyle\int \frac{du}{u^n\sqrt{a + bu}} = -\frac{\sqrt{a + bu}}{(n - 1)au^{n-1}} - \frac{(2n - 3)b}{2(n - 1)a}\int \frac{du}{u^{n-1}\sqrt{a + bu}}$ $n \neq 1$

D. Integrals Involving $\sqrt{a^2 - u^2}$

42. $\displaystyle\int \sqrt{a^2 - u^2} \, du = \frac{u}{2}\sqrt{a^2 - u^2} + \frac{a^2}{2} \operatorname{Arcsin}\left(\frac{u}{a}\right) + C$

43. $\displaystyle\int u\sqrt{a^2 - u^2} \, du = -\frac{1}{3}(a^2 - u^2)^{3/2} + C$

44. $\displaystyle\int u^n\sqrt{a^2 - u^2} \, du = -\frac{u^{n-1}(a^2 - u^2)^{3/2}}{n + 2} + \frac{(n - 1)a^2}{n + 2}\int u^{n-2}\sqrt{a^2 - u^2} \, du$ $n \neq -2$

45. $\displaystyle\int \frac{\sqrt{a^2 - u^2}}{u} \, du = \sqrt{a^2 - u^2} - a \ln \left| \frac{a + \sqrt{a^2 - u^2}}{u} \right| + C$

46. $\displaystyle\int \frac{\sqrt{a^2 - u^2}}{u^2} \, du = -\frac{\sqrt{a^2 - u^2}}{u} - \operatorname{Arcsin}\left(\frac{u}{a}\right) + C$

47. $\displaystyle\int \frac{du}{u\sqrt{a^2 - u^2}} = -\frac{1}{a} \ln \left| \frac{a + \sqrt{a^2 - u^2}}{u} \right| + C$

48. $\displaystyle\int \frac{du}{u^2\sqrt{a^2 - u^2}} = -\frac{\sqrt{a^2 - u^2}}{a^2 u} + C$

49. $\displaystyle\int \frac{u^2 \, du}{\sqrt{a^2 - u^2}} = -\frac{u}{2}\sqrt{a^2 - u^2} + \frac{a^2}{2} \operatorname{Arcsin}\left(\frac{u}{a}\right) + C$

50. $\displaystyle\int \frac{u^n \, du}{\sqrt{a^2 - u^2}} = -\frac{u^{n-1}\sqrt{a^2 - u^2}}{n} + \frac{(n - 1)a^2}{n}\int \frac{u^{n-2}\, du}{\sqrt{a^2 - u^2}}$ $n \neq 0$

51. $\displaystyle\int \frac{du}{u^n\sqrt{a^2 - u^2}} = \frac{-\sqrt{a^2 - u^2}}{(n - 1)a^2 u^{n-1}} + \frac{n - 2}{(n - 1)a^2}\int \frac{du}{u^{n-2}\sqrt{a^2 - u^2}}$ $n \neq 1$

52. $\displaystyle\int (a^2 - u^2)^{3/2} \, du = \frac{u}{4}(a^2 - u^2)^{3/2} + \frac{3a^2 u}{8}\sqrt{a^2 - u^2} + \frac{3a^4}{8} \operatorname{Arcsin}\left(\frac{u}{a}\right) + C$

53. $\displaystyle\int \frac{(a^2 - u^2)^{3/2}}{u} \, du = \frac{1}{3}(a^2 - u^2)^{3/2} - a^2\sqrt{a^2 - u^2} + a^3 \ln \left| \frac{a + \sqrt{a^2 - u^2}}{u} \right| + C$

54. $\displaystyle\int \frac{du}{(a^2 - u^2)^{3/2}} = \frac{u}{a^2\sqrt{a^2 - u^2}} + C$

55. $\displaystyle\int \frac{u^2 \, du}{(a^2 - u^2)^{3/2}} = \frac{u}{\sqrt{a^2 - u^2}} - \text{Arcsin}\left(\frac{u}{a}\right) + C$

56. $\displaystyle\int \frac{du}{u(a^2 - u^2)^{3/2}} = \frac{1}{a^2\sqrt{a^2 - u^2}} - \frac{1}{a^3} \ln \left| \frac{a + \sqrt{a^2 - u^2}}{u} \right| + C$

E.   Integrals Involving $\sqrt{u^2 + a^2}$ or $\sqrt{u^2 - a^2}$

57. $\displaystyle\int \sqrt{u^2 \pm a^2} \, du = \frac{1}{2}\left[u\sqrt{u^2 \pm a^2} \pm a^2 \ln \left| u + \sqrt{u^2 \pm a^2} \right| \right] + C$

58. $\displaystyle\int \frac{\sqrt{u^2 + a^2}}{u} \, du = \sqrt{u^2 + a^2} - a \ln \left| \frac{a + \sqrt{u^2 + a^2}}{u} \right| + C$

59. $\displaystyle\int \frac{\sqrt{u^2 - a^2}}{u} \, du = \sqrt{u^2 - a^2} - a \, \text{Arccos}\left(\frac{a}{u}\right) + C$

60. $\displaystyle\int \frac{\sqrt{u^2 \pm a^2}}{u^2} \, du = -\frac{\sqrt{u^2 \pm a^2}}{u} + \ln \left| u + \sqrt{u^2 \pm a^2} \right| + C$

61. $\displaystyle\int u^2 \sqrt{u^2 \pm a^2} \, du = \frac{u}{8}(2u^2 \pm a^2)\sqrt{u^2 \pm a^2} - \frac{a^4}{8} \ln \left| u + \sqrt{u^2 \pm a^2} \right| + C$

62. $\displaystyle\int \frac{du}{u\sqrt{u^2 + a^2}} = \frac{1}{a} \ln \left| \frac{u}{a + \sqrt{u^2 + a^2}} \right| + C$

63. $\displaystyle\int \frac{du}{u\sqrt{u^2 - a^2}} = \frac{1}{a} \text{Arccos}\left(\frac{a}{u}\right) + C$

64. $\displaystyle\int \frac{u^2 \, du}{\sqrt{u^2 \pm a^2}} = \frac{u}{2}\sqrt{u^2 \pm a^2} \mp \frac{a^2}{2} \ln \left| u + \sqrt{u^2 \pm a^2} \right| + C$

65. $\displaystyle\int \frac{du}{u^2\sqrt{u^2 \pm a^2}} = \frac{\mp\sqrt{u^2 \pm a^2}}{a^2 u} + C$

66. $\displaystyle\int \frac{du}{(u^2 \pm a^2)^{3/2}} = \frac{\pm u}{a^2\sqrt{u^2 \pm a^2}} + C$

67. $\displaystyle\int \frac{u^2 \, du}{(u^2 \pm a^2)^{3/2}} = -\frac{u}{\sqrt{u^2 \pm a^2}} + \ln \left| u + \sqrt{u^2 \pm a^2} \right| + C$

68. $\displaystyle\int (u^2 \pm a^2)^{3/2} \, du = \frac{u}{8}(2u^2 \pm 5a^2)\sqrt{u^2 \pm a^2} + \frac{3a^4}{8} \ln \left| u + \sqrt{u^2 \pm a^2} \right| + C$

F.   Integrals Involving Trigonometric Expressions

69. $\displaystyle\int \sin^2 u \, du = \frac{1}{2}u - \frac{1}{4}\sin 2u + C$

70. $\displaystyle\int \sin^n u \, du = -\frac{1}{n}\sin^{n-1}u \cos u + \frac{n-1}{n} \int \sin^{n-2}u \, du$

71. $\displaystyle\int \cos^2 u \, du = \frac{1}{2}u + \frac{1}{4}\sin 2u + C$

72. $\displaystyle\int \cos^n u \; du = \frac{1}{n}\cos^{n-1}u \sin u + \frac{n-1}{n}\int \cos^{n-2}u \; du$

73. $\displaystyle\int \tan^2 u \; du = \tan u - u + C$

74. $\displaystyle\int \tan^n u \; du = \frac{\tan^{n-1}u}{n-1} - \int \tan^{n-2}u \; du$

75. $\displaystyle\int \cot^2 u \; du = -\cot u - u + C$

76. $\displaystyle\int \cot^n u \; du = -\frac{\cot^{n-1}u}{n-1} - \int \cot^{n-2}u \; du$

77. $\displaystyle\int \sec^n u \; du = \frac{\sec^{n-2}u \tan u}{n-1} + \frac{n-2}{n-1}\int \sec^{n-2}u \; du$

78. $\displaystyle\int \csc^n u \; du = -\frac{\csc^{n-2}u \cot u}{n-1} + \frac{n-2}{n-1}\int \csc^{n-2}u \; du$

79. $\displaystyle\int u \sin u \; du = \sin u - u \cos u + C$

80. $\displaystyle\int u \cos u \; du = \cos u + u \sin u + C$

81. $\displaystyle\int u^n \sin au \; du = -\frac{u^n}{a}\cos au + \frac{n}{a}\int u^{n-1}\cos au \; du \qquad$ n a positive integer

82. $\displaystyle\int u^n \cos au \; du = \frac{u^n}{a}\sin au - \frac{n}{a}\int u^{n-1}\sin au \; du \qquad$ n a positive integer

83. $\displaystyle\int \frac{\sin u}{u^n}\; du = -\frac{\sin u}{(n-1)u^{n-1}} + \frac{1}{n-1}\int \frac{\cos u}{u^{n-1}}\; du \qquad n \neq 1$

84. $\displaystyle\int \frac{\cos u}{u^n}\; du = -\frac{\cos u}{(n-1)u^{n-1}} - \frac{1}{n-1}\int \frac{\sin u}{u^{n-1}}\; du \qquad n \neq 1$

85. $\displaystyle\int \frac{du}{\sin u \cos u} = \ln \left| \tan u \right| + C$

86. $\displaystyle\int \sin au \sin bu \; du = -\frac{\sin (a+b)u}{2(a+b)} + \frac{\sin (a-b)u}{2(a-b)} + C$

87. $\displaystyle\int \cos au \cos bu \; du = \frac{\sin (a+b)u}{2(a+b)} + \frac{\sin (a-b)u}{2(a-b)} + C$

88. $\displaystyle\int \sin au \cos bu \; du = -\frac{\cos (a+b)u}{2(a+b)} - \frac{\cos (a-b)u}{2(a-b)} + C$

89. $\displaystyle\int \sin^m u \cos^n u \; du = \frac{\sin^{m+1}u \cos^{n-1}u}{m+n} + \frac{n-1}{m+n}\int \sin^m u \cos^{n-2}u \; du \qquad m, n > 0$

$\displaystyle\qquad\qquad\qquad = -\frac{\sin^{m-1}u \cos^{n+1}u}{m+n} + \frac{m-1}{m+n}\int \sin^{m-2}u \cos^n u \; du \qquad m, n > 0$

G. Integrals Involving Exponential Functions

90. $\displaystyle\int u e^{au}\; du = \left(\frac{au-1}{a^2}\right)e^{au} + C$

91. $\displaystyle \int u^2 e^{au}\, du = \frac{e^{au}}{a^3}(a^2 u^2 - 2au + 2) + C$

92. $\displaystyle \int u^n e^{au}\, du = \frac{1}{a}u^n e^{au} - \frac{n}{a}\int u^{n-1} e^{au}\, du \qquad n > 0$

93. $\displaystyle \int e^{au} \ln u\, du = \frac{e^{au}\ln u}{a} - \frac{1}{a}\int \frac{e^{au}}{u}\, du$

94. $\displaystyle \int e^{au} \sin bu\, du = \frac{e^{au}(a \sin bu - b \cos bu)}{a^2 + b^2} + C$

95. $\displaystyle \int e^{au} \cos bu\, du = \frac{e^{au}(a \cos bu + b \sin bu)}{a^2 + b^2} + C$

96. $\displaystyle \int u^n a^u\, du = \frac{a^u u^n}{\ln a} - \frac{n}{\ln a}\int u^{n-1} a^u\, du$

97. $\displaystyle \int \frac{e^{au}}{u^n}\, du = \frac{1}{n-1}\left(-\frac{e^{au}}{u^{n-1}} + a\int \frac{e^{au}}{u^{n-1}}\, du\right) \qquad n \text{ an integer} > 1$

98. $\displaystyle \int \frac{du}{a + be^u} = \frac{u - \ln|a + be^u|}{a} + C$

H. Integrals Involving Logarithmic Functions

99. $\displaystyle \int \ln u\, du = u \ln u - u + C$

100. $\displaystyle \int u^n \ln u\, du = \frac{u^{n+1}\ln u}{n+1} - \frac{u^{n+1}}{(n+1)^2} + C \qquad n \neq -1$

101. $\displaystyle \int \frac{\ln u}{u}\, du = \frac{1}{2}(\ln u)^2 + C$

102. $\displaystyle \int \frac{du}{u \ln u} = \ln|\ln u| + C$

I. Integrals Involving Inverse Trigonometric Functions

103. $\displaystyle \int \text{Arcsin } u\, du = u \text{ Arcsin } u + \sqrt{1 - u^2} + C$

104. $\displaystyle \int \text{Arccos } u\, du = u \text{ Arccos } u - \sqrt{1 - u^2} + C$

105. $\displaystyle \int \text{Arctan } u\, du = u \text{ Arctan } u - \frac{1}{2}\ln(1 + u^2) + C$

106. $\displaystyle \int u^n \text{ Arcsin } u\, du = \frac{u^{n+1}\text{ Arcsin } u}{n+1} - \frac{1}{n+1}\int \frac{u^{n+1}\, du}{\sqrt{1 - u^2}} \qquad n \neq -1$

107. $\displaystyle \int u^n \text{ Arccos } u\, du = \frac{u^{n+1}\text{ Arccos } u}{n+1} + \frac{1}{n+1}\int \frac{u^{n+1}\, du}{\sqrt{1 - u^2}} \qquad n \neq -1$

108. $\displaystyle \int u^n \text{ Arctan } u\, du = \frac{u^{n+1}\text{ Arctan } u}{n+1} - \frac{1}{n+1}\int \frac{u^{n+1}\, du}{\sqrt{1 + u^2}} \qquad n \neq -1$

J. Miscellaneous Integrals

109. $\displaystyle\int \frac{du}{au^2 + bu + c} = \frac{2}{\sqrt{4ac - b^2}} \text{Arctan}\left(\frac{2au + b}{\sqrt{4ac - b^2}}\right) + C \qquad 4ac - b^2 > 0$

110. $\displaystyle\int \frac{u\,du}{au^2 + bu + c} = \frac{1}{2a} \ln|au^2 + bu + c| - \frac{b}{a\sqrt{4ac - b^2}} \text{Arctan}\left(\frac{2au + b}{\sqrt{4ac - b^2}}\right) + C \qquad 4ac - b^2 > 0$

111. $\displaystyle\int \frac{du}{(au^2 + bu + c)^n} = \frac{2au + b}{(n - 1)(4ac - b^2)(au^2 + bu + c)^{n-1}} + \frac{2a(2n - 3)}{(n - 1)(4ac - b^2)} \int \frac{du}{(au^2 + bu + c)^{n-1}}$

$$n \neq 1,\ 4ac - b^2 > 0$$

112. $\displaystyle\int \frac{du}{\sqrt{au^2 + bu + c}} = \frac{1}{\sqrt{a}} \ln\left|2au + b + 2\sqrt{a}\sqrt{au^2 + bu + c}\right| \qquad a > 0$

$$= \frac{1}{\sqrt{-a}} \text{Arcsin}\left(-\frac{2au + b}{\sqrt{b^2 - 4ac}}\right) \qquad a < 0,\ b^2 - 4ac > 0$$

113. $\displaystyle\int \sqrt{au^2 + bu + c}\,du = \frac{2au + b}{4a}\sqrt{au^2 + bu + c} + \frac{4ac - b}{8a} \int \frac{du}{\sqrt{au^2 + bu + c}}$

114. $\displaystyle\int \sqrt{2au - u^2}\,du = \frac{u - a}{2}\sqrt{2au - u^2} + \frac{a^2}{2} \text{Arcsin}\left(\frac{u - a}{a}\right) + C$

115. $\displaystyle\int \frac{du}{\sqrt{2au - u^2}} = \text{Arccos}\left(\frac{a - u}{a}\right) + C$

Answers to Odd-Numbered Problems

Chapter 1

Exercises 1-1, p. 8

1. Rational **3.** Irrational **5.** Rational **7.** Rational **9.** Integer, rational **11.** Integer, rational
13. < **15.** < **17.** < **19.** > **21.** < **23.** > **25.** 13 **27.** 7.2 **29.** -3
31. 33 **33.** $-\dfrac{1}{4}$ **35.** $\dfrac{8}{3}$ **37.** $\dfrac{1}{\sqrt{7}}$ **39.** -5 **41.** 56 lb **43.** 50 km/h **45.** 5 m/s

47.

49. (a) Potential energy equals mass times acceleration due to gravity times height.
   (b) E, m, and h are variables; g is a constant
51. (a) Yes (b) No **53.** (a) 260 mm (b) 360 mm **55.** 2.5 seconds

Exercises 1-2, p. 16

1. -8 **3.** -17 **5.** -9 **7.** -9 **9.** -16 **11.** -51 **13.** 4 **15.** -7 **17.** $\dfrac{1}{6}$ **19.** 1.1

21. 0.96 **23.** $-\dfrac{21}{8}$ **25.** 34 **27.** -140 **29.** 24 **31.** -6.913 **33.** -3.67

35. Distributive law **37.** Commutative law of multiplication **39.** Commutative law of addition
41. Commutative law of multiplication **43.** 2710 V **45.** $-13°$F **47.** \$758,000 **49.** Negative

Exercises 1-3, p. 20

1. 17 **3.** 24 **5.** 6 **7.** 0 **9.** 0 **11.** Undefined **13.** 13 **15.** 23 **17.** 62 **19.** 53
21. 10 **23.** -7 **25.** -17.2692 **27.** -8.18 **29.** 49 **31.** $12(16) + 20(8.40) + 50(2.50) = \$485.00$
33. $0.86 + 0.52(8 - 1) = \$4.50$ **35.** $372°$C

Exercises 1-4, p. 28

1. 5^{11} **3.** y^{12} **5.** $2^{15} \cdot 5^6$ **7.** $2y^8$ **9.** 3^{12} **11.** m^{14} **13.** a^4b^4 **15.** 10^3 **17.** $2^4 \cdot x^8 = 16x^8$
19. a^6b^{12} **21.** $\dfrac{2^3}{5^3} = \dfrac{8}{125}$ **23.** $\dfrac{w^4}{z^4}$ **25.** $\dfrac{2^2}{y^{12}} = \dfrac{4}{y^{12}}$ **27.** 7^3 **29.** 1 **31.** $\dfrac{1}{y^4}$ **33.** 1 **35.** $\dfrac{1}{5^3}$

37. 4 **39.** y^3 **41.** $\dfrac{5}{x^4}$ **43.** $\dfrac{c^4}{a^6b^2d^7}$ **45.** 8^{-7} **47.** 10^{-4} **49.** $p^{-3}q^{-3}$ **51.** 10^{-10} **53.** c^{-3}

55. -25 **57.** x^8 **59.** $-x^{15}$ **61.** 65 **63.** 45,000 lb **65.** $\dfrac{1}{p}$ **67.** $\dfrac{2x^3w}{3}$

Exercises 1-5, p. 33

1. 7×10^3 **3.** 5.3×10^{-4} **5.** 4.07×10^4 **7.** 6.3×10^1 **9.** 7×10^{-1} **11.** 6×10^{-2} **13.** 10^5
15. 5×10^6 **17.** 50,000,000 **19.** 0.00065 **21.** 7,250,000 **23.** 4.9 **25.** 0.0002389
27. 0.00000001 **29.** 8×10^7 **31.** 7.8×10^2 **33.** 2.0×10^2 **35.** 8×10^{-11} **37.** 8.4×10^{-10}
39. 6.6×10^{13} **41.** 4.04 **43.** 1.3×10^1 cal/s **45.** 4×10^{-10} W **47.** 1.53×10^{-4} μF
49. -1.1 V **51.** 2.7×10^{-26} kg

Exercises 1-6, p. 39

1. 2 **3.** -10 **5.** 2 **7.** -5 **9.** 4 **11.** $2\sqrt{3}$ **13.** $5\sqrt{3}$ **15.** $3\sqrt{5}$ **17.** $14\sqrt{2}$
19. Imaginary **21.** $\dfrac{14}{15}$ **23.** $\dfrac{3}{5}$ **25.** $\dfrac{\pm\sqrt{3}}{2}$ **27.** $\dfrac{2\sqrt{6}}{7}$ **29.** $4\sqrt{5}$ **31.** 46 **33.** 37 **35.** 43.08
37. 24 **39.** 17 **41.** 33.90 **43.** 2.31, -0.53 **45.** 12 cm **47.** $1856 \div 29 = 64$ **49.** 89,000 lb/in.²
51. 19,000 Ω

Exercises 1-7, p. 41

1. (b) **3.** (c), (e) **5.** (g) **7.** (a) $\dfrac{5}{16}$ (b) 17 (c) 5π (d) 2.5 (e) $\sqrt{5}$ (f) $0.\overline{37}$
9. Distributive law **11.** Associative law of addition **13.** Commutative law of addition **15.** -15
17. -6 **19.** 27 **21.** -88 **23.** 7 **25.** 0 **27.** -19 **29.** 180 **31.** -20 **33.** 20
35. -51 **37.** 261 **39.** 6 **41.** 4 **43.** $4\sqrt{2}$ **45.** $\dfrac{2}{5}$ **47.** -1 **49.** 13 **51.** 57 **53.** 35
55. -20 **57.** 3^{13} **59.** $3^3x^3 = 27x^3$ **61.** $2^3x^6y^9 = 8x^6y^9$ **63.** $\dfrac{1}{m^4}$ **65.** 1 **67.** $\dfrac{7}{y^3}$ **69.** y^8
71. x **73.** 2^{-5} **75.** 10^{-5} **77.** $a^{-4}x^{-4}$ **79.** 9.4×10^{11} W **81.** 10,000 years
83. 2×10^{-4} kcal/s·m·°C **85.** 0.000057 Wb/m² **87.** 8.5×10^9 **89.** 1.6×10^{-1} **91.** 5.4 V
93. -748 kcal **95.** 1.1 **97.** 8.2×10^{-7} erg **99.** 68 ft **101.** 9.7×10^6 m/s **103.** 1290

Chapter 2

Exercises 2-1, p. 49

1. $7x$ **3.** $11m + 2n$ **5.** $-12v + 9w$ **7.** $-2x + 5y - 4z$ **9.** $12x^2 - 18x$ **11.** $8ab + 3a$
13. $11x - 5y - 12z$ **15.** $2m + 7$ **17.** $-6a - 10b$ **19.** $-5x^2 - 20x + 8$ **21.** $6m - 2mn$
23. $-4a - 2b - 2c$ **25.** $-13 + 3x$ **27.** $-10m^2 + 3m$ **29.** $2p$

Exercises 2-2, p. 53

1. $-15x^3$ **3.** m^4n^6 **5.** $-70x^4y^3$ **7.** $18a^8b$ **9.** $-12m + 28p$ **11.** $6y^3 - 21y^2$
13. $-30w^5 + 42w^3$ **15.** $6x^3y + 10x^2y - 16xy$ **17.** $3m^3n - 5m^2n^2 + 6m^2n$ **19.** $45x^2 + 105xy$
21. $x^2 + 10x + 21$ **23.** $x^2 - 36$ **25.** $16m^2 + 28m - 30$ **27.** $6x^2 + 13x - 28$ **29.** $10a^2 + 39ab - 27b^2$
31. $56m^4 + 75m^2 + 25$ **33.** $x^2 - 10x + 25$ **35.** $25a^2 + 70ab + 49b^2$ **37.** $x^3 - 2x^2 - 41x - 56$
39. $2p^4 - p^3 + 8p^2 + 21p$ **41.** $8x^3 - 20x^2 - 252x$ **43.** $15x^3 - 8x^2 - 69x + 14$
45. $m^3 - 27m^2 + 243m - 729$ **47.** $A = D + \dfrac{Dpt}{100}$ **49.** $m_1v_a^2 - m_1v_b^2 = m_2v_c^2 - m_2v_d^2$ **51.** $2n^2 + 6n$
53. $100 + 200r + 100r^2$

Exercises 2-3, p. 58

1. $-4x^3$ **3.** $\dfrac{2}{m^3}$ **5.** $4x^2y^2$ **7.** $\dfrac{8p^3}{q}$ **9.** $\dfrac{14y^3}{5}$ **11.** $\dfrac{-5x}{2z^2}$ **13.** $6x - 2$ **15.** $-2mn - \dfrac{3m^2}{n}$

17. $2c^2 - 3c + 1$ **19.** $-2xy^2 + 3y + \dfrac{7y^3}{2x}$ **21.** $\dfrac{ax^{r-s}}{c} - \dfrac{bx^{r+1-s}}{c}$ **23.** $x + 5$ **25.** $2x - 9$

27. $3x^2 - 5x - 7$ **29.** $4x^3 - 3x^2 - 2x - 4 + \dfrac{1}{3x - 2}$ **31.** $x^2 + 2x + 4$ **33.** $3x^2 + 5$ **35.** $\dfrac{1}{z_c} + \dfrac{\mu}{z_o} - \dfrac{\mu}{z_r}$

Exercises 2-4, p. 64

1. -8 **3.** 15 **5.** -8 **7.** -21 **9.** 36 **11.** 6 **13.** -12 **15.** 12 **17.** 2 **19.** -7

21. $\dfrac{4}{3}$ **23.** -5 **25.** 2 **27.** -5 **29.** $\dfrac{5}{9}$ **31.** $\dfrac{20}{17}$ **33.** No solution **35.** -3.3

37. $t = 4$ s **39.** $A = 4$

Exercises 2-5, p. 70

1. 200 V **3.** 20 lx **5.** $104°$F **7.** $27°$C **9.** 314 ft **11.** 114 lb **13.** 1000 ft **15.** $\$22,000$

17. 76 ft **19.** $a = \dfrac{F}{m}$ **21.** $H_2 = H_1 - H_R$ **23.** $D = \dfrac{CA}{LU}$ **25.** $L = \dfrac{RD^2}{r}$ **27.** $T = \dfrac{Mv^2}{2R}$

29. $v_o = \dfrac{2s - at^2}{2t}$, or $v_o = \dfrac{s}{t} - \dfrac{at}{2}$ **31.** $P_1 = \dfrac{2L}{s} - P_2$ **33.** $i_e = \dfrac{i_b}{1 - \alpha}$ **35.** $D = \dfrac{R^2N^2 - 9RL}{10L}$ **37.** 12

39. 13.4 **41.** 11.3 ft **43.** 140 mi/h

Exercises 2-6, p. 81

1. 2105 **3.** 128 and 256 kbytes **5.** 17 cm by 25 cm **7.** 15 in., 15 in., 35 in. **9.** 360 mi/h **11.** $\dfrac{1}{4}$ h

13. 530 mi/h **15.** 10 mi **17.** 76%: 21 L; 12%: 7 L **19.** 10 gal **21.** 3 L **23.** 30 L

25. 12 min **27.** 10 h

Exercises 2-7, p. 89

1. $\dfrac{3}{1}$ **3.** $\dfrac{9}{4}$ **5.** $\dfrac{1}{2}$ **7.** $\dfrac{7}{10}$ **9.** $\dfrac{4}{1}$ **11.** $\dfrac{2}{3}$ **13.** $\dfrac{30}{1}$ **15.** 0.12 **17.** $\dfrac{9}{4}$ **19.** 9 **21.** $\dfrac{297}{17}$

23. 0.05 **25.** 320 rev/min **27.** 880 mg/dL **29.** $\$48.44$ **31.** 230 mg **33.** 37 gal **35.** 15 lb

37. 350 rev/min **39.** 140 lb cement, 560 lb gravel **41.** 12 **43.** $13\dfrac{1}{3}$

Exercises 2-8, p. 97

1. $x = ky$ **3.** $a = kbc$ **5.** $m = \dfrac{k}{n^3}$ **7.** $e = \dfrac{kfg^3}{h}$ **9.** (a) $k = 3$ ft/s (b) $x = 3y$

11. (a) $k = 48$ (b) $m = \dfrac{48}{n^3}$ **13.** (a) $k = \dfrac{1}{6}$ (b) $w = \dfrac{1}{6}xz^2$ **15.** (a) $k = 7$ (b) $x = \dfrac{7y}{z^2}$

17. 20 A **19.** $6\frac{3}{4}$ **21.** 240 **23.** 2.7 **25.** 8 A **27.** 2100 lb **29.** 313 W **31.** 2.4×10^{-2} cm

33. 65 hp **35.** 3.2×10^{-8} W/m^2 **37.** 0.69 A **39.** $56,000$ tons

Exercises 2-9, p. 100

1. $-5x$ **3.** $-3ab - 4ab^2 - 6a^2b$ **5.** $-3c + 8$ **7.** $n + 13$ **9.** $-90a^3b^4$

11. $-15m^3n + 21m^2n^2 + 27mn^3$ **13.** $3y^4 - 19y^2 + 30$ **15.** $25x^2 + 70x + 49$ **17.** $-6y^3 + 62y^2 - 112y$

19. $-34x - 55y$ **21.** $-7x^2 + 41x - 89$ **23.** $\dfrac{2c^3}{3a^3b^5}$ **25.** $4x^2 + 1$ **27.** $5x - 4$

29. $2x^3 + 3x^2 - x - 5$ **31.** $-\dfrac{3}{4}$ **33.** 7 **35.** $-\dfrac{8}{3}$ **37.** -14 **39.** 4 **41.** -1 **43.** $88.0 \ \Omega$

45. 9.4×10^{-8} F   **47.** 289 ft   **49.** 816 cm²   **51.** $C_2 = C - C_1 - C_3$ **53.** $L = \dfrac{RA}{\rho}$ **55.** $R = \dfrac{I - P}{PT}$

57. $L = \dfrac{8m}{G' - G}$ **59.** 16   **61.** $\dfrac{3}{1}$ **63.** $\dfrac{1}{6}$ **65.** $\dfrac{150}{1}$ **67.** $\dfrac{24}{1}$ **69.** 5   **71.** 28.6   **73.** $5\dfrac{1}{4}$ in.

75. 0.4 Ω   **77.** 20   **79.** 825   **81.** 8 lb   **83.** 5 h   **85.** 70%: 20 kg; 40%: 40 kg

87. $48(m + n) = 48m + 48n$ **89.** $\dfrac{5}{2}\mu RT$ **91.** $v_{10} - v_{20} = -v_{1f} + v_{2f}$ **93.** $\dfrac{MG}{r^2} - \dfrac{3L}{16\pi\rho scr^2}$ **95.** 160 kg

97. 150   **99.** 18   **101.** 560 km   **103.** 200 Btu/h   **105.** 649 g/mol

Chapter 3

Exercises 3-1, p. 109

1. $P(L) = 16 + 2L$ **3.** $s(P) = \dfrac{P}{3}$ **5.** $W(g) = 1.5 + 11.4g$ **7.** $L(P) = \dfrac{P - 8}{2}$

9. (a) 2 (b) -16 (c) -7 **11.** (a) 6 (b) 0 (c) 1.4   **13.** (a) 0 (b) 10 (c) 70

15. (a) 2 (b) $-\dfrac{5}{2}$ (c) $\dfrac{7}{8}$ **17.** (a) $\dfrac{38}{5}$ (b) $-\dfrac{11}{5}$ (c) $\dfrac{79}{5}$ **19.** (a) 24.57 (b) 236.8352 (c) 63,220

21. (a) $4t + 5$ (b) $12t - 3$ (c) $4t - 7$ **23.** (a) $x^4 - 7x^2 + 2$ (b) $4x^2 - 14x + 2$ (c) $4x - 10$

25. All real numbers except 5   **27.** All real numbers $a \geq 3$ **29.** All real numbers except -5 and 4

31. All real numbers except 0 and 2   **33.** 280 in.²   **35.** 40.4 lb/in.²   **37.** 0.198 W

39. (a) $\dfrac{1}{3}$ (b) $\eta = \dfrac{T + x}{40 - T - x}$ **41.** (a) -7 (b) 10

Exercises 3-2, p. 114

1. -2 **3.** 4   **5.** IV, III, II, I   **7.** $(2, -1)$ **9.** $(0, 3)$ **11.** $(-5, 0)$

13.   **15.**   **17.**

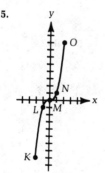

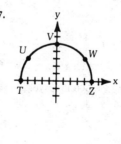

19. $(8, -2)$ **21.** II and III   **23.** II and IV   **25.** III and IV   **27.** III

Exercises 3-3, p. 120

1.   **3.**   **5.** $f(a)$ **7.**

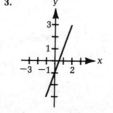

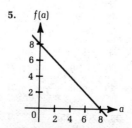

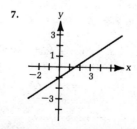

9.
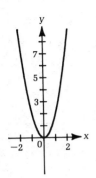

11.
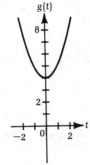

13.
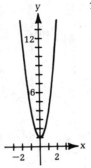

15.
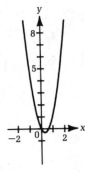

17.
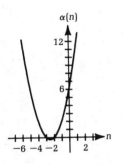

19.
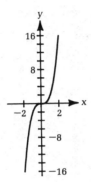

21.
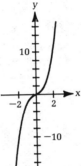

23.
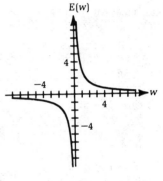

25.
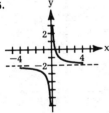

27.
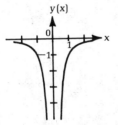

29.
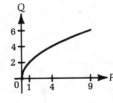

31.
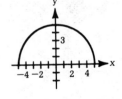

33.
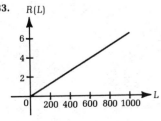

35.
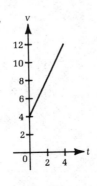

37.
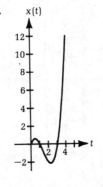

39.
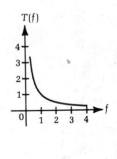

41.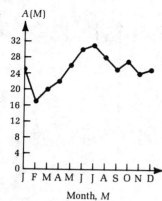

43.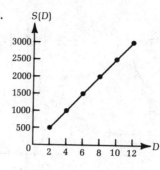

45.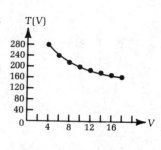

Exercises 3-4, p. 126

1. 2   **3.** -2.7 **5.** -0.5 **7.** 5   **9.** 10   **11.** -2.5 **13.** 0, 1   **15.** 4, 1   **17.** 2.7, -0.4
19. 5.8, 0.2   **21.** 0, 3, -3 **23.** 1.6, -1.6, 0.6, -0.6 **25.** 0.5   **27.** 6   **29.** $L = 60$ ft   **31.** $t = 2.2$ s
33. $M = 0.5$

Exercises 3-5, p. 127

1. $A(d) = \dfrac{\pi d^2}{4}$ **3.** $d(v) = 7v$ **5.** $r(C) = \dfrac{C}{2\pi}$ **7.** (a) -3 (b) -17 (c) -6

9. (a) -15 (b) 9 (c) $-3\frac{3}{4}$ **11.** (a) -308 (b) 5.72 (c) 2.2   **13.** (a) $5y - 9$ (b) $-10y + 1$ (c) 20
15. All real numbers except 0.   **17.** All real numbers between and including -6 and 6.

19.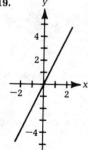

21.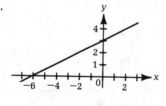

23.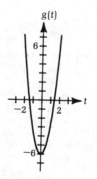

25.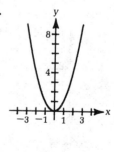

27.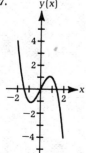

29.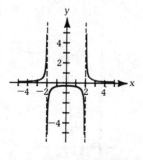

31.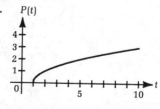

33. 3   **35.** −1.3   **37.** 1.8   **39.** 0, 3   **41.** 2.4, −0.9   **43.** 0, 2.2, −2.2   **45.** −2.4, 2.4

47. (a) 5 (b) 7   **49.** (a) I (b) IV (c) III   **51.** (1, 13)   **53.** (a) $5.62 (b) $5r^2 + 11r + 6.05$

55. q = 2.5

57. t = 3 s

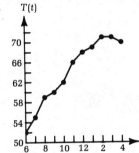

59. F

61. L(x)

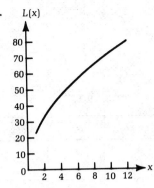

63. T(t)

65. E(L)

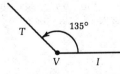

Chapter 4

Exercises 4-1, p. 136

1.

3.

5.

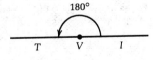

7.

9.

11.

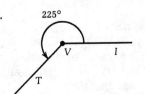

13. 420°, −300°   **15.** 525°, −195°   **17.** 402°15′, −317°45′   **19.** 497.9°, −222.1°   **21.** 52.3°

23. 238.75°   **25.** 65.72°   **27.** 98.43°   **29.** 58°12′   **31.** 227°15′   **33.** 8°37′   **35.** 164°50′

37. III   **39.** IV

41. 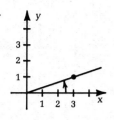   **43.**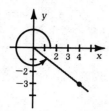

Exercises 4-2, p. 141

	sin	cos	tan	csc	sec	cot
1.	$\dfrac{4}{5}$	$\dfrac{3}{5}$	$\dfrac{4}{3}$	$\dfrac{5}{4}$	$\dfrac{5}{3}$	$\dfrac{3}{4}$
3.	$\dfrac{5}{13}$	$\dfrac{12}{13}$	$\dfrac{5}{12}$	$\dfrac{13}{5}$	$\dfrac{13}{12}$	$\dfrac{12}{5}$
5.	$\dfrac{5}{\sqrt{29}}$	$\dfrac{2}{\sqrt{29}}$	$\dfrac{5}{2}$	$\dfrac{\sqrt{29}}{5}$	$\dfrac{\sqrt{29}}{2}$	$\dfrac{2}{5}$
7.	$\dfrac{20}{\sqrt{409}}$	$\dfrac{3}{\sqrt{409}}$	$\dfrac{20}{3}$	$\dfrac{\sqrt{409}}{20}$	$\dfrac{\sqrt{409}}{3}$	$\dfrac{3}{20}$
9.	0.919	0.394	2.333	1.088	2.539	0.429
11.	0.986	0.168	5.857	1.014	5.942	0.171

13. (a) $\dfrac{\sqrt{7}}{4}$ (b) $\dfrac{\sqrt{7}}{3}$ **15.** (a) $\sqrt{5}$ (b) $\dfrac{2}{\sqrt{5}}$ **17.** (a) 0.43 (b) 2.07 **19.** (a) 0.768
(b) 1.201 **21.** (a) 0.5358 (b) 1.1843 **23.** $x = 10.5$

Exercises 4-3, p. 148

1. 0.899 **3.** 0.810 **5.** 0.530 **7.** 3.072 **9.** 1.036 **11.** 0.386 **13.** 0.935 **15.** 0.332
17. 1.096 **19.** 0.989 **21.** 0.987 **23.** 2.455 **25.** 1.338 **27.** 1.936 **29.** 0.760 **31.** 0.161
33. 59.3° **35.** 35.0° **37.** 53.1° **39.** 34.6° **41.** 33.9° **43.** 79.1° **45.** 64.5° **47.** 49.6°
49. 25°47′ **51.** 70°15′ **53.** 67°25′ **55.** 81°8′ **57.** 22°55′ **59.** 56°34′ **61.** 251.6 ft
63. 815.7 Ω

Exercises 4-4, p. 156

1. $A = 38.0°, b = 10.2, c = 13.0$ **3.** $c = 18.6, A = 53.7°, B = 36.3°$ **5.** $A = 45.8°, a = 15.1, c = 21.1$
7. $b = 20.1, A = 43.1°, B = 46.9°$ **9.** $a = 6.887, c = 21.15, B = 71°0′$ **11.** $a = 42.37, A = 27°50′, B = 62°10′$
13. $A = 18°58′, a = 114.1, b = 331.9$ **15.** $c = 8.751, A = 72°35′, B = 17°25′$ **17.** 410 ft **19.** 5.2 in.
21. $x = 2.07$ in., $y = 0.688$ in. **23.** 81° **25.** 25 ft **27.** 158° **29.** 58°0′ **31.** 14 ft **33.** 15°
35. 5.48 cm **37.** 35 ft **39.** 29.3 ft

Exercises 4-5, p. 161

1. 400°, −320° **3.** 584°18′, −135°42′ **5.** 24.4° **7.** 256.13° **9.** 94°30′ **11.** 285°47′

13. $\sin \theta = 0.6$, $\cos \theta = 0.8$, $\tan \theta = 0.75$, $\csc \theta = 1.667$, $\sec \theta = 1.25$, $\cot \theta = 1.333$

15. $\sin \theta = 0.894$, $\cos \theta = 0.447$, $\tan \theta = 2$, $\csc \theta = 1.118$, $\sec \theta = 2.236$, $\cot \theta = 0.5$

17. (a) $\dfrac{24}{25}$ (b) $\dfrac{7}{24}$ **19.** (a) $\dfrac{1}{3}$ (b) $\dfrac{1}{\sqrt{10}}$ **21.** (a) 0.275 (b) 0.962 **23.** 0.616 **25.** 1.664

27. 0.941 **29.** 1.183 **31.** 0.599 **33.** 0.564 **35.** 23.6° **37.** 72.8° **39.** 71.5° **41.** 48°10′

43. 47°25′ **45.** 33°1′ **47.** $B = 63.0°$, $a = 4.59$, $c = 10.1$ **49.** $c = 19.7$, $A = 36.1°$, $B = 53.9°$

51. $A = 72.2°$, $a = 3.74$, $c = 3.93$ **53.** $b = 18.9$, $A = 37.1°$, $B = 52.9°$ **55.** $b = 5.901$, $a = 6.623$, $B = 41°42′$

57. $a = 2.601$, $A = 55°31′$, $B = 34°29′$ **59.** 44° **61.** 2.1 mm **63.** 48° **65.** 22°19′ **67.** 450 ft

69. 32°

Chapter 5

Exercises 5-1, p. 168

1. (a) Yes (b) Yes (c) No **3.** (a) Yes (b) Yes (c) Yes **5.** (a) Yes (b) No (c) Yes

7. (a) $y = -4$ (b) $y = -18$ (c) $x = 2$ **9.** (a) $x = 2$ (b) $x = -2$ (c) $y = -\dfrac{9}{4}$

11. (a) $y = -26$ (b) $y = -59$ (c) $x = 142$ **13.** Yes **15.** No **17.** No **19.** Yes **21.** Yes

23. No **25.** Yes

Exercises 5-2, p. 174

1. $(4, 0)$ **3.** Dependent **5.** $(1, 2)$ **7.** $(0, 5)$ **9.** $(1.5, -0.5)$ **11.** $(1, -1)$ **13.** Inconsistent

15. $(0, -5)$ **17.** $(5.7, 1.3)$ **19.** $(4, -1)$ **21.** $(5, 1)$ **23.** $(2.9, -1.4)$ **25.** ($1.50, $0.90)

27. $S = 4$ lb, $L = 6$ lb

Exercises 5-3, p. 181

1. $(6, 12)$ **3.** $(1, 1)$ **5.** $\left(-\dfrac{14}{5}, \dfrac{2}{5}\right)$ **7.** $(3, -2)$ **9.** Dependent **11.** $\left(\dfrac{22}{23}, -\dfrac{28}{23}\right)$ **13.** $(8, 3)$

15. $\left(2, -\dfrac{2}{3}\right)$ **17.** $\left(2, -\dfrac{1}{3}\right)$ **19.** Inconsistent **21.** $(0, -3)$ **23.** $(0, -3)$ **25.** $\left(48, \dfrac{35}{2}\right)$ **27.** $(5, 2)$

29. $(12, 29)$ **31.** $(2, 2)$ **33.** Dependent **35.** $(3, 4)$ **37.** 20 mL of 20% acid, 4 mL of 50% acid

39. Plane: 231 mi/h, wind: 21 mi/h **41.** $m = 0.005$, $b = 0.15$ **43.** 35 ft, 105 ft

Exercises 5-4, p. 188

1. 1 **3.** -6 **5.** 28 **7.** -32 **9.** 17 **11.** -55.86 **13.** $(-7, 2)$ **15.** $(0, -6)$ **17.** $\left(\dfrac{3}{2}, 3\right)$

19. $\left(\dfrac{35}{8}, -\dfrac{55}{8}\right)$ **21.** Inconsistent **23.** $\left(\dfrac{7}{2}, 4\right)$ **25.** $\left(\dfrac{1}{2}, \dfrac{5}{2}\right)$ **27.** $(-3, -2)$ **29.** $(27, 38)$

31. 15 cm, 9 cm **33.** $0.24 (first minute), $0.16 (each additional) **35.** $8000 at 16%, $2000 at 6%

37. 8 L of 65%, 12 L of 15% **39.** 27 black, 19 two-color

Exercises 5-5, p. 194

1. $(1, 2, 3)$ **3.** $(-1, -1, 3)$ **5.** $(5, -3, 4)$ **7.** $\left(\dfrac{1}{2}, -\dfrac{2}{3}, 4\right)$ **9.** Dependent **11.** $(-4, 5, 1)$

13. $\left(\dfrac{1}{5}, -3, 6\right)$ **15.** $(6, -8, 7)$ **17.** $\left(\dfrac{1}{2}, -\dfrac{1}{3}, -\dfrac{3}{4}\right)$ **19.** 40 mL 25%, 10 mL 50%, 50 mL 60%

21. A produces 6 pages, B produces 3 pages, C produces 8 pages   **23.** $X = 4, Y = 6, Z = 2$
25. 6000 of model A, 1500 of model B, 4000 of model C

Exercises 5-6, p. 201

1. 18   **3.** 243   **5.** -222 **7.** 17   **9.** -330 **11.** -20.114 **13.** $(1, 2, -1)$ **15.** $(3, 1, 0)$

17. $(-1, -3, 2)$ **19.** $\left(\dfrac{3}{2}, 3, -4\right)$ **21.** Inconsistent   **23.** $\left(\dfrac{5}{2}, \dfrac{2}{3}, -\dfrac{3}{4}\right)$ **25.** $(31, 27, 53)$

27. $i_1 = \dfrac{25}{34}, i_2 = -\dfrac{6}{17}, i_3 = -\dfrac{13}{34}$ **29.** $s_0 = 8$ ft, $v_0 = 6$ ft/s, $a = 3$ ft/s^2

31. Invest \$8000 at 6%, \$8000 at 10%, \$4000 at 18%   **33.** Run X for 5 h, Y for 3 h, Z for 4 h

Exercises 5-7, p. 205

1. $(-2, -3)$ **3.** $(3, 0)$ **5.** $(3, 2.5)$ **7.** $(2.3, 1.7)$ **9.** $(7, 2)$ **11.** $\left(\dfrac{10}{3}, -\dfrac{1}{3}\right)$ **13.** $(0, -2)$

15. Inconsistent   **17.** $\left(\dfrac{5}{2}, -3\right)$ **19.** $(-2, -5)$ **21.** $(-3, -2, 1)$ **23.** $(4, 7, -2)$ **25.** $\left(\dfrac{1}{2}, 2, -\dfrac{2}{3}\right)$

27. Inconsistent   **29.** 22   **31.** 3   **33.** -2 **35.** -376 **37.** $(1, -4)$ **39.** $(-3, 8)$ **41.** $(-2, -4)$

43. Inconsistent   **45.** $\left(\dfrac{3}{2}, -\dfrac{2}{3}\right)$ **47.** $(43, 38)$ **49.** $(-3, -3, -2)$ **51.** $(2, 4, 3)$ **53.** $\left(-\dfrac{2}{3}, \dfrac{5}{6}, 4\right)$

55. Dependent   **57.** $T_1 = 250$ lb, $T_2 = 433$ lb   **59.** $v_0 = 4$ ft/s, $a = 2$ ft/s^2 **61.** 3 L of 30%, 9 L of 10%
63. \$0.63 (first pound), \$0.23 (each additional)
65. 60 lb weight is 16 in. from fulcrum; 80 lb weight is 12 in. from fulcrum
67. X for 3 days, Y for 4 days, Z for 2 days   **69.** $A = 36, B = 18, C = 28$ **71.** 40 kg A, 40 kg B, 20 kg C

Chapter 6

Exercises 6-1, p. 213

1. $15x + 60$ **3.** $2m^2 - 10m$ **5.** $-10c^3 - 25c^2$ **7.** $x^2 - 16$ **9.** $4a^2 - 9$ **11.** $4x^2 - 49y^2$
13. $x^2 + 12x + 36$ **15.** $4f^2 - 28f + 49$ **17.** $16x^2 - 24xy + 9y^2$ **19.** $w^2 + 10w + 24$
21. $x^2 + 3x - 40$ **23.** $m^2 + 8m + 15$ **25.** $x^2 + 3x - 10$ **27.** $56 + 15a + a^2$ **29.** $x^2 - 2xy - 15y^2$
31. $3x^2 - 108$ **33.** $36m^3 - 4m$ **35.** $5a^2 + 20a + 20$ **37.** $-16v^3 + 48v^2 - 36v$ **39.** $7x^3 - 35x^2 + 28x$
41. $x^2 + 2xy + y^2 + 8x + 8y + 16$ **43.** $x^2 + 2xy + y^2 - 25$ **45.** $5c^2 + 10cd + 5d^2 + 20c + 20d + 20$

47. $\dfrac{V}{1 + 2p + p^2}$ **49.** $2L^2 + 3Lb + b^2$

Exercises 6-2, p. 218

1. $2(x + 4)$ **3.** $x(x - 5)$ **5.** $4x(x + 2)$ **7.** $4m(2m - 3n)$ **9.** Unfactorable   **11.** $2(2x - 3y + 4z)$
13. $6x^2(2y - 4x - 5x^2y^2)$ **15.** $2a^2b(9a^2bc - 14ab^2c^2 - 18)$ **17.** $(x + 5)(x - 5)$ **19.** $(9 + x)(9 - x)$
21. $(3a + 4)(3a - 4)$ **23.** $(5m + 6n)(5m - 6n)$ **25.** $3(y + 4)(y - 4)$ **27.** $x(x + 2)(x - 2)$
29. Unfactorable   **31.** $9(2 + y)(2 - y)$ **33.** $5y(x^2 + 3)(x^2 - 3)$ **35.** $(x^4 + 9)(x^2 + 3)(x^2 - 3)$
37. $s = \sqrt{4\varepsilon(2R - \varepsilon)}$ **39.** $d^2 = (L + H)(L - H)$ **41.** $8y(4y^4 - 20y^2 + 15)$

Exercises 6-3, p. 224

1. $(x + 8)(x + 1)$ **3.** $(a + 7)(a - 6)$ **5.** $(x - 5)^2$ **7.** $(x + 11y)(x + 3y)$ **9.** Unfactorable
11. $(3x + 1)(x + 3)$ **13.** $(5x - 7y)(x + y)$ **15.** $(2n + 9)(n + 3)$ **17.** $(3x + 7y)(x - 3y)$ **19.** $(3x + 5)^2$
21. $(3x + 1)(2x + 3)$ **23.** $(8x - 3)(x - 4)$ **25.** $(5y - 2)(3y + 2)$ **27.** $3(4m^2 - 11mn + 3n^2)$

29. $(-x + 10)(x + 2)$ or $(x - 10)(-x - 2)$ **31.** $2(x - 4)(x - 2)$ **33.** $y^2(y - 9)(y + 4)$
35. $(x + 2)(x - 2)(x + 3)(x - 3)$ **37.** $16(t + 3)(t - 1)$ **39.** $4(I - 3)(I + 33)$ **41.** $(2Q + 7)(Q - 3)$

Exercises 6-4, p. 229

1. $\dfrac{15}{24}$ **3.** $\dfrac{6xy}{3y^2}$ **5.** $\dfrac{15mn^2 + 3n^3}{6n^3}$ **7.** $\dfrac{15x}{3x(x + 3)}$ **9.** $\dfrac{2x^2 + 8x}{2x^2 - 10x}$ **11.** $\dfrac{m^2 - m - 6}{m^2 + 6m + 8}$

13. $\dfrac{6y^2 + 13y + 6}{9y^2 - 4}$ **15.** $\dfrac{3}{7}$ **17.** $\dfrac{2}{3x}$ **19.** $\dfrac{3a}{5b}$ **21.** $\dfrac{1}{2}$ **23.** $\dfrac{x + 3y}{x - 3y}$ **25.** Irreducible

27. $\dfrac{1}{x - 5}$ **29.** $\dfrac{x}{x + 5}$ **31.** $\dfrac{x - 4}{x + 2}$ **33.** $\dfrac{y - 9}{3y + 9}$ **35.** $\dfrac{x + 4}{x - 4}$ **37.** $\dfrac{-3 - x}{x + 7}$ **39.** $\dfrac{x - 4}{x + 1}$

41. $\dfrac{3x - 1}{3x + 1}$ **43.** $\dfrac{-4 - 2x}{x - 2}$

Exercises 6-5, p. 233

1. $\dfrac{15}{28}$ **3.** $\dfrac{1}{2}$ **5.** $\dfrac{5x^2y}{14}$ **7.** $\dfrac{1}{3m}$ **9.** $2x^2 - 8y^2$ **11.** $\dfrac{4x^2 - 8x}{3x + 9}$ **13.** -1 **15.** $x^2 - xy - 6y^2$

17. $\dfrac{11}{15}$ **19.** $\dfrac{27}{16}$ **21.** $\dfrac{3q^2}{4mp^2}$ **23.** $\dfrac{4}{v}$ **25.** $\dfrac{12}{10y - 15}$ **27.** $\dfrac{1}{y^2 + 8y + 16}$ **29.** 1 **31.** $\dfrac{1}{a}$

33. $\dfrac{9m}{2x^2}$ **35.** $\dfrac{3(m - 3)^2(m + 3)}{(m - 4)^2(m + 2)^2}$ **37.** $\dfrac{q}{\varepsilon_0}$

Exercises 6-6, p. 239

1. $\dfrac{13}{9}$ **3.** $-\dfrac{1}{x}$ **5.** 2 **7.** $\dfrac{-2}{x + 3}$ **9.** $\dfrac{3}{2}$ **11.** $\dfrac{29}{144}$ **13.** $\dfrac{2 - 3y}{4y}$ **15.** $\dfrac{5m + a^2m}{a^3}$ **17.** $\dfrac{10y - 5x}{12x^2}$

19. $\dfrac{8a - 15b + 7a^2}{12a^2b}$ **21.** $\dfrac{y + 3}{y - 6}$ **23.** $\dfrac{x^2 + 3x + 28}{x^2 - 16}$ **25.** $\dfrac{16 - 15x}{10x + 20}$ **27.** $\dfrac{6m - 25}{2m^2 + 10m}$ **29.** $\dfrac{12x + 21}{x^2 - 9}$

31. $\dfrac{-y^2 + 22y - 7}{9y^2 - 1}$ **33.** $\dfrac{-2x^2 + 15x - 46}{(2x - 7)(x + 2)(2x + 5)}$ **35.** $\dfrac{x^2 + 13x - 2}{x^2 + 3x - 10}$ **37.** $\dfrac{10}{x}$ **39.** $\dfrac{x^2 + x}{8x + 6}$

41. $\dfrac{x^2 - 3}{1 - 3x}$ **43.** $\dfrac{-x^4 + 6x^2 + 2x - 4}{x^3 - 5x}$ **45.** $\dfrac{4mD^2}{L^2}$ **47.** $\dfrac{4L^2d^2 + d^4}{8L^3}$ **49.** $\dfrac{2rd}{\left(r - \dfrac{d}{2}\right)^2\left(r + \dfrac{d}{2}\right)^2}$

51. $\dfrac{nC}{s + nC}$

Exercises 6-7, p. 245

1. -4 **3.** -6 **5.** $\dfrac{50}{9}$ **7.** 32 **9.** $-\dfrac{39}{2}$ **11.** $\dfrac{2}{5}$ **13.** 10 **15.** 23 **17.** 2 **19.** $\dfrac{5}{11}$

21. $\dfrac{1}{3}$ **23.** No solution **25.** 9 **27.** No solution **29.** $-\dfrac{20}{3}$ **31.** $-\dfrac{15}{2}$ **33.** $y = \dfrac{9 - 10x}{10}$

35. $c = \dfrac{22 - 8a}{2a - 11}$ **37.** $R_1 = \dfrac{VR_2 + VR_3}{E - V}$ **39.** $S = \dfrac{V}{1 - G}$ **41.** $L_1 = \dfrac{M^2 - 2ML + LL_2}{L_2 - L}$ **43.** 6.0 V

45. 6 h **47.** 40 min

Exercises 6-8, p. 247

1. $-8x + 32y$ **3.** $20ac - 35c^2$ **5.** $y^2 - 25$ **7.** $m^2 - 8m + 16$ **9.** $x^2 + 12x + 35$
11. $15w^2 + 34w - 16$ **13.** $4x^2 - 4$ **15.** $-8y^3 + 56y^2 - 98y$ **17.** $6(x - 4)$ **19.** $2x^2(4x + 5)$
21. $4(2x - 3y + 4z)$ **23.** $(x + 6)(x - 6)$ **25.** $(4a + 5)(4a - 5)$ **27.** $7x(x + 2)(x - 2)$ **29.** $4(4x^2 + 9)$
31. $(x + 6)(x - 2)$ **33.** $(m + 7n)(m + 2n)$ **35.** $(3x + 5)(x - 1)$ **37.** $(3x - 4)^2$ **39.** $(2x + 5)(x - 3)$

41. $(3x + 4y)(2x - 5y)$ **43.** $x(4x - 3)(3x + 8)$ **45.** $\dfrac{5}{x}$ **47.** $\dfrac{1}{a - b}$ **49.** $\dfrac{2x + 3}{3x + 2}$ **51.** $\dfrac{3x - 3}{4x}$

53. 1 **55.** $\dfrac{8}{x}$ **57.** $-a - 1$ **59.** 2 **61.** $\dfrac{ab - 18b^2 + 15a^2}{6a^2 b^2}$ **63.** $\dfrac{7y + 35}{49 - y^2}$

65. $\dfrac{6x^2 - 11x}{(2x - 1)(x + 3)(2x + 1)}$ **67.** $\dfrac{22}{5}$ **69.** 12 **71.** $-\dfrac{4}{7}$ **73.** $\dfrac{3}{14}$ **75.** -4 **77.** $a = \dfrac{-35}{2x - 17}$

79. $S = \dfrac{BR_1}{R_1^2 + 2R_1 R_T + R_T^2}$ **81.** $F = Q(2Q + 1)$ **83.** $c^2(p^2 + m^2 c^2)$ **85.** $\dfrac{R^2(9a^2 + L^2)}{4L^4}$ **87.** $\dfrac{2\pi f x j}{v j + T t}$

89. $P = \dfrac{Ak^2 r^2}{kr^2 + L^2}$ **91.** $f_2 = \dfrac{f f_1}{f_1 - f}$ **93.** $z_0 = \dfrac{-z\mu z_c z_r}{z z_r - \mu z z_c - z_c z_r}$ **95.** $13\frac{1}{3}$ h

Chapter 7

Exercises 7-1, p. 256

1. $5x^2 - 9x - 4 = 0$; $a = 5, b = -9, c = -4$ **3.** $2x^2 - 5 = 0$; $a = 2, b = 0, c = -5$ **5.** Not quadratic
7. $-8x^2 + 3x = 0$; $a = -8, b = 3, c = 0$ **9.** Not quadratic   **11.** $x^2 - 6x + 9 = 0$; $a = 1, b = -6, c = 9$

13. $8, 4$ **15.** $7, -5$ **17.** $9, -5$ **19.** $\dfrac{5}{2}, 2$ **21.** $-\dfrac{1}{5}, -3$ **23.** $4, -4$ **25.** $\dfrac{7}{3}, -\dfrac{7}{3}$ **27.** $0, -9$

29. $0, 3$ **31.** -3 **33.** $\dfrac{1}{2}, -\dfrac{3}{2}$ **35.** $\dfrac{3}{2}, 4$ **37.** $7, 8$ **39** $\dfrac{4}{5}, -1$ **41.** $-11, 10$ **43.** 10 s

45. 250 gal/min   **47.** 9 m by 13 m   **49.** 2 ft high by 6 ft wide or 3 ft high by 4 ft wide

Exercises 7-2, p. 263

1. ± 4 **3.** $\pm\dfrac{11}{3}$ **5.** $\pm 2\sqrt{2}$ **7.** $\pm\dfrac{\sqrt{7}}{2}$ **9.** $9, 1$ **11.** $-2 \pm 2\sqrt{2}$ **13.** $-6 \pm \sqrt{5}$ **15.** $-8, -2$

17. $-7, 6$ **19.** $-3 \pm \sqrt{3}$ **21.** $6.275, -1.275$ **23.** $3, \dfrac{5}{3}$ **25.** $2.081, -0.481$,   **27.** $\dfrac{-1 \pm \sqrt{73}}{12}$

29. $\dfrac{1 \pm \sqrt{145}}{2}$ **31.** $-2m \pm \sqrt{4m^2 - n}$ **33.** Fast outlet: 6.8 h; slow outlet: 9.8 h
35. 4.7 ft wide by 16.7 ft long

Exercises 7-3, p. 268

1. $-1, -4$ **3.** $7, -4$ **5.** -3 **7.** $\dfrac{-5 \pm \sqrt{61}}{2}$ **9.** $4, -\dfrac{7}{2}$ **11.** $\dfrac{3 \pm \sqrt{89}}{10}$

13. No real solution   **15.** $\dfrac{7 \pm \sqrt{89}}{4}$ **17.** $\dfrac{\pm 2\sqrt{15}}{5}$ **19.** $0, \dfrac{7}{2}$ **21.** $0.189, -3.523$

23. $-1.102, 0.227$ **25.** $\dfrac{3 \pm \sqrt{15}}{2}$ **27.** $\dfrac{-5t \pm \sqrt{25t^2 - 12}}{2}$ **29.** $150°C$ or $71°C$

31. $t = \dfrac{-v_0 + \sqrt{v_0^2 + 2as}}{a}$ **33.** 990 km/h   **35.** 18 in. by 16 in.

Exercises 7-4, p. 269

1. $9, -9$ **3.** $0, -\dfrac{10}{3}$ **5.** -5 **7.** $-6, -3$ **9.** $8, -4$ **11.** $5, -\dfrac{3}{5}$ **13.** ± 5 **15.** $\dfrac{\pm 2\sqrt{3}}{7}$

17. $13, -7$ **19.** $8, -6$ **21.** -6 **23.** $-5 \pm 2\sqrt{7}$ **25.** $\dfrac{-5 \pm \sqrt{17}}{2}$ **27.** $\dfrac{1 \pm \sqrt{1 + 3a^2}}{a}$

29. 1.290, 0.310   **31.** 0.393, −3.393   **33.** No real roots   **35.** 1.175, −0.425   **37.** 1.472, −7.472
39. 10.179, −1.179   **41.** 1.372, −4.372   **43.** 25 A, −40 A   **45.** Approx 12 s   **47.** 170°C, 50°C
49. 6 Ω and 12 Ω   **51.** 9 in. by 12 in. or 6 in. by 18 in.   **53.** A: 7.5 h; B: 8.5 h   **55.** $R = \dfrac{4X}{3}$
57. 0.019 mole/L

Chapter 8

Exercises 8-1, p. 277

1. +   **3.** +   **5.** −   **7.** +   **9.** −   **11.** +   **13.** −   **15.** −   **17.** −   **19.** −   **21.** +
23. +   **25.** I, IV   **27.** I, II   **29.** I, III   **31.** IV   **33.** III   **35.** IV   **37.** 90°   **39.** 0°, 180°
41. 0°, 180°   **43.** 270°

	sin	cos	tan	csc	sec	cot
45.	0.3162	0.9487	0.3333	3.1623	1.0541	3
47.	−0.6	−0.8	0.75	−1.6667	−1.25	1.3333
49.	−0.4472	0.8944	−0.5	−2.2361	1.1180	−2
51.	0.3162	−0.9487	−0.3333	3.1623	−1.0541	−3

Exercises 8-2, p. 284

1. −0.469   **3.** −0.682   **5.** 0.987   **7.** 0.423   **9.** −0.748   **11.** 0.833   **13.** 1.691   **15.** 6.596
17. −1.235   **19.** −0.734   **21.** 0.515   **23.** −1.257   **25.** 0.985   **27.** 0.737   **29.** −sin 80°
31. −tan 61°23′   **33.** −cos 67°   **35.** sec 41°   **37.** −cot 24.4°   **39.** $-\dfrac{1}{2}$ **41.** 1   **43.** $-\dfrac{\sqrt{3}}{2}$
45. $\theta_I = 23.6°$, $\theta_{II} = 156.4°$ **47.** $\theta_I = 55.7°$, $\theta_{IV} = 304.3°$ **49.** $\theta_{II} = 125.5°$, $\theta_{IV} = 305.5°$
51. $\theta_{II} = 113.3°$, $\theta_{III} = 246.7°$ **53.** $\theta_I = 17.9°$, $\theta_{III} = 197.9°$ **55.** $\theta_{II} = 134.4°$, $\theta_{III} = 225.6°$
57. $\theta_{III} = 233°31′$, $\theta_{IV} = 306°29′$ **59.** $\theta_I = 49°59′$, $\theta_{IV} = 310°1′$ **61.** $\theta_I = 61°30′$, $\theta_{III} = 241°30′$
63. $\theta_{II} = 165°3′$, $\theta_{IV} = 345°3′$ **65.** $\theta_I = 30°$, $\theta_{II} = 150°$ **67.** $\theta_{II} = 150°$, $\theta_{III} = 210°$ **69.** $\theta_{II} = 150°$, $\theta_{IV} = 330°$
71. −0.8997   **73.** 2.246   **75.** 292°   **77.** 150°   **79.** 18.4 mA   **81.** 6103 ft

Exercises 8-3, p. 291

1. IV   **3.** III   **5.** I   **7.** II   **9.** II   **11.** I   **13.** $\dfrac{\pi}{3}$ **15.** $\dfrac{55\pi}{36}$ **17.** $\dfrac{2\pi}{5}$ **19.** $\dfrac{10\pi}{9}$ **21.** $\dfrac{5\pi}{18}$
23. $\dfrac{7\pi}{4}$ **25.** 0.98   **27.** 2.02   **29.** 3.92   **31.** 5.10   **33.** −0.74 or 5.54   **35.** 1.31   **37.** 225°
39. 210°   **41.** 105°   **43.** 112.5°   **45.** 247.5°   **47.** 255°   **49.** 80.2°   **51.** 264.7°   **53.** 297.9°
55. 175.3°   **57.** 230°44′   **59.** 345°57′   **61.** $-\dfrac{\sqrt{3}}{2}$ **63.** $-\dfrac{\sqrt{3}}{2}$ **65.** $-\sqrt{3}$ **67.** −3.732   **69.** −2
71. −0.707   **73.** 0.878   **75.** −0.763   **77.** −0.616   **79.** 1.073   **81.** −6.308   **83.** 0.270
85. $\theta_I = \dfrac{\pi}{6}$, $\theta_{IV} = \dfrac{11\pi}{6}$ **87.** $\theta_{II} = \dfrac{3\pi}{4}$, $\theta_{IV} = \dfrac{7\pi}{4}$ **89.** $\theta_{III} = \dfrac{7\pi}{6}$, $\theta_{IV} = \dfrac{11\pi}{6}$ **91.** $\theta_I = 1.16$, $\theta_{IV} = 5.12$
93. $\theta_{III} = 3.68$, $\theta_{IV} = 5.74$ **95.** $\theta_I = 1.31$, $\theta_{III} = 4.45$ **97.** $\theta_{II} = 2.75$, $\theta_{IV} = 5.89$ **99.** $\theta_I = 0.28$, $\theta_{II} = 2.86$
101. $\theta_{II} = 1.79$, $\theta_{III} = 4.50$ **103.** −4.7 A   **105.** −0.093 m   **107.** 0.75 s

Exercises 8-4, p. 297

1. 3.40 in. **3.** 9.0 m **5.** 3.68 **7.** 91.8 mm **9.** 187 in.² **11.** 229 m² **13.** 2.20 **15.** 42.7 cm
17. 3.01 in. **19.** 358 ft **21.** 180 ft **23.** 376 in.² **25.** 113.9° **27.** 510 in./s **29.** 5.40 ft/s
31. 66 rad/s

Exercises 8-5, p. 301

	sin	cos	tan	csc	sec	cot
1.	0.3846	−0.9231	−0.4167	2.6	−1.0833	−2.4
3.	−0.7593	0.6508	−1.1667	−1.3171	1.5366	−0.8571

5. $-\cos 28°$ **7.** $-\tan 36°$ **9.** $\sin 78°45'$ **11.** $-\sec 42°$ **13.** $\dfrac{55\pi}{36}$ **15.** $\dfrac{11\pi}{9}$ **17.** 0.86 **19.** 3.78

21. 45° **23.** 292.5° **25.** 63.0° **27.** 140.4° **29.** −0.839 **31.** −1.111 **33.** −0.784 **35.** 1.274

37. −1.956 **39.** 0.924 **41.** 0.577 **43.** 0.383 **45.** 1.238 **47.** 0.130 **49.** $-\dfrac{1}{2}$ **51.** $-\sqrt{3}$

53. $\dfrac{\sqrt{3}}{2}$ **55.** 131.5°, 228.5° **57.** 20.9°, 200.9° **59.** 204°12′, 335°48′ **61.** 113°35′, 246°25′

63. 1.94, 5.08 **65.** 1.06, 5.22 **67.** 3.43, 6.00 **69.** 1.26, 4.40 **71.** 60°, 120° **73.** 120°, 300°

75. $\dfrac{\pi}{4}, \dfrac{7\pi}{4}$ **77.** 24.9 mA **79.** 78.1 ft **81.** 10.6 in. **83.** 3.5 **85.** 0.586 m/s **87.** 0.107 mi/h

89. 5.8 ft/s **91.** 21.5 cm **93.** 68 in.² **95.** 18 mi²

Chapter 9

Exercises 9-1, p. 312

1. Scalar **3.** Vector **5.** Scalar **7.** Vector **9.** Vector **11.** Scalar

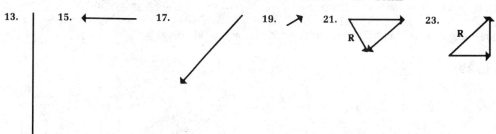

27. $R = 3, \theta = 330°$ **29.** $R = 3, \theta = 105°$ **31.** $R = 6, \theta = 35°$

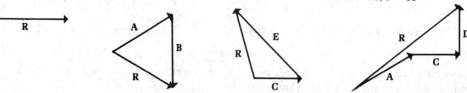

33. $R = 2.0, \theta = 35°$ **35.** $R = 1, \theta = 90°$ **37.** $R = 5, \theta = 270°$ **39.** $R = 6, \theta = 165°$

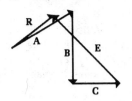
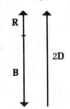
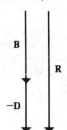
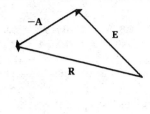

41. $R = 8, \theta = 70°$ **43.** $R = 15, \theta = 345°$

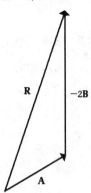
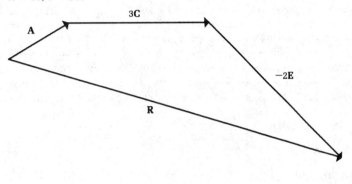

Exercises 9-2, p. 319

1. $A_x = 1.55$ ft/s, $A_y = 0.790$ ft/s   **3.** $C_x = -172$ mi, $C_y = 228$ mi   **5.** $E_x = -0.99$ lb, $E_y = -5.6$ lb
7. $G_x = 995$ N, $G_y = -757$ N   **9.** $I_x = -67, I_y = 52$ **11.** $K_x = -229, K_y = -114$ **13.** $R = 6.6$ m/s, $\theta_R = 29°$
15. $T = 33.8$ ft/s², $\theta_T = 327.4°$ **17.** $V = 897$ lb, $\theta_V = 162.2°$ **19.** $L = 91.4$ mi, $\theta_L = 214.4°$
21. $R = 78.4$ lb, $\theta = 37.7°$ **23.** $R = 429$ mi/h, $\theta = 81.8°$ **25.** $R = 2.21$ m/s², $\theta = 264.5°$
27. $R = 92.0$ cm, $\theta = 23.9°$ **29.** $R = 330, \theta = 61.0°$ **31.** $R = 164, \theta = 162.7°$ **33.** $R = 29.7, \theta = 175.3°$
35. $R = 397, \theta = 37.2°$

Exercises 9-3, p. 325

1. 9.8 mi, 52° north of west   **3.** 400 N, 62.5° from the 185 N force   **5.** 677 mi/h, 4.7° north of east
7. 3120 km, 19.5° north of west   **9.** 540 mi/h, 53° north of west   **11.** 733 lb   **13.** 3.85 ft   **15.** 18.7 lb
17. 5.83 lb   **19.** 9.6 in. at 29°   **21.** Rope 1: 41 lb, rope 2: 77 lb, rope 3: 65 lb

Exercises 9-4, p. 332

1. $C = 75°, c = 9.2, b = 8.4$ **3.** $A = 39.0°, a = 202, c = 301$
5. $A = 55.9°, C = 94.1°, c = 116$; or $A = 124.1°, C = 25.9°, c = 50.6$ **7.** $C = 23.4°, A = 44.6°, a = 13.2$
9. $A = 112.0°, b = 36.0, c = 52.1$ **11.** $B = 27.9°, b = 3.79, a = 5.87$
13. $A = 83.3°, B = 19.3°, b = 174$; or $A = 96.7°, B = 5.9°, b = 53.5$ **15.** $A = 14.2°, C = 127.0°, c = 12.4$
17. $C = 64°30', a = 0.0432, b = 0.0706$ **19.** $C = 65°20', A = 70°0', a = 7.75$; or $C = 114°40', A = 20°40', a = 2.92$
21. 34° west of south   **23.** $v_c = 1.8$ knots, $v_b = 7.3$ knots   **25.** 41 ft   **27.** 36.3 ft

Exercises 9-5, p. 337

1. $c = 5.6$, $A = 92°$, $B = 47°$ **3.** $b = 690$, $A = 62.4°$, $C = 15.6°$ **5.** $C = 94.2°$, $A = 37.5°$, $B = 48.3°$
7. $C = 140.1°$, $B = 17.8°$, $A = 22.1°$ **9.** $b = 78$, $A = 65°$, $C = 49°$ **11.** $c = 5490$, $A = 136.2°$, $B = 16.0°$
13. $B = 149.9°$, $A = 7.1°$, $C = 23.0°$ **15.** $A = 94.3°$, $B = 61.6°$, $C = 24.1°$ **17.** $a = 0.865$, $B = 23°0'$, $C = 38°30'$
19. $B = 141.5°$, $C = 26.7°$, $A = 11.8°$ **21.** 8.7 mi   **23.** 52.1°   **25.** 182 ft   **27.** 6.6 knots, 33° north of east

Exercises 9-6, p. 339

1. $A_x = 9.8$ m/s, $A_y = 11$ m/s   **3.** $C_x = -102$ lb, $C_y = -223$ lb   **5.** $v_x = 8.6$, $v_y = -4.4$
7. $E = 31.5$ lb, $\theta = 62.4°$ **9.** $G = 0.64$ m/s², $\theta = 210°$ **11.** $R = 89.6$ lb, $\theta = 33.2°$
13. $R = 4.59$ ft/s, $\theta = 248.5°$ **15.** $R = 1.22$ in., $\theta = 296.2°$ **17.** $R = 313$, $\theta = 215.0°$
19. $C = 76.0°$, $c = 19.2$, $b = 12.4$ **21.** $c = 9.8$, $A = 53°$, $B = 31°$ **23.** $b = 190$, $A = 39.8°$, $C = 93.2°$
25. $c = 53.1$, $A = 107.0°$, $a = 95.9$ **27.** $A = 155.6°$, $B = 14.8°$, $C = 9.6°$ **29.** $A = 24.9°$, $B = 38.9°$, $b = 1730$
31. $C = 52°$, $B = 91°$, $b = 430$; or $C = 128°$, $B = 15°$, $b = 110$ **33.** $C = 94.7°$, $A = 22.0°$, $B = 63.3°$
35. 8.4 mi/h, 17° downstream from the perpendicular direction   **37.** 100 lb at 3° clockwise from the dashed line
39. 4.4 lb   **41.** 710 mi/h at 44° north of west   **43.** 860 ft/s, 7° above horizontal
45. $v_c = 1.53$ knots, $v_b = 12.6$ knots   **47.** 44 in.   **49.** 13 knots, 44° south of west   **51.** 1.08 in.

Chapter 10

Exercises 10-1, p. 350

1.
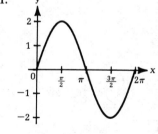

3.
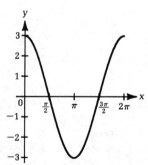

5.
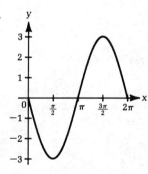

7.
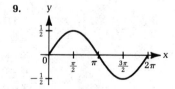

9.

11.
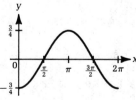

13.
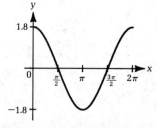

15.

Exercises 10-2, p. 355

1. Period: $\frac{\pi}{2}$; Amplitude: 1   **3.** Period: 3π; Amplitude: 1   **5.** Period: $\frac{\pi}{3}$; Amplitude: 3

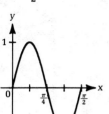

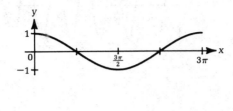

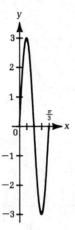

7. Period: 1; Amplitude: 2   **9.** Period: $\frac{\pi}{4}$; Amplitude: 1   **11.** Period: 4; Amplitude: 3

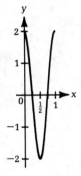

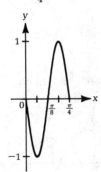

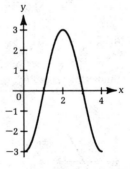

13. Period: $\frac{2\pi}{5}$; Amplitude: $\frac{1}{2}$ **15.** Period: $\frac{1}{2}$; Amplitude: $\frac{3}{4}$ **17.** Period: $\frac{8\pi}{3}$; Amplitude: 2.8

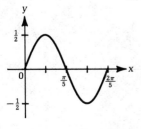

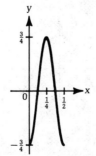

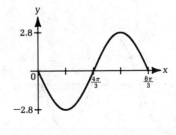

19. Period: 0.2; Amplitude: 0.1

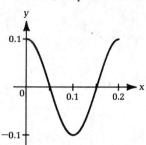

21.

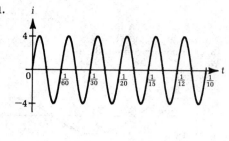

23.

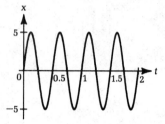

Exercises 10-3, p. 359

1. Period: 2π; Amplitude: 1; Phase shift: $-\dfrac{\pi}{3}$

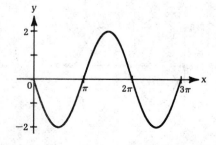

3. Period: 2π; Amplitude: 1; Phase shift: $\dfrac{\pi}{8}$

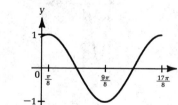

5. Period: 2π; Amplitude: 2; Phase shift: π

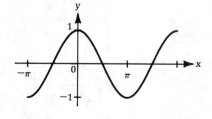

7. Period: 2π; Amplitude: 1; Phase shift: $-\pi$

9. Period: 4π; Amplitude: $\dfrac{3}{2}$; Phase shift: $-\dfrac{\pi}{4}$

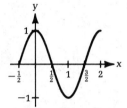

11. Period: 6π; Amplitude: 4; Phase shift: $\dfrac{\pi}{2}$

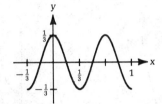

13. Period: 2; Amplitude: 1; Phase shift: $-\dfrac{1}{2}$

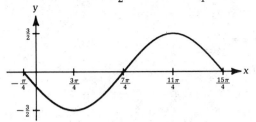

15. Period: $\dfrac{2}{3}$; Amplitude: $\dfrac{1}{3}$; Phase shift: $-\dfrac{1}{3}$

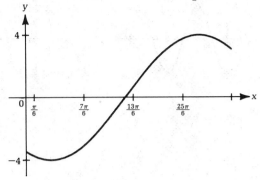

17. Period: 4; Amplitude: $\dfrac{2}{3}$; Phase shift: $\dfrac{2}{\pi} \approx 0.64$

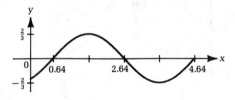

19. Period: 1; Amplitude: 2; Phase shift: $\dfrac{2}{\pi} \approx 0.64$

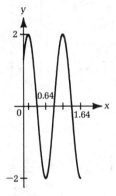

21. Period: $\dfrac{2\pi}{3}$; Amplitude: 0.4; Phase shift: $-\dfrac{\pi}{24}$

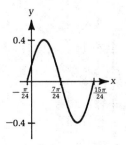

23. Period: 4; Amplitude: 2.6; Phase shift: $-\dfrac{2}{\pi} \approx -0.64$

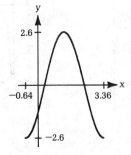

25.

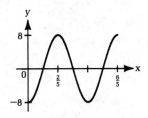

27.

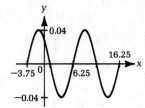

Exercises 10-4, p. 365

1.

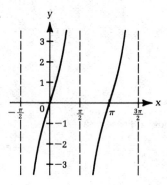

3.

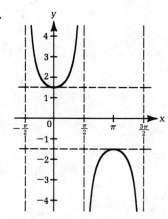

5.

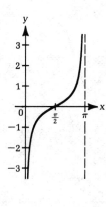

7.

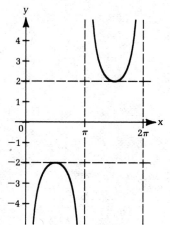

9.

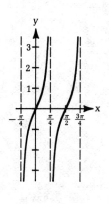

11.

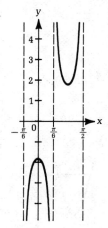

13.

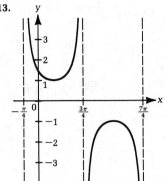

15.

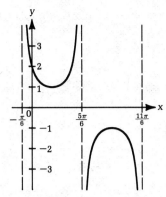

17.

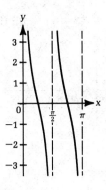

19.

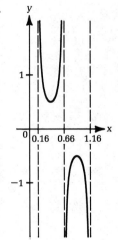

Exercises 10-5, p. 371

1. (a) $|a| = 6.0$ ft, $f = 1$ Hz, $T = 1$ s, $\phi = -\dfrac{\pi}{4}$ (b) -4.2 ft (c) At $t = \dfrac{3}{8}$ s, $1\dfrac{3}{8}$ s, ...

(d)

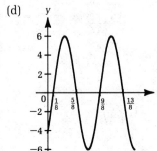

3. (a) $|a| = 0.80$ ft, $f = 1.5$ Hz, $T = \frac{2}{3}$ s, $\phi = \pi$ (b) 0   (c) At $t = \frac{1}{2}$ s, $1\frac{1}{6}$ s, ...

(d)

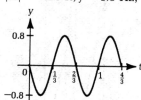

5. $y = 3 \sin\left(\pi t + \dfrac{\pi}{3}\right)$ **7.** $y = 0.60 \sin\left(\dfrac{8\pi t}{3} - \dfrac{\pi}{6}\right)$

9.

11.

13.

15.

Exercises 10-6, p. 377

1.

3.

5.

7.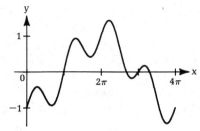

9.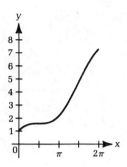

11.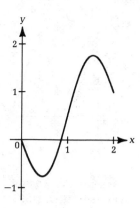

13.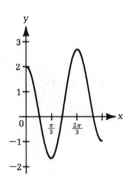

15.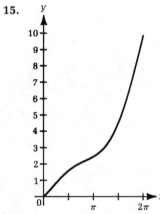

17.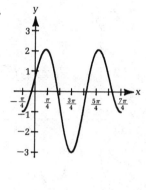

19.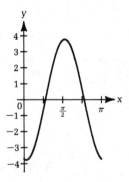

21.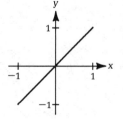

23.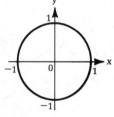

25.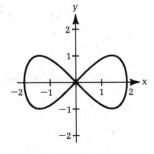

27.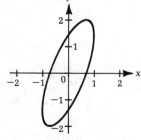

29.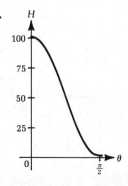

31.

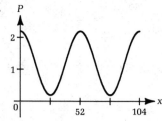

33.

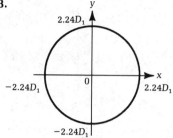

Exercises 10-7, p. 379

1.

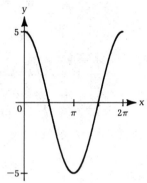

3.

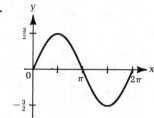

5.

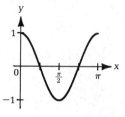

7.

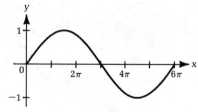

9.

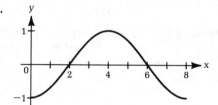

11.

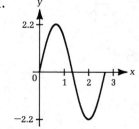

13.

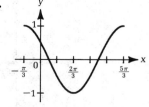

15.

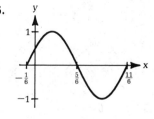

17.

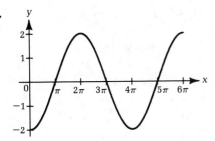

19.

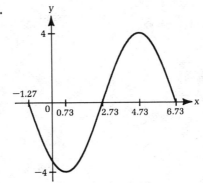

21.

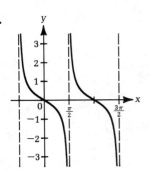

23.

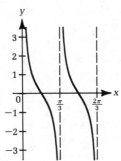

25.

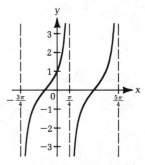

27.

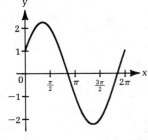

29.

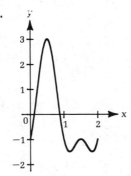

31.

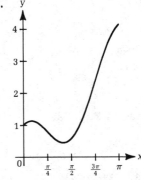

33.

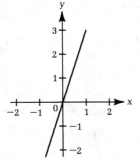

35.

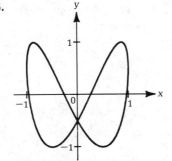

37. $y \approx 0.38$ cm at $t = 2.0$ s

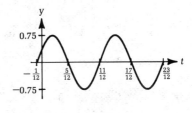

39. $i \approx -2.2$ A at $t = \dfrac{1}{90}$ s

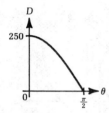

41. $y = 0.8 \sin \left(\dfrac{\pi t}{2} + \dfrac{\pi}{4} \right)$

43.

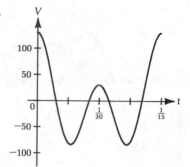

45.

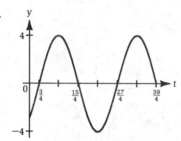

47.

Chapter 11

Exercises 11-1, p. 385

1. a^4 **3.** $\dfrac{1}{y^4}$ **5.** $\dfrac{8y^3}{x^6}$ **7.** $\dfrac{n^4}{16m^2}$ **9.** $\dfrac{1}{a^4}$ **11.** $\dfrac{c^2}{d^3}$ **13.** $\dfrac{ab^6}{c^5}$ **15.** $\dfrac{y^{12}}{x^{10}}$ **17.** $\dfrac{x^4}{4}$ **19.** $\dfrac{1}{w^{15}v^{10}}$

21. 5   **23.** 1   **25.** $\dfrac{1}{x+2}$ **27.** $\dfrac{y+3}{3y}$ **29.** $\dfrac{4}{a^5b^4}$ **31.** $\dfrac{y^3+xy}{x^2}$ **33.** $\dfrac{x^5-2y^2}{2y^2x^2}$ **35.** $\dfrac{mn}{m+n}$

37. $\dfrac{b^2}{(3ab+1)^2}$ **39.** $\dfrac{x+y}{y-x}$ **41.** $\dfrac{2n^2+m^2}{mn(n-m)}$ **43.** $\dfrac{4x-1}{(x-4)^3}$ **45.** $\dfrac{\text{kg}\cdot\text{m}^2}{\text{s}^3}$ **47.** $e^{1.8}$

Exercises 11-2, p. 390

1. 4   **3.** 2   **5.** 48   **7.** $\dfrac{1}{10}$ **9.** 25   **11.** 64   **13.** $\dfrac{1}{10}$ **15.** $-\dfrac{1}{8}$ **17.** 7.48   **19.** 31.29

21. 1.95   **23.** 16.40   **25.** 3.03   **27.** 3.32   **29.** 0.74   **31.** 2.01   **33.** 0.46   **35.** 1.44   **37.** $x^{3/4}$

39. $\dfrac{1}{a^{1/2}}$ **41.** $w^{3/2}$ **43.** m **45.** $y^{1/4}$ **47.** $m^{5/3}$ **49.** $x^{7/12}$ **51.** $b^{1/4}$ **53.** $\dfrac{9}{x^{1/2}}$ **55.** $m^{11/18}$

57. $\dfrac{y^{1/6}}{x}$ **59.** $\dfrac{1+32x}{8x^3}$ **61.** $\dfrac{-3}{(x+5)^{1/2}}$ **63.** 0.67 or 67%   **65.** \$17,900,000, rounded   **67.** 3.0 ft³/s

Exercises 11-3, p. 396

1. $3\sqrt{2}$ **3.** $5\sqrt{5}$ **5.** $x^2y\sqrt{x}$ **7.** $v^2w^2\sqrt{vw}$ **9.** $c^3\sqrt{7}$ **11.** $3m^2n^3\sqrt{3m}$ **13.** $2\sqrt[3]{3}$ **15.** $-2\sqrt[5]{2}$

17. $5m\sqrt[3]{m}$ **19.** $2b\sqrt[4]{3a^3b^2}$ **21.** $\sqrt[6]{3}$ **23.** $\sqrt[8]{9x}$ **25.** $\dfrac{3\sqrt{2y}}{2y}$ **27.** $\dfrac{\sqrt[3]{18m^2}}{3m}$ **29.** $\dfrac{\sqrt{15}}{5}$ **31.** $\dfrac{\sqrt{5x}}{5}$

33. $\dfrac{\sqrt[3]{18}}{3}$ **35.** $\dfrac{\sqrt[5]{2401x^3}}{7}$ **37.** $\sqrt[3]{6}$ **39.** $\sqrt[3]{2}$ **41.** $\sqrt{xy}$ **43.** $\dfrac{\sqrt{7}}{2}$ **45.** $x+2$ **47.** $\dfrac{(2x-1)\sqrt{2}}{2}$

49. $\dfrac{y\sqrt{21xy}}{9x^3z}$ **51.** $\dfrac{\sqrt{x^2-1}}{x-1}$ **53.** $\dfrac{\sqrt{5}}{\pi} \approx 0.71$ **55.** $1 + \dfrac{\sqrt{m^2+2mh}}{m}$ or $\dfrac{m+\sqrt{m^2+2mh}}{m}$

Exercises 11-4, p. 398

1. $6\sqrt{2}$ **3.** $-\sqrt{5}$ **5.** $5\sqrt{x}$ **7.** $\sqrt{3}-2\sqrt{2}$ **9.** $2\sqrt{5}-7\sqrt[3]{5}$ **11.** $3\sqrt{y}+\sqrt{2y}$ **13.** $3\sqrt{2}$

15. $-4\sqrt{5}$ **17.** $\sqrt{6}$ **19.** $7\sqrt{3}+30\sqrt{2}$ **21.** $\dfrac{19\sqrt{10}}{10}$ **23.** $5\sqrt{6}$ **25.** $\dfrac{68\sqrt{3}}{3}$ **27.** $5\sqrt[3]{2}$

29. $-\sqrt[4]{3}$ **31.** $(xy+2y^2)\sqrt{y}$ **33.** $(2w^2+3v-5)\sqrt{3v}$ **35.** $-\dfrac{\sqrt{3y}}{3}$ **37.** $\dfrac{b\sqrt{a^2+ab}}{a^2+ab}$ **39.** $\dfrac{\sqrt{3}}{2}x$

Exercises 11-5, p. 401

1. $\sqrt{10}$ **3.** $\sqrt{6xy}$ **5.** $12\sqrt{35}$ **7.** $5\sqrt{2}$ **9.** $2\sqrt[3]{3}$ **11.** $3m$ **13.** 180 **15.** $54x^4$

17. $4x+20$ **19.** $2\sqrt{3}$ **21.** $\sqrt{15}-\sqrt{10}$ **23.** $16\sqrt{5}+30\sqrt{2}$ **25.** $5x\sqrt{y}-10\sqrt{xy}$ **27.** -4

29. $x-4y$ **31.** $16+8\sqrt{5}$ **33.** $3-5\sqrt{14}$ **35.** $6x-19\sqrt{x}+15$ **37.** $21-8\sqrt{5}$

39. $13+4x+12\sqrt{x+1}$ **41.** $\sqrt[6]{1125}$ **43.** $\sqrt[4]{45x^3}$ **45.** 1 **47.** $2+\dfrac{x}{3}+\dfrac{2\sqrt{30x}}{5}$ **49.** $\dfrac{3KA\sqrt[3]{4L^2}}{16L}$

Exercises 11-6, p. 405

1. $\sqrt{3}$ **3.** y^2 **5.** $2\sqrt{2}$ **7.** $x\sqrt[3]{x}$ **9.** $\dfrac{\sqrt{2}}{4}$ **11.** $\dfrac{\sqrt{5y}}{2y^2}$ **13.** $\dfrac{\sqrt{66}}{6}$ **15.** $\dfrac{2\sqrt{15xy}}{5y}$ **17.** $\sqrt[6]{18}$

19. $\dfrac{\sqrt{5x}+\sqrt{10}}{5}$ **21.** $\dfrac{3\sqrt{2}-\sqrt{30}}{3}$ **23.** $\dfrac{3\sqrt{5}+3}{4}$ **25.** $\dfrac{5\sqrt{m}-5\sqrt{n}}{m-n}$ **27.** $-12-5\sqrt{6}$

29. $\dfrac{6x-3\sqrt{x}}{4x-1}$ **31.** $\dfrac{21+4\sqrt{5}}{19}$ **33.** $\dfrac{6-5\sqrt{2y}+2y}{4-2y}$ **35.** $\dfrac{11-\sqrt{2}}{7}$ **37.** $\dfrac{2x-3\sqrt{xy}+y}{4x-y}$

39. $d=\dfrac{2f\sqrt{3}}{3}$

Exercises 11-7, p. 406

1. $\dfrac{1}{m^2}$ **3.** $\dfrac{1}{x}$ **5.** $\dfrac{3b^4}{a^2}$ **7.** $\dfrac{y^2}{x^6}$ **9.** $\dfrac{-3}{x^5}$ **11.** $\dfrac{1}{m^{1/9}n^{4/9}}$ **13.** $\dfrac{1}{(x+y)^2}$ **15.** $\dfrac{1-8x}{2x^2}$ **17.** $\dfrac{a}{1+2ab}$

19. $\dfrac{1}{x+2}$ **21.** 3 **23.** 54 **25.** 7.29 **27.** 11.39 **29.** 1953.13 **31.** 0.91 **33.** $7\sqrt{2}$

35. $4x^2\sqrt{5xy}$ **37.** $2n^2\sqrt[3]{2n}$ **39.** $\sqrt[4]{25y}$ **41.** $\dfrac{\sqrt{55x}}{5x}$ **43.** $\dfrac{\sqrt[3]{7x^2}}{x}$ **45.** $\sqrt[4]{5xy^2}$ **47.** $6\sqrt{7}$

49. $3\sqrt{y}+3\sqrt{3y}$ **51.** $24\sqrt{3}-12\sqrt{2}$ **53.** $(2a^2b^2-a^2)\sqrt{b}$ **55.** $\sqrt{15mn}$ **57.** $2n\sqrt[4]{2n}$ **59.** $12x+4$

61. $3\sqrt{6}-12\sqrt{5}$ **63.** 3 **65.** $2x+3\sqrt{xy}-2y$ **67.** $x-1+4\sqrt{x-5}$ **69.** $2\sqrt{6}$ **71.** $\dfrac{3\sqrt{5}-\sqrt{30}}{3}$

73. $-8-5\sqrt{3}$ **75.** $\dfrac{kg}{m \cdot s^2}$ **77.** $5.2 \text{ ft}^3/\text{s}$ **79.** $s=\dfrac{c\sqrt{2}}{2}$ **81.** $4\sqrt[4]{450} \approx 18 \text{ ft}$

Chapter 12

Exercises 12-1, p. 414

1. $3j$ **3.** $j\sqrt{5}$ **5.** $2j\sqrt{3}$ **7.** $\dfrac{j}{2}\sqrt{3}$ **9.** $0.2j$ **11.** $29.2j$ **13.** $-2\sqrt{5}$ **15.** -2 **17.** -10

19. -1 **21.** $-j$ **23.** 1   **25.** j **27.** -2 **29.** $-4j$ **31.** Real: 2, Imaginary: $-3j$
33. Real: -31, Imaginary: $0j$ **35.** Real: 3, Imaginary: $4j$ **37.** Real: -2, Imaginary: $-3j$ **39.** $5 + j$
41. $-6 - 2j$ **43.** $5j$ **45.** 17   **47.** $x = 1, y = -11$ **49.** $x = -5, y = -2$ **51.** $x = \dfrac{2}{3}, y = 2$

53. $x = -4, y = 4$ **55.** $Z = (24 - 24j)\ \Omega$ **57.** $Z = \dfrac{j\omega L}{1 - \omega^2 LC}$

Exercises 12-2, p. 417

1. $8 + 5j$ **3.** $-12 + 8j$ **5.** $19 - 2j$ **7.** $19 - 6j$ **9.** $-2 + 5j$ **11.** $11 - 17j$ **13.** $-30j$
15. $-120j\sqrt{6}$ **17.** $-3j\sqrt{3}$ **19.** $40 + 15j$ **21.** $42 + 70j$ **23.** $18 + j$ **25.** 34   **27.** $-21 + 20j$
29. $88 - 16j$ **31.** $366 - 53j$ **33.** $-1 + 5j\sqrt{2}$ **35.** $-6 - 13j$ **37.** $-\dfrac{15}{34} + \dfrac{9}{34}j$ **39.** $\dfrac{35}{26} - \dfrac{7}{26}j$
41. $\dfrac{6}{5} + \dfrac{2}{5}j$ **43.** $-\dfrac{1}{5} - \dfrac{2}{5}j$ **45.** $\dfrac{46}{85} + \dfrac{37}{85}j$ **47.** $-\dfrac{4}{5} + \dfrac{1}{15}j$ **49.** $\dfrac{11}{17} - \dfrac{27}{17}j$ **51.** $60.72 + 62.6j$
53. -10.147 **55.** $-413{,}504 + 203{,}904j$ **57.** $2a$ **59.** $a^2 + b^2$
61. (a) $2a^2 - 2b^2$ (b) $2a^3 - 6ab^2$

Exercises 12-3, p. 420

1.   **3.**   **5.**

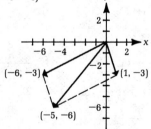

7. $8 - j$

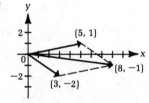

9. $-5 - 6j$

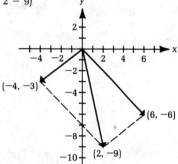

11. $7 - j$

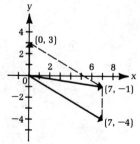

13. $2 - 9j$

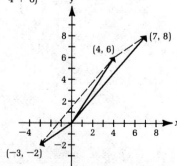

15. $4 + 6j$

17. $6 + j$

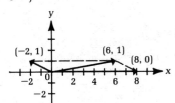

19. $-2 + 6j$

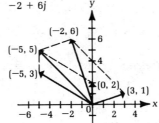

21. $4 - j$

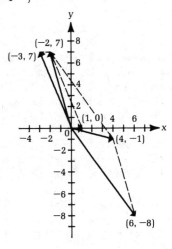

23. 6

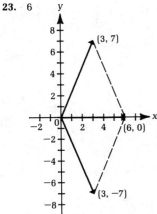

25.

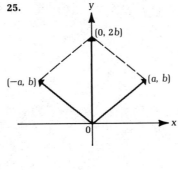

27.

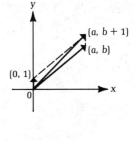

Exercises 12-4, p. 424

1. $5(\cos 306.9° + j \sin 306.9°)$ **3.** $17(\cos 118.1° + j \sin 118.1°)$ **5.** $8.06(\cos 60.3° + j \sin 60.3°)$
7. $9.43(\cos 212.0° + j \sin 212.0°)$ **9.** $4.22(\cos 323.7° + j \sin 323.7°)$ **11.** $2.55\underline{/45°}$ **13.** $7.03\underline{/251.9°}$
15. $6\underline{/0°}$ **17.** $3\underline{/270°}$ **19.** $0.845 + 1.81j$ **21.** $2.56 - 2.39j$ **23.** $4.54 - 16.6j$
25. $-19.0 + 134j$ **27.** 9   **29.** $-12j$ **31.** $0.949 - 0.316j$ **33.** $7.46 - 11.5j$
35. (a) $r = a^b\sqrt{a^2 + b^2}$; $\tan (\text{Ref } \theta) = \left|\dfrac{b}{a}\right|$ (b) $r = ab\sqrt{2(a + b)^2}$; $\tan (\text{Ref } \theta) = 1$

Exercises 12-5, p. 427

1. $3.60e^{0.960j}$ **3.** $12.7e^{4.33j}$ **5.** $0.640e^{1.69j}$ **7.** $485e^{5.44j}$ **9.** $10e^{2.21j}$ **11.** $3.61e^{3.73j}$
13. $23.1e^{0.995j}$ **15.** $2360e^{5.62j}$ **17.** $15.0(\cos 91.7° + j \sin 91.7°)$ **19.** $12.6(\cos 26.0° + j \sin 26.0°)$
21. $2.60(\cos 183.9° + j \sin 183.9°)$ **23.** $-3.62 + 1.71j$ **25.** $-4.55 + 1.37j$ **27.** $1.66 + 1.06j$
29. $0.472e^{5.27j}$, $i_0 \approx 0.472$ A   **31.** -93 V

Exercises 12-6, p. 436

1. $15e^{6j}$ **3.** $9.5e^{4.6j}$ **5.** $3e^{j}$ **7.** $1.81e^{5.14j}$ **9.** $9(\cos 60° + j \sin 60°)$ **11.** $5.55\underline{/88°}$
13. $2\underline{/65°}$ **15.** $2.5(\cos 111° + j \sin 111°)$ **17.** $27(\cos 105° + j \sin 105°)$ **19.** $8.35\underline{/92°}$
Note: For Problems 21–29 the unrounded values of r and θ were used to calculate the rectangular solution.
21. $130(\cos 329.5° + j \sin 329.5°)$ or $112 - 66j$ **23.** $47.3(\cos 222.0° + j \sin 222.0°)$ or $-35.2 - 31.7j$
25. $2.50(\cos 56.3° + j \sin 56.3°)$ or $1.38 + 2.08j$
27. $0.272(\cos 195.5° + j \sin 195.5°)$ or $-0.262 - 0.0727j$
29. $2030(\cos 106.3° + j \sin 106.3°)$ or $-567 + 1940j$
31. $1.41(\cos 15° + j \sin 15°), 1.41(\cos 195° + j \sin 195°)$
33. $2(\cos 45° + j \sin 45°), 2(\cos 165° + j \sin 165°), 2(\cos 285° + j \sin 285°)$
35. $1.59 + 1.45j, -2.05 + 0.655j, 0.459 - 2.11j$
37. $1, 0.309 + 0.951j, -0.809 + 0.588j, -0.809 - 0.588j, 0.309 - 0.951j$
39. $0.924 + 0.383j, -0.383 + 0.924j, -0.924 - 0.383j, 0.383 - 0.924j$
41. $1 + j\sqrt{3}, -2, 1 - j\sqrt{3}$ **43.** $2.82 + 0.815j, -2.82 - 0.815j$

Exercises 12-7, p. 444

1. (a) 157Ω (b) 21.2Ω (c) 136Ω (d) 0.110 A (e) $87.2°$ (f) 0.739 V (g) 17.3 V (h) 2.34 V
3. $Z = 32.5 \Omega, i = 3.7$ A   **5.** $Z = 4.12 \Omega, \phi = -14.0°$ **7.** 64.0 V   **9.** 1.49 A   **11.** 368 Hz
13. (a) 4.7×10^{-13} F (b) 4.5×10^{-6} A (c) 1.2 V   **15.** 2.6 F

Exercises 12-8, p. 446

1. $x = 2, y = -3$ **3.** $x = -\dfrac{9}{5}, y = 5$ **5.** $5 - 4j$ **7.** $-4 + 3j$ **9.** $10 + 16j$ **11.** $-70j$ **13.** 49

15. $-6 + 43j$ **17.** $117 - 44j$ **19.** $-\dfrac{6}{29} + \dfrac{14}{29}j$ **21.** $-\dfrac{60}{149} + \dfrac{42}{149}j$ **23.** $-41 - 4j$

25. $-1 + 5j$ 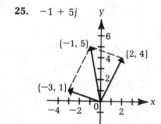   **27.** $3 - j$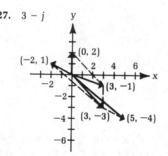

29. $15(\cos 53.1° + j \sin 53.1°), 15e^{0.927j}$ **31.** $3.61(\cos 303.7° + j \sin 303.7°), 3.61e^{5.30j}$
33. $31.9(\cos 299.9° + j \sin 299.9°), 31.9e^{5.23j}$ **35.** $7(\cos 0° + j \sin 0°), 7e^{0j} = 7$ **37.** $2.30 + 1.93j$
39. $-0.965 + 0.262j$ **41.** $-4.00 + 4.47j$ **43.** $1.69 - 4.97j$ **45.** $20e^{1.72j}$ **47.** $6e^{2j}$
49. $20(\cos 65° + j \sin 65°)$ **51.** $4\underline{/60°}$ **53.** $125(\cos 165° + j \sin 165°)$
55. $50(\cos 16.3° + j \sin 16.3°) = 48 + 14j$ **57.** $1.94(\cos 33.7° + j \sin 33.7°) = 1.62 + 1.08j$
59. $100(\cos 286.3° + j \sin 286.3°) = 28 - 96j$ **61.** $4(\cos 60° + j \sin 60°), 4(\cos 240° + j \sin 240°)$
63. $2.01 + 0.594j, -1.52 + 1.44j, -0.491 - 2.04j$ **65.** $1.19, -1.19, 1.19j, -1.19j$
67. $-3, \dfrac{3}{2} + j\dfrac{3\sqrt{3}}{2}, \dfrac{3}{2} - j\dfrac{3\sqrt{3}}{2}$ **69.** $2.92e^{5.25j}, i_0 \approx 2.92$ A

71. (a) 15.7Ω (b) 18.2Ω (c) 8.85Ω (d) 2.71 A (e) $-16.3°$ (f) 23.0 V (g) 42.6 V (h) 49.3 V
73. 54.7 V   **75.** 1080 Hz   **77.** $Z = 8.60 \Omega, \phi = -35.5°$

Chapter 13

Exercises 13-1, p. 453

1. 19,400   **3.** 36.6   **5.** 0.347   **7.** 0.0638   **9.** 13.5   **11.** 1.47

13.
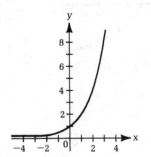

15.
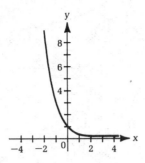

17.
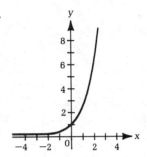

19.
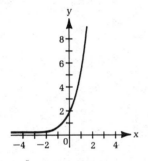

21.
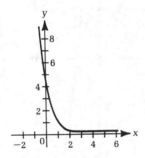

23.
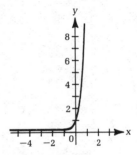

25.
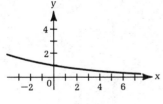

27.
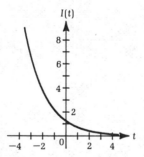

Exercises 13-2, p. 458

1. $4 = \log_2 16$ **3.** $-3 = \log_3\left(\dfrac{1}{27}\right)$ **5.** $\dfrac{1}{4} = \log_{256} 4$ **7.** $2 = \log_{1/2}\left(\dfrac{1}{4}\right)$ **9.** $x = \log_4 y$ **11.** $5 = \log_a c$

13. $2^5 = 32$ **15.** $8^1 = 8$ **17.** $5^{-1} = \dfrac{1}{5}$ **19.** $\left(\dfrac{1}{2}\right)^{-2} = 4$ **21.** $10^n = m$ **23.** $a^y = 12$ **25.** $x = 2$

27. $x = -3$ **29.** $x = 2$ **31.** $N = 10,000$ **33.** $N = 2$ **35.** $N = \dfrac{1}{128}$ **37.** $b = 10$ **39.** $b = 27$

41. $b = \dfrac{1}{7}$

43.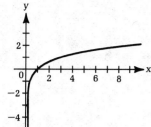

45.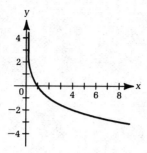

47.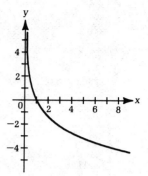

49.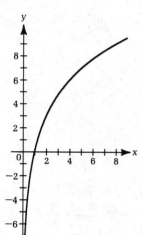

51.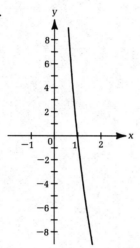

53. $t = -\dfrac{L}{R} \log_e\left(\dfrac{LI}{A}\right)$ **55.** $v = \dfrac{e^{-ct/m}(mg + cv_0) - mg}{c}$

Exercises 13-3, p. 464

1. $\log_2 3 + \log_2 x$ **3.** $\log_5 a - \log_5 b$ **5.** $3 \log_6 x$ **7.** $\dfrac{1}{2} \log_4 11$ **9.** $\log_{16} 3 + \log_{16} x + \log_{16} y$

11. $\log_5 m + \dfrac{1}{4} \log_5 n$ **13.** $\log_3 wy$ **15.** $\log_7\left(\dfrac{3}{p}\right)$ **17.** $\log_6 x^2$ **19.** $\log_2 81$ **21.** $\log_8 15$

23. $\log_c\left(\dfrac{1}{2}\right)$ **25.** $\log_2\left(\dfrac{x^3 y}{5}\right)$ **27.** $\log_2 75$ **29.** 3 **31.** -2 **33.** 3.2 **35.** $\dfrac{1}{2}$ **37.** $\dfrac{4}{3}$ **39.** $-\dfrac{5}{4}$

41. $1 + \log_2 5$ **43.** $\log_4 3 - 1$ **45.** $\dfrac{1}{2} + \dfrac{1}{2} \log_5 3$ **47.** $-2 + \log_6\left(\dfrac{1}{2}\right)$ or $-2 - \log_6 2$ **49.** $2 + \log_{10} 5$

51. $\dfrac{5}{2}$

Exercises 13-4, p. 469

1. 1.58 **3.** 2.74 **5.** 2.71 **7.** 7.81 **9.** −1.08 **11.** 13.6 **13.** −3.76 **15.** −11.8
17. −0.207 **19.** Undefined **21.** 14.8 **23.** 126 **25.** 5.95 **27.** 4.30 **29.** −1.76
31. 2.10 **33.** 603 **35.** 6.99×10^{-17} **37.** 49.5 **39.** 0.564 **41.** 1.64 **43.** 0.298 **45.** 2.9%
47. 1.49 **49.** 32%

Exercises 13-5, p. 476

1. −2 **3.** 1.90 **5.** 1.99 **7.** 8 **9.** 1.52 **11.** 0.441 **13.** 0.957 **15.** 1.92 **17.** 4.95
19. 11.0 **21.** 9 **23.** −0.838 **25.** $\frac{7}{3} \approx 2.33$ **27.** No solution **29.** 1.70 **31.** −1

33. No solution **35.** $\frac{5}{2} = 2.5$ **37.** $\frac{2}{3} \approx 0.667$ **39.** 8.19

Note: If you have been getting answers in addition to ours, you may have forgotten to reject those that do not check. Remember, the logarithm of a negative number is undefined.

41. 8 digits **43.** 294 years **45.** (a) 0.106 (b) 17 min **47.** 47 years **49.** $n = \dfrac{g}{e^{a+bE}}$

Exercises 13-6, p. 483

1.
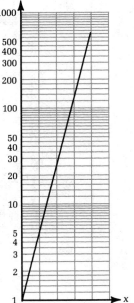

3.
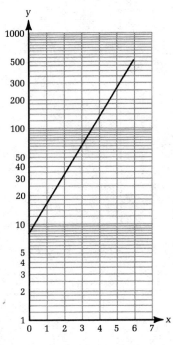

5.
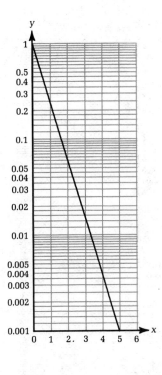

7.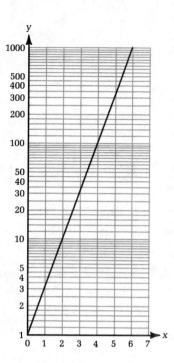

9.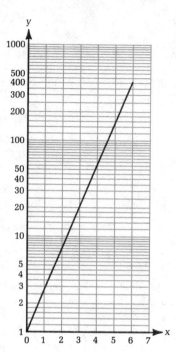

11.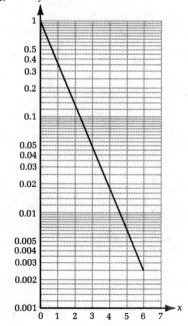

13.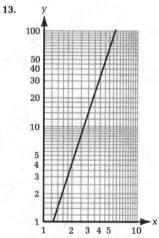

15.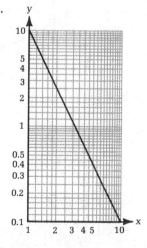

17.

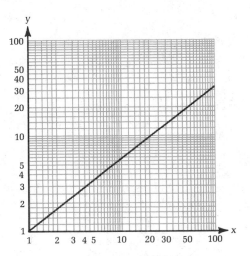

19.

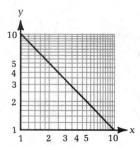

21.

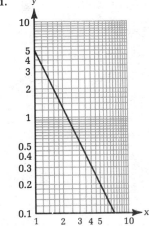

23.

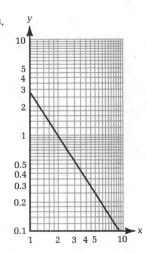

25.

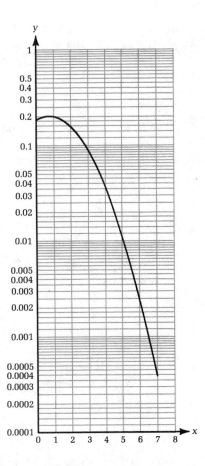

27.

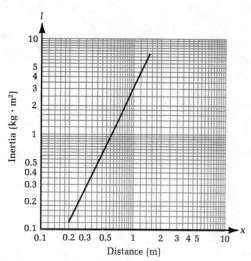

29.

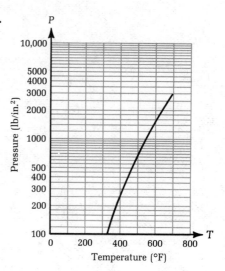

Exercises 13-7, p. 485

1. 4   **3.** 3.2   **5.** 64   **7.** 5   **9.** 9   **11.** 25   **13.** 1.76   **15.** 6.17   **17.** 6.32   **19.** 174,000
21. 0.0000651   **23.** −2.24   **25.** 1.79   **27.** 7.75   **29.** 5.01×10^{14} **31.** 0.169
33. $\log_3 x + \log_3 y + \log_3 z$ **35.** $4 \ln c$ **37.** $3 \log a - 2 \log b$ **39.** 4   **41.** −0.738   **43.** −0.0794
45. 2.27   **47.** 8   **49.** $-\dfrac{17}{6} \approx -2.83$ **51.** $\dfrac{2}{3} \approx 0.667$ **53.** 6

55.

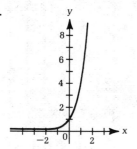

57.

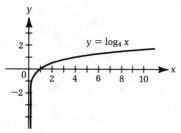

59.

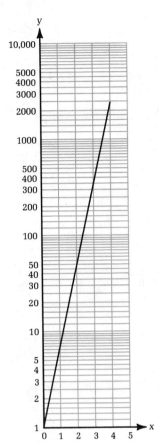

61.

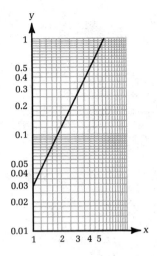

63. $t = -RC \ln\left(\dfrac{q}{Q}\right)$ **65.** 150 dB **67.** 25 **69.** 9.5 mi **71.** 408 years **73.** $a = be^{-2\pi\epsilon_0 V/\lambda}$

75.

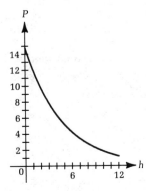

77.

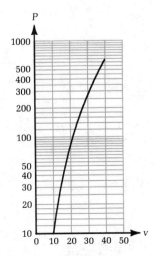

Chapter 14

Exercises 14-1, p. 497

1. Hyperbola   **3.** Circle   **5.** Parabola   **7.** Ellipse   **9.** $(1, 1)$ **11.** No solution   **13.** $(-3.3, 1.3)$
15. $(1.8, 1.2), (-1.8, 1.2), (0.7, -1.5), (-0.7, -1.5)$ **17.** $(1.6, 1.2), (-1.6, 1.2)$
19. $(3.2, 2.4), (3.2, -2.4), (-3.2, 2.4), (-3.2, -2.4)$ **21.** $(2.9, 1.3), (2.9, -1.3), (-2.9, 1.3), (-2.9, -1.3)$
23. $(0.9, 0.8)$ **25.** $(1, 0)$ **27.** $(1.4, 1.4), (-1.4, -1.4)$ **29.** 2.3 in. by 8.7 in.   **31.** No solution
33. $i_1 = 1.5$ A, $i_2 = 0.7$ A or $i_1 = 0.3$ A, $i_2 = 1.9$ A

Exercises 14-2, p. 501

1. $(2, 1), (-1, -2)$ **3.** $(0, 1), \left(-\dfrac{8}{5}, -\dfrac{3}{5}\right)$ **5.** $\left(\sqrt{2}, 1\right), \left(-\sqrt{2}, 1\right), \left(\sqrt{2}, -1\right), \left(-\sqrt{2}, -1\right)$

7. $\left(-\dfrac{3}{2}, -\dfrac{8}{3}\right), (2, 2)$ **9.** $\left(\sqrt{2}, 2\right), \left(-\sqrt{2}, 2\right)$ **11.** $(2.55, 1.58), (-2.55, 1.58), (2.55, -1.58), (-2.55, -1.58)$
13. $(0, 0), (9, 3)$ **15.** $(2.41, 1.79), (-2.41, 1.79), (1.10, -2.79), (-1.10, -2.79)$ **17.** No solution
19. $(1.47, 2.42), (-0.27, -2.82)$ **21.** 45 ft by 32 ft   **23.** $x = 0.40$ in., $y = 0.18$ in.   **25.** 19 in. by 15 in.
27. $T_1 = 303$ K, $T_2 = 553$ K   **29.** $x = 360$ m, $y = 90$ m (at $t = 7$ s)

Exercises 14-3, p. 507

1. 7   **3.** $\dfrac{19}{2}$ **5.** 4   **7.** 3   **9.** -2 **11.** 2, 1   **13.** 6.65   **15.** 16   **17.** 29.42   **19.** 4

21. 10.29   **23.** $B = \dfrac{4\pi^2 I}{T^2 M}$ **25.** $J = \dfrac{4EV^{3/2}}{9d^2}\sqrt{\dfrac{2e}{m}}$ **27.** 48 cm by 14 cm   **29.** $r = \dfrac{d^2 - h^2}{2h}$

Exercises 14-4, p. 512

1. $3, -1$ **3.** $\dfrac{1}{7}, -\dfrac{1}{6}$ **5.** $\pm 2.10, \pm 1.26$ **7.** $0.52, -0.69$ **9.** ± 0.92 **11.** 4, 1   **13.** $\dfrac{1}{81}$
15. $216, -8$ **17.** $-1, -3, 1$ **19.** $\pm 1.87, \pm 2$ **21.** 5 in. by 12 in.

Exercises 14-5, p. 513

1. $(1.4, 2.8), (-1.4, -2.8)$ **3.** $(-3.8, 2.9), (2.6, 0.8)$ **5.** $(4.6, 18.8), (0.4, -1.8)$
7. $(1.3, 1.0), (1.3, -1.0), (-1.3, 1.0), (-1.3, -1.0)$ **9.** $(1.0, 0.5)$ **11.** $(0, 0), (4, 16)$ **13.** No solution
15. $\left(2, \sqrt{2}\right), \left(2, -\sqrt{2}\right), \left(-2, \sqrt{2}\right), \left(-2, -\sqrt{2}\right)$ **17.** $(5, 3), (5, -3), (-2, 1.41), (-2, -1.41)$ **19.** $(0, -2), (3, 7)$
21. $\pm 4, \pm 1$ **23.** $\pm 1, \pm 2$ **25.** 81   **27.** $27, \dfrac{1}{8}$ **29.** 19   **31.** 5, 2   **33.** 5.19   **35.** 0.06

37. $D = \dfrac{4H^2 + W^2}{4H}$ **39.** 12 cm by 21 cm   **41.** $t \approx 5$ s   **43.** 186 ft by 248 ft
45. $v_1 = 8.9$ m/s, $v_2 = 2.2$ m/s

Chapter 15

Exercises 15-1, p. 519

1. 2   **3.** 0   **5.** 1918   **7.** 5   **9.** 5512   **11.** 0   **13.** -27 **15.** 4   **17.** -36 **19.** Yes
21. No   **23.** Yes   **25.** No   **27.** No   **29.** No   **31.** Yes   **33.** Yes   **35.** No

Exercises 15-2, p. 524

1. $x^2 - 3, R = 2$ **3.** $x^3 + x^2 + x - 1$ **5.** $x^2 + 2x - 6, R = 14$ **7.** $x^4 + 3x^3 + 4x^2 + 12x + 27, R = 83$
9. $2x^2 - 8x + 3, R = 6$ **11.** $2x^4 - 3x^3 - 5x^2 - 2x - 2, R = 2$ **13.** $2x^3 + 12, R = -43$
15. $14x^2 + 87x + 845, R = 1295$ **17.** No   **19.** Yes   **21.** No   **23.** Yes   **25.** Yes   **27.** No
29. Yes   **31.** No

Exercises 15-3, p. 529

1. $-2, -3$ **3.** $-4, 3$ **5.** $4, -1$ **7.** $\dfrac{3}{2}, -2$ **9.** $0.61, -3.28$ **11.** $1 \pm j\dfrac{\sqrt{6}}{2}$ **13.** $4, -j$ **15.** $5, 1$

17. $0.32, -6.32$ **19.** $\dfrac{1}{2} \pm \dfrac{1}{2}j$ **21.** $\dfrac{7}{3}, -2, j$ **23.** $38, -24$ **25.** $5, 8, -8, 7, -7,$

27. $(x + 1)(x - 3)(x - 4) = 0$ or $x^3 - 6x^2 + 5x + 12 = 0$ (Any nonzero multiple of the left side of the equation also results in an equation with these roots.)

Exercises 15-4, p. 535

1. (a) $\pm 4. \pm 2, \pm 1$ (b) 2 (c) 1
3. (a) $\pm 20, \pm 10, \pm\dfrac{20}{3}, \pm 5, \pm 4, \pm\dfrac{10}{3}, \pm 2, \pm\dfrac{5}{3}, \pm\dfrac{4}{3}, \pm 1, \pm\dfrac{2}{3}$ (b) 2 (c) 2   **5.** $1, -2, 2$ **7.** $-1, 2.56, -1.56$

9. $4, -2, -\dfrac{1}{2}$ **11.** $\dfrac{1}{2}, -\dfrac{1}{4} \pm j\dfrac{\sqrt{31}}{4}$ **13.** $-1, 1, 2, -3$ **15.** $5, -2, 0.62, -1.62$ **17.** $-\dfrac{3}{2}, -2, 2, -1$

19. $-\dfrac{5}{2}, -\dfrac{1}{2}, j, -j$ **21.** $7, -2, 1, -1$ (double)   **23.** $-\dfrac{1}{3}, -2, 1$ (double), $-\dfrac{1}{2} \pm j\dfrac{\sqrt{3}}{2}$ **25.** $x = 3$

27. $t = 2$ s   **29.** 0.5 in. or 3.9 in.

Exercises 15-5, p. 541

1. 0.78   **3.** 0.71   **5.** -2.89 **7.** 1.55   **9.** 1.71   **11.** -0.83 **13.** 3.91   **15.** 1.29
17. $3.45, -1.45$ **19.** 5.76 years   **21.** 6.02 m by 11.02 m by 3.02 m

Exercises 15-6, p. 543

1. -16 **3.** -4 **5.** Yes   **7.** No   **9.** No   **11.** Yes   **13.** Yes   **15.** No
17. $x^2 - 3x - 4, R = -21$ **19.** $4x^2 - 7x + 19, R = -79$ **21.** $x^3 - 2x^2 - 10x - 3$
23. $x^4 + 4x^3 + 7x^2 + 15x + 40, R = 162$ **25.** $6, -4$ **27.** $-\dfrac{3}{4} \pm j\dfrac{\sqrt{39}}{4}$ **29.** -2 (double)

31. $-2j, \pm 1.73$ **33.** $1, -1, -3$ **35.** $-5, \dfrac{1}{4} \pm j\dfrac{\sqrt{7}}{4}$ **37.** $6, 2, -0.44, -4.56$ **39.** $-\dfrac{1}{4}, 8, -1$ (double)

41. 0.31   **43.** 1.74   **45.** 2.38 s   **47.** 1.28 ksi   **49.** 0.8 in.

Chapter 16

Exercises 16-1, p. 553

1. 10   **3.** -12 **5.** 40   **7.** 873,040   **9.** -32 **11.** -66 **13.** 210   **15.** 2,868,907
17. $x = 2, y = 1, z = -1$ **19.** $a = 3, b = 4, c = -2$ **21.** $w = 1, x = 3, y = -1, z = 0$

23. $a = \frac{4}{3}, b = -\frac{1}{2}, c = 1, d = 2$ **25.** $I_1 = 0.4$ A, $I_2 = 0.2$ A, $I_3 = 0$ A

27. 30 g A, 40 g B, 20 g C, 10 g D

Exercises 16-2, p. 560

1. 0   **3.** 24   **5.** 0   **7.** -24 **9.** -20 **11.** -12 **13.** 3   **15.** 20   **17.** -168 **19.** -78

21. -23 **23.** $x = -3, y = 1, z = 3$ **25.** $a = \frac{1}{2}, b = \frac{1}{3}, c = 2$ **27.** $w = 1, x = 2, y = -3, z = 3$

29. $a = -\frac{1}{2}, b = -\frac{2}{3}, c = 1, d = 2$ **31.** $F_1 = 3, F_2 = 1, F_3 = 4, F_4 = -2$

33. 10 kg of 30%, 30 kg of 40%, 60 kg of 60%

Exercises 16-3, p. 566

1. $x = 1, y = -3, z = -5, u = -2, v = 27, w = 2$ **3.** $p = -3, q = 10, r = 18$ **5.** $\begin{bmatrix} -1 & 3 & 5 \\ -1 & -5 & 3 \end{bmatrix}$

7. $\begin{bmatrix} 8 & 1 \\ -3 & -11 \end{bmatrix}$ **9.** $\begin{bmatrix} 6 & 4 & 8 \\ -2 & -6 & 10 \end{bmatrix}$ **11.** $\begin{bmatrix} 1 & 3 & -6 \\ 0 & -4 & -8 \end{bmatrix}$ **13.** $\begin{bmatrix} 4 & 5 & -2 \\ -1 & -7 & -3 \end{bmatrix}$

15. $\begin{bmatrix} 1 & -4 & 16 \\ -1 & 5 & 21 \end{bmatrix}$ **17.** Cannot be added   **19.** $M + N = N + M = \begin{bmatrix} 2 & -2 & 6 & 3 \\ 0 & 7 & -2 & 7 \end{bmatrix}$

21. $M + 0 = M = \begin{bmatrix} -1 & -3 & 2 & 4 \\ 2 & 5 & -2 & 3 \end{bmatrix}$ **23.** $\begin{bmatrix} 28 & 30 & 24 \\ 17 & 25 & 16 \\ 7 & 11 & 14 \\ 12 & 3 & 4 \end{bmatrix}$ **25.** $\begin{bmatrix} \frac{1}{2} & -\frac{1}{2} \\ -\frac{1}{2} & \frac{1}{2} \end{bmatrix}$

Exercises 16-4, p. 572

1. No   **3.** No   **5.** Yes   **7.** Yes   **9.** Yes   **11.** Yes   **13.** $\begin{bmatrix} -1 & 7 \end{bmatrix}$

15. $\begin{bmatrix} -21 & -7 \\ 28 & -8 \end{bmatrix}$ **17.** $\begin{bmatrix} 77 & 10 \\ -37 & 14 \end{bmatrix}$ **19.** Not possible   **21.** $\begin{bmatrix} -27 \end{bmatrix}$

23. $\begin{bmatrix} 6278 & 8376 \\ 8254 & 10{,}200 \end{bmatrix}$ **25.** $\begin{bmatrix} -3 & 10 & -24 \\ -9 & 11 & -27 \\ 12 & -21 & 7 \end{bmatrix}$ **27.** $\begin{bmatrix} 5 & 38 & -2 \\ -2 & 28 & -1 \\ -2 & 5 & 15 \end{bmatrix}$

33. $\begin{bmatrix} 0 & 1 & 0 & 0 & 0 \\ 0 & 0 & 1 & 0 & 0 \\ 0 & 0 & 0 & 1 & 0 \\ 0 & 0 & 0 & 0 & 1 \\ 1 & 0 & 0 & 0 & 0 \end{bmatrix}$ **35.** $\begin{bmatrix} -1 & -3 \\ -2 & -1 \\ -6 & -3 \\ -3 & -2 \end{bmatrix}$

Exercises 16-5, p. 578

1. $\begin{bmatrix} \frac{2}{7} & -\frac{1}{7} \\ \frac{1}{7} & \frac{3}{7} \end{bmatrix}$ **3.** $\begin{bmatrix} 0 & \frac{1}{2} \\ \frac{1}{5} & \frac{3}{10} \end{bmatrix}$ **5.** $\begin{bmatrix} -8 & -5 \\ -3 & -2 \end{bmatrix}$ **7.** $\begin{bmatrix} -0.597 & -0.840 \\ 0.372 & 0.369 \end{bmatrix}$ **9.** $\begin{bmatrix} -\frac{1}{5} & \frac{4}{5} \\ \frac{2}{5} & -\frac{3}{5} \end{bmatrix}$

11. $\begin{bmatrix} \frac{2}{3} & 1 \\ 1 & 2 \end{bmatrix}$ **13.** $\begin{bmatrix} \frac{3}{7} & \frac{1}{7} \\ \frac{8}{7} & \frac{5}{7} \end{bmatrix}$ **15.** $\begin{bmatrix} -7 & 8 \\ 1 & -1 \end{bmatrix}$ **17.** $\begin{bmatrix} \frac{1}{7} & -\frac{6}{7} & \frac{2}{21} \\ \frac{1}{7} & \frac{1}{7} & \frac{2}{21} \\ \frac{1}{7} & \frac{1}{7} & -\frac{5}{21} \end{bmatrix}$

19. $\begin{bmatrix} \frac{1}{2} & \frac{5}{2} & -8 \\ \frac{1}{2} & \frac{5}{2} & -7 \\ -\frac{1}{2} & -\frac{7}{2} & 10 \end{bmatrix}$ **21.** $\begin{bmatrix} \frac{4}{9} & \frac{4}{3} & \frac{13}{9} \\ \frac{1}{9} & \frac{1}{3} & \frac{1}{9} \\ -\frac{1}{9} & \frac{2}{3} & \frac{8}{9} \end{bmatrix}$ **23.** $\begin{bmatrix} 4 & 11 & 4 \\ 12 & 30 & 11 \\ -3 & -8 & -3 \end{bmatrix}$ **25.** $D_J = 1 \cdot 1 - 1 \cdot 1 = 0$

Exercises 16-6, p. 583

1. $x = 2, y = -1$ **3.** $x = 4, y = 1$ **5.** $x = \frac{1}{2}, y = -2$ **7.** $x = -\frac{2}{3}, y = 5$ **9.** $x = -5, y = 2, z = 4$

11. $x = 6, y = 2, z = -3$ **13.** $x = -\frac{1}{3}, y = -6, z = 5$ **15.** $x = 86.5, y = 120, z = 38.5$

17. $V = \frac{P}{2}, M = -\frac{PL}{4}$ **19.** $s_0 = 1$ ft, $v_0 = 5$ ft/s, $a = 2$ ft/s^2

Exercises 16-7, p. 585

1. 13   **3.** 85   **5.** -4 **7.** -1714.6772 **9.** 0   **11.** 0   **13.** -24 **15.** 40   **17.** -140
19. 387   **21.** $x = 1, y = 2, z = -1$ **23.** $a = -2, b = 0, c = 3, d = -5$

25. $\begin{bmatrix} 1 & 1 \\ 6 & -1 \\ 2 & -1 \end{bmatrix}$ **27.** $\begin{bmatrix} 3 & -7 \\ -6 & 3 \\ 8 & -3 \end{bmatrix}$ **29.** $\begin{bmatrix} 4 & -6 \\ 0 & 2 \\ 10 & -4 \end{bmatrix}$ **31.** $\begin{bmatrix} 4 & -1 \\ 12 & -1 \\ 9 & -4 \end{bmatrix}$

33. $\begin{bmatrix} 8 & -32 & -13 \\ -11 & 56 & 20 \\ -7 & 20 & 11 \end{bmatrix}$ **35.** Cannot be multiplied   **37.** $[-8 \quad 12 \quad 10]$

39. $\begin{bmatrix} -18 & 10 & -4 \\ -9 & 21 & 15 \\ -1 & -19 & -21 \end{bmatrix}$ **41.** $\begin{bmatrix} 1 & -2 \\ 1 & -3 \end{bmatrix}$ **43.** $\begin{bmatrix} -\frac{1}{2} & -\frac{1}{4} \\ \frac{3}{8} & \frac{5}{16} \end{bmatrix}$ **45.** $\begin{bmatrix} 3 & \frac{5}{2} & 1 \\ 5 & \frac{9}{2} & 2 \\ 10 & \frac{17}{2} & 4 \end{bmatrix}$

47. $x = -3, y = -5$ **49.** $x = 9.8, y = 6.2$ **51.** $x = 1, y = -3, z = 4$ **53.** $x = 3, y = 4, z = -2$

55. $n_1 = -\frac{1}{2}, n_2 = \frac{1}{2}$ **57.** $x = 0.181k, y = 0.480k, z = 2.07k$ **59.** $\begin{bmatrix} 3 & 3 & 2 \\ 6 & 3 & 2.3 \\ 6 & 6 & 3.2 \end{bmatrix}$
61. $a = 0.01, b = 0.1$ **63.** A, 5 days; B, 4 days; C, 4 days

Chapter 17

Exercises 17-1, p. 599

1. $x > -5$

3. $y \le -2$

5. $x < -3$

7. $m \ge -5$

9. $x < -6$

11. $a \le 12$

13. $x \le \dfrac{9}{2}$

15. $w \le 2$

17. $x \le -4$

19. $x > -2$

21. $-1 < x < 3$

23. $x \ge 2$ or $x < -3$

25. $0 < x < 1$

27. $-4 \le x \le -3$

29. $(6 < x < 11)$ or $(x > 15$ or $x < 1)$

31. $(5 < x \le 7)$ or $(10 \le x < 12)$

33. $x < 4.76$

35. $x < -87.13$

37. $1100 \text{ mi/h} < v < 1600 \text{ mi/h}$, rounded

39. $0.15(x - 4000) > 1200$
$$x > \$12{,}000$$

Exercises 17-2, p. 604

1. $x > 2$ or $x < -2$

3. $-\dfrac{5}{2} < x < \dfrac{5}{2}$

5. All real numbers

7. $-7 < x < 0$

9. $x \le -4$ or $x \ge -2$

11. $\dfrac{5}{2} < x < 3$

13. $-7 \le x \le 0$ or $x \ge 4$

15. $-3.4 < x < -0.6$

17. $-3 < x < 3$

19. $x \le 4.8$

21. $x < -4$ or $x > 3$

23. $x \le 1$ or $x > 2$

25. $-3 < x < -2$

27. $-5 < x \le -4$ or $x > 7$

29. $-2 \le x < 5$

31. $-3 < x < -2$ or $0 < x < 5$

33. $t > 1 \text{ ms}$ **35.** $v_y \ge 55 \text{ ft/s}$

Exercises 17-3, p. 609

1. $x > 4$ or $x < -10$ **3.** $4 \le x \le 6$ **5.** $x \ge 2$ or $x \le 1$ **7.** $-1 < x < \dfrac{13}{3}$ **9.** $x < 0$ or $x > 4$

11. $2 < x < 5$ **13.** $x \ge -2$ or $x \le -8$ **15.** $x > 15$ or $x < -3$ **17.** $(-4 < x < -3)$ or $(2 < x < 3)$

19. $(x \ge 5$ or $x \le 2)$ or $(3 \le x \le 4)$ **21.** $-3.5 < x < 8.5$ **23.** Minimum is 50°F, maximum is 120°F

25. $9.8\ \Omega < R < 10.6\ \Omega$

Exercises 17-4, p. 615

1.

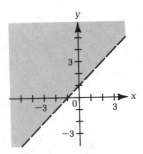

3.

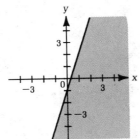

5.

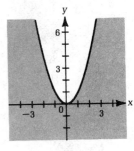

7.

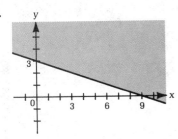

9.

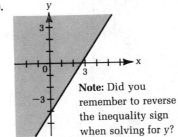

Note: Did you remember to reverse the inequality sign when solving for y?

11.

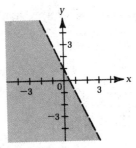

13.

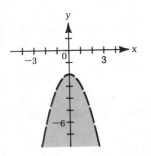

15.

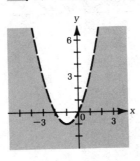

17.

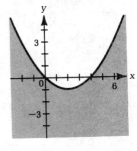

19.

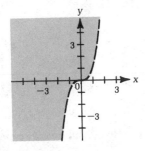

21.

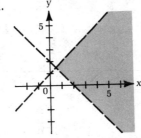

23.

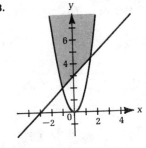

25.

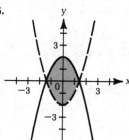

27. 50 tables, 37 chairs   **29.** 20 days of Y, 0 days of X

Exercises 17-5, p. 616

1. $x < -4$ **3.** $x \leq 6$ **5.** $x < -2$ **7.** $x > 2$ **9.** $x \geq -16$ **11.** $x \geq 1$ or $x < -9$

13. $x \leq -4$ or $x \geq 4$ **15.** $0 < x < 1$ **17.** $x < -\dfrac{7}{2}$ or $x > 2$ **19.** $x \leq -3$ or $-2 \leq x \leq 2$ **21.** $3 < x \leq 5$

23. $x < -5$ or $x > 8$ **25.** $3 < x < 11$ **27.** $x \leq 2$ or $x \geq 3$

29.

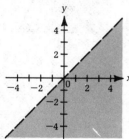

31.

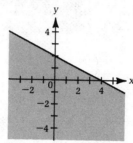

33.

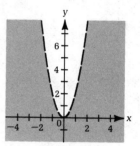

35.

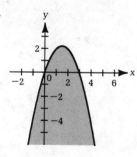

37.

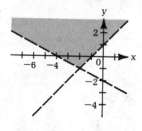

39. $23°F \leq x \leq 50°F$ **41.** $w < 25$ oz   **43.** $d > 6$ ft   **45.** Minimum is 1500 parts, maximum is 2500 parts

Chapter 18

Exercises 18-1, p. 625

1. 20   **3.** -13 **5.** $\dfrac{59}{2}$ **7.** 25   **9.** -56 **11.** 42   **13.** -2016 **15.** -1270

17. $n = 10, S_{10} = 200$ **19.** $d = -2, S_{24} = -360$ **21.** $a_{40} = \dfrac{179}{2}$ or 89.5, $S_{40} = 1630$ **23.** $n = 14, d = -3$

25. $n = 15, S_{15} = 165$ **27.** $d = \dfrac{7}{16}, a_{45} = 20$ **29.** $a_{26} = 88, a_1 = 138$ **31.** $n = 18, a_{18} = 38$

33. $a_1 = -3, d = 2, S_{12} = 96$ **35.** 160,400   **37.** 38.2 m/s   **39.** 380.8 m   **41.** After slightly over 95 months

Exercises 18-2, p. 632

1. 13,122   **3.** $\dfrac{3}{16} = 0.1875$ **5.** 972   **7.** $\dfrac{15}{32} = 0.46875$ **9.** $-\dfrac{127}{8} = -15.875$

11. $-\dfrac{1261}{32} = -39.40625$ **13.** 378   **15.** $S_5 = -484$ if $r = 3$; $S_5 = -244$ if $r = -3$ **17.** $n = 4, S_4 = 255$

19. $a_1 = -2, a_7 = -1458$ **21.** $a_1 = -5, S_6 = -315$ **23.** $n = 8, a_8 = 2187$ **25.** 32   **27.** \$33,626

29. 81.0 ppb   **31.** About 30 weeks   **33.** 110 cd   **35.** $n = 1 + \dfrac{\log(a_n/a_1)}{\log r}$

Exercises 18-3, p. 636

1. 24   **3.** $\dfrac{8}{3}$ **5.** 25   **7.** 54   **9.** $-\dfrac{192}{7}$ **11.** $\dfrac{-27 + 9\sqrt{3}}{2} \approx -5.7$ **13.** $\dfrac{16a^3}{2a - 1}$ **15.** $\dfrac{7}{9}$ **17.** $\dfrac{13}{33}$

19. $\dfrac{127}{999}$ **21.** $\dfrac{78}{185}$ **23.** $2\dfrac{173}{495}$ **25.** 166.7 mm   **27.** 400.0 m

Exercises 18-4, p. 641

1. $m^4 + 16m^3 + 96m^2 + 256m + 256$ **3.** $y^5 - 10y^4 + 40y^3 - 80y^2 + 80y - 32$

5. $64x^6 - 192x^5 + 240x^4 - 160x^3 + 60x^2 - 12x + 1$

7. $1024m^5 + 3840m^4n + 5760m^3n^2 + 4320m^2n^3 + 1620mn^4 + 243n^5$

9. $x^5 + 25x^4 + 250x^3 + 1250x^2 + 3125x + 3125$

11. $x^6 - 6x^5y + 15x^4y^2 - 20x^3y^3 + 15x^2y^4 - 6xy^5 + y^6$

13. $m^6 - 12m^5n + 60m^4n^2 - 160m^3n^3 + 240m^2n^4 - 192mn^5 + 64n^6$ **15.** $512a^3 + 1344a^2b + 1176ab^2 + 343b^3$

17. $x^{12} + 18x^{10}y + 135x^8y^2 + 540x^6y^3 + 1215x^4y^4 + 1458x^2y^5 + 729y^6$ **19.** $y^9 - 27y^8 + 324y^7 - 2268y^6$

21. $1 - 3x + 6x^2 - 10x^3$ **23.** $3240x^7$ **25.** $330x^7y^4$ **27.** $1 - \dfrac{R}{F} + \dfrac{R^2}{F^2}$

Exercises 18-5, p. 643

1. 44   **3.** 19,683   **5.** -55 **7.** $-\dfrac{1,953,125}{256} \approx -7629$ **9.** 39,364   **11.** $\dfrac{45}{2}$ **13.** 231   **15.** -210

17. -73 **19.** -65 **21.** 2   **23.** 10   **25.** $\dfrac{2}{9}$ **27.** $\dfrac{254}{333}$

29. $x^5 - 20x^4 + 160x^3 - 640x^2 + 1280x - 1024$ **31.** $81x^4 + 540x^3y + 1350x^2y^2 + 1500xy^3 + 625y^4$

33. $x^{11} + 22x^{10} + 220x^9 + 1320x^8$ **35.** $1 - 4x + 10x^2 - 20x^3$ **37.** 45,300   **39.** $163,296x^5$

41. 11.4 cm   **43.** \$137,435   **45.** 3.20 ppb   **47.** 30 ft

Chapter 19

Exercises 19-1, p. 651

1. $\cot 71°$: 71 [tan] [1/x] → ‎ 0.3443276

$\dfrac{\cos 71°}{\sin 71°}$: 71 [cos] [÷] 71 [sin] [=] → ‎ 0.3443276

3. $1 + \tan^2\left(\dfrac{5\pi}{6}\right)$: [RAD] 5 [×] [π] [÷] 6 [=] [STO] [tan] [x²] [+] 1 [=] → ‎ 1.3333333

$\sec^2\left(\dfrac{5\pi}{6}\right)$: [RCL] [cos] [1/x] [x²] → ‎ 1.3333333

Note: In the proofs in this chapter, R designates the given right member and L designates the given left member of the stated problem. The basic intermediate steps for each proof are shown; however, in some cases your instructor may want you to show additional detail.

5. $R = \dfrac{\cos \theta}{\dfrac{\cos \theta}{\sin \theta}} = L$ **7.** $R = \cos x\left(\dfrac{1}{\sin x}\right) = \dfrac{\cos x}{\sin x} = L$ **9.** $L = \cos \theta \left(\dfrac{\sin \theta}{\cos \theta}\right)\left(\dfrac{\cos \theta}{\sin \theta}\right)\left(\dfrac{1}{\cos \theta}\right) = R$

11. $R = \cos^2 x - 2 \sin x \cos x + \sin^2 x = L$ **13.** $R = \sin x + \cos x\left(\dfrac{\cos x}{\sin x}\right) = \dfrac{\sin^2 x + \cos^2 x}{\sin x} = \dfrac{1}{\sin x} = L$

15. $L = \dfrac{\dfrac{\sin \theta}{\cos \theta} + 1}{\dfrac{\sin \theta}{\cos \theta}} = \dfrac{\dfrac{\sin \theta + \cos \theta}{\cos \theta}}{\dfrac{\sin \theta}{\cos \theta}} = R$ **17.** $L = \dfrac{1 - \dfrac{1}{\sin x}}{2 \cdot \dfrac{1}{\sin x}} = \dfrac{\dfrac{\sin x - 1}{\sin x}}{\dfrac{2}{\sin x}} = R$

19. $R = \dfrac{\dfrac{\cos \theta}{\sin \theta} + \dfrac{\sin \theta}{\cos \theta}}{\dfrac{1}{\cos \theta}} = \dfrac{\dfrac{\cos^2 \theta + \sin^2 \theta}{\sin \theta \cos \theta}}{\dfrac{1}{\cos \theta}} = \dfrac{1}{\sin \theta \cos \theta} \cdot \cos \theta = \dfrac{1}{\sin \theta} = L$

21. $R = \left(\dfrac{\sin y}{\cos y} - \dfrac{1}{\cos y}\right)^2 = \left(\dfrac{\sin y - 1}{\cos y}\right)^2 = \dfrac{(\sin y - 1)^2}{1 - \sin^2 y} = \dfrac{-1(1 - \sin y)(\sin y - 1)}{(1 - \sin y)(1 + \sin y)} = L$

23. $L = \sin \theta \csc \theta - \sin^2 \theta = 1 - \sin^2 \theta = R$ **25.** $R = \dfrac{\csc x}{\sec x} + \dfrac{1}{\sec x} = \dfrac{\cos x}{\sin x} + \cos x = L$

27. $R = \dfrac{\cos \theta}{\dfrac{1}{\cos \theta} - 1} - \dfrac{\cos \theta}{\dfrac{\sin^2 \theta}{\cos^2 \theta}} = \dfrac{\cos^2 \theta}{1 - \cos \theta} - \dfrac{\cos^3 \theta}{1 - \cos^2 \theta} = \dfrac{\cos^2 \theta + \cos^3 \theta - \cos^3 \theta}{1 - \cos^2 \theta} = \dfrac{\cos^2 \theta}{\sin^2 \theta} = L$

29. $R = \dfrac{1}{\cot \theta + \csc \theta} \cdot \dfrac{\csc \theta - \cot \theta}{\csc \theta - \cot \theta} = \dfrac{\csc \theta - \cot \theta}{\csc^2 \theta - \cot^2 \theta} = \dfrac{\csc \theta - \cot \theta}{1 + \cot^2 \theta - \cot^2 \theta} = L$

31. $R = \dfrac{\dfrac{\tan x}{\tan x} - \dfrac{\sin x}{\dfrac{\sin x}{\cos x}}}{} = 1 - \cos x = (1 - \cos x)\left(\dfrac{1 + \cos x}{1 + \cos x}\right) = \dfrac{1 - \cos^2 x}{1 + \cos x} = L$

33. $L = \dfrac{1 + 2 \sin n + \sin^2 n + \cos^2 n}{\cos n(1 + \sin n)} = \dfrac{2 + 2 \sin n}{\cos n(1 + \sin n)} = \dfrac{2(1 + \sin n)}{\cos n(1 + \sin n)} = \dfrac{2}{\cos n} = R$

35. $L = \cot m - \dfrac{\sin m}{1 + \cos m} \cdot \dfrac{1 - \cos m}{1 - \cos m} = \cot m - \dfrac{\sin m(1 - \cos m)}{\sin^2 m} = \cot m - \dfrac{1 - \cos m}{\sin m}$

$= \cot m - \dfrac{1}{\sin m} + \dfrac{\cos m}{\sin m} = R$

37. $L = \dfrac{1 + 2 \cos y + \cos^2 y + \sin^2 y}{\sin y(1 + \cos y)} = \dfrac{2(1 + \cos y)}{\sin y(1 + \cos y)} = \dfrac{2}{\sin y} = R$

39. $L = \dfrac{\dfrac{\sin \theta}{\cos \theta}}{\dfrac{2 \sin \theta}{\cos \theta} - \sin \theta} = \dfrac{\dfrac{\sin \theta}{\cos \theta}}{\dfrac{2 \sin \theta - \sin \theta \cos \theta}{\cos \theta}} = \dfrac{\sin \theta}{\cos \theta} \cdot \dfrac{\cos \theta}{\sin \theta(2 - \cos \theta)} = R$

41. Let $x = \dfrac{2sE}{MG}$. We now have $\sin \theta = (1 - x^2)^{-1/2}$

Therefore, $\sin^2 \theta = \dfrac{1}{1 + x^2}$ and $\cos^2 \theta = 1 - \dfrac{1}{1 + x^2} = \dfrac{x^2}{1 + x^2}$.

So $\cot^2 \theta = \dfrac{\cos^2 \theta}{\sin^2 \theta} = \dfrac{\dfrac{x^2}{1 + x^2}}{\dfrac{1}{1 + x^2}} = x^2$ and $\cot \theta = x = \dfrac{2sE}{MG}$.

43. $g \sin^4 \theta - Lh^2 \cos \theta = 0$

$Lh^2 \cos \theta = g \sin^4 \theta$

$h^2 = \dfrac{g \sin^4 \theta}{L \cos \theta} = \dfrac{g(1 - \cos^2 \theta)^2}{L \cos \theta}$

Exercises 19-2, p. 658

1. $\dfrac{\sqrt{6} + \sqrt{2}}{4}$ 3. $2 + \sqrt{3}$ 5. $\dfrac{63}{65}$ 7. $\dfrac{56}{65}$ 9. $\dfrac{3}{5}$ 11. $-\dfrac{44}{125}$ 13. $\sin(-A)$ 15. $\cos 2y$

17. $\tan 3m$ 19. $L = \cos 90° \cos x - \sin 90° \sin x = 0 - 1 \cdot \sin x = R$

21. $L = \sin \dfrac{3\pi}{2} \cos x + \cos \dfrac{3\pi}{2} \sin x = -1 \cdot \cos x + 0 = R$

23. $L = \cos x \cos \dfrac{\pi}{4} + \sin x \sin \dfrac{\pi}{4} = \dfrac{\sqrt{2}}{2} \cdot \cos x + \dfrac{\sqrt{2}}{2} \cdot \sin x = R$

25. $L = \sin \dfrac{5\pi}{6} \cos x - \cos \dfrac{5\pi}{6} \sin x = \dfrac{1}{2} \cdot \cos x - \left(-\dfrac{\sqrt{3}}{2}\right) \sin x = R$ 27. $L = \dfrac{\tan x + \tan 45°}{1 - \tan x \tan 45°} = R$

29. $L = \dfrac{\cos x \cos y - \sin x \sin y}{\cos x \sin y} = \dfrac{\cos x \cos y}{\cos x \sin y} - \dfrac{\sin x \sin y}{\cos x \sin y} = \dfrac{\cos y}{\sin y} - \dfrac{\sin x}{\cos x} = R$

31. $R = \dfrac{1}{2}[\sin x \cos y + \cos x \sin y + \sin x \cos y - \cos x \sin y] = \dfrac{1}{2}(2 \sin x \cos y) = L$

33. $L = \dfrac{\sin(90° - x)}{\cos(90° - x)} = \dfrac{\sin 90° \cos x - \cos 90° \sin x}{\cos 90° \cos x + \sin 90° \sin x} = \dfrac{\cos x}{\sin x} = R$

35. $L = \sin(0 - x) = \sin 0 \cos x - \cos 0 \sin x = 0 - \sin x = R$

37. $V = -E\left[\sin 2\pi ft \cos \dfrac{5\pi}{6} - \cos 2\pi ft \sin \dfrac{5\pi}{6} + \sin 2\pi ft \cos \dfrac{7\pi}{6} - \cos 2\pi ft \sin \dfrac{7\pi}{6}\right]$

$= -E\left[-\dfrac{\sqrt{3}}{2} \sin 2\pi ft - \dfrac{1}{2} \cos 2\pi ft - \dfrac{\sqrt{3}}{2} \sin 2\pi ft + \dfrac{1}{2} \cos 2\pi ft\right]$

$= -E\left[-\sqrt{3} \sin 2\pi ft\right] = E\sqrt{3} \sin(2\pi ft)$

39. Denominator $= \sin[\pi - (\alpha + \beta)] = \sin \pi \cos(\alpha + \beta) - \cos \pi \sin(\alpha + \beta) = \sin(\alpha + \beta)$

Solving for a, we get $a \sin(\alpha + \beta) = a \sin \alpha + h \sin \alpha$

$$a = \dfrac{h \sin \alpha}{\sin(\alpha + \beta) - \sin \alpha}$$

41. $\dfrac{e^{j\alpha}}{e^{j\beta}} = \dfrac{\cos \alpha + j \sin \alpha}{\cos \beta + j \sin \beta} = \dfrac{\cos \alpha + j \sin \alpha}{\cos \beta + j \sin \beta} \cdot \dfrac{\cos \beta - j \sin \beta}{\cos \beta - j \sin \beta}$

$= \cos \alpha \cos \beta + \sin \alpha \sin \beta + j(\sin \alpha \cos \beta - \cos \alpha \sin \beta)$

But, $\dfrac{e^{j\alpha}}{e^{j\beta}} = e^{j(\alpha-\beta)} = \cos(\alpha - \beta) + j \sin(\alpha - \beta)$

Equating the real and imaginary parts,

$\cos(\alpha - \beta) = \cos \alpha \cos \beta + \sin \alpha \sin \beta$ and $\sin(\alpha - \beta) = \sin \alpha \cos \beta - \cos \alpha \sin \beta$

Exercises 19-3, p. 664

1. $\cos 2\alpha = -\dfrac{7}{25},\ \sin 2\alpha = \dfrac{24}{25}$ 3. $\cos 2\alpha = -\dfrac{7}{25},\ \sin 2\alpha = \dfrac{24}{25}$ 5. $\cos \dfrac{\beta}{2} = \dfrac{3\sqrt{10}}{10},\ \sin \dfrac{\beta}{2} = \dfrac{\sqrt{10}}{10}$

7. $\cos \dfrac{\beta}{2} = -\dfrac{\sqrt{17}}{17},\ \sin \dfrac{\beta}{2} = \dfrac{4\sqrt{17}}{17}$ 9. $2 \sin 4x$ 11. $-2 \cos 6x$ 13. $R = \dfrac{1 + 2\cos^2 x - 1}{2} = \dfrac{2\cos^2 x}{2} = L$

15. $R = \dfrac{1 - \dfrac{\sin^2 x}{\cos^2 x}}{1 + \dfrac{\sin^2 x}{\cos^2 x}} = \dfrac{\dfrac{\cos^2 x - \sin^2 x}{\cos^2 x}}{\dfrac{\cos^2 x + \sin^2 x}{\cos^2 x}} = \cos^2 x - \sin^2 x = L$

17. $L = \dfrac{\sin x}{\cos x} + \dfrac{\cos x}{\sin x} = \dfrac{\sin^2 x + \cos^2 x}{\cos x \sin x} = \dfrac{1}{\cos x \sin x} = \dfrac{2}{2 \cos x \sin x} = R$

19. $L = \sin^2 x + \cos^2 x + 2 \sin x \cos x = 1 + 2 \sin x \cos x = R$

21. $L = 2\left(\sqrt{\dfrac{1 + \cos x}{2}}\right)^2 = 1 + \cos x = (1 + \cos x)\left(\dfrac{1 - \cos x}{1 - \cos x}\right) = \dfrac{1}{1} R$

23. $R = \dfrac{\cos\frac{x}{2} + \sin\frac{x}{2}}{\cos\frac{x}{2} - \sin\frac{x}{2}} \cdot \dfrac{\cos\frac{x}{2} + \sin\frac{x}{2}}{\cos\frac{x}{2} + \sin\frac{x}{2}} = \dfrac{\cos^2\left(\frac{x}{2}\right) + \sin^2\left(\frac{x}{2}\right) + 2\cos\frac{x}{2}\sin\frac{x}{2}}{\cos^2\left(\frac{x}{2}\right) - \sin^2\left(\frac{x}{2}\right)}$

$= \dfrac{1 + 2\left(\sqrt{\dfrac{1+\cos x}{2}}\right)\left(\sqrt{\dfrac{1-\cos x}{2}}\right)}{\dfrac{1+\cos x}{2} - \dfrac{1-\cos x}{2}} = \dfrac{1 + 2\left(\dfrac{\sqrt{1-\cos^2 x}}{2}\right)}{\dfrac{2\cos x}{2}} = L$

25. By Example 25, $L = \dfrac{\sin x}{1 + \cos x}$. Multiply by $\dfrac{1 - \cos x}{1 - \cos x}$ to get $\dfrac{\sin x - \sin x\cos x}{1 - \cos^2 x}$

$= \dfrac{\sin x - \sin x \cos x}{\sin^2 x} = \dfrac{1}{\sin x} - \dfrac{\cos x}{\sin x} = R$

27. $L = \dfrac{1}{\sin\frac{x}{2}} = \dfrac{1}{\pm\sqrt{\dfrac{1-\cos x}{2}}} = R$ **29.** $R = \dfrac{(4\sin^2\beta - 1)^2}{(3 - 4\sin^2\beta)^2}$ **31.** $x = 2r(1 - \cos\theta)$

Exercises 19-4, p. 670

1. $\dfrac{7\pi}{6}, \dfrac{11\pi}{6}$ **3.** π **5.** $\dfrac{\pi}{3}, \dfrac{2\pi}{3}$ **7.** $\dfrac{\pi}{4}, \dfrac{3\pi}{4}, \dfrac{5\pi}{4}, \dfrac{7\pi}{4}$ **9.** $0.88, 2.26, 4.03, 5.40$ **11.** $0, \pi$

13. $0, \pi, \dfrac{3\pi}{2}$ **15.** $\dfrac{2\pi}{3}, \pi, \dfrac{4\pi}{3}$ **17.** $\dfrac{3\pi}{2}, 3.31, 6.12$ **19.** $\dfrac{\pi}{6}, \dfrac{5\pi}{6}$ **21.** $0, \pi, \dfrac{\pi}{3}, \dfrac{5\pi}{3}$ **23.** $0, \dfrac{2\pi}{3}, \dfrac{4\pi}{3}$

25. $\dfrac{\pi}{3}, \pi, \dfrac{5\pi}{3}$ **27.** $0, \dfrac{\pi}{3}, \dfrac{2\pi}{3}, \pi, \dfrac{4\pi}{3}, \dfrac{5\pi}{3}$ **29.** 0 **31.** $\dfrac{3\pi}{4}, \dfrac{7\pi}{4}, 0.93, 4.07$ **33.** $\dfrac{\pi}{4}, \dfrac{3\pi}{4}, \dfrac{5\pi}{4}, \dfrac{7\pi}{4}$

35. 6.88 min and 9.12 min

Exercises 19-5, p. 680

1. $\dfrac{\pi}{3}$ **3.** π **5.** $\dfrac{\pi}{4}$ **7.** $-\dfrac{\pi}{2}$ **9.** $\dfrac{5\pi}{6}$ **11.** $-\dfrac{\pi}{6}$ **13.** 0.83 **15.** 1.4 **17.** 1.0 **19.** -0.384

21. 2.64 **23.** -1.26 **25** $\dfrac{1}{2}$ **27.** $\dfrac{\sqrt{3}}{2}$ **29.** 0.472 **31.** -0.972 **33.** $\sqrt{1 - x^2}$ **35.** $\dfrac{\sqrt{1-x^2}}{x}$

37. x **39.** $x^2 - 1$ **41.** $\alpha = 2\,\text{Arctan}\,\dfrac{y}{x}$ **43.** 0.559

Exercises 19-6, p. 682

1. $\dfrac{44}{125}$ **3.** $\dfrac{3}{5}$ **5.** $\dfrac{24}{25}$ **7.** 5 **9.** $3\sin 8x$ **11.** $4\cos 6x$ **13.** $\sin(-4x)$

15. $L = \dfrac{\dfrac{1}{\sin x}}{\dfrac{1}{\cos x}} = \dfrac{\cos x}{\sin x} = R$ **17.** $\sin x\left(\dfrac{\sin x}{\cos x}\right)\left(\dfrac{1}{\sin x}\right)\left(\dfrac{\cos x}{\sin x}\right) = R$

19. $L = \dfrac{1}{\cos\theta} - \dfrac{\sin\theta}{\sin\theta} = \dfrac{1 - \sin\theta}{\cos\theta} \cdot \dfrac{\sin\theta}{\sin\theta} = \dfrac{1 - \sin^2\theta}{\cos\theta(1 + \sin\theta)} = \dfrac{\cos^2\theta}{\cos\theta(1 + \sin\theta)} = R$

21. $L = \dfrac{2\cos x}{1 - \sin^2 x} = \dfrac{2\cos x}{\cos^2 x} = \dfrac{2}{\cos x}$

23. $R = \dfrac{\dfrac{\sin^4 x}{\cos^2 x} \cdot \sin^2 x}{\dfrac{\sin x}{\cos x} - \sin x} = \dfrac{\dfrac{\sin^4 x}{\cos^2 x}}{\dfrac{\sin x - \cos x \sin x}{\cos x}} = \dfrac{\sin^3 x}{\cos x(1 - \cos x)}$

25. $L = \cos x\cos\dfrac{2\pi}{3} + \sin x\sin\dfrac{2\pi}{3}$ $= \dfrac{\sin x(1 - \cos^2 x)}{\cos x(1 - \cos x)} = \dfrac{\sin x(1 + \cos x)}{\cos x} = \dfrac{\sin x}{\cos x} + \dfrac{\sin x\cos x}{\cos x} = L$

$= -\dfrac{1}{2}\cos x + \dfrac{\sqrt{3}}{2}\sin x = R$

27. $R = \dfrac{1}{2}(\sin x \cos y + \cos x \sin y - \sin x \cos y + \cos x \sin y) = \dfrac{1}{2}(2 \cos x \sin y) = L$

29. $R = \dfrac{1 - (1 - 2\sin^2\theta)}{2} = \dfrac{2\sin^2\theta}{2} = L$

31. $L = \cos(2x + x) = \cos 2x \cos x - \sin 2x \sin x = (2\cos^2 x - 1)\cos x - 2\sin^2 x \cos x$
$\quad = 2\cos^3 x - \cos x - 2(1 - \cos^2 x)\cos x = 2\cos^3 x - \cos x - 2\cos x + 2\cos^3 x = R$

33. $L = \dfrac{1}{\cos\dfrac{x}{2}} \cdot \dfrac{1}{\sin\dfrac{x}{2}} = \sqrt{\dfrac{2}{1 + \cos x}} \cdot \sqrt{\dfrac{2}{1 - \cos x}} = \dfrac{2}{\sqrt{1 - \cos^2 x}} = R$

35. $\dfrac{5\pi}{6}, \dfrac{7\pi}{6}$ **37.** $0, \pi$ **39.** $\dfrac{\pi}{6}, \dfrac{5\pi}{6}, 3.87, 5.55$ **41.** $0.23, 2.91$ **43.** $\dfrac{\pi}{2}$ **45.** $\dfrac{2\pi}{3}$ **47.** $\dfrac{\pi}{3}$

49. $\dfrac{\sqrt{2}}{2}$ **51.** -0.35 **53.** 1.37 **55.** -1.07 **57.** 0.944 **59.** x **61.** $2x\sqrt{1 - x^2}$

63. $\sin^2\alpha = \dfrac{p^{-2} - n^{-2}}{p^{-2} - m^{-2}}, \quad \tan^2\alpha = \dfrac{p^{-2} - n^{-2}}{n^{-2} - m^{-2}}$ **65.** -1.2

67. $n = \dfrac{\sin 2\theta}{\cos\dfrac{\theta}{2}} = \pm\dfrac{2\sin\theta\cos\theta}{\sqrt{\dfrac{1 + \cos\theta}{2}}} = \pm\dfrac{2\sqrt{2}\sin\theta\cos\theta}{\sqrt{1 + \cos\theta}} = \pm\dfrac{2\sqrt{2 + 2\cos\theta}\,\sin\theta}{\cos\theta}1 + \cos\theta$

69. $6.3°$ **71.** 0.17 s, 1.2 s

Chapter 20

Exercises 20-1, p. 694

1. 4.12 **3.** 5 **5.** 5 **7.** 4.47 **9.** 4 **11.** $-\dfrac{4}{3}$ **13.** Undefined   **15.** -2 **17.** $(11, 2)$

19. $\left(\dfrac{21}{2}, \dfrac{9}{2}\right)$ **21.** Perpendicular   **23.** Neither   **25.** Parallel   **27.** $k = -9$ or $k = -5$ **29.** $k = -12$

31. $k = -3$ **33.** $m_{AB} \cdot m_{BC} = 1(-1) = -1$ **35.** $m_{GH} = m_{IJ} = \dfrac{4}{3}, m_{HI} = m_{GJ} = -\dfrac{2}{7}$ **37.** 12 sq units

39. $\sqrt{3}$ **41.** $112.1°$

Exercises 20-2, p. 701

1. $-6x + y - 23 = 0$ **3.** $3x + 4y + 18 = 0$ **5.** $-4x + y + 7 = 0$ **7.** $x + 3y - 9 = 0$
9. $-2x + y - 14 = 0$ **11.** $-2x + 3y + 15 = 0$ **13.** $y = 5$ **15.** $x = -2$ **17.** $-5x + y + 16 = 0$
19. $3x + y - 22 = 0$

21. $m = 3, b = -3, a = 1$ **23.** $m = -3, b = \dfrac{1}{2}, a = \dfrac{1}{6}$ **25.** $m = \dfrac{7}{3}, b = 0, a = 0$

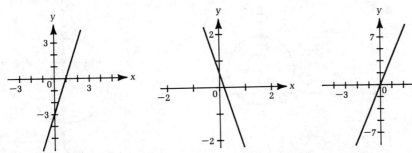

27. $m = 0$, $b = -2$, no x-intercept (see figure below)   **29.** Neither   **31.** Neither   **33.** Parallel

35. $k = -1$ **37.** $k = -2$ **39.** $v = 2t + 8$ (see figure below)   **41.** $R = 0.008T + 4$; $R = 4.104$ Ω at 13°C

43. $L = \dfrac{W}{2} + 8$; $L = 10.5$ in. at $W = 5$ lb   **45.** $C = 1.5N + 650$; $15,650   **47.** V

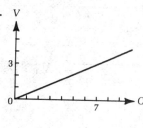

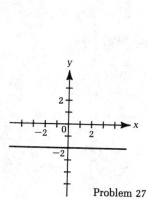

Problem 27

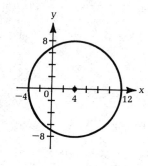

t Problem 39

Exercises 20-3, p. 706

1. $(0, 0)$, $r = 9$

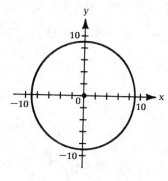

3. $(4, 0)$, $r = 8$

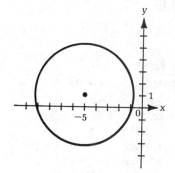

5. $(-5, 1)$, $r = 4.2$

7. $(3, 5)$, $r = 3$

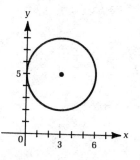

9. $(0, 2)$, $r = 2.8$

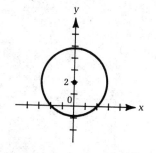

11. $(-1, -3)$, $r = 7$

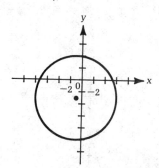

13. $(-3, 2)$, $r = 1.9$

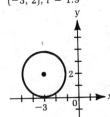

15. $x^2 + y^2 = 121$ **17.** $x^2 + (y - 5)^2 = 16$ **19.** $(x - 3)^2 + (y + 8)^2 = 1$

21. $(x + 8)^2 + (y + 9)^2 = 11$ **23.** $(x + 3)^2 + (y + 4)^2 = 34$

25. $(x + 8)^2 + (y - 4)^2 = 16$ **27.** $(x - 9)^2 + (y - 4)^2 = 13$

29. $x^2 + (y - 2.25)^2 = 0.04$

Exercises 20-4, p. 716

	Vertex	**Focus**	**Directrix**
1.	$(0, 0)$	$(3, 0)$	$x = -3$
3.	$(0, 0)$	$(0, -3)$	$y = 3$
5.	$(0, 0)$	$\left(-\frac{1}{2}, 0\right)$	$x = \frac{1}{2}$
7.	$(0, 0)$	$\left(0, \frac{3}{4}\right)$	$y = -\frac{3}{4}$

	Vertex	**Focus**	**Directrix**
9.	$(3, 5)$	$(2, 5)$	$x = 4$
11.	$(0, -3)$	$\left(-\frac{3}{2}, -3\right)$	$x = \frac{3}{2}$
13.	$(0, -5)$	$(0, -7)$	$y = -3$
15.	$(-1, 6)$	$\left(-1, \frac{13}{2}\right)$	$y = \frac{11}{2}$

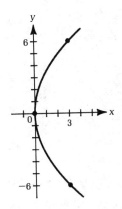

Problem 1

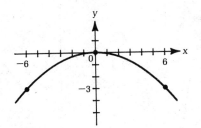

Problem 3

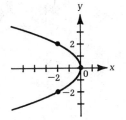

Problem 5

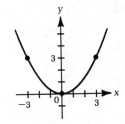

Problem 7

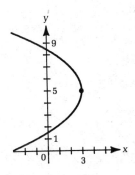

Problem 9

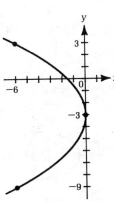

Problem 11

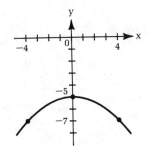

Problem 13

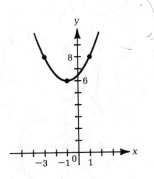

Problem 15

17. Vertex at (4, 2)

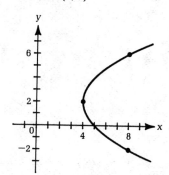

19. Vertex at (0, −8)

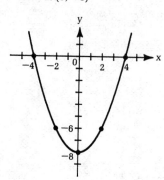

21. Vertex at (6, 5)

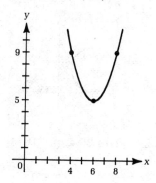

23. Vertex at (0, −2)

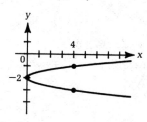

25. $x^2 = 20y$ **27.** $(y - 4)^2 = -12(x + 1)$ **29.** $(x + 3)^2 = -12(y + 8)$

31. $(y - 5)^2 = \dfrac{49}{2}(x - 3)$ **33.** See figure below.

35. 32 ft   **37.** At $\left(0, \dfrac{25}{16}\right)$ **39.**

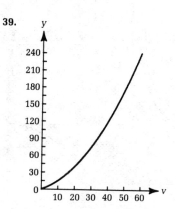

Problem 33

Exercises 20-5, p. 725

	Center	Vertices	Foci
1.	(0, 0)	(0, 5), (0, −5)	(0, 3), (0, −3)

	Center	Vertices	Foci
3.	(0, 0)	(4, 0), (−4, 0)	(3.5, 0), (−3.5, 0)

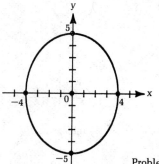

Problem 1

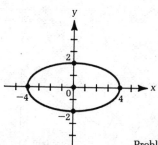

Problem 3

	Center	**Vertices**	**Foci**
5.	(0, 0)	(0, 6), (0, −6)	(0, 5.3), (0, −5.3)
7.	(3, 2)	(3, 5), (3, −1)	(3, 4.2), (3, −0.8)
9.	(−4, 1)	(4, 1), (−12, 1)	(2.9, 1), (−10.9, 1)

	Center	**Vertices**	**Foci**
11.	(1, −5)	(4, −5), (−2, −5)	(3.8, −5), (−1.8, −5)
13.	(−3, 0)	(−3, 4.9), (−3, −4.9)	(−3, 3.5), (−3, −3.5)

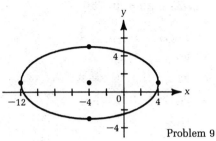

Problem 9

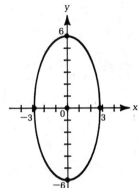

Problem 5

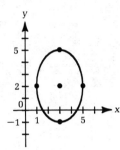

Problem 7

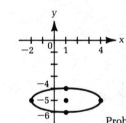

Problem 11

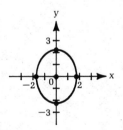

Problem 13

15. Center at (0, 0), vertices at (2, 0) and (−2, 0)   **17.** Center at (0, 0), vertices at (0, 2.2) and (0, −2.2)

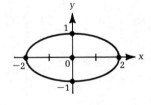

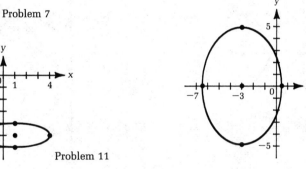

19. Center at (3, −5), vertices at (3, −1) and (3, −9)   **21.** Center at (−4, −2), vertices at (−4, 0.8) and (−4, −4.8)

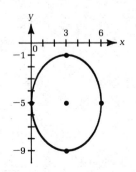

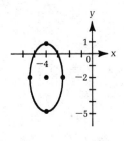

23. $\dfrac{x^2}{7} + \dfrac{y^2}{16} = 1$ **25.** $\dfrac{(x+5)^2}{4} + (y-2)^2 = 1$ **27.** $\dfrac{(x+3)^2}{49} + \dfrac{(y-7)^2}{24} = 1$ **29.** $\dfrac{(x+4)^2}{36} + \dfrac{(y+1)^2}{81} = 1$

31. $\dfrac{x^2}{2.64 \times 10^7} + \dfrac{y^2}{2.59 \times 10^7} = 1$ **33.** $(2.2, 0), (-2.2, 0)$ **35.** 11.8 ft

Exercises 20-6, p. 736

	Center	**Vertices**	**Foci**		**Center**	**Vertices**	**Foci**
1.	(0, 0)	(4, 0), (−4, 0)	(5, 0), (−5, 0)	**9.**	(4, −1)	(4, 2), (4, −4)	(4, 2.2), (4, −4.2)
3.	(0, 0)	(0, 8), (0, −8)	(0, 10.2), (0, −10.2)	**11.**	(3, 0)	(6.5, 0), (−0.5, 0)	(9.1, 0), (−3.1, 0)
5.	(0, 0)	(1, 0), (−1, 0)	(3.2, 0), (−3.2, 0)	**13.**	(−1, 7)	(−1, 9.4), (−1, 4.6)	(−1, 10.5), (−1, 3.5)
7.	(2, 5)	(9, 5), (−5, 5)	(14.2, 5), (−10.2, 5)				

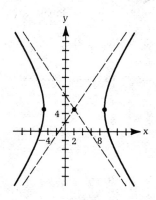

Problem 1

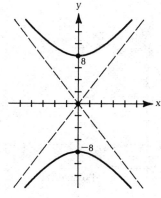

Problem 3

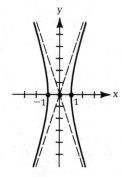

Problem 5

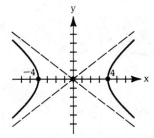

Problem 7

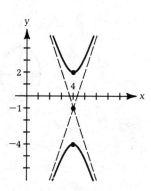

Problem 9

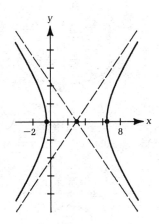

Problem 11

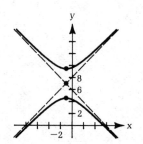

Problem 13

15. Center at (0, 0), vertices at (2, 2) and (−2, −2)

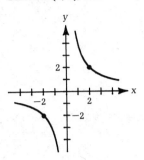

17. Center at (0, 0), vertices at (1.4, −1.4) and (−1.4, 1.4)

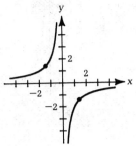

19. Center at (0, 0), vertices at (0, 4) and (0, −4)

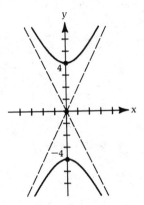

21. Center at (0, 0) vertices at $\left(\dfrac{5}{3}, 0\right)$ and $\left(-\dfrac{5}{3}, 0\right)$

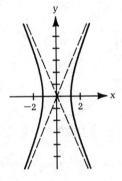

23. Center at (2, 6), vertices at (2, 9) and (2, 3)

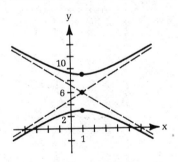

25. Center at (3, −2), vertices at (6.2, −2) and (−0.2, −2)

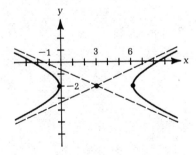

27. Center at $\left(-1, \dfrac{5}{2}\right)$, vertices at (−1, 5) and (−1, 0)

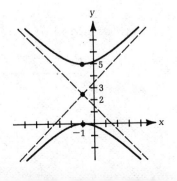

29. $\dfrac{y^2}{25} - \dfrac{x^2}{24} = 1$ **31.** $\dfrac{(x+4)^2}{25} - \dfrac{(y-3)^2}{49} = 1$

33. $\dfrac{(y-\sqrt{3})^2}{9} - \dfrac{(x+2)^2}{3} = 1$ **35.** $\dfrac{y^2}{16} - \dfrac{(x+4)^2}{\frac{576}{29}} = 1$

37.

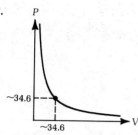

39.

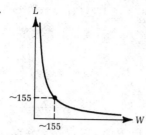

Exercises 20-7, p. 742

1. Circle   **3.** Hyperbola   **5.** Ellipse   **7.** Parabola   **9.** Circle   **11.** Hyperbola   **13.** Parabola
15. Ellipse   **17.** Center at $(0, 3)$

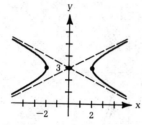

19. Imaginary ellipse   **21.** Degenerate parabola: imaginary parallel lines   **23.** Center at $(3, -6)$

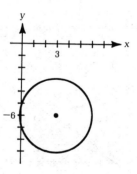

Exercises 20-8, p. 749

1.

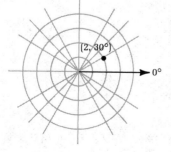

3.

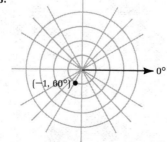

5.

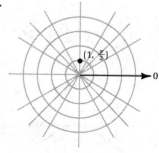

7.

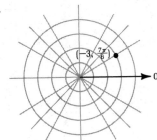

$\left(-3, \frac{7\pi}{6}\right)$

0

9.

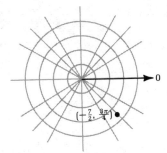

$\left(-\frac{7}{2}, \frac{3\pi}{4}\right)$

0

11.

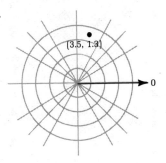

(3.5, 1.3)

0

13. $\left(2, \dfrac{\pi}{6}\right)$ **15.** $(3.6, 2.6)$ **17.** $(2\sqrt{2}, 225°)$ **19.** $(5, 323°)$ **21.** $(\sqrt{2}, \sqrt{2})$ **23.** $(-4.1, -1.4)$

25. $(0, 2)$ **27.** $(-2.8, -2.1)$ **29.** $r = \dfrac{4}{\cos \theta}$ **31.** $r^2 = 25$ **33.** $r^2 = \dfrac{4 \cos \theta}{\sin^2 \theta}$

35. $r^2 = \dfrac{9}{\cos^2 \theta + 2 \sin^2 \theta}$ **37.** $x^2 + y^2 = 3y$ **39.** $x = -6$ **41.** $x^2 + y^2 = 3\sqrt{x^2 + y^2} + y$

43. $(x^2 + y^2)^2 = x^2 - y^2$ **45.** $r^2 = \dfrac{E^2}{E^2 \cos^2 \theta + \sin^2 \theta}$ **47.** $(x^2 + y^2)\sqrt{x^2 + y^2} = 2x^2$

Exercises 20-9, p. 758

1.

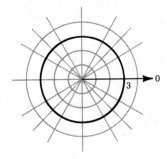

0

3

3.

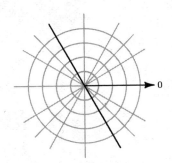

0

5.

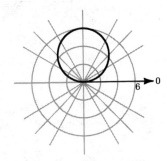

0

6

7.

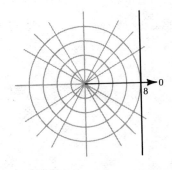

0

8

9.

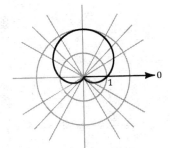

0

1

11.

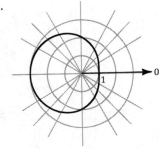

0

1

13.

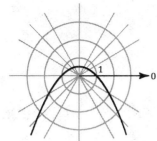

15.

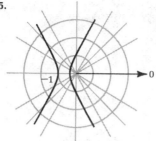

17.

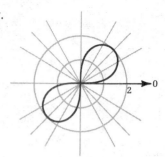

19.

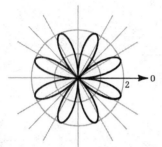

21.

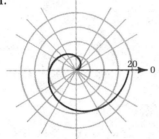

23.

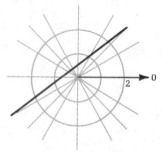

25.

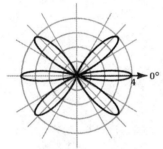

27.

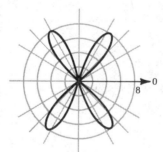

29.

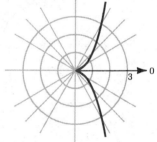

Exercises 20-10, p. 761

1. $d = 6.32$, $M(-3, 10)$, $m = -3$ **3.** $d = 6.32$, $M(8, -4)$, $m = -\dfrac{1}{3}$

5. Slope is undefined, no y-intercept   **7.** $m = -2$, $b = 5$ **9.** Vertex at $(0, 0)$, focus at $(0, -2)$, directrix: $y = 2$

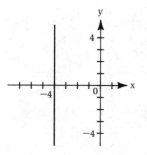

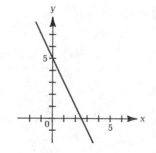

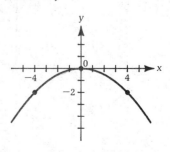

11. Center at $(-1, 5)$, vertices at $(1, 5)$ and $(-3, 5)$, foci at $(0.7, 5)$ and $(-2.7, 5)$

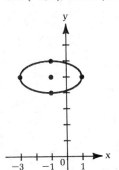

13. Center at $(-4, 5)$, vertices at $(-2, 5)$ and $(-6, 5)$, foci at $(-1.4, 5)$ and $(-6.6, 5)$

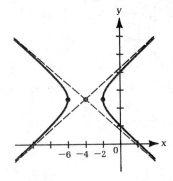

15. Center at $(-4, -2)$, $r = 5$

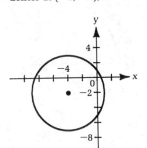

17. Vertex at $(-5, 1)$, focus at $\left(-5, \dfrac{3}{4}\right)$, directrix: $y = \dfrac{5}{4}$

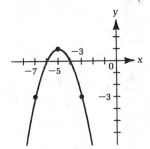

19. Center at $(0, 5)$, vertices at $(0, 6)$ and $(0, 4)$, foci at $(0, 8.2)$ and $(0, 1.8)$ **21.** $y = -3x - 9$ **23.** $x + 5 = 0$

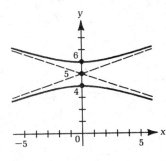

25. $(x - 6)^2 + (y + 2)^2 = 64$ **27.** $(x + 2)^2 = 8(y - 5)$ **29.** $\dfrac{x^2}{36} + \dfrac{y^2}{20} = 1$

31. $\dfrac{y^2}{33} - \dfrac{x^2}{16} = 1$ **33.** $r^2 = 81$ **35.** $r^2 = \dfrac{12}{3\cos^2\theta + \sin^2\theta}$

37. $x^2 + y^2 = 5y$ **39.** $x^2 + y^2 = (1 - x)\sqrt{x^2 + y^2}$

41.

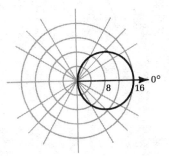

43.

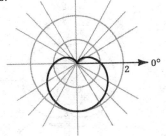

45.

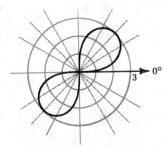

47.

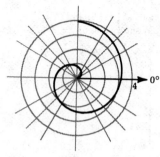

49. $k = -11$ or $k = -5$ **51.** Perpendicular   **53.** $\dfrac{29}{2}$ square units

55. $v = 180 - 32t$

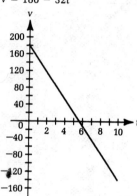

57. $I = 15t + 110$

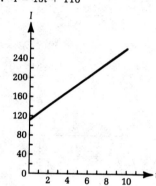

59.

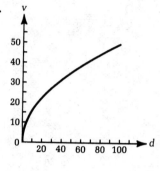

61. $(0, 9)$ **63.** Approx. 5.7 s

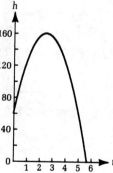

65. $\dfrac{x^2}{6.75} + \dfrac{y^2}{6.25} = 1$ **67.** 22 ft

69. Major axis: 2.8; minor axis: 1.6

71.

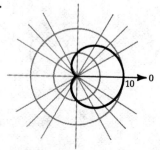

73.

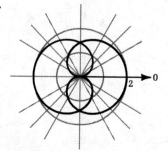

75. $\dfrac{x^2}{16} - \dfrac{y^2}{9} = 1$

Chapter 21

Exercises 21-1, p. 771

1. $\dfrac{1}{6}$ **3.** $\dfrac{1}{2}$ **5.** 0   **7.** $\dfrac{1}{36}$ **9.** 0.5   **11.** 0.78   **13.** 0.24   **15.** 0.05   **17.** 0.067   **19.** 0.2

21. 0.018   **23.** $\dfrac{1}{4}$ **25.** $\dfrac{3}{13}$ **27.** $\dfrac{1}{16}$ **29.** 0.98   **31.** 0.00001   **33.** 0.000003   **35.** 0.78

Exercises 21-2, p. 778

1.

Number of Defective Light Bulbs	3	4	5	6	7	8
Frequency	2	3	4	2	1	1

3.

Weekly Fuel Consumption (gal)	220–224	225–229	230–234	235–239	240–244	245–249	250–254
Frequency	1	4	2	7	4	1	1

5.

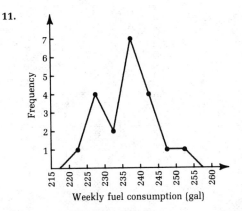

7.

9.

11.

13. 5   **15.** 236 gal   **17.** 235.9 gal   **19.** 5   **21.** 237 gal   **23.** 5   **25.** 237 gal

27.

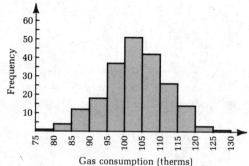

29. 102.6 therms

31.

Acceleration Times (s)	10.5	10.6	10.7	10.8	10.9	11.0	11.1
Frequency	1	2	3	4	3	1	1

33. 10.8 s   **35.** $566   **37.** Approx. 61,900 lb

Exercises 21-3, p. 787

1. 5.453 in.   **3.** 10.2   **5.** 73%   **7.** 67%   **9.** 67%   **11.** 8.8 therms   **13.** 0.0018 in.
15. ±0.18 cm

Exercises 21-4, p. 794

1. $y = 1.5x + 3.6$

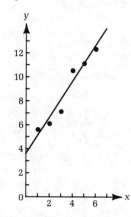

3. $y = -0.92x + 35$

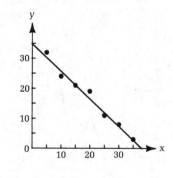

5. $y = 7.5x^2 - 6.9$

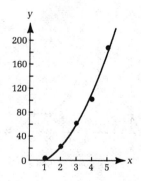

7. $y = 9.9(10^x) + 16$

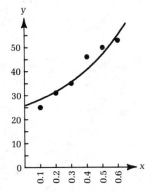

9. $v = 1.052T + 1044$; $v \approx 1040$ ft/s at $0°F$

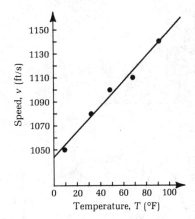

11. $s = 55.97t + 14.27$; $s \approx 154$ mi after 2.50 h

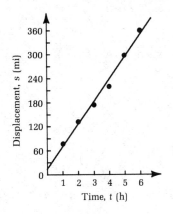

13. $T = 82{,}000\left(\dfrac{1}{f}\right) - 2.7$; $T \approx 66$ lb · ft at $f = 1200$ rev/min

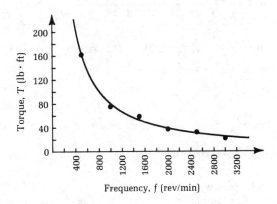

15. $L = 21.7 \cos \theta + 60.4$

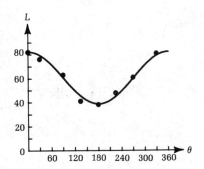

Exercises 21-5, p. 799

1. 0.47   **3.** 0.32   **5.** 0.22   **7.** 0.6   **9.** 0.7   **11.** 0.35   **13.** 78   **15.** 9.43

17.

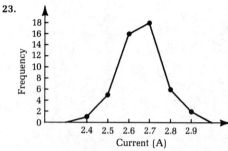

19. 386.05 N · m   **21.** 1.37 N · m

23.

25. 2.66 A   **27.** 0.02; 10,204   **29.** 71,500   **31.** $y = 6.7x - 7.5$ **33.** $p = 0.18w + 4.6$; 50 lb

35. $i = 493 \left(\dfrac{1}{d^2} \right) + 7.22$; 12 lx

Chapter 22

Exercises 22-1, p. 818

1. 3   **3.** Does not exist   **5.** $\frac{1}{3}$ **7.** 1   **9.** ∞ **11.** 9   **13.** 8   **15.** 1   **17.** 0   **19.** 1   **21.** $\frac{1}{2}$
23. ∞ **25.** Does not exist   **27.** 5   **29.** Does not exist   **31.** 0
33. (a) 0   (b) -2 (c) Does not exist   **35.** Continuous   **37.** Not continuous; $\lim_{x \to 2} g(x) \neq g(4)$
39. Not continuous   **41.** (a) 3   (b) 0   (c) $-\infty$ **43.** (a) 5   (b) 5   (c) Does not exist
45. (a) ∞ (b) ∞ (c) 0   (d) 1   (e) ∞ **47.** 5.5×10^{-7} H   **49.** $2\frac{1}{3}$
51. $\lim_{s \to 100^-} C(s) \approx \lim_{s \to 100^+} C(s) \approx 30.00$ ksi

Exercises 22-2, p. 833

1. 8   **3.** 0.25   **5.** 2.18   **7.** 8   **9.** 0   **11.** 4   **13.** $2x$ **15.** $8x$ **17.** $-1/x^2$ **19.** $-3/x^2$
21. $1/2\sqrt{x}$ **23.** $1/\sqrt{2x}$ **25.** $-2x/(x^2 + 2)^2$ **27.** $3x^2 + 2x - 1$
29. 0 when $t = 0$; -4 when $t = 1$; 4 when $t = -1$
31. -1 when $x = 0$; not differentiable when $x = 1$; -1 when $x = 2$; $-\frac{1}{4}$ when $x = -1$
33. 0 when $x = 0$; 3 when $x = 1$; 0 when $x = 2$
35. Not differentiable when $x = 0$; 0.244 when $x = 1.15$; 0.360 when $x = -1.25$ **37.** 8 cm/s

Exercises 22-3, p. 843

1. 0   **3.** $7x^6$ **5.** $15x^2$ **7.** $-55x^{10}$ **9.** $2x^3$ **11.** $-x/3$ **13.** $-6x$ **15.** $1 + 4x^3$
17. $20x^3 - 20x^4$ **19.** $21x^6 - 10x^9$ **21.** $6x^5 - 5x^4 + 4x^3 - 6x^2$ **23.** 0.06   **25.** -13 **27.** 0.002528

29. 116   **31.** 35   **33.** 1.06   **35.** (1, 3)   **37.** (1.41, 11.31); (−1.41, −11.31)   **39.** (1, 0) and $(-\frac{1}{3}, -\frac{4}{27})$

41. $\frac{1}{2}$ **43.** $\frac{3}{10}$ **45.** 2.230, −0.897   **47.** 0.211, 0.789   **49.** $8x - y + 5 = 0$ **51.** $4\pi r^2$

53. (a) $2r - 2h$ (b) $3r/2$ **55.** 52π, 163.4 cm   **57.** $0.32\ \mu\Omega \cdot \text{cm/°C}$ **59.** $6s^2$

Exercises 22-4, p. 852

1. $12x^2 - 16x$ **3.** $6x^5 + 3x^2$ **5.** $12x^3 + 6$ **7.** $-5x^4 + 4x^3 - 2x + 1$ **9.** $-48x^5 - 10x^4 + 12x^3$

11. $5x^4 - 24x^3 + 27x^2 - 8x$ **13.** $-5x^4 + 4x^3 - 3x^2 + 2x$ **15.** $8x^3 - 14x + 3$

17. $-35x^6 + 60x^5 + 25x^4 - 40x^3 - 15x^2 + 20x$ **19.** $-\dfrac{3}{(x-3)^2}$ **21.** $\dfrac{x^2 - 1}{2x^2}$ **23.** $\dfrac{3}{x^2}$ **25.** $\dfrac{x-6}{2x^3}$

27. $\dfrac{-8x^2 + 24}{(x^2 + 3)^2}$ **29.** $\dfrac{-4x^2 - 12x + 15}{(2x+3)^2}$ **31.** $\dfrac{4x^2 - 6x - 2}{(2x^2 + 1)^2}$ **33.** $-\dfrac{5}{x^6}$ **35.** $\dfrac{18}{x^7}$ **37.** $\dfrac{-12x^3 - 5x^2 + 4x}{(2x+1)^2}$

39. $2x - \dfrac{4x}{(2x^2 + 1)^2}$ **41.** (a) -1 (b) 7   (c) $\dfrac{5 \pm \sqrt{43}}{18}$ **43.** (a) 2   (b) 18   (c) $-\frac{1}{8}$

45. (a) 1   (b) Not defined   (c) $\dfrac{2 \pm \sqrt{2}}{2}$ **47.** 48   **49.** $2x + 9y - 7 = 0$ **51.** $\dfrac{6600}{(R+60)^2}$

53. $\dfrac{a - b + ay^2}{(a - b - ay^2)^2}$ **55.** $\dfrac{6}{V^3} - \dfrac{RT}{(V-24)^2}$

Exercises 22-5, p. 859

1. $8(1 + 2x)^3$ **3.** $10x(x^2 - 2)^4$ **5.** $27x^2(4 + 3x^3)^2$ **7.** $10(1 - x)(1 + 2x - x^2)^4$

9. $10(2x - 3x^2)(x^2 - x^3)^9$ **11.** $16(4x^3 - 5x^4)(1 - x^4 + x^5)^3$ **13.** $-\dfrac{2x}{(x^2 - 1)^2}$ **15.** $-\dfrac{4}{(2x+3)^3}$

17. $\dfrac{-24x^2 + 16x}{(x^3 - x^2 + 1)^5}$ **19.** $(9x^2 + 2)(x^2 + 2)^3$ **21.** $2(7x^2 - 2)(x^2 - 2)^2$ **23.** $-(10x^2 - 8x + 4)(1 - x)(2 + x^2)^3$

25. $2x + 15x^2(1 + x^3)^4$ **27.** $10x(1 + x^2)^4 - (2x - x^2)^3(8 - 8x)$ **29.** $-\dfrac{3}{x^4} - 12x^2(x^3 - 2)^3$ **31.** $\dfrac{-4x^2 - 8}{(x^2 - 2)^2}$

33. $\dfrac{-42x}{(2x - 3)^3}$ **35.** $\dfrac{-3x(x^3 - 6x - 2)(1 + x^3)^2}{(2 - x^2)^4}$ **37.** $\dfrac{8(5x^2 - 2)^3(5x^2 + 2x + 2)}{(2x + x^2)^5}$ **39.** 0, ±2   **41.** 300

43. $32ki/9\pi d^2$ **45.** 78 kW

Exercises 22-6, p. 866

1. $-\frac{4}{5}$ **3.** $-\dfrac{y+1}{x}$ **5.** $\dfrac{y+1}{4y - x}$ **7.** $\dfrac{3x^2}{4y^3}$ **9.** $\dfrac{x}{y}$ **11.** $-\dfrac{y + 2x}{2y + x}$ **13.** $\dfrac{6x^2 + 2xy}{2y - x^2}$ **15.** $\dfrac{5 - 3x^2}{3y^2}$

17. $\dfrac{2x - y}{x + 2y}$ **19.** $-\dfrac{y}{x}$ **21.** $-\dfrac{2(1 + 3y)}{3(2x + 3)}$ **23.** $\dfrac{-x - 1}{2(2y - 3)}$ **25.** $\dfrac{1}{3x^{2/3}}$ or $\dfrac{1}{3\sqrt[3]{x^2}}$ **27.** $-\dfrac{1}{2x^{3/2}}$ or $-\dfrac{1}{2x\sqrt{x}}$

29. $-\dfrac{2}{3x^{4/3}}$ or $-\dfrac{2}{3x\sqrt[3]{x}}$ **31.** $\dfrac{3\sqrt{x}}{2}$ **33.** $-\dfrac{(x+1)\sqrt{x}}{2x(x - 1)^2}$ **35.** $-\dfrac{x}{\sqrt{x^2 + 1}}$ **37.** $\dfrac{3}{4}x^{-1/4} - \dfrac{1}{4}x^{-3/4}$

39. $\frac{2}{3}(2x - 1)^{-2/3}$ **41.** $\dfrac{8x^3}{3\sqrt[3]{(2x^4 - 1)^2}}$ **43.** $\dfrac{-6x - 9}{(x^2 + 3x)^{5/2}}$ **45.** $\sqrt{\dfrac{y}{x}}$ **47.** -1 **49.** -8 **51.** 2.5

53. 13 m/s   **55.** -5.5×10^{-7} gauss/cm   **57.** $-\dfrac{1}{4\pi L}\sqrt{\dfrac{g}{L}}$ **59.** $-\left(\dfrac{i}{o}\right)^2$

Exercises 22-7, p. 869

1. 13   **3.** ∞ **5.** -1 **7.** -4 **9.** $\frac{1}{2}$ **11.** 2   **13.** 6   **15.** $6x^2$ **17.** $\dfrac{x}{\sqrt{x^2 + 4}}$ **19.** 0

21. $4x^7$ **23.** $4x^3 - 6x^2 + 14x - 4$ **25.** $-10x^4 + 16x$ **27.** $-\dfrac{15}{(x-5)^2}$ **29.** $\dfrac{15x^2 - 52x + 10}{(3x^2 - 2)^2}$

31. $-18x(4 - 3x^2)^2$ **33.** $-\dfrac{12x}{(3x^2 - 2)^3}$ **35.** $4(7x^3 + 11x^2 - 6x - 6)(x^2 - 2)^3(4x + 2x^2)^2$ **37.** $\dfrac{4}{3}$

39. $\dfrac{3(x - y)^2 - 2x}{2y + 3(x - y)^2}$ **41.** $\dfrac{4}{(x^2 + 4)\sqrt{x^2 + 4}}$ **43.** 0 **45.** $5x + y - 8 = 0$ **47.** $\dfrac{1}{2}$

49. $C(0.002) = \lim\limits_{q \to 0.002} C(q) = 140$ **51.** $\lim\limits_{r \to R^-} E(r) = \lim\limits_{r \to R^+} E(r) = -0.01$ **53.** $\lim\limits_{f \to 10^-} A(f) > 0$ but $\lim\limits_{f \to 10^+} A(f) < 0$

55. $-3.1 \times 10^{-6}\ \text{s}^{-2}$ **57.** $-4.2\ \text{V/s}$ **59.** $\dfrac{2\pi R\sqrt{2\pi k\rho L}}{k}$ **61.** $-\dfrac{Q}{\pi\varepsilon_0 r^5}$ **63.** $-4.26 \times 10^{-4}\ \text{g/cm}^3 \cdot {}^\circ\text{C}$

Chapter 23

Exercises 23-1, p. 882

	Tangent	**Normal**
1.	$6x + y + 3 = 0$	$x - 6y + 19 = 0$
3.	$x - 4y + 1 = 0$	$8x + 2y - 9 = 0$
5.	$7x - 2y - 3\sqrt{3} = 0$	$2x + 7y - 16\sqrt{3} = 0$
7.	$x + 2y - 1 = 0$	$2x - y + 3 = 0$
9.	$x\sqrt{3} + 3y - 4\sqrt{3} = 0$	$x\sqrt{3} - y = 0$
11.	$x + 4y + 2 = 0$	$8x - 2y - 1 = 0$
13.	$x - 4y + 7 = 0$	$4x + y - 6 = 0$

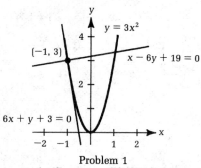

Problem 1

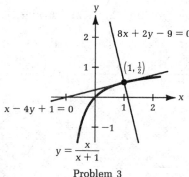

Problem 3

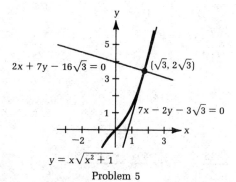

Problem 5

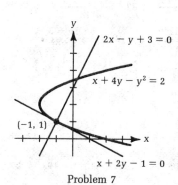

Problem 7

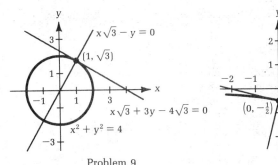

Problem 9

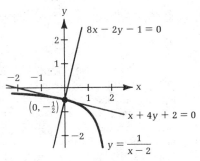

Problem 11

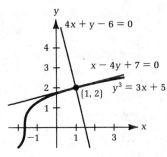

Problem 13

15. $60x^3, 180x^2$ **17.** $2 - \dfrac{6}{x^4}, \dfrac{24}{x^5}$ **19.** $24x^2 - 2, 48x$ **21.** $2, 0$ **23.** $-\dfrac{1}{4}x^{-3/2}, \dfrac{3}{8}x^{-5/2}$

25. $-\dfrac{1}{(x^2 - 1)\sqrt{x^2 - 1}}$ **27.** $\dfrac{10}{9}x^{-1/3}$ **29.** $12(x^3 - 2x)^2(11x^4 - 16x^2 + 4)$ **31.** -4 **33.** $\dfrac{3}{4}$ **35.** -1

37. $\dfrac{6kq_1q_2}{r^4}$ **39.** $-\dfrac{2B}{p}$ or $\dfrac{-2BV}{AT - BT^2}$ **41.** $8x - 3y - 29 = 0$

Exercises 23-2, p. 893

1. -2 m/s, 2 m/s^2 **3.** -20 m/s, -22 m/s^2 **5.** -4.65 m/s, 9 m/s^2 **7.** -3 m/s, 0 m/s^2
9. -3.75 m/s, -2.25 m/s^2 **11.** 9.7 m/s, 0.09 m/s^2 **13.** 4.5 ft/s, 6.75 ft/s^2
15. 160 ft/s at $-83°$, 32 ft/s^2 at $-90°$ **17.** 0.80 ft/s at $83°$, 0.01 ft/s^2 at $-45°$
19. (a) 0.011 m (b) 306 m/s at $-11°$ **21.** 2.0×10^4 m at $-27°$
23. (a) 240 ft (b) 3.9 s (c) 0 ft/s (d) 120 m/s, -120 m/s

Exercises 23-3, p. 899

1. (a) 315 (b) 2800 km/min at $-89°$ **3.** 3 **5.** -0.006 m/s^3 **7.** -1.3×10^6 N/C **9.** 4×10^{-3} J/s
11. 12π ft^3/min **13.** 0.03 cm/s **15.** 1100 m^2/s **17.** 8.5×10^{-4} Hz/s **19.** -8.5 ft^3/s **21.** 0.08 Ω/min
23. 7.9 mi/h **25.** 0.016 ft/min **27.** -4.5 cm^2/s **29.** (a) 0.027 ft/min (b) 0.031 ft/min

Exercises 23-4, p. 922

1. (a) Increasing for $x > 2$ (b) Concave up for all x (c) Minimum when $x = 2$ (d) No inflection
points
3. (a) Increasing for $x > -\frac{1}{2}$ (b) Concave up for all x (c) Minimum when $x = -\frac{1}{2}$ (d) No inflection
points

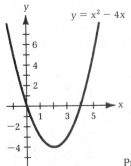

Problem 1

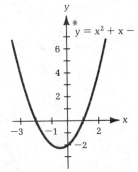

Problem 3

5. (a) Increasing for x > 2   (b) Concave up for all x   (c) Minimum when x = 2   (d) No inflection points

7. (a) Increasing for all x   (b) Concave up for x > 0   (c) No local minimum or maximum
   (d) Inflection point when x = 0

9. (a) Increasing for $x < -2\sqrt{3}/3$ and $x > 2\sqrt{3}/3$ (b) Concave up for x > 0   (c) Maximum when $x = -2\sqrt{3}/3$, minimum when $x = 2\sqrt{3}/3$ (d) Inflection point when x = 0

11. (a) Increasing for x < 0 and x > 4   (b) Concave up for x > 2   (c) Maximum when x = 0, minimum when x = 4   (d) Inflection point when x = 2

13. (a) Increasing for x > 1   (b) Concave up for all x < 0 and $x > \frac{2}{3}$ (c) Minimum when x = 1
   (d) Inflection points when $x = \frac{2}{3}$ and x = 0

15. (a) Increasing for all x   (b) Concave up for $x > \frac{2}{3}$ (c) No local maximum or minimum   (d) Inflection point when $x = \frac{2}{3}$

17. (a) Increasing for $-\sqrt{2} < x < 0$ and $x > \sqrt{2}$ (b) Concave up for $x < -\sqrt{6}/3$ and $x > \sqrt{6}/3$ (c) Minimum when $x = -\sqrt{2}$ and $x = \sqrt{2}$, maximum when x = 0   (d) Inflection points when $x = -\sqrt{6}/3$ and $x = \sqrt{6}/3$

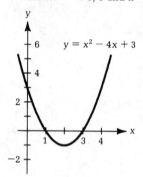

Problem 5

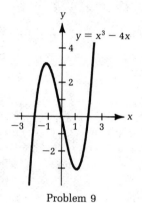

Problem 7

Problem 9

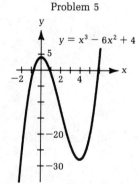

Problem 11

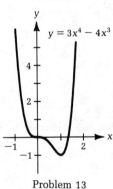

Problem 13

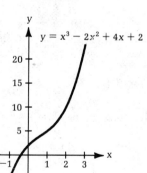

Problem 15

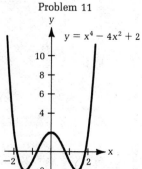

Problem 17

19.

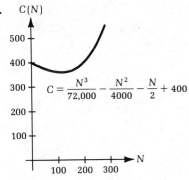

$$C = \frac{N^3}{72,000} - \frac{N^2}{4000} - \frac{N}{2} + 400$$

21.

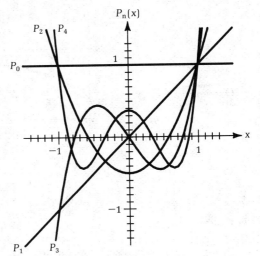

Exercises 23-5, p. 931

1.

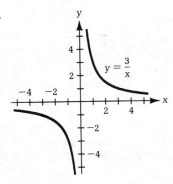

$$y = \frac{3}{x}$$

3.

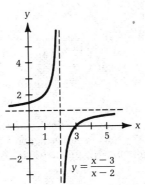

$$y = \frac{x - 3}{x - 2}$$

5.

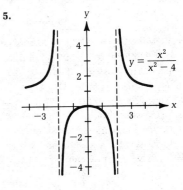

$$y = \frac{x^2}{x^2 - 4}$$

7.

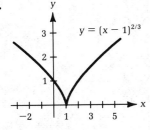

$$y = (x - 1)^{2/3}$$

9.

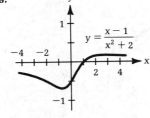

$$y = \frac{x - 1}{x^2 + 2}$$

11.

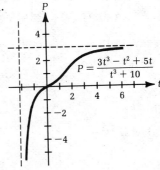

$$P = \frac{3t^3 - t^2 + 5t}{t^3 + 10}$$

13.

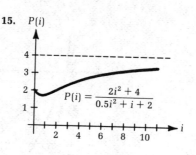

$$U(r) = \frac{1}{r} - \frac{1}{r^3}$$

15.

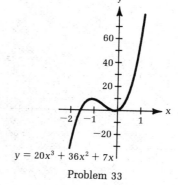

$$P(i) = \frac{2i^2 + 4}{0.5i^2 + i + 2}$$

Exercises 23-6, p. 940

1. 0.4 μm   **3.** $\frac{1}{2}$ **5.** 12.6 g   **7.** 290   **9.** (0, 0)   **11.** 4.9 in.   **13.** 2.4 in.   **15.** 6 cm by 10.5 cm
17. $r \approx 7.6$ ft, $h \approx 19$ ft   **19.** 16π in.^{3} **21.** 9.2 in. by 9.8 in.   **23.** (3, 2)   **25.** $R = r/2$
27. $\sqrt{10}$ **29.** 37.5 ft from the right end

Exercises 23-7, p. 945

1. Tangent: $2x + y - 4 = 0$; normal: $x - 2y - 2 = 0$ **3.** Tangent: $x + 3y - 10 = 0$; normal: $3x - y = 0$
5. $\dfrac{d^2u}{dr^2} = \dfrac{5q^2}{8\pi^2 \varepsilon_0 r^6}$ **7.** $v = -6$ m/s, $a = 2$ m/s^2 **9.** $v \approx 0.31$ m/s, $a \approx 0.16$ m/s^2
11. $v \approx 155$ m/s at $\theta \approx -73°$, $a \approx 32$ m/s^2 at $\theta \approx -90°$ **13.** (a) $v \approx 190$ ft/s at $\theta \approx 10°$
   (b) $a = 32$ ft/s^2 at $\theta = -90°$ **15.** (a) 41 yd   (b) 20.5 yd   (c) 2.8 yd/s at $\theta = -45°$
17. 12 cm/s at 81°   **19.** 990 cm^2/s   **21.** 135 mi/h   **23.** 345 mi/h   **25.** 2.4×10^{-3} ft/s
27. 5 in.^{2}/s   **29.** (a) Increasing for $x < 3$ (b) Concave down for all x   (c) Maximum when x = 3
   (d) No inflection points
31. (a) Decreasing for all x   (b) concave up for x < 0
   (c) No local maximum or minimum   (d) Inflection point when x = 0
33. (a) Increasing for $x < -1.1$ and $x > -0.1$ (b) Concave up for $x > -0.6$ (c) Maximum when
   $x \approx -1.1$, Minimum when $x \approx -0.1$ (d) Inflection point when x = -0.6

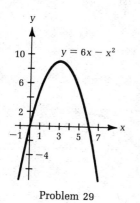

Problem 29

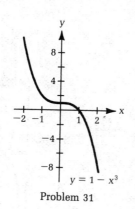

Problem 31

Problem 33

35.

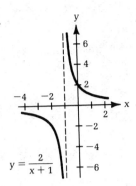

$$y = \frac{2}{x+1}$$

37.

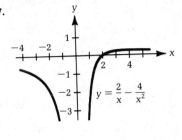

$$y = \frac{2}{x} - \frac{4}{x^2}$$

39.

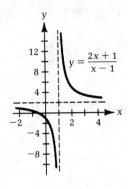

$$y = \frac{2x+1}{x-1}$$

41.

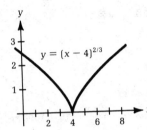

$$y = (x-4)^{2/3}$$

43.

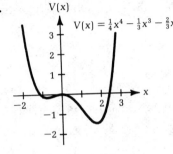

$$V(x) = \tfrac{1}{4}x^4 - \tfrac{1}{3}x^3 - \tfrac{2}{3}x^2$$

45. 20 and 20   **47.** (0, 0)   **49.** 4.8 s   **51.** 1.4 s   **53.** Height $= 8\sqrt{3}$ cm; radius $= 4\sqrt{6}$ cm
55. 0.1 mi from A **57.** 18 in. by 18 in. by 36 in.

Chapter 24

Exercises 24-1, p. 958

1. $(3x^2 - 4x)\,dx$ **3.** $-6x(2 - x^2)^2\,dx$ **5.** $2x(4x^2 - 1)(x^2 - 1)^2\,dx$ **7.** $\dfrac{(x^2 - 2x)\,dx}{(x - 1)^2}$ **9.** $\dfrac{3\sqrt{x}\,dx}{2}$

11. $-2x\,dx$ **13.** $-0.1, -0.11$ **15.** $1.35, 1.39$ **17.** $5.88, 5.91$ **19.** $0.063, 0.059$
21. 91.86 ± 21.49 cm³, $0.234, 23.4\%$ **23.** 0.44 cm³ or 1.5% **25.** (a) 0.002 ft   (b) 6.4 ft
27. $\Delta P/P = \Delta T/T$ **29.** 3.12×10^{-7}

Exercises 24-2, p. 971

1. $3x + C$ **3.** $\dfrac{3x^2}{2} + C$ **5.** $-\dfrac{1}{3t^3} + C$ **7.** $\dfrac{2x^{7/2}}{7} + C$ **9.** $m^2 - 5m + C$ **11.** $x^4 + \dfrac{3}{x} + C$

13. $\dfrac{2x^{3/2}}{3} - \dfrac{2x^{5/2}}{5} + \dfrac{2x^{7/2}}{7} + C$ **15.** $\tfrac{1}{9}(3x + 2)^3 + C$ **17.** $\tfrac{1}{4}(x^2 + 5)^4 + C$ **19.** $\tfrac{1}{18}(2x^3 + 1)^6 + C$

21. $\tfrac{2}{9}(3x^2 + 2)^{3/2} + C$ **23.** $\tfrac{1}{8}(6x - 2)^{4/3} + C$ **25.** $-\tfrac{1}{2}(x^2 + 4)^{-2} + C$ **27.** $6x\sqrt{x - 6} + C$

29. $\tfrac{1}{10}(x^4 - 2x^2)^5 + C$ **31.** $f(x) = \dfrac{4x^3}{3} - \dfrac{2}{3}$ **33.** $f(x) = \tfrac{1}{3}(2x + 1)^{3/2} - 2$

35. $N = 150t + t^2 - \dfrac{0.1t^3}{3} + C, 1247$ **37.** $v = 2t + \dfrac{t^3}{3} - \dfrac{2}{t + 1} + 2$ **39.** $W = 2\sqrt{t^2 + 1} - 2$

Exercises 24-3, p. 980

1. (a) 10 (b) 8.8 **3.** (a) 1.25 (b) 0.80 **5.** (a) 8.75 (b) 7.48 **7.** (a) 1.05 (b) 1.25
9. (a) 44 (b) 36.78 **11.** (a) 1.78 (b) 1.92 **13.** (a) 0.48 (b) 0.51 **15.** 6 **17.** $41\frac{1}{3}$
19. $14\frac{2}{3}$ **21.** 8

Exercises 24-4, p. 988

1. 4 **3.** 80 **5.** $12\frac{2}{5}$ **7.** $-\frac{2}{3}$ **9.** $2\frac{8}{15}$ **11.** -4 **13.** $\frac{1}{15}$ **15.** $312\frac{2}{5}$ **17.** $\frac{7}{36}$
19. 37.5 sq units **21.** 45 sq units **23.** $8\frac{2}{3}$ sq units **25.** $3\frac{5}{6}$ sq units **27.** $2\sqrt{a}$ **29.** $k\left(\sqrt{r^2 + a^2} - r\right)$

Exercises 24-5, p. 996

1. 12.13 **3.** 0.52 **5.** 17.05 **7.** 14.54 **9.** 10.14 **11.** 0.20 **13.** 2.1 **15.** 1.8 **17.** 1800 rev
19. 865.82 V **21.** 21.3 sq units

Exercises 24-6, p. 1009

1. 14.5 ft/s **3.** 44.4 m/s **5.** 34.5 m **7.** $55\frac{1}{3}$ m **9.** 1.3 s, 2.4 s **11.** 800 lb·in.
13. 7.5 lb/in., 77 ft·lb **15.** 144 ft·lb **17.** 1.5k J **19.** -0.36 J **21.** (a) $\omega = \dfrac{t^3}{3} + \dfrac{t^2}{2}$ (b) 1000 rad
23. 2 **25.** 1 **27.** Diverges **29.** 9 **31.** 2 **33.** -0.54 J

Exercises 24-7, p. 1013

1. $(12x^3 - 5)\,dx$ **3.** $\dfrac{x\,dx}{(4 - x^2)^{3/2}}$ **5.** $dy = 4.8$, $\Delta y = 4.96$ **7.** $2x + C$ **9.** $11\frac{2}{3}$ **11.** $\dfrac{2x^{9/2}}{9} + C$
13. $\frac{1}{9}(6x - 5)^{3/2} + C$ **15.** $x^3 - \dfrac{x^2}{2} + 4x + C$ **17.** $-6\frac{2}{3}$ **19.** $\frac{1}{20}(5x - 2)^4 + C$ **21.** $\dfrac{2x^{9/2}}{9} - \dfrac{8x^{5/2}}{5} + C$
23. 4.57 **25.** $y = \dfrac{2x^3}{3} - 3x^2 + 4$ **27.** 4.25 sq units **29.** $5\frac{1}{3}$ sq units **31.** $12\frac{4}{9}$ sq units
33. 12.94 sq units **35.** 0.30 **37.** 44.1 ± 14.7 m **39.** $M = \dfrac{20}{\sqrt{t + 1}} - 15$ **41.** 14,700 units
43. 675 ft·lb **45.** -0.6 J **47.** 96.6 ft/s **49.** (a) 0.02 m (b) 0.06% **51.** 3 **53.** $-\frac{5}{6}$

Chapter 25

Exercises 25-1, p. 1027

1. $58\frac{2}{3}$ **3.** $64\sqrt{2}/3$ **5.** $1\frac{1}{3}$ **7.** $\frac{1}{6}$ **9.** 40.5 **11.** $\frac{1}{2}$ **13.** $20\frac{5}{6}$ **15.** $50\frac{5}{6}$ **17.** 36 **19.** 2.79
21. $\frac{5}{12}$ **23.** $7\frac{1}{3}$ **25.** 14.54 **27.** $2\frac{2}{3}$ **29.** $\frac{5}{12}$ **31.** $21\frac{1}{12}$ **33.** $28\frac{1}{2}$ **35.** 2630 Btu
37. 6 in.², \$1050 **39.** $21\frac{1}{3}$ cm²

Exercises 25-2, p. 1045

1. $64\pi/3$ **3.** 14.1 **5.** $25\frac{1}{15}\pi$ **7.** 176.2 **9.** 8π **11.** 54.33 **13.** $8\pi/3$ **15.** 516π
17. $125\pi/12$ **19.** $16\pi/15$ **21.** 2844.4 **23.** 8.4 **25.** $17\frac{1}{15}\pi$ **27.** 77.9 **29.** $51\frac{11}{15}\pi$ **31.** 23.7
33. 8π **35.** 56π in.³ **37.** 18π **39.** 7.2 kg

Exercises 25-3, p. 1060

1. 0.8 cm   **3.** -0.9 m   **5.** $(-0.5, 0.6)$ cm   **7.** $(2.3, 1.6)$ m   **9.** $(2.1, 2.4)$ cm   **11.** $(-1.3, -1.1)$ cm
13. $(2.7, 2.7)$ **15.** $(2, 2)$ **17.** $(0, 0.9)$ **19.** $(2.2, 0.9)$ **21.** $(3, 6.6)$ **23.** $(1.5, 3.6)$ **25.** $(5, 2.9)$
27. $(0, 1.4)$ **29.** $(1.1, 0)$ **31.** $(0, 1)$ **33.** $(0.75, 0)$ **35.** $(0, 2)$ **37.** $(1, 1.2)$ in.   **39.** $(0, 4/3\pi)$

Exercises 25-4, p. 1076

1. (a) 2990, 3.9   (b) 1700, 2.9   **3.** (a) 66.8, 2.7   (b) 97, 3.3   **5.** (a) 13.5m   (b) 3.7
7. (a) m/5   (b) 0.4   **9.** (a) 4m/5   (b) 0.9   **11.** (a) 6m   (b) 2.4   **13.** (a) 2.7m   (b) 1.6
15. (a) 6m   (b) 2.4   **17.** (a) 17.2m   (b) 4.2   **19.** (a) 6.4m   (b) 2.5   **21.** $\frac{1}{3}a^2$m   **23.** 0.2m
25. 80m

Exercises 25-5, p. 1090

1. 5.33   **3.** 2.17   **5.** 0.04   **7.** 2.92   **9.** 1.41   **11.** 4.02   **13.** 46 mi/h
15. (a) 0.078 W   (b) 0.11 W   **17.** 72°F   **19.** 0.62 W   **21.** 4.7×10^7 J   **23.** 2.8×10^6 ft·lb
25. 2.7×10^5 ft·lb   **27.** 2.7×10^7 N·m   **29.** 230 lb   **31.** (a) 11.7 lb   (b) 38.0   (c) 26.3 lb
33. 45,000 lb   **35.** 2080 lb

Exercises 25-6, p. 1094

1. $2\frac{2}{3}$ **3.** 3.81   **5.** $2\frac{2}{3}$ **7.** $125\pi/6$ **9.** $384\pi/5$ **11.** 18.15   **13.** 0.3 cm   **15.** $(2.6, 2.7)$ cm
17. $(\frac{2}{3}, \frac{2}{3})$ **19.** $(2.3, 2.3)$ **21.** $(0, \frac{5}{3})$ **23.** (a) $\dfrac{12m}{7}$ (b) 1.3   **25.** (a) 2.4m   (b) 1.5
27. 46.7   **29.** 3.0 solar units   **31.** 4.2×10^6 ft·lb   **33.** 44.5 N   **35.** 7.1 sq units   **37.** 64π
39. 27.4m

Chapter 26

Exercises 26-1, p. 1109

1. $2 \cos 2x$ **3.** $8 \cos (4x - 3)$ **5.** $-12 \sin 3t$ **7.** $-5 \sin (5x - 2)$ **9.** $-\cos\left(\dfrac{\pi}{2} - x\right)$

11. $(2x + 2) \cos (x^2 + 2x)$ **13.** $6 \sin 3x \cos 3x$ **15.** $-18 \cos^2(3\theta - 1) \sin (3\theta - 1)$ **17.** $\dfrac{\cos 2x}{\sqrt{\sin 2x}}$

19. $2x \cos 2x + \sin 2x$ **21.** $-4x^3 \sin 4x + 3x^2 \cos 4x$ **23.** $\dfrac{x \cos \sqrt{x^2 + 1}}{\sqrt{x^2 + 1}}$ **25.** $2x + 5 \cos 5x$

27. $-2 \sin 2x + 2 \cos x$ **29.** $-x \sin x + \cos x - 4x$ **31.** $\dfrac{2x \cos 2x - \sin 2x}{x^2}$ **33.** $\dfrac{3x^2 \cos x + 2x^3 \sin x}{\cos^3 x}$

35. $\dfrac{-3 \sin x^2 \sin 3x - 2x \cos x^2 \cos 3x}{\sin^2 x^2}$ **37.** $-2x \sin 3x \sin x^2 + 3 \cos 3x \cos x^2$

39. $-8 \sin x \sin 2x + 4 \cos x \cos 2x$ **41.** $\dfrac{3 - 3x^2 \sin x^3}{\pm 2\sqrt{3x - \cos x^3}}$ **43.** 0   **45.** $\pi/4$ **47.** $-A\omega \cos (kx - \omega t)$

49. $\dfrac{-T_m F \sin (\theta/2) \cos (\theta/2)}{[1 + F \sin^2(\theta/2)]^2}$ **51.** $-\dfrac{16 I_m \pi d}{9\lambda} \cos \dfrac{\pi d \sin \theta}{\lambda} \sin \dfrac{\pi d \sin \theta}{\lambda} \cos \theta$

Exercises 26-2, p. 1117

1. $4 \sec^2 4x$ **3.** $-24x^3 \csc^2 2x^4$ **5.** $-\dfrac{\csc \sqrt{x} \cot \sqrt{x}}{2\sqrt{x}}$ **7.** $5 \sec (x + \pi/4) \tan (x + \pi/4)$

9. $6 \tan 3\theta \sec^2 3\theta$ **11.** $-18x \cot^2 x^2 \csc^2 x^2$ **13.** $-\dfrac{\csc 2x \cot 2x}{\sqrt{\csc 2x}}$ **15.** $36t \sec^2(3t)^2 \tan (3t)^2$

17. $6x^3 \sec^2 x^3 + 2 \tan x^3$ **19.** $t \csc^3 t (2 - 3t \cot t)$ **21.** $2x \cos 2x \sec^2 x^2 - 2 \tan x^2 \sin 2x$

23. $6 \cot \theta \cos 2\theta - 3 \sin 2\theta \csc^2 \theta$ **25.** $\dfrac{8x^4 \sec^2 x^4 - 2 \tan x^4}{4x^2}$

27. $\dfrac{\csc 3x(-3 \cot 3x - 3 \cos x^2 \cot 3x + 2x \sin x^2)}{(1 + \cos x^2)^2}$ **29.** $2x \sec^2 x^2 + 2 \sec 2x \tan 2x$

31. $6 \csc^2 3x(\cot 3x - 1)$ **33.** $\dfrac{1 - \csc^2 x}{\pm 2\sqrt{x + \cot x}}$ **35.** -4.9×10^{-4} rad/s

37. 24 **39.** 7.8 m/s

Exercises 26-3, p. 1125

1. $\dfrac{6}{\sqrt{1 - 4x^2}}$ **3.** $-\dfrac{1}{(x + 1)\sqrt{x^2 + 2x}}$ **5.** $-\dfrac{3x^2}{\sqrt{1 - x^6}}$ **7.** $\dfrac{2}{\sqrt{x - x^2}}$ **9.** $\dfrac{12x^3}{1 + 9x^8}$ **11.** $\dfrac{6}{(1 + 2x)\sqrt{2x}}$

13. $-\dfrac{2x^2}{\sqrt{1 - x^4}} + \text{Arccos } x^2$ **15.** $\dfrac{3x^2 - 3}{1 + 9x^2} + 2x \text{ Arctan } 3x$ **17.** $\dfrac{4 \text{ Arcsin } 2x}{\sqrt{1 - 4x^2}}$ **19.** $\dfrac{6(\text{Arctan } x)^2}{1 + x^2}$

21. $\dfrac{(\sqrt{1 - x^2}) \text{ Arcsin } x - x}{\sqrt{1 - x^2} \,(\text{Arcsin } x)^2}$ **23.** $\dfrac{x - (1 + x^2) \text{ Arctan } x}{2x^2(1 + x^2)}$ **25.** $\dfrac{4x^4 + 4x + 1}{1 + 4x^4}$ **27.** $\dfrac{3 - 9x}{\sqrt{1 - 9x^2}}$

29. 0.444 rad $\pm 1.5\%$ **31.** 1.5 **33.** 5.3 rad/s   **35.** 3.3 m/s

Exercises 26-4, p. 1131

1. $\dfrac{3 \log e}{x}$ **3.** $\dfrac{15 \log e}{5x - 1}$ **5.** $\dfrac{2x}{x^2 - 3}$ **7.** $-3 \tan 3x$ **9.** $\dfrac{\log e}{x(x + 1)}$ **11.** $\dfrac{9(x^2 + 2)}{x(3x^2 + 2)}$ **13.** $\dfrac{3x^2}{2(5 + x^3)}$

15. $\dfrac{2 \sec^2 x \tan x + 1}{\sec^2 x + x}$ **17.** $\dfrac{2}{x + 1}$ **19.** $\dfrac{2 \ln 2x}{x}$ **21.** $2 + \ln x^2$ **23.** $\dfrac{4(\ln 3x - 1)}{(\ln 3x)^2}$ **25.** $\frac{3}{5}$ **27.** $\frac{1}{2}$

29. 0.054 dB $\cdot$ W$^{-1} \cdot$ m^2 **31.** 0.24

Exercises 26-5, p. 1147

1. $4 \cdot 3^{4x} \ln 3$ **3.** $2x(4^{x^2} \ln 4)$ **5.** $2xe^{x^2}$ **7.** $(4 + 8x)e^{x + x^2}$ **9.** $-\dfrac{\ln 3}{3^x}$ **11.** $\dfrac{e^{\sqrt{x}}(2 + \sqrt{x})}{2}$

13. $e^{\sin x}(\cos x)$ **15.** $\dfrac{e^{2x}(2x - 1)}{2x^2}$ **17.** $\ln 2(2^x + 3 \cdot 2^{3x})$ **19.** $3x^2 + 3e^{3x}$ **21.** $\dfrac{e^{\ln 2x}}{x} = 2$ **23.** $-\dfrac{2e^{2/x}}{x^2}$

25. $-2e^{2x} \tan (e^{2x})$ **27.** $\dfrac{3e^{3x}}{\sqrt{1 - e^{6x}}}$ **29.** $\dfrac{e^x}{2\sqrt{1 + e^x}}$ **31.** $2xe^{x^2} \cos e^{x^2}$ **33.** $2e^{2 + 2x}$ **35.** $\dfrac{kabc^{-kx} \ln c}{(1 + bc^{-kx})^2}$

37. -1 **39.** 1 **41.** $kabe^{-kt - be^{-kt}}$ **43.** 1.9 cases/week   **45.** -7.0×10^{-4} lb/in.$^2 \cdot$ mi   **47.** 1.5

49. (a) $k = 1.2 \times 10^{-4}$ per year   (b) 1618 years   (c) 5.0×10^{-3} dis/min $\cdot$ year

51. (a) 146 s   (b) $-0.24°$C/s   **53.** $\$116.80$ per year   **55.** 0.03 A   **57.** 0.23 g/s

Exercises 26-6, p. 1150

1. $6 \cos 6x$ **3.** $\dfrac{3 \sec^2 \sqrt{x}}{\sqrt{x}}$ **5.** $8x^3 \sec^2 x^4 \tan x^4$ **7.** $-24x^2 \cos x^3 \sin x^3$

9. $\sec x^2(2x \sin 3x \tan x^2 + 3 \cos 3x)$ **11.** $8x^3 + 2 \csc (2x + 5) \cot (2x + 5)$

13. $\dfrac{-2x \sin x^2 - 2x \cot x \sin x^2 + \cos x^2 \csc^2 x}{(1 + \cot x)^2}$ **15.** $\dfrac{6x^2}{\sqrt{1 - x^6}}$ **17.** $\dfrac{1}{2\sqrt{x}(1 + x)}$

19. $\dfrac{\sqrt{1-x^4}\,\text{Arccos}\,x^2 + 2x^2}{\sqrt{1-x^4}(\text{Arccos}\,x^2)^2}$ **21.** $\dfrac{2\log e}{x}$ **23.** $2x^2 + 3x^2\ln x^2$ **25.** $\dfrac{2}{(1+4x^2)\,\text{Arctan}\,2x}$

27. $3x^2(5^{x^3}\ln 5)$ **29.** $(-3\sin 3x)e^{\cos 3x}$ **31.** $\dfrac{e^{\sqrt{x+1}}(x+2\sqrt{x+1})}{2\sqrt{x+1}}$ **33.** $\dfrac{e^{x^2}(1+2x^2\ln x)}{x}$

35. $32e^{8x}$ **37.** $\dfrac{(\tan 2x)^2}{\sqrt{2x-\tan 2x}}$ **39.** $y=2$ **41.** $4x-2y-\pi=0$ **43.** $-2\omega Y_m\sin kx\sin\omega t$

45. 1.12 gauss/rad   **47.** 0.70%   **49.** -0.098 rad/s   **51.** 2.9 V/s   **53.** 0.011 C/ms

55. \$120.51 per year

Chapter 27

Exercises 27-1, p. 1158

1. $-\dfrac{1}{2(2x-1)}+C$ **3.** $\dfrac{2(x^2+2)^{3/2}}{3}+C$ **5.** $\dfrac{1}{4}\sin^4 x+C$ **7.** $\dfrac{2}{3}$ **9.** $-2\cos^2 x-\dfrac{1}{5}\cos^5 x$

11. $\dfrac{1}{2(1-\sin x)^2}+C$ **13.** $-\dfrac{2}{\sqrt{e^x+2}}+C$ **15.** $\dfrac{1}{3}(\text{Arcsin}\,x)^3+C$ **17.** $\dfrac{1}{3}(\sin 2x)^{3/2}+C$

19. $\dfrac{1}{2}(\ln x)^2+C$ **21.** $\dfrac{1}{2(1-e^{2t})}+C$ **23.** $\dfrac{1}{3}$ **25.** 0.170   **27.** $\dfrac{\sin^4 x-3}{3\sin x}+C$

29. 0.011 C   **31.** 0.364   **33.** $-5(1+e^{-t})^2+C$

Exercises 27-2, p. 1166

1. $-\ln|1-x|+C$ **3.** $-\dfrac{1}{4}\ln|1-2x^2|+C$ **5.** $\ln|\tan\theta|+C$ **7.** $\ln|\ln x|+C$ **9.** $e^{3x}+C$

11. $2e^{x/2}+C$ **13.** $-\dfrac{1}{2}e^{-2x}+C$ **15.** $\dfrac{1}{2}e^{2x-3}+C$ **17.** 1.03   **19.** $\dfrac{1}{2}x^2+\dfrac{1}{3}e^{3x}+C$

21. $x-2e^{-x}-\dfrac{1}{2}e^{-2x}+C$ **23.** 4.65   **25.** 5400   **27.** 1.07   **29.** $\dfrac{2^{4x}}{4\ln 2}+C$ **31.** 3.64

33. $\ln|x+2x^2|+C$ **35.** 1.19   **37.** 0.20   **39.** 7.3 N/C·s   **41.** 45.23   **43.** $2(1-e^{-3x^2})$

45. 1.20 sq units   **47.** 9.06 cu units

Exercises 27-3, p. 1173

1. $-\dfrac{1}{3}\cos 3x+C$ **3.** $-\dfrac{1}{2}\ln|\sec(-2t)+\tan(-2t)|+C$ **5.** $2\tan\dfrac{x}{2}+C$ **7.** -0.217 **9.** 0.48

11. $\sin(x+2)+C$ **13.** $\dfrac{1}{2}\sec 2t+C$ **15.** $-\dfrac{1}{2}\ln|\cos x^2|+C$ **17.** $-\dfrac{1}{\sin x}+C$ **19.** $\dfrac{1}{2\cos^2 x}+C$

21. $-\cos(\ln x)+C$ **23.** 0.11   **25.** 1.4×10^{-3} N·m   **27.** 0.44   **29.** π cu units   **31.** 5.43 sq units

33. $V=-\dfrac{1}{2}\sin(1-2t)+0.42$

Exercises 27-4, p. 1185

1. $\dfrac{x}{2}-\dfrac{1}{4}\sin 2x+C$ **3.** $2\sin\dfrac{x}{2}-\dfrac{4}{3}\sin^3\dfrac{x}{2}+\dfrac{2}{5}\sin^5\dfrac{x}{2}+C$ **5.** $\dfrac{3x}{8}+\dfrac{1}{4}\sin(2x+2)+\dfrac{1}{32}\sin(4x+4)$

7. $\dfrac{1}{\cos x}+C$ **9.** $-\dfrac{3}{4}\cos^{4/3}t+\dfrac{3}{10}\cos^{10/3}t+C$ **11.** $\dfrac{1}{5}\sin^5 x-\dfrac{1}{7}\sin^7 x+C$ **13.** $-\dfrac{\cos^5 2x}{10}+\dfrac{\cos^7 2x}{14}+C$

15. $\frac{1}{6}\cos^6\theta - \frac{1}{4}\cos^4\theta + C$ **17.** $\frac{1}{2}\tan^2 x + \ln|\cos x| + C$ **19.** $-\frac{1}{\pi}\cot \pi x - x + C$

21. $\frac{1}{5}\tan^5 t + \frac{2}{3}\tan^3 t + \tan t + C$ **23.** $\frac{1}{3}\sec^3 x - \sec x + C$ **25.** $\frac{1}{5\pi}\tan^5 \pi\theta + C$

27. $-\frac{1}{14}\csc^7 x^2 + \frac{1}{5}\csc^5 x^2 - \frac{1}{6}\csc^3 x^2 + C$ **29.** $\frac{\pi}{4}$ **31.** $\frac{20}{21}$ **33.** $0.02 \text{ kg} \cdot \text{m}^2$ **35.** $1 - \frac{\pi}{4}$ sq unit

37. 24.5 W

Exercises 27-5, p. 1195

1. $\text{Arcsin } \frac{x}{3} + C$ **3.** 1.89   **5.** $\frac{\sqrt{3}}{6}\text{Arctan } \frac{2q\sqrt{3}}{3} + C$ **7.** $\frac{2}{3}\text{Arcsin } \frac{3x}{4} + C$ **9.** $\frac{1}{3}(x^2 - 1)^{3/2} + C$

11. $2\text{ Arcsin } \frac{x}{2} - \frac{x}{2}\sqrt{4 - x^2} + C$ **13.** $\frac{\sqrt{2}}{4}\text{Arcsin } 2x\sqrt{2} + C$ **15.** 282.4   **17.** $\text{Arcsin } \frac{t^2}{3} + C$

19. $\text{Arctan } x + \frac{1}{2}\ln(1 + x^2) + C$ **21.** $\frac{x}{4\sqrt{4 - x^2}} + C$ **23.** $\frac{\sqrt{x^2 - 16}}{16x} + C$

25. $\frac{9}{2}\text{Arcsin } \frac{x + 1}{3} + \frac{1}{2}(x + 1)\sqrt{9 - (x + 1)^2} + C$ **27.** $\frac{\sqrt{3}}{3}\text{Arctan } \frac{\sqrt{3}(x + 3)}{3} + C$ **29.** 0.812

31. $\frac{1}{2}\left(\text{Arcsin } e^x + e^x\sqrt{1 - e^{2x}}\right) + C$ **33.** $v = 20 \text{ Arctan } 10t \text{ cm/s}$ **35.** 1.62 cu units   **37.** 0.13 N

39. 0.85 sq unit

Exercises 27-6, p. 1202

1. $-x \cos x + \sin x + C$ **3.** $-\frac{1}{4}e^{-2x}(2x + 1) + C$ **5.** $\frac{x^2}{2}\ln x(\ln x - 1) + \frac{1}{4}x^2 + C$

7. $y(\ln y)^3 - 3y(\ln y)^2 + 6y \ln y - 6y + C$ **9.** $\frac{2\pi - 1}{4}$ **11.** $\frac{1}{4}e^{2t}(\sin 2t - \cos 2t) + C$

13. $\sin 2x\left(\frac{x^2}{2} - \frac{1}{4}\right) + \frac{x}{2}\cos 2x + C$ **15.** $y \tan y + \ln|\cos y| - \frac{1}{2}y^2 + C$

17. $\frac{2q}{3}(q + 2)^{3/2} - \frac{4}{15}(q + 2)^{5/2} + C$ **19.** 1.56   **21.** $\frac{y}{5}(y - 2)^5 - \frac{1}{30}(y - 2)^6 + C$

23. $\frac{1}{4}\sec^3\theta \tan\theta + \frac{3}{8}\sec\theta \tan\theta + \frac{3}{8}\ln|\sec\theta + \tan\theta| + C$ **25.** $\frac{1}{2}e^{x^2}(x^2 - 1) + C$ **27.** 0.29 sq unit

29. $\frac{Ve^{kt}}{L(\omega^2 + k^2)}(k \sin \omega t - \omega \cos \omega t) + C$ **31.** $s = \frac{4}{15}(3t - 2)(1 + t)^{3/2} + \frac{8}{15}$

Exercises 27-7, p. 1207

1. $\frac{1}{12}(2 + 3x)^4 + C$ **3.** $-\ln\left|\frac{1 - 2t}{t}\right| + C$ **5.** $\frac{2(5x - 4)}{75}\sqrt{2 + 5x} + C$ **7.** $\ln|\tan\theta| + C$ **9.** 0.144

11. $\frac{\sqrt{2}}{2}\ln\left|\frac{\sqrt{2 + x} - \sqrt{2}}{\sqrt{2 + x} + \sqrt{2}}\right| + C$ **13.** 0.324   **15.** $-\frac{1}{3}\ln\left|\frac{3 + 5x}{x}\right| + C$

17. $\frac{x}{8}(2x^2 - 25)\sqrt{x^2 + 5} + \frac{75}{8}\ln\left|x + \sqrt{x^2 + 5}\right| + C$ **19.** 4.728   **21.** $\frac{2}{5}x^2\sqrt{2 + x} - \frac{16}{15}(x - 4)\sqrt{2 + x} + C$

23. $-\frac{1}{6}\cot^3 2x + \frac{1}{2}\cot 2x + x + C$ **25.** $V(t) = \frac{e^{kt}(k \sin \omega t - \omega \cos \omega t) + \omega}{k^2 + \omega^2}$ **27.** 0.19   **29.** 2.92×10^8

31. 33 rev

Exercises 27-8, p. 1210

1. $\dfrac{1}{5}\sin^5 x + C$ 3. $\dfrac{1}{2}\tan^2\theta + C$ 5. $\dfrac{1}{1-e^x} + C$ 7. $-\dfrac{1}{4}(\text{Arccos } x)^4 + C$ 9. $\dfrac{1}{2}(\ln 3t)^2 + C$

11. $-\dfrac{1}{2}\ln|3 - 2x| + C$ 13. 0.231 15. $e^{\sin x} + C$ 17. 8.835 19. $\dfrac{4^{2x}}{2\ln 4} + C$ 21. $\dfrac{1}{2}\sin x^2 + C$

23. $-\ln|\cos(\ln 2x)| + C$ 25. $\dfrac{1}{3}\sec x^3 + C$ 27. $\dfrac{1}{3}\sin 3x - \dfrac{1}{9}\sin^3 3x + C$

29. $\dfrac{1}{3}\sin^3 x - \dfrac{2}{5}\sin^5 x + \dfrac{1}{7}\sin^7 x + C$ 31. $\dfrac{1}{8}\sin^4 2x - \dfrac{1}{12}\sin^6 2x + C$ 33. $\dfrac{1}{15}\tan^5 3x + \dfrac{1}{9}\tan^3 3x + C$

35. $-\dfrac{1}{18}\cot^3 6x + \dfrac{1}{6}\cot 6x + x + C$ 37. $\text{Arcsin }\dfrac{t}{6} + C$ 39. 0.148

41. $\dfrac{x}{2}\sqrt{x^2 + 6} - 3\ln\left|x + \sqrt{x^2 + 6}\right| + C$ 43. $\text{Arcsin } e^x + C$ 45. $\dfrac{x}{2}\sin 2x + \dfrac{1}{4}\cos 2x + C$

47. $-e^{-x}(x^2 + 2x + 2) + C$ 49. $\dfrac{(9x - 2)(2 + 3x)^3}{108}$ 51. $\dfrac{2}{3}(x - 2)\sqrt{1 + x} + C$ 53. 0.2

55. 0.364 cu unit   57. $(2.27, 0.40)$ 59. 3.01×10^{-3} C   61. 50.3 V   63. $V(t) = -\dfrac{4}{3}\cos\left(3t + \dfrac{\pi}{2}\right)$

65. 18.8 N

Chapter 28

Exercises 28-1, p. 1221

1. $4 - 2x^2 + \dfrac{x^4}{6}$ 3. $x - x^2 + \dfrac{x^3}{3}$ 5. $-x - \dfrac{x^2}{2} - \dfrac{x^3}{3}$ 7. $1 + x - \dfrac{x^2}{2}$ 9. $1 - x^2 + x^4$

11. $1 - 2x^2 + x^4$ 13. $1 - \theta^2 + \dfrac{\theta^4}{3}$ 15. $x + 3x^2 + \dfrac{9}{2}x^3$ 17. $1 + \dfrac{x}{3} - \dfrac{x^2}{9}$ 19. $\theta + \dfrac{\theta^3}{3}$ 21. $x - \dfrac{x^3}{3}$

23. $1 + x$ 25. $x^2 - \dfrac{x^4}{3}$ 27. $2x - 2x^2$ 29. $1 + \dfrac{x^2}{2} + \dfrac{x^4}{24} + \dfrac{x^6}{720}$ 31. $1 + ax - \dfrac{a^2 x^2}{2} - \dfrac{a^3 x^3}{6}$

Exercises 28-2, p. 1229

1. $1 + 2x + 2x^2 + \dfrac{4x^3}{3}$ 3. $-2x - 2x^2 - \dfrac{8x^3}{3} - 4x^4$ 5. $1 - \dfrac{9\theta^2}{2} + \dfrac{27\theta^4}{8} - \dfrac{81\theta^6}{80}$ 7. $x^2 - \dfrac{x^6}{3!} + \dfrac{x^{10}}{5!} - \dfrac{x^{14}}{7!}$

9. $1 - 3t + 9t^2 - 27t^3$ 11. $\dfrac{1}{2} + \dfrac{3x}{4} + \dfrac{9x^2}{8} + \dfrac{27x^3}{16}$ 13. $\dfrac{x}{2} - \dfrac{x^3}{48} + \dfrac{x^5}{3840} - \dfrac{x^7}{645,120}$ 15. $1 - t + \dfrac{t^3}{3} - \dfrac{t^4}{6}$

17. $x^2 - \dfrac{x^4}{2!} + \dfrac{x^6}{4!} - \dfrac{x^8}{6!}$ 19. $\dfrac{1}{x} + 1 + \dfrac{x}{2} + \dfrac{x^2}{6}$ 21. $1 - \dfrac{\theta^2}{3!} + \dfrac{\theta^4}{5!} - \dfrac{\theta^6}{7!}$ 23. $C + x - \dfrac{x^3}{3} + \dfrac{x^5}{10}$

25. $C + x - \dfrac{x^2}{4} + \dfrac{x^3}{9}$ 27. 0.0198 29. $10 + 80x + 240x^2$ 31. $-\dfrac{1}{x^2} - \dfrac{1}{6} + \dfrac{x^2}{40}$ 33. $-2.945P_0$

35. 1.031 37. 0.982 cu unit

Exercises 28-3, p. 1234

1. 0.53125 3. 0.0871557 5. 0.6970667 7. 0.264 9. 1.625 11. 0.5 13. 1.1875

15. $2.705, 0.743$ 17. $0.3420203, 1.3 \times 10^{-7}$ 19. $0.5302917, 0.0446$ 21. 3.13906

23. $0.872832V_0, 0.0853138V_0$ 25. -0.356

Exercises 28-4, p. 1240

1. $-\left(x - \dfrac{\pi}{2}\right) + \dfrac{1}{6}\left(x - \dfrac{\pi}{2}\right)^3 - \dfrac{1}{120}\left(x - \dfrac{\pi}{2}\right)^5$ **3.** $e^3\left[1 + (x - 3) + \dfrac{1}{2}(x - 3)^2\right]$ **5.** $\dfrac{1}{2} - \dfrac{1}{16}(x - 4) + \dfrac{3}{256}(x - 4)^2$

7. $2 + \dfrac{1}{12}(x - 8) - \dfrac{1}{288}(x - 8)^2$ **9.** $\ln 2 + \dfrac{1}{2}(x - 2) - \dfrac{1}{8}(x - 2)^2$ **11.** 0.4668574 **13.** 1.9617969

15. 1.1441901 **17.** 0.7547490 **19.** 3.1416782

Exercises 28-5, p. 1253

1. $\dfrac{4}{\pi} \sin x + \dfrac{4}{3\pi} \sin 3x + \dfrac{4}{5\pi} \sin 5x$ **3.** $\dfrac{1}{2} + \dfrac{2}{\pi} \sin \pi x + \dfrac{2}{3\pi} \sin 3\pi x + \dfrac{2}{5\pi} \sin 5\pi x$

5. $-\dfrac{1}{2} + \dfrac{4}{\pi^2} \cos \pi t + \dfrac{4}{9\pi^2} \cos 3\pi t$ **7.** $\dfrac{\pi}{4} - \dfrac{2}{\pi} \cos x - \dfrac{2}{9\pi} \cos 3x + \sin x - \dfrac{1}{2} \sin 2x + \dfrac{1}{3} \sin 3x$

9. $\dfrac{2}{\pi} \cos x + \dfrac{2}{9\pi} \cos 3x + \dfrac{2}{25\pi} \cos 5x + \dfrac{1}{2} \sin 2x + \dfrac{1}{4} \sin 4x + \dfrac{1}{6} \sin 6x$

11. $1 + \dfrac{4}{\pi} \cos \dfrac{\pi t}{2} - \dfrac{4}{3\pi} \cos \dfrac{3\pi t}{2} + \dfrac{4}{5\pi} \cos \dfrac{5\pi t}{2}$ **13.** $\dfrac{2}{\pi} - \dfrac{4}{3\pi} \cos 2t - \dfrac{4}{15\pi} \cos 4t - \dfrac{4}{35\pi} \cos 6t$

15. $a_0 = \dfrac{2P + e^{-2P} - 1}{2P}$, $a_n = \dfrac{P}{P^2 + n^2\pi^2}(e^{-2P} - 1)$, $b_n = \dfrac{n\pi}{P^2 + n^2\pi^2}(e^{-2P} - 1)$

Exercises 28-6, p. 1256

1. $1 - 3x + \dfrac{9}{2}x^2$ **3.** $2x - x^3 + \dfrac{1}{12}x^5$ **5.** $2x - 2x^2 + \dfrac{8}{3}x^3$ **7.** $2 - 5x + 2x^3$ **9.** $x + x^2 + \dfrac{1}{3}x^3$

11. $x + \dfrac{x^2}{2} + \dfrac{x^4}{8}$ **13.** $3x^2 - \dfrac{5x^4}{6} + \dfrac{7x^6}{120}$ **15.** 1.48 **17.** 0.9925462 **19.** 0.2013573 **21.** 4.6×10^{-9}

23. $e^{-4}\left[1 - (x - 4) + \dfrac{1}{2}(x - 4)^2\right]$ **25.** $\ln \dfrac{1}{2} + \sqrt{3}\left(x - \dfrac{\pi}{6}\right) - 2\left(x - \dfrac{\pi}{6}\right)^2$ **27.** 0.2680169 **29.** 4.417205

31. $-\dfrac{4}{\pi} \sin x - \dfrac{4}{3\pi} \sin 3x - \dfrac{4}{5\pi} \sin 5x$ **33.** 0.96 sq unit **35.** 0.342 **37.** $1 + 2x + 3x^2 + 4x^3$

39. $f_0(1 - \beta)\left(1 + \dfrac{\beta^2}{2} + \dfrac{3\beta^4}{8}\right)$ **41.** $1 - \dfrac{\theta^2}{2!} + \dfrac{\theta^4}{4!} + \cdots = 1 - \dfrac{x}{2} + \dfrac{3}{8}x^2 + \cdots; \dfrac{\theta^2}{2} \approx \dfrac{x}{2}; \theta \approx x = \dfrac{1}{\sqrt{L}}$ **43.** $\dfrac{5\pi + 8}{8}$

Chapter 29

Exercises 29-1, p. 1269

1. (a) 1 (b) 1 **3.** (a) 2 (b) 1 **5.** (a) 1 (b) 1 **7.** (a) 2 (b) 1 **9.** (a) 1 (b) 3

11. Yes **13.** No **15.** Yes **17.** No **19.** $y = \dfrac{2x^3}{3} + x + C$ **21.** $y = x + \cos x + C$

23. $y = -\dfrac{1}{3}(4 - x^2)^{3/2} + C$ **25.** $y^2 = x^2 + C$ **27.** $y = Ce^{1/x}$ **29.** $y = 1 - Ce^{-x^2/2}$ **31.** $y = 1 + \dfrac{1}{C - e^{-x}}$

33. $y = \text{Arcsin}\left(\dfrac{2}{x} + C\right)$ **35.** $y = C - e^{x+2}$ **37.** $y = x + \dfrac{x^2}{2} + \ln x + C$ **39.** $y^3 = \dfrac{3x^2}{2} + 1$

41. $y = \text{Arccos } x$ **43.** $q = VC(1 - e^{-t/RC})$ **45.** 4 h

Exercises 29-2, p. 1279

1. $y^2 = 2x - x^2 + C$ **3.** $y = 2x \ln |x| + Cx$ **5.** $y^3 + Cy - x = 0$

7. $y^4 - 2x^2y^2 - 2y^2 - 2x^2 + x^4 + C = 0$ **9.** $y = \dfrac{x}{\sin x + C}$ **11.** $\ln(x^2 + y^2) = -\dfrac{2y^3}{3} + C$

13. $\left(\dfrac{y}{x}\right)^{3/2} = \dfrac{3x^2}{4} + C$ **15.** $e^{xy} = x + C$ **17.** $y = \dfrac{x^3}{2} + 4x \ln x + Cx$ **19.** $y = \left(\dfrac{x-1}{x}\right)e^x + \dfrac{C}{x}$

21. $y = e^{-4x}(x^3 + C)$ **23.** $y = e^{-2x}(x \ln x - x + C)$ **25.** $y = Ce^{-2x} - e^{-3x}$ **27.** $y = x + \dfrac{C}{x}$

29. $y = -2x^4 + Cx^2$ **31.** $y = \dfrac{x^3}{4} + \dfrac{2x^2}{3} - \dfrac{x}{2} + \dfrac{C}{x}$ **33.** $y = \dfrac{1}{x^2 - 1}\left(\dfrac{x^3}{3} + \dfrac{x^2}{2} + C\right)$

35. $y = -e^{-x} \cos x + e^{-x}$ **37.** $y = \cos x\,(1 - \ln|\cos x|)$ **39.** $y = -\dfrac{1}{2}\cos 2x + \dfrac{1}{4x}\sin 2x - \dfrac{\pi}{4x}$

Exercises 29-3, p. 1296

1. $y = 3x + 2$ **3.** $y = 2e^{x^2} - 1$ **5.** $y = x - 1 + e^{-x}$ **7.** $y = 1 - \cos x$ **9.** $y^2 = 1 - 2 \ln x$

11. $y^2 + \dfrac{x^2}{2} = C$ **13.** $y - \dfrac{y^2}{2} = \dfrac{x^2}{6} + C$ **15.** $\dfrac{dL}{dt} = kL,\ L = L_0e^{kt}$ **17.** $26{,}000$

19. $N = N_0 + \dfrac{1}{k}(e^{kt} - 1)$, where $k = \dfrac{1 - a}{N_0}$ **21.** (a) $y = \left(\dfrac{kt}{2} + 2\right)^2$ (b) 289 **23.** $P = \dfrac{P_0}{2P_0 + (1 - 2P_0)e^{-kt}}$

25. 2.2 years   **27.** 22.7 days   **29.** $x = A + (1 - A)e^{-kt}$ **31.** $i = 2.4(1 - e^{-50t})$ **33.** $i = \dfrac{55}{3}(e^{-2t} - e^{-5t})$

35. 470 ft/s   **37.** $0.27\ \text{lb} \cdot \text{s}^2 \cdot \text{ft}^{-2}$ **39.** 33 lb   **41.** 37 lb

Exercises 29-4, p. 1307

1. $y = \dfrac{x^4}{12} + c_1x + c_2$ **3.** $y = -\sin x + c_1x + c_2$ **5.** $y = |x|\ln|x| - |x| + c_1x + c_2$

7. $y = -\dfrac{x^2}{2} - x + c_1e^x + c_2$ **9.** $y = c_1x^2 + c_2$ **11.** $y = \dfrac{x}{2}\sqrt{c_1^2 - x^2} + \dfrac{c_1^2}{2}\operatorname{Arcsin}\left(\dfrac{x}{c_1}\right) + c_2$

13. $y = c_1e^x + c_2$ **15.** $y^2 = c_1x + c_2$ **17.** $y = c_1e^{-2x} + c_2e^x$ **19.** $y = c_1e^{-2x} + c_2e^{-x}$

21. $y = c_1e^{3x/2} + c_2e^{-x}$ **23.** $y = c_1e^{3x} + c_2e^{-3x}$ **25.** $y = c_1 + c_2e^x$ **27.** $y = c_1e^{-x} + c_2xe^{-x}$

29. $y = c_1e^{x/2} + c_2xe^{x/2}$ **31.** $y = e^x\!\left(c_1e^{x\sqrt{3}} + c_2e^{-x\sqrt{3}}\right)$ **33.** $y = e^{-x/2}\!\left(c_1e^{x\sqrt{5}/2} + c_2e^{-x\sqrt{5}/2}\right)$

35. $y = e^x(c_1 \cos x + c_2 \sin x)$ **37.** $y = e^{-2x}(c_1 \cos x + c_2 \sin x)$ **39.** $y = e^{-2x}\!\left(c_1 \cos x\sqrt{2} + c_2 \sin x\sqrt{2}\right)$

41. $y = e^{2x} - e^{-4x}$ **43.** $y = \dfrac{e^x}{2}\!\left(e^{x\sqrt{2}} - e^{-x\sqrt{2}}\right)$

Exercises 29-5, p. 1320

1. $y = c_1e^{-3x} - c_2e^{2x} - \dfrac{1}{9} - \dfrac{2x}{3}$ **3.** $y = c_1 + c_2e^{2x} - \dfrac{3x}{2}$ **5.** $y = c_1e^{4x} + c_2e^{-2x} + \dfrac{5}{16} - \dfrac{x}{2} + x^2$

7. $y = c_1e^{2x} + c_2e^{-x} - \dfrac{3}{10}\sin x + \dfrac{1}{10}\cos x$ **9.** $y = c_1e^x + c_2xe^x + \dfrac{1}{2}e^{3x}$

11. $y = c_1e^{x/2} + c_2xe^{x/2} - \dfrac{6}{25}\sin x + \dfrac{8}{25}\cos x + x + 4$ **13.** $y = c_1e^{4x} + c_2xe^{4x} + \dfrac{1}{25}e^{-x} + \dfrac{x}{8} + \dfrac{1}{8}$

15. $y = e^{-x}(c_1 \cos x + c_2 \sin x) - \dfrac{1}{10}\sin 2x - \dfrac{1}{5}\cos 2x$ **17.** $y = c_1 \cos 2x + c_2 \sin 2x + \dfrac{x^2}{4} - \dfrac{1}{8}$

19. $y = e^x(c_1 \cos 2x + c_2 \sin 2x) - \dfrac{1}{16}e^{3x} + \dfrac{1}{8}xe^{3x}$ **21.** $y = c_1e^x + c_2xe^x - \dfrac{1}{2}\sin x - \dfrac{1}{2}\cos x - \dfrac{1}{2}x \sin x$

23. $y = c_1e^{5x} + c_2e^{-x} - \dfrac{2}{5} + \dfrac{1}{8}e^x$ **25.** $y = -xe^{-x/2} + e^x$ **27.** $y = 2 \cos 2x + \dfrac{1}{3}\sin 2x + x + \dfrac{1}{3}\sin x$

29. $x = -4 \cos \dfrac{t}{2} + 6 \sin \dfrac{t}{2}$ **31.** $x = 2 \cos 4t\sqrt{3}$ **33.** $x = e^{-2t}\!\left(20 \cos 2t\sqrt{3} + \dfrac{20\sqrt{3}}{3}\sin 2t\sqrt{3}\right)$

35. $x = e^{-4t}(20 + 80t)$ **37.** $q \approx e^{-2.1t}(c_1 \cos 5.4t + c_2 \sin 5.4t) + \dfrac{3}{80}$

39. $q \approx e^{-2.1t}(c_1 \cos 5.4t + c_2 \sin 5.4t) + 0.49 \sin t + 3.80 \cos t$

Exercises 29-6, p. 1325

1. $y = -\dfrac{5x^2}{2} + c$ **3.** $y = x + e^x + c$ **5.** $y = \dfrac{2x^{3/2}}{3} + x + c$ **7.** $y = \dfrac{c}{1-x} - 1$ **9.** $y^{3/2} = \dfrac{3x^2}{4} + \dfrac{3x}{2} + 1$

11. $y = 4 + \dfrac{c}{x}$ **13.** $y = xe^x + cx$ **15.** $y = \dfrac{1}{3x}(x^2 + 1)^{3/2} + \dfrac{c}{x}$ **17.** $y = -1 + x \cot x - c \csc x$

19. $y = \dfrac{x^3 + 3x + c}{x^2 + 1}$ **21.** $y = \dfrac{(x-1)(e^x + 1)}{x}$ **23.** $y = ce^{-x}$ **25.** $y = \dfrac{x^4}{12} + \dfrac{x^3}{3} + c_1 x + c_2$

27. $y = -\dfrac{1}{c_1} + c_2 e^{-c_1 x}$ **29.** $y = e^{3x}\left(c_1 e^{x\sqrt{2}} + c_2 e^{-x\sqrt{2}}\right)$ **31.** $y = e^{2x}(c_1 \cos 2x + c_2 \sin 2x)$

33. $y = e^{3x}(2 \cos 2x - \sin 2x)$ **35.** $y = c_1 + c_2 e^{-x} - 2 \sin x - 2 \cos x$ **37.** $V = 12e^{-2t} + 4e^{-4t}$

39. 236 lb **41.** 2.4 s **43.** 6440 years **45.** $N \approx \dfrac{52.5}{1 + 42.5e^{-0.656t}}$ **47.** $x \approx 0.132e^{-2.61t} - 0.032e^{-10.72t}$

49. $q \approx c_1 e^{-4t} + c_2 t e^{-4t} - 0.16 \cos 5t + 0.71 \sin 5t$

Appendix B   Exercises B-1, p. A11

Note: All answers have been rounded according to the rules outlined in Section B-3.

1. 4500 g **3.** 211 yd **5.** 430 kHz **7.** 34 lb/in.^{2} **9.** 6.28 gal **11.** 125 mL
13. 4.6 tons **15.** 3150 Btu **17.** 11.2 in. **19.** 13.2 qt **21.** 7.8 hp **23.** 156 lx
25. 70 ft/s **27.** 10 mi/h **29.** 24 yd^3 **31.** 0.302 m^2 **33.** 0.792 g/cm^3 **35.** 189 L
37. 4.63 yd^3 **39.** 4800 slugs **41.** 232°C

Exercises B-2, p. A14

1. A **3.** E **5.** A **7.** E **9.** E **11.** A
13. (a) 3 significant digits   (b) Hundredths **15.** (a) 1 significant digit   (b) Ten-thousandths
17. (a) 3 significant digits   (b) Units **19.** (a) 4 significant digits   (b) Hundredths
21. 570 **23.** 7.94 **25.** 16.4 **27.** 3.20 **29.** 6.5 **31.** 0.08 **33.** 1,573,000

Exercises B-3, p. A18

1. 6.8 **3.** 5 **5.** 36 **7.** 6.1 **9.** 2400 **11.** 226 **13.** 1.8 **15.** 2200
17. 0.7928 **19.** 3.52 **21.** 0.450 **23.** 2.81 in. **25.** 70 m **27.** 34.77° or 34°46′
29. 26 Ω **31.** 5.5 lx **33.** 120 W

Appendix C   Exercises, p. A28

1. 115° **3.** 65° **5.** 115° **7.** 65° **9.** $\angle t, \angle u, \angle x, \angle y$
11. $\angle s$ and $\angle z$; $\angle t$ and $\angle y$ **13.** 63° **15.** 48.4° **17.** 122° **19.** 61°20′
21. Equilateral; acute (also equiangular, oblique) **23.** 6 in.; 6 in. **25.** 37.5°
27. Scalene; right **29.** 5.8 in. **31.** 58 m **33.** (a) Triangle   (b) 54.0 in.   (c) 126 in.^{2}
35. (a) Rectangle   (b) 126.2 m   (c) 853 m^2 **37.** (a) Square   (b) 900 yd   (c) 50,600 yd^2
39. (a) Circle   (b) 8.17 in.   (c) 5.31 in.^{2} **41.** (a) Right circular cylinder   (b) 6670 ft^2 (c) 41,700 ft^3
43. (a) Sphere   (b) 17,700 cm^2 (c) 221,000 cm^3 **45.** (a) Square pyramid   (b) 2400 mm^2 (c) 6000 mm^3
47. 11 in.^{2} **49.** $A = 389$ m^2; $c = 42.5$ m **51.** $A = 6080$ yd^2; $C = 276$ yd
53. $A = 42.3$ ft^2; $P = 26.00$ ft **55.** $V = 796.3$ cm^3; $S = 415.5$ cm^2
57. $S = 135,000$ mm^2; $V = 3,375,000$ mm^3 **59.** 0.327 in.^{3} **61.** 1345 cm^2 **63.** 330 gal
65. 200 km **67.** (a) 11.4 ft   (b) 9.16 ft **69.** 122 cm^3 **71.** 163 in.^{2} **73.** 42 ft^2

Index to Applied Problems